Handbook of
Analytical Derivatization Reactions

Handbook of
Analytical Derivatization Reactions

Daniel R. Knapp

Department of Pharmacology,
Medical University of South Carolina,
Charleston, South Carolina

A Wiley-Interscience Publication

JOHN WILEY & SONS

New York • Chichester • Brisbane • Toronto • Singapore

Library of Congress Cataloging in Publication Data:

Knapp, Daniel R

Handbook of analytical derivatization reactions.

"A Wiley-Interscience publication."
Includes indexes.
1. Chemistry, Analytic—Technique—Handbooks, manuals, etc. 2. Chromatographic analysis—Handbooks, manuals, etc. 3. Mass spectrometry—Handbooks, manuals, etc. I. Title. II. Title: Analytic derivatization reactions.

QD78.K63 543 78-12944
ISBN 0-471-03469-X

Printed in the United States of America

10

Preface

Much of the analytical work involving gas chromatography, mass spectrometry, combined gas chromatography-mass spectrometry, and, more recently, high performance liquid chromatography involves the use of analytical derivatization prior to the analysis. Analytical derivatization reactions may find application in virtually any area of work involving analysis of organic compounds, particularly in work involving biological samples. The numerous and diverse analytical derivatization methods have been published in widespread areas and, until recently, no effort had been made to consolidate this methodology into a single reference volume. Encyclopedic- and compendium-type complications of reagents and methodology have proven to be extremely useful in the area of synthetic chemistry. Since analytical derivatization is in reality a special type of (usually) microscale synthetic chemistry, it was felt that a similar presentation of analytical derivatization methods would be of considerable value to those whose work involves analytical chemistry, whether in biomedical, environmental, forensic, industrial, or any of numerous other areas. This book presents such a collection of analytical derivatization methods for both chromatographic (GC and HPLC) and mass spectral analysis.

The scope of this volume has been restricted to methods involving covalent derivatives of organic compounds formed prior to the analysis; ion pairs, derivatives of inorganic ions, and postcolumn derivatives are excluded. An effort was made to include a wide and representative range of examples, but the coverage is, by no means, intended to be exhaustive. Due to the magnitude of existing literature on derivatization, omissions in particular areas have likely occurred; for these oversights, the author apologizes.

Much of the discussion of analytical derivatization in existing references is organized according to derivative or reagent type. In practice however, analytical derivatization problems are usually approached from a different point of view, that is, with a sample in need of derivatization. Accordingly, the methods in this book are grouped into sections by sample type. Each section covering a given sample type includes an overview of derivatization of that sample type followed by specific methods for derivatization. Within each major compound type, the derivatization methods are further organized into subsections of compound type and within each subsection according to derivative type. Each method description includes structural formulas for the derivatization reaction, experimental directions, pertinent comments, and references to the original literature. In addition to the sections covering derivatizations of specific sample compound types, sections on derivatives for substituent location by mass spectrometry and derivatives for chromatographic separation of optical isomers are included. A general overview, including a discussion of practical aspects of derivatization and a description of commonly used equipment, is presented in the introduction. Appendices include a table of brand name reagents and their compositions and addresses of both U.S. and foreign suppliers of reagents and apparatus.

A comprehensive system of indexes is essential to the utility of this type of reference. In addition to both general subject and author indexes, an index of derivatives and reagents is presented to aid in the location of methods used for one sample type that may be applicable to other sample types. Cross references to methods are included within the sections as well.

This book is intended to serve both as a self-contained methodology reference and as an efficient entry point to the original literature for workers in biomedical, industrial, forensic, environmental, and numerous other areas employing analytical derivatization methodology. In particular, this work should be useful for those who, like the author, are confronted with a wide variety of sample types. For those involved with more narrowly defined sample types, it is hoped that this work will serve as a stimulus to explore derivatization methods applied in other areas.

The author strongly believes that this type of reference is most useful in the laboratory rather than in the library. Therefore, every effort was made to minimize the cost of the book to make it affordable for the individual. The abbreviations used in the text and the shortened reference format were an effort to conserve space. Apologies are extended to the authors whose names are buried in the *et al.*'s; all names do appear in the author index, however.

If the readers find the book useful, it is hoped to issue supplementary volumes as sufficient new methods become available. The author welcomes suggestions of methods to be included and any other general suggestions for the improvement of this work.

I would like to express my appreciation to the staff of the Medical University of South Carolina Library and to our Department of Pharmacology office staff for assistance in various ways during the preparation of the book. I would also like to thank T. Gaffney for his support and encouragement in this work and T. Walle for helpful discussion and criticism. Most of all I would like to express my appreciation to Rebecca Knapp who was a partner in this project and without whose many hours spent in typing, editing, drawing formulas, and tending to the many other details, the book could not have been completed. I, however, accept all responsibility for any errors and would appreciate readers calling my attention to them.

DANIEL R. KNAPP

April
1979
Charleston, South Carolina

Contents

Abbreviations

abs	absolute (anhyd)
Abu	α-aminobutyric acid
Ac	acetyl
AcCl	acetyl chloride
Ac_2O	acetic anhydride
AcOH	acetic acid
ADMS	allyldimethylsilyl
ADMSI	allyldimethylsilyl imidazole
Ala	alanine
i-AmOH	isoamyl alcohol
n-AmOH	normal amyl alcohol
AMP	adenosine monophosphate
amt	amount
amu	atomic mass units (Daltons)
anhyd	anhydrous
aq	aqueous
Arg	arginine
Asp	aspartic acid
Asn	asparagine
BANS	5-dibutylaminonapthalene-1-sulfonyl
BDSA	bis(dimethylsilyl)acetamide
BHT	2,6-di-*t*-butylhydroxytoluene ("butylated hydroxytoluene")
BMDCS	bromomethyldimethylchlorosilane
BMDMS	bromomethyldimethylsilyl
BOC	butoxycarbonyl
BSTFA	bis(trimethylsilyl)trifluoroacetamide
BTFA	bis(trifluoracetamide)
Bu	butyl
BuOAc	butyl acetate
BuOH	butanol
t-BuOK	potassium *y*-butoxide
ca	*circa*, approximately
calcd	calculated
CDF	chlorodifluoroacetyl
CDI	*N*,*N*′-carbonyldiimidazole
CFA	*sym*-trichlorotrifluoroacetone
CHD	cyclohexanedione
CI	chemical ionization
Cit	citrulline
cm	centimeter
CMDMS	chloromethyldimethylsilyl
CMP	cytidine monophosphate
cont	continue, continued
cryst	crystal, crystallize, crystalline
CSF	cerebrospinal fluid

CySH	cysteine
Cys	cystine
DAG	dianhydrogalactitol
DANS	dansyl, 5-dimethylaminonaphthalene-1-sulfonyl
dansyl	see DANS
DCC	dicyclohexylcarbodiimide
DCTFA	1,3-dichloro-1,1,3,3-tetrafluoroacetone
DDTC	sodium diethyldithiocarbamate
DEA	diethylamine
deriv(s)	derivative, derivatives, derivatize
dil	dilute, diluted
diss	dissolve, dissolved
dist	distill, distilled
DMA	*N*,*N*-dimethylacetamide
DMAP	dimethylaminopyridine
DMB	dimethylaminobenzaldehyde
DMCS	dimethyldichlorosilane
DME	dimethoxyethane
DMES	dimethylethylsilyl
DMESI	dimethylethylsilyl imidazole
DMF	dimethylformamide
DMMCS	dimethylmonochlorosilane
DMP	dimethoxypropane
DMPS	dimethyl-*n*-propylsilyl
DMPSI	dimethyl-*n*-propylsilyl imidazole
DMS	dimethylsilyl
DMSO	dimethylsulfoxide
DNB-Cl	dinitrobenzoyl chloride
DNBS	dinitrobenzenesulfonic acid (or dinitrobenzenesulfonate)
DNCB	dinitrochlorobenzene
DNFB	dinitrofluorobenzene
DNP	dinitrophenyl
DNPH	dinitrophenylhydrazine
DNT	4-trifluoromethyl-2,6-dinitrophenyl
DNT-Cl	4-trifluoromethyl-2,6-dinitrochlorobenzene
DOPA	dihydroxyphenylalanine
DOPAC	dihydroxyphenylacetic acid
DTPA	diethylenetriamine pentaacetic acid
ECD	electron capture detection, detector
EDTA	ethylenediamine tetraacetic acid
EFA	ethoxyformic anhydride
EFU	eicosafluoroundecanoate
EI	electron impact
EO	ethoxime
equiv	equivalent
est	estimate, estimated, estimation
Et	ethyl
Et_3N	triethylamine
Et_2O	diethyl ether
EtOAc	ethyl acetate

EtOH	ethanol
ETSA	ethyl trimethylsilylacetate
evap	evaporate
ext	extract, extracted
FD	field desorption
fg	femptogram (10^{-15} gram)
FID	flame ionization detection
g	gram
GABA	γ-aminobutyric acid
GC	gas chromatography, gas chromatographic, gas chromatograph
GC-MS	gas chromatography-mass spectrometry gas chromatograph-mass spectrometer
Gln	glutamine
Glu	glutamic acid
Gly	glycine
GSH	glutathione
HCl	hydrochloric acid, hydrochloride, hydrogen chloride
HFAA	hexafluoracetylacetone
HFB	heptafluorobutyryl
HFBA	heptafluorobutyric anhydride
HFB-Cl	heptafluorobutyryl chloride
HFBI	heptafluorobutyrylimidazole
HFIP	hexafluoroisopropanol
HFN	hexadecafluorononanoate
His	histidine
HMDS	hexamethyldisilazane
HOLys	hydroxylysine
HOPro	hydroxyproline
HPLC	high performance liquid chromatography (chromatographic, chromatograph)
hr	hour
HSN	*N*-hydroxysuccinimido β-napthoate
HVA	homovanillic acid
id	inside diameter
IDA	iminodiacetic acid
IDMS	iodomethyldimethylsilyl
Ile	isoleucine
int stand	internal standard
IPOTMS	isopropenyloxytrimethylsilane
IR	infrared
μl	microliter (10^{-6} liter)
Leu	leucine
lt	light
Lys	lysine
m	meter
M	molar (gram molecular weight per liter)
MBTFA	*N*-methyl-bis(trifluoroacetamide)
Me	methyl (CH_3-)
MDPF	2-methoxy-2,4-diphenyl-3(2H)-furanone
Me_2O	dimethyl ether

MeOAc	methyl acetate
MeOH	methanol
MeOTFA	methyl trifluoroacetate
meq	milliequivalent
Met	methionine
mg	milligram
min	minute, minimum
mixt	mixture
ml	milliliter
mm	millimeter (mm Hg, Torr)
MN	metanephrine
MNT	4-trifluoromethyl-2-nitrophenyl
MNT-F	4-trifluoromethyl-2-nitrofluorobenzene
MO	methoxime
mod	modify, modified
mol	mole, gram molecular weight, molecular
μmol	micromole, 10^{-6} gram molecular weight
MOPEG	3-methoxy-4-hydroxyphenyl ethylene glycol
MS	mass spectrum, mass spectral, mass spectrometer
MSA	*N*-methyl-*N*-trimethylsilylacetamide
MSHFBA	*N*-methyl-*N*-trimethylsilylheptafluorobutyramide
MsOH	methanesulfonic acid
MSTFA	*N*-methyl-*N*-trimethylsilyltrifluoroacetamide
3-MT	3-methoxytyramine
MTFA	*N*-methyl-trifluoroacetamide
MTP-AcCl	α-methoxy-α-trifluoromethylphenylacetyl chloride
MU	methylene units
N	normal (gram-equivalent weight per liter)
NaOBu	sodium *n*-butoxide
NaOMe	sodium methoxide
NBD	4-nitrobenzo-2,1,3-oxadiazole
NBS	*N*-bromosuccinimide
ng	nanogram (10^{-9} gram)
nm	nanometer
NMN	normetanephrine
NMR	nuclear magnetic resonance
NTA	nitrilotriacetic acid
NTFB	2-nitro-4-trifluoromethylfluorobenzene
od	outside diameter
org	organic
Orn	ornithine
pet ether	petroleum ether
PFB	pentafluorobenzyl (not pentafluorbenzoyl)
PFBA	pentafluorobenzaldehyde
PFB-Br	pentafluorobenzyl bromide
PFBC	pentafluorobenzyl chloroformate
PFBHA	pentafluorobenzyl hydroxylamine
PFO	pentadecafluorooctanoyl
PFOA	pentadecafluorooctanoic anhydride
PFP	pentafluoropropionyl

PFPA	pentafluoropropionic anhydride
PFPI	pentafluoropropionylimidazole
PFPT	pentafluoropropionyltriazole
pg	picogram (10^{-12} gram)
PG	prostaglandin
PGE_2	prostaglandin E_2
Phe	phenylalanine
PHPA	*p*-hydroxyphenylacetic acid
pmol	picomole (10^{-6} gram molecular weight)
PNB	*p*-nitrobenzyl (not *p*-nitrobenzoyl)
PNBA	*p*-nitrobenzyloxyamine
PNBDI	*O-p*-nitrobenzyl-*N,N'*-diisopropyl isourea
PNBTT	1-*p*-nitrobenzyl-3-*p*-tolyltriazene
ppb	parts per billion (10^9)
ppm	parts per million (10^6)
ppt	precipitate
prep	prepare, preparative
press	pressure
Pro	proline
$(i\text{-Pr})_2O$	di-isopropyl ether
i-PrOH	isopropyl alcohol (2-propanol)
n-PrOH	*n*-propyl alcohol (1-propanol)
q.v.	*quode vide*, which see
quant	quantitative, quantity
RaNi	Raney nickel
red	reduce, reduced
redist	redistill, redistilled
rel S.D.	relative standard deviation
rxn	reaction
satd	saturated
SCOT	surface coated open tubular
SD	standard deviation
SE	standard error
sep	separate, separatory
Ser	serine
SIM	Selected ion monitoring (multiple ion detection, mass fragmentography, etc.)
SNPA	*N*-succinimidyl-*p*-nitrophenylacetate
soln	solution
TBAF	tetrabutylammonium fluoride
TBAH	tetrabutylammonium hydroxide
$TBA\text{-}HSO_4$	tetrabutylammonium hydrogen sulfate
TBDMS	*t*-butyldimethylsilyl
TBDMS-Cl	*t*-butyldimethylchlorosilane
TCE	trichloroethanol
TEAH	tetraethylammonium hydroxide
temp	temperature
TFA	trifluoroacetyl, trifluoroacetic acid
TFAA	trifluoroacetic anhydride
TFAI	trifluoroacetylimidazole

THAH	tetrahexylammonium hydroxide
THA-HSO_4	tetrahexylammonium hydrogen sulfate
THASS	tetrahexylammonium sodium sulfate
THC	tetrahydrocannabinol
THF	tetrahydrofuran
Thr	threonine
TLC	thin layer chromatography
TMA	trimethylamine
TMAH	tetramethylammonium (*not* trimethylanilinium) hydroxide
TMBAH	trimethylbenzylammonium hydroxide
TMPAH	trimethylphenyl ammonium hydroxide (trimethylanilinium hydroxide)
TMBS	trimethylbromosilane
TMCS	trimethylchlorosilane
TMDS	tetramethydisilazane
TMIPS	cyclotetramethyleneisopropylsilyl
TMIPS-Cl	cyclotetramethyleneisopropylchlorosilane
TMS	trimethylsilyl, $(CH_3)_3Si$-
TMSDEA	trimethylsilyldiethylamine
TMSDMA	trimethylsilyldimethylamine
TMSI	trimethylsilylimidazole (Note: TSIM is often seen used as abbreviation)
TMSM	trimethylsilylmorphine
TMSPI	trimethylsilylpiperidine
TMSPY	trimethylsilylpyrrolidine
TMTBS	cyclotetramethylene-*t*-butylsilyl
TMTBS-Cl	cyclotetramethylene-*t*-butylchlorosilane
TNBS	trinitrobenzenesulfonic acid
TNP	trinitrophenyl
TPAH	tetrapentylammonium hydroxide
TPhA	tetraphenylammonium
TPrAH	tetrapropylammonium hydroxide
Trp	tryptophan
TSIM	see TMSI
TsCl	*p*-toluenesulfonyl chloride
TsOH	*p*-toluenesulfonic acid
Tyr	tyrosine
UMP	uridine monophosphate
unsatd	unsaturated
UV	ultraviolet
vac	vacuum
Val	valine
vol	volume
VMA	vanillylmandelic acid
v/v	volume/volume
w/w	weight/weight
μ (Greek mu)	micro (prefix), 10^{-6}

Organization of the Book and Suggestions for Use

Information in this book is organized into three major components: a general introduction to the subject (Part I), specific derivatization methods (Part II), and indexes and appendices (Part III). Part I presents an overview of analytical derivatization in general, including a section on apparatus. The reasons for using analytical derivatization and the major classes of derivatization reactions and reagents are reviewed. References are given for the reader who wishes to pursue individual subjects in more depth.

Part II, the major emphasis of the book, is a collecton of analytical derivatization methods organized in sections according to the nature of the sample to be derivatized. The reader is referred to the table of contents for the sample types covered in the sections. Each section begins with an overview of derivatization of the particular type of sample, with the remainder of the section consisting of descriptions of analytical derivatization methods that have been applied to the given sample type. Each method description follows a standard format:

Derivative Type
Reagent
Sample Type
Reaction
Typical Procedure
Comments
References

For methods that utilize more than one derivatizing group, e.g., *N*-acetyl, methyl esters of amino acids, the derivative types are listed in the order of attachment of the derivative groups. Reagents are listed in the order of use with those used in different steps separated by commas and multiple reagents within a step separated by shilling marks (/). The sample type is given followed by structural formulas for the derivatization reaction. Structural formulas are given for all reactions. To have done otherwise would have required a decision of where to draw the line on triviality, a difficult task in a book intended for a wide range of users. The specific examples shown in the structural formulas were selected wherever possible to illustrate the presence of other functional groups in addition to those being derivatized. These illustrations of nonaffected functional groups are intended to give additional information on the scope of the reactions. In the procedure section, experimental details of the derivatization method, as extracted from the original literature, are presented. Abbreviations are used extensively, as in the experimental sections of many journals, to conserve space and, hence, cost. A complete list of abbreviations appears at the front of the book. In many cases, the procedure includes the isolation of the analyzed compound from the sample matrix since the isolation is often closely linked to the derivatization. Following the method description is a list of references to the original literature sources. In all cases the first reference is the principal source for the given method, although it may not actually be cited by number in the text. References with more than two authors are abbreviated with *et al.* All of the author names appear in the index, however.

Part III contains the appendices and indexes. The indexes are an integral and an indispensible part of this book. Many of the derivatization methods have been applied to multiple sample types, but often the references to these additional applications appear only in the comments sections. Furthermore, many derivatized compounds contain multiple functional groups making any classification scheme somewhat arbitrary. The result is that information on derivatization of a given functional group or compound class may appear in sections other than the primary section for that compound class. Therefore, when seeking information about a derivatization method, the reader should consult, in addition to the primary sample type section, both the Derivative and Reagent Index and the General Subject Index. Note, however, that the General Subject Index lists only those references to derivative types that appear in the introductory and overview sections since the references to derivative types for the methods sections are covered in the Derivative and Reagent Index. In summary, retrieval of information from this book may be approached systematically in several ways: (1) via the Table of Contents and the General Subject Index for information on derivatization of a given *sample*, (2) via the Derivative and Reagent Index and the General Subject Index for information on a type of *derivative* or a given *reagent*, or (3) via the Author Index for information on the work of a given *author*. It is suggested also that the user take time to simply peruse the book, for in doing so he is likely to happen upon ideas for new approaches, which may likely not be found solely by systematic reference.

Handbook of

Analytical Derivatization Reactions

Part I. Introduction

Uses of Analytical Derivatization

Analytical derivatization finds application in a wide range of disciplines. In addition to the traditional areas of chemistry and biochemistry, derivatization methods are an integral part of much of the chemical analysis practiced in the medical, forensic, and environmental. sciences to name but a few. Derivatization has enabled the use of analytical methods such as gas chromatography (GC) and mass spectrometry (MS) on samples not otherwise amenable to these techniques and has thus extended the application of these powerful tools to new fields. In studies ranging from analysis of cellular components to examination of lunar samples, analytical derivatization plays a key role.

In spite of numerous applications of analytical derivatizations, few books have been devoted to the subject. One of these, *Silylation of Organic Compounds* by Pierce,[1] was a significant factor in the transfer of silylation reactions from the relatively esoteric field of organosilicon chemistry to the status of perhaps the most widely practiced of derivatization methods. Only quite recently have two other books appeared: one covering derivatives for chromatography,[2] and another dealing specifically with derivatives for liquid chromatography.[3] Derivatization for gas[4–6] and liquid[7–10] chromatography have also been the subject of review papers. This book extends the coverage to include derivatives for mass spectrometry as well as gas and liquid chromatography.

In a broad sense, analytical derivatization is employed for two reasons: (1) to permit analysis of compounds not directly amenable to analysis due to, for example, inadequate volatility or stability, and (2) to improve the analysis by, for example, improving chromatographic behavior or detectability. Beyond this, the specific reasons for derivatization differ with the analytical technique in use.

Derivatives for GC Analysis

The foremost reason for employing analytical derivatization in gas chromatographic analysis is to impart volatility to otherwise nonvolatile compounds. Low volatility of a compound may be due to sheer size of the molecules and the resultant large dispersion forces holding the molecules together. In smaller molecules low volatility may result from strong intermolecular attractions between polar groups. In the former case derivatization cannot increase volatility, but in the latter case masking of polar groups by derivatization can yield dramatic increases in volatility. Polar N-H, O-H, and S-H groups that can undergo hydrogen bonding contribute significantly to intermolecular attraction. Replacement of hydrogen in these groups by alkylation, acylation, or silylation significantly increases volatility, especially in compounds with multiple polar groups. The monosaccharides are a prime example of a relatively low molecular weight group of compounds that exhibit little volatility even up to temperatures at which they begin to decompose. Replacement of the active hydrogens with trimethylsilyl groups gives volatile products that readily undergo GC analysis.

In some cases, such as the analysis of very low molecular weight amines, excessive volatility is a problem. Chemical derivatization may then be used to yield less volatile derivatives to minimize losses in manipulation and to help separate the gas chromatographic sample peaks from the solvent front.

Some sensitive compounds, although able to be volatilized, undergo partial thermal decomposition in the gas chromatograph. Derivatization may be employed to yield a more stable product and thereby improve chromatographic performance and peak shape. Unsymmetrical peaks may also result from adsorption effects in the column. Polar samples tend to adsorb on active surfaces of both the column walls and the solid support. Reduction of these adsorption effects accounts for a large proportion of the applications of derivatization in gas chromatographic analysis.

Closely related compounds that are poorly separated by gas chromatography may yield much improved separations by use of an appropriate derivative. For example, chloromethyldimethylsilyl derivatives gave much improved separations of isomeric dihydroxynaphthalenes (p. 59). Optical isomers may be separated by gas chromatography as diastereomeric derivatives. The derivatization of enantiomeric samples with chiral derivatizing agents is discussed in Section 9. Derivatization may also be employed to effect group separation of complex mixtures. In steroid analysis formation of benzyloximes allows separation of ketosteroids from hydroxy steroids by shifting the ketosteroid derivatives to much longer retention times (p. 485). In all these cases, the derivatization serves to accentuate the differences in the sample compounds to facilitate the chromatographic separation.

Derivatization for Detectability Enhancement in GC Analysis

Derivatization may be employed to confer improved detectability to sample compounds. Electron capture detection (ECD) is an extremely sensitive detection technique, but relatively few compounds are sensitive to ECD in their native state. Halogenated derivatives in particular have been used to confer ECD sensitivity for trace level analysis. The most commonly employed ECD derivatizations utilize halogenated acyl groups, but substituted alkyl, aryl, silyl, and other groups have also been employed. Poole has surveyed the types of derivatives used for ECD-GC.[11]

In general the halogen substituents increase electron affinity in the order I > Br > Cl > F. Nitro groups are similar in effect to Cl. Multiple subsitutions upon the same carbon have a synergistic effect. Aromatic ring substituents affect ECD sensitivity in accordance with their ability to withdraw electrons from the aromatic system; electron donating substituents, particularly those acting via resonance effects, tend to reduce electron affinity. To explain the ECD response of various halogenated acetate derivatives, Landowne and Lipsky[12] proposed that electron capture initially occurred on the carbonyl group. Clarke et al.,[13] on the other hand, reasoned that the increased sensitivity of heptafluorobutyryl derivatives compared to trifluoroacetyl indicated that electron capture occurred on the halogenated alkyl chain. Matin and Rowland[14] studied ECD derivatives of amines and showed that a double bond (either C=O or C=N) adjacent to the halogenated group was necessary for optimal ECD response. These results lend support to the Landowne and Lipsky[12] theory.

Tables 1 and 2 list comparative ECD responses of halogenated derivatives of hydroxyl compounds and amines. Bromo and iodo compounds are not included in these listings because practical considerations of volatility and peak shape render them of little use. The trends in increasing sensitivity of the derivatives are accompanied by increasing molecular weights and resultant decreasing volatility. The choice of an optimal derivative is thus limited by volatility requirements.

Eaborn *et al.* introduced the use of halomethyldimethylsilyl ethers for ECD analysis.[22, 23] The sensitivity of the derivatives increases in the order I > Br > Cl > F. For comparison to the haloacyl derivatives, the chloromethyldimethylsilyl derivatives are comparable to chloroacetyl derivatives, and the iodomethyldimethylsilyl derivatives are comparable to the heptafluorobutyryl derivatives. The volatility of the derivatives is in the reverse order and limits the utility of the more sensitive derivatives. Morgan and Poole[24,25] have examined series of silyl derivatives with larger fluorinated substituents (Table 3). The pentafluc phenyldimethylsilyl (flophemesyl) derivatives appear to be particularly promising, but increased sensitivity is accompanied by a significant decrease in volatility.

Other halogenated derivatives for ECD include pentafluorobenzy pentafluorobenzyloximes, and pentafluorophenylhydrazones. For a discuss theoretical aspects of the ECD, the reader is referred to a review by Pelizarri.[26] D for sensitivity enhancement has also been employed in conjunction wi detection. Methods have been described primarily for introduction

Table 1. Comparative Electron Capture Response of Some Derivatives of Alcohols and Phenols

Derivative	Compound			
	Thymol[a]	Cholesterol[b]	Testosterone[b]	Testosterone[c]
Acetyl	—	—	—	0.1
Monofluoroacetyl	0.007	—	—	—
Monochloroacetyl	0.3	—	0.20	4.0
Chlorodifluoroacetyl	—	2.08	1.41	34.
Trifluoroacetyl	—	0.09	0.01	0.4
Pentafluoropropionyl	1.3	—	0.28	5.
Heptafluorobutyryl	1.0	1.00	1.00	19.
Diheptafluorobutyryl	—	—	2.30	—
Perfluorooctanoyl	—	1.85	2.10	60.
Pentafluorobenzoyl	6.9	—	—	50.
Pentafluorobenzyl	5.9	—	—	—
Dinitrophenyl	0.3	—	—	—

[a]Data from Reference 15.
[b]Data from Reference 16.
[c]Data from Reference 17.

containing groups, but it has also been shown that boronate derivatives are sensitive to alkali flame detection.[27]

Derivatives for MS Analysis

As with GC volatility enhancement is also a consideration in derivatization for MS analysis. Formation of chemical derivatives permits analysis of compounds not otherwise amenable even to direct probe analysis. Chemical derivatization may also be employed to reduce the contribution of thermal or catalytic decomposition to the mass spectrum.

The use of chemical derivatives to assist in structure determination is an important tool in organic mass spectrometry. Numbers and types of functional groups present in a compound may often be inferred from mass shifts observed after formation of suitable derivatives. The use of both unlabeled and heavy isotope labeled derivatives gives further information. Derivatives formed with mixtures of labeled and unlabeled reagents or mixtures of homologous reagents yield ion clusters that facilitate identification of certain compound types within a complex mixture or certain fragments within a compound.

Derivatives are also used to stabilize ions formed in the mass spectrometer to favor structurally informative fragmentation modes. This approach has been extensively studied in relation to peptide sequence analysis and in relation to location of unsaturated or otherwise substituted sites in long alkyl chains. The increasing use of quantitative GC-MS in conjunction with selected on monitoring has led to development of derivatization reactions yielding products that form intense high mass ions. By use of such derivatives, the resulting concentration of ion current into fewer peaks increases sensitivity, whereas shifting the monitored peaks to the less background obscured higher mass region enhances selectivity. Derivatization methods are therefore useful for both qualitative and quantitative mass spectrometry.

Table 2. Comparative Electron Capture Responses of Acyl, Schiff Base, and Alkyl Derivatives of Some Amines

Derivative	Compound					
	Amphetamine[a]	Amphetamine[b]	Benzylamine[c]	Phenethylamine[d]	Aniline[e]	Cyclohexylamine[e]
Acyl						
Trifluoroacetyl	<0.1		1.0			
Monochloroacetyl	0.1		37.5			
Trichloroacetyl	54					
Pentafluoropropionyl	4		286			
Heptafluorobutyryl	9	2.2	894		670	1
Perfluorooctanoyl	23					
Pentafluorophenylacetyl				11		
Pentafluorobenzoyl	77	225		78	133	500
p-Trifluoromethylbenzoyl	12					
p-Nitrobenzoyl		11				
Schiff Base						
Heptafluorobutylidene		2.1				
Pentafluorobenzylidene		43		100		
Alkyl						
Heptafluorobutyl		0.41				
Pentafluorobenzyl		3.7		4.1		
Dipentafluorobenzyl				34		
Dinitrophenyl		2.8				

[a]Data from Reference 18.
[b]Data from Reference 14.
[c]Data from Reference 19.
[d]Data from Reference 20.
[e]Data from Reference 21.

Table 3. Comparison of the Relative Volatility and ECD Sensitivity of a Series of R—$Si(CH_3)_2^-$ Silyl Derivatives of Cholesterol[a]

R=	Relative Retention Time[b]	Least Detectable *Amount (ng)*
CF_3–	1.17	—
$CF_3(CH_2)_2$–	1.47	1500
$CF_3(CF_2)_2(CH_2)_2$–	1.60	115
$ClCH_2$–	2.50	75
C_6F_5–	3.68	4
$C_6F_5CH_2CH_2$–	—	200

[a]Data from Reference 11.
[b]1% OV–101, 250°.

Derivatives for Liquid Chromatographic Analysis

Although high performance liquid chromatography has undergone explosive growth in recent years, a major bottleneck in the use of this technique has been the limited sensitivity of available detection methods. Ultraviolet detection has been most widely applied, but many compounds have only very weak ultraviolet (UV) absorption. Increasing use is being made of chromogenic derivatization to impart or increase UV detection sensitivity. With increasing availability of suitable fluorometric detectors, fluorogenic derivatization is also becoming increasingly applied.

Detection enhancement is by far the most common use of derivatization in liquid chromatography, but derivatives may also be used to alter separation properties. Ion pair derivatives are an important example of such use, but, not being covalent derivatives, are outside the scope of this book. Examples are presented, however, of derivatives used for liquid chromatographic separation of enantiomers.

Derivatization may assist the analyst in other ways as well. Several examples are cited that employ acetylation of sample compounds in aqueous solution thereby assisting their isolation by solvent extraction in addition to forming a useful derivative for GC.

Derivative Types and Reagents

A variety of reactions are employed in the preparation of derivatives, but most of the resulting products may be classified into a few major categories. Replacement of active hydrogen in polar groups (OH, SH, NH) accounts for a large portion of analytical derivatization and involves principally alkylation, acylation, silylation, and condensation (defined as a reaction in which two molecules are joined with elimination of water) reactions. These derivative types and other derivatization reactions are discussed below.

Alkyl Derivatives

Alkylation of OH (including COOH), SH, and NH (including (CONH) groups yield less polar derivatives by replacement of the active hydrogens with alkyl groups. The principal reaction employed for preparation of these derivatives is nucleophilic displacement of the type shown in equation 1, where X is a halogen or other good leaving group.

$$\text{Sample-OH} \xrightarrow[\text{base}]{\text{R'-X}} \text{Sample-O-R'} + \text{HX} \qquad (1)$$

The resulting products are ethers, esters, thioethers, thiolesters, N-alkyl amines, and N-alkyl amides. Alkylation of weakly acidic groups requires strongly basic catalysts; methylsulfinyl carbanion, for example, is employed in permethylation of compounds containing alcohol OH and amide NH groups. More acidic OH groups, as in phenols and carboxylic acids, require less strongly basic catalysts. The reagents employed are principally alkyl bromides or iodides, but nitro substituted chloro- or fluorobenzenes are used to prepare nitrophenyl derivatives.

Catalytic methods have been utilized in analytical alkylation of acidic compounds to render the anions of the compounds soluble in nonprotic organic solvents. The "naked" anions that result in these solvents are extremely reactive toward alkyl halide reagents enabling rapid, high yield derivatization. Crown ethers yield an organic soluble complex of a cation by binding the ion in the center of the ring (see Reference 28 for a review of the chemistry of crown ethers). These complexing agents have been used for analytical derivatization of carboxylic acids (equation 2). Anions may be also be rendered soluble in

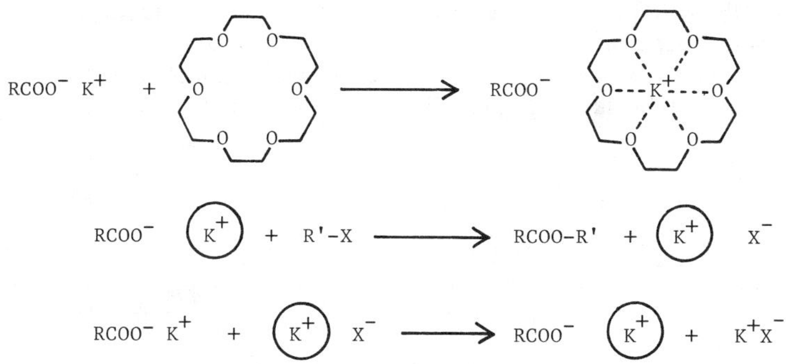

(2)

nonprotic organic solvents by forming salts with tetraalkylammonium ions. Phase transfer alkylation employing such quaternary ammonium salts has been reviewed.[29–31] Equation 3 illustrates the phase transfer alkylation of a phenol (ArOH) with an alkyl halide (R′X). This approach, which enables simultaneous extraction and analytical derivatization of the sample compound, has also been referred to as "extractive alkylation."

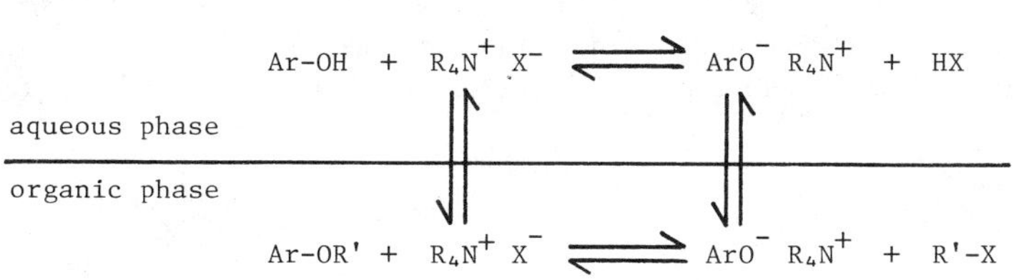

(3)

Other alkylating agents commonly used for acidic OH groups include the tetraalkylammonium hydroxide reagents (equation 4), the dimethylformamide dialkyl acetals (equation 5), and the diazoalkanes (equation 6).

The tetraalkylammonium reagents form salts with acidic groups; pyrolysis in the hot GC injector yields the alkylated product plus a tertiary amine. Substitution of phenyl for one of the four alkyl groups yields a reagent with a better leaving group (a dialkylaniline) permitting use of lower pyrolysis temperatures. An electron withdrawing group (e.g. CF_3) on the ring

$$\text{Sample-OH} \xrightarrow[\Delta]{R_4N^+} \text{Sample-OR} + R_3N \qquad (4)$$

$$\text{Sample-OH} \xrightarrow{(CH_3)_2N\text{-}CH(OR)_2} \text{Sample-OR} + \text{ROH} + \text{DMF} \qquad (5)$$

$$\text{Sample-OH} \xrightarrow{CH_2N_2} \text{Sample-OCH}_3 + N_2 \qquad (6)$$

yields a still better leaving group. The tetraalkylammonium reagents will alkylate carboxyls, phenolic hydroxyls, and acidic NH groups such as in barbiturates.

The dimethylformamide dialkyl acetals are readily available with alkyl groups up to butyl. These reagents are primarily used for alkylation (esterfication) of carboxyl groups. Diazoalkanes likewise are used principally for esterfications although phenolic groups are alkylated by these reagents at a slower rate. With Lewis acid catalysts, diazoalkanes will alkylate the less acidic alcohol hydroxyls.

Less commonly used alkylation reagents include the 3-alkyl-p-tolyltriazenes (equation 7) and O-alkyl isourea reagents (equation 8). The latter reagents are prepared by allowing one mole of alcohol to react with a carbodiimide. The triazenes are applicable to both carboxylic acids and phenols, but the isourea reagents have only been demonstrated on carboxylic acids.

$$\text{Sample-COOH} + \text{R-NH-N=N-}C_6H_5CH_3 \longrightarrow \text{Sample-COOR} + N_2 + H_2N\text{-}C_6H_5CH_3 \qquad (7)$$

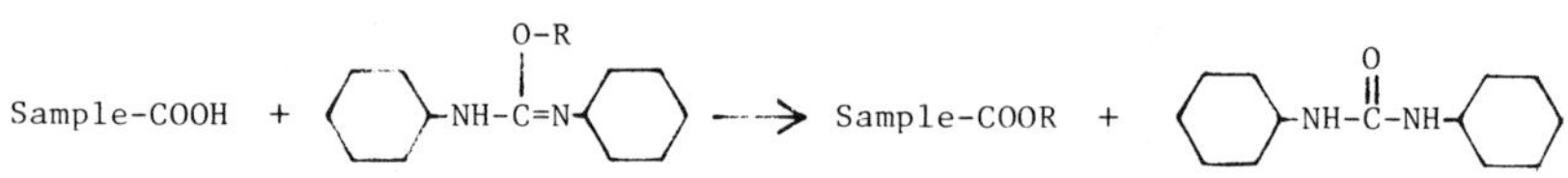
(8)

Safety Note. Any potent alkylating reagent should be handled with care and treated as a potential mutagen or carcinogen. Of the reagents discussed, the diazoalkanes and the triazenes especially deserve particular respect due to their toxic potential.

Silyl Derivatives

Silyl derivatives are also formed by replacement of active hydrogen on OH, SH, and NH groups. The general reaction for silylation is shown in equation 9. Trimethylsilyl derivatives are most commonly employed but other substituted silyl derivatives have been prepared for specific purposes.

$$\text{Sample-OH} + R_3Si\text{-X} \longrightarrow \text{Sample-O-Si}R_3 + \text{HX} \qquad (9)$$

The mechanism appears to involve nucleophilic attack upon silicon (equation 10).[32] Thus the properties desired of X, the leaving group, for a good silylating reagent are low basicity, ability to stabilize a negative charge in the transition state, and little or no (p→d) π back-bonding between X and the silicon atom.[33]

The early work involving formation of trimethysilyl derivatives utilized trimethylchlorosilane and hexamethyldisilazane. These reagents are still used, but a variety of newer reagents have been developed. The more common trimethylsilylating reagents are listed in Table 4 in the approximate order of increasing silyl donor ability. TMSI and the bis (amide)

Table 4. Common Trimethylsilylating Reagents Listed in Order of Increasing Silyl Donor Strength

Reagent	Abbreviation	Structure
Hexamethyldisilazane	HMDS	$(CH_3)_3Si\text{-}N(H)\text{-}Si(CH_3)_3$
Trimethylchlorosilane (with base catalyst)	TMCS	$(CH_3)_3Si\text{-}Cl$
N-Methyl-*N*-trimethylsilyl-acetamide	MSA	$CH_3\text{-}C(=O)\text{-}N(CH_3)\text{-}Si(CH_3)_3$
N-Trimethylsilyldiethylamine	TMSDEA	$(CH_3)_3Si\text{-}N(C_2H_5)_2$
N-Trimethylsilyldimethylamine	TMSDMA	$(CH_3)_3Si\text{-}N(CH_3)_2$
N-Methyl-*N*-trimethylsilyl-trifluoroacetamide	MSTFA	$CF_3\text{-}C(=O)\text{-}N(CH_3)\text{-}Si(CH_3)_3$
N,*O*-Bis(trimethylsilyl)-acetamide	BSA	$CH_3\text{-}C(O\text{-}Si(CH_3)_3)=N\text{-}Si(CH_3)_3$
N,*O*-Bis(trimethylsilyl)-trifluoroacetamide	BSTFA	$CF_3\text{-}C(O\text{-}Si(CH_3)_3)=N\text{-}Si(CH_3)_3$
N-Trimethylsilylimidazole	TMSI	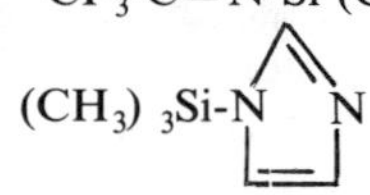

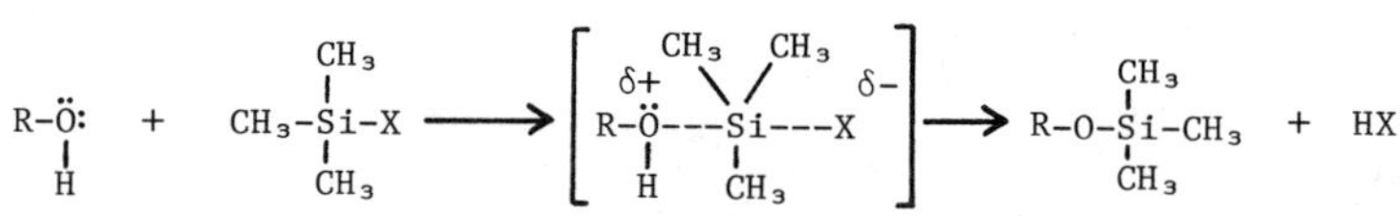

(10)

reagents are particularly good donors due to the ability of the leaving groups, imidazole, acetamide, and trifluoroacetamide, to stabilize through resonance the partial negative charge in the transition state. TMSI is unusual in that, unlike the somewhat less potent BSA and BSTFA, it reacts readily with hydroxyls but not at all with aliphatic amines. It does, however, silylate indole nitrogen. The silyl donor potencies of the reagents are markedly influenced by the solvent systems and added catalysts. TMCS if often added to the other reagents to increase the silyl donor strength. Basic catalysts such as potassium acetate have been used to promote silyl enol ether formation. The mixture BSA/TMSI/TMCS (1:1:1, v/v/v) appears to be the most potent general silylating agent that has been described.[34] The ease of reactivity of the functional groups toward silylation follows the order:

alcohol hydroxyl > phenol hydroxyl > carboxyl > amine > amide

Within each group, reactivity appears to be primarily a function of steric hindrance with the order for alcohols being:

primary > secondary > tertiary

Silyl derivatives in general tend to be sensitive to moisture and are normally prepared in sealed vessels immediately prior to use. Methods have been described, however, where silyl derivatives were isolated by solvent extraction with aqueous washing. Thus all silyl derivatives, particularly of alcohols, are not as hydrolytically labile as is commonly assumed.

In addition to the extensively utilized trimethylsilyl derivatives, other alkyl substituted silyl groups have been studied. Dimethyl derivatives are more volatile than trimethysilyl derivatives and may offer some advantages for analysis of polyfunctional compounds. The dimethylsilyl derivatives, however, tend to be less stable toward hydrolysis and the reactive Si-H can lead to extraneous reactions. Dimethylethyl- and dimethylpropylysilyl derivatives have been examined as a means of determining the number of hydroxyls in a steroid from retention index shifts (p. 46). Halogenated silyl groups were discussed in the section on derivatization for ECD analysis.

The *t*-butyldimethylsilyl group was first employed as a more water stable alternative to the trimethylsilyl group to be used as a protecting group for synthesis.[35] The *t*-butyldimethylsilyl derivatives were subsequently found to have useful mass spectral properties in that they tended to give intense high mass ions due to facile loss of the *t*-butyl radical. Quilliam and Westmore have examined a series of other "sterically crowded" trialkylsilyl ethers (p. 465). The cyclotetramethylene-*t*-butylsilyl and other similar derivatives gave improved separations of epimeric steroids and also exhibited useful mass spectral properties. Allyldimethylsilyl derivatives have been examined but appear to be of limited utility.

Silyl derivatives have the undesirable property of fouling flame ionization detectors with silica deposits, especially when samples are injected in excess reagent. These deposits are reduced by use of fluorine containing reagents such as BSTFA. *N*-Methyl-*N*-trimethylsilyl-heptafluorobutyramide (MSHFBA) was prepared as a reagent incorporating higher fluorine content and was claimed to cause less fouling of the detector (p. 58).

Acyl Derivatives

Acylation is another commonly used derivatization for polar OH, SH, and NH groups. Acyl derivatives are formed with acyl anhydride, acyl halide, and activated acyl amide reagents. The anhydrides and acyl halide reagents form acid by-products, which must be removed in GC analysis to present destructive effects on the column. Acylations with anhydride reagents are normally performed in pyridine, tetrahydrofuran, or another solvent capable of accepting the acid by-product. Amine bases may also be used as catalysts/acid acceptors. A few analytical applications of the potent acylation catalyst dimethylaminopyridine have been reported. This catalyst appears to be very useful and deserves further examination.

The activated amide reagents such as trifluoroacetylimidazole (TFAI) or *N*-methyl-bis (trifluoroacetamide) (MBTFA) have the advantage of yielding no acid by-products giving only imidazole and *N*-methyltrifluoroacetamide respectively. The novel acylating agent TFA Nylon 66 (p. 77) is an activated amide reagent in which the by-product is an insoluble solid. This reagent could be the harbinger of further useful applications of solid phase chemistry in analytical derivatization reactions.

The acetyl group is the principal nonhalogenated acyl group employed analytically. Higher acyl groups give increasingly less volatile and usually less useful derivatives, but have been employed in some cases where a derivative is needed to shift the gas chromatography peak to a longer retention time to avoid interferences. Fluorinated acyl groups, in contrast, exhibit little increase in volatility in going from trifluoracetyl to heptafluorobutyryl. The use of the halogenated acyl groups was discussed under ECD derivatives. The acyl derivatives in general exhibit good GC behavior and account for a large part of the derivatization methods in use. Acylations have also been utilized to form fragmentation directing derivatives for MS and chromogenic derivatives for high performance liquid chromatography.

Derivatives from Condensation Reactions

Condensations have been defined to include those reactions in which two molecules are joined with loss of water. Under this definition, the esterfication of carboxylic acids with alcohols is included. Esterfication with alcohols has been performed on the analytical scale using both Brönsted acid (e.g., HCl) and Lewis acid (e.g., BF_3) catalysts. Difunctional alcohols have been esterfied with alkane boronic acids to form cyclic boronate derivatives (equation 11). Boronate derivatives may also be formed with other suitably positioned functional groups (equation 12).

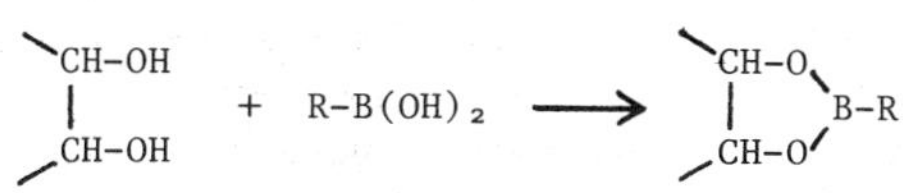

(11)

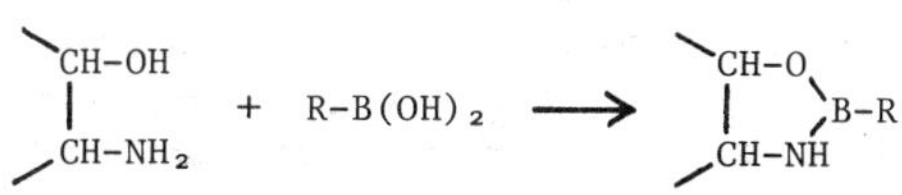

(12)

Condensations with carbonyl groups are another important subset of analytical condensations. Oxime (equation 13) and hydrazone (equation 14) derivatives have been formed for both gas and liquid chromatographic analysis. Pentafluorobenzyloximes are among the most sensitive ECD derivatives known. Condensation of carbonyl groups with primary amines give Schiff base derivatives (equation 15). In most cases these derivatives have been formed with carbonyl compounds as reagents to give volatile derivatives for GC. A similar condensation of primary amines with the "thiocarbonyl" compound carbon disulfide yields

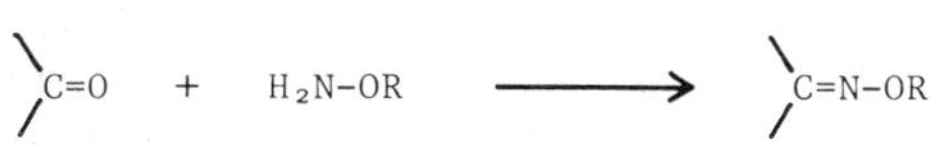

(13)

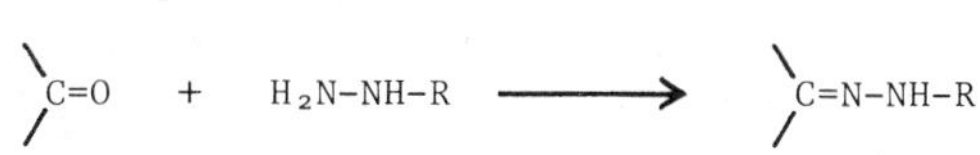

(14)

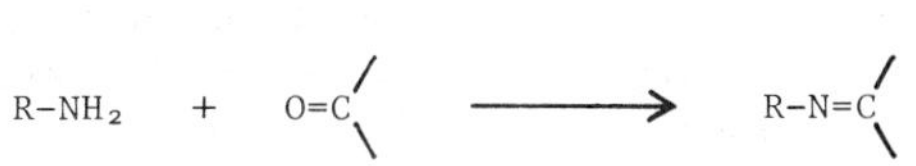

(15)

isothiocyanate derivatives (equation 16). These derivatives have received considerable study for both GC and MS analysis of primary amines. Amine-carbonyl condensation may also be employed to form cyclic derivatives (equation 17).

$$\text{R-NH}_2 \quad + \quad \text{S=C=S} \longrightarrow \text{R-N=C=S} \tag{16}$$

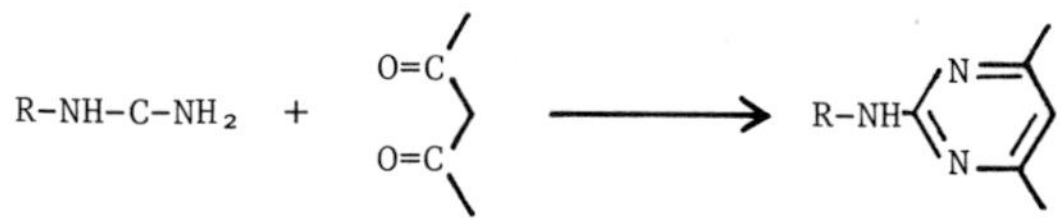

(17)

Other Derivatives

A large number of other organic reactions have been employed on the micro scale in selected cases to produce analytical derivatives. These include formation of such products as phosphoryl and sulfonyl derivatives, ketals, ureas, oxazolidine, oxazolidinones, and many others. The reader is referred to the Derivatives and Reagents Index for specific examples.

General Considerations in Analytical Derivatization

Analytical derivatization reactions are a form of synthetic organic chemistry in which the goal is a 100% yield of a single pure product. Although this goal is rarely achieved in macro scale organic synthesis, the use of high purity highly reactive reagents, often in large excess, makes many microscale analytical derivatization reactions essentially quantitative. For those derivatization reactions that are not quantitative, it is important for consistent results to know the rate of the reaction under the conditions being used. The use of an internal standard compound with similar reactivity is extremely advantageous. A properly selected standard, i.e., as similar as possible to the sample compound in physical and chemical behavior, added to the sample at the beginning of the analysis enables correction for losses in all steps of the analysis. An ideal solution is the use of a heavy isotope labeled variant of the sample compound in conjunction with analysis by GC-MS with selected ion monitoring.

The reagent qualities of high reactivity and very high purity are a difficult combination, since highly reactive reagents tend to react readily with extraneous materials and otherwise decompose. The analyst must therefore take care to insure the purity of the reagents and of the solvents employed. Although most reagents are less expensive in bulk, such purchase is false economy if a large part of the quantity is rendered useless due to deterioration. Small quantities of reagents, packaged in sealed ampules, for example, offer the most convenient solution to the need for pure reagents. Such packages may be purchased or prepared by the user but are without a doubt the best way to store and use reactive derivatization reagents. The highest quality solvents should be used for analytical derivatizations and the quantities used held to the minimum. Repeated evaporations tend to concentrate even trace impurities. The reader is reminded that "spectral grade" solvents are not necessarily the most pure since they may contain significant amounts of spectrophotometrically invisible contaminants.

Reagents used in excess must often be removed prior to the analysis. This is especially crucial in ECD or chromogenic derivatization where the excess reagents may "swamp" the detection system. Volatile reagents may be removed simply by evaporation but keeping in mind that losses of volatile derivatives may also occur. Solvent extraction of excess reagent may also be employed provided the derivative is stable to the extraction conditions. A useful strategy for organic soluble derivatives is to add an excess of another difunctional substrate that yields an aqueous extractable derivative. For example, in a derivatization method for carboxylic acids, excess pentafluorobenzyl bromide reagent was removed by reaction with an aminophenol (p. 184). The products formed with excess reagent, as well as excess aminophenol, were removed by extraction into aqueous acid.

Apparatus for Analytical Derivatization

The need for manipulation of small samples in analytical derivatization work has led to the development of specialized apparatus for this purpose, as well as the adaptation of apparatus originally intended for other uses. Analytical derivatizations require suitable containers for the samples and reagents; reaction vessels for the derivatization reactions; means of transferring and measuring samples and reagents; and means of heating, mixing, and evaporating the samples. This section describes some of the apparatus used and some of the practical aspects of analytical derivatization. Names and addresses of companies supplying apparatus for analytical derivatization are given in Appendix B.

Sample Containers and Reaction Vessels

The small samples commonly used in analytical derivatizations require proper containers for convenient manipulations without excessive losses. The vessels in which the derivatization reactions are carried out are often also the containers in which the derivative samples are transported for analysis. Reaction vials such as those shown in Figure 1 have been widely accepted as sample containers and analytical derivatization reaction vessels. These vials, which are available from several suppliers, come in sizes ranging from 0.1 to 10.0 ml, and all feature a tapered inside to allow access to even a few microliters of sample solution via a syringe needle. The vials are supplied with open top screw caps and can be sealed with rubber septum stoppers or Teflon® lined discs. The heavy walls and the good seal afforded by the Teflon® lined discs allow heating of samples to moderately high temperatures with safety. Thermostatically controlled heating units are available with aluminum blocks drilled to exactly fit the vials. Although Teflon® is quite inert, it must be remembered that some samples

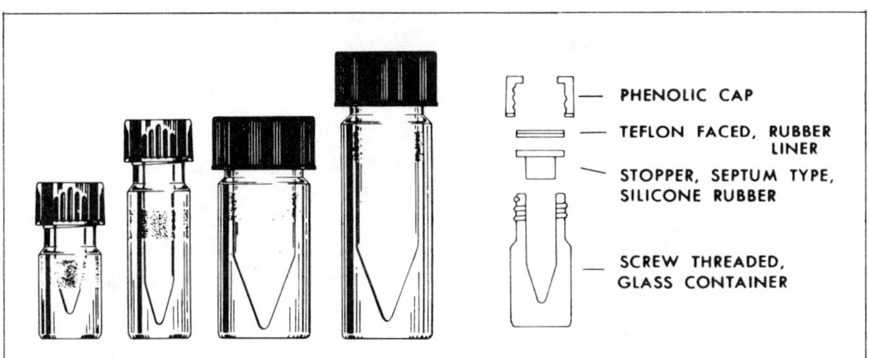

Figure 1. Tapered reaction vials. Vials may be sealed with either Teflon® faced rubber discs or silicone rubber septum stoppers. (Illustration curtesy of Regis Chemical Co.)

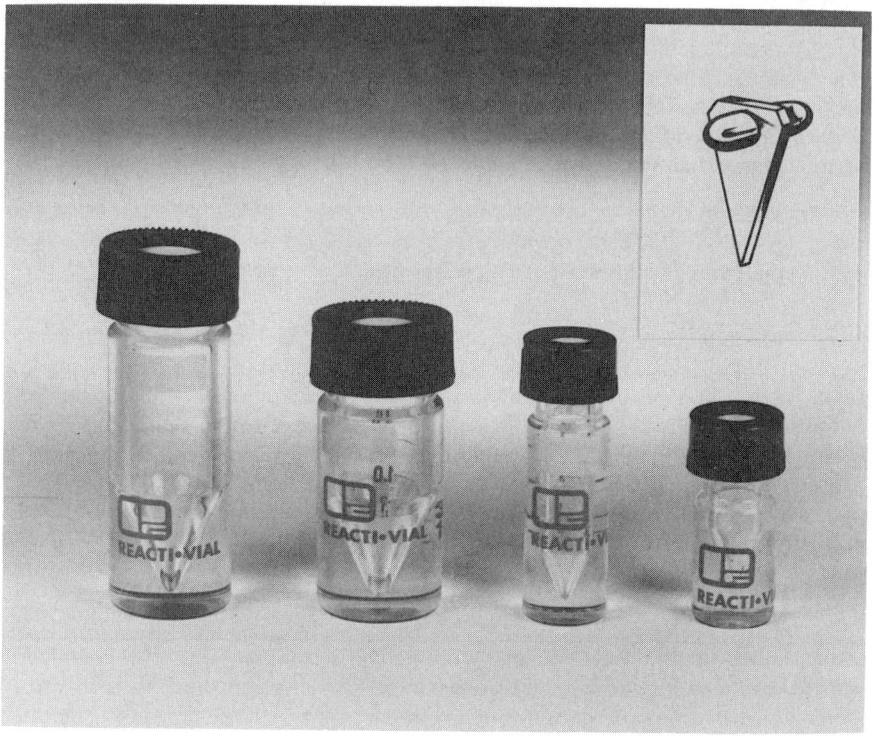

Figure 2. Tapered reaction vial with Teflon® magnetic stirring paddle (inset, upper right). (Illustration courtesy of Pierce Chemical Co.)

and reagents will dissolve in Teflon®, thus these containers, although extremely useful, do have limitations.

The ground bottoms of the vials permit them to stand upright on a flat surface and form a convenient area for pencil markings. Miniature Teflon® magnetic stirring paddles are made to fit the vials (Figure 2). These vials are also available with an inside standard taper glass joint and with sealable ampule tops which, when broken off, allow use of the regular sealing cap. The latter type is particularly useful for storage of air or moisture sensitive reagents.

Unfortunately, the tapered reaction vials are rather expensive, and this can become a significant disadvantage when large numbers are needed. A similar but somewhat less expensive container is the Silli-Vial® (Figure 3). Carrier blocks that can be used for heating the vials and a matching evaporator unit are also available.

Centrifuge tubes with Teflon® lined screw caps and tapered bottoms are used extensively for sample extractions, evaporations, and derivatizations. The larger volume compared to tapered reaction vials or Silli-Vials® can be a disadvantage when sample solutions must be heated. The large volume also results in exposure of the sample to a larger air space. The problems associated with materials dissolving in the Teflon® cap liners can be overcome by use of centrifuge tubes with glass stoppers.

Several other less expensive analytical derivatization vessels can be used with some improvisation. Screw capped culture tubes are considerably less expensive than centrifuge tubes but are generally less strong mechanically. The tubes can be pulled apart in a torch

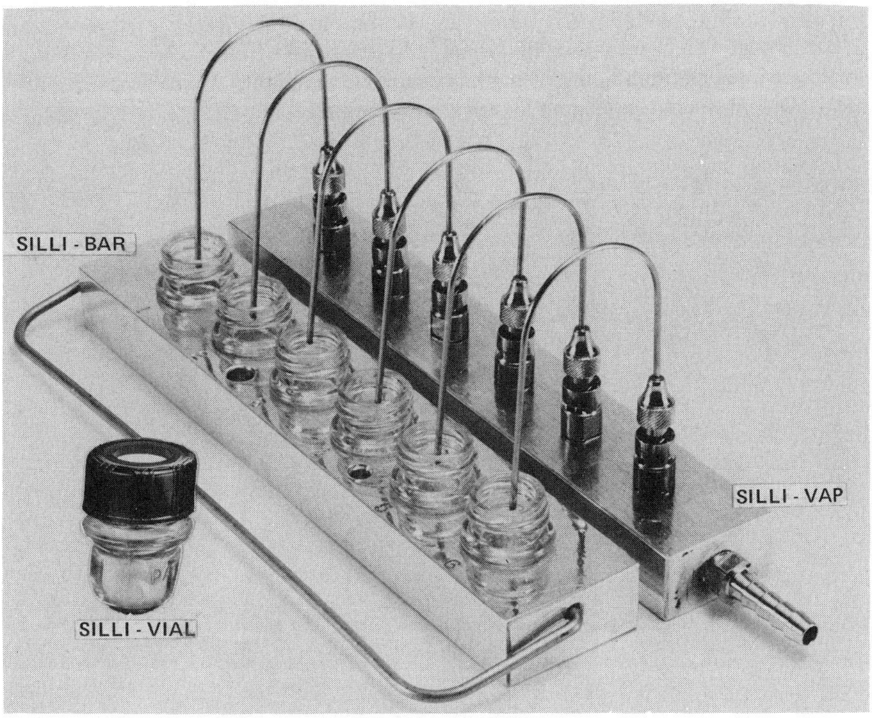

Figure 3. Silli-Vials® with matching heating bar and evaporator unit. (Illustration courtesy of Pierce Chemical Co.)

flame to form smaller vessels with tapered bottoms similar to more expensive containers (see Figure 4). It is somewhat difficult to achieve uniformity of size and shape in these homemade vessels, but they offer an inexpensive alternative for room temperature derivatizations. An even less expensive container can be prepared from small common screw capped bottles or vials. Small Teflon® cap liners can be cut from the readily available larger liners or sheet stock with a cork borer. These vials do not seal as well as the more expensive containers and cannot be safely heated more than a few degrees, but are quite satisfactory for room temperature use.

Analytical procedures requiring vessels sealed under vacuum have classically been carried

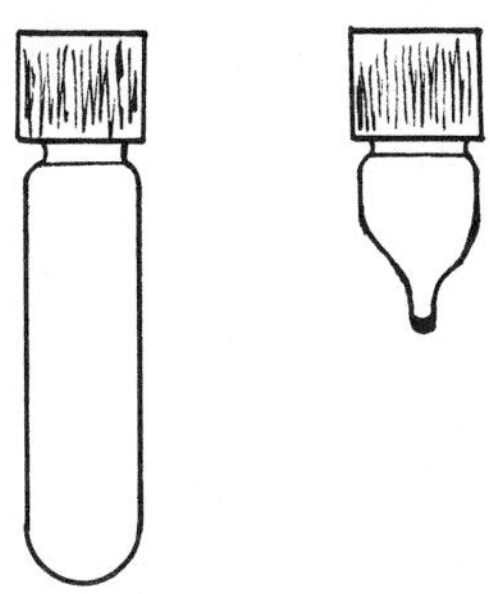

Figure 4. Screw capped culture tube (left) can be drawn out in a flame to form a tapered bottom vessel (right).

out in glass tubes requiring torch sealing and breakage of the tube to retrieve the sample. Sealable vacuum reaction tubes with Teflon® plunger seals are now commercially available (Figure 5). Matching aluminum heating blocks are also available. These reuseable tubes are particularly convenient for vacuum hydrolysis procedures.

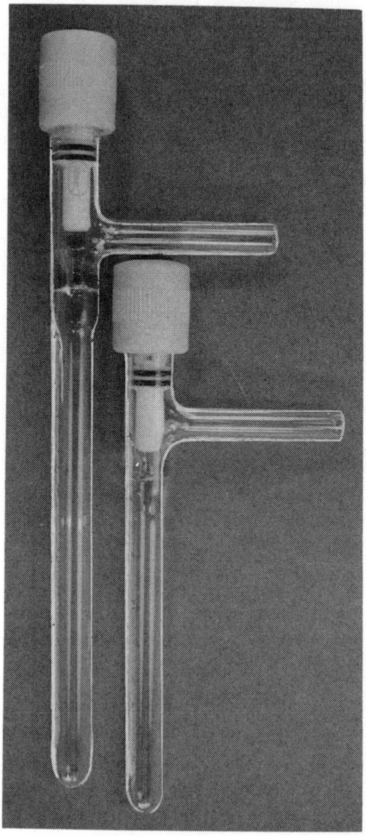

Figure 5. Vacuum reaction tubes. (Illustration courtesy of Regis Chemical Co.)

Sample and Reagent Transfer

Many workers prefer to transfer samples and reagents with miniature pipettes. For sensitive reagents the use of a microliter syringe gives less exposure to atmospheric moisture. Examples of several types of these syringes are shown in Figure 6. The syringes with Teflon tipped plungers are more convenient than conventional all metal plunger types particularly for volatile reagent transfers. The Teflon® plunger tip forms a better seal and facilitates withdrawal of reagents from sealed vials. Any of the syringes will retain some reagent in the barrel and the all metal plunger types are particularly prone to corrosion and freezing if not properly cleaned. The best cleaning method is to completely remove and wash the plunger and pull solvent through the bore of the syringe with a vacuum. A frozen plunger, resulting from

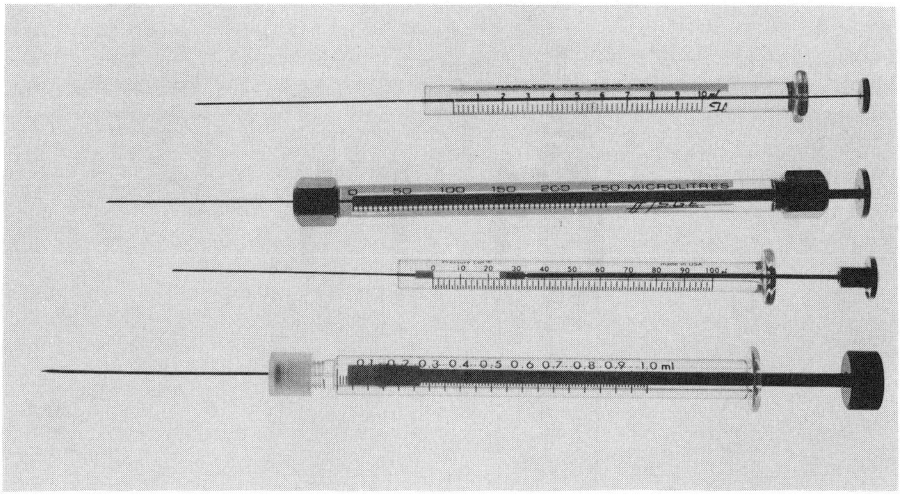

Figure 6. Microliter syringes used for sample and reagent transfer with all metal plungers (top) and Teflon® tipped plungers (bottom).

failure to clean the syringe, can usually be freed by soaking the entire syringe in a graduate cylinder filled with methanol.

Many of the derivatization reagents are sensitive to oxygen or atmospheric moisture. Such reagents are best kept in small containers such as sealed ampules and opened as needed. Reagents kept in regular screw capped vials are susceptible to atmospheric exposure with repeated openings. For reagents compatible with silicone rubber, the use of septum stoppered vials allows removal of reagent via a syringe without opening the vial. A device called the Mininert® valve is available to fit the tapered reaction vials and other containers (Figure 7).

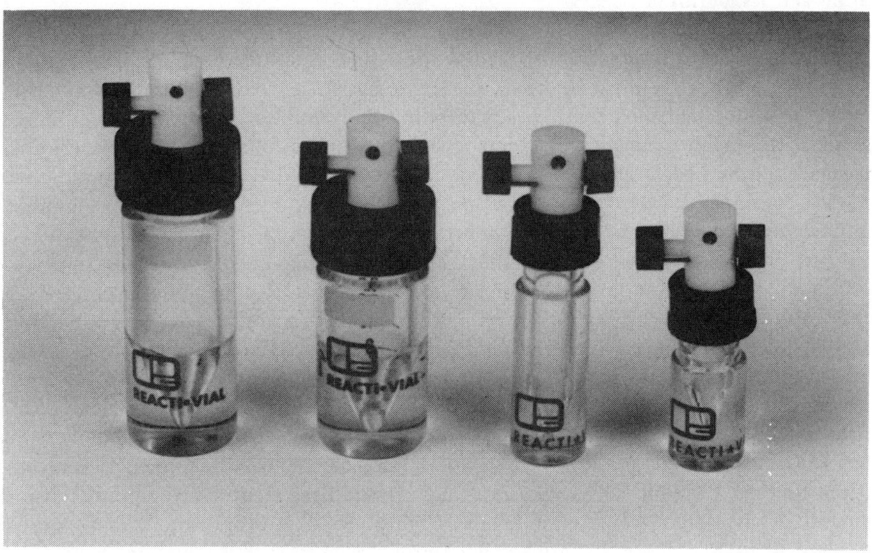

Figure 7. Tapered reaction vials with Mininert® valves attached. (Illustration courtesy of Pierce Chemical Co.)

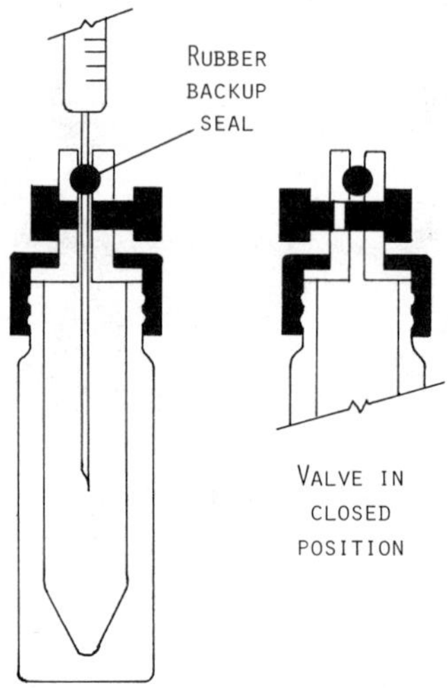

Figure 8. Operation of the Mininert® valve. In the open position (left) a syringe needle can be inserted through the rubber backup seal. In the closed position (right) the contents of the vial contact only glass and Teflon®.

This valve, like the rubber septum, allows removal of reagent via a syringe but the reagent stored in the container contacts only Teflon® and glass (Figure 8). Reagents purchased in small sealed glass ampules may be quickly transferred to a vial sealed with a Mininert® valve. Subsequent withdrawals of reagent can then be made without exposure of remaining reagent to atmospheric moisture.

Heating and Evaporation

In addition to thermostatically controlled heating blocks, sample tubes may be heated in a water or sand bath. Care must be taken to exclude moisture from the samples when using a water bath. With either the heating block or bath small samples are usually heated in sealed containers. Refluxing quantities of liquid less than 50 μl is difficult. The surface forces of such small volumes are such that bumping cannot be avoided even with a boiling stone. Dünges[36] developed a micro refluxing apparatus, shown in Figure 9, which is suitable for quantities down to 5 μl. The small volume samples are placed in 2 mm i.d. tubes with the upper part of the tubes held in an aluminum cooling block. The bottoms of the tubes are heated in an air bath. Bumping is avoided by a continuous vibrating action produced with a solenoid attached to the cooling block. With this apparatus, quantities as small as 5 μl could be refluxed without disturbance for over an hour. This microrefluxer is now available commercially.[37] A refluxer for 0.2–5 ml samples was also described by Dünges.[38]

It is often necessary to evaporate solvents and excess reagents from analytical derivatization reaction mixtures. Evaporation can be accomplished with a stream of inert gas or under vacuum. Several devices have been developed for simultaneous evaporation of multiple samples. The Silli-Vap® evaporator has already been mentioned (Figure 3). Another commercially available device is shown in Figure 10. This evaporator can accomodate containers of various sizes and features an integral thermostatically controlled heating bath. Analytical chemists have been ingenious in devising less expensive evaporating devices for multiple samples. A variety of types of manifolds can be assembled and connected to

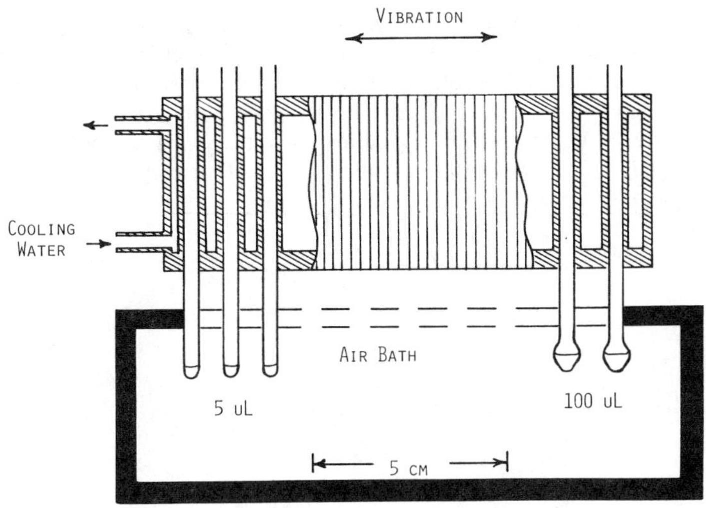

Figure 9. Microrefluxing apparatus. (Reproduced from Reference 36 with permission.)

Figure 10. Multiple sample evaporator. (Illustration courtesy of Organomation Associates.)

disposable Pasteur pipettes for delivery of gas streams to small containers. A particularly useful homemade device can be constructed from gang valves such as those used for air lines on an aquarium. The individual needle valves on each stream permit precise adjustment of gas flow for each sample.

Whelan[39] reported a simple device for evacution and nitrogen backfilling of septum sealed vials (Figure 11). The three-way stopcock allows evacuation of the vial and subsequent nitrogen filling via the same needle.

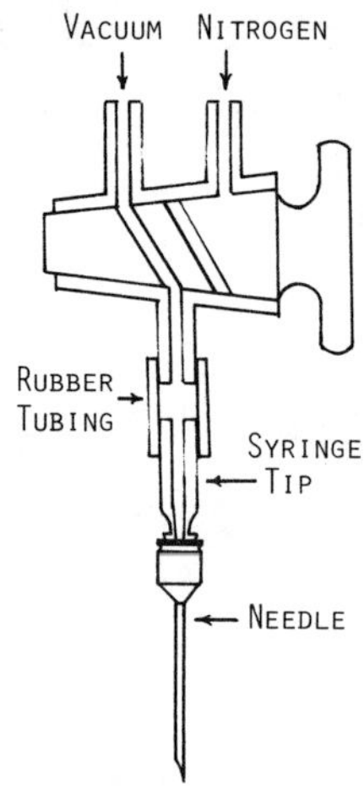

Figure 11. Device used for evacuation and nitrogen backfilling of septum sealed vials.[39]

Vapor Phase Derivatization

It is also possible to perform derivatization without solvents by depositing the sample on a solid support such as paper or stainless steel gauze and exposing the sample to reagent vapors.[40–43] The sample is then analyzed via a GC solid injector system. The technique offers the particular advantage for GC–MS of no solvent peak. Engel and Orr[44] summarized the applications of vapor phase derivatization and solid sample injection in the GC–MS analysis of steroids.

Deactivation of Glassware Surfaces

In very low level analysis, losses due to adsorption on glassware surfaces can be significant. Amines are particularly prone to adsorptive loss on the slightly acidic silica surface of glass. The effect with amines is especially marked with acid washed glassware. To prevent such losses, glassware used for low level analysis is usually silanised. The object of the treatment is to mask the polar Si–OH groups on the glass surface, by chemically binding a nonadsorptive silicone layer, in effect "derivatizing" the glass. The most common silanization technique is to treat the glassware with a solution of 5–10% dimethyldichlorosilane (DMCS) in toluene. After standing in contact with the solution for 30 min. (some workers use longer and others claim adequate silylation in shorter times) the glassware is rinsed with dry toluene followed immediately with absolute methanol. An alternative method is to treat the glassware with DMCS vapor. In this method a small amount of DMCS is placed in the vessel to be silanized, the vessel is then sealed and allowed to stand overnight. Excess DMCS is rinsed away with toluene and the process completed with a methanol rinse. The chemistry of the silanization procedure is shown below.

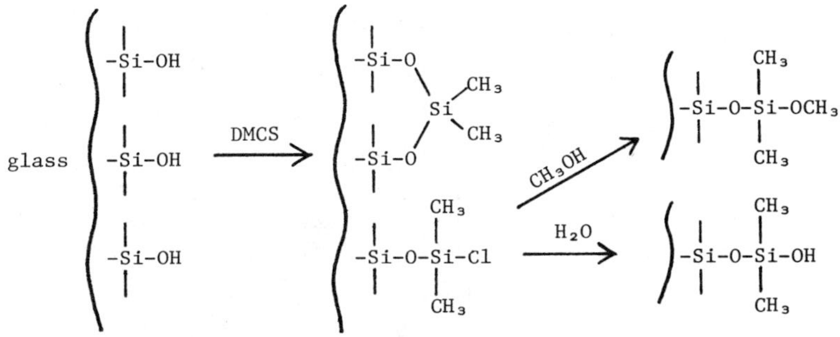

A molecule of DMCS may react with a single Si—OH or with two Si—OH groups to give a cyclic structure. The methanol treatment converts any remaining Si—Cl groups to Si—OCH_3. Exposure to moisture before conversion of the Si—Cl groups gives Si—OH groups, defeating the purpose of silanization.

It has been observed that glassware used for diazomethane esterfication appears to become less adsorptive. This effect could be due to methylation of the weekly acidic surface Si–OH groups. In experiments comparing adsorption of compounds upon untreated, silanized, and diazomethane treated glassware, it appeared that diazomethane treatment significantly reduced adsorption of amino compounds.[45] Diazomethane may offer another method of dectivating glass surfaces, but due to the hazards of diazomethane any advantages of the method must be weighed carefully against the potential risks.

In addition to deactivation of glassware surfaces, another method of reducing adsorption losses in trace analysis is the addition of a compound that competes for the adsorptive sites. Amines are particularly prone to losses from nonpolar extraction solvents (e.g., heptane) by adsorption on glass. Addition of a small amount (often less than 1%) of an alcohol such as butanol to the solvent significantly reduces these losses.

Apparatus for Diazomethane Generation

Diazomethane, and the other diazoalkanes, are usually prepared by the analyst rather than purchased. The classical technique for preparation of diazomethane consists of generation from any of several precursers and codistillation with diethyl ether. The reader is reminded that diazomethane is both toxic and explosive and must be handled with due care. Explosions have most often occured during distillation and can be initiated photolytically or by exposure of the gas to sharp edges. Ground joints must therefore be avoided. Glassware kits made especially for diazomethane generation featuring Clearseal® joints are available commercially. Frequently only small amounts of diazomethane are needed; such amounts can be prepared conveniently with apparatus designed specifically for small scale generation. One such apparatus was described by Schlenk and Gellerman[46] (Figure 12). This apparatus is constructed from sidearm test tubes. A stream of nitrogen is saturated with diethyl ether in the first tube and then passed through a diazomethane generating solution containing *N*-methyl-*N*-nitroso-*p*-toluenesulfonamide (Diazald®, Aldrich Chemical Company) and potassium hydroxide. The generated diazomethane is carried into the sample tube via the nitrogen stream. Samples are treated until excess diazomethane is present as evidenced by development of a yellow color in the sample. Similar methods were described by Levitt[47] and

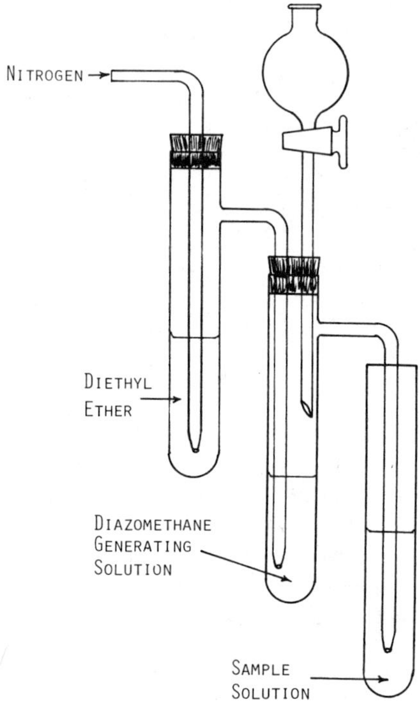

Figure 12. Apparatus for small scale generation of diazomethane.[46]

Crotte *et al.*[48] In the latter case diazomethane was generated from hydrazine, chloroform, and potassium hydroxide.

Fales *et al.*[49] described an apparatus for generating small quantities of diazomethane solution without the need for codistillation (Figure 13). In this apparatus, diazomethane is generated from *N*-Methyl-*N*-nitroso-N^1-nitroguanidine in the center tube. *N*-Methyl-*N*-nitroso-*p*-toluenesulfonamide, generally the preferred reagent for diazomethane preparation, reacted too slowly at room temperature to be useful in this apparatus. The generating reaction is initiated by injection of sodium hydroxide solution via the septum, and the generated diazomethane dissolves in the ether by gaseous diffusion. The dissolving process can be speeded by magnetic stirring of the ether. The method is also applicable to diazoethane preparation from *N*-ethyl-*N*-nitroso-N^1-nitroguanidine. The apparatus is commercially available with either Clearseal® tapered or O-ring joints. It must be emphasised that these micro scale devices do not eliminate the hazards of diazomethane, and the proper precautions (i.e., fume hood, safety screen, avoidance of strong light, and sharp edges) must still be observed. It should also be noted that many people become allergic to *N*-methyl-*N*-nitroso-N^1-nitroguanidine and that this diazomethane precursor has been shown to be a potent mutagen.[50] The preparation and reactions of diazomethane have been reviewed.[51,52]

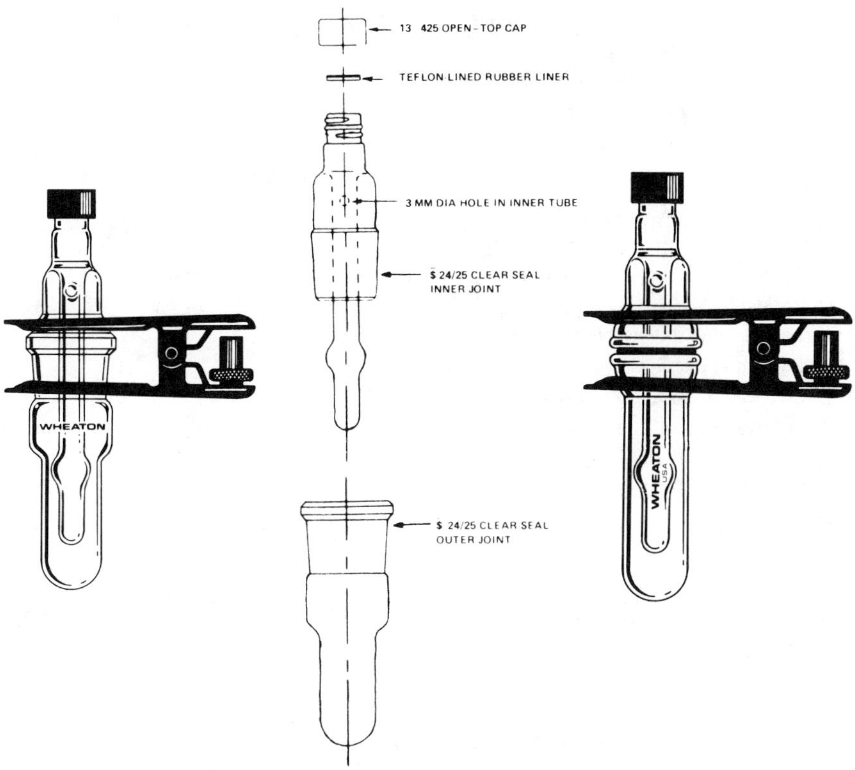

Figure 13. Apparatus for generation of small quantities of diazomethane without the need for codistillation.[49] (Illustration courtesy of Wheaton Glass Co.)

References

1. A. E. Pierce, *Silylation of Organic Compounds*, Pierce Chemical Co., Rockford, IL. 1968.
2. K. Blau and G. King (editors), *Handbook of Derivatives for Chromatography*, Heyden, London, 1977.
3. J. F. Lawrence and R. W. Frei, *Chemical Derivatization in Liquid Chromatography*, Elsevier, Amsterdam, 1976.
4. J. Drozd, *J. Chromatogr.* **113**, 303 (1975).
5. S. Ahuja, *J. Pharm. Sci.* **65**, 163 (1976).
6. W. P. Cochrane, *J. Chromatogr. Sci.* **13**, 246 (1975).
7. R. W. Frei and W. Santi, *Z. Anal. Chem.* **277**, 303 (1975).
8. T. H. Jupille, *Am. Lab.* **8**(5), 85 (1976).
9. R. W. Frei, *Res./Development* **28**(2), 42 (1977).
10. M. S. F. Ross, *J. Chromatogr.* **141**, 107 (1977).
11. C. F. Poole, *Chem. Ind.* 479 (1976).
12. R. A. Landowne and S. R. Lipsky, *Anal. Chem.* **35**, 532 (1963).

13. D. D. Clarke *et al., J. Gas Chromatogr.* **5**, 307 (1967).
14. S. B. Matin and M. Rowland, *J. Pharm. Sci.* **61**, 1235 (1972).
15. N. K. McCallum and R. J. Armstrong, *J. Chromatogr.* **78**, 303 (1973).
16. K. Nakagawa *et al., Steroids* **7**, 329 (1966).
17. L. Dehennin *et al., J. Chromatogr. Sci.* **10**, 224 (1972).
18. E. Anggard and A. Hankey, *Acta Chem. Scand.* **23**, 3110 (1969).
19. D. D. Clarke *et al., J. Gas Chromatogr.* **4**, 310 (1966).
20. A. C. Moffat *et al., J. Chromatogr.* **66**, 255 (1972).
21. L. M. Cummins, in *Recent Advances in Gas Chromatography,* I. I. Domsky and J. A. Perry (editors), Dekker, New York, 1971, p. 313.
22. C. Eaborn *et al., Chem. Ind.,* 827 (1967).
23. C. Eaborn *et al., J. Chem. Soc. C,* 2502 (1969).
24. E. D. Morgan and C. F. Poole, *J. Chromatogr.* **89**, 225 (1974).
25. E. D. Morgan and C. F. Poole, *J. Chromatogr.* **104**, 351 (1975).
26. E. D. Pellizari, *J. Chromatogr.* **98**, 323 (1974).
27. R. Greenhalgh and P. J. Wood, *J. Chromatogr.* **82**, 410 (1973).
28. C. J. Pederson and H. K. Frensdorff, *Angew. Chem. Int. Ed. Eng.* **11**, 16 (1972).
29. J. Dockx, *Synthesis,* 441 (1973).
30. E. V. Dehmlow, *Angew. Chem. Int. Ed. Eng.* **13**, 170 (1974).
31. R. A. Jones, *Aldrichim. Acta* (Aldrich Chemical Company) **9**(3), 35 (1966).
32. See Reference 1, p. 54.
33. See Reference 1, p. 55.
34. N. Sakauchi and E. C. Horning, *Anal. Lett.* **4**, 41 (1971).
35. E. J. Corey and A. Venkateswarlu, *J. Am. Chem. Soc.* **94**, 6190 (1972).
36. W. Dünges, *Anal. Chem.* **45**, 963 (1973).
37. Regis Chemical Co., Morton Grove, IL.
38. W. Dünges, *Chromatographia* **9**, 571 (1976).
39. J. K. Whelan, *J. Chromatogr.* **111**, 337 (1975).
40. E. Menini and J. K. Norymberski, *Biochem. J.* **95**, 1 (1965).
41. J. K. Norymberski and A. Riondel, *Experientia* **23**, 318 (1967).
42. F. L. Rigby *et al., Steroids* **16**, 703 (1970).
43. B. Tuewissen and A. Darbre, *J. Chromatogr.* **49**, 298 (1970).
44. L. L. Engel and J. C. Orr, in *Biomedical Applications of Mass Spectrometry,* G. R. Waller, ed., Wiley, New York, 1972, p. 546.
45. D. R. Knapp and T. Walle, unpublished work.
46. H. Schlenk and J. L. Gellerman, *Anal. Chem.* **32**, 1412 (1960).
47. M. J. Levitt, *Anal. Chem.* **45**, 618 (1973).
48. C. Crotte *et al., Bull. Soc. Chim. Biol.* **52**, 108 (1970).
49. H. M. Fales *et al., Anal. Chem.* **45**, 2302 (1973).
50. J. D. Mandell and J. Greenburg, *Biochem. Biophys. Res. Commun.* **3**, 575 (1960).
51. H. B. Hopps, *Aldrichim. Acta* (Aldrich Chemical Company) **3**, 9 (1970).
52. J. Pizey, *Synthetic Reagents*, Vol. 2, Halsted, New York, 1974, p. 65.

Part II. Derivatization of Particular Compound Types

1. Hydroxyl, Sulfhydryl, and Epoxy Compounds

Derivatives used in the analysis of alcohols, thiols, phenols, and epoxides are discussed in this section. Derivatization of the hydroxyl group, however, is often applied to compounds other than simple alcohols and phenols; therefore, in addition to the derivatives of alcohols, thiols, and phenols contained in this section, examples of derivatization of hydroxyl and sulfydryl groups will be found in several other sections, particularly:

- 2.5 Amino alcohols and amino thiols, p. 119
- 2.6 Phenolic amines, p. 126
- 3.4 Hydroxy acids, p. 191
- 3.5 Ketoacids and Krebs cycle acids, p. 196
- 3.6 Phenolic acids, p. 206
- 9.1 Enantiomeric alcohols, p. 408
- 10. Fatty lipids, p. 437
- 11. Steroids, p. 449
- 12. Prostaglandins, p. 515
- 13. Carbohydrates, p. 539
- 14. Nucleotides, etc., p. 599
- 15. Drugs, p. 621

It is suggested that the reader consult these other sections as well as the indexes in locating information on derivatives of hydroxyl and sulfhydryl groups. Derivatives of the epoxide group are all contained in this section (epoxides formed *as* derivatives of unsaturated compounds are discussed on p. 231).

Alcohols. Most of the underivatized monofunctional alcohols have sufficient volatility for GC and MS analysis, however, derivatization may be employed to improve GC behavior or to impart detectability. In MS analysis, derivatives may be prepared to obtain a desired fragmentation behavior. Derivatives of alcohols are also used to impart detectability in HPLC analysis. The derivatives of alcohols fall primarily into the three classes of alkyl, acyl, and silyl derivatives. Alcohols are insufficiently acidic (pK_a 16-19) to undergo alkylation by the common alkylation reagents but may be alkylated by use of very strong bases and alkyl halides (see permethylation of carbohydrates, p. 543) or by diazomethane with suitable catalysis. Boron trifluoride catalyzed methylation of alcohol groups by diazomethane has been used for prostaglandins (see p. 517). Methylation of alcohols with diazomethane catalyzed by fluoboric acid has been used preparatively (p. 30) and may be adaptable for analytical use.

Acyl derivatives of alcohols are most commonly prepared with acid anhydrides, acid chlorides, or reactive amide reagents. The classical technique for preparing acetyl derivatives involves acetylation with acetic anhydride, which may be base (frequently pyridine) catalyzed (see p. 30) or acid catalyzed (see p. 31). A novel acetylation technique utilizes the mixed anhydride with methanesulfonic acid (p. 31). Acylation with perfluoro anhydrides is also commonly used. The perfluoroacyl derivatives are more volatile than the corresponding nonfluorinated derivatives and may be used to impart electron capturing ability. The fluorinated and nonfluorinated derivatives of a series of alcohols were analyzed by GC on three stationary phases. The data plotted on triangular graphs could be used for identification of unknown alcohols (p. 33).

The acid chlorides, less commonly utilized reagents for analytical acylation of alcohols, are used primarily in the form of benzoyl chloride reagents. Pentafluorobenzoyl chloride has been used to prepare derivatives of alcohols for ECD-GC (p. 32). and dinitrobenzoyl derivatives have been prepared with the acid chloride reagent for HPLC analysis (p. 41).

The reactive amide reagents include the acyl imidazole reagents and the bis (amide) reagents MBTFA and BTFA (p. 32). These reagents are particularly useful for acid sensitive compounds since, unlike the anhydrides and acid chlorides, they do not yield acid by-products.

The silyl derivatives of alcohols are also valuable analytical derivatives. Although the trimethylsilyl ethers of the simple monofunctional alcohols have higher boiling points, in general, than the alcohols themselves, they are useful in GC analysis to reduce adsorption and improve GC behavior. The derivatives of the polyfunctional alcohols, in contrast, show significantly enhaced volatility. Alcohols are among the most easily silylated functions and yield some of the most stable silyl derivatives. The less powerful silylating reagents such as HMDS are usually adequate for silylation of alcohols although tertiary and other hindered alcohols may require more vigorous conditions (p. 33). The ease of formation of TMS derivatives of the alcohols is primarily a function of steric factors with the order of reactivity of the alcohols being primary > secondary > tertiary. In addition to the common reagents, other reagents such as trimethylsilylpiperidine (p. 34), (isopropenyloxy) trimethylsilane (p. 34), *N*-trimethylsilylacetanilide (p. 35), and bis (trimethylsilyl) urea (p. 35) may be used to silylate alcohols. Deuterated trimethylsilyl derivatives of alcohols have been employed for MS analysis (p. 36).

Dimethylsilyl derivatives of alcohols have been prepared with tetramethyldisilazane/dimethylmonochlorosilane (p. 36) or bis(dimethylsilyl)acetamide (p. 36). These dimethylsilyl derivatives are less volatile than corresponding trimethylsilyl compounds and offer a possible alternative for compounds that are inseparable as TMS derivatives. The dimethylsilyl derivatives are useful for MS analysis of alcohols since they tend to fragment in a manner that is structurally more informative than the trimethylsilyl derivatives. The *t*-butyldimethyl silyl derivative of alcohols, which was introduced as a synthetic protecting group, has also been useful in MS analysis since its primary fragmentation (loss of the *t*-butyl group) yields intense high mass ions (p. 37).

Halogenated silyl derivatives have been used for ECD-GC analysis of alcohols. Bromomethyldimethylsilyl derivatives have been prepared with the chlorosilane reagent and converted to the iodomethyldimethylsilyl derivatives with sodium iodide (p. 37). The iodo derivatives tend to be more sensitive to ECD than the bromo derivatives. Pentafluorophenyldimethylsilyl (flophemesyl) derivatives have been prepared from flophemesylamine/flophemesyl chloride (p. 38). These derivatives useful both for ECD-GC and MS analysis are unusual in their MS behavior because they tend to give predominant molecular ions, which are sometimes even the base peak. Heterocyclic phosphorus containing derivatives have been used for GC analysis of alcohols using an alkali flame ionization detector (p. 38).

Glycols. The 1,2-diols constitute a special class of difunctional alcohols in that they are amenable to formation of cyclic derivatives specific for the 1,2-dihydroxy function. These cyclic derivatives include the acetonide (p. 39) and alkylboronate (p. 39) derivatives. One example is given of a chromogenic derivative, the 3,5-dinitrobenzoate, for HPLC analysis (p. 41).

The phenylethylene glycols are a particular group of 1,2-diols of considerable analytical importance since they are one type of metabolite of biogenic amines. These compounds have been subjected both to derivatizations used generally for alcohols and to those used specifically for glycols. These compounds often include phenolic groups, and the differences in reactivity between the aliphatic and aromatic hydroxyls permit the formation of mixed derivatives. Trifluoroacetyl (p. 41), pentafluoropropionyl (p. 42), and heptafluorobutyryl (p. 42) derivatives have been prepared with the corresponding anhydrides. Specific acylation of the phenolic groups with acetic anhydride and bicarbonate permit subsequent conversion of the alcohol groups to heptafluorobutyryl (p. 42) or trimethylsilyl (p. 43) derivatives. Boronate derivatives have been prepared from the phenylethylene glycols (p. 43), and mixed boronate, trimethylsilyl derivatives from phenylethylene glycols with single phenolic groups (p. 44).

The dihydrodiols are of analytical interest because of their role in foreign compound metabolism. These compounds, which easily lose water to revert to an aromatic system, must be derivatized with care. Methods for preparation of trimethylsilyl (p. 45) and chloromethyldimethylsilyl (p. 45) derivatives have been described. The structurally similar dihydrodiol metabolite of iminostilbene has been analyzed as a cyclic boronate (p. 46). The menthyloxyacetyl derivative has been used for enantiomeric analysis of a dihydrodiol by GC (p. 411).

Thiols. The aliphatic thiols ($pK_a \sim 12$) are more acidic than alcohols (pK_a 16–19) and are more similar to phenols (pK_a 9–11) in acidity. Accordingly, they are easier to alkylate than alcohols but form silyl and acyl derivatives with more difficulty. Pentafluorobenzyl derivatives of thiols have been prepared for ECD-GC analysis (p. 47). Benzoyl and 2,4-dinitrophenyl derivatives were used in the analysis of thiols in food aromas (p. 47). The N-dansylaminoethyl derivative (p. 48), which exhibits specificity for thiols, is potentially useful as a chromogenic derivative for HPLC analysis. Other derivatives of thiols are discussed in the sections on aminothiols (p. 119) and amino acids (p. 288).

Phenols. Many derivatives used in the analysis of alcohols are also applicable in the analysis of phenols. The increased acidity of the phenols (pK_a9–11 for simple phenols) as compared to alcohols also permits use of additional derivatives. Unlike alcohols, the phenols react with diazomethane without a catalyst (p. 48). Phenols substituted with electron withdrawing groups (which increase the acidity) react more rapidly than simple unsubstituted phenols. Phenols may be alkylated with alkyl halides in the presence of base; potassium carbonate has been used in single phase systems (p. 49) and tetraalkylammonium bases in phase transfer systems (p. 49). Alkylation of the phenolic group with heptyl iodide and tetramethylammonium hydroxide, followed by flash methylation of the amide group has been used in the GC analysis of acetaminophen (p. 50). Alkylation with pentafluorobenzyl bromide in the presence of potassium carbonate has been used to prepare derivatives of phenols for ECD-GC analysis (p. 50).

In addition to the alkyl derivatives, phenols have been analyzed as acyl derivatives. The anhydride reagents have been used to prepare acetyl (p. 51), trifluoroacetyl (p. 52) heptafluorobutyryl (p. 52), perfluorooctanoyl (p. 53), and chloroacetyl (p. 53) derivatives. The use of the acylation catalyst 4-dimethylaminopyridine (p. 51) gave improved acylation with acetic anhydride compared to the more commonly used pyridine. Acid chloride reagents have been used to prepare dichloroacetyl (p. 54) and pentafluorobenzoyl (p. 54) derivatives for ECD-GC analysis. The *O*-benzoyl derivatives of acetaminophen was prepared with benzoyl chloride in aqueous solution in the presence of potassium carbonate (p. 55). The anisoyl (*p*-methoxybenzoyl) derivative, prepared from the acid chloride, has been employed to enhance the ultraviolet absorption for HPLC analysis of a phenolic compound (p. 55).

Phenolic hydroxyl groups are easily silylated. Trimethylsilyl derivatives can be prepared from unhindered phenols with any of the usual reagents such as BSTFA (p. 56) and BSA (p. 57). Hindered phenols require somewhat more vigorous conditions; trifluoroacetic acid was found to effectively promote silylation of hindered phenols with BSTFA (p. 58). A relatively new reagent, MSHFBA, first applied to phenols (p. 58), was designed to contain more fluorine than BSTFA to reduce silica fouling of flame ionization detectors. Of several derivatives examined, chloromethyldimethylsilyl derivatives were found to give the best separation of dihydroxynaphthols (p. 59).

Dinitrophenyl (p. 60) and trifluoromethyldinitrophenyl (p. 61) derivatives have been used for ECD-GC analysis of phenols. These derivatives were prepared from the halide reagents in the presence of base. The dinitrophenyl derivatives could be prepared from samples on TLC plates, and then eluted for GC analysis (p. 60).

Diethyl phosphate derivatives of phenols were analyzed by GC with flame photometric detection but the method was less sensitive than ECD-GC analysis of pentafluorobenzoyl derivatives (p. 61). DANS derivatives of phenols have been widely employed for MS and HPLC analysis (p. 62). Boronate derivatives can be prepared from catechols (p. 63) and also from single phenolic hydroxyls when there exists a suitable substituent on an adjacent sidechain (p. 63). Thus formation of a cyclic boronate cannot be considered absolutely specific for 1,2-dihydroxy functions.

Epoxides. Epoxides react with trimethylchlorosilane to yield trimethylsilyl chlorohydrins (p. 63). This derivatization has been applied to epoxide metabolites of allyl barbituric acids for GC-MS analysis (p. 64). The epoxide group has also been derivatized with sodium diethyldithiocarbamate to form bis(dithiocarbamoyl) esters for HPLC analysis (p. 64).

1.1. DERIVATIVES OF ALCOHOLS

1.1.1. Alkyl Derivatives of Alcohols

Derivative Type. Methyl ether.

Reagent. Diazomethane/fluoboric acid.

Compound Type. Alcohols.

Reaction.

$$R\text{—}OH \xrightarrow[HBF_4]{CH_2N_2} R\text{—}OCH_3$$

Typical Procedure (Preparative). Diss 10.0 g *n*-octyl alcohol in 20 ml CH_2Cl_2 and add 23 mg fluoboric acid (*ca* 16 *N*). Cool to below 0° and slowly add 120 mmol CH_2N_2 in CH_2Cl_2. Then add a further 17 mg of fluoboric acid and subsequently 0.5 g of the alcohol (removes the remaining CH_2N_2). Neutralize with KOH, filter to remove polymethylene, evap solvent, then warm residue with a sm amt of Na to react with excess alcohol. Isolate the ether by distillation.

Comments. This preparative method could be adaptable to analytical methylation of alcohols. It was applied preparatively to several alcohols and steroids and to ascorbic acid.[1] Diazomethane has also been used with BF_3 to prepare methyl ethers.[2]

References

[1]M. Neeman *et al.*, *Tetrahedron* **6**, 36 (1959).

[2]E. Müller and W. Rundel, *Angew. Chem.* **70**, 105 (1958).

1.1.2. Acyl Derivatives of Alcohols

Derivative Type. Acetyl.

Reagent. Acetic anhydride (Ac_2O)/pyridine.

Compound Type. Terpene alcohols.

Reaction

$$R\text{—}OH \xrightarrow[\text{pyridine}]{Ac_2O} R\text{—}OCOCH_3$$

Typical Procedure. Diss the alcohol sample in Ac_2O/pyridine (1:1, v/v). Allow to react at room temp (time not specified) then analyze the deriv by GC.

Comments. TMS derivs were also prepared using BSA at room temp. Attempts to prepare derivs from TFAA were unsuccessful yielding mainly degradation products. This problem was attributed to the presence of double bonds and active methylene groups. GC data were reported for the derivs of five terpene alcohols.

References

[1]R. B. Watts and R. G. O. Kekwick, *J. Chromatogr.* **88**, 15 (1974).

Derivative Type. Acetyl.

Reagent. Acetic anhydride (Ac_2O)/perchloric acid.

Compound Type. Alcohols.

Reaction

$$R\text{—}OH \xrightarrow[HClO_4]{Ac_2O} R\text{—}OCOCH_3$$

Typical Procedure. Prepare reagent by carefully adding 4 g (2.35 ml) 72% $HClO_4$ (see caution below) to 150 ml EtOAc. Add 8 ml Ac_2O and allow to stand at room temp at least 30 min. Cool to 5° and add 42 ml cold Ac_2O. Maintain at 5° for 1 hr, then allow to warm to room temp.

Diss alcohol sample containing 3–4 mmol of hydroxyl in 5 ml of the 2 *M* Ac_2O soln in EtOAc. Allow to stand at least 5 min at room temp for complete reaction.

Comments. The acetylation reaction was used in a procedure for quant anaylsis of hydroxyl groups but may be useful for GC or GC-MS analysis. The perchloric acid catalyzed acetylation was much faster (5 min at room temp) than other procedures. Sugars were acetylated using 2 *M* Ac_2O in pyridine containing 0.15 *M* TsOH and heating 5–10 min at 50°.

Caution! Mixtures of perchloric acid and organic compds are potentially explosive.

References

[1]J. S. Fritz and G. H. Schenk, *Anal. Chem.* **31**, 1808 (1959).

Derivative Type. Acetyl.

Reagent. Acetyl methanesulfonate on Celite/phosphorous pentoxide.

Compound Type. Alcohols.

Reaction

$$R\text{—}OH \xrightarrow[\substack{\text{Celite} \\ P_2O_5}]{CH_3COOSO_2CH_3} R\text{—}OCOCH_3$$

Typical Procedure. Prepare the impregnated Celite by first grinding 4 g Celite 545 (dried 18 hr at 500–600°) and 1 g P_2O_5 together (under anhyd conditions) and storing overnight in a dessicator over P_2O_5. Grind this mixt with 1 ml acetyl methanesulfonate until a homogenous tan color is obtained. This material can be stored at −18° over P_2O_5 for at least 1 yr.

Prepare a column of the impregnated Celite by tamping 1.5 cm into a 50mm × 1.5–2.0 mm melting point capillary and compressing (with the end of a paper clip) to 1 cm. Add a sample of 0.2–200 μg of the alcohol in 1–10 μl CCl_4 to the capillary and force it through the column with light air or N_2 pressure. Collect the effluent and analyze directly by GC or GC-MS.

Comments. The method was successfully applied to primary, secondary, and tertiary alcohols although some sensitive alcohols (e.g., cinnamyl, furfuryl) did not emerge from the column. Ac_2O and AcOH also emerged from the column with the derivs. These could be removed by addition of a layer of K_2HPO_4 impregnated Celite on the exit end of the column.

References

[1]D. P. Schwartz, *Anal, Biochem.* **71**, 24 (1967).

Derivative Type. Pentafluorobenzoyl.

Reagent. Pentafluorobenzoyl chloride.

Compound Type. Alcohols, phenols, and amines.

Reaction

$$R{-}OH \xrightarrow{C_6F_5COCl} R{-}OCOC_6F_5$$

Typical Procedure (Preparative). Diss 100 mg of the sample compd in CH_2Cl_2 and add an excess of the acid chloride. For amines, add pyridine also. Evap soln under N_2 and diss residue in 5 ml Et_2O. Wash with 1 ml satd $NaHCO_3$ and 2 × 1 ml H_2O. Dry over Na_2SO_4 and evap under vac. Analyze the deriv by ECD-GC.

Comments. The derivs were examined along with pentafluorophenylacetyl and pentafluorophenoxyacetyl derivs (also formed from the acid chlorides) in a study of the temp dependence of ECD response. It was noted that the retention times of the latter two derivs were impractically long. The temp dependence of ECD sensitivity could be exploited to maximize response to desired components and minimize response to interfering compds.

References

[1]A. Zlatkis and B. C. Pettitt, Chromatographia **2**, 484 (1969).

Derivative Type. Trifluoroacetyl.

Reagent. *N*-Methyl-bis(trifluoroacetamide) (MBTFA), or bis(trifluoroacetamide) (BTFA). (BTFA).

Compund Type. Alcohols, thiols, carboxylic acids, and amines.

Reaction

$$R{-}OH + (CF_3CO)_2NCH_3 \longrightarrow R{-}OCOCF_3 + CF_3CONHCH_3$$

Typical Procedure. Treat approx 2 mg compd with 0.5 ml MBTFA. If no reaction occurs at room temp, heat reaction mixt to 120° for 1 hr. Those compounds not reacting after 1 hr at 120°C can be made to react by using a mixt of MBTFA and an aprotic solvent like acetonitrile, pyridine, DMF, or THF (1:4, v/v).

Comments. BTFA is a solid is easily soluble in many aprotic organic solvents. Adding 30 g BTFA to 10 g THF forms a clear soln that can be measured with a syringe. By-products of derivatization are neutral trifluoroacetamides that elute with the solvent front.

References

[1]M. Donike, *J. Chromatogr.* **78**, 273 (1973).

Derivative Type. Esters.

Reagent. Acetic, propionic, butyric anhydrides/toluenesulfonic acid (TsOH) and perfluoroacids.

Compound Type. Alcohols.

Reaction

$$R{-}OH \xrightarrow[TsOH]{Ac_2O} R{-}OCOCH_3$$

$$R{-}OH \xrightarrow{CF_3COOH} R{-}OCOCF_3$$

Typical Procedure. Mix 50 μl of the alcohol with 100 μl of the anhydride containing 2% TsOH. Allow to stand 15 min then analyze by GC.

Comments. The perfluoroesters could be prepared simply by using the perfluoro acid. TMS derivs were also prepared with HMDS/TMCS. Retention indices were reported for the derivs of a large number of alcohols on SE30, Carbowax 20M, and polydiethylene glycol phases. Plotting the retention data from the three phases on a triangular diagram facilitated identification of unknown alcohols.[1]

References

[1] D. Zarazir *et al., Chromatographia* **3**, 180 (1970).

1.1.3. Silyl Derivatives of Alcohols

Derivative Type. Trimethylsilyl.

Reagent. Hexamethyldisilazane (HMDS).

Compound Type. Alcohols.

Reaction

$$R{-}OH \xrightarrow{HMDS} R{-}OTMS$$

Typical Procedure. Diss the alcohol sample in excess HMDS. Anaylze by GC.

Comments. The order of reactivity of alcohols is primary > secondary > tertiary. Tertiary alcohols do not react in HMDS alone but can be silylated with HMDS in DMF or DMSO with heating. HMDS with TMCS or any of the more reactive silyl donors can also be used but are generally unnecessary since alcohols are the most easily silylated functions.

References

[1] A. E. Pierce, *Silylation of Organic Compounds*, Pierce Chemical Co., Rockford, IL., 1968, p.72.

Derivative Type. Trimethylsilyl.

Reagent. *N*-Methyl-*N*-trimethysilyltrifluoroacetamide (MSTFA).

Compound Type. Hydroxyl, carboxyl, and amino compounds.

Reaction

$$R{-}OH \xrightarrow{MSTFA} R{-}OTMS$$

Typical Procedure. Warm sample with MSTFA for a short time. Analyze soln directly by GC.

Comments. The reagent was synthesized from TMCS and *N*-methyltrifluoroacetamide

(MTFA). The by-product of silylation was the volatile MTFA. If necessary, TMCS could be used to catalyze silylation with MSTFA.

References

[1]M. Donike, *J. Chromatogr.* **42**, 103 (1969).

Derivative Type. Trimethylsilyl.

Reagent. *N*-Trimethylsilylpiperidine (TMSPI).

Compound Type. Hydroxyl, carboxyl, amino, imino, and mercapto.

Reaction

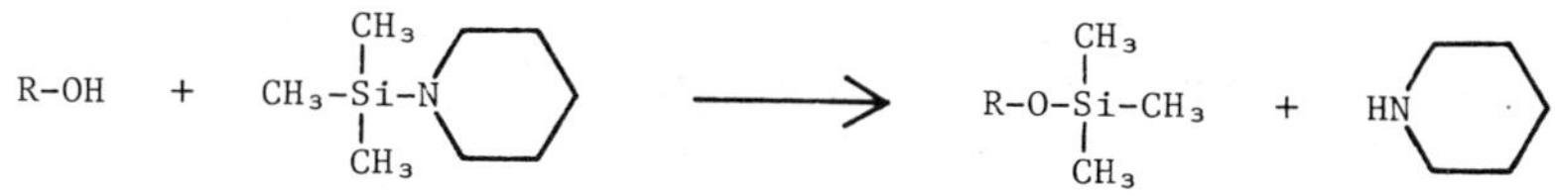

Typical Procedure. To a 0.1–0.5 ml soln of the org compd in pyridine, DMF, or CH_3CN add 0.1–0.5 ml of TMSPI. Shake for 30 sec and allow to stand a variable amount of time dependent upon the type of compound and degree of steric hindrance (see comments). Analyze an aliquot of the soln by GC.

Comments. TMSPI[2] was compared with trimethylsilylpyrrolidine (TMSPY)[2] and trimethylsilylmorpholine (TMSM)[2] as well as the conventional agents BSA, TMSDEA, and TMSI. In general TMSPI and TMSPY were stronger silyl donors than TMSDEA and TMSI and in some cases were as effective as BSA.

References

[1]R. Piekos *et al.*, *J. Chromatogr.* **116**, 315 (1976).

[2]L. Birkofer *et al.*, *Chem. Ber.* **93**, 2804 (1960).

Derivative Type. Trimethylsilyl.

Reagent. (Isopropenyloxy) trimethylsilane (IPOTMS).

Compound Type. Alcohols, phenols, carboxylic acids.

Reaction

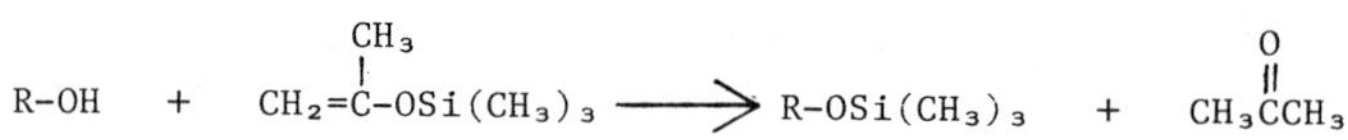

Typical Procedure. A. To 5 mg cholesterol in 500 μl tetrahydrofuran, add 200 μl IPOTMS and 1 drop HOAc. Heat to boiling under reflux for 2 hr. Analyze by GC.

B. To 5 mg cholesterol plus 500 μl IPOTMS, add 2 drops HOAc. Boil 30 min under reflux. Analyze by GC.

Comments. The reagent was immiscible in and only slowly hydrolyzed by H_2O. It could also be used in CH_3CN. HOAc or TMCS was usually necessary as catalyst. The reagent could not be used with primary amines (yielded Schiff bases) or α-hydroxyamines (yielded oxazolidines). The reagent has also been used to synthesize acetonides from 1,2-diols[3] and hemithioketals from β-mercapto alcohols[4].

References

[1]Appl. Sci. Newsl., **16** (6) (Nov./Dec. 1975).

[2]M. Donike and L. Jaenicke, *Angew. Chem. Int. Ed.* **8**, 974 (1969).
[3]G. L. Larson and A. Hernandez, *J. Org. Chem.* **38**, 3935 (1973).
[4]G. L. Larson and A. Hernandez, *Synth. Commun.* **4**, 61 (1974).

Derivative Type. Trimethylsilyl.

Reagent. *N*-Trimethylsilylacetanilide.

Compound Type. Alcohols and phenols.

Reaction

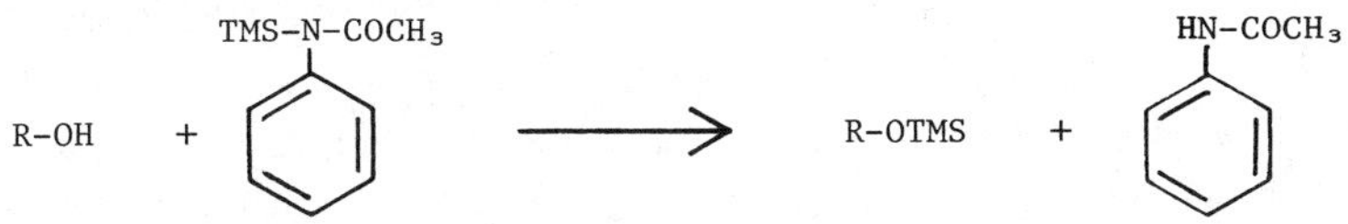

Typical Procedure. To 0.1–0.5 ml of a 0.2–1.0 *M* soln of the sample compd in pyridine, CH_3CN, or DMF add 0.1–0.5 ml of the reagent. Mix vigorously 30 secs; then allow to stand at 20–30°C for a few min to several hr (dependent upon steric hindrance of sample hydroxyls). Analyze by GC.

Comments. *N*-Trimethylsilyl-*p*-ethoxyacetanilide was also examined.[1] The reagents were evaluated on a series of alcohols, phenols, and other compds with replaceable hydrogen; and compared with BSA, TMSI, and TMSDEA. The reagents were simple to prepare (the syntheses were reported) and had good solubility in org solvents. Both reagents and their desilylation products gave single sharp GC peaks. The trimethylsilylacetanilide was useful for compds whose derivs eluted in the ranges 40–140°C and 160–350°C on OV—1. The *p*-ethoxy reagent was useful in the ranges 40–160°C and 180–350°C. The reagents appeared to have their greatest utility for compds whose derivs elute early and would be obscured by tailing reagent fronts with other reagents.

References

[1]R. Piekos *et al.*, *J. Chromatogr.* **117**, 431 (1976).

Derivative Type. Trimethysilyl.

Reagent. *N, N'*-Bis (trimethylsilyl) urea.

Compound Type. Hydroxyl and amino compounds.

Reaction

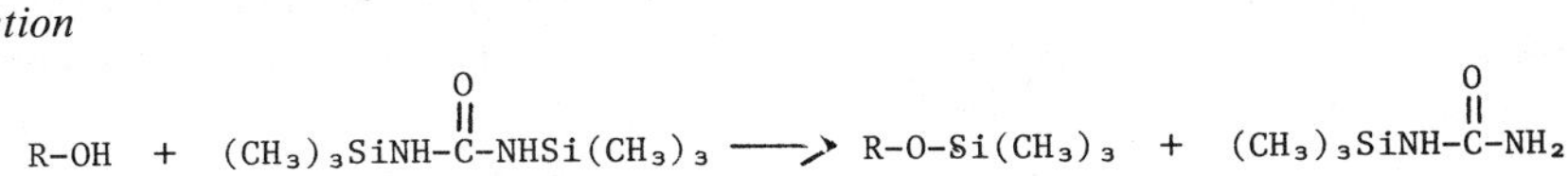

Typical Procedure. No details given.

Comments. The method was said to be useful for carrying out silylations under neutral conditions. The reagent was soluble in ether and acetone but insoluble in hexane. The by-product urea was insoluble in most organic solvents.[1]

References

[1]PCR Reports (PCR Inc., Gainesville, Fla.), Dec. 1975.

Derivative Type. d_9-Trimethylsilyl.

Reagent. d_{18}-Bis(trimethylsilyl)acetamide (d_{18}-BSA).

Compound Type. Hydroxylated compounds.

Reaction

$$HO\text{-}CH_2(CH_2)_8CH_2OH \xrightarrow[Et_3N]{d_{18}\text{-}BSA} (CD_3)_3SiOCH_2(CH_2)_8CH_2OSi(CD_3)_3$$

Typical Procedure. Treat sample with a soln of the reagent in Et_3N (see below) and analyze soln directly by GC-MS.

Comments The derivs were used for MS fragmentation interpretation. The reagent was prepared by refluxing a mixt of 2.6 mmol acetamide, 14 mmol Et_3N, and 7.2 mmol d_9-TMCS for 16 hr while protected from moisture. Filtration of the ppt Et_3N-HCl left a soln of d_{18}-BSA and d_9-trimethylsilylacetamide which was used without further purification. d_9-TMS derivs were also prepared by addition of d_9-TMCS to a soln of the sample in pyridine or Et_3N. d_{18}-BSA is now commercially available.

References

[1]J. A. McCloskey *et al.*, *Anal. Chem.* **40**, 233 (1968).

Derivative Type. Dimethylsilyl (DMS).

Reagent. Tetramethyldisilazane (TMDS) dimethylmonochlorosilane (DMMCS).

Compound Type. Alcohols and phenols.

Reaction

$$R\text{-}OH \xrightarrow[(CH_3)_2HSiCl]{[(CH_3)_2HSi]_2NH} R\text{-}O\text{-}\underset{CH_3}{\overset{CH_3}{Si}}\text{-}H$$

Typical Procedure. Diss approx. 50 mg sample in 1 ml TMDS/DMMCS/pyridine v/v/v) and allow to react 10–30 min. Analyze by GC.

Comments. Derivs were prepared from alcohols, glycols, phenols, aldoses, alditols, and steroids and compared with the corresponding TMS derivs. The DMS derivs gave shorter retention times and could be analyzed by GC at lower temp. In some cases the DMS derivs allowed separation of compds not separable as TMS derivs, but in other cases the converse was also true.

References

[1]W. R. Supina *et al.*, *J. Am. Oil Chem. Soc*, **44**, 74 (1967).

Derivative Type. Dimethylsilyl (DMS).

Reagent. Bis (dimethylsilyl) acetamide (BDSA).

Compound Type. Aliphatic alcohols.

Reaction

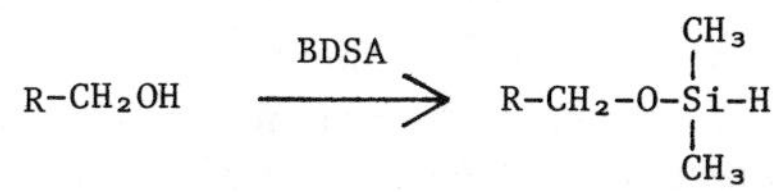

Typical Procedure. Diss sample in BDSA and analyze by GC-MS.

Comments. The derivs of five primary unbranched, two primary branched, and nine secondary unbranched alcohols from C_5 to C_{10} were examined. Unlike the TMS derivs which showed predominant cleavage of Si—CH_3 bonds, the DMS derivs showed pronounced C—C cleavage of the bond adjacent to the oxygen to give structurally informative ions.

References.

[1]W. J. Richter and D. H. Hunnemann, *Helv. Chim. Acta* **57**, 1131 (1974).

Derivative Type. *t*-Butyldimethylsilyl (TBDMS).

Reagent. *t*-Butyldimethylchlorosilane (TBDMS-Cl)/imidazole.

Compound Type Alcohols.

Reaction

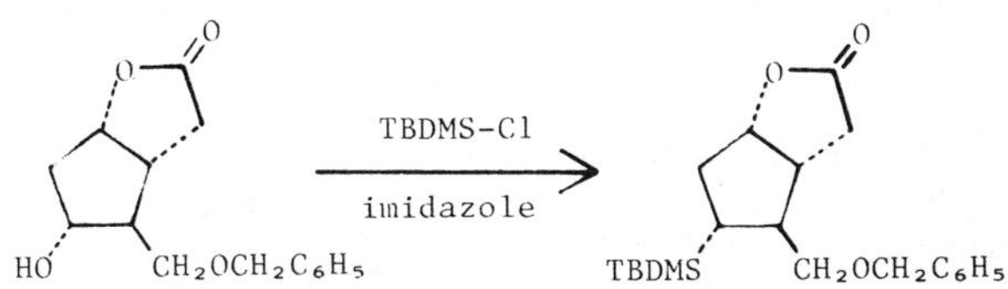

Typical Procedure. Treat hydroxylactone with TBDMS-Cl (1.2 eq) and imidazole (2.5 eq) in DMF (2 ml/g of lactone) at 35°C for 10 hr.

Comments. The deriv was introduced as a relatively H_2O stable protecting group for alcohols in synthetic work but has since become quite useful in MS work. Although the mixt of TBDMS-Cl and imidazole formed TBDMS-imidazole, the use of TBDMS-imidazole alone gave a much slower reaction.[2] The reaction was apparently catalyzed by the HCl formed when TBDMS-Cl was used. Phenols, on the other hand, react quite readily with TBDMS-imidazole at room temp.

References.

[1]E. J. Corey and A. Venkateswarlu, *J. Am. Chem. Soc.* **94**, 6190 (1972).

[2]J. L. Kutschinski, Ph. D. Thesis, University of Calfornia, 1975; *Diss. Abstr. Int. B* **35**, 4838 (1975).

Derivative Type. Iodomethyldimethylsilyl ether.

Reagent Bromomethyldimethylchlorosilane (BMDCS)/diethylamine (DEA), Sodium iodide.

Compound Type. Alcohols.

Reaction.

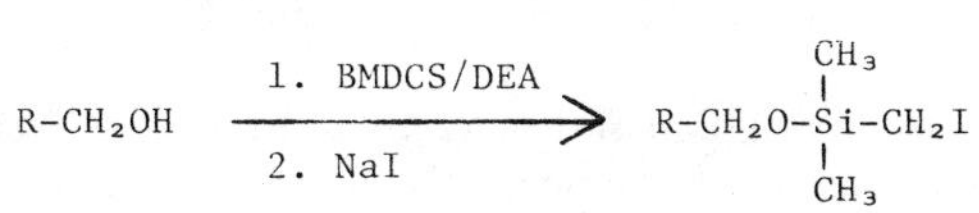

Typical Procedure. Ext aq sample with 20 ml Et_2O. Evap ext under an air stream to 1 ml and dry with approx. 100 mg $MgSO_4$. Evap soln to near dryness. Add 1 drop DEA/$CHCl_3$ (1:11, v/v) and 1 drop BMDCS/$CHCl_3$ (1:11, v/v). Seal and heat at 80° for 1 hr. Evap to dryness and add 0.1 ml satd (at 25°) NaI in acetone. Warm 30 min at 35°C. Add 0.9 ml $CHCl_3$, shake, centrifuge, and discard supernatant. Evap $CHCl_3$ to near dryness, add 1 ml xylene, and analyze 0.5 μ by GC.

Comments. The iodine deriv was more highly electron-capturing than the bromine compd. Carboxylic acids could be analyzed concurrently as iodomethyl tetramethyldisiloxane esters.[1] The method was useful for short chain acids, hydroxyacids, and alcohols.

References

[1]J. B. Brooks *et al.*, *Anal. Chem.* **47**, 1960 (1975).

Derivative Type. Pentafluorophenyldimethylsilyl (flophemesyl).

Reagent. Flophemesylamine/flophemesyl chloride.

Compound Type. Alcohols.

Reaction

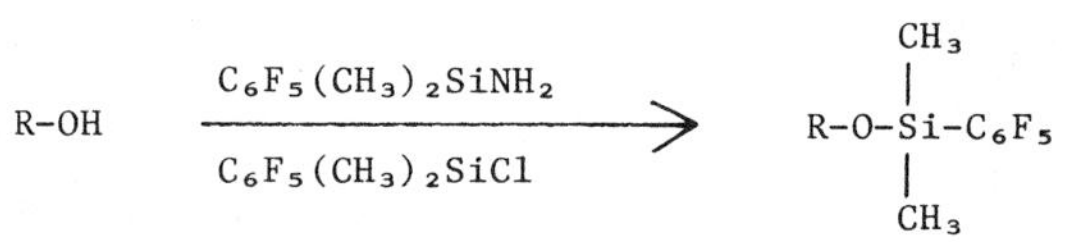

Typical Procedure. Diss alcohol sample in pyridine and add a 1:1 mixt of the flophemesylamine-chloride. Analyze immediately by GC or GC-MS for primary alcohols or allow to react 10 min at 25° C for tertiary alcohols.

Comments. GC data were reported for the derivs of 23 low mol wt alcohols.[1] All the derivs gave prominent mol ions (sometimes the base peak) in the MS. The temp-response relationship was investigated for ECD quantitation of the derivs; maximum sensitivity was achieved by using the maximum ECD operating temp. Linear response was observed from 25 fg to 2.5 pg for the deriv of 2, 2-dimethylpropanol.

References

[1]P. M. Burkinshaw and E. D. Morgan, *J. Chromatogr.* **132**, 548 (1977).

1.1.4. Other Derivatives of Alcohols

Derivative Type. 2-Alkoxy-1, 3, 2-dioxaphospholanes.

Reagent. 2-Chloro-1, 3, 2-dioxaphospholane/triethylamine (Et_3N).

Compound Type. Alcohols.

Reaction

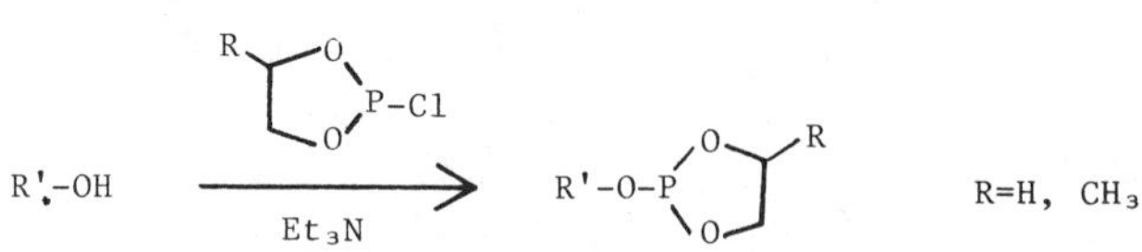

Typical Procedure. To a soln of 1–2 μl of the alcohol in 1 ml benzene add 30 μl Et_3N and 30 μl of the reagent. The reaction occurs immediately. Analyze an aliquot of the soln by GC.

Comments. The analogous 2-alkoxy-1, 3, 2-dioxaphosphorinanes (6-membered rings) were also prepared. The derivs were used for analysis of low mol wt alcohols using an alkali flame ionization detector.

References

[1]R. Vilceanu and P. Schulz, *J. Chromatogr.* **82**, 279 (1973).

1.2. DERIVATIVES OF GLYCOLS

1.2.1. Derivatives of Aliphatic Glycols

Derivative Type. Acetonide.

Reagent. Acetone/*p*-toluenesulfonic acid (TsOH).

Compound Type. Diols.

Reaction

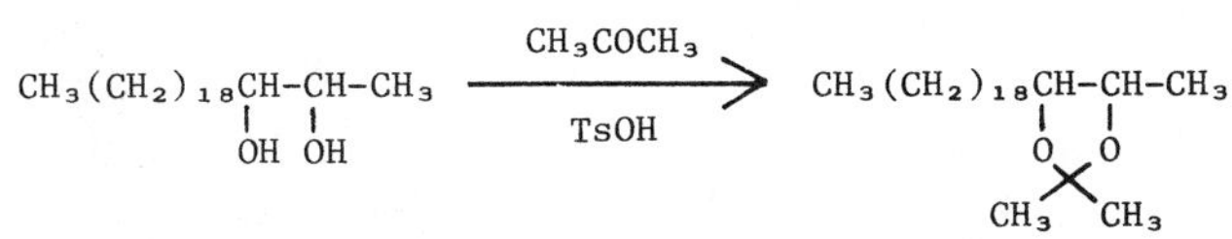

Typical Procedure. Diss 1–2 mg of the diols in 1 ml acetone, and add 1 mg TsOH. Allow to stand at room temp 10–20 min. Cool in ice and add 1 ml hexane and 3 ml H_2O. Mix well, remove hexane phase, dry over Na_2SO_4, and analyze by GC or MS.

Comments. The method was used to identify diols in the unsaponifiable fraction of a wax. The derivs of the *threo* and *erythro* isomers of the diols were GC-separable.[1]

References

[1]E. O. A. Haahti and H. M. Fales, *J. Lipid Res.* **8**, 131 (1967).

Derivative Type. *n*-Butylboronate.

Reagent. *n*-Butaneboronic acid.

Compound Type. Diols, hydroxyacids, hydroxyamines, ketoacids, amino acids.

Reaction

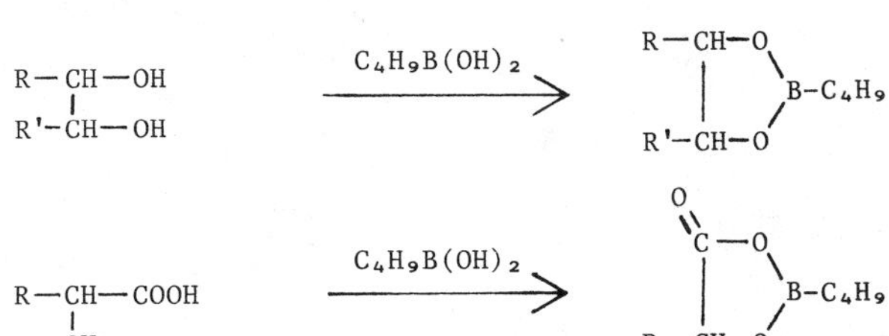

Typical Procedure. Diss 1 mg sample in 100 μl anhyd DMF. Add the boronic acid and allow to stand at room temp. Analyze by GC or GC-MS.

Comments. The derivs of the *erythro* and *threo* 9, 10-dihydroxystearates were separable by GC.[1] Both α-hydroxy and β-hydroxy acids yielded derivs. The hydroxyamine derivs were formed by mixing the reagent and sample without solvent. Phenylpyruvic acid was the only keto acid and anthranilic the only amino acid that formed a deriv. MS data were reported for the *n*-butyl boronates of 35 compds and for the phenyl, methyl, and cyclohexyl boronates of mandelic and salicylic acids.

References

[1]C. J. W. Brooks and I. Maclean, *J. Chromatogr. Sci.* **9**, 18 (1971).

Derivative Type. *n*-Butylboronate.

Reagent. *n*-Butaneboronic acid.

Compound Type. Menthoglycols.

Reaction

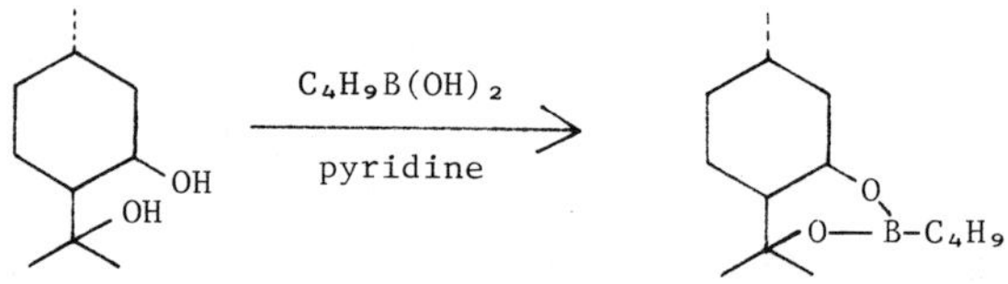

Typical Procedure. Diss 1.7 mg menthoglycol or neomenthoglycol and 1 mg *n*-butaneboronic acid (1 :1 mol ratio) in pyridine. After 3 min analyze by GC.

Comments. The reaction was shown to be complete in 3 min at room temp. TMS and HFB derivs were also prepared. MS data were reported for the derivs.[1] The deriv has also been used to separate and identify diastereomeric duvatrienediols in tobacco.[2] For this application, excess boronic acid was used with heating at 100°C for 15 min.

References

[1]P. Bournot, *J. Chromatogr.* **57**, 55 (1971).

[2]S. Y. Chang and C. Grunwald, *J. Lipid Res.* **17**, 7 (1976).

Derivative Type. *n*-Butylboronate.

Reagent. *n*-Butaneboronic acid.

Compound Type. Dianhydrogalactitol (DAG).

Reaction

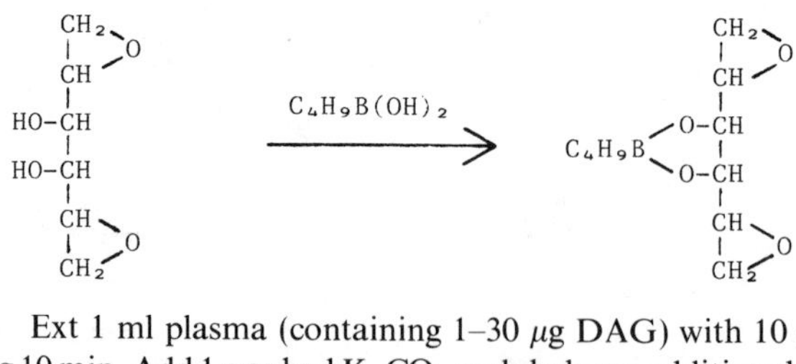

Typical Procedure. Ext 1 ml plasma (containing 1–30 μg DAG) with 10 ml *i*-PrOH/$CHCl_3$ (9:1, v/v) by shaking 10 min. Add 1 g anhyd K_2CO_3 and shake an additional 10 min. Centrifuge and remove 9 ml of the org layer. Evap to dryness at 40°C.

Diss the residue in 100 μl acetone containing 250 μg butaneboronic acid and 5.0 μg tetra-(OTMS)-*meso*-erythritol int stand. Allow to stand at room temp 5 min then analyze an aliquot by GC.

Comments. Acetyl and TMS derivs were also examined but the boronate deriv was found to be simpler to prepare and adequate for the analysis.[1]

References

[1]T. Kimura *et al.*, *Clin. Chem.* **22**, 1639 (1976).

Derivative Type. 3, 5-Dinitrobenzoate.

Reagent. 3, 5-Dinitrobenzoyl chloride (DNB-Cl).

Compound Type. Glycols.

Reaction

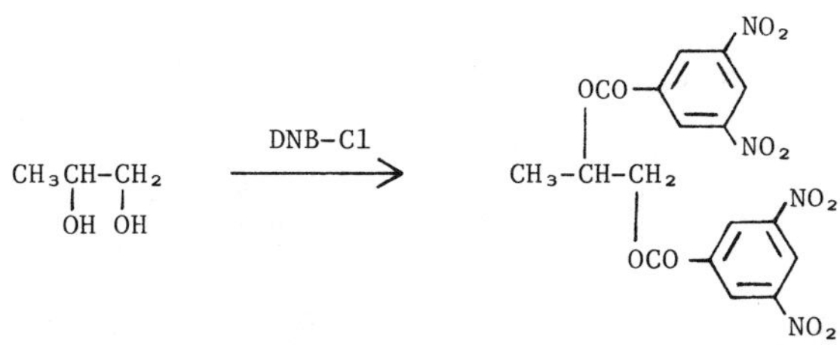

Typical Procedure. Diss 0.5 g DNB-Cl in 30 ml pyridine (use 20 ml for polyethylene glycol 300–1000). Add 1 meq of glycol, mix, and heat at 60°C for 15 min. Adjust to pH 2.5 with HCl. Cool to room temp and add 25 ml BuOAc. Sep BuOAc and wash with 50 ml 1% Na_2CO_3, 25 ml 0.5 *N* H_2SO_4, and 10 ml H_2O. Analyze the BuOAc soln HPLC.

Comments. The method was used for chromogenic deriv for analysis of monoethylene and diethylene glycols in polyethylene glycol using propylene glycol as int stand.

References

[1]M. A. Carey and A. F. Persinger, *J. Chromatogr. Sci.* **10**, 537 (1972).

1.2.2. Derivatives of Phenylethylene Glycols

Derivative Type. Trifluoroacetyl.

Reagent. Trifluoroacetic anhydride (TFAA).

Compound Type. 3-Methoxy-4-hydroxyphenylethylene glycol (MOPEG).

Reaction

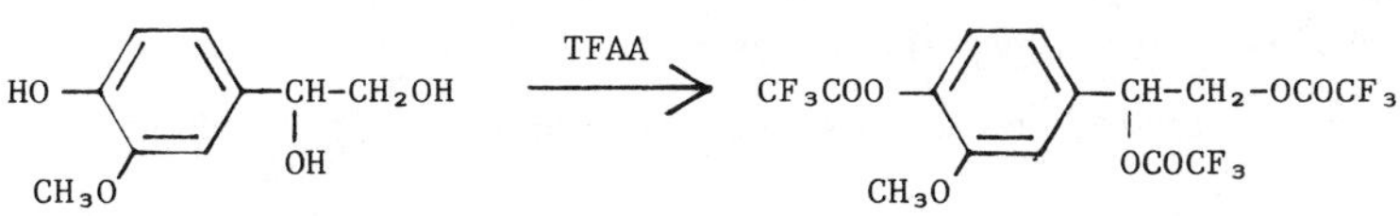

Typical Procedure. Acidify a urine sample (corresponding to 2 mg creatinine) with 0.1 vol 3 *N* HCl and ext with 2× 2 vol EtOAc. Adjust the aq phase to pH 5.2 and incubate overnight with 0.1 ml Glusulase.[2] Dil to 5 ml, centrifuge, and pass supernatant over a 5× 1.2 cm column of Biorad AG1X4 resin (Cl form). Wash column with 10 ml H_2O and elute with 20 ml H_2O. Collect the eluate, sat with NaCl, and ext with 20, 10, and 10 ml EtOAc. Evap ext to dryness and transfer residue with 2× 1 ml EtOAc. Add 0.25 ml TFAA and allow to stand 1 hr at room temp. Evap to dryness under N_2 and diss residue in 1 ml EtOAc. Analyze by ECD-GC.

Comments. The method was adapted for analysis of CSF.[3] Tyramine was added to the CSF samples and the size of the TFA-tyramine peak used as a check for deriv hydrolysis.[3] Dekirmenjian and Maas[4] used the deriv in their reported improved procedure for MOPEG. Bertilsson[5] used the deriv for a GC-MS-SIM assay of MOPEG in CSF.

References.

[1]S. Wilk, *et al.*, *Clin. Chin. Acta* **16**, 403 (1967).
[2]Glucuronidase-sulfatase, Endo Corporation, Garden City, NY.
[3]S. Wilk, *et al.*, *Anal. Biochem.* **39**, 498 (1971).
[4]H. Dekirmenjian and J. W. Maas, *Anal. Biochem.* **35**, 113 (1970).
[5]L. Bertilsson, *J. Chromatogr.* **87**, 147 (1973).

Derivative Type. Pentafluoropropionyl.

Reagent. Pentafluoropropionic anhydride

Compound Type. Alcoholic metabolites of biogenic amines.

Reaction.

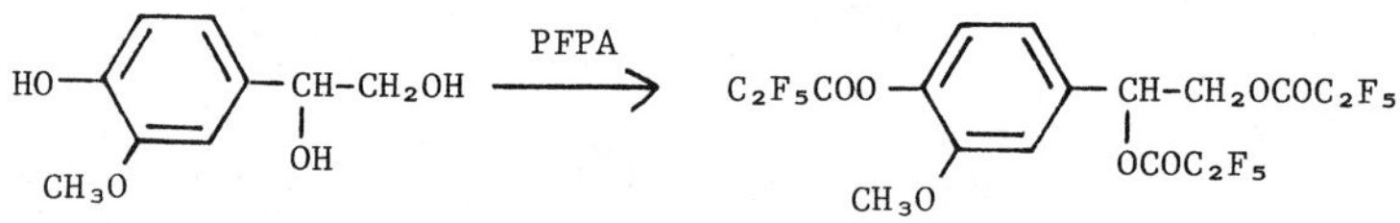

Typical Procedure.[1] Diss sample in 0.1 ml EtOAc (dried and stored over $CaCl_2$) and add 50 μl PFPA. Heat at 60°C for 10 min. Evap under N_2 and diss residue in 0.2 ml EtOAc. Analyze by GC or GC-MS.

Comments. Complete MS of the derivs of 4-hydroxy-3-methoxyphenylethanol, 4-hydroxy-3-methoxyphenylglycol, and tryptophol were reported.[2] The deriv was used for quantitation of the alcoholic metabolites in urine,[2] brain tissue,[3-5] and CSF.[3]

References.

[1]F. Karoum, *et al.*, *Biochem. Med.* **5**, 505 (1971).
[2]F. Karoum, *et al.*, *Clin. Chim. Acta* **43**, 127 (1973).
[3]F. Karoum, *et al.*, *Biomed. Mass Spectrom.* **2**, 183 (1975).
[4]F. Karoum, *et al.*, *J. Neurochem.* **25**, 653 (1975).
[5]F. Karoum, *et al.*, *Brit. J. Pharmacol.* **56**, 403 (1976).

Derivative Type. Acetyl, heptafluorobutyryl.

Reagent. Acetic anhydride (Ac_2O) / sodium bicarbonate, heptafluorobutyric anhydride (HFBA).

Compound Type. 4-Hydroxy-3-methoxyphenylethylene glycol (MOPEG) and 3,4,-dihydroxyphenylethylene glycol.

Reaction.

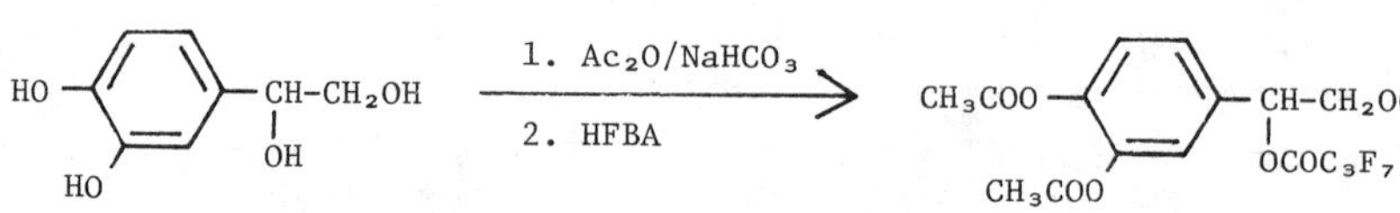

Typical Procedure. Homogenize tissue (up to 0.5 g) in 4 ml 0.1 *N* HCl. Sat with Solid KCl and centrifuge. Mix supernatant with 0.3 ml Ac_2O and add a slight excess of $NaHCO_3$ portionwise. Ext with 2× 10 ml CH_2Cl_2. Dry ext by filtering through Na_2SO_4 and evap under N_2 at 60°C. Diss residue in 20–200 μl THF and add 0.25 vol of HFBA. Allow to stand at room temp 5–10 min then evap to dryness under N_2 at 60°C. Diss deriv in hexane and analyze by ECD-GC.

Comments. Norepinephrine and normetanephrine could be measured by the same method in pure soln but not in crude tissue ext.[1] MOPEG from serum has been analyzed as the acetyl, TFA deriv by GC-MS-SIM.[2]

References.

[1]D. F. Sharman, *Brit. J. Pharmacol.* **36**, 523 (1969).

[2]S. Takahashi, *et al.*, *Clin. Chim. Acta* **81**, 183 (1977).

Derivative Type. Diacetyl, bis-Trimethylsilyl.

Reagent. Acetic anhydride (Ac_2O), Hexamethyldisilazane (HMDS) / trimethylchlorosilane (TMCS) / pyridine.

Compound Type. 3, 4-Dihydroxyphenylglycol.

Reaction.

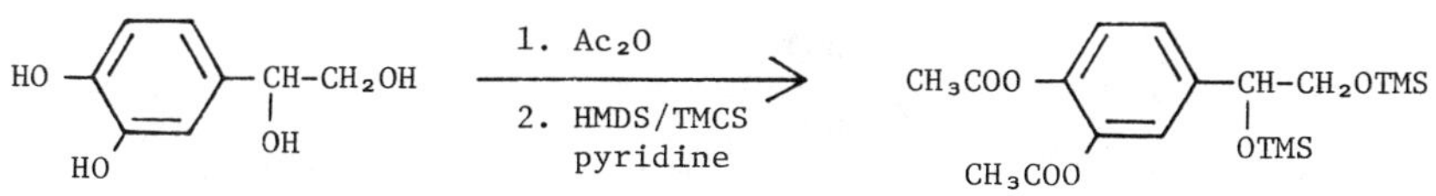

Typical Procedure. To 5 ml urine add 0.5 mg ascorbic acid (antioxidant) and dil with 5 ml H_2O. Adjust to pH 5.8 and add 0.2 ml Glusulase.[2] Flush tube with N_2, seal, and incubate at 37°C for 18 hr. Dil to 20 ml with H_2O, add 0.5 ml 4% ascorbic acid soln and 0.5 ml 8% EDTA soln, and adjust to pH 7.1. Pass through a column (5 mm i.d.) of 2 g acid alumina (washed previously with 5 ml 0.2 *M* NaOAc, pH 8.5). Wash column with 10 ml H_2O and elute with 10 ml 0.25 *N* HCl.

Add 0.2 g ascorbic acid and 5 ml acetone to the eluate. Add carefully 2.25 g $KHCO_3$ followed by 1ml Ac_2O. Stir 30 min at room temp, adjust to pH 8.5 with 5 *N* NaOH and ext with 5 × 10 ml $CHCl_3$. Dry ext over Na_2SO_4 and evap under N_2. Add 100 μl HMDS/TMCS/pyridine (3:1:9, V/V/V) and heat 1 hr at 40°.C Analyze by GC.

Comments. Acetylation in aq soln improved the extractability of the compd. MS data were reported for the deriv.

References.

[1]Z. Kahane, *et al.*, *Clin. Chim. Acta* **73**, 203 (1976).

[2]Sulphatase-glucuronidase, Endo Laboratories, Garden City, NY.

Derivative Type. *n*-Butylboronate.

Reagent. *n*-Butaneboronic acid.

Compound Type. Phenylethylene glycol.

Reaction.

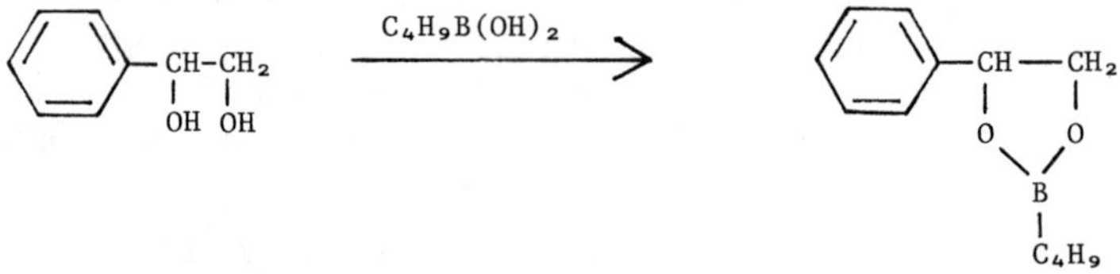

Typical Procedure. Diss sample in 100 μl of 0.5 μg/μl *m*-dinitrobenzene (int stand) in acetone. Add 2.5 μl of a 100 μg/μl soln of *n*-butaneboronic acid in DMF. Analyze a 1 μl aliquot by GC.

Comments. The method was used in assays for styrene monooxygenase and styrene epoxide hydratase. MS data were given for the deriv.

References.

[1]G. Belvedere, *et al.*, *J. Chromatogr.* **118**, 387 (1976).

Derivative Type. Methane or *n*-butaneboronates, trimethylsilyl ether.

Reagent. Methane or *n*-butaneboronic acid, bis(trimethylsilyl)trifluoroacetamide (BSTFA).

Compound Type. 3-Methoxy-4-hydroxyphenylethylene glycol (MOPEG).

Reaction

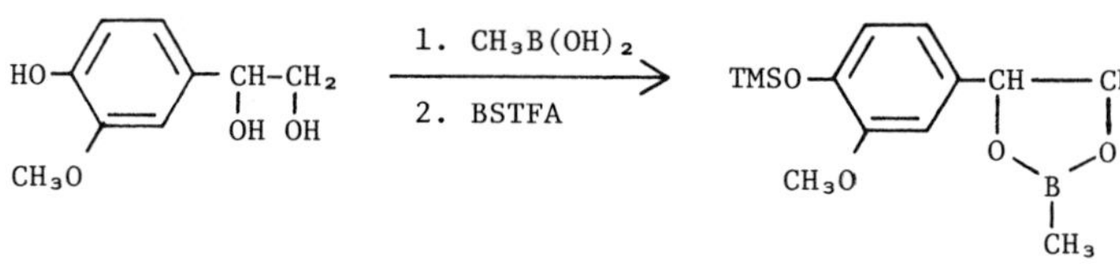

Typical Procedure. Diss 5–40 μg MOPEG in 0.3 ml dimethoxypropane (DMP) containing 300 μg of CH_3-$B(OH)_2$ or C_4H_9-$B(OH)_2$ and 15 μg phenanthrene int stand. Allow to stand 15 min at room temp, then inject 1 μl together with 0.2 μl BSTFA into GC.

Comments. DMF, pyridine, and acetone were also examined as solvents, but DMP was found to be best in terms of both deriv formation and stability. Direct silylation in the injection step was used because mixing BSTFA with the reaction mixt led to decomp of the boronate and formation of the tri-TMS deriv. The acetyl boronate deriv was prepared by acylation of the phenolic hydroxyl followed by reaction with the boronic acid.[1] MS data were reported for the derivs.

References

[1]M. Cagnasso and P. A. Biondi, *Anal. Biochem.* **71**, 597 (1976).

[2]P. Biondi and M. Cagnasso, *Anal. Lett* **9**, 507 (1976).

1.2.3. Derivatives of Dihydrodiols

Derivative Type. *N, N*-Dimethyl, *O*-trimethylsilyl.

Reagent. Diazomethane, bis(trimethylsilyl)acetamide(BSA)/trimethylchlorosilane (TMCS).

Compound Type. Dihydrodiol metabolite of phenobarbital and mephobarbital.

Reaction

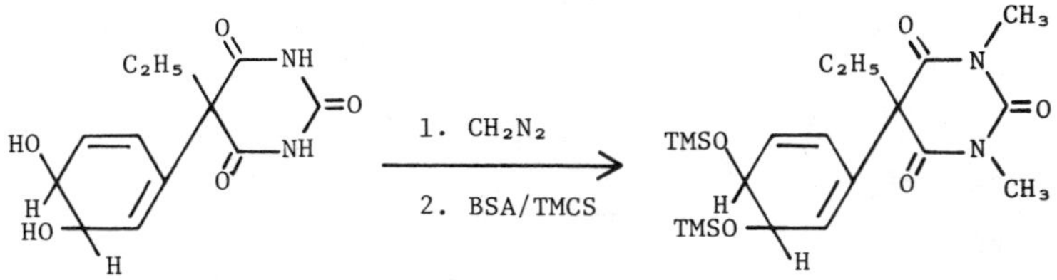

Typical Procedure. Diss metabolites in 0.5 ml MeOH, and add excess CH_2N_2 in Et_2O. After 15 min evap under N_2, diss residue in 10 μl pyridine, and add 10 μl BSA and 5 μl TMCS. Allow to stand overnight at room temp; then anaylze by GC-MS.

Comments. The method yielded a dimethyl deriv of the barbiturate moiety (primarily *N, N*- but with some of the *N,O*- and *O,O*-isomers). Phenolic hydroxyls were converted to a mixt of methyl and TMS ethers. The d_9-TMS analog was also prepared. A full MS of the deriv was given. The same deriv was used to identify a dihydrodiol metabolite of phenytoin.[2]

References

[1]D. J. Harvey *et al.*, *Res. Commun. Chem. Pathol. Pharmacol.* **3**, 557 (1972).

[2]M. G. Horning *et al.*, *Anal. Lett.* **4**, 537 (1971).

Derivative Type. Trimethylsilyl.

Reagent. Hexamethyldisilazane(HMDS)/trimethylchlorosilane(TMCS)/pyridine, bis(trimethylsilyl)trifluoroacetamide(BSTFA).

Compound Type. 3-Methylcholanthrene-11,12-dihydrodiol.

Reaction

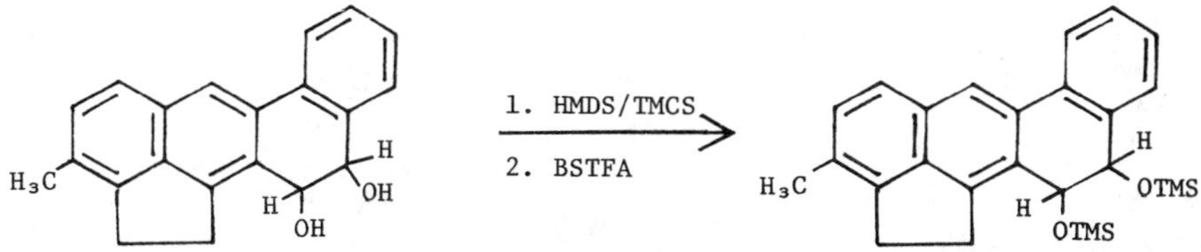

Typical Procedure. To a 1 mg sample add 0.4 ml Tri-Sil[3] and heat at 60° C for 30 min. Add 0.2 ml BSTFA and dil to 1 ml with CH_3CN. Store at 5°C prior to analysis by GC.

Comments. The method was used in a GC assay of epoxide hydrase.[2] 3-Methylcholanthrene-11,12-oxide formed a mixt of 11-TMSO and 12-TMSO derivs under the same conditions. These mono-TMS derivs were separable by GC from the dihydrodiol di-TMS deriv.

References

[1]T. Stoming *et al.*, *Life Sci. II* **12**, 425 (1973).

[2]T. Stoming and E. Bresnick, *Science* **181**, 951 (1973).

[3]HMDS-TMCS-pyridine, Pierce Chemical Co., Rockford, IL.

Derivative Type. Chloromethyldimethylsilyl.

Reagent. Bis(chloromethyl)tetramethyldisilazane/chloromethyldimethylchlorosilane.

Compound Type. Polycyclic aromatic hydrocarbon dihydrodiols.

Reaction

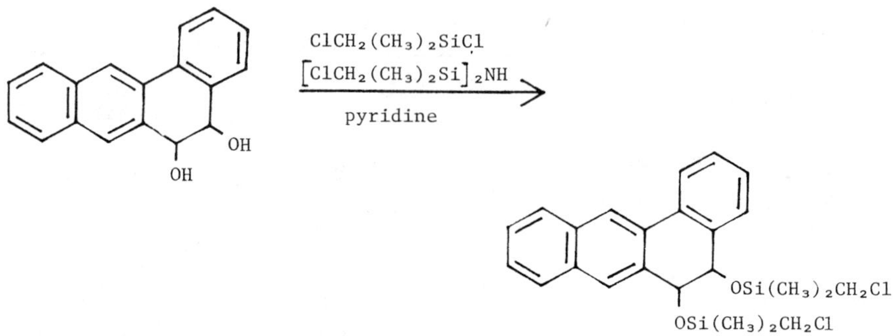

Typical Procedure. To a 0.4 ml microsomal incubation mixt (to which was added 62.5–1250 ng arene oxide) add 2 ml cold EtOAc containing 50 ng aldrin-6, 7-dihydrodiol int stand. Shake 2 min, centrifuge, and remove org layer. Ext aq layer with another 1 ml EtOAc and combine ext. Dry over $MgSO_4$ and evap to dryness at 40° C under N_2. Add 20 μl bis (chloromethyl)tetramethyldisilazane/chloromethyldimethylchlorosilane/pyridine (1:0.5:2, v/v/v) and heat at 60°C for 30 min. Evap to dryness and diss residue in 50 μl hexane. Analyze 1–2 μl by ECD-GC.

Comments. The method was used for an assay of epoxide hydrase activity. TMS derivs were also prepared using 20 μl Tri-Sil®[2] and 10 μl BSTFA under the same conditions. The TMS derivs were quantitated by GC-MS-SIM using deuterated dihydrodiol int stand. Conc as low as 1 ng/ml could be estimated.

References

[1]A. Bettencourt *et al.*, *J. Chromatogr.* **134**, 323 (1977).

[2]HMDS-TMCS-pyridine, Pierce Chemical Company, Rockford, IL.

Derivative Type. Butylboronate.

Reagent. *n*-Butaneboronic acid.

Compound Type. Dihydrodiol metabolites.

Reaction

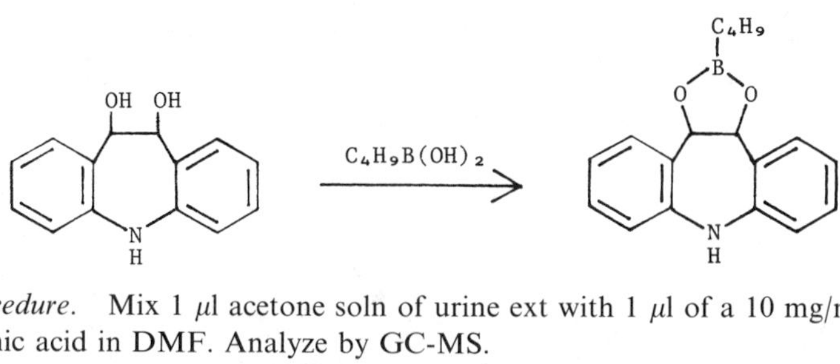

Typical Procedure. Mix 1 μl acetone soln of urine ext with 1 μl of a 10 mg/ml soln of *n*-butaneboronic acid in DMF. Analyze by GC-MS.

Comments. The method was applied to dihydrodiol metabolites of iminostilbene (shown above),[1] cytenamide,[2] and carbamazepine.[3] The underiv compds underwent rearrangement to the corresponding acridine- or anthracenecarboxaldehydes in the GC.

References

[1]C. Pantarotto *et al.*, *J. Chromatogr.* **134**, 307 (1977).

[2]A. Frigerio *et al.*, *J. Chromatogr.* **134**, 299 (1977).

[3]P. L. Morselli and A. Frigerio, *Drug Metab. Rev.* **4**, 97 (1975)

1.3. DERIVATIVES OF THIOLS

Derivative Type. Pentafluorobenzyl.

Reagent. Pentafluorobenzyl bromide (PFB-Br)/potassium carbonate.

Compound Type. Mercaptans, phenols, and organic acids.

Reaction

$$\begin{matrix} R\text{-}SH \\ + \\ Ar\text{-}OH \\ + \\ R\text{-}COOH \end{matrix} \quad \xrightarrow[K_2CO_3]{PFB\text{-}Br} \quad \begin{matrix} R\text{-}SCH_2C_6F_5 \\ + \\ Ar\text{-}OCH_2C_6F_5 \\ + \\ R\text{-}COOCH_2C_6F_5 \end{matrix}$$

Typical Procedure. Diss water ext sample in approx 4 ml acetone per mg sample and add approx 10 mg PFB-Br per mg sample (as approx 40 mg/ml soln in EtOH). Add approx 4 mg K_2CO_3 per mg sample and reflux 2 hr. (Use both reagents in at least five-fold excess). Evap at 50° C under vac (50mm). Diss residue in hexane/Et_2O/acetone (97:1:1, v/v/v) and dil as necessary for ECD-GC analysis.

Comments. The method used for analysis of water pollutants extracted by carbon filtration and $CHCl_3$ ext.[1] The reagent has also been used for analysis of mixt of phenols and anilines.[2] Deriv without K_2CO_3 gave only aniline derivs while addn of K_2CO_3 gave only phenol derivs.

References

[1]F. K. Kawahara, *Environ. Sci. Technol.* **5**, 235 (1971).

[2]D. E. Bradway and T. Shafik, *J. Chromatogr. Sci.* **15**, 322 (1977).

Derivative Type. 2, 4-Dinitrophenylthioethers and sulfones.

Reagent. 2, 4-Dinitrochlorobenzene (DNCB), potassium permanganate.

Compound Type. Thiols.

Reaction

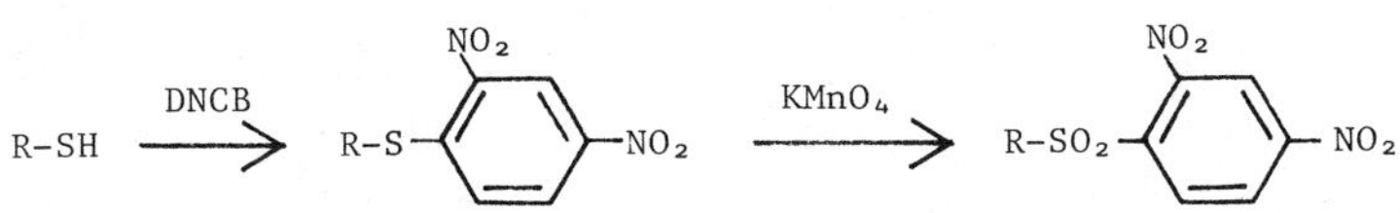

Typical Procedure. Diss conc of volatiles in 3 ml $CHCl_3$ and add 1 ml 0.5 *M* H_2SO_4. Dist under vac and collect distillate in a trap cooled with liquid N_2. Diss the frozen distillate in 10 ml MeOH and add 4 drops 6 *M* NaOH. Add 1 ml MeOH containing 4 mg/ml DNCB. Reflux 10 min, and then evap under vac. Ext the residue with $CHCl_3$ until the ext is no longer yellow. Evap under vac and diss the residue in a known vol of $CHCl_3$ prior to GC analysis.

To prepare the sulfones diss the residue in a min vol of HOAc and add a few drops of aq

$KMnO_4$ until the soln remains pink. Remove excess oxidant by adding aq Na_2SO_3 until the color just disappears. Dil with H_2O, and ext with $CHCl_3$. Wash ext with 10% aq Na_2CO_3 then evap under vac. Diss the residue in a known vol of $CHCl_3$ for GC analysis.

Comments. The method was used to identify thiols in food aromas. Benzoyl derivs were also prepared using benzoyl chloride in pyridine.

References

[1]L. Gasco and R. Barrera, *Anal. Chim. Acta* **61**, 253 (1972).

Derivative Type. *N*-Dansylaminoethyl.

Reagent. *N*-Dansylaziridine.

Compound Type. Sulfhydryl compounds (e.g., proteins, amino acids).

Reaction

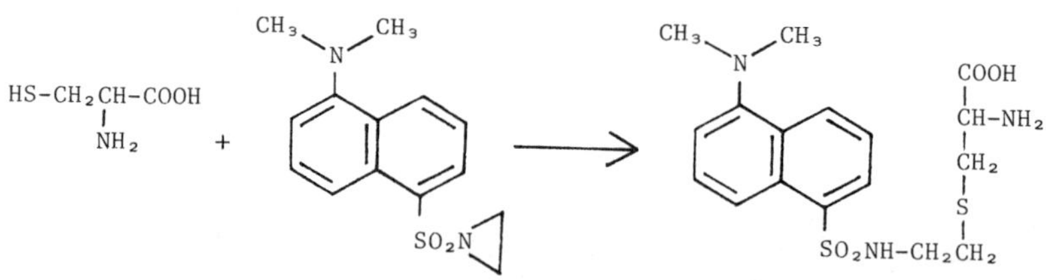

Typical Procedure (Preparative). Diss 279 mg of reagent in 20 ml EtOH and add dropwise with stirring a soln of 135 mg Cys in 30 ml H_2O adjusted to pH 8.6 with Na_2CO_3. Stir overnight at room temp then evap to dryness under vac. Recryst from EtOH-BuOH-H_2O.

Comments. The reagent reacted with free protein sulfhydryl groups to produce fluorescent derivs. The reagent was developed for determining sulfhydryl content of proteins but could be useful as a specific fluorogenic deriv reagent for HPLC. The synthesis of the reagent from DANS-Cl and ethyleneimine was reported.

References

[1]W. H. Scouten *et al.*, *Biophys. Acta* **336**, 421 (1974).

1.4. DERIVATIVES OF PHENOLS

1.4.1. Alkyl Derivatives of Phenols

Derivative Type. Methyl ethers and esters.

Reagent. Diazomethane.

Compound Type. Dinitrophenols and chlorophenoxyacetic acids.

Reaction

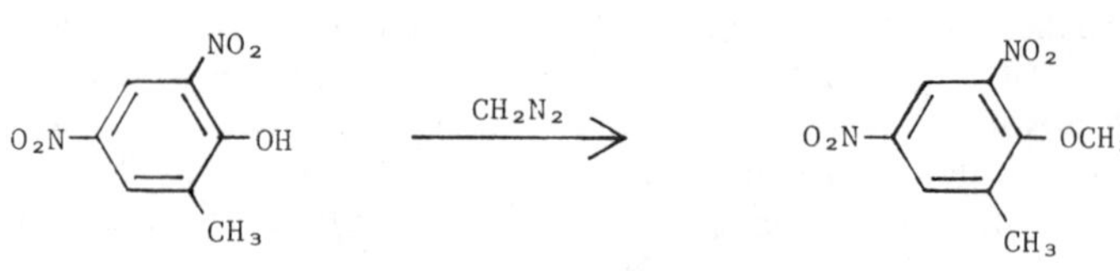

Typical Procedure. To a sample containing 2–8 μg each of 8 phenols and acids add 2 ml CH_2N_2 in Et_2O and 1 ml isooctane. Mix, add a boiling chip, and immediately evap Et_2O from the sample in a 70° H_2O bath until only the isooctane remains. Analyze by GC.

Comments. The use of CH_2N_2 alone, with MeOH, and with MeOH and HCl were compared. The completeness of methylation was variable with the latter two methods. Isooctane was added to minimize losses of the volatile derivs.

References

[1]S. F. Howard and G. Yip, *J. Assoc. Off. Anal. Chem.* 54, 970 (1971).

Derivative Type. Methyl ether.

Reagent. Methyl iodide/potassium carbonate.

Compound Type. Hydroxylated 4,4′ -dichlorobiphenyl.

Reaction

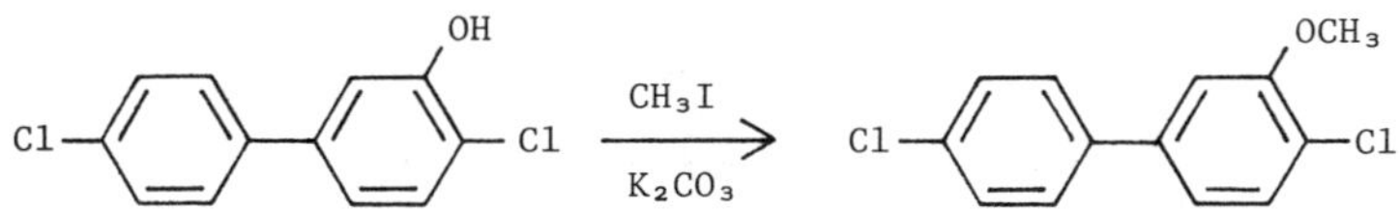

Typical Procedure. Collect urine in 4 *N* H_2SO_4. Heat at 80° for 2 hr then ext 3 x with Et_2O hexane (1 : 1, v/v). Evap the ext and diss the residue in acetone. Treat with excess CH_3I and K_2CO_3 at reflux overnight. Evap to dryness and diss the residue in hexane. Isolate the derivs by TLC (silica gel, hexane) and analyze by GC-MS.

Comments. The derivs were used in identification of metabolites of chlorinated biphenyls. Methylation was also performed with CH_2N_2. Cyclic *n*-butylboronate derivs of dihydroxy compounds were formed with butaneboronic acid in DMF with dimethoxypropane.

References

[1]M. T. M. Tulp and O. Hutzinger, *J. Chromatogr.* **139**, 51 (1977).

[2]M. T. M. Tulp *et al., Chemosphere* **5**, 425 (1976).

Derivative Type. Methyl.

Reagent. Tetrabutylammonium hydrogen sulfate/methyl iodide.

Compound Type. 8-Hydroxyquinolines.

Reaction

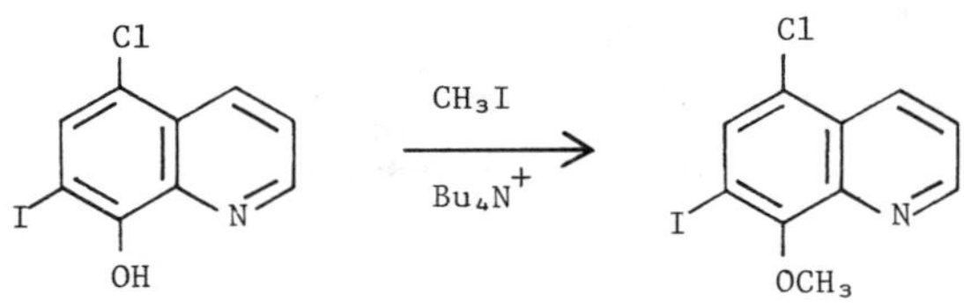

Typical Procedure. Alkalinize 0.1 or 0.5 ml plasma or urine with 2 drops 5 *M* NaOH and add 0.1 ml int stand soln. Add 1 ml 0.1 *M* tetrabutylammonium in pH 11 phosphate buffer and dil with H_2O to 2 ml. Add 1 ml CH_2Cl_2 and 100 μl CH_3I and shake 30 min. Remove org phase and add to 0.5 ml heptane. Evap CH_3I and CH_2Cl_2 under N_2. Add 1 ml heptane, and wash soln with 1 ml satd Ag_2SO_4 soln. Analyze the org phase by ECD-GC.

Comments. The method was used to quantitate clioquinol (shown above) and cloroquinaldol (5, 7-dichloro-2-methyl-8-quinolinol) for each using the other as int stand (1 μg/ml soln in pH 11 phosphate buffer). Tetrabutylammonium was reported to cause less decomposition than tetrahexylammonium which had previously been used[2] for clioquinol.

References

[1]P. Hartvig and C. Fagerlund, J. Chromatogr. **140**, 170 (1977).

[2]P. H. Degen *et al.*, *J. Chromatogr.* **117**, 407 (1976).

Derivative Type. *O*-Heptyl, *N*-methyl.

Reagent. Iodoheptane/tetramethylammonium hydroxide (TMAH), trimethylphenylammonium hydroxide (TMPAH).

Compound Type. Acetaminophen.

Reaction

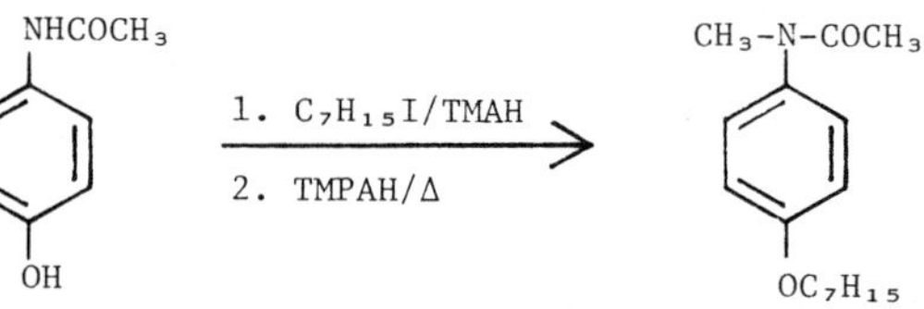

Typical Procedure. Combine 1 ml serum or saliva with 1 ml dist H_2O, 1 g NaCl, and 10 ml Et_2O/toluene (40:60, v/v) containing 10 μg of int stand (*N*-propionyl-*p*-aminophenol). Shake 15 min, then combine with 2 ml of 0.1 *M* NaOH. Shake 10 min, centrifuge, and discard org phase. Add 0.5 ml 1 *M* pH 6.8 phosphate buffer, 1 g NaCl and 10 ml Et_2O. Shake 10 min, centrifuge, remove Et_2O phase, and evap under N_2 in 40° H_2O bath. Add 80 μl DMA and 20 μl TMAH soln (24 g/1 MeOH). Vortex 5 sec after each addition and add 20 μl iodoheptane. Vortex 5 sec, allow to stand at room temp 10 min, then centrifuge to remove tetramethylammonium iodide. Transfer supernatant to 1 ml vial and evap under N_2 at 40°. Add 25 μl TMPAH soln (25 g/1 MeOH), mix on vortex for 10 sec, and inject 1 μl into GC.

Comments. The deriv was analyzed on a 3% OV-17 column (150–260°) with the injector at 250°.

References

[1]W. A. Dechtiaruk *et al.*, *Clin. Chem.* **22**, 879 (1976).

Derivative Type. Pentafluorobenzyl.

Reagent. Pentafluorobenzyl bromide (PFB-Br).

Compound Type. Phenol and carboxylic acid pesticide residues.

Reaction

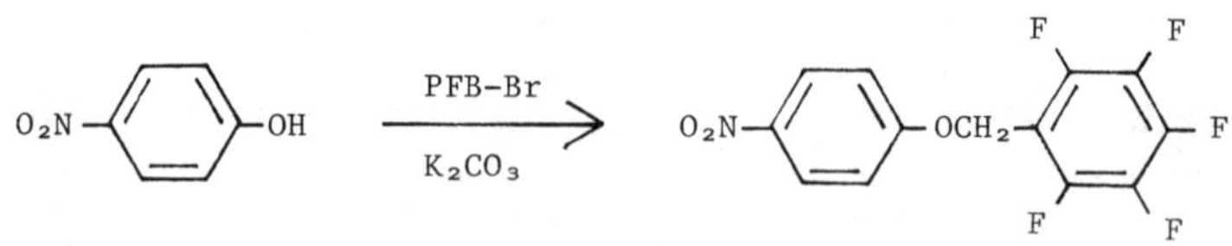

Typical Procedure. To a soln of $\leqslant$50 μg pesticide in acetone add *ca* 50 mg K_2CO_3 and evap to 0.5 ml. Add 1 ml PFB-Br (20 mg/ml) in acetone and heat at 50° for 15 min. Add 1.5 ml isooctane and evap to 1.5ml. Add to a column of 1 g silica gel prewet with 5 ml hexane. Rinse the tube with 1 ml hexane and add rinse to the column. Elute excess reagent with 8 ml 5% benzene in hexane. Elute with 6 ml 25% benzene in hexane and collect. Elute with 8 ml 75% benzene in hexane and collect separately. Conc each eluate and analyze by ECD-GC.

Comments. The method was applied to 1-naphthol, *p*-nitrophenol, 2,4-dichlorophenoxyacetic acid (2,4-D),*p,p'*-dichlorodiphenylacetic acid (*p,p'*-DDA), and *sec*-amyl phenols.[1] The less polar 1-naphthol and *sec*-amyl phenol derivs were eluted with 25% benzene. The method was an improvement over the earlier method, which used distillation to remove excess reagent. In this work the derivs of 16 phenols and mercaptans were separated by GC.[2]

References

[1]L. G. Johnson, *J. Assoc. Off. Anal. Chem.* **56**, 1503 (1973).

[2]F. K. Kawahara, *Anal. Chem.* **40**, 1009, 2073 (1968).

1.4.2. Acyl Derivatives of Phenols

Derivative Type. O-Acetyl.

Reagent. Acetic anhydride (Ac_2O)/pyridine.

Compound Type. Acetaminophen.

Reaction.

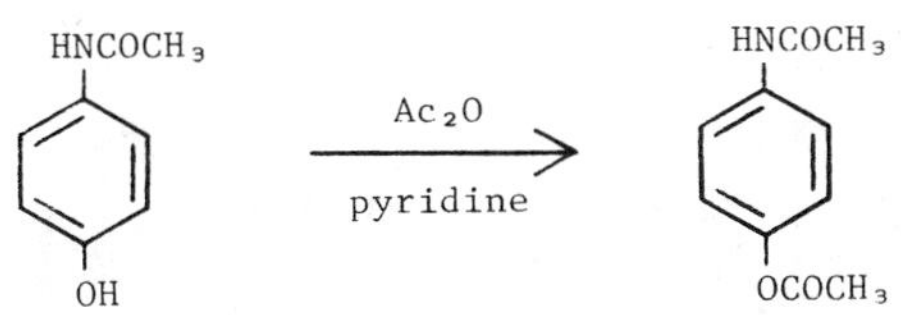

Typical Procedure.[1] Mix 2.0 ml plasma or urine with 1.0 ml 1 *M* pH 7.4 phosphate buffer. Ext with 5.0 ml EtOAc containing 5 μg/ml *N*-butyryl-*p*-aminophenol. Evap ext to dryness and diss residue in 5 μl pyridine and 15 μl Ac_2O. Heat at 45° for 20 min then analyze 1–3 μl aliquots by GC.

Comments. Acetaminophen had earlier been analyzed as the *O*-TMS (from TMSI) or *N,O*-di-TMS (from BSA) derivs but the derivs were quite moisture labile.[2] The acetyl deriv was more stable and gave better sensitivity.

References.

[1]L. F. Prescott, *J. Pharm. Pharmacol.* **23**, 807 (1971).

[2]L. F. Prescott, *J. Pharm. Pharmacol.* **23**, 111 (1971).

Derivative Type. Acetyl.

Reagent. Acetic anhydride (Ac_2O)/4-dimethylaminopyridine (DMAP).

Compound Type. Phenols.

Reaction.

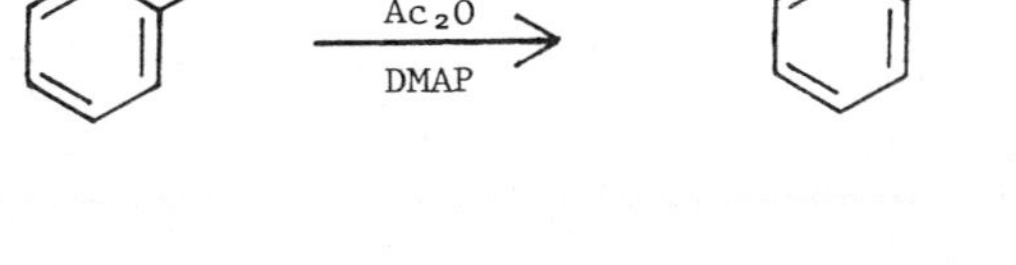

Typical Procedure. To an aq phenol soln or hydrolyzed urine sample add 2 ml pH 7 (0.5 *M*) phosphate buffer, sat with NaCl, and ext with 3 × 10 ml Et_2O and 3 × 10 ml EtOAc. Pass ext through a column (1 × 10 cm) of Na_2SO_4 then wash column with 3 × 10 ml EtOAc. To the combined ext and wash add 0.5 ml Ac_2O and 10 mg DMAP in EtOAc. Allow to react 1 hr then wash with 3 × 20 ml satd $NaHCO_3$. Evap to *ca* 1 ml and analyze by GC.

Comments. The DMAP was a much more powerful acylation catalyst than pyridine alone and gave rapid and quant deriv of phenols in dil soln.[1] Connors and Albert[2] used DMAP for analytical acetylation of alcohols and phenols. Under the reaction conditions using DMAP, it was observed that amino acids were converted to azalactones at room temp.[1]

References.

[1]V. Fell and C. R. Lee, *J. Chromatogr.* **121**, 41 (1976).

[2]K. A. Connors and K. S. Albert, *J. Pharm. Sci.* **62**, 845 (1973).

Derivative Type. Trifluoroacetyl.

Reagent. Trifluoroacetic anhydride (TFAA).

Compound Type. Phenols.

Reaction.

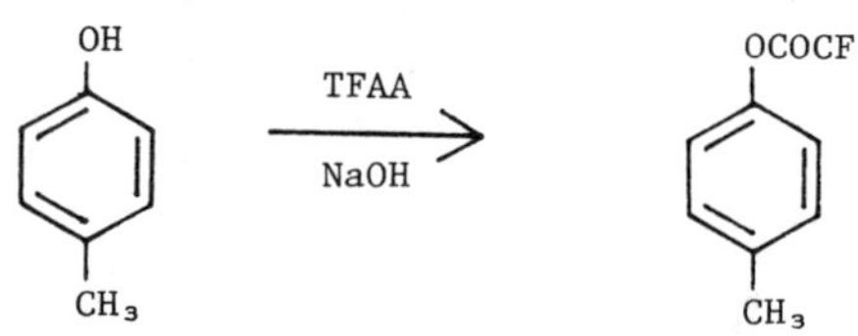

Typical Procedure. (Preparative). Suspend 0.1 mol of the phenol in 0.11 mol TFAA and add a trace of 50 % aq NaOH. Reflux until the mixt is homogeneous. Distil off excess TFAA and the by-product acid. Distil the deriv. Analyze by GC.

Comments. Derivs were prepared from 19 methylated phenols.

References.

[1]A. T. Shulgin, *Anal. Chem.* **36**, 920 (1964).

Derivative Type: Heptafluorobutyryl

Reagent. Heptafluorobutyric anhydride (HFBA)/trimethylamine (TMA).

Compound Type: Phenols.

Reaction.

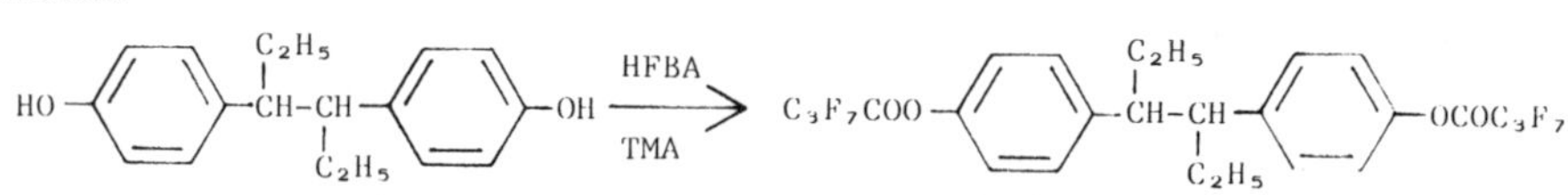

Typical Procedure: Diss ng amts of the phenol in 0.5 ml benzene. Add 100 μl 0.1 *M* TMA in benzene and 10 μl HFBA. Let stand 10 min at room temp and shake with 1 ml pH 6.0 phosphate buffer for 30 sec. Centrifuge and analyze benzene phase by ECD-GC.

Comments: Ten phenols (hexestrol shown) were studied with the derivs detectable to *ca* 1 pg by ECD.[1] The deriv was used for analysis of several "agricultural chemicals" in foods.[2] These included diethylstilbestrol, hexestrol, estradiol, clopidol, carbofuran, 3-ketocarbofuran,

3-hydroxycarbofuran, baygon, and linuron, In comparing ECD with electrolytic conductivity detection it was found that ECD had better sensitivity but poorer selectivity.[2]

References.

[1]H. Ehrsson, *et al.*, *Acta Pharm. Suecica.* **8**, 319 (1971).

[2]J. F. Lawrence and J. J. Ryan, *J. Chromatogr.* **130**, 97 (1977).

Derivative Type. Pentadecafluorooctanoyl (PFO).

Reagent. Pentadecafluorooctanoic anhydride (PFOA)/trimethylamine (TMA).

Compound Type. Diethylstilbestrol and clopidol.

Reaction.

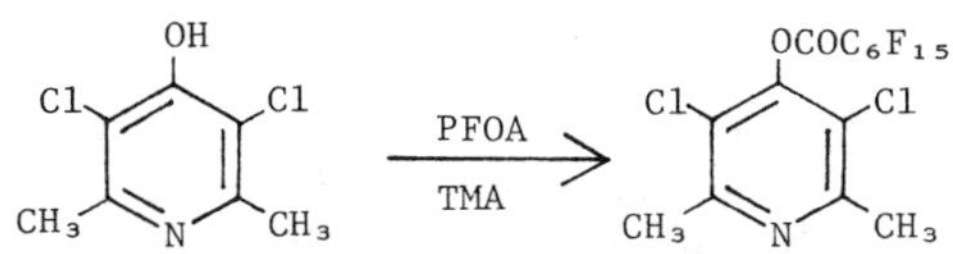

Typical Procedure. Dis 1–10 μg of sample in 0.1 ml 0.1 *M* TMA in benzene. Add 20 μl PFOA and dil to 0.5 ml with benzene. Allow to stand at room temp 1 hr. Add 0.5 ml benzene and 0.5 ml H_2O and shake 30 sec. Centrifuge and analyze an aliquot of the benzene layer by ECD-GC. Alternatively, wash with an additional 2 × 5 ml H_2O and analyze by GC with electrolytic conductivity detection.

Comments. The PFO derivs were compared with TFA, PFP, and HFB derivs (prepared in the same manner using TFAA, PFPA, and HFBA, respectively) in terms of detection sensitivity by ECD and electrolytic conductivity detection (halogen mode). ECD sensitivities were found to be *ca* 10–100 times greater. The HFB derivs showed the greatest ECD sensitivity and the PFO derivs the greatest sensitivity by electrolytic conductivity detection.

References.

[1]J. J. Ryan and J. F. Lawrence, *J. Chromatogr.* **135**, 117 (1977).

Derivative Type Chloroacetate.

Reagent. Chloroacetic Anhydride.

Compound Type. Phenols.

Reaction.

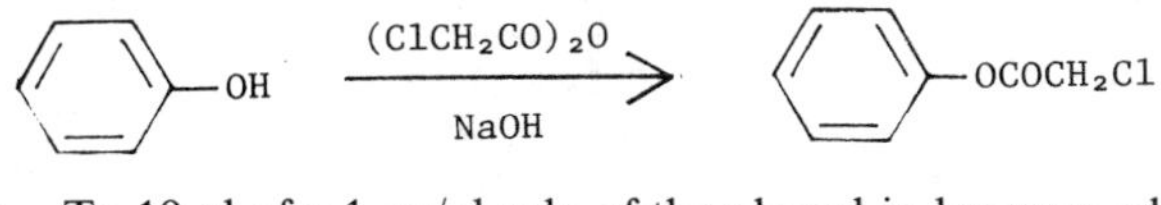

Typical Procedure. To 10 μl of a 1 μg/μl soln of the phenol in benzene add 15 ml of 0.25 *N* NaOH and 10 ml of a 5 mg/ml soln of chloroacetic anhydride in benzene. Shake 2 min and analyze a 5 μl aliquot of the benzene by ECD-GC.

Comments. A linear response was demonstrated for 0.1–3 ng of injected deriv. GC data were reported for 32 phenols, 12 of which were phenols that result from hydrolysis of pesticides.

References.

[1]R. J. Argauer, *Anal. Chem.* **40**, 122 (1968).

Derivative Type. Dichloroacetyl ester.

Reagent. Dichloroacetyl chloride.

Compound Type. Diethystilbestrol (DES).

Reaction

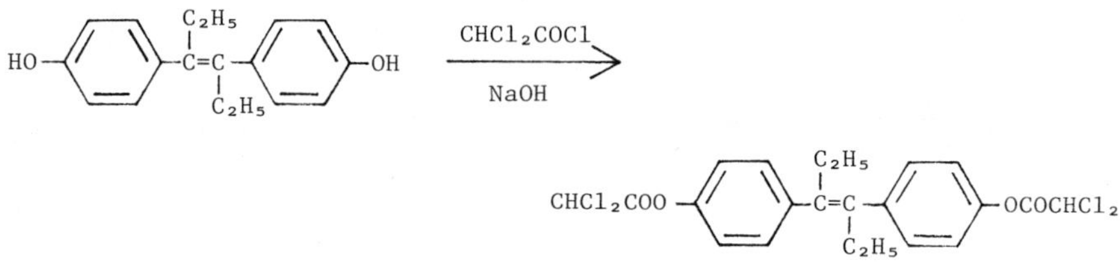

Typical Procedure. Ext 2 ml of a 0.25 μg/ml soln of DES in $CHCl_3$ with 10 ml 1 *N* NaOH. to the aq ext add 7 ml $CHCl_3$ and 0.1 ml dichloroacetyl chloride. Shake 10–15 sec; then allow phases to separate. Drain $CHCl_3$ through *ca* 5 g $CHCl_3$ washed anhyd Na_2SO_4. Repeat addition of 7 ml $CHCl_3$ and 0.1 ml dichloroacetyl chloride, shake, and again drain through the Na_2SO_4. Rinse the Na_2SO_4 with 2–3 ml $CHCl_3$ and evap the combined $CHCl_3$ to dryness. Diss residue in benzene and analyze by ECD-GC.

Comments. A method was given for ext of DES and its glucuronide from meat samples. McGregor *et al.*[2] used the dimethyl ether deriv formed with dimethyl sulfate to measure DES by GC. DES in feeds has been measured as the TMS deriv from BSA.[3]

References

[1] A. L. Donoho *et al.*, *J. Assoc. Off Anal. Chem.* **56**, 785 (1973).

[2] R. F. McGregor *et al.*, *Anal. Biochem.* **2**, 441 (1961).

[3] I. E. Smiley and E. D. Schall, *J. Assoc. Off. Anal. Chem.* **52**, 107 (1969).

Derivative Type. Pentafluorobenzoate.

Reagent. Sodium hydride, pentafluorobenzoyl chloride.

Compound Type. Phenols.

Reaction.

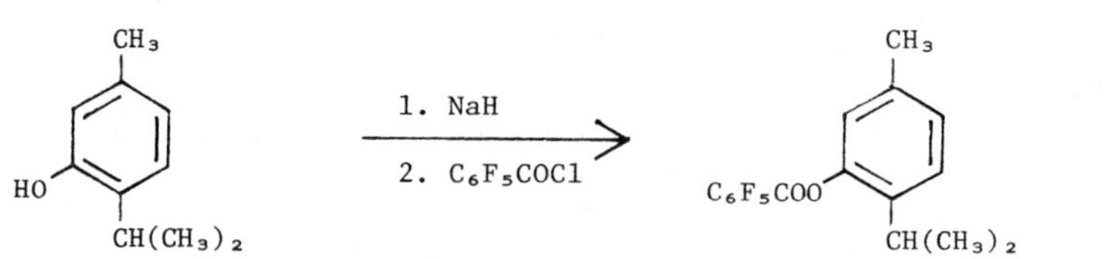

Typical Procedure. Add excess NaH to a soln of thymol in Et_2O. After the bubbling subsides, add the supernatant to a stirred soln of the reagent (1.1 mol equiv) in anhyd Et_2O. Stir 4 hr, then filter and conc to yield the product. Analyze by ECD-GC.

Comments. The pentafluorobenzoyl deriv was found to give the greatest sensitivity of 7 derivs examined. The derivs, reagents used, and relative sensitivities were as follows:

Pentafluorobenzoate	Pentafluorobenzoyl chloride	6.9
Pentafluorobenzyl ether	Pentafluorobenzyl bromide	5.9

Pentafluoropropionate	Pentafluoropropionic anhydride	1.3
Heptafluorobutyrate	Heptafluorobutyric anhydride	1.0
Monochloroacetate	Monochloroacetyl chloride	0.3
2, 4-Dinitrophenyl ether	2,4-Dinitrofluorobenzene	0.3
Monofluoroacetate	Monofluoroacetyl chloride	0.007

An attempt to prepare the iodosilane[2] was unsuccessful.

References

[1]N. K. McCallum and R. J. Armstrong, *J. Chromatogr.* 78, 303 (1973).

[2]H. Van der Molen *et al., Clin. Chim. Acta* 34, 169 (1971).

Derivative Type. *O*-Benzoyl, *N*-trimethylsilyl.

Reagent. Benzoyl chloride, bis (trimethylsilyl) acetamide (BSA).

Compound Type. Acetaminophen.

Reaction

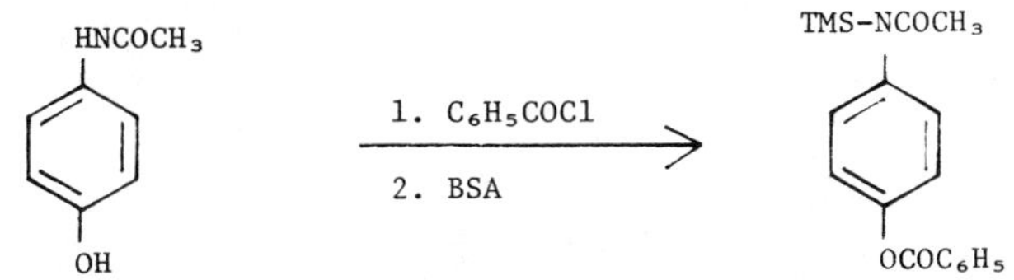

Typical Procedure. Mix 1 ml plasma with ml H_2O and add 50 μl of a soln of N-butyryl-*p*-aminophenol int stand (20 μg) in acetone. Add 2 g NaCl and ext with 2 × 50 ml Et_2O. Ext the combined Et_2O with 4 × 3 ml 5% aq Na_2CO_3. To the combined Na_2CO_3 ext add 20 μl benzoyl chloride and shake vigorously for 2 min. Ext with 50 ml Et_2O. Dry ext over Na_2SO_4 and evap to dryness on a boiling H_2O bath. Diss residue in 200 μl acetone. Evap 30 μl to dryness in an injector capsule and analyze by GC for quantitation. For qualitative confirmation, evap 50 μl of the soln to dryness and treat residue with 40μl BSA. Evap to dryness in the injector capsule and analyze by GC.

Comments. Analysis of two derivs gave confirmation of the identity of the compd and permitted distinguishing acetaminophen from interfering barbiturates.

References

[1]H. V. Street, *J. Chromatogr.* **109**, 29 (1975).

Derivative Type. Anisoyl ester.

Reagent. Anisoyl chloride (*p*-methoxybenzoyl chloride).

Compound Type. Hexachlorophene [2, 2-methylenebis (3, 4, 6-trichlorophenol)]

Reaction

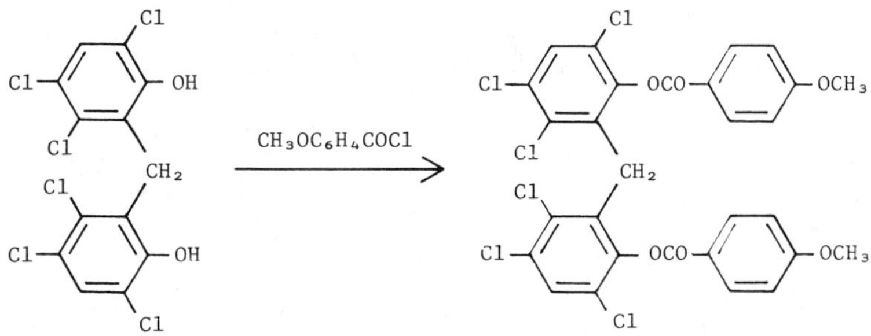

Typical Procedure. To a 4–20 μg sample of hexachlorophene add 1 ml 5 % NaOH and 30 μl anisoyl chloride. Vortex 1 min, and allow to stand at room temp 20 min. Add 9 ml H_2O and mix. Ext with 3 × 10 ml hexane. Evap hexane ext to dryness under N_2 on a steam bath. Diss residue in 1ml *n*-butyl chloride and analyze 10 μl aliquots by HPLC.

Comments. The deriv was used for HPLC quantitation of hexachlorophene using 254 nm UV detection. The deriv gave 64-fold increase in absorptivity compared to the parent compd.

References

[1]P. J. Porcaro and P. Shubiak, *Anal. Chem.* **44**, 1865 (1972).

1.4.3. Silyl Derivatives of Phenols

Derivative Type. Trimethylsilyl ether.

Reagent. Bis(trimethylsilyl)trifluoroacetamide (BSTFA).

Compound Type. Phenols.

Reaction

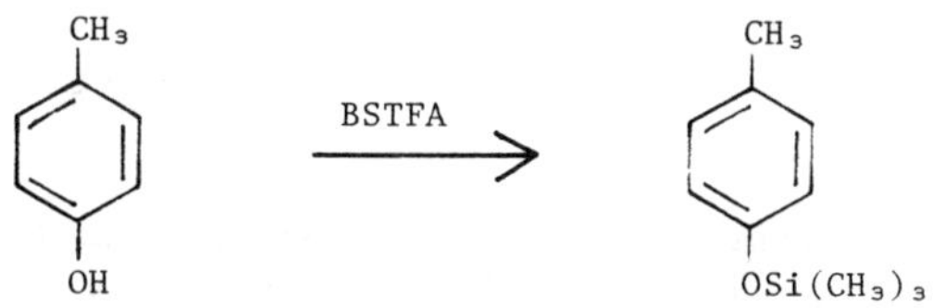

Typical Procedure. Add *o*-cresol int stand to one 2 ml serum sample, and analyze a duplicate 2 ml sample without int stand. To each sample add 13 ml H_2O and 1.5 ml 12 *M* HCl. Distil until 5 ml of distillate is collected, then add 10 ml H_2O and distil another 10 ml. Repeat addition and 10 ml distillation. Ext the combined distillate with 2 vol of $CHCl_3$. Evap ext to *ca* 100 μl under N_2 and add 20 μl BSTFA. Allow to stand overnight at room temp, and analyze by GC.

Comments. Phenols were isolated from serum by codistillation with H_2O. The glucuronide conjugates were determined after hydrolysis with β-glucuronidase, but phenol sulfates were cleaved upon heating in the HCl soln.

References

[1]B. Wengle and K. Hellström, *Scand. J. Clin. Lab Invest.* **28**, 477 (1977).

Derivative Type. Trimethylsilyl.

Reagent. Bis(trimethylsilyl)trifluoroacetamide (BSTFA) or *N,O*-bis(trimethylsilyl)acetamide (BSA).

Compound Type. Acetaminophen, phenacetin, acetanilide.

Reaction

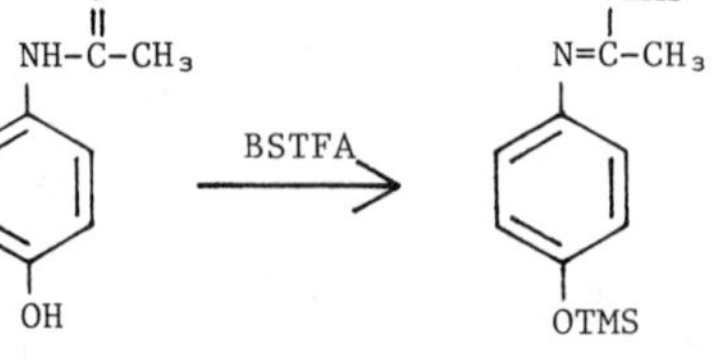

Typical Procedure. Add 2–10 μg of labeled standard to 1 ml plasma. Buffer to pH 5 with 0.4 ml 3 *M* NaH_2PO_4. Ext with 10 ml EtOAc, centrifuge, and remove org phase. Evap to dryness under N_2 and treat residue with 50 μl of BSA or BSTFA at 80° for 30 min. Inject 2 μl into GC-MS.

Comments. Compounds were quantitated by GC-MS using d_3-labeled internal standards. BSA formed only the mono-TMS derivative.

References

[1]J. D. Baty *et al.*, *Biomed. Mass Spectrosc.* **3**, 60 (1976).

Derivative Type. Trimethylsilyl.

Reagent. Bis(trimethylsilyl)acetamide (BSA).

Compound Type. Sterically hindered phenols.

Reaction

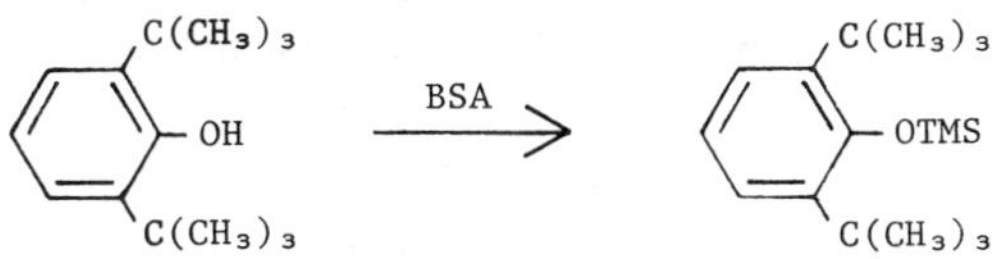

Typical Procedure (Preparative). Diss 2,6-di-*t*-butylphenol in CH_3CN, and add BSA. Heat at 90° for 15 hr.

Comments. Refluxing the same phenol with TMCS and Et_3N for 5 days yielded only 10% conversion to the TMS deriv. 2,6-Diphenylphenol was converted to the TMS deriv by BSA in $CHCl_3$ at room temp in 30 min.

References

Derivative Type. Trimethylsilyl.

Reagent. Bis(trimethylsilyl)acetamide (BSA).

Compound Type. Phentolamine.

Reaction

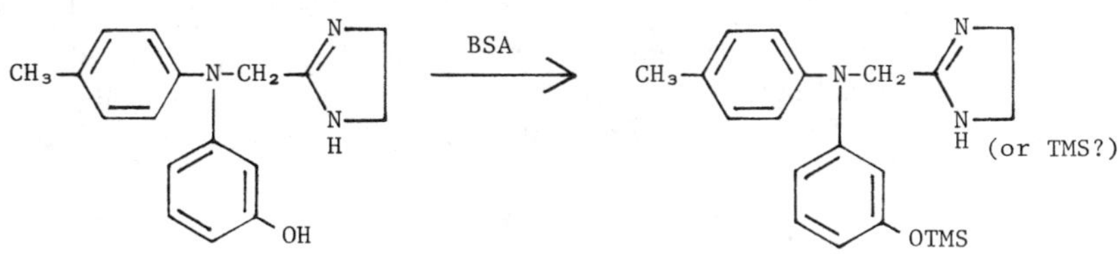

Typical Procedure. Treat sample with BSA and analyze soln by GC.

Comments. The preparation of the TMS deriv with TMCS/HMDS was reported to be unsuccessful. The structure of the deriv was not confirmed by MS, and thus there is some question of whether the deriv contained an *N*-TMS group.

References

[1]P. F. G. Boon and W. Sudds, *J. Pharm. Pharmacol.* **19**, 88S (1967).

Derivative Type. Trimethylsilyl.

Reagent. Bis(trimethylsilyl)trifluoroacetamide (BSTFA)/trifluoroacetic acid.

Compound Type. Sterically hindered phenols.

Reaction

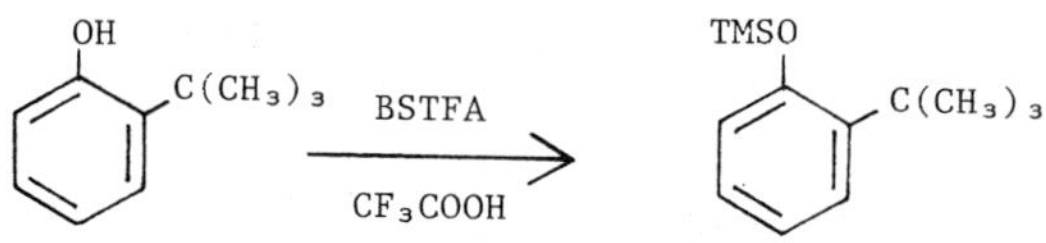

Typical Procedure. Dil a 10 mg/ml soln of the phenol 1 : 10 with isooctane. Mix 200 μl of this soln with 200 μl BSTFA and 10 μl CF_3COOH. After 20 min analyze by GC.

Comments. It was found that CF_3COOH promoted silylation of hindered phenols and was superior to TMCS for this purpose. The use of CF_3COOH with HMDS, TMCS, BSA, BSTFA, and TMSI was studied. It was speculated that trimethylsilyl trifluoroacetate was formed as the active silylation agent. The method has been applied to deriv of Δ^1-tetrahydrocannabinol.[2]

References

[1] N. E. Hoffman and K. A. Peteranetz, *Anal. Lett.* **5**, 589 (1972).

[2] N. E. Hoffman and R. Y. Yang, *Anal. Lett.* **5**, 7 (1972).

Derivative Type. Trimethylsilyl.

Reagent. *N*-Methyl-*N*-trimethylsilyl heptafluorobutyramide (MSHFBA)/trimethylchlorosilane (TMCS).

Compound Type. Phenolic compounds.

Reaction

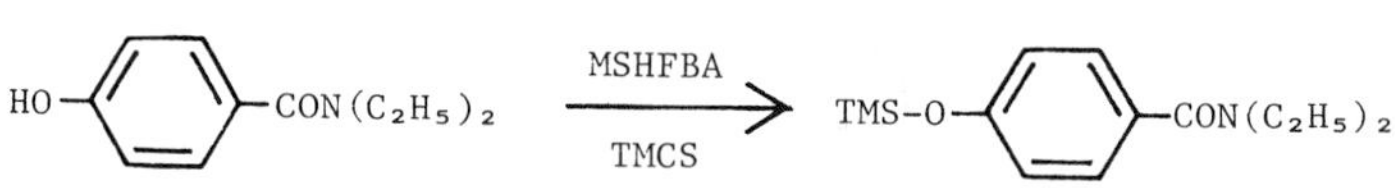

Typical Procedure.[1,2] Diss a sample containing 30 ppm each of *p*-hydroxyamphetamine and ethamivan (shown above) in MSHFBA/TMCS/BuOAc (40 : 1 : 59). Analyze by FID-GC.

Comments. The reagent, and the pentafluoropropionamide and pentafluorobenzamide analogs, were developed as reagents that, due to increased fluorine content, would cause less SiO_2 fouling of the FID. Thies[3] used the reagent for analysis of glucosinolates in seeds and leaves.

References

[1] A. Radmacher, German Pat. 2,262,842 (30 May 1974); *Chem. Abstr.* **81**, 114256r (1974).

[2] M. Donike, U.S. Pat. 3,954,651 (4 May 1976); *Chem. Abstr.* **85**, 103518j (1976).

[3] W. Thies, *Fette Seifen Anstrichem.* **78**, 231 (1976); *Chem. Abstr.* 85, 76445z (1976).

Derivative Type. Chloromethyldimethylsilyl (CMDMS).

Reagent. Chloromethyldimethylchlorosilane/bis(chloromethyl)tetramethyldisilazane.

Compound Type. Naphthalenediols.

Reaction

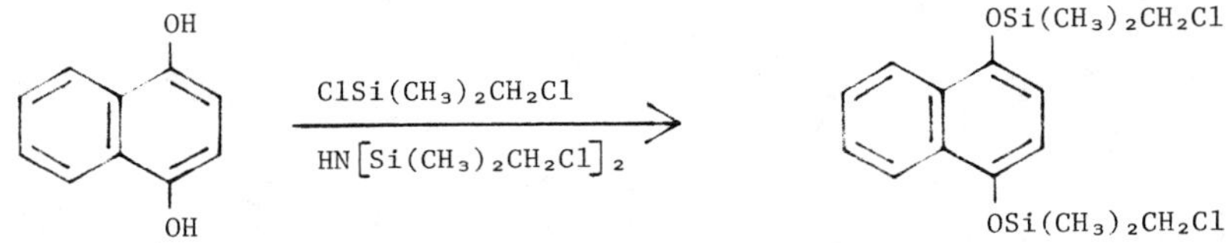

Typical Procedure. Diss 1 mg sample in 6 ml benzene, and add 0.2–0.4 ml anhyd pyridine. Add 0.3 ml of a soln of 1 part chloromethyldimethylchlorosilane/bis (chloromethyl)tetramethyldisilazane (1 :3, v/v) in 10 parts (v/v) benzene. Allow to stand 15 min at 25°. Wash with 3 × 5 ml H_2O. Analyze by GC.

Comments. CMDMS, TMS, TFA, HFB, and chloroacetyl derivs were prepared from 1-naphthol and 1, 4-, 1, 5-, and 1, 6-naphthalenediols.[1] The CMDMS derivs gave the best GC separation. The same reagents were used to prepare CMDMS derivs of four hydroxystilbenes.[2].

References

[1]S. Khalifa and R. O. Mumma, *J. Agr. Food Chem.* **20**, 632 (1972).

[2]J. E. Sinsheimer and R. V. Smith, *J. Pharm. Sci.* **56**, 1280 (1967).

Derivative Type. Trimethylsilyl.

Reagent. Hexamethyldisilazane (HMDS)/trimethylchlorosilane (TMCS).

Compound Type. Anthocyanins.

Reaction

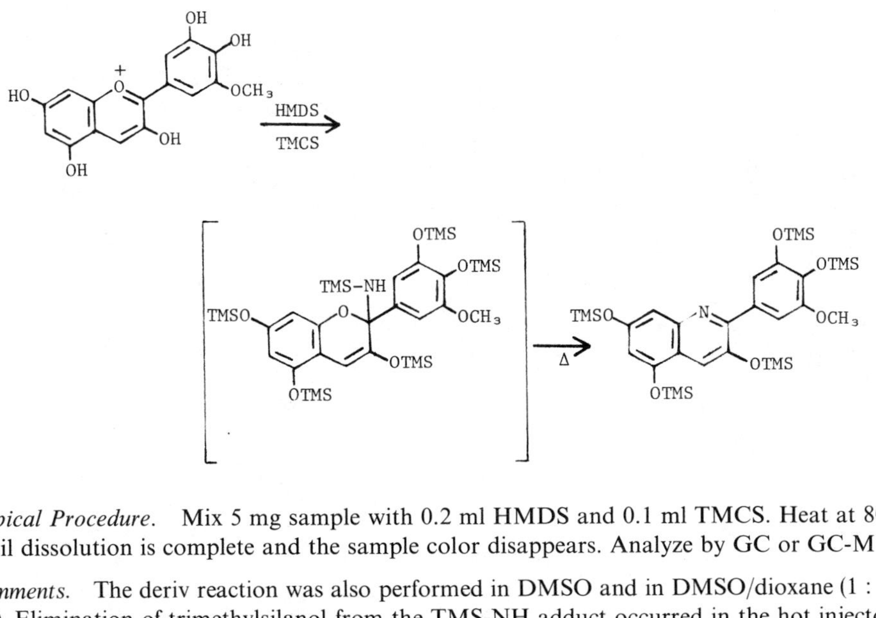

Typical Procedure. Mix 5 mg sample with 0.2 ml HMDS and 0.1 ml TMCS. Heat at 80° until dissolution is complete and the sample color disappears. Analyze by GC or GC-MS.

Comments. The deriv reaction was also performed in DMSO and in DMSO/dioxane (1 : 1, v/v). Elimination of trimethylsilanol from the TMS-NH adduct occurred in the hot injector to give the quinoline deriv.

References

[1]E. Bombardelli *et al.*, *J. Chromatogr.* **120**, 115 (1976).
[2]E. Bombardelli *et al.*, *J. Chromatogr.* **139**, 111 (1977).

1.4.4. Dinitrophenyl Derivatives of Phenols

Derivative Type. 2,4-Dinitrophenyl (DNP).

Reagent. 2,4-Dinitrofluorobenzene (DNFB).

Compound Type. Phenols.

Reaction

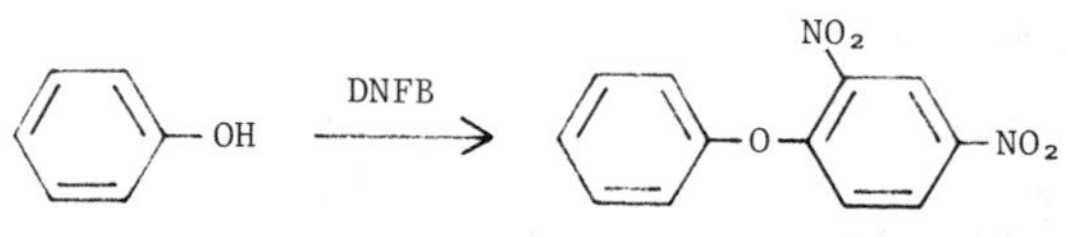

Typical Procedure. A. Diss 10 μg of the phenol in 4 ml acetone. Add 0.1 ml of a satd soln of NaOMe in MeOH and 1 ml of a 1 % (w/v) soln of DNFB in acetone. Reflux 30 min then add to 25 ml of 2.5 % (w/v) aq NaOH. Ext with 25 ml $CHCl_3$. Dry ext over Na_2SO_4. Evap to a small vol, diss residue in acetone, and analyze by GC.

B. Mix 25 ml H_2O soln of 10 μg phenols with 0.1 ml of a satd soln of NaOMe in MeOH and 1 ml of 1 % (w/v) DNFB in acetone. Shake 2 min then add 25 ml 2.5 % (w/v) aq NaOH and 10 ml hexane. Shake 1 min, separate hexane, and dry with a short column of Na_2SO_4. Analyze 5 μl aliquot by GC.

C. Spot 10 μg of phenol in acetone soln on a TLC plate. Spray with a satd soln of NaOMe in MeOH (7 ml per 400 cm^2) followed by a similar quantity of 4 % (w/v) DNFB in acetone. Cover the plate with another glass plate and heat at 190° for 40 min. Cool and scrape areas of interest. Elute using a short column and 2 × 5 ml acetone. Analyze eluate by GC.

Comments. GC retention and ECD sensitivity data were reported for 36 phenols. The method for deriv formation on TLC plates was used for analysis of phenol hydrolysis products of pesticides.

References

[1]I. C. Cohen *et al.*, *J. Chromatogr.* **44**, 251 (1969).

Derivative Type. 2,4-Dinitrophenyl (DNP).

Reagent. 2,4-Dinitrofluorobenzene (DNFB)/triethylamine (Et_3N).

Compound Type. Phenolic metabolites of carbofuran.

Reaction

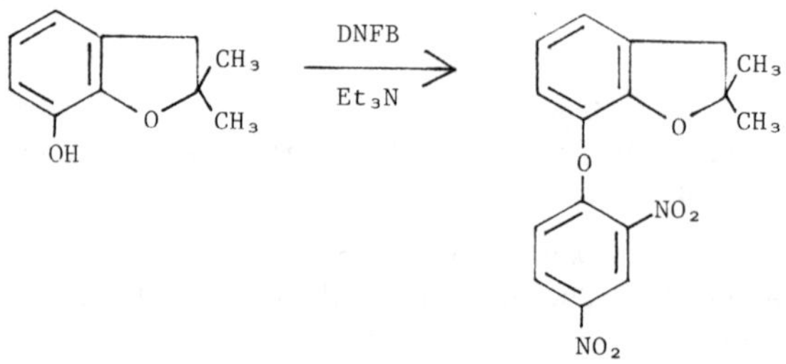

Typical Procedure. To the dried ext of a 100 g potato sample (or 50g meat sample) add 100 ml acetone, 25 ml pH 11.0 phosphate buffer and 1 ml 1% DNFB in acetone. Conc at 90–95° then heat another 30 min after all solvent has been removed. Ext with 2× 20 ml hexane and dry ext over Na_2SO_4. Conc ext to 5 ml. Purify the deriv on a charcoal column and analyze by GC.

Comments. The method was applied to the phenol shown and its 3-keto and 3-hydroxy metabolites. The 3-hydroxy compd was converted to the 3-ethoxy deriv by refluxing in EtOH/HCl prior to treatment with DNFB. A microcoulometric nitrogen detector was used.

References

[1]R. F. Cook *et al., J. Agr. Food Chem.* **25**, 1013 (1977).

Derivative Type. 2,6-Dinitro-4-trifluoromethylphenyl (DNT).

Reagent. 4-Chloro-α,α,α-trifluoro-3,5-dinitrotoluene (DNT-Cl).

Compound Type. Phenols (from pesticides).

Reaction

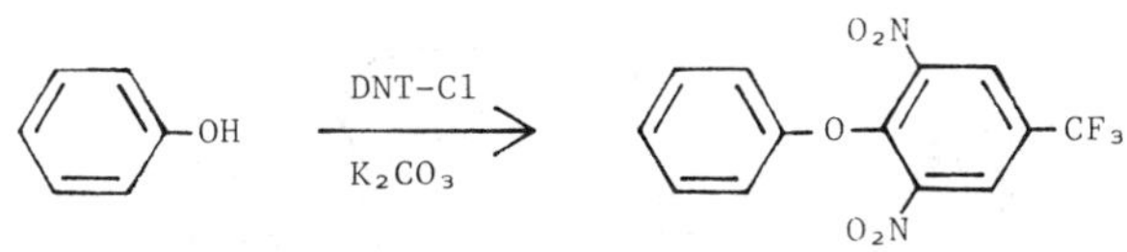

Typical Procedure. To a soln of 10–25 μg of the phenol in 0.2 ml acetone add 20 μl 5% aq K_2CO_3 (or KOH) and mix briefly. Add 9 ml acetone and 0.25 ml DNT-Cl (5 mg/ml) in acetone. Dil to 10 ml with acetone, shake vigorously 30 sec, then allow to stand 2 hr in the dark. Analyze 1–3 μl aliquots by GC.

Comments. The DNT derivs were compared with DNP (from DNFB) and PFB (from PFB-Br) derivs prepared in the same manner. The three types of derivs were found to be similar in terms of ease of preparation, stability, and ECD response. GC and MS data were reported for the DNT derivs of 28 phenols.

References

[1]J. N. Seiber *et al., J. Chromatogr.* **73**, 89 (1972).

1.4.5. Other Derivatives of Phenols

Derivative Type. Diethyl phosphate.

Reagent. Sodium *t*-butoxide/diethyl phosphorochloridate.

Compound Type. Phenols.

Reaction.

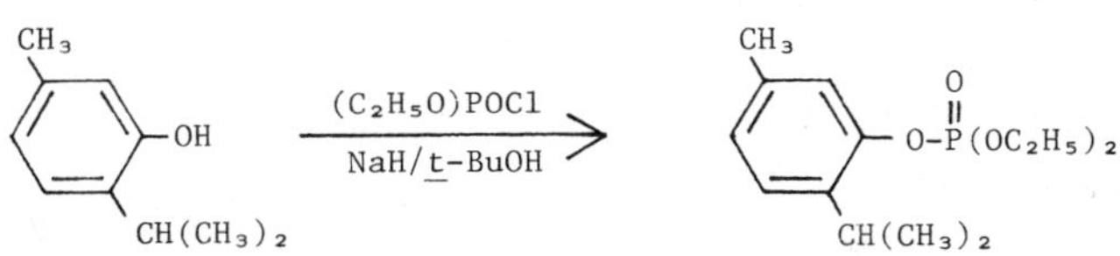

Typical Procedure. Diss submicrogram phenol sample in 0.5 ml benzene/*t*-BuOH (9 : 1, v/v). Add *ca* 300 mg NaH under anhyd conditions. Add soln (and tube washings, total *ca* 1

ml) dropwise to 2 drops of redist diethyl phosphorochloridate. Shake and heat at 50° for 30 min. Evap benzene and add 2 ml 5% aq Na_2CO_3. Shake and allow to stand 2 min. Ext with benzene and wash ext with 0.1 *N* HCl then with H_2O. Dry benzene (method not stated), evap to dryness, and rediss in benzene prior to GC analysis.

Comments. The method was used with flame photometric GC detection. The detectability limit was about 5 pg for the deriv of thymol. ECD-GC of the pentafluorobenzoyl deriv was about five times more sensitive.

References.

[1]M. P. Heenan and N. K. McCallum, *J. Chromatogr. Sci.* **12**, 89 (1974).

Derivative Type. 5-Dimethylaminonaphthalene-1-sulfonyl (DANS).

Reagent. 5-Dimethylaminonaphthalene-1-sulfonyl chloride (DANS-Cl)

Compound Type. Phenols.

Reaction.

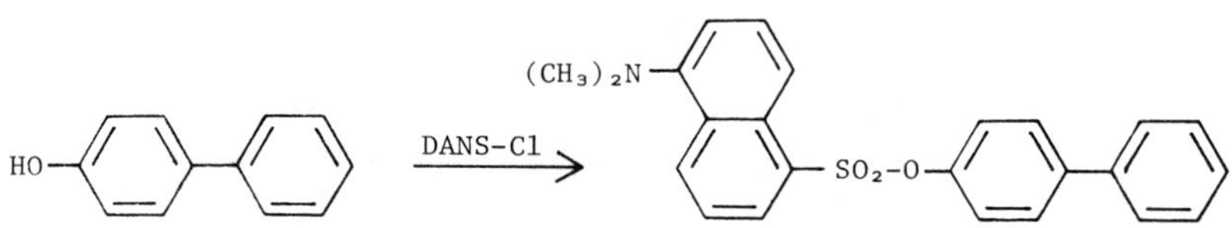

Typical Procedure. To 10 μl of a 1m*M* soln of the sample in acetone add 200 μl of 0.1% DANS-Cl in acetone and 30 μl 0.1 *M* aq Na_2CO_3. Heat 15–20 min at 43–45°. Cool to room temp and add 2 drops 1 *N* NaOH to hydrolyze excess reagent. Ext deriv with 500 μl *n*-hexane. Purify deriv by TLC and analyze by MS.

Comments. The method was applied to hydroxybiphenyls,[1] chlorophenols,[2] and hydroxychlorobiphenyls.[3] DANS derivs have been used for numerous other phenols including estrogens[4] and cannabinols.[5] The DANS derivs have also been used to quantitate phenols in urine by HPLC.[6]

References.

[1]M. Frei-Hausler, *et al., J. Chromatogr.* **79**, 209 (1973).

[2]M. Frei-Hausler, *et al., J. Chromatogr.* **84**, 214 (1973).

[3]O. Hutzinger, *et al., J. Chromatogr* **97**, 233 (1974).

[4]L. P. Penzes and G. W. Oertel, *J. Chromatogr.* **51**, 325 (1970).

[5]J. Friedrich-Fiechtl, *et al., Naturwissenschaften* **60**, 207 (1973).

[6]R. M. Cassidy, *et al., J. Chromatogr. Sci.* **12**, 85 (1974).

Derivative Type. Butaneboronate.

Reagent. Butaneboronic acid.

Compound Type. Hydroxysalicylanilides.

Reaction.

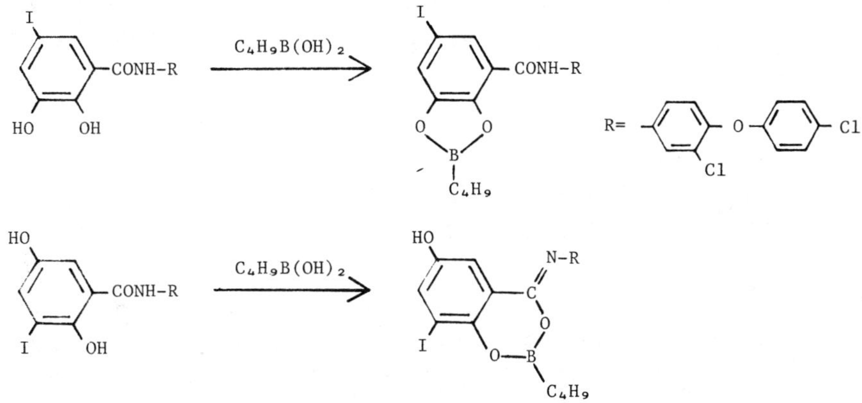

Typical Procedure. Diss 25 μg of sample and a severalfold excess of butaneboronic acid in dry acetone. Allow to stand 30 min at 25°. Analyze by MS.

Comments. The benzeneboronate was also prepared. The results demonstrated that caution must be exercised in MS interpretation when basing structure determination upon wt increases after deriv. Salicylanilide was also observed to form a cyclic boronate but benzanilide and *p*-hydroxybenzanilide did not. TMS derivs of the hydroxysalicylanilides were also prepared using BSA at 75°.

References.

[1]W. J. A. VandenHeuval, *Anal. Lett.* **6**, 51 (1973).

1.5. DERIVATIVES OF EPOXIDES

Derivative Type. Trimethylsilyl chlorohydrin.

Reagent. Trimethylchlorosilane (TMCS).

Compound Type. Epoxides.

Reaction.

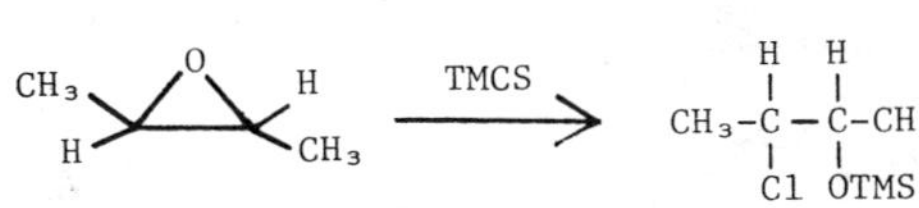

Typical Procedure. Mix 25 μl of epoxide and 0.2 ml of TMCS in a tube. Seal and heat at 100° for a period ranging from 5 min to 150 hr. Cool, open tube, and evap under vac. Add 2 ml dry pentane and centrifuge. Conc pentane soln and analyze by GC or GC-MS.

Comments. The reactivity varied considerably for seven different epoxides studied. It was demonstrated that all epoxides do not necessarily react rapidly with TMCS. The derivs were also prepared with HMDS/TMCS/pyridine.

References.

[1]O. Ceder and B. Hansson, *Acta Chem. Scand. B* **30**, 574 (1976).

Derivative Type. *N, N*-Dimethyl, trimethylsilyl chlorohydrin.

Reagent. Diazomethane, trimethylchlorosilane (TMCS), bis(trimethylsilylacetamide) (BSA).

Compound Type. Aliphatic epoxide metabolites of allyl barbituric acids.

Reaction.

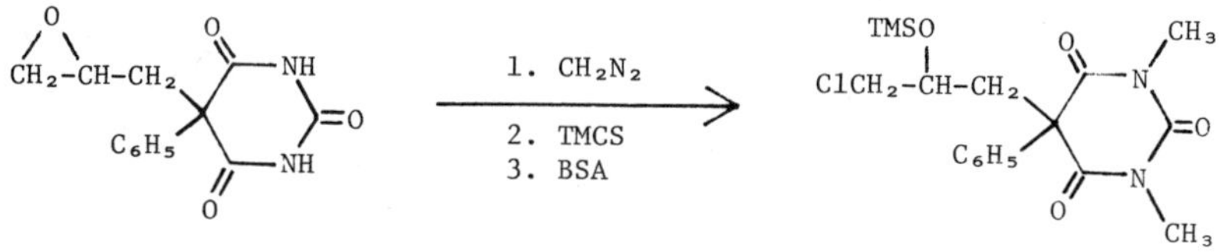

Typical Procedure. Diss isolated metabolites in MeOH and add excess CH_2N_2 in Et_2O. After 15 min at room temp evap under N_2. Diss residue in 10 μl pyridine and add 5 μl TMCS. Heat at 60° for 30 min, then add 10 μl BSA to insure silylation of all hydroxyl groups. Allow to stand overnight at room temp and then analyze by GC-MS.

Comments. The procedure converted the barbiturate moiety to the dimethyl deriv (primarily *N,N*-, but with small amts of *N, O*-, and *O, O*-isomers), aromatic hydroxyls to a mixt of methyl and TMS ethers, and aliphatic hydroxyls to TMS ethers. Epoxides were converted to TMS ethers of the chlorohydrin. MS showed the deriv to be the 3-chloro-2-TMSO deriv.

References.

[1]D. J. Harvey, *et al.*, *Res. Commun. Chem. Pathol. Pharmacol.* **4**, 247 (1972).

[2]D. J. Harvey, *et al.*, *Anal. Lett.* **5**, 745 (1972).

Derivative Type. Bis(dithiocarbamoyl) ester.

Reagent. Sodium diethyldithiocarbamate (DDTC).

Compound Type. Dianhydrogalactitol (DAG).

Reaction.

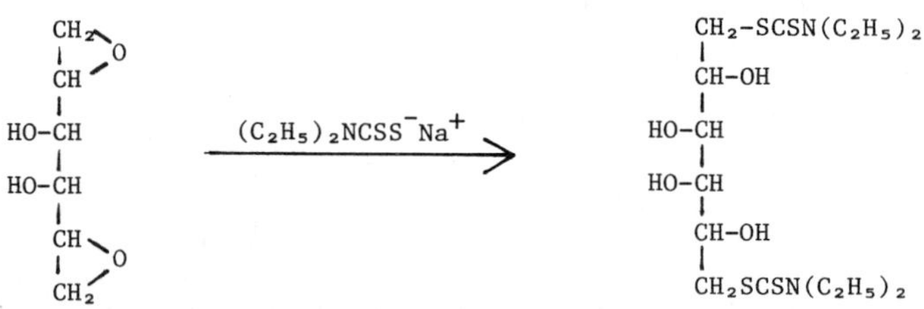

Typical Procedure. Mix 1 ml plasma (containing 0.05–50 μg DAG) with 0.5 ml 0.1 *M* phosphate buffer (pH 7.0), 0.5 ml of a 5 % (w/v) aq soln of DDTC, and 2 ml H_2O. Cap vial and allow to stand at room temp 1 hr. Ext with 10 ml $CHCl_3$ and wash ext with 2–5 ml portions of 33 % (w/v) aq NaCl. Remove 8 ml of the $CHCl_3$ and evap to dryness at 40°. Diss residue in 200 μl heptane/$CHCl_3$ (7 : 3, v/v) and analyze 10 μl of the soln by HPLC.

Comments. The method permitted detection of DAG to below 50 ng/ml using UV detection at 245 nm. The deriv method was shown to yield a single product quantitatively by conversion of the deriv to the diacetonide and analysis by HPLC and NMR.

References.

[1]D. Munger, *et al.*, *J. Chromatogr.* **143**, 375 (1977).

2. Amino Compounds

Derivatization of the amino group constitutes one of the largest areas of application of analytical derivatization reactions. Derivatives are utilized in GC, MS, and HPLC analysis of amino compounds. Although many amino compounds are sufficiently volatile for direct GC and MS analysis, the polar N–H groups and unshared electrons on the nitrogen can cause adsorption and tailing in the GC. Indeed, in very low level analyses such adsorption can cause total loss of the sample. Derivatization methods have been employed to reduce the polarity of the amino group and improve GC behavior of the compounds. Derivatization of the amino group may also be employed to achieve a separation of compounds that are otherwise inseparable or to confer detectability as in ECD-GC or HPLC analysis. This section contains the various derivatives that have been used in the analysis of amino compounds. Amino acids and peptides are covered separately in Section 5. The derivatives of other nitrogen containing functional groups are discussed in Section 7.

Primary and Secondary Amines. The most commonly used derivatives for GC analysis are the acyl derivatives. In the normal acyl series (acetyl, propionyl, butyryl, etc.), the volatility decreases with increasing molecular weight. The fluorinated acyl derivatives are more volatile than the corresponding nonfluorinated compounds, but there is generally little difference in retention times for trifluoroacetyl versus heptafluorobutyryl derivatives. Apparently the effect on volatility of the increased molecular weight is offset by the volatility enhancement of the fluorine substitution. Pentafluoropropionyl derivatives have similar retention times but have in some cases been reported to be the most stable of the three. In addition to improving GC behavior, the fluorinated acyl groups are used to impart electron capture detectability to amines. The ECD sensitivity generally increases with increasing fluorine content. Trichloroacetyl derivatives tend to have higher ECD sensitivities than even the perfluorooctanoyl derivatives but the fluorinated aromatic derivatives (e.g., pentafluorobenzoyl) have still higher sensitivities. The trichloroacetyl and pentafluorobenzoyl derivatives, however, tend to be less volatile than the smaller fluoroacyl derivatives. The acyl derivatives thus have a wide range of volatility and permit selection of a derivative for optimization of desired retention and/or separation. In the case of the electron capturing derivatives, selection is on the basis of sensitivity within the constraint of usable volatility.

The acyl derivatives are generally prepared from the anhydride, acid chloride, or acyl imidazole reagents. In the case of the anhydride and acid chloride reagents, it is usually necessary to remove excess reagent and by-product acid since these agents can have detrimental effects upon the GC column. Derivatives prepared with the acyl imidazole reagents can normally be analyzed without removal of excess reagent. A related reagent, pentafluoropropionyltriazole (PFPT), has been used to prepare pentafluoropropionyl derivatives (p. 131). Trifluoroacetyl derivatives have also been prepared from acyl amide reagents. The reagent *N*-methyl-bis(trifluoroacetamide) is very volatile and yields the also volatile and innocuous (with respect to column damage) *N*-methyl-trifluoroacetamide as by-product. The novel trifluoroacetyl nylon reagent (p. 77) solves the problem of removal of excess reagent and by-product simply by filtration.

Whereas the *N*-acyl derivatives tend to be more stable than *O*-acyl derivatives, the *N*-trimethylsilyl derivatives are much less stable than the *O*-trimethylsilyl derivatives. Indeed, several *N*-TMS compounds are used as trimethylsilylating reagents. The *N*-*t*-butyldimethylsilyl derivatives would be expected to be more stable than the corresponding *N*-trimethylsilyl derivatives, but little work has been done on these derivatives (see pp. 10 and 87).

Permethylated derivatives of amines have been prepared for GC analysis using formaldehyde/borohydride (p. 87); formaldehyde/formic acid (p. 88), and methyl iodide/sodium hydride (p. 88). These derivatives eliminate the polar N—H groups but retain the troublesome adsorptive properties of tertiary amines. Schiff base derivatives of primary amines (p. 89) prepared using a variety of aldehydes and ketones are useful for GC analysis. The derivative from pentafluorobenzaldehyde is among the most sensitive yet found for ECD-GC (pp. 11, and 90). Dimethylformamide dimethyl acetal also forms a Schiff base type derivative with primary amines. (p. 91).

Nitrophenyl derivatives have been used for analysis of primary and secondary amines. These derivatives impart ECD sensitivity but have also been used for FID-GC analysis. In the latter case the derivatives are particularly useful for low molecular weight amines that have inconveniently short retention times as other (e.g., trifluoroacetyl) derivatives. The 2, 4-dinitrophenyl derivatives have been prepared with dinitrofluorobenzene (p. 92) or 2,4-dinitrobenzenesulfonic acid (p. 93). Addition of a trifluoromethyl group to a nitrophenyl or dinotrophenyl derivative increases the ECD sensitivity and decreases the GC retention time. 4-Trifluoromethyl-2-nitrophenyl and -2,6-dinitrophenyl derivatives have been prepared from the fluoride and chloride reagents respectively (p. 94). Surprisingly, the dinitrophenyl derivatives were less sensitive to ECD than the mononitro derivative.

Several other amine derivatives have been employed for GC and MS analysis. Reaction of primary and secondary amines with isocyanates yields urea derivatives (p. 94) that have been used for GC and GC-MS analysis of amines. The 3-trifluoromethylphenyl urea derivatives, prepared from the corresponding isocyanate, were useful for ECD-GC analysis (p. 95). Amines react with diethyl pyrocarbonate to form volatile carbamates that have been analyzed by GC (p. 95).

Reaction of primary amines with the γ-diketone 2,5-hexanedione gave dimethylpyrrole derivatives that could be analyzed by GC (p. 95). This method was also applicable to amino acid methyl esters. Primary amines undergo a Schiff base type condensation with the "thiocarbonyl" compound carbon disulfide to yield volatile isothiocyanate derivatives (pp. 12, 96, and 139). These derivatives, which are formed only from the primary amines, have been used for GC and GC-MS analysis, and were reported to be more volatile than the *N*-trimethylsilyl derivatives. It should be noted that, since carbon disulfide is sometimes used as a very volatile injection solvent for GC analysis, isothiocyanate derivatives will be formed with any underivatized primary amines in the sample.

The primary and secondary amines been the object of considerable work toward preparation of chromogenic derivatives for HPLC analysis. The nitrophenyl group, in addition to conferring ECD sensitivity, also confers UV absorption to the amine derivatives. The nitrophenyl acyl derivatives in particular have been employed for HPLC analysis. These derivatives were prepared with the acid chloride reagents (p. 96) or the ester from *N*-hydroxy-succinimide (p. 97). Sulfonamide derivatives have been prepared from 5-dimethylamino- (p. 98) and 5-di-*n*-butylamino- (p. 100) naphthalene-1-sulfonyl chlorides (DANS-Cl and BANS-Cl). In chromogenic derivatization as in ECD analysis, it is usually necessary to remove excess reagent prior to the analysis to avoid "swamping" the detector with reagent absorption. The use of an amino acid to react with excess reagent permits separation of the desired derivative (organic soluble) from the excess reagent product (water soluble) by solvent extraction. This strategy, employing proline, was used in the preparation of the DANS derivatives of polyamines (p. 98). The DANS and BANS derivatives have been used in MS analysis, exploiting both their favorable MS behavior and their relatively low volatility, which minimizes losses of volatile amines. The BANS derivatives (p. 100) are preferred for MS analysis, since their major fragmentation involves loss of a butyl group but retention of the derivatized amine moiety. DANS derivatives, in contrast, tend to lose the sulfonamide group by C—S cleavage in their principal fragmentation.

Reaction of primary amines with the nonfluorescent reagent fluorescamine yields intensely fluorescent derivatives that are useful in HPLC analysis (p. 100). The development of the reagent resulted from studies of the mechanism of formation of fluorescent derivatives from ninhydrin.[1-3] Although the majority of work with this reagent has employed post column derivatization, it has been used for precolumn derivatization in the analysis of polyamines (p. 101) and catecholamines (p. 143). The fluorescamine derivatives have also been used for MS analysis of amines (p. 101).

o-Phthalaldehyde, which likewise forms a fluorescent derivative with primary amino groups (p. 102), has been used primarily with post-column derivatization in the analysis of amino acids. The use of *o*-phthalaldehyde has two possible advantages: (1) it is water soluble while fluorescamine requires organic solvents; (2) the resulting derivative potentially has wider utility, for example in MS analysis, since it is among the smallest of any fluorescent derivative formed with amino groups.

Another fluorogenic reagent for amines is 4-chloro-7-nitrobenzo-2,1,3-oxadiazole (NBD-Cl) (p. 102). Although the majority of the work with this reagent has involved TLC analysis, the derivatives should be amenable to HPLC.

Indole Amines. The indole amines represent a biologically important subset of the amines. The indole nitrogen is much less basic than an aliphatic amine, but can be acylated under forcing conditions. With mild conditions it is possible to selectively acylate a more basic sidechain amine (pp. 103 and 110). In a study of acylation of the indole nitrogen, it was found that the acylation with trifluoroacetic anhydride was faster in nonpolar solvents (p. 104). Care must be taken in acylation of indoles with the anhydride reagents since the indole ring system is acid sensitive. Use of acyl indole alkyl amines can be complicated by a cyclization reaction. This cyclization was originally thought to yield β-carboline type compounds (p. 106) but more recent evidence suggests formation of a spiro product (p. 105).

The indole nitrogen has been silylated by a variety of reagents (p. 108). The use of the common catalyst TMCS tended to decrease the silylation rate, whereas basic catalysts were effective in increasing the rate (p. 108). TMSI appears to be the best reagent for silylation of the indole nitrogen. Mixed derivatives with silylated indole nitrogen have also been prepared. Sidechain acetylated (p. 110) and trifluoroacetylated (p. 111) derivatives have been prepared with acetic anhydride and MBTFA, respectively. Primary indoleamines containing other functional groups (e.g., phenolic hydroxyls) have been converted to mixed TMS and Schiff base derivatives (p. 111), although the HMDS reagent used in this method apparently did not silylate the indole nitrogen. DANS derivatives have been prepared from indoleamines for MS analysis (p. 112). The DANS-Cl reagent likewise did not react with the indole nitrogen.

Tertiary Amines. The tertiary amines are devoid of the polar N—H groups of the primary and secondary amines but still suffer from adsorptive effects giving undesirable GC behavior. Lacking any replaceable hydrogen on the amine nitrogen, the tertiary amines are not directly acylable. One approach with acyl amines has been to acylate the aromatic ring with trifluoroacetic anhydride (p. 112). This method, while it does confer ECD sensitivity, does not directly attack the problem of poor GC behavior due to the tertiary amine nitrogen. The tertiary amines can be dealkylated using chloroformate reagents to yield acylated (p. 113) or carbamate (p. 113) derivatives. Another approach to derivatization of tertiary amines has been to completely eliminate the amine nitrogen. This has been accomplished by Hofmann degradation (p. 115) or by oxidative degradations of amenable structures (p. 117).

Quaternary Ammonium Compounds. Quaternary ammonium compounds represent a more difficult situation with regard to volatility for GC or MS analysis. Jenden and co-workers (p. 118) introduced the analytical application of benzenethiolate dealkylation of quaternary ammonium compounds to yield the volatile tertiary amines. Even though, as noted above, the tertiary amines do not have ideal GC behavior, they do have adequate volatility, and this derivatization procedure has enabled the use of GC and GC-MS for analysis of choline and its esters. Pyridinium type quaternary amines can be reduced with borohydride to tertiary piperidine type derivatives. This approach has been applied to the analysis of the dipyridinium compound paraquat in plasma (p. 118). In some cases the ammonium compound cleaves in the analysis to a volatile fragment as in the Hofmann elimination (p. 119).

Amino Alcohols and Thiols. In the amino alcohols and amino thiols, the alcohol or thiol group as well as the amino group may require derivatization. Anhydride reagents can be used to simultaneously acylate both functions. Trifluoroacetyl derivatives of aliphatic amines or alcohols do not normally exhibit good ECD sensitivity, but the presence of two adjacent trifluoroacetyl groups, as in the derivatives of β-hydroxyamines, have significantly better ECD response (pp. 120 and 128). By selection of proper conditions, only *N*-acylated derivatives can be prepared (pp. 121).

Amino thiols have been converted to trimethylsilyl derivatives with BSTFA (p. 122). In an analytical procedure for bufuralol, treatment with BSTFA followed by TFAA gave the *O*-TMS, *N*-TFA derivative (p. 123). Substitution of PFPA for TFAA gave a derivative suitable for ECD-GC analysis. Preparation of dinitrophenyl derivatives of amino alcohols followed by trimethylsilylation of the hydroxyl has also been used for ECD-GC analysis (p. 123).

Reaction of aldehydes and ketones with primary amines normally yields Schiff bases. In amino alcohols and thiols where the hydroxyl or sulfyldryl is on the β-carbon, the carbonyl compounds yield oxazolidine (p. 124) or thiazolidine (p. 125) derivatives. These cyclic derivatives are useful for GC analysis. The oxazolidine derivative formed with pentafluorobenzaldehyde, however, has been found to have much less ECD sensitivity than the Schiff base derivatives emphasizing the importance of a double bond adjacent to the aromatic ring for good ECD response (p. 124). The oxazolidine derivatives, unlike Schiff bases, can be formed from secondary amines. Secondary amines without the requisite β-substituent for oxazolidine formation can give extraneous reactions with pentafluorobenzaldehyde. In at least one example (p. 124) displacement of fluorine was observed leading to aryl tertiary amines.

Phenolic Amines. The phenolic amines include the important subgroup of catecholamines and their metabolites, which largely accounts for the extensive derivatization efforts that have been directed toward the group. Like the amino alcohols, the phenolic amines can be *N*-acylated and *O*-acylated (including any alcohol hydroxyls) in a single step by proper application of previously discussed acylation reagents. Acetyl (p. 126), fluoroacyl (pp. 126–131), and pivalyl (p. 131) derivatives have been reported. The catecholamine compounds with benzylic hydroxyls exhibit unusual reactivity of these alcohol hydroxyls. Methyl ethers can be prepared from these compounds by treatment of initially formed trifluoroacetyl derivatives with methanol (p. 132) or by treating the compounds themselves with HCl in methanol prior to acylation (p. 133). In the case of norepinephrine, the methyl ether, pentafluoropropionyl derivative showed a gain in ECD sensitivity compared to the fully acylated derivative (p. 133). With epinephrine, however, there was only a slight difference in sensitivity. The analogous ethyl ether derivatives have been used in conjunction with isothiocyanate formation for analysis of catecholamines (p. 141).

Trimethylsilyl derivatives have been prepared from phenolic amines using HMDS, BSA, BSTFA, and TMSI (p. 133). Both *O*-TMS and *N*-TMS derivatives are produced except with TMSI, which is specific for hydroxyls (p. 135), or in the case of hindered amino groups (p. 134). Sulfate esters of phenolic amines can be simultaneously cleaved and converted to trimethyl silyl derivatives with BSTFA or BSA (p. 403). The *t*-butyldimethylsilyl derivative of a phenolic amine has also been reported (p. 135).

Since *O*-TMS derivatives tend to be more stable than *N*-TMS, and *N*-acyl derivatives tend to be more stable than *O*-acyl, it would be predicted that the mixed *O*-TMS, *N*-acyl derivatives should be more stable than either of the single types of derivatives. This appears to be the case and several methods have been reported for their preparation (p. 136). A sequence of reactions involving acetylation, methylation, and trimethylsilylation has been used to prepare derivatives to distinguish dopamine 3- and 4-*O*-sulfate isomers (p. 404).

Mixed trimethylsilyl, Schiff base derivatives have been prepared from phenolic amines (p. 138). Use of pentafluorobenzaldehyde results in derivatives that may be quantitated by ECD-GC (p. 138). *N*-Dinitrophenyl, O-trimethylsilyl derivatives have also been used for ECD-GC analysis of monophenolic amines, but the high pH used in the method was incompatible with catecholamines (p. 139).

The isothiocyanate derivatives of primary amines, discussed previously, have been applied to phenolic amines both with and without further derivatization of the phenolic groups (p. 139). One step preparation of the *O*-trimethylsilyl isothiocyanate derivatives was reported using a mixture of TMSI and carbon disulfide in dimethylformamide (p. 140).

Use of chloromethyldimethylsilyl (CMDMS) reagents gave the expected CMDMS derivatives of phenolic hydroxyls but yielded cyclic dimethylsilylmethylene derivatives from β-hydroxyamine sidechains (p. 141). A similar result was obtained with an amino acid. Cyclic derivatives have also been prepared with the boronic acids (p. 142). In this case both the catechol function and the β-hydroxyamine sidechain gave cyclic boronate derivatives. Treatment of phenolic amines with BANS-Cl resulted in reaction of both the amino acid and the phenol groups to yield a mixture of totally and partially derivatized products (p. 143). The derivatives were used for quantitative MS analysis, with the totally derivatized product deemed most useful.

DANS-Cl has been used to prepare chromogenic derivatives for HPLC analysis of catecholamines and their *O*-methyl metabolites (p. 143). Derivatization of both amino and phenolic groups was obtained. Fluorescamine derivatives have also been applied to catecholamines and metabolites (p. 144).

Hydrazines. Hydrazines were converted to Schiff base derivatives with 5-nitro-2-hydroxybenzaldehyde for GC analysis (p. 145). These nitroaromatic derivatives should also be applicable to HPLC analysis. Treatment of hydrazines with acetylacetone gave pyrazole derivatives, which could be analyzed by GC (p. 145).

2.1. DERIVATIVES OF PRIMARY AND SECONDARY AMINES

2.1.1. Acyl Derivatives

2.1.1.1. *Acetyl Derivatives*

Derivative Type. Acyl.

Reagent. Anhydrides/pyridine.

Compound Type. Primary amines.

Reaction

$$R{-}NH_2 \xrightarrow[\text{pyridine}]{(R'CO)_2O} R{-}NHCO{-}R' \qquad R' = CH_3, C_2H_5, C_3H_7, CF_3, C_2F_5, C_3F_7$$

Typical Procedure (Details Not Given). Diss the amine in EtOAc and treat with the anhydride using pyridine as catalyst. Analyze by GC.

Comments. Acetyl, propionyl, butyryl, TFA, PFP, and HFB derivs were prepared from a series of long chain and alicyclic monoamines, aliphatic diamines, and aromatic amines. The elution order for the ordinary amides was acetamide < propionamide < butyramide. The fluorinated derivs showed no large increase in retention time for the TFA versus the HFB. On the phases examined (CNSi, SE-52, F-60-Z) the PFP derivs had the shortest retention times.

References

[1]W. J. A. VandenHeuval *et al., Anal. Chem.* **36**, 1550 (1964).

Derivative Type. Acetyl.

Reagent. Acetic anhydride (Ac_2O).

Compound Type. 4-Methoxy- and 3,4-dimethoxyphenethylamine.

Reaction.

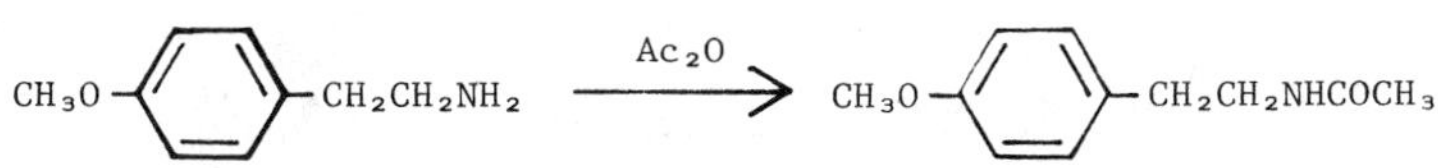

Typical Procedure. Ext 400 ml urine at pH 1 with 3 × 1 vol Et_2O and discard ext. Add 1 g EDTA, adjust to pH 9 with 3 *N* NaOH, and ext with 800 ml $CHCl_3$. Add 10 ml 3 *N* NaOH and continue ext. Separate $CHCl_3$ and ext aq phase with another 2 × 600 ml $CHCl_3$. Dry combined $CHCl_3$ over Na_2SO_4, evap to a small vol, and isolate the amines by paper chromatography. Elute spots with 0.01 *N* HCl, alkalinize eluates, and ext with $CHCl_3$. Dry ext over Na_2SO_4 and evap to dryness. Add 0.5 ml Ac_2O to the residue and reflux 3 min. Cool, add 6 ml 3 *N* NaOH, and ext with Et_2O. Dry ext over Na_2SO_4 and conc to 0.1 ml. Analyze by GC.

Comments. Benzoyl derivs were also prepared by adding 2 ml 10 *N* NaOH and 0.5 ml benzoyl chloride to the eluate and shaking 1 hr. The benzoyl derivs were extd with Et_2O after adjusting the pH to 11.8. The methods were used to identify the compds in urines from schizophrenic patients.

References.

[1]N. P. Sen and P. L. McGeer, *Biochem. Biophys. Res. Commun.* **14**, 227 (1964).

Derivative Type. Acetyl.

Reagent. Acetic anhydride (Ac_2O)/pyridine.

Compound Type. Desmethyldoxepin.

Reaction.

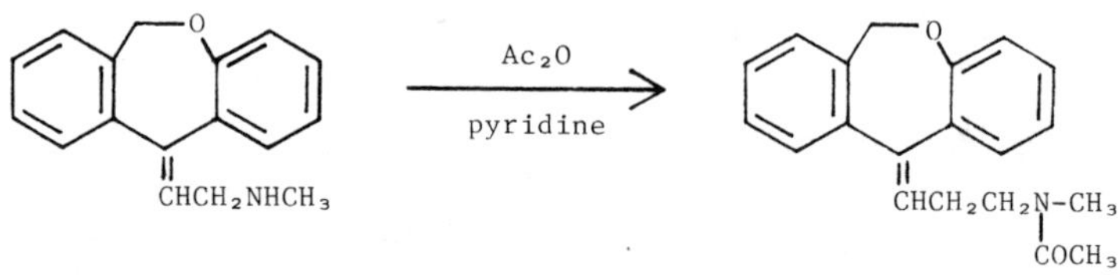

Typical Procedure. Alkalinize a 1 ml volume of heparinized plasma (previously acidified with 0.5 ml 0.1 *N* HCl and stored at −20°) with 0.5 ml 1 *N* NaOH and ext with 3 ml *n*-hexane by shaking 30 min. Centrifuge and remove 2.5 ml of the org phase and transfer to a silanized tube that has just been rinsed with isoamyl alcohol.

Add 50 μl pyridine and 100 μl Ac_2O and heat at 70° (sand bath) for 30 min. Evap under N_2 at 70°. Cool to room temp and diss residue in 100 μl abs EtOH containing 1 μg/ml promazine int stand. Analyze 1–3 μl by GC-MS.

Comments. The method was used to quantitate doxepin and its desmethyl deriv in plasma to 10 ng/ml using GC-MS-SIM. The acetyl deriv was chosen for its convenient retention time.

References.

[1]A. Frigerio, *et al., J. Chromatogr.* **130**, 354 (1977).

Derivative Type. *N*-Acetyl.

Reagent. Acetyl chloride (AcCl).

Compound Type. Amphetamine.

Reaction.

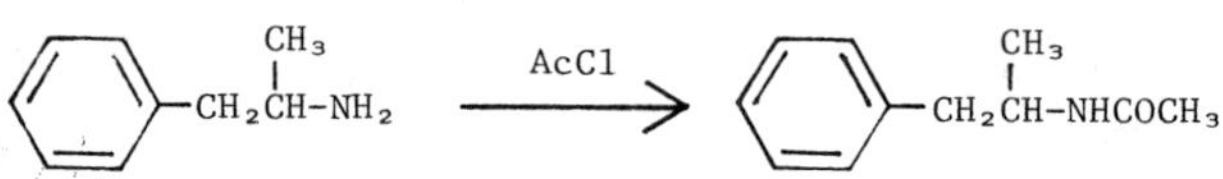

Typical Procedure. Acidify 10 ml urine to pH 1 with 5 *N* HCl and ext with 3 × 15 ml Et_2O. Add 5 g NaCl to the aq phase, adjust to pH 10 with 5 *N* NaOH, and ext with 3 × 5 ml Et_2O. Dry the Et_2O ext over Na_2SO_4 and add 0.5 ml AcCl. Allow to stand at room temp 30 min then evap to dryness under N_2. Diss the residue in 100 μl benzene and analyze by GC.

Comments. The method permitted detection of as little as 100 ng amphetamine.

References

[1]D. A. Toseland and P. H. Scott, *Clin. Chim. Acta* **25**, 75 (1969).

2.1.1.2. *Trifluoroacetyl Derivatives*

Derivative Type. Trifluoroacetyl (TFA).

Reagent. Trifluoroacetylimidazole (TFAI).

Compound Type. Desipramine.

Reaction.

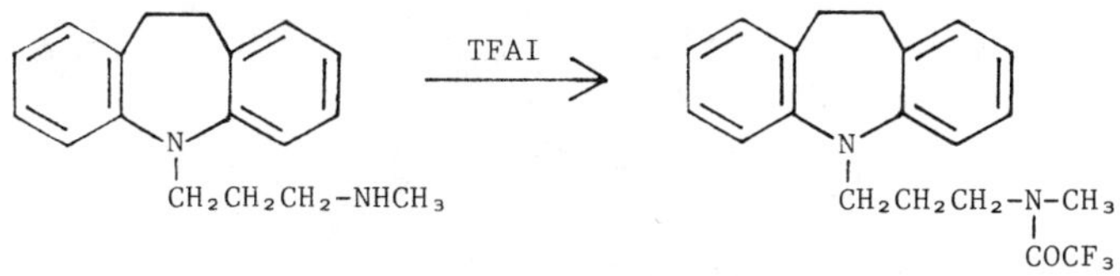

Typical Procedure. Add 1 ml 0.1 *M* NaOH and 100 μl of d_4-imipramine and d_4-desipramine int stand solns (0.6 μg and 1 μg respectively) to 2 ml plasma and ext with 2 × 3 ml hexane. Evap exts to dryness under N_2.

Add 20 μl hexane and 2 μl TFAI. Allow to stand at room temp 30 min then analyze 1–2 μl by GC-MS-SIM.

Comments. The method was used to quantitate imipramine and desipramine in plasma to low ng levels using methane CI-MS. Use of TFAA for deriv yielded a bis-TFA deriv of desipramine[2] and totally degraded imipramine.

References

[1]M. Claeys *et al.*, *Biomed. Mass Spect.* **3**, 110 (1976).
[2]T. Walle, *et al.*, *Acta Pharm. Suecica* **9**, 509 (1972)

Derivative Type. Trifluoroacetyl (TFA).

Reagent. Trifluoroacetic anhydride (TFAA).

Compound Type. Desipramine.

Reaction

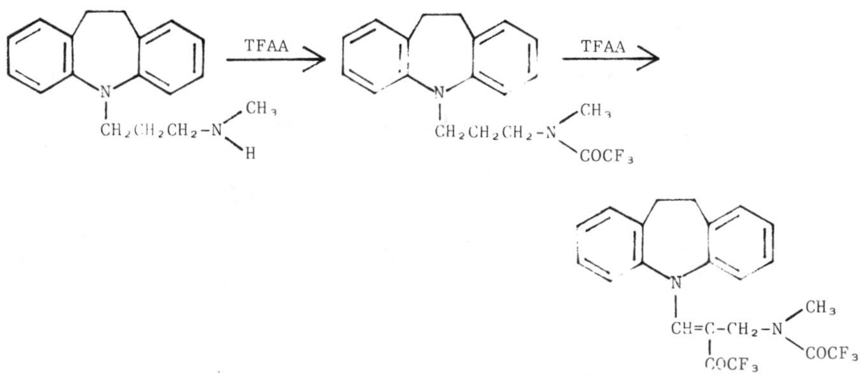

Typical Procedure. Diss 10 ng desipramine in 50 μl EtOAc and add 100 μl of TFAA. Allow to stand 45 min at 35° then evap to dryness under N_2. Diss residue in benzene and analyze by ECD-GC.

Comments. The method was used for quantitation of desipramine in serum.[1] The deriv was detectable to 0.5 pg by ECD-GC. The mono-TFA deriv was formed rapidly with the di-TFA deriv requiring 45 min for quant conversion. In the absence of EtOAc, quant conversion to the di-TFA took 4 hr. The di-TFA was synthesized on a larger scale for structure proof.[2] A di-

HFB deriv was also formed with HFBA but at a slower rate.[1] The di-HFB deriv has been used for quantitation of desipramine by GS-MS-SIM.[3]

References

[1]M. Ervik *et al.*, *Acta Pharm. Suecica* **7**, 625 (1972).

[2]T. Walle *et al.*, *Acta Pharm. Suecica* **9**, 509 (1972).

[3]K. Matsubayashi *et al.*, *J. Chromatogr.* **143**, 571 (1977).

Derivative Type. Trifluoroacetyl (TFA).

Reagent. Trifluoroacetic anhydride (TFAA).

Compound Type. Polyamines.

Reaction

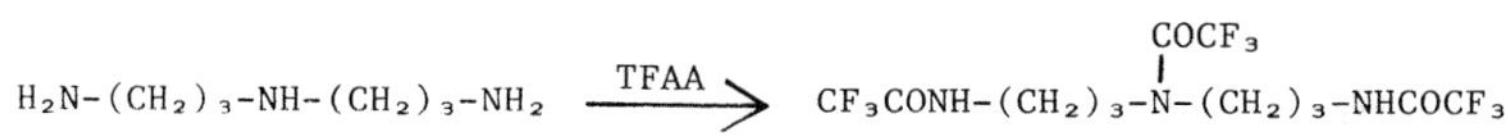

Typical Procedure. Evap aliquots of polyamine soln in 1.0 *N* HCl to dryness in a 0.3 ml vial. Add deuterated int stand and evap to dryness under N_2. Add 100 μl TFAA, seal, and sonicate 10–15 min. Evap to dryness under N_2 and diss residue in CH_2Cl_2 prior to GC-MS analysis.

Comments. The method was used to quantitate polyamines down to the 50 pmol level using d_4-putrescine, d_4-cadaverine, d_6-spermidine, and d_8-spermine as int stand.

References

[1]T. Walle, in *Polyamines in Normal and Neoplastic Growth*, ed by D. H. Russell, Raven Press, New York, 1973, pp. 355–365.

[2]R. G. Smith and G. Davies, *Biomed. Mass Spect.* **4**, 146 (1977).

Derivative Type. Trifluoroacetyl (TFA).

Reagent. Trifluoroacetic anhydride (TFAA).

Compound Type. Aryl amines (anilines and alkylated derivs).

Reaction

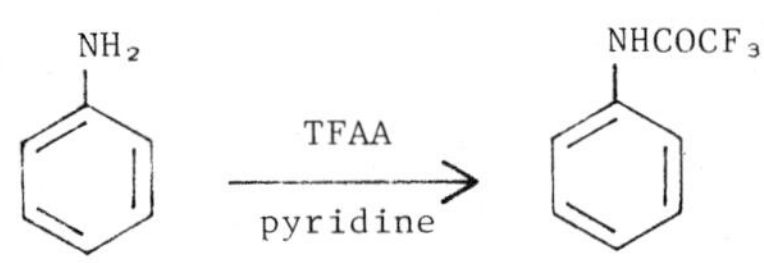

Typical Procedure. Diss 200 mg mixed amine sample in 2 ml THF and add 5 drops pyridine. Cool in an ice bath and add 2 ml TFAA. Heat at 50° for 10 min then cool and add 5 ml H_2O. Ext with 8 ml and 4 ml CH_2Cl_2. Wash the combined CH_2Cl_2 with satd aq $NaHCO_3$ and 5 ml H_2O. Dry over Na_2SO_4 and analyze 5 μl by GC.

Comments. The method was used to quantitate 19 aryl amines using *n*-dodecane as int stand.

References

[1]R. A. Dove, *Anal. Chem.* **39**, 1188 (1967).

Derivative Type. Trifluoroacetyl (TFA).

Reagent. Trifluoroacetic anhydride (TFAA).

Compound Type. Amines.

Reaction

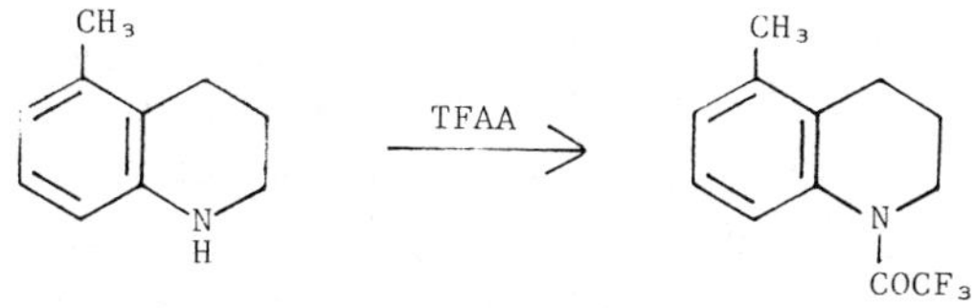

Typical Procedure. Diss the amine in Et_2O and add TFAA in slight excess. Allow the soln to stand a few min then wash with aq $NaHCO_3$. Wash with H_2O and dry over Mg SO_4. Evap to dryness and analyze by GC or GC-MS.

Comments. The TFA derivs of a series of isomers of tetrahydroquinolines, tetrahydroisoquinolines, and tetrahydronaphthylamines were examined by GC-MS. [2] The deriv procedure was also applied to a series of 48 primary and secondary amines with GC data reported.[1]

References

[1]W. J. Irvine and M. J. Saxby, *J. Chromatogr.* **43**, 129 (1969).
[2]W. J. Irvine and M. J. Saxby, *J. Chromatogr.* **70**, 141 (1972).

Derivative Type. Trifluoroacetyl (TFA).

Reagent. Trifluoroacetic anhydride (TFAA).

Compound Type. Amphetamine and phentermine.

Reaction

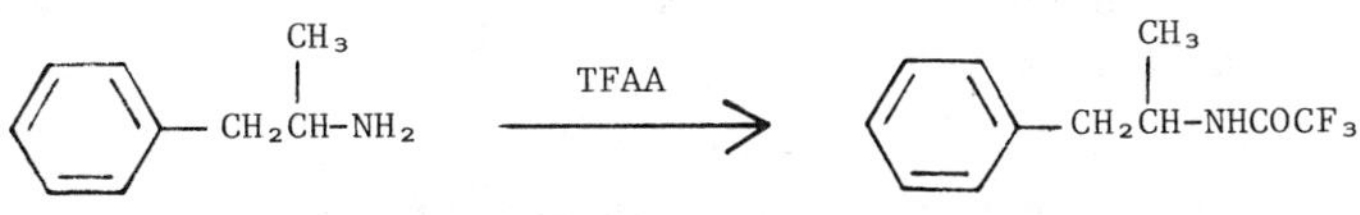

Typical Procedure. Alkalinize 5 ml blood with 0.5 ml 20% NaOH and ext with 10 ml benzene. Centrifuge and decant benzene into Teflon® coated tube. Add 1 μl diethylamine and form HCl salts by bubbling HCl into soln. Add 10μl of a 0.05% soln of 1-adamantanamine int stand soln in chloroform. Centrifuge to conc ppt at bottom of tube then evap to dryness under N_2 at 90°. Wash down walls of tube with 1 ml CH_2Cl_2 and evap to dryness under N_2. Add 50μl hexane and 50μl TFAA. After 30–45 min, analyze by GC.

Comments. The method was used for quantitation of amphetamine and phentermine in blood and urine. Formation of the HCl salt was necessary to prevent loss of the volatile amine during evap of the ext.

References

[1]J. E. O'Brien *et al.*, *J. Chromatogr. Sci.* **10**, 336 (1972).

Derivative Type. Di(trifluoroacetyl).

Reagent. Trifluoroacetic anhydride (TFAA)/trimethylamine (TMA).

Compound Type. Primary amines.

Reaction

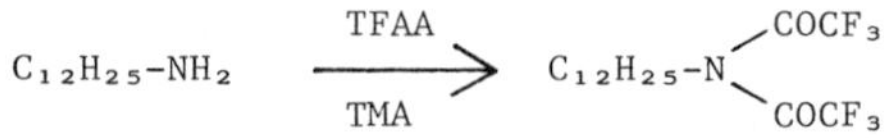

Typical Procedure. Mix 100 μl of a soln of dodecylamine in benzene (0.97–12.2 μg/ml) with 200 μl 0.3 *M* TMA in benzene. Add 25 μl TFAA, mix, and allow to stand at room temp for 15 min. Add 2 ml pH 6.0 phosphate buffer, shake 15 sec, and centrifuge. Analyze 1 μl aliquot of benzene phase by ECD-GC.

Comments. The conc of TMA was critical to the yield of diacyl deriv. Di-HFB derivs were also prepared with HFBA. The di-HFB derivs gave higher ECD response.

References

[1]H. Ehrsson and H. Brötell, *Acta Pharm. Suecica* **8**, 591 (1971).

Derivative Type. Trifluoroacetyl (TFA).

Reagent. Trifluoroacetic anhydride (TFAA).

Compound Type. Fatty amines.

Reaction

$$\text{R-NH}_2 \xrightarrow{\text{TFAA}} \text{R-NHCOCF}_3$$

Typical Procedure. Mix 5 drops melted fatty amines with 5 drops $CHCl_3$ and add 6–10 drops TFAA. Allow to stand a few min at room temp then add 5–8 drops H_2O or *n*-BuOH to destroy excess TFAA. Mix, allow to settle, then analyze an aliquot of the lower phase by GC.

Comments. The method was applied to mixt of C_8–C_{22} fatty amines including unsatd amines.

References

[1]R. A. Morrissette and W. E. Link, *J. Gas Chromatogr.* **3**, 67 (1965).

Derivative Type. Trifluoroacetyl.

Reagent. *N*-Methyl bis(trifluoroacetamide) (MBTFA).

Compound Type. Lofexidine.

Reaction

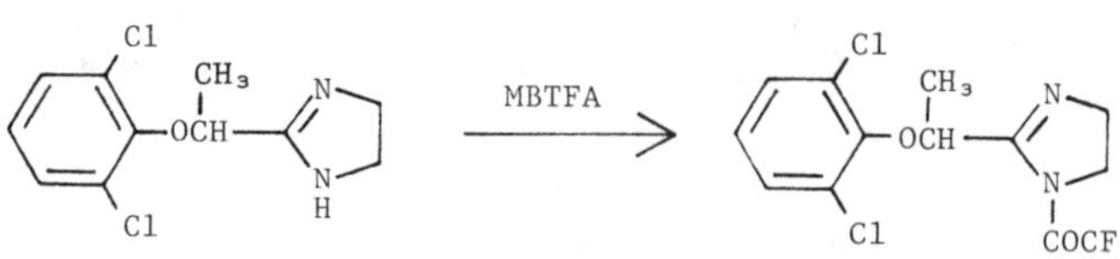

Typical Procedure. Adjust a 2 ml aq sample (urine or plasma) to pH 9–10 with 10% (w/v) Na_2CO_3 and ext with 2 × 5 ml EtOAc. Back ext the EtOAc with 2 ml 0.1 *M* HCl. Adjust the aq soln to pH 10 with aq Na_2CO_3 and ext with 5 ml EtOAc. Evap the org phase to dryness under

N_2 and treat the residue with 10μl MBTFA. Allow to stand 1 hr at room temp then analyze by GC-MS-SIM.

Comments. The use of TFAA was avoided because the aq wash which was needed to remove excess reagent caused problems with hydrolytic opening of the imidazoline ring. The drug could alternatively be intentionally hydrolyzed to the ring-opened form after trifluoroacetylation.

References

[1]D. R. Knapp, unpublished work.

Derivative Type. Trifluoroacetyl (TFA).

Reagent. Trifluoroacetyl Nylon 66.

Compound Type. Amines and alcohols.

Reaction

$$R_2NH \xrightarrow{\text{TFAA Nylon 66}} R_2NCOCF_3$$

Typical Procedure. Add 230 mg TFA Nylon 66 to a soln of 1 mmol of amine in 5 ml CH_3CN. Stir at room temp 12 hr, centrifuge, and wash ppt with 2 × 5 ml CH_3CN. Combine CH_3CN supernatant and washes and evap to yield the deriv.

Comments. The preparation of the reagent from poly(hexamethyleneadipamide) (Nylon 66) and TFAA was reported. Analytical data supported the following structure for the reagent:

$$[-\underset{\substack{|\\ COCF_3}}{N}-(CH_2)_6-\underset{\substack{|\\ COCF_3}}{N}-CO-(CH_2)_4-CO-]_n$$

References

[1]H. Schuttenberg and R. C. Schulz, *Angew. Chem. Int. Ed. Eng.* **15**, 777 (1976).

2.1.1.3. *Pentafluoropropionyl Derivatives*

Derivative Type. Pentafluoropropionyl (PFP).

Reagent. Pentafluoropropionic anhydride (PFPA).

Compound Type. Anilines (metabolites of pesticides).

Reaction

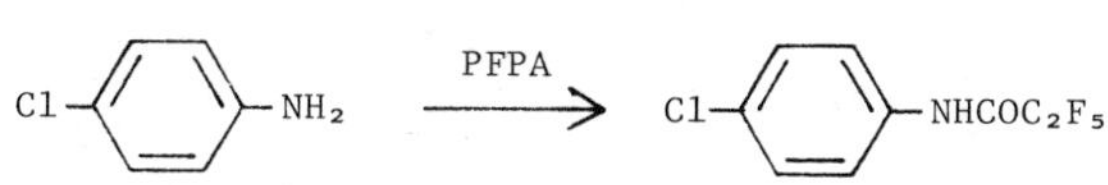

Typical Procedure. To 2.0 ml urine or H_2O add 1 ml 10% NaOH and 1 g NaCl, and ext with 4 ml and 2 × 2 ml benzene. Evap ext under N_2 to 0.2 ml. Add 0.2 ml EtOAc and 0.1 ml PFPA. Allow to stand 1 hr at room temp then add 2 ml H_2O and ext with 5 ml benzene. Wash ext with 2 × 2 ml H_2O and dry over Na_2SO_4. Evap to 1 ml and dil to 5 ml with hexane. Purify deriv on a 1 g column of silica gel topped by 1 g Na_2SO_4. Prewash column with hexane, add ext, then elute deriv with 15 ml 60% benzene in hexane. Analyze eluate by GC or GC-MS.

Comments. A variety of derivs were examined for analysis of the anilines with the PFP deemed the best in terms of "clean" chromatograms at low levels. The HFB derivs (prepared with HFBA or HFBI) were nearly as clean. Other deriv reagents examined included *p*-bromobenzoyl bromide, pentafluorobenzoyl chloride and bromide, 1-fluoro-2,4-dinitrobenzene, 4-chloro-α,α,α-trifluoro-3,5-dinitrotoluene, and α,α,α,4-tetrafluoro-3-nitrotoluene; all gave low yields, multiple products, or unuseable chromatograms. MS data were reported for the PFP derivs of six anilines.

References

[1]D. E. Bradway and T. Shafik, *J. Chromatogr. Sci.* **15**, 322 (1977).

Derivative Type. Pentafluoropropionyl (PFP).

Reagent. Pentafluoropropionic anhydride (PFPA).

Compound Type. Amines.

Reaction

$$R\text{-}NH_2 \xrightarrow{\text{PFPA}} R\text{-}NHCOC_2F_5$$

Typical Procedure. React one equiv of amine with 2 equiv of PFPA in CH_2Cl_2 at 0° for 1 hr. Evap under vac and add ice H_2O to the residue. Separate by filtration (solid derivs) and recryst from EtOH by precipitation with ice H_2O. Dry in a dessicator over Drierite. Analyze by GC.

Comments. The TFA, PFP, and HFB derivs were compared for use in separating enantiomeric amines on an optically active stationary phase (the ureide of Val isopropyl ester). The separation was best with the PFP derivs. The method is given here also as a method for preparative scale synthesis of the pure derivs.

References

[1]J. A. Corbin and L. B. Rogers, *Anal. Chem.* **42**, 974 (1970).

2.1.1.4. *Heptafluorobutyryl Derivatives*

Derivative Type. Heptafluorobutyryl (HFB.)

Reagent. Heptafluorobutyric anhydride (HFBA)/dimethylformamide (DMF)/pyridine.

Compound Type. Nortriptyline and desmethylnortriptyline.

Reaction

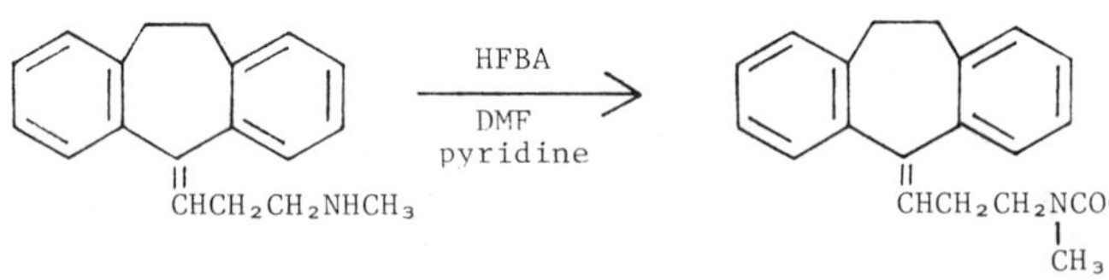

Typical Procedure. To 3 ml plasma add 3.26 μg int stand (CIBA 34276 HCl) in 0.1 ml 0.01 *M* HCl. Adjust to pH 10.5 by addition of 1 ml 2 *M* $NaHCO_3$ and ext with 10 ml hexane. Back ext 9 ml of the hexane with 2 ml 0.05 *M* H_2SO_4. Wash the acid phase with 10 ml hexane then alkalinize with 2 ml 0.1 *M* NaOH and 0.5 ml 2 *M* $NaHCO_3$. Ext with 4.5 ml hexane. To 4 ml of the ext add 20 μl DMF, 20 μl pyridine, and 100 μl HFBA. Heat at 63° for 1 hr. Add 1.5 ml H_2O-MeOH (2:1, v/v), shake, and separate hexane phase. Evap hexane to dryness and diss residue in 50 μl hexane. Analyze by GC-MS-SIM.

Comments. The method permitted quantitation to a few ng/ml.

References

[1]O. Borga *et al., Anal. Lett.* **4**, 837 (1971).

Derivative Type. Heptafluorobutyryl (HFB)

Reagent. Heptafluorobutyrylimidazole (HFBI)

Compound Type. Protriptyline.

Reaction

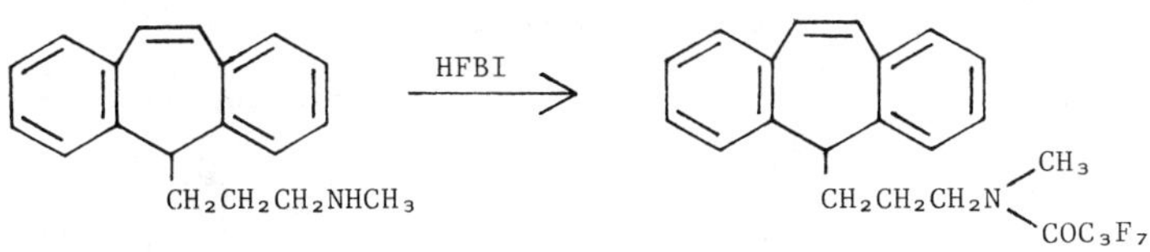

Typical Procedure. Mix 2 ml plasma, urine, or erythrocyte specimen with 2 ml H_2O and 0.2 ml 5*N* NaOH. Ext with 5 ml Et_2O, remove 3 ml of the Et_2O phase, and ext with another 3 ml Et_2O. Back ext the combined Et_2O phases with 1 ml 1 *N* H_2SO_4 and discard Et_2O. Heat the aq phase 5 min at 60° to expel residual Et_2O then mix with 1 ml *n*-hexane and 0.4 ml 5 *N* NaOH. Mix well, remove 0.6 ml hexane, and repeat ext with 2 × 0.6 ml hexane. Add 0.1 ml HFBI to the combined hexane ext and heat 15 min at 60°. Evap under N_2 and diss residue in 0.3 ml toluene. Wash with 1 ml 1 *N* NaOH and 0.5 ml 0.2 *N* H_2SO_4; then refrigerate prior to GC analysis.

Comments. The deriv was detectable to 50 pg permitting quantitation in biological samples to *ca* 10 ng/ml.

References

[1]S. F. Sisenwine *et al., Anal. Lett.* **2**, 315 (1969).

Derivative Type. Heptafluorobutyryl (HFB).

Reagent. Heptafluorobutyric anhydride (HFBA).

Compound Type. Primary and secondary amines.

Reaction

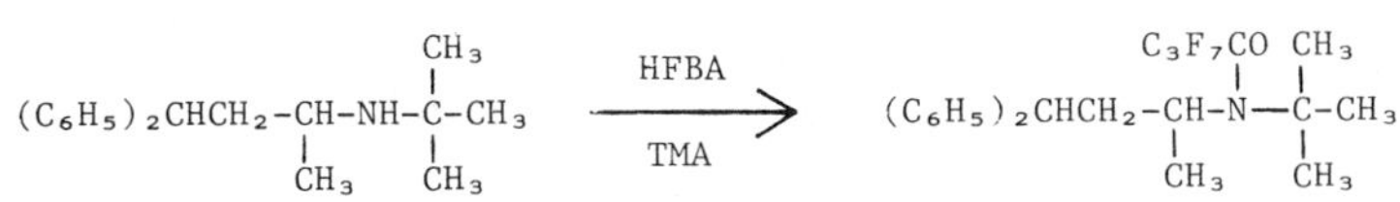

Typical Procedure. Mix 4 ml of plasma or urine with 15 ml benzene. Add 0.5 ml of aq int stand soln and 0.5 ml 5 *N* NaOH. Shake, centrifuge, and remove benzene layer. Evap under N_2 at 80° in a glycerine bath. Add 50 μl of 0.05 *M* TMA in benzene and 10 μl of HFBA. Warm 10 min at 50°. Add 1 ml H_2O and shake 1 min. Add 1 ml 5% aq NH_3 and shake 5 min. Centrifuge and analyze 1–5 μl by ECD-GC.

Comments. A back ext into 0.4 ml 0.1 *M* H_2SO_4 followed by basification with 0.1 ml 5 *M* NaOH and ext into 2 ml benzene could be performed for additional cleanup. The method was applied to 14 amines, 6 alcohols, and the amino alcohol ephedrine.[2] The applicability to highly hindered amines was illustrated with terodiline (shown above). The minimum detectable quantity was 0.3 pg (HFB aniline). Commercial HFBA gave low yields of deriv presumably due to contaminating HFB acid. The purification of HFBA by distillation from P_2O_5 was described.

References

[1]T. Walle and H. Ehrsson, *Acta Pharm. Suecica* **8**, 27 (1971).
[2]T. Walle and H. Ehrsson, *Acta Pharm. Suecica* **7**, 389 (1970).

Derivative Type. Heptafluorobutyryl (HFB).

Reagent. Heptafluorobutyric anhydride (HFBA)/trimethylamine (TMA).

Compound Type. Terodiline (*N*-*t*-butyl-l-methyl-3, 3-diphenylpropylamine).

Reaction

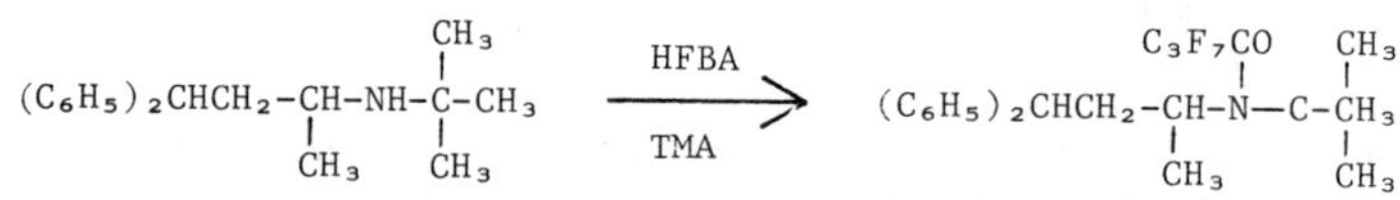

Typical Procedure. Mix serum sample containing 25–100 ng of Terodiline (2 ml or less) with 100 μl of int stand (butylamine analog, 2 μg/ml in H_2O). Add 0.5 ml 2 *M* NaOH and H_2O to make 10 ml. Ext 30 min with 10 ml heptane and discard aqueous phase. Add 1.0 ml 0.1 *M* H_2SO_4 and extract 10 min. Remove aq phase and add to it 0.5 ml 2 *M* NaOH and 0.1 ml heptane. Ext 10 min, centrifuge, and completely remove aq phase. To the heptane soln add 100 μl 1.5 *M* TMA in heptane and transfer entire phase to another ext tube. Add 15 μl HFBA and hold at 50° for 30 min. Add 2 ml 0.1 *M* NaOH, ext 10 min, and centrifuge. Anaylze 2–5 μl of org phase by GC.

Comments. The method formed an ECD sensitive deriv of a highly hindered amine.

References

[1]P. Hartvig *et al., Acta Pharm. Suecica* **11**, 97 (1974).

Derivative Type. Heptafluorobutyryl (HFB).

Reagent. Heptafluorobutyric anhydride (HFBA).

Compound Type. Amphetamine and related amines.

Reaction

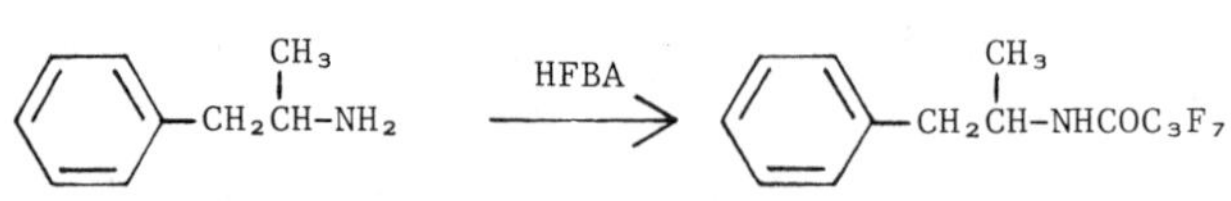

Typical Procedure. Add int stand to 0.5 ml blood, mix, and add 1.0 ml 2 *N* NaOH and 5.0 ml H_2O. Ext with 10 ml followed by 5 ml pentane. Dry ext by passing through Na_2SO_4 then wash Na_2SO_4 with 2 ml pentane. Evap combined pentane under a gentle stream of N_2 and add 100 μl HFBA to the residue. Allow to react 30 min with occasional mixing. Evap excess HFBA under a gentle stream of N_2. Cool the tube in ice and add 0.5 ml 2 *N* NaOH. Keeping the mixt cold, ext with 4 ml, 3 ml and 3 ml portions of pentane. Pass pentane through Na_2SO_4, and wash Na_2SO_4 with 2 ml pentane. Allow combined pentane to stand overnight at room temp. Evap pentane under gentle stream of N_2, diss residue in 0.2 ml pentane, and anaylze 2–4 μl aliquots by ECD-GC.

Comments. The 5% OV-1 column used required conditioning by injection of HFBA in pentane prior to use. Chlorphentermine was used as int stand for amphetamine, methamphetamine, and fenfluramine; amphetamine was used as int stand for chlorphentermine.

References

[1]R. B. Bruce and W. R. Maynard, *Anal. Chem.* **41**, 977 (1969).

2.1.1.5. *Trichloroacetyl Derivatives*

Derivative Type. Trichloroacetyl.

Reagent Trichloroacetyl chloride.

Compound Type. 1-α-Acetylmethadol metabolites.

Reaction

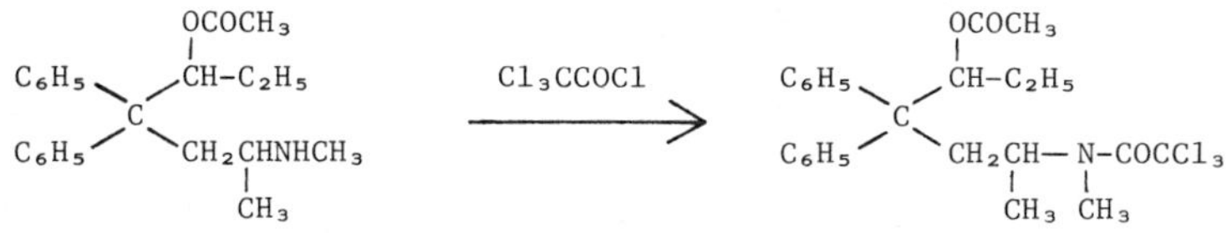

Typical Procedure. Dil 2 ml plasma with 2 ml H_2O and adjust to pH 9.5 with 0.1 *N* NaOH. Ext with 2 × 10 ml 1-chlorobutane. Evap to dryness under N_2 and add 0.5 ml 0.25 m*M* trichloroacetyl chloride in toluene. Heat 2 hr at 70° then evap under N_2. Diss residue in hexane and analyze by ECD-GC.

Comments. Derivs were prepared from noracetylmethadol, dinoracetylmethadol and normethadol. Billings *et al.*[2] used the same reagent to prepare the trichloroacetyl derivs of the first two but not of normethadol. Kaiko *et al.*[3] used TFAI to prepare TFA derivs of the first two but the calibration curves obtained were nonlinear and inconsistent. Trichloroacetic anhydride, pentafluorobenzoyl chloride, HFB-Cl, HFBA, and TFAA were also evaluated for deriv of the three metabolites but trichloroacetyl chloride was found to give the best results.[1] ECD and FID response data were tabulated for all of the derivs.

References

[1]D. H. Lau and G. H. Henderson, *J. Chromatogr.* **129**, 329 (1976).

[2]R. E. Billings *et al., J. Med. Chem.* **16**, 305 (1973).

[3]R. F. Kaiko *et al., J. Chromatogr.* **109**, 247 (1975).

Derivative Type. *N*-Trichloroacetyl.

Reagent. Trichloroacetyl chloride.

Compound Type. Sympathomimetic amines.

Reaction

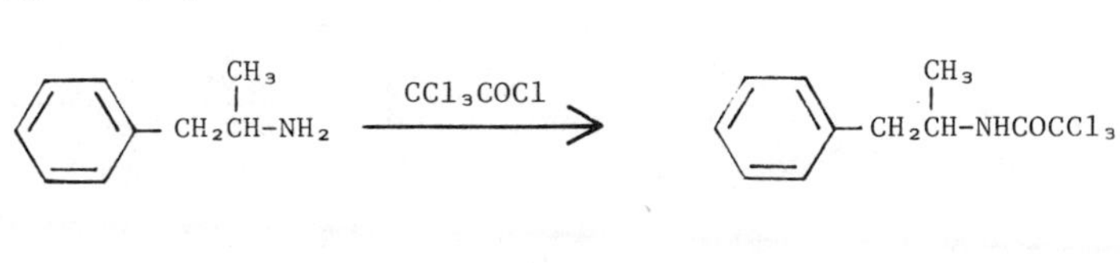

Typical Procedure. To a soln of 1 μg/ml of the amine in hexane or EtOAc add a 50–100 fold excess of trichloroacetyl chloride. Allow to react 30 min at room temp then analyze by ECD-GC. Alternatively, remove excess reagent, by washing briefly with 0.1 *N* NaOH before analysis.

Comments. A series of derivs of amphetamine and related compds were prepared and their ECD response measured. The *N*-trichloroacetyl derivs were found to have good GC properties and favorable ECD response as well as being easy to prepare. The ECD response of the various derivs of amphetamine and phenmetrazine are tabulated below.

		ECD Response (Lindane = 100)	
Derivative[a]	Reagent	Amphetamine Derivative	Phenmetrazine Derivative
TFA	anhydride	<0.1	<0.1
PFP	anhydride	4	6
HFB	anhydride	9	8
Perfluorooctanoyl	anhydride	23	16
Monochloroacetyl	anhydride	0.1	1
Trichloroacetyl	acid chloride	54	47
Maleamide	anhydride/CH_2N_2	13	7
p-Trifluoromethylbenzoyl	acid chloride	12	16
Pentafluorobenzoyl	acid chloride	77	52
Isothiocyanate	CS_2	0.2	—

[a]MS data were reported for the derivs.

References

[1]E. Anggard and A. Hankey, *Acta Chem. Scand.* **23**, 3110 (1969).

Derivative Type. Trichloroacetyl.

Reagent. Trichloroacetyl chloride.

Compound Type. Amphetamine and phenmetrazine.

Reaction

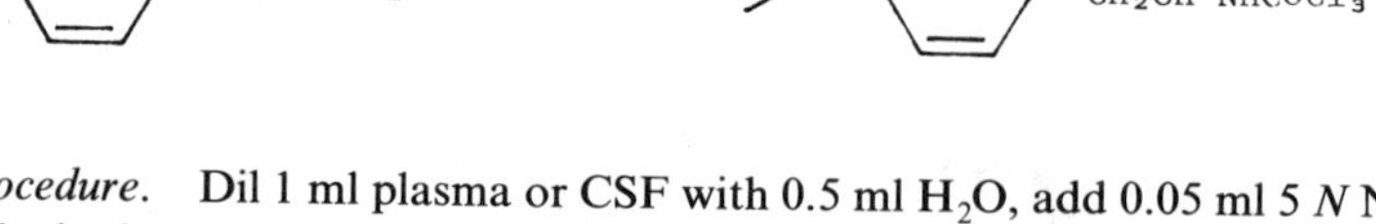

Typical Procedure. Dil 1 ml plasma or CSF with 0.5 ml H_2O, add 0.05 ml 5 *N* NaOH, and ext with 2.5 ml toluene. Remove 2 ml of the toluene ext and add to it 10 μl 1 % trichloroacetyl chloride in toluene. Heat at 40° for 30 min. Cool to room temp and analyze a 4 μl aliquot by ECD-GC.

Comments. The trichloroacetyl deriv was used because it was less volatile and had *ca* 5 × the sensitivity of the HFB deriv. The method permitted blood level measurements to 1 ng/ml. For the higher levels in urine, TFA derivs (from TFAA, 60°, 15 min) were used with FID-GC.

References

[1]E. Anggard *et al., Scand. J. Clin. Lab. Invest.* **26**, 137 (1970).

Derivative Type. Trichloroacetyl.

Reagent. Trichloroacetyl chloride.

Compound Type. Amphetamine.

Reaction

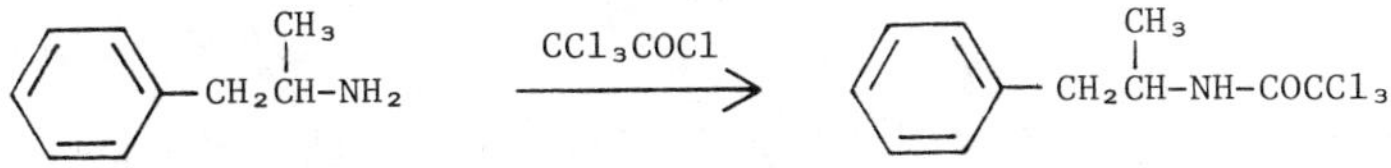

Typical Procedure. Mix 4 ml oxalated plasma with 0.5 ml 10 *N* NaOH and ext with 4 ml hexane/isooctane (3:1, v/v). Remove 2 ml org phase and add to it 50 μl trichloroacetyl chloride. Mix and allow to stand at room temp 20 min. Add 2 ml 1 *N* NaOH and mix. Centrifuge and analyze 4 μl aliquots of the hexane phase by ECD-GC.

Comments. If the deriv soln was not to be analyzed immediately, the hexane soln was separated from the final NaOH wash soln. The method was used to quantitate amphetamine in plasma, urine, and saliva of horses.

References

[1]J. S. Noonan *et al., J. Pharm. Exp. Therap.* **168**, 205 (1969).

2.1.1.6. *Other Acyl Derivatives*

Derivative Type. *N*-Propionyl.

Reagent. Propionic anhydride.

Compound Type. 1, 3-Dihydro-3-phenylspiro [isobenzofuran-1, 4-piperidine].

Reaction

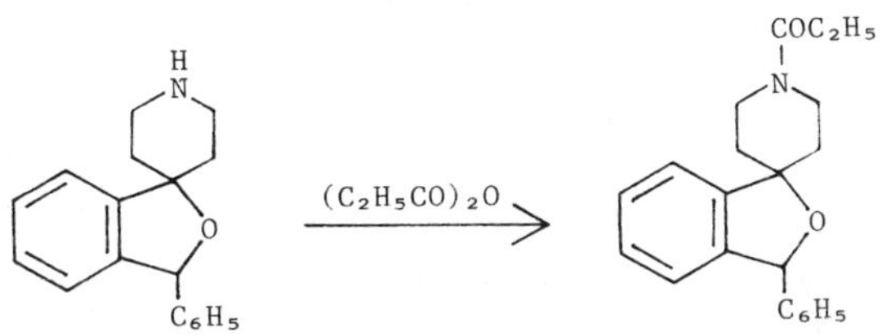

Typical Procedure. To 2 ml plasma (or 1 ml urine) add 0.5 ml 1 *M* NaOH and 100 ng int stand (*p*-methyl analog) and ext with 10 ml hexane. Separate hexane and back-ext with 1 ml 1 *M* HOAc. Wash aq phase with 1 ml hexane then basify with 1.5 ml 1 *M* NaOH. Ext with 5 ml hexane and to the ext add 0.5 ml 0.5% (v/v) propionic anhydride and 1% (v/v) pyridine in hexane. Heat 1 hr at 60°. Add 1 ml 1 *M* HOAc and mix. Separate hexane and evap to dryness under N_2 at 40°. Diss residue in 25μl cyclohexane containing 5% (v/v) EtOAc. Analyze by GC.

Comments. The TFA, HFB, and acetyl derivs were also prepared, but the *N*-propionyl gave the longest retention time and best separation from other compds in the ext.

References

[1]T. A. Bryce and J. L. Burrows, *J. Chromatogr.* **136**, 401 (1977).

Derivative Type. Chlorodifluoroacetyl.

Reagent. Chlorodifluoroacetic anhydride/triethylamine (Et_3N).

Compound Type. 1-(2-Phenyladamant-1-yl)-2-methylaminopropane.

Reaction

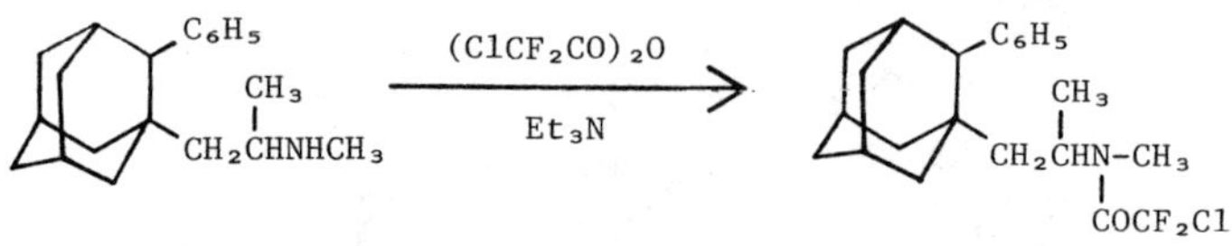

Typical Procedure. Mix 2–3 ml plasma with 0.5 ml satd aq Na_2HPO_4 and ext with 4 ml benzene. Back ext the benzene with 2 ml 0.1 *N* HCl. Neutralize the HCl soln with 0.2 ml 1 *N* NaOH and ext with $CHCl_3$. Evap the $CHCl_3$ to dryness, diss the residue in 20 μl $CHCl_3$, and add 3μl of the anhydride reagent and 1.5μl Et_3N. Heat 30 min at 50° then evap to dryness. Neutralize the residual reagent with excess $NaHCO_3$, then ext the deriv with $CHCl_3$. Evap the ext, diss the deriv in toluene, and analyze by ECD-GC.

Comments. Urine (5 ml) was treated similarly except that solid Na_2HPO_4 was added to raise the pH. The deriv permitted GC separation of the two diastereoisomers of the drug. Various acyl (RCO-) derivs were examined with the following order of ECD response observed: $R = CF_2Cl \approx n\text{-}C_3F_7 > CHCl_2 \approx C_2F_5 > CCl_3 > CF_3 \gg CH_3$. The chlorodifluoroacetyl deriv was preferred to the HFB deriv because it gave very similar ECD response for the two diastereoisomers whereas the HFB derivs gave a two- to threefold differential response. The HFB deriv solns also showed more reagent peak interferences.

References

[1]A. F. Cockerill *et al., J. Chromatogr.* **114**, 151 (1975).

Derivative Type. 4′-Nitroazobenzene-4-carboxamides.

Reagent. 4′-Nitroazobenzene-4-carboxylic acid chloride.

Compound Type. Amines.

Reaction

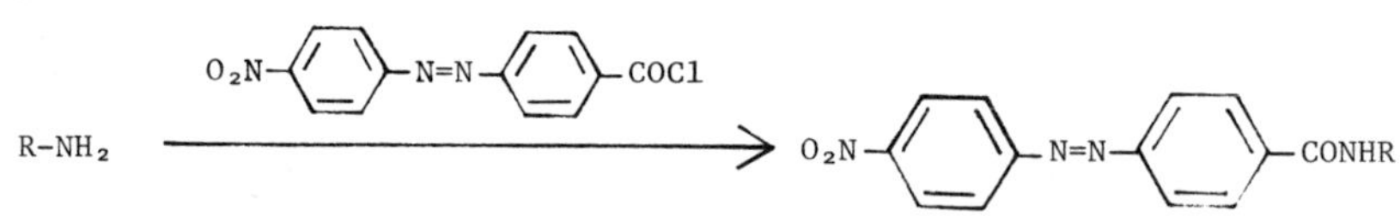

Typical Procedure. Prepare the derivs by heating the amines with the acid chloride in pyridine. Purify the derivs by preseparation on a silica column followed by TLC on silica. Analyze by GC.

Comments. GC data were reported for the derivs of an extensive list of primary and secondary amines. It was reported that primary amines in mixts could be identified by disappearance of their deriv peaks in the GC after treatment of the deriv mixt with "nitrous gases". The deriv has also been used for GC ident of hydroxyl compds. GC retention data have been reported for the derivs of over 100 alcohols and phenols.[2]

References

[1]G. Neurath and W. Luttich, *J. Chromatogr.* **37**, 205 (1968).

[2]G. Neurath and W. Luttich, *J. Chromatogr.* **34**, 253 (1968).

Derivative Type. Pentafluorobenzoyl.

Reagent. Pentafluorobenzoyl chloride.

Compound Type. Phenethylamine.

Reaction

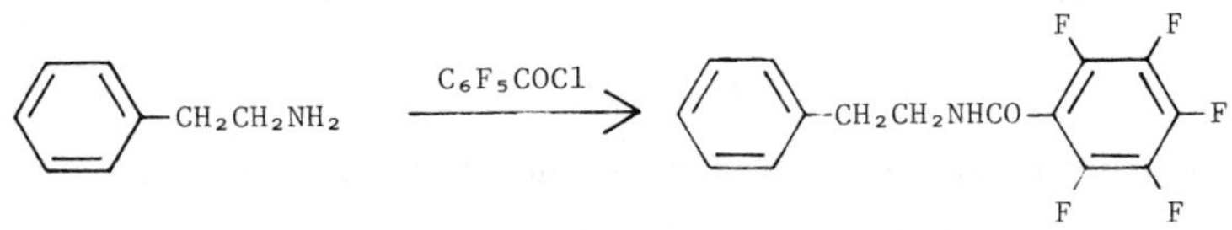

Typical Procedure. Diss 1 mg amine in 0.2 ml CH_3CN and add 0.1 ml pentafluorobenzoyl chloride. Heat at 60° for 1 hr. Dil to 1 ml with hexane and analyze 1 μl by FID-GC or further dil and analyze by ECD-GC.

Comments. The deriv was compared with other pentafluorophenyl substituted derivs for relative ECD response with the following results:

Deriv	MU Value (5% SE-30)	Relative ECD Response
Pentafluorobenzoyl	18.81	78
Pentafluorophenylacetyl	19.69	11
Pentafluorobenzyl	16.81	4.1
N, N-Di-pentafluorobenzyl	21.38	34
Pentafluorobenzaldehyde Schiff base	16.81	100

A double bond to the carbon adjacent to the fluorinated ring was deemed to be a structural requirement for good ECD response.

References

[1]A. C. Moffat *et al., J. Chromatogr.* **66**, 255 (1972).

Derivative Type. *N*-Trifluoroacetyl-L-prolyl.

Reagent. *N*-Trifluoroacetyl-L-prolyl chloride.

Compound Type. Amines.

Reaction

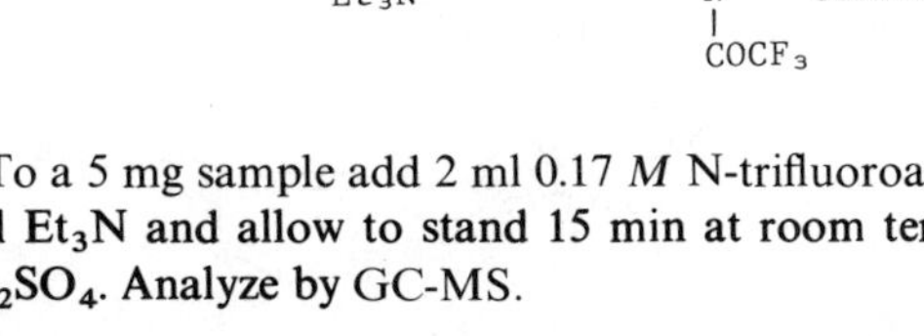

Typical Procedure.[2] To a 5 mg sample add 2 ml 0.17 *M* N-trifluoroacetyl-L-prolyl chloride[3] in $CHCl_3$. Add 0.1 ml Et_3N and allow to stand 15 min at room temp. Wash with 2–3 ml H_2O and dry over Na_2SO_4. Analyze by GC-MS.

Comments. The MS fragmentations for the derivs of 21 biologically significant amines were examined to determine the suitability of the derivs for GC-MS-SIM analysis.[1] Hydrolylated amines were converted to TMS derivs in a second step using BSTFA with 1% TMCS.

References

[1]H. Iwase and A. Murai, *Chem. Pharm. Bull.* **25**, 1215 (1977).
[2]H. Iwase and A. Murai, *Chem. Pharm. Bull.* **22**, 8 (1974).
[3]J. C. Dabrowiak and D. W. Cooke, *Anal. Chem.* **43**, 791 (1971).

2.1.2. Silyl Derivatives

Derivative Type. Trimethylsilyl (TMS).

Reagent. Bis(trimethylsilyl)acetamide (BSA)/trimethylchlorosilane (TMCS).

Compound Type. Phenylethylamines and tryptamines.

Reaction

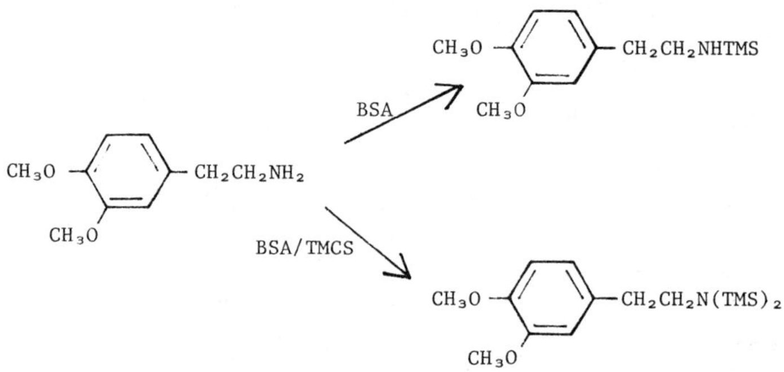

Typical Procedure. A. Diss 50μg of 3, 4-dimethoxyphenylethylamine (DMPEA) in 40μl BSA and 20μl pyridine and allow to stand at room temp for 10 min or at 60° for 5 min. Analyze by GC-MS.

B. Diss 50 μg of DMPEA in 40μl BSA, 20 μl pyridine, and 10 μl TMCS. Heat at 60° for 20 min then analyze by GC-MS.

Comments. The same procedures applied to tryptamine yielded the mono- and di-TMS derivs of the sidechain amino group. Procedure B yielded a small amount of tri-TMS deriv. Increasing the reaction time to 45 min yielded *ca* 50% deriv of the indole-N (i.e., the tri-TMS deriv).

References

[1]W. J. A. Vandenheuval, *J. Chromatogr.* **36**, 354 (1968).

Derivative Type. Trimethysilyl.

Reagent. Bis(trimethylsilyl)acetamide (BSA)/trimethylchlorosilane (TMCS).

Compound Type. Histamine and metabolites.

Reaction

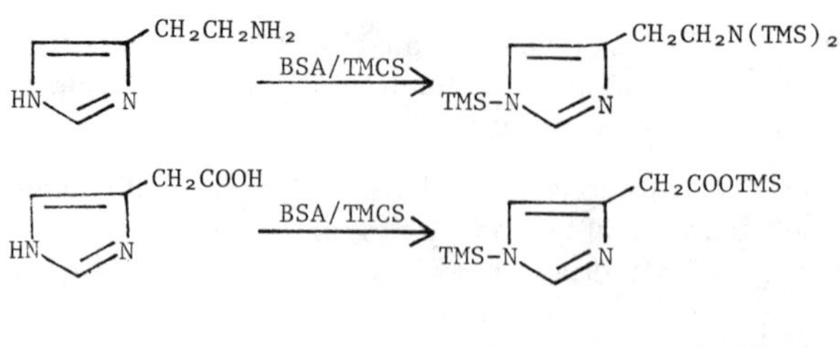

Typical Procedure. To a mixt of 20μg of each compd add 40 μl 4% TMCS in BSA. Heat 30 min at 60° then analyze by GC.

Comments. Histamine and 4-imidazoleacetic acid (shown above) and their 1-methyl derivs were analyzed simultaneously. The derivs of the *N*-methylated compds contained one less TMS group. Detection limits were on the order of 10 ng for the acids and 100 ng for the amines. MS data were given for the derivs.

References

[1]N Mahy and E. Gelpi, *J. Chromatogr.* **130**, 237 (1977).

Derivative Type. *t*-Butyldimethylsilyl (TBDMS).

Reagent. *N*-TBDMS-*N*-methyltrifluoroacetamide.

Compound Type. Amines.

Reaction

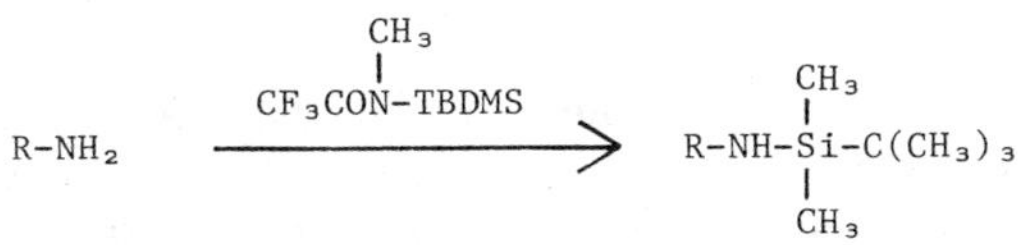

Typical Procedure. Treat sample with 2 mol equiv of the reagent in DMF or CH_3CN. Analyze by GC.

Comments. The reagent was evaluated for synthetic use but could be useful for analytical deriv formation. The reaction with primary amines in DMF or CH_3CN was immediate, but was considerably slower in pyridine. Secondary amines did not form the deriv in DMF but instead catalyzed the formation of di-TBDMS siloxane. TBDMS derivs of secondary amines (e.g., BuMeNH) were preparable in CH_3CN. TBDMS-Cl and TBDMS-dimethylamine were also useful for preparation of TBDMS derivs of amines.

References

[1]J. L. Kutschinski, Ph.D. Thesis, University of California 1975; Diss. Abstr. Int. B **35**, 4838 (1975).

2.1.3. Permethyl Derivatives

Derivative Type. Permethyl.

Reagent. *N*-TBDMS-*N*-methyltrifluoroacetamide.

Compound Type. Polyamines.

Reaction

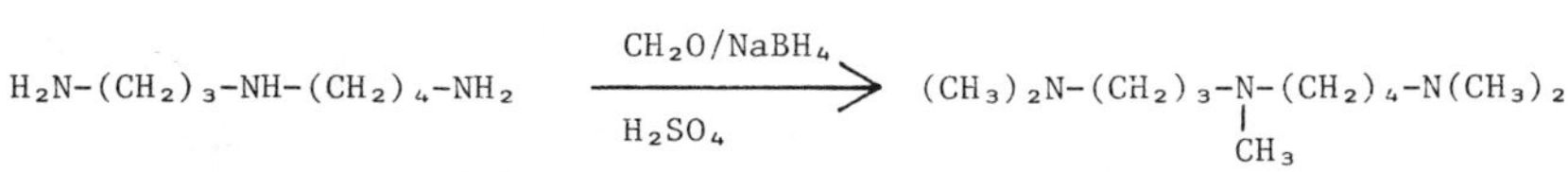

Typical Procedure Diss polyamine sample in 3 *M* aq H_2SO_4 (0.55 mol per NH group) and 40% aq formaldehyde (3.5 mol/NH). Add solid $NaBH_4$ (2.1 mol/NH) portionwise with stirring while maintaining the temp at 0–20°. At the end of the addition, acidify strongly with H_2SO_4 and wash with Et_2O. Alkalinize with KOH and ext deriv with Et_2O. Analyze by GC or GC-MS.

Comments Use of excess H_2O depressed the deriv yield. HCl should not be used in place of H_2SO_4 since HCl would lead to production of toxic di(chloromethyl) ether. Derivs were prepared from 9 di- and triamines including spermine and spermidine.

References

[1]A. G. Giumanini *et al., Anal. Chem.* **48**, 484 (1976).

Derivative Type. Tetramethyl.

Reagent. Formaldehyde/formic acid.

Compound Type. Diamines.

Reaction

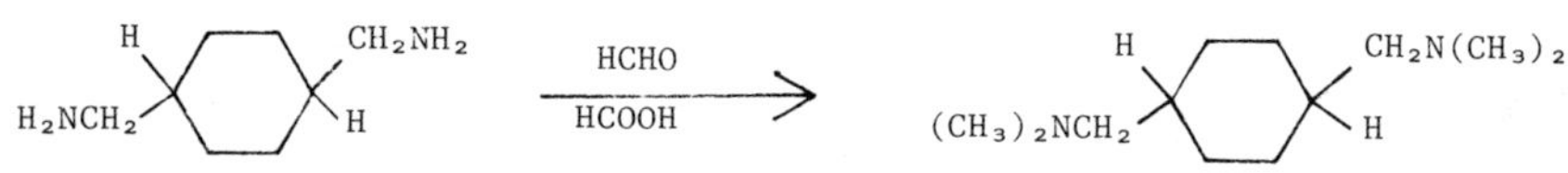

Typical Procedure. Mix 0.01 mol of diamines with 15 ml (0.50 mol) 37% aq formaldehyde and 20 ml (0.50 mol) formic acid. Reflux 6 hr; then add 2 ml conc HCl. Evap to a yellow syrup. Ext with Et_2O and discard ext. Basify residue and ext with Et_2O. Back ext Et_2O with 4 *N* HCl. Basify HCl soln and ext with Et_2O. Dry Et_2O soln and analyze by GC.

Comments. The deriv procedure was a modification of the Leuckart Reaction.[2] The method was used to separate geometric isomers of cyclohexane diamines.

References

[1]M. W. Scoggins *et al., J. Chromatogr. Sci.* **10**, 678 (1972).

[2]M. L. More, *Org. Reactions* **5**, 301 (1949).

Derivative Type. *N*-Methyl.

Reagent. Methyl iodide/sodium hydride.

Compound Type. Triazine herbicides.

Reaction

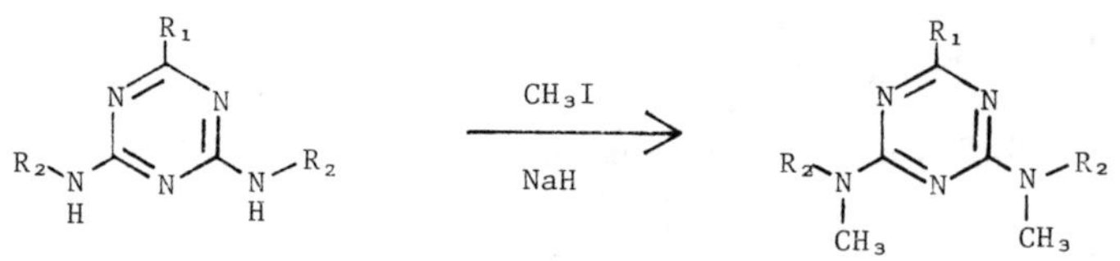

R_1 = Cl, OCH_3, SCH_3

R_2 = *i*-Pr, Et

Typical Procedure. To an herbicide sample add 0.5 ml dry DMSO, *ca* 10 mg NaH, and 0.1 ml CH_3I. Heat at 45° for 10 min. Cool to room temp and add 1 ml benzene. Carefully add 2 ml H_2O dropwise with care to release the rapidly evolved H_2. When H_2 evolution ceases shake well and remove benzene phase. Ext with 2 × 1 ml benzene and combine the three exts. Conc under N_2 and analyze by GC.

Comments. The method was used as a confirmatory test for the herbicides. MS data supported the dimethyl deriv structure. For chlorotriazines, methoxylation using NaOMe/MeOH was also used.

References

[1]J. F. Lawrence, *J. Agr. Food Chem.* **22**, 936 (1974).

2.1.4. Schiff Base Derivatives

Derivative Type. Schiff bases.

Reagent. Ketones.

Compound Type. Primary amines.

Reaction

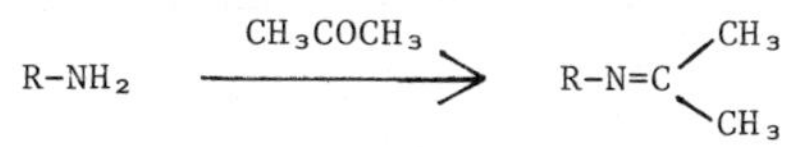

Typical Procedure. Diss the amine sample in acetone (or another ketone) and analyze by GC.

Comments. Derivs were prepared from a series of long chain and alicyclic monoamines, aliphatic diamines, and aromatic amines using acetone, 1,1,1-trifluoroacetone, cyclopentanone, cyclohexanone, and cycloheptanone.[1] Hoshika analyzed C_1–C_4 alkyl primary amines as the Schiff bases from benzaldehyde.[2]

References

[1]W. J. A VandenHeuval *et al.*, *Anal. Chem.* **36**, 1550 (1961).

[2]Y. Hoshika, *J. Chromatogr.* **115**, 596 (1975).

Derivative Type. Schiff base.

Reagent. Acetone.

Compound Type. Sympathomimetic amines.

Reaction.

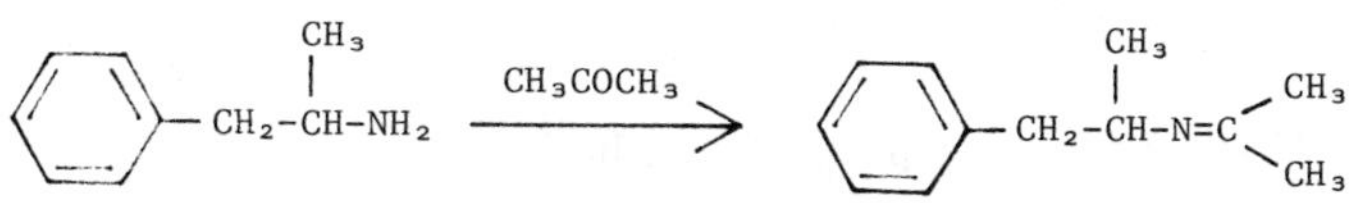

Typical Procedure. Diss amine in acetone and allow to stand at room temp 2–3 hr. Analyze by GC.

Comments. The derivs of 11 primary amines were examined. Monophenolic amines could be analyzed as the Schiff bases, but the derivs of catecholamines decomposed. Butanone derivs were also examined.

References

[1]E. Brochmann-Hanssen and A. B. Svendson, *J. Pharm. Sci.* **51**, 938 (1962).

Derivative Type: Schiff base.

Reagent. Cyclohexanone.

Compound Type: Amphetamine.

Reaction

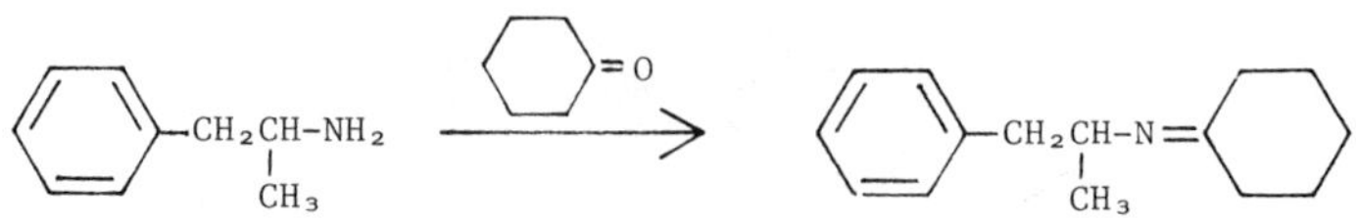

Typical Procedure: Grind tablets and sieve. To an equivalent of 10 mg amphetamine salt add 2.0 ml int stand soln (1.25 g phenethylamine-HCl plus 5 ml conc HCl dil to 200 ml with H_2O) and heat on steam bath 15 min. Cool to room temp, add 5.0 ml deriv reagent (10 ml cyclohexanone dil to 100 ml with MeOH) and 1.0 ml 5 *N* NaOH, mix, and let stand 30 min. Add 10 ml hexane and shake 1 min. Analyze hexane layer by GC as soon as possible.

Comments: The method was used for analysis of amphetamine in solid dosage forms.

References

[1]C. Clark, *J. Assoc. Off. Anal. Chem.* **58**, 1174 (1975).

Derivative Type. Schiff base.

Reagent. 2-Thiophene aldehyde.

Compound Type. Aliphatic primary amines (C_1–C_7).

Reaction

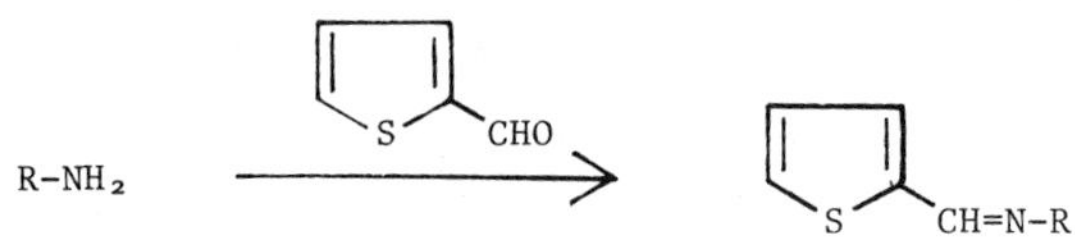

Typical Procedure. Mix 10 μmol amine and 20 μmol of 2-thiophene aldehyde in 1 ml EtOH and allow to react at 60° for 1 hr. Analyze by GC.

Comments. The method was used for complete separation of the derivs of 13 lower aliphatic amines using a glass capillary column.

References

[1]Y. Hoshika, *J. Chromatogr.* **136**, 253 (1977).

Derivative Type. Pentafluorophenyl Schiff base.

Reagent. Pentafluorobenzaldehyde (PFBA).

Compound Type. Low molecular weight aliphatic amines.

Reaction

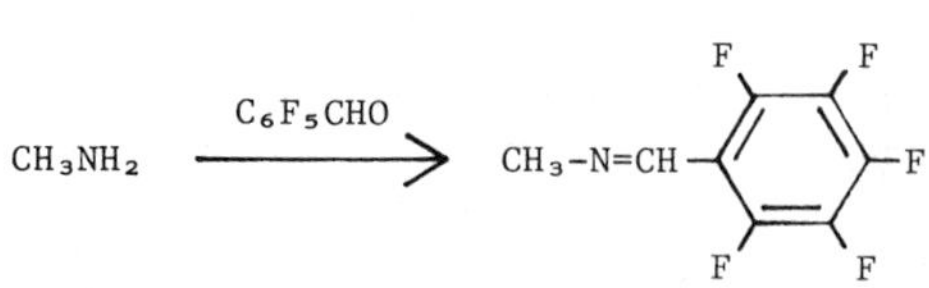

Typical Procedure. Mix 10–100 μl of amine soln in EtOH (10^{-5} mol/ml) with 1 ml hexane and add 10–100μl of PFBA in EtOH (10^{-4} mol/ml). Heat at 60° for 1–3 hr then remove excess PFBA by adding 5 ml of 0.1 *N* NaOH and shaking vigorously for 1 min. Analyze the aliquot of hexane soln by GC.

Comments. The minimum detectable quantity of deriv was 0.02 ng using ECD-GC. The method was applied to C_1–C_5 alkyl amines.

References

[1]Y. Hoshika, *Anal. Chem.* **49**, 541 (1977).

Derivative Type. Schiff base.

Reagent. Pentafluorobenzaldehyde (PFBA).

Compound Type. Primary amines.

Reaction

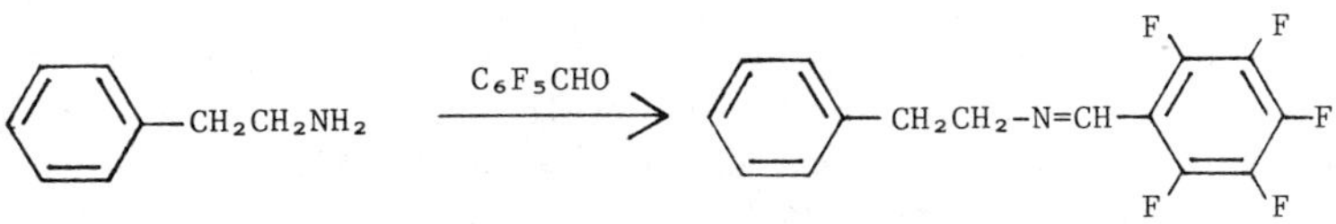

Typical Procedure. Diss 1 mg phenethylamine in 0.2 ml CH_3CN. Add 0.1 ml PFBA and heat at 60° for 1 hr. Dil with hexane or CH_3CN prior to analysis by GC-ECD.

Comments. The deriv was compared to 12 other derivs for ECD detection of primary amines and found to give the greatest sensitivity (measurable to 10 pg). The Schiff base from perfluorooctanaldehyde had similar sensitivity but multiple products were formed. The HFB deriv was the most sensitive acylated deriv that had detectability comparable to the 2,4-dinitrophenyl deriv (both an order of magnitude lower than the PFBA Schiff base).

References

[1]A. C. Moffat and E. C. Horning, *Anal. Lett.* **3**, 205 (1970).

Derivative Type. Dimethylaminomethylene.

Reagent. Dimethylformamide (DMF) dimethyl acetal.

Compound Type. Diastereomeric diamines.

Reaction

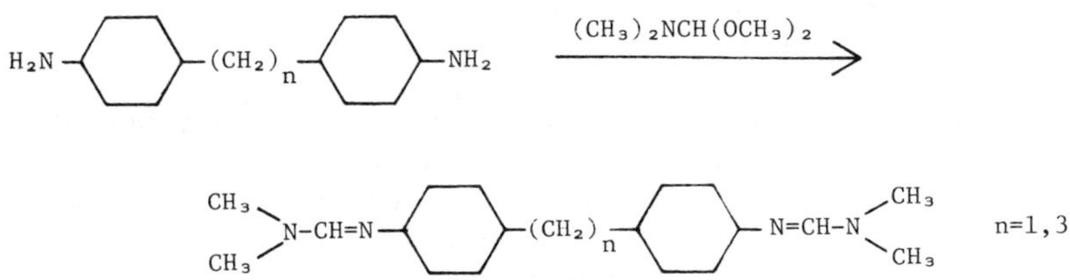

Typical Procedure. Mix 25 mg diamine with 0.25 ml DMF dimethylacetal and heat at 60° for 10 min. Analyze by GC.

Comments. The method permitted GC separation of the *cis-cis, cis-trans,* and *trans-trans* isomers.

References

[1]M. Scoggins, *J. Chromatogr. Sci.* **13**, 146 (1975).

2.1.5. Nitrophenyl Derivatives

Derivative Type. 2,4-Dinitrophenyl (DNP).

Reagent. 2,4-Dinitrofluorobenzene (DNFB).

Compound Type. Primary and secondary amines.

Reaction

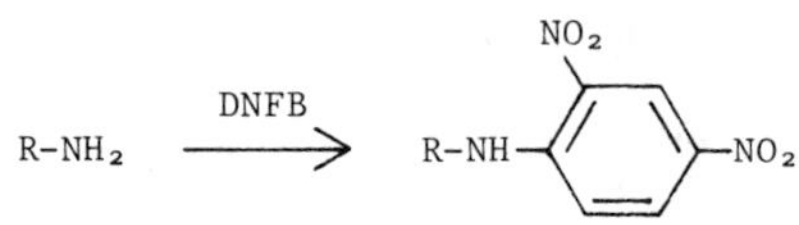

Typical Procedure. Mix 10 ml of a 1.0 μg/ml aq soln of the amine(s) and 5 ml of a 2.5% (w/v) aq soln of $Na_2B_4O_7$-$10H_2O$. Add 2 ml of a soln of 2 ml DNFB in 25 ml *p*-dioxane. Heat at 60° for 20 min. Add 2 ml 2 *N* NaOH and heat an additional 30 min. Cool and ext with 10 ml cyclohexane. Wash ext with 3 × 15 ml 0.1 *N* Na_2CO_3. Dry over Na_2SO_4. Analyze by GC or TLC on silica (hexane/Et_2O, 70 : 30, v/v) and analyze three "zones" obtained by GC.

Comments. The method was used to separate the 11 C_1–C_4 primary and secondary amines by a combination of TLC and GC. Three "zones" were obtained on TLC with all of the derivs within each "zone" separable by GC.

References

[1]E. W. Day *et al., Anal. Chem.* **38**, 1053 (1966).

Derivative Type. 2,4-Dinitrophenyl (DNP).

Reagent. 2,4-Dinitrofluorobenzene (DNFB).

Compound Type. Amines.

Reaction

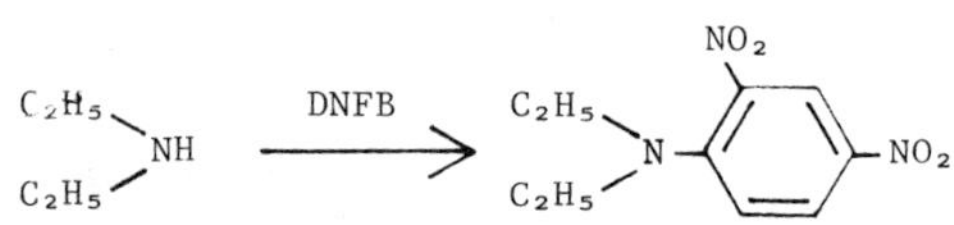

Typical Procedure. To 3 ml of aq amine sample soln or urine, add 1 ml of amine int stand soln, 1 ml 5 *M* K_2CO_3 and 30 ml benzene. Shake and separate benzene phase. To 25 ml of the benzene phase add a fivefold excess of DNFB diss in 1 ml benzene. Allow to stand 5 min at room temp, then heat at 60° for 15 min. Analyze 2 – 5μl by ECD-GC. For amine conc below 5 × 10^{-8} *M*, conc benzene soln under N_2 before analysis.

Comments. The method was evaluated on 13 primary and secondary aliphatic and aromatic amines. Amphetamine was quantitated in urine to 5ng/ml and diethylamine to 50 ng/ml.

References

[1]T. Walle, *Acta Pharm. Suecica* **5**, 367 (1968).

Derivative Type. 2, 4-Dinitrophenyl (DNP).

Reagent. 2,4-Dinitrofluorobenzene (DNFB).

Compound Type. Çyclohexylamine.

Reaction

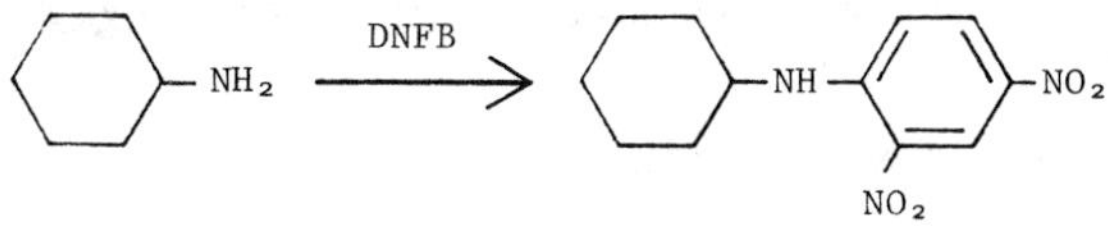

Typical Procedure. To a 1–10 μg sample in 10 ml 2.5% $Na_2B_4O_7$ buffer, add 0.5 ml 4% (w/v) DNFB) in dioxane and heat at 60° for 20 min. Ext with 5 ml hexane and dry ext over Na_2SO_4.Analyze by ECD-GC.

Comments. The method was applied to the analysis of cyclohexylamine in cyclamates [1, 2] and in soft drinks[2] and was usable down to 1 ppm and 0.1 ppm, respectively. Gunner and O'Brien[3] analyzed cyclohexylamine as the trinitrophenyl deriv (from 2,4,6-trinitrochlorobenzene) and reported sensitivity to 1 ppm.

References

[1]R. E. Weston and B. B. Wheals, *Analyst* **95**, 680 (1970).

[2]M. D. Soloman *et al., Anal. Lett.* **4**, 301 (1971).

[3]S. W. Gunner and R. C. O'Brien, *J. Assoc. Off. Anal. Chem.* **52**, 1200 (1969).

Derivative Type. 2, 4-Dinitrophenyl (DNP).

Reagent. Sodium 2, 4-dinitrobenzenesulfonate (DNBS).

Compound Type. Amines.

Reaction.

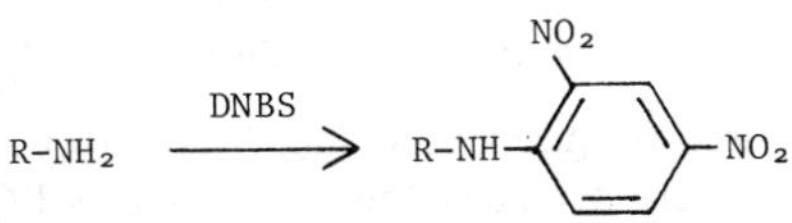

Typical Procedure.[1] Dil a 10 ml oxalated blood sample with 40 ml H_2O and add 5 ml 20% $ZnSO_4$ and 5 ml 1 *N* NaOH. Stir thoroughly, centrifuge, and remove supernatant. Wash ppt 2x with 35 ml H_2O and combine washes with supernatant. Add 1 ml 5% Na_2EDTA and adjust to neutral pH if necessary with 0.1 *N* NaOH. Pass through a 2.5 × 10 cm column of Dowex 1-X8 (OH form) and wash column with H_2O until the eluate is no longer alkaline. Add effluent including washings to a stirred suspension of Amberlite CG-50 (4 g, H form) in H_2O. Stir 1 hr. Pour resin into a column and wash with 2 × 20 ml H_2O. Elute with 40 ml 0.5 *N* HCl. Evap eluate to dryness under vac over a mixt of solid NaOH and granular $CaCl_2$.

Diss the residue in 3 ml H_2O. Add 3 ml 0.25 *M* DNBS in satd aq $Na_2B_4O_7$ and 1.5 ml satd aq $Na_2B_4O_7$. Heat 1 hr at 100°, cool, and dil with 40 ml half satd $Na_2B_4O_7$. Ext with 40, 35, and 30 ml benzene. Wash ext with 2 × 50 ml NaCl; then dil to 100 ml with benzene. Dry over Na_2SO_4 and evap under vac. Diss residue in 0.3 ml THF containing 75μg DNP-*t*-butylamine int stand and analyze 3μl by FID-GC.

Comments. The method was used for separation and identification of aliphatic amines in whole blood.[1] Piperidine was not deriv under these conditions. GC data for DNP derivs of 26 amines were reported by the same authors.[2]

References

[1]S. Baba *et al., J. Chromatogr.* **88**, 373 (1974).

[2]Y. Ishitoya *et al., Clin Chim. Acta* **46**, 55 (1973).

Derivative Type. 4-Trifluoromethyl-2,6-dinitroaniline (DNT) or 4-trifluoromethyl-2-nitroaniline (MNT).

Reagent. 4-Chloro-α,α,α-trifluoro-3,5-dinitrotoluene (DNT-Cl) or α,α,α,4-tetrafluoro-3-nitrotoluene (MNT-F).

Compound Type. Amines from pesticides.

Reaction

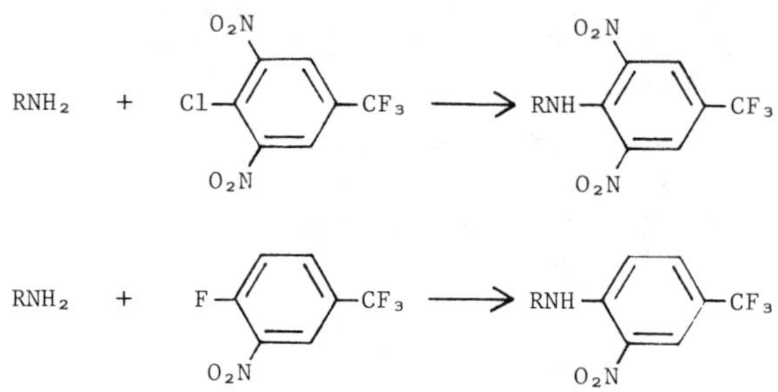

Typical Procedure (for Molinate in H_2O). Acidify 1 liter H_2O sample with a few drops of conc HCl and ext with 4 × 50 ml hexane. Transfer 20 ml of combined ext to a 40 ml glass stoppered tube and add 0.5 ml conc H_2SO_4. After shaking, remove most of the hexane and heat the remaining phase at 85° for 20 min. Cool in ice, dil to 5 ml, and neutralize (phenolphthalein end point) with 30% aq KOH. Add two additional drops of base and ext with 5 ml benzene. Ext benzene with 1 ml 0.1 *M* HCl. To HCl ext add 0.15 ml of a 15 mg/ml soln of either DNT-Cl of MNT-F in acetone, 2 ml of 0.1 *M* $Na_2B_4O_7$, 0.2 ml of 0.1 *M* KOH, and 5 ml acetone. Heat at 55° for 1–2 hr. Wash into 100 ml flask with H_2O and add 1.0 ml hexane. Shake, allow phases to separate, and analyze an aliquot of the hexane by GC.

Comments. The method was used to quant pesticides by hydrolysis and derivatization of the resulting amines. The derivs were detected down to 50 pg by ECD-GC. Dinitrophenyl derivs were also examined, but the DNT and MNT derivs provided greater detector response, better resolution, and shorter retention times. In most instances, the MNT derivs gave better sensitivity than the DNT. Physical data and retention times were given for a series of aliphatic, dicyclic, and aromatic amines derivable from pesticides.

References

[1]D. G. Crosby and J. B. Bowers, *J. Agric. Food Chem.* **16**, 839 (1968).

2.1.6. Other Derivatives for GC or MS Analysis

Derivative Type. *N-t*-Butylurea.

Reagent. *t*-Butylisocyanate.

Compound Type. Primary and secondary amines.

Reaction

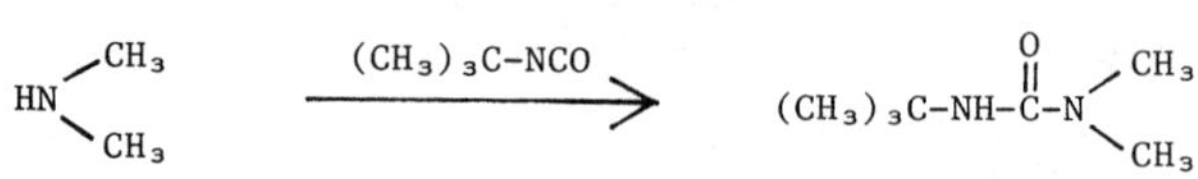

Typical Procedure. Prepare derivs by mixing equimolar amts of the amine and the isocyanate in hexane or benzene. On a prep scale the urea product precipitates. Diss the product in acetone and analyze by GC or GC-MS.

Comments The 3-trifluoromethylphenyl urea derivs were also prepared from 3-trifluoromethylphenyl isocyanate and were detectable to 100 pg using ECD.[1] The *t*-butyl ureas were detectable to 500 pg using GC-MS-SIM.[2] GC and MS data were reported for the derivs of several low mol wt amines. The method was examined for use in measuring amine metabolites of pesticides.

References

[1]I. Nitsche *et al., J. Chromatogr.* **94**, 65 (1974).

[2]J. Slemrova and I. Nitsche, *J. Chromatogr.* **135**, 401 (1977).

Derivative Type. Carbamates.

Reagent. Diethyl pyrocarbonate.

Compound Type. Amines.

Reaction

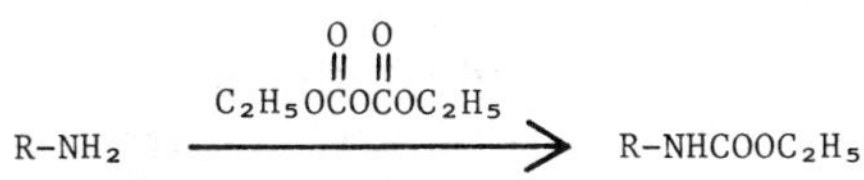

Typical Procedure. Diss 0.05–0.7 mg amine in 5 ml H_2O and adjust the pH to 9.5 with NaOH. Add 10 mg diethyl pyrocarbonate and allow to react 30–40 min at room temp. Analyze the deriv by GC.

Comments. GC data were reported for the derivs of ammonia, eight aliphatic, and two aromatic amines.

References

[1]T. Gejvall, *J. Chromatogr.* **90**, 157 (1974).

Derivative Type. 2,5-Dimethylpyrrole.

Reagent. 2,5-Hexanedione.

Compound Type. Primary amines and amino acid methyl esters.

Reaction

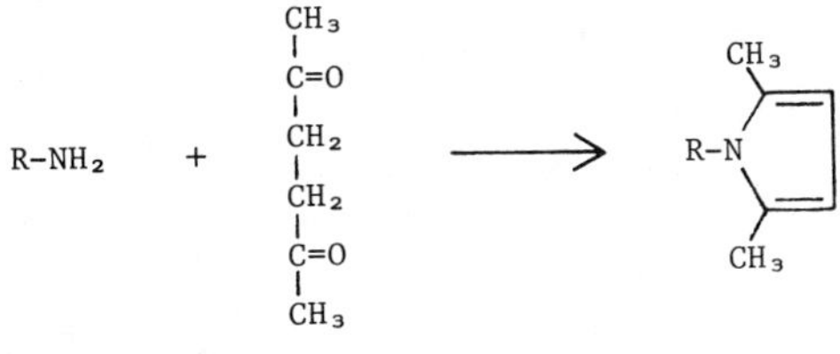

Typical Procedure. Diss amine in Et_2O or MeOH and add a fivefold excess of 2,5-hexanedione. Heat at 60° for 30 min protected from light with N_2 slowly bubbling through the soln. During the procedure the solvent slowely evaporates. Diss the remainder in Et_2O and analyze an aliquot by GC.

Comments Amino acid methyl esters were deriv by the same procedure using 5 % pyridine in MeOH as solvent. The esters were prepared with MeOH/HCl.

References

[1]T. Walle, *Acta Pharm. Suecica* **5**, 353 (1968).

Derivative Type. Isothiocyanate.

Reagent. Carbon disulfide.

Compound Type. Primary amines.

Reaction

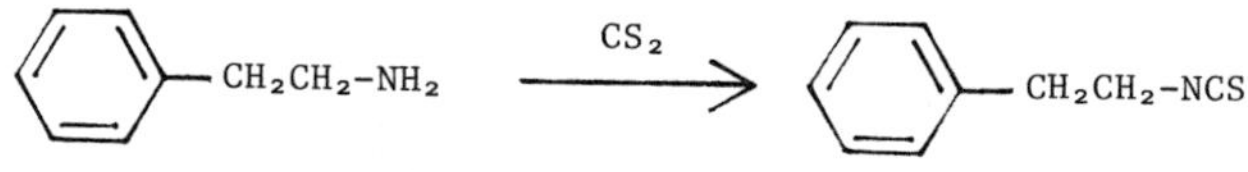

Typical Procedure. Diss 1 ml free amine in 5 ml EtOAc and add 0.5 ml CS_2. Shake 30 min; then evap to dryness under vac. Diss residue in 1ml EtOAc and analyze by GC or GC-MS.

Comments. The salts of fully methylated (methoxy) amines were extd at pH 12 with EtOAc. Salts of phenolic amines were neutralized with $NaHCO_3$ or NH_3 and extd into EtOAc. Phenolic or indole amines could be converted to TMS derivs (after treatment with CS_2) by heating with 1 % TMCS in BSTFA at 90° for 15 min. The isothiocyanate derivs were found to be more volatile than the corresponding *N*-TMS derivs. GC and MS data were reported for the derivs and compared with the TMS derivs.[1,2] It was later shown that the acetylated amines (which have the same mass) could be used as ext stand for quant of the isothiocyanate derivs by GC-MS-SIM.[3] The isothiocyanate deriv of amphetamine had been reported earlier with MS data.[4]

References

[1]N. Narasimhachari and P. Vouros, *Anal. Biochem.* **45**, 154 (1972).

[2]N. Narasimhachari and P. Vouros, *Biomed. Mass Spectrosc.* **1**, 367 (1974).

[3]N. Narasimhachari and J. M. Davis, Presented at the 25th Annual Conference on Mass Spectrometry and Related Topics, Washington, D.C., May 29–June 3, 1977.

[4]H. Brandenberger and E. Hellbach, *Helv. Chim. Acta* **50**, 958 (1967).

2.1.7. Chromogenic Derivatives

2.1.7.1. *Nitrobenzoyl and Nitrophenylacetyl Derivatives*

Derivative Type. 4-Nitrobenzamide.

Reagent. 4-Nitrobenzoyl chloride.

Compound Type. Arylalkylamines.

Reaction

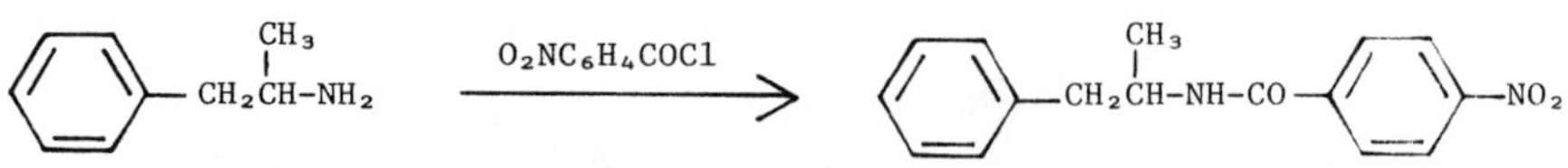

Typical Procedure. To 1 ml of aq amine soln (20–300 μg/ml) add 2 ml of a freshly prepared 4-nitrobenzoyl chloride soln in THF (50 mg/ml) and 1 ml of 1 *M* NaOH. Seal vial and heat at 65° for 1hr. Analyze by HPLC or GC.

Comments. The derivs could be isolated by adjusting soln to pH 12 with 1 *M* NaOH and ext with 2× 10 ml $CHCl_3$. Wash ext with 2× 20 ml 10% K_2CO_3, wash with H_2O, dry over $MgSO_4$, and evap to dryness. Diss deriv in MeOH for analysis. The derivs could be formed in the aq sample and were less prone to loss in isolation due to volatility than the free amines.

References

[1]C. R. Clark *et al., Anal. Chem.* **49**, 912 (1977).

Derivative Type. 3, 5-Dinitrobenzoyl.

Reagent. 3, 5-Dinitrobenzoyl chloride.

Compound Type. Diethylamine.

Reaction

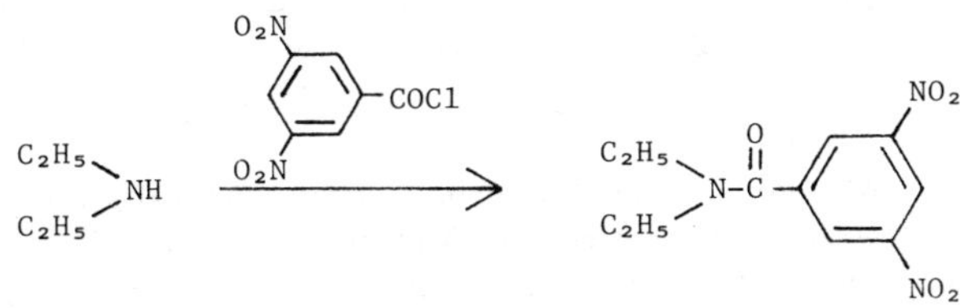

Typical Procedure. Mix 2 ml urine (previously acidified to pH 4 with HOAc) with 15 ml Et_2O, 0.2 ml pyridine, 2 ml 5% aq K_2CO_3, and 0.5 ml of 3,5-dinitrobenzoyl chloride in Et_2O (0.25 g/ml) in a 125 ml sep funnel. Shake periodically for 20 min; then remove and discard aq layer. Wash Et_2O with 2 × 5 ml 5% $NaCO_3$, 2 × 5 ml 5% HCl, and 2 × 5 ml H_2O. Dry over Na_2SO_4 and evap to dryness. TLC residue on silica gel with $CHCl_3$/MeOH (99:1). Locate spots with I_2 vapor then allow I_2 to evap overnight. Elute spots with MeOH and read UV absorbance at 300 nm.

Comments. Method was used for spectrophotometric quantitation but could be adaptable for HPLC or GC analysis.

References

[1]D. H. Neiderhiser *et al., J. Chromatogr.***117**, 187 (1976).

Derivative Type. *p*-Nitrophenylacetate.

Reagent. *N*-Succinimidyl-*p*-nitrophenylacetate (SNPA).

Compound Type. Amines and amino acids.

Reaction

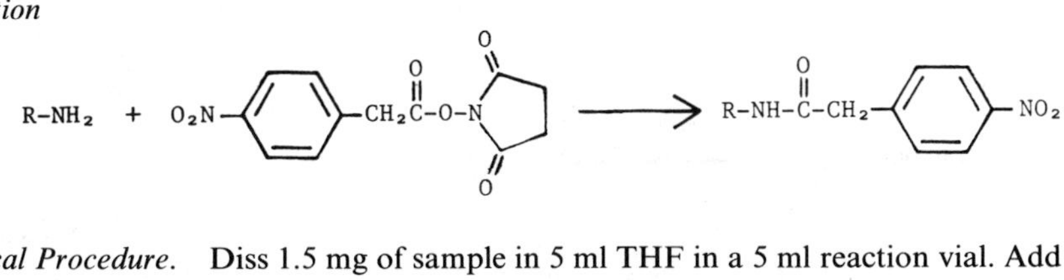

Typical Procedure. Diss 1.5 mg of sample in 5 ml THF in a 5 ml reaction vial. Add 50 mg SNPA and heat at 60° for 1 hr. Cool and analyze by HPLC.

Comments. The deriv was used to enhance UV absorption for HPLC. The suggested HPLC solvent for a silica column was heptane/$CHCl_3$/MeOH (85:14:1, v/v/v).

References

[1]Regis Lab Notes No. 17, Regis Chemical Company, Nov. 1974

2.1.7.2. *5-Dialkylaminonaphthalene-l-sulfonyl Derivatives*

Derivative Type. 5-Dimethylaminonaphthalene-l-sulfonyl (DANS).

Reagent. 5-Dimethylaminonaphthalene-1-sulfonyl chloride (DANS-Cl).

Compound Type. Polyamines.

Reaction

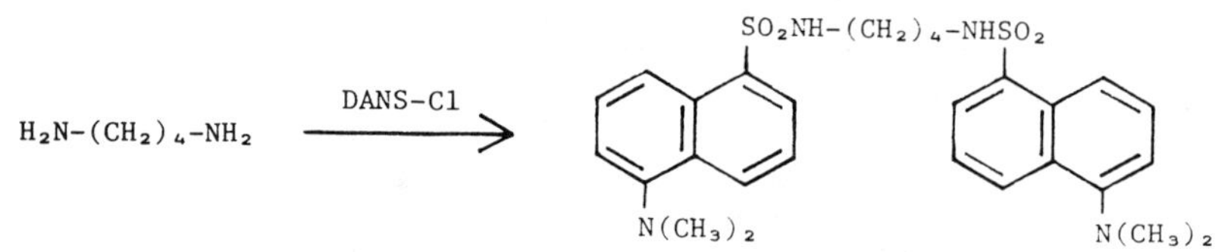

Typical Procedure. Mix 100 μl of tissue ext in 0.3 *N* $HClO_3$ with 100 μl H_2O, or dil 100 μl of ref aq polyamine soln with 100 μl 0.3 *N* $HClO_3$. Add 100 μl satd aq $NaHCO_3$, 20 μl of 0.2 m*M* 1, 2-diaminoethane—HCl (int stand), and 600 μl DANS-Cl (7 mg/ml) in acetone. Shake gently overnight at room temp in the dark. Evap under N_2 at 40°; then partition residue between 0.5 ml H_2O and 5 ml benzene. Evap benzene phase under N_2 at 40° and diss residue in 50 μl CH_2Cl_2 for HPLC analysis.

Comments. HPLC of the pure derivs had been reported earlier.[2] The method was extended to the analysis of tissues permitting detection to 40 pmol for putrescine and 20 pmol for spermine and spermidine. Satisfactory separation was also achieved for N^1- and N^8-acetyl di-DANS spermidines.[2] Pro has been used to react with excess DANS-Cl permitting solvent ext of DANS-polyamines leaving DANS-Pro in the aq phase.[3]

References

[1]N. E. Newton *et al., J. Chromatogr.* **124**, 277 (1976).

[2]M. M. Abdel-Monem and K. Ohno, *J. Chromatogr.* **107**, 416 (1975).

[3]N. Seilar and A. Askar, *J. Chromatogr.* **62**, 121 (1971).

Derivative Type. 5-Dimethylaminonaphthalene-l-sulfonyl (DANS).

Reagent. 5-Dimethylaminonaphthalene-1-sulfonyl chloride (DANS-Cl).

Compound Type. Tocainamide.

Reaction

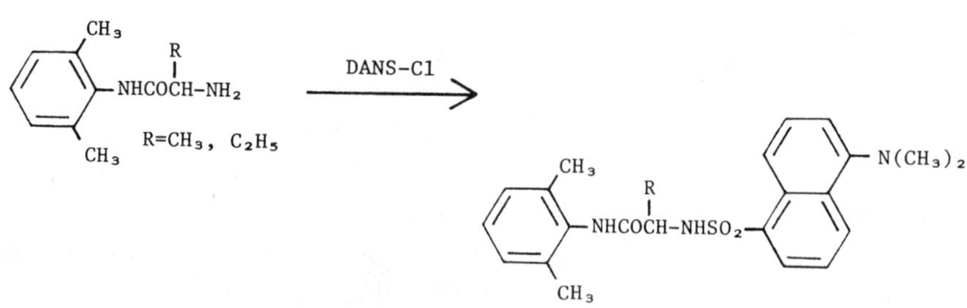

Typical Procedure. To a sample (0.05–2.0 ml) of plasma or blood containing 0.1–5.0 μg tocainamide (R = CH_3), add 0.675 μg of int stand (R = C_2H_4) and 200 μl 2 *N* NaOH. Ext with 3 ml EtOAc. Take 2 ml of the EtOAc ext and back ext with 50 μl 1 *N* H_2SO_4. Discard the EtOAc, wash the acid phase with 1 ml hexane (freeze the aq phase in dry ice/acetone to achieve the final separation). To the thawed acid phase add 50 μl 1 *M* $NaHCO_3$ and 200 μl 1 mg/ml DANS-Cl in acetone. Heat 30 min at 60°; then add 100 μl of 0.5 *N* NaOH to convert excess reagent to the Na salt. Remove acetone under vac at 40°. Add 1 ml pH 6.0 citrate buffer and ext with 200 μl CCl_4. Analyze an aliquot of the CCl_4 by HPLC.

Comments. The method was used with fluorescence detection to quantitate the drug to 0.1 μg/ml. MS data were given for the derivs.

References

[1]P. J. Meffin *et al., J. Pharm. Sci.* **66**, 583 (1977).

Derivative Type. 5-Dimethylamino-l-naphthanlene sulfonyl (DANS).

Reagent. 5-Dimethylamino-l naphthalene sulfonyl chloride (DANS-Cl).

Compound Type. Amphetamine.

Reaction

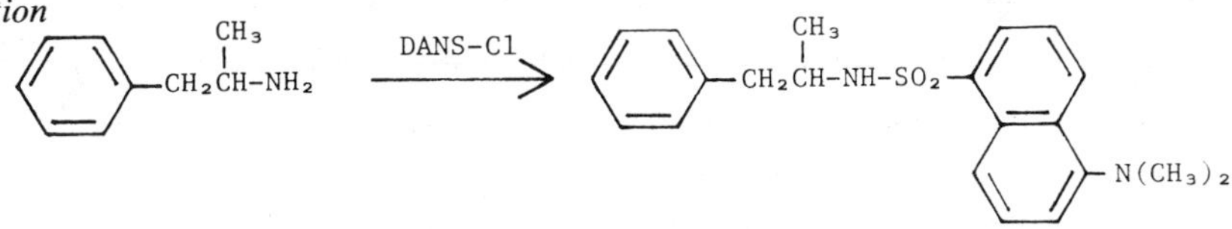

Typical Procedure. Adjust 5 ml urine to pH 10 with 2 *N* NaOH and centrifuge. Add 112 ng of d_5-amphetamine int stand to the supernatant and ext with 2 × 5 ml benzene. Add 5 drops 3 *N* HCl and evap to dryness under N_2 at 40°.

Diss residue in 1 ml satd $NaHCO_3$ then add 2 ml acetone and 200 μl of DANS-Cl in acetone (5 mg/ml). Shake and let stand overnight at room temp. Evap to dryness under N_2 and ext residue with 2 ml benzene. Evap ext and TLC residue on silica gel using benzene/Et_3N (8:1, v/v). Locate deriv by 365 nm UV and elute with EtOAc. Evap to dryness and TLC residue on silica gel using the upper phase of pet ether (100–120°)/toluene/AcOH/H_2O (133:67:170:30, v/v/v/v). Elute deriv with 500 μl EtOAc and analyze by direct probe MS.

Comments. The deriv method was also used to quantitate amphetamine in tissue samples. Quantitation was achieved by integrating the ion current of masses characteristic of the labeled and unlabeled amphetamine derivs.

References

[1]T. J. Danielson and A. A. Boulton, *Biomed. Mass Spectrosc.* **1**, 159 (1974).

Derivative Type. 5-Dimethylaminonaphthalene-l-sulfonyl (DANS).

Reagent. 5-Dimethylaminonaphthaline-1-sulfonyl chloride (DANS-Cl).

Compound Type. β-Phenylethylamine, tryptamine, tyramine.

Reaction

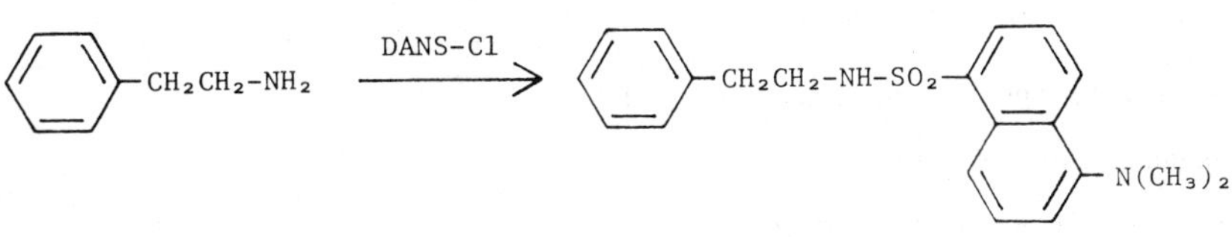

Typical Procedure. Homogenize tissue sample at 0° in 20 ml 0.4 *N* $HClO_4$. Add 25 – 100 ng d_2-phenylethylamine as int stand. Mix, centrifuge, and separate supernatant. Add Triton X-100 to conc of 0.2% (w/v) and adjust pH to 7.0. Pass through 1.5 × 3 cm column of Biorad AG50W-X2 (H^+). Wash column with 10 ml H_2O, 20 ml 0.1 *M* NaOAc, and 10 ml H_2O. Elute with 10 ml MeOH-HCl (73:27) and evap eluate on rotary evaporator.

Diss residue in 1 ml 10% Na_2CO_3 and add 1.5 ml DANS-Cl soln in acetone (1 mg/ml). Heat at 37° for 2.5 hr. Add 9 ml acetone to ppt Na_2CO_3. Remove supernatant and evap under vac at 45°. TLC residue on silica gel, elute deriv, and analyze by MS.

Comments. The method was used to quantitate biogenic amines in rat tissue.[1] DANS derivs have been used for both qualitative and quantitative analysis of noncatecholic biogenic amines by MS.[2] The derivs can also be prepared using $NaHCO_3$ instead of Na_2CO_3.[3]

References

[1]A. A. Boulton *et al., J. Chromatogr.* **82**, 137 (1973).

[2]D. A. Durden *et al., Biomed. Mass Spectr.* **1**, 83 (1974).

[3]C. R. Creveling *et al., Clin. Chem.* **14**, 302 (1968).

Derivative Type. Di-*n*-butylaminonaphthalene-l-sulfonyl (BANS).

Reagent. 5-Di-*n*-butylaminonaphthalene-l-sulfonyl chloride (BANS-Cl).

Compound Type. Piperidine.

Reaction

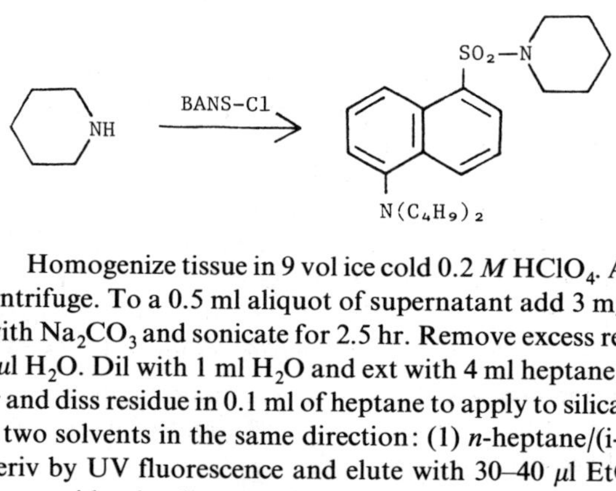

Typical Procedure. Homogenize tissue in 9 vol ice cold 0.2 *M* $HClO_4$. Allow to stand 1 hr at room temp then centrifuge. To a 0.5 ml aliquot of supernatant add 3 mg BANS-Cl in 1.5 ml acetone. Sat soln with Na_2CO_3 and sonicate for 2.5 hr. Remove excess reagent by addition of 5 mg proline in 50 μl H_2O. Dil with 1 ml H_2O and ext with 4 ml heptane. Evap ext to dryness under stream of air and diss residue in 0.1 ml of heptane to apply to silica gel TLC plate. TLC with the following two solvents in the same direction: (1) *n*-heptane/(i-Pr)$_2$O (2:1, v/v), (2) benzene. Locate deriv by UV fluorescence and elute with 30–40 μl EtOAc. Evap eluate in capillary and analyze residue by direct probe MS.

Comments. The method was used to quantitate piperidine in tissues in pmol amts using BANS-pyrrolidine as int stand.

References

[1]N. Seiler and H. H. Schneider, *Biomed. Mass Spectrom.* **1**, 381 (1974).

2.1.7.3. *Fluorescamine Derivatives*

Derivative Type. Fluorescamine.

Reagent. Fluorescamine.

Compound Type. Primary amines.

Reaction

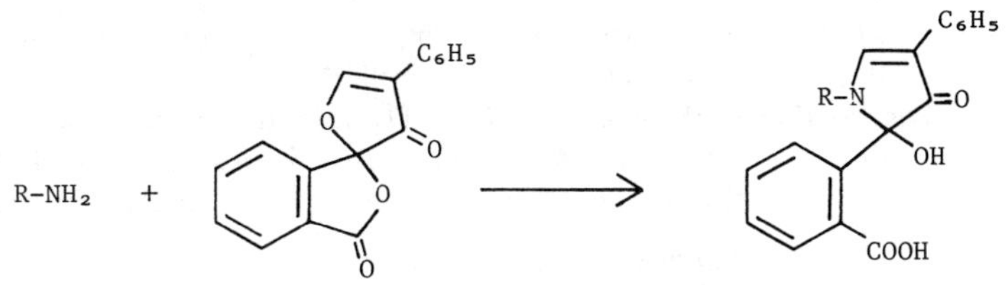

Typical Procedure.[1] Dil 100 μl of a 1 mg/ml aq soln of the amine HCl with 0.5 ml pH 9.0 borate buffer. Add 0.5 ml of a 30 mg/ml soln of fluorescamine in EtOH and stir 2 min. Acidify with 0.5 ml 0.5 *N* HCl and ext with 10 ml EtOAc. Dry ext over Na_2SO_4, evap to dryness, and analyze by direct probe MS.

Comments. The reagent, which is not fluorescent itself, yielded fluorescent derivs of primary amines and amino acids. The reagent was developed as a result of studies of the structure of fluorescent derivs formed with ninhydrin.[2] Excess reagent hydrolyzed ($t_{1/2}$5–10 sec) at pH 9 to yield a nonfluorescent hydrolysis product. Use of acetone as solvent (as in the original work[2]) for fluorescamine led to some formation of the Schiff base deriv. Derivs of amino acid ethyl esters were also examined. The derivs were found to be useful for MS analysis of primary amines and were said to be superior to DANS derivs.

References

[1]N. Narasimhachari, *Biochem, Biophys. Res. Commun.* **55**, 231 (1973).

[2]S. Udenfriend *et al., Science* **178**, 871 (1972).

Derivative Type. Fluorescamine.

Reagent. Fluorescamine.

Compound Type. Diamines and polyamines.

Reaction

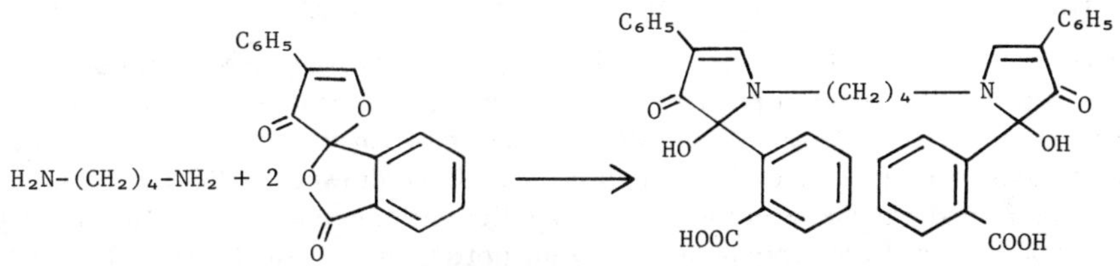

Typical Procedure. Homogenize tissue sample in 1 ml/100 mg chilled 0.3 *N* $HClO_4$ and centrifuge. Use the resulting clear supernatant as the sample soln. Deproteinize serum (0.5–1.0 ml) with 1.5 *N* $HClO_4$ to a final conc of 0.3 *N*. Centrifuge and use the supernatant as the sample soln. Mix urine (0.5–1.0 ml) with an equal vol of conc HCl and heat at 115° for 15 hr. Filter the hydrolyzed urine and remove the HCl under vac to yield the sample soln. In each case add the appropriate int stand to the initial mixt.

Isolate diamine and polyamine fractions by chromatography on a carboxymethylcellulose column using pyridine-HOAc buffers. Place the eluted fractions in a dessicator over 30% aq HOAc (to maintain an acidic atmosphere to retard evap loss of amines) and remove the solvent by overnight evap at 0.3 mm Hg using a −80° cold trap. Treat the residue by one of two methods depending upon the amt of amines present.

Method I. Diss residue in 100 μl of 0.1 M $Na_2B_4O_7$ buffer (pH 8) and add 20 μl of fluorescamine (2 mg/ml) in acetone with rapid mixing. For polyamines add 5 μl 1 M NaOAc buffer (pH 5.4). After 10 min, analyze an aliquot by HPLC.

Method II. Diss residue in 100 μl 0.01 N HCl and transfer to a 0.3 ml micro vial. Carefully evap to dryness under vac over KOH in a dessicator. Diss the residue in 20 μl of the borate buffer and add 10 μl of the fluorescamine soln as in Method I. Analyze by HPLC.

Comments. Fluorescence detection was used in the HPLC analysis. 1, 6-Hexanediamine and *N*-3-aminopropyl-1,6-hexanediamine were used as int stand. The method was useful for samples containing down to 100 pmol of the amines.

References

[1]K. Samejima *et al., Anal. Biochem.* **76**, 392 (1976).

[2]K. Samejima, *J. Chromatogr.* **96**, 250 (1974).

2.1.7.4. *Other Chromogenic Derivatives*

Derivative Type. 1-Alkylthio-2-alkylisoindole.

Reagent. *o*-Phthalaldehyde/2-mercaptoethanol or ethanethiol.

Compound Type. Primary amines and amino acids.

Reaction

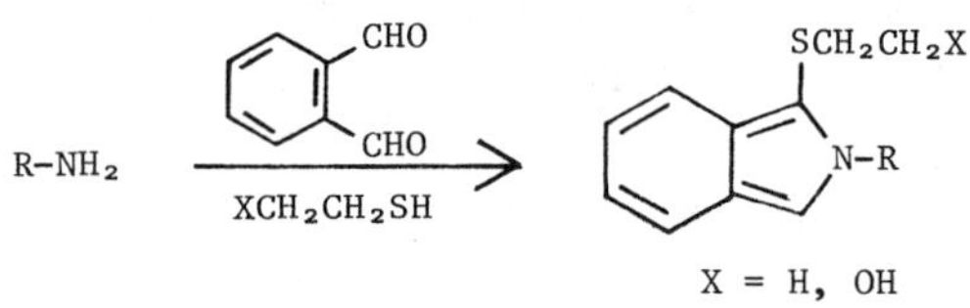

Typical Procedure.[1] Prepare reagent by mixing 1.5 ml 10 mg/ml *o*-phthalaldehyde in EtOH with 90 ml pH 9.5 $Na_2B_4O_7$ buffer (0.05 M). To this soln add 1.5 ml 5 μl/ml 2-mercaptoethanol in EtOH and mix. For analysis mix 100 μl sample soln (10^{-4}–10^{-3} M) with 3 ml freshly prepared reagent and read fluorescence within 5–25 min (excitation, 340 nm; fluorescence, 455 nm.

Comments. The reagent was used initially for fluorescence analysis of amino acids.[1] It reacted with all the common primary amino acids except Cys (the secondary amines Pro HOPro did not react). Benson and Hare[2] showed pmol sensitivity in amino acid analysis using postchromatographic deriv formation. The deriv showed 5–10× the sensitivity of fluorescamine and was stable and soluble in aq buffers. Simons and Johnson[3] determined the structure of the deriv from propylamine but could not actually isolate the deriv. The deriv was among the smallest of those described for amino acid analysis and could be of more general utility for analysis of primary amines.

References

[1]M. Roth, *Anal. Chem.* **43**, 880 (1971).

[2]J. R. Benson and P. E. Hare, *Proc. Natl. Acad. Sci.* (USA) **72**, 619 (1975).

[3]S. S. Simons and D. F. Johnston, *J. Am. Chem. Soc.* **98**, 7098 (1976).

Derivative Type. 7-Nitrobenzo-2,1,3-oxadiazole (NBD).

Reagent. 4-Chloro-7-nitrobenzo-2,1,3-oxadiazole (NBD-Cl).

Compound Type. Amphetamine and analogs.

Reaction

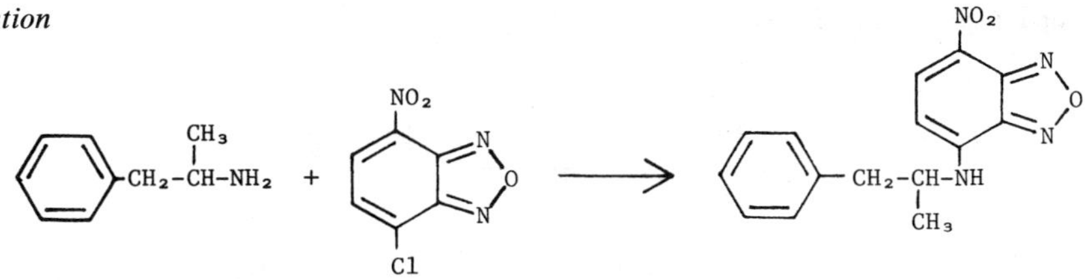

Typical Procedure. To 0.1 mg sample add 0.2 ml of 0.1 *M* $NaHCO_3$ and 0.2 ml of a 1 % soln of NBD-Cl in methyl isobutyl ketone. Heat at 80° for 30 min. Cool and analyze upper layer by TLC.

Comments The method was used for fluorescent deriv for TLC analysis but should be adaptable for HPLC. Fluorescence maxima (482 nm excitation) were reported for the derivs of amphetamine and 13 related amines.

References

[1]F. VanHoof and A. Heyndrickx, *Anal. Chem.* **46**, 286 (1974).

2.2 DERIVATIVES OF INDOLE AMINES

2.2.1. Acyl Derivatives

Derivative Type. *N*-Acetyl.

Reagent. Acetic anhydride (Ac_2O)/pyridine.

Compound Type. Tryptamines

Reaction

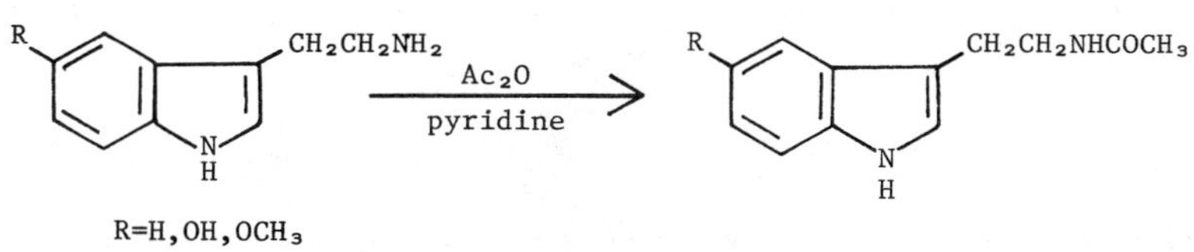

Typical Procedure. Mix 100 mg of the amine with 0.2 ml Ac_2O and 0.2 ml dry pyridine and heat at 75° for 30 min in a sealed vial. Evap to dryness under an air stream and diss residue in MeOH. Isolate the deriv by prep GC and analyze by MS.

Comments. Acetylation of the indole-*N* of *N, N*-dimethyltryptamines was achieved by heating 3 hr at 135°. The acetyl derivs were preferred to the HFB or TMS derivs for MS analysis since they underwent fewer rearrangements and gave prominent ions from the ring system.

References

[1]M. W. Couch and C.M. Williams, *Anal. Biochem.* **50**, 612 (1972).

Derivative Type. Trifluoroacetyl (TFA).

Reagent. Trifluoroacetic anhydride (TFAA)/trimethylamine (TMA).

Compound Type. Indoles.

Reaction

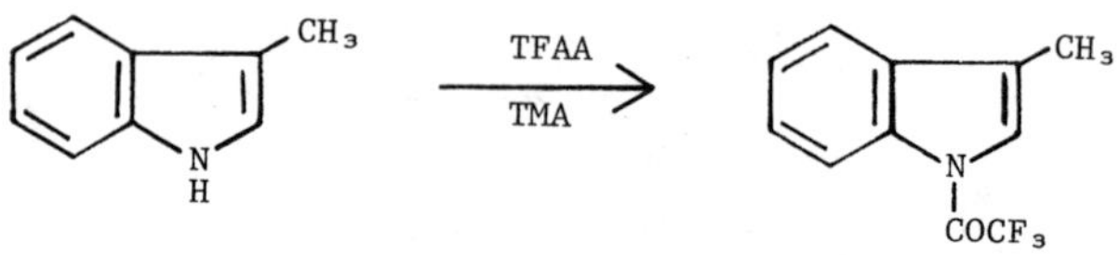

Typical Procedure. Mix 0.5 ml of a soln of 3-methylindole (20 ng/ml–1 μg/ml) and the int stand (*p*-dibromobenzene) in hexane with 200μl 0.9 *M* TMA in hexane and 10μl TFAA. Allow to react 30 min at room temp then wash with 2 ml pH 6 phosphate buffer. Analyze 2 μl of the hexane soln by ECD-GC.

Comments. The acylation of small amts of indoles was studied using 3-methylindole as a model compd. The reaction was very slow in EtOAc (only 10% reaction after 2 hr). In benzene the reaction was 95% complete in 1 hr. The reaction in hexane was quantitative in less than 30 min. The presence of TMA was necessary to prevent TFA catalyzed destruction of the indole. A linear standard curve was shown for the conc range 20 ng/ml to 1 μg/ml.

References

[1]H. Ehrsson, *Acta Pharm. Suecica* **9**, 419 (1972).

Derivative Type. Trifluoroacetyl.

Reagent. Trifluoroacetic anhydride (TFAA).

Compound Type. 5-Hydroxytryptamine (5-HT), norepinephrine, dopamine.

Reaction

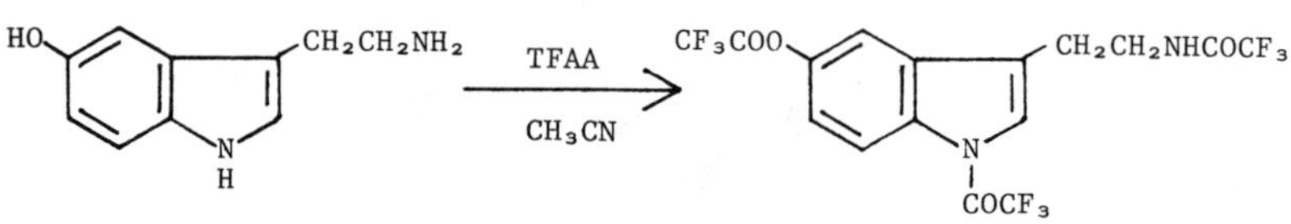

Typical Procedure. Diss the extracted amine sample (as the formic acid salt) in 0.5 ml CH_3CN and add 50μl TFAA. After 5 min at room temp, analyze 1 μl aliquots by ECD-GC.

Comments. Norepinephrine formed a tetra-TFA deriv and dopamine a tri-TFA deriv. The structures of the derivs were confirmed by MS. The method was used for estimation of the compds in rat brain. The catecholamines were ext as ion pairs using di-(2-ethylhexyl)phosphate. The method permitted estimation of as little as 5 ng of dopamine and norepinephrine and 10 ng of 5-HT in a single tissue sample.

References

[1]I. L. Martin and G. B. Ansell, *Biochem. Pharmacol.* **22**, 521 (1973).

Derivative Type. Pentafluoropropionyl.

Reagent. Pentafluoropropionic anhydride (PFPA).

Compound Type. Indole alkyl amines.

Reaction

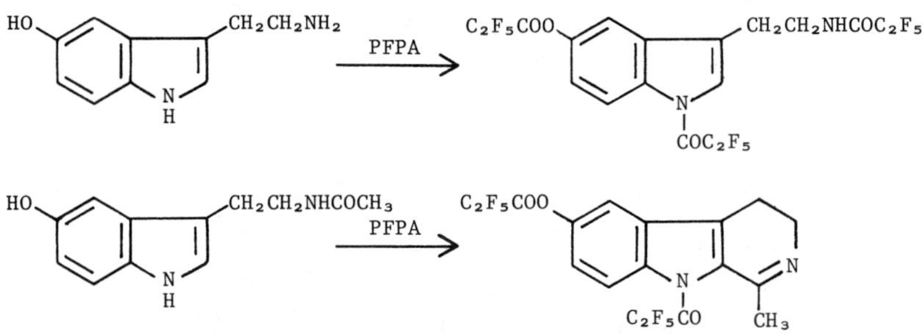

Typical Procedure Treat 1–600 pmol of sample and 20 pmol int stand with 100μl PFPA and 20 μL EtOAc at 60° for 3 hr. Evap to dryness under N_2 and diss residue in 10μl EtOAc. Analyze by GC-MS-SIM.

Comments. The method was used to quantitate the compds in rat pineals. *N*-Acetyltryptamine was used as int stand for *N*-acetylserotonin and melatonin and α-methylserotonin for serotonin and 5-methoxytryptamine. The *N*-acetylated compds reportedly formed β-carboline type cyclized products upon treatment with PFPA (see however the method of Blau *et al.*[2] which follows).

References

[1]F. Cattabeni *et al., Science* **178**, 166 (1972).

[2]K. Blau *et al., Biomed Mass Spectrom.* **4**, 232 (1977).

Derivative Type. 3, 3-Spirocyclic pentafluoropropionyl.

Reagent. Pentafluoropropionic anhydride (PFPA).

Compound Type. Melatonin and related acyl tryptamines.

Reaction

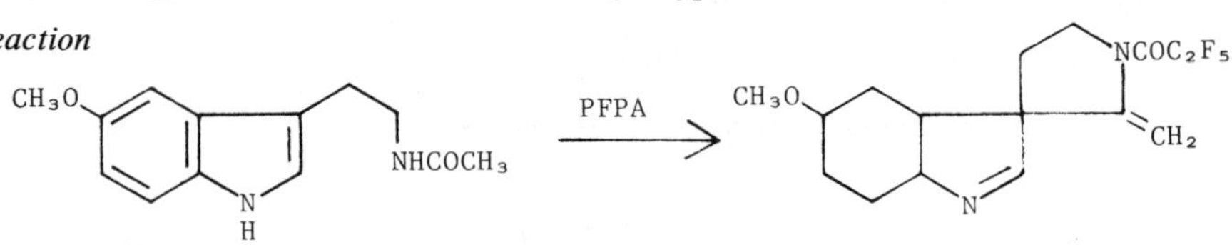

Typical Procedure. Diss 72 mg melatonin in 50 ml 2% PFPA in dry benzene at 0° C. After 10 min evap to dryness under vac. Purify by chromatography on a 20 × 0.9 cm column of silica gel using $CHCl_3$ as eluant. Discard the first 10 ml and collect the next 40 ml of eluant to give 108 mg (97%) of the deriv after evap of the $CHCl_3$. Analyze by GC-MS.

Comments. Cattabeni *et al.*[2] reported the formation of *1* from heating melatonin with PFPA. In this work[1] preparation of *1* from *2* using the PFPI gave a different product than that from melatonin and PFPA. MS and NMR analysis using isotopic labeling confirmed the spirocyclic structure for the melatonin deriv. Derivs were also prepared from several analogs of melatonin.

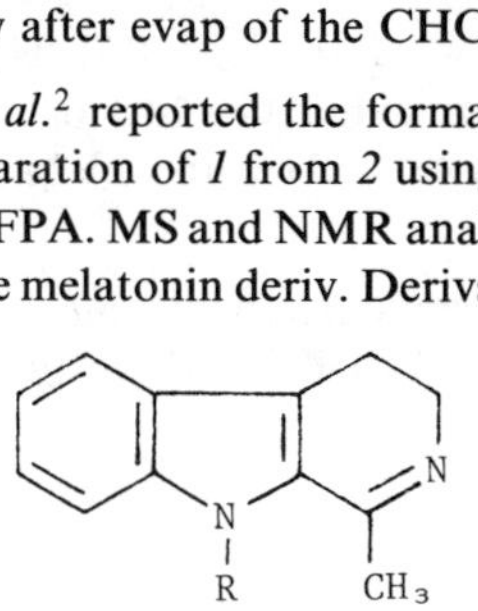

1 R = COC_2F_5

2 R = H

References

[1]K. Blau *et al.*, *Biomed. Mass Spectrom.* **4**, 232 (1977).

[2]F. Cattabeni *et al.*, *Science* **178**, 166 (1962).

Derivative Type. β-Carboline, [1]*N*-trifluoroacetyl.

Reagent. Acetic anhydride (Ac_2O), trifluoroacetic anhydride (TFAA).

Compound Type. Tryptamine.

Reaction

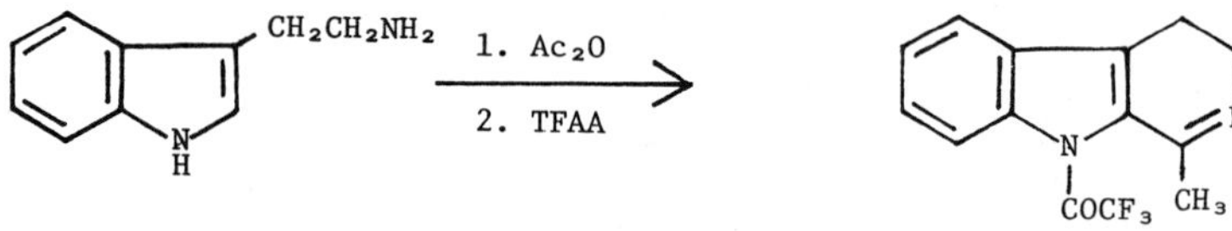

Typical Procedure. Adjust 2 ml urine to pH 5.0–5.5 with 1 *N* HCl. Centrifuge and transfer 1.5 ml supernatant to another tube containing 2 g NaCl and 1 ml H_2O. Add 25 ng d_2-tryptamine int stand. Ext with 2×5 ml Et_2O and discard ext. Adjust pH of aq soln to 10.3 ± 0.2 by adding 2 ml 1 *M* $NaHCO_3$ buffer (pH 10.5) and a few drops of 1 *N* NaOH. Add 5 ml 1% Ac_2O in Et_2O, shake 5 min and centrifuge. Remove org phase and dry over Na_2SO_4. Evap 3–4 ml to dryness under N_2 and add TFAA/EtOAc (5:1,v/v). Allow to react 1 hr at room temp. Evap to dryness under N_2 and diss in 25 μl EtOAc. Analyze by GC-MS-SIM.

Comments. The method was also used for analysis of brain tissue. The assay was sensitive to 500 pg. (See comments in preceding method regarding deriv structure.)

References

[1]J. J. Warsh *et al.*, *Biochem. Med.* **18**, 10 (1977).

Derivative Type. Heptafluorobutyryl.

Reagent. Heptafluorobutyryl imidazole (HFBI).

Compound Type. Indole amines and indole alcohols.

Reaction

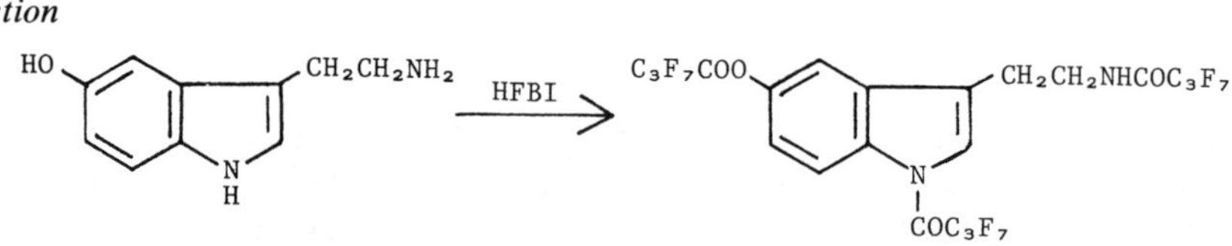

Typical Procedure. Diss 1–2 mg sample in 0.1–0.2 ml HFBI. Heat at 80° for 2–3 hr in a sealed tube. Ext with 3×5 ml hexane. Chill hexane ext (−14°) for 1 hr to ppt reagent extd into hexane. Remove hexane and wash ppt with 2×5 ml hexane. Evap combined hexane to 0.5 ml and analyze 1–3 μl by GC.

Comments. Derivs were prepared from tryptamine, serotonin, tryptophol, and several *N*- and *O*-alkyl analogs. The derivs were useful for either FID or ECD-GC. MS data and fragmentation mechanisms were reported.

References

[1]J. Vessman *et al.*, *Anal. Lett.* **2**, 81 (1969).

Derivative Type. Heptafluorobutyryl.

Reagent. Heptafluorobutyrylimidazole (HFBI).

Compound Type. Indolealkylamines.

Reaction

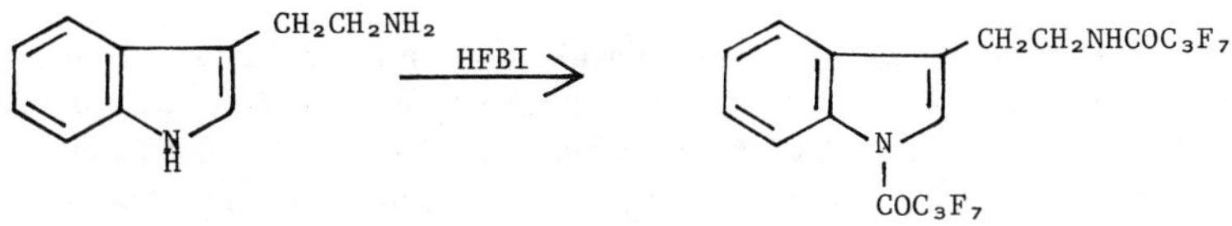

Typical Procedure. Deproteinize a 5 ml sample of CSF with $HClO_4$; then adjust to pH 11 with NaOH. Ext 2× with an equal vol of CH_2Cl_2. Dry the ext over Na_2SO_4 and evap to dryness. Add 0.1 ml HFBI and heat at 85° for 1 hr. Add 2 ml toluene and 1 ml H_2O. Separate the toluene layer, dry over Na_2SO_4, and analyze by ECD-GC.

Comments. The method was demonstrated with tryptamine added to CSF. GC data were reported for the derivs of five indolealkylamines.

References

[1]F. Bennington *et al., J. Chromatogr.* **106**, 435 (1975).

Derivative Type. Heptafluorobutyryl.

Reagent. Heptafluorobutyrylimidazole (HFBI).

Compound Type. Biological amines.

Reaction

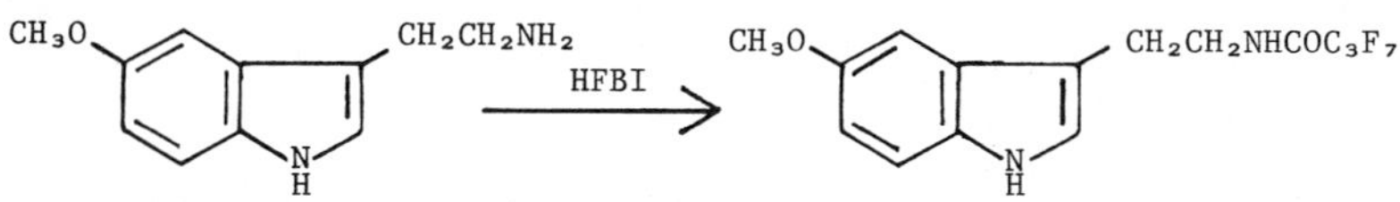

Typical Procedure. Treat the sample with a slight excess of 3% (v/v) HFBI in EtOAc containing 1% (v/v) Et_3N. Allow to react in the dark at room temp for 30 min. Analyze by GC.

Comments. It was reported that the reaction with HFBI should be carried out in the dark since the presence of light gave production of a green-yellow color and irreproducible conversion. The indole nitrogen did not react. Et_3N was added to inhibit acid catalyzed dimerization of the indole compds.[2]

References

[1]J. J. Franken and M. M. F. Trijbels, *J. Chromatogr.* **91**, 425 (1974).

[2]H. Ehrsson, *Acta Pharm. Suecica* **9**, 419 (1972).

Derivative Type. Heptafluorobutyryl.

Reagent. Heptafluorobutyrylimidazole (HFBI).

Compound Type. Melatonin.

Reaction

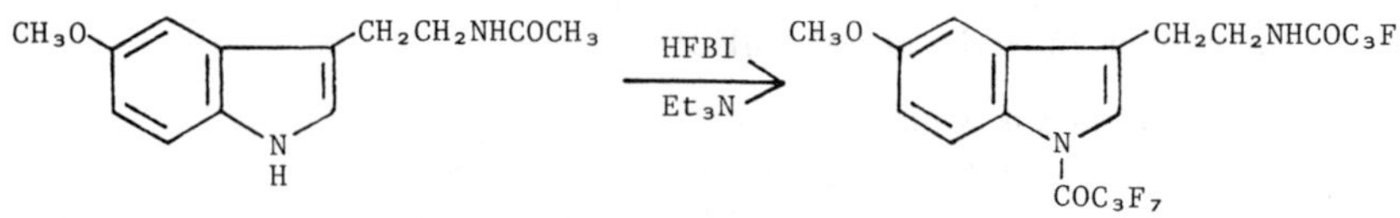

Typical Procedure. Homogenize two rat pineals in 0.2 ml $CHCl_3$ by sonication with ice cooling. Centrifuge and discard top pellet. Cool $CHCl_3$ in ice and ext with 0.2 ml 0.05 *N* NaOH. Centrifuge and discard NaOH layer. Ext with 0.2 ml 0.1 *N* HCl, centrifuge, and discard HCl layer. Dry ext by transferring to a 2.5 cm column of anhyd Na_2SO_4 with 2 ml $CHCl_3$ and eluting with another 2 ml $CHCl_3$. Evap eluates to dryness under N_2.

Treat residue with 2 μl triethylamine and 20 μl HFBI and heat at 90° for 2.5 hr. Cool, add 20 μl heptane, and mix. Add 20 μl ice cold H_2O and shake 10 sec. Centrifuge and add 1 μl (442 pg) of a soln of dieldrin int stand in heptane. Cool, centrifuge, and chill H_2O layer in solid CO_2/acetone. Remove heptane and analyze by ECD-GC.

Comments. The method was used to quantitate melatonin to 35 pg. The structure of the deriv was confirmed by MS. GC properties of the derivs of 11 other indoleamines were also examined.

References

[1]P. H. Degen *et al., Anal. Biochem.* **45**, 634 (1972).

2.2.2. Silyl Derivatives

Derivative Type. Trimethylsilyl.

Reagent. *N*-Methyl-*N*-trimethylsilyltrifluoroacetamide (MSTFA)/trimethylsilylimidazole (TMSI).

Compound Type. Indole compounds.

Reaction

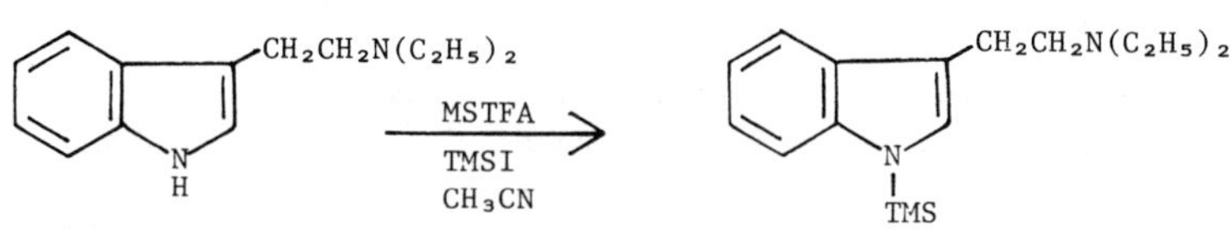

Typical Procedure. Diss sample in 1 ml MSTFA/CH_3CN (1:1, v/v) containing 5% (v/v) TMSI and heat at 60° for 5 min. Analyze by GC.

Comments. It was found that the commonly used TMCS catalyst actually decreased the rate of [1]*N*-TMS formation from indoles whereas basic catalysts (pyridine, triethylamine) increased the rate. TMSI was the most effective catalyst giving complete [1]*N*-silylation of *N,N*-diethyltryptamine, tryptamine, serotonin, and tryptophan in 5 min at 60°.

References

[1]M. Donike, *Chromatographia* **9**, 440 (1976).

Derivative Type. Trimethylsilyl.

Reagent. Bis(trimethylsilyl)trifluoroacetamide(BSTFA)/trimethylchlorosilane (TMCS).

Compound Type. Tryptamines.

Reaction

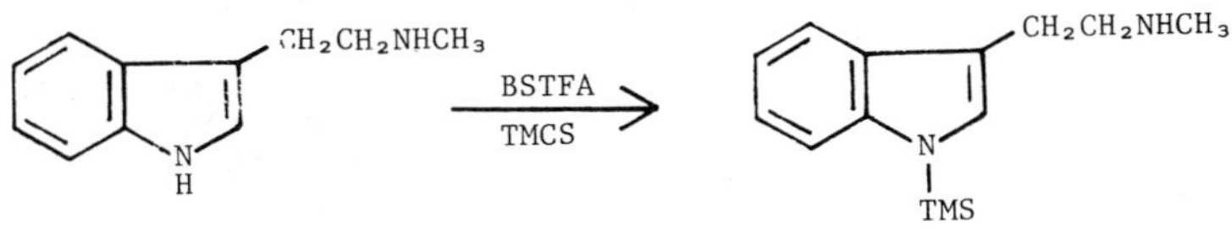

Typical Procedure. Heat 100 μg of free amine to 100° for 15 min in a sealed tube with 100μl of BSTFA containing 1 % TMCS. Inject soln directly into GC-MS.

Comments. The method yielded selective introduction of TMS at the indole N. Serotonin yielded a di-TMS deriv. The method was also applicable to primary amines. The free primary amino group was confirmed by conversion to the isothiocyanate with CS_2.[3]

References

[1]N. Narasimhachari *et al., J. Chromatogr. Sci.* **9**, 502 (1971).

[2]N. Narasimhachari and P. Vouros, *Anal. Biochem.* **45**, 154 (1972).

[3]N. Narasimhachari and K. Leiner, *J. Chromat. Sci.* **15**, 181 (1977).

Derivative Type. Trimethylsilyl.

Reagent. Bis(trimethylsilyl)acetamide (BSA)/trimethylchlorosilane (TMCS)/trimethylsilylimidazole (TMSI).

Compound Type. Melatonin.

Reaction

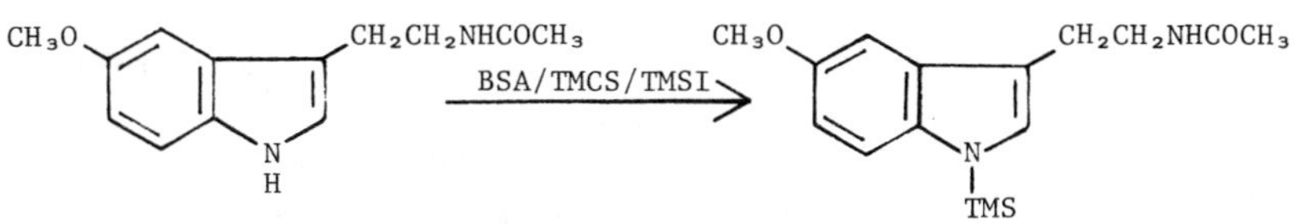

Typical Procedure. To 2 ml of plasma or CSF add 50 ng ω-N-*n*-hexanoyl 5-methoxytryptamine int stand and ext with 4 ml $CHCl_3$. Evap to dryness under Ar and add BSA/TMCS/TMSI (3:2:3, w/w/w). Allow to stand 15 min at room temp then analyze by GC-MS-SIM.

Comments. The method permitted detection of melatonin in plasma to 10 pg/ml.

References

[1]B. W. Wilson *et al., Anal. Biochem.* **81**, 283 (1977).

Derivative Type. Trimethylsilyl.

Reagent. Bis(trimethylsilyl)trifluoroacetamide (BSTFA).

Compound Type. Psilocin and psilocybin.

Reaction

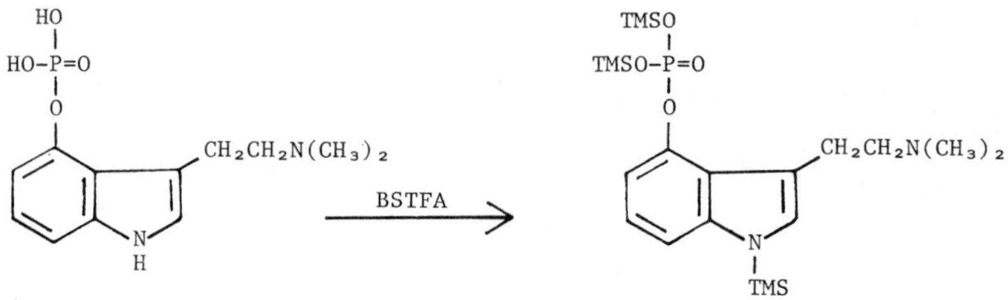

Typical Procedure. Grind 50 mg of freeze-dried mushroom tissue to a fine powder with sand and mix with 5 ml MeOH. Shake 24 hr at room temp then filter and wash the solids with MeOH. Evap the combined MeOH to dryness under vac and treat the residue with 100 μl BSTFA under dry N_2. Heat 15 min at 140°. Cool to room temp and analyze 1 μl aliquots by GC or GC-MS.

Comments. MS data were given for the derivs.

References

[1]D. B. Repke *et al., J. Pharm. Sci.* **66**, 743 (1977).

2.2.3. Mixed Silyl and Other Derivatives

Derivative Type. *N*-Acetyl-[1]N, *O*-trimethylsilyl.

Reagent. Acetic anhydride (Ac_2O), *N*-methyl-*N*-trimethylsilyltrifluoroacetamide (MSTFA).

Compound Type. Serotonin, tryptamine.

Reaction

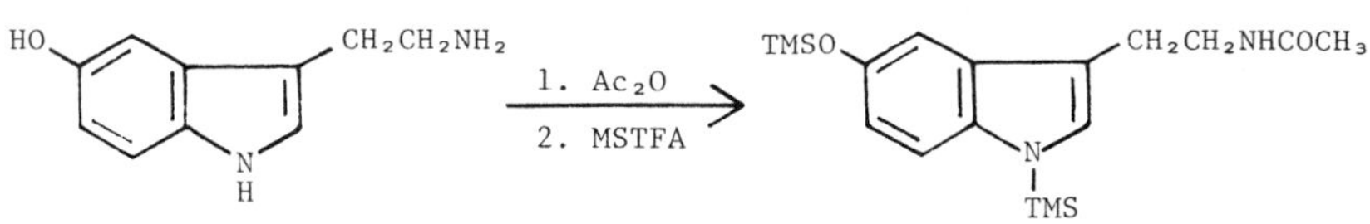

Typical Procedure. To 5 ml of H_2O soln containing 1–10 μg each of serotonin or tryptamine and *N*-acetyldiphenylamine (int stand), add $NaHCO_3$ (1 g) to saturation. Add dropwise with constant stirring 500 μl of Ac_2O (or 1.5 ml of heptafluorobutyric anhydride). Cont stirring for 5 min after evolution of CO_2 ceases and all $NaHCO_3$ is dissolved. Transfer soln to a 25 ml pear-shaped flash and ext 4 × with 0.5 ml CH_2Cl_2. Transfer CH_2Cl_2 with 500 μl syringe into a 1 ml tapered reaction vial and evap instantly under N_2. Add 25 μl of MSTFA to the dry residue and heat the mixt in a closed vial for 30 min at 80°. Inject an aliquot of this mixt directly into the GC-MS.

Comments. This method avoided the necessity of isolating the free amines from aqueous soln by acylating in the aqueous phase. The amides were then more easily extractable by organic solvents. BSTFA tended to give more reaction with the amide N that MSTFA, therefore MSTFA was the reagent of choice.

References

[1]J. Eyem and L. Bergstedt, *Adv. Mass Spect. Biochem. Med.* **1**, 497 (1976).

Derivative Type. [1]N-Trimethylsilyl-[ω]N-trifluoroacetyl.

Reagent. N-Methyl-N-trimethylsilyltrifluoroacetamide (MSTFA)/N-trimethylsilylimidazole (TMSI), N-methyl bis(trifluoroacetamide) (MBTFA).

Compound Type. Indolalkylamines.

Reaction

Typical Procedure. Diss 2–3 mg of indolalkylamine in 1 ml of MSTFA/N-methyltrifluoroacetamide (MTFA)/imidazole (1.2:2.0:0.2, mol/mol/mol) and heat at 80° for 15 min. Allow to cool and add 50 μl of MBTFA. Heat 10 min at 80° or allow to stand at room temp for 1 hr. Analyze an aliquot of the soln by GC or GC-MS-SIM.

Comments. Selective deriv of [1]N and [ω]N was possible because [1]N-silylation was kinetically controlled while reaction at [ω]N was thermodynamically controlled. The method was compared with pertrimethylsilylation for quantitation of serotonin in tissue samples by GC-MS-SIM and found to give better specificity. Quantitation was possible in the femptomole range.

References

[1]M. Donike *et al., J. Chromatogr.* **134**, 385 (1977).

Derivative Type. Trimethylsilyl, Schiff base.

Reagent. Hexamethyldisilazane (HMDS)/acetone.

Compound Type. Indoleamines.

Reaction

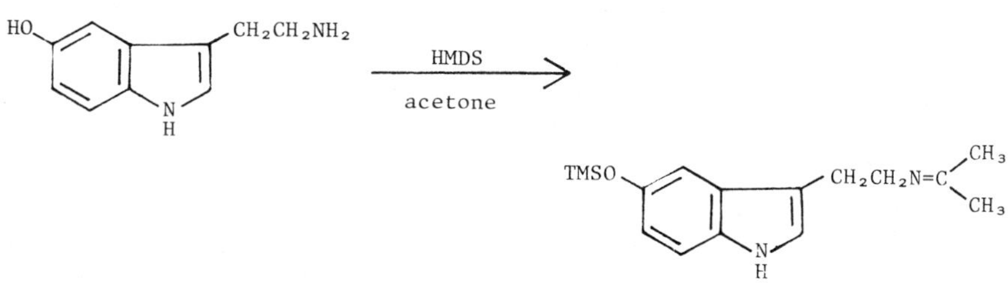

Typical Procedure. Diss sample in a soln of HMDS in acetone. Allow to stand overnight at room temp then analyze by GC.

Comments. The GC behavior of the derivs of a series of indoleamines was studied.[1] The acetone Schiff base formation provided a convenient method of distinguishing the primary amines. Data on the derivs of a series of 10 tryptamines, N,N-dimethyltryptamines, the N,N-diethyltryptamines and hydroxy and methoxy subsituted compds were reported.[2]

References

[1]B. Holmstedt *et al., Anal. Biochem.* **8**, 151 (1964).
[2]E. C. Horning *et al., Anal. Chem.* **36**, 1546 (1964).

2.2.4. Other Derivatives

Derivative Type. 5-Dimethylamino-l-naphthalenesulfonyl (DANS).

Reagent. 5-Dimethylamino-l-naphthalenesulfonyl chloride (DANS-Cl).

Compound Type. Tetrahydroisoquinolines, tetrahydro-β-carbolines.

Reaction

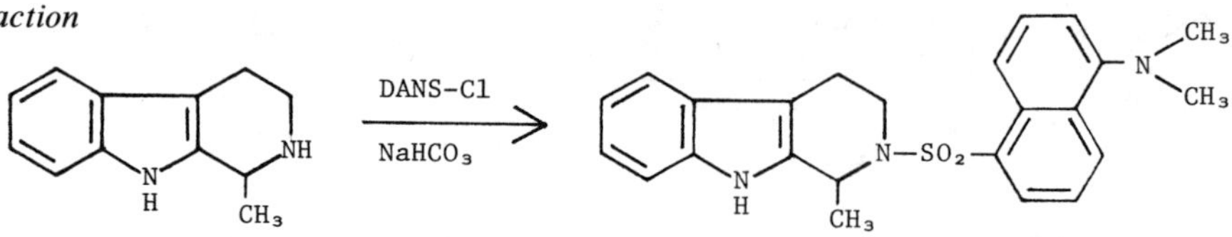

Typical Procedure. Diss alkaloid in minimum amt of H_2O and add an equal vol of a N_2 satd soln containing 1 vol of satd $NaHCO_3$ and 2 vol of DANS-Cl in acetone (4 mg/ml) so as to have *ca* 25 % excess of DANS-Cl. Seal under N_2 and allow to stand at 20° for 16 hr. Ext deriv into EtOAc, wash ext with H_2O, and recryst from EtOH-H_2O. Analyze by MS.

Comments. Derivs were examined for use in quantitation of the alkaloids using deuterated int stand. Phenolic alkaloids were also examined as *O*-methyl, *N*-DANS derivs. The *O*-methylation was performed with diazomethane at 25° for 4 hr. The mixed derivs gave greater sensitivity but would be indistinguishable from the derivs of the corresponding methoxy alkaloids.

References

[1]D. A. Durden *et al., Adv. Mass Spect. Biochem. Med.* **2**, 597 (1976).

2.3. TERTIARY AMINES

2.3.1. Acyl Derivatives

Derivative Type. Trifluoroacyl.

Reagent. Trifluoroacetic anhydride (TFAA).

Compound Type. Tertiary aromatic amines.

Reaction

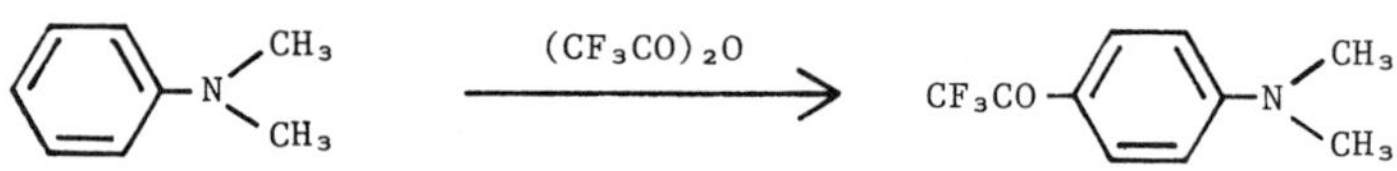

Typical Procedure. Diss sample in 200 μl DMF or EtOAc. Add 100 μl 1 *M* trimethylamine in benzene, 25 μl (0.2 mmol) TFAA, and heat at 50° for 60 min. Cool and shake with 0.5 ml 0.5 *M* NaOH. Centrifuge and analyze an aliquot of the benzene layer by GC.

Comments. The deriv was detectable to less than 1 pg by ECD. A similar ring acylation occurred upon treating methoxybenzene with TFAA and benzoic acid forming the ECD sensitive *p*-benzoyl deriv.

References

[1]T. Walle, *J. Chromatogr.* **111,** 133 (1975).

[2]H. Brink *et al., Acta Pharm. Suecica* **12**, 235 (1975).

Derivative Type. Heptafluorobutyramide.

Reagent. Methylchloroformate, hydrogen bromide/acetic acid (HOAc), heptafluorobutyric anhydride (HFBA)/trimethylamine (TMA).

Compound Type. Tertiary amines.

Reaction

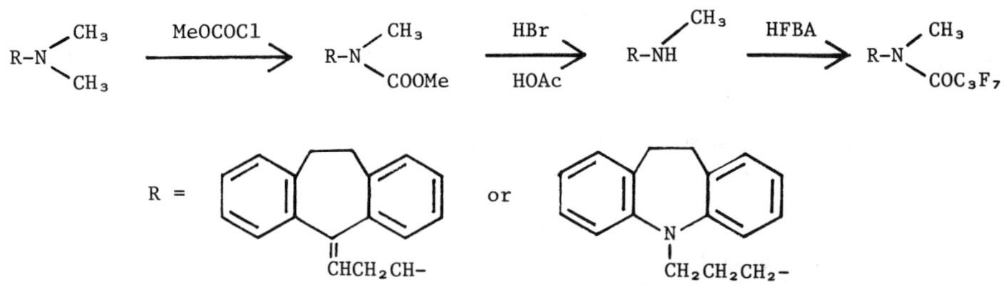

Typical Procedure. Alkalinize a 1 ml aq sample (containing 0.5 mg amine) with 0.2 ml 2 *N* NaOH and ext with 1 ml heptane. Remove heptane phase and mix with 200 μl methyl chloroformate and 10 mg Na_2CO_3. Heat 30 min at 100°. Evap to dryness and add 0.5 ml 30% HBr in HOAc. Heat at 100° for 10 min. Add 1 ml conc NH_3 soln (13 *M*) and ext with 0.5 ml heptane. Remove heptane phase and add 50 μl 0.1 *M* TMA in heptane and 5 μl HFBA. Allow to stand 10 min at room temp then wash with 2 ml 0.1 *M* NaOH. Anaylze 5 μl of heptane phase by GC.

Comments. This method[2] was an improvement of the original method[1] which was more time consuming. Derivs were prepared from imipramine, amitriptyline, and a series of diphenylmethane alkyl tertiary amines. Methyl chloroformate was found to be the best of several alkyl chloroformates examined.

References

[1]J. Vessman *et al., Anal. Lett.* **6**, 699 (1973).

[2]P. Hartvig and J. Vessman, *Acta Pharm. Suecica* **11**, 115 (1974).

2.3.2. Carbamate Derivatives

Derivative Type. Pentafluorobenzyl carbamate.

Reagent. Pentafluorobenzyl chloroformate.

Compound Type. Tertiary amines.

Reaction

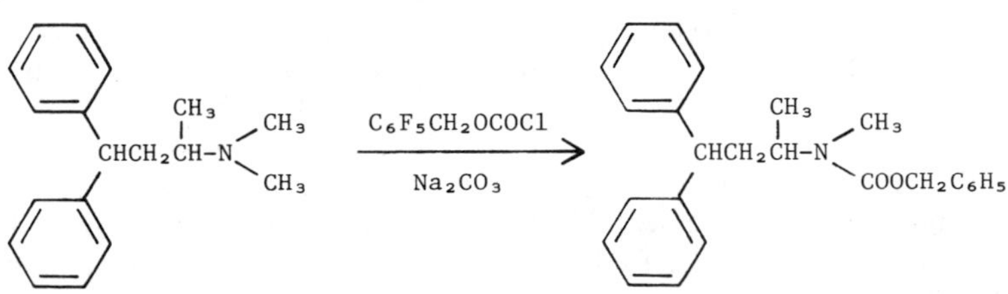

Typical Procedure.[1] To 0.2 ml of heptane soln of the amine (0.5–1 mg/ml) add *ca* 50 μl of the chloroformate and *ca* 10 mg of anhyd Na_2CO_3. Attach an air condenser and heat for 30 min on a boiling water bath. Allow to cool to room temp and shake with 1 ml 1 *M* NaOH. Analyze an aliquot of the heptane phase by ECD-GC.

Comments. The method was developed for ECD-GC determination of tertiary amines.[1,2] The benzyl carbamate and nitrobenzyl carbamate were also examined but gave lower detector response. The method was applied to quantitation of imipramine using chloroimipramine as int stand.[3] Some compds (e.g., methadone, promethazine) were deaminated to neutral products.[2]

References

[1] P. Hartvig and J. Vessman, *J. Chromatogr. Sci.* **12**, 722 (1974).

[2] P. Hartvig and J. Vessman, *Anal. Lett.* **7**, 223 (1974).

[3] P. Hartvig *et al.*, *Anal. Chem.* **48**, 390 (1976).

Derivative Type. Trichloroethyl carbamate.

Reagent. Trichloroethyl chloroformate.

Compound Type. Amitriptyline and nortriptyline.

Reaction

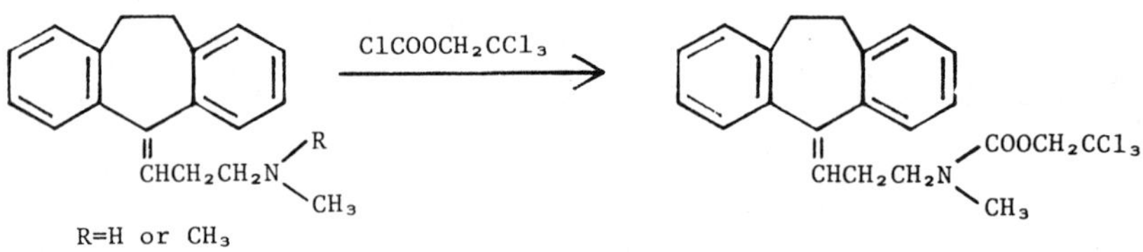

Typical Procedure. Mix 0.5 ml plasma, 1.0 ml of int stand soln (see below), and 2 g cellulose. Pack this mixt on top of a column (30 × 1 cm) prep by mixing 0.2 g cellulose, 0.1 ml 1 *M* citrate buffer (pH 5.0), and 0.05 ml blank plasma. Elute amitriptyline and 2-chloroamitriptyline with 10 ml *n*-heptane and ext with 1.0 ml 2 *M* H_3PO_4. Alkalinize the aq phase and ext with 0.25 ml toluene. Mix the toluene ext with 25 μl trichloroethyl chloroformate and add 10 mg Na_2CO_3. Attach air condenser and heat for 1 hr at 110°. Shake reaction mixt with 1.0 ml alcoholic alkali (see below) for 10 min. Add 1.0 ml H_2O and shake 5 min. Remove aq phase and add 1.0 ml of satd KOH in MeOH. Shake 10–15 sec, add 1.0 ml H_2O, and shake 10 min. Discard aq phase and analyze 5 μl of the org phase by ECD-GC.

Elute nortriptyline and its standard with 10 ml of CH_2Cl_2/BuOH (4 : 1, v/v). Add 10 ml *n*-heptane to the eluate and ext with 1 ml 2 *M* H_3PO_4. Alkalinize the aq phase and ext with 0.25 ml toluene. Add 20 μl trichloroethyl chloroformate to the toluene ext and allow to stand 30 min at room temp. Remove excess reagent and analyze as for amitriptyline.

Comments. 2-Chloroamitriptyline was used as int stand for amitriptyline and 2-chlorodemethylimipramine for nortriptyline. The two stand were diss together in pH 5.0 citrate soln to a conc of 40 ng/ml. The alcoholic alkali was prep by diss 2.8 g KOH in 22 g H_2O and adding 75 g MeOH.

Since both amitriptyline and nortriptyline formed the same deriv it was necessary to sep the two compds prior to deriv. The method was used to determine the compds in plasma to 1 ng/ml. The method was also applied to imipramine, 2-chloroimipramine, and their *N*-demethylmetabolites,[1] and to a series of diphenylmethane alkyl tertiary amines.[2] The trichloroethyl derivs gave slightly lower ECD response than the corresponding pentafluorobenzyl derivs but gave fewer interferences and less pronounced solvent fronts.[2]

References

[1]P. Hartvig and B. Nasland, *J. Chromatogr.* **133**, 367 (1977).
[2]P. Hartvig *et al.*, *Acta Pharm. Suecica* **13**, 181 (1976).

Derivative Type. Trichloroethyl carbamate.

Trichloroethyl chloroformate/sodium carbonate.

Compound Type. Pethidine and norpethidine.

Reaction

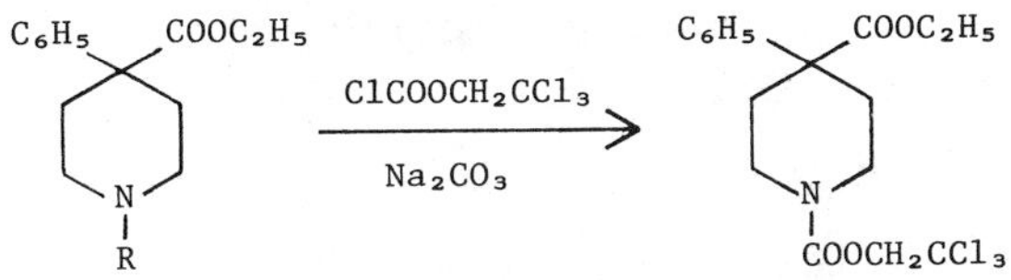

R = H, CH_3

Typical Procedure. Mix 0.1 ml plasma and 0.1 ml int stand soln (see below) with 0.3 g cellulose and pack on the top of a column of 2 g cellulose prepared with 1.3 ml pH 7.4 phospate buffer. Elute pethidine and its int stand with 10 ml heptane and ext into 1.0 ml pH 2 phosphate buffer. Alkalinize aq phase and ext with 0.25 ml toluene. Add 25 μl trichloroethyl chloroformate and 10 mg Na_2CO_3 to the toluene ext and heat 1 hr at 125°. Add 1 ml alcoholic alkali (2.8 g KOH, 75 g MeOH, 22 g H_2O) and shake 10 min. Add 1 ml H_2O and shake 5 min. Remove aq phase and add 1ml satd KOH in MeOH. Shake 10–15 sec, add 1 ml H_2O, and mix 10 min. Discard aq phase and analyze 1–2 μl of the org phase by ECD-GC.

Elute norpethidine and its int stand with 10 ml CH_2Cl_2/*n*-BuOH (4:1, v/v). Ext into 0.5 ml 0.1 *M* H_3PO_4, alkalinize, and ext with 0.25 ml toluene. Add 10 μl reagent to the toluene ext and allow to stand 15 min at room temp. Remove excess reagent as above and analyze by ECD-GC.

Comments. Butyl 1-methyl-4-phenyl-4-piperidinecarboxylate (for pethidine) and propyl 4-phenyl-4-piperidinecarboxylate (for norpethidine) were used as int stands. The int stand soln consisted of 500 ng/ml each in pH 7.4 phosphate buffer.

References

[1]P. Hartvig *et al.*, *Europ. J. Clin. Pharm.* **11**, 65 (1977).
[2]P. Hartvig *et al.*, *J. Chromatogr.* **121**, 235 (1976).

2.3.3. Derivatives with the Amino Group Removed

Derivative Type. Hofmann degradation product.

Reagent. Methyl iodide, silver oxide.

Compound Type. Amitriptyline and other tertiary amines.

Reaction

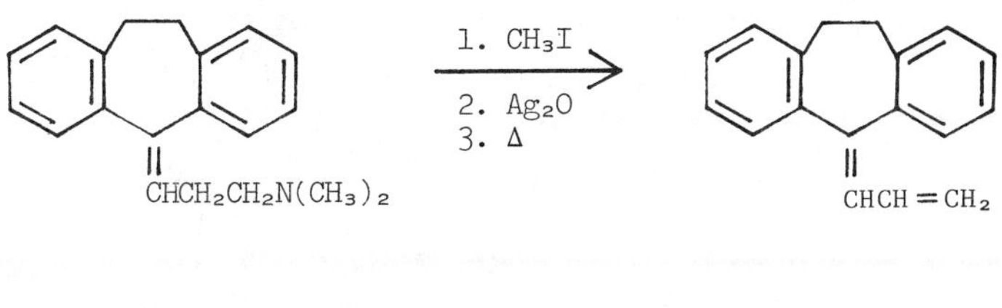

Typical Procedure. Diss 1–10 μg of the tertiary amine in 10 μl benzene and add 4 μl methyl iodide. Allow to stand at room temp 5 min. Add a small particle of moist silver oxide then heat at 70° until the benzene evaporates. Remove remaining traces of H_2O under a stream of N_2. Diss residue in 5–10 μl CS_2 and analyze by GC.

Comments. Hofmann degradation of the quaternary ammonium hydroxide intermediate occurred in the hot injector. Some of the compds gave two isomeric olefin products.[2] The method was shown to be useful for quantitation of amitriptyline[1] as well as for qualitative identification of a variety of other tertiary amines.

References

[1]H. B. Hucker and J. K. Miller, *J. Chromatogr.* **32**, 408 (1968).
[2]H. V. Street, *J. Chromatogr.* **73**, 73 (1972).

Derivative Type. Hofmann degradation product.

Reagent. Methyl iodide, silver oxide.

Compound Type. Butyrophenones.

Reaction

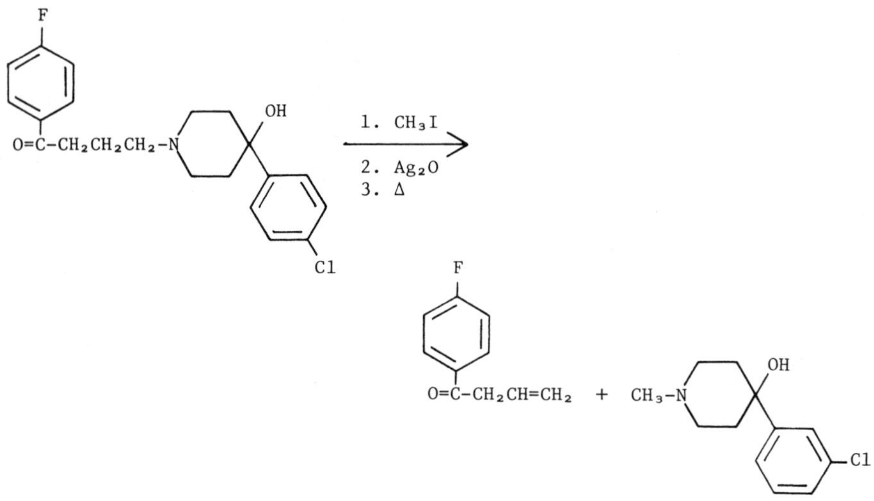

Typical Procedure. To the drug soln in benzene add 1 ml CH_3I and allow to stand at room temp 15 min. Evap under N_2 and diss residue in MeOH. Add a small amt of solid Ag_2O, shake well, then allow to settle. Analyze an aliquot of the supernatant by GC.

Comments. The method was applied to haloperidol (shown above), droperidol, and penfluoridol. The Hofmann degradation of the quaternary ammonium hydroxide intermediate was accomplished in the hot (300°) injector. MS data were given for the derivs.

References

[1]J. Rosenfeld *et al., J. Chromatogr.* **129**, 387 (1976).

Derivative Type. *p*-Chlorobenzaldehyde.

Reagent. Sodium borohydride, sodium periodate.

Compound Type. Lofepramine.

Reaction

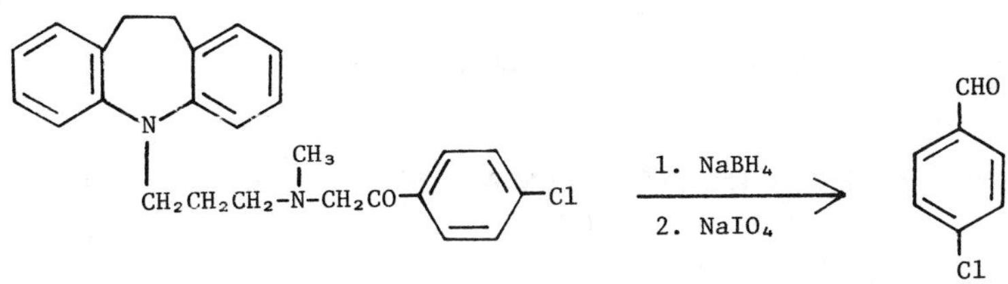

Typical Procedure. Adjust 2 ml plasma to pH 7–8 with 0.25 *M* NaOH and add *ca* 75 ng int stand (bromo analog of lofepramine). Add 1 ml freshly prepared soln of 250 mg $NaBH_4$ in 25 ml 0.04 *M* NaOH. Dil to 4 ml and allow to react 15 min. Ext with 2 × 5 ml hexane and add 10 ml Et_2O to the combined ext. Back ext into 5 ml 0.25 *M* H_2SO_4 and evap any Et_2O in the aq phase under a stream of N_2. Adjust to pH 9 with 5 *M* NaOH and add 10 mg $NaIO_4$ and 10 mg Na_2HPO_4. Distill *ca* 10 ml into a receiver containing 3.3 g NaCl. Ext distillate with 500 μl hexane and analyze by ECD-GC.

Comments. The amino ketone drug was reduced to an amino alcohol which was cleaved by $NaIO_4$ to the aldehyde. The lofepramide metabolite, desmethylimipramine, was analyzed as the HFB deriv after removal of the parent drug by reduction. The drug has also been measured as the ethyl *p*-chlorobenzoate deriv formed by H_2O_2 oxidation and esterfication with EtOH/HCl.[2]

References

[1]R. Lundgren *et al.*, *Acta Pharm. Suecica* **14**, 81 (1977).

[2]K. Matsubayashi *et al.*, *J. Chromatogr.* **143**, 571 (1977).

Derivative Type. Anthraquinone.

Reagent. Ceric sulfate/sulfuric acid.

Compound Type. Amitriptyline and nortriptyline.

Reaction

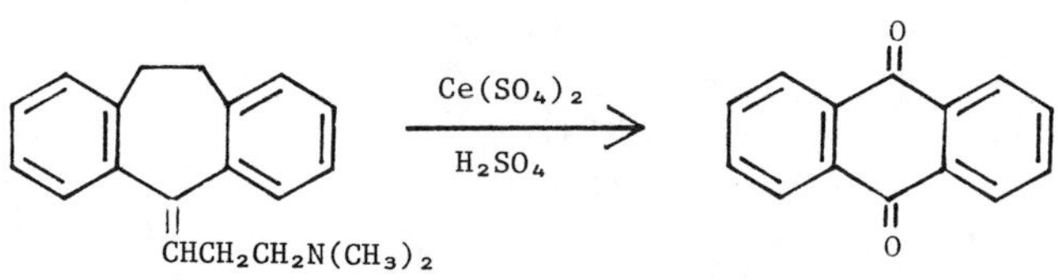

Typical Procedure. Add 1 drop 1 *N* NaOH to 0.5 ml plasma and ext with 2.5 ml hexane. Transfer 2.3 ml of the hexane to another tube containing 2.0 ml of 5.5 *M* H_2SO_4 containing 25 mg/ml $Ce(SO_4)_2$. Mix well then remove 1.8 ml of the aq phase and heat at 95° for 10 min. Allow to cool then ext with 0.1 ml hexane containing 100 ng/ml ethylanthraquinone int stand. Analyze 5 μl of the ext by ECD-GC.

Comments. The method permitted quantitation of the combined compds down to a few ng/ml. Hartvig *et al*[2]. employed a column ext procedure to specifically determine amitriptyline as the same deriv. It was reported[3] that 3,3′-spirobiphthalide was a significant side product in the oxidation. This product would be removed by the NaOH wash in the Hartvig *et al.* method. Protriptyline and cyproheptadine were potential interfering compds yielding the same deriv.

References

[1]J. E. Wallace *et al., Anal. Chem.* **47**, 1516 (1975).

[2]P. Hartvig *et al., J. Chromatogr.* **118**, 65 (1976).

[3]A. W. Missen and H. M. Stone, *Anal. Chem.* **49**, 1864 (1977).

2.4. DERIVATIVES OF QUATERNARY AMMONIUM COMPOUNDS

2.4.1. Dealkylated Derivatives

Derivative Type. Demethyl.

Reagent. Sodium benzenethiolate.

Compound Type. Acetylcholine.

Reaction

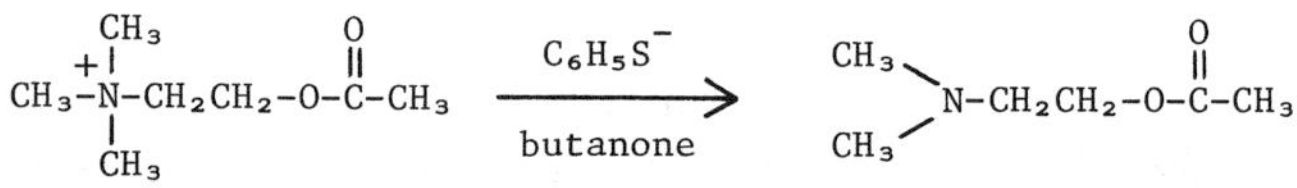

Typical Procedure.[1] To a sample containing 0.5–25 nmol of acetylcholine add *ca* 10 nmol hexyldimethylamine HCl int stand. Evap to dryness and add 0.5 ml of 6 mg/ml (50 m*M*) sodium benzenethiolate in butanone. Purge tube with N_2, cap, and heat at 80° for 30 min (shake every 5 min). Cool tube and add 0.1 ml 0.5 *M* citric acid and 2 ml pentane. Shake, centrifuge, and discard upper layer. Wash aq layer with 2 × 1 ml pentane. Evap traces of pentane from aq phase under N_2. Add 50 μl $CHCl_3$ and 0.1 ml of 2 *M* ammonium citrate-7.5 *M* ammonium hydroxide. Shake, centrifuge, and analyze 5 μl of $CHCl_3$ phase by GC.

Comments. The benzenethiolate reagent was freshly prepared each day and stored under N_2. The procedure should be generally useful for quantitation of quaternary amines. The method was adapted for simultaneous analysis of choline by acylating the choline with propionyl chloride.[3]

References

[1]D. L. Jenden *et al., Anal. Chem.* **40**, 125 (1968).

[2]D. L. Jenden *et al., Anal. Biochem.* **35**, 209 (1970).

[3]D. J. Jenden *et al., Anal. Chem.* **44**, 1879 (1972).

2.4.2. Reduced Derivatives

Derivative Type. Hexahydro deriv.

Reagent. Sodium borohydride.

Compound Type. Paraquat.

Reaction

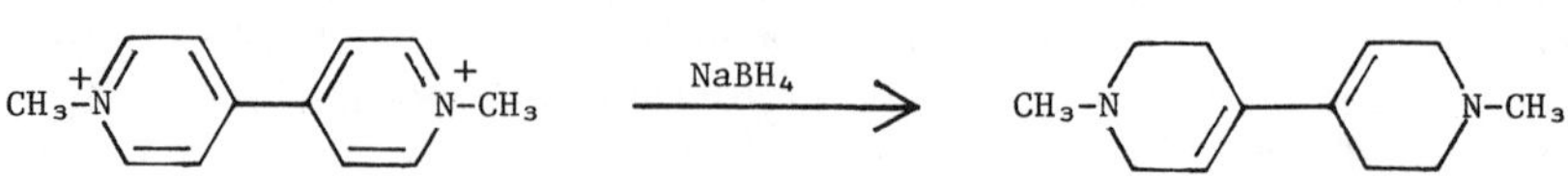

Typical Procedure. To 3 ml plasma add 1 ml of 1 μg/ml aq soln of 1,1′-diethyl-4,4′-dipyridyl (int stand) and 1 ml 25% aq TCA. Centrifuge and wash ppt with 5 ml 5% aq TCA. Centrifuge and combine supernatants. Adjust to pH 10 with 1 ml 5 *N* NaOH and add 150 mg $NaBH_4$. Allow to stand at room temp with intermittent shaking for 60–90 min. Add 8 g NaOAc and ext with 9 ml Et_2O. Back-ext Et_2O with 0.5 ml 1 *N* HCl. Remove traces of Et_2O under N_2, alkalinize with 0.15 ml 5 *N* NaOH, and ext with 30 μl CH_2Cl_2. Analyze an aliquot of the CH_2Cl_2 by GC or GC-MS.

Comments. The method was used for quantitation of paraquat in plasma to 0.1 μg/ml with FID-GC and 0.025 μg/ml with nitrogen selective detection. GC-MS-SIM was used for enhanced selectivity. MS data were given for the derivs of paraquat, d_6(methyl)-paraquat, and the int stand.

References

[1]G. H. Draffan *et al.*, *J. Chromatogr.* **139**, 311 (1977).

2.4.3. Cleavage Products

Derivative Type. Trifluoroacetyl.

Reagent. Trifluoroacetic anhydride (TFAA).

Compound Type. Thiamine.

Reaction

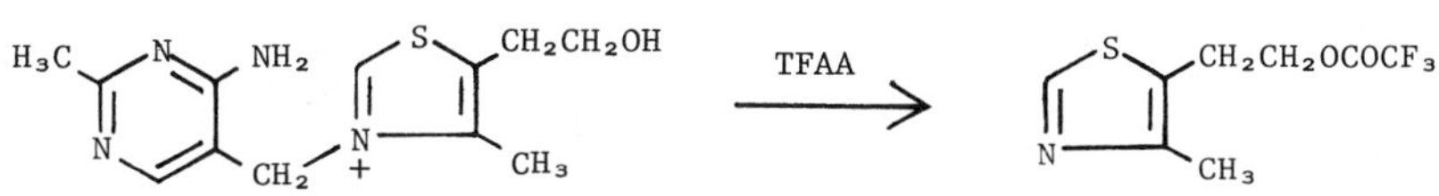

Typical Procedure. Evap 20 μl plasma to dryness under N_2 at 70°. Add 100 μl of TFAA/CH_2Cl_2 (1:3, v/v). Sonicate 1 min then heat at 150° for 5 min. Analyze by GC using an electrolytic conductivity detector.

Comments. The method was demonstrated on submicrogram quantities of thiamine and thiamine pyrophosphate in rat plasma using the nitrogen specific detector. No structural data were reported for the derivs,[1] but MS data indicate the structure shown above for the deriv of thiamine.[2]

References

[1]D. M. Hilker and J. M. L. Mee, *J. Chromatogr.* **76**, 239 (1973).

[2]T. D. Eller and D. R. Knapp, unpublished work.

2.5 DERIVATIVES OF AMINO ALCOHOLS AND AMINO THIOLS

2.5.1. Acyl Derivatives

Derivative Type. Acetyl.

Reagent. Acetic anhydride (Ac_2O).

Compound Type. Phenylethylamines.

Reaction

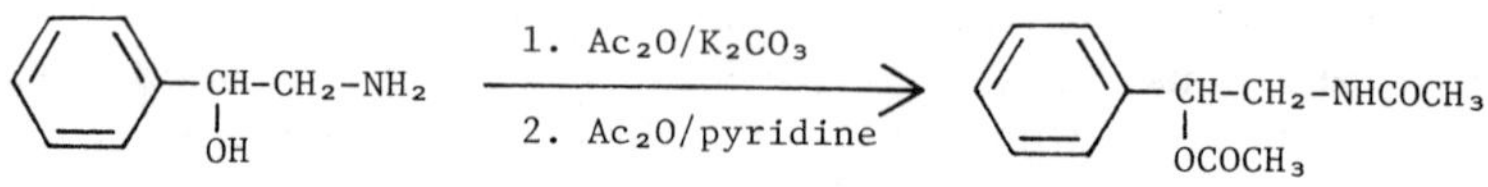

Typical Procedure. Prepare soln of 20–60 μg of amine in 100 ml H_2O or urine. Adjust to pH > 8 with solid K_2CO_3. Add portionwise 10, 5, and 5 ml of Ac_2O over a period of 2 hr while adding K_2CO_3 to maintain pH > 8. Allow to stand overnight; then ext with a total of 300 ml EtOAc. Wash ext with aq NaOAc, dry over Na_2SO_4, and evap to dryness. Treat residue with 20 μl pyridine and 30 μl Ac_2O. Evap to dryness under vac, diss residue in EtOAc, and analyze by GC.

Comments. Acetylation in aq medium was examined as a means of deriv as well as facilitating extraction. Reacetylation under anhyd conditions was performed to insure acetylation of benzylic hydroxyls. β-Hydroxyphenethylamine and tyramine showed 95% recovery from aq media while octopamine and normetanephrine recovery was only 30–40%.

References

[1]C. J. W. Brooks and E. C. Horning, *Anal. Chem.* **36**, 1540 (1964).

Derivative Type. Trifluoroacetyl (TFA).

Reagent. Trifluoroacetic anhydride (TFAA)/trimethylamine (TMA).

Compound Type. Beta blocker drugs.

Reaction

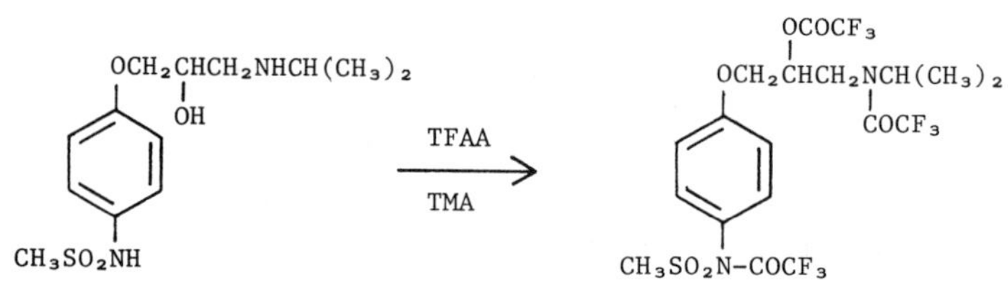

Typical Procedure. Diss sample in 50–200μl of benzene or use similar volume of conc ext. Add 75 μl 1 *M* TMA in benzene and 75 μl TFAA. Heat 5 min at 50°. Wash with 2 ml pH 6.0 (0.5 *M*) phosphate buffer and analyze benzene layer by ECD-GC.

Comments. Method was used for analysis of propranolol, oxprenolol, alprenolol, pronethalol, dichloroisoproterenol, practolol, and sotalol. Sotalol (shown above) and practolol (4-acetamido compd) formed tri-TFA derivs and the other compds di-TFA derivs. Determination methods in plasma, urine, and tissues were given as well as MS data for the derivs.

References

[1]T. Walle, *J. Pharm. Sci.* **63**, 1885 (1974).

Derivative Type. Trifluoroacetyl (TFA).

Reagent. Trifluoroacetic anhydride (TFAA)/trimethylamine (TMA).

Compound Type. Propranolol and metabolites.

Reaction

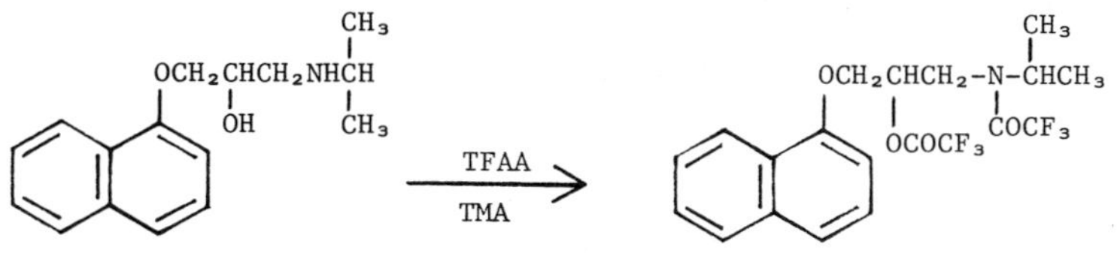

Typical Procedure. Homogenize 500 mg sample of brain tissue with 1.5 ml CH_3CN, 1.5 ml 0.8 *N* H_3PO_4, and int stand oxprenolol in an ice-cooled Potter-Elvehjem homogenizer. Centrifuge and transfer supernate to a conical centrifuge tube. Adjust to pH > 12 with 5 *N* NaOH and ext with 5 ml benzene. Ext benzene with 2 ml 1 *N* H_2SO_4 leaving neutral metabolite (propranolol glycol) in benzene. Adjust acidic soln to pH > 12 with 5 *N* NaOH and ext with 5 ml benzene to yield ext of basic compds. Evap each ext to 500 μl at 50–60° under N_2. Add 25 μl of 1 *M* TMA in benzene and 25 μl of TFAA to each sample and heat 5 min at 50–60°. Before analysis, wash each sample with 3 ml H_2O and analyze an aliquot of the org phase by GC.

Comments. The method was used for quantitation of propranolol and the metabolites 3-(α-naphthoxy)-1,2-propanediol and *N*-desisopropylpropranolol in brain tissue. Levels of 10–250 ng/g were measured by ECD-GC. MS of the deriv were given.

References

[1]D. A. Saelens *et al.*, *J. Chromatogr.* **123**, 185 (1976).

Derivative Type. Heptafluorobutyryl (HFB).

Reagent. Heptafluorobutyric anhydride (HFBA)/pyridine.

Compound Type. Pseudoephedrine and related ephedrines.

Reaction

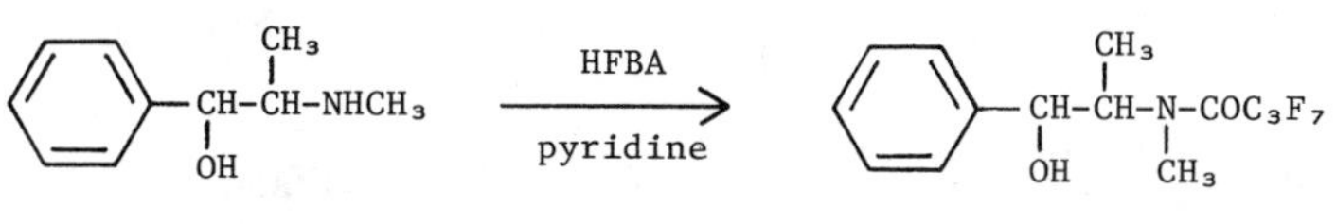

Typical Procedure. Mix 2 ml serum with 2 ml 0.5 *N* HCl and ext with 2 × 5 ml benzene. Centrifuge after each ext and discard 4 ml of the benzene phase. Adjust to pH 11 with 2 ml 0.1 *M* Na_3PO_4 buffer (pH 12.0). Ext with 5 ml benzene and remove 4 ml of benzene phase. To the benzene ext add 0.1 ml pyridine and 0.1 ml HFBA. Mix, stopper, and allow to stand at room temp at least 4 hr. Cool in an ice bath and ext with 3 × 10 ml cold 0.01 *N* HCl. Remove 3 ml of the benzene phase and analyze 2 μl aliquots by ECD-GC.

Comments. The method was used to measure therapeutic levels of pseudoephedrine in plasma. The ECD response was linear from 0–1000 ng for the derivs of pseudoephedrine, ephedrine, and norephedrine. Only *N*-acylated derivs were observed under the deriv conditions used.

References

[1]L. M. Cummins and M. J. Fourier, *Anal. Lett.* **2**, 403 (1969).

Derivative Type. Pentafluorobenzamide.

Reagent. Pentafluorobenzoyl chloride.

Compound Type. Amphetamine and ephedrines.

Reaction

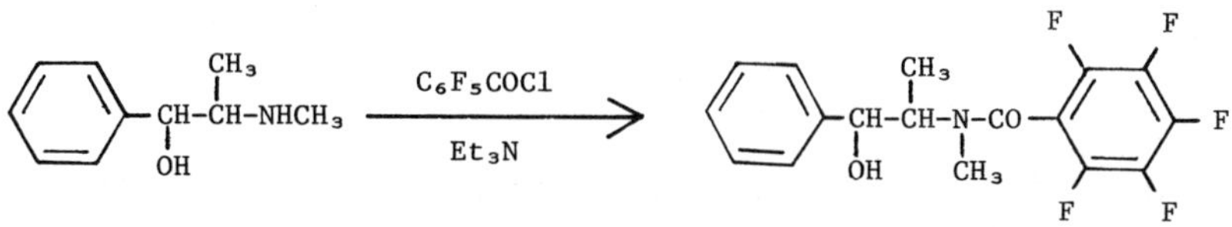

Typical Procedure. Diss 10^{-3} mol sample and 10^{-3} mol Et_3N in 20 ml EtOAc. Add 1.5×10^{-3} mol pentafluorobenzoyl chloride dropwise with shaking. Allow to stand 2 hr then evap to dryness under vac at 50°. Diss residue in 100 ml 95% EtOH. Analyze by ECD-GC.

Comments. The TFA, PFP, HFB, and pentadecafluorooctanamide derivs were also examined but the pentafluorobenzamide derivs had the highest ECD response. GC data were reported for β-phenylethylamine, amphetamine, and the four ephedrine isomers. MS data indicated that only amide derivs were obtained (i.e., the hydroxyl groups were not deriv).

References

[1]G. R. Wilkinson, *Anal. Lett.* **3**, 289 (1970).

2.5.2. Silyl and Mixed Silyl Derivatives

Derivative Type. Trimethylsilyl.

Reagent. Bis(trimethylsilyl)trifluoroacetamide (BSTFA).

Compound Type. Aminothiols.

Reaction

$$\text{HS-(CH}_2)_n\text{-NH}_2 \xrightarrow[\text{DMF}]{\text{BSTFA}} \text{TMS-S-(CH}_2)_n\text{-N(TMS)}_2$$

Typical Procedure. Diss 0.5–10 mg sample in 0.2 ml dry DMF and add 0.2 ml BSTFA. Allow to stand in a sealed vial overnight. Analyze by GC.

Comments. Derivs were prepared from a series of C_2—C_6 unsubstituted and *N*-alkyl aminothiols and aminodisulfides. Although no rigorous structural proof was given, the number and time course of partially silylated products indicated formation of di-TMS primary amines.

References

[1]R. T. Lofberg, *Anal. Lett.* **4**, 77 (1971).

Derivative Type. *O*-Trimethylsilyl (TMS), *N*-Trifluoroacetyl (TFA).

Reagent. Bis(trimethylsilyl)trifluoroacetamide (BSTFA), trifluoroacetic anhydride (TFAA).

Compound Type. Bufuralol and metabolites.

Reaction

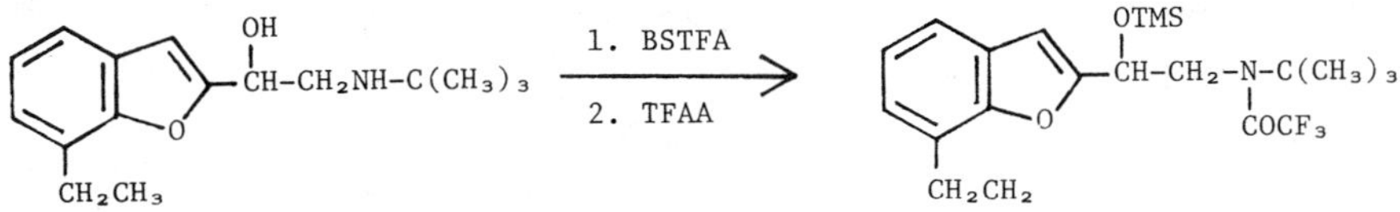

Typical Procedure. To 1 ml plasma in 11 ml glass centrifuge tube add 0.1 ml int stand soln (100 ng of stand), 2 ml satd Na_2HPO_4, and 5 ml EtOAc. Ext on rotator 20 min. Centrifuge and transfer org layer to another tube containing 1 ml 0.1 *M* H_2SO_4. Rotate 10 min and centrifuge. Discard org layer and add 0.2 ml 0.880 *M* NH_3 and 1 ml EtOAc to aq layer. Vortex 1 min, centrifuge, and transfer org layer to third tube containing *ca* 250 mg $MgSO_4 \cdot 3H_2O$. Vortex briefly, allow to stand 10 min, then centrifuge and transfer supernatant to a 1 ml conical vial. Evap to dryness under N_2 and add 30 μl cyclohexane and 5 μl BSTFA. Cap vial and heat at 60° for 10 min. Cool and add 5 μl TFAA. Cap vial and reheat at 60° for 5 min. Evap to dryness under N_2 and diss in 10 μl cyclohexane. Analyze 5 μl aliquots by GC.

Comments. The method was used to quantitate bufuralol (1 ng/ml) and metabolites (250 pg/ml) in plasma by GC-MS. The method was also used for GC-ECD quantitation to about 10 ng/ml by substituting PFPA for TFAA.

References

[1]R. J. Francis *et al.*, *Biomed. Mass Spectrom.* **3**, 281 (1976).

2.5.3. Dinitrophenyl Derivatives

Derivative Type. 2,4-Dinitrophenyl (DNP).

Reagent. 2,4-Dinitrobenzenesulfonic acid (DNBS).

Compound Type. Amino alcohols and amines.

Reaction

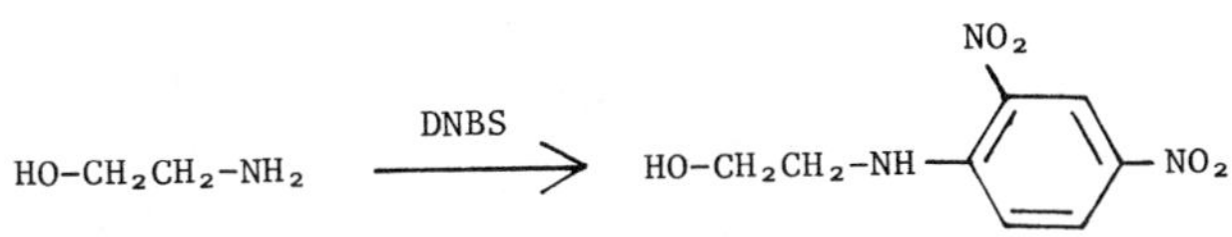

Typical Procedure. Prepare reagent soln by making a 0.25 *M* soln of the Na salt of DNBS in satd aq $Na_2B_4O_7$. Decant the soln from the ppt formed and wash the soln 4 × with $CHCl_3$. Mix 1 ml of amine soln with 1 ml of reagent soln and heat at 100° for 40 min. Ext with 4 ml benzene. Evap benzene to yield deriv.

Comments. The method was used for TLC and column chromatography of amines and amino alcohols in urine. The method has subsequently been used for ECD-GC analysis of phenylethylamines after silylation of hydroxyl groups.[2] DNT derivs were similarly prepared.[2] The reagent should be useful for prep of chromogenic derivs for HPLC.

References

[1]A. D. Smith and J. B. Jepson, *Anal. Biochem.* **18**, 36 (1967).

[2]D. J. Edwards and K. Blau, *Anal. Biochem.* **45**, 387 (1972).

Derivative Type. 2,4-Dinitrophenyl (DNP).

Reagent. 2,4-Dinitrobenzenesulfonic acid (DNBS).

Compound Type. Amino alcohols.

Reaction

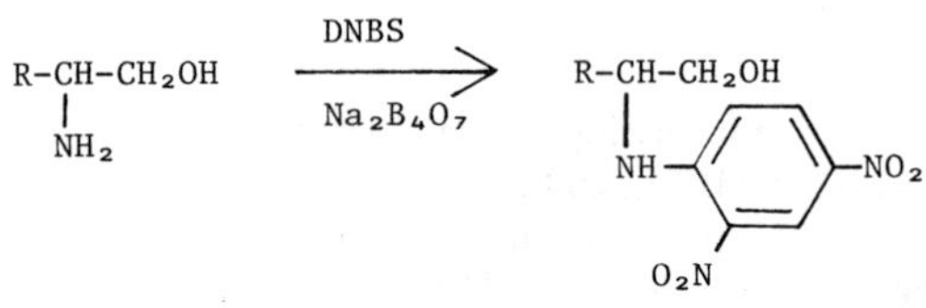

Typical Procedure. Diss 20 mg protein hydrolyzate and 1.5 mg amino alcohols in 1 ml H_2O and adjust pH to 9.5–10.0 with 1 *N* NaOH. Add 1 g $Na_2B_4O_7$ and 200 mg DNBS. Heat in a sealed tube at 95° for 20 hr. Dil with 10 ml H_2O and ext with Et_2O. Purify the derivs by chromatography on Hyflo Super Cel.

Comments. The derivs were used for identification of amino alcohols obtained from hydrolysis of C-terminal reduced peptides. The reagent formed only derivs of the amino group and did not react with hydroxyls.

References

[1]J. C. Crawhall and D. F. Elliot, *Biochem. J.* **61**, 264 (1955).

2.5.4. Derivatives Formed with Aldehydes and Ketones

Derivative Type. Pentafluorophenyl oxazolidine.

Reagent. Pentafluorobenzaldehyde (PFBA).

Compound Type. Pseudoephedrine.

Reaction

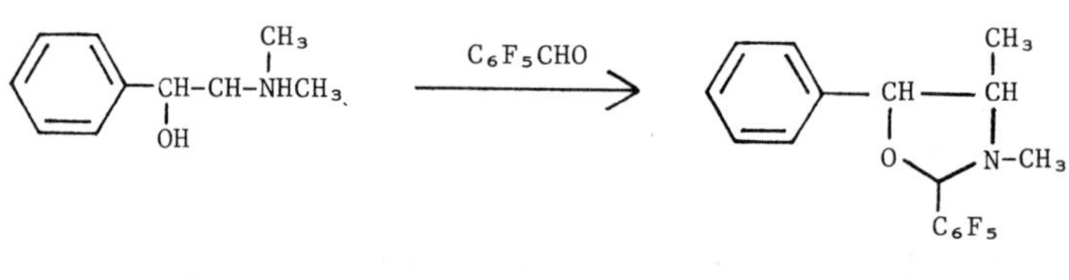

Typical Procedure. Diss 1 mg of amine in a soln of 20 mg PFBA in 0.1 ml CH_3CN and heat at 60° for 1 hr. Dil to 1 ml with hexane and analyze 1 μl by FID-GC or further dil and analyze by ECD-GC.

Comments. The deriv reaction was observed in a study of ECD sensitivity of various pentafluorophenyl derivs. The pentafluorophenyl oxazolidine had only 3% of the ECD sensitivity of the PFBA Schiff base of phenethylamine. *N*-Methylphenethylamine reacted with PFBA by N displacement of F to give *o*- and *p*-isomers of a tertiary amine deriv.

References

[1]A. C. Moffat *et al., J. Chromatogr.* **66**, 255 (1972).

Derivative Type. Schiff base or oxazolidine.

Reagent. Acetone.

Compound Type. Phenylalkylamines or indolealkylamines.

Reaction

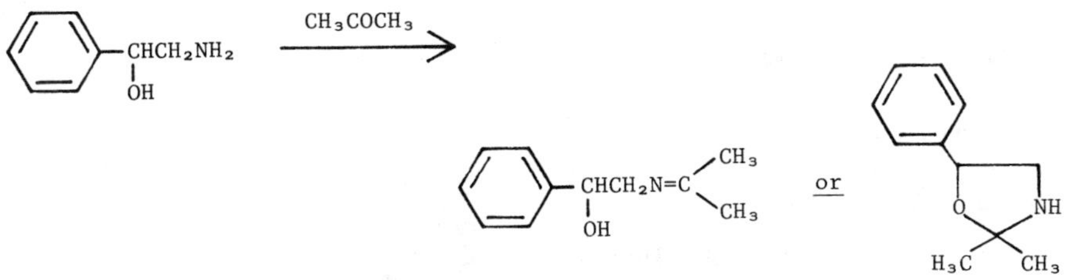

Typical Procedure. Diss free amine in dry acetone. Analyze soln by GC.

Comments. The β-hydroxy amines presumably formed oxazolidines rather than Schiff bases; the deriv structures were not proven, however. The derivs of the catecholamines showed substantial decomp upon GC analysis. Retention data were reported for seven phenylalkylamines and two indolealkylamines.

References

[1]C. J. W. Brooks and E. C. Horning, *Anal. Chem.* **36**, 1540 (1964).

Derivative Type. Neopentylidene.

Reagent. Pivaldehyde (2,2-dimethylpropanal).

Compound Type. Aminothiols and aminodisulfides.

Reaction

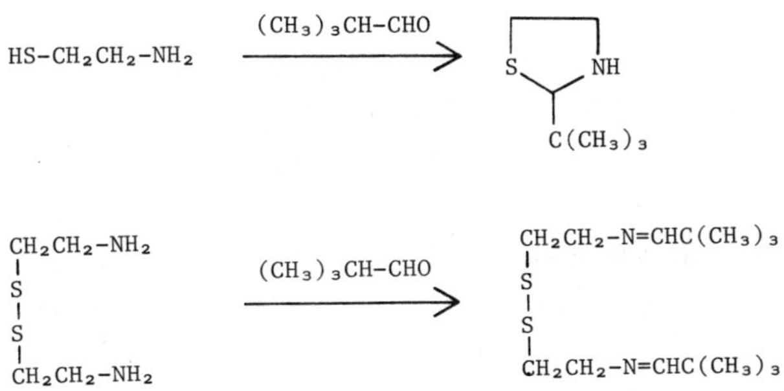

Typical Procedure. Ext 100 μl biological fluid (or tissue sample) with trichloroacetic acid or aq EtOH and evap ext to dryness. Add 1 ml 1.25 *N* HCl in MeOH and allow to stand 30 min. Evap under vac and diss residue in 0.1 ml MeOH. Add 0.1 ml pivaldehyde and AG1-X8 ion exchange beads (HCO_3 form) until the soln is pH 7–8. Add 3A mol sieve to absorb H_2O. After 10 min, analyze 1 μl aliquots by GC.

Comments. Sulfur containing amino acids [Cys, homo-Cys, *S*-methyl-Cys, Met, $(Cys)_2$] were converted first to the methyl esters with the MeOH/HCl. This step was omitted for cysteamine and cystamine determinations. The aldehyde condensation reaction required basic conditions but the use of pyridine gave a slow reaction and Et_3N caused decomposition in the GC. The anion exchange resin was found to give the best results. MS data for the derivs were reported.

References

[1]E. Jellum *et al.*, *Anal. Biochem.* **31**, 339 (1969).

2.6. PHENOLIC AMINES

2.6.1. Acyl Derivatives

Derivative Type. Acetyl.

Reagent. Acetic anhydride (Ac_2O)/pyridine.

Compound Type. Phenylalkylamines, imidazolealkylamines, and indolealkylamines.

Reaction

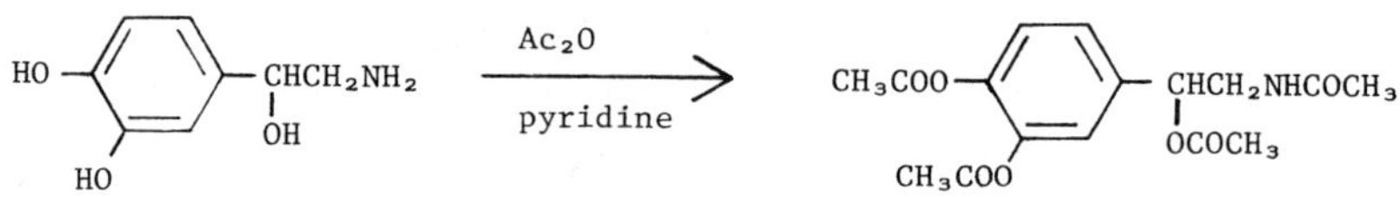

Typical Procedure. Diss 10 mg of amine in 0.3 ml CH_3CN. Add 0.1 ml Ac_2O and 0.1 ml pyridine. Heat to reflux. Anaylze by GC.

Comments. GC retention data were reported for the derivs of 22 phenylalkylamines, 10 indolealkylamines, and 2 imidazolealkylamines.[1] Propionyl and pentafluoropropionyl (PFP) derivs of the catecholamines were also prep by the same procedure.[2] The PFP derivs gave erratic results when the reaction mixt was analyzed directly which required isolation of the derivs by addition of $NaHCO_3$ soln and ext with EtOAc.

References

[1]C. J. W. Brooks and E. C. Horning, *Anal. Chem.* **36**, 1540 (1964).
[2]E. C. Horning *et al., Anal. Chem.* **36**, 1546 (1964).

Derivative Type. Trifluoroacetyl.

Reagent. Trifluoroacetic anhydride (TFAA).

Compound Type. Phenolic amines.

Reaction

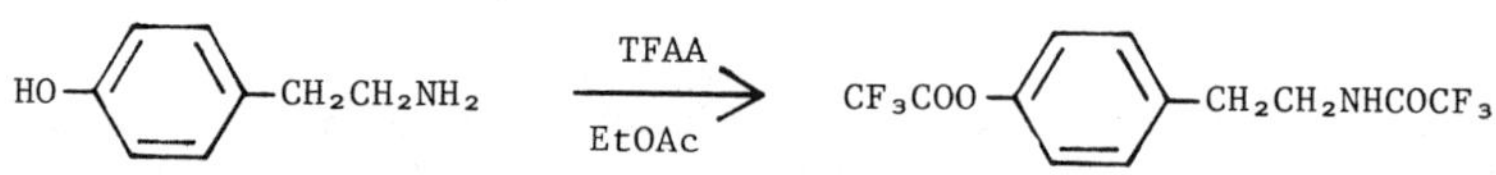

Typical Procedure. Diss 1 mg sample in 1 ml EtOAc and add 0.5 ml TFAA. Heat at 50° for 1 hr then evap to dryness under vac. Diss residue in 1 ml EtOAc and dil 0.1 ml to 10 ml for ECD-GC analysis.

Comments. The TFA derivs of various amines were compared to the PFP, HFB, and chloroacetyl derivs in terms of ECD sensitivity. For nonhydroylated amines the order of sensitivity was HFB > PFP > chloroacetyl > TFA. The *N*-TFA group was relatively insensitive to ECD. The *N,O*-di-TFA tyramine (shown above) was 10,000 times more sensitive to ECD than the *N*-mono-TFA deriv.

References

[1]D. D. Clarke *et al., J. Gas. Chromatogr.* **4**, 310 (1966).

Derivative Type. Trifluoroacetyl.

Reagent. Trifluoroacetic anhydride (TFAA).

Compound Type. Phenolic amines.

Reaction

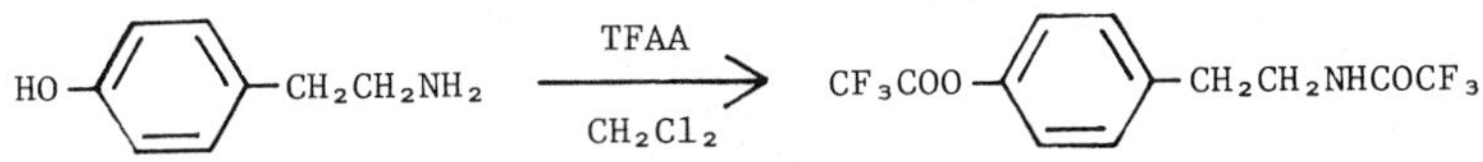

Typical Procedure. Pass 10 ml of the aq sample soln through a 1.0 × 2.5 cm column of Dowex 50-X2 (H form). Wash column with 20 ml H_2O, 20 ml 0.1 *N* NaOAc, and 20 ml H_2O. Elute with 15 ml 1.0 *N* NH_4OH in 65 % EtOH. Evap eluate to dryness under vac at 60°. To the residue add 1 ml CH_2Cl_2 and 1 ml TFAA and allow to stand at room temp 1 hr. Evap to dryness under vac, diss residue in 1 ml CH_2Cl_2 and analyze an aliquot by FID-GC.

Comments. The method was used to analyze for tyramine; octopamine, synephrine, and their β-methoxy derivs; and hordenine in fruit juices.[1] Sen used the TFA derivs to analyze tyramine, octopamine, and synephrine by ECD-GC.[2]

References

[1]D. E. Coffin, *J. Assoc. Off. Anal. Chem.* **52**, 1044 (1969).
[2]N. P. Sen, *J. Food Sci.* **34**, 22 (1969).

Derivative Type. Trifluoroacetyl.

Reagent. Trifluoroacetic anhydride (TFAA).

Compound Type. Phenolic amines.

Reaction

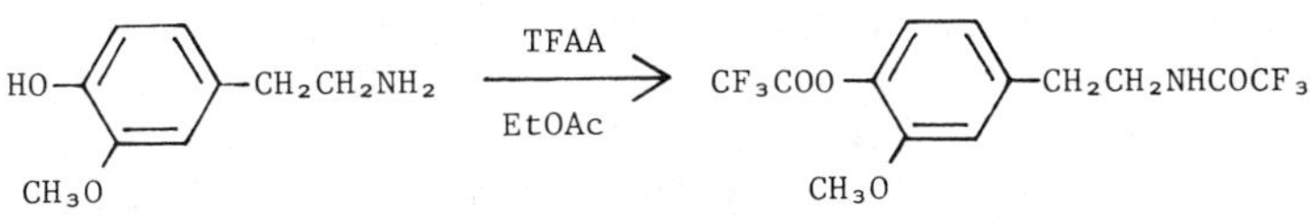

Typical Procedure. Hydrolyze 25 ml urine at pH < 1 for 30 min then adjust pH to 4–6. Absorb amines on a 1 × 5 cm column of Dowex 50W-X2 (H form). Wash with 20 ml 0.1 *N* NaOAc and 10 ml H_2O then elute with 8 ml 5 *N* NH_4OH. Evap under vac. Add 1.5 ml EtOAc and 0.75 ml TFAA and allow to stand at room temp 1 hr. Evap under N_2 and diss residue in 1 ml dry EtOAc. Analyze by ECD-GC.

Comments. The derivs of metanephrine and normetanephrine were shown to be separable from the derivs of tyramine, 3-methoxytyramine, octopamine, tryptamine, and serotonin.

References

[1]L. M. Bertani *et al., Clin. Chim. Acta.* **30**, 227 (1970).

Derivative Type. Trifluoroacetyl.

Reagent. Trifluoroacetic anhydride (TFAA).

Compound Type. Catecholamines.

Reaction

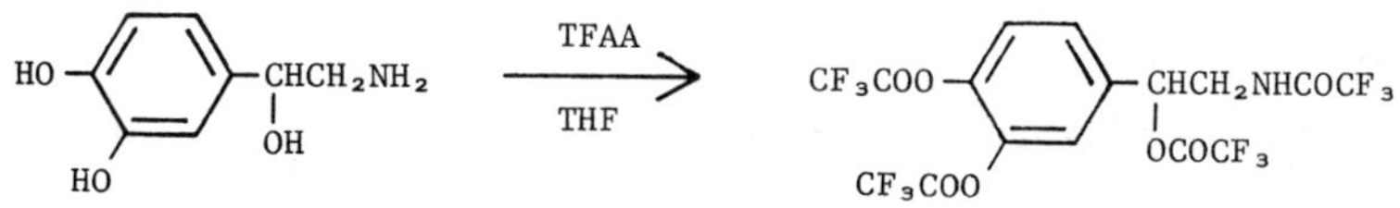

Typical Procedure.[1] Evap 1 ml of a soln of catecholamines (1 μg each) in 0.1 *N* HOAc to dryness at 50° under vac. Add 50 μl THF and 1 drop of TFAA. Allow to react 10 min at room temp. Dil with hexane and analyze by ECD-GC.

Comments. The deriv method has been applied to the analysis of catecholamines in urine,[2] tissues,[2,3] and serum.[3] Imai *et al.*,[3] used EtOAc in place of THF.

References

[1]S. Kawai and Z. Tamura, *Chem. Pharm. Bull.* **16**, 699 (1968).

[2]S. Kawai and Z. Tamura, *Chem. Pharm. Bull.* **16**, 1091 (1968).

[3]K. Imai *et al., Chem. Pharm. Bull.* **19**, 409 (1971).

Derivative Type. Trifluoroacetyl.

Reagent. Trifluoroacetic anhydride (TFAA).

Compound Type. Dopamine.

Reaction

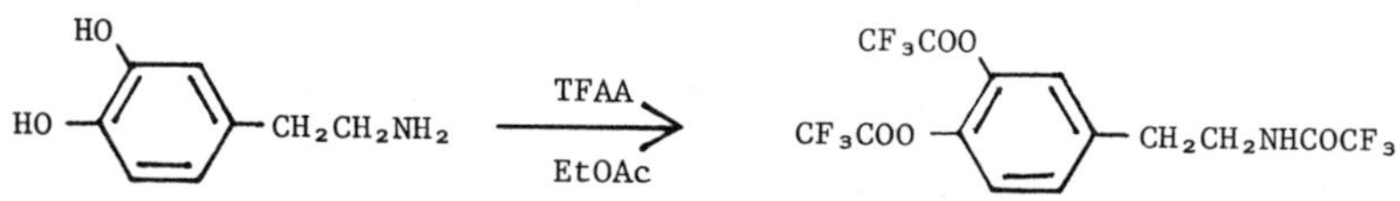

Typical Procedure. Mix 25 ml urine with 1.5 ml 70% $HClO_4$, centrifuge, and decant supernatant into a beaker containing 0.5 g alumina. Adsorb dopamine on the alumina by adjusting pH to 8–8.5 with NaOH. Stir additional 3 min, then allow to settle and decant supernatant. Transfer alumina to a 15 ml centrifuge tube and wash with 3 × 13 ml H_2O. Elute dopamine by shaking alumina 15 min with 3 ml 0.25 *N* HOAc. Centrifuge and remove 0.5 ml of the eluate. Evap to dryness under vac. To the residue add 1.5 ml EtOAc and 0.25 ml TFAA. Allow to stand 1 hr at room temp. Evap to dryness under N_2. Diss residue in 1 ml EtOAc and analyze by ECD-GC.

Comments. The derivs permitted GC separation of dopamine, epinephrine, and norepinephrine. The method was used for quantitation of dopamine in urine.

References

[1]D. D. Clarke *et al., J. Gas Chromatogr.* **5**, 307 (1967).

Derivative Type. Trifluoroacetyl.

Reagent. Trifluoroacetic anhydride (TFAA)/trimethylamine (TMA).

Compound Type. Propranolol and 4-hydroxypropranolol.

Reaction

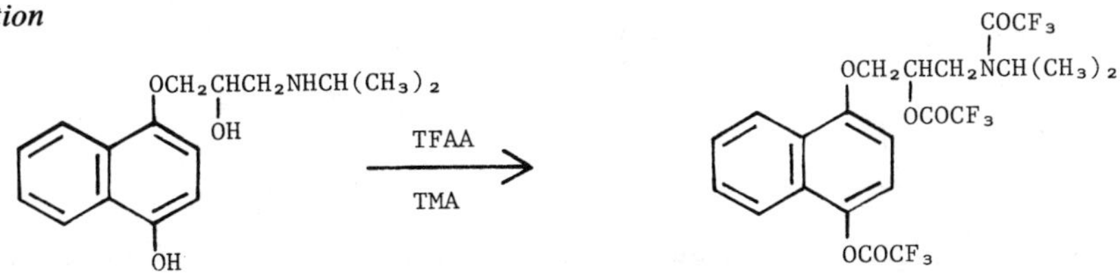

Typical Procedure. To 1 ml plasma add 20 mg $NaHSO_3$, 25 ng oxprenolol HCl, 1 ml 1 *M* pH 9.6 carbonate buffer, and 10 ml EtOAc. Shake, separate EtOAc, and evap to dryness at 60° under N_2. Add 200 μl benzene, 25 μl 1 *M* TMA in benzene, and 50 μl TFAA. After 15 min at 50° wash with 1 *M* pH 6.0 phosphate buffer and analyze 2–4 μl aliquots by GC-MS.

Comments. The method was used for quantitation in plasma of propranolol to 1 ng/ml and 4-hydroxypropranolol to 5 ng/ml using GC-MS-SIM with oxprenolol as int stand. $NaHSO_3$ was used as antioxidant to prevent loss of the 4-hydroxypropranolol.

References

[1]T. Walle *et al.*, *J. Chromatogr.* **114**, 351 (1975).
[2]T. Walle, *J. Pharm. Sci.* **63**, 1885 (1974).

Derivative Type. Trifluoroacetyl.

Reagent. Trifluoroacetic anhydride (TFAA).

Compound Type. *p*-Hydroxyamphetamine and *p*-hydroxynorephedrine.

Reaction

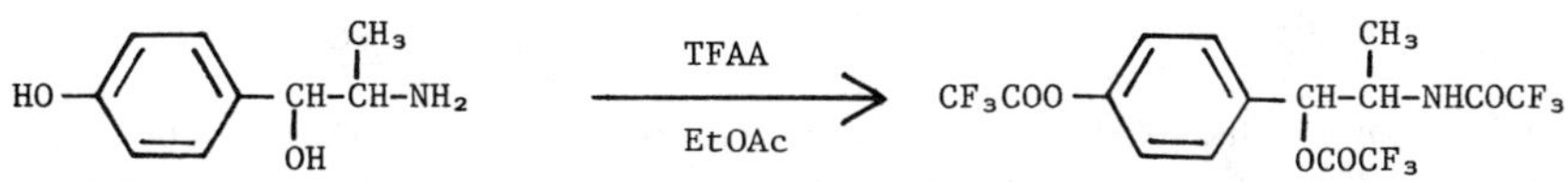

Typical Procedure. Homogenize rat brain with 6 ml/g of a cold mixt of acetone/1 *N* HCOOH (85 : 15, v/v). Centrifuge and ext supernatant twice with heptane/$CHCl_3$ (4 : 1, v/v) Discard org ext and dil aq phase to 1.5 ml. Acidify to pH < 2 with 85% HCOOH and wash with 3 ml benzene. Adjust to pH 10 with 5 *N* NaOH and wash with 3 ml benzene. Add 0.5 g NaCl and ext with 2 ml EtOAc.

Evap ext to dryness under N_2 and treat residue with 50 μl EtOAc and 30 μl TFAA. Heat at 65° for 10 min then evap to dryness under N_2. Diss residue in 300μl EtOAc containing 1 ng/μl *p*-nitroanisole int stand and analyze by ECD-GC.

Comments. MS data for the derivs were reported.

References

[1]G. Belvedere *et al.*, *J. Chromatogr.* **84**, 355 (1973).

Derivative Type. Trifluoroacetyl.

Reagent. Trifluoroacetic anhydride (TFAA).

Compound Type. Tetrahydropapaveroline.

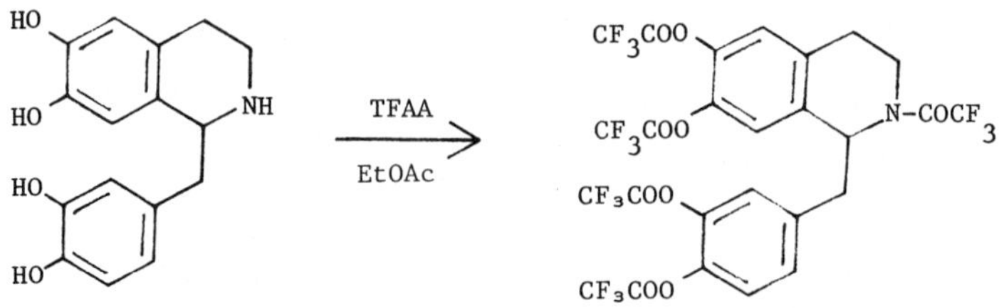

Typical Procedure. Homogenize tissue sample in 4 volumes of 0.4 *M* $HClO_4$ containing 0.1% (w/v) of ascorbate and EDTA. Centrifuge and rehomogenize ppt as above. Combine supernatants and ppt $KClO_4$ by careful addition of 1 *M* KOH. Adsorb catechols on alumina at pH 8.3. Wash alumina with H_2O and elute with 2 × 2 ml 1 *M* HOAc. Lyophilize eluate and treat powder with TFAA/EtOAc (1 : 3) at 60° for 3 hr. Analyze directly by GC-MS.

Comments. The method was used to determine tetrahydropapaveroline in rat brain by GC-MS-SIM. No int stand was used. The deriv was detectable to *ca* 50 pg. Solvent extraction could not be used because the free base was very unstable.

References

[1]A. J. Turner *et al.*, *Mass Spectrometry in Biochemistry and Medicine*, Raven Press, New York, NY, 1974, p. 99.

Derivative Type. Pentafluoropropionyl (PFP).

Reagent. Pentafluoropropionic anhydride (PFPA).

Compound Type. Biogenic amines.

Reaction

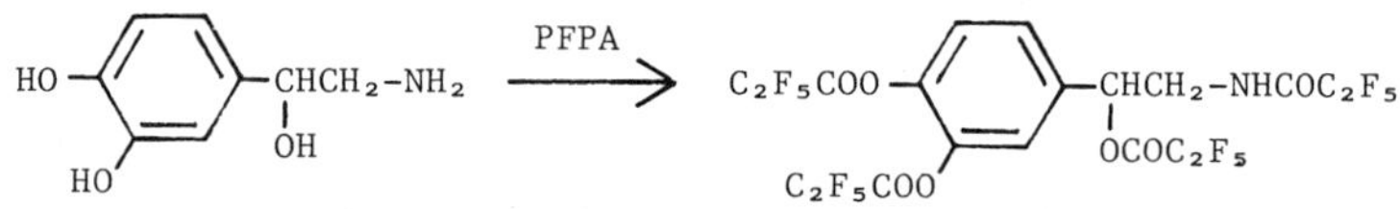

Typical Procedure. Diss 0.1–1 μg amine salt in 40 μl PFPA. Add 20 μl redist EtOAc previously dried and stored over $CaCl_2$. Seal and heat at 60° for 10 min. Cool and evap to dryness under N_2. Redissolve in EtOAc for analysis by ECD-GC or GC-MS.

Comments. The PFP derivs of 13 amines were examined and their GC properties compared to the HFB and TMS, HFB derivs.[1] The highest GC resolution was obtained for the PFP derivs. MS data for the derivs were given. Gelpi *et al.*[2] reported GC behaviour and optimum deriv conditions for the preparation of PFP derivs of 28 catecholamines and related compds. MS data were also reported. The PFP derivs of metanephrine, normetanephrine, and 3-methoxytyramine were compared with TFA and HFB derivs.[3] The PFP and HFB derivs were more stable and had better ECD response than the TFA derivs. The deriv stability was better in EtOAc than in hexane. MS data for the fluoroacyl derivs, as well as TMS and acetyl derivs, were reported.[3] The PFP derivs have been used to quantitate catecholamines in 0.1 mg tissue samples by GC-MS-SIM.[4]

References

[1]F. Karoum *et al.*, *Anal. Biochem.* **47**, 550 (1972).

[2]E. Gelpi, *et al.*, *J. Chromatogr. Sci.* **12**, 701 (1974).

[3]E. Anggard and G. Sedall, *Anal Chem.* **41**, 1250 (1969)

[4]S. H. Koslow *et al.*, *Science* **176**, 177 (1972).

Derivative Type. Pentafluoropropionate.

Reagent. Pentafluoropropionic anhydride (PFPA).

Compound Type. Dopamine.

Reaction

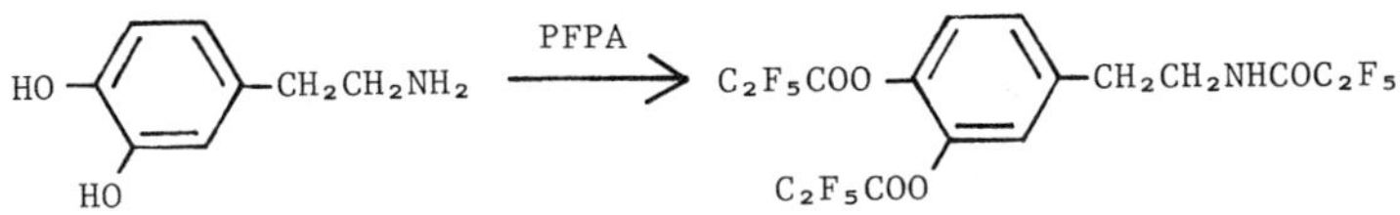

Typical Procedure. Homogenize tissue sample with 150 μl of a soln of 0.1 *N* HCOOH and 0.05 *M* ascorbic acid. Add 500 μl acetone containing 1 % pentafluoropropionic acid and 1 % deionized H_2O then dry under N_2 to a gel. Add 100 μl fresh PFPA reagent. (20 % PFPA, 10 % EtOAc, 70 % CH_3CN, v/v/v). Prior to analysis, remove the excess reagent under N_2. Diss the residue in 200 μl of benzene and analyze by GC-MS.

Comments The method was used with a deuterated int stand for GC-MS quantitation of dopamine in mouse brain. PFP derivs were found to be more stable than the TFA or HFB derivs.

References

[1]H. Ko *et al., Anal. Lett.* 7, 243 (1974).

Derivative Type Pentafluoropropionyl.

Reagent. Pentafluoropropionyl-1,2,4-triazole (PFPT).

Compound Type. Biogenic amines and metabolites.

Reaction

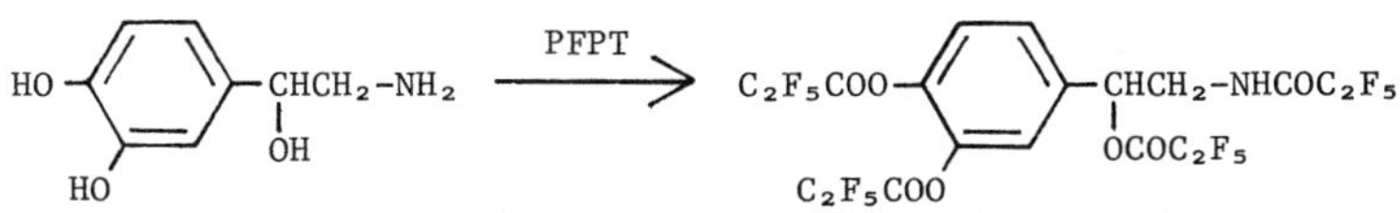

Typical Procedure. Diss 1–2 μg sample of the amine in 0.1 ml PFPT and allow to stand at room temp 30 min. Analyze by GC-MS.

Comments. The derivs of 10 biogenic amines and four acid metabolites were prepared and analyzed by CI-MS. It was found that CI gave 10 × higher sensitivity than EI permitting detection of the norepinephrine deriv to 1 pg. The acid metabolites were also examined by first converting them to the methyl esters with 1 *N* PFPA in MeOH (60°, 1 hr). After removal of the excess reagents under vac, the esters were treated with PFPT as for the amines.

References

[1] H. Miyazaki *et al., J. Chromatogr.* **99**, 575 (1974).

Derivative Type Trimethylacetyl (Pivalyl).

Reagent. Trimethylacetic anhydride/triethylamine (Et_3N).

Compound Type. Pirbuterol.

Reaction

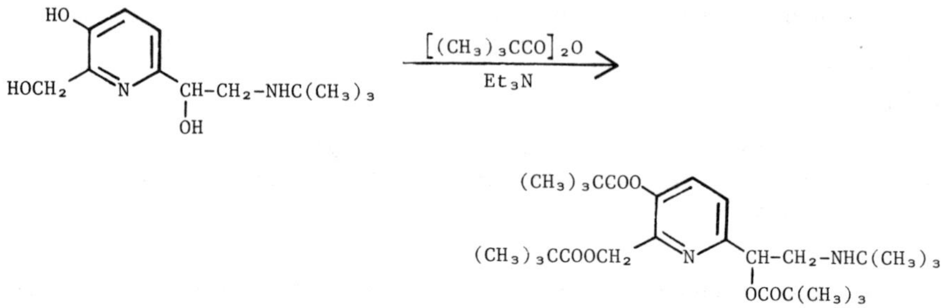

Typical Procedure. Ext drug from plasma with acetone/Na_2CO_3. Evap acetone ext to dryness under N_2. Add 0.5 ml of a soln of trimethylacetic anhydride, Et_3N, DMF, and EtOAc (1 : 1 : 1 : 7, v/v/v/v). Stopper tube and heat at 100° in a sand bath for 30 min. Evap soln to *ca* 50μl under N_2 and analyze by GC-MS.

Comments. d_9-Pirbuterol was used as int stand for GC-MS quantitation. The method was useable at levels of 1–20 ng/ml in plasma.

References

[1]F. C. Falkner and H. M. McIlhenny, *Biomed. Mass Spectrom.* **3**, 207 (1976).

2.6.2. Acyl Methyl Ether Derivatives

Derivative Type. *N*-Trifluoroacetyl, *β*-*O*-Methyl.

Reagent. Trifluoroacetic anhydride (TFAA), methanol (MeOH).

Compound Type. Metanephrine (MN) and normetanephrine (NMN).

Reaction

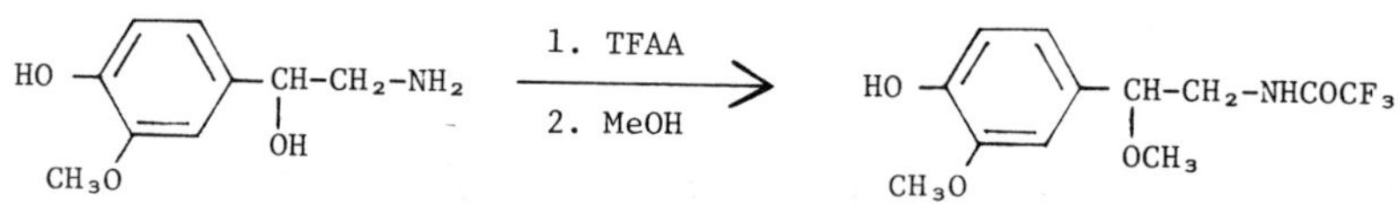

Typical Procedure. To a dried sample add 1.0 ml TFAA and heat 10 min at 75°. Evap to dryness and diss residue in MeOH. Allow to stand 30 min then analyze by GC.

Comments. The method was used to measure MN, NMN, and 3-methoxytyramine in urine. The initially formed *N,O*-TFA derivs apparently reacted with MeOH to give the free phenolic hydroxyl and methylation of the *β*-hydroxyl. MS data were reported for the derivs.

References

[1]M. Greer *et al., Clin Chim. Acta* **21**, 247 (1968).

Derivative Type. Methyl ester/ether, pentafluoropropionyl.

Reagent. Methanol (MeOH)/HCl, pentafluoropropionic anhydride (PFPA).

Compound Type. Catecholamines and related compounds.

Reaction

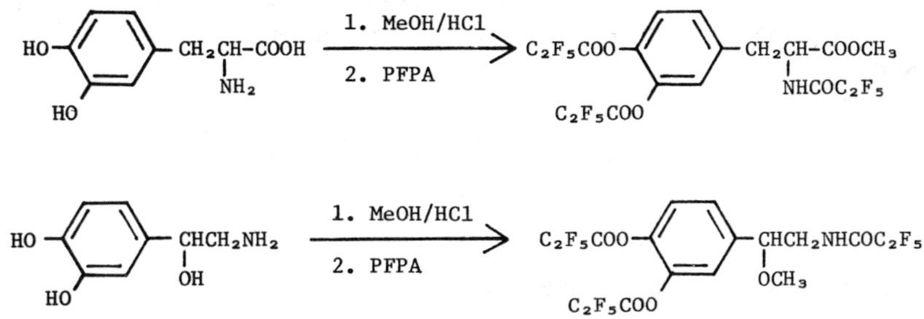

Typical Procedure. Homogenize 500 mg tissue in 2 ml 0.4 *N* $HClO_4$ and wash homogenizer with 0.4 *N* $HClO_4$ to yield a total of 4.5 ml. Add 0.5 ml 50 mg/ml $NaHSO_3$, centrifuge, and add the supernatant to 200 mg of activated alumina in a 10 ml freeze drying vial along with 1 ml of 100 mg/ml Na_2EDTA. Quickly adjust pH of mixture to 8.6 with 0.5 *N* NaOH and stir magnetically for 5 min. Allow alumina to settle by gravity, remove supernatant with pipette, and wash alumina with 4 × 3 ml double dist H_2O. Add 0.2 ml 0.1 *N* HCl, freeze in dry acetone bath, and lyophilize. Add 3 ml 2 *N* anhyd HCl in MeOH, stir 30 min, and remove 2 ml of soln. Dry under vac and add 1 ml of 20 % (v/v) PFPA in CH_3CN. Allow to stand at room temp for 10 min then dry under N_2. Diss residue in benzene for GC.

Comments. TFA derivs were formed by the same method with TFAA but were less stable. Compds examined included DOPA, dopamine, DOPAC (3, 4-dihydroxyphenylacetic acid), epinephrine, and norepinephrine. The methyl ether, PFP deriv of norepinephrine showed a gain in ECD sensitivity over the fully acylated deriv but the methyl ether/PFP deriv of epinephrine showed only a slight increase. The method has been adapted to measure dopamine, norepinephrine, and their α-methyl analogs simultaneously by CI-MS using deuterated int stands.[2]

References

[1]E. L. Arnold and R. Ford, *Anal. Chem.* **45**, 85 (1973).

[2]C.R. Freed *et al., Anal Biochem.* **78**, 319(1977).

2.6.3. Silyl Derivatives

Derivative Type Trimethylsilyl.

Reagent. Hexamethyldisilazane (HMDS)/pyridine.

Compound Type. Phenolic and catecholic amines.

Reaction

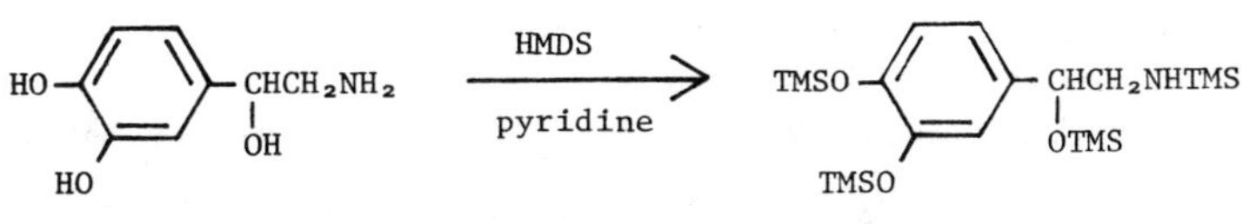

Typical Procedure. To 10 mg of amine or HCl salt add 0.5 ml pyridine and 1.5 ml HMDS. Shake 20 hr at room temp then heat 30 min at 95°. Analyze an aliquot of the soln by GC.

Comments. GC data were given for TMS derivs of five catecholamines and methylated metabolites relative to the deriv of 3-methoxytyramine. Although no structural data was reported, *N*-trimethylsilylation was inferred from the fact that phenethylamine yielded a TMS deriv.

References

[1] N. P. Sen and P. L. McGeer, *Biochem. Biophys. Res. Commun.* **13**, 390 (1963).

Derivative Type. Trimethylsilyl.

Reagent. Bis(trimethylsilyl)acetamide (BSA).

Compound Type. Biogenic amines and narcotic alkaloids.

Reaction

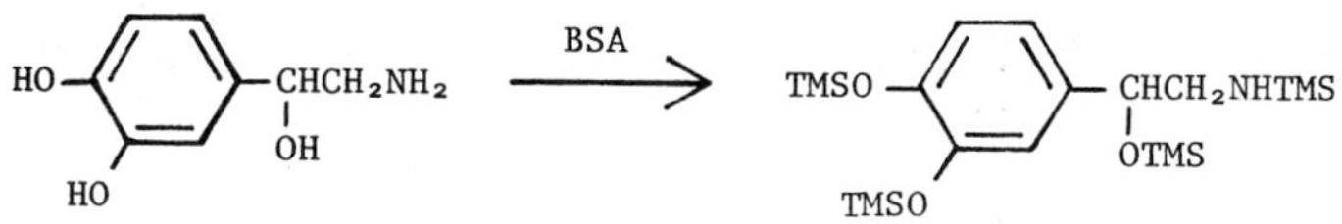

Typical Procedure. Dry a 1–30 μg sample in a vial in a vac dessicator at 50° for 1–2 hr. Seal vial with a septum cap and inject 100 μl BSA. Heat 3 hr at 76°. Analyze by GC.

Comments. The method was applied to biogenic amines, tetrahydroisoquinoline alkaloids, and phenanthrene (opium) alkaloids. The *N*-TMS derivs were susceptible to hydrolysis by traces of moisture to give multiple peaks.

References

[1] J. L. Cashaw *et al., J. Chromatogr. Sci.* **9**, 98 (1971).

Derivative Type. Trimethylsilyl.

Reagent. Bis(trimethylsilyl)trifluoroacetamide (BSTFA).

Compound Type. Terbutaline.

Reaction

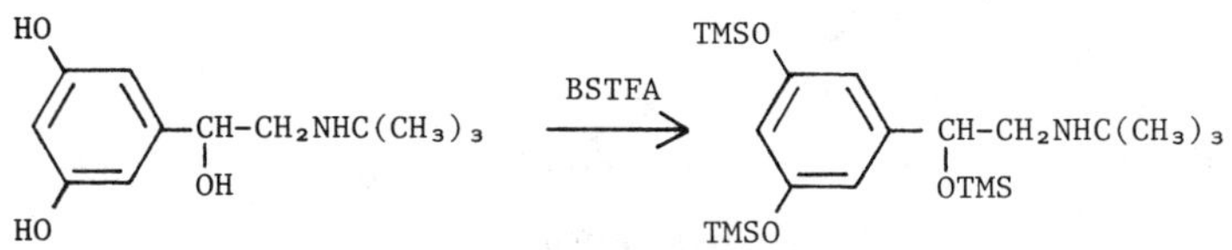

Typical Procedure. To 1 ml serum add 20 ng of d_6-terbutaline int stand (as aq soln). Allow to equilibrate 15 min, then add 8 ml of 0.015% (w/v) soln of bis(2-ethylhexyl) hydrogen phosphate in EtOAc. Mix on a Vortex mixer 5 min then centrifuge and remove org phase. Evap to dryness under N_2 at 55°.

Add 20 μl BSTFA and heat at 80° for 15 min. Analyze 1 μl aliquots by GC-CI-MS-SIM.

Comments. The hydrophilicity of terbutaline and its instability at high pH necessitated the use of ion pair ext. The method was used to quant terbutaline in serum to 1 ng/ml. For quant in urine, 0.1 ml of urine was dil with 1 ml pH 7.3 phosphate buffer, 100 ng of int stand was added, and the sample extd as above.

References

[1]J. G. Leferink *et al., J. Chromatogr.* **143**, 299 (1977).
[2]J. G. Leferink *et al., Biomed. Mass Spectrom.* **3**, 255 (1976).

Derivative Type. *O*-Trimethylsilyl.

Reagent. Trimethylsilylimidazole (TMSI).

Compound Type. Catecholamines.

Reaction

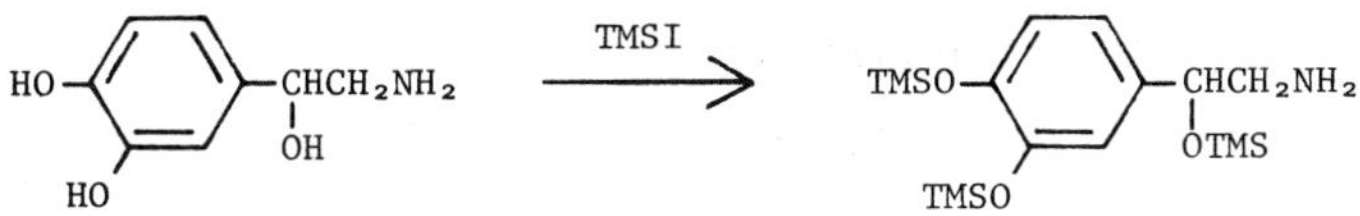

Typical Procedure. Diss sample in a soln of TMSI in CH_3CN (or pyridine) and heat 2–3 hr at 60°. Analyze by GC.

Comments. The TMSI formed only *O*-TMS derivs. BSA/TMCS was used to form *N,N*-di-TMS derivs of the primary amines. Norepinephrine was diss in 100 μl CH_3CN or pyridine and 200 μl BSA and 100 μl TMCS added. At 60°, the reaction was complete in 5 hr in CH_3CN and 50 hr in pyridine. With 2 μl H_2O added as catalyst, reaction times were shortened to 2 and 5 hr, respectively. Secondary (*N*-methyl) amines did not form *N*-TMS derivs under these conditions.

References

[1]M. Horning *et al., Biochim. Biophys. Acta* **148**, 597 (1967).

Derivative Type. *t*-Butyldimethylsilyl (TBDMS).

Reagent. *t*-Butyldimethylchlorosilane (TBDMS-Cl)/imidazole.

Compound Type. Salbutamol.

Reaction

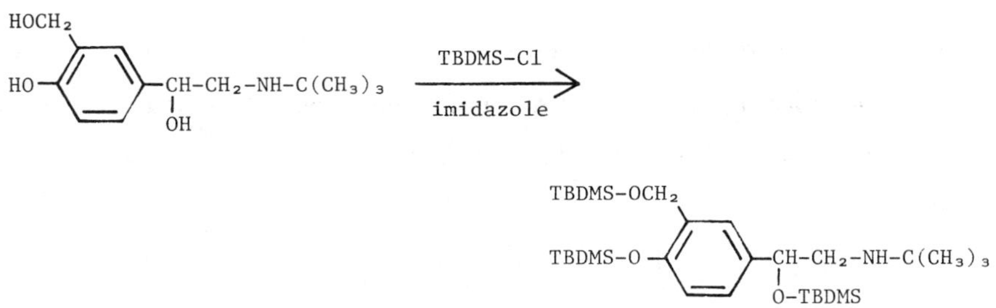

Typical Procedure. Add deuterated int stand to 2 ml plasma and ext with 3 ml H_2O satd Et_2O for 10 min on rotary mixer. Centrifuge and discard org phase. Dil aq phase with 1 ml buffer (pH 8.5, 0.5 *M*, $NaHCO_3$-$NaCO_3$) and ext 10 min with 2 × 2 ml heptan-3-one containing 10 mg/ml sodium tetraphenylboron. Centrifuge to remove org layers. Evap combined org phases under vac with heating and dry residue overnight over P_2O_5 in a vac dessicator. To the residue add 100 μl of 0.6 *M* imidazole in pyridine and 100 μl of 0.6 *M*

TBDMS in pyridine. Seal tube and heat at 100° for 4 hr. Cool and add 2 ml heptane. Mix, centrifuge, and remove the heptane layer. Evap under N_2 and diss residue in 200 μl heptane. Analyze by GC-MS.

Comments. The method used an ion pair extraction with tetraphenylboron anion.

References

[1]L. E. Martin *et al.*, *Biomed. Mass Spectrom.* **3**, 184 (1976).

2.6.4. Mixed Silyl and Other Derivatives

Derivative Type. Triacetyl, trimethylsilyl.

Reagent. Acetic anhydride (Ac_2O)/sodium bicarbonate, hexamethyldisilazane (HMDS).

Compound Type. Catecholamines.

Reaction

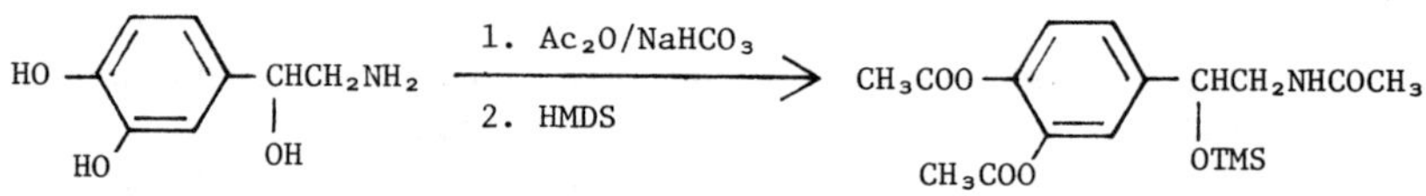

Typical Procedure. Add solid $NaHCO_3$ to an aq soln of the catecholamine followed by Ac_2O. Shake until evolution of CO_2 ceases then allow to stand 5 min. Ext with $CHCl_3$ and evap ext to dryness. Treat residue with HMDS in THF. Allow to stand overnight at room temp or reflux 2–3 hr. Analyze by GC.

Comments. The acetylation procedure was adapted from the USP procedure[2] for gravimetric analysis of epinephrine. The derivs of epinephrine and norepinephrine were separated on an SE-30 column.

References

[1]E. Brochmann-Hanssen and A. B. Svendson, *J. Pharm. Sci.* **51**, 938 (1962).

[2]United States Pharmacopeia, Sixteenth Ed., Mack Publ. Co., Easton, PA, 1960, p. 263.

Derivative Type. Trifluoroacetyl, trimethylsilyl.

Reagent. Trifluoroacetic anhydride (TFAA), bis(trimethylsilyl)trifluoroacetamide (BSTFA).

Compound Type. Amines (hydroxylated and nonhydroxylated).

Reaction

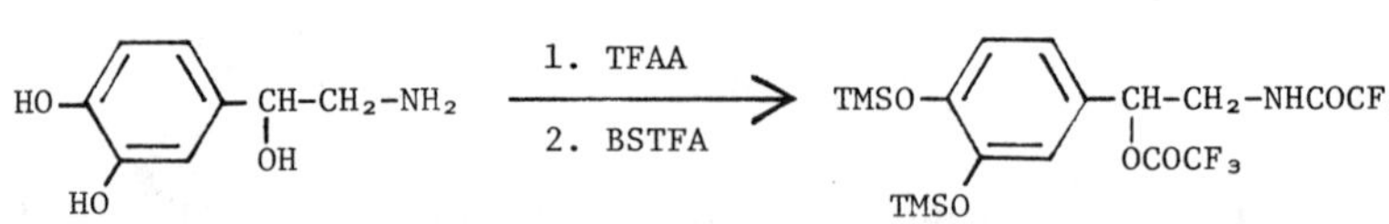

Typical Procedure. To a dried amine sample add 50 μl CH_2Cl_2 and 25 μl TFAA. Heat at 140° for 7 min. Cool quickly and evap under vac. Add 50 μl pyridine and 50 μl BSTFA. Heat at 104° for 25 min. Evap under vac. Diss residue in 10 μl acetone (if sample contains more than 5 nmol of each amine) or 10 μl $CHCl_3$ (less than 5 nmol). Analyze by GC.

Comments. Amine groups and primary and secondary alcohols gave TFA derivs while phenolic hydroxyls were converted to TMS derivs. The mixed deriv gave better stability than either the per-TFA or the per-TMS deriv.

References

[1]P. Cancalon and J. D. Klingman, *J. Chromatogr. Sci.* **10**, 253 (1972).

Derivative Type. Trimethylsilyl ether, *N*-trifluoroacetyl.

Reagent. *N*-Methyl-*N*-trimethylsilyltrifluoroacetamide (MSTFA), *N*-Methylbis(trifluoroacetamide) (MBTFA).

Compound Type. Catecholamines and related compounds.

Reaction

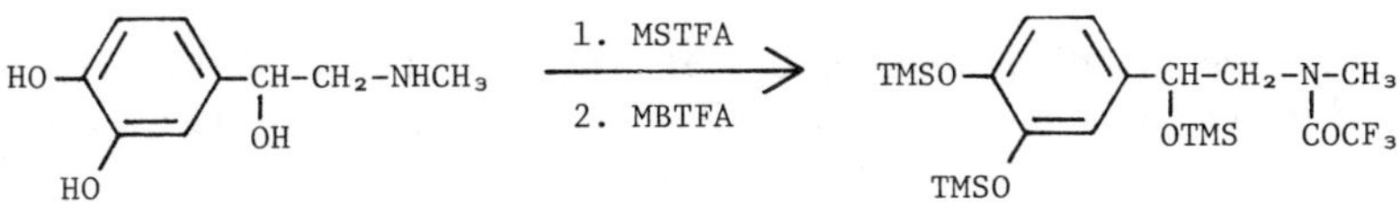

Typical Procedure. Mix an aliquot of a catecholamine soln in 0.1 *N* HOAc with 0.2 ml 1% glycine soln and 0.1 ml 0.1 *N* HCl. Evap to dryness under vac then dry in vac dessicator over P_2O_5/KOH for 4 hr. Diss the residue in 0.1 ml of a reagent soln consisting of 5 ml CH_3CN, 1.5 ml MSTFA, and 0.1 ml mercaptoethanol. Heat at 80° for 30 min. Cool and add 10 μl MBTFA. After 10 min, heat at 80° for 5 min. Cool and analyze by GC-MS.

Comments. Addition of 10 μl HMDS to the final soln served to maintain the silylation capacity and permitted multiple openings of the tube without loss of the deriv by hydrolysis. The glycine served to reduce adsorption on the glass surface and to maintain a constant "silylation potential."[2] Derivs were detectable to the femtomole region by GC-MS-SIM. MS data for the derivs was reported.

References

[1]M. Donike, *Chromatographia* **7**, 651 (1974).
[2]M. Donike, *J. Chromatogr.* **103**, 91 (1975).

Derivative Type. Trimethylsilyl ether, *N*-heptafluorobutyryl.

Reagent. Trimethylsilyl imidazole (TMSI), heptafluorobutyrylimidazole (HFBI).

Compound Type. Catecholamines and related compounds.

Reaction

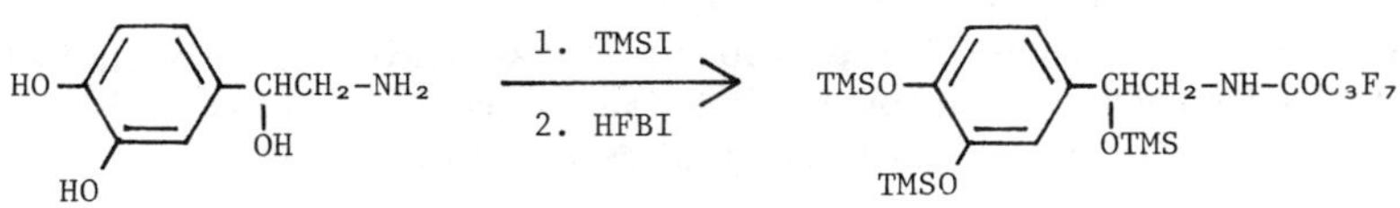

Typical Procedure. Diss 1 mg amine HCl in 0.1 ml CH_3CN. Add 0.2 ml TMSI and heat at 60° for 3 hr. Add 0.1 ml HFBI and heat 30 min at 60°. Analyze soln directly by GC.

Comments. Derivs were investigated for ECD-GC analysis of catecholamines. Derivs showed shorter retention times than the corresponding TMS, acetyl derivs.

References

[1]M. G. Horning *et al.*, *Anal. Lett.* **1**, 311 (1968).

Derivative Type. Pentafluorobenzylimine, trimethylsilyl ether.

Reagent. Pentafluorobenzaldehyde (PFBA), bis(trimethylsilyl)acetamide (BSA).

Compound Type. Catecholamines.

Reaction

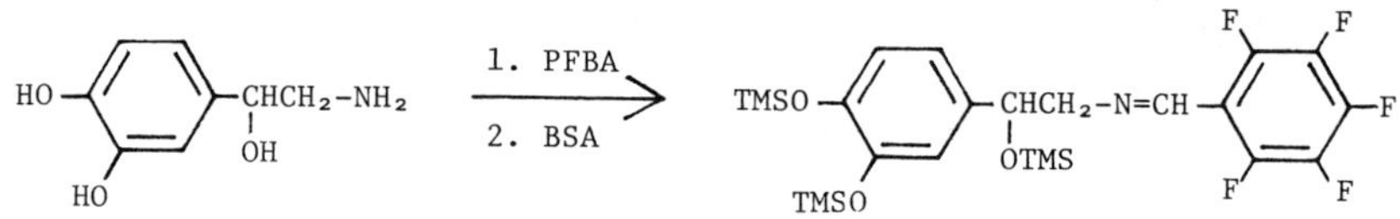

Typical Procedure. Add 1 μg amine in 50 μl DMF to 100 μl of a soln of PFBA (100 μg/ml) in CH_3CN. Heat at 85° for 5 min then add 2 μl BSA. After 15 min at room temp ext with 850 μl hexane and analyze 1 μl of the hexane by GC.

For ng level deriv, add 10 μl of a CH_3CN soln of PFBA (4 μg/ml) to 1–10 ng amine. Heat at 85° for 15 min then add 1 μl BSA and 9 μl hexane. Seal, heat at 60° for 5 min, and analyze 1 μl of the hexane phase.

Comments. The method was a modification of that of Moffat and Horning.[2] The derivs were quantitated by ECD-GC and GC-MS-SIM. Pulsed ECD allowed quantitation to 10 pg.

References

[1]J.-C. Lhuguenot and B. F. Maume, *J. Chromatogr. Sci.* **12**, 411 (1974).
[2]A. C. Moffat and E. C. Horning, *Biochem. Biophys. Acta* **222**, 248 (1970).

Derivative Type. Trimethylsilyl, Schiff base.

Reagent. Hexamethyldisilazane (HMDS), acetone.

Compound Type. Catecholamines and other biologic amines.

Reaction

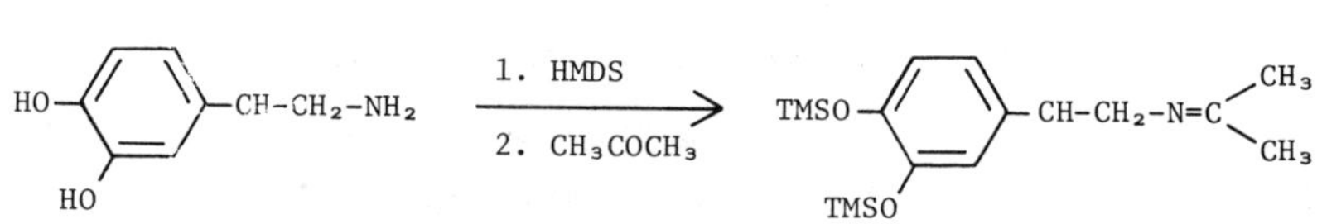

Typical Procedure. Diss 0.5–1 mg of amine or HCl salt in 0.05 ml DMF. Add 0.15 ml of HMDS and allow to stand at room temp 30 min. Prepare HMDS-acetone soln by heating 1 ml HMDS in 10 ml of acetone to boiling and then allowing to cool. Add 0.4 ml of this soln to the DMF soln and allow to stand 12 hr. Analyze 1–2 μl by GC-MS.

Comments. Cyclobutanone was also used in place of acetone. GC data were reported for the derivs of 16 amines using both ketones. MS data were also reported. Cyclopentanone and cyclohexanone formed carbinolamines with some of the amines and thus were eliminated as useful reagents. Kawai et al reported similar derivs using HMDS in DMSO and 4-methyl-2-pentanone[2,3] and using HMDS in pyridine and 2-pentanone.[4] In the latter study[4] acetone and 3-pentanone were also examined, but only 2-pentanone allowed separation of epinephrine, norepinephrine, dopamine, metanephrine, and normetanephrine as single derivs (the secondary amines did not form Schiff bases).

References

[1]P. Capella and E. C. Horning, *Anal. Chem.* **38**, 316 (1966).
[2]S. Kawai and Z. Tamura, *J. Chromatogr.* **25**, 471 (1966).
[3]S. Kawai and Z. Tamura, *Chem. Pharm. Bull.* **15**, 1493 (1967).
[4]S. Kawai *et al.*, *Chem. Pharm. Bull.* **14**, 618 (1966).

Derivative Type. *N*-Dinitrophenyl, *O*-trimethylsilyl.

Reagent. 2,4-Dinitrobenzenesulfonic acid (DNBS), bis (trimethylsilyl) acetamide (BSA).

Compound Type. Phenylethylamines.

Reaction

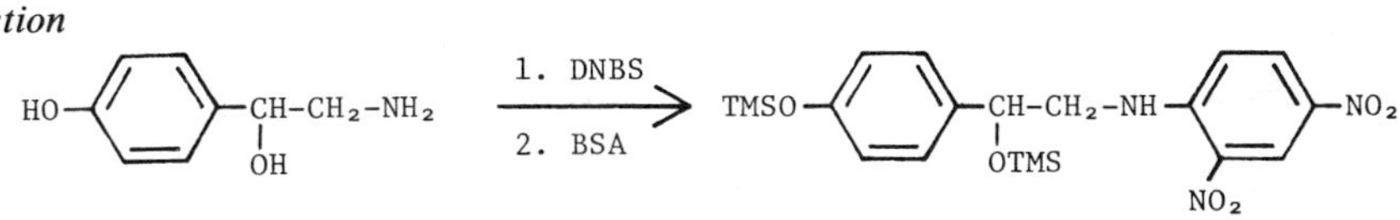

Typical Procedure. Homogenize tissue in 4 vol of cold 0.01 *N* HCl containing 50 mg/l pargyline HCl. Centrifuge at 10,000 g for 15 min under refrigeration. Mix 1.0 ml of the supernatant with 0.6 g NaCl and 10 μl thioglycolic acid (antioxidant). Adjust to pH 8 with 2.5 *N* NaOH and ext with 2 × 1 ml *n*-BuOH. Mix *n*-BuOH ext with 0.5 ml 0.01 *N* HCl and 4 ml *n*-heptane. Sep aq layer and evap to dryness under vac over P_2O_5.

Diss residue in 0.05 ml H_2O and add 0.05 ml of 0.25 *M* DNBS in satd aq $Na_2B_4O_7$. Heat at 100° for 15 min. Cool and ext with 0.4 and 0.2 ml benzene. Evap ext to dryness and diss residue in 5 μl BSA. Heat at 60° for 30 min, cool, and evap to dryness under vac for 1 hr. Add 1 μl BSA and make up to 10 μl with benzene. Analyze 1 μl aliquots by GC-ECD.

Comments. The use of DNBS in alkaline soln permitted sep of the derivs from excess reagent by benzene ext. DNBS did not react with hydroxyl groups. The method did not deriv secondary amines. Catecholamines were not recovered at the high pH used in the method. The method permitted quantitation of phenylethylamine, octopamine, tyramine, normetanephrine, and 3,4-dimethoxyphenylethylamine, in tissues to 20 ng/g. The DNT derivs were also prep and found to be somewhat more sensitive to ECD, however the reagent reacted also with the hydroxyl groups and could not be separated from the derivs by ext.

References

[1]D. J. Edwards and K. Blau, *Anal. Biochem.* **45**, 387 (1972).

2.6.5. Isothiocyanate Derivatives.

Derivative Type. Isothiocyanate.

Reagent. Carbon disulfide.

Compound Type. 3-Methoxy-4-hydroxy- and 4-methoxy-3-hydroxydopamine.

Reaction

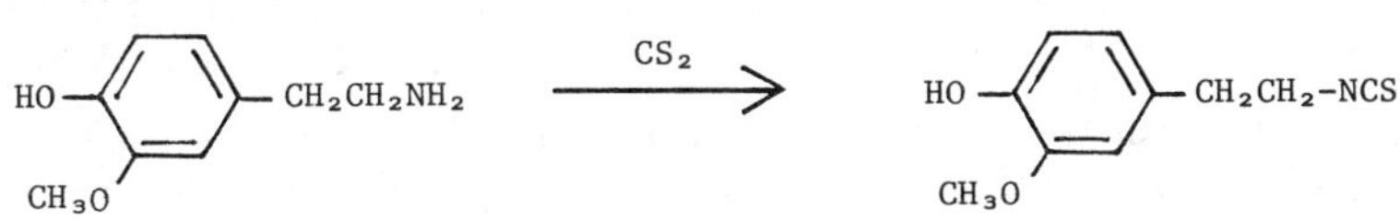

Typical Procedure. Adjust 1 ml of a 1 mg/ml aq soln of the amine HCl to pH 8 with $NaHCO_3$ soln. Ext with 10 ml EtOAc and evap to dryness under vac. Diss the residue in 1 ml EtOAc, add 0.1 ml CS_2 and shake 30 min. Evap to dryness under vac and diss residue in 1 ml EtOAc. Analyze by GC or GC-MS.

Comments. The derivs of the two isomers were separable by GC on OV-101 or OV-225. The TMS, isothiocyanate derivs were not separable on either type of column. MS data were reported for the derivs.

References.

[1]N. Narasimhachari and P. Vouros, *J. Chromatogr.* **70**, 135 (1972).

Derivative Type. Isothiocyanate, trimethylsilyl ether.

Reagent. Carbon disulfide, bis(trimethylsilyl) trifluoroacetamide (BSTFA)/trimethylchlorosilane (TMCS).

Compound Type. 3-*O*-Methyldopamine and 4-*O*-methyldopamine.

Reaction

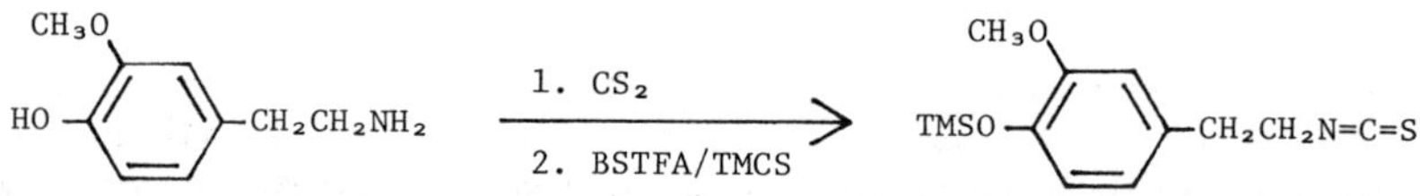

Typical Procedure. Adjust 1 ml of a 1 mg/ml soln of the amine HCl to pH 8 with $NaHCO_3$ and ext with 10 ml EtOAc. Dry ext over Na_2SO_4 and evap to dryness under vac. Diss residue in 1 ml EtOAc, add 0.1 ml CS_2, and shake for 30 min. Evap to dryness and diss the residue in 1 ml pyridine. Mix 5 μl of the pyridine soln with 5 μl of BSTFA/TMCS and allow to stand for 5 min. Analyze by GC or GC-MS.

Comments. The BSTFA/TMCS ratio was not stated but a 5:1 mixt of d_{18}-BSA and d_9-TMCS was used to prepare the deuterated derivs. The isothiocyanate, trifluoroacetyl derivs were also prepared using TFAA. MS data were reported for the derivs.

References

[1]N. Narasimhachari and P. Vouros, *J. Chromatogr.* **70**, 135 (1972).

Derivative Type. Isothiocyanate, Trimethylsilyl.

Reagent. Carbon disulfide, trimethylsilylimidazole (TMSI).

Compound Type. Hydroxylated phenylethylamines.

Reaction

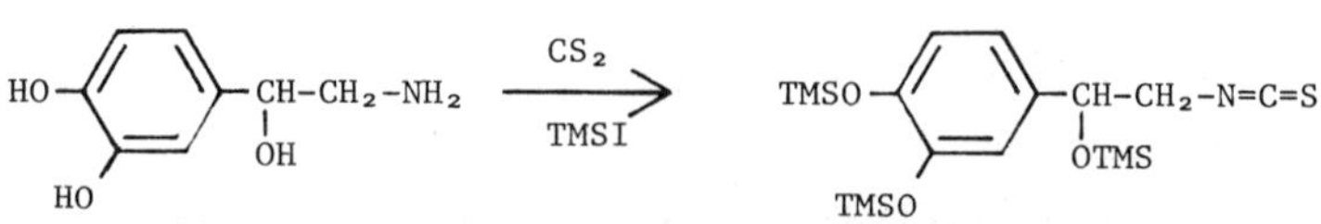

Typical Procedure. Mix a soln of the sample in DMF with an equal vol of TMSI and of CS_2. Warm for 3 hr under reflux then evap to a small vol and analyze by GC-MS-SIM.

Comments. The method was applied to amphetamine and substituted analogs, mescaline, dopamine, norephedrine, and norepinephrine. The derivs were detectable down to *ca* 10 pg.

References

[1]H. Brandenberger and D. Schnyder, *Z. Anal. Chem.* **261**, 297 (1972).

Derivative Type. *O*-Ethyl, Isothiocyanate (primary amines), ethyl ester (carboxylic acids).

Reagent. Ethanol (EtOH)/HCl, carbon disulfide.

Compound Type. Metanephrine (MN), normetanephrine (NMN), and vanillylmandelic acid (VMA).

Reaction

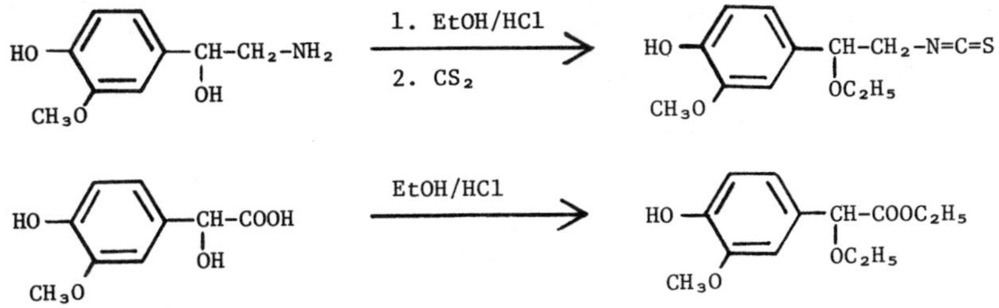

Typical Procedure. Diss 3 mg of compd in 2 ml of 3 *N* HCl in EtOH. Heat in a stoppered tube for 2 hr at 90–100°. Evap to dryness under vac, then add EtOH and evap again to insure complete removal of H_2O. Diss residues of ethyl esters in EtAOc and analyze by GC-MS.

For the residues from MN and NMN, shake 15 min with 10 ml EtOAc and 1 ml 10% aq NH_3 soln. Dry EtOAc phase over Na_2SO_4 and evap to dryness under vac. Diss in EtOAc and shake 30 min with 0.5 ml CS_2. Analyze by GC-MS.

Comments. The free phenolic hydroxyl groups permitted GC separation of the isomeric 3- and 4-methoxy compds. MN and NMN gave quantitative conversion of the alcohol hydroxy to the ethyl ether. The VMA deriv showed 5–10% of the underiv alcoholic hydroxyl. 3-Methoxy-4-hydroxyphenylglycol formed the epoxide under the deriv conditions.

References

[1]N. Narasimhachari, *J. Chromatogr.* **90**, 163 (1974).

2.6.6. Other Derivatives for GC and MS Analysis

Derivative Type. Cyclic dimethylsilylmethylene.

Reagent. 1,3-Bis(chloromethyl)-1,1,3,3-tetramethyldisilazane/chloromethyldimethylchlorosilane.

Compound Type. β-Hydroxyamines.

Reaction

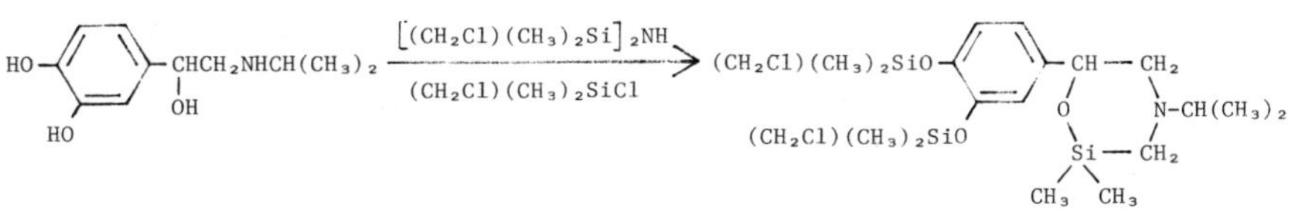

Typical Procedure. To a soln of the sample compd (1 mg/ml) in CH_2Cl_2 or DMF add the silazane-silane reagents in a ratio of 2:1 (v/v). Allow to react at room temp or heat 5–15 min at 60°. Analyze by GC-MS.

Comments. The deriv method was applicable to primary secondary, tertiary, and quaternary amines. It was also applied to one amino acid (Phe). MS fragmentation data were given for the derivs.

References

[1]C. G. Hammar, *Biomed. Mass Spectrom.* **5**, 25 (1978).

Derivative Type. *n*-Butylboronates.

Reagent. *n*-Butaneboronic acid.

Compound Type. Catecholamines.

Reaction

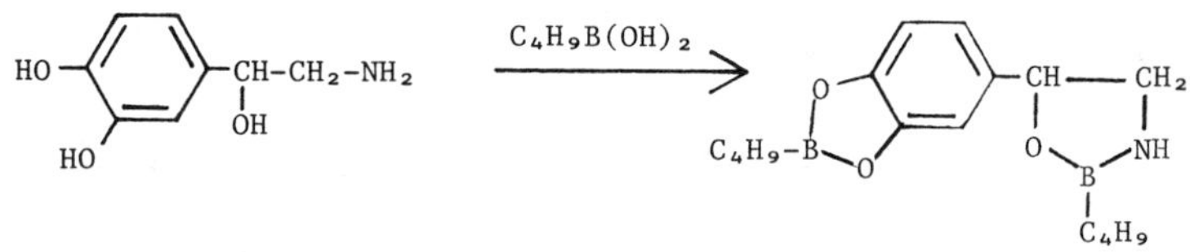

Typical Procedure. Diss 1 mg sample in 1 ml pyridine or DMF. Add 1.5–2 mol equiv of the boronic acid and allow to stand at 22° overnight. Analyze soln directly by GC-MS, or evap to dryness, ext residue with Et_2O, and analyze Et_2O soln by GC-MS.

Comments. The method was applicable to amine salts or free bases as long as the compd was soluble.[1] Acetone was a suitable solvent for α- and β-diols but was unsuitable for β-hydroxyamines. Cyclohexyl,[1,2] phenyl,[1,2] *t*-butyl[2], and methyl[2] boronates were also prepared. GC and MS data were reported for the derivs of a series of catecholamines and related compds.[1–3] GC data were also reported for the boronate derivs of ephedrine and ψ-ephedrine.[2]

References

[1]G. M. Anthony *et al., J. Chromatogr. Sci.* **7**, 623 (1969).

[2]G. M. Anthony *et al., J. Pharm Pharmacol.* **22**, 205 (1970).

[3]C. J. W. Brooks *et al., Org. Mass Spectrom.* **2**, 1023 (1969).

Derivative Type. 5-Dibutylamino-1-naphthalenesulfonyl.

Reagent. 5-Dibutylamino-1-naphthalenesulfonyl chloride (BANS-Cl).

Compound Type. Dopamine and metabolites.

Reaction

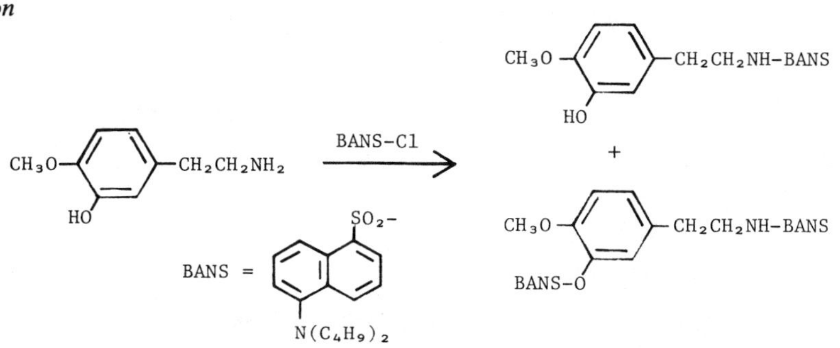

Typical Procedure. Acidify (H_2SO_4 to pH 1) an aq soln of the amine (*ca* 10 μl/ml) and mix with 6 ml of a soln of BANS-Cl in acetone (*ca* 1 mg/ml). Sat soln with Na_2CO_3 and allow to react at room temp 3 hr. Decant soln from the salt and wash salt once with 1 ml acetone. Sat the combined soln with KH_2PO_4 and evap acetone under N_2. Add 3 ml MeOH to residual aq phase and ext with 2 ml *n*-heptane or toluene. Evap to dryness under N_2 and diss residue in 100 μl EtOAc. TLC on silica with cyclohexane-EtOAc (4:1). Locate derivs by fluoroscence and ext with 500 μl EtOAc. Conc eluate under N_2 and analyze by direct probe EI or FD-MS.

Comments. The derivs were used to quantitate dopamine, 3-*O*-methyldopamine, and 4-*O*-methyldopamine by MS. A mixt of totally and partially deriv products were obtained; the totally deriv product was most useful for quantitation.

References

[1]W. D. Lehmann *et al., Anal. Chem.* **48**, 1572 (1976).

2.6.7. Chromogenic Derivatives

Derivative Type. 5-Dimethylamino-l-naphthalenesulfonyl (DANS).

Reagent. 5-Dimethylamino-1-naphthalenesulfonyl chloride (DANS-Cl).

Compound Type. Catecholamines.

Reaction

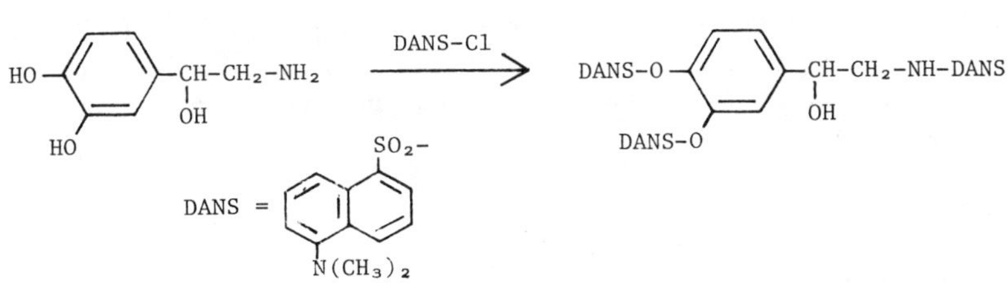

Typical Procedure. Dil 10 μl of catecholamine soln (containing up to 10 μg) to 0.1 ml with H_2O. Add 0.2 ml of 0.1 % DANS-Cl soln in acetone, 0.4 ml 0.1 *M* aq $NaHCO_3$ (pH 9.0), and 0.4 ml acetone. Seal vial and heat at 60° for 20 min. Allow to cool at room temp and analyze an aliquot by reversed phase HPLC. Alternatively, ext with 3 ml Et_2O or benzene and analyze an aliquot of the ext by adsorption HPLC.

Comments. A systematic study was carried out to determine the optimum deriv cond for epinephrine, norepinephrine, and dopamine.[1] The derivs were detectable to 50 pg using fluorescence detection.[2] Similar methodology was used for catecholamine *O*-methyl metabolites.[3]

References

[1]G. Schwedt and H. H. Bussemas, *Z. Anal. Chem.* **283**, 23 (1977).
[2]G. Schwedt and H. H. Bussemas, *Chromatographia* **9**, 17 (1976).
[3]G. Schwedt and H. H. Bussemas, *Z. Anal. Chem.* **285**, 381 (1977).

Derivative Type. Fluorescamine.

Reagent. Fluorescamine (4-phenylspiro[furan-2-(3H),1′-phthalan]-3,3′-dione).

Compound Type. Catecholamines and *O*-methyl metabolites.

Reaction

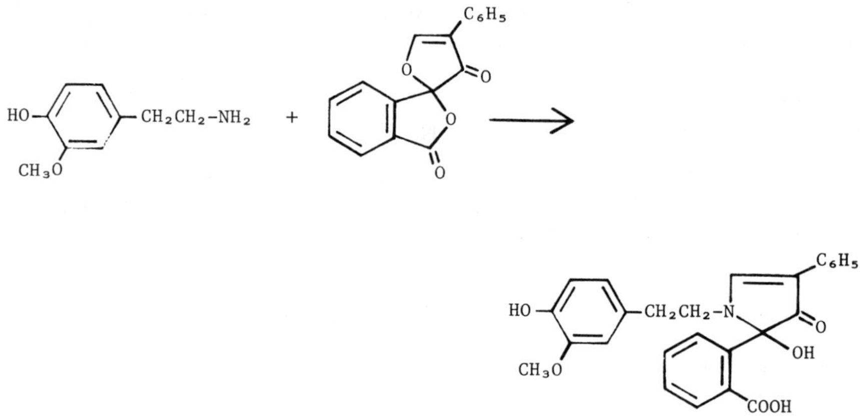

Typical Procedure. Dil a few μl of the amine soln (less than 1 m*M*) to 20 μl with pH 8.0 (0.05 *M*) phosphate or borate buffer. Add 10 μl of a 2 mg/ml soln of fluorescamine in acetone. Mix by shaking then analyze an aliquot by HPLC.

Comments. The derivs were used to quantitate the amines to the 100 pmol level using fluorescence detection (excitation 390 nm, emission 475 nm).[1] Schwedt also used the derivs for dopamine and norepinephrine analysis by HPLC and reported similar sensitivity with fluorescence detection.

References

[1]K. Imai, *J. Chromatogr.* **105**, 135 (1975).
[2]G. Schwedt, *J. Chromatogr.* **118**, 429 (1976).

2.7. DERIVATIVES OF HYDRAZINES

Derivative Type. 5-Nitro-2-hydroxybenzal.

Reagent. Nitro-2-hydroxybenzaldehyde.

Compound Type. Hydrazines.

Reaction

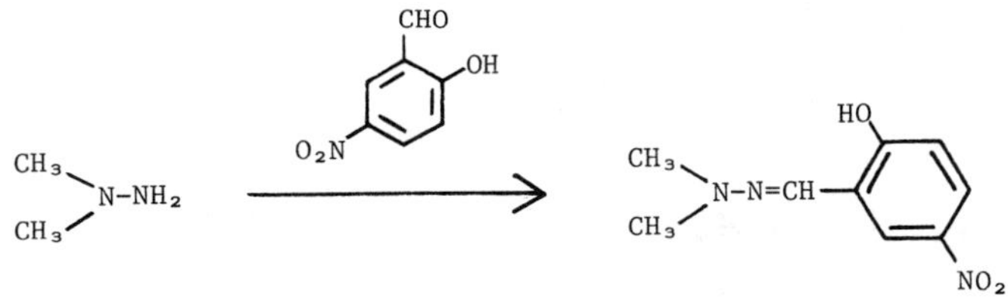

Typical Procedure.[1] To 10 μg of the hydrazine in 5 ml EtOH add 20 μg of the reagent. Heat on a H_2O bath then analyze by GC.

Comments. The method was also applicable to *N*-nitroso compds after $LiAlH_4$ reduction to the hydrazines.[1] GC data were reported for the derivs of 25 hydrazines.[2] The derivs were originally analyzed by TLC[1] and could be adaptable to HPLC analysis.

References

[1]G. Neurath *et al.*, *Chem. Ber.* **97**, 1631 (1964).
[2]G. Neurath and W. Lüttich, *J. Chromatogr.* **34**, 257 (1968).

Derivative Type. Pyrazole.

Reagent. Acetylacetone (2,4-pentanedione).

Compound Type. Hydrazine and methylhydrazine.

Reaction

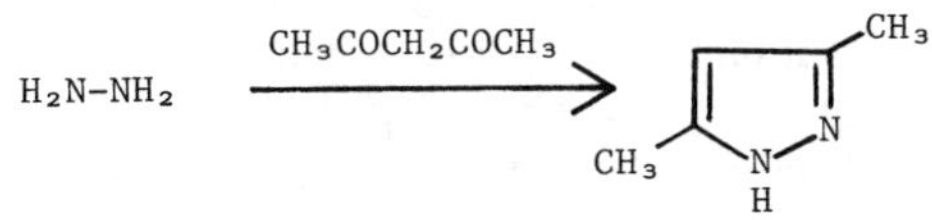

Typical Procedure. Adjust 100 ml aq sample (containing 0.1–50 ppm hydrazines) to pH 6–9 with 1 *N* NaOH or 1 *N* H_2SO_4 as required. Add 50 μl acetylacetone, shake well, and allow to stand at room temp 1 hr. Analyze 5 μl of the soln by GC.

Comments. The method was used for simultaneous determination of hydrazine and methylhydrazine in dil aq soln.

References

[1]L. A. Dee, *Anal. Chem.* **43**, 1416 (1971).

3. Carboxylic Acids

Derivatization of carboxylic acids has been employed for GC, MS, and HPLC analysis. Although many of the monocarboxylic acids are sufficiently volatile for GC analysis, derivatization yields improved peak shape and reduced adsorption losses. For dicarboxylic acids and many multifunctional carboxylic acids, derivatization is essential for successful GC analysis. Various derivatives of carboxylic acids have been prepared to impart specific detectability in GC analysis. Derivatization is also of value in improving volatility and stability of carboxyli. acid compounds for MS analysis. For HPLC analysis, derivatives have been used to confer UV absorption or fluorescence to improve detectability. This section contains derivatives that have been applied in the analysis of carboxylic acids including acids which contain additional functional groups. Derivatives used for location of substituents (double bonds, branching, and cyclopropane rings) are covered separately in Section 4. Derivatization of carboxyl groups in amino acids and peptides are covered in Section 5, and derivatives of carboxyl groups for enantiomeric analysis in Section 6. The prostaglandins are covered in Section 12 and the sugar carboxylic acids in Section 13.

Esterification of the Carboxyl Group. The classical derivatives of both fatty acids and carboxylic acids in general have been the methyl esters. Several other esters have also been used, and a wide variety of methods have been employed for such esterification. The esterification of fatty acids in particular was recently reviewed.[1] Methyl esters can be prepared from the fatty acids by direct esterification in methanol with a suitable catalyst. Catalysts that have been used include hydrogen chloride (p. 151), boron trichloride (p. 153), and boron trifluoride (p. 154). The catalyst must be chosen with care to avoid isomerization and artifact production with unsaturated or cyclopropane substituted acids. In this respect, boron trichloride appears to be preferable to the trifluoride (p. 154). Cupric acetate has been used used in conjunction with hydrogen chloride catalyzed esterification to avoid undesired alteration of polyunsaturated fatty acids (p. 153). Methyl ester formation using *N,N'*-carbonyldiimidazole was employed to minimize transesterification of lipids present in plasma samples being analyzed for free fatty acids (p. 155). Fatty acid methyl esters have also been prepared with methyl iodide and base (p. 155).

Diazomethane is a particularly useful reagent for preparation of methyl esters. It gives a rapid reaction, with nitrogen as essentially the only by-product (some polymethylene may also be formed). The reagent must be handled with care, however, due to its toxicity, explosion hazard, and potential to form artifacts with unsaturated and keto acids (discussed further below). The esterification can be performed either in solution (p. 156) or by exposure of the acids on Celite to the gaseous reagent (p. 157).

Flash methylation by pyrolysis of tetraalkylammonium salts of the carboxylic acids in the injector of the gas chromatograph has also been used for methyl ester formation. The tetramethylammonium (p. 157), trimethylphenylammonium (trimethylanilinium) (p. 158), and (*m*-trifluoromethyl)trimethylammonium (p. 158) hydroxides have been used for this purpose. The aromatic reagents, having better leaving groups (the aryldimethylamines), can be used with lower injector temperatures than the tetraalkylammonium reagents. Methyl esters, as well as higher esters, can be prepared from the corresponding dialkyl acetals of dimethylformamide (p. 159). The deuterated reagent is available for preparation of d_3-methyl esters (p. 159).

The methods used for methyl esterification of fatty acids have also been applied to other carboxylic acids (p. 169). In addition, other methyl esterification methods have been reported for these acids. Methyl esters of several acids were prepared by coupling with methanol using *N,N'*-dicyclohexylcarbodiimide (p. 171). Methylation with methyl sulfate in the presence of potassium carbonate was used for the sulfonamido benzoic acid probenecid (p. 172). Heptafluorobutyric anhydride formed a mixed anhydride with some dicarboxylic acids derived from a substituted pyridine compound; subsequent treatment with methanol gave the methyl esters (p. 172). Methyl esterification with thionyl chloride and methanol, via the intermediate dimethyl sulfite, was first applied to amino acids but has subsequently been used for other carboxylic acids (p. 172). The carboxylic acid groups of hemins could be esterfied with trimethyloxonium tetrafluoroborate without altering hydroxyl, vinyl, or formyl substituents (p. 173).

Higher alkyl esters have been prepared both from fatty acids (p. 159) and from other carboxylic acids (p. 174). In particular, the higher esters are useful for the short chain fatty acids whose methyl esters are too volatile for easy manipulation. Examples include isopropyl (p. 160), *n*-propyl (p. 159), *n*-butyl (p. 160), and decyl (p. 161) esters of fatty acids prepared from the alcohols by acid catalysis. *n*-Butyl (p. 174) and butoxyethyl (p. 174) esters were prepared from chlorinated phenoxyacetic acids using the alcohols and mineral acids.

Diazoalkanes are also useful reagents for preparation of higher esters. Benzyl esters of fatty acids (p. 161) and trifluoroacetic acid (p. 175) have been prepared with phenyldiazomethane. Diazoethane, rather than the more common diazomethane, was used in the preparation of the ethyl ester of indomethacin since the demethylated (phenolic) metabolite would have been converted to the same derivative as the parent drug with diazomethane (p. 175).

Substituted phenacyl (p. 162) and substituted benzyl (p. 162) esters were prepared from fatty acids using the corresponding bromide reagents in the presence of potassium hydroxide. Alkyl iodide reagents and tetraalkylammonium bases were used to prepare the ester derivatives from fatty acids (p. 163) and indomethacin (p. 176). The coupling reagent, *N,N'*-dicyclohexylcarbodiimide can be coupled with an alcohol to form a reactive isourea such as *N,N'*dicyclohexyl-*O*-benzylisourea, which was used to prepare benzyl esters from fatty acids (p. 163).

Glycerides. Fatty acid analysis often involves analysis of fatty acids in triglycerides. Such analysis may be done by saponification of the fat, isolation of the acids, and use of any of the preceding derivatization methods. Direct conversion of the triglyceride to the fatty acid derivatives, however, is much more convenient. The methyl ester derivatives, commonly used for analysis of fatty acids in triglycerides, may be prepared by sequential saponification and esterification (p. 164) or by base catalyzed (p. 164) or acid catalyzed (p. 165) transesterfication procedures. To avoid possible base catalyzed isomerizations, a novel method was developed involving formation of a complex with sulfuric acid at $-60°$ followed by treatment with methanol (p. 167). As stated earlier, the short chain methyl esters can be troublesome due to excessive volatility. To circumvent the problem of excessive volatility of the short chain methyl esters, methods have been developed for formation of butyl esters from triglycerides (p. 168). While the lower volatility of the butyl esters simplifies handling of the short chain acids, the derivatives are still sufficiently volatile for analysis of the longer acids as well.

Other Carboxylic Acid Derivatives. Silylated derivatives of carboxylic acids, although less widely applied than the alkyl esters, can be useful. Trimethylsilyl derivatives are formed readily from carboxylic acids but are somewhat less stable toward hydrolysis than the same derivatives of hydroxyl groups. Any of the usual trimethylsilyl reagents may be used with carboxylic acids; examples include HMDS/TMCS (p. 176), HMDS/TMCS/BSA (p. 177), BSTFA (p. 178), and TMSI (p. 178). The latter reagent is particularly useful for the very volatile trimethylsilyl derivatives of short chain acids since the excess reagent elutes after the derivatives. The *t*-butyldimethylsilyl derivatives are more stable to hydrolysis than the trimethylsilyl derivatives and have the added advantage of more favorable MS fragmentation. These derivatives have been prepared with the silyl chloride and imidazole (p. 179). The silyl amide reagent from *N*-methyltrifluoroacetamide, although generally less reactive, does react readily with carboxylic acids (p. 179). This reagent, which yields the neutral and quite volatile *N*-methyltrifluoroacetamide as by-product, could be useful for analytical derivatization of carboxylic acids for both GC and MS analysis.

The analytical use of amide derivatives of carboxylic acids has been reported (p. 180). These derivatives were prepared by treating the acids with thionyl chloride and reaction of the resulting acid chlorides with aniline.

Various alkyl esters, silyl esters, and an amide derivative have been reported in methods designed to enhance detectability for either GC or HPLC analysis. Esterification of herbicidal acids with 2-chloroethanol has been accomplished with boron trifluoride (p. 180), boron trichloride (p. 181) and *N,N′*-dicyclohexylcarbodiimide (p. 181). These derivatives were analyzed by ECD-GC, but the chloroethanol probably made little increase in their electron capturing ability. The trichloroethyl esters, on the other hand, significantly enhance ECD sensitivity and have been prepared utilizing the mixed anhydride approach with trifluoroacetic (p. 182) and heptafluorobutyric (p. 182) anhydrides. Pentafluorobenzyl esters were prepared from carboxylic acids using pentafluorobenzyl bromide in the presence of potassium carbonate (p. 183) or tetraalkylammonium bases (p. 184). Halogenated silyl derivatives have also been used to enhance detectability by ECD-GC. Bromomethyl- or chloromethyldimethylsilyl esters were formed with the silyl chloride in the presence of triethylamine (p. 185). These derivatives could be further converted to the more electron capturing iodomethyltetramethyldisiloxane esters by treatment with sodium iodide (p. 186). α-Hydroxyphosphonate esters were prepared from dimethyl α-hydroxyalkanephosphonates and *N,N′*-dicyclohexylcarbodiimide for alkali flame detection of carboxylic acids (p. 186).

Chromogenic esters for HPLC analysis were prepared from the halide reagents and base yielding phenacyl (p. 187), *p*-bromophenacyl (p. 187), 2-naphthacyl (p. 188), *p*-nitrobenzyl (p. 188), benzyloxymethyl (p. 189), and 4-methyl-7-methoxycoumarin (p. 189) esters. The methods for the *p*-bromophenacyl and *p*-nitrobenzyl esters employed a crown ether to sequester the cation from the potassium salts yielding very reactive "naked" carboxylate anions in the nonprotic solvent. Benzyl (p. 189) and *p*-nitrobenzyl (p. 190) esters have been prepared from the triazene reagents. An alternative method for preparing *p*-nitrobenzyl esters employs an *O*-(*p*-nitrobenzyl) isourea (p. 190) to replace the possibly carcinogenic triazene reagent. *p*-Methoxyanilide derivatives of fatty acids were prepared by forming acid chlorides with triphenylphosphine and carbon tetrachloride and then allowing them to react with *p*-methoxyaniline (p. 191).

Hydroxy Acids. The hydroxy acids have been analyzed as a series of mixed derivatives employing esterification methods discussed in this section and hydroxyl derivatives discussed in Section 1. Examples include methyl and butyl esters with *O*-acetyl, *O*-trifluoroacetyl, and *O*-heptafluorobutyryl groups (p. 191). In all of these cases, the esterification was performed prior to derivatization of the hydroxyl groups. A simultaneous esterification, acylation method for analysis of hydroxy and amino acids employed heptafluorobutyric anhydride, ethanol, and pyridine (p. 193). Simultaneous derivatization of hydroxy and carboxyl groups can also be achieved with silylation as exemplified by formation of trimethylsilyl derivatives using HMDS/TMCS (p. 194). The presence of the hydroxyl group in the α-position allows formation of cyclic boronate derivatives. Butaneboronates have been prepared from both aromatic and aliphatic α-hydroxy acids (p. 195).

Keto Acids and Krebs Cycle Acids. The keto acids and other acids of the Krebs cycle (citric, *cis*-aconitic, isocitric, α-ketoglutaric, succinic, fumaric, malic, and oxalacetic) present particular problems in analytical derivatizations. Their propensity for multiple product and artifact formation has complicated the development of suitable derivatization methods. A series of alcohol esterifications employing acid catalysis (p. 196) have met with varying degrees of success suggesting that the "ideal" method is yet to be devised. Diazomethane esterification is particularly prone to formation of adducts with both the keto (p. 200) and the unsaturated (p. 199) acids. Use of diazomethane at a very low temperature, however, has met with some success (p. 199).

Silylation of keto acids is complicated by formation of derivative mixtures containing enol ether as well as ester derivatives. The non-enolizable mesoxalic acid partially forms a reagent adduct (p. 201). These problems have been circumvented by formation of oxime derivatives prior to silylation. Both *O*-trimethylsilyl (p. 201) and *O*-alkyl (p. 203) oxime trimethylsilyl esters have been reported. Keto acids have also been converted to quinoxalone derivatives with *o*-phenylenediamine and subsequently trimethylsilylated for GC analysis (p. 204). The quinoxalones are UV absorbing and fluorescent and thus may possibly be employed for HPLC analysis (p. 205). A cyclic imidazolidine-2-carboxylic acid was prepared from glyoxylic acid and subsequently trimethylsilylated for GC analysis (p. 205).

Phenolic Acids. The phenolic acids include the acid metabolites of the phenolic biogenic amines. Studies on the analysis of these metabolites have been largely responsible for the volume of work on phenolic acids. The majority of the derivatizations of the phenolic acids have involved alkyl esters of the carboxyl group. In one method, the ester was analyzed by GC without further derivatization of the phenolic hydroxyl (p. 206). Most methods, however, have included derivatization of the phenolic hydroxyl, especially formation of acyl derivatives. Methyl esterification followed by acetylation was used for vanillylmandelic and homovanillic acids (p. 207). Acetylation followed by methyl esterification with diazomethane has also been reported (p. 207). Aspirin, an example of an already acetylated phenolic acid, has been analyzed by esterification in methanol with boron trifluoride (p. 208). Methyl esterification followed by perfluoroacylation yields derivatives which have been used for ECD-GC and GC-MS analysis. Examples include pentafluoropropionyl derivatives (p. 208) and heptafluorobutyryl derivatives (p. 205). Fluorinated acyl derivatives with halogenated alkyl ester groups such as pentafluoropropionyl derivatives of pentafluoropropyl (p. 209) and chloropentafluoroisopropyl (p. 210) esters, and trifluoroacetyl derivatives of hexafluoroisopropyl esters (p. 210) have been employed both for ECD-GC and GC-MS analysis.

Ether derivatives of phenolic acid esters have also been utilized analytically. Diazomethane will methylate both carboxylic acid and phenol groups (the latter at a much slower rate), and has been used to prepare methyl ester, methyl ether derivatives (p. 211). Both functional groups are also sufficiently acidic to be alkylated with alkyl halides. This approach has been used to prepare pentafluorobenzyl ester, ether derivatives for ECD-GC analysis (p. 212).

Silylated derivatives of phenolic acids have been used to a lesser extent. Methyl esters of phenolic acids have been treated with HMDS/TMCS for silylation of the phenolic groups (p. 213). Trimethylsilyl ester, ether derivatives have been prepared with HMDS (p. 213), BSA (p. 214), and BSTFA/TMCS (p. 214). The sodium salts of phenolic acids were converted to trimethylsilyl ester, ether derivatives with BSA/TMCS (p. 215). It has been reported that improved resolution can be obtained if the phenolic group is converted to the sulfate ester prior to silylation (p. 215). Chloromethyldimethylsilyl ester, ether derivatives have been used for ECD-GC analysis (p. 216).

Vanillylmandelic acid was oxidatively cleaved to the aldehyde derivative and subsequently trifluoroacetylated for ECD-GC analysis (p. 216). A modification of this technique added borohydride reduction of the aldehyde and formation of a di-trifluoroacetyl derivative by acylation of both the phenols and alcohol hydroxyls (p. 217).

Indole Acids. Indole acids are also of considerable analytical interest as metabolites of biogenic amines. Phenolic indole acids have, in addition to the carboxyl and hydroxyl groups, the indole nitrogen as a possible site of derivatization. Phenolic indole carboxylic acids have been methylated at all three sites using trimethylphenylammonium hydroxide (p. 217). Methyl esters prepared with diazomethane (p. 218) or methanol/boron trichloride (p. 218) have been subsequently both *N*- and *O*-acylated with perfluoroacyl anhydride reagents. Pentafluoropropyl ester, *N,O*-pentafluoropropionyl derivatives have been prepared in a single step method for GC-MS analysis (p. 219). Phenolic indole acids have also been analyzed as the per-trimethylsilyl derivatives formed with TMSI (p. 219).

Polyamides and Polyesters. Polyesters are somewhat analogous to the triglycerides in that it is often necessary to analyze the carboxylic acid components. Similar techniques involving base catalyzed transesterification have been employed to prepare the methyl esters (p. 220). Cleavage with tetramethylammonium hydroxide followed by treatment with methyl iodide has also been used to prepare the methyl ester derivatives (p. 221). Polyester prepolymer mixtures have been analyzed as trimethylsilyl derivatives (p. 221). Polyamides were cleaved with aqueous hydrochloric acid prior to analysis as trimethylsilyl and *N*-trifluoroacetyl, methyl ester derivatives (p. 222).

Vinylogous Acids. The vinylogous acids are those compounds that contain a "carboxyl group" in which the carbonyl is separated from the hydroxyl by a vinyl group. These compounds behave similarly to the carboxylic acids because the reasonance effect, which stabilizes the carboxylate anion, is transmitted through the vinyl group. The 4-hydroxycoumarin drugs constitute an important group of vinylogous acid type compounds and several methods have been employed for their analytical derivatization. Methylated derivatives have been prepared with diazomethane (p. 222), pentafluorobenzyl derivatives with the bromide reagent (p. 223), and trimethylsilyl derivatives with HMDS/TMCS (p. 223). Whereas the mixed anhydrides of carboxylic acids have been used only as reactive intermediates for ester formation, the trifluoroacetyl derivatives of the 4-hydroxycoumarins (vinylogous mixed anhydrides) have been used analytically along with the corresponding acetyl and trichloroacetyl derivatives (p. 224). The vinylogous acids thus appear to behave somewhat like phenols also. The hop and beer bitter acids, which contain both vinylogous acid groups and alcohol hydroxyls, have been analyzed as trimethylsilyl derivatives prepared with HMDS (p. 224).

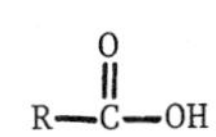

carboxyl

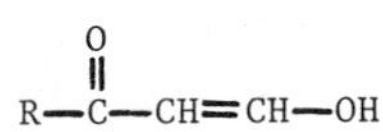

vinylogous carboxyl

References

1. A. J. Sheppard and J. L. Iverson, *J. Chromatogr. Sci.* **13**, 448 (1975).

3.1. ALKYL ESTER DERIVATIVES OF CARBOXYLIC ACIDS

3.1.1. Derivatives of Fatty Acids

3.1.1.1. Fatty Acid Methyl Esters

Derivative Type. Methyl ester.

Reagent. Methanol (MeOH)/HCl.

Compound Type. Fatty acids.

Reaction

$$\text{R-COOH} \xrightarrow[\text{HCl}]{\text{MeOH}} \text{R—COOCH}_3$$

Typical Procedure. Diss 1–10 mg fatty acids in 4 ml 0.5 *N* HCl in MeOH and 0.5 ml benzene. Heat at 80° for 1 hr with frequent shaking. Cool to room temp and add 10 ml H_2O. Ext with 3 × 3 ml Et_2O. Dry Et_2O over a mixt of $Na_2SO_4/NaHCO_3$ (4 : 1, w/w) and analyze by GC.

Comments. The MeOH/HCl was prepared by bubbling dry HCl gas into dry MeOH under anhydr conditions.[1] If the presence of methyl acetate is not a problem, MeOH/HCl can be prepared by adding AcCl to anhyd MeOH.

References

[1]N. Pelick and V. Mahadevan, in *Analysis of Lipids and Lipoproteins*, ed by E. G. Perkins, Am. Oil Chem. Soc., Champaign, IL, 1975, p. 24.

Derivative Type. Methyl ester.

Reagent. Methanol (MeOH)/HCl.

Compound Type. Formic and acetic acids (in cigarette smoke).

Reaction

$$\text{HCOOH} \xrightarrow[\text{HCl}]{\text{MeOH}} \text{HCOOCH}_3$$

Typical Procedure. Pass cigarette smoke through H_2O in a bubbler trap. Pass the resulting H_2O soln through a 10 g column of Rexyn 202 ion exchange resin (F^- form). Wash column with 100 ml H_2O, 100 ml H_2O/MeOH (1 : 1, v/v), and 100 ml MeOH. Elute with 7–10% HCl in MeOH. Discard first 5 ml of eluate and collect the following 35 ml. Analyze collected eluate by GC.

Comments. Use of the strongly basic anion exchange resin in the fluoride form permitted collection of the free acids without hydrolysis of formate or acetate esters. Elution with HCl/MeOH gave simultaneous esterfication.

References

[1]E. T. Oakley *et al.*, *Anal. Chem.* **37**, 380 (1965).

Derivative Type. Methyl ester.

Reagent. Strong base anion exchange resin, Methanol (MeOH)/HCl.

Compound Type. Free fatty acids (in fats).

Reaction

$$\text{R—COOH} \xrightarrow[\text{2. MeOH/HCl}]{\text{1. IRA-400}} \text{R—COOCH}_3$$

Typical Procedure. Diss 0.1–1 g fat in 40 ml pet ether and add 10–25 mg *n*-heptadecanoic acid int stand. Stir 5 min with 10 g Amberlite IRA-400 (OH form), separate resin, and wash with 3 × 25 ml pet ether. Add 25 ml 5–10% (w/w) HCl in anhyd MeOH and stir 25 min. Filter and wash the resin with 2 × 15 ml of the HCl/MeOH. Add 10 ml of H_2O to the combined filtrate and washings, and ext with 50 ml and 2 × 20 ml pet ether. Wash combined org phases with H_2O until all HCl is removed then dry over Na_2SO_4. Conc under N_2 and analyze by GC.

Comments. The ion exchange resin was pretreated with 1 *N* NaOH (25 ml/10 g) and washed with H_2O, EtOH, and pet ether.

References

[1]I. Hornstein *et al.*, *Anal. Chem.* **32**, 540 (1960).

Derivative Type. Methyl ester.

Reagent Methanol/HCl/chloroform/cupric acetate [Cu(OAc)$_2$].

Compound Type. Fatty acids.

Reaction

$$\text{R-COOH} \xrightarrow[\text{Cu(OAc)}_2]{\text{MeOH/HCl}} \text{R-COOCH}_3$$

Typical Procedure. Diss 1 mg acid in 0.2 ml $CHCl_3$. Add 0.2 ml 0.02 *M* $Cu(OAc)_2$ in MeOH and 1 ml of 0.5 *N* HCl in MeOH. After 10–60 min at room temp or at 37°, add 0.4 ml H_2O and ext with 3 × 2 ml hexane. Alternatively add 2 ml H_2O and ext with 2 ml $CHCl_3$. Wash ext with H_2O and evap to dryness under N_2. Analyze by GC.

Comments. Cupric acetate monohydrate was used to prepare the soln in MeOH. The method did not cause alteration of and was suitable for use with polyunsaturated fatty acids.

References

[1]M. Hoshi *et al.*, *J. Lipid Res.* **14**, 599 (1973).

Derivative Type. Methyl ester.

Reagent Methanol (MeOH)/boron trichloride.

Compound Type. Fatty acids.

Reaction

$$\text{R-COOH} \xrightarrow[\text{BCl}_3]{\text{MeOH}} \text{R-COOCH}_3$$

Typical Procedure. Diss 4 mg of fatty acid in 1 ml 10% BCl_3 in MeOH in a glass ampoule. Flush ampoule with N_2 and seal. Heat at 120° for 90 min. Open ampoule and ext esters with hexane. Wash ext with H_2O then dry over Na_2SO_4. Analyze by GC.

Comments. BCl_3/MeOH esterification was compared with BF_3/MeOH esterification in sealed ampoules and screw-capped vials. The procedure above was optimum with respect to loss of unsatd acids and artifact formation. BF_3 caused some loss of unsatd acids in Teflon® lined screw-cap vials and in sealed ampoules. Losses were more marked in vinyl lined screw-cap vials. Long term studies at 80° showed BCl_3 to cause less losses than BF_3.

References

[1]W. E. Klopfenstein, *J. Lipid Res.* **12**, 773 (1971).

Derivative Type. Methyl esters.

Reagent Boron trichloride/methanol (MeOH).

Compound Type. Cyclopropane fatty acids.

Reaction

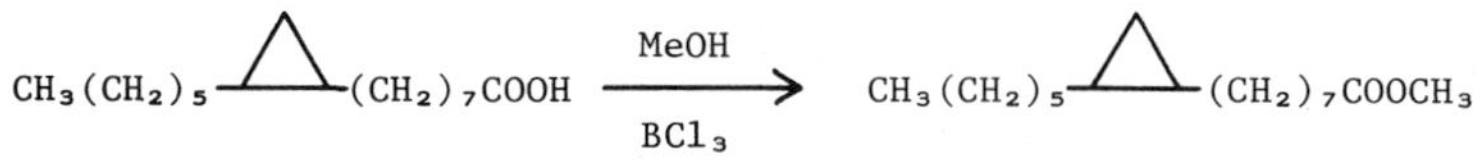

Typical Procedure. Diss a sample containing 1 mg of each acid in 2 ml of 10% (w/v) BCl_3 in MeOH. Heat 2 min at 100°, dil with 4 ml $CHCl_3$, and wash with 1 ml H_2O. Remove the $CHCl_3$ phase and evap under N_2. Diss the esters in 1 ml $CHCl_3$ and analyze by GC.

Comments. Esterification of cyclopropane fatty acids with BCl_3/MeOH was compared with the use of BF_3/MeOH and CH_2N_2.[1] Both BCl_3/MeOH and CH_2N_2 gave satisfactory results but BF_3/MeOH gave extraneous peaks. BF_3/MeOH has been shown to react with disubstituted cyclopropanes to give methoxy derivs and olefins.[2]

References

[1] B. L. Brian *et al.*, *J. Chromatogr.* **66**, 138 (1972).

[2] D. E. Minnikin and N. Polgar, *Chem. Commun.*, 312 (1967).

Derivative Type. Methyl esters.

Reagent Methanol (MeOH)/boron trifluoride.

Compound Type. Fatty acids and lipids.

Reaction

$$\text{R—COOH} \xrightarrow[\text{BF}_3]{\text{MeOH}} \text{R—COOCH}_3$$

Typical Procedure. To 150 mg lipid add 4 ml 0.5 *M* NaOH in MeOH. Heat on a steam bath until the lipid diss (*ca* 5 min). Add 5 ml 12.5% (w/v) BF_3 in MeOH and boil 2 min. Add satd NaCl soln and ext with pet ether. Analyze the ext by GC.

Comments. Free fatty acids (150 mg) were treated directly with 3 ml of the BF_3/MeOH reagent. Kleiman et al.[3] compared the method with the use of MeOH/HCl and NaOMe/MeOH and found it suitable for fats containing unusual structures such as cyclopropane rings, vicinal hydroxyls, oxirane oxygen, conjugated dienols, and conjugated trienes. In some cases it was less destructive than the mineral acid methods. Lough[4] had reported methoxy substituted artifacts from the use of BF_3/MeOH on fatty acids.

References

[1] L. D. Metcalfe and A. A. Schmitz, *Anal. Chem.* **33**, 363 (1961).

[2] L. D. Metcalfe and A. A. Schmitz, *Anal. Chem.* **38**, 514 (1966).

[3] R. Kleiman *et al.*, *Lipids* **4**, 118 (1969).

[4] A. K. Lough, *Biochem. J.* **90**, 4C (1964).

Derivative Type. Methyl esters.

Reagent. *N,N′*-Carbonyldiimidazole (CDI), methanol (MeOH)/triethylamine (Et_3N).

Compound Type. Fatty acids.

Reaction

$$\text{R-COOH} \xrightarrow[\text{2. MeOH/Et}_3\text{N}]{\text{1. CDI}} \text{R-COOCH}_3$$

Typical Procedure. Mix 0.1–1.0 ml plasma or serum with 0.033 *N* H_2SO_4 to make 3 ml and ext with 4 ml heptane/*i*-PrOH (3 : 7, v/v) containing pentadecanoic acid int stand. Add 1 ml of the ext to 20–50 mg silicic acid (100–150 mesh, washed with heptane and dried at 110° overnight). Mix well then decant heptane phase and evap under N_2. Diss residue in 0.1 ml 0.325 *M* CDI in $CHCl_3$. Add 1 ml 10% Et_3N in abs MeOH. After 1 min or more add 3 ml 1.0 *N* aq NaOH satd with CCl_4. Cap vials and shake 2 min. Puncture caps with a needle to relieve the pressure then centrifuge to separate the bottom $CHCl_3$ phase. Analyze an aliquot of the $CHCl_3$ by GC.

Comments. The method was used for analysis of free fatty acids (C_{14}–C_{18}) in serum and plasma. The gentle esterification did not transesterfy triglycerides or cholesteryl esters and caused less than 20% transesterification of phospholipids. Other alcohols could be used for esterification but MeOH was found to be optimal for this analysis.

References

[1]H. Ko and M. E. Royer, *J. Chromatogr.* **88**, 253 (1974).

Derivative Type. Methyl esters.

Reagent Methyl iodide/potassium carbonate.

Compound Type. Fatty acids.

Reaction

$$\text{R—COOH} \xrightarrow[\text{K}_2\text{CO}_3]{\text{CH}_3\text{I}} \text{R—COOCH}_3$$

Typical Procedure. Mix 50 μl of serum or plasma with 10 μl 1 *N* H_2SO_4 and ext with 1.1 ml $CHCl_3$/MeOH (2 : 1, v/v) containing heptadecanoic acid int stand. Evap 1 ml of the ext under vac and treat the residue with 20 μl CH_3I/EtOAc (1:1, v/v) and 1–2 mg K_2CO_3 at 90° for 1 hr. Analyze the resulting derivs by GC.

Comments. The method was used for analysis of long chain (C_{10}–C_{18}) fatty acids in serum.

References

[1]A. Grünert and K. H. Bassler, *Z. Anal. Chem.* **267**, 342 (1973).

Derivative Type. Methyl ester.

Reagent. Silver nitrate, methyl iodide.

Compound Type. Fatty acids.

Reaction

$$\text{R-COOH} \xrightarrow[\text{3. CH}_3\text{I}]{\text{1. KOH} \;\; \text{2. AgNO}_3} \text{R-COOCH}_3$$

Typical Procedure. To a sample of up to 50 mg fatty acids or their K salts add 5 ml of 0.3 *N* KOH in 95% EtOH. Warm on a steam bath and add EtOH to diss K salts. Add about a two-fold excess of aq $AgNO_3$ and evap EtOH in a stream of N_2 on a steam bath. Add 3 ml of a satd soln of $AgNO_3$ in 95% EtOH, mix, and chill to $-20°$. Remove EtOH by filtration and mix residual salts with 50 mg clean sand. Add 1 ml *n*-pentane containing 0.1 ml CH_3I. Shake and allow to stand 8 hr under inert atmosphere. Analyze by GC.

Comments. GC data were reported for C_6–C_{18} fatty acids. The method was also applied to dicarboxylic- and hydroxy-acids.

References

[1]C. W. Gehrke and D. F. Goerlitz, *Anal. Chem.* **35**, 76 (1963).

Derivative Type. Methyl esters.

Reagent. Diazomethane.

Compound Type. Fatty acids.

Reaction

$$R\text{—}COOH \xrightarrow{CH_2N_2} R\text{—}COOCH_3$$

Typical Procedure. Diss 5–30 mg fatty acids in 2–3 ml Et_2O containing 10% MeOH. Generate CH_2N_2 from *ca* 2 mmol precursor per mmol fatty acids and pass through the soln via a stream of Et_2O satd N_2. When the sample soln develops a slight yellow color, stop the esterfication and destroy the excess CH_2N_2 by adding a dil soln of HOAc in Et_2O or by evap under N_2. Analyze by GC.

Comments. The reaction was slow or incomplete when the 10% MeOH was omitted from the Et_2O. The apparatus[1] consisted of three test tubes with side arms (see p. 22). A stream of N_2 was satd with Et_2O in the first tube and bubbled through the CH_2N_2 generating soln in the second tube carrying the CH_2N_2 into the sample soln in the third tube. The N_2 flow was adjusted to *ca* 6 ml/min. The second tube contained 0.7 ml Et_2O, 0.7 ml carbitol, and 0.6 g KOH in ml H_2O, to which was added *N*-methyl-*N*-nitroso-*p*-toluenesulfonamide to generate CH_2N_2. Vorbeck *et al.*[2] esterfied fatty acids with CH_2N_2 from a microgenerator[3] (CH_2N_2 made from *N*-nitrosomethylurea and KOH) and compared the method with methylation by MeOH/HCl[4,5] and MeOH/BF_3.[6] Esters of fatty acid mixts containing low mol wt fatty acids were best prepared by use of CH_2N_2. No significant loss of polyunsaturated fatty acids was found as a result of using CH_2N_2.[2] It had been reported that freshly dist CH_2N_2 was needed to avoid spurious peaks in the C_{10}–C_{12} region.[7]

References

[1]H. Schlenk and J. L. Gellerman, *Anal. Chem.* **32**, 1412 (1960).

[2]M. L. Vorbeck *et al., Anal. Chem.* **33**, 1512 (1961).

[3]R. Roper and T. S. Ma, *Microchem. J.* **1**, 245 (1957).

[4]W. Stoffel *et al., Anal. Chem.* **31**, 307 (1959).

[5]I. Hornstein *et al., Anal. Chem.* **32**, 540 (1960).

[6]L. D. Metcalf and A. A. Schmitz, *Anal. Chem.* **33**, 363 (1961).

[7]W. R. Morrison *et al., Chem. Ind.* 1534 (1961).

Derivative Type. Methyl ester.

Reagent Diazomethane.

Compound Type. Carboxylic acids.

Reaction

$$R\text{—}COOH \xrightarrow{CH_2N_2} R\text{—}COOCH_3$$

Typical Procedure. Prepare a 1.5 cm column of compacted Celite in a piece of melting point capillary. Apply 0.5–5 μg of the acid(s) in 2 μl Et_2O to the column. Insert the capillary in a rubber septum which is then attached to the top of a 1 ml vial containing a stirred mixt (10 μl each) of Diazald[2] in Et_2O (1 g/3 ml) and aq KOH (50% w/w). After 2 min remove capillary and elute esters with 10–15 μl of CH_2Cl_2, CCl_4, Et_2O, or CS_2. Analyze by GC.

Comments. The method was used for microscale preparation of methyl esters.

References

[1]D. P. Schwartz and R. S. Bright, *Anal. Biochem.* **61**, 271 (1974).

[2]*N*-Methyl-*N*-nitroso-*p*-toluenesulfonamide, Aldrich Chemical Co., Milwaukee, WI.

Derivative Type. Methyl ester.

Reagent Tetramethylammonium hydroxide (TMAH).

Compound Type. Carboxylic acids.

Reaction

$$R\text{–}COOH \xrightarrow{TMAH} R\text{—}COO^-(CH_3)_4N^+ \xrightarrow[\Delta]{} R\text{–}COOCH_3 + (CH_3)_3N$$

Typical Procedure. Diss acid in MeOH and titrate to phenolphthalein end point with TMAH in MeOH. Analyze an aliquot by GC with injector at 365°

Comments. Methylation was achieved by pyrolysis of the ammonium salt in the hot injector. Seven aliphatic acids and benzoic, cinnamic, succinic, levulinic, glycolic, and lactic acids were successfully derivatized.[1] Oxalic, malonic, malic, and citric did not yield methyl esters by this procedure.[1] The deriv method was subsequently used in various applications including analysis of aliphatic[2,3] and aromatic[2] acids, rosin acids,[4,5] indole acids,[6] and Krebs cycle acids.[7,8] The reagent has also been used to convert polyunsaturated fatty acids in lipids to conjugated unsaturated fatty acid methyl esters for analysis of degree of unsaturation.[9].

References

[1]E. W. Robb and J. J. Westbrook III, *Anal. Chem.* **35**, 1644 (1963).

[2]J. J. Bailey, *Anal. Chem.* **39**, 1485 (1967).

[3]D. T. Downing and R. S. Greene, *Anal. Chem.* **40**, 827 (1968).

[4]N. E. Hetman *et al.*, *J. Am. Oil Chem. Soc.* **42**, 255 (1965).

[5]N. M. Joye *et al.*, *J. Am. Oil Chem. Soc.* **51**, 195 (1974).

[6]R. H. Leonard, *J. Gas Chromatogr.* **5**, 323 (1967).

[7]A. Kuksis and P. Vishwakarma, *Can. J. Biochem. Physiol.* **41**, 2353 (1963).

[8]A. Kuksis and P. Prioreschi, *Anal. Biochem.* **19**, 468 (1967).

[9]E. D. Bitner *et al.*, *J. Am. Oil Chem. Soc.* **48**, 633 (1971).

Derivative Type. Methyl ester.

Reagent. Tetramethylammonium hydroxide (TMAH).

Compound Type. Carboxylic acids.

Reaction

$$\text{R–COOH} \xrightarrow[\Delta]{\text{TMAH}} \text{RCOOCH}_3$$

Typical Procedure. Prepare tetramethylammonium salts of the acids by treatment with a twofold mol excess of 25 % aq TMAH. Dry an aliquot of the solution in a capillary probe and analyze by GC with the vaporizor at 300°.

Comments. It was found that analysis of dried samples via capillary probes gave higher and more reproducible yields of ester derivs than by direct injection of TMAH solns.

References

[1]D. T. Downing, *Anal. Chem.* **39**, 218 (1967).

Derivative Type. Methyl ester.

Reagent. Trimethylphenylammonium hydroxide (TMPAH).

Compound Type. Fatty acids.

Reaction

$$\text{R-COOH} \xrightarrow[\Delta]{\text{TMPAH}} \text{R-COOCH}_3$$

Typical Procedure. Diss 1 mg fatty acid mixt in 0.5 ml 0.2 *M* TMPAH in MeOH. Analyze 1 μl aliquots by GC with injector at 265°.

Comments. The production of methyl esters appeared to be quantitative and no side products were noted.

References

[1]B. S. Middleditch and D. M. Desiderio, *Anal. Lett.* **5**, 605 (1972).

Derivative Type. Methyl esters.

Reagent. (*m*-Trifluoromethylphenyl) trimethylammonium hydroxide.

Compound Type. Fatty acids.

Reaction

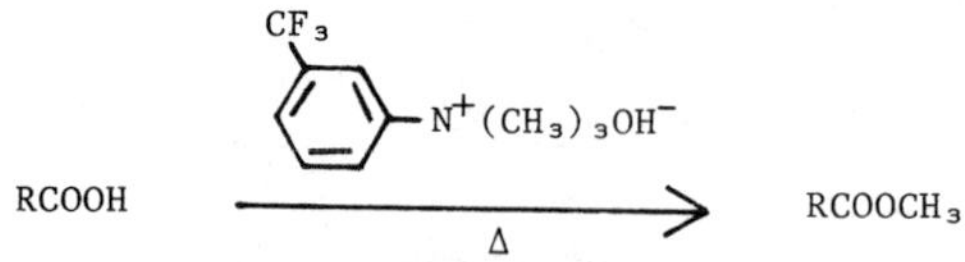

Typical Procedure. Hydrolyze a sample of tissue (3–5 mg) or serum (5–100 μl) with 0.5 ml

KOH (150 mg/ml) in MeOH, acidify (0.7 ml 1 *M* H_3PO_4), then ext into 5 ml hexane. Ext hexane with 10 μl aq 0.5 *M* (*m*-trifluoromethylphenyl) trimethylammonium hydroxide and inject the resulting ammonium salts (lower phase) into the GC.

Comments. The reagent caused less destruction of unsatd fatty acids than either TMAH or TMPAH. Having a better leaving group, the reagent could be used at lower injector temp (240°) than that required for either TMAH or TMPAH.

References

[1]J. MacGee and K. G. Allen, *J. Chromatogr.* **100**, 35 (1974).

Derivative Type. Alkyl ester.

Reagent. Dimethylformamide (DMF) dialkyl acetal.

Compound Type. Fatty acids.

Reaction

$$\text{R–COOH} \xrightarrow{(\text{CH}_3)_2\text{N–CH(OCH}_3)_2} \text{R–COOCH}_3$$

Typical Procedure.[1] Diss fatty acid sample in DMF dialkyl acetal and inject soln directly into the GC. A "sandwich" technique can also be used. Draw successive solns into a syringe in the following order: 1 μl of DMF dialkyl acetal, 1 μl of pyridine containing 2 μg of docosane (standard), and 1 μl of pyridine containing 2 μg of fatty acid. Inject into the GC.

Comments. The method was used for methyl, ethyl, propyl, *n*-butyl, and *t*-butyl acetals. Orthoesters would not give a similar result. The deuterated methyl reagent is commercially available for preparation of d_3-methyl esters.[4]

References

[1]J.-P. Thenot *et al., Anal. Lett.* **5**, 217 (1972).
[2]A. Vorbrüggen, *Angew Chem. Int. Ed.* **2**, 211 (1963).
[3]H. Brechbuhler *et al., Angew. Chem. Int. Ed.* **2**, 212 (1963).
[4]Tri-Deuter-8®, Pierce Chemical Co., Rockford, IL.

3.1.1.2. *Other Fatty Acid Esters*

Derivative Type. *n*-Propyl esters.

Reagent *n*-Propanol (*n*-PrOH)/ortho-phosphoric acid/boron trifluoride.

Compound Type. Mono- and dicarboxylic fatty acids (as sodium salts).

Reaction

$$\text{R–COO}^-\text{Na}^+ \xrightarrow[\text{2. }\underline{n}\text{-PrOH/BF}_3]{\text{1. }\underline{n}\text{-PrOH/H}_3\text{PO}_4} \text{R–COOC}_3\text{H}_7$$

Typical Procedure. Grind a sample of the sodium salts with 1 ml *n*-PrOH containing 3–4 drops H_3PO_4. Add 3 ml *n*-PrOH and 2 ml 25 % BF_3 in *n*-PrOH. Heat 2 min at 90° then cool in ice. Add 30 ml ice H_2O and ext with pet ether (30-60°) or cyclopentane. Analyze by GC.

Comments. The method was suitable for analysis of C_4–C_{20} monocarboxylic acids and C_3–C_{12} dicarboxylic acids. It could be extended to formic, acetic, propionic, and oxalic acids but these required calibration curves since they were less efficiently extracted from aq phase.

References

[1]A. J. Appleby and J. E. O. Mayne, *J. Gas Chromatogr.* **5**, 266 (1967).

Derivative Type. Isopropyl esters.

Reagent. Isopropanol (*i*-PrOH)/boron trifluoride.

Compound Type. Carboxylic acids.

Reaction

$$\text{R—COOH} \xrightarrow[\text{BF}_3]{i\text{-PrOH}} \text{R—COOCH(CH}_3)_2$$

Typical Procedure. To 25–100 μg of acids add 150 μl of 2 *N* BF_3 in *i*-PrOH and heat at 100° for 20 min. Cool in an ice bath and add 0.2 ml *n*-pentane, 0.2 ml Et_2O and 0.15 ml *n*-pentane/Et_3N (1:1, v/v) (pyridine may be used in place of Et_3N). Mix well, centrifuge, and analyze an aliquot of the supernatant by GC.

Comments. *n*-Butyl esters were also prepared in the same manner. The method was developed as an improvement over earlier methods that employed an aq wash to remove BF_3 and was reported to be simpler and give better recoveries. The Et_3N or pyridine reacted with BF_3 to give an insoluble adduct which was removed by centrifugation.

References

[1]P. A. Biondi and M. Cagnasso, *J. Chromatogr.* **109**, 389 (1975).

Derivative Type. *n*-Butyl ester.

Reagent. *n*-Butanol (*n*-BuOH)/boron trifluoride.

Compound Type. Short chain fatty acids.

Reaction

$$\text{R—COOH} \xrightarrow[\text{BF}_3]{n\text{-BuOH}} \text{R—COOC}_4\text{H}_9$$

Typical Procedure. Evap an Et_2O soln containing *ca* 0.5 mg of each acid to 0.2 ml and add 0.1 ml 14% (w/v) BF_3 in *n*-BuOH. Allow to stand a few min for the remaining Et_2O to evap; then cap and heat at 100° for 5–10 min. Cool to room temp, add 0.1 ml $CHCl_3$, then wash with 0.2 ml H_2O. Remove org layer and analyze by GC.

Comments. The method was used to analyze C_1–C_8 acids including branched and hydroxy acids. The hydroxy acid esters were converted to trifluoroacetyl derivs with TFAA (10 min at room temp). MS data were reported for the derivs.[2]

References

[1]M. A. Lambert and C. W. Moss, *J. Chromatogr.* **74**, 335 (1972).

[2]C. Choudhary and C. W. Moss, *J. Chromatogr.* **128**, 261 (1976).

Derivative Type. Decyl ester.

Reagent. Decyl alcohol/HCl.

Compound Type. Short chain fatty acids.

Reaction

$$R\text{—}COOH \xrightarrow[HCl]{C_{10}H_{21}OH} R\text{—}COOC_{10}H_{21}$$

Typical Procedure. Mix 5–100 mg of fatty acid K salts with 1 ml decyl alcohol containing 10% (w/w) HCl and 4 ml 60–80° pet ether. Shake 1 hr at room temp then wash several times with H_2O. Remove excess alcohol by passing the soln through a column of 25 g alumina (Activity I) prepared with Et_2O. Elute the column with Et_2O collecting the first 300 ml. Evap the Et_2O under slight vac at 45° to conc the soln. Analyze by GC.

Comments. Butyl esters were also prepared from diazobutane and phenacyl esters from α-bromoacetophenone and KOH. The three methods were compared for propionic through pelargonic acids and the decyl ester deemed the best.

References

[1]B. M. Craig *et al.*, *J. Am. Oil Chem. Soc.* **40**, 61 (1963).

Derivative Type. Benzyl ester.

Reagent. Phenyldiazomethane.

Compound Type. Fatty acids.

Reaction

$$R\text{—}COOH \xrightarrow{C_6H_5CHN_2} R\text{—}COOCH_2C_6H_5 + N_2$$

Typical Procedure. Prepare 1% (w/v) soln of fatty acids in pentane and add phenyldiazomethane in pentane until the red color persists. Allow to react 10–15 min at room temp then destroy excess reagent by adding 1% H_3PO_4 in Et_2O. Analyze 1 μl of the soln by GC.

Comments. Longer chain fatty acids required the use of $CHCl_3$/MeOH (2:1,v/v) in place of pentane for solubility.[2] The reaction was also carried out at 0° allowing a 30 min reaction time. The preparation of the reagent was reported.[1]

References

[1]H.-P. Klemm *et al.*, *J. Chromatogr.* **75**, 19 (1973).
[2]U. Hintze *et al.*, *J. Chromatogr.* **87**, 481 (1973).

Derivative Type. Benzyl ester.

Reagent. Phenyldiazomethane.

Compound Type. Carboxylic acids.

Reaction

$$R\text{—}COOH \xrightarrow{C_6H_5CHN_2} R\text{—}COOCH_2C_6H_5$$

Typical Procedure. Ext a 1 ml aq soln of the acids with 3 × 6 ml Et_2O. Add 0.5–1.0 ml of a 50% (v/v) soln of phenyldiazomethane in pet ether (30–40°) and allow to stand 15 min at room temp. Cool to 0° and evap under N_2 at −10–0° to *ca* 50 μl. If the red color does not persist, add additional phenyldiazomethane. Add a few pellets of 4A mol sieve and allow to dry for 1 hr. Analyze the resulting pale yellow soln by GC.

Comments. The method was used initially for analysis of short chain (C_2–C_{10}) acids in bacterial cultures[1] but was extended to a wide variety of carboxylic acids.[2] Doms[3] used phenyldiazomethane for analysis of formic acid[3] and C_1–C_5 fatty acids (including lactic acid)[4]. Hexanoic acid was added to destroy excess reagent.[4]

References

[1]D. L. Corina and P. M. Dunstan, *Anal. Biochem.* **53**, 571 (1973).

[2]D. L. Corina, *J. Chromatogr.* **87**, 254 (1973).

[3]E. K. Doms, *J. Chromatogr.* **105**, 79 (1975).

[4]E. K. Doms, *J. Chromatogr.* **140**, 29 (1977).

Derivative Type. *p*-Bromophenacyl ester.

Reagent. *p*-Bromophenacyl bromide/potassium hydroxide.

Compound Type. Carboxylic acids (C_2–C_{10}).

Reaction

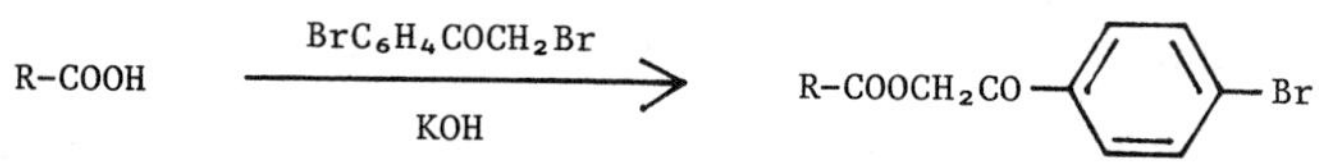

Typical Procedure. Mix 2.5 ml of an EtOH soln of the acids (0.1 *M* each) with 5 ml H_2O and one drop of 0.5% phenolphthalein. Neutralize with a few drops of 5 *N* KOH then make just acid with 3 drops of 1 *N*HCl. Add 0.63 g *p*-bromophenacyl bromide and reflux 10 min. Add a few drops of EtOH if necessary to diss solids. Cool and add 25 ml EtOAc. Add cold H_2O until the EtOAc sep. Remove EtOAc, dry over Na_2SO_4, and analyze by GC.

Comments. *p*-Phenylphenacyl esters were prepared in the same manner from *p*-phenylphenacyl bromide. The derivs were used for qualitative and quantitative analysis of C_2–C_{10} acids using FID–GC.

References

[1]E. O. Umeh, *J. Chromatogr.* **56**, 29 (1971).

Derivative Type. *p*–Bromobenzyl ester.

Reagent. Bromobenzyl bromide/potassium hydroxide.

Compound Type. Carboxylic acids (C_1 to C_3).

Reaction

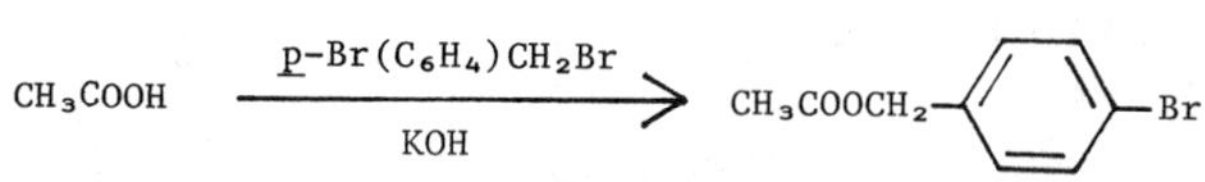

Typical Procedure. Mix 10 μl of a 1 μg/μl soln of the acid in EtOH with 3 μl of a 3 μg/μl soln of KOH in EtOH and 5 μl of the reagent (3 μg/μl) in EtOH in a melting point capillary. Flame seal the tube and heat at 110° for 1 hr. Cool and analyze a 2 μl aliquot by GC.

Comments. *p*-Nitrobenzyl and *p*-methylbenzyl esters were prepared by the same method using *p*-nitrobenzyl iodide and *p*-methylbenzyl bromide respectively. The *p*-bromobenzyl derivs gave optimum results.

References

[1]R. Watson and P. Crescuolo, *J. Chromatogr.* **52**, 63 (1970).

Derivative Type. Butyl ester.

Reagent. 1-Iodobutane/trimethylphenylammonium hydroxide (TMPAH).

Compound Type. Fatty acids.

Reaction

$$\text{R—COOH} \xrightarrow[\text{TMPAH}]{C_4H_9I} \text{R—COOC}_4\text{H}_9$$

Typical Procedure. A. Mix 0.05 mmol stearic acid (14.2 mg) with 4 ml DMA and 1 ml 0.1 *M* TMPAH in MeOH. Add 0.10 ml 1-iodobutane (0.88 mmol) and mix (98 % reacted in 10 min). Analyze by GC.

B. Mix 0.088 mmol stearic acid (25 mg), 4 ml DMA, 0.95 ml MeOH, 0.05 ml. TMAH soln (0.104 mmol), and 0.10 ml 1–iodobutane. Allow the tetramethylammonium iodide to ppt to yield 99 % conversion. Analyze by GC.

Comments. This rapid esterfication method was also used with methyl, ethyl, and 2-butyl iodides and with either TMPAH or TMAH. DMF was also a suitable solvent.

References

[1]R. H. Greeley, *J. Chromatogr.* **88**, 229 (1974).

Derivative Type. Benzyl ester.

Reagent. *N,N'*-Dicyclohexyl-O-benzylisourea.

Compound Type. Fatty Acids.

Reaction

$$\text{R-COOH} + C_6H_{11}\text{-N=}\overset{\overset{\displaystyle OCH_2C_6H_5}{|}}{C}\text{-NH-}C_6H_{11} \longrightarrow \text{R-COOCH}_2C_6H_5 + C_6H_{11}\text{-NH-}\overset{\overset{\displaystyle O}{||}}{C}\text{-NH-}C_6H_{11}$$

Typical Procedure. Diss 10 mg fatty acids per 1 ml benzene and reflux 1 hr with slight excess of *N,N'*-dicyclohexyl-O-benzylisourea. Allow to cool to room temp (dicyclohexylurea sediments) and analyze 1 μl of the supernatant by GC.

Comments. Other solvents (e.g., dioxane, THF, CCl_4) could be used and did not have to be anhydrous. The derivs, being less volatile than methyl esters, offered some advantages for short chain acids. The preparation of the reagent was reported.

References

[1]H.-P Klemm *et al.*, *J. Chromatogr.* **75**, 19 (1973).

3.1.2. Fatty Acid Esters from Glycerides

3.1.2.1. Methyl Esters from Glycerides

Derivative Type. Fatty acid methyl esters.

Reagent. Sodium hydroxide/methanol (MeOH), methanol/ammonium chloride/sulfuric acid.

Compound Type. Lipids.

Reaction

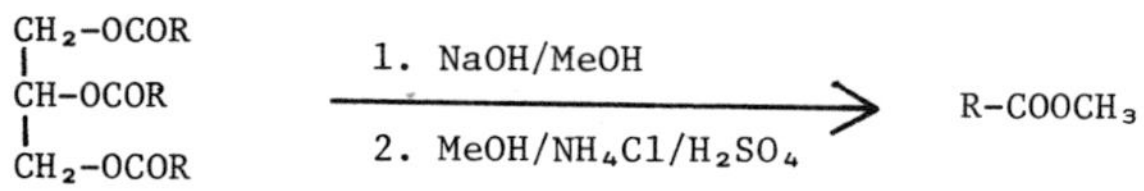

Typical Procedure. Boil 200–500 mg lipid with 5 ml 0.5 *N* NaOH or KOH in MeOH for 3–5 min. Add 15 ml of the esterfication mixt (see below) and reflux 3 min. Transfer to a sep funnel with 25 ml lt pet ether and 50 ml H_2O. Discard the aq phase and wash the org phase with 2× 25 ml H_2O. Analyze by GC.

Comments. The NH_4Cl/H_2SO_4/MeOH soln was prepared by adding 2 g NH_4Cl to 60 ml MeOH followed by 3 ml conc H_2SO_4 and refluxing 15 min. The method gave comparable results to those obtained with the older methods of alkali catalyzed methanolysis and saponification with ethanolic KOH followed by esterfication with 1% H_2SO_4 in MeOH.

References

[1]L. Hartman and R. C. A. Lago, *Lab. Pract.* **22**, 475 (1973).

Derivative Type Fatty acid methyl esters.

Reagent. Methanol (MeOH)/sodium methoxide (NaOMe).

Compound Type. Lipids.

Reaction

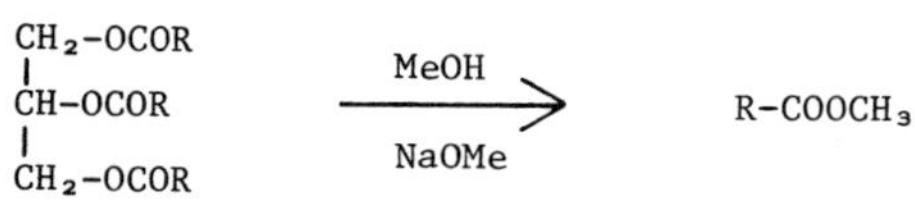

Typical Procedure. Diss 10–30 mg lipid in 1 ml benzene and add 1 ml 0.5 *N* NaOMe in MeOH. Heat at 80° for 15–20 min. Cool to room temp and add 3 ml H_2O and 3 ml Et_2O. Mix well, separate org layer, and wash with 2–3 ml H_2O. Dry over Na_2SO_4 and analyze by GC.

Comments. The method was applicable to triglycerides, cholesteryl esters, and phospholipids. The reagent was prepared by diss metallic Na in anhyd MeOH.

References

[1]N. Pelick and V. Mahadevan, in *Analysis of Lipids and Lipoproteins,* ed by E. G. Perkins, Am. Oil Chem. Soc., Champaign. IL., 1975, p. 29.

Derivative Type. Fatty acid methyl esters.

Reagent. Methanol (MeOH)/potassium methoxide (KOMe).

Compound Type. Lipids

Reaction

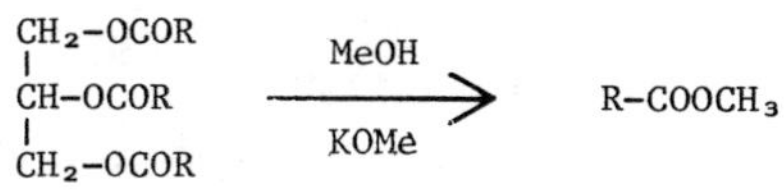

Typical Procedure. To 1–30 mg of oil add 0.25 ml 0.4 *N* KOMe in MeOH and seal in a vial under N_2. Heat at 65° for 30 sec with shaking then another 90 sec without shaking. Cool to room temp and add 0.6 g of a 1 : 1 mixt of silica gel and $CaCl_2$.Stir thoroughly then ext with 3 ml CS_2 by shaking 1–2 min. Centrifuge and analyze an aliquot of the CS_2 soln by GC.

Comments. The method was developed to circumvent losses of volatile short chain methyl esters.

References

[1]F. E. Luddy *et al., J. Am. Oil Chem. Soc.* **45**, 549 (1968).

Derivative Type. Methyl esters.

Reagent. Methanol (MeOH)/potassium hydroxide.

Compound Type. Triglycerides.

Reaction

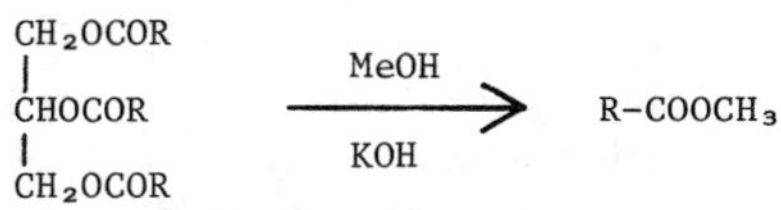

Typical Procedure. To 10 vol of a 10% (w/v) soln of milk fat in pet ether add 1 vol of 2 *N* KOH in MeOH. Mix until soln becomes clear (*ca* 20 sec). The soln will rapidly become turbid due to ppt of glycerol. Allow to settle then analyze an aliquot of the soln by GC.

Comments. The method could also be used with 2 *N* NaOMe in place of KOH.

References

[1]S. W. Christopherson and R. L. Glass, *J. Dairy Sci.* **52**, 1289 (1969).

Derivative Type. Methyl ester.

Reagent. Methanol (MeOH)/chloroform/sulfuric acid.

Compound Type. Triglycerides.

Reaction

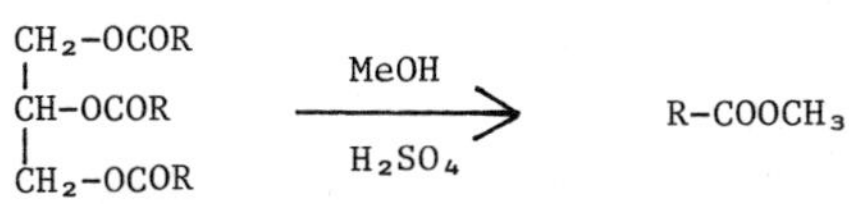

Typical Procedure. Diss 30 mg sample in *ca* 0.75 ml $MeOH/CHCl_3/H_2SO_4$ (100:100:1,v/v/v). Heat in a sealed tube at 170° for 5 min. Cool under cold water, open, add a small piece of Zn and evap in a hot H_2O bath. Add 1 ml H_2O and 1 ml pet ether. Mix well then separate pet ether soln. Dry over Na_2SO_4 and analyze by GC.

Comments. A simple pressure sealing device for glass tubes was described. The Zn was added to prevent H_2SO_4 charring during the evap step.

References

[1]K. V. Peisker, *J. Am. Oil Chem. Soc.* **41**, 87 (1964).

Derivative Type. Methyl esters and dimethyl acetals.

Reagent. Methanol (MeOH)/boron trifluoride.

Compound Type. Lipids.

Reaction

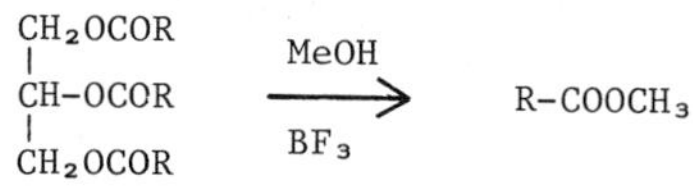

Typical Procedure. To 4–16 mg lipid add under N_2 1 ml of a mixt of 5 parts 14% (w/v) BF_3 in MeOH, 4 parts benzene, and 11 parts MeOH. Heat at 100° for 30 min (other times for other lipid types — see below). Cool, add 2 ml pentane and 1 ml H_2O, and shake. Separate the org layer and analyze by GC.

Comments. The ext method was modified for analysis of fatty aldehydes to prevent hydrolysis of the acetal derivs. The reaction mixt was cooled and 4 ml pentane and 1 ml 5 *N* NaOH added at 0°. The ext was then analyzed as above. Sterol esters were treated with a mixt of 7 parts BF_3/MeOH, 6 parts benzene, and 7 parts MeOH for 45 min at 100°. Other lipids were treated with just the 14% BF_3/MeOH. Fatty acids required only 2 min heating. Mono- and diglycerides; phosphatidyl ethanolamines, -serines, and -cholines; monophosphoinositides; and monogalactosyl glycerides required 10 min. Phosphatidal cholines and digalactosyl glycerides required 30 min while sphingomyelins were heated 90 min. Use of BCl_3 in place of BF_3 gave slower reaction rates. The BF_3/MeOH method has been shown to produce artifacts from unsatd[2] and cyclopropane[3,4] fatty acids.

References

[1]W. R. Morrison and L. M. Smith, *J. Lipid Res.* **5**, 600 (1964).
[2]W. K. Fulk and M. S. Shorb, *J. Lipid Res.* **11**, 276 (1970).
[3]E. A. Dawidowicz and T. E. Thompson, *J. Lipid Res.* **12**, 636 (1971).
[4]D. E. Minnikin and N. Polgar, *Chem. Commun.* 312 (1967).

Derivative Type. Fatty acid methyl esters, isopropylidene glycerol.

Reagent. Dimethoxypropane (DMP)/methanol (MeOH)/HCl.

Compound Type. Triglycerides.

Reaction

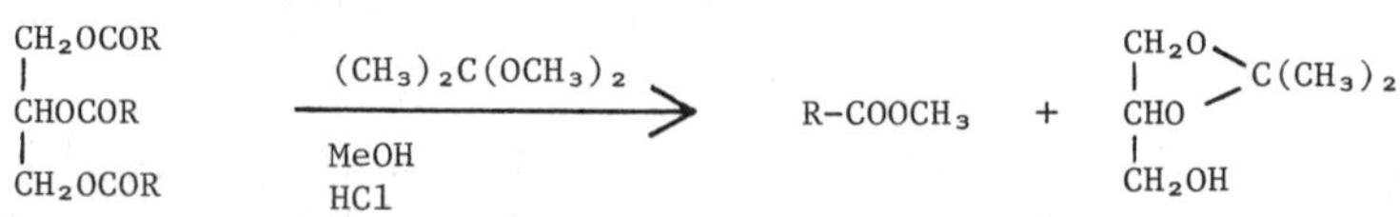

Typical Procedure. Place *ca* 0.2 g fat or oil in a 50 ml flask and add 14 ml dry benzene, 1 ml DMP, and 5 ml 10% HCl in MeOH. Allow to stand overnight at 22° then neutralize with *ca* 2

g of $NaHCO_3/Na_2CO_3/Na_2SO_4$ (2:1:2, w/w/w, dried overnight at 110°). After 30 min, analyze an aliquot by GC.

Comments. The method permitted esterfication without the use of elevated temp. The DMP formed polymeric by-products which gave extraneous GC peaks. Simmonds and Zlatkis[2] found that 0.25% DMSO effectively inhibited by-product formation. The use of DMP/MeOH/HCl for esterfication of free fatty acids had been reported earlier.[3] TsOH was used in place of HCl to esterfy the acids in castor oil with DMP/MeOH.[4]

References

[1]M. E. Mason and G. R. Waller, *Anal. Chem.* **36**, 583 (1964).

[2]P. G. Simmonds and A. Zlatkis, *Anal. Chem.* **37**, 302 (1965).

[3]N. S. Radin *et al., J. Lipid Res.* **1**, 250 (1960).

[4]N. B. Lorette and J. H. Brown, *J. Org. Chem.* **24**, 261 (1959).

Derivative Type. Methyl esters.

Reagent. Methanol (MeOH)/HCl.

Compound Type. Fatty acids in lipids.

Reaction

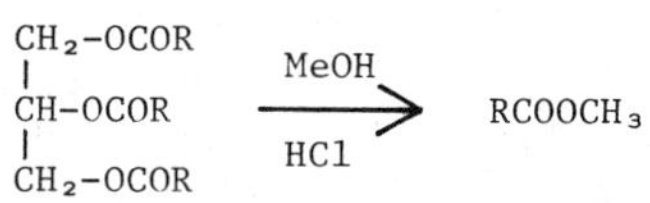

Typical Procedure. Mix a 1–10 mg sample with 4 ml 5% HCl in "super dry" MeOH and 0.5 ml benzene in a tube protected with a $CaCl_2$ moisture trap. Reflux 2 hr in an 80–100° oil bath then cool to room temp. Add 2 vol H_2O and ext with 3 × 3 ml pet ether. Dry and neutralize simultaneously over a mixt of Na_2SO_4 and $NaHCO_3$. Analyze by GC.

Comments. The method gave methyl esters in nearly quant yield from glycerides, phosphatides, and cholesterol esters.

References

[1]W. Stoffel *et al., Anal. Chem.* **31**, 307 (1959).

Derivative Type. Fatty acid methyl esters.

Reagent. Sulfuric acid, methanol (MeOH).

Compound Type. Fats.

Reaction

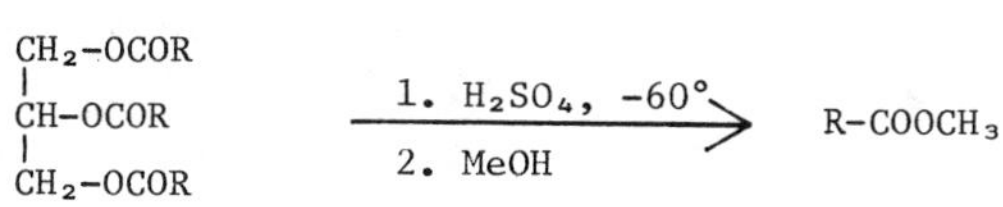

Typical Procedure. Diss 200 mg fat in 20 ml peroxide free Et_2O and cool to −60° in a dry ice-acetone bath. Add 2 ml conc H_2SO_4 gradually over a 2 min period. Allow to warm to −10° over a 10 min period then recool to −60°. Add 15 ml anhyd MeOH followed by 13 ml 35% KOH in MeOH. Remove from bath and allow to warm to room temp with stirring. Dil with 150 ml H_2O and ext with 30 ml and 2 × 15 ml pentane. Dry ext over Na_2SO_4. Analyze by GC.

Comments. The H_2SO_4 complex with the fatty acids was decomposed by MeOH to the methyl esters. The method was developed as a low temp acid catalysis alternative to basic interesterfication to avoid possible base catalyzed isomerizations.

References

[1]G. W. McGinnis and L. R. Dugan, *J. Am. Oil Chem. Soc.* **42**, 305 (1965).

3.1.2.2. *Butyl Esters from Glycerides*

Derivative Type. Butyl ester.

Reagent. Di-*n*-butyl carbonate/sodium butoxide (NaOBu).

Compound Type. Fatty acids (in milk fat).

Reaction

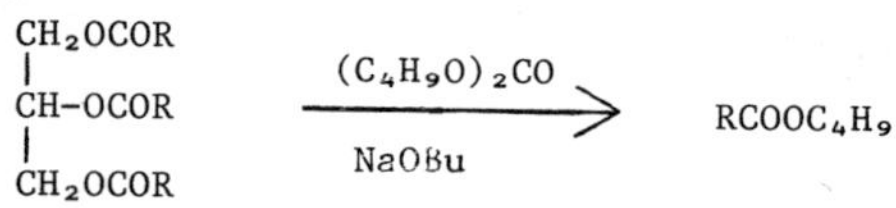

Typical Procedure. To a sample containing up to 100 mg glycerides add 2–5 drops of di-*n*-butyl carbonate and 0.5 ml 0.2 *N* NaOBu. Cap tube and heat on a stream bath 2 min. Cool and add 0.2 ml 1 *N* HCl. Transfer tube contents to a Babcock milk fat test bottle and rinse tube with 0.5 ml H_2O, 0.5 ml acetone, and 0.5 ml H_2O. Add 10 ml satd NaCl and sufficient H_2O to half fill the bottle. Shake then add H_2O to bring level within 2 cm of bottle top. Centrifuge and analyze an aliquot of the top layer by GC.

Comments. The method permitted analysis of both long and short chain acids in a single GC run. It was a modification of that used by Glass *et al.*[2] to analyze milk fat fatty acids as methyl and ethyl esters. The procedure was not applicable to free fatty acids.

References

[1]J. Sampugna *et al.*, *J. Dairy Sci.* **49**, 1462 (1966).
[2]R. L. Glass *et al.*, *J. Dairy Sci.* **48**, 1106 (1965).

Derivative Type. *n*-Butyl ester.

Reagent. *n*-Butanol (*n*-BuOH)/boron trifluoride.

Compound Type. Fatty acids in triglycerides.

Reaction

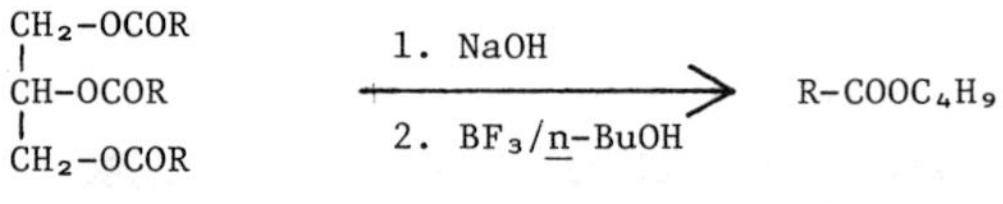

Typical Procedure. To 100 mg triglyceride add 2 ml 0.5 *N* NaOH in *n*-BuOH and reflux 2 min on steam bath. Cool, add 3 ml 12.5% (w/v) BF_3 in *n*-BuOH and 2 ml hexane, and boil for 3 min. Cool, add 40 ml H_2O, and shake. Separate hexane and wash 2 × with H_2O. Analyze an aliquot of the hexane soln by GC.

Comments. The method was used in preference to methyl ester formation for butter analyses due to problems of losses of volatile short chain methyl esters.

References

[1]J. L. Iverson and A. J. Sheppard, *J. Assoc. Off. Anal. Chem.* **60**, 284 (1977).

3.1.3. Derivatives of Other Carboxylic Acids

3.1.3.1. Methyl Esters of Carboxylic Acids

Derivative Type. Methyl ester.

Reagent. Methanol (MeOH)/boron trifluoride.

Compound Type. 2,4-Dichlorophenoxyacetic acid (2,4-D).

Reaction

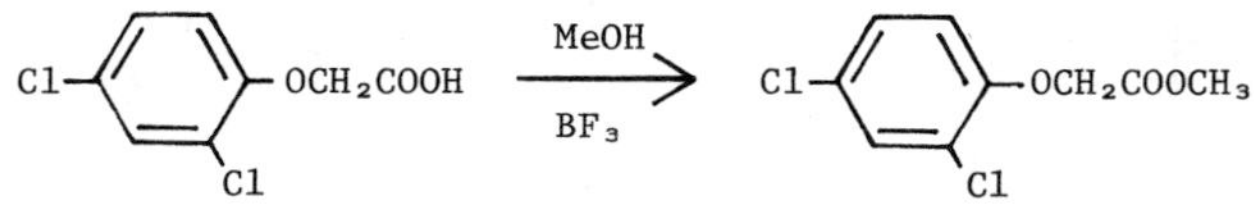

Typical Procedure (Preparative). Diss 1.7 g of 2,4-D in 3 ml MeOH and add 3 ml MeOH containing 0.2 g/ml BF_3. Reflux at 100° for 10 min then evap under vac. Add 20 ml H_2O, shake 5 min, and allow to separate. Remove ester layer and ext the aq phase with 3 × 10 ml Et_2O. Evap the ext and combine the residue with the ester layer. Analyze the product by GC.

Comments. Esterfication of 2, 4-D with BF_3/alcohol, diazoalkane, and alcohol/mineral acid were compared in terms of yield and production of impurities. The BF_3/alcohol was deemed superior to diazoalkanes, which yielded excessive impurities and mineral acid catalysis which was inefficient. The use of BF_3/MeOH for 2,4-D analysis had been reported earlier.[2]

References.

[1]J. Horner *et al., Anal. Chem.* **46**, 110 (1974).
[2]W. H. Gutenmann and D. J. Lisk, *J. Agr. Food Chem.* **11**, 304 (1963).

Derivative Type. Methyl ester.

Reagent. Diazomethane.

Compound Type. Hippuric acid.

Reaction

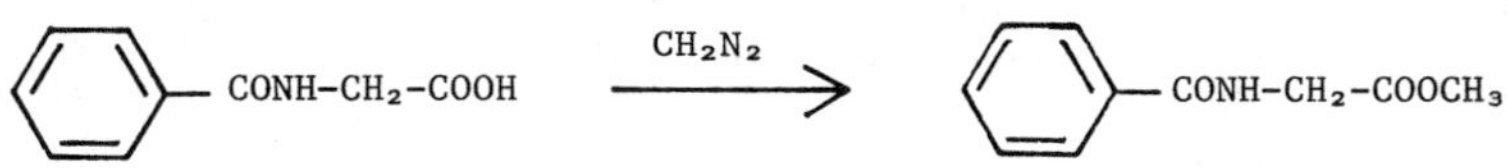

Typical Procedure. Diss *ca* 5 mg hippuric acid in 2 ml of 10% MeOH in Et_2O. Pass CH_2N_2 in N_2 through the soln until a slight yellow color persists (see p. 22 for apparatus).[2] Remove excess CH_2N_2 with a stream of N_2 and analyze by GC.

Comments. Silylation with BSA/TMCS or BSTFA yielded a mixt of mono- and di-TMS derivs. Methylation with BF_3/MeOH yielded a mixt of hippuric and benzoic acid methyl esters. On-column methylation with TMPAH yielded methyl hippurate and its *N*-methyl deriv. The deriv from CH_2N_2 yielded a single GC peak.

References

[1]U. Langenbeck and J. E. Seegmiller, *J. Chromatogr.* **80**, 81 (1973).
[2]H. Schlenk and J. L. Gellerman, *Anal. Chem.* **32**, 1412 (1960).

Derivative Type. Methyl ester.

Reagent. Diazomethane.

Compound Type. 4,4-Dimethyloxazolidine-N-oxyl (doxyl) stearic acids.

Reaction

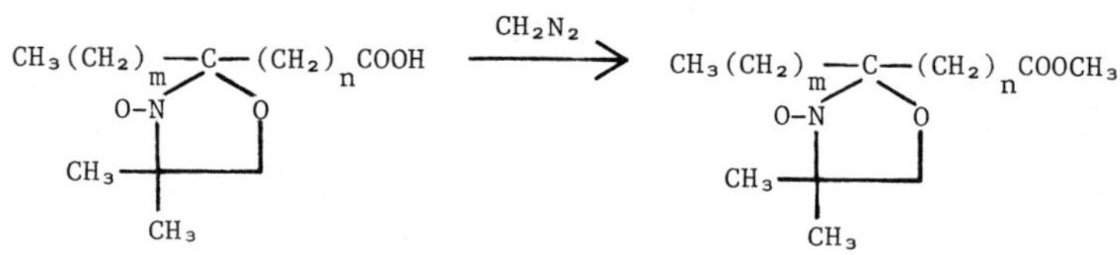

Typical Procedure. Prepare methyl esters using CH_2N_2 in Et_2O and analyze by GC-MS or direct probe MS.

Comments. TMS derivs were also prepared using BSA and TBDMS derivs using TBDMS-Cl/imidazole. The derivs could be analyzed by direct probe MS in the presence of large amts of fatty acids since the fatty esters evaporated at lower temps than the doxyl derivs.

References

[1]L. Murai *et al., Chem. Phys. Lipids* **17**, 213 (1976).

Derivative Type. Methyl or ethyl ester.

Reagent. Diazomethane or diazoethane.

Compound Type. Pesticide related acids and phenols.

Reaction

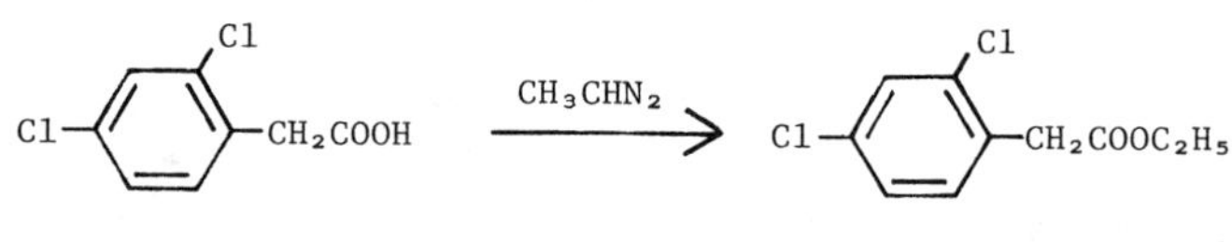

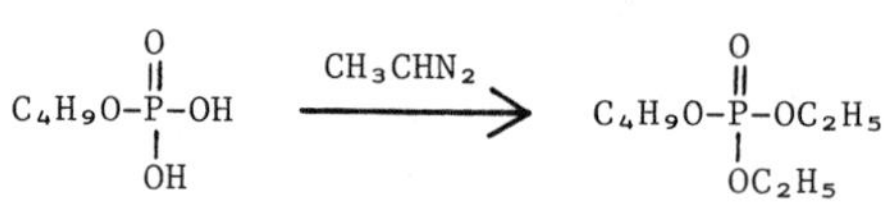

Typical Procedure Transfer up to 0.5 ml of sample ext to a small vial and add 1 drop of EtOAc/conc HCl (1:1, v/v). Add 2 ml of diazoalkane in Et_2O followed by additional aliquots until the color persists. Allow to stand several min then evap to the original vol under N_2. Analyze by GC.

Comments. Diazomethane was prepared from *N*-methyl-*N'*-nitro-*N*-nitrosoguanidine and diazoethane from *N*-ethyl-*N'*-nitro-*N*-nitrosoguanidine by addition of the precursor to the aq KOH and ext of the diazoalkane into Et_2O.[2] Chlorinated phenoxyacetic, propionic, and butyric acids, pentachlorophenol, *o*-phenylphenol, and 4, 6-dinitro-*o*-*sec*-butylphenol were separable as well as the methyl, ethyl, and butyl phosphoric acids.

References

[1]C. W. Stanley, *J. Agric. Food Chem.* **14**, 321 (1966).
[2]A. F. McKay *et al., Can. J. Res.* **28B**, 683 (1950).

Derivative Type. Methyl ester.

Reagent. *N,N*-Dimethylformamide dimethyl acetal.

Compound Type Trifluoroacetic acid (TFA).

Reaction

$$CF_3COOH \xrightarrow{(CH_3)_2NCH(OCH_3)_2} CF_3COOCH_3$$

Typical Procedure. Adjust 20 ml of aq sample to pH 7.0 with 10 *N* NaOH and ext with 3 × 40 ml Et_2O. Adjust remaining aq phase to pH 1.0 with 6 *N* HCl and ext with 3 × 40 ml Et_2O. Combine these extracts and conc to *ca* 0.5 ml under dry N_2. Add 50 μl of the conc ext to 50 μl DMF dimethyl acetal in a vial. Seal and heat at 60° for 15 min. Analyze by GC or GC–MS.

Comments. The method was used to detect TFA from halothane metabolism. TFA has also been analyzed as the 2, 2, 2-trichloroethyl ester (from Cl_3CCH_2OH/H_2SO_4; 50°, 15 min) using headspace sampling and ECD-GC.[2]

References

[1]D. Karashima *et al., Biomed. Mass Spectrom.* **3**, 41 (1976).
[2]L. Witte *et al., J. Chromatogr.* **143**, 329 (1977).

Derivative Type. Methyl ester.

Reagent. Methanol (MeOH)/dicyclohexylcarbodiimide (DCC)/pyridine.

Compound Type. Carboxylic acids.

Reaction

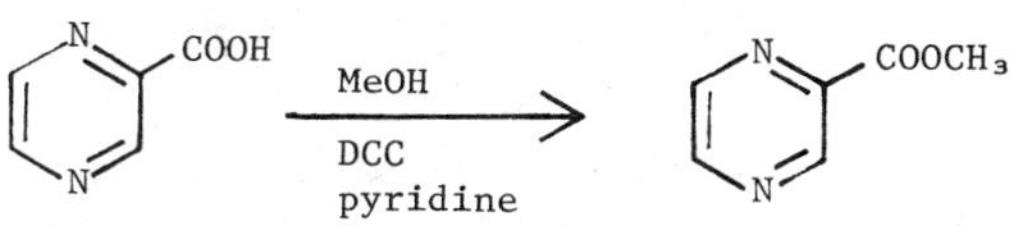

Typical Procedure. Dis 10 meq of the acid in 25 ml MeOH and 4 ml pyridine. Add 12 meq of DCC (more if the acid is wet or contains crystallization H_2O). Allow to react 15 min at room temp then analyze an aliquot by GC.

Comments. The method was initially used with pyrazinoic acid (shown) but was extended to a series of aliphatic, aromatic, and heterocyclic acids using a variety of alcohols. For solid alcohols, the amt of pyridine was increased to serve as solvent and 20 meq of the alcohol was used. Higher alcohols required heating at 40–80° for 30–120 min. The methyl esters, prepared by this method, have also been used to separate monohalogenated benzoic acids.[2]

References

[1]E. Felder *et al., J. Chromatogr.* **82**, 291 (1973).
[2]E. Felder *et al., J. Chromatogr.* **82**, 390 (1973).

Derivative Type. Methyl ester.

Reagent. Dimethyl sulfate/potassium carbonate.

Compound Type. Probenecid.

Reaction

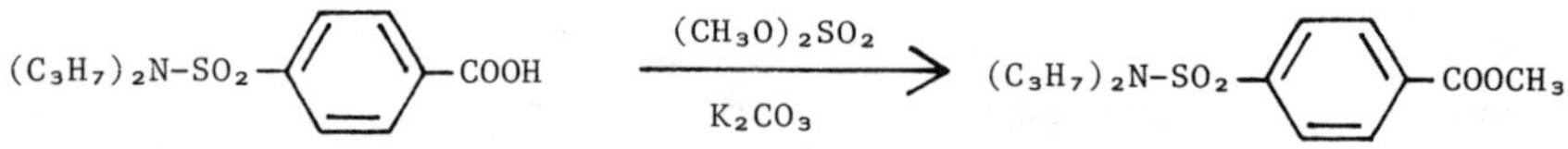

Typical Procedure. Acidify 0.2 ml plasma, bile, or urine sample with 1 ml 1 *N* HC1 and ext with 5 ml $CHCl_3$. Sep layers and adjust aq layer to pH 5 with 1 *N* NaOH. Add 2 ml *β*-glucuronidase soln (1 mg/ml in pH 5.0 acetate buffer) and incubate overnight. Acidify and ext with $CHC1_3$. Evap each ext (representing free and conjugated drug respectively) to dryness under an air stream.

To each residue add 1 ml K_2CO_3 in MeOH (prepared by mixing 1 ml 5% aq K_2CO_3 with 9 ml MeOH) and 0.1 ml $(CH_3O)_2SO_2$ and heat at 70° for 5 min. Evap MeOH under a stream of air, add 1 ml pH 5.6 acetate buffer, and ext with 5 ml $CHC1_3$. Evap ext to dryness and dis residue in 100 μl $CHCl_3$. Analyze by GC.

Comments. MS data were reported for the deriv.

References

[1]K. Sabih *et al., J. Pharm. Sci.* **60**, 745 (1971).

Derivative Type. Methyl ester.

Reagent. Heptafluorobutyric anhydride (HFBA)/methanol (MeOH).

Compound Type. Dicarboxylic acid metabolites of 5-(4′-Chloro-*n*-butyl) picolinic acid.

Reaction

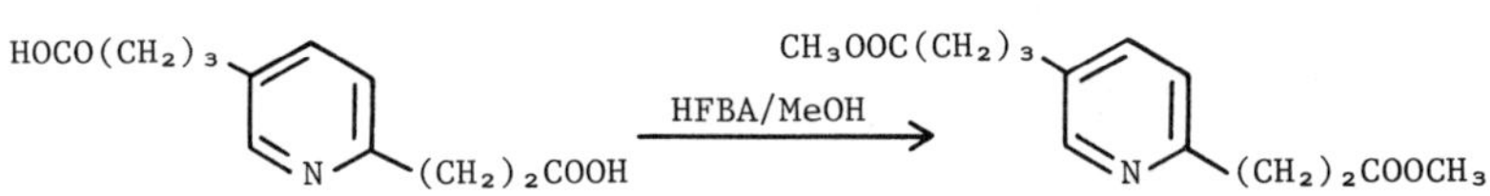

Typical Procedure. Treat compd with HFBA at 100° for 30 min. Add MeOH and reflux 1 hr. Analyze by GC-MS.

Comments. The trideuteromethyl esters were prepared in the same manner from CD_3OD. The dicarboxylic acid metabolites examined included α, *β*-unsatd and *β*-keto acids.

References

[1]H. Miyazaki *et al., Biomed. Mass Spectrom.* **3**, 140 (1976).

Derivative Type. Methyl ester.

Reagent. Methanol (MeOH)/thionyl chloride.

Compound Type. Amino acids.

Reaction

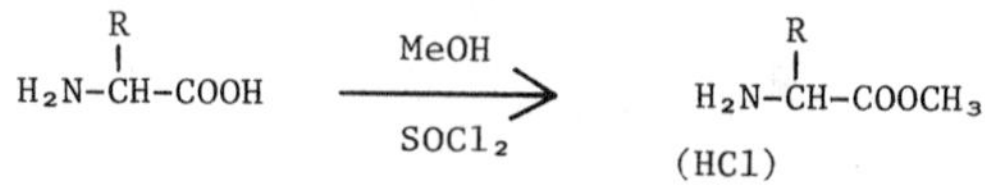

Typical Procedure (*Preparative*). Cool 30 ml (0.78 mol) MeOH in an ice-NaCl bath and add dropwise 7.9 ml (0.11 mol) $SOCl_2$ with stirring. Add portionwise with stirring 13.1 g (0.10 mol) Leu maintaining the temp below $-5°$. Allow the mixt to warm slowly to 40° and maintain at 40° for 2 hr with stirring. Evap the MeOH under vac and dry the residue under vac at 100°. Add 6–8 ml H_2O and shake with 250 ml Et_2O. Add conc aq NH_3 with cooling until alkaline to phenolphthalein then filter the NH_4Cl. Wash the Et_2O with Na_2SO_4 and evap to yield the deriv.

Comments. This esterfication method has been adapted for several analytical deriv methods.

References

[1]M. Brenner and W. Huber, *Helv. Chim. Acta* **36**, 1109 (1953).

Derivative Type. Methyl ester.

Reagent. Trimethyloxonium tetrafluoroborate (Me_3OBF_4).

Compound Type. Hemins.

Reaction

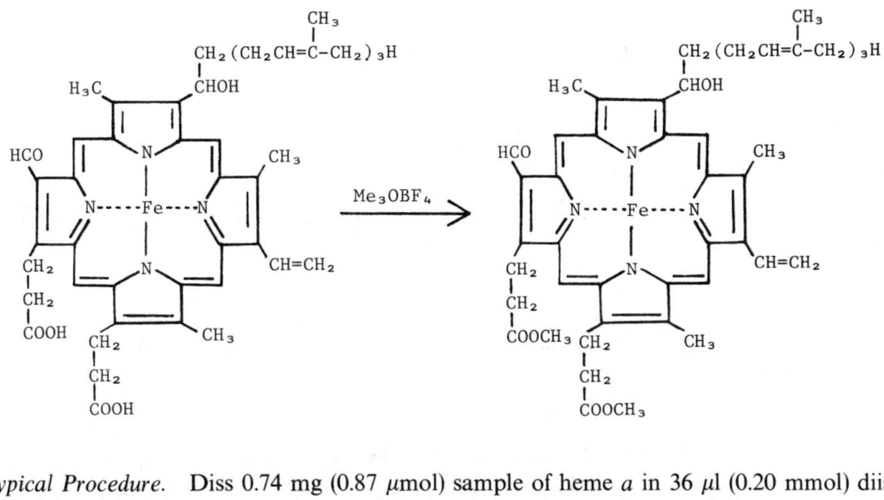

Typical Procedure. Diss 0.74 mg (0.87 μmol) sample of heme *a* in 36 μl (0.20 mmol) diisopropylethylamine, 0.4 ml H_2O, 9.0 ml 95% EtOH, and 0.6 ml 1.0 *M* aq $NaHCO_3$. Add 30, 27, and 22 mg of Me_3OBF_4 at 5 min intervals (total 0.53 mmol). (During the reaction the $NaHCO_3$ dissolves and CO_2 is evolved.) Allow to stand 10 min then add 25 ml $CHCl_3$ and 50 ml H_2O. Shake, separate $CHCl_3$ phase, and wash with 2×50 ml H_2O. Dry over NaCl then over Na_2SO_4. Evap to dryness under vac. Analyze by direct probe MS.

Comments. The method was used for selective methylation of carboxyl groups in heme *a* (shown above), protohemin IX, and hematohemin. The hydroxyl, formyl, and vinyl groups of heme *a* were unaffected. Benzilic acid was studied as a model compd for esterfication of the carboxyl group without affecting the hydroxyl group.

References

[1]R: T. Dean *et al.*, *Anal. Biochem.* **76**, 1 (1976).

3.1.3.2. *Other Alkyl Esters of Carboxylic Acids*

Derivative Type. *n*-Butyl ester.

Reagent. *n*-Butanol (*n*-BuOH)/sulfuric acid.

Compound Type. 2,4,5-Trichlorophenoxyacetic acid

Reaction

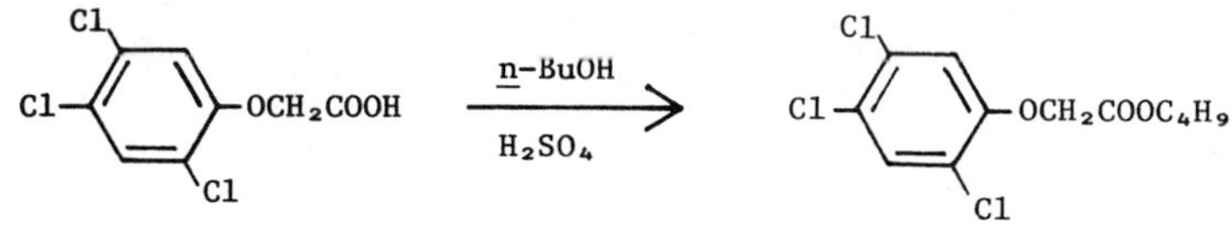

Typical Procedure. Grind 10 g soil sample with 10 g anhyd Na_2SO_4 and shake 1 hr with 50 ml $Et_2O/CHCl_3/HOAc$ (25:25:1, v/v/v). Allow to settle and evap 5 ml supernatant to dryness under an air stream at 35°. Add 1 ml *n*-BuOH and 3 drops conc H_2SO_4. Heat at 100° for 30 min. Add 20 ml H_2O and 5 ml 2,2,4-trimethylpentane and shake. Analyze an aliquot of the org phase by GC.

Comments. The butyl ester was used to avoid problems of methyl and ethyl esters coeluting with other soil constituents.

References

[1]C. E. McCone and R. J. Hance, *J. Chromatogr.* **69**, 204 (1972).

Derivative Type. 2-Butoxyethyl ester.

Reagent. 2-Butoxyethanol/HCl.

Compound Type. 2,4-Dichlorophenoxyacetic acid and 2-(2,4,5-trichlorophenoxy)propionic acid.

Reaction

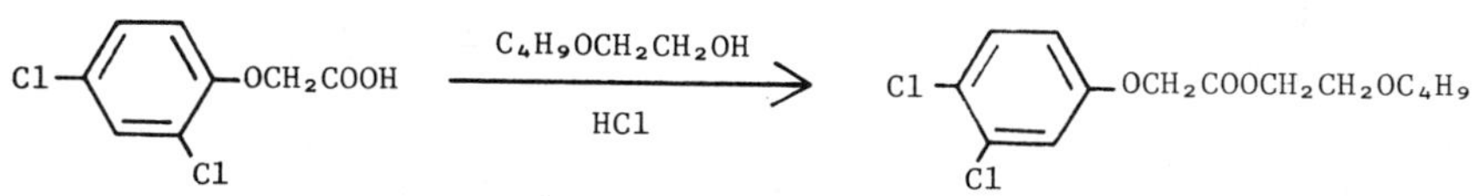

Typical Procedure. To a residue ext in $CHCl_3$ add 1 ml 2-butoxyethanol satd with HCl. Heat on a steam bath 5 min. Add 50 ml hexane and wash with 7 × 50 ml 0.2 *M* K_2HPO_4 and 3 × 50 ml H_2O. Purify the deriv on a 2 × 12 cm column of Florisil containing 2.5 cm Na_2SO_4 on the top. Wash column with 50 ml hexane then elute deriv with 100 ml Et_2O. Evap the eluate under vac and diss the residue in hexane prior to ECD-GC analysis.

Comments. The butoxyethyl ester was used because its relatively long retention time gave resolution of the derivs from interfering substances in citrus fruit samples.

References

[1]W. R. Meagher, *J. Agr. Food Chem.* **14**, 374 (1966).

Derivative Type. Benzyl ester.

Reagent. Phenyldiazomethane.

Compound Type. Trifluoroacetic acid (TFA).

Reaction

$$CF_3COOH \xrightarrow{C_6H_5CHN_2} CF_3COOCH_2C_6H_5$$

Typical Procedure. Ext TFA from biological sample 3–9 times with 10 ml Et_2O. Combine ext and dry over $MgSO_4$. Add phenyldiazomethane soln in Et_2O until the red color persists. Stir overnight at room temp, conc to *ca* 1 ml on a H_2O bath, and keep at −20° for over 12 hr. Analyze the soln directly by GC or GC-MS.

Comments. The method was used to analyze TFA in halothane treated microsomal suspensions. The reagent was prepared via the following route[2,3]:

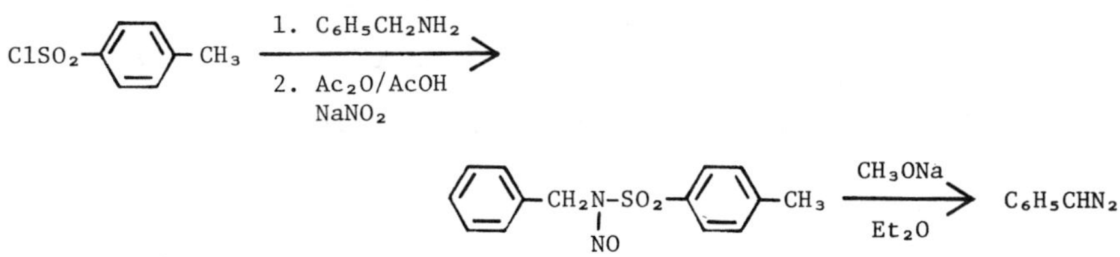

References

[1]D. Karashima *et al., J. Chromatogr.* **130**, 77 (1977).

[2]C. G. Overberger and J. P. Anselme, *J. Org. Chem.* **28**, 592 (1963).

[3]H.-P. Klemm *et al., J. Chromatogr.* **75**, 19 (1973).

Derivative Type. Ethyl ester.

Reagent. Diazoethane.

Compound Type. Indomethacin.

Reaction

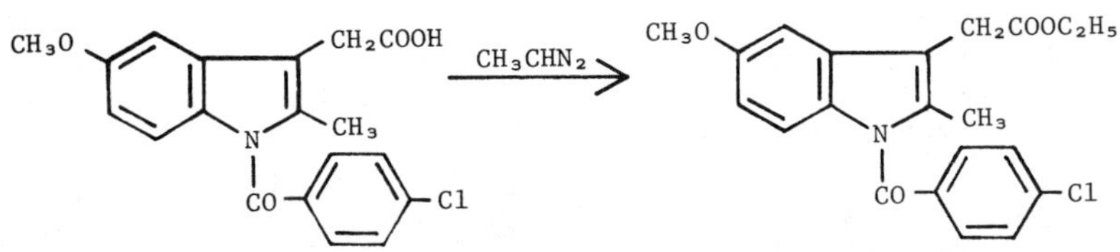

Typical Procedure. Mix 0.5 ml of serum or urine with 0.5 ml pH 5.0 citrate buffer and ext with 7.0 ml 1,2-dichloroethane. Transfer 6 ml of the org phase to another tube and evap to dryness under N_2 at 37°. Rinse tube sides with 0.5 ml Et_2O and add 0.5 ml diazoethane in Et_2O. Let stand 30 min at 30° then evap to dryness and diss residue in benzene. Analyze by ECD-GC.

Comments. The method was used for quantitation of indomethacin down to 50 ng/ml. Diazomethane was not used because it could convert the *O*-desmethyl metabolite to the same deriv. It has been shown, however, that esterfication can be achieved without methylation of the phenolic hydroxyl by reaction with diazomethane at 0°.[2] The methyl ester was also prepared with DMF dimethyl acetal and the TMS ester with BSA.[2] Diazoethane was prepared from 1-ethyl-1-nitrosourea using the method for diazomethane reported by Arndt.[3]

References

[1]L. Helleberg, *J. Chromatogr.* **117**, 167 (1976).

[2]B. Plazonnet and W. J. A. VandenHeuval, *J. Chromatogr.* **142**, 587 (1977).

[3]F. Arndt, *Org. Syn. Coll.* **2**, 165 (1944).

Derivative Type. Propyl ester.

Reagent. *n*-Propyl iodide/tetrabutylammonium hydrogen sulfate (TBA-HSO_4)/sodium bicarbonate.

Compound Type. Indomethacin.

Reaction

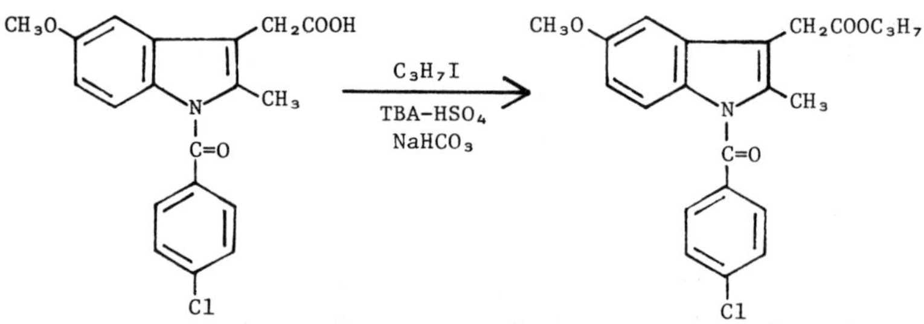

Typical Procedure. Mix 0.5–1.0 ml plasma with 2.0 ml 1 *M* NaH_2PO_4 and 0.1 ml int stand soln (methyl ester in 5% *n*-AmOH in heptane), and ext with 5.0 ml *n*-heptane containing 5% *n*-AmOH. Evap ext to dryness and add 1.0 ml 0.01 *M* TBA-HSO_4 in CH_2Cl_2, 0.3 g $NaHCO_3$, and 50 μl C_3H_7I. Shake 30 min then add 5.0 ml H_2O. Shake 5 min, centrifuge, remove the org phase, and evap to dryness. Diss residue in heptane and analyze by ECD-GC.

Comments. The method was compared with propylation by diazopropane and by extractive alkylation and gave the lowest rel. S.D. of the three methods presumably due to less decomposition of indomethacin under the mild conditions. The method was useable down to 5 ng per sample.

References

[1]A. Arbin, *J. Chromatogr.* **144**, 85 (1977).

3.2. OTHER DERIVATIVES OF CARBOXYLIC ACIDS

3.2.1. Silyl Derivatives of Carboxylic Acids

Derivative Type. Trimethylsilyl ester.

Reagent. Hexamethyldisilazane (HMDS)/trimethylchlorosilane (TMCS).

Compound Type. Fatty acids and resin acids.

Reaction

$$\text{R–COOH} \xrightarrow[\text{TMCS}]{\text{HMDS}} \text{R–COOTMS}$$

Typical Procedure. Diss 10 mg fatty or resin acid in 2 ml 30–80° pet ether and add 100 mg HMDS and 100 mg TMCS. Heat 5–10 min on a steam bath then evap under vac. Ext the TMS ester from the by-product NH_4Cl with dry pet ether. Filter through a sintered glass funnel and conc for GC analysis.

Comments. Pyridine was found to be detrimental to formation of the TMS esters but they could be readily prepared in hydrocarbon solvents.

References

[1]D. F. Zinkel *et al., J. Gas Chromatogr.* **6**, 158 (1968).

Derivative Type. Trimethylsilyl.

Reagent. Hexamethyldisilazane (HMDS)/trimethylchlorosilane (TMCS)/bis(trimethylsilyl) acetamide (BSA)/pyridine.

Compound Type. Fatty acids.

Reaction

$$\text{R—COOH} \xrightarrow[\text{pyridine}]{\text{HMDS/TMCS/BSA}} \text{R–COOTMS}$$

Typical Procedure.[1] Diss fatty acid sample in a soln of HMDS/TMCS/pyridine (2:1:10, v/v/v) and BSA (2.5 meq/ml) in pyridine. (The relative amts of each soln were not stated.[1]) Allow to stand 1–2 hr at room temp (the reaction is usually complete in 5 min but the longer time yields complete silylation of any neutral alcohols present[2]). Analyze by GC.

Comments. The method was applied to the analysis of lipids[1] and free fatty acids[2] in plasma. GC and MS data were given for the derivs.[2] There has been disagreement in the literature as to the value of pyridine in fatty acid silylation; some workers found it to be detrimental,[3,4] whereas others reported it to be beneficial as long as it was completely dry.[5,6] In the plasma analysis work[2] it was included to insure silylation of the other lipid materials present in the sample.

References

[1]A. Kuksis *et al., J. Lipid Res.* **10**, 660 (1969).
[2]A. Kuksis *et al., Anal. Biochem.* **70**, 303 (1976).
[3]Z. Horii *et al., Chem. Ind.* 1494 (1965).
[4]D. F. Zinkel *et al., J. Gas Chromatogr.* **6**, 158 (1968).
[5]C. E. Dalglish *et al., Biochem. J.* **101**, 792 (1966).
[6]G. E. Martin and J. S. Swinehart, *J. Gas, Chromatogr.* **6**, 533 (1968).

Derivative Type. Trimethylsilyl.

Reagent. Bis(trimethylsilyl) trifluoroacetamide (BSTFA).

Compound Type. Fatty acids.

Reaction

$$\text{R—COOH} \xrightarrow[\text{pyridine}]{\text{BSTFA}} \text{R—COOTMS}$$

Typical Procedure. Mix 1 mg fatty acid with 50 μl BSTFA and 50 μl pyridine. Heat at 50–60° for several hr. Analyze by GC-MS.

Comments. The method was used for analysis of nonvolatile fatty acids in cigarette smoke. Pyridine was particularly advantageous as solvent since it gave complete soln of the smoke condensate.

References

[1]M. R. Guerin *et al.*, *Anal. Chem.* **46**, 761 (1974).

Derivative Type. Trimethylsilyl.

Reagent. Bis(trimethylsilyl) trifluoroacetamide (BSTFA)/trimethylchlorosilane (TMCS).

Compound Type. Bilirubins.

Reaction

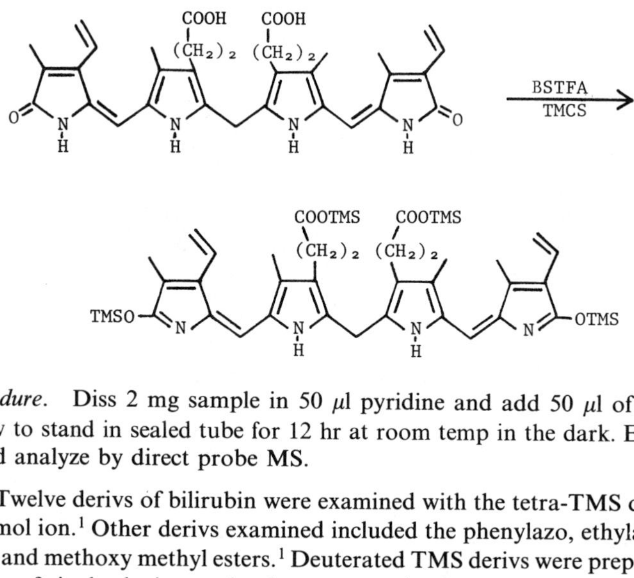

Typical Procedure. Diss 2 mg sample in 50 μl pyridine and add 50 μl of 1% TMCS in BSTFA. Allow to stand in sealed tube for 12 hr at room temp in the dark. Evap to dryness under vac and analyze by direct probe MS.

Comments. Twelve derivs of bilirubin were examined with the tetra-TMS deriv giving the most intense mol ion.[1] Other derivs examined included the phenylazo, ethylanthranilylazo, methyl esters, and methoxy methyl esters.[1] Deuterated TMS derivs were prepd with d_9-BSA. Hydrogenation of vinyl substituents has been reported to increase the stability of the derivs.

References

[1]M. Salmon and C. Fenselau, *Biomed. Mass Spectrom.* **1**, 219 (1974).
[2]F. Compernolle, *Biomed. Mass Spectrom.* **3**, 155 (1976).
[3]N. Blanckaert *et al.*, *Biochem. J.* **155**, 405 (1976).

Derivative Type. Trimethylsilyl.

Reagent. Trimethylsilylimidazole (TMSI).

Compound Type. Low molecular weight acids (C_1–C_5).

Reaction

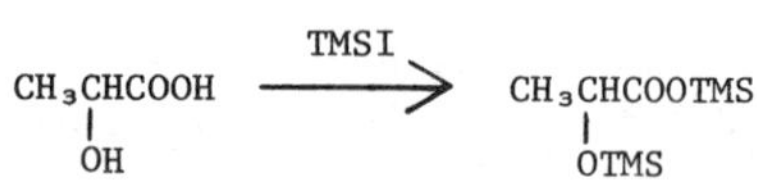

Typical Procedure. Alkalinize (pH 13) 2 ml urine with 5 *M* NaOH satd with NaCl and ext with 3 × 2 ml Et_2O. Discard Et_2O and acidify (pH 1–2) aq phase with 5 *M* HCl. Ext with 3 × 2 ml Et_2O, dry over Na_2SO_4, and evap half of the Et_2O to near dryness under N_2 (the other half is silylated with BSA for analysis of the higher acids). Diss the residue in 50 μl TMSI and analyze by GC.

Comments. The method was applied to C_1–C_5 fatty acids and lactic, pyruvic, and β-hydroxybutyric acids. The excess reagent eluted after the lactic acid deriv on OV-1 and after β-hydroxybutyric acid deriv on OV-17 permitting analysis of the acids without interferences due to solvent or reagent peak tailing.

References

[1]O. A. Mamer and B. F. Gibbs, *Clin. Chem.* **19**, 1006 (1973).

Derivative Type. *t*-Butyldimethylsilyl ester.

Reagent. *t*-Butyldimethylsilyl chloride (TBDMS-Cl).

Compound Type. Fatty acids.

Reaction

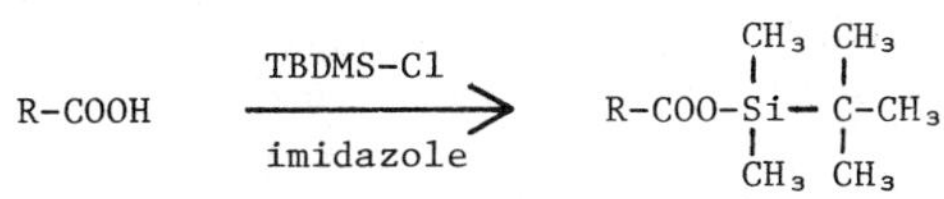

Typical Procedure. Diss 500 μg of the acid in 100 μl of a soln of 2 mmol imidazole and 1 mmol TBDMS-Cl in 10 ml DMF. Heat 15 min at 60°. Add aq NaCl soln and ext deriv into Et_2O. Analyze by GC or GC-MS.

Comments. The MS of the derivs showed intense M-57 ions which facilitated mol wt determination while also giving other structurally useful ions.

References

[1]G. Phillipou *et al., Lipids* **10**, 714 (1975).

Derivative Type. *t*-Butyldimethylsilyl (TBDMS) ester.

Reagent. *N*-TBDMS-*N*-methyltrifluoroacetamide.

Compound Type. Carboxylic acids.

Reaction

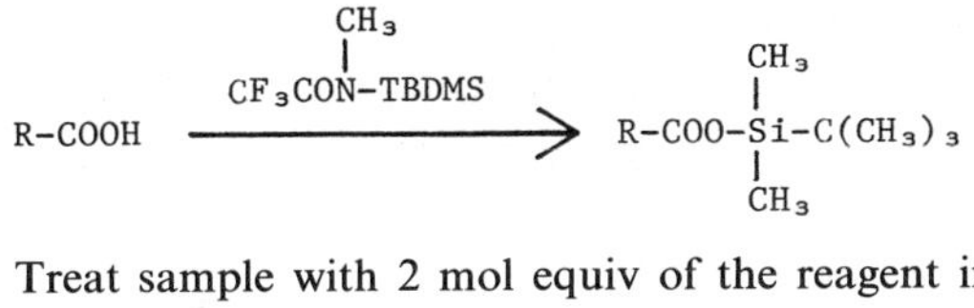

Typical Procedure. Treat sample with 2 mol equiv of the reagent in DMF or CH_3CN. Analyze by GC.

Comments. The reaction with acids was virtually instantaneous. The reagent was evaluated for synthetic use but could be useful for analytical deriv formation since the by-product of the reaction is the neutral and volatile *N*-methyltrifluoroacetamide. The preparation of the reagent from TBDMS-Cl and the Na salt of the amide was reported. The derivs were also prepared with TBDMS imidazole.

References

[1]J. L. Kutschinski, Ph.D. Thesis, University of California 1975; *Diss. Abstr. Int. B.* **35**, 4838 (1975).

3.2.2. Anilide Derivatives of Carboxylic Acids.

Derivative Type. Anilide.

Reagent. Thionyl chloride, aniline.

Compound Type. C_1–C_8 carboxylic acids.

Reaction

$$\text{R—COOH} \xrightarrow[\text{2. } C_6H_5NH_2]{\text{1. } SOCl_2} \text{R—CONH—}C_6H_5$$

Typical Procedure. Mix 0.5 ml of an EtOAc soln of the acids (0.5 *M* each) with 0.15 ml $SOCl_2$ then add 0.5 ml aniline dropwise. Heat 5 min at 60–80° then cool 2 min in ice. Add 10 ml 1% (w/v) methyl myristate (int stand) in EtOAc and mix. Wash soln with 10 ml 1 *M* aq $NaHCO_3$ and analyze 2 μl of the soln by GC.

Comments. Derivs of acetic through *n*-octanoic acids were formed quant in EtOAc but in very low yield in aq medium. Conversely the deriv of formic acid was formed quant in aq medium but hardly at all in EtOAc due to decomp to CO upon $SOCl_2$ treatment. Formyl and acetyl derivs were also investigated for analysis of anilines.[2]

References

[1]E. O. Umeh, *J. Chromatogr.* **51**, 147 (1970).

[2]E. O. Umeh, *J. Chromatogr,* **51**, 139 (1970).

3.3. CARBOXYLIC ACID DERIVATIVES FOR ENHANCEMENT OF DETECTABILITY

3.3.1. Derivatives for Electron Capture Detection

Derivative Type. 2-Chloroethyl ester.

Reagent. 2-Chloroethanol/boron trifluoride.

Compound Type. 2-Methyl-4-chlorophenoxyacetic acid.

Reaction

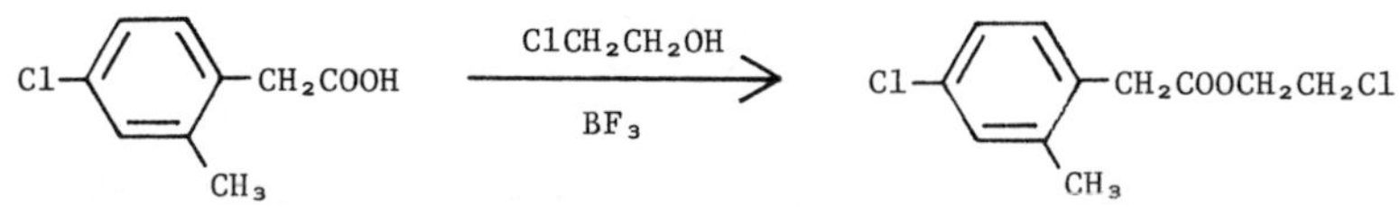

Typical Procedure. To the residue extd from a 10 g soil sample add 1 ml of a soln of 20 g BF_3 in 100 ml redist 2-chloroethanol. Heat in a 138° oil bath for 10 min. Cool and add 1 ml hexane and 5 ml 2% aq Na_2SO_4. Shake vigorously, allow to separate, then analyze 5 μl of the hexane phase by ECD-GC.

Comments. The method was sensitive to about 0.5 ppm. Several other chloro- and bromo-alcohols were also examined but all except 2-chloroethanol gave tars with BF_3.

References

[1]W. H. Gutenmann and D. J. Lisk, *J. Assoc. Off. Anal. Chem.* **47**, 353 (1964).

Derivative Type. 2-Chloroethyl ester.

Reagent. 2-Chloroethanol/boron trichloride.

Compound Type. 2,4-Dichlorophenoxyacetic acid and other acid herbicides.

Reaction

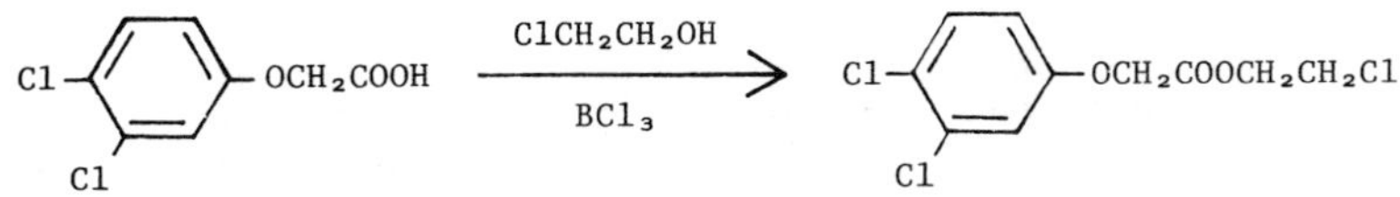

Typical Procedure. Mix a 75 g soil sample with 30 ml H_2O and 150 ml Et_2O. Add 50% (v/v) H_2SO_4 to make pH 3 or less then mix 4 hr on a torator. Filter through Na_2SO_4-Celite and mix with 50 ml of the Et_2O and 50 ml 0.05 *N* NaOH. Add more NaOH if necessary to make the soln pH10 or greater. Remove the aq soln, adjust to pH 3 or less with 50% H_2SO_4, and ext with 50 ml Et_2O. Add 1 ml 0.01% Nujol in hexane and evap to dryness. Add 1 ml 10% (w/w) BCl_3 in 2-chloroethanol and heat at 90° for 10 min. Cool and add 5 ml hexane and 10 ml 7% (w/v) aq Na_2SO_4. Mix well, allow to separate, and analyze an aliquot of the hexane soln by GC.

Comments. The back extraction into alkali removed interferences from other soil constituents. Retention and sensitivity data were given also for the derivs of 2,4,5-trichlorophenoxyacetic; 2,3,6-trichlorophenylacetic, 3-amino-2, 5-trichlorobenzoic, and 4-amino-3, 5, 6-trichloropicolinic acids.

References

[1]D. W. Woodham *et al., J. Agric. Food Chem.* **19**, 186 (1971).

Derivative Type. 2-Chloroethyl ester.

Reagent. 2-Chloroethanol/dicyclohexylcarbodiimide (DCC).

Compound Type. Herbicidal carboxylic acids.

Reaction

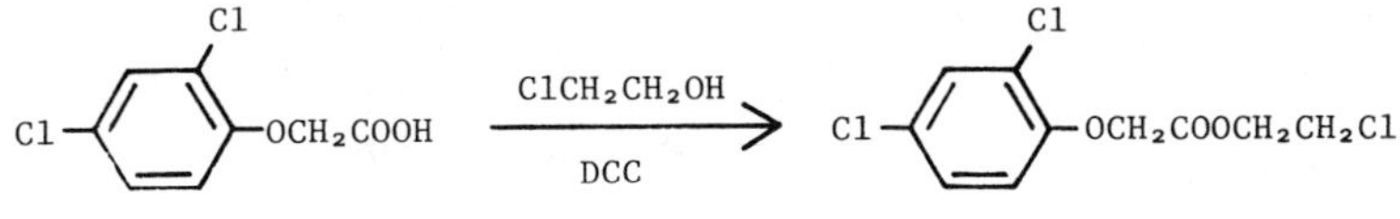

Typical Procedure To 50–500μl stand soln (1 ng/μl in benzene) or an aliquot of a sample ext add 500μl DCC soln (2g/100 ml pyridine) and 200 μl 2-chloroethanol. Mix, stopper, and allow to react overnight at room temp. Add 10 ml benzene and 3–4 ml dil HCl, shake, and discard aq phase. Wash 2× with H_2O and dry over Na_2SO_4. Analyze aliquot directly or conc under N_2 and analyze by ECD-GC.

Comments. The esterfication using DCC was found to be superior to the method using boron trichloride/2-chloroethanol. The method was applied to 10 herbicidal acids including both aliphatic and aromatic compds.

References

[1]A. S. Y. Chau and K. Terry, *J. Assoc. Off Anal. Chem.* **58**, 1294 (1975).

Derivative Type. 2,2,2-Trichloroethyl ester.

Reagent. 2,2,2-Trichloroethanol/trifluoroacetic anhydride (TFAA)/sulfuric acid.

Compound Type. Phenoxyacetic acid herbicides.

Reaction

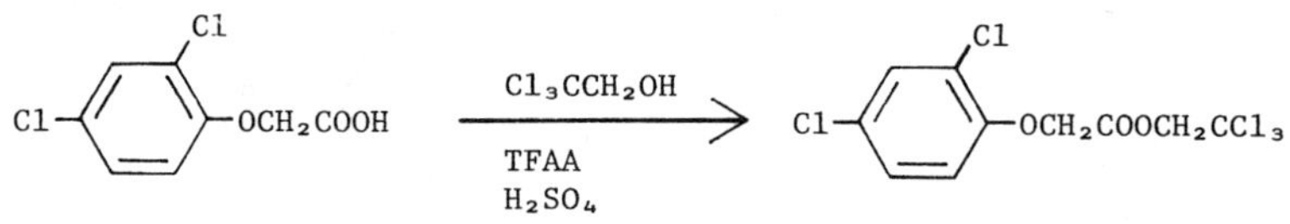

Typical Procedure To the residue ext from 1000 ml H_2O add 1 ml freshly prepared 20 % (v/v) trichloroethanol in TFAA and 5μl conc H_2SO_4. Reflux at 100° for 15 min or allow to stand at room temp overnight. Evap to *ca* 0.1 ml under an air stream at room temp then add 2 ml isooctane and shake. Add 6 ml 0.5 *M* KOH and shake. Analyze 2 μl of the org phase by ECD-GC.

Comments. The method was used for analysis of 2,4-dichlorophenoxyacetic acid and 4-chloro-2-methylphenoxyacetic acid in H_2O samples by sorption on XAD-4 resin, elution with benzene, and esterfication. For actual surface water samples, cleanup of the isooctane soln on a small silica gel column was required. A similar procedure without the H_2SO_4 was used to prepare the derivs of 10 aromatic acids.[2] A 10 mg sample was heated 10 min on a steam bath in 1 ml 10% trichloroethanol in TFAA and the deriv purified on a silica gel column eluted with EtOAc.

References

[1]S. Mierzwa and S. Witek, *J. Chromatogr.* **136**, 105 (1977).
[2]R. V. Smith and S. L. Tsai, *J. Chromatogr.* **61**, 29 (1971).

Derivative Type. 2,2,2-Trichloroethyl ester.

Reagent. 2,2,2-Trichloroethanol (TCE) and heptafluorobutyric anhydride (HFBA).

Compound Type. Short chain aliphatic acids (C_2–C_8).

Reaction

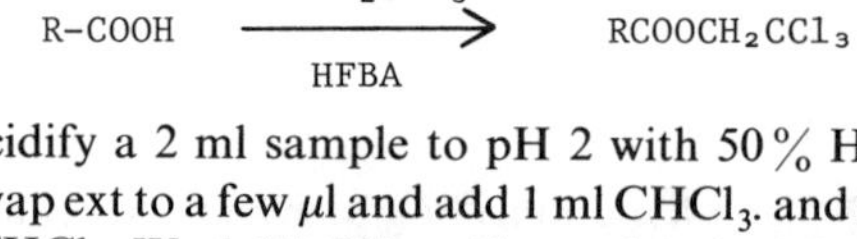

Typical Procedure. Acidify a 2 ml sample to pH 2 with 50% H_2SO_4 and ext with 20 ml $CHCl_3$/EtOH (94:6). Evap ext to a few μl and add 1 ml $CHCl_3$. and *ca* 100 mg $MgSO_4$. Shake, centrifuge, and decant $CHCl_3$. Wash $MgSO_4$ with another 1 ml $CHCl_3$. Evap to a few μl and add 1 drop fresh soln of TCE in $CHCl_3$ (1:99) and 3 drops HFBA. After 32 min at room temp add 1 drop of palmitic acid in $CHCl_3$ (0.25 g/ml) and 1 drop HFBA. After 16 min add 100 μl $CHCl_3$ and wash soln with 2 drops 0.1 *N* HCl allowing to stand 3 min. Remove aq layer and add 2 drops 0.1 *N* NaOH. After 30 min remove organic layer and evap to dryness. Diss residue in 100μl Et_2O for analysis.

Comments. The deriv was detectable to the low ng range.

References

[1]C. C. Alley *et al., Anal. Chem.* **48**, 387 (1976).

Derivative Type. Pentafluorobenzyl ester.

Reagent. Pentafluorobenzyl bromide (PFB-Br).

Compound Type. Carboxylic acids.

Reaction

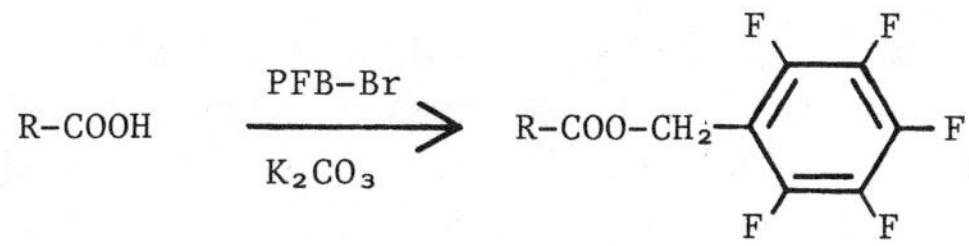

Typical Procedure. Diss 0.8 mg of the acid in 100 ml acetone and add 250 mg PFB-Br and 50 mg K_2CO_3. Reflux 3 hr then add 500 ml Et_2O and 20 ml EtOAc. Wash briefly with H_2O then dry over Na_2SO_4. Evap to dryness and diss residue in hexane containing 1 % acetone and 1 % Et_2O. Analyze aliquot by ECD-GC.

Comments. The deriv was examined for use in analysis of water for trace amts of carboxylic acids. The derivs of the low mol wt acids encountered some loss in the evap of the ext due to high volatility. Retention data were reported for the derivs of 14 acids.

References

[1]F. K. Kawahara, *Anal. Chem.* **40**, 2073 (1968).

Derivative Type. Pentafluorobenzyl (PFB) ester.

Reagent. Pentafluorobenzyl bromide (PFB-Br)/potassium carbonate.

Compound Type. Flurbiprofen [2-(2-fluoro-4-biphenyl) propionic acid].

Reaction

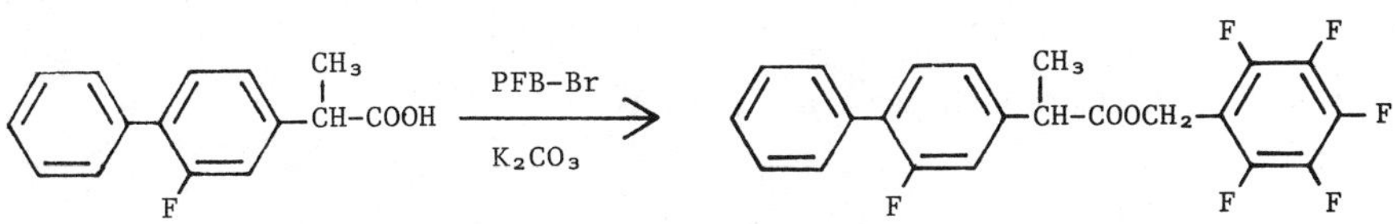

Typical Procedure. Evap 1 ml of a 4 μg/ml soln of 2-(2-methoxy-4-biphenyl) propionic acid int stand in 2-butanone to dryness in a tube. Add 1 ml plasma and acidify with 0.25 ml 1 *N* H_2SO_4. Ext with 5 ml benzene and transfer 4 ml to another tube. Evap to dryness under N_2. Diss residue in 50 μl $CHCl_3$ and purify by TLC on silica gel (10% HOAc in toluene). Locate compd by UV and elute with MeOH. Evap eluate to dryness under N_2.

Add 25 mg anhyd K_2CO_3 and 2 ml 25 μg/ml PFB-Br in acetone. Heat at 60° for 90 min. Evap to dryness under N_2 and 1 ml H_2O and 1 ml cyclohexane. Mix well then analyze 2 μl of org phase by ECD-GC.

Comments. The method was used to measure flurbiprofen in plasma to 50 ng/ml.

References

[1]D. G. Kaiser *et al., J. Pharm. Sci.* **63**, 567 (1964).

Derivative Type. Pentafluorobenzyl ester.

Reagent. Tetrabutylammonium hydrogen sulfate (TBA-HSO_4)/pentafluorobenzyl bromide (PFB-Br).

Compound Type. Fatty acids.

Reaction

$$\text{R-COOH} \xrightarrow[\text{TBA-HSO}_4]{\text{PFB-Br}} \text{R-COO-CH}_2\text{C}_6\text{F}_5$$

Typical Procedure. Acidify 10–50 μl serum with 500 μl 1 *M* H_2SO_4 and ext with 500 μl heptane containing 4 μg margaric acid int stand. Ext heptane with 50 μl 0.1 *M* NaOH. Discard heptane and add 250 μl of 0.1 *M* TBA-HSO_4 in 0.2 *M* NaOH and 250 μl of 0.066 *M* PFB-Br in CH_2Cl_2. Shake 25 min then add 50 μl of 0.5 *M* hordenine sulfate in 1*M* NaOH. Shake 45 min then sep CH_2Cl_2 phase and evap to dryness under N_2 at 30°. Add 200 μl heptane and 2.5 ml of 0.1 *M* H_2SO_4. Shake 15 min then centrifuge. Add 2.0 ml heptane and analyze 1 μl of the heptane phase by ECD-GC.

Comments. The method was used to quantitate 6.2 μg of palmitic acid in 50 μl serum with a rel S.D. of 6.1%. Excess PFB-Br was removed by reaction with hordenine (bottom reaction below) and ext of the deriv into aq acid.

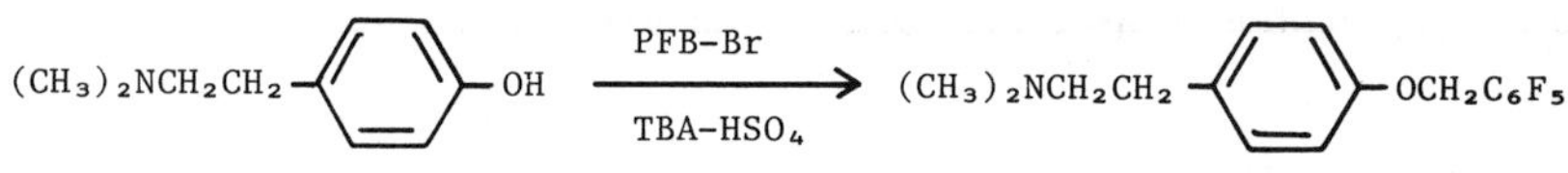

References

[1]O. Gyllenhaal *et al., J. Chromatogr.* **129**, 295 (1976).

Derivative Type. Pentafluorobenzyl ester or ether.

Reagent. Pentafluorobenzyl bromide (PFB-Br)/tetrapentylammonium hydroxide (TPAH).

Compound Type. Carboxylic acids and phenols.

Reaction

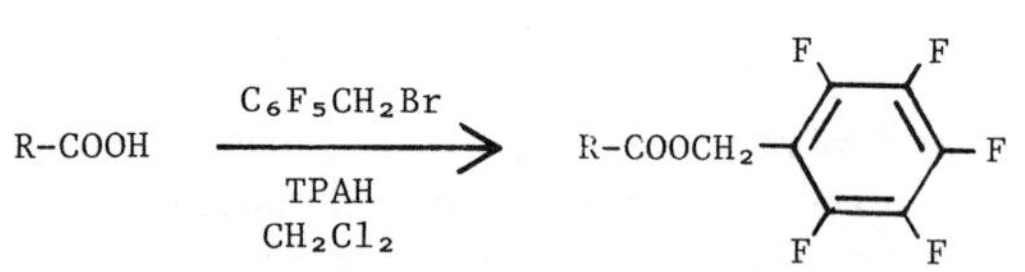

Typical Procedure. Diss 2 mmol carboxylic acid in 1 ml CH_2Cl_2 and add 1 ml pH 9 carbonate buffer containing 0.1 *M* TPAH. Shake for 10 min at 25° and analyze CH_2Cl_2 phase by GC.

Comments. TBAH gave slower reactions. No deriv formation was seen in 120 min using benzene, EtOAc, or Et_2O in place of CH_2Cl_2.

References

[1]H. Ehrsson, *Acta Pharm. Suecica* **8**, 113 (1971).

Derivative Type. Pentafluorobenzyl ester (of hydrolysis product).

Reagent. Aqueous sodium hydroxide, tetraphenylammonium (TPhA)/pentafluorobenzyl bromide (PFB-Br).

Compound Type. Oxyphenonium bromide.

Reaction

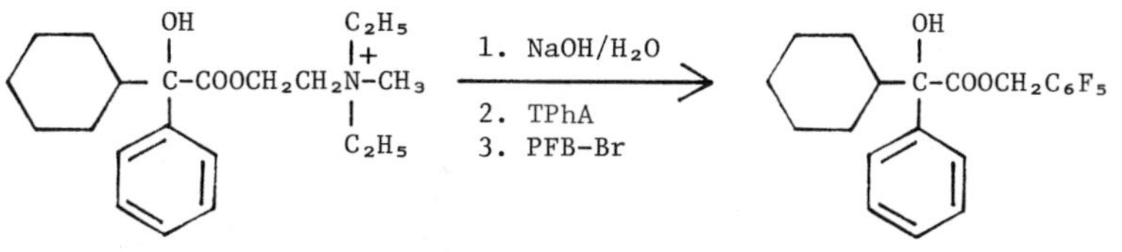

Typical Procedure. To 1.0 ml plasma add 50 μl of aq int stand (benactyzine methiodide, 50 ng) soln and 100 μl 1 *N* $NaClO_4$; then ext with 4 ml dichloroethane. Remove 3 ml of the org phase and mix with 1 ml aq 0.002 *M* (pH 7.5) tetraphenylammonium (TPhA) soln. Mix well, centrifuge, then remove 800 μl of aq phase and mix with 1 ml CH_2Cl_2. Mix well, centrifuge, then transfer 600 μl of the aq phase to another tube containing 50 μl 1 *N* NaOH. Hydrolyze by heating 30 min at 40°. Cool, add 100 μl 1 *M* NaH_2PO_4, and ext the TPhA-carboxylate ion pair with 1 ml CH_2Cl_2. Evap 500 μl of the CH_2Cl_2 phase to dryness under N_2, add 50 μl 0.1% (v/v) PFB-Br in CH_2Cl_2, and heat 3 min at 65°. Evap under N_2, diss residue in 50 μl *n*-heptane-EtOAc (98:2, v/v), and analyze 2.5 μl by ECD-GC.

Comments. The method permitted quantitation to 2 ng/ml in plasma. The preparation of the TPhA soln from TPhA iodide was given.

References

[1]J. E. Greving *et al.*, *J. Chromatogr.* **142**, 611 (1977).

Derivative Type. Bromomethyldimethylsilyl.

Reagent. Bromomethyldimethylchlorosilane (BMDCS)/diethylamine (Et_3N).

Compound Type. Carboxylic acids and phenols (herbicides and metabolites).

Reaction

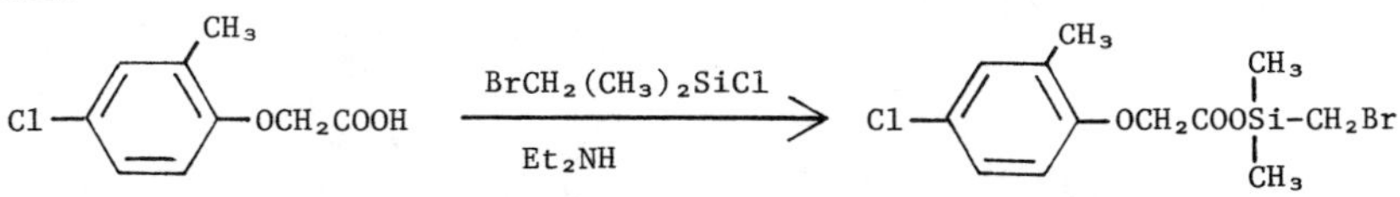

Typical Procedure. Mix 1 ml hexane, 0.075 ml Et_2NH, and 0.09 ml BMDCS. Shake, centrifuge, and transfer 0.4 ml of the supernatant to another tube containing the sample compound (up to 100 μg) in 0.1 ml EtOAc. Attach an air condenser and reflux at 65° for 30 min. Analyze soln by GC.

Comments. The chloromethyldimethylsilyl derivs were prep in the same manner. The derivs were used to enhance the response of the compds to ECD and emission spectrometric detection. The method was applied to six carboxylic acids and six phenols permitting detection in the range of 1–100 ng.

References

[1]C. A. Bache *et al.*, *Anal. Chem.* **40**, 1241 (1968).

Derivative Type. Iodomethyltetramethyldisiloxane ester.

Reagent. Bromomethyldimethylchlorosilane (BMDCS)/diethylamine (Et_2NH), sodium iodide.

Compound Type. Carboxylic acids.

Reaction

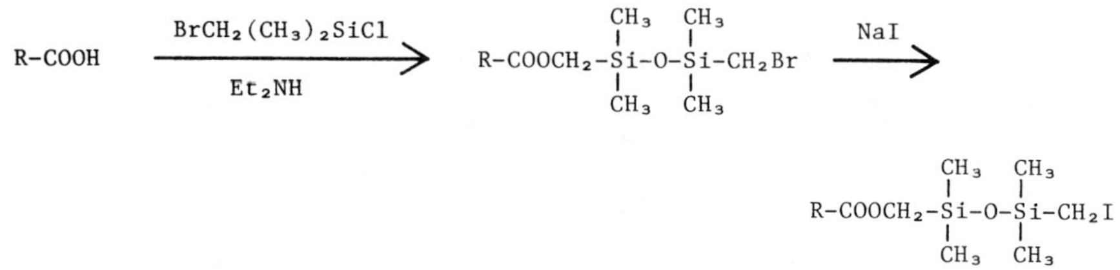

Typical Procedure. Ext a 2 ml aq sample with 20 ml Et_2O. Evap under an air stream to 1 ml and dry with *ca* 100 mg $MgSO_4$. Evap soln to near dryness. Add 1 drop $Et_2NH/CHCl_3$ (1:11, v/v) and 1 drop $BMDCS/CHCl_3$ (1:11, v/v). Seal and heat at 80° for 1 hr. Evap to dryness and add 0.1 ml satd (at 25°) NaI in acetone. Warm for 30 min at 35°. Add 0.9 ml $CHCl_3$, shake, centrifuge, and discard supernatant. Evap $CHCl_3$ to near dryness, add 1 ml xylene, and analyze 0.5 μl by ECD-GC.

Comments. The iodine deriv was more highly electron capturing than the bromine deriv. Alcohols could be analyzed concurrently as iodomethyl dimethylsilyl ethers. The method was reported to be useful for short chain acids, hydroxy acids, and alcohols.

References

[1]J. B. Brooks *et al., Anal. Chem.* **47**, 1960 (1975).

3.3.2. Derivatives for Alkali Flame Detection

Derivative Type. α-Hydroxyphosphonate ester.

Reagent. Dimethyl α-hydroxymethanephosphonate/dicyclohexylcarbodiimide (DCC).

Compound Type. Carboxylic acids.

Reaction

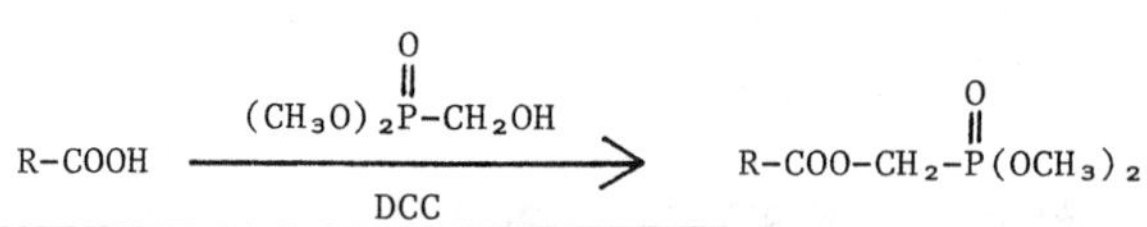

Typical Procedure. To a 100 μl soln of the acids in benzene add 150 μl of a soln of 25 g DCC and 12.5 g pyridine in 100 ml benzene. Add 50 μl of the phosphonate ester and heat 1 hr at 50°. Cool to room temp analyze by GC.

Comments. The method permitted detection of the acids to the low pg range using a alkali flame detector. Derivs were prepared from C_1–C_7 fatty acids using methyl and ethyl α-hydroxymethane, through butanesulfonates.

References

[1]P. Schulz and R. Vîlceanu, *J. Chromatogr.* **111**, 105 (1975).

3.3.3. Chromogenic Derivatives for HPLC

Derivative Type. Phenacyl ester.

Reagent. Phenacyl bromide/triethylamine (Et_3N).

Compound Type. Fatty acids.

Reaction

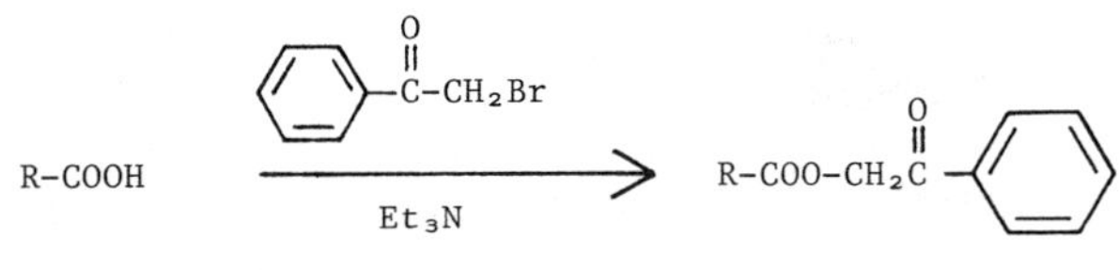

Typical Procedure. Mix 100 μg fatty acid with 10 μl phenacyl bromide soln (12 mg/ml acetone) and 10 μl triethylamine soln (10 mg/ml acetone) and allow to stand overnight at room temp. Inject aliquot of soln directly into HPLC.

Comments. Alternatively mixt was heated for 2 hr at 50° rather than standing overnight. C_{12}–C_{24} fatty acid derivs (both satd and unsatd) were separated on an alkyl bonded phase column using CH_3CN—H_2O. Detection limit was *ca* 100 ng at 254 nm (ε 6,000) but operation at the λ_{max}(ε 14,000) would give greater sensitivity.

References

[1]R. F. Borch, *Anal. Chem.* **47**, 2437 (1975).

Derivative Type. *p*-Bromophenacyl ester.

Reagent. *p*-Bromophenacyl bromide/18-crown-6(1,4,7,10,13,16-hexaoxacyclooctadecane).

Compound Type. Carboxylic acids.

Reaction

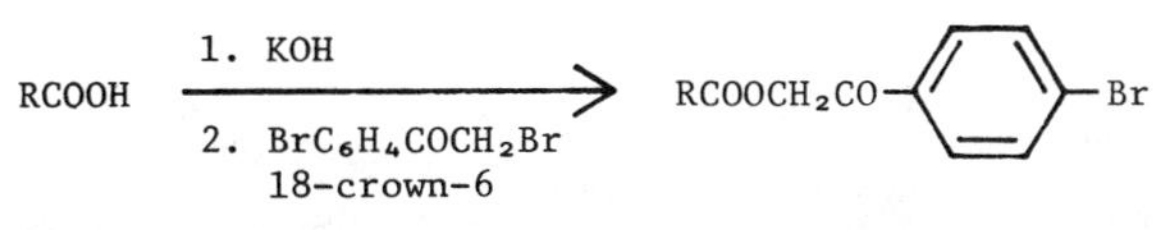

Typical Procedure. Neutralize the acid sample with KOH-MeOH, then evap MeOH. Add an excess of the phenacyl bromide/crown ether (20:1 for larger samples; 10:1 for smaller samples). Heat with stirring at 80° for 15 min. Cool soln and analyze by HPLC.

Comments. The method utilized a soln of alkylating agent-crown ether in benzene or acetonitrile in molar ratio 20:1 or 10:1. A soln of 85 %KOH in MeOH was used to prepare K salts. It was possible to add 3–5 mol equiv of solid $KHCO_3$ to the free acid sample and then add alkylating agent-crown ether soln. This method required extending the heating time to 30 min. The method yielded a chromogenic deriv for HPLC analysis. Quantitative yields (>97%) were obtained without excess reagent. The HPLC system used was an ankyl phase column with heptane-$CHCl_3$ or MeOH—H_2O. Sensitivity was reported in the low ng range. It was subsequently shown that positional and geometrical isomers of unsatd acids could be separated by HPLC as the *p*-bromophenacyl derivs.[2]

References

[1]H. D. Durst *et al., Anal. Chem.* **47**, 1797 (1975).

[2]P. T.-S. Pei *et al., Lipids* **11**, 814 (1976).

Derivative Type. 2-Naphthacyl esters.

Reagent. 2-Naphthacyl bromide/*N*,*N*-diisopropylethylamine.

Compound Type. Long chain fatty acids.

Reaction

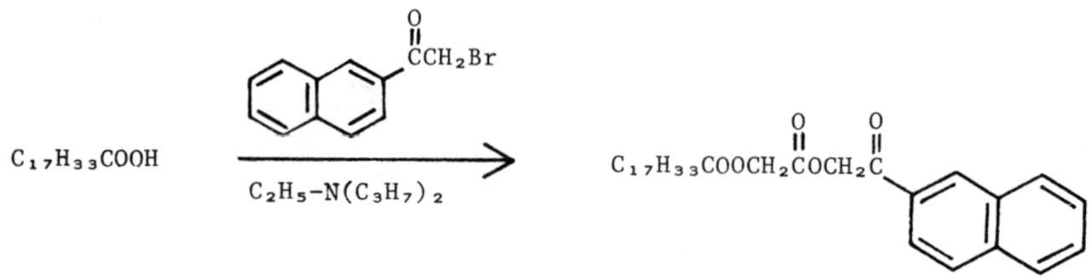

Typical Procedure. Diss 10 μmol of the acid, 20 μmol of 2-naphthacyl bromide, and 40 μmol of *N*,*N*-diisopropylethylamine in 1 ml DMF. Heat at 60° for 10 min then analyze an aliquot by HPLC.

Comments. The UV absorbing deriv was detectable to the ng range. MS data (EI and CI) for the derivs were given.

References

[1]M. J. Cooper and M. W. Anders, *Anal. Chem.* **46**, 1849 (1974).

Derivative Type. *p*-Nitrobenzyl ester.

Reagent. *p*-Nitrobenzyl bromide/dicyclohexyl-18-Crown-6.

Compound Type. Dicarboxylic acids.

Reaction

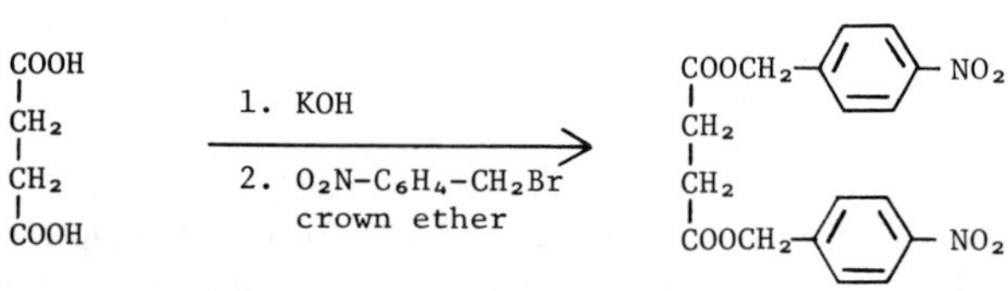

Typical Procedure. Neutralize a MeOH or H_2O soln of the acid to phenolphthalein end point with KOH-MeOH soln. Evap under vac to give the K salts. If necessary, dry by azeotropic removal of H_2O with benzene. For each 0.1 mmol of acid add 1.2 ml of a soln of 0.2 mmol *p*-nitrobenzyl bromide and 0.15 nmol dicyclohexyl-18-Crown-6 per ml CH_3CN. Reflux 1.5 hr. Analyze by HPLC.

Comments. Phenacyl and *p*-bromophenacyl esters were prepared by the same method using α-bromoacetophenone and α, *p*-dibromoacetophenone. These derivs, especially the *p*-bromo ester, suffered from inadequate solubility in the HPLC solvents. A phenacyl ester deriv was also prepared from citric acid.

References

[1]E. Grushka *et al., J. Chromatogr.* **112**, 673 (1975).

Derivative Type. Benzyloxymethyl ester.

Reagent. Benzyl chloromethyl ether.

Compound Type. Carboxylic acids.

Reaction

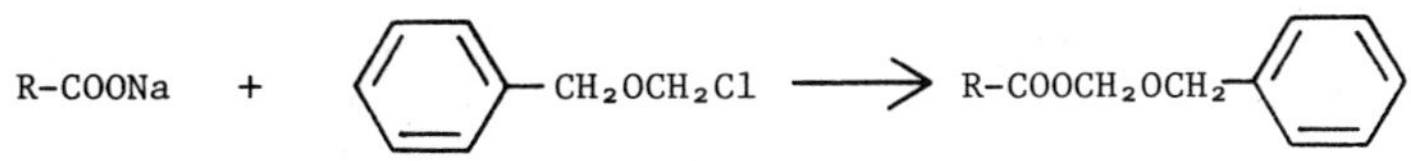

Typical Procedure (Synthetic Scale). Stir sodium hexanoate (13.8 g, 0.1 mol) and benzylcholoromethyl ether (15.7 g, 0.1 mol) in 80 ml hexamethylphosophoramide at room temp for 2 days. Pour into 700 ml H_2O and ext with 2 × 500 ml hexane. Wash hexane ext with 5% $NaHCO_3$ (500 ml) and H_2O (500 ml). Dry over $MgSO_4$ and evap to yield product.

Comments. This synthetic reaction could possibly be adaptable to chromogenic derivatization.

References

[1] P. A. Zoretic *et al., J. Org. Chem.* **40**, 2962 (1975).

Derivative Type. 4-Methyl-7-methoxycoumarin ester.

Reagent. 4-Bromomethyl-7-methoxycoumarin.

Compound Type. Acidic herbicides.

Reaction

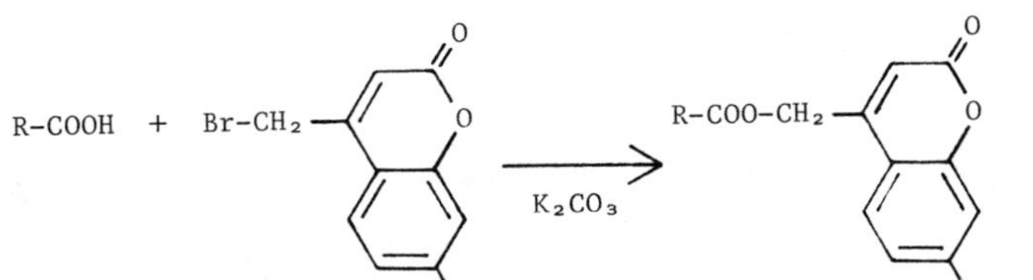

Typical Procedure. Diss acid in acetone with a 2–3 × mol excess of the reagent and add solid K_2CO_3. Reflux for 1 hr in a microrefluxer.[2] Apply acetone soln directly to TLC plate and measure spots, after separation, by fluorescence.

Comments. The method was suggested as suitable for HPLC analysis.

References

[1]W. Dinges, *Chromatographia* **9**, 624 (1976).
[2]W. Dünges, *Anal. Chem.* **45**, 963 (1973).
[3]W. Dünges, *Anal. Chem.* **49**, 442 (1977).

Derivative Type. Benzyl ester.

Reagent. 1-Benzyl-3-*p*-tolyltriazene.

Compound Type. Fatty acids.

Reaction

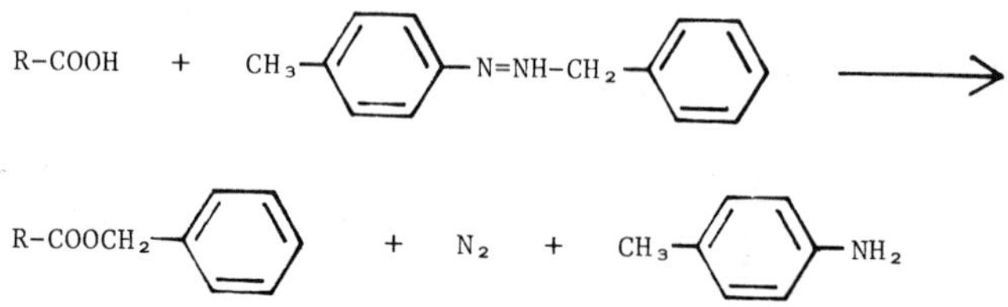

Typical Procedure. Diss 2 mmol fatty acid in 10 ml Et_2O. Add 10 mmol 1-benzyl-3-*p*-tolyltriazene in 5 ml Et_2O. Heat at 36° with stirring for 3 hr. Cool and wash with 5 ml 10% HCl and 2 × 5 ml 10% Na_2CO_3. Dry over $MgSO_4$ and evap to dryness. Analyze residue by HPLC.

Comments. The deriv was detectable to 1–2 μg using 254 nm detector. The commercial reagent was recryst from Et_2O *n*-hexane (1:1) prior to use. MS data for the deriv was given.

References

[1]I. R. Politzer *et al., Anal. Lett.* **6**, 539 (1973).

Derivative Type. *p*-Nitrobenzyl esters.

Reagent. 1-*p*-Nitrobenzyl-3-*p*-tolytriazene (PNBTT).

Compound Type. Carboxylic acid.

Reaction

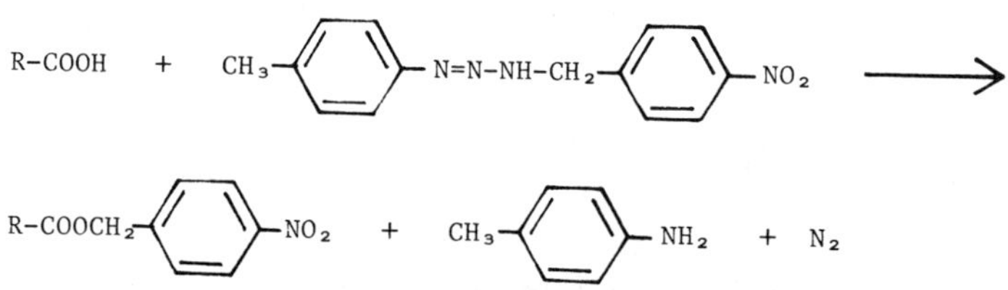

Typical Procedure. To 2–3 mg sample in 5 ml reaction vial add 3 ml EtOH and 50 mg PNBTT. Cover *loosely* and heat at 65° for 1 hr. Cool and analyze by HPLC; or evap to dryness under N_2, diss in 2–3 ml Et_2O, wash with dil HCl, then H_2O, and analyze Et_2O soln by HPLC.

Comments. The chromogenic deriv was used for HPLC analysis with UV detection. The optional acid wash removed excess reagent and the by-product toluidine.

References

[1]Regis Lab Notes, No. 17 Nov. 1974.

Derivative Type. *p*-Nitrobenzyl ester.

Reagent. *O*-*p*-Nitrobenzyl-*N*,*N*′-diisopropylisourea (PNBDI).

Compound Type. Carboxylic acid.

Reaction

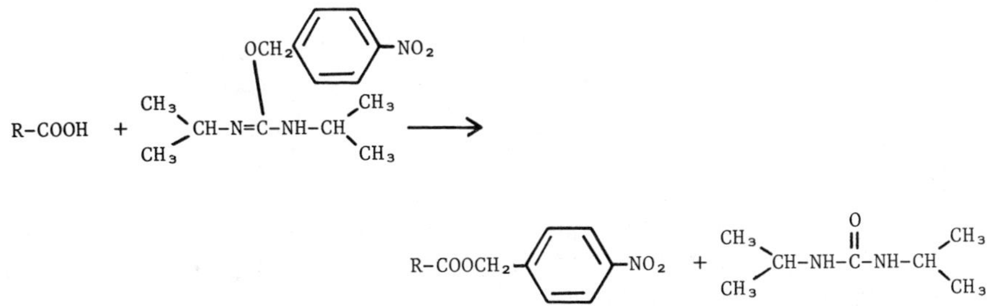

Typical Procedure. To a soln of 3 μmol of fatty acid in 125 μl of CH_2Cl_2, add 9 μmol of PNBDI in 125 μl of CH_2Cl_2. Seal vial and heat at 80° for 2 hr. Inject aliquot into HPLC.

Comments. The chromogenic deriv was used for HPLC analysis. Detectability was shown for 4 pmol of stearate with a 254 mm detector.

References

[1]D. R. Knapp and S. A. Krueger, *Anal. Lett.* **8**, 603 (1975).

Derivative Type. *p*-Methoxyanilide.

Reagent. Triphenylphosphine/carbon tetrachloride, *p*-methoxyaniline.

Compound Type. Fatty acids.

Reaction

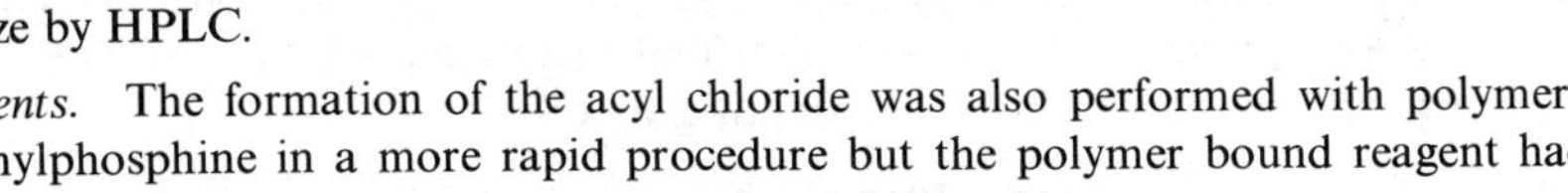

Typical Procedure. Mix fatty acid sample (total *ca* 0.7 mmol) with 0.5 g (1.9 mmol) triphenylphosphine and 2 ml dry CCl_4. Seal and heat at 80° for 5 min or 140° for 2 min. Cool and add 0.5 g (4.1 mmol) *p*-methoxyaniline in 8 ml dry EtOAc. Heat at 80° for 1 hr or 140° for 20 min. Analyze by HPLC.

Comments. The formation of the acyl chloride was also performed with polymer bound triphenylphosphine in a more rapid procedure but the polymer bound reagent had to be synthesized. The derivs showed maximum absorbance at 254 nm.

References

[1]N. E. Hoffman and J. C. Liao, *Anal. Chem.* **48**, 1104 (1976).

3.4. DERIVATIVES OF HYDROXYACIDS

3.4.1. *O*-Acyl Alkyl Ester Derivatives

Derivative Type. Methyl ester, *O*-acetyl.

Reagent. Methanol (MeOH)/boron trifluoride, acetic anhydride (Ac_2O)/toluenesulfonic acid (TsOH)/pyridine.

Compound Type. Hydroxy fatty acids.

Reaction

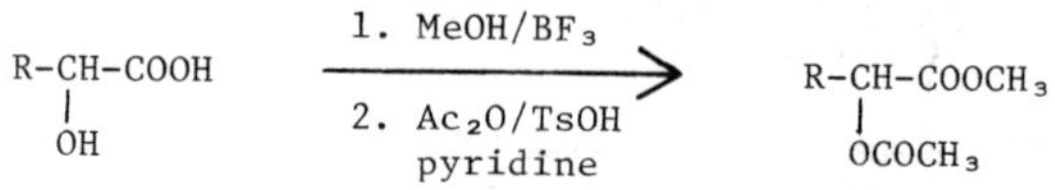

Typical Procedure Prepare the methyl esters by treatment with 12.5% (w/v) BF_3 in MeOH.[2] Diss 2 mg of the acid in 2 ml Ac_2O-pyridine (1:2, v/v) containing 60 mg TsOH. Heat 3 min at 100°, add 1 ml H_2O, and ext with pet ether. Evap under vac and add Et_2O and evap until all of the pyridine is removed. Analyze the derivs by GC.

Comments. Satd fatty acid methyl esters were acetylated by $HClO_4$ catalysis in Ac_2O-EtOAc[3] but this method produced extraneous products with unsatd esters. GC data were listed for derivs of all of the isomers of hydroxypalmitic acid.

References

[1]J. S. O'Brien and G. Rouser, *Anal. Biochem.* **7**, 288 (1964).

[2]L. D. Metcalfe and A. A. Schmitz, *Anal. Biochem.* **33**, 363 (1961).

[3]J. S. Fritz and G. H. Schenk, *Anal. Chem.* **31**, 1808 (1959).

Derivative Type. Methyl ester, *O*-acetyl.

Reagent. Dimethoxypropane (DMP)/methanol (MeOH)/HCl, isopropenyl acetate.

Compound Type. 2-Hydroxy fatty acids.

Reaction

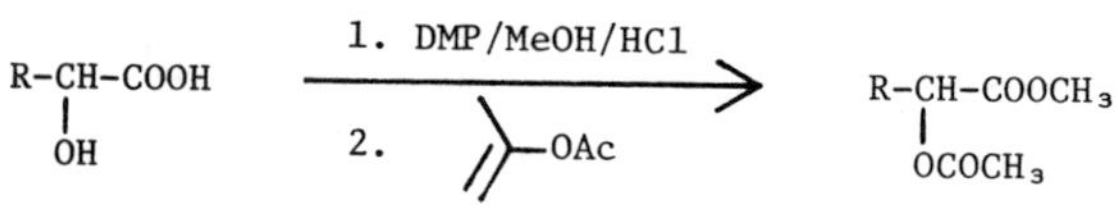

Typical Procedure. Diss the hydroxyacid sample (0.6 μmol) in 1 ml freshly mixed MeOH/conc HCl/DMP (10:1:25, v/v/v) and allow to stand 1 hr. Add 5 ml toluene and evap on a rotary evaporator. Add 50 μl freshly mixed isopropenyl acetate/conc H_2SO_4 (500:1, v/v). Heat at 60° for 30 min. Add 0.3 ml H_2O and ext with 0.3 ml Et_2O/toluene (1:1, v/v). Dry ext over Na_2CO_3 then evap. Diss residue in CS_2 and analyze by GC.

Comments. A procedure was given for specific isolation of 2-hydroxy fatty acids from brain lipids using saponification, silica gel chromatography, and copper chelation.

References

[1]Y. Kishimoto and N. S. Radin, *J. Lipid Res.* **4**, 130 (1963).

Derivative Type. *n*-Butyl ester, trifluoroacetyl.

Reagent. *n*-Butanol (*n*-BuOH)/boron trifluoride, Trifluoroacetic anhydride (TFAA).

Compound Type. Hydroxy acids.

Reaction

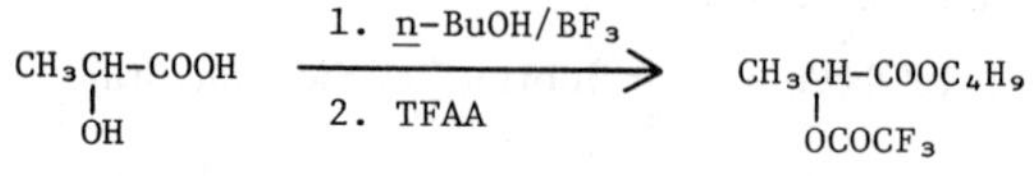

Typical Procedure. To a 5–10 μmol sample in *ca* 0.1 ml $CHCl_3$ add 2 drops (*ca* 0.1 ml) BF_3 in *n*-BuOH. Remove most of the $CHCl_3$ by evap under an air stream; then heat at 100° for 4 min. Cool and ext esters with 5 drops (*ca* 0.1 ml) $CHCl_3$/MeOH (14:1, v/v). Wash with 2 drops (*ca* 0.1 ml) 0.1 *N* NaOH, separate, and wash with 2 drops 4% HCl. Evap $CHCl_3$ soln to *ca* 0.1 ml and add 8 drops (*ca* 26μl) of a cold (4°) soln of TFAA/pyridine (9:1, v/v). Allow to stand 10 min then wash with 2 drops 4% HCl. Allow to stand 30 min then analyze by GC.

Comments. The method was used for analysis of hydroxy acids produced by bacteria.[1] The *n*-butyl esters have been used for analysis of fatty acids from bacteria.[2]

References

[1]J. B. Brooks *et al., Can. J. Microbiol.* **18**, 157 (1972).

[2]J. B. Brooks *et al., Can. J. Microbiol.* **17**, 531 (1971).

Derivative Type. Methyl ester, heptafluorobutyryl (HFB).

Reagent. Diazomethane, heptafluorobutyrylimidazole (HFBI).

Compound Type. Gibberellins.

Reaction

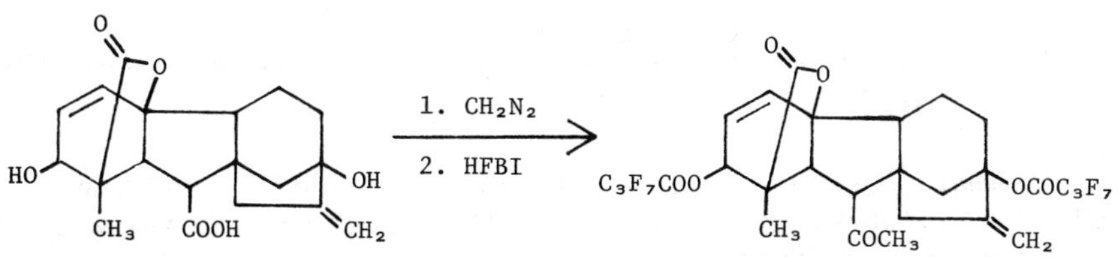

Typical Procedure. Methylate a 0.1–1 μg sample with CH_2N_2 in Et_2O. Evap to dryness under N_2 and add (in a P_2O_5 dried N_2 atmosphere) HFBI. Seal vial and heat at 60° for 3 hr. Evap under vac and diss residue in anhyd hexane. Analyze by ECD-GC.

Comments. The TFA derivs were prepared by the same method using TFAI and showed 49.5% of the ECD sensitivity of the HFB deriv. The derivs were prepared in an N_2 atmosphere with P_2O_5 drying which was reported to be necessary for deriv submicrogram samples. The method was used for analysis of gibberellins in apple seed.

References

[1]S. D. Seeley and L. E. Powell, *Anal. Biochem.* **58**, 38 (1974).

Derivative Type. Ethyl ester, heptafluorobutyryl (HFB).

Reagent. Ethanol (EtOH), heptafluorobutyric anhydride (HFBA)/pyridine.

Compound Type. Hydroxyacids and amino acids.

Reaction

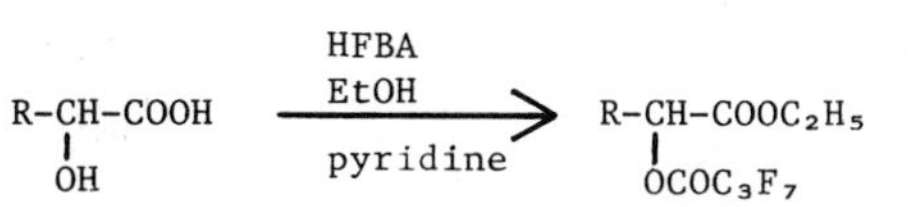

Typical Procedure. Diss extracted hydroxyacids in 0.1 ml $CHCl_3$. Add 0.01 ml EtOH/$CHCl_3$ (1:4) and mix by shaking. Add 0.01 ml pyridine/$CHCl_3$ (1:3, v/v) and mix by shaking. Add

0.03 ml HFBA and mix by shaking. Evap $CHCl_3$ to *ca* 1/2 drop and heat in boiling H_2O bath 4 min. Cool under tap H_2O and dil with 0.14 ml $CHCl_3$. Add *ca* 0.09 ml 0.1 *N* HCl and sep $CHCl_3$ layer. Wash with 0.1 *N* NaOH and allow to stand 30 min. Sep $CHCl_3$ layer and evap solvent almost to dryness. Add 0.1 ml Et_2O or acetone for GC analysis.

Comments. The method achieved simultaneous esterification of carboxyl groups and formation of HFB derivs of hydroxyl and amine groups. The procedure was demonstrated for analysis of synovial fluid using ECD-GC.

References

[1]J. B. Brooks *et al., Anal. Chem.* **46**, 1930 (1974).

3.4.2. Other Hydroxy Acid Derivatives

Derivative Type. Trimethylsilyl.

Reagent. Hexamethyldisilazane (HMDS)/trimethylchlorosilane (TMCS)/pyridine.

Compound Type. Hydroxyfatty acids (as methyl esters) and monoglycerides.

Reaction

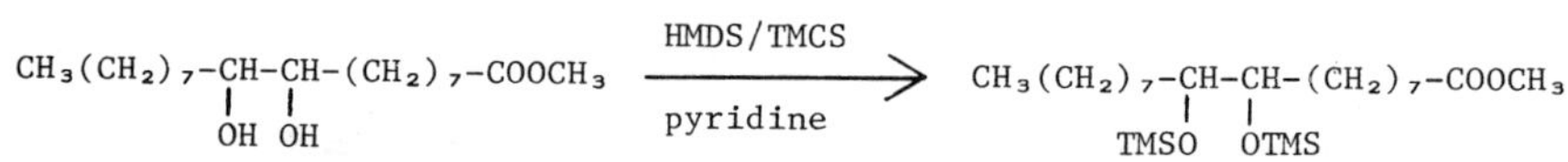

Typical Procedure. To 1 ml sample add 1.0 ml pyridine, 0.2 ml HMDS, and 0.1 ml TMCS. Shake 15–30 sec and allow to stand 5 min. Add 5 ml hexane and 5 ml H_2O. Mix well, remove hexane, and ext aq phase with another 5 ml hexane. Dry ext over $CaSO_4$ and evap under N_2. Diss in hexane and evap until pyridine odor is gone. Analyze by GC.

Comments. The method was used for analysis of di-, tri-, and tetrahydroxystearates[1] and 1-, and 2-monoglycerides.[2]

References

[1]R. D. Wood *et al., J. Am. Oil Chem. Soc.* **42**, 81 (1965).
[2]R. D. Wood *et al., J. Am. Oil Chem. Soc.* **42**, 161 (1965).

Derivative Type. Trimethylsilyl.

Reagent. Hexamethyldisilazane (HMDS)/trimethylchlorosilane (TMCS).

Compound Type. Shikimic acid and related compounds.

Reaction

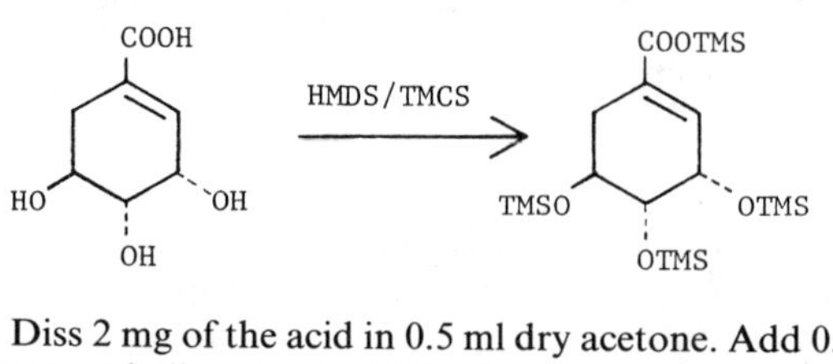

Typical Procedure. Diss 2 mg of the acid in 0.5 ml dry acetone. Add 0.2 ml HMDS and 0.1 ml TMCS. Shake 30 sec and allow to stand 10 min at room temp. Analyze 1 μl aliquots by GC.

Comments. The method was applied to shikimic, 5-dehydroshikimic, quinic, and 5-dehydroquinic acids.

References

[1]J. P. Shyluk *et al.*, *J. Chromatogr.* **26**, 268 (1976).

Derivative Type. *n*-Butylboronate.

Reagent. *n*-Butylboronic acid.

Compound Type. Proximal difunctional compounds (α- and β-hydroxyacids; *o*-phenolic acids; enolizable α-oxo acids; 1,2- and 1,3-diols; enolizable 1,2- and 1,3-ketols; β- and γ-hydroxyamines).

Reaction

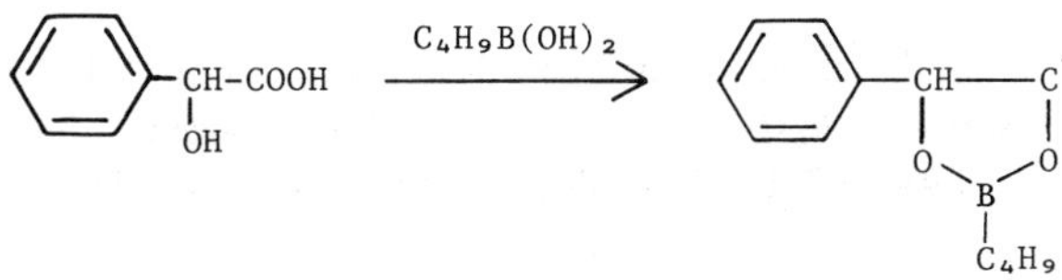

Typical Procedure. Diss 1–2 mg sample in 200 μl anhyd DMF or acetone and add an equivalent wt of *n*-butylboronic acid. Seal and allow to stand at room temp. Analyze by GC-MS.

Comments. GC and MS data were reported for 15 compds with various proximal difunctional groupings.[1] Work was reported earlier on the derivs of 1,2- and 1,3-diols.[2]

References

[1]G. M. Anthony *et al., J. Chromatogr. Sci.* **7**, 623 (1969).

[2]C. J. W. Brooks and J. Watson, *Chem. Commun.* 952 (1967).

Derivative Type. *n*-Butyl boronate.

Reagent. *n*-Butaneboronic acid.

Compound Type. α-Hydroxy fatty acids.

Reaction

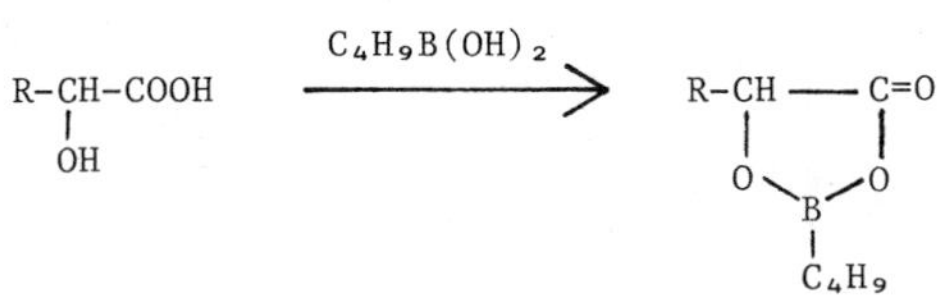

Typical Procedure. Diss *ca* 10 mg sample in a minimum amt of $CHCl_3$. Add *ca* 10 mg *n*-butaneboronic acid and allow to stand at room temp 15 min. Analyze an aliquot by GC.

Comments. The derivs of α-hydroxy-lauric, -myristic, -palmitic, -stearic, and -arachidic acids were separated on OV-17.

References

[1]Applied Science Laboratories, State College, PA, Technical Bulletin 25.

[2]R. F. Kruppa *et al.*, presented at the Eastern Anal. Symp., New York, Nov. 20, 1970.

3.5. DERIVATIVES OF KETO ACIDS AND KREBS CYCLE ACIDS

3.5.1. Alkyl Ester Derivatives

Derivative Type. Methyl ester.

Reagent. Methanol (MeOH)/HCl/thionyl chloride.

Compound Type. Carboxylic acids.

Reaction

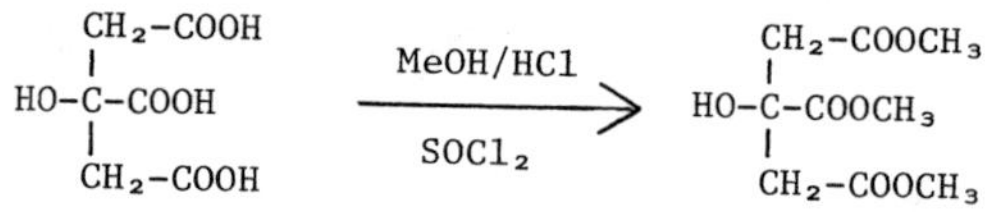

Typical Procedure. To a 50 mg sample of free acids or ammonium salts add 10 ml 5–10% HCl in anhyd MeOH. Carefully add 0.1 ml $SOCl_2$ and reflux 10 min in a hot H_2O bath. Evap under vac, add MeOH, and repeat evap to remove volatile nonesters. Diss residue in MeOH and analyze by GC:

Comments. The method was applied to nonvolatile acids in plant materials. The reagent was reported as particularly useful for those acids that are difficult to esterfy by other methods, e.g., diazomethane or mineral acid in alcohol. MeOH/$SOCl_2$ has also been used for α-keto acids.[2]

References

[1]M. Gee, *Anal. Chem.* **37**, 926 (1965).

[2]C. R. Atkins and K. T. Canvin, *Can. J. Biochem.* **49**, 949 (1971).

Derivative Type. Dimethyl ketal, methyl ester.

Reagent. Methanol (MeOH)/HCl/trimethyl orthoformate.

Compound Type. Oxalacetic and α-ketoglutaric acids.

Reaction

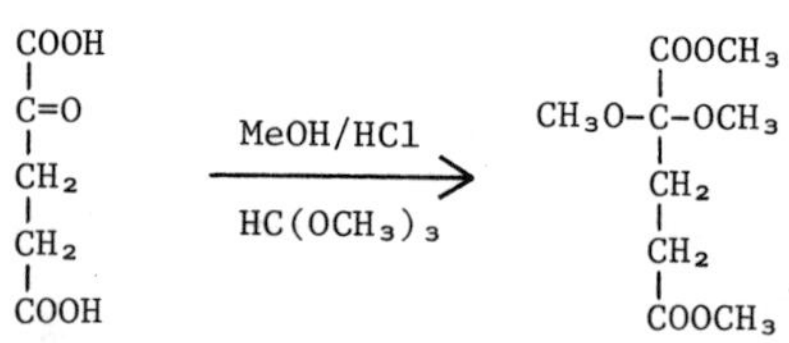

Typical Procedure. Diss 200 mg of the acid in 20 ml 10% HCl in MeOH. Add 2 ml redist trimethyl orthoformate and reflux 6 hr on a 70° oil bath. Neutralize by addition of satd $NaHCO_3$ soln and conc to *ca* 10 ml under vac. Ext with 4 × 5 ml Et_2O. Dry ext over $MgSO_4$ and analyze by GC or GC-MS.

Comments. The derivs were formed in 80% and 68% yield respectively for α-ketoglutaric and oxalacetic acids.

References

[1]P. G. Simmonds *et al.*, *Anal. Chem.* **39**, 163 (1967).

Derivative Type. *n*-Butyl ester.

Reagent. *n*-Butyl alcohol (*n*-BuOH)/HCl.

Compound Type. Citric and nitrilotriacetic acids.

Reaction

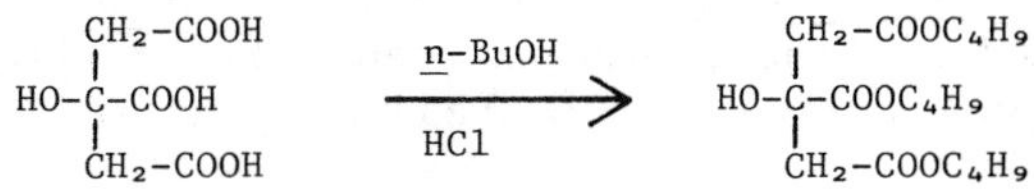

Typical Procedure. Add 2 ml 4 *M* HCOOH to a 50 ml H_2O sample and adjust pH to 2.5 with additional HCOOH. Bubble N_2 through sample for 30 min at 60°. Cool and pass through 5 × 0.5 cm diameter bed of Bio-Rad AG1-X2, 50–100 mesh anion exchange resin followed by 2 × 5 ml 0.1 *M* HCOOH. Elute column with 16 *M* HCOOH collecting 8 ml. Evap under N_2(2 hr at 85° and ambient temp overnight). Add 2 ml 3 *N* HCl in *n*-BuOH and place in ultrasonic bath at 75° for 25 min (sonicate only first 5 min). Evap under N_2 and diss residue in 100 μl acetone. Analyze 2 μl aliquots by GC.

Comments. The method was reported to be useful for analysis at 1–10,000 ppb in aq samples.

References

[1]W. A. Aue *et al.*, *J. Chromatogr.* **72**, 259 (1972).

Derivative Type. Ethyl ester.

Reagent. Ethanol (EtOH)/HCl.

Compound Type. Krebs cycle acids.

Reaction

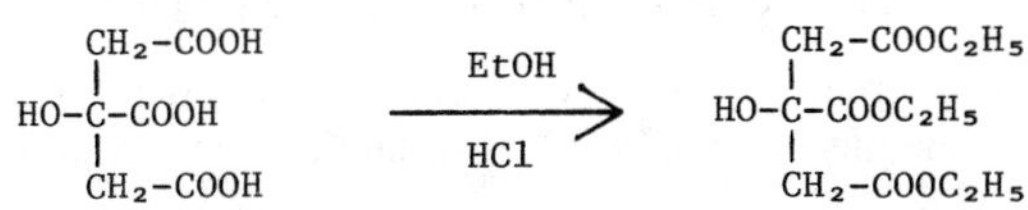

Typical Procedure. Reflux the acid sample in 5% HCl in EtOH for 2 hr. Evap almost to dryness under vac, add 5 ml H_2O, and ext with 2 × 10 ml Et_2O. Wash ext with aq $NaHCO_3$, dry over Na_2SO_4 and evap to dryness under vac at room temp. Analyze by GC.

Comments. Methyl esters were prepared in the same manner, and with CH_2N_2 and BF_3/MeOH. CH_2N_2 gave by products with the keto acids. Isocitric acid gave a lactone diester deriv. GC data were reported for ethyl and methyl esters of 12 acids.

References

[1]H. M. Kellogg *et al.*, *J. Pharm. Sci.* **53**, 420 (1964).

Derivative Type. Methyl ester.

Reagent. (MeOH)/boron trifluoride.

Compound Type. Carboxylic acids.

Reaction

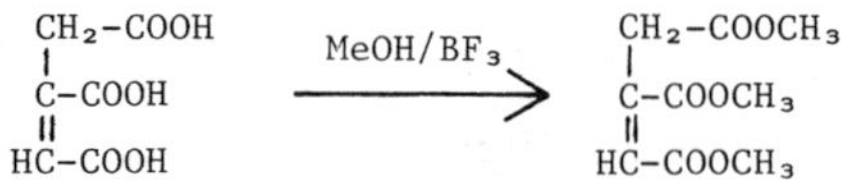

Typical Procedure. Diss 10–100 mg of acid in 3 ml of 51 % BF_3 in MeOH. Shake overnight at 28° or heat at 100° for 2–3 min. Cool in an ice bath and add 5 ml H_2O. Ext with 2 × 2 ml $CHCl_3$ (ext must be done within 5 min of H_2O addn). Dry ext over Na_2SO_4 and analyze by GC.

Comments. The method was useful for quant determination of 18 mono-, di-, and tricarboxylic acids. A separation of the derivs of the seven tricarboxylic acid cycle acids was reported. The derivs could be estimated to as low as 0.1 μg using FID.

References

[1]M. A. Harmon and H. W. Doelle, *J. Chromatogr.* **42**, 157 (1969).

Derivative Type. Methyl ester.

Reagent. Methanol (MeOH)/boron trifluoride.

Compound Type. Organic acids.

Reaction

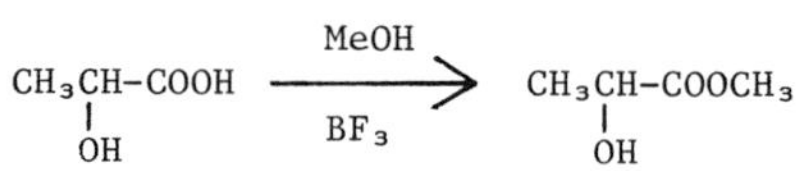

Typical Procedure. Pass an Et_2O soln of the acids through a 1 × 10 cm column of acid alumina (activated at 120° for 16 hr). Apply suction to column to remove traces of solvent. Elute the column with 10 ml 14% (w/v) BF_3 in MeOH. Allow eluate to stand at room temp overnight. Dil with an equal vol of H_2O and ext with 3 × 2 ml $CHCl_3$. Analyze the ext by GC.

Comments. The column method was used to avoid losses encountered in evap of an Et_2O soln of volatile acids. The method was applied to lactic, pyruvic, and six Krebs cycle acids.

References

[1]N. W. Alcock, *Anal. Biochem.* **71**, 335 (1965).

Derivative Type. Methyl 2,2-dimethoxypropionate.

Reagent. Methanol/boron trifluoride.

Compound Type. Pyruvic acid.

Reaction

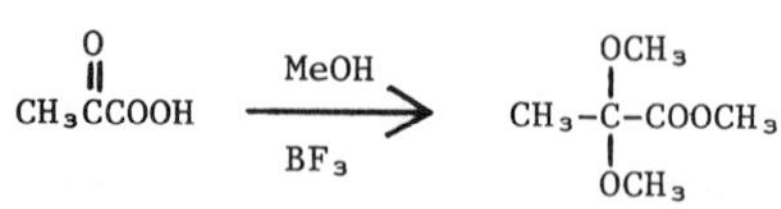

Typical Procedure. Mix 0.5–5.0 ml of a 1 mg/ml EtOH soln of the acid with 1.0 ml 2 *N* NH_3 and evap under vac 20 °. Add 2 ml 12.5 % (w/v) BF_3 in MeOH and allow to stand 16 hr at room temp. Add 4 ml 52 % satd aq $(NH_4)_2SO_4$ and ext with 2.0 ml $CHCl_3$. Dry the ext over Na_2SO_4 and analyze by GC.

Comments. The ammonium salt was formed to reduce the volatility of the acid during the evap step. The method affected essentially quant conversion of pyruvate to the 2,2-dimethoxypropionate and allowed simultaneous analysis of lactic, fumaric, succinic, malic, and citric acids.

References

[1]E. Hautala and M. L Weaver, *Anal. Biochem.* **30**, 32 (1969).

Derivative Type. *n*-Propyl ester.

Reagent. *n*-Propanol (*n*-PrOH)/boron trichloride.

Compound Type. Lactic and succinic acids.

Reaction

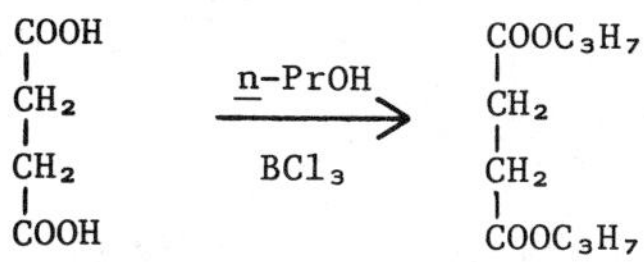

Typical Procedure. Reflux sample for 15 min in 1 ml *n*-PrOH satd with BCl_3 (*ca* 12%). Cool, add 4 ml satd $(NH_4)_2SO_4$ soln, and ext with 5 ml CH_2Cl_2. Dry ext by filtering through *ca* 0.5 g Na_2SO_4. Analyze by GC.

Comments. The method was used for analysis of the acids in eggs. The acids were isolated from an aq soln by continuous ext with Et_2O.

References

[1]P. Daemens and L. Laruelle, *J. Assoc. Off. Anal. Chem.* **59**, 613 (1976).

Derivative Type. Methyl ester.

Reagent. Diazomethane.

Compound Type. Unsaturated dibasic acids.

Reaction

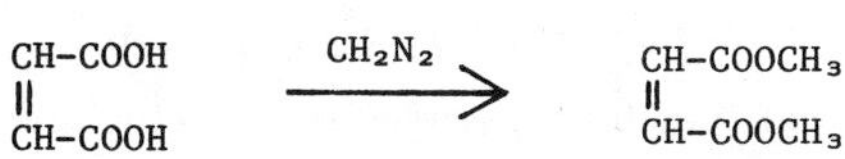

Typical Procedure. Dil 10 μl of a 50 mg/ml MeOH soln of the acid with 50 μl Et_2O. Cool to $-60°$ (solid CO_2), and saturate with CH_2N_2.[2] Remove excess CH_2N_2 under a stream of N_2 at $-60°$ until the soln is colorless. Analyze by GC.

Comments. Esterfication of the unsaturated diabasic acids at $-60°$ was found to give predominantly the diesters with a minimum formation of CH_2N_2 addition side products. It was also reported that the substitution of dioxane for Et_2O in the Schlenk and Gellerman[2] method was advantageous for quantitative analysis of satd aliphatic and aromatic acids. It had previously been shown that maleic, fumaric, and *cis*-aconitic acids could be esterfied with CH_2N_2 at $-70°$ but that they yielded pyrazolines at 25°.[3] Other workers reported successful esterfication of pyruvic, lactic, malic, succinic, and fumaric acids with CH_2N_2 at "low" temp.[4]

References

[1]O. Mlejnek, *J. Chromatogr.* **70**, 59 (1972).

[2]H. Schlenk and J. L. Gellerman, *Anal. Chem.* **32**, 1412 (1960).

[3]G. G. McKeown and S. I. Read, *Anal. Chem.* **37**, 1780 (1965).

[4]F. L. Estes and R. C. Bachmann, *Anal. Chem.* **38**, 1178 (1966).

Derivative Type. 1-Pyrazoline, trimethyl ester.

Reagent. Diazomethane.

Compound Type. *cis*-Aconitic acid.

Reaction

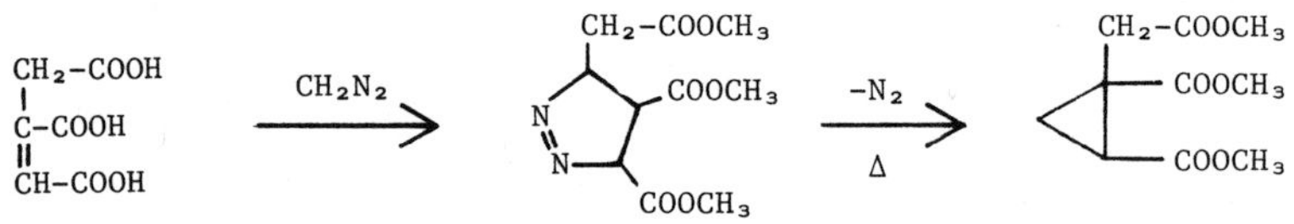

Typical Procedure. Treat acid with excess CH_2N_2 in Et_2O. Analyze by GC.

Comments. The pyrazoline was found to undergo varying degrees of decomp depending upon the GC column used. It was concluded that diazomethane was an unsuitable deriv reagent for analysis of Krebs Cycle acids.

References

[1]S. Dilli and J. Szyszka, *J. Chromatogr.* **132**, 148 (1977).

Derivative Type. Methyl ester, epoxide.

Reagent. Diazomethane.

Compound Type. α-Keto acids (Krebs cycle acids).

Reaction

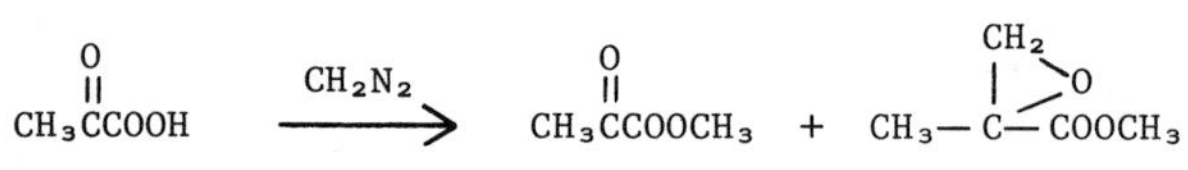

Typical Procedure. Treat acid with excess CH_2N_2 in Et_2O. Analyze by GC or GC-MS.

Comments. It was observed that both pyruvic and α-ketoglutaric acids gave increasing proportion of the glycidic ester with increasing reaction time. Oxalacetic acid gave predominantly the enol ether ester rather than the epoxide ester presumably due to the greater tendency toward enolization of the acid. The keto esters and enol ether ester were sensitive to ECD. An epoxide ester deriv was also reported for α-ketoisovaleric acid.[2]

References

[1]P. G. Simmonds *et al.*, *Anal. Chem.* **39**, 163 (1967).

[2]D. Gompertz and G. H. Draffan, *Clin. Chim. Acta* **40**, 5 (1972).

3.5.2. Silyl and Oxime Silyl Derivatives

Derivative Type. Trimethylsilyl, BSTFA addition product.

Reagent. Bis(trimethylsilyl)trifluoroacetamide (BSTFA).

Compound Type. Mesoxalic acid.

Reaction

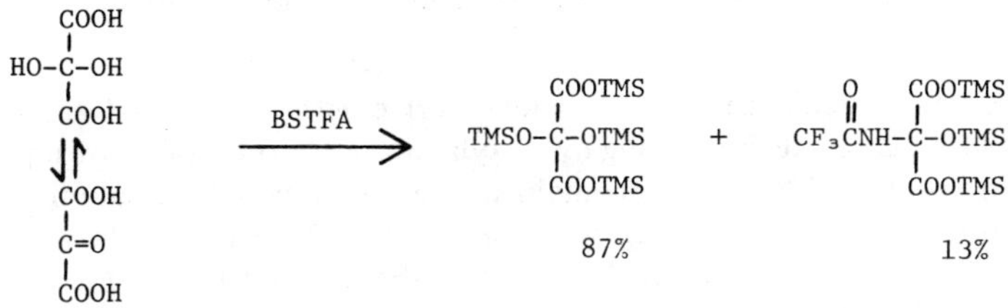

Typical Procedure. Treat 1 mg of mesoxalic acid with 100-200 μl BSTFA containing 1% TMCS at room temp for 1 hr. Analyze by GC-MS.

Comments. The addition product was an artifact in silylation of mesoxalic acid. BSA gave 2–5% of the acetamide deriv.

References

[1]S. P. Markey, *J. Chromatogr. Sci.* **11**, 417 (1973).

Derivative Type. Trimethylsilyl oximes.

Reagent. Hydroxylamine, hexamethyldisilazane (HMDS)/trimethylchlorosilane (TMCS).

Compound Type. Krebs cycle acids.

Reaction

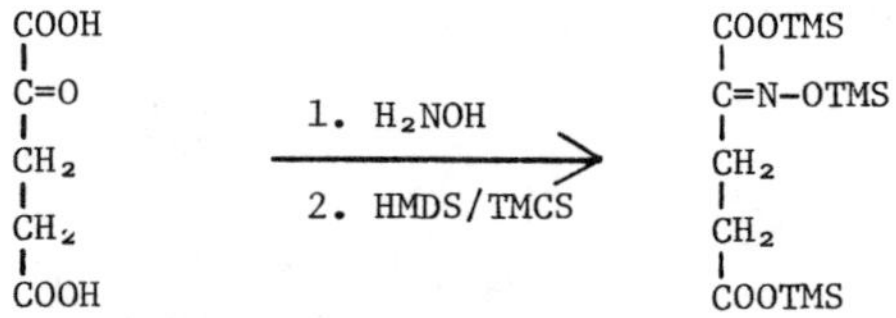

Typical Procedure. Diss 10 mg acids and 10 mg hydroxylamine HCl in 1 ml dry pyridine and allow to stand 10 min at room temp. Add 0.1 ml HMDS and 0.1 ml TMCS. After a few min at room temp, analyze by GC.

Comments. GC separation of 10 Krebs cycle acid was demonstrated.[1] For samples containing no keto acids, the hydroxylamine was omitted. BSTFA was used instead of HMDS/TMCS to prepare the derivs of 15 Krebs cycle acids and related compds.[2] The structures of the derivs were confirmed by MS. TMS-ester, TMS-enol ether derivs were also prepared using BSTFA/TMCS (1:4, v/v; room temp, 2 hr).[2]

References

[1]Z. Horii *et al., Chem. Ind.* **34**, 1494 (1965).

[2]M. G. Horning *et al., Anal. Lett.* **1**, 713 (1968).

Derivative Type. Oxime, trimethylsilyl.

Reagent. Hydroxylamine, bis (trimethylsilyl) acetamide (BSA).

Compound Type. α-Ketoacids.

Reaction

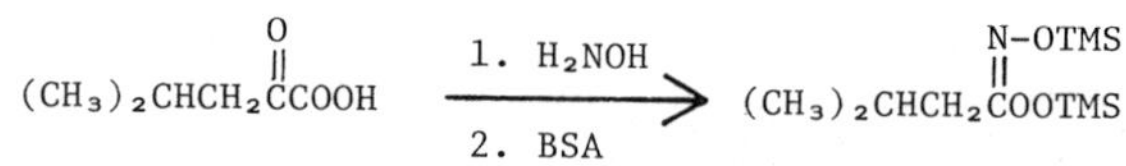

Typical Procedure. Alkalinize a 4 ml aliquot of urine with 30% NaOH and ext 3 × with equal vol Et_2O; discard ext. Add 25 mg hydroxylamine HCl to 1/2 of above aq phase. Make further additions of 30% NaOH to maintain strongly alkaline soln. Heat mixt to 60° for 30 min and acidify with 5 *N* HCl. Saturate with NaCl and ext 3 × with equal vol Et_2O. Conc ext to 2 ml. Evap an aliquot of conc ext to dryness, cap tube, and add a vol of Tri-Sil-BSA®[3] equal to 2% original vol of urine. Heat at 50–60° for 30 min. Withdraw aliquots through serum cap and inject into GC.

Comments. Plasma or serum was deproteinized by addition of equal vol of 95% EtOH and the supernatant was treated in the manner described for urine.[1]

References

[1]G. Lancaster *et al., Clin. Chim. Acta* **48**, 279 (1973).

[2]O. A. Mamer *et al., Clin. Chim. Acta* **32**, 171 (1971).

[3]Pierce Chemical Co., Rockford, IL.

Derivative Type. Oxime, trimethylsilyl.

Reagent. Hydroxylamine, bis(trimethylsilyl)acetamide (BSA)/trimethylchlorosilane (TMCS).

Compound Type. α-Keto acids.

Reaction

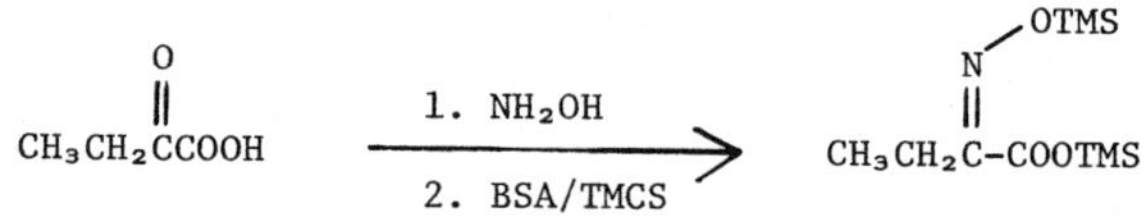

Typical Procedure. To 2.0 ml urine add 25 μl of a soln of 1 mg *n*-eicosane int stand in 1 ml hexane. Add 2 ml satd aq NaCl and 2 drops conc HCl. Ext with 3 × 3 ml EtOAc. Evap exts to dryness under N_2 at room temp.

Diss residue in 50 μl of hydroxylamine HCl soln (20 mg/ml) in pyridine. After 30 min at room temp add 50 μl BSA containing 10% TMCS. After 30 min at room temp, analyze 1 μl by GC.

Comments. Nine α-keto acids and 15 phenolic acids and related compds could be determined in a single analysis. An ext method for serum was also reported.

References

[1]H. J. Stermowsky *et al., Clin. Chim. Acta* **47**, 371 (1973).

Derivative Type. Oxime, Trimethylsilyl.

Reagent. Hydroxylamine HCl, hexamethyldisilazane (HMDS)/trimethylchlorosilane (TMCS)/pyridine.

Compound Type. Phenylpyruvic acid.

Reaction

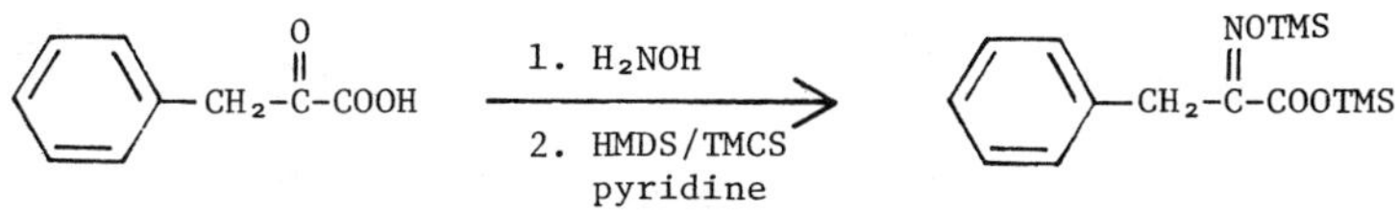

Typical Procedure. Treat urine with a large excess (*ca* 300 mg/10–20 mg creatinine) $H_2NOH{\cdot}HCl$ immediately after voiding and adjust to pH 11 with NaOH. Allow to stand 30 min at room temp then acidify to pH 1 with HCl. If not to be analyzed immediately, freeze at $-20°$. If sample vol exceeds 10 ml, conc to *ca* 10 ml at 30° on a rotary evaporator. Sat with NaCl and ext with 4 × 1 ml EtOAc and 7 × 3 vol Et_2O. Dry combined ext over Na_2SO_4 and evap under vac. Diss residue in 6 ml pyridine and add 0.8 ml HMDS and 0.8 ml TMCS. Allow to stand at room temp a few min, centrifuge, analyze an aliquot of the supernatant by GC.

Comments. The method was used to measure the Phe metabolites phenylpyruvic, phenyllactic, phenylacetic and *o*-hydroxyphenylacetic acids in urine. Phenylpyruvic acid was unstable in frozen urine ($-20°$) but the oxime was stable for at least 2 weeks. Thompson[2] also used the deriv to circumvent the problem of partial decarboxylation in the ext of phenylpyruvic acid from urine.[2]

References

[1]J. M. Vavich and R. R. Howell, *J. Lab. Clin. Med.* **77**, 159 (1971).

[2]R. M. Thompson, *Clin. Chim. Acta* **61**, 367 (1975).

Derivative Type. *O*-Ethyloxime, trimethylsilyl.

Reagent. *O*-Ethylhydroxylamine HCl, bis(trimethylsilyl)acetamide (BSA).

Compound Type. Oxocarboxylic acids.

Reaction

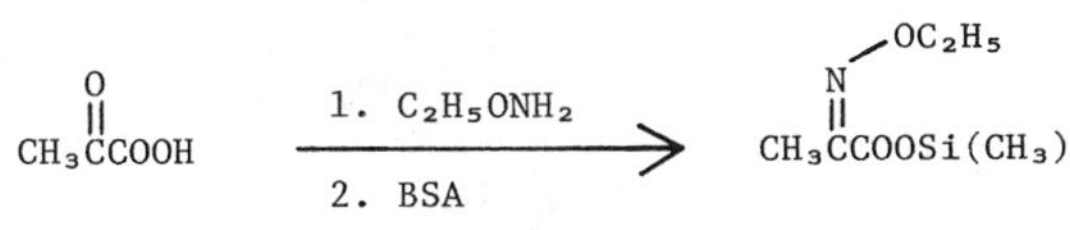

Typical Procedure. Neutralize 0.25–10.0 ml of urine, add 1.0 ml of undecanedioic acid int stand soln (1.00 mg), and apply to a 4.5 × 0.7 cm column of A25 DEAE-Sephadex. Wash column with 2 × 5 ml H_2O and eluate with 15 ml 1.5 *M* aq pyridinium acetate.

Add eluate to 50 mg *O*-ethylhydroxylamine HCl, mix, and allow to stand at room temp at least 4 hr. Freeze dry and add 900 μl BSA in pyridine (TriSil BSA®[2]). Allow to stand at least 4 hr. Add BSA as necessary to give a clear soln. Analyze by GC.

Comments. The method was used to quantitate organic acids (including non-oxo acids) in urine. The *O*-ethyloxime derivs were found to be superior to the *O*-methyl and *O*-benzyl derivs for GC analysis. *O*-TMS oximes were found to be unstable.[3] MS data for methyl-, ethyl-, and benzyloxime TMS ester derivs of 16 metabolically important oxocarboxylic acids were also reported.[4]

References

[1]R. A. Chalmers and R. W. E. Watts, *Analyst* **97**, 958 (1972).
[2]Pierce Chemical Company, Rockford, IL.
[3]R. A. Chalmers and R. W. E. Watts, *Analyst* **97**, 951 (1972).
[4]A. M. Lawson *et al., Biomed. Mass Spectrom.* **1**, 199 (1974).

3.5.3. Quinoxalone Derivatives

Derivative Type. Trimethylsilyl quinoxalone.

Reagent. *o*-Phenylenediamine, bis(trimethylsilyl) trifluoroacetamide (BSTFA).

Compound Type. α-Keto acids.

Reaction

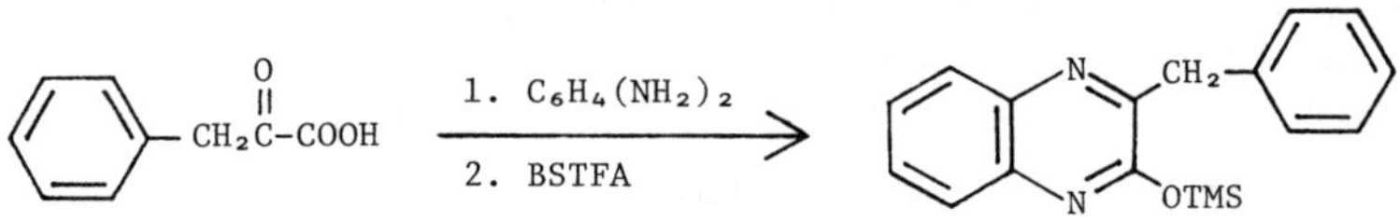

Typical Procedure.[1] Adjust 50 ml urine to pH > 11 with 12% NaOH. Evap to dryness below 90° under vac. Add a 10 ml soln containing 0.5 g *o*-phenylenediamine and 3.8 ml conc HCl. Heat 15 min at 100° then 45 min at 40–45°. Ext with 3 × 10 ml EtOAc. Ext combined EtOAc with 20 ml 6% NaOH. Add solid CO_2 to the NaOH ext to give pH 7. Ext the resulting soln with 3 × 10 ml EtOAc. Evap the combined EtOAc to dryness under vac. Add 0.1 ml pyridine and 0.6 ml BSTFA to the residue. Analyze by GC.

Comments. The method was used to measure pyruvic, phenylpyruvic, and α-ketoisocaproic acids in urine.[1] A rubidium sulfate flame detector was used to enhance the selectivity and sensitivity for the derivs.[1] Frigerio *et al.* showed the deriv to be the *O*-TMS isomer using MS data.[5] Langenbeck *et al* used the derivs for GC[6] and GC-MS-SIM[7] analysis of α-keto acids and have published an extended study[8] of their behavior.

References

[1]N. E. Hoffman and K. M. Gooding, *Anal. Lett.* **2**, 479 (1969).
[2]N. E. Hoffman and T. A. Killinger, *Anal. Chem.* **41**, 162 (1969).
[3]N. E. Hoffman and T. A. Killinger, *Clin. Chem.* **14**, 807 (1968).
[4]N. E. Hoffman *et al., Res. Commun. Pathol. Pharmacol.* **2**, 87 (1971).
[5]A. Frigerio *et al., J. Chromatogr.* **81**, 139 (1973).
[6]U. Langenbeck, *et al., J. Chromatogr.* **115**, 65 (1975).
[7]U. Langenbeck *et al., J. Chromatogr.* **143**, 39 (1977).
[8]U. Langenbeck *et al., Biomed. Mass Spectrom.* **4**, 197 (1977).

Derivative Type. Quinoxalone, trimethylsilyl.

Reagent. *o*-Phenylenediamine, bis (trimethylsilyl) acetamide.

Compound Type. Pyruvic and α-ketoglutaric acids.

Reaction

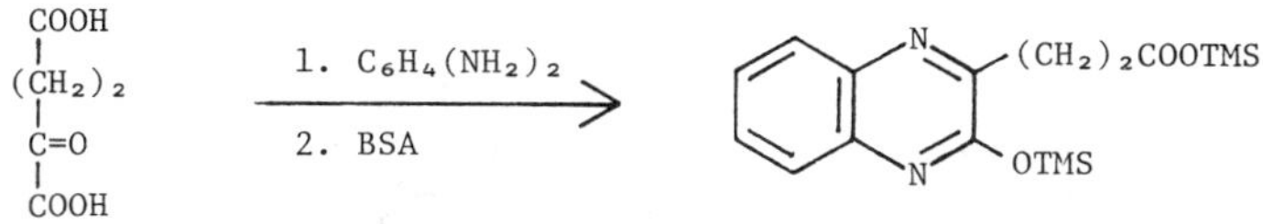

Typical Procedure. Mix soln of 60 mmol of sublimed *o*-phenylenediamine in 100 ml 10% AcOH with a soln of 30 mmol of keto acid in 30 ml H_2O. After 15 min filter quinoxalone ppt. Diss quinoxalone in dry pyridine to yield 10 mg/ml soln. Mix 5–50 μl of quinoxalone soln with 20 μl of a 20 mg/ml pyridine soln of int stand (6-methyl-2-naphthol, or *p*-nitrophenyl phenyl ether). Add 200 μl BSA and 50 μl pyridine. Analyze by GC.

Comments. The quinoxalones alone gave poor GC performance,[2] but silylation resulted in good GC behavior. IR and NMR data were reported to indicate that the derivs existed in the keto form,[1] but Frigerio *et al.*[3] presented MS data supporting the *O*-TMS form shown above.

References

[1]N. E. Hoffman and T. A. Killinger, *Anal. Chem.* **41**, 162 (1969).

[2]P. G. Simmonds *et al.*, *Anal. Chem.* **39**, 163 (1967).

[3]A. Frigerio *et al.*, *J. Chromatogr.* **81**, 139 (1973).

Derivative Type. Quinoxalinols (quinoxalones).

Reagent. *o*-Phenylenediamine.

Compound Type. α-Ketoacids.

Reaction

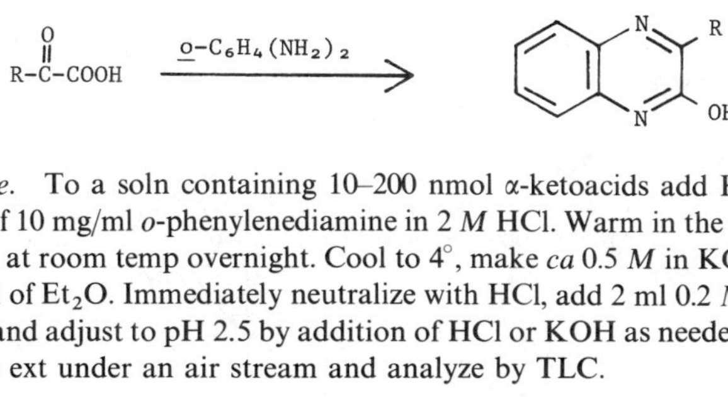

Typical Procedure. To a soln containing 10–200 nmol α-ketoacids add HCl to make 2 *M* then add 0.2 ml of 10 mg/ml *o*-phenylenediamine in 2 *M* HCl. Warm in the dark 1–3 hr at 37° or allow to stand at room temp overnight. Cool to 4°, make *ca* 0.5 *M* in KOH, and wash 3 × with an equal vol of Et_2O. Immediately neutralize with HCl, add 2 ml 0.2 *M* sodium formate buffer (pH 2.5), and adjust to pH 2.5 by addition of HCl or KOH as needed. Ext with 2 × 10 ml $CHCl_3$. Conc ext under an air stream and analyze by TLC.

Comments. The derivs were UV absorbing and fluorescent. Both absorption and emission spectra were given. The derivs could be useable for HPLC.

References

[1]J. Mowbray and J. H. Ottaway, *Biochem. J.* **120**, 171 (1970).

3.5.4. Other Derivatives of Keto Acids

Derivative Type. 1, 3-Diphenylimidazolidine-2-carboxylic acid, trimethylsilyl ester.

Reagent. *N,N′*-Diphenylethylenediamine, bis(trimethylsilyl)acetamide (BSA).

Compound Type. Glyoxylic acid.

Reaction

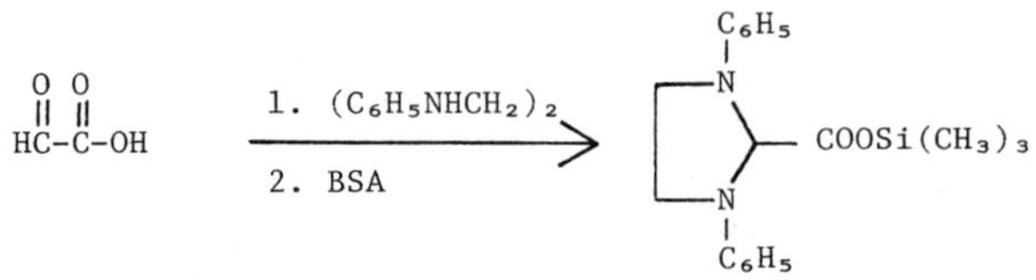

Typical Procedure. Mix 2 mg glyoxylic acid with 10 mg *N,N′*-diphenylethylenediamine and diss in 0.4 ml pyridine. Allow to stand 2 hr at room temp; then add 0.5 ml BSA. Allow to stand at least 2 hr and then analyze by GC.

Comments. The deriv was found to be suitable for GC analysis of glyoxylic acid and was suggested to be possibly useful for other aldehydes. Dimedone derivs were found to be insufficiently volatile for GC analysis.

References

[1]R. A. Chalmers and R. W. E. Watts, *Analyst* **97**, 951 (1972).

3.6. DERIVATIVES OF PHENOLIC ACIDS

3.6.1. Alkyl Ester Derivatives of Phenolic Acids

Derivative Type. Ethyl ester.

Reagent. Ethanol (EtOH)/HCl.

Compound Type. Homovanillic (HVA), iso-homovanillic (iso-HVA), and 3, 4-dimethoxyphenylacetic acids.

Reaction

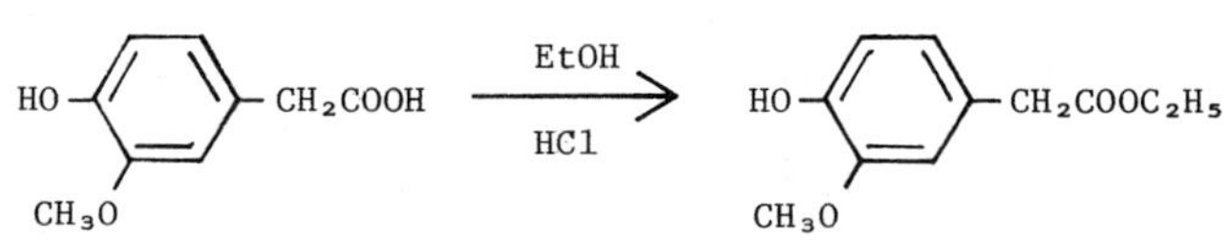

Typical Procedure. Adjust a sample of urine to pH 2 with 6 *M* HCl and ext with 2 × 1 vol EtOAc. Back ext the combined EtOAc with 3 ml 7% aq $NaHCO_3$. Acidify 0.5 ml of the aq phase with HCl, sat with NaCl, and ext with 10 ml EtOAc. Dry ext over Na_2SO_4 and evap to dryness. Diss the residue in 2 ml 3 *M* HCl in EtOH and heat 1 hr at 100°. Evap under vac, add abs EtOH, and evap again to insure dehydration. Diss the residue in EtOAc and analyze by GC.

Comments. The method was used to determine the ratio of HVA to iso-HVA in urine of patients taking L-DOPA.

References

[1]N. Narasimhachari *et al.*, *Clin. Chim. Acta* **50**, 337 (1974).

Derivative Type. Methyl ester, acetate.

Reagent. Diazomethane, acetic anhydride (Ac_2O).

Compound Type. Vanillylmandelic acid (VMA), homovanillic acid (HVA).

Reaction

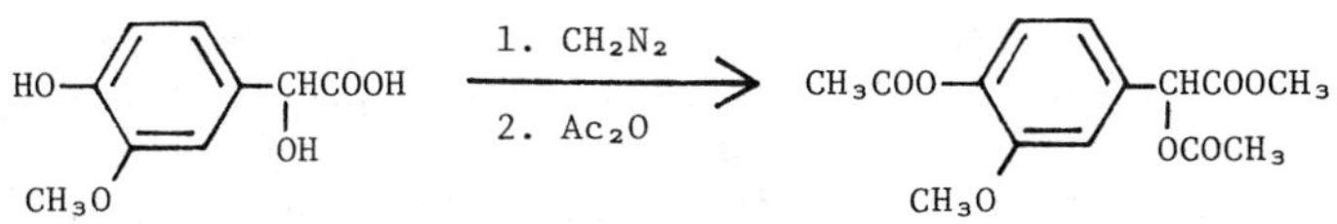

Typical Procedure. Diss the acid in MeOH and add CH_2N_2 in Et_2O to prepare the methyl ester. Remove excess CH_2N_2 immediately in stream of N_2 to prevent methylation of phenolic hydroxyls. Diss the ester in 50 μl pyridine and add 50 μl Ac_2O. Evap excess reagents under N_2, diss residue in EtOAc, and analyze by GC-MS.

Comments. MS data were reported for the derivs and also the d_3-acetates. Methyl ester, TMS ethers were also prepared by treating the esters with BSA in THF (1 : 1, v/v) for 15 min at 37°.

References

[1]E. Anggärd and G. Sedvall, *Anal. Chem.* **41**, 1250 (1969).

Derivative Type. Acetyl, methyl ester.

Reagent. Acetic anhydride (Ac_2O)/trifluoroacetic acid (TFA), diazomethane.

Compound Type. Dihydroxy aromatic acids.

Reaction

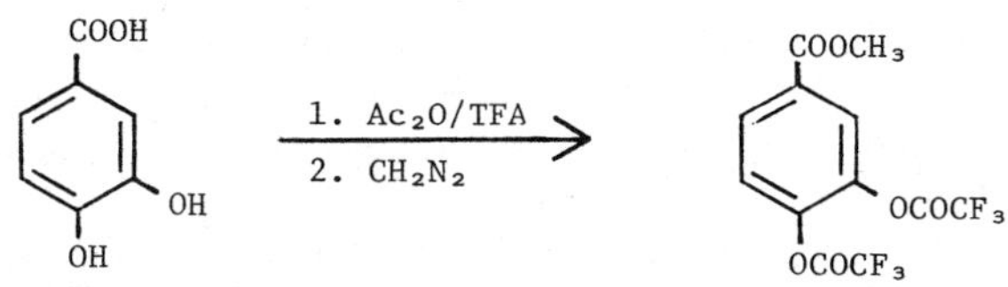

Typical Procedure. Diss 10–15 mg of the acid in 1.0 ml Ac_2O and 0.5 ml TFA. Heat 30 min at 75° then evap under an air stream. Treat residue with CH_2N_2 in Et_2O until the bubbling ceases and the yellow color persists. Evap under air and diss residue in acetone or THF prior to GC analysis.

Comments. The method was applied to the analysis of aromatic acids in urine. Deriv of dihydroxy aromatic acids with MeOH/BF_3 or MeOH/HCl was reported to yield mixts.[2]

References

[1]C. M. Williams and R. H. Leonard, *Anal. Biochem.* **5**, 362 (1963).

[2]C. M. Williams, *Anal. Biochem.* **4**, 423 (1962).

Derivative Type. Methyl ester.

Reagent. Methanol (MeOH)/boron trifluoride.

Compound Type. Acetylsalicylic acid.

Reaction

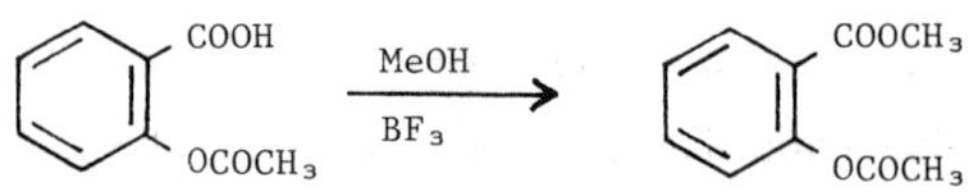

Typical Procedure. Grind dry sample (tablet, dry tissue, or soln residue) and ext with 25 ml hot anhyd MeOH for 5 to 10 min in a modified (siphon tube replaced by a run-off tube for minimum holdup) 50 ml Soxhlet extractor. To the hot MeOH soln add 20 ml of 20% (w/w) BF_3 in MeOH. Change extractor to a condenser and reflux 5–10 min. Transfer soln to a 50 ml vol flask, rinse, and make up to 50 ml with MeOH. Analyze 5 μl aliquot by GC.

Comments. The method was reported to be rapid and convenient for analysis of aspirin especially in single component products.

References

[1]R. C. Crippen and H. C. Freimuth, *Anal. Chem.* **36**, 273 (1964).

Derivative Type. Methyl ester, pentafluoropropionyl.

Reagent. Methanol (MeOH)/HCl, pentafluoropropionyl imidazole (PFPI).

Compound Type. Acid metabolites of biogenic amines.

Reaction

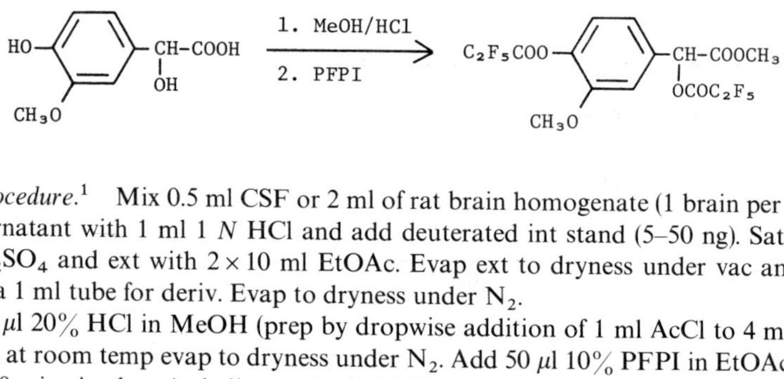

Typical Procedure.[1] Mix 0.5 ml CSF or 2 ml of rat brain homogenate (1 brain per 10 ml 1 *N* HCl) supernatant with 1 ml 1 *N* HCl and add deuterated int stand (5–50 ng). Sat soln with anhyd Na_2SO_4 and ext with 2 × 10 ml EtOAc. Evap ext to dryness under vac and transfer residue to a 1 ml tube for deriv. Evap to dryness under N_2.

Add 100 μl 20% HCl in MeOH (prep by dropwise addition of 1 ml AcCl to 4 ml MeOH). After 5 min at room temp evap to dryness under N_2. Add 50 μl 10% PFPI in EtOAc and heat at 70° for 10 min. Analyze 1 μl aliquots by GC-MS.

Comments. The use of PFPI was found to give higher yields of PFP derivs of the acidic metabolites than did PFPA. The opposite was true for alcoholic metabolites. The method was used for ng level quantitation of the metabolites in human CSF and rat brain.[1] MS data were reported for the derivs of VMA, DOPAC, and PHPA.[1]

References

[1]F. Karoum *et al., Biomed. Mass Spectrom.* **2**, 183 (1975).

[2]F. Karoum *et al., J. Neurochem.* **25**, 653 (1975).

[3]F. Karoum *et al., Brit. J. Pharmacol.* **56**, 403 (1976).

Derivative Type. Methyl ester, heptafluorobutyrate (HFB).

Reagent. Diazomethane, heptafluorobutyric anhydride (HFBA).

Compound Type. Homovanillic acid (HVA) and vanillylmandelic acid (VMA).

Reaction

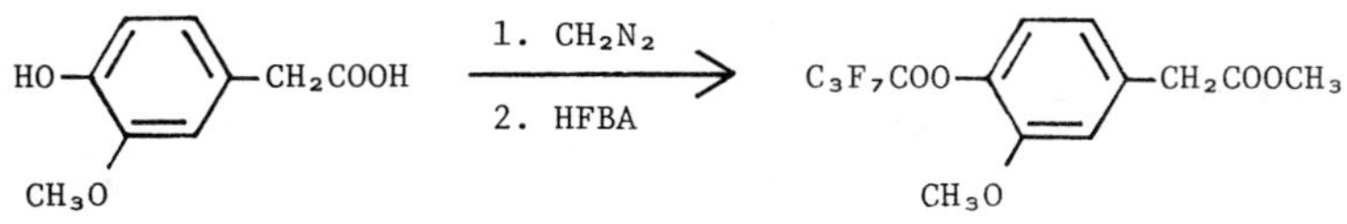

Typical Procedure. Diss sample in MeOH and add CH_2N_2 in Et_2O. Remove excess CH_2N_2 immediately under stream of N_2 (longer exposure leads to methylation of phenolic hydroxyls). Diss residue in redist EtOAc and add an equal vol of HFBA. Let stand 15 min at room temp and evap to dryness under N_2. Diss deriv in EtOAc and analyze by ECD-GC.

Comments. PFP and TFA derivs were also prepared using PFPA and TFAA, HFB and PFP derivs were more stable and had higher ECD response than TFA derivs. MS data were reported for the derivs. The HFB methyl ester was used to quantitate HVA in CSF by ECD-GC and GC-MS-SIM.[2] The PFP methyl ester deriv has been used to quantitate HVA in serum by GC-MS-SIM.[3]

References

[1]E. Anggärd and G. Sedvall, *Anal. Chem.* **41**, 1250 (1969).

[2]B. Sjöquist and E. Anggärd, *Anal. Chem.* **44**, 2297 (1972).

[3]S. Takahashi *et al., Clin. Chim. Acta* **81**, 183 (1977).

Derivative Type. Pentafluoropropionyl, pentafluoropropyl ester.

Reagent. Pentafluoropropionic anhydride (PFPA)/2,2,3,3,3-pentafluoro-1-propanol.

Compound Type. Phenolic carboxylic acids.

Reaction

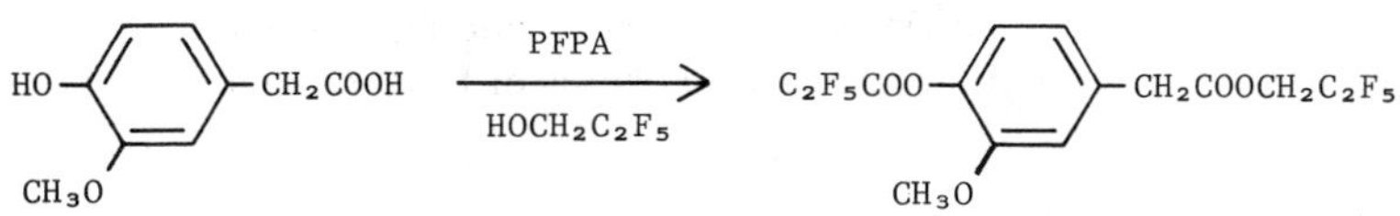

Typical Procedure. To 100 ng of phenolic acid add 40 μl PFPA and 10 μl $HOCH_2C_2F_5$. Seal and heat at 75° for 10 min. Evap to dryness under N_2 and add 25 μl PFPA. Heat at 75° for 5 min. Cool to room temp and evap under N_2. Diss residue in 1 ml EtOAc for GC or GC-MS analysis.

Comments. Derivs were examined from *p*-hydroxyphenylacetic acid, 3, 4-dihydroxyphenylacetic acid, vanillylmandelic acid, homovanillic acid, 5-hydroxy-indoleacetic acid, and 2-pyrrolidone-5-carboxylic acid.[1] Other halogenated propyl and pentyl alcohols were also examined. The derivs could be detected to 5 pg by ECD-GC. The method has also been used for GC-MS analysis in studies where the trifluoroethyl esters were also examined.[2] The procedure above was shown to produce a mixt of mono- and di-PFP derivs of 5-HIAA, but the partial acylation could be minimized by extending the heating time to 3 hr.[2] The method caused partial alkylation of the side chain hydroxyl in *p*-hydroxy substituted mandelic acids.[2] The derivs have been used for analysis of 3, 4-dihydroxy- and 3-methoxy-4-hydroxyphenylacetic acids in brain tissue.[3]

References

[1]E. Watson *et al., Anal. Biochem.* **59**, 441 (1974).

[2]D. D. Godse *et al., Anal. Chem.* **49**, 915 (1977).

[3]F.-A. Wiesel *et al., J. Neurol. Transm.* **35**, 319 (1974).

Derivative Type. 1-Chloro-1,1,3,3,3-pentafluoro-2-propyl ester, pentafluoropropionyl (PFP).

Reagent. 1-Chloro-1,1,3,3,3-pentafluoro-2-propanol/pentafluoropropionic anhydride (PFPA), PFPA.

Compound Type. 3,4-Dihydroxyphenylacetic acid (DOPAC) and homovanillic acid (HVA).

Reaction

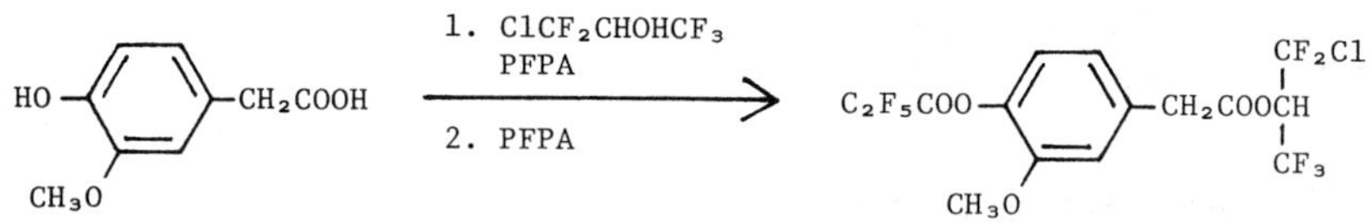

Typical Procedure. Homogenize 6–9 mg tissue in 1 ml 1 *N* HCl and centrifuge. Add 16 ng of each int stand to 100 μl of supernatant and ext with 1 ml toluene. Discard ext, cool aq phase in ice H_2O, and ext with 1 ml cold Et_2O. Evap ext to dryness and add 10 μl of 1-chloro-1,1,3,3,3-pentafluoro-2-propanol and 50 μl PFPA. Heat 15 min at 75°. Cool to room temp and evap under N_2. Add 25 μl PFPA and heat 5 min at 75°. Evap under N_2 and diss residue in 200 μl toluene. Analyze by ECD-GC.

Comments. The propionic acid analogs of DOPAC and HVA were used as int stand. Derivs were also prepared using 1,1,1,3,3,3-hexafluoro-2-propanol and 2,2,3,3,3-pentafluoro-1-propanol, but the deriv from the chlorofluoro alcohol permitted simultaneous determination of DOPAC and HVA on a single column without interferences.[1] The PFP, 2,2,3,3,3-pentafluoro-1-propyl ester deriv was used for analysis of DOPAC and HVA in CSF.[2]

References

[1]E. Watson *et al., Life Sci.* **15**, 2167 (1974).

[2]E. Watson and S. Wilk, *Psychopharmacologia* **42**, 57 (1975).

Derivative Type. Trifluoroacetyl, hexafluoroisopropyl ester.

Reagent. Trifluoroacetic anhydride (TFAA), hexafluoroisopropanol/boron trifluoride.

Compound Type. Homovanillic (HVA), iso-homovanillic (*i*-HVA), and dihydroxyphenylacetic acids.

Reaction

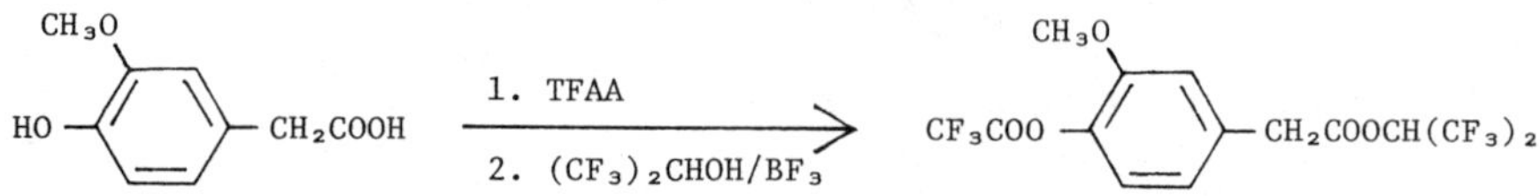

Typical Procedure.[1] Acidify 1 ml urine with 0.3 ml 6 *N* HCl and dil to 3 ml with H_2O. Sat with NaCl (1 g) and ext with 2 × 6 ml EtOAc. Evap ext to dryness under N_2 at 40°. Transfer residue to a small vial with 0.5 ml EtOH and 2 × 0.75 ml EtOAc. Evap to dryness and add 0.5 ml TFAA. Allow to react 30 min, then evap to dryness. Add 0.5 ml of a mixt of hexafluoroisopropanol and satd BF_3 in Et_2O (10:1, v/v). Heat at 100° for 15 min; then evap to a thick oil. Diss in 2 ml EtOAc and analyze by ECD-GC.

Comments. The method was used to quantitate HVA and *i*-HVA in urine[1,2] and CSF.[3] The deriv method was extended to include dihydroxyphenylacetic acid for analysis of the three compds in nervous tissue.[4] Davis *et al.*,[5] in applying a similar method (with PFPA) to analysis of *m*- and *p*-hydroxyphenylacetic acids in brain tissue, encountered losses of deriv in the evap step and therefore used a wash with 1 *M* (pH 6) phosphate buffer to remove excess acylating agent and by-product acid.

References

[1] S. W. Dziedzic *et al., Anal. Biochem.* **47**, 592 (1972).

[2] S. W. Dziedzic *et al., J. Lab. Clin. Med.* **82**, 829 (1973).

[3] S. W. Dziedzic and S. E. Gitlow, *J. Neurochem.* **22**, 333 (1974).

[4] J. D. M. Pearson and D. F. Sharman, *Br. J. Pharmacol.* **53**, 143 (1975).

[5] B. A. Davis *et al., J. Chromatogr.* **142**, 517 (1977).

Derivative Type. Methyl ester, methyl ether.

Reagent. Diazomethane.

Compound Type. Phenolic acids.

Reaction

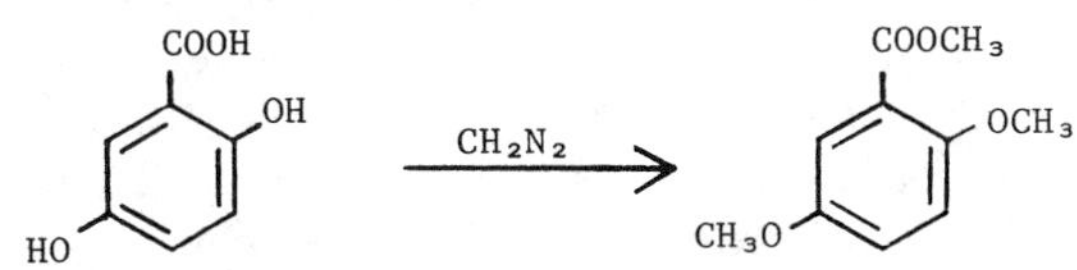

Typical Procedure. Diss sample in a minimum amt of EtOH, add CH_2N_2 until the yellow color persists, and then add *ca* 20% excess. Allow to stand at room temp 48 hr, adding CH_2N_2 if the soln becomes clear. Evap under N_2 and diss residue in pyridine. Analyze by GC.

Comments. Watson *et al.*[2] also used CH_2N_2 (in THF) for analysis of aspirin and salicylic acid but suppressed ether formation by cooling to −15°.

References

[1] C. H. Morris *et al., J. Pharm. Sci.* **59**, 270 (1970).

[2] J. R. Watson *et al., J. Pharm. Sci.* **60**, 455 (1971).

Derivative Type. Methyl ester, methyl ether.

Reagent. Diazomethane/methanol (MeOH).

Compound Type. Anacardic acids.

Reaction

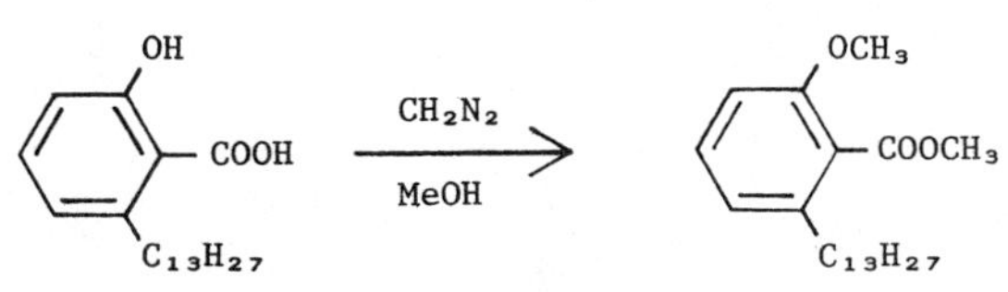

Typical Procedure. Diss sample in a soln of CH_2N_2 in Et_2O containing 20% MeOH. Allow to stand overnight at room temp. Analyze by GC.

Comments. The method was used for analysis of anacardic acids in cashew nuts and ginkgo leaves. Treatment of ext lipids with 3% H_2SO_4 in MeOH esterfied the fatty acids but not the anacardic acids. Mild treatment with CH_2N_2 in Et_2O gave esterfication only. The methyl esters were acetylated with Ac_2O in pyridine at 100° for 45 min.

References

[1]J. L. Gellerman and H. Schlenk, *Anal. Chem.* **40**, 739 (1968).

Derivative Type. Methyl ester, methyl ether.

Reagent. Diazomethane.

Compound Type. Dihydroxybenzoic acids.

Reaction

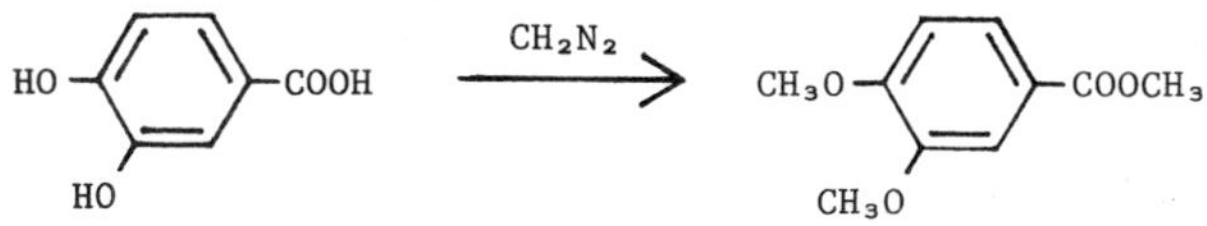

Typical Procedure. Diss acid in MeOH and add excess CH_2N_2 in Et_2O. Allow to react 16 hr at room temp then analyze by GC.

Comments. Reaction with CH_2N_2 for 1 min gave esterfication with negligible ether formation. Attempted esterfication with BF_3/MeOH was successful. Attempts to form diacetyl (with Ac_2O), ditrifluoroacetyl (with TFAA), and ditrimethylsilyl (with HMDS/TMCS) methyl esters via two step deriv yielded multiple derivs. GC data were reported for the derivs of six isomeric dihydroxybenzoic acids.[1] The deriv reaction was applied to the analysis of HVA and VMA in urine.[2]

References

[1]C. M. Williams, *Anal. Biochem.* **11**, 224 (1965).
[2]C. M. Williams and M. Greer, *Clin. Chim. Acta* **7**, 880 (1962).

Derivative Type. Pentafluorobenzyl.

Reagent. Pentafluorobenzyl bromide (PFB-Br)/18-crown-6 ether/base (see comments).

Compound Type. Carboxylic acids and phenols.

Reaction

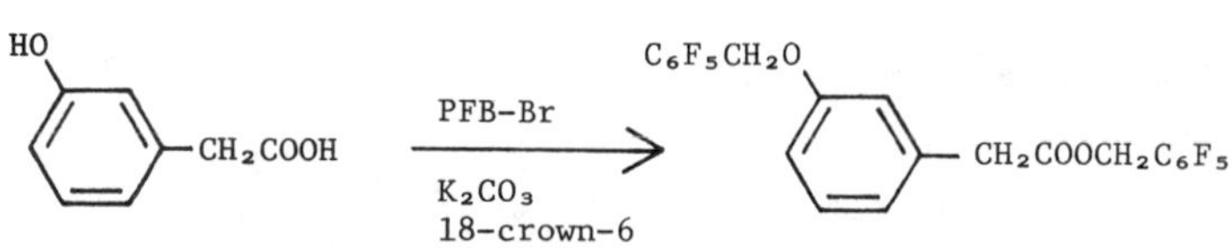

Typical Procedure. To up to 10 μg of sample add 50 μl of a soln of PFB-Br (0.3 μl/ml) and 18-crown-6 (3.0 mg/ml) in benzene. Add a few grains of powdered base (see below) and heat at 40° for 90 min. Analyze an aliquot of soln directly by GC.

Comments. The high reactivity of the sample anion when crown ether was used obviated the necessity of a large excess of reagent and subsequent removal of the excess before ECD-GC. Use of a strong base (K_2CO_3) resulted in deriv of both carboxylic acids and phenols, while weaker base (KOAc, $KHCO_3$, or KCN) gave selective deriv of carboxylic acids. The presence of moisture led to production of artifacts. Derivs were detectable to 0.3 pmol by ECD-GC.

References

[1]B. Davis, *Anal. Chem.* **49**, 832 (1977).

3.6.2. Silyl Derivatives of Phenolic Acids

Derivative Type. Methyl ester, trimethylsilyl ether.

Reagent. Diazomethane, hexamethyldisilazane (HMDS)/trimethylchlorosilane (TMCS).

Compound Type. Phenolic acids.

Reaction

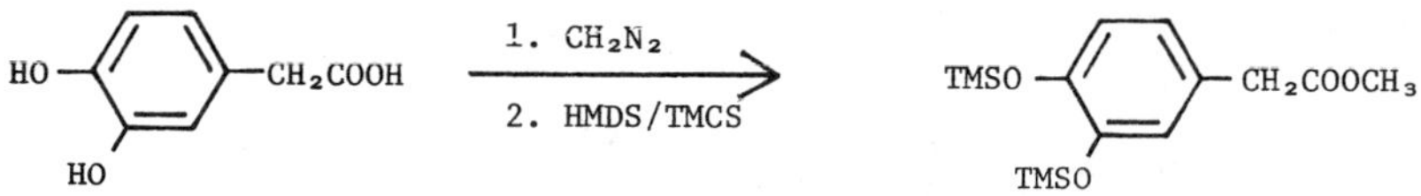

Typical Procedure. Esterfy acid sample (0.5 mg) or urine ext residue by diss in 0.2 ml MeOH and adding excess CH_2N_2 in Et_2O. After 1 min evap under N_2. Diss residue in 0.2 ml dioxane. Add 0.2 ml HMDS and 0.1 ml TMCS. After 30 min at room temp, centrifuge and analyze an aliquot of the supernatant by GC.

Comments. Trimethylsilyl ether, ester derivs were prepared by the same method omitting the diazomethane step. GC data were reported for the derivs of 48 biologically important phenolic acids and alcohols.[1] The derivs have also been prepared using HMDS/TMCS in pyridine.[2]

References

[1]F. Karoum *et al., Clin. Chim. Acta* **20**, 427 (1968).
[2]M. G. Horning *et al., Anal. Biochem.* **17**, 224 (1966).

Derivative Type. Trimethylsilyl.

Reagent. Hexamethyldisilazane (HMDS).

Compound Type. Salicylic acid and aspirin.

Reaction

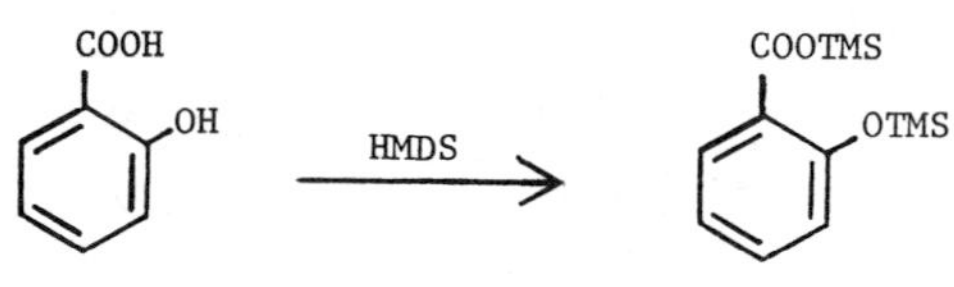

Typical Procedure. Add 2.5 mg KF (as a 50% w/w aq soln) to plasma samples immediately after collection and freeze until analyzed. To 1 ml of the plasma sample add 1 ml 10% (w/v) aq $KHSO_4$ and ext with 5 ml $CHCl_3$. Transfer 4.0 ml of the ext to a tube containing 25 μg *n*-butyl benzoate and 0.5 μg ethyl anisate int stands. Evap to dryness under vac and add 100 μl HMDS and 100μl acetone. Allow to stand 60 min at room temp, then analyze by GC.

Comments. The KF was used to inhibit hydrolysis of aspirin. Of several ext solvents examined, $CHCl_3$ offered the best compromise between volatility and ext of moisture. BSA, TMCS, or HMDS/TMCS were reported to offer no advantage over HMDS alone for silylation.[1] Rowland *et al.*[2, 3] had previously used 2% HMDS/0.1% TMCS in CS_2 for silylation of aspirin. Blakley[4] had used HMDS/TMCS/dioxane (1:1:8, v/v/v, 15 min, room temp) for silylation of phenolic acids.

References

[1]L. J Walter *et al., J. Pharm. Sci.* **63**, 1754 (1974).

[2]M. Rowland and S. Riegelman, *Anal. Biochem.* **20**, 463 (1967).

[3]M. Rowland and S. Riegelman, *J. Pharm. Sci.* **56**, 717 (1964).

[4]E. R. Blakley, *Anal. Biochem.* **15**, 350 (1966).

Derivative Type. Trimethylsilyl.

Reagent. Bis(trimethylsilyl) acetamide (BSA).

Compound Type. Aromatic acids.

Reaction

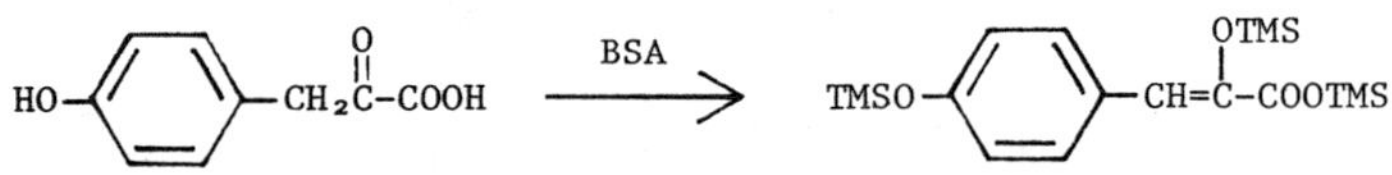

Typical Procedure. Diss acid in pyridine (1 ml/g acid) and add 2 × mol excess of BSA. Warm on a hot plate (temp not given) 15 min then isolate deriv by prep GC.

Comments. Derivs were prepared of 38 aromatic acids and their GC, IR, and NMR properties studied.[1] The deriv procedure converted all carboxyl, hydroxyl, and enol groups to the TMS derivs. BSTFA was used in place of BSA for analysis of *o*-hydroxyphenylacetic, *β*-phenyllactic, and phenylpyruvic acids in urine.[2]

References

[1]N. E. Hoffman *et al., Anal. Biochem.* **32**, 386 (1969).

[2]N. E. Hoffman and K. M. Gooding, *Anal. Biochem.* **31**, 471 (1969).

Derivative Type. Trimethylsilyl.

Reagent. Bis(trimethylsilyl) trifluoroacetamide (BSTFA)/trimethylchlorosilane (TMCS).

Compound Type. Homovanillic acid (HVA) and vanillylmandelic acid (VMA).

Reaction

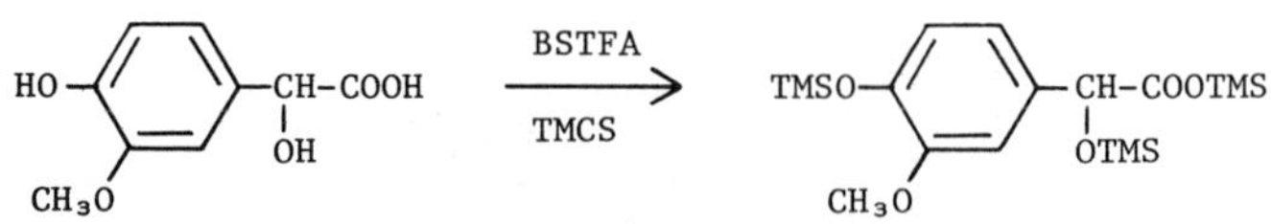

Typical Procedure. Acidify a urine sample (corresponding to 10 mg creatinine) to pH 1 with HCl, dil with equal vol of H_2O, and sat with NaCl. Ext with 5 × 10 ml Et_2O. Evap to dryness under N_2 then under vac. Add 500 μl BSTFA containing 1% TMCS and heat at 70° for 20 min. Analyze by GC or GC-MS.

Comments. The method was more rapid than previous multistep deriv methods. MS data were given for the derivs.[1] Another method for urinary HVA and VMA using the same reagents employed a 1 hr reaction time at 60°.[2]

References

[1]T. J. Sprinkle *et al., Clin. Chim. Acta* **25**, 409 (1969).

[2]S. Addanki *et al., Clin. Chem.* **22**, 310 (1976).

Derivative Type. Trimethylsilyl.

Reagent. Bis(trimethylsilyl) acetamide (BSA)/trimethylchlorosilane (TMCS).

Compound Type. Carboxylic acid sodium salts.

Reaction

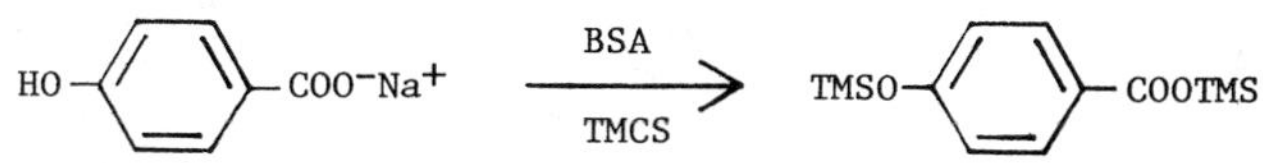

Typical Procedure. Diss 2 mg of the Na salt in 300 μl BSA/TMCS (9:1, v/v). After 30 min at room temp, analyze by GC or GC-MS.

Comments. BSA alone did not form TMS derivs of the Na salts but did deriv the compds when either TMCS or hydroxylamine hydrochloride was present. The method offered a way of preventing losses of carboxylic acids during lyophilization by converting them to the sodium salts. Keto acids (e.g., phenylpyruvic) were converted to oxime, TMS derivs in the presence of hydroxylamine·HCl, pyridine, BSA, and TMCS.

References

[1]A. J. Poole *et al., Clin. Chim. Acta* **73**, 527 (1976).

Derivative Type. Sulfonyl trimethylsilyl esters.

Reagent. Chlorosulfonic acid, bis(trimethylsilyl) trifluoroacetamide (BSTFA).

Compound Type. Phenolic acids.

Reaction

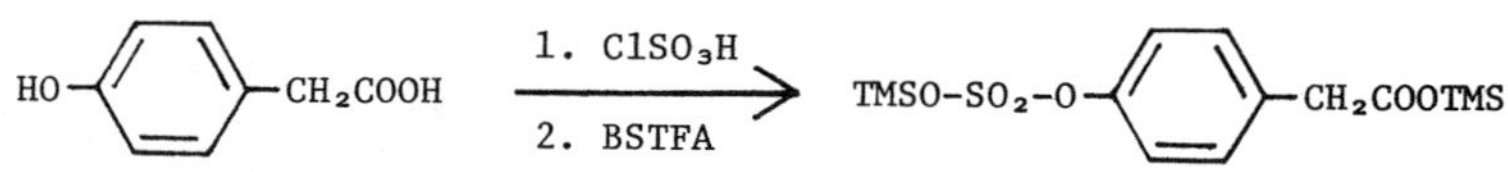

Typical Procedure. To a soln of a few mg of the acid in 2 ml EtOAc add 0.3 ml chlorosulfonic acid. Heat 5 min at 60°, cool, and add 2 ml H_2O. Ext with 3 × 3 ml EtOAc and back ext the combined EtOAc into 2 × 2 ml 1 *M* Na_2CO_3. Acidify the aq soln with 5 *M* HCl and ext with 3 × 4 ml Et_2O. Dry ext over Na_2SO_4 and evap to dryness. Add 0.1 ml BSTFA and heat at 60° for 10 min. Analyze a 5 μl aliquot by GC.

Comments. The deriv was used to obtain better resolution in the GC of phenolic acids.

References

[1]P. H. Scott, *J. Chromatogr.* **70**, 67 (1972).

Derivative Type. Chloromethyldimethylsilyl.

Reagent. Di(chloromethyl) tetramethyldisilazane/chloromethyldimethylchlorosilane.

Compound Type. Phenolic acids.

Reaction

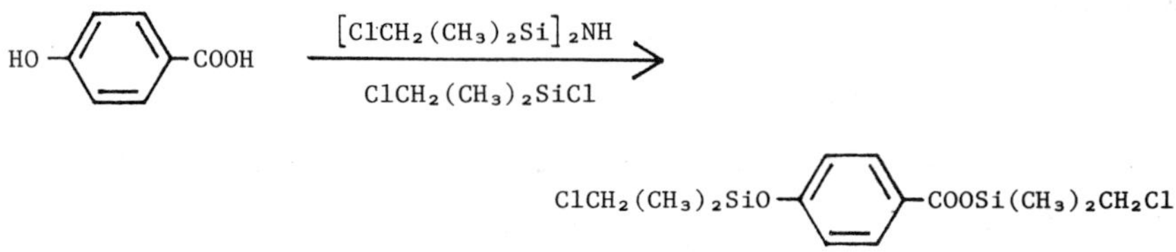

Typical Procedure. Diss mg sample of phenolic acid in 600 μl pyridine. Add 200 μl di(chloromethyl) tetramethyldisilazane and 100 μl chloromethyldimethylchlorosilane. After 30 min dil with hexane and analyze by GC.

Comments. The deriv had a detection limit of 4 ng using ECD-GC. Derivs of some of the compds tested appeared to be unstable and gave multiple GC peaks. MS data were given for the derivs of 2-hydroxy, 4-hydroxy, and 3,4-dihydroxybenzoic acids.

References

[1]H. Morita and W. Montgomery, *J. Chromatogr.* **123**, 454 (1976).

3.6.3. Other Derivatives of Phenolic Acids

Derivative Type. Trifluoroacetyl vanillin.

Reagent. Sodium periodate, trifluoroacetic anhydride (TFAA).

Compound Type. Vanillylmandelic acid (VMA).

Reaction

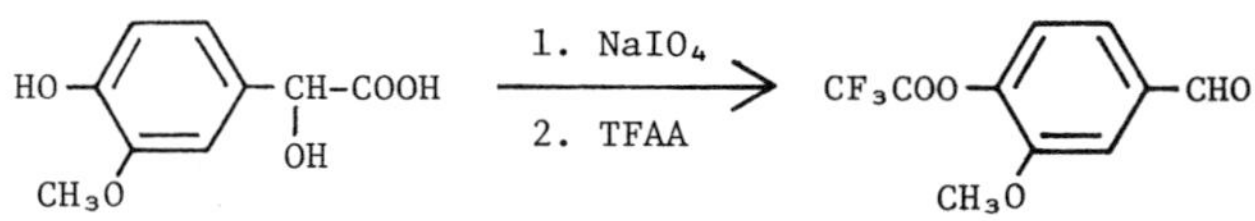

Typical Procedure. Acidify a urine sample (corresponding to 3 mg creatinine) with 0.1 vol 3 *N* HCl.Ext with 2, 1, and 1 vol of EtOAc. Ext the combined EtOAc with 1 ml 1 *M* K_2CO_3. Add 0.2 ml 2% $NaIO_4$ and heat 30 min at 50°. Cool and neutralize with 0.4 ml 5 N HOAc and 0.6 ml pH 6.2 phosphate buffer. Ext with 20 ml toluene and evap ext under vac. Transfer residue with 2 × 1 ml dry EtOAc and add 0.5 ml TFAA. Allow to stand at room temp 1 hr then evap under vac. Diss in 1 ml EtOAc and analyze by ECD-GC.

Comments. Trifluoroacetylation at higher temp in the presence of traces of H_2O led to a less stable deriv in which the aldehyde group was converted to a di-TFA deriv.

References

[1]S. Wilk *et al., Anal. Biochem.* **13**, 544 (1965).

Derivative Type. Trifluoroacetyl vanillyl alcohol.

Reagent. Sodium periodate, potassium borohydride, trifluoroacetic anhydride (TFAA).

Compound Type. Vanillylmandelic acid.

Reaction

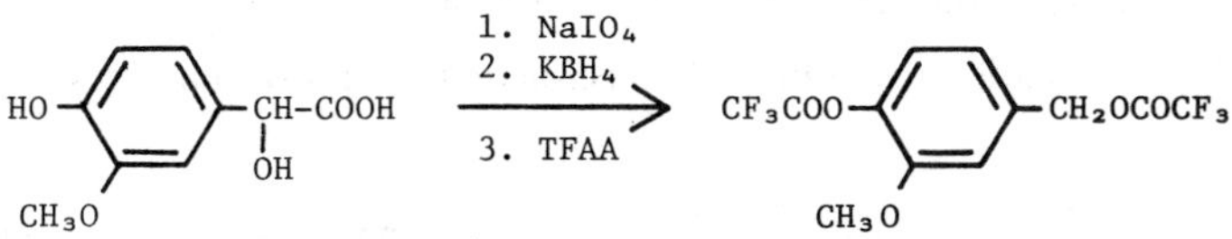

Typical Procedure. Acidify 3 ml urine with 3 ml 3 *N* HCl and ext with 3 × 10 ml EtOAc. Ext the combined EtOAc with 1 ml 1.0 *M* K_2CO_3. Add 0.2 ml 2% aq $NaIO_4$ to the aq phase and heat 30 min at 50°. Add 0.2 ml 10% aq $NaHSO_3$, cool, and add 100 mg KBH_4. Allow to react 10 min at room temp then adjust to pH 7.0 with 5 *N* HOAc. Add 0.6 ml pH 7.2 phosphate buffer and ext with 2 × 10 ml EtOAc. Evap to dryness under vac; then add MeOH and evap twice to remove borate as methyl borate. Diss the residue in 2 ml EtOAc and add 0.5 ml TFAA. Allow to react 1 hr at room temp; then evap under vac. Diss residue in 1.0 ml EtOAc and analyze an aliquot by ECD-GC.

Comments. The method was developed as an improvement over the method of Wilk *et al.*[2] The initially produced vanillin was reduced to the less volatile vanillyl alcohol to minimize losses during evap.

References

[1]H. Derkirmenjian and J. W. Maas, *Clin. Chim. Acta* **32**, 310 (1971).

[2]S. Wilk *et al., Anal. Biochem.* **13**, 544 (1965).

3.7. DERIVATIVES OF INDOLE CARBOXYLIC ACIDS

Derivative Type. Methyl.

Reagent. Trimethylphenylammonium hydroxide (TMPAH).

Compound Type. Aromatic acids.

Reaction

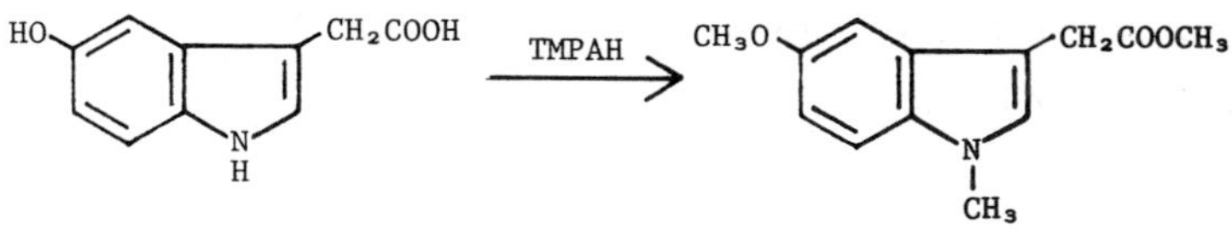

Typical Procedure. Acidify 1 ml urine with 0.1 ml 6 *N* HCl and add 2 ml satd NaCl soln. Ext with 2 × 2 ml EtOAc and 2 × 2 ml Et_2O. Dry combined ext over $MgSO_4$ and evap to dryness under N_2. Diss residue in 0.1 ml of 0.4 *M* TMPAH in MeOH and add 3A mol sieve. Allow to stand 10 min at room temp then analyze by GC or GC-MS.

Comments. The method methylated carboxyl groups, indole nitrogen, phenolic hydroxyls, and α-hydroxyls in α-hydroxyphenylacetic acids. MS data were reported for the derivs of eight aromatic acids found in urine.

References

[1]I. Gan *et al.*, *J. Chromatogr.* **92**, 435 (1974).

Derivative Type. Methyl ester, pentafluoropropionyl.

Reagent. Methanol/boron trichloride, pentafluoropropionic anhydride (PFPA).

Compound Type. Acidic metabolites of catecholamines and tryptamines.

Reaction

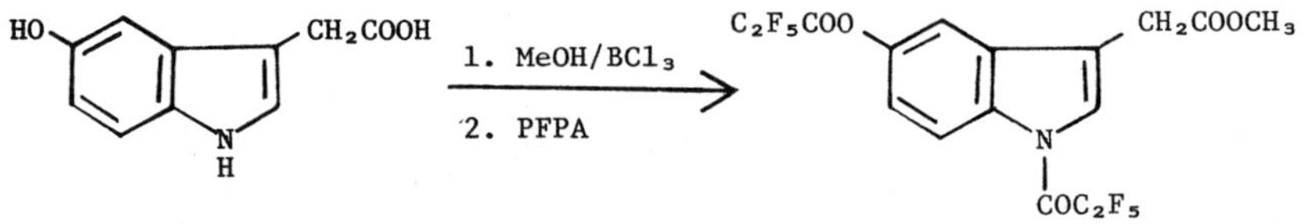

Typical Procedure. Diss sample in 10% (w/v) BCl_3 in MeOH and heat at 80–100° for 5 min (indole acids and indole amino acids). Evap to dryness under N_2 and diss residue in 20–100 μl of EtOAc. Add 20–60 μl of PFPA and heat at 100° for 3 hr (indoleacetic and 5-hydroxyindoleacetic acids). Evap to dryness under N_2 and diss residue in benzene prior to GC or GC-MS analysis.

Comments. No compromise conditions were found to maximize the yield of fully acylated derivs of Trp and 5-hydroxy-Trp. Acylation for 2 min at 100° gave 90% yield of the mono- and diacylated derivs, respectively. Although the structure was not firmly established, it is assumed that the indole N remains underiv in the latter deriv.

Dihydroxyphenylacetic, homovanillic, and dihydroxymandelic acids were esterfied at room temp for 15 min to minimize methylation of hydroxyl groups. Perfluoroacylation was carried out at 60° for 1 hr.

Picogram level sensitivity for the derivs was demonstrated using ECD-GC. It was found subsequently that acylation in CH_3CN satd with $NaHSO_4$ gave improved yields.[2]

References

[1]E. Gelpi *et al.*, *J. Chromatogr. Sci.* **12**, 701 (1974).
[2]J. Segura *et al.*, *Biomed. Mass Spectrom.* **3**, 91 (1971).

Derivative Type. Methyl ester, *N*-trifluoroacetyl.

Reagent. Diazomethane, trifluoroacetic anhydride (TFAA).

Compound Type. Indole acids.

Reaction

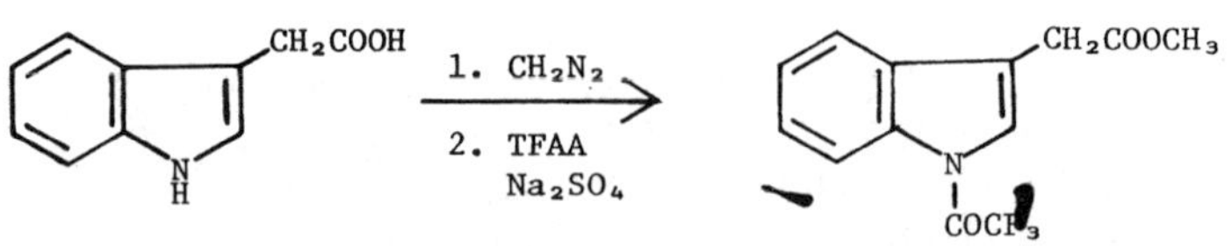

Typical Procedure. Diss 1 mg sample in 1 ml Et_2O and add CH_2N_2 in Et_2O until the yellow color persists. After 15 min evap under N_2 and diss residue in 3 ml EtOAc. Add 0.5 ml TFAA and a small amt of Na_2SO_4. Stir 1 hr at room temp, then evap under vac. Diss residue in EtOAc and analyze by GC.

Comments. The use of Na_2SO_4 in the trifluoroacetylation was reported to shorten the deriv time. The deriv method has been used for analysis of indoleacetic acid in apple seeds.

References

[1]J. L. Brook *et al., Anal. Biochem.* **18**, 453 (1967).

[2]S. D. Seeley and L. E. Powell, *Anal. Biochem.* **58**, 39 (1974).

Derivative Type. Pentafluoropropyl ester, pentafluoropropionyl.

Reagent. Pentafluoropropanol, pentafluoropropionic anhydride (PFPA).

Compound Type. Homovanillic acid (HVA) and 5-hydroxyindoleacetic acid (5-HIAA).

Reaction

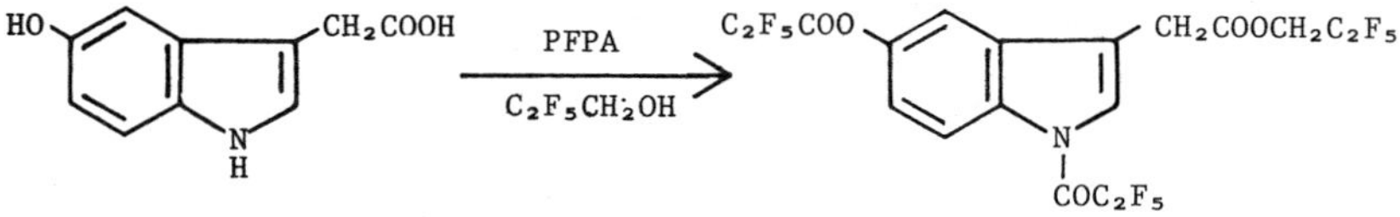

Typical Procedure. To 2 ml CSF add *ca* 5.5 nmol d_2-HVA and 1.5 nmol d_2-5-HIAA int stand. Acidify to pH 3 with HCOOH, add 1.5 g NaCl, and ext with 4 × 2 ml EtOAc. Evap ext to dryness under N_2 and transfer residue in MeOH (2 × 0.3 ml) into small tubes and evap to dryness under N_2.Add 50 μl of a mixt of 20% $C_2F_5CH_2OH$ in PFPA. Heat at 75° for 15 min then evap to dryness under N_2. Add 30 μl PFPA, heat 5 min at 75°, then evap to dryness under N_2. Diss residue in 30 μl 1% PFPA in EtOAc and analyze by GC-MS-SIM.

Comments. The method permitted simultaneous determination of HVA and 5-HIAA in CSF.

References

[1]C.-G. Fri *et al., Life Sci.* **14**, 2469 (1974).

Derivative Type. Trimethylsilyl.

Reagent. Trimethylsilylimidazole (TMSI).

Compound Type. 5-Hydroxytryptamine (5-HT) and 5-hydroxyindoleacetic acid (5-HIAA).

Reaction

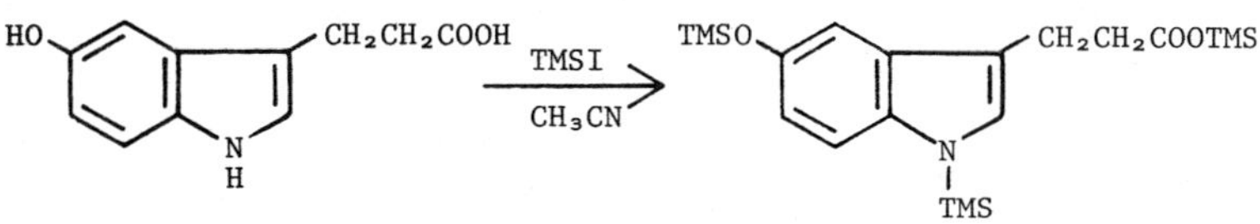

Typical Procedure. Diss sample in 50 μl (1:1, v/v) TMSI/CH_3CN. Heat at 60° for 1 hr (2 hr for 5-HT). Analyze by GC.

Comments. The derivs were used for GC analysis of 5-HT and 5-HIAA in brain tissue. The structures of the derivs were not established. Immediate injection of the deriv mixt (i.e., no heating) gave a deriv for 5-HT with a shorter retention time but gave a nonlinear standard curve and less sensitivity. It is presumed that the deriv after heating was an *N, O*-TMS deriv, while that from immediate analysis was the *O*-TMS only.

References

[1]Y. Maruyama and A. E. Takemori, *Biochem. Pharmacol* **20**, 1833 (1971).

3.8. DERIVATIVES OF CARBOXYLIC ACIDS FROM POLYAMIDES AND POLYESTERS

Derivative Type. Methyl ester.

Reagent. Methanol (MeOH)/sodium methoxide (NaOMe).

Compound Type. Polyester resin.

Reaction

$$\int \text{-OCO-}(CH_2)_3\text{-COO-} \int \xrightarrow[\text{NaOMe}]{\text{MeOH}} CH_3OCO\text{-}(CH_2)_3\text{-}COOCH_3$$

Typical Procedure. Reflux 10 g resin with 100 ml MeOH containing 0.1 g NaOMe for 18 hr. Analyze an aliquot by GC.

Comments. The constituents of polyester resins were analyzed by basic methanolysis followed by GC analysis of the resulting esters and free glycols.

References

[1]D. F. Percival, *Anal. Chem.* **35**, 236 (1963).

Derivative Type. Methyl esters.

Reagent. Lithium methoxide (LiOMe)/methanol (MeOH).

Compound Type. Carboxylic acids in polyester resins.

Reaction

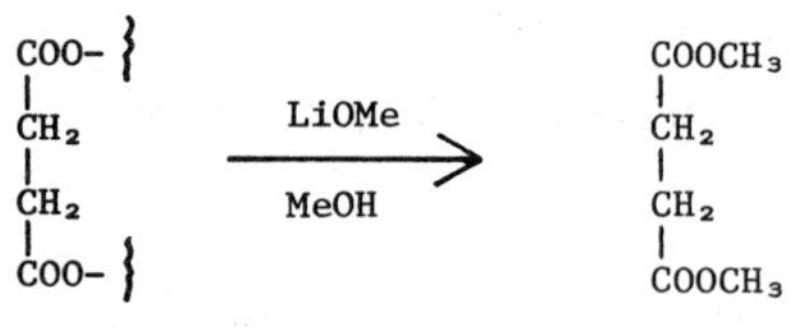

Typical Procedure. Reflux a 0.3 g resin sample in 15 ml 0.5 *N* LiOMe in MeOH for 2 min. add 5 ml 6 *N* H_2SO_4, dil to 50 ml with H_2O and ext with 35 ml CH_2Cl_2. Wash ext with H_2O to remove H_2SO_4; then analyze esters by GC.

Comments. The method was used to identify 19 of the most commonly encountered acids in synthetic resins.

References

[1]G. G. Esposito and M. H. Swann, *Anal. Chem.* **34**, 1048 (1962).

Derivative Type. Methyl ester.

Reagent. Tetramethylammonium hydroxide (TMAH), methyl iodide.

Compound Type. Carboxylic acids from lipids, alkyd resins, polyester resins, and ester plasticizers

Reaction

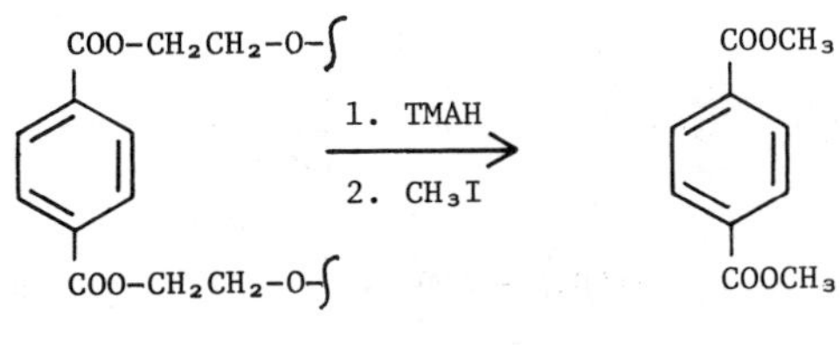

Typical Procedure. Mix 100–200 μg sample with 0.5 ml MeOH and 0.5 ml 24% TMAH in MeOH. Heat on steam bath for 10–15 min or until all solids diss. Add 5 ml DMF, mix, and add 0.5 ml CH_3I. Mix, then allow the tetramethylammonium iodide ppt to settle. Analyze an aliquot of the supernatant by GC.

Comments. The methyl esters were alternatively extd with hexane before analysis. For phthalates, the more polar butyl acetate was required for ext. Since some methylation occurred during saponification, use of higher alkyl iodides did not yield quantitative conversion to the higher esters.

References

[1]J. C. West, *Anal. Chem.* **47**, 1708 (1975).

Derivative Type. Trimethylsilyl.

Reagent. Bis(trimethylsilyl) trifluoroacetamide (BSTFA)/pyridine.

Compound Type. Terephthalate prepolymer mixture.

Reaction

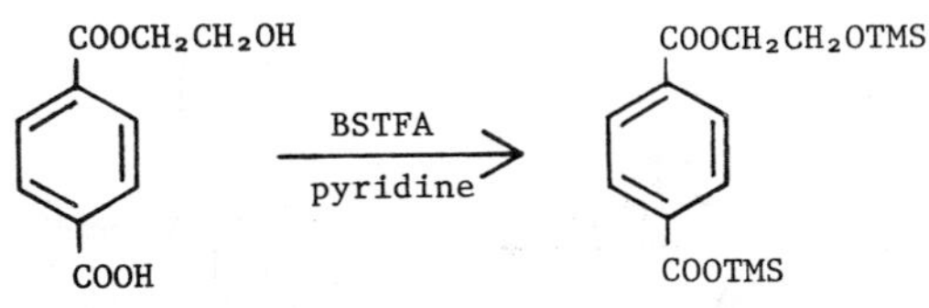

Typical Procedure. Mix 20 mg sample with 0.50 ml 2 mg/ml dibenzyl int stand soln in pyridine and 0.50 ml BSTFA. Heat at 80° for 10 min. Analyze 2 μl aliquots by GC.

Comments. The method was used to quantitate bis(2-hydroxyethyl) terephthalate, 2-hydroxyethylterephthalate, terephthalic acid, ethylene glycol, and diethylene glycol in polyester prepolymers. The use of BSA was also examined, but chromatographic interferences were encountered for the ethylene glycol deriv.

References

[1]E. R. Atkinson and S. I. Calouche, *Anal. Chem.* **43**, 460 (1971).

Derivative Type. Trimethylsilyl.

Reagent Hydrochloric acid, bis(trimethylsilyl) acetamide (BSA)/triethylamine (Et_3N).

Compound Type. Polyamide resins.

Reaction

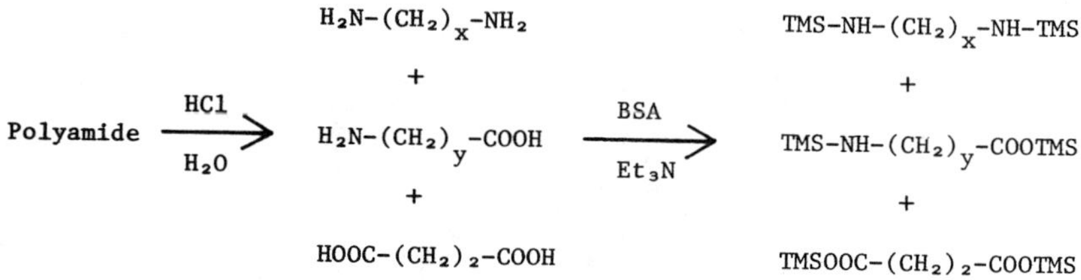

Typical Procedure. Hydrolyze a 0.1 g sample in 10 ml 6 *N* HCl at 130° for 2–8 hr (depending upon sample type) in a sealed tube. Evap to dryness on a steam bath. Diss 10 mg of the residue in 0.15 ml CH_3CN and add 0.03 ml Et_3N and 0.15 ml BSA. Heat 30 min at 90° under reflux with protection from moisture. Add 0.15 ml each of CH_3CN and BSA and heat another 30 min. Dil to 2 ml with CH_3CN and analyze by GC.

Comments. The hydrolysis products of polyamides have also been analyzed as *N*-TFA, methyl ester derivs.[2] The TMS deriv procedure was simpler. The Et_3N was used to scavenge HCl from the amine HCl salts obtained from the hydrolysis.

References

[1]S. Mori *et al.*, *Anal. Chem.* **42**, 959 (1970).
[2]S. Mori *et al.*, *Anal. Chem.* **42**, 138 (1970).

3.9. DERIVATIVES OF VINYLOGOUS ACIDS

Derivative Type. *O*-Methyl.

Reagent. Diazomethane.

Compound Type. Acenocoumarol.

Reaction

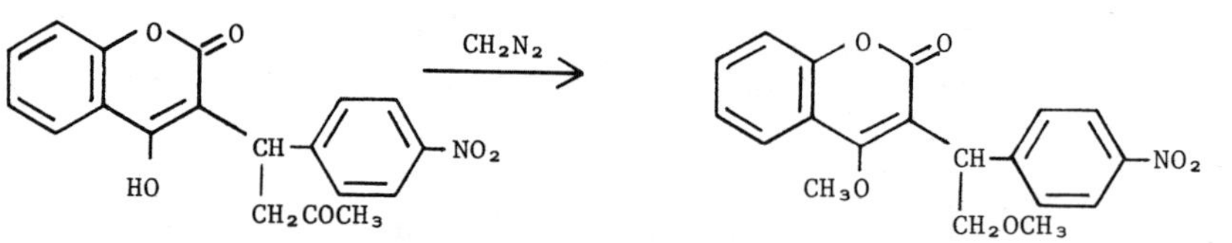

Typical Procedure. To 2 ml plasma add 1 ml int stand soln (15 μg) and 3 ml 3 *N* HCl. Ext with 20 ml $ClCH_2CH_2Cl$. Transfer 18 ml of ext to a tube containing 10 ml 1 *N* NaOH. Shake well, centrifuge, and transfer 9 ml of the aq phase to another tube containing 3 ml 5 *N* HCl. Ext with 20 ml $ClCH_2CH_2Cl$. Evap 18 ml of the ext to *ca* 4 ml under N_2 at 85°. Add a crystal of anhyd $CaCl_2$ and evap to dryness. Diss residue in 100μl CH_2N_2 in Et_2O and allow to stand at room temp 10 min. Evap to dryness at room temp; then diss residue in 25 μl CS_2. Analyze by GC.

Comments. The γ-oxo deriv of phenylbutazone was used as int stand. Warfarin and phenprocoumon also gave derivs by this method, but they eluted too near the solvent front to be useful. Methylation with trimethylphenylammonium hydroxide was unsuccessful. The formation of vinylogous esters with CH_2N_2 should be generally useful for the 4-hydroxy-coumarin compounds. MS data were given for the acenocoumarol deriv.

References

[1]K. K. Midha and J. K. Cooper, *J. Pharm. Sci.* **66**, 799 (1977).

Derivative Type. Pentafluorobenzyl.

Reagent. Pentafluorobenzyl bromide (PFB-Br)/potassium carbonate.

Compound Type. Warfarin [3-(α-acetonylbenzyl)-4-hydroxycoumarin].

Reaction

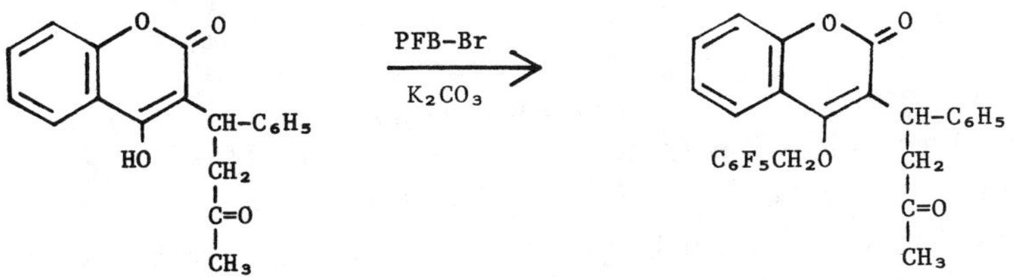

Typical Procedure. To a tube containing 2 μg 3-(α-acetonyl-*p*-chlorobenzyl)-4-hydroxy coumarin int stand add 0.5 ml 0.01 *N* NaOH and 1 ml plasma. Mix well; then add 0.5 ml 3 *N* H_2SO_4 and ext with 10 ml 1,2-dichloroethane. Evap ext under N_2 and diss residue in 100 μl dichloroethane. TLC on silica gel using 10% (v/v) acetone and dichloroethane. Visualize the drug zone under UV and elute with 2 ml MeOH. Evap 1 ml of the eluate under N_2. Add 25 mg anhyd K_2CO_3 and 0.5 ml PFB/Br (0.1 ml/100 ml) in acetone. Heat at 60° for 1 hr and then evap under N_2. Partition between 1 ml H_2O and 2 ml hexane, analyze an aliquot of the hexane soln by ECD-GC.

Comments. The method was used for quantitation of warfarin in plasma to 20 ng/nl.[1] The same deriv has been used for analysis of acenocoumarin [3-(α-acetonyl-*p*-nitrobenzyl)-4-hydroxycoumarin].[2]

References

[1]D. G. Kaiser and R. S. Martin, *J. Pharm. Sci.* **63**, 1579 (1974)

[2]G. Bianchetti *et al., J. Chromatogr.* **124**, 331 (1976).

Derivative Type. Trimethylsilyl.

Reagent. Hexamethyldisilazane (HMDS)/trimethylchlorosilane (TMCS).

Compound Type. 4-Hydroxycoumarins.

Reaction

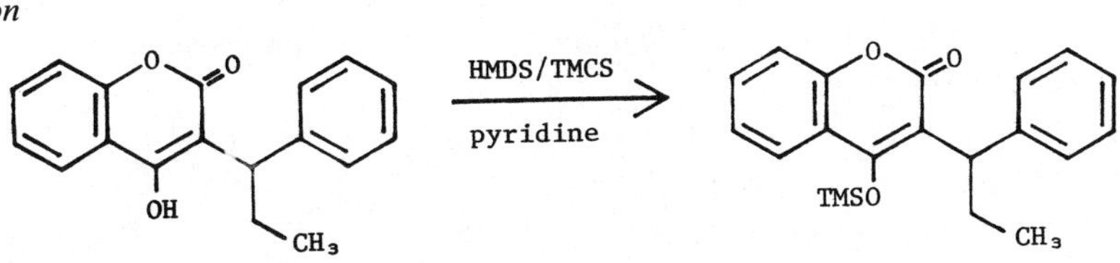

Typical Procedure. Diss 1–2 mg sample in 100 μl anhyd pyridine and add 100 μl HMDS and 50 μl TMCS. Mix 1 min then centrifuge. Analyze an aliquot of the supernatant by GC.

Comments. GC data were given for the derivs of 6 anticoagulant 4-hydroxycoumarins.

References

[1]F. W. Deckert, J. Chromatogr. **64**, 355 (1972).

Derivative Type. Trifluoroacetate.

Reagent. Trifluoroacetic anhydride (TFAA).

Compound Type. 4-Hydroxycoumarins.

Reaction

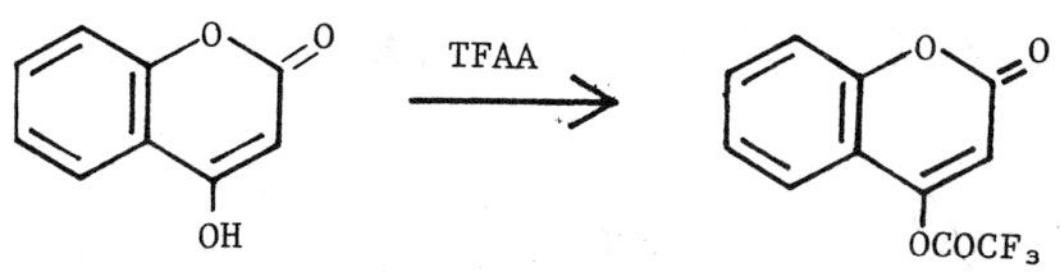

Typical Procedure. Diss 250 μg sample in 0.5 ml benzene and add 10 μl anhyd pyridine and 10–50 μl TFAA. Allow to stand 20–30 min at room temp then shake with 1.0 ml H_2O. Centrifuge and analyze 1 μl of the benzene layer by GC.

Comments. The method was applied to 4-hydroxycoumarin, warfarin, and phenprocoumin. Acetyl and trichloroacetyl derivs were also prepared by similar methods but the reactions required heating. The GC behavior of the derivs was compared with that of the TMS derivs.[2]

References

[1]F. Deckert, *J. Chromatogr.* **69**, 201 (1972).
[2]F. Deckert, *J. Chromatogr.* **64**, 355 (1972).

Derivative Type. Trimethylsilyl.

Reagent. Hexamethyldisilazane (HMDS)/dimethoxypropane (DMP).

Compound Type. Hop and beer bitter acids.

Reaction

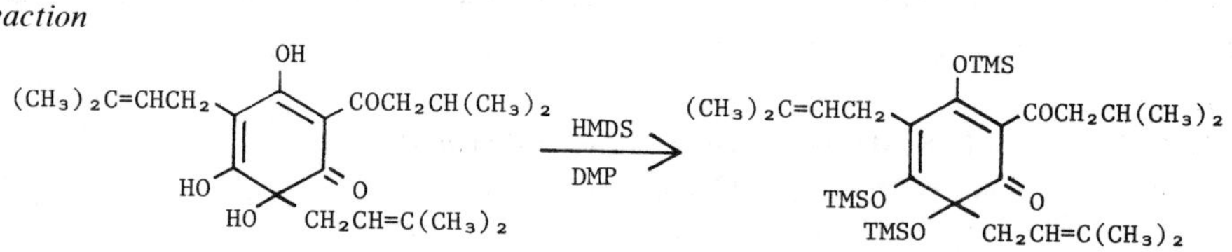

Typical Procedure. Diss 0.2–2 mg of the acid in 150 μl DMP and 100 μl HMDS. Heat at 50° for 3 hr. Analyze by glass capillary GC.

Comments. BSA gave isomerization and multiple derivs. TMSI apparently did not react. Butylboronates and phenylboronates (reported earlier by Shaw[2]) were also examined but deemed not useful for analysis. The TMS derivs deteriorated on packed columns but could be analyzed on glass capillaries.

References

[1]M. Verzele *et al., Anal. Chem.* **45**, 1549 (1973).
[2]S. Shaw, *Tetrahedron Lett.* **26**, 3033 (1968).

4. Unsaturated, Cyclopropane, and Methyl Branched Fatty Acids: Derivatives for Determination of Substituent Location

Mass spectrometric analysis of compounds containing long chain unsaturated hydrocarbon groups is complicated by isomerizations that make it difficult to determine the original position or geometry of the unsaturation. Several approaches have been used for chemical modification of these compounds to facilitate such analysis. The majority of the work in this area has been applied to analysis of unsaturated fatty acids, but the techniques are also applicable to other unsaturated compounds. The classical technique for double bond location has been oxidative cleavage followed by identification of the resulting fragments. A variety of derivatives of the intact compounds have also been examined. The first such derivative reported was the deuterated reduction product prepared with deuterohydrazine,[1] but this approach met with limited success.[2] Other derivatives that introduced charge stabilizing substituents at the double bond locus were subsequently developed. The double bonds have been converted to epoxides and ketones by oxidation with peracids (p. 229). The epoxide derivatives of *cis* and *trans*-olefins could be separated by GC (p. 231). Oxidation with osmium tetroxide and reduction of the resulting intermediate has been employed to convert the double bonds to glycols, which were then converted to methyl or trimethylsilyl ethers (p. 231). Amino alcohol derivatives have been prepared by oxidation with peracid followed by reaction of the resulting epoxide with dimethylamine.

Most of these methods of derivatizing the double bond yield difunctional compounds in which there is free rotation of the carbon-carbon bond between the two derivative functional groups. These derivatives are useful for determining the position of unsaturation but are of limited utility for determining double bond geometry. By contrast, stereospecific oxidation with osmium tetroxide followed by formation of a cyclic ketal with acetone yields a derivative in which the geometry of the original double bond is preserved (*cis→erythro*; *trans→threo*). (p. 233). These derivatives therefore provide information on both the location and geometry of the unsaturated site. The analogous hexafluoroacetone ketal derivative has been examined and found to have potential utility (p. 234). Boronate derivatives of glycols have been prepared (p. 39) but have not been explored for analysis of unsaturated compounds.

Treatment of unsaturated compounds with mercuric acetate and methanol, followed by reduction, yields methoxy and cyclic ether derivatives which are useful for determining double bond location (p. 235). Treatment of the adducts with sodium chloride instead of reduction yields chloromercurimethoxy alkanes, which are also useful for double bond location (p. 237).

An alternate approach to derivatization employs, instead of modification of the double bond, attachment of a charge stabilizing group at the end of the molecule. Pyrrolidide derivatives of unsaturated fatty acids allow MS determination of double bond position by yielding stabilized ions resulting from cleavage at successive positions of the chain (p. 237). These derivatives have been applied to mono- and polyunsaturated acids. A similar approach employed *N*-alkyl-2-pyrrolidone derivatives from long chain primary alcohols (p. 238). Derivatives with terminal charge stabilizing groups have also been utilized to determine the location of branching points in branched long chain compounds.

Determination of the location of cyclopropane rings in long chain compounds presents problems similar to those of unsaturated chains. Several approaches have been used to modify the cyclopropane ring to facilitate locus determination by MS. Treatment with boron trifluoride and methanol yields a mixture of ring opened methoxy derivatives whose structures can be used to deduce the original ring position (p. 238). This mixture of as many as six derivatives from the cyclopropane ring also illustrates potential artifacts that may be encountered in boron trifluoride/methanol esterfication of fatty acids containing cyclopropane rings (p. 154). The cyclopropane ring can be opened by catalytic reduction and the resulting products used to deduce the original locus (p. 239). Oxidation of the cyclopropane compounds to yield cyclopropyl ketones gives the smallest number of derivatives since oxidation on either side of the ring gives only a pair of ketones (p. 240). These derivatives, which have been applied to both fatty acids and hydrocarbons, were deemed preferable to the pyrrolidide derivatives.

The MS characteristics of the derivatives used for substituent location have been discussed in reviews of the MS of fatty acids.[3,4]

References

1. N. Dinh-Nguyen, *et al.*, *Ark. Kemi* **15**, 433 (1960).
2. N. Dinh-Nguyen, *et al.*, *Ark. Kemi* **18**, 393 (1961).
3. J. A. McCloskey, in *Topics in Lipid Chemistry*, ed. by F. D. Gunstone, Logos Press, London, 1970, p. 369.
4. G. Odham and E. Stenhagen, in *Biomedical Applications of Mass Spectrometry*, ed. by G. R. Waller, Wiley Interscience, New York, NY 1972, p. 211.

4.1. DERIVATIVES FOR DETERMINATION OF UNSATURATION LOCATION

4.1.1. Double Bond Cleavage Derivatives

Derivative Type. Ethyl ester.

Reagent. Sodium metaperiodate/potassium permanganate, ethanol/boron trifluoride.

Compound Type. Unsaturated fatty acids.

Reaction

$$CH_3(CH_2)_7\text{-}CH{=}CH\text{-}(CH_2)_7COOH \xrightarrow[\text{2. EtOH/BF}_3]{\text{1. NaIO}_4/\text{KMnO}_4} \begin{array}{c} CH_3(CH_2)_7\text{-}COOC_2H_5 \\ + \\ C_2H_5OCO\text{-}(CH_2)_7\text{-}COOC_2H_5 \end{array}$$

Typical Procedure. To 500 ml H_2O containing 1.5 mmol KOH add 0.5 mmol (141 mg) oleic acid and shake under N_2. Add 85 ml (3.97 mmol) 1% aq soln $NaIO_4$ and adjust to pH 8 with 1 ml 0.1 *M* $KMnO_4$ in 400 ml H_2O. Stir *ca* 2 hr then adjust to pH 9 and allow to react for a total of 24 hr. Acidify with HCl and reduce excess oxidant with SO_2. Alkalinize with KOH and evap to near dryness on a steam bath under an air stream. Add a few ml 95% EtOH and gaseous CO_2 to react with excess alkali. Evap under vac at or below room temp. Add 2–4 ml 15% (w/w) HCl in abs EtOH and *ca* 2 ml 15% (w/w) BF_3 in EtOH. Reflux *ca* 10 min, cool, and remove most of the EtOH under vac at or below room temp. Add 5–10 ml H_2O and neutralize with satd aq $NaHCO_3$. Sat with NaCl and ext with 2 × 5–10 ml Et_2O. Wash ext with H_2O, dry over Na_2SO_4, and analyze by GC.

Comments. The method was used to determine the position and amt of unsaturation in fatty acids. Butyl esters were prepared for analysis of short chain acids to circumvent losses of the more volatile ethyl esters.

References

[1]E. P. Jones and V. L. Davison, *J. Am. Oil Chem. Soc.* **42**, 121 (1965).

Derivative Type. Carboxylic acid methyl ester.

Reagent. Sodium metaperiodate/potassium permanganate, tetramethylammonium hydroxide (TMAH).

Compound Type. Unsaturated fatty acid methyl esters.

Reaction

$$CH_3(CH_2)_nCH{=}CH(CH_2)_mCOOCH_3 \xrightarrow[\text{2. TMAH, }\Delta]{\text{1. NaIO}_4/\text{KMnO}_4}$$

$$CH_3(CH_2)_nCOOCH_3 \; + \; CH_3OCO(CH_2)_mCOOCH_3$$

Typical Procedure. Diss 1 mg of fatty acid methyl ester or mixt in 1 ml *t*-BuOH. Add 1 ml 2.5 mg/ml aq K_2CO_3 and 1 ml of a soln of $NaIO_4$ (20.86 g/l) and $KMnO_4$ (395 mg/l). Shake 1 hr at room temp. Add 3 ml hexane and 1 drop conc H_2SO_4. Shake vigorously 20 sec; then sep hexane phase. Shake remaining aq phase with 3 ml hexane that has been equilibrated with 33% *t*-BuOH. Shake combined hexane with 0.3 ml 1% aq TMAH. Dry a 3 μl aliquot of the TMAH on a capillary probe and analyze by GC with the vaporizor at 280°.

Comments. The method (an adaptation of the von Rudloff oxidation method[2]) was used for location of double bonds in unsaturated fatty acids. It was shown that even the short chain acids were efficiently extracted by the hexane washes.

References

[1]D. T. Downing and R. S. Greene, *Lipids* **3**, 96 (1968).

[2]E. von Rudloff, *Can. J. Chem.* **34**, 1413 (1950).

Derivative Type. Aldehyde cleavage products of monoenes.

Reagent. Hydrazine, ozone, hydrogen/Lindlar catalyst.

Compound Type. *cis, trans*-Polyunsaturated fatty acids.

Reaction

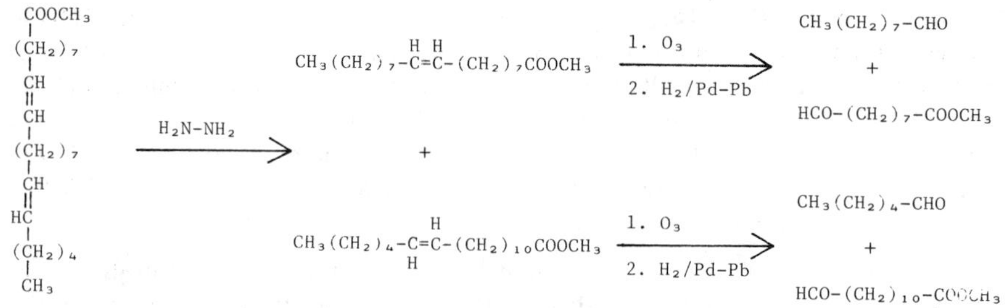

Typical Procedure. Diss 5–100 mg of unsatd fatty acid in 100 vol of 10% (v/v) hydrazine hydrate in MeOH. Stir 1.5–2 hr at 36° (for dienes, longer for more unsatd compds). Add to *ca* 3 ml of 6% (w/w) HCl in MeOH and seal in a tube under N_2. Heat at 100° for 1 hr. Evap and separate *cis* and *trans* monoenes by TLC on $AgNO_3$-silica gel. Elute fractions with Et_2O/pet ether (1 : 1, v/v) and treat the eluted monoenes with 10 ml pentane satd with O_3 at −65°.[2] After 1 min evap under vac and purify ozonides by TLC. Reduce with hydrogen over Lindlar catalyst.[3] Analyze the resulting aldehyde cleavage products by GC.

Comments. The method was used to determine positions and geometric configuration of double bonds in polyunsaturated fatty acids.

References

[1]O. S. Privett and E. C. Nickell, *Lipids* **1**, 98 (1966).

[2]O. S. Privett and E. C. Nickell, *J. Am. Oil Chem. Soc.* **39**, 414 (1962).

[3]H. Lindlar, *Helv. Chim. Acta* **35**, 446 (1952).

Derivative Type. Methyl ester.

Reagent. Ozone, methanol (MeOH)/boron trifluoride.

Compound Type. Unsaturated fatty acids (monoenoic).

Reaction

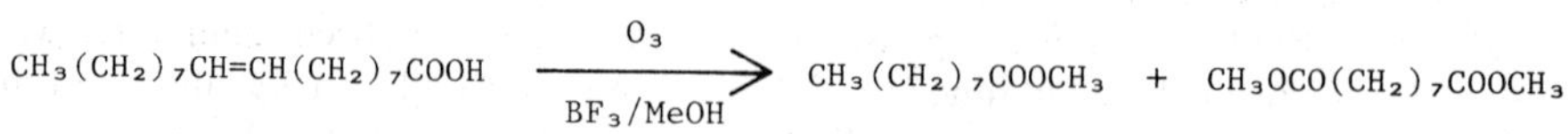

Typical Procedure. Diss 1–10 mg unsatd acid in 2 ml 14% BF_3 in MeOH and bubble in ozone at a brisk rate for 1 min. Heat at 100° for 1 hr. Cool, add 2 ml H_2O and 2 ml of methylcyclohexane, and agitate. Transfer 1.9 ml of org phase to another tube and re-ext

H_2O/MeOH layer with 2 ml methylcyclohexane. Transfer 1.9 ml of org phase to other tube and wash combined ext with 1 ml H_2O. Conc org phase under N_2 to smaller volume for GC or GC-MS analysis.

Comments. No side products were detected by GC analysis.

References

[1]R. G. Ackman, *Lipids* **12**, 293 (1977).

Derivative Type. Methyl ester.

Reagent. Chromic acid, methanol (MeOH)/H_2SO_4.

Compound Type. Unsaturated glycerol ether (as dimethoxy deriv).

Reaction

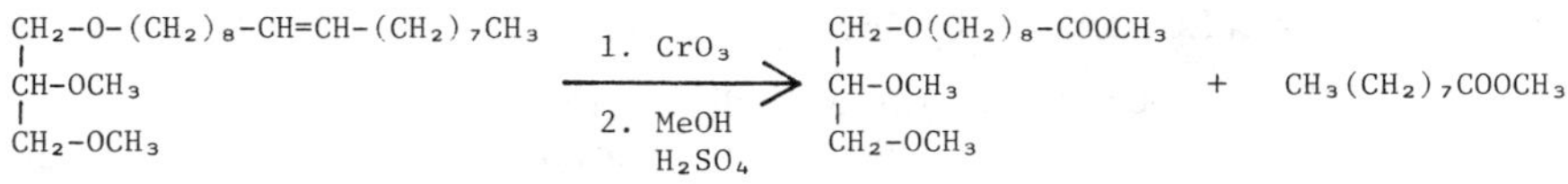

Typical Procedure. Mix 7.5 mg of dimethoxy glycerol ether with 7.5 mg CrO_3, 0.5 ml HOAc, and 0.1 ml H_2O. Reflux on steam bath for 1 hr. Add 1 ml MeOH to reduce excess CrO_3 followed by 4 ml 2% H_2SO_4 in MeOH. Reflux for 1.5 hr on steam bath. Add 5 ml H_2O and ext with Et_2O. Conc Et_2O under N_2 at less than 20° and analyze residue by GC.

Comments. The deriv was used to locate the double bond position in the unsatd alkyl groups.

References

[1]B. Hallgren and S. Larsson, *J. Lipid Res.* **3**, 31 (1962).

4.1.2. Oxidation Derivatives

Derivative Type. Ketones.

Reagent. Monoperphthalic acid, sodium iodide/methyl iodide.

Compound Type. Olefins.

Reaction

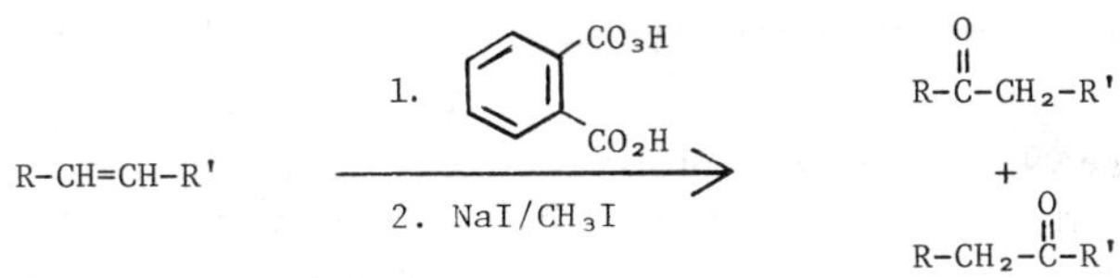

Typical Procedure (Complete Details Not Reported). Epoxidize the olefin with monoperphthalic acid. Isomerize the epoxide (5 mg) to the ketones by mixing with 40 mg NaI, 2 ml CH_3I, and 2 ml DMF and refluxing 4 hr. Analyze the resulting ketones by MS.

Comments. The deriv was used to determine the location of the double bond by MS. The isomerization could be done in 2-butanone or CH_3CN but went at a slower rate.

References

[1]G. Kenner and E. Stenhagen, *Acta Chem. Scand.* **18**, 1551 (1964).

Derivative Type. Epoxide.

Reagent. Perlauric acid.

Compound Type. Unsaturated fatty acids (as methyl esters).

Reaction

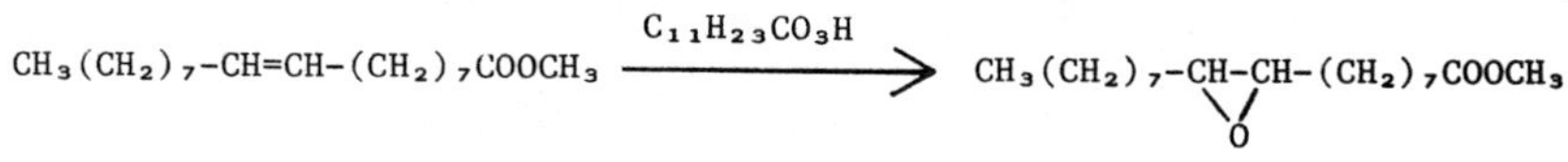

Typical Procedure (Details Not Given). Prepare the epoxides by treatment with perlauric acid and purify by TLC. Analyze by direct probe MS.

Comments. MS cleavage of the epoxide and adjacent to the epoxide permitted assignment of the position of unsaturation in unsatd aliphatic chains of fatty acids, alcohols, and hydrocarbons.

References

[1]R. T. Alpin and L. Coles, *Chem. Commun.* 858 (1967).

Derivative Type. Epoxide.

Reagent. *m*-Chloroperbenzoic acid.

Compound Type. Olefins.

Reaction

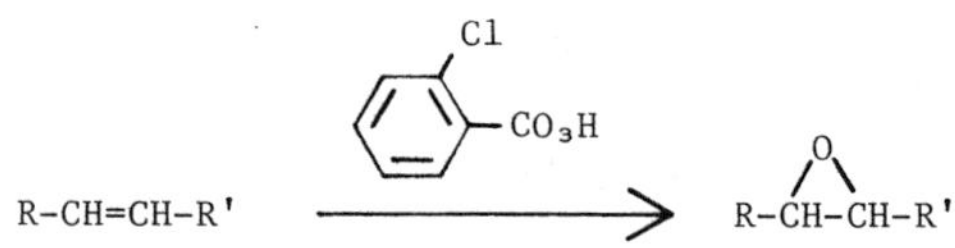

Typical Procedure. Diss olefin in CH_2Cl_2 containing a molar excess of 99+ % *m*-chloroperbenzoic acid and allow to stand at room temp from 30 min to several hr (time dependent upon olefin structure). Analyze aliquot of mixt directly by GC-chemical ionization MS.

Comments. The 99+ % pure *m*-chloroperbenzoic acid was prepared by washing the 85 % commercial material with pH 7.5 phosphate buffer.[2] By sequential sampling of the reaction mixt monoepoxides of diolefins could be examined. Chemical ionization MS using isobutane or methane gave characteristic ions which facilitated location of the epoxide groups. MS data was given for derivs of a series of C_{10} olefins and a series of olefinic alcohols (as acetate derivs).

References

[1]J. H. Tumlinson *et al., Anal. Chem.* **46**, 1309 (1974).
[2]N. M. Schwartz and J. H. Blumbergs, *J. Org. Chem.* **29**, 1976 (1964).

Derivative Type. Epoxides.

Reagent. *m*-Chloroperbenzoic acid.

Compound Type. Olefins.

Reaction

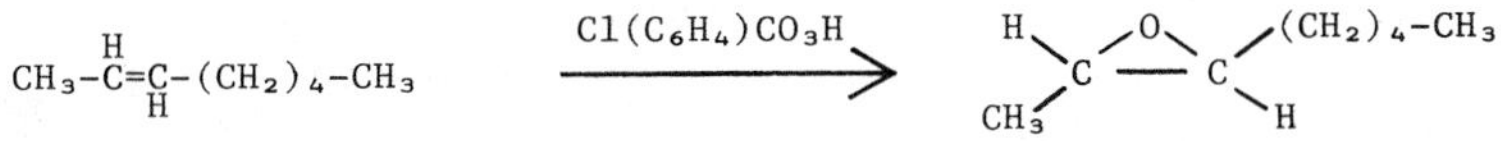

Typical Procedure. Diss 0.1 g olefin and an equimolar quant of *m*-chloroperbenzoic acid in 25 ml CH_2Cl_2 and refrigerate overnight at 6°. Ext with 5 × 25 ml 1 *N* NaOH and 5 × 25 ml H_2O. Dry over Na_2SO_4 and evap under vac at 30°. Diss the residue in hexane and analyze by GC.

Comments. The method was used to analyze the *cis-trans* isomer content of olefins. The epoxide derivs from *cis* olefins showed longer retention times than those from *trans* isomers, and the difference in retention times increased with increasing polarity of the stationary phase.

References

[1]L. M. McDonough and D. A. George, *J. Chromatogr. Sci.* **8**, 158 (1970).

4.1.3. Glycol Type Derivatives

Derivative Type. Polymethoxy methyl ester.

Reagent. Osmium tetroxide, sodium sulfite, methylsulfinyl carbanion/methyl iodide.

Compound Type. Unsaturated fatty acids.

Reaction

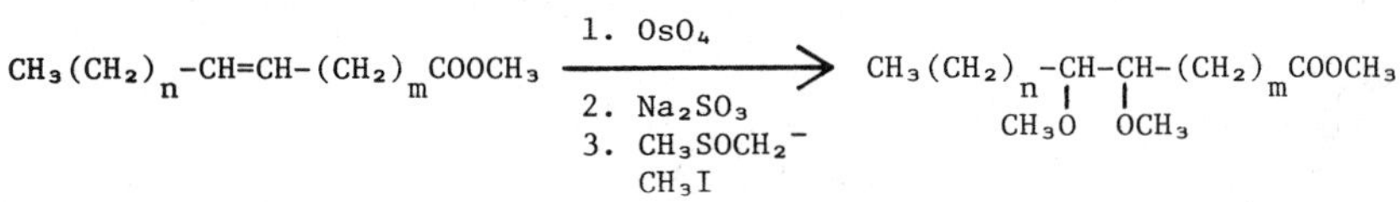

Typical Procedure. Diss fatty acid (0.1–100 μmol) in 1 ml dioxane-pyridine (8:1) and add 0.1 ml of 5% OsO_4 in dioxane. After 1 hr at room temp add 2.5 ml MeOH and 8.5 ml 16% Na_2SO_3 in H_2O and allow mix to stand 1 more hr. Centrifuge, dil supernatant with 4 vol MeOH, and filter. Evap filtrate to dryness, suspend in 2 ml MeOH, add 4 ml $CHCl_3$, filter suspension, and evap solvent. Diss residue in 1 ml dry DMSO, add 1 ml methylsulfinyl carbanion soln,[2] flush with N_2, and let stand at room temp 15–20 min. Add 0.2 ml CH_3I dropwise, flush with N_2, then let stand 10 min. Add 10 ml Et_2O and 5 ml H_2O, shake twice and remove Et_2O. Ext 2 more times with Et_2O, combine extracts, dry over $MgSO_4$ and evap to dryness. Diss residue in hexane for GC-MS analysis.

Comments. Alkaline permanganate oxidation could be substituted for OsO_4. [Diss 0.1–100 μmol fatty acid in 0.2 ml 0.25 *N* NaOH and dil with 1 ml cold H_2O. Add 0.2 ml 0.05 *M* $KMnO_4$. After 5 min decolorize soln with SO_2. Ext derivs with $CHCl_3/CH_3OH$ (2:1). Dry ext over $MgSO_4$ and evap to dryness.] The methylsulfinyl carbanion soln (*ca* 2 *M*) was prepared from dry DMSO and NaH[3] and stored frozen under N_2 until used. The method was applicable to acids containing 1–5 double bonds.

References

[1]W. G. Niehaus, Jr. and R. Ryhage, *Anal. Chem.* **40**, 1840 (1968).

[2]S. Hakomori, *J. Biochem. (Tokyo)* **55**, 205 (1964).

[3]E. Corey and M. Chakovsky, *J. Am. Chem. Soc.* **84**, 254 (1962).

Derivative Type. Trimethylsilyloxy.

Reagent. Osmium tetroxide, sodium sulfite, bis (trimethylsilyl) acetamide (BSA).

Compound Type. Unsaturated fatty acid methyl esters.

Reaction

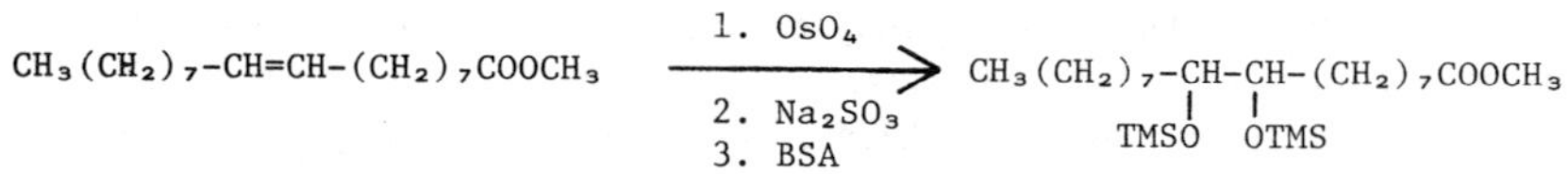

Typical Procedure. Diss 1 mg of unsatd ester and 2 mg OsO_4 in pyridine/dioxane (1:8, v/v) and allow to stand 1 hr at room temp. Add 6 ml of suspension of Na_2SO_3 made by addition of 8.5 ml 16% Na_2SO_3 in H_2O to 2.5 ml MeOH. Allow to stand 1.5 hr at room temp. Add 2 ml H_2O and ext with 5 ml $CHCl_3$. Dry ext over Na_2SO_4 and evap under N_2. Add 0.5 ml BSA to the residue and heat under reflux for 1 hr. Evap under vac and diss residue in 50 μl isooctane prior to GC-MS analysis.

Comments. The method was used to determine double bond position in unsatd fatty acids by GC-MS.[1] It was subsequently extended to include dienoic and trienoic acids.[2] The derivs have been used to locate double bonds in the identification of unsatd long chain hydroxy acids in apple cuticle.[3] MS data for the derivs were reported along with data on an *O*-TFA methyl ester deriv which illustrated that not all derivs of the vicinal dihydroxy compds are structurally useful.[4] Et_2O could be used in place of dioxane.[3,4] The Et_2O then served to ext the derivs after addition of the H_2O/MeOH obviating the need for the $CHCl_3$ ext.

References

[1]C. J. Argoudelis and E. G. Perkins, *Lipids* **3**, 379 (1968).

[2]E. G. Perkins and C. J. Argoudelis, *Lipids* **4**, 619 (1969).

[3]G. Eglington and D. H. Hunneman, *Phytochemistry* **7**, 313 (1968).

[4]G. Eglington *et al.*, *Org. Mass Spectrom.* **1**, 593 (1968).

Derivative Type. Trimethylsilyloxy.

Reagent. Osmium tetroxide, sodium sulfite, hexamethyldisilazane (HMDS)/trimethylchlorosilane (TMCS).

Compound Type. Unsaturated fatty acid esters.

Reaction

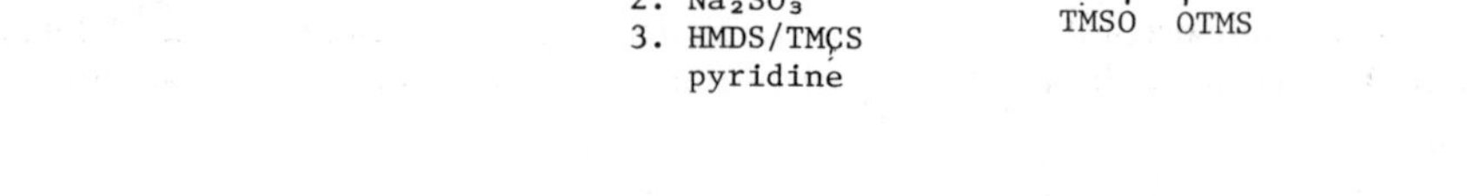

Typical Procedure. Mix 1 ml unsatd fatty ester in 0.2 ml pyridine/dioxane (1 : 8) with 2 mg OsO_4 (fresh 2% soln in dioxane) and allow to stand at room temp for 2 hr. Add 6 ml of Na_2SO_3 suspension (8.5 ml 16% Na_2SO_3 in H_2O added to 2.5 ml MeOH). Let stand 1.5 hr with occasional shaking. Centrifuge and evap under vacuum. Diss residue in Et_2O and dry over Na_2SO_4. Evap Et_2O and treat residue with 0.3 ml pyridine/HMDS/TMCS (10:2:1, v/v/v). After 5 min inject directly into GC-MS.

Comments. The deriv was used for localization of double bonds in unsatd fatty acids. Araki, *et al.* employed a similar method for analysis of unsatd fatty acids in serum lipids using CI-MS.[2]

References

[1]P. Capella and C. M. Zorzut, *Anal. Chem.* **40**, 1458 (1968).
[2]E. Araki *et al., Biomed. Mass Spectrom.* **3**, 261 (1976).

Derivative Type. Polytrimethylsilyloxy.

Reagent. Osmium tetroxide, hydrogen sulfide, hexamethyldisilazane (HMDS)/trimethylchlorosilane (TMCS).

Compound Type. Polyunsaturated fatty acids (or methyl esters).

Reaction

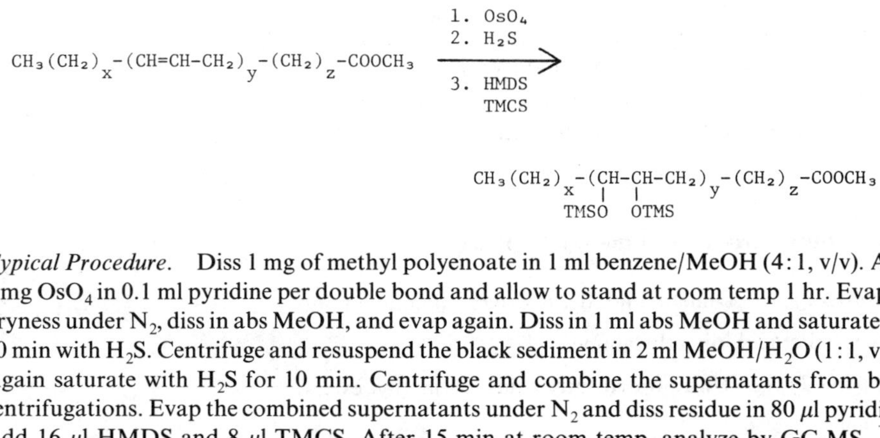

Typical Procedure. Diss 1 mg of methyl polyenoate in 1 ml benzene/MeOH (4:1, v/v). Add 2 mg OsO_4 in 0.1 ml pyridine per double bond and allow to stand at room temp 1 hr. Evap to dryness under N_2, diss in abs MeOH, and evap again. Diss in 1 ml abs MeOH and saturate for 10 min with H_2S. Centrifuge and resuspend the black sediment in 2 ml MeOH/H_2O (1:1, v/v). Again saturate with H_2S for 10 min. Centrifuge and combine the supernatants from both centrifugations. Evap the combined supernatants under N_2 and diss residue in 80 μl pyridine. Add 16 μl HMDS and 8 μl TMCS. After 15 min at room temp, analyze by GC-MS.

Comments. When analyzing free fatty acids, the H_2S treatment in MeOH was omitted due to the higher polarity of the polyhydroxy derivs. The use of H_2S reduction avoided the problems of large amts of salts when using Na_2SO_3. Isopropylidene, alkylboronate, and methyl ether derivs were also examined but only the TMS derivs gave all the information necessary for determination of chain length as well as number and position of double bonds.

References

[1]V. Dommes *et al., J. Chromatogr. Sci.* **14**, 360 (1976).

Derivative Type. Isopropylidine.

Reagent. Osmium tetroxide, acetone/cupric sulfate.

Compound Type. Unsaturated fatty acid esters.

Reaction

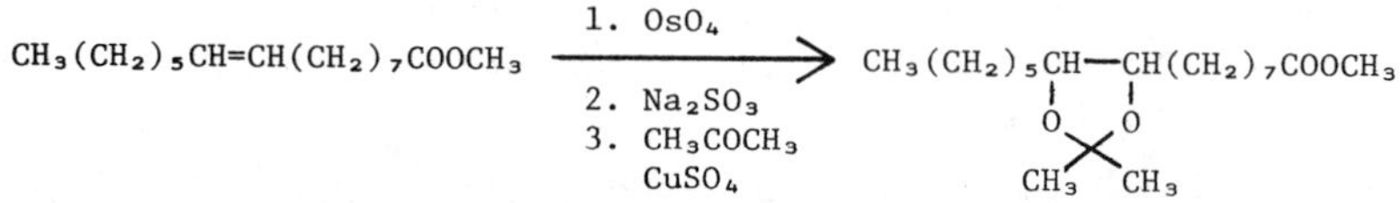

Typical Procedure. Diss 2 mg OsO_4 and 1 mg of unsatd ester in pyridine/dioxane (1:8) and allow to stand 1 hr at room temp. Add 6 ml of a suspension of 8.5 ml 16% aq $NaSO_3$ in 2.5 ml MeOH and allow to stand 1.5 hr at room temp. Centrifuge and evap supernatant under vac. Diss residue in Et_2O and again evap to dryness. Add 1 ml or less of acetone, 50 mg anhyd $CuSO_4$ and heat 2 hr at 50°. Inject mixt directly into GC-MS.

Comments. The method was used to locate double bonds by MS. The geometry of the double bond was also reflected in the deriv. The method was subsequently applied to unsatd hydrocarbons[2] and perdeuterated fatty acid esters.[3] A hexafluoroacetone ketal deriv has also been prepared for use in MS analysis of double bond location.[4] The preparative scale method reported involved preparation of the bromohydrin with *N*-bromosuccinimide in wet DMSO and reaction with hexafluoroacetone in a sealed tube. The deriv appeared to be potentially useful but an analytical scale preparation was not given.

References

[1]J. A. McCloskey and M. J. McClelland, *J. Am. Chem. Soc.* **87**, 5090 (1965).

[2]R. E. Wolff *et al., Tetrahedron* **22**, 3093 (1966).

[3]G. Wendt and J. A. McCloskey, *Biochemistry* **9**, 4854 (1970).

[4]B. M. Johnson and J. W. Taylor, *Anal. Chem.* **44**, 1438 (1972).

4.1.4. Other Derivatives of the Double Bond

Derivative Type. Dimethyl amide, hydroxy, dimethylamino.

Reagent. Peracid, dimethylamine.

Compound Type. Unsaturated fatty acids (as methyl esters).

Reaction

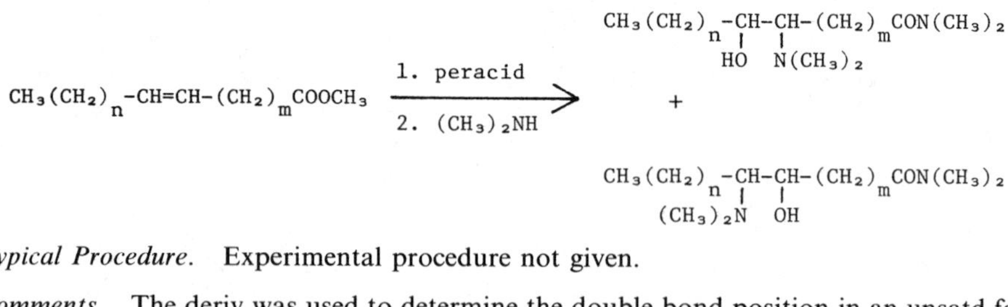

Typical Procedure. Experimental procedure not given.

Comments. The deriv was used to determine the double bond position in an unsatd fatty acid by MS.

References

[1]H. Audier *et al., Bull. Soc. Chim. Fr.* 3034 (1964).

Derivative Type. Methoxy.

Reagent. Mercuric acetate/methanol, sodium borohydride.

Compound Type. Unsaturated fatty acid methyl ester.

Reaction

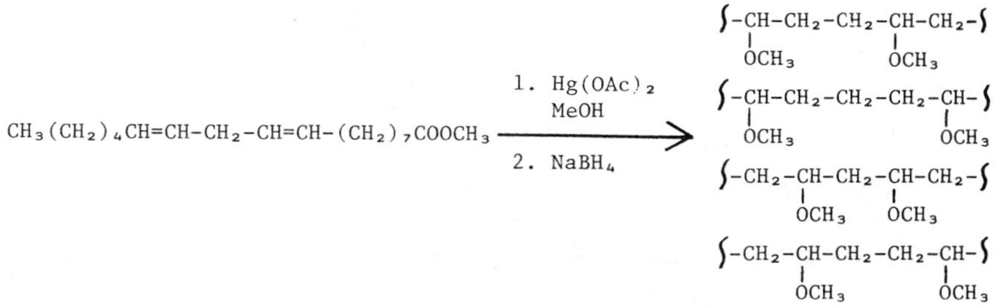

Typical Procedure. Reflux soln of unsatd fatty acid methyl ester in MeOH with 5–10% excess of $Hg(OAc)_2$ for 1 hr. Add solid $NaBH_4$ until no more Hg precipitates followed by a few drops of HOAc to decompose any remaining borohydride. Evap to dryness and partition between Et_2O and H_2O. Dry Et_2O over Na_2SO_4 and evap to dryness. Purify methoxy esters by TLC on silica gel with hexane/Et_2O (80:20, v/v). Analyze purified methoxy esters by GC-MS.

Comments. The mixtures of methoxy derivs were only partially separable by TLC or GC, but the mixt mass spectra were still suitable for assignment of double bond location. MS data were given for derivs of oleic, elaidic, linoleic, linolenic, and erucic acids and two monounsaturated acids from *Lactobacillus casei.*[1] Cyclopropane acid methyl esters were not affected by the $Hg(OAc)_2$-MeOH reagent.[2] Di-and poly- unsatd compds could be converted to partially or totally deriv products (separable by TLC) by adjustment of reaction conditions.[2] The presence of hydroxy substituents β or γ to the double bond gave tetrahydrofurans (1,4-epoxides) and δ to the double bond gave tetrahydropyrans (1,5-epoxides) upon oxymercuration.[3]

References

[1] D. E. Minnikin *et al., Lipids* **9**, 135 (1974).

[2] P. Abley *et al., Chem. Commun.* 348 (1970).

[3] F. D. Gunstone and R. P. Inglis, *Chem. Phys. Lipids* **10**, 73 (1973).

Derivative Type. Monomethoxy, methyl ester.

Reagent. Mercuric acetate/methanol, sodium borohydride, hydrogen/palladium.

Compound Type. Polyenoic fatty esters.

Reaction

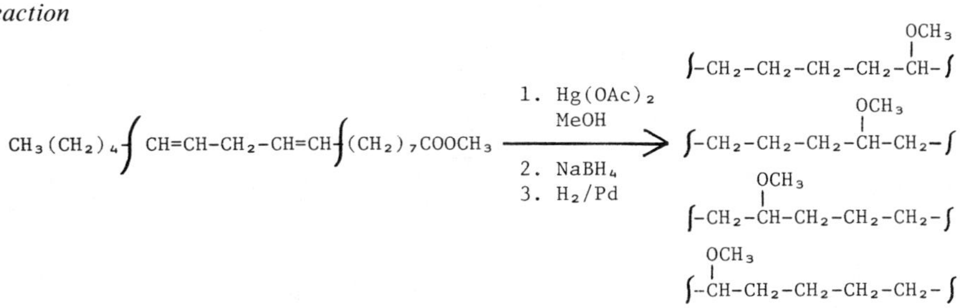

Typical Procedure. Reflux 1–5 mg of polyenoic fatty acid methyl ester with equimolar amount of mercuric acetate in 1–2 ml methanol for 30 min. Add solid sodium borohydride until no further precipitation of mercury. Analyze by hydrogenation GC-MS (H_2 carrier gas, with precolumn containing 1% palladium chloride on Chomosorb P[2]).

Comments. The deriv was used to locate double bonds in polyenoic fatty acids containing up to four double bonds. The method yielded a mixt of monomethoxy derivs for each acid. Hydrogenation in the GC reduced the analysis time.

References

[1]R. D. Plattner *et al.*, *Lipids* **11**, 222 (1976).

[2]M. Beroza and R. Sermiento, *Anal. Chem.* **38**, 1042 (1966).

Derivative Type. 1,4-Epoxides (tetrahydrofurans) and 1,5-epoxides (tetrahydropyrans).

Reagent. Mercuric acetate/sodium borohydride.

Compound Type. Dimethylene interrupted dienoic esters.

Reaction

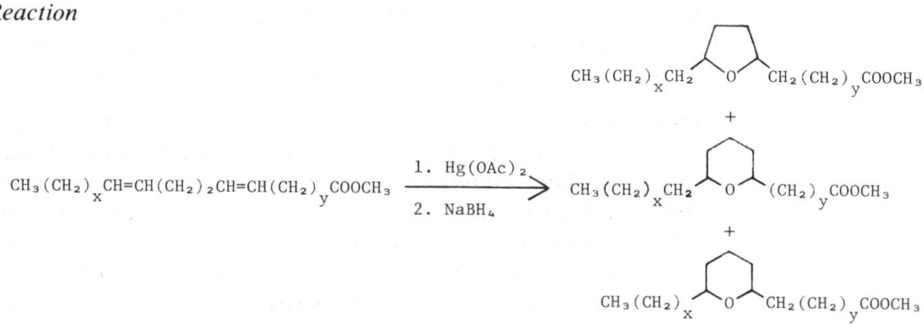

Typical Procedure. Diss 40 mg of methyl octadecadienoate and 100 mg $Hg(OAc)_2$ in 4 ml THF and 6 ml H_2O and shake for 4 days. Add 40 mg $NaBH_4$ in 10 ml 3 *M* NaOH. Sat with NaCl and ext with Et_2O. Sep the 1,4- and 1,5-epoxides by prep TLC on silica gel using Et_2O/lt pet ether (1:9). Analyze products by GC or MS.

Comments. Derivs were prepared from 15 dimethylene interrupted methyl octadecadienoates. GC and MS data were reported.

References

[1]M. S. F. Lie Ken Jie and C. H. Lam, *J. Chromatogr.* **129**, 181 (1976).

Derivative Type. 1,4-Epoxides (tetrahydrofurans) and 1,5-epoxides (tetrahydropyrans).

Reagent. Lithium aluminum hydride, mercuric acetate/sodium borohydride.

Compound Type. Δ3, Δ4, and Δ5 alcohols, aldehydes, acids, and esters.

Reaction

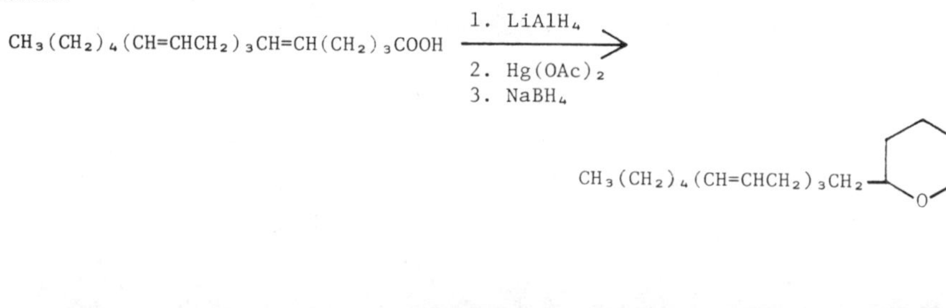

Typical Procedure. Diss 100 mg of acid, ester, or glyceride in 5 ml dry Et_2O and add to a stirred susp of 20 mg $LiAlH_4$ in 2 ml Et_2O. Stir 10 min, cool to 0°, then destroy excess hydride by careful addition of wet Et_2O then H_2O. Add 20 ml 2 *M* H_2SO_4 and ext with 2× 20 ml Et_2O. Evap ext to yield the alcohols.

Diss 50 mg of alcohol and 100–150 mg of $Hg(OAc)_2$ in DMF and allow to stand 2–4 days at room temp. Cool to 0° and add dropwise a soln of 20–80 mg $NaBH_4$ in 10 ml H_2O. Stir 30 min at room temp then sat with NaCl and ext with Et_2O. Analyze by GC or MS.

Comments. Acids, esters, and glycerides were reduced to alkenols first with $LiAlH_4$. It was assumed the method would also be applicable to aldehydes. The alkenols were then converted to 1,4-epoxides (Δ3 and Δ4 compds) or 1,5-epoxides (Δ5 compds). The other double bonds did not react when the reaction was run in DMF. The method was used to identify Δ3, Δ4, and Δ5 fatty acids in lipids.

References

[1]F. D. Gunstone and R. P. Inglis, *Chem. Phys. Lipids* **10**, 105 (1973).

Derivative Type. Chloromercuri-methoxyalkane.

Reagent. Mercuric acetate/methanol (MeOH), sodium chloride.

Compound Type. Olefins.

Reaction

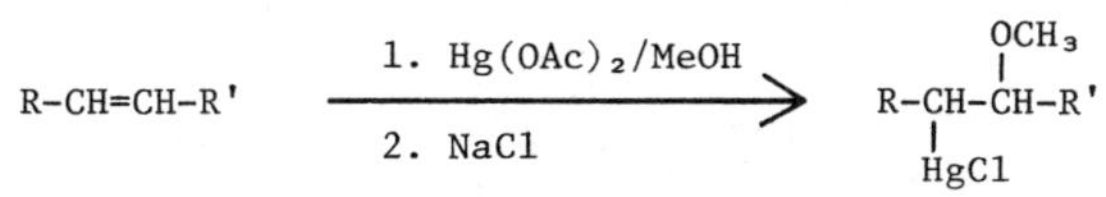

Typical Procedure. Mix 0.1 mol olefin in 100 ml MeOH with a suspension of 0.1 mol $Hg(OAc)_2$ in 250 ml MeOH and shake until dissolution is complete. Add 150 ml CH_2Cl_2 and then mix with aq 1.5 *M* NaCl. Separate the org phase, wash with H_2O, and dry over Na_2SO_4. Analyze by MS.

Comments. The method was used to determine the double bond position by MS.

References

[1]U. Basler and K. Doerffel, *Z. Chem.* **17**, 26 (1977).

4.1.5. Derivatives with Unchanged Double Bonds

Derivative Type. Pyrrolidide.

Reagent. Pyrrolidine.

Compound Type. Unsaturated fatty acid methyl esters.

Reaction

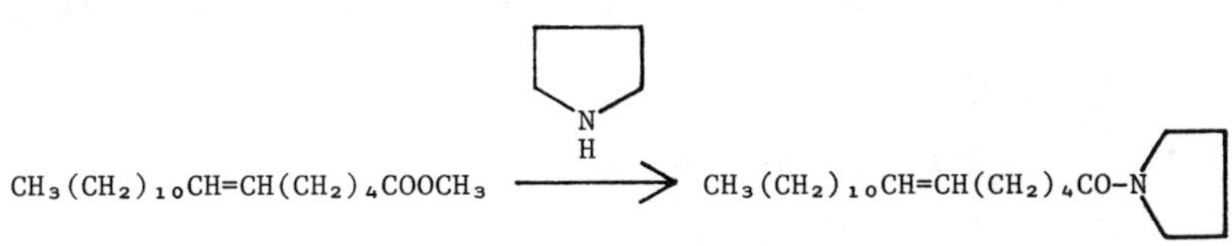

Typical Procedure. Diss 10 μl fatty acid methyl ester in 1 ml freshly dist pyrrolidine and 0.1 ml HOAc. Heat 30 min at 100° in a sealed tube. Cool, diss in CH_2Cl_2, and wash with dil HCl and H_2O. Dry over $MgSO_4$ and analyze aliquot by MS. Alternatively analyze the pyrrolidine soln directly by GC-MS.

Comments. The method was used to locate double bonds by MS analysis of derivs of monounsatd[1] and polyunsatd[3] fatty acids. Several amides were examined to determine the best deriv for double bond location.[2] Tertiary amides gave the best structural information with the pyrrolidide being the most useful. The derivs of deuterated and C-13 labeled acids were also studied.[4,5]

References

[1]B. A. Andersson and R. T. Holman, *Lipids* **9**, 185 (1974).

[2]B. A. Andersson *et al.*, *Lipids* **9**, 443 (1974).

[3]B. A. Andersson *et al.*, *Lipids* **10**, 215 (1975).

[4]B. A. Andersson *et al.*, *Chem. Scripta* **8**, 200 (1975).

[5]B. A. Andersson *et al.*, *Chem. Scripta* **9**, 155 (1976).

Derivative Type. *N*-Alkyl-2-pyrrolidone.

Reagent. Methanesulfonyl chloride, pyrrolidone/potassium *t*-butoxide (*t*-BuOK).

Compound Type. Long chain primary alcohols.

Reaction

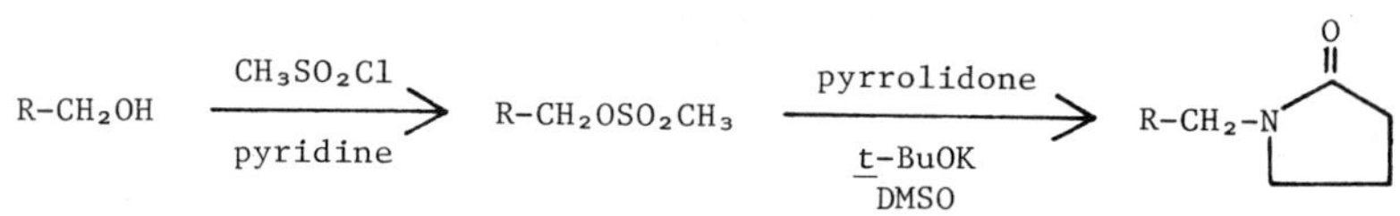

Typical Procedure. Diss *ca* 1 μmol of alcohol in 0.5 ml pyridine containing 2 μmol CH_3SO_2Cl. Allow to stand 1 hr at room temp then add a few drops of glycerol. After a few min add hexane and H_2O. Sep the hexane and dry over Na_2SO_4. Conc the dried hexane soln to a sm vol and add to 0.5 ml of reagent soln prepared from 19 mg *t*-BuOK and 25 μl 2-pyrrolidone in 10 ml DMSO (dist over CaH_2). After 1 hr at room temp add hexane and 1 *N* HCl. Sep the hexane soln, dry over Na_2SO_4, and analyze by GC-MS.

Comments. The derivs were used to locate double bond positions and branching points in long chain alcohols.

References

[1]W. Vetter *et al.*, *Helv. Chim. Acta* **60**, 1203 (1977).

4.2. DERIVATIVES FOR DETERMINATION OF CYCLOPROPANE LOCATION

Derivative Type. Methoxy.

Reagent. Methanol (MeOH)/boron trifluoride.

Compound Type. Cyclopropane fatty acid methyl esters.

Reaction

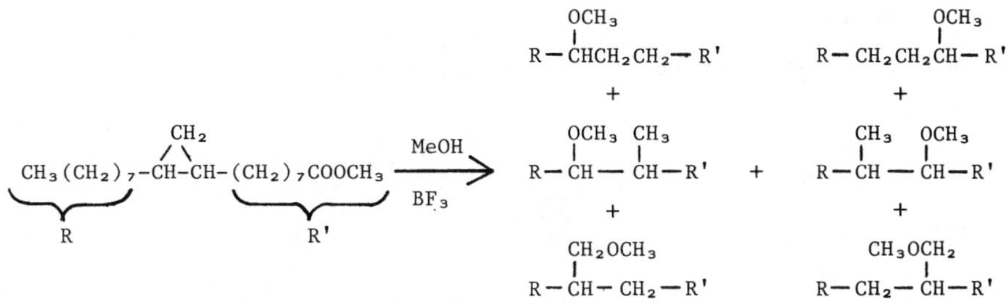

Typical Procedure. Diss 35 mg of the ester in 3 ml CH_2Cl_2 and add 1 ml 50% BF_3 in MeOH. Allow to stand 3 hr. Pour into H_2O and sep CH_2Cl_2 layer. Dry over Na_2SO_4 and isolate methoxy esters by prep TLC. Analyze by GC or MS.

Comments. The mixt of derivs was reported to be useful for location of the cyclopropane group by MS. Cleavage adjacent to the methoxy substituted carbons gave prominent MS peaks indicative of the original location of the cyclopropane ring. Unsatd fatty acid esters were also partially attacked by the reagent to form a mixt of methoxy esters. The derivs illustrate types of artifacts that may be produced using MeOH/BF_3 for methanolysis of lipids.

References

[1]D. E. Minnikin, *Lipids* **7**, 398 (1972).

Derivative Type. Ring-opened reduction products.

Reagent. Hydrogen/platinum.

Compound Type. Cyclopropane fatty acid methyl esters.

Reaction

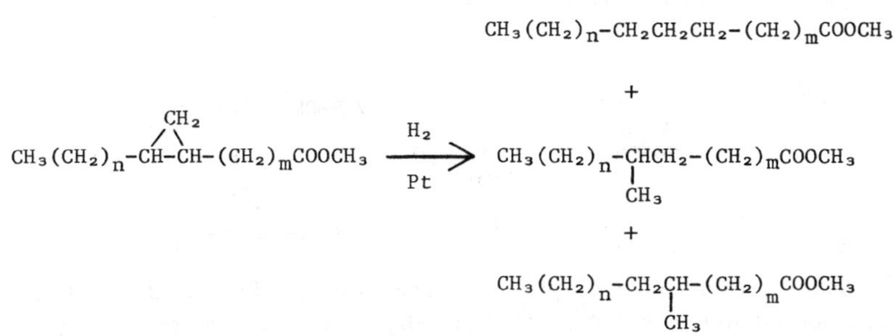

Typical Procedure. To 100 μg of the ester sample add 2–3 mg PtO and 2.0 ml HOAc. Hydrogenate at 20 psi with shaking for 20 hr. Filter to remove the catalyst and analyze by GC-MS.

Comments. Identification of the methyl branched fatty acid esters permitted determination of the cyclopropane position in the parent compd.

References

[1]J. A. McCloskey and J. H. Law, *Lipids* **2**, 225 (1967).

Derivative Type. Cyclopropane keto acids.

Reagent. Chromic acid.

Compound Type. Cyclopropane fatty acids (as methyl esters).

Reaction

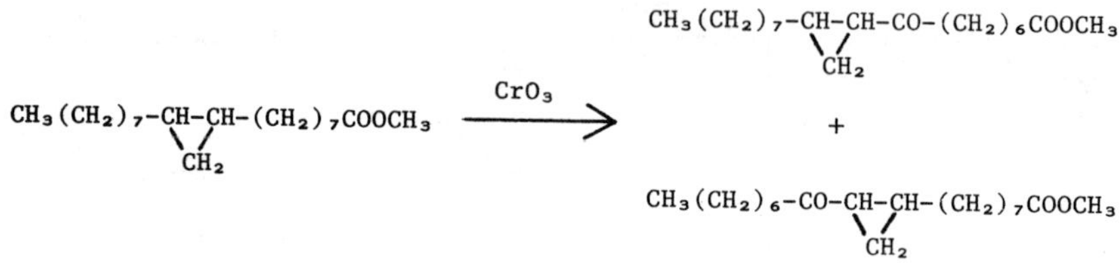

Typical Procedure.[1] To 24 mg of the cyclopropane fatty acid methyl ester add 0.1 ml H_2O and 1 ml 4% chromic acid in HOAc containing 3% H_2O. Allow to react 24 hr then stop the reaction by addition of 25 ml satd aq $NaHCO_3$. Ext the deriv and analyze by GC and MS.

Comments. The oxidation of cyclopropane fatty acids and hydrocarbons with chromic acid was studied. For the hydrocarbons, CCl_4 was used in the oxidation mixt for solubility. The derivs could be used to locate the position of the cyclopropane ring. In a comparison of the keto derivs with pyrrolidide derivs, the keto derivs were deemed preferable.[3]

References

[1]C. Asselineau *et al., Bull. Soc. Chim. France* 1911 (1969).

[2]J. -C. Promé, *Bull Soc. Chim. France* 655 (1968).

[3]W. J. Gensler and J. P. Marshall, *J. Org. Chem.* **42**, 126 (1977).

Derivative Type. Cyclopropyl ketones.

Reagent. Ozone.

Compound Type. Cyclopropyl hydrocarbons.

Reaction

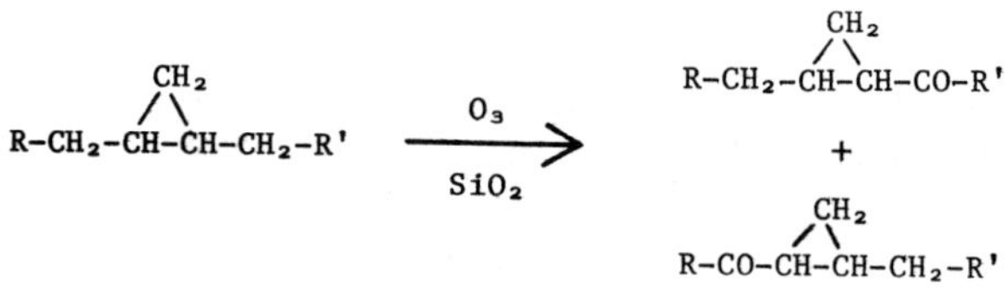

Typical Procedure. Adsorb the sample on silica gel by evap a soln on sufficient silica gel to give *ca* 1% adsorbed sample. Pass O_3 through the dry silica gel at −78° (see cautionary note in Ref. 2) until it is satd. Allow to warm to room temp over *ca* 30 min then elute the products from the silica gel.

Comments. The "dry ozonization"[2] of alkylcyclopropanes was shown to yield cyclopropyl ketones.[1] This reaction could be useful for location of the cyclopropyl ring in alkyl chains using MS.

References

[1]E. Proksch and A. de Meijere, *Angew Chem. Int. Ed.* **15**, 761 (1976).

[2]Z. Cohen *et al., J. Org. Chem.* **40**, 2141 (1975).

4.3. DERIVATIVES FOR LOCATION OF METHYL BRANCHING

Derivative Type. Pyrrolidide.

Reagent. Pyrrolidine.

Compound Type. Methyl branched fatty acids (as methyl esters).

Reaction

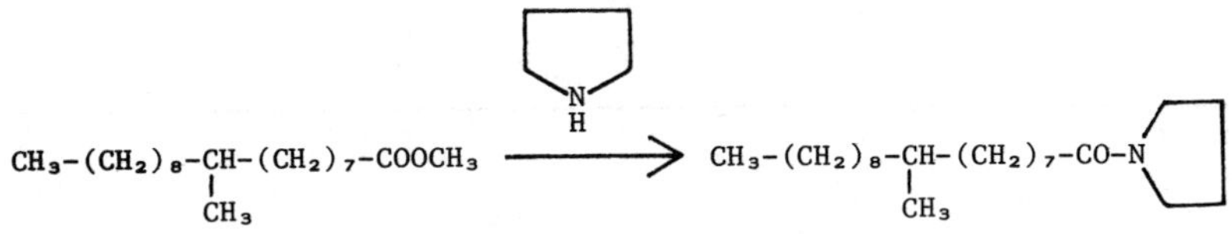

Typical Procedure. Diss 10 mg methyl ester in 1 ml pyrrolidine and 0.1 ml HOAc. Heat in sealed tube at 100° for 30 min. Analyze aliquot by GC-MS.

Comments. The deriv was used to determine the location of methyl branching in the fatty acid.

References

[1]B. Andersson and R. Holman, *Lipids* **10**, 716 (1975).

5. Amino Acids and Peptides

The key role of proteins and their constituent amino acids in the structure and function of living matter has stimulated extensive study of the analysis of these compounds. For the analysis of the individual protein amino acids, the use of ion exchange chromatography has held the preiminent position. Stemming from the classical work of Moore and Stein,[1,2] the modern automated amino acid analyzer remains the major tool for amino acid analysis. In the last two decades, the application of gas chromatography to amino acid analysis has been extensively investigated. Significant effort has also been directed toward the utilization of mass spectrometry for peptide analysis. The low volatility of the amino acids and peptides has necessitated derivatization for the application of gas chromatography and most applications of mass spectrometry. This section presents derivatives used for analysis of amino acids and peptides by gas chromatography and mass spectrometry, and chromogenic derivatives used for HPLC analysis. In addition to the protein amino acids, derivatives of the iodoamino acids and other compounds containing amino and carboxyl groups are included.

Protein Amino Acids. The analysis of amino acids by gas chromatography has been reviewed by Husek and Macek.[3] Due to the presence of both the amino and carboxyl groups, the amino acids must be derivatized to achieve sufficient volatility for GC analysis. The derivatization of amino acids is complicated, however, by the presence of other functional groups in addition to the always present amino and carboxyl groups. These additional groups and the corresponding amino acids in which they occur are summarized below:

alcohol hydroxyl	serine (Ser), threonine (Thr)
phenol hydroxyl	tyrosine (Tyr)
second carboxyl	aspartic acid (Asp), glutamic acid (Glu)
carboxamide	asparagine (Asn), glutamine (Gln)
second amino	lysine (Lys)
guanidine	arginine (Arg)
indole	tryptophan (Trp)
imidazole	histidine (His)
disulfide	cystine (Cys)
thiol	cysteine (CySH)

The remaining protein amino acids, which contain only single amino and carboxyl groups, are glycine (Gly), alanine (Ala), valine (Val), leucine (Leu), isoleucine (Ile), proline (Pro), methionine (Met), and phenylalanine (Phe). Additional biologically important amino acids, not usually considered among the "twenty protein amino acids," include hydroxyproline (HOPro), hydroxylysine (HOLys), ornithine (Orn), and others. Many of this latter group also contain additional functional groups. A generally useful derivatization method must not only derivatize the carboxyl and amino groups but the other polar groups as well. Furthermore, sensitive structures such as the indole ring system must not be destroyed in the process. The fulfillment of these requirements for a broad spectrum of functional groups has made derivatization of the amino acids a considerable analytical challenge.

Since the amino acid mixtures to be analyzed are often the constituent amino acids of a peptide, hydrolysis of the peptide linkage is necessary. The use of 6 *N* hydrochloric acid at 110° for 24 hr has been the most commonly used[4] of several methods applied. These methods are summarized in the review by Husek and Macek.[3] The range of structures present in the amino acids complicates the hydrolysis techniques. Some of the peptide bonds, for example those involving Val, Leu, or Ile, require more rigorous hydrolysis, while particularly sensitive structures such as Trp are increasingly destroyed by the more rigorous conditions. Thus the hydrolytic cleavage of the amino acids, as well as the subsequent derivatization, requires considerable care.

The first published application of GC to amino acids involved analysis of the aldehydes liberated upon reaction with ninhydrin.[5] Amino acids were subsequently analyzed as aldehydes produced by other reagents, as nitriles, as hydroxy and chloro acids, as decarboxylated amines, and as the reduced amino alcohols.[3] The most useful derivatives, however, have been those in which the existing polar groups have not been eliminated or changed to new groups, but rather have been rendered less polar via alkylation, acylation, or similar techniques. It is these derivatives which are emphasized in this section.

Most derivatizations of amino acids have included esterification of the carboxyl group. Esters of some of the amino acids have been analyzed by GC and are sufficiently volatile for direct MS analysis (p. 265). More generally useful GC derivatives have included derivatization, most commonly acylation, of the amino group as well. In general, the esters are prepared by acid catalyzed esterification in the anhydrous alcohol; hydrogen chloride in methanol has been used extensively (p. 249). Acid catalysis by the use of acidic ion exchange resin has also been employed (p. 251). Other, less widely utilized methods include esterification with dimethyl sulfite (p. 251) and with thionyl chloride in methanol, which forms dimethyl sulfite as an intermediate (p. 252). Diazomethane has been used in esterification following derivatization of the amino group (p. 252) or amino and phenolic groups (p. 252), but the highly volatile derivatives of some amino acids are subject to losses during evaporation of excess reagent. Higher diazoalkanes, such as diazobutane, have proven useful, however (p. 263). Methanol with acetic anhydride was used to form the methyl ester and *N*-acetyl derivatives in a single step (p. 253). The high volatility of the methyl ester derivatives of some of the amino acids, and the resulting losses incurred in their manipulation, severely limit their utility. This problem has led to examination of higher esters, especially propyl (p. 253), butyl (p. 257), and amyl (p. 263) ester derivatives. Although the amino acids and their hydrochloride salts are adequately soluble in methanol/hydrogen chloride, solubility problems arise in the use of higher alcohols. The amino acids with more than one site for hydrochloride formation, in particular, Lys, His, and Cys, present significant difficulties. These solubility difficulties can be circumvented by formation of methyl esters and then conversion to higher esters by acid catalyzed transesterification (p. 257 and 263). Direct esterification of the amino acids in *n*-butanol/hydrogen chloride is facilitated by use of ultrasonic mixing (p. 258). Addition of 10% methylene chloride to the *n*-butanol/hydrogen chloride allows preparation of the butyl esters in five minutes with sonication (p. 259).

The esterfied amino acids have been converted to various *N*-acyl derivatives of the alkyl esters. Fu and Mak[6] reported data on a series of *N*-acyl alkyl ester derivatives of amino acids and analyzed the effect of substitution upon GC behavior. The *N*-acyl substituents that have received the most attention have been the acetyl and the two, three, and four carbon perfluoroacyl groups. These derivatives have been prepared principally by use of the anhydride reagents although methyl trifluoroacetate was used in early studies to prepare *N*-trifluoroacetyl derivatives (p. 249). *N*-Benzoyl (p. 261), *N*-pentafluorobenzoyl (p. 261), and *N*-(trifluoroacetylprolyl) (p. 262) derivatives were prepared with the acid chloride reagents. The latter derivatives, although used predominantly for enantiomeric analysis, have also been found useful for GC-MS-SIM analysis.

Of the many combinations of *N*-acyl and alkyl ester groups, the derivative which has emerged as the most generally useful is the *N*-trifluoroacetyl, *n*-butyl ester (p. 257). The application of this derivative was developed principally by Gehrke and coworkers; an extensive discussion was presented in a monograph by Gehrke *et al.* published in 1968.[7] The derivative has been widely applied to amino acid analysis by GC.[3] In a recent application, *N*-trifluoroacetyl butyl esters were used for picomole level analysis by GC-MS-SIM.[8]

Other ester derivatives of amino acids have not been found to be as useful, in general, as the

N-trifluoroacetyl, butyl ester derivatives. Some of these do, however, offer the advantage of single step derivatization but none are generally applicable.

The most generally useful single step derivative is the trimethylsilyl derivative (p. 269). The use of TMS derivatives of amino acids in GC was reported as early as 1960,[9] but it was not until later in the 1960s that Smith and co-workers began systematic studies of the derivatives for amino acid analysis (p. 271). A large part of the recent work in this area has been carried out by Gehrke and co-workers. The major problems in the use of the TMS derivatives are the difficulty of sylylation of the amide group and the instability of the resulting *N*-TMS derivatives. The difficulty of the *N*-silylation and the instability of the derivatives lead to formation of multiple GC peaks, complicating the analysis. The most generally applicable method for preparation of the TMS derivatives appears to be the use of 1:1 BSTFA–CH_3CN (p. 270). While all 20 protein amino acids can be analyzed by this technique, the long (2.5 hr) reaction time and the extreme instability of some of the derivatives makes *N*-acyl, alkyl ester derivatives preferable. Trimethylsilyl derivatives have been used in combination with *N*-trifluoroacetyl (p. 273) and with *n*-butyl ester (p. 273) derivatives, but neither method has become generally applied. Trp was converted to a β-carboline with formaldehyde and then analyzed as the trimethylsilyl derivative (p. 274).

The formation of phenylthiohydantoin derivatives of *N*-terminal amino acids is the basis of the Edman degradation method of peptide sequence determination (p. 274). The phenylthiohydantoin derivatives, prepared from phenylisothiocyanate have normally been associated with this method, but other substituted thiohydantoins have also been used. Methyl isothiocyanate (p. 276) and pentafluorophenyl isothiocyanate (p. 278) have been used to prepare more volatile thiohydantoin derivatives. Use of bromophenyl isothiocyanate gave bromophenylthiohydantoin derivatives in which the bromine isotopic doublet facilitated mass spectral analysis (p. 277). Colored thiohydantoin derivatives were prepared using *p*-phenylazophenyl isothiocyanate (p. 278). The thiohydantoin derivatives are usually prepared by a two step procedure in which the initially formed thiourea is subsequently cyclized under acid catalysis. Fairwell *et al.* showed that the intermediate thioureas could be cyclized on the probe of a mass spectrometer for direct mass spectral analysis (p. 275). The thiohydantoins, while principally employed in Edman degradation analysis, have been applied to free amino acids. In one example, the phenylthiohydantoin derivatives were used for mass spectral isotope ratio analysis of Phe and Tyr in plasma (p. 275). Many of the amino acid thiohydantoin derivatives, especially the more volatile methylthiohydantoins, can be analyzed directly by GC. Those containing additional amino acid functional groups give improved GC behavior after trimethylsilylation (p. 278). GC analysis of phenylthiohydantoins of some of the amino acids as acetyl and trifluoroacetyl derivatives has also been reported (p. 279).

2,4-Dinitrophenyl derivatives, used primarily for analysis of *N*-terminal amino acids in peptides, have also been used for analysis of free amino acids in serum (p. 281). These derivatives can be analyzed by GC as the methyl esters (p. 280), or, for the derivatives of the hydroxylated amino acids, as the trimethylsilyl methyl esters (p. 281). The dinitrophenyl derivatives are normally prepared with dinitrofluorobenzene; trinitrophenyl derivatives have been prepared with trinitrobenzenesulfonic acid (p. 282).

Amino acids treated with trifluoroacetic anhydride without prior esterification yield cyclic trifluoromethyloxazolin-5-one derivatives (p. 282). These derivatives are more volatile than the corresponding *N*-trifluoroacetyl *n*-butyl esters and can be analyzed by GC. A series of other substituted oxaxolin-5-ones have been prepared from a group of acyl leucine derivatives and analyzed by GC (p. 283). Volatile oxazolidin-5-one derivatives prepared by condensation of amino acids with 1,3-dichloro-1, 1, 3, 3-tetrafluoroacetone (p. 283) also show good ECD response. This method has been applied to iodoamino acids and to all protein amino acids and is potentially a generally useful technique. Similar derivatives had been prepared with hexafluoroacetone on a preparative scale (p. 284).

Post column derivatization is usually employed in the analysis of amino acids by high performance ion exchange chromatography in modern automated analyzers. The resulting postcolumn derivatives are not within the defined scope of this book; however, application of preformed derivatives of amino acids for HPLC analysis are reported. Phenylthiohydantoins have been analyzed by HPLC, and the nitrophenyl derivatives possess a chromophore suitable for UV detection. Other derivatives analyzed by HPLC include the 5-dimethylamino- (p. 284) and 5-dibutylamino- (p. 285) naphthalene-1-sulfonamides and the pyridoxyl derivatives (p. 286) prepared by a reductive amination reaction. Other potentially useful derivatives for HPLC analysis are also described (p. 286). The fluorescamine derivatives of amino acids, normally used as postcolumn fluorogenic derivatives, were analyzed by MS after esterification with diazomethane (p. 287).

Some of the amino acids, in particular Cys, are troublesome to analyze in peptide hydrolyzates. Methods have been devised to chemically alter Cys and CySH for protection during hydrolysis. Both *S*-carboxymethyl (p. 288) and *S*-carboxyethyl (p. 288) derivatives can be prepared and analyzed as the *N*-acetyl esters.

Potentially useful miscellaneous acylation methods are presented (p. 289), although specific analytical applications are not cited.

Iodoamino Acids. A series of volatile derivatives have been prepared from six iodinated tyrosine and thyronine amino acids with particular interest in the analysis of the clinically significant triiodothyronine (T_3) and thyroxine (T_4). Although the methylated derivatives formed with diazomethane can be analyzed by GC (p. 290), most of the work on these compounds has also employed *N*-acylation. *N*-Trifluoroacetyl (p. 291), heptafluorobutyryl (p. 291), and pivalyl (p. 292) methyl esters have been investigated. Efforts to utilize trimethylsilyl derivatives for GC analysis of the iodinated amino acids have met with only limited success (p. 293).

Nonprotein Amino Acids. Many of the derivatization methods applied to the protein amino acids have also been applied to nonprotein amino acids. Tranexamic acid (p. 294) and baclofen (p. 293) have both been analyzed as *N*-heptafluorobutyryl, *n*-butyl esters. The derivatives, however, were prepared with different reagents and utilized opposite sequences of esterification and acylation. The amino acid methyl ester, methylphenidate, was quantitated by GC-MS as the *N*-trifluoroacetyl derivative (p. 295). Another *N*-acylated derivative not included among the derivatives of the protein amino acids is the *N*-pentafluoropropionyl, hexafluoroisopropyl ester, which was utilized in the measurement of Glu and γ-aminobutyric acid in tissue samples (p. 295). The *N*-pentafluoropropionyl mixed anhydride was used for GC analysis of mefenamic acid (p. 296).

Nonprotein amino acids have been analyzed as trimethylsilyl derivatives utilizing BSA (p. 296) and mixed silylating reagents (p. 297). Nitrilotriacetic acid and the related tertiary amine polyacids have been analyzed as trimethylsilyl as well as alkyl esters (p. 298). α-Methyl amino acids were converted to 2-phenyl-oxazolin-5-one derivatives by *N*-benzylation and cyclization with dicyclohexylcarbodiimide (p. 301). Derivatives of nonprotein amino acids have also been prepared by chemical conversion of either the amino or the carboxyl group to a different functional group. For example, anthranilic acid was converted to salicylic acid by diazotization and analyzed as the methyl ester (p. 301). The carboxyl group of methylphenidate was reduced to amino alcohol, which was analyzed as the pentafluoropropionyl derivative (p. 302). This method permitted formation of a diacylated derivative for improved ECD sensitivity compared to the monoacylated parent compound.

Peptides. The high polarity and resulting low volatility of the amino acids is also shared by peptides likewise necessitating their derivatization for GC and most applications of MS analysis. Although a few applications of GC analysis of peptides are reported, the major emphasis of this section is on derivatives for MS and GC-MS analysis. As with the individual amino acids, peptides have been converted to more volatile *N*-acyl ester derivatives including *N*-trifluoroacetyl, methyl (p. 302) and butyl (p. 303) derivatives that have been used for GC analysis of tripeptides. The use of *N*-pentafluoropropionyl pentafluoropropyl ester derivatives permitted analysis by ECD-GC (p. 304). In recently reported work a dimethyl derivative of tripeptide amide was prepared with diazomethane for GC-MS-SIM quantitation (p. 304). Derivatives used for GC and LC separations of peptides were surveyed by Deyl.[10] The application of MS to peptide analysis has been reviewed.[11-14]

Two general approaches have been employed in the sequence analysis of peptides. The first, and classical, approach has been to successively cleave and identify single amino acid residues. The thiohydantoin derivatives as used in the Edman degradation and the nitrophenyl derivatives for *N*-terminal end group analysis have been discussed. The use of these derivatives in analysis of peptides has been reviewed by Rosmus and Deyl.[15] The same authors have also reviewed a variety of other derivatives for peptide end group analysis.[16]

The second and more recently applied technique for peptide sequence analysis involves partial hydrolysis of the peptide, sequence determination of the overlapping fragments by mass spectrometry, and reassembly of these fragment sequences to determine the sequence of the starting peptide. It has been noted that this second approach to peptide sequencing using mass spectrometry, rather than replacing the older methods such as Edman degradation, should be used as a complement to them. Particular sequences, which are poorly defined by the Edman method, may be well defined by mass spectrometry, and *vice versa*. The complementary nature of the two general methods is illustrated by Hudson and Biemann in a study of sequence determination, using mass spectrometry, of the 44 amino acid subunit I of monellin.[17] Application of the Edman degradation to this peptide provided most of the sequence, but some areas of ambiguity remained. These areas were well defined by the mass spectral approach allowing determination of the complete sequence.

The simplest derivative for mass spectral analysis of peptides is the methyl ester (p. 305). This derivative has been used for field desorption mass spectrometry which does not have the same volatility requirements as convention MS. The volatility requirement of conventional MS usually necessitates *N*-derivatization as well. Such *N*-derivatization has been performed primarily via either acylation (p. 305) or Schiff base formation (p. 312). In an early study in this area, Prox and Sun examined 14 different *N*-substituted derivatives of the tripeptide Val-Ile-Ala methyl ester.[18] These included the following

formyl	benzoyl
acetyl	*p*-chlorobenzoyl
trifluoroacetyl	pentafluorobenzoyl
chlorodifluoroacetyl	benzyloxycarbonyl
decanoyl	phthaloyl
octadecanoyl	benzenesulfonyl
cyclohexancarbonyl	2,4-dinitrophenyl

Methods for preparing *N*-substituted derivatives are given beginning on pages 305 (acyl derivatives), 312 (Schiff base derivatives), and 314 (other derivatives).

In addition to volatility, there are other features that affect the utility of the different *N*-derivatives. Paramount among these is the nature of the mass spectral fragmentation. Derivatives with charge stabilizing properties that yield prominent "sequence peaks" in the mass spectrum are desired. Strategies used to further facilitate identification of sequence peaks have included use of equimolar mixtures of different acyl groups (p. 307) or of isotope labeled and unlabeled acyl groups (p. 334) to produce sequence peaks as visually conspicuous equal intensity ion pairs. Usefulness of a derivative can also be improved by an upward shift in molecular weight, which "moves" the sequence peaks to the less background obscured higher mass region. The increased molecular weight, however, generally results in decreased volatility, thus limiting the usefulness of this approach. Finally, other properties, such as color, can also facilitate the isolation process, yielding more useful derivatives.

Much of the effort in peptide derivitization has been aimed at obtaining mass spectral sequence information from as large a peptide as possible. An alternate approach to peptide sequencing involves analysis only of dipeptides, thus simplifying the derivative requirements. This approach utilizes dipeptidylaminopeptidase digestion of the peptide to yield a mixture of dipeptides. Analysis of this mixture, and analysis of the mixture obtained from the peptide minus one amino acid, gives a series of overlapping dipeptides, which are "reassembled" to yield the peptide sequence. Derivatization methods that are useful for this approach are listed on pages 315 (acyl, ester derivatives), 317 (Schiff base, ester derivatives), and 318 (other derivatives).

Many of the previously discussed derivatives, both for dipeptides and larger peptides, are not applicable to all amino acid residues. The presence of certain multifunctional amino acids in the peptides precludes the preparation of a volatile product and leads to undesired reactions. Arginine is particularly troublesome in this respect. A series of derivatives designed for modification of specific amino acid residues are presented beginning on page 321.

Even though the *N*-derivatives of peptide esters that have been discussed may allow mass spectral analysis of peptides containing up to seven or eight amino acids, these derivatives still have limited volatility. This limited volatility is likely due to the peptide bond N—H groups that remain free to form hydrogen bonds. Methylation of the amide nitrogens yields more volatile derivatives. The use of these permethyl derivatives has been reviewed by Das and Lederer.[19] The permethylation of *N*-acetylated derivatives was performed initially with methyl iodide and silver oxide (p. 327), but most permethyl derivitizations have employed "Hakomori" methylation using methyl iodide and methylsulfinyl carbanion (p. 329). The method is complicated by the formation of methyl ammonium and sulfonium salts from the basic and sulfur containing amino acids and by *C*-methylation of certain (particularly Gly) amino acid residues. An "equimolar method" using limited amounts of reagent, introduced to circumvent these problems, is limited by the need for exact knowledge of the amount of peptide in the sample (p. 331). A simpler approach used a short (60 sec) reaction time to minimize formation of the nonvolatile salts (p. 332). The problem of C-methylation can be minimized by use of deuterated methyl iodide, which permits distinguishing *C*-methylated residues from higher homologs (p. 333). Equimolar mixtures of deuterated and unlabeled acetyl derivatives have been used to form equal intensity mass spectral ion pairs that facilitate sequence determination (p. 334). Permethylated *N*-trifluoroacetyl derivatives have also been examined (p. 335).

A different type of modification of the peptide chain, introduced by Biemann and coworkers, employed reduction of the *N*-acetyl peptide esters to yield polyamino alcohols (p. 335). These derivatives were more volatile than the permethyl derivatives and could be analyzed by GC-MS. The derivatives with more than one hydroxyl were trimethylsilylated prior to analysis to improve their GC behaviour. The basic approach has been modified by forming *N*-heptafluorobutyryl derivatives prior to the reduction and trimethylsilylation (p. 337). This modification yields the most volatile peptide derivatives yet devised. The derivatives are amenable to GC-MS analysis permitting use of both GC and MS data for their identification. This derivatization method applied to partial hydrolysis mixtures from peptides and employing GC-MS analysis and computerized reassembly of peptide fragments constitutes a very sophisticated mass spectral approach to peptide sequencing.[13]

References

1. S. Moore and W. H. Stein, *J. Biol. Chem.* **192**, 663 (1951); **211**, 895 (1954).
2. S. Moore *et al.*, *Anal. Chem.* **30** 1185, 1190 (1958).
3. P. Husek and K. Macek, *J. Chromatogr.* **113**, 139 (1975).
4. S. Moore and W. H. Stein, *Meth. Enzymol.* **6**, 819 (1963).
5. I. R. Hunter *et al.*, *Chem. Ind.* **16**, 294 (1956).
6. S.-C. J. Fu and D. S. H. Mak, *J. Chromatogr.* **54**, 205 (1971); **78**, 211 (1973).
7. C. W. Gehrke *et al.*, *Quantitative Gas-Liquid Chromatography of Amino Acids in Proteins and Biological Substances*, Anal. Biochem. Labs., Columbia, MO, 1968.
8. W. Frick *et al.*, *Anal. Chem.* **49**, 1241 (1977).
9. R. Birkofer and A. Ritter, *Chem. Ber.* **93**, 424 (1960).
10. Z. Deyl, *J. Chromatogr.* **127**, 91 (1976).
11. M. M. Shemyakin *et al.*, in *Mass Spectrometry: Techniques and Applications*, ed by G. W. A. Milne, Wiley, New York, 1971, p. 289.
12. K. Biemann, in *Biomedical Applications of Mass Spectrometry*, ed. by G. R. Waller, Wiley, New York, 1972, p. 405.
13. H. Nau, *Angew. Chem. Int. Ed. Eng.* **15**, 75 (1976).
14. H. Falter, *Mol. Biol. Biochem. Biophys.* **25**, 123 (1977).
15. J. Rosmus and Z. Deyl, *J. Chromatogr.* **70**, 221 (1972).
16. J. Rosmus and Z. Deyl, *Chromatogr. Rev.* **13**, 163 (1971).
17. G. Hudson and K. Biemann, *Biochem. Biophys. Res. Commun.* **71**, 212 (1976).
18. A. Prox and K. K. Sun, *Z. Naturforsch.* **21b**, 1028 (1966).
19. B. C. Das and E. Lederer, in *New Techniques in Amino Acid, Peptide, and Protein Analysis*, Ann Arbor Science Publ., Ann Arbor, MI, 1971, p. 175.

5.1. DERIVATIVES OF PROTEIN AMINO ACIDS

5.1.1. N-Acyl Methyl Ester Derivatives

Derivative Type. Methyl ester, *N*-trifluoroacetyl.

Reagent. Methanol (MeOH)/HCl, methyl trifluoroacetate.

Compound Type. Amino acids and dipeptides.

Reaction

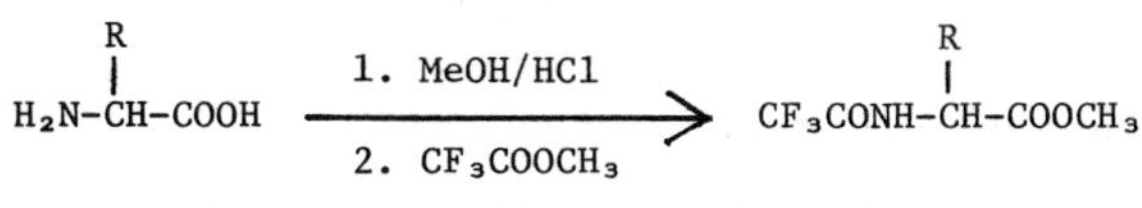

Typical Procedure. Convert 1 g amino acid to the HCl salt by diss in 5 ml conc HCl and evap under vac. Diss in 20 ml 0.2 *N* HCl in MeOH and reflux 2 hr. Evap under vac and diss residue in 15 ml MeOH. Add 2.3 ml Et_3N and 5 ml methyl trifluoroacetate. Let stand 2 hr at 20° then evap under vac. Partition the residue between EtOAc and H_2O and evap the EtOAc phase under vac. Analyze the residual deriv by GC.

Comments. Group separation of amino acids and dipeptides as well as separation of members within each group was possible. The derivs of the diastereomeric pair L-Ala-L-Phe, L-Ala-D-Phe were also separable by GC. The MS behaviour of *N*-TFA peptide methyl esters was studied extensively,[2,3] but the derivs were not prepared directly from the parent peptides. Bayer *et al.*[4] studied *N*-TFA methyl ester derivs of Cys peptides as *S*-benzyl derivs.[4]

References

[1]F. Weygand *et al.*, *Z. Physiol. Chem.* **322**, 38 (1960).

[2]F. Weygand *et al.*, *Z. Naturforsch* **18B**, 1924 (1963).

[3]F. Weygand *et al.*, *Z. Naturforsch* **20B**, 1169 (1965).

[4]E. Bayer *et al.*, *Z. Naturforsch* **22B**, 924 (1967).

Derivative Type. Methyl ester, *N*-TFA.

Reagent. Methanol (MeOH)/HCl, trifluoroacetic anhydride (TFAA).

Compound Type. Amino acids.

Reaction

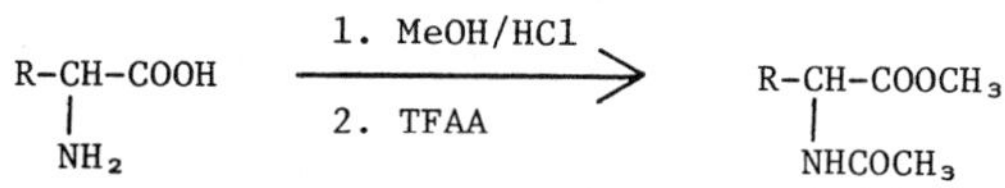

Typical Procedure.[1,2] Diss 2 mg amino acids in 2 ml dry MeOH and bubble in dry HCl gas for 30 min while heating at 70°. Evap to dryness under vac. Add 0.2 ml TFAA and allow to stand at room temp at least 30 min. Evap under vac while cooling tube in an ice bath (limit to 4 min to minimize loss of volatile derivs). Diss residue in 100 μl CH_2Cl_2 and analyze by GC.

Comments. His gave a low yield of deriv. For deriv of Arg, it was necessary to acylate in a sealed tube at 140° for 10 min. GC separation of the derivs of 22 amino acids was demonstrated.[1] The deriv method was adapted for prep of the derivs on a platinum support utilizing a solid injection GC technique.[3] GC of the methyl ester, TFA derivs of amino acids had been reported earlier.[4] Di-TFA derivs were formed from Thr, Ser, Cys, HO-Pro, His, Tyr, Orn, Lys, and Trp; Arg formed a tri-TFA deriv.[4] The diastereomers of β-hydroxy-α-amino acids have been separated by GC of the *N,O*-TFA methyl esters.[5]

References

[1]A. Darbre and A. Islam, *Biochem. J.* **106**, 923 (1968).

[2]A. Islam and A. Darbre, *J. Chromatogr.* **43**, 11 (1969).

[3]B. Teuwissen and A. Darbre, *J. Chromatogr.* **49**, 298 (1970).

[4]S. Makisumi and H. A. Saroff, *J. Gas Chromatogr.* **3**, 21 (1965).

[5]M. Y. Karpeisky *et al.*, *Tetrahedron Lett.* 3885 (1965).

Derivative Type. Methyl ester, trifluoroacetyl.

Reagent. Methanol (MeOH)/HCl, trifluoroacetic anhydride (TFAA).

Compound Type. Amino acids.

Reaction

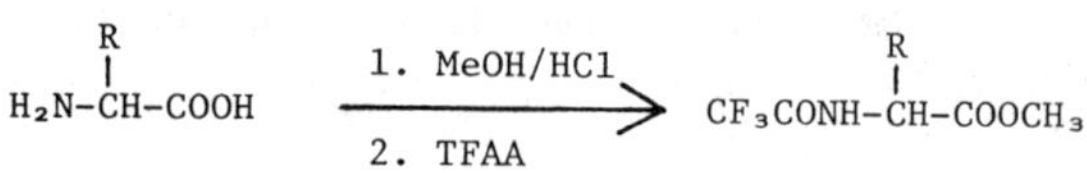

Typical Procedure. Dry under vac (0.1 – 1 Torr) over P_2O_5 and NaOH an aq sample containing 0.5 μmol of each amino acid. Add 1 ml 4 *M* HCl in MeOH and heat at 65° for 90 min. Dry as before and diss residue in 200 μl of 20% TFAA in CH_2Cl_2 containing 1% (w/v) methyl stearate int stand. Seal and heat at 120° for 20 min. Apply 10 μl aliquot to a precolumn and evap reagents under vac. Analyze by GC.

Comments. The reproducibility of analyses using the above method were studied and it was found that quite small changes in the conditions of deriv prep gave erratic results. Using standard mixtures it was found that Met, Tyr, Arg, and Cys had exceptionally large (> 20%) coefficients of variation. Those for the remaining 18 amino acids ranged from 0.7 to 8.0%.

References

[1]A. J. Cliffe *et al.*, *J. Chromatogr.* **78**, 333 (1973).

Derivative Type. Methyl ester, trifluoroacetyl.

Reagent. Methanol/Dowex 50, methyl trifluoroacetate.

Compound Type. Amino acid.

Reaction

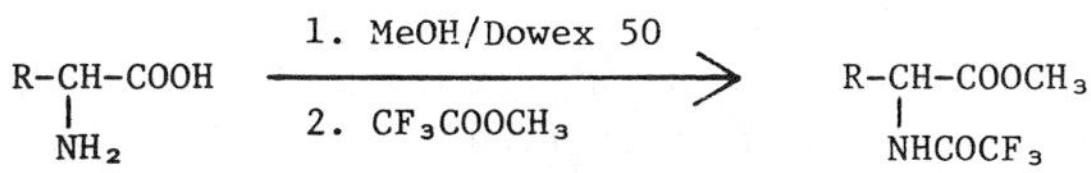

Typical Procedure. Add 2 mmol amino acid to 10 ml dry MeOH in which is suspended 2.5 g of Dowex 50 resin (H form, 50–100 mesh). Reflux 30–45 min, decant MeOH, and wash resin with dry MeOH. To the washed resin add 10 ml dry MeOH and 2.1 g methyl trifluoroacetate. Alkalinize by adding 3.3 g Et_3N (or by bubbling in Et_3N for 10 min). Reflux 15 min, cool, and decant from the resin. Analyze an aliquot of the soln by GC.

Comments. Use of the sulfonic acid type ion exchange resin was an improvement over the use of MeOH/HCl, which yielded side products (e.g., reaction of Gly with other amino acids). GC data were reported for the derivs of 14 common amino acids.

References

[1]H. A. Saroff and A. Karmen, *Anal. Biochem.* **1**, 344 (1960).

Derivative Type. Methyl ester, trifluoroacetyl.

Reagent. Dimethyl sulfite, trifluoroacetic anhydride (TFAA).

Compound Type. Amino acids.

Reaction

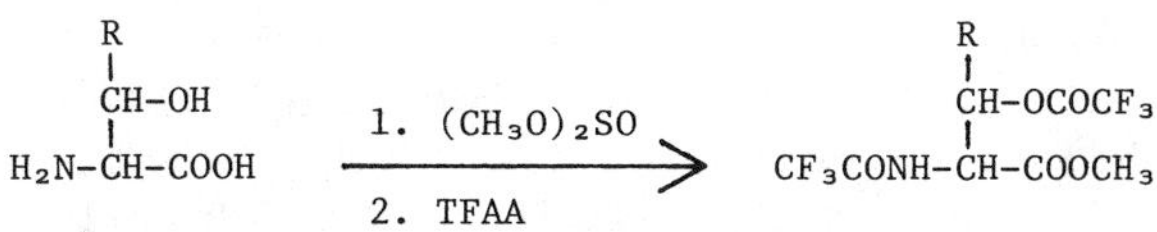

Typical Procedure. Suspend 100 μmol or less of amino acids in 5 ml MeOH and saturate with anhyd HCl (ice bath cooling). Add 1 ml dimethyl sulfite and reflux 30 min under a $CaCl_2$ drying tube. Evap to dryness under vac and dry residue under high vac. Add 1 ml TFAA and reflux 10 min (with $CaCl_2$ drying tube). Evap to dryness under N_2. Redissolve in TFAA prior to GC analysis.

Comments. Cystine was reduced to CySH by diss sample in 4 ml H_2O, neutralizing with a few drops of satd aq $NaHCO_3$, and refluxing 5 min with 0.1 ml ethanethiol. The deriv method gave single derivs of all 21 amino acids except Arg. The reverse order of deriv was examined using TFAA acylation followed by CH_2N_2 esterfication. The hydroxyl groups in the hydroxy amino acids remained unacylated in the reverse procedure.

References

[1]P. A. Cruickshank and J. C. Sheehan, *Anal. Chem.* **36**, 1191 (1964).

Derivative Type. Methyl ester, *N*-trifluoroacetyl.

Reagent. Methanol (MeOH)/thionyl chloride, trifluoroacetic anhydride (TFAA).

Compound Type. Amino acids.

Reaction

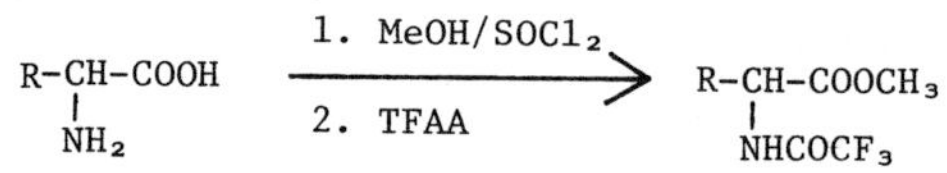

Typical Procedure. Diss amino acid mixt in 20 ml dry MeOH. Chill in dry ice bath and add 2–4 ml $SOCl_2$. Maintain at 40° for 2 hr under a $CaCl_2$ drying tube. Evap to dryness under vac at room temp; then add 1–2 ml TFAA. Allow to stand at room temp 2 hr then evap to dryness under vac at room temp. Diss the residue in 5 ml dry MeOH and analyze by GC.

Comments. GC separation of the derivs of 19 protein amino acids was achieved with two GC columns.

References

[1]P. B. Hagen and W. Black, *Can. J. Biochem.* **43**, 309 (1965).

Derivative Type. *N, O,*-Trifluoroacetyl, methyl ester.

Reagent. Trifluoroacetic anhydride (TFAA), diazomethane.

Compound Type Phenylalanine and tyrosine.

Reaction

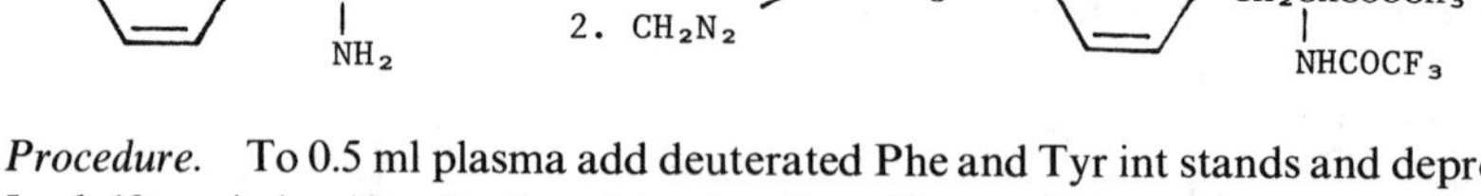

Typical Procedure. To 0.5 ml plasma add deuterated Phe and Tyr int stands and deproteinize with 0.5 ml 40 mg/ml sulfosalicylic acid soln. Centrifuge twice to remove ppt then apply supernatant to a 20 × 0.7 cm column of Dowex 50W-X8 (H form). Rinse column with 100 ml H_2O and elute with 1 *N* NH_3 soln. Collect the UV positive (240 nm) fractions and evap to dryness under vac. Dry the residue by evap of CH_2Cl_2.

Add 200 μl TFAA and 200 μl CH_3CN, allow to react 10 min at room temp, then evap under N_2. Add excess CH_2N_2 in Et_2O then evap under N_2. Diss residue in 200 μl CH_3CN, and 40 μl TFAA. Analyze by GC-MS-SIM.

Comments. The method was used to quant Phe and Tyr in plasma. Several derivs were examined including methyl, isobutyl, and TMS esters, and *N,O*-HFB and -TMS derivs. MS data were given for the derivs. In the method above, use of TFAA first eliminated possible methylation of the phenolic hydroxyl in Tyr by CH_2N_2.

References

[1]M.-J. Zagalak *et al.*, *J. Chromatogr.* **142**, 523 (1977).

Derivative Type. *N*-Acetyl, methyl ester.

Reagent. Acetic anhydride (Ac_2O)/methanol (MeOH).

Compound Type. Amino acids.

Reaction

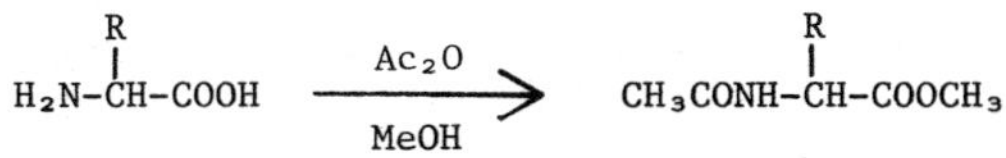

Typical Procedure. To a 0.5 ml vial add 200 μl 80% (v/v) EtOH, *ca* 12 nmol labeled amino acid standard, and a 4 mm filter paper disk containing a dry blood spot sample (*ca* 5 μl). Seal and sonicate 5 min. Remove paper disk (which retains denatured protein) and evap soln to dryness under N_2 at 80°.

Add 10 μl Ac_2O, 50 μl anhyd MeOH, and 1 stick of 3 A mol sieve. Seal, sonicate 2 min, then heat at 100° for 5 min. Transfer 10 μl aliquot to probe tip sample holder, evap under vac, and analyze by direct probe CI-MS.

Comments. The deriv could be used to quantitate amino acids down to the 25 ng level. The method was applied to Phe determination in microdrop blood samples from newborns.

References

[1]J. M. L. Mee *et al.*, *Biomed. Mass Spectrom.* **4**, 178 (1977).

5.1.2. N-Acyl Propyl Ester Derivatives

Derivative Type. *n*-Propyl, *N*-acetyl.

Reagent. *n*-Propanol (*n*-PrOH)/HCl, acetic anhydride (Ac_2O)/triethylamine (Et_3N).

Compound Type. Amino acids.

Reaction

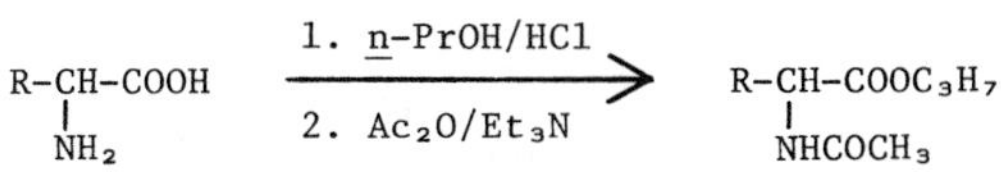

Typical Procedure. Ext amino acids from 0.1 ml serum or 0.5 ml urine on a 50 mg column of Dowex 50X8 (H form) and elute with 2 ml 2 *M* NH_4OH. Evap eluate to dryness under vac. To the dried amino acid sample add 50 μl DMP (H_2O scavenger) and 1 ml 8 *M* HCl in *n*-PrOH. Heat at 110° for 20 min. Cool tube and open. Evap to dryness at 110° under N_2. Cool and add 1 ml acetone/Et_3N/Ac_2O (5 : 2 : 1, v/v/v). Warm 30 sec at 60°. Evap excess reagents at 60° under N_2. Diss residue in EtOAc and analyze by GC.

Comments. Norleucine was used as int stand to quantitate amino acids to the nmol level in protein hydrolyzates, plant samples, serum, and urine. The method resolved all 21 common amino acids. The basic acylation reagent allowed deriv of Arg and His obviating the need for separate procedures for these compds. Asn and Gln were converted to the acids in the esterfication step. Cys was analyzed as CySH following reduction with stannous chloride. Dried protein hydrolyzates could be deriv directly.

References

[1]R. F. Adams, *J. Chromatogr.* **95**, 189 (1974).

Derivative Type. *n*-Propyl ester, *N*-acetyl.

Reagent. *n*-Propanol (*n*-PrOH)/HCl, acetic anhydride (Ac_2O)/pyridine.

Compound Type. Amino acids.

Reaction

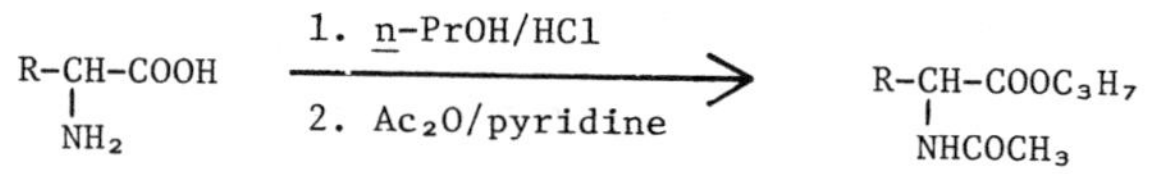

Typical Procedure. Evap 0.1 ml protein hydrolyzate (*ca* 10^{-6} mol of each amino acid) quickly to dryness at 100° under dry N_2. Add 0.4 ml 8 *M* HCl in *n*-PrOH and heat at 100° for 10 min. Evap to dryness under N_2 and repeat propylation and evap. Treat residue with 0.4 ml pyridine/Ac_2O (4:1, v/v). Let stand 5 min at room temp and carefully evap to dryness under N_2 removing heat and N_2 just as the last pyridine is evapd. Diss the residual oil in EtOAc and analyze by GC.

Comments. The method was reported to be satisfactory for all amino acids except Arg and His. Arg was analyzed by conversion to Orn with arginase and His by conversion to Asp by ozonolysis.. Trp was destroyed by the acid esterfication in a time-dependent manner necessitating strict adherence to the time schedule. McGregor[2] employed essentially the same method for analysis of 16 amino acids in urine hydrolyzates using norleucine as int stand.

References

[1]J. R. Coulter and C. S. Hann, *J. Chromatogr.* **36**, 42 (1968).
[2]R. F. McGregor, *Clin. Chim. Acta* **48**, 65 (1973).

Derivative Type. Isopropyl ester, *N*-trifluoroacetyl.

Reagent. Isopropanol (*i*-PrOH)/HCl, trifluoroacetic anhydride (TFAA).

Compound Type. Amino acids.

Reaction

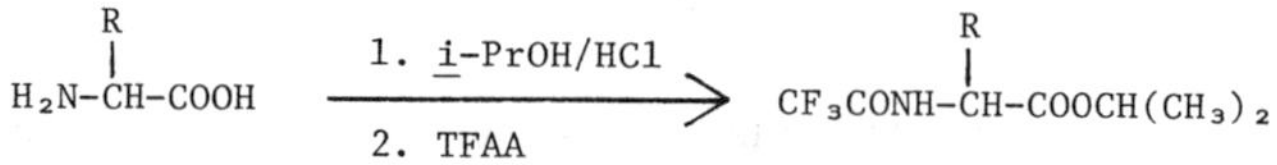

Typical Procedure. Diss a 5 mg amino acid sample in 20 ml 3 *N* HCl in *i*-PrOH and heat 30 min at 100°. Evap under vac and diss residue in 20 ml CH_2Cl_2. Cool in dry ice-acetone and add 4 ml TFAA. Allow to warm to room temp then let stand 1 hr. Evap under vac and diss residue in $CHCl_3$ for GC analysis.

Comments. The derivs were used in studies of separation of enantiomers on optically active stationary phases but it was noted that the esterification procedure could cause racemization.

References

[1]W. Parr and P. Howard, *Chromatographia* **4**, 162 (1971).

Derivative Type. Isopropyl ester, *N*-pentafluoropropionyl (PFP).

Reagent. Isopropyl alcohol (*i*-PrOH)/HCl, pentafluoropropionic anhydride (PFPA).

Compound Type. Amino acids.

Reaction

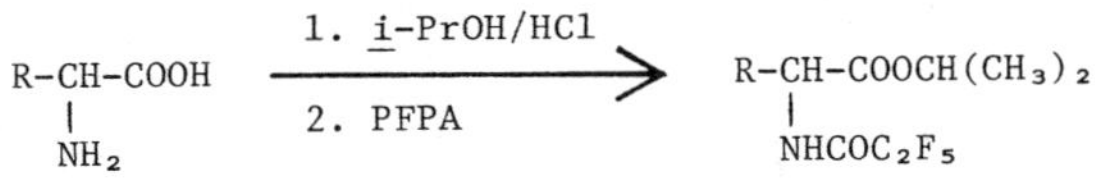

Typical Procedure.[2] Diss 10 mg of the amino acid in 30 ml 1.25 *N* HCl in *i*-PrOH. Heat in a sealed tube at 100° for 3 hr. Evap to dryness under vac. Add PFPA and allow to react at room temp in a sealed tube.[3] Analyze by GC.

Comments. *N*-PFP isopropyl esters were compared with the corresponding *N*-TFA, *N*-HFB, and *N*-pentadecafluorooctanoic isopropyl esters for use in GC resolution of amino acid enantiomers on optically active stationary phases. The PFP derivs were most volatile. The methyl and *t*-butyl esters were also examined.

References

[1]W. Parr *et al., J. Chromatogr.* **50**, 510 (1970).

[2]W. A. Koenig *et al., J. Chromatogr. Sci.* **8**, 183 (1970).

[3]C. W. Gehrke and D. L. Stalling, *Separ. Sci.* **2**, 101 (1967).

Derivative Type. *n*-Propyl ester, *N*-heptafluorobutyryl.

Reagent. *n*-Propanol (*n*-PrOH)/HCl, heptafluorobutyric anhydride (HFBA).

Compound Type. Amino acids.

Reaction

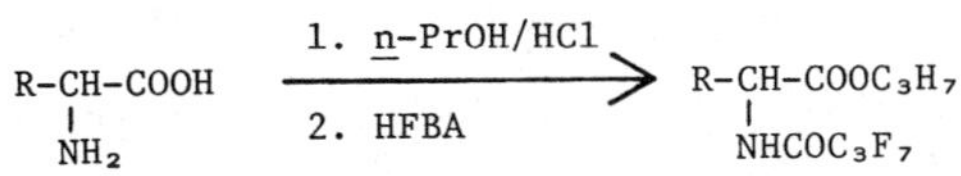

Typical Procedure.[1,2] Esterfy amino acid sample (*ca* 10^{-6} mol each) by heating with 0.4 ml 8 *M* HCl in *n*-PrOH at 100° for 10 min.[3] Evap to dryness under N_2 and repeat esterification and evap. Add 0.2 ml HFBA and 0.1 ml EtOAc. Heat in the sealed tube at 150° for 10 min; then allow to cool to room temp. Evap just to dryness under N_2; then diss residue in 0.1 ml EtOAc. Inject 3 μl of this soln simultaneously with 2 μl Ac_2O into the GC.

Comments. The method permitted GC resolution of all 20 protein amino acids on an OV 1 column in a 43 min GC run. By comparison the 20 *N*-TFA, *n*-propyl derivs were incompletely resolved. Omission of the coinjection of Ac_2O yielded a broad peak for His, probably the mono-HFB *n*-propyl deriv. On-column acylation with the Ac_2O yielded a sharp His deriv peak. The method was later applied to the analysis of hydroxy-Pro and hydroxy-Lys.[4]

References

[1]C. W. Moss *et al., J. Chromatogr.* **60**, 134 (1971).
[2]C. W. Moss and M. S. Lambert, *J. Chromatogr.* **57**, 209 (1971).
[3]J. R. Coulter and C. S. Hann, *J. Chromatogr.* **36**, 42 (1968).
[4]C. W. Moss and M. A. Lambert, *Anal. Biochem.* **59**, 259 (1974).

Derivative Type. *n*-Propyl ester, *N*-heptafluorobutyryl (HFB).

Reagent. *n*-Propanol (*n*-PrOH)/HCl, heptafluorobutyric anhydride (HFBA).

Compound Type. Amino acids.

Reaction

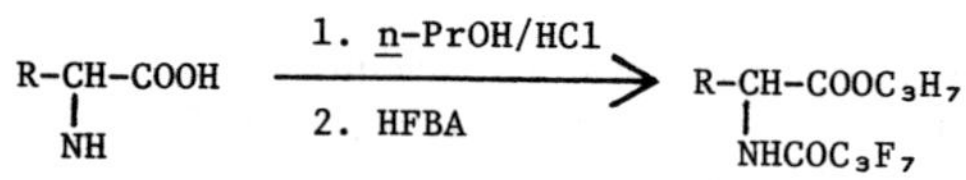

Typical Procedure. To an amino acid HCl mixt (from hydrolysis of *ca* 10 mg protein) add 2 ml 3.5 *M* HCl in *n*-PrOH. Heat at 110° for 30 min. Evap 1 ml of the soln just to dryness under N_2 at 90–100°. Add 0.2 ml HFBA and 0.1 ml CH_3CN. Heat at 150° for 5 min with the tube immersed just to the liquid level in an oil bath. Draw 3 μl Ac_2O into a 10 μl syringe followed by 3 μl of the acylation mixt and inject into the GC.

Comments. The derivs permitted a 25 min GC separation of all the protein amino acids on a 3% SE-30 column. Quantitative results were obtained for all but His. Amino acid sample preparation methods were given for food plant samples. Jönsson *et al.*[2] analyzed the same derivs on glass capillary columns with good quantitative results except for CySH (CySH could be analyzed as S-methyl CySH).

References

[1]M. A. Kirkman, *J. Chromatogr.* **97**, 175 (1974).
[2]J. Jönsson *et al., Anal. Biochem.* **51**, 204 (1973).

Derivative Type. *n*-Propyl ester, *N*-heptafluorobutyryl (HFB).

Reagent. *n*-Propanol (*n*-PrOH)/HCl, heptafluorobutyric anhydride (HFBA).

Compound Type. Amino acids.

Reaction

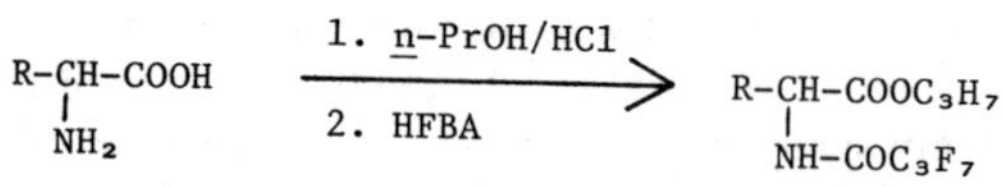

Typical Procedure. Evap amino acid (*ca* 3μmol) sample soln to dryness under N_2 and diss residue in 0.4 ml 6 *M* HCl in *n*-PrOH. Heat at 150° for 3.5 min then cool rapidly. Evap under vac at 35°. Add 0.1 ml of a soln of 2,6-di-*t*-butyl-*p*-cresol (BHT) (0.03%, w/v) in EtOAc and 0.1 ml HFBA. Heat at 150° for 12 min then cool rapidly. Evap at room temp under N_2. Diss residue in 0.05 ml EtOAc and 0.05 ml Ac_2O. Store at −20° under dessication prior to GC analysis.

Comments. The method was used for analysis of the 20 protein amino acids. BHT was used as an antioxidant and was necessary for reliable results. Removal of HFBA was necessary to insure the conversion of the di-HFB deriv of histidine (which elutes at the same position as the glutamic acid deriv) to the mono-HFB deriv.

References

[1]J. F. March, *Anal. Biochem.* **69**, 420 (1975).

5.1.3. N-Acyl, Butyl Ester Derivatives

Derivative Type. *n*-Butyl ester, *N*-trifluoroacetyl.

Reagent. *n*-Butanol (*n*-BuOH)/HCl/dibutoxypropane, trifluoroacetic anhydride (TFAA).

Compound Type. Amino acids.

Reaction

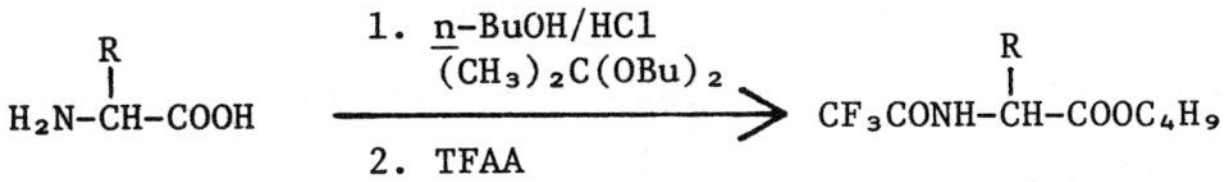

Typical Procedure. Diss up to 10 mg amino acids in 20 ml 5% HCl in *n*-BuOH and 2 ml DMF. Add 3 mol equiv (based on carboxyl groups) of dibutoxypropane to scavenge H_2O and heat under a $CaCl_2$ drying tube for 3 hr at 55–60°. Evap under vac at 60° and neutralize residue with 1 *N* Na_2CO_3. Ext with 3 × 10 ml CH_2Cl_2 and evap under vac. Add 10 ml CH_2Cl_2 and evap again to remove H_2O. Diss residue in 2–5 ml CH_2Cl_2 containing 2% DMF. Add 0.2–0.5 ml TFAA and heat 30 min at 28°. Evap under vac and diss residue in acetone for GC analysis.

Comments. Separation of the derivs of 19 protein amino acids on a single column was demonstrated.

References

[1]C. Zomzely *et al., Anal. Chem.* **34**, 1414 (1962).

Derivative Type. *n*-Butyl, *N*-trifluoroacetyl.

Reagent. Methanol (MeOH)/HCl, *n*-Butanol (*n*-BuOH)/HCl, Trifluoroacetic anhydride (TFAA)

Compound Type. Amino acids.

Reaction

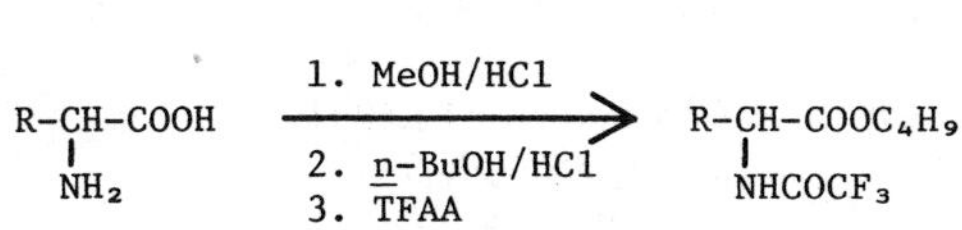

Typical Procedure.[1,2] Diss the amino acid mixt (< 60 mg) in 10 ml 1.2 *N* HCl in MeOH. Stir 30 min at room temp then evap to dryness under vac at 60°. Add 10 ml 1.2 *N* HCl in *n*-BuOH and stir 2.5 hr at 100°. Evap to dryness under vac at 60°. Add 4 ml CH_2Cl_2 and 1.0 ml TFAA. Stir 15 min at room temp. Transfer 1.0 ml of the soln to a heavy wall tube, seal, and heat at 150° for 5 min. Cool and analyze solns directly by GC.

Comments. The method was used for quant analysis of the 20 protein amino acids.[2] Pollack[3] prepared the corresponding PFP and HFB derivs using the same method and showed the retention times to be HFB < PFP < TFA. Comparable separations to those obtained with TFA derivs could be obtained in a shorter time with HFB derivs. In addition, the HFB and PFP derivs had much better ECD sensitivity. Other workers have further studied experimental conditions for use of TFA butyl esters for amino acid analyses in immunochemistry.[4] The derivs have also been used for amino acid analysis by GC-MS-SIM.[5]

References

[1]W. M. Lamkin and C. W. Gehrke, *Anal. Chem.* **37**, 383 (1965).

[2]C. W. Gehrke and D. L. Stalling, *Separation Sci.* **2**, 101 (1967).

[3]G. E. Pollack, *Anal. Chem.* **39**, 1194 (1967).

[4]N. Lachavitzki and B. Bjorklund, *Anal. Biochem.* **38**, 446 (1970).

[5]W. E. Pereira *et al.*, *Anal. Biochem.* **55**, 236 (1973).

Derivative Type. *n*-Butyl ester, trifluoroacetamide.

Reagent. *n*-Butanol (*n*-BuOH)/HCl, trifluoroacetic anhydride (TFAA).

Compound Type. Amino acids.

Reaction

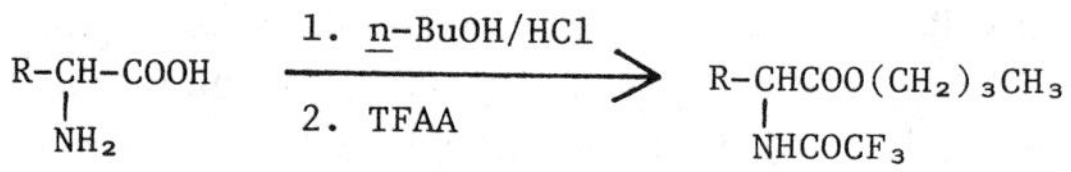

Typical Procedure. Evap the sample soln to dryness under N_2 at 100° in a screw capped tube. For each 100 μg of amino acids add 150 μl of 3 *N* HCl in *n*-BuOH. Mix in an ultrasonic bath and heat 15 min at 100°. Evap to dryness at 100° under N_2 and remove H_2O azeotropically with an added 150 μl CH_2Cl_2. To the dry residue add 60 μl of CH_2Cl_2 and 20 μl TFAA per 100 μg of amino acids. Heat 5 min at 150°, then analyze by GC.

Comments. A 35 min esterfication time was required for Ile; some Trp was destroyed in this time. All 20 protein amino acids could be separated using a mixed phase of 2% OV-17 and 1% OV-210[2] or on 10% Apiezon M.[3] Stearic acid, tranexamic acid, and Orn were used as int stand. GC data have also been reported for derivs of nonprotein amino acids.[4] Sample preparation techniques for protein hydrolyzates, physiological fluids, and other types of samples for use with the deriv method have been reported.[5,6] The deriv has also been used for GC-MS-SIM analysis of 12 amino acids using deuterated int stands.[7] The complete MS of 48 amino acid derivs have been reported.[8]

References

[1]D. Roach and C. W. Gehrke, *J. Chromatogr.* **44**, 269 (1969).

[2]C. W. Gehrke *et al.*, *J. Chromatogr.* **57**, 209 (1971).

[3]C. W. Gehrke and H. Takeda, *J. Chromatogr.* **76**, 63 (1973).

[4]D. J. Casagrande *et al.*, *J. Chromatogr.* **49**, 537 (1970).

[5]F. E. Kaiser *et al.*, *J. Chromatogr.* **94**, 113 (1974).

[6]C. W. Gehrke *et al.*, *J. Agr. Food Chem.* **19**, 605 (1971).

[7]R. E. Summons *et al.*, *Anal. Chem.* **46**, 582 (1974).

[8]K. R. Leimer *et al.*, *J. Chromatogr.* **141**, 121 (1977).

Derivative Type. *n*-Butyl ester, *N*-trifluoroacetyl.

Reagent. *n*-Butanol (*n*-BuOH)/HCl, trifluoroacetic anhydride (TFAA).

Compound Type. Amino acids.

Reaction

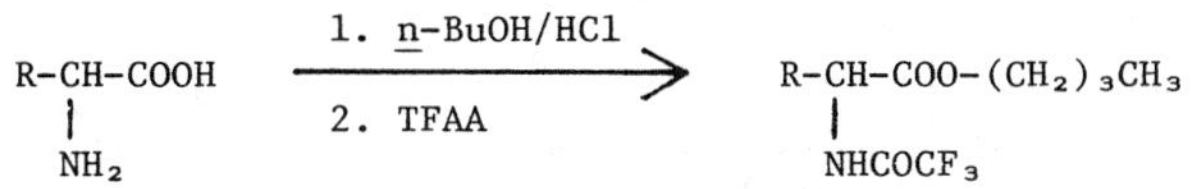

Typical Procedure. Take an aliquot of an aq soln of amino acids (5–100 ng each) and evap to dryness under an IR lamp. Cap vial and allow to cool. Add 50 μl CH_2Cl_2 and evap under the IR lamp to remove the last traces of H_2O azeotropically. Add 100 μl 3 *N* HCl in *n*-BuOH and sonicate 30 sec. Heat in 100° sand bath for 30 min immersing tube just to the liquid level and allowing to reflux in the tube. Allow to cool, open vial, and evap just to dryness under the IR lamp. Add 100 μl of TFAA (0.5–10 μl/100 μl) in CH_2Cl_2. Cap vial, sonicate 30 sec, and heat at 100° in the sand bath for 10–20 min (depending upon wall thickness of the vial used). Allow to cool to room temp and analyze by GC.

Comments. Detection of the derivs to the pg level was achieved using ECD. *N*-Heptafluorobutyryl butyl ester derivs of Met and Cys were prep using HFBA and were detectable at lower levels than the corresponding TFA derivs. The method was used for amino acid analysis of returned lunar samples.

References

[1]R. W. Zumwalt *et al.*, *J. Chromatogr.* **57**, 193 (1971).

Derivative Type. *n*-Butyl ester, *N*-trifluoroacetyl.

Reagent. *n*-Butanol (*n*-BuOH)/HCl, trifluoroacetic anhydride (TFAA).

Compound Type. Amino acids.

Reaction

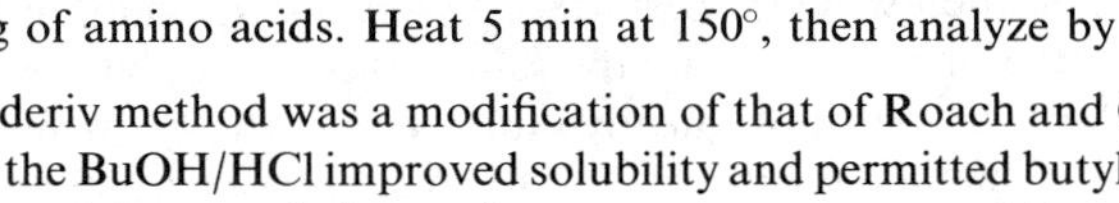

Typical Procedure. To each 100 μg of amino acids add 150 μl of a mixt of 9:1 (v/v) 3.5 *M* HCl in *n*-BuOH and CH_2Cl_2. Heat 5 min at 100° with sonication. Evap to dryness at 100° under N_2, add 150 μl CH_2Cl_2, and evap again. To the residue add 60 μl of CH_2Cl_2 and 20 μl TFAA per 100 μg of amino acids. Heat 5 min at 150°, then analyze by GC.

Comments. The deriv method was a modification of that of Roach and Gehrke.[2] Addition of 10% CH_2Cl_2 to the BuOH/HCl improved solubility and permitted butyl ester formation in 5 min of sonication. A better relative molar response was reported for the hydroxy amino acids, Orn, and Trp. The use of carbonyl free *n*-BuOH was necessary to prevent formation of extraneous GC peaks.

References

[1]P. Cancalon and J. D. Klingman, *J. Chromatogr. Sci.* **12**, 349 (1974).
[2]D. Roach and C. W. Gehrke, *J. Chromatogr.* **44**, 269 (1969).

Derivative Type. N^{α}-Trifluoroacetyl-N^{im}-carbethoxy, *n*-butyl ester.

Reagent. *n*-Butanol/HCl, trifluoroacetic anhydride (TFAA), ethoxyformic anhydride (EFA).

Compound Type. Histidine.

Reaction

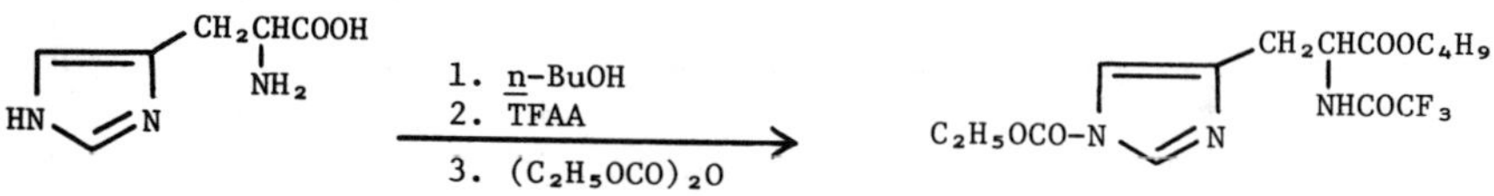

Typical Procedure. First deriv by the procedure of Roach and Gehrke[2] (q.v.). Evap almost to dryness under N_2 to yield the N^{α}-TFA His *n*-Butyl ester.[3] Add 1 ml CH_2Cl_2 and 1 μl EFA. Heat at 150° for 20 min then analyze by GC.

Comments. The method yielded a more stable deriv of His which could be analyzed concurrently with TFA butyl ester derivs of the other amino acids. The EFA step affected the response of Met, Arg, and Tyr and thus the usual analysis[2] had to be run also.

References

[1]I. M. Moodie, *J. Chromatogr.* **99**, 495 (1974).

[2]D. Roach and C. W. Gehrke, *J. Chromatogr.* **44**, 269 (1969).

[3]C. W. Gehrke *et al., J. Chromatogr.* **43**, 311 (1969).

Derivative Type. Isobutyl ester, *N*-heptafluorobutyryl.

Reagent. Isobutyl alcohol (*i*- BuOH)/HCl, heptafluorobutyric anhydride (HFBA).

Compound Type. Amino acids.

Reaction

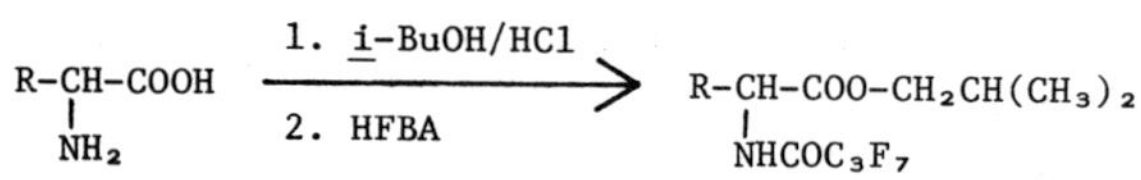

Typical Procedure. Dry a sample containing 0.05–0.25 μmol of each amino acid either by freeze drying or evap under N_2. Remove traces of H_2O azeotropically by evap of added CH_2Cl_2. Diss residue in 200 μl 1.25 *N* HCl in *i*-BuOH and heat at 120° for 20 min. Allow to cool and evap to room temp. Diss residue in 50 μl EtOAc, add 20 μl HFBA and heat 10 min at 150°. Cool to room temp, evap just to dryness and diss residue in EtOAc. Inject an aliquot of the soln along with Ac_2O (sample-Ac_2O ratio 2:1, v/v) into the GC.

Comments. The esterfication was also performed by interesterfication of the methyl esters (formed with 1.25 *N* HCl in MeOH) using the same *i*-BuOH/HCl soln.[1] The initially used esterfication at 110° for 1 hr[1] was subsequently shortened to 120° for 20 min.[2] GC conditions were reported use of this method for Pro and HO-Pro analysis.[3] The structures of the derivs were confirmed by GC-MS.[4]

References

[1]S. L. MacKenzie and D. Tenaschuk, *J. Chromatogr.* **97**, 19 (1974).

[2]S. L. MacKenzie and D. Tenaschuk, *J. Chromatogr.* **111**, 413 (1975).

[3]S. L. MacKenzie and D. Tenaschuk, *J. Chromatogr.* **104**, 176 (1975).

[4]S. L. MacKenzie and L. R. Hogge, *J.Chromatogr.* **132**, 485 (1977).

Derivative Type. Isobutyl ester, *N*-heptafluorobutyryl.

Reagent. Isobutanol (*i*-BuOH)/HCl, heptafluorobutyric anhydride (HFBA).

Compound Type. Amino acids.

Reaction

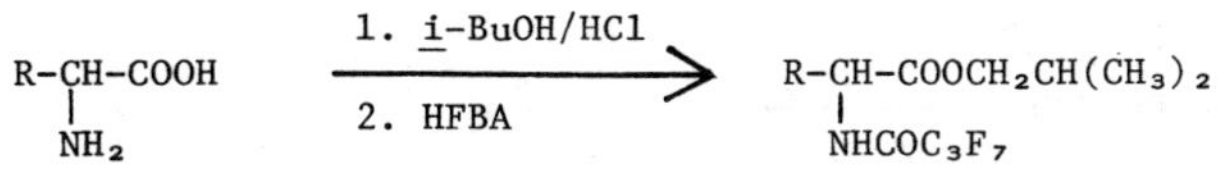

Typical Procedure. Freeze dry sample containing 25–250 nmol amino acids; then add 100 μl CH_2Cl_2 and evap to remove traces of H_2O. Add 100 μl 3 *N* HCl in *i*-BuOH, flush vial with N_2, seal, and heat at 110° for 30 min. Cool to room temp, evap, then remove H_2O with CH_2Cl_2 as above. Add 25 μl EtOAc containing 2,6-di-*t*-butylhydroxytoluene (BHT, antioxidant) in a *ca* equimolar (to each amino acid) amt. Add 10 μl HFBA, flush with N_2, seal, and heat at 110° for 10 min. If His determination is not required, evap almost to dryness and diss residue in 30 μl EtOAc. If His determination is required, evap as above and add 5 μl ethoxyformic anhydride (EFA) and 30 μl EtOAc. Flush with N_2, seal, and heat at 110° for 5 min. Evap almost to dryness and diss residue in 30μl EtOAc. Analyze by GC.

Comments. The method permitted analysis of all the common amino acids on a SCOT column in 30 min. Use of BHT antioxidant and heating with EFA gave a volatile His deriv. HFB derivs have also been prepared with HFBA in CH_3CN in a study reporting GC data for 50 biologically interesting amino acids.[2]

References

[1] R. J. Pearce, *J. Chromatogr.* **136**, 113 (1977).

[2] R. J. Siezen and T. H. Mague, *J. Chromatogr.* **130**, 151 (1977).

Derivative Type. *n*-Butyl ester, *N*-pentafluorobenzoyl.

Reagent. *n*-Butanol (*n*-BuOH)/HCl, pentafluorobenzoyl chloride.

Compound Type. Amino acids.

Reaction

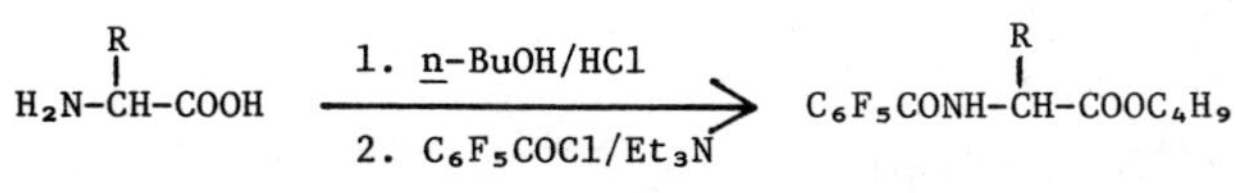

Typical Procedure. For each 100 μg of amino acid add 150 μl of 3 *N* HCl in *n*-BuOH.[2] Mix ultrasonically then heat 15 min at 100°. Evap to dryness at 100° under N_2. Remove traces of H_2O by addition and evap of 150 μl CH_2Cl_2. Add pentafluorobenzoyl chloride and neutralize the mixt with Et_3N. Allow to stand 15 min at room temp then wash with H_2O and dry over Na_2SO_4. Analyze by GC-MS.

Comments. *N*-Pentafluorobenzoyl and *N*-benzoyl amino acid *n*-butyl esters were found to be useful derivs for ultramicrodetermination of amino acids by GC-MS-SIM.[1] MS data were reported for the derivs of Ala, Val, Leu, Pro, and Glu.

References

[1] H. Iwase and A. Murai, *Anal. Biochem.* **78**, 340 (1977).

[2] D. Roach and C. W. Gehrke, *J. Chromatogr.* **44**, 269 (1969).

Derivative Type. *n*-Butyl ester, *N*-trifluoroacetyl-L-prolyl.

Reagent. *n*-Butanol (*n*-BuOH)/HCl, *N*-trifluoroacetyl-L-prolyl chloride.

Compound Type. Amino acids.

Reaction

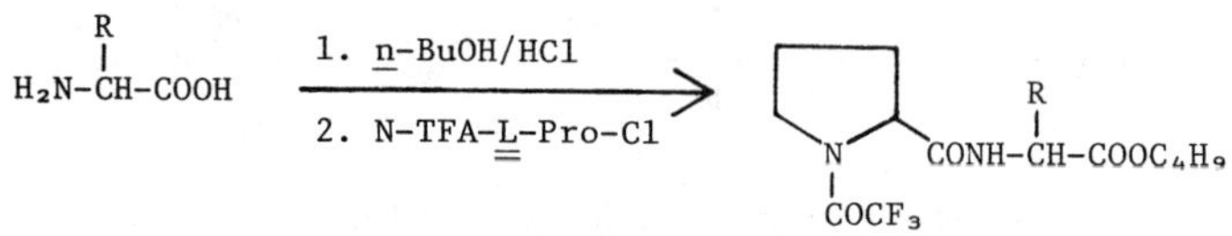

Typical Procedure.[3,4] For each 100 μg amino acids add 150 μl 3 *N* HCl in *n*-BuOH. Mix ultrasonically; then heat at 100° for 15 min. Evap to dryness at 100° under N_2. Remove traces of H_2O by addition and evap of 150 μl CH_2Cl_2. Add 200 μl 0.17 *M* N-trifluoroacetyl-L-prolyl chloride[5] in $CHCl_3$ and 10 μl Et_3N. Allow to stand at room temp then wash with H_2O and dry over Na_2SO_4. Analyze by GC-MS.

Comments. The derivs were examined for their utility for GC-MS-SIM analysis.[1,2] MS data were given for the derivs of 46 amino acids.[1] Hydroxylated amino acids were converted to TMS derivs with BSTFA/TMCS in CH_3CN.[4] The derivs were also used for GC resolution of amino acid enantiomers.[6-9]

References

[1]H. Iwase and A. Murai, *Che. Pharm. Bull.* **25**, 285 (1977).

[2]H. Iwase and A. Murai, *Anal. Biochem.* **78**, 340 (1977).

[3]D. Roach and C. W. Gehrke, *J. Chromatogr.* **44**, 269 (1969).

[4]H. Iwase and A. Murai, *Chem. Pharm. Bull.* **22**, 8 (1974).

[5]J. C. Dabrowiak and D. W. Cooke, *Anal. Chem.* **43**, 791 (1971).

[6]H. Iwase and A. Murai, *Chem. Pharm. Bull* **22**, 1455 (1974).

[7]H. Iwase, *Chem. Pharm. Bull.* **22**, 1663 (1974).

[8]H. Iwase, *Chem. Pharm. Bull.* **22**, 2075 (1974).

[9]H. Iwase, *Chem. Pharm Bull.* **23**, 217 (1975).

Derivative Type. Propyl ester.

Reagent. Diazopropane.

Compound Type. *N*-Acetylaspartic acid.

Reaction

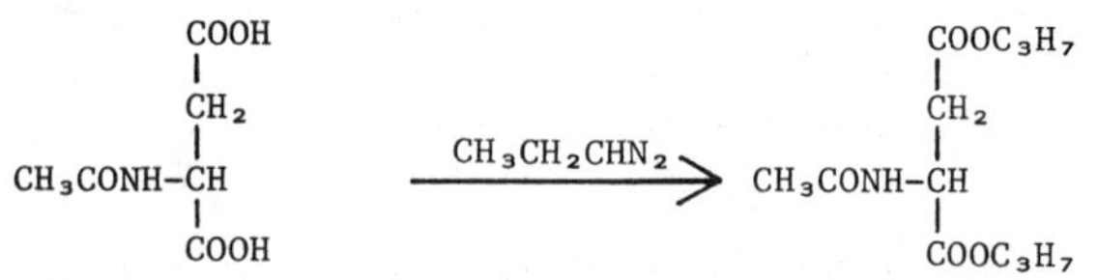

Typical Procedure. Homogenize rat brain in 10 vols 75% aq EtOH (ice cold), maintain 1 hr at 4°, then centrifuge. Evap 2.5 ml of supernatant to dryness under vac and diss residue in 2 ml MeOH. Add 5 ml diazopropane in Et_2O (prepared by adding 50% aq KOH to *N*-propyl-*N'*-nitrosonitroquanidine in Et_2O). After 5 min evap to dryness and diss residue in 0.25 ml dichloroethane. Analyze 2 μl (corresponding to 1.6 mg brain) by GC.

Comments. The propyl esters and butyl esters (prepared from diazobutane[2]) were used to reduce losses incurred in isolation of the more volatile methyl esters.

References

[1]F. Marcucci and E. Mussini, *J. Chromatogr.* **25**, 11 (1966).

[2]E. Mussini and F. Marcucci, *J. Chromatogr.* **26**, 481 (1967).

5.1.4. *N*-Acyl Amyl Ester Derivatives

Derivative Type. Isoamyl ester, heptafluorobutyryl.

Reagent. Isoamyl alcohol (*i*-AmOH)/HCl, heptafluorobutyric anhydride (HFBA).

Compound Type. Amino acids.

Reaction

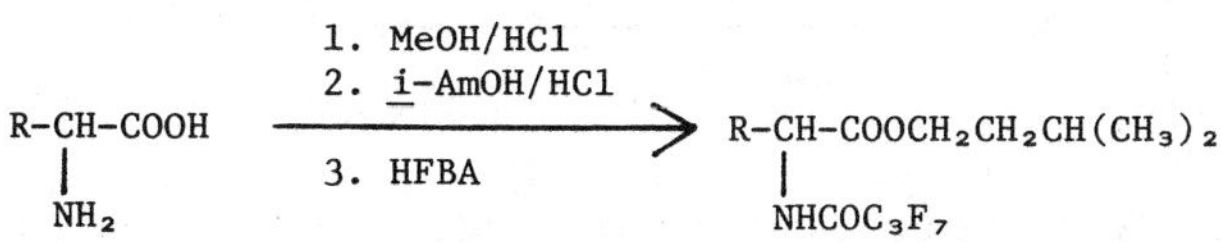

Typical Procedure. Freeze dry a sample containing 1–50 μg of amino acid and diss residue in 200 μl of 1.25 *M* HCl in MeOH. After 30 min at room temp evap to dryness under N_2 at 50°. Diss residue in 200 μl of 1.25 *M* HCl in *i*-AmOH and heat 110° for 150 min. Cool and evap to dryness under N_2 at 80°. Diss residue in 100 μl of CH_3CN and add 20 μl HFBA. Seal and heat at 150° (sand bath) for 10 min. Cool and evap under N_2 until the sample is just dry. Diss in EtOAc and analyze by GC.

Comments. It was possible to conc the sample without loss of the derivs permitting quantitation to 0.1 ng. All of the common protein amino acids could be sepd on an SE-30 column. Pipecolic acid was used as int stand.

References

[1]J. P. Vanetta and G. Vincendon, *J. Chromatogr.* **76**, 91 (1973).

Derivative Type. Isoamyl ester, heptafluorobutyryl.

Reagent. Methanol (MeOH)/acetyl chloride (AcCl), isoamyl alcohol (*i*-AmOH)/AcCl, heptafluorobutyric anhydride (HFBA).

Compound Type. Amino acids.

Reaction

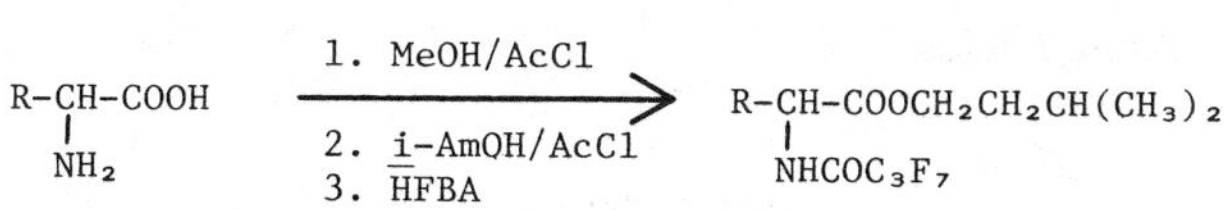

Typical Procedure. Dry amino acid soln (up to 50 μg amino acids) under N_2 at 70° and add 0.4 ml freshly prepared 20% (v/v) AcCl in MeOH. Sonicate 15–20 sec and heat at 70° for 30 min. Dry at 50° under N_2 and add 0.4 ml 20% (v/v) AcCl in *i*-AmOH. Flush vial with N_2, evacuate, sonicate 15–20 sec, and heat at 100° for 2.5 hr. Dry at 70° under N_2 and add 100 μl EtOAc and 20 μl PFBA. Flush vial with N_2, evacuate, sonicate, and heat at 150° for 5 min.

Cool to room temp and carefully evap under N_2 until sides of vial are just wet (critical step). Rediss in EtOAc and analyze by GC.

Comments. His, Cys, CySH, and Trp could not be measured by this method. GC retention data and MS data were reported for derivs of 15 amino acids.

References

[1]P. Felker and R. S. Bandurski, *Anal. Biochem* **67**, 245 (1975).

Derivative Type. Pentyl (*n*-amyl) ester, *N*-trifluoroacetyl.

Reagent. l-Pentanol (*n*-amyl alcohol, *n*-AmOH)/HCl, trifluoroacetic anhydride (TFAA).

Compound Type. Amino acids.

Reaction

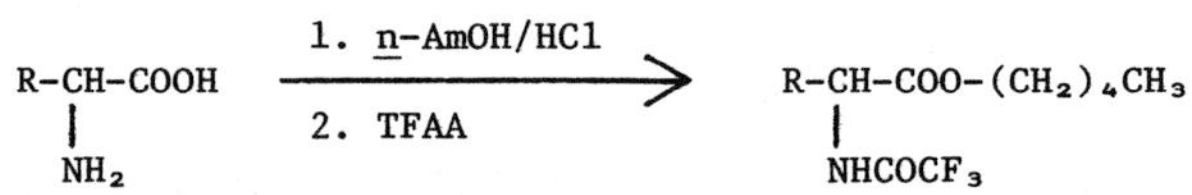

Typical Procedure. Mix 2 mg of amino acid with 2 ml *n*-AmOH and place in an oil bath at 108°. Pass a continuous stream of dry HCl through the alcohol via a Pasteur pipette while heating at 108° for 25 min. Evap to dryness under vac. Add 0.1 ml TFAA and allow to stand at room temp 1 hr. Evap to dryness under vac and diss residue in 100 μl butanone or CH_3NO_2. Analyze by GC.

Comments. GC retention data for the derivs were reported for a large number of stationary phases but complete resolution of the derivs of all the amino acids examined could not be accomplished on a single column. The hydroxy amino acids formed *N,O*-di-TFA derivs and CySH formed an *N,S*-di-TFA deriv.

References

[1]A. Darbre and K. Blau, *J. Chromatogr.* **17**, 31 (1965).

[2]K. Blau and A. Darbre, *J. Chromatogr.* **17**, 445 (1965).

[3]A. Darbre and K. Blau, *Biochem. Biophys. Acta* **100**, 298 (1965).

[4]A. Darbre and K. Blau, *Biochem. Biophys. Acta* **126**, 591 (1966).

[5]K. Blau and A. Darbre, *J. Chromatogr.* **26**, 35 (1967).

[6]A. Darbre and K. Blau, *J. Chromatogr.* **29**, 49 (1967).

Derivative Type. *n*-Amyl ester, *N*-acetyl.

Reagent. *n*-Amyl alcohol (*n*-AmOH)/hydrogen bromide, acetic anhydride (Ac_2O).

Compound Type. Amino acids.

Reaction

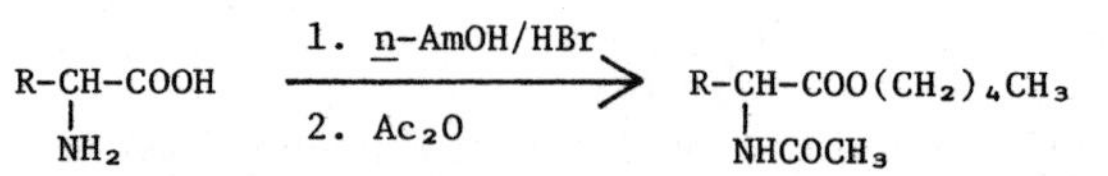

Typical Procedure. Suspend 1–10 mg amino acids in 20 ml *n*-AmOH and sat with anhyd HBr (2–4 min required). Heat in a 165° oil bath (*ca* half the *n*-AmOH dist in 30 min). Evap under vac at 60° and mix residue with 8 ml Ac_2O. Allow to stand at 26° for 5 min. Evap under vac at 60°. Diss residue in *n*-AmOH or benzene and analyze by GC.

Comments. Preliminary studies were done with *n*-butyl, isobutyl, and isoamyl esters also. The method was used to separate 35 amino acids including 18 protein amino acids. Poor yields were obtained for Arg and His. Trp did not give a GC peak. Met was converted to the sulfoxide in the esterification step.

References

[1]D. E. Johnson *et al., Anal. Chem.* **33**, 669 (1961).

5.1.5. Other Ester Derivatives of Protein Amino Acids

Derivative Type. Ethyl ester.

Reagent. Ethanol (EtOH)/HCl.

Compound Type. Amino acids.

Reaction

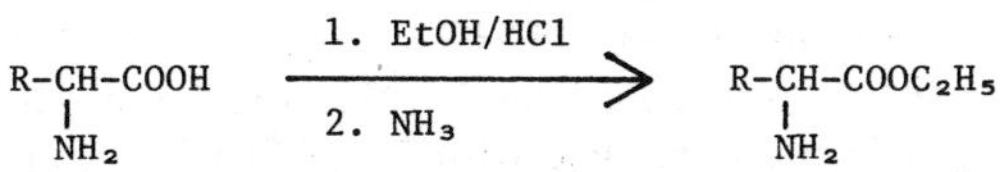

Typical Procedure. Reflux 0.5–5.0 μmol amino acid with 2 ml 2% HCl in EtOH for 2 hr. Evap in a vac dessicator over NaOH. Diss residue in 2 ml CH_2Cl_2 and pass dry NH_3 through the soln. Filter and evap the solvent at 60°. Analyze the residue by MS.

Comments. The MS behavior of the derivs of 24 amino acids were studied.

References

[1]K. Biemann *et al., J. Am. Chem. Soc.* **83**, 3795 (1961).

Derivative Type. *N*-Isobutyloxycarbonyl, methyl ester.

Reagent. Isobutyl chloroformate, diazomethane.

Compound Type. Amino acids.

Reaction

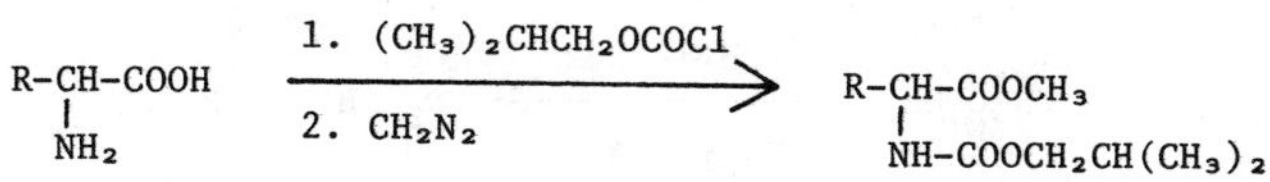

Typical Procedure. Mix aq sample containing up to 2 mg total amino acids with 0.5 ml norleucine int stand soln (100 μg/ml in 0.1% HCl). Add 0.5 ml 10% Na_2CO_3 and dil to 2 ml with H_2O. Add 0.1 ml isobutyl chloroformate and shake 10 min at room temp. Ext with 2 × 3 ml Et_2O and discard exts. Saturate aq phase with NaCl, adjust to pH 1–2 with 10% H_3PO_4 and ext with 3 × 3 ml Et_2O. Add 1 ml MeOH to the combined exts and bubble in CH_2N_2 until the soln becomes yellow. Allow to stand 5 min then evap to dryness under N_2 at 50°. Diss in EtOAc, dry over Na_2SO_4, and analyze by GC.

Comments. All of the common protein amino acids were derivatized except Arg. Arg was converted to Orn by arginase and derivatized. The derivs were resolved by dual column GC. The method was also applied to nonprotein amino acids.[2]

References

[1]M. Makita *et al.*, *J. Chromatogr.* **120**, 129 (1976).
[2]M. Makita *et al.*, *J. Chromatogr.* **124**, 92 (1976).

Derivative Type. Propyl ester, isothiocyanate.

Reagent *n*-Propanol (*n*-PrOH)/HCl, carbon disulfide/triethylamine (Et_3N), methyl chloroformate.

Compound Type. Amino acids.

Reaction

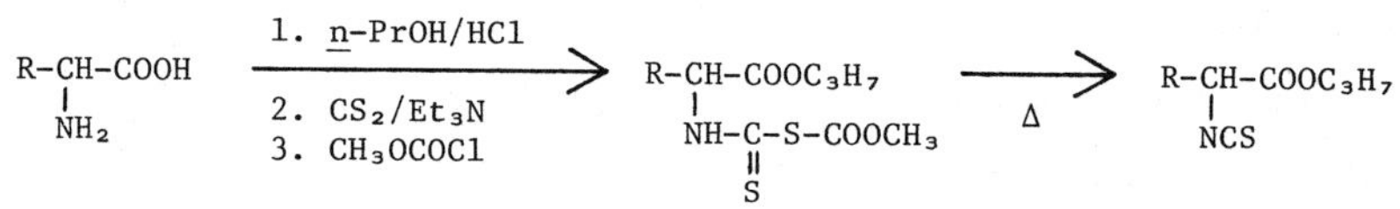

Typical Procedure. Esterfy 0.2–0.5 mg amino acids with 1 ml 25–30% (w/w) HCl in *n*-PrOH at 80°. Evap and repeat. Evap and add 100 μl CH_2Cl_2, 10 μl CS_2, and 8 μl Et_3N at −5°. Allow to stand 1 hr at 20° then cool to −5° and add 2 μl methyl chloroformate. Allow to stand 1 hr at 20° then wash into 1 ml 2% (w/w) aq citric acid with 5 ml CH_2Cl_2. Separate CH_2Cl_2 and wash with 1 ml H_2O. Conc to 0.2 ml and analyze by GC.

Comments. The derivs of all the amino acids (except Leu and Ile) were separated in 60 min on a 5% QF-1 column. Leu and Ile were separable on SE-52 or XE-60. MS data were given for the derivs.

References

[1]B. Halpern *et al.*, *Tetrahedron Lett.* 3119 (1968).

Derivative Type. Isopropyl.

Reagent. 2-Bromopropane/sodium hydride/dimethylsulfoxide (DMSO).

Compound Type. Amines and amino acids.

Reaction

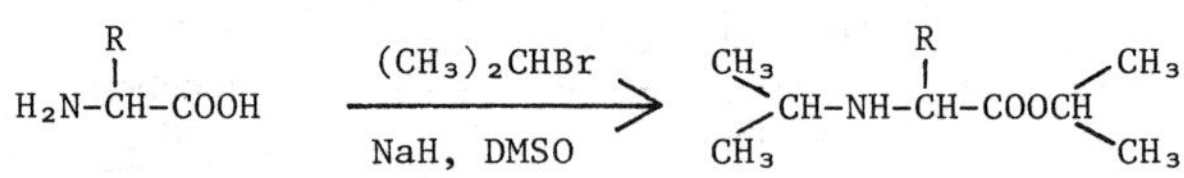

Typical Procedure. Suspend 1–10 mg of amino acid in 2 ml DMSO (dried over 4 A mol sieve). Wash NaH with 4 × 5 ml hexane in a separate tube then add to the DMSO soln. Add excess 2-bromopropane, seal tube, and allow to stand overnight. Add 3 ml satd aq NaCl and ext with $CHCl_3$ (for ECD use benzene). Wash ext 4× with satd NaCl, dry over Na_2SO_4, and analyze by GC.

Comments. Derivs were obtained from all the amino acids (including mono- and diiodotyrosine) except Arg.[1] CySH formed the *N, O, S*-triisopropyl deriv. The method has also been applied to urinary amines and amino acids.[2] Additional alkylation of N was apparently inhibited by the bulkiness of the isopropyl group.

References

[1]B. C. Pettit and J. E. Stouffer, *J. Chromatogr. Sci.* **8**, 735 (1970).

[2]B. Blessington and N. I. Y. Fiagbe, *J. Chromatogr.* **78**, 343 (1973).

Derivative Type. Dimethylaminomethylene, methyl ester.

Reagent. *N, N*-Dimethylformamide (DMF) dimethylacetal.

Compound Type. Amino acids.

Reaction

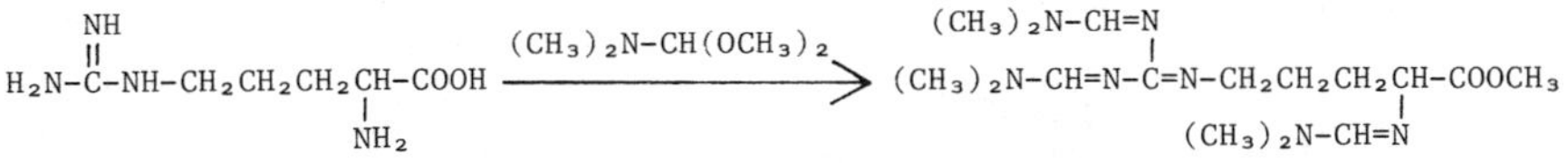

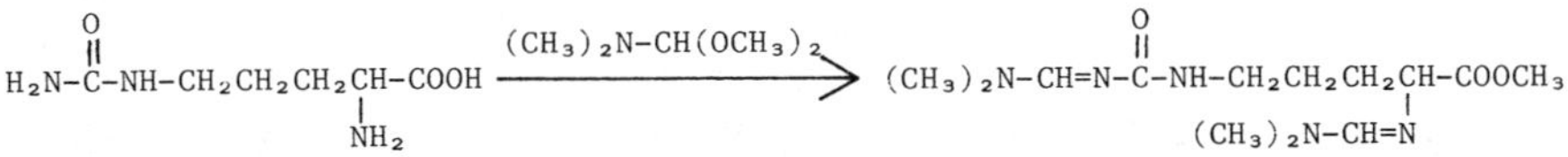

Typical Procedure. To 1–3 mg of amino acid (free base or HCl) add 0.2 ml of DMF dimethyl acetal/CH_3CN/MeOH (3:2:1, v/v/v). Heat at 100° for 15 min. Analyze 1 μl aliquots by direct probe MS.

Comments. The method was used to examine derivs of Arg. Cit, Cys, CysH, His, HOPro, Met, Orn, Pro, Ser, Thr, Trp, and Tyr. It was shown that derivs of amino acids that required further deriv (e.g. TMS) for GC analysis[2] could be analyzed by direct probe MS after this single step deriv. The reagent did not react with alcoholic or phenolic OH, SH, thioether, disulfide, or indole nitrogen. The ethyl, *n*-propyl, and *n*-butyl acetals gave analogous derivs.[2]

References

[1]I. Horman and F. J. Hesford, *Biomed. Mass Spectrom.* **1**, 115 (1974).

[2]J. P. Thenot and E. C. Horning, *Anal. Lett.* **5**, 519 (1972).

Derivative Type. Methyl ester, neopentylidene.

Reagent. Methanol/thionyl chloride, pivaldehyde.

Compound Type. Phenylalanine.

Reaction

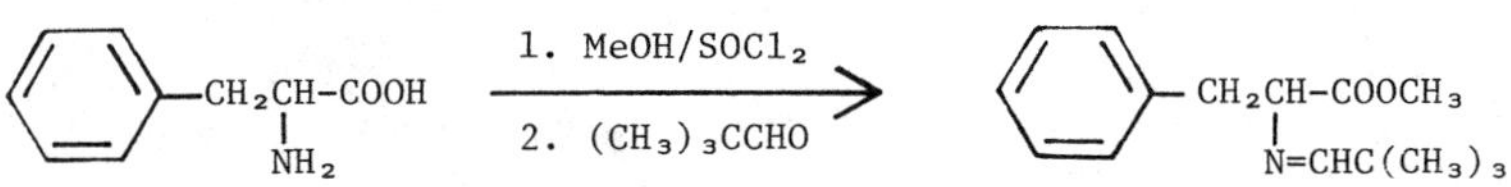

Typical Procedure. Mix 50 μl serum, 5 μl 4 mg/ml aq norleucine int stand, and 200 μl EtOH. Centrifuge, separate supernatant, and evap to dryness under vac. Add 0.5 ml reagent (prepared by adding 1 ml $SOCl_2$ to 10 ml anhyd MeOH at –25°) and heat 15 min at 85°. Evap under vac, add 25 μl MeOH and 5 μl pivaldehyde, and alkalinize (pH 7–8) with 5–10 mg ion exchange resin (HCO_3 form). Add 1 pellet of 3A mol sieve to scavenge H_2O. After 5 min analyze 1–2 μl by GC.

Comments. The method was also suitable for 5 μl serum samples using 1/10 amts of the reagents.

References

[1]E. Jellum *et al.*, *Anal. Biochem.* **31**, 227 (1969).

Derivative Type. Neopentylidene, methyl ester

Reagent. Trimethylphenylammonium hydroxide (TMPAH)/pivaldehyde/methanol (MeOH).

Compound Type. Amino acids.

Reaction

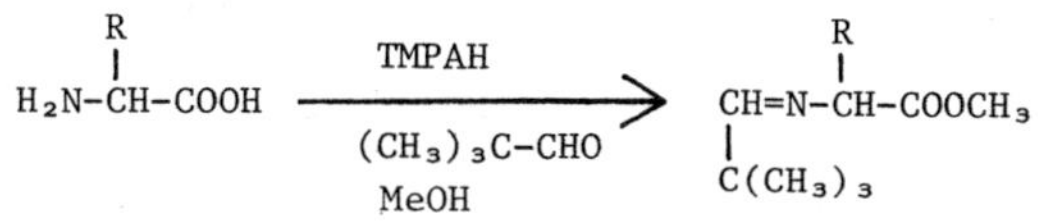

Typical Procedure. Diss a sample containing 0.1 mg of each amino acid in 0.5 ml 0.2 *M* TMPAH in MeOH and add 20 μl pivaldehyde and 3A mol sieve. Heat 15 min at 80°; then analyze by GC.

Comments. Neutral, basic, and acidic amino acids were chromatographed as the neopentylidene methyl esters. The phenolic group of Tyr and the ring nitrogens of Pro and Trp were also methylated. No data were given for Arg or His however.

References

[1]K. M. Williams and B. Halpern, *Anal. Lett.* **6**, 839 (1973).

Derivative Type. Methyl ester, *N*-diethylphosphoryl.

Reagent. Methanol (MeOH)/HCl, diethylchlorophosphate/triethylamine (Et_3N).

Compound Type. Amino acids.

Reaction

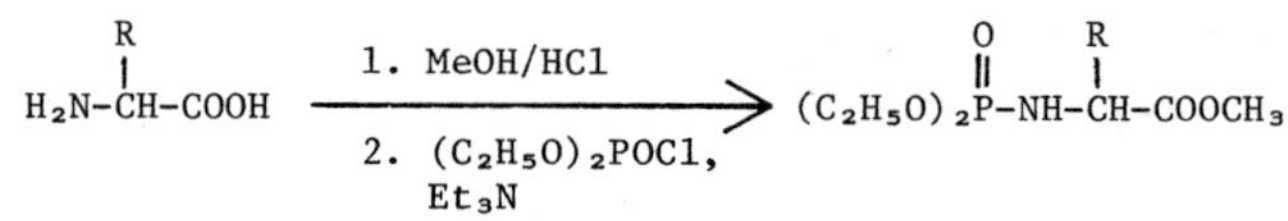

Typical Procedure. Diss 1.5 mg peptide hydrolyzate in 5 ml anhyd MeOH containing 1.2 ± 0.1 meq/ml anhydr HCl and stir 30 min at room temp. Evap to dryness and add 10 mg Et_3N in 1 ml anhyd Et_2O. Sonicate 10 min, add 6 mg diethylchlorophosphate, and allow to stand at room temp 30 min. Analyze an aliquot by GC.

Comments. The derivs were used with an alkali flame detector. The deriv of Ala was detectable to 5 pg.

References

[1]G. Ertingshausen *et al.*, *Sep. Sci.* **2**, 681 (1967).

5.1.6. Trimethylsilyl Derivatives

Derivative Type. Trimethylsilyl.

Reagent. Bis (trimethylsilyl) trifluoroacetamide (BSTFA).

Compound Type. Sulfur amino acids.

Reaction

Typical Procedure. To a 10 mg sample add 0.5 ml BSTFA and 1.5 ml CH_3CN. Heat at 150° for 5 min. Allow to cool and analyze by GC.

Comments. Derivs were prepared from 12 sulfur containing amino acids. BSTFA was the recommended reagent except for Met, Met sulfoxide, and *S*-methyl CySH for which BSA was preferred. Analytical data indicated silylation of NH_2, SO_3H, SH, and COOH groups; both hydrogens on the amino group of taurine were replaced.

References

[1]F. Shahrokhi and C. W. Gehrke, *J. Chromatogr.* **36**, 31 (1968).

Derivative Type. Trimethylsilyl.

Reagent. Bis(trimethylsilyl) trifluoroacetamide (BSTFA).

Compound Type. Amino acids.

Reaction

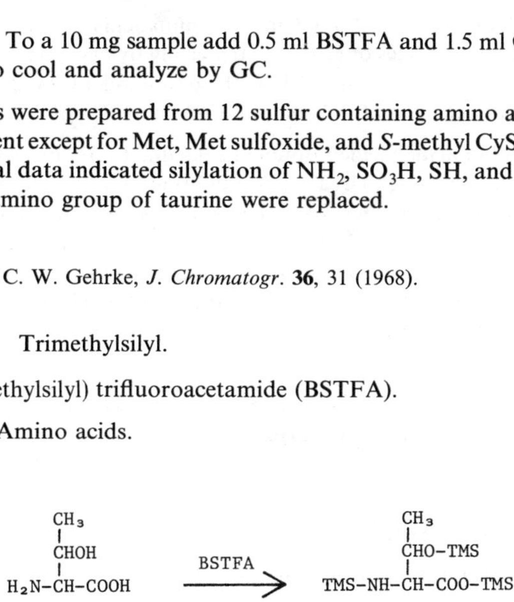

Typical Procedure. Diss 1.0 mg of dry amino acid or HCl salt and 1.0 mg of phenanthrene (int stand) in 0.2 ml CH_3CN and 0.2 ml BSTFA. Cap tube and place in 125°(±1°) oil bath just to the liquid level. Heat 15 min then allow to cool. Analyze aliquot by GC.

Comments. The synthesis of the new silylating agent BSTFA was reported.[1] BSTFA and its by product MSTFA had increased volatility compared to BSA and appeared in the solvent front. The fluorine content helped to reduce SiO_2 deposits in the FID. The method gave single reproducible products for 18 of the 20 protein amino acids. CySH was deriv in 30 min at 150°. No reproducible GC peak was obtained for Arg. Asn was converted to the TMS deriv in 30 min at 150° and Gln in 30 min at 70°. BSTFA and MSTFA, being more volatile than BSA, did not interfere with the GC separation of Ala and Gly. Laseter *et al.*[2] used the method to analyze amino acids in plants and reported MS data for the derivs.

References

[1]D. L. Stalling *et al., Biochem. Biophys. Res. Commun.* **31**, 616 (1968).

[2]J. L. Laseter *et al., Anal. Lett.* **4**, 671 (1971).

Derivative Type. Trimethylsilyl.

Reagent. Bis(trimethylsilyl)trifluoroacetamide (BSTFA).

Compound Type. Amino acids.

Reaction

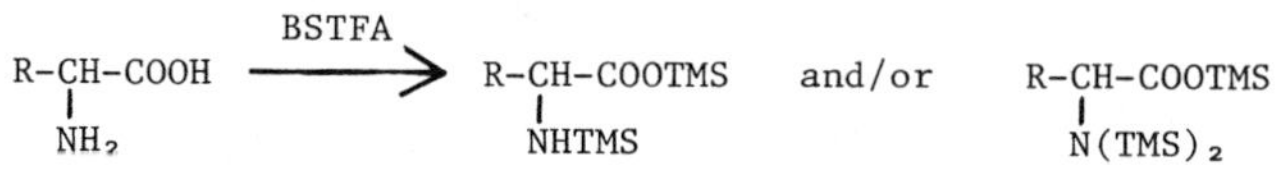

Typical Procedure. Evap aq soln of 0.5–6 mg of amino acids to dryness in a tube under a stream of N_2 at 70°. Remove traces of H_2O azeotropically by two 0.5 ml portions of CH_2Cl_2. For each mg of amino acids add 250 μl of CH_3CN (containing int stand) and 250 μl BSTFA. Seal tube, place in ultrasonic bath for 1 min, and heat at 150° for 2.5 hr. Analyze by GC.

Comments. Some amino acids yielded double peaks due to di- and tri-TMS derivs. TMS derivs of amino acids are unstable to hydrolytic conditions and therefore should be analyzed immediately if possible. In earlier work, 14 protein amino acids were reproducibly derivatized in 15 min at 135°.[2] Further heating at 135° for 4 hr was necessary to silylate Glu (2 TMS), Arg (4 TMS), Lys (4 TMS), His (3 TMS), Tyr (3 TMS), and CySH (4 TMS). Derivs were found to be stable at room temp for several days if kept sealed except that Gly deteriorated from the di-TMS to the tri-TMS within 3 hr.

References

[1]C. W. Gehrke and K. Leimer, *J. Chromatogr.* **57**, 219 (1971).

[2]C. W. Gehrke *et al.*, *J. Chromatogr.* **45**, 24 (1969).

Derivative Type. Trimethylsilyl.

Reagent. Bis(trimethylsilyl) trifluoroacetamide (BSTFA).

Compound Type. Glycine, lysine, arginine.

Reaction

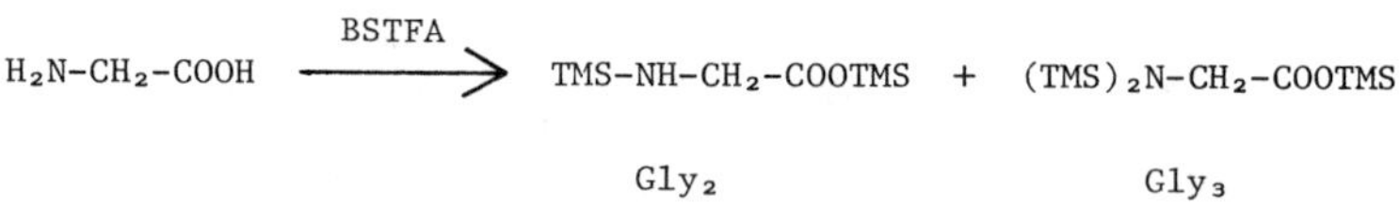

Typical Procedure.[1] Evap 2 ml of a 0.1 mg/ml amino acid soln to dryness under N_2 at 75°. Add 1 ml CH_2Cl_2 and evap under N_2 to remove traces of H_2O. Add 1 ml CH_3CN (or CH_2Cl_2) and 1 ml BSTFA. Heat at 135° for 15 min. Analyze by GC.

Comments. Trimethylsilylation in CH_3CN and other polar solvents yielded multiple derivs for Gly and Lys but a single peak for Arg. Replacement of CH_3CN with CH_2Cl_2 or other nonpolar solvents yielded only Gly_2 and Lys_3. No peak was seen for Arg, however. In samples containing large amts of Gly, the Gly_3 peak interferred with the Ile and Pro peaks. MS data were reported for the Gly and Lys derivs and for the TMS derivs of 6-aminocaproic acid.[2] Butts[3] used BSTFA with 1% TMCS (60°, 16 hr) but still obtained only partially disubstituted amino groups for some amino acids.

References

[1]C. W. Gehrke and K. Leimer, *J. Chromatogr.* **53**, 201 (1970).
[2]K. Bergstrom *et al.*, *Anal. Biochem.* **34**, 74 (1970).
[3]W. C. Butts, *Anal. Biochem.* **46**, 187 (1972).

Derivative Type. Trimethylsilyl.

Reagent. Bis(trimethylsilyl)acetamide (BSA)/trimethylchlorosilane (TMCS)/pyridine.

Compound Type. Amines and amino acids.

Reaction

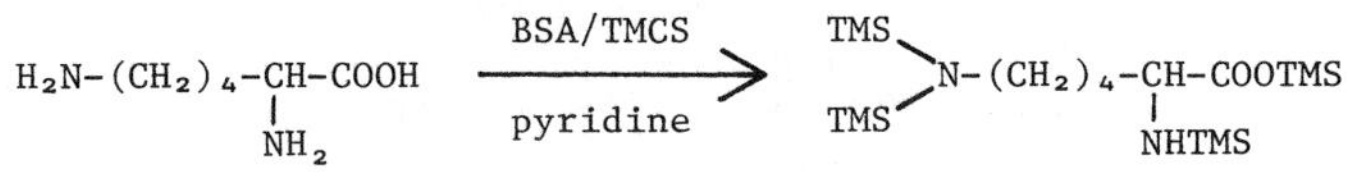

Typical Procedure. Homogenize 10 mg tissue sample in 100 μl cold 1.5 *N* HCOOH/acetone (15:85,v/v) and centrifuge at 2000 g for 10 min. Lyophilize 10 μl of the supernatant and diss the residue in 10 μl BSA/TMCS/pyridine (5:1:6, v/v/v). Cap tube and heat at 60° for 12–14 hr. Analyze 1 μl aliquots by GC-MS-SIM.

Comments. The method permitted detection of the compds to 10–100 fmol levels in micro samples of tissue. SIM of m/e 174 permitted detection of amines and ω-amino acids. SIM of m/e 218 was used for α-amino acids. GC and MS data were reported for the derivs of 43 amines and amino acids. The primary amines and ω-amino groups formed di-TMS derivs while α-amino groups in the α-amino acids (except glycine) formed mono-*N*-TMS derivs.

References

[1]F. P. Abramson *et al.*, *Anal. Biochem.* **57**, 482 (1974).

Derivative Type. Trimethylsilyl.

Reagent. Trimethylsilyldiethylamine (TMSDEA).

Compound Type. Amino acids.

Reaction

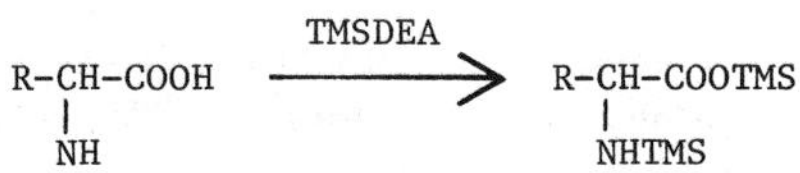

Typical Procedure. Reflux the amino acid with excess reagent for 1 hr under N_2. Analyze by GC.

Comments. TMSDEA was compared with TMSDMA; BSA, BSTFA, *N*-methyl-*N*-TMS acetamide, and TMSI for deriv of Phe, Tyr, and Lys. Under the test conditions the amine reagents gave superior yields to the amide reagents. TMSI was completely ineffective. Use of amino acid HCl salts gave no significant difference in results. HMDS, HMDS/TMCS, and TMSDEA were examined for amino acid deriv in previous work,[2] and the use of TMSDEA for preparative scale synthesis of TMS amino acid derivs (isolated by distillation) had been reported earlier.[3]

References

[1]E. D. Smith and K. L. Shewbart, *J. Chromatogr. Sci.* **7**, 704 (1969).

[2]E. D. Smith and H. Sheppard, *Nature* **208**, 878 (1965).

[3]K. Rühlmann, *Chem. Ber.* **94**, 1876 (1961).

Derivative Type. Trimethylsilyl.

Reagent. *N*-Trimethylsilyl-*N*-methyl acetamide (MSA) or -formamide.

Compound Type. Amino acids, fatty acids, polyols, sugars, amines, phenolalkylamines.

Reaction

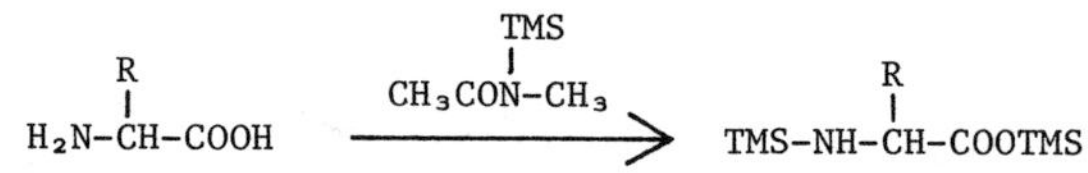

Typical Procedure. Add 1–2 mg of sample to 100 μl MSA and shake 5 min at room temp. If the compd has not diss, heat at 60–100° until soln is complete. Analyze by GC.

Comments. Hydrochlorides of amines and amino acids reacted with the reagents to give TMCS and the *N*-methyl amides.

References

[1]L. Birkhofer and M. Donike, *J. Chromatogr.* **26**, 270 (1967).

Derivative Type. Trimethylsilyl.

Reagent. *N*-Methyl-*N*-trimethylsilyltrifluoroacetamide (MSTFA).

Compound Type. Amines, amino acids, nucleosides.

Reaction

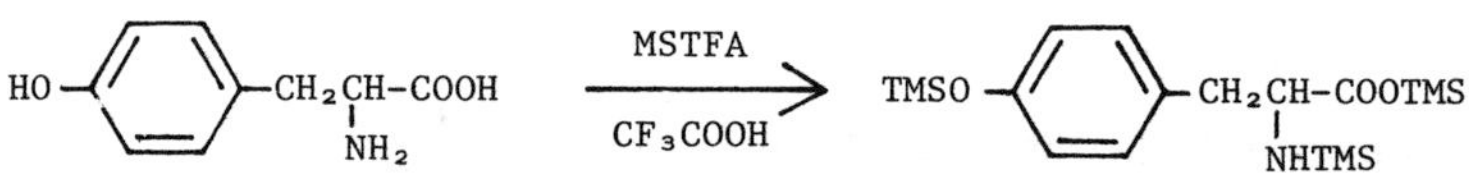

Typical Procedure. Diss 1–2 mg sample in 100 μl CF_3COOH and add dropwise 0.9 ml MSTFA. Analyze by GC.

Comments. The silylation of polar compds with MSTFA was examined in formic, acetic, trifluoroacetic, pentafluoropropionic, and heptafluorobutyric acids as solvents. It was found that silylation was virtually instantaneous in the homogeneous state. Esterification of hydroxyl groups, a possible side reaction, did not occur. Use of these solvents permitted silylation of samples in very small volumes without the necessity of heating. The silylated solvents and reagent were both extremely volatile.

References

[1]M. Donike, *J. Chromatogr.* **85**, 1 (1973).

Derivative Type. Trimethylsilyl.

Reagent. *N*-Methyl-*N*-trimethylsilyltrifluoroacetamide (MSTFA)/trimethylchlorosilane (TMCS)/acetonitrile.

Compound Type. Phenolalkylamines, amino acids, nucleoside bases, and sugars (on TLC plates.

Reaction

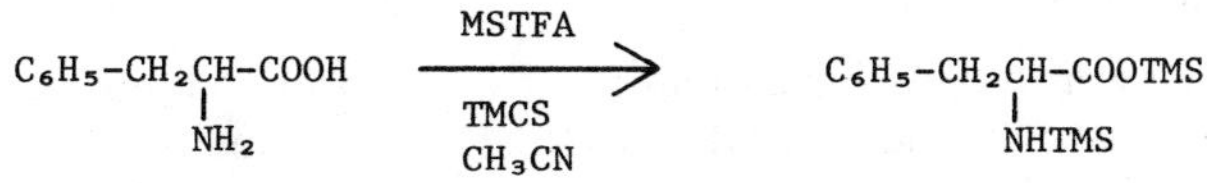

Typical Procedure. Scrape silica gel area containing the sample compd from the TLC plate and add to a soln of 0.5 ml MSTFA, 0.5 ml CH_3CN, and 0.05 ml TMCS. Heat 1 hr at 120°. Centrifuge and analyze an aliquot of the supernatant by GC.

Comments. It was found that the compds could be eluted from the TLC scrapings by the silylation mixt without the necessity of a separate elution step. The recovery of compds with vicinal functional groups that tend to form metal complexes was poor. Removal of contaminating heavy metals such as Fe^{+3} prevented these losses.

References

[1]M. Donike *et al.*, *J. Chromatogr.* **85**, 9 (1973).

Derivative Type. Trimethylsilyl, trifluoroacetyl.

Reagent. *N*-Methyl-*N*-trimethylsilyltrifluoroacetamide (MSTFA), *N*-methyl-bis (trifluoroacetamide) (MBTFA).

Compound Type. Amino acids, phenol alkylamines, hydroxyamines.

Reaction

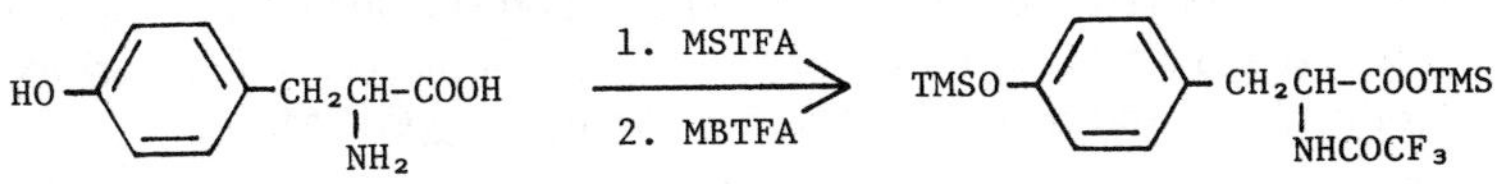

Typical Procedure. Diss 1.8 mg Tyr in 100 μl CF_3COOH containing 20 μg of methyl orange. Add MSTFA until the soln turns yellow (360 μl) then add 10% more MSTFA. Add 20 μl MBTFA. After 20 min analyze by GC.

Comments. Methyl orange was used as an indicator of "trimethylsilylation potential" allowing adjustment to form TMS derivs of hydroxyls and carboxyls while allowing amino groups to form TFA derivs. The indicator also permitted visual determination of sufficient TMS capacity in stored samples to prevent attack of TMS derivs by MBTFA.

References

[1]M. Donike, *J. Chromatogr.* **115**, 591 (1975).

Derivative Type. *n*-Butyl ester, *N*-trimethylsilyl.

Reagent. *n*-Butanol (*n*-BuOH)/HCl, bis (trimethylsilyl) trifluoroacetamide (BSTFA).

Compound Type. Amino acids.

Reaction

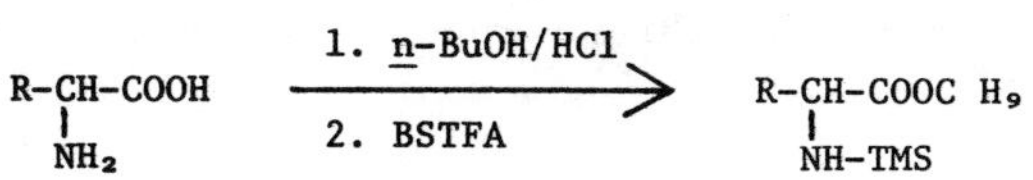

Typical Procedure. Evap 1 ml of a soln of 20 amino acids (2.5 mM each in 1 N HCl) to dryness under N_2 at 50°. Add 3 N HCl in n-BuOH and heat 15 min at 150°. Evap to dryness under N_2 at 50°. Add 0.5 ml BSTFA and 0.5 ml of CH_3CN or CH_2Cl_2 and heat at 150° for 90 min. Analyze 1 μl aliquots by GC.

Comments. Derivs were prepd from 20 amino acids. Gly formed mono- and di-TMS derivs, and Lys and Arg formed di- and tri-TMS derivs. CH_2Cl_2 favored the formation of the less silylated derivs while CH_3CN favored the more silylated deriv. It was observed that significant amts of H_2O were formed in the esterfication by the reaction.

$$\mathrm{BuOH} + \mathrm{HCl} \longrightarrow \mathrm{BuCl} + \mathrm{H_2O}$$

References

[1]J. P. Hardy and S. L. Kerrin, *Anal. Chem.* **44**, 1497 (1972).

Derivative Type. β-Carboline, trimethylsilyl.

Reagent. Formaldehyde, bis (trimethylsilyl) trifluoroacetamide (BSA).

Compound Type. Tryptophan and tryptamines.

Reaction

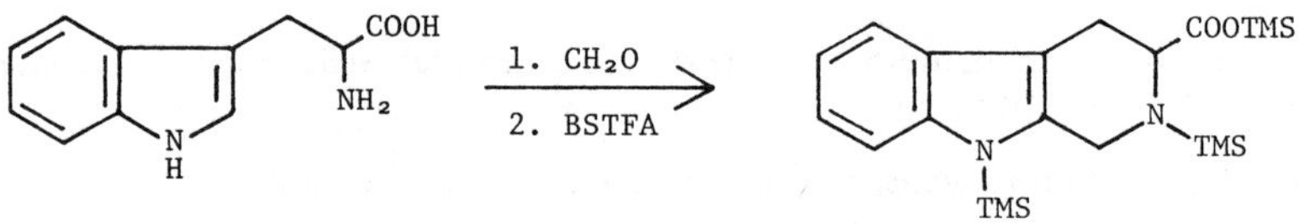

Typical Procedure. Diss Trp in 0.1 N NaOH containing formalin and allow to stand at 40° overnight.[2] Adjust to pH 6.5 by addition of 0.1 N HCl to ppt the β-carboline carboxylic acid. Collect the product, dry, and treat with BSTFA. Analyze by GC-MS.

Comments. The method was also applied to tryptamines.[3] Silylation of the β-carbolines with TMSI yielded only O-TMS derivs while BSA was used to form both O- and N-TMS derivs. Other aldehydes yielded homologous derivs.

References

[1]B. S. Middleditch, *Anal. Lett.* **8**, 397 (1975).

[2]D. G. Harvey *et al.*, *J. Chem. Soc.* 153 (1941).

[3]B. S. Middleditch, *J. Chromatogr.* **126**, 581 (1976).

5.1.7. Thiohydantoin Derivatives

Derivative Type. Phenylthiohydantoin (PTH).

Reagent. Phenylisothiocyanate.

Compound Type. Amino acids.

Reaction

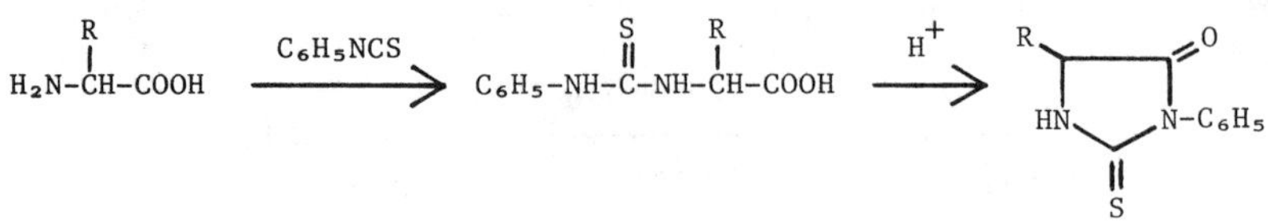

Typical Procedure.[1] Diss 1.5 μmol peptide in 1 ml 60% aq pyridine containing 15 mg phenylisothiocyanate. Warm at 40° for 1 hr then dil with 1 ml H_2O. Ext excess reagent with 4 × 2 ml benzene. Evap aq soln to dryness then add 1.5 ml of 1 : 1 (v/v) 3 *N* HCl–60% HOAc. Warm at 40° for 30 min. Dil with 2 ml H_2O and ext with 2 ml EtOAc and 2 ml benzene. Analyze the extd derivs by HPLC.

Comments. The reaction is the basis of the Edman degradation procedure for sequencing peptides.[2] The methylthiohydantoin (MTH) derivs result from the same reaction using methylisothiocyanate. The PTH and MTH derivs have been analyzed by adsorption,[3,4] reversed-phase,[5] and ion exchange[6] chromatography. Fales *et al.* identified the PTH derivs by CI-MS.[7]

References

[1]J. F. Lawrence and R. W. Frei, *Chemical Derivatization in Liquid Chromatography*, Elsevier, New York, NY, 1976, p. 113.

[2]P. Edman and G. Begg, *Eur. J. Biochem.* **1**, 80 (1967).

[3]G. Frank and W. Strubert, *Chromatographia* **6**, 522 (1973).

[4]A. P. Graffeo *et al., Anal. Lett.* **6**, 505 (1973).

[5]A. Haag and K. Langern, *Chromatographia* **7**, 659 (1974).

[6]V. M. Stepanov *et al., Anal. Biochem.* **43**, 209 (1971).

[7]H. M. Fales *et al., Anal. Biochem.* **43**, 288 (1971).

Derivative Type. *N*-Methyl- or *N*-phenylthiourea.

Reagent. Methyl or phenylisothiocyanate.

Compound Type. Amino acids or peptides.

Reaction

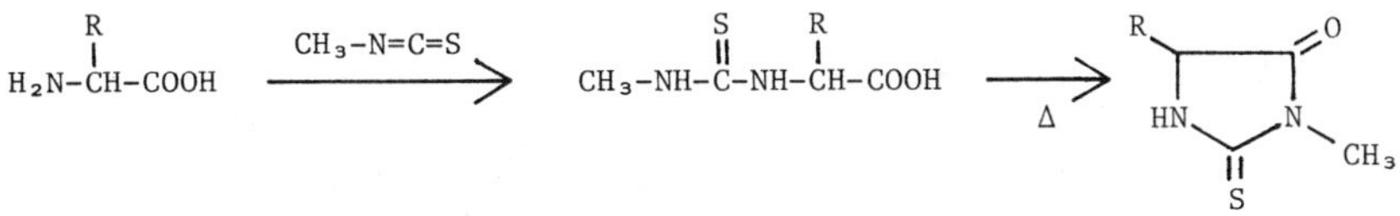

Typical Procedure. Treat amino acid or peptide sample with methyl (or phenyl) isothiocyanate in pyridine/H_2O (80 : 20) for 2 hr at 50°. Evap to dryness under vac and analyze residue by direct probe MS.

Comments. The thiourea derivs apparently formed the thiohydantoins upon heating on the direct probe. The method was shown to be applicable for *N*-terminal analysis of peptides by cleavage of the *N*-terminal amino acid deriv on the probe to give the corresponding thiohydantoin.

References

[1]T. Fairwell *et al., Anal. Biochem.* **53**, 115 (1973).

Derivative Type. Phenylthiohydantoin (PTH).

Reagent. Phenylisothiocyanate.

Compound Type. Phenylalanine and tyrosine.

Reaction

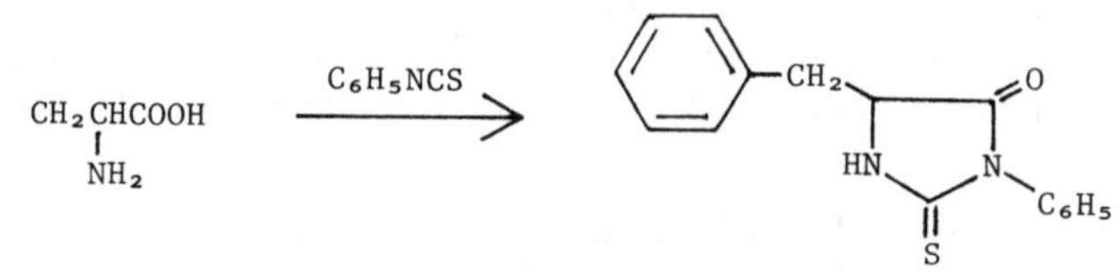

Typical Procedure. Mix 1 ml plasma with 1 ml acetone and centrifuge. Mix clear supernatant with 1 ml triethylammonium acetate (Et_3NHOAc) buffer (see below) and 5 μl phenylisothiocyanate. Seal tube and heat at 40° for 1 hr. Ext with 2 ml benzene and discard ext. Add 1 ml 1 *N* HCl in HOAc, flush tube with N_2, seal, and heat at 80° for 50 min. Ext derivs with 2 × 10 ml CH_2Cl_2. Evap ext to dryness under N_2. Diss residue in 270 μl HPLC solvent (see below), centrifuge, and apply supernatant to a silica HPLC column. Collect the product eluate, evap to dryness, and diss residue in 10 μl acetone for direct probe MS analysis.

Comments. The buffer was prepared by mixing 2 ml 2 *N* HOAc with 1.2 ml Et_3N and diluting with 96.8 ml H_2O/acetone (1 : 1, v/v). The HPLC solvent was prepared by stirring 885 ml CH_2Cl_2, 75 ml EtOH, and 40 ml H_2O for 2 hr, allowing to separate for 2 hr, and mixing 100 ml of the org phase with 275 ml CH_2Cl_2 and 125 ml *n*-hexane. The method was used for determination of deuterium content in plasma Phe and Tyr after administration of labeled Phe.

References

[1]F. K. Trefz, *Clin. Chim. Acta* **73**, 431 (1976).

Derivative Type. Methyl thiohydantoin (MTH).

Reagent. Methyl isothiocyanate.

Compound Type. Amino acids.

Reaction

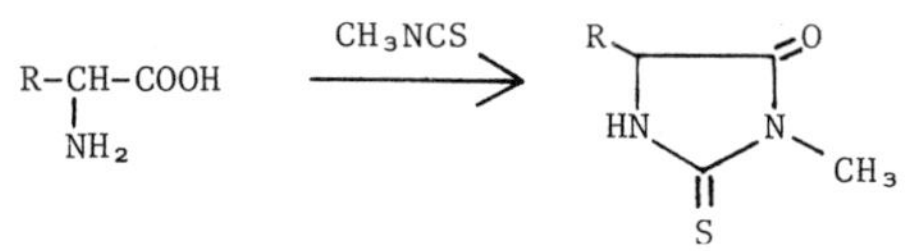

Typical Procedure (Preparative).[2–4] Treat amino acid in H_2O with CH_3NCS at 40° for 15 min maintaining pH 9 by addition of KOH. Adjust to pH 1 with HCl and reflux 10 min to give the MTH derivs. Analyze derivs by GC.[1]

Comments. The MTH derivs of Asp, Ser, Arg, carboxymethyl-CySH, and cysteic acid gave decomposition leading to a common GC peak. Trimethylsilylation of these derivs with BSTFA/TMCS gave unique GC peaks.[1] The MTH deriv of CySH was not successfully analyzed. TMS MTH derivs of all the common protein acids except Arg have been prepared with BSA/CH_3CN (1 : 3, v/v; room temp, 3 min).[5] Other workers[6] used equal parts of CH_3CN and BSTFA with 2% TMCS to prepare TMS derivs of MTH amino acids.

References

[1]J. E. Attrill *et al.*, *Anal. Lett.* **3**, 59 (1970).

[2]V. M. Stepanov and V. F. Krivtsov, *Z. Obshch. Khim.* **35**, 53 (1965); *Chem. Abstr.* **62**, 13137 (1965).

[3]V. F. Krivtsov and V. M. Stepanov, *Z. Obshch. Khim.* **35**, 556 (1965); *Chem. Abstr.* **63**, 669 (1965).

[4]V. M. Stepanov and V. F. Krivtsov, *Z. Obshch. Khim.* **35**, 982 (1965); *Chem. Abstr.* **63**, 9934 (1965).

[5]D. E. Vance and D. S. Feingold, *Anal. Biochem.* **36**, 30 (1970).
[6]J. Eyem and J. Sjöquist, *Anal. Biochem.* **52**, 255 (1973).

Derivative Type. Methylthiohydantoin (MTH).

Reagent. Methyl isothiocyanate, trifluoroacetic acid, HCl.

Compound Type. Amino acids from peptides.

Reaction

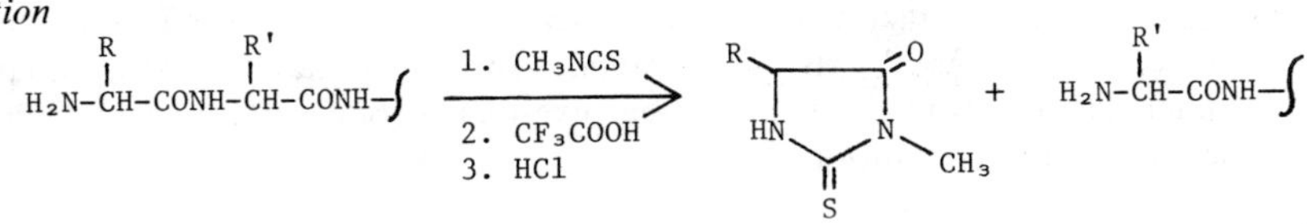

Typical Procedure. Diss 0.5–1.0 μmol peptide in 250 μl 0.4 *M* dimethylallylamine (in 40% aq pyridine adjusted to pH 9.5 with trifluoroacetic acid). Add 10 μl methyl isothiocyanate (melted by warming at 40°) and heat at 50° for 30 min. Ext with 2 × 2 ml benzene and discard ext. Evap traces of benzene under N_2 and lyophilize. Add 50–100 μl CF_3COOH and heat at 50° for 10 min. Evap under N_2 and ext residue with 2 ml dichloroethane. Evap ext to dryness and add 200 μl 1 *N* HCl. Heat at 80° for 10 min. Ext with 3 ml EtOAc and evap ext to dryness. Diss residue in 1–5 μl EtOAc and analyze by GC.

Comments. The method was a modification of the Edman degradation[2] using methyl isothiocyanate in place of phenyl isothiocyanate. The derivs were silylated with BSA (50% in EtOAc, 50°, 10 min) to resolve Ser, Thr, and *S*-carboxymethyl-Cys and to volatilize Asp and Glu. The derivs of all of the amino acids could be resolved sufficiently for quantitative analysis except Arg. The deriv of Arg could be chromatographed after blocking the guanidine group with cyclohexanedione.[3] The MTH derivs have also been used for isotope dilution quantitation by MS using ^{15}N labeled int stand.[4]

References

[1]M. Waterfield and E. Haber, *Biochemistry* **9**, 832 (1970).
[2]P. Edman and G. Begg, *Eur. J. Biochem.* **1**, 80 (1967).
[3]K. Toi *et al.*, *J. Biol. Chem.* **240**, 3455 (1965).
[4]F. F. Richards *et al.*, *Nature* **221**, 1241 (1970).

Derivative Type. *p*-Bromophenylthiohydantoin.

Reagent. *p*-Bromophenylisothiocyanate.

Compound Type. Peptides.

Reaction

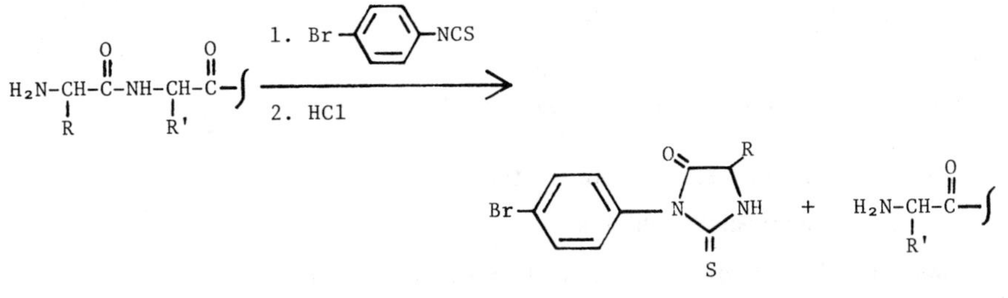

Typical Procedure.[2] To a soln of 1 mmol amino acid (or peptide) in 5 ml pyridine/H_2O (1 : 1, v/v) add 1 *N* NaOH to adjust pH to 9. Add 1.1 mmol bromophenylisothiocyanate and observe consumption of NaOH with an autotitrator. When the reaction is finished, ext 4 × with benzene to remove *N,N′*-bis(bromophenyl)thiourea. Evap soln to dryness and redissolve residue in H_2O. Add 1–1.5 ml 1 *N* HCl and stir 24 hr to convert to the bromophenylthiohydantoin which precipitates. Analyze by MS.

Comments. The method was used in the Edman degradation of peptides. The *p*-bromophenylisothiocyanate reacted almost twice as rapidly as the unbrominated reagent. The bromine isotope doublet facilitated MS identification of the products.[3] Tetradeuterated (phenyl) derivs were used as int stand for quantitative MS analysis.[2] Pentafluorophenylisothiocyanate has also been investigated but was found to hydrolyze rapidly under the reaction conditions.[4]

References

[1]H. Tschesche and E. Wachter, *Eur. J. Biochem.* **16**, 187 (1970).
[2]M. Schneider and H. Tschesche, *Z. Physiol. Chem.* **357**, 1339 (1976).
[3]F. Weygand and R. Obermeier, *Eur. J. Biochem.* **20**, 72 (1971).
[4]R. M. Lequin and H. D. Niall, *Biochim. Biophys. Acta* **257**, 76 (1972).

Derivative Type. *p*-Phenylazophenylthiohydantoin.

Reagent. *p*-Phenylazophenyl isothiocyanate.

Compound Type. Peptides.

Reaction

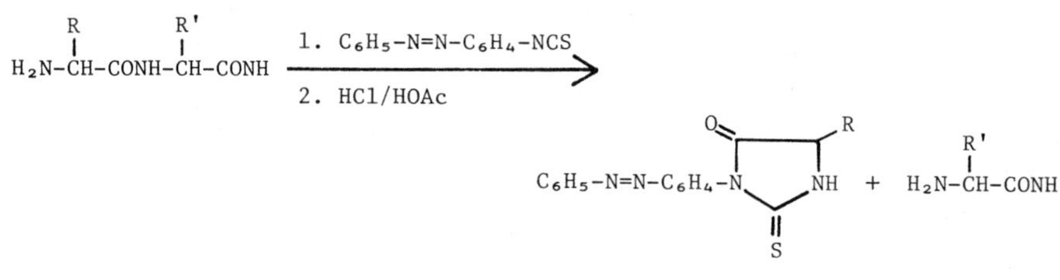

Typical Procedure (Preparative). Diss 1 mmol amino acid or peptide in 3 ml dioxane/H_2O (2 : 1) and adjust to pH 9.7 with 0.1 *N* NaOH. Add 48 mg of the isothiocyanate and stir at 2–3° for 1 hr. After an additional 2 hr standing at 2–3°, ext 6–7 × with benzene; then freeze dry the aq phase. Diss residue in cold 1 : 1 conc HCl and HOAc, store in refrigerator overnight, then freeze dry. Ext residue 2–3 × with EtOAc and evap EtOAc to yield red cryst product.

Comments. The method was used to obtain colored thiohydantoin derivs of *N*-terminal amino acids. HCl was omitted in the cyclization of tryptophan thiohydantoin. The method could be useful for HPLC analysis.

References

[1]S. Datta *et al.*, *Biochem. Biophys. Res. Commun.* **72**, 1296 (1976).

Derivative Type. Trimethylsilyl phenylthiohydantoins.

Reagent. Bis(trimethylsilyl)acetamide (BSA).

Compound Type. Phenylthiohydantoin (PTH) derivs of amino acids.

Reaction

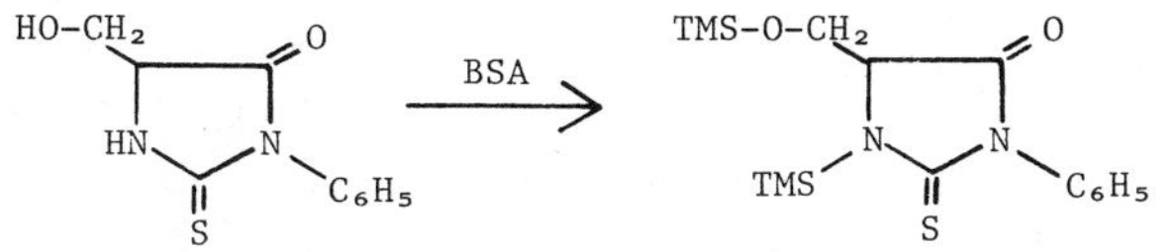

Typical Procedure. Diss sample in BSA and heat at 60° for 2 min. Analyze by GC or GC-MS.

Comments. The derivs of 17 PTH amino acids were examined.[1] GC-MS showed that di-TMS derivs were formed from PTH-Ser (shown above),-Thr, - Glu, -Gln, -Tyr, -His, and -Trp.[1] The MS data indicated that the PTH ring contained an N-TMS group in the deriv.[1] Guerin and Shults used BSTFA to prepare TMS derivs of 26 PTH amino acids and suggested the use of sulfur specific photometric detection.[2] It was reported that PTH derivs could be analyzed directly by GC except for Asp, Glu, and cysteic acid which required TMS formation.[3] The PTH derivs of Ser, Thr, Lys, and *S*-carboxymethyl-CySH showed improved response after TMS deriv. Methylthiohydantoin derivs were also analyzed by GC.[4]

References

[1]R. E. Harman *et al., Anal. Biochem.* **25**, 452 (1968).

[2]M. R. Guerin and W. D. Shults, *J. Chromatogr. Sci.* **7**, 701 (1969).

[3]J. J. Pisano and T. J. Bronzert, *J. Biol. Chem.* **244**, 5597 (1969).

[4]J. J. Pisano *et al., Anal. Biochem.* **45**, 43 (1972).

Derivative Type. Trimethylsilyl thiohydantoins.

Reagent. Bis(trimethylsilyl)trifluoroacetamide (BSTFA).

Compound Type. Thiohydantoins of amino acids.

Reaction

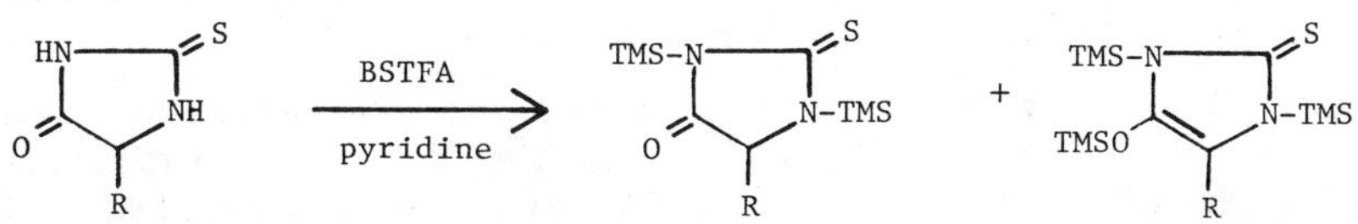

Typical Procedure. To 1 mg thiohydantoin in a 2.5 ml tube with a glass stopper add 500 μl BSTFA/pyridine (1:1, v/v). Heat at 50° for 10 min. Inject 1 μl aliquots directly into GC.

Comments. All of the derivs except the Arg deriv gave satisfactory GC peaks, but some gave two peaks. Identification of the derivs was done on OV-17 or Dexsil 300. MS data were given.

References

[1]A. Rangarajan *et al., J. Chromatogr.* **87**, 499 (1973).

Derivative Type. Acetyl or trifluoroacetyl phenylthiohydantoins.

Reagent. Acetic anhydride (Ac_2O) or trifluoroacetic anhydride (TFAA).

Compound Type. Phenylthiohydantoin (PTH) amino acids.

Reaction

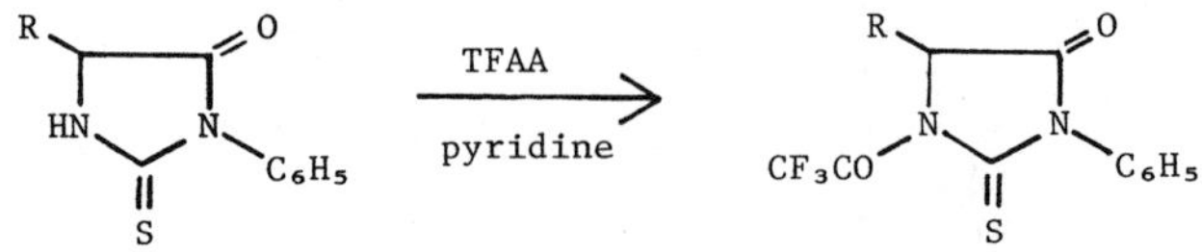

Typical Procedure. Diss 2 mg PTH amino acid in 2 ml $CHCl_3$ and add 0.2 ml Ac_2O or TFAA. Let stand 30 min at room temp and analyze by GC.

Comments. GC data were reported for the derivs of Ala, Gly, Val, Leu, and Ile (not sepd), Met (acetyl deriv only), and Phe.[1] The structures of the derivs were not established. Acetyl derivs have also been prepared in Ac_2O/pyridine (4:1, v/v); most of the amino acids required a 6–14 hr reaction time but 3 hr–6 days were needed for His and Arg.[2]

References

[1]B. L. Brian *et al.*, *J. Chromatogr.* **63**, 386 (1971).

[2]A. S. Inglis and P. W. Nicholls, *J. Chromatogr.* **86**, 117 (1973).

Derivative Type. Trifluoroacetyl (TFA) phenylthiohydantoins.

Reagent. Trifluoroacetic anhydride (TFAA).

Compound Type. Amino acid phenylthiohydantoin (PTH) derivs.

Reaction

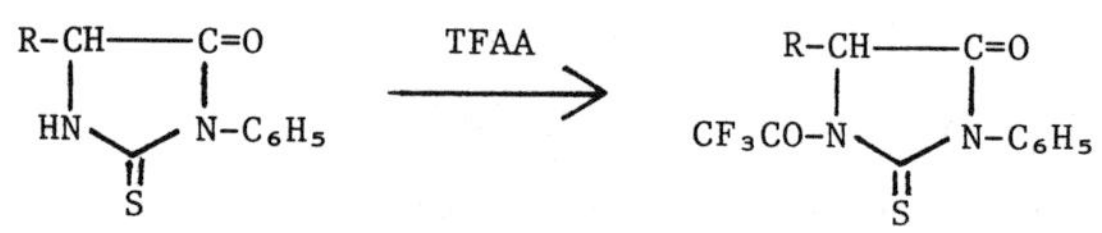

Typical Procedure. Diss 3 mg of each PTH amino acid in 2 ml CH_2Cl_2 and add 0.3 ml TFAA. Allow to stand at room temp 30 min then analyze by GC.

Comments. The derivs were examined as a possible means of enabling complete analysis of all the amino acid PTH derivs as the more volatile TFA PTH derivs. Derivs were prepared from PTH-Gly, Ala, Val, Leu, Ile, and Pro. All were GC separable except Leu and Ile. The PTH derivs alone had been analyzed by GC earlier[2] but the derivs of Ser, Thr, Asn, Gln, and the basic amino acids presented problems.

References

[1]G. Roda and A. Zamorani, *J. Chromatogr.* **46**, 315 (1970).

[2]J. J. Pisano *et al.*, *Biochem. Biophys. Res. Commun.* **7**, 82 (1962).

5.1.8. Nitrophenyl Derivatives

Derivative Type. Methyl ester.

Reagent. Diazomethane.

Compound Type. Dinitrophenyl (DNP) amino acids.

Reaction

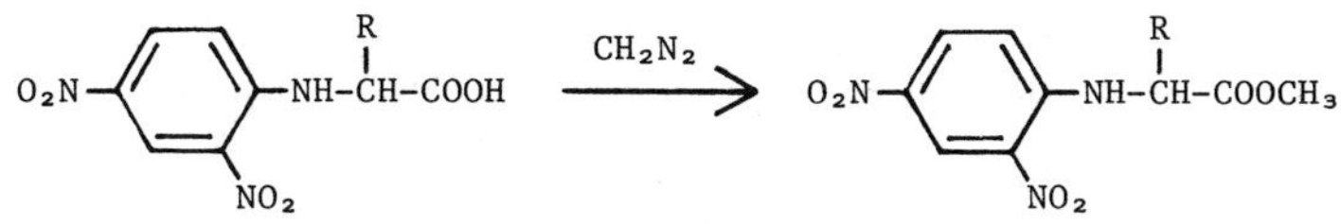

Typical Procedure. Treat DNP amino acid with CH_2N_2 in Et_2O and MeOH as needed for solubility. Evap soln, redissolve residue in MeOH, and analyze by GC.

Comments. The ECD sensitivities of DNP amino acid methyl esters were reported. The investigators used commercially available DNP amino acids rather than derivatizing free amino acids.

References

[1]R. A. Landowne and S. R. Lipsky, *Nature* **199**, 141 (1963).

Derivative Type. Dinitrophenyl (DNP), methyl ester, trimethylsilyl ether.

Reagent. Dinitrofluorobenzene (DNFB), diazomethane, hexamethyldisilazane (HMDS)/trimethylchlorosilane (TMCS).

Compound Type. Amino acids.

Reaction

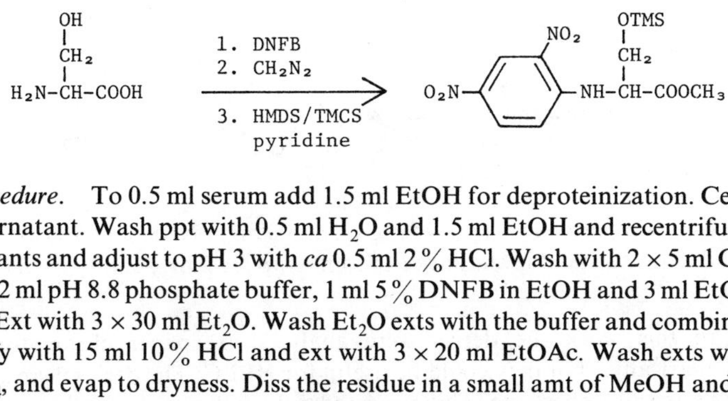

Typical Procedure. To 0.5 ml serum add 1.5 ml EtOH for deproteinization. Centrifuge and remove supernatant. Wash ppt with 0.5 ml H_2O and 1.5 ml EtOH and recentrifuge. Combine the supernatants and adjust to pH 3 with *ca* 0.5 ml 2% HCl. Wash with 2 × 5 ml $CHCl_3$. to the aq layer add 2 ml pH 8.8 phosphate buffer, 1 ml 5% DNFB in EtOH and 3 ml EtOH. Shake at 40° for 2 hr. Ext with 3 × 30 ml Et_2O. Wash Et_2O exts with the buffer and combine buffer and wash. Acidify with 15 ml 10% HCl and ext with 3 × 20 ml EtOAc. Wash exts with H_2O, dry over Na_2SO_4, and evap to dryness. Diss the residue in a small amt of MeOH and add CH_2N_2 in Et_2O. Evap to dryness and add 0.5 ml pyridine, 0.2 ml HMDS, and 0.1 ml TMCS. Shake and allow to stand at room temp 20 min. Evap under N_2 and diss residue in THF. Analyze by GC.

Comments. Methyl ester DNP derivs of Ala, Gly, Val, Leu, Ile, Pro, Asp, Glu, Met, and Phe were separated by GC. Thr, Ser, and HO-Pro required trimethylsilylation.

References

[1]N. Ikekawa *et al., Anal. Biochem.* **17**, 16 (1966).

Derivative Type. 2, 4-Dinitrophenyl (DNP).

Reagent. 2, 4-Dinitrofluorobenzene (DNFB).

Compound Type. Amino acids.

Reaction

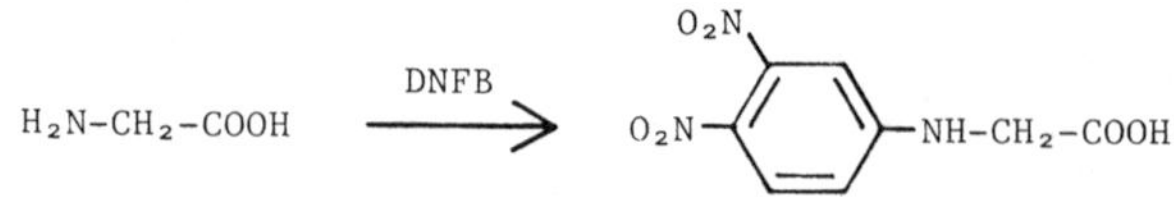

Typical Procedure (Preparative). Diss 1 g $NaHCO_3$ in 10 ml H_2O and add 0.5 g amino acid. Add 20 ml DMF, DMSO, or DMAC followed by 1 ml DNFB and stir at room temp 15 min. Transfer to a separatory funnel with 100 ml H_2O. Ext with 2×20 ml Et_2O to remove excess DNFB. Acidify aq layer with 6 *N* HCl to the congo red end point. Chill if necessary to induce crystallization and collect product by filtration.

Comments. The method showed a shorter reaction time and increased yield of deriv in aprotic solvents compared to EtOH. Pisano *et al.*[2] have reported GC analysis of DNP amino acids.

References

[1]J. A. Vinson and L. D. Pepper, *Anal. Chim Acta* **58**, 245 (1972).

[2]J. J. Pisano *et al., Biochem. Biophys. Res. Commun.* **7**, 82 (1962).

Derivative Type. 2, 4, 6-Trinitrophenyl (TNP).

Reagent. 2, 4, 6-Trinitrobenzenesulfonic acid (TNBS).

Compound Type. Amino acids and peptides.

Reaction

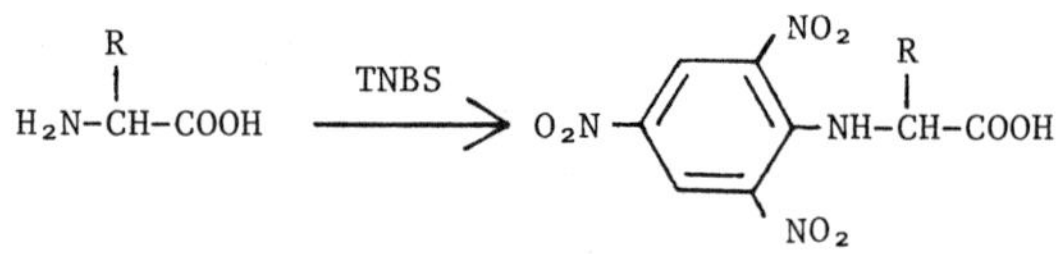

Typical Procedure (Preparative). Diss 75 mg Gly, 200 mg TNBS, and 200 mg $NaHCO_3$ in 10 ml H_2O and allow to stand in the dark at room temp for 2 hr. Acidify to pH 1 and collect the deriv ppt.

Comments. In this work the derivs were analyzed by paper chromatography[1] and spectrophotometrically,[2] but they could be useful for HPLC. The derivs were also prepared from 2, 4, 6-trinitrochlorobenzene. TNBS reacted only with primary amino groups and not with H_2O, phenolic hydroxyls, nor imidazole imino groups. The derivs could be cleaved to the parent amino acids or peptides with ammonia.[2]

References

[1]T. Okuyama and K. Satake, *J. Biochem.* **47**, 454 (1960).

[2]K. Satake *et al., J. Biochem.* **47**, 654 (1960).

[3]K. Satake *et al., J. Biochem.* **50**, 6 (1961).

5.1.9. Oxazolinone and Oxazolidinone Derivatives

Derivative Type. Trifluoromethyloxazolin-5-one.

Reagent. Trifluoroacetic anhydride (TFAA).

Compound Type. Amino acids.

Reaction

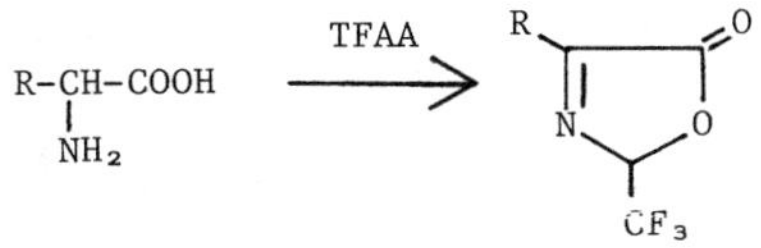

Typical Procedure. Mix 10 mg of amino acids with 2 ml TFAA and heat in a sealed tube at 150° for 10 min. Cool to room temp and evap under vac. Diss residue in EtOAc and analyze by GC.

Comments. Derivs were prepared from 10 amino acids and separated by GC.[1] The derivs were more volatile than the corresponding *N*-TFA, *n*-butyl esters. The 2-H, 2-methyl, and 2-phenyloxazolin-5-one derivs were prepared by refluxing the *N*-formyl, -acetyl, or -benzoyl derivs with dicyclohexylcarbodiimide.[2] Trifluoromethyloxazolin-5-one derivs had been reported earlier.[3]

References

[1]O. Grahl-Nielsen and E. Solheim, *J. Chem. Soc. Chem. Commun.*, 1092 (1972).
[2]O. Grahl-Nielsen and E. Solheim, *J. Chromatogr.* **69**, 366 (1972).
[3]F. Weygand *et al.*, *Ann.* **658**, 128 (1962).

Derivative Type. Oxazolin-5-one.

Reagent. Dicyclohexylcarbodiimide (DCC).

Compound Type. Acyl leucine.

Reaction

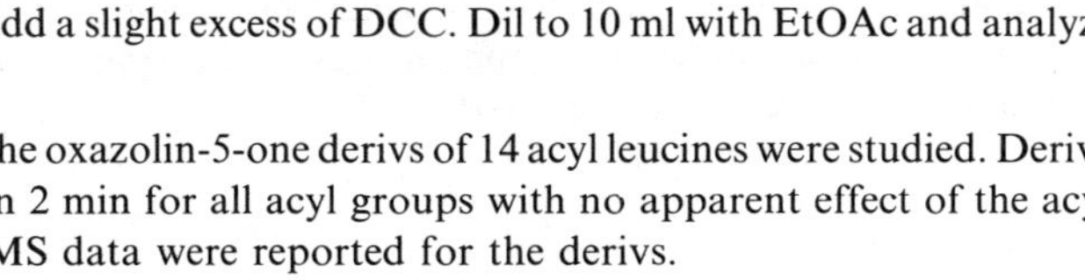

Typical Procedure. Diss *ca* 50 μmol of acyl Leu (prepared by acyl chloride treatment of Leu) in EtOAc and add a slight excess of DCC. Dil to 10 ml with EtOAc and analyze 1 μl by GC or GC-MS.

Comments. The oxazolin-5-one derivs of 14 acyl leucines were studied. Deriv formation was complete within 2 min for all acyl groups with no apparent effect of the acyl group on the reaction rate. MS data were reported for the derivs.

References

[1]O. Grahl-Nielsen and S. Solheim, *J. Chromatogr.* **105**, 89 (1975).

Derivative Type. 2, 2-Di (difluorochloromethyl) oxazolidin-5-one.

Reagent. 1, 3-Dichloro-1, 1, 3, 3-tetrafluoroacetone (DCTFA).

Compound Type. Amino acids.

Reaction

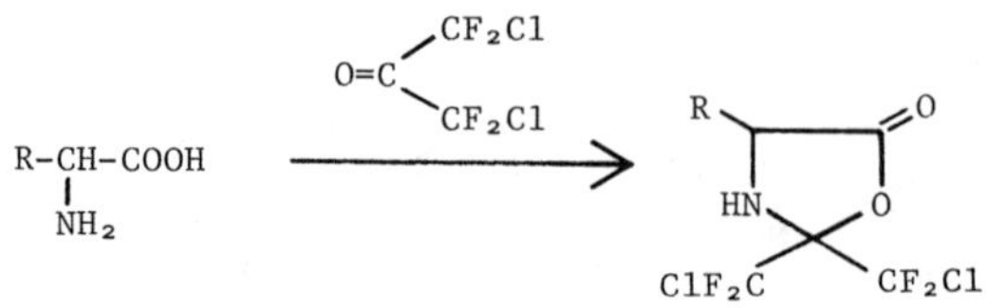

Typical Procedure. Diss 1–3 μg of amino acid in 120 μl of 1% pyridine in CH_3CN. Add 30 μl of DCTFA and heat at 50° for 5 min. Evap to dryness and diss residue in EtOAc. Analyze by FID-GC or ECD-GC.

Comments. Derivs have been prepared from all 20 protein amino acids and the iodoamino acids. Hydroxylated aromatic amino acids were further derivatized with TFAA, HFBA, Ac_2O, TMAA, or HMDS in the same reaction medium. The oxazolidine derivs showed good ECD response.

References

[1]P. Husek, *J. Chromatogr.* **91**, 475 (1974).

[2]P. Husek, *J. Chromatogr.* **91**, 483 (1974).

Derivative Type. 2, 2-Bis (trifluoromethyl)-5-oxazolidones.

Reagent. Hexafluoroacetone.

Compound Type. Amino acids.

Reaction

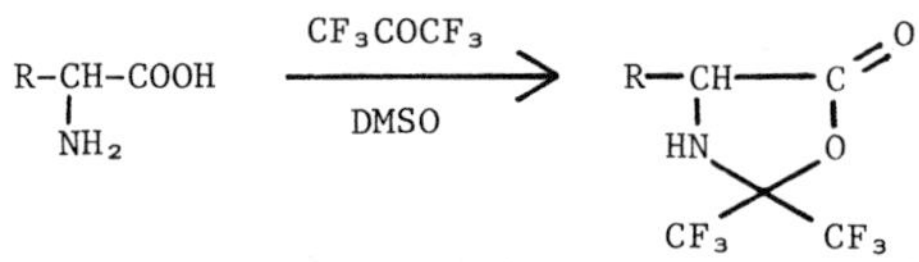

Typical Procedure (Preparative). Diss 20–30 mmol amino acid in 15–20 ml DMSO and slowly add hexafluoroacetone under anhyd conditions. Warm 20–60 min with stirring until the amino acid dissolves. Stir 4 hr then distil off excess reagent. Add soln to H_2O and ext with CH_2Cl_2. Wash ext 2 × with H_2O, dry over Na_2SO_4, and evap to yield deriv. Analyze by GC.

Comments. GC separation of the derivs of 7 amino acids was demonstrated. Physical data were reported on the derivs of 26 amino acids and analogs. The preparation of derivs from *sym*-dichlorotetrafluoroacetone was also reported.

References

[1]F. Weygand *et al., Chem. Ber.* **99**, 1461 (1966).

5.1.10. Chromogenic Derivatives

Derivative Type. 5-Dimethylaminonaphthalene-l-sulfonyl (DANS).

Reagent. 5-Dimethylaminonaphthalene-l-sulfonyl chloride (DANS-Cl).

Compound Type. Amino acids.

Reaction

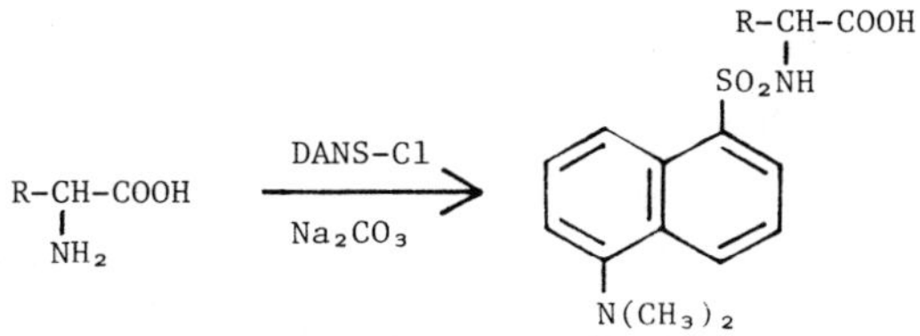

Typical Procedure. Diss amino acid sample in 1 ml 0.1% DANS-Cl in acetone. Add 2–3 drops 0.5 *M* Na_2CO_3 soln and heat at 45° for 1 hr. Evap to dryness and diss residue in benzene. Analyze by HPLC.

Comments. The derivs were used for fluorescence detection HPLC. Detection limits were in the low ng range. The derivs have also been prepared in pH 9.0 aq borate buffer by adding DANS-Cl in acetone.[2]

References

[1]R. W. Frei and J. F. Lawrence, *J. Chromatogr.* **83**, 321 (1973).

[2]E. Bayer *et al., Anal. Chem.* **48**, 1106 (1976).

Derivative Type. 5-Di-*n*-butylaminonaphthalene-l-sulfonyl (BANS).

Reagent. 5-Di-*n*-butylaminonaphthalene-l-sulfonyl chloride (BANS-Cl).

Compound Type. Amines, amino acids, peptides.

Reaction

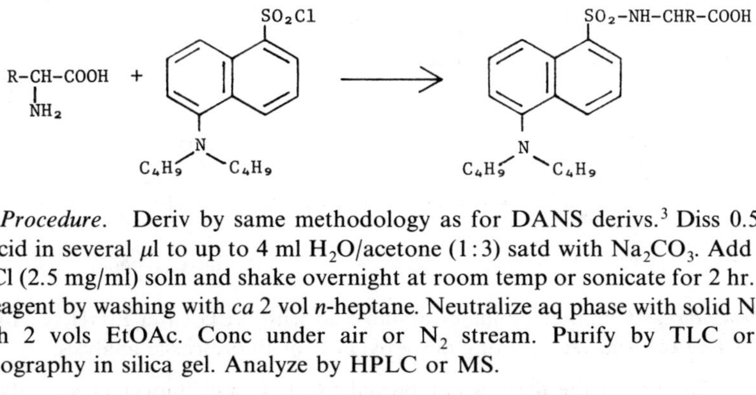

Typical Procedure. Deriv by same methodology as for DANS derivs.[3] Diss 0.5–5 nmol amino acid in several μl to up to 4 ml H_2O/acetone (1:3) satd with Na_2CO_3. Add 10–15 μl BANS-Cl (2.5 mg/ml) soln and shake overnight at room temp or sonicate for 2 hr. Remove excess reagent by washing with *ca* 2 vol *n*-heptane. Neutralize aq phase with solid NaH_2PO_4. Ext with 2 vols EtOAc. Conc under air or N_2 stream. Purify by TLC or column chromatography in silica gel. Analyze by HPLC or MS.

Comments. The derivs could be used for HPLC analysis with fluorescence detection or for MS analysis. The BANS deriv does not split off dimethylamine (as does the DANS) and thus can be used for quantitation of dimethylamine. Fluorescence yields of BANS derivs were *ca* 15% greater (in EtOAc) than DANS derivs. The BANS derivs also gave characteristic intense high mass MS peaks.

References

[1]N. Seiler *et al., J. Chromatogr.* **84**, 95 (1973).

[2]N. Seiler *et al., J. Chromatogr.* **97**, 286 (1973).

[3]N. Seiler *et al., Meth. Biochem. Anal.* **18**, 259 (1970).

Derivative Type. Pyridoxyl.

Reagent. Pyridoxal/sodium borohydride.

Compound Type. Amino acids.

Reaction

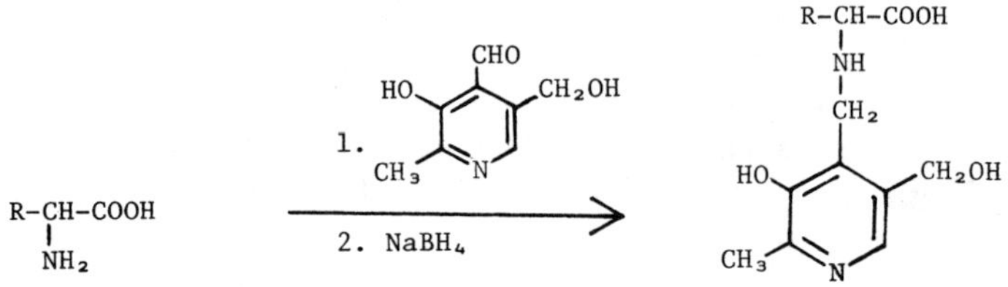

Typical Procedure. Mix the amino acid soln (in 0.5 *M* pH 9.5 phosphate buffer) with an equal volume of 0.4 *M* pyridoxal freshly adjusted to pH 9.3 and allow to stand 30 min at 8°. Add 100 mg $NaBH_4$ in 1 ml 0.1 *N* NaOH. Acidify with HCl and analyze by LC.

Comments. The derivs could be detected in the pmol range by fluorescence detection. Use of $NaBT_4$ in place of $NaBH_4$ and radiochemical detection permitted sensitivity to 10^{-13} mol. α-Keto acids could be determined by an analogous reaction using pyridoxamine.

References

[1]H. -W. Lange *et al., Z. Anal. Chem.* **261**, 337 (1972).

Derivative Type. 4-Nitrobenzo-2,1,3-oxadiazole (NBD).

Reagent. 7-Chloro-4-nitrobenzo-2,1,3-oxadiazole (NBD-Cl).

Compound Type. Amines and amino acids.

Reaction

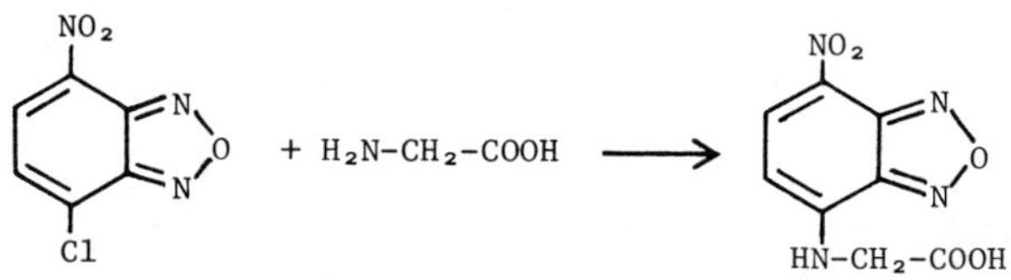

Typical Procedure. Combine amine or amino acid with NBD-Cl in EtOH or H_2O in the presence of KOAc. Analyze resulting deriv by TLC or fluorescence analysis.

Comments. The use of the deriv was reported for TLC and fluorometric analysis. The reagent itself was not fluorescent. CySH methyl ester and glutathione in aq phosphate buffer (pH 7) formed *N, S*-bis-NBD derivs but the strong fluorescence was reported to arise from the *N*-NBD portion of the derivs. Fluorescence data were reported for a series of derivs using 464 nm excitation. The derivs should be useful for HPLC analysis.

References

[1]P. B. Ghosh and M. W. Whitehouse, *Biochem. J.* **108**, 155 (1968).

Derivative Type. 3,5-Diphenyl-5-hydroxy-2-pyrrolin-4-one.

Reagent. 2-Methoxy-2,4-diphenyl-3 (2*H*)-furanone.

Compound Type. Peptides.

Reaction

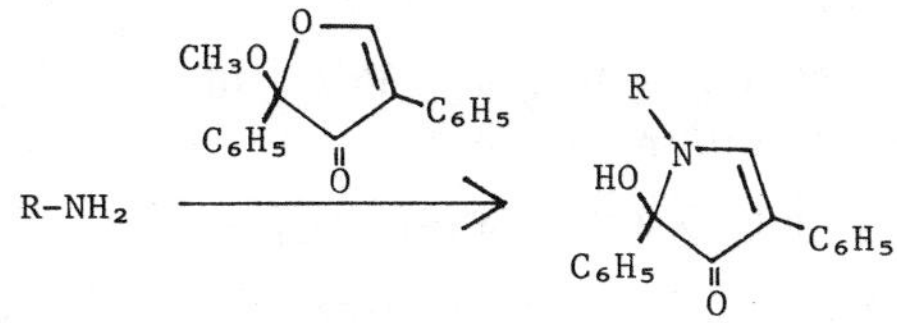

Typical Procedure. Dil 5 ml aq peptide soln (30–50 mg/ml) with 5 ml 0.1 *M* borate buffered saline (pH 9.5). Cool soln in ice and add 2–4 mg reagent in 1 ml acetone over a period of 60 min. Continue stirring 60 min then dialyze at 4° against 100 vol 0.025 *M* phosphate buffered saline (pH 6.85) followed by 12 hr dialysis against fresh buffer.

Comments. The reagent itself was nonfluorescent and hydrolyzed at a slower rate than the rate of reaction with amino groups. The reagent was used for fluorescent labeling of peptides for immunological staining but could be useful for other fluorescent labeling of amino groups in aq soln

References

[1]M. Weigele *et al.*, *Biochem. Biophys. Res. Commun.* **54**, 899 (1973).

Derivative Type. Fluorescamine deriv, methyl ester.

Reagent. Fluorescamine, diazomethane.

Compound Type. Amino acids.

Reaction

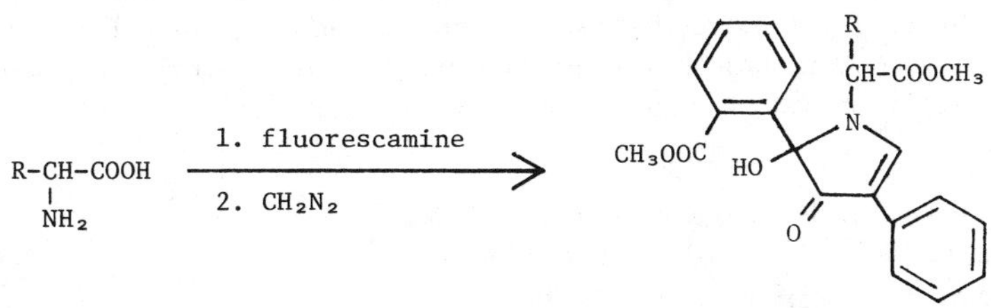

Typical Procedure. Diss amino acid in H_2O and adjust to pH 9.0 with 0.2 *M* borate buffer. Add excess fluorescamine in acetone and mix. Acidify with 0.1 *N* HCl and ext with $CHCl_3$-MeOH (MeOH content was adjusted as a function of the amino acid sidechain polarity). Purify deriv on a silica column then convert to the methyl ester with diazomethane. Analyze by MS.

Comments. EI[2] and CI[2] MS of the derivs of 17 natural amino acids were studied. MS studies had previously been reported for the derivs of Gly and Phe[3] and of dipeptides.[4]

References

[1]J. J. Shieh *et al.*, *Org. Mass Spectrom.* **11**, 479 (1976).
[2]J. J. Shieh *et al.*, *Anal. Lett.* **10**, 575 (1977).
[3]N. Narasimhachari, *Biochem. Biophys. Res. Commun.* **55**, 231 (1973).
[4]D. G. Pritchard *et al.*, *Biochem. Biophys. Res. Commun.* **65**, 312 (1975).

5.1.11. Derivatives of Cystine and Cysteine

Derivative Type. *N*-Acetyl, *n*-Propyl, *S*-carboxymethylcysteinate.

Reagent. Mercaptoethanol, iodoacetic acid, *n*-Propanol (*n*-PrOH)/HCl/dimethoxypropane (DMP), acetic anhydride (Ac_2O)/triethylamine (Et_3N).

Compound Type. Cystine and cysteine (in peptides).

Reaction

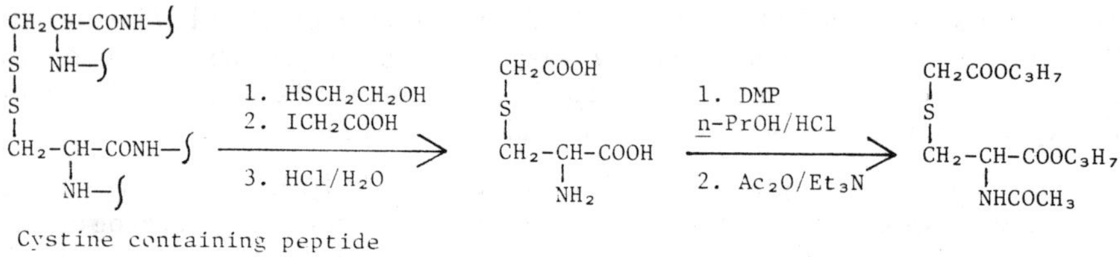

Typical Procedure. Reduce peptide with $HSCH_2CH_2OH$ in 8 *M* urea under N_2.[2] Add ICH_2COOH and 1 *N* NaOH soln.[2] After 15 min purify the deriv peptide on Sephadex G-75.[2] Hydrolyze peptide and lyophilize.

Add 50 μl dimethoxypropane and 1 ml 8 *M* HCl in *n*-PrOH. Heat at 110° for 20 min. Evap to dryness under N_2 at 60°. Add 50 mg anhyd Na_2CO_3 and 1 ml freshly prepared acetone/Et_3N/Ac_2O (5:2:1, v/v/v). Sonicate 15 sec then heat at 60° for 3 min. Evap to dryness under N_2 at 60° and diss residue in 0.5 ml dry EtOAc. Filter through glass wool then analyze by GC.

Comments. The procedure was developed as a means of obviating errors in CySH (or Cys) analysis arising from losses during peptide hydrolysis. The method was demonstrated on a reference soln of *S*-carboxymethylcysteine and 20 amino acids.[1] The reduction, S-carboxymethylation, and hydrolysis of peptides had been reported by other workers.[2] The esterfication-acylation was essentially as described by Adams.[3]

References

[1]I. M. Moodie and R. D. George, *J. Chromatogr.* **124**, 315 (1976).

[2]A. M. Crestfield *et al.*, *J. Biol. Chem.* **238**, 622 (1963).

[3]R. F. Adams, *J. Chromatogr.* **95**, 189 (1974).

Derivative Type. *S*-Carboxyethyl (cysteine), methyl ester, *N*-acetyl.

Reagent. Acrylonitrile, methanol/acetyl chloride, acetic anhydride (Ac_2O)/triethylamine (Et_3N).

Compound Type. Cystine and cysteine (in peptides).

Reaction

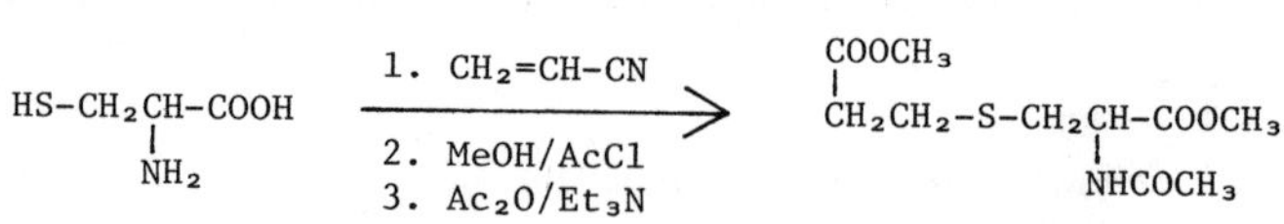

Typical Procedure. Treat 300 μg of protein with acrylonitrile and hydrolyze in the presence of 1,2-ethanethiol. Treat the dried protein hydrolysate with 0.4 ml freshly prepared acetyl chloride in MeOH for 30 min at 70°. Evap to dryness under N_2 and add 80 μl Ac_2O and 20 μl Et_3N. Heat at 100° for 30 min. Cool and evap almost to dryness under N_2. Add 30 μl Ac_2O to the residue and analyze by GC.

Comments. CySH was protected as the cyanoethyl deriv, and Trp was protected by 1,2-ethanethiol during protein hydrolysis. The structures of the derivs were confirmed by MS and His and Trp were both found to be converted to the diacyl derivs. Quantitation was achieved by radioisotope dilution utilizing ^{14}C labeled amino acids and splitting the GC effluent for counting.

References

[1]P. Felker, *Anal. Biochem.* **76**, 192 (1976).

5.1.12. Miscellaneous Acylation Methods

Derivative Type. Acetyl.

Reagent. *N*-Acetoxysuccinimide.

Compound Type. Amino acids.

Reaction

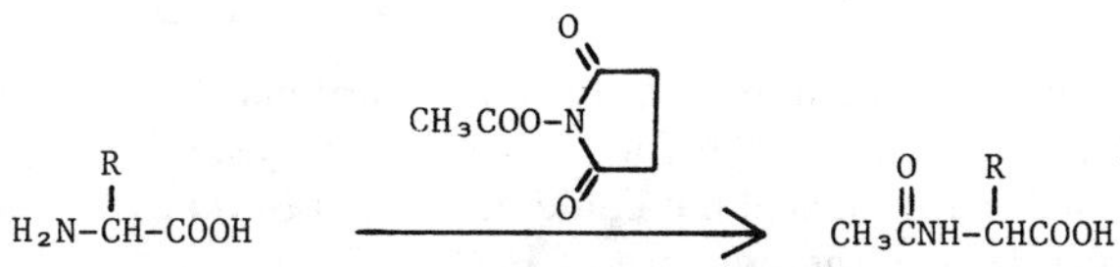

Typical Procedure (Preparative). Add a soln of *N*-acetoxysuccinimide (300 mg, 1 mmol) in 10 ml THF to a soln of the amino acid (1 mmol) and $NaHCO_3$ (84 mg, 1 mmol) in 10 ml H_2O. After 16 hr, acidify soln to pH 2 with 1 *N* HCl. Remove org solvent under vac. Add 50 ml H_2O, filter, dry, and crystallize from $CHCl_3$-pet ether.

Comments. The reagent was shown to be a highly specific acylating agent for free amino acids and did not react with hydroxyl groups or with the amino groups of adenine or cytosine. The reagent could be useful for analytical deriv preparation.

References

[1]Y. Lapidot *et al.*, *J. Lipid Res.* **8**, 142 (1967).

[2]N. DeGroot *et al.*, *Biochem. Biophys. Res.* **25**, 17 (1966).

Derivative Type. Trifluoroacetyl.

Reagent. *sym*-Trichlorotrifluoroacetone (CFA)/dimethylsulfoxide (DMSO).

Compound Type. Amino acids and peptides.

Reaction

Typical Procedure. Stir amino acid or peptide in DMSO with a threefold excess of CFA for 24 hr at room temp. Pour into ice H_2O and ext with *n*-BuOH.

Comments. The reagent trifluoroacetylated peptides without cleavage of the peptide bond.[1] Hexachloroacetone yielded the trichloroacetyl deriv under similar conditions.[2] The reaction could be useful for analytical purposes.

References

[1]C. A. Panetta and T. G. Casanova, *J. Org. Chem.* **35**, 4275 (1970).

[2]C. A. Panetta and T. G. Casanova, *J. Org. Chem.* **35**, 2423 (1970).

[3]C. A. Panetta, *Org. Syn.* **56**, 122 (1977).

Derivative Type. Trifluoroacetyl.

Reagent. Ethyl thioltrifluoroacetate.

Compound Type. Amino acids and peptides.

Reaction

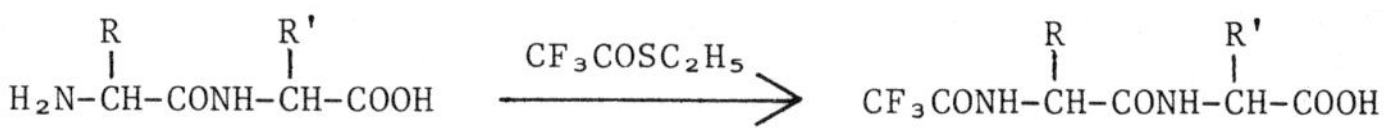

Typical Procedure. Diss amino acid or peptide in 1.0 equiv of 1 *N* NaOH. Add 1.6 equiv of ethyl thioltrifluoroacetate (0.20 ml/mmol of sample). Shake for 24 hr then acidify with conc HCl. Collect product ppt.

Comments. The deriv was used as a protecting group[1] and for selective modification of peptides.[2,3] The reagent and deriv were found to be relatively stable in aq media compared to TFAA. The deriv was stable in acid but could be cleaved by aq base. The reagent might be useful for analytical deriv in the presence of H_2O.

References

[1]E. E. Schallenberg and M. Calvin, *J. Am. Chem. Soc.* **77**, 2779 (1955).

[2]D. Levy and R. A. Paselk, *Biochim. Biophys. Acta* **310**, 398 (1973).

[3]R. A. Paselk and D. Levy, *Biochim. Biophys. Acta* **359**, 215 (1974).

5.2. DERIVATIVES OF IODOAMINO ACIDS

Derivative Type. Methyl ether, ester.

Reagent. Diazomethane/HCl.

Compound Type. Thyroxine (T_4).

Reaction

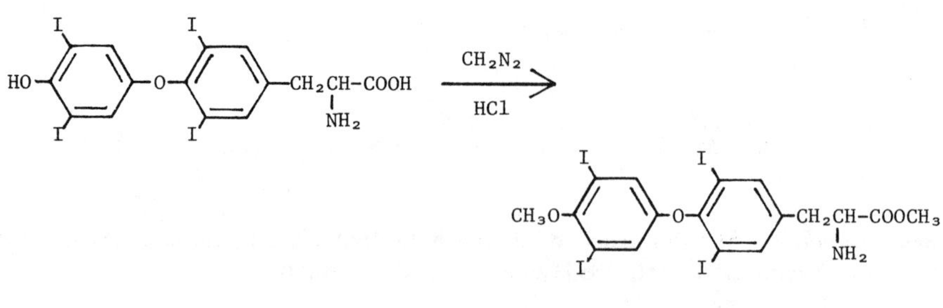

Typical Procedure. Diss 8 mg sodium thyroxine in 0.1 ml DMSO containing 5% aq HCl (v/v). Add 1 ml CH_2Cl_2 containing 1 mg estradiol benzoate int stand followed by 1 ml CH_2N_2 (*ca* 0.6 *M*) in Et_2O/CH_2Cl_2 (2:1, v/v). Allow to react 30 min then evap under N_2 to *ca* 0.2 ml. Add 1 ml CH_2Cl_2, mix, and allow ppt to settle. Analyze 1.5 μl by GC.

Comments. Methylation of T_4 gave two GC peaks but the major peak accounted for 93%. Although the deriv was not rigorously identified, it was stated that NMR evidence suggested methylation of both the carboxyl and phenolic hydroxyl. Higher conc of DMSO or substitution of formic or acetic acids for HCl gave inferior results.

References

[1]R. O. Zimmerer and L. T. Grady, *J. Pharm. Sci.* **60**, 493 (1971).

Derivative Type. Methyl ester, trifluoroacetyl.

Reagent. Methanol (MeOH)/HCl, trifluoroacetic anhydride (TFAA).

Compound Type. Iodoamino acids.

Reaction

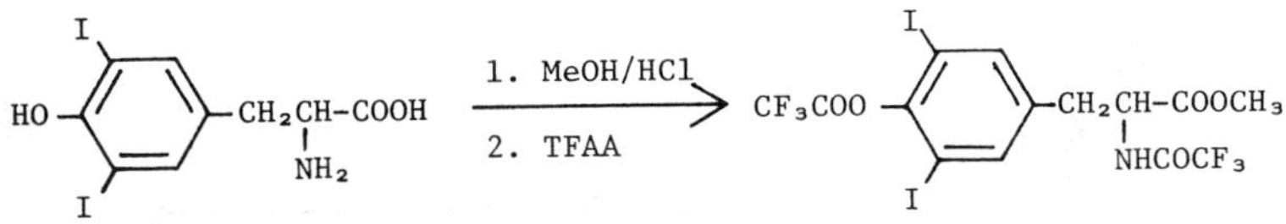

Typical Procedure. Mix 10 μmol sample with 10 ml anhyd MeOH and sat with dry HCl. After 30 min evap to dryness under vac. Dry under vac over NaOH and P_2O_5 for 2 hr. Diss residue in 1 ml TFAA, 0.9 ml CH_2Cl_2, and 0.1 ml DMF. After 30 min, analyze by GC.

Comments. In a comparison[2] of the TFA methyl ester derivs[1] and the pivalyl methyl ester derivs,[3] the latter were deemed superior for ECD-GC quantitation of thyroid hormones.

References

[1]A. H. Richards and W. B. Mason, *Anal. Chem.* **38**, 1751 (1966).
[2]R. Docter and G. Henneman, *Clin. Chim. Acta* **34**, 297 (1971).
[3]J. E. Stouffer *et al., Biochim. Biophys. Acta* **127**, 261 (1966).

Derivative Type. Methyl ester, heptafluorobutyryl.

Reagent. Methanol (MeOH)/HCl, heptafluorobutyric anhydride (HFBA).

Compound Type. Thyroid hormones and analogs.

Reaction

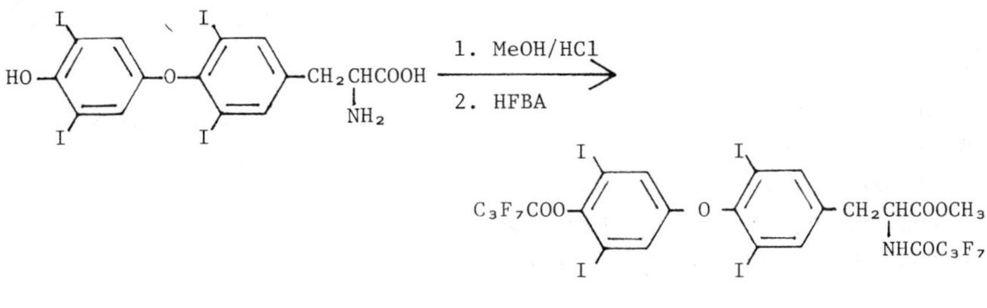

Typical Procedure. Treat 1 mg of hormone with 1 ml 25% (w/w) HCl in MeOH (freshly prepared) at 60° for 1 hr. Evap to dryness under N_2. Add a freshly prepared soln of 400 μl HFBA in 500 μl CH_3CN. Heat at 60° for 1 hr. Evap to dryness under N_2. Diss residue in CH_3CN and analyze by ECD-GC or GC-MS-SIM.

Comments. The method was applied to 3,5-diiodo-, 3,3′,5-triiodo-, 3,3′,5,5′-tetraiodo-, 3′,5′-dichloro-3,5-diiodo-, and 3′,5′-dibromo-3,5-diiodothyronine. The derivs were used for ECD-GC[1] (detection limit 0.2 pg) and GC-MS-SIM[2] (detection limit 0.5 pg) quantitation. MS data for the derivs were reported.[2] The method has been applied to quantitation of the hormones in serum after isolation by dialysis and ion exchange.[1]

References

[1]B. A. Petersen *et al., J. Chromatogr.* **126**, 503 (1976).

[2]B. A. Petersen and P. Vouros, *Anal. Chem.* **49**, 1304 (1977).

Derivative Type. Methyl ester, *N,O*-pivalyl.

Reagent. Methanol/thionyl chloride, pivalic anhydride/triethylamine (Et_3N).

Compound Type. Thyroid hormones.

Reaction

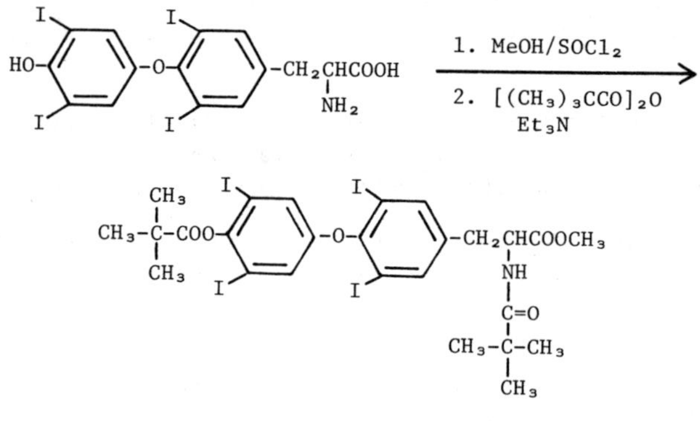

Typical Procedure. Convert the sample (0.5–1.0 mg) to the methyl ester by diss in 1.0 ml dry MeOH containing a few drops of $SOCl_2$,[2] or by diss in 1 ml 25% (w/w) HCl in MeOH, and heating 1 hr at 70°. Evap to dryness under N_2. Add 0.2 ml pivalic anhydride and 0.05 ml Et_3N and heat 30 min. Analyze by GC or GC-MS.

Comments. The method was applied to Tyr, mono- and diiodo-Tyr, triiodothyronine (T_3), and thyroxine (T_4). MS data were reported for the derivs.[1] Volpert *et al.*[4] prepared and isolated the pure derivs. The deriv has been used for analysis of T_3 and T_4 in serum.[5]

References

[1]J. E. Stouffer *et al., Biochim. Biophys Acta* **127**, 261 (1966).

[2]M. Brenner and W. Huber, *Helv. Chim. Acta* **36**, 1109 (1953).

[3]P. I. Jaakonmaki and P. I. Stouffer, *J. Gas Chromatogr.* **5**, 303 (1967).

[4]E. M. Volpert *et al., J. Chromatogr.* **50**, 507 (1970).

[5]N. N. Nihei, *Anal. Biochem.* **43**, 433 (1971).

Derivative Type. Trimethylsilyl.

Reagent. Bis (trimethylsilyl) acetamide (BSA).

Compound Type. Iodoamino acids (and other iodinated compounds).

Reaction

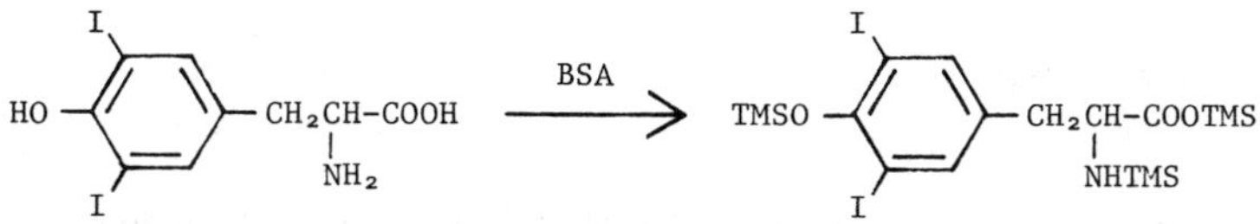

Typical Procedure. Mix 1 mg sample with 100 μl BSA. Allow to stand at room temp or heat at 80°. Analyze by FID or ECD-GC.

Comments. For qualitative work, use of BSA with 10 % CH_3CN was more rapid (5–7 min at 80°) but the derivs were less stable in the presence of CH_3CN. The method was applied to a variety of iodinated compds including the iodoamino acids, their sidechain and position analogs, intermediates and degradation products, and related compds.[1] The derivs were very prone to hydrolysis by moisture but could be used successfully if prepared in sealed vials in an excess of BSA. MS data showed that the derivs of both the iodothyronines and the iodothyroxines contained three TMS groups.[1] Other workers prepared the derivs using BSA/CH_3CN (1:1, v/v, 5 min, 50°),[2] BSA/CH_3CN (1:3, v/v, 30 min, 150°),[3] HMDS/TMCS/pyridine (2:1:7, v/v/v, 90 min, 125°),[3] BSA/5 % TMCS (20 min, 50°),[4] and BSA (5 min, 60°),[5] Lawson *et al.*[5] prepared derivs of 16 iodoamino acids and related compds and reported MS data for the derivs. Problems were encountered in GC of the derivs due to loss of the *N*-TMS. It was concluded that the TMS derivs were not optimum for quant application.

References

[1]K. Funakoshi and H. J. Cahnmann, *Anal. Biochem.* **27**, 150 (1969).

[2]N. M. Alexander and R. Scheig, *Anal. Biochem.* **22**, 187 (1968).

[3]F. Shahrokhi and C. W. Gehrke, *Anal. Biochem.* **24**, 281 (1968).

[4]E. T. Backer and V. J. Pileggi, *J. Chromatogr.* **36**, 351 (1968).

[5]A. M. Lawson *et al., Biomed. Mass Spectrom.* **1**, 374 (1974).

5.3. DERIVATIVES OF NONPROTEIN AMINO ACIDS

5.3.1. *N*-Acyl Derivatives

Derivative Type. *n*-Butyl ester, *N*-heptafluorobutyryl.

Reagent. Methanol (MeOH)/HCl, *n*-Butanol (*n*-BuOH)/HCl, heptafluorobutyrylimidazole (HFBI).

Compound Type. Baclofen [γ-amino-β-(*p*-chlorophenyl) butyric acid].

Reaction

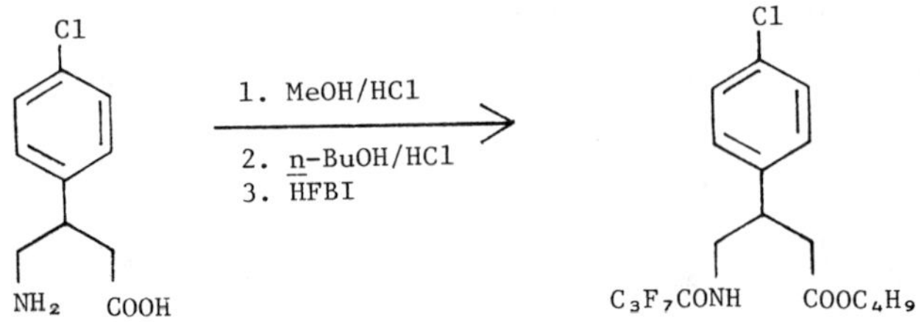

Typical Procedure. Diss sample in 0.5 ml MeOH/HCl and heat at 70° for 45 min. Evap under N_2, add 0.5 ml *n*-BuOH/HCl, and heat at 100° for 15 min. Evap under N_2 at 60° and add 25 μl HFBI. Heat at 80° for 2 hr. Cool and add 2 ml H_2O and 5 ml heptane. Shake and immediately place in a bath of solid CO_2 for 10 min. Remove org phase and analyze by ECD-GC.

Comments. The alcohol/HCl solns were prepared by adding 0.25 ml AcCl to 5 ml of the alcohol. Baclofen was measured in plasma or urine using the dichloro analog as int stand. The drug and stand were isolated by adsorption on charcoal, converted to butyl esters, and purified by base specific extraction. The esters were then acylated with HFBI, the derivs isolated by solvent ext, and analyzed by ECD-GC.

References

[1]P. H. Degen and W. Riess, *J. Chromatogr.* **117**, 399 (1976).

Derivative Type. *N*-Heptafluorobutyryl, *n*-butyl ester.

Reagent. Heptafluorobutyric anhydride (HFBA), *n*-butanol (*n*-BuOH)/HFBA.

Compound Type. Tranexamic acid.

Reaction

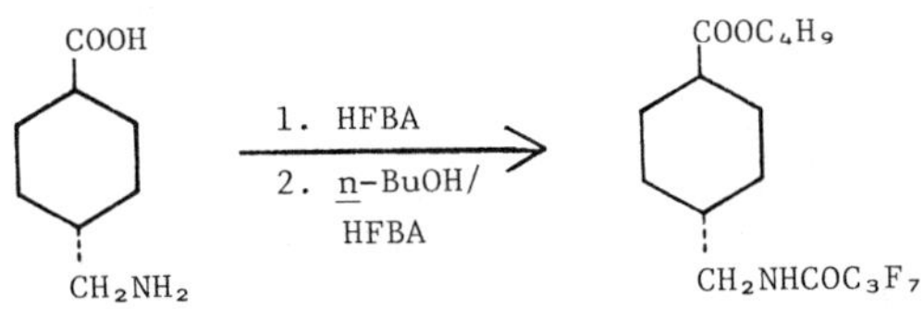

Typical Procedure. Add 4.5 μg d_6 int stand to 1 ml serum then add 3 ml H_2O and mix. Add 100 μl heptafluorobutyric acid gradually. Shake, allow to stand 10 min, and centrifuge. Evap 2 ml supernatant under reduced pressure. Add 0.5 ml hexane to the residue, cool in an ice bath, and add 0.5 ml HFBA. Heat at 60° for 5 min. Add 0.5 ml HFBA and 0.5 ml *n*-BuOH. Heat at 60° for 10 min. Evap under reduced pressure. Ext residue with 4 × 1 ml EtOAc. Wash combined ext with 1.5 ml H_2O. Dry EtOAc layer with Na_2SO_4 and conc to *ca* 0.1 ml. Chromatograph on 0.9 × 5 cm silica column with hexane/Et_2O (85:15). Collect first 50 ml and conc to 1 ml. Analyze by GC-MS.

Comments. The method was used with d_6-tranexamic acid as int stand for quantitation by GC-MS. Heptafluorobutyric acid was used for deproteination because it was more easily removed than the more commonly used trichloroacetic acid. Of several esters, examined the Bu ester gave the most favorable MS behavior for quantitation. Esterfication using BuOH/HCl was incomplete. Some *trans-cis* conversion was seen in the amidation step.

References

[1]H. Miyazaki *et al.*, *Chem. Pharm. Bull.* **23**, 1806 (1975).

Derivative Type. Trifluoroacetyl.

Reagent. Trifluoroacetic anhydride (TFAA).

Compound Type. Methylphenidate.

Reaction

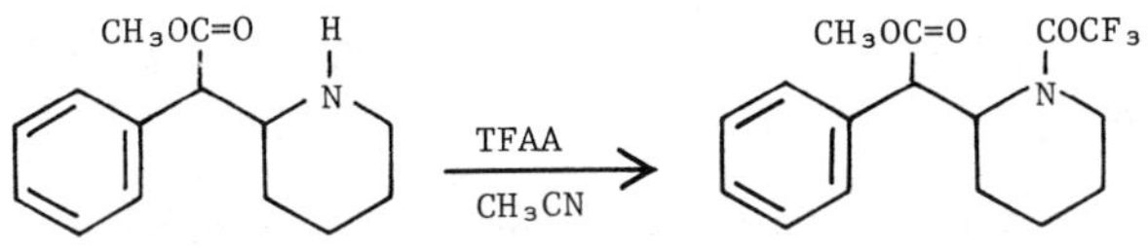

Typical Procedure. To 2 ml plasma add 500 pmol int stand and 0.6 g $(NH_4)_2CO_3$. Ext with 6 ml hexane. Evap 4.5 ml of the ext to dryness under N_2. To the residue add 50 μl CH_3CN and 50 μl TFAA. Heat at 70° for 15 min. Evap to dryness under N_2 and diss residue in 50 μl CH_3CN. Analyze by GC-MS-SIM.

Comments. The ethyl ester analog was used as int stand. The method permitted quantitation down to 1.2 ng/ml in plasma or brain homogenate.

References

[1]J. Gal *et al.*, *J. Pharm. Sci.* **66**, 866 (1977).

Derivative Type. Hexafluoroisopropyl ester, *N*-pentafluoropropionyl.

Reagent. Hexafluoroisopropanol (HFIP)/pentafluoropropionic anhydride (PFPA).

Compound Type. Glutamic acid and γ-aminobutyric acid (GABA).

Reaction

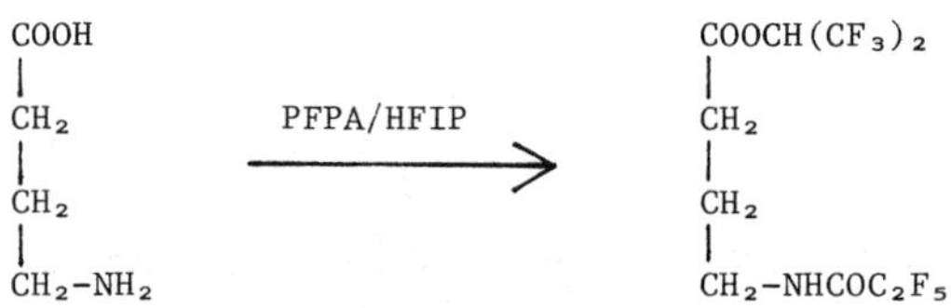

Typical Procedure. Homogenize tissue sample in 80% aq EtOH containing int stand (d_5-Glu, d_2-Glu, d_2-GABA, 5-amino-*n*-valeric acid). Centrifuge and transfer the supernatant or a 50 μl aliquot to another tube. Evap to dryness under N_2.

Add 50 μl HFIP and 100 μl PFPA. Seal tube and heat at 60° for 1 hr. Store at 4°. Just before analysis, evap to dryness and diss residue in 10–100 μl EtOAc. Analyze 1–3 μl by GC-MS-SIM.

Comments. The method was used to quantitate Glu and GABA in micro samples (*ca* 50 μg protein) of rat brain.[1] Other workers have used the same deriv for GABA in brain tissue.[2,3]

References

[1]L. Bertilsson and E. Costa, *J. Chromatogr.* **118**, 395 (1976).

[2]J. D. M. Pearson and D. F. Sharman, *J. Neurochem.* **24**, 1225 (1975).

[3] R. Schmid and M. Karobath, *J. Chromatogr.* **139**, 101 (1977).

Derivative Type. Pentafluoropropionyl.

Reagent. Pentafluoropropionic anhydride (PFPA).

Compound Type. Mefenamic acid.

Reaction

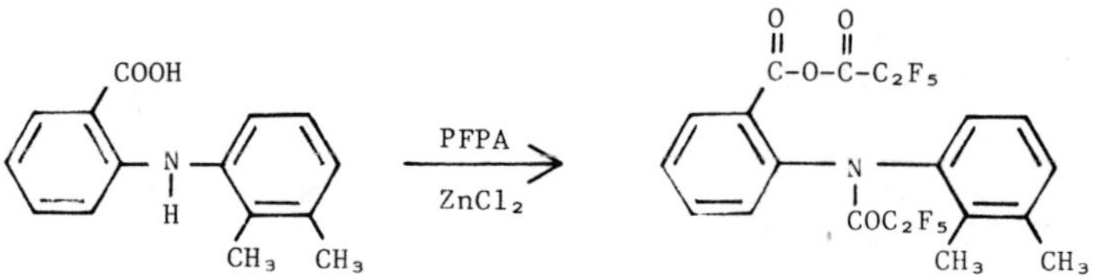

Typical Procedure. Combine 2.0 ml aq sample with 1.0 ml satd aq KH_2PO_4 and ext with 2.0 ml CH_2Cl_2. Evap ext to dryness under N_2 and add 50 μl PFPA and 50 μl EtOAc satd with $ZnCl_2$. Cap sample tube and heat at 60° for 5 min. Add 1.0 ml cyclohexane and wash soln with 8.0 ml satd $Na_2B_4O_7$. Centrifuge and analyze an aliquot of the cyclohexane phase by GC.

Comments. $ZnCl_2$ was used as a Lewis acid catalyst. IR and UV data were reported to support the assigned structure of the deriv. The deriv was detectable to 50 ng/ml.

References

[1]S. A. Bland *et al.*, *J. Chromatogr. Sci.* **14**, 201 (1976).

5.3.2. Trimethylsilyl Derivatives

Derivative Type. Trimethylsilyl.

Reagent. Bis(trimethylsilyl)acetamide (BSA).

Compound Type. ω-Amino acids.

Reaction

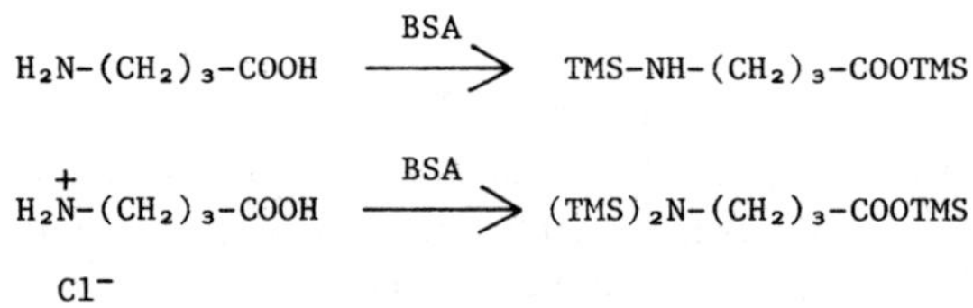

Typical Procedure. Diss 5–10 mg ω-amino acid or the HCl in 0.5 ml BSA. Heat at 90° for 30–60 min then let stand at room temp for 24 hr. Analyze by GC-MS.

Comments. The conditions described yielded 100% di-TMS deriv for free amino acids and 100% tri-TMS deriv for the hydrochlorides. The use of CH_3CN or Et_3N tended to yield mixt of di- and tri-TMS derivs. MS data were reported for the derivs of C_4,C_6,C_7,C_8, and C_{11} ω-amino acids.

References

[1]J. Marik *et al.*, *J. Chromatogr.* **128**, 1 (1976).

Derivative Type. Trimethylsilyl.

Reagent. Hexamethyldisilazane (HMDS)/trimethylchlorosilane (TMCS)/bis(trimethylsilyl) trifluoroacetamide (BSTFA).

Compound Type. γ-Aminobutyric acid (GABA).

Reaction

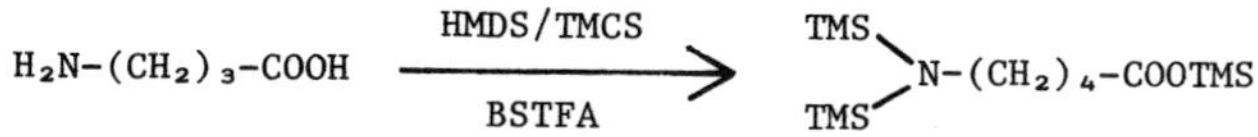

Typical Procedure. Homogenize rat brain tissue in 30 vol 0.1 *N* HCOOH containing 5-aminovaleric acid int stand. Centrifuge, remove 5 μl supernatant, and evap under N_2. Add 30 μl HMDS/TMCS/BSTFA/pyridine (3:1:3:9, v/v/v/v). Allow to react 90 min at room temp then analyze an aliquot by GC-MS-SIM.

Comments. The deriv was measurable to the low picomole range. Full MS were given for the derivs of GABA and the int stand.

References

[1]F. Cattabeni *et al., Anal. Biochem.* **72**, 1 (1976).

Derivative Type. Trimethysilyl.

Reagent. Bis(trimethylsilyl)trifluoroacetamide (BSTFA)/trimethylchlorosilane (TMCS)/ trimethylsilyldiethylamine (TMSDEA)/pyridine.

Compound Type. Tyrosine, tryptophan, and metabolites.

Reaction

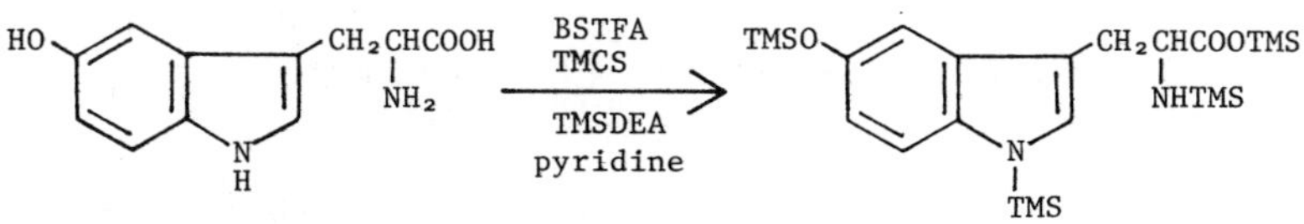

Typical Procedure. Add *n*-docosane int stand to a mixt of BSTFA/ TMSDEA/TMCS/pyridine (99:30:1:100, v/v/v/v) to give 50 μg/ml. To up to 2 mg of sample compds add 0.30 ml reagent and heat in a sealed vial at 100° for 10 min. Allow to stand overnight at room temp. Analyze 3 μl aliquots by GC.

Comments. The method was used for simultaneous analysis of acidic, neutral, and basic metabolites of Tyr and Trp. Derivs of 16 Trp metabolites, 12 Tyr metabolites, and several related compds were examined. Eleven silylation reagent mixt were evaluated. TMSDEA/TMCS (8:1, v/v) gave quantitative deriv but did not react with indole nitrogen; without indole deriv all of the metabolites were not separable by the GC methods used.

References

[1]P. W. Albro and L. Fishbein, *J. Chromatogr.* **55**, 297 (1971).

5.3.3. Derivatives of Nitrilotriacetic Acid and Related Compounds

Derivative Type. Methyl ester.

Reagent. Methanol (MeOH)/boron trifluoride.

Compound Type. Nitrilotriacetic acid (NTA), ethylenediamine tetraacetic acid (EDTA), diethylenetriamine pentaacetic acid (DTPA).

Reaction

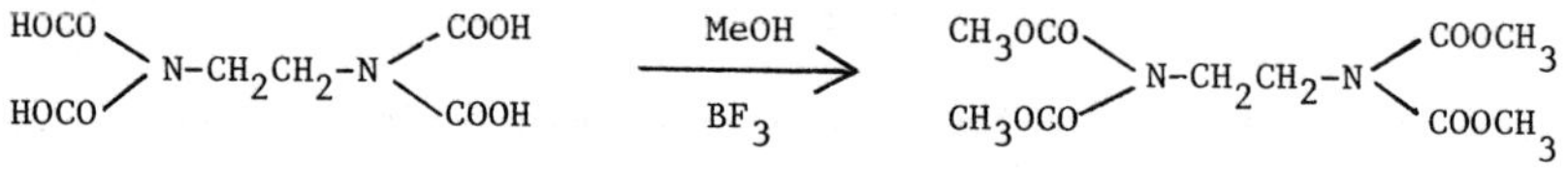

Typical Procedure. Filter a 10 ml aq sample and adjust to pH 7.0. Ext with an equal vol of $CHCl_3$ and transfer 5 ml of aq layer to another tube. Add 25 μl of int stand soln (cyclohexane-1,2-diamine tetraacetic acid, 0.5 g/1) and immerse tube in 100° oil bath. Evap to dryness under N_2, cool, and add 1 ml BF_3/MeOH (10%, w/v). Seal and place in an ultrasonic bath at 100° for 40 min. Cool, add 1 ml $CHCl_3$, and transfer to a tube containing 3 ml pH 7.0 KH_2PO_4 buffer (1 *M*). Shake 1 min, transfer an aliquot (0.6–0.8 ml) of the $CHCl_3$ to a dessicator, and evap under vacuum at 20°. Diss residue in 50 μl $CHCl_3$ and analyze 5 μl by GC.

Comments. The method was used to determine EDTA and DTPA down to 10 μg/1 and NTA down to 200 μg/1.

References

[1]L. Rudling, *Water Res.* **6**, 871 (1972).

Derivative Type. Methyl ester.

Reagent. Methanol (MeOH)/HCl/dimethyl sulfite.

Compound Type. Nitrilotriacetic acid (NTA).

Reaction

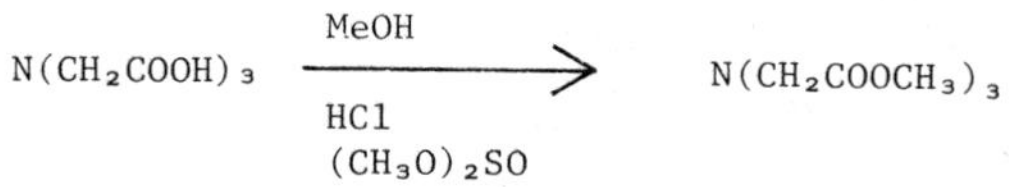

Typical Procedure. Filter H_2O sample, acidify with 2 ml conc HCl per 250 ml sample, and store frozen. Prior to analysis, thaw, and aliquot 10 ml. Add 0.1 ml conc HCl. Lyophilize and add 1 ml of 2 *N* HCl in MeOH to residue. Add 0.2 ml dimethyl sulfite, seal vial, and heat at 68° for 3 hr. Evap on rotary evaporator, add 20 μl int stand (4.65 mg methyl heptadecanoate in 25 ml $CHCl_3$), and 150 μl $CHCl_3$. Analyze by GC.

Comments. The method was used to quantitate NTA to 25 μg/1 with FID-GC.

References

[1]D. Morray and D. Povoledo, *J. Fish Res. Board Can.* **28**, 1043 (1971).

Derivative Type. Propyl ester.

Reagent. *n*-Propanol (*n*-PrOH)/HCl.

Compound Type. Nitrilotriacetic acid.

Reaction

$$N(CH_2COOH)_3 \xrightarrow[\text{HCl}]{\text{PrOH}} N(CH_2COOC_3H_7)_3$$

Typical Procedure. Filter a 200 ml H_2O sample and acidify with HNO_3 to pH 3–4. Pass sample through a Dowex-1 column (in formate form) and wash column with 300 ml H_2O. Elute with 100 ml 3 *M* HCOOH. Collect the eluate and conc to 1–2 ml in a 90–100° sand bath. Transfer concentrate to a 10 ml ampule and evap to dryness under N_2. Add 2 ml HCl satd *n*-PrOH and 0.1 ml *n*-PrOH containing 20 μg of heptadecanoic acid int stand. Seal and place in boiling H_2O bath for 30 min. Cool, open, evap to dryness, and diss residue in 50 μl benzene. Analyze 5 μl aliquots by GC.

Comments. The methyl, *n*-propyl, and *n*-butyl esters were investigated with the *n*-propyl giving the most suitable GC volatility and separation from interfering esters of fatty acids and amino acids. BF_3 was also examined but HCl gave superior results. The lower detection limit of the deriv was given as 10 ng using FID-GC.

References

[1]Y. K. Chau and M. E. Fox, *J. Chromatogr. Sci.* **9**, 271 (1971).

Derivative Type. *n*-Butyl ester.

Reagent. *n*-Butyl alcohol (*n*-BuOH)/HCl.

Compound Type. Nitrilotriacetic acid (NTA), *N*-methyl-iminodiacetic acid and *N*-oxalyl-iminodiacetic acid.

Reaction

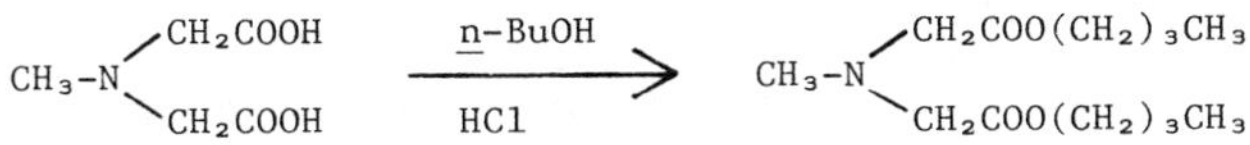

Typical Procedure. Freeze dry the H_2O sample (with int stand) in a screw cap test tube and add 2 ml 3 *N* *n*-BuOH/HCl. Seal and place in 75° ultrasonic bath for 5 min followed by a 65° H_2O bath for 45 min. Remove BuOH/HCl on rotary evaporator and diss residue in Freon 113 (1,1,2-trichlorofluoroethane) prior to GC analysis.

Comments. The method was useful for H_2O sample concentrations from 20 to 0.025 mg/liter. This method combined with subsequent trifluoroacetylation (TFAA in CH_2Cl_2, room temp, 30 min) allowed quantitation of iminodiacetic acid (IDA), glycine, and sarcosine also using glutamic acid as int stand. *N*-Nitroso-IDA was converted to IDA by BuOH/HCl.

References

[1]C. B. Warren and E. J. Malec, *J. Chromatogr.* **64**, 219 (1972).

Derivative Type. Tri (2-chloroethyl) ester.

Reagent. 2-Chloroethanol/boron trifluoride.

Compound Type. Nitrilotriacetic acid (NTA).

Reaction

$$N(CH_2COOH)_3 \xrightarrow[BF_3]{ClCH_2CH_2OH} N(CH_2COOCH_2CH_2Cl)_3$$

Typical Procedure. Filter 100 ml H_2O sample and adjust pH to 5.0 with acetate buffer (1 *M* NaOAc + 1 *M* HOAc). Pass soln through a column containing 3 g of Dowex 1X8 (50–100 mesh, H form). Wash column with 3 × 20 ml H_2O and elute with 1.2 *M* HCl collecting the first 30 ml. Evap 2 ml of eluate in an ampule at 100°. Add 1 ml of BF_3 in 2-chloroethanol (10% w/v), seal, and maintain at 112° for 2 hr. Open vial and add 1 ml benzene. Transfer the mixt to a tube containing 5 ml H_2O and shake for 1 min. Transfer 0.5 ml of org phase to another tube containing 2 ml H_2O. Analyze aliquots of benzene phase by ECD-GC.

Comments. Ion exchange was used to separate NTA from interfering ions. MS data were given for the deriv. The deriv was also analyzed by GC-MS using selected ion monitoring.

References

[1]L. Rudling, *Water Res.* **5**, 831 (1971).

Derivative Type. Trimethylsilyl.

Reagent. Bis(trimethylsilyl)trifluoroacetamide (BSTFA).

Compound Type. Nitrilotriacetate (NTA).

Reaction

$$N(CH_2COOH)_3 \xrightarrow{BSTFA} N(CH_2COOTMS)_3$$

Typical Procedure. Pipet a 1.00 ml sample into a precleaned and dried 6 ml screw cap vial and add 1 ml of 2 mg/ml ammonium EDTA soln. Evap just to dryness in a 150° sandbath under a stream of dry N_2. Add and evap under N_2 in a 70° sandbath, 2 × 1.5 ml portions of dry CH_2Cl_2. Immediately seal vial with septum cap and add 10 μl DMF containing 2 mg/ml of *n*-octadecane int stand. Add 200 μl each of BSTFA and DMF and heat at 70° for 30 min. Cool and analyze 1.0 μl aliquots by GC.

Comments. The method was shown to be suitable for analysis of samples containing 1–100 μg/ml NTA. EDTA was added to remove ions from NTA chelates which inhibited deriv of NTA.

References

[1]R. J. Stolzberg and D. N. Hume, *Anal. Chem.* **49**, 374 (1977).

5.3.4. Other Derivatives of Nonprotein Amino Acids

Derivative Type. *N*-(2-Nitro-4-trifluoromethylphenyl)/ethyl ester.

Reagent. 2-Nitro-4-trifluoromethylfluorobenzene (NTFB), Ethyl iodide/tetrabutylammonium hydrogen sulfate (TBA-HSO_4).

Compound Type. Tranexamic acid (*trans*-4-aminomethylcyclohexanecarboxylic acid).

Reaction

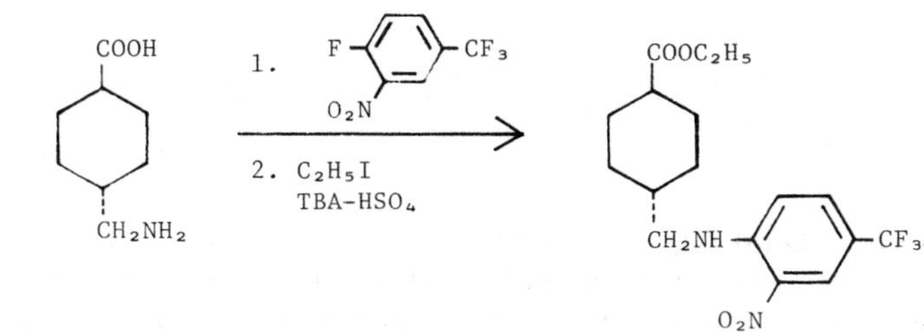

Typical Procedure. Mix up to 200 μl of plasma (containing 0.2–1 μg tranexamic acid) with 100 μl int stand soln (4-aminomethylbicyclo[2.2.2]octane-l-carboxylic acid, 250 mg/1). Add 200 μl of 0.2 *M* pH 9.4 borate buffer and 1.0 ml NTFB soln (0.4 ml NTFB in 100 ml DMSO). Shake gently and allow to stand at least 5 min. Wash mixt with 3 ml toluene and discard toluene. Add 3 ml C_2H_5I soln (3 *M* in CH_2Cl_2) and 3 ml TBA-HSO_4 soln (0.15 *M*, pH 9.5). Shake 5 min, then filter org phase through silanized glass wool in a Pasteur pipette into a 12 ml glass stoppered centrifuge tube. Evap to dryness on a boiling H_2O bath, cool, and diss residue in 3 ml toluene. Wash with 5 ml 0.1 *M* NaOH then evap toluene to approx 0.5 ml. Analyze 1 μl aliquots by ECD-GC.

Comments. The deriv was detectable to about 40 pg.

References

[1]J. Vessman and S. Stromberg, *Anal. Chem.* **49**, 369 (1977).

Derivative Type. 2-Phenyl-oxazolin-5-one.

Reagent. Benzoyl chloride/dicyclohexylcarbodiimide (DCC).

Compound Type. α-Methyl amino acids.

Reaction

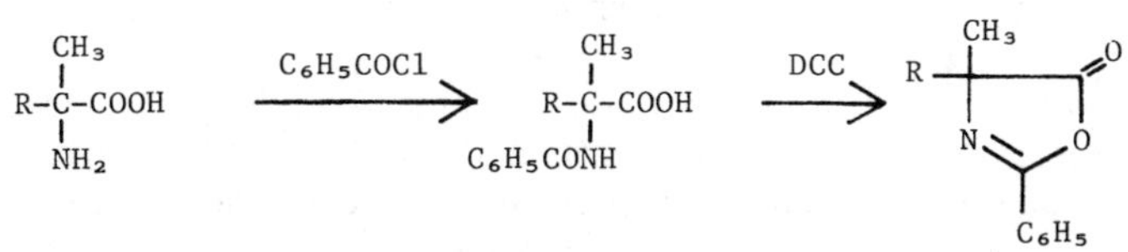

Typical Procedure. Diss amino acid HCl in H_2O and alkalinize slightly with NaOH. Add benzoyl chloride with vigorous stirring keeping the soln alkaline with NaOH. After 15 min additional stirring, remove excess benzoyl chloride by ext with Et_2O. Acidify with HCl, ext with EtOAc and dry ext over Na_2SO_4. Add DCC, mix, and analyze an aliquot by GC-MS.

Comments. MS data were reported for the derivs of 12 α-methyl amino acids.

References

[1]O. Grahl-Nielsen and E. Solheim, *Anal. Chem.* **47**, 333 (1975).

Derivative Type. Methyl salicylate.

Reagent. Nitrous acid, diazomethane.

Compound Type. Anthranilic acid.

Reaction

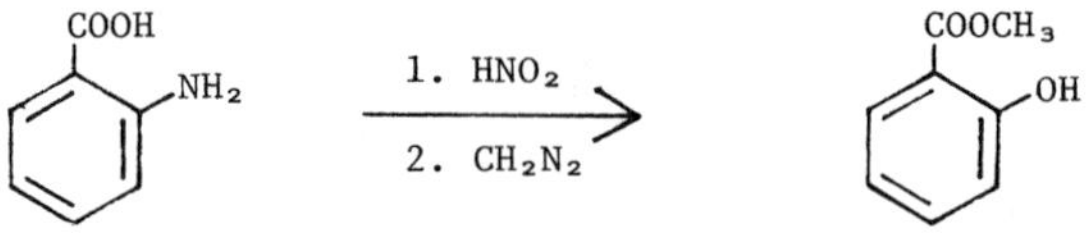

Typical Procedure. Dil 10 ml urine with 1 ml conc HCl and 9 ml 0.1 *N* HCl. Add 1 ml 0.2 *M* $NaNO_2$ and allow to stand 20 min in an ice bath. Destroy excess HNO_2 by addition of 1 ml 0.2 *M* sulfamic acid and heating at 80° for 35 min. Cool in an ice bath and add phenylpropionic acid int stand in 0.5 ml $CHCl_3$. Ext with 5 ml portions of $CHCl_3$. To the combined ext add 0.5 ml dimethoxypropane (dehydrating agent) and evap to just dryness under vac. Add 1 ml CH_2N_2 in Et_2O and allow to stand at room temp 10 min. Add a small vol $CHCl_3$ and evap under vac to *ca* 0.1 ml. Analyze 1.0 μl of the soln by GC.

Comments. GC on a mixed stationary phase (OV-17, XF-1105) was necessary to separate the deriv from methyl phenylacetate.[2] MS data were reported for the deriv.[2]

References

[1]K. Hirano *et al.*, *J. Chromatogr.* **64**, 174 (1972).

[2]K. Hirano *et al.*, *J. Chromatogr.* **70**, 53 (1972).

Derivative Type. Pentafluoropropionyl amino alcohol.

Reagent. Lithium aluminum hydride, pentafluoropropionic anhydride (PFPA).

Compound Type. Methylphenidate.

Reaction

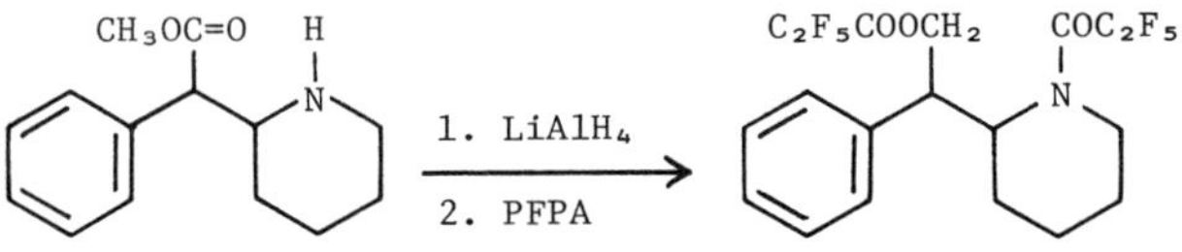

Typical Procedure. To 4.0 ml plasma add 2.0 ml satd aq $Na_2B_4O_7$ and ext with 4 ml cyclohexane. Separate the ext and add to it 50 μl of a satd soln of $LiAlH_4$ in Et_2O. Allow to react 3 min then add 50 μl H_2O and shake. Add 50 μl PFPA and allow to react 3–5 min. Wash with 6 ml satd aq $Na_2B_4O_7$ and analyze an aliquot of the org phase by ECD-GC.

Comments. Longer reduction and acylation times gave no increase in yield of the deriv. The diacyl deriv was five times more sensitive to ECD than the *N*-HFB deriv of the parent drug. The method was also applied to cocaine.

References

[1]R. Huffman *et al.*, *J. Chromatogr. Sci.* **12**, 382 (1974).

5.4. DERIVATIVES OF PEPTIDES

5.4.1. Derivatives for Quantitation of Peptides by GC

Derivative Type. Methyl ester, trifluoroacetyl.

Reagent. Diazomethane, methyl trifluoroacetate (MeOTFA).

Compound Type. Peptides.

Reaction

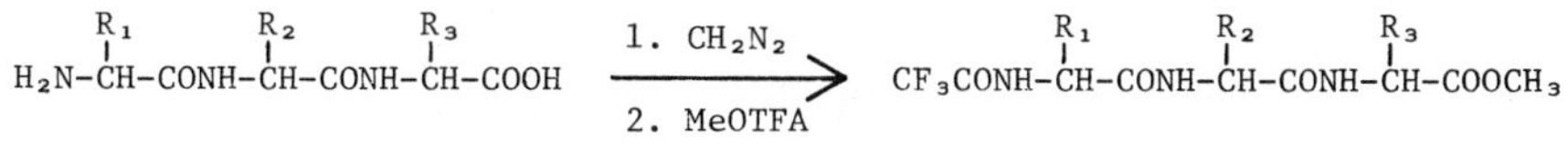

Typical Procedure. Diss 100 mg of peptide in 5 ml conc HCl and maintain at 37° for 72 hr. Evap to dryness under vac and diss residue in MeOH. Add excess CH_2N_2 in Et_2O and after several min destroy excess by addition of a small amt of AcOH. Evap to dryness under vac and diss residue in 5 ml MeOH. Add 2 ml MeOTFA and adjust soln to pH 7.5–8.0 with Et_3N. After 5 hr at room temp evap solvent to yield deriv. Analyze by GC.

Comments. The *S*-benzyl group was used to protect CySH thiol groups.[2] The use of MeOTFA allowed trifluoroacetylation under mild conditions to prevent decomposition of the peptides.

References

[1]E. Bayer and W. A. Koenig, *J. Chromatogr.* **7**, 95 (1969).
[2]E. Bayer *et al.*, *Z. Naturforsch.* **22b**, 924 (1967).

Derivative Type. *n*-Butyl ester, *N*-trifluoroacetyl.

Reagent. *n*-Butanol (*n*-BuOH)/HCl, trifluoroacetic anhydride (TFAA).

Compound Type. Glutathione (GSH).

Reaction

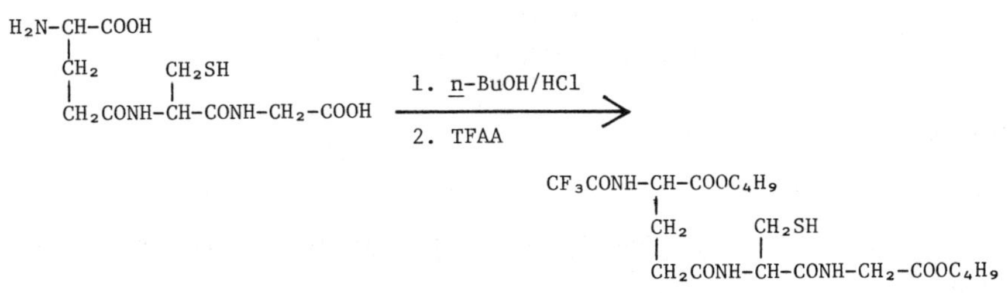

Typical Procedure.[3] Mix 10 μl of plasma (or 2.5 m*M* GSH standard) with 10 μl 2.5 m*M* α-aminocaprylic acid int stand soln. Add 150 μl 3 *N* HCl in BuOH, sonicate 15 sec at room temp, then heat at 100° for 15 min. Evap to dryness at 100° under N_2. Add 150 μl CH_2Cl_2 and evap to remove traces of H_2O. Add 60 μl CH_2Cl_2 and 20 μl TFAA and heat at 150° for 5 min. Analyze by GC using an electrolytic conductivity detector.

Comments. The esterfication and acylation were performed on a heating block apparatus.[2] The method was applied to GSH analysis in swine blood. Oxidized glutathione (GSSG) gave a deriv which eluted at the same GC temp, but it was reported that MS analysis did not indicate formation of the same deriv. The deriv structure shown above is assumed to be correct, but the actual MS data were not reported.

References

[1]J. M. L. Mee, *J. Chromatogr.* **87**, 258 (1973).
[2]J. M. L. Mee and C. C. Brooks, *J. Chromatogr.* **62**, 138 (1971).
[3]D. Roach and C. W. Gehrke, *J. Chromatogr.* **44**, 269 (1969).

Derivative Type. 2,2,3,3,3-Pentafluoro-l-propyl ester, *N*-pentafluoropropionyl.

Reagent. 2,2,3,3,3-Pentafluoro-l-propanol/pentafluoropropionic anhydride (PFPA), PFPA.

Compound Type. γ-Glutamyl amino acids.

Reaction

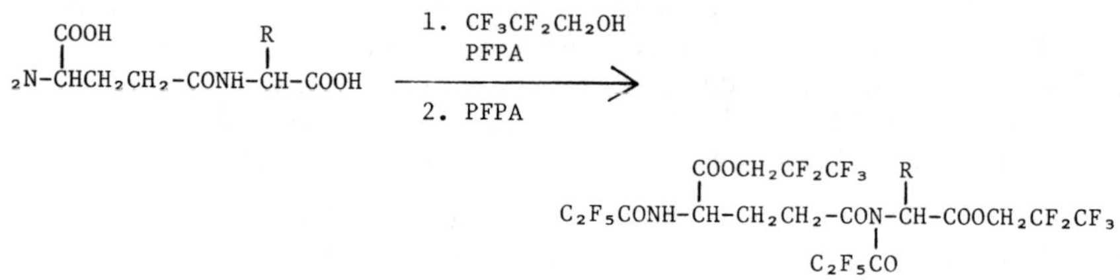

Typical Procedure. Mix sample with 0.25 ml of 20% (v/v) 2,2,3,3,3-pentafluoro-l-propanol in PFPA. Heat 15 min at 75°. Evap to dryness under N_2. Add 0.1 ml PFPA and heat 5 min at 75°. Evap to dryness under N_2 and diss residue in 1 ml EtOAc. Analyze by ECD-GC.

Comments. The method was used to quantitate γ-glutamyl amino acids and pyrrolidone carboxylate in urine, CSF, and tissue samples. No structural data were given for the derivs, but it was stated that GC properties suggested that all functional groups had been derivatized. Since the amide N in pyrrolidone carboxylate was acylated[2] it was assumed that the amide N was also acylated in the γ-glutamyl amino acid derivs.

References

[1]S. Wilk and M. Orlowski, *Anal. Biochem.* **69**, 100 (1975).

[2]E. Watson *et al.*, *Anal. Biochem.* **59**, 441 (1974).

Derivative Type. Dimethyl.

Reagent. Diazomethane.

Compound Type. Thyrotropin releasing hormone (TRH, pyro-Glu-His-Pro-NH_2).

Reaction

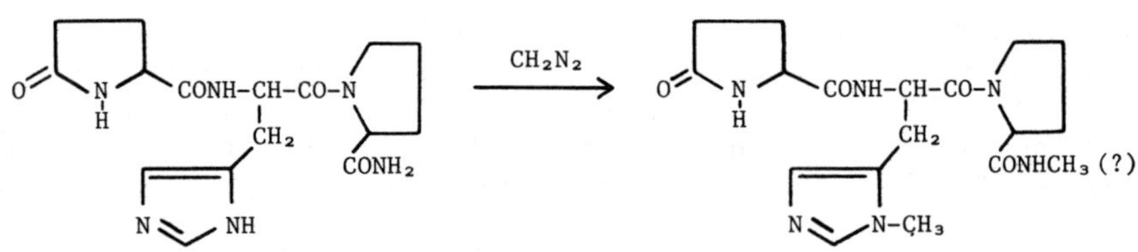

Typical Procedure. Treat TRH sample with a soln of CH_2N_2 in Et_2O and EtOH. Analyze by GC-MS-SIM.

Comments. The method was used to quantitate TRH in 2 ml urine or serum samples. The methyl group on the His ring could be on either N due to tautomerism of the imidazole structure. The location of the second methyl group was uncertain, but the MS data given could be rationalized to support the structure shown above. This method is one of the few examples of quant of a peptide by GC-MS-SIM.

References

[1]N. Heki *et al.*, *Fol. Endocrinol. Jap.* **53**, 690 (1977).

5.4.2. Peptide Esters.

Derivative Type. Methyl ester.

Reagent. Methanol (MeOH)/acetyl chloride (AcCl).

Compound Type. Peptide.

Reaction

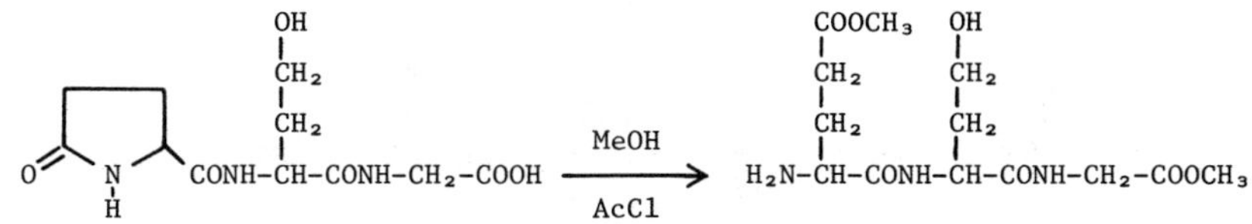

Typical Procedure. Diss μg or smaller peptide sample in 25 μl anhyd MeOH (or CD_3OH) and 5 μl AcCl. Heat at 50° for 5–30 min. Remove HCl by twice evap MeOH to dryness under N_2. Diss residue in MeOH and apply to emitter. Analyze by FD-MS.

Comments. The derivs were used to determine the number of carboxyl groups present in a peptide, to identify the presence of a *pyro*-glutamyl N-terminus, and to obtain sequence information. Paired experiments using MeOH and deutero MeOH facilitated MS interpretation.

References

[1]W. Frick *et al.*, *Biomed. Mass Spectrom.*, **4**, 152 (1977).

5.4.3. N-Acyl Peptide Esters

Derivative Type. Methyl ester, *N*-acetyl.

Reagent. Methanol (MeOH)/HCl, acetic acid (HOAc)/acetic anhydride (Ac_2O).

Compound Type. Peptides.

Reaction

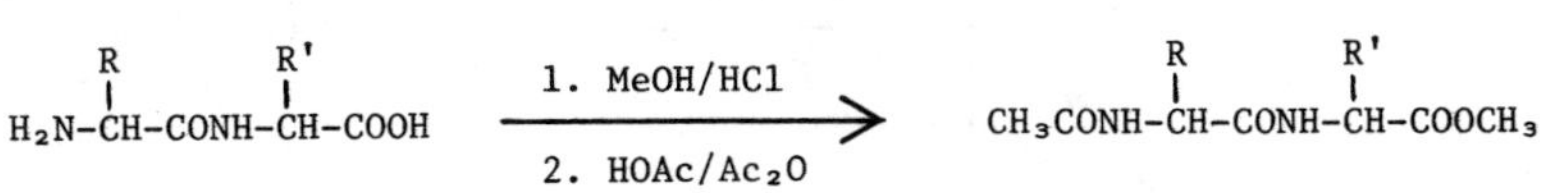

Typical Procedure. Diss 0.1–0.6 μmol peptide in 40 ± 10 μl 2–2.5 *N* HCl in abs MeOH and allow to stand overnight in a sealed tube. Evap under vac and diss residue in 40 ± 10 μl HOAc/Ac_2O (1 : 1, v/v). Allow to stand 4–6 hr at room temp then evap under vac. Analyze by MS.

Comments. The derivs were used for sequencing by MS. His residues appeared to have been methylated. Deuteroacetyl derivs were prepared using CD_3COOH/CH_2Cl_2 (1 : 1, v/v) and trifluoroacetyl derivs using CF_3COOH/TFAA (1 : 1, v/v).[1] Alpin *et al.*[2] compared the acetyl, trifluoroacetyl, benzyloxycarbonyl, methoxycarbonyl, ethoxycarbonyl, methylamino carbonyl (from methylisocyanate), phenylaminocarbonyl (from phenylisocyanate), phthaloyl, and stearyl derivs of peptide methyl esters and concluded that the acetyl derivs provided the best combination of volatility and abundance of sequence ions. They also pointed out that MeOH/HCl gave some peptide cleavage in the esterfication step.

References

[1]M. Senn *et al., J. Am Chem. Soc.* **88**, 5593 (1966).
[2]R. T. Alpin *et al., Org Mass Spectrom.* **2**, 795 (1969).

Derivative Type. Ethyl ester.

Reagent. Triethyloxonium fluoroborate.

Compound Type. *N*-acetyl amino acids or peptides.

Reaction

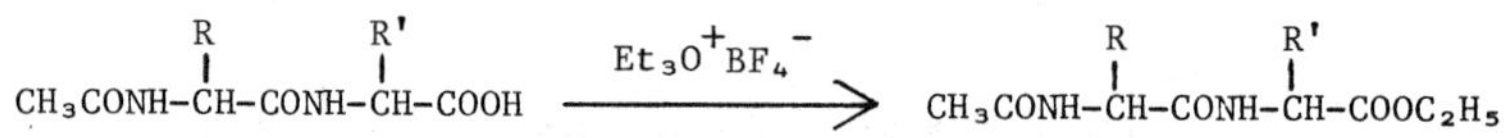

Typical Procedure. To 1 mmol acyl amino acids or peptides in 10 ml H_2O containing 100 mg $NaHCO_3$ add 3.42 g (18 mmol) of the reagent in 1 ml CH_3CN and 1.7 g (20 mmol) $NaHCO_3$ over a period of 10 min. Stir 20 min then ext with EtOAc. Wash ext with 10% $NaCO_3$, and satd NaCl, then dry over Na_2SO_4. Evap to yield the ester deriv.

Comments. Although not used analytically, this aq soln esterfication method could be useful for analytical applications. His and Met containing peptides were converted to ammonium and sulfonium salts, respectively.

References

[1]T. Hamada and O. Yonemitsu, *Chem. Pharm. Bull.* **19**, 1444 (1971).

Derivative Type. *N*-Pivalyl, methyl ester.

Reagent. *N*-Pivalyl chloride, methanol (MeOH)/HCl, diazomethane.

Compound Type. *N*-Terminal amino acids in peptides.

Reaction

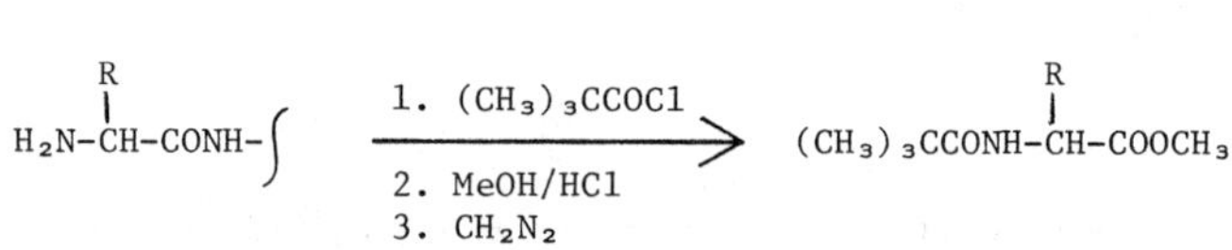

Typical Procedure. Diss 10–50 nmol peptide in 100 μl 40% *N*-ethylmorpholine in H_2O and add 20 μl pivalyl chloride. Allow to react 30 min at room temp then evap under N_2. Add 300 μl MeOH satd with HCl and heat 30 min at 70°. Evap the MeOH, add 200 μl 0.1 *N* HCl, and ext with 2 $\times$ 600 μl EtOAc. Evap the ext and treat the residue with excess CH_2N_2 in Et_2O for a few min. Evap and diss the residue in 5 ml CH_2Cl_2. Analyze by GC.

Comments. The method was used for *N*-terminal analysis of polypeptide chains. Analogous results were obtained with benzoyl chloride. *N*-Terminal acylation with pivalyl or benzoyl groups promoted acid catalyzed cleavage of the adjacent peptide bond presumably via an oxazolinone intermediate. The treatment with CH_2N_2 was to methylate ω-carboxyls of Asp and Glu and phenolic groups of Tyr.

References

[1]J. C. Cavadore *et al., Anal. Biochem.* **60**, 608 (1974).

Derivative Type. *N*-Heptafluorobutyryl (HFB).

Reagent. Heptafluorobutyric anhydride (HFBA).

Compound Type. Peptide methyl esters.

Reaction

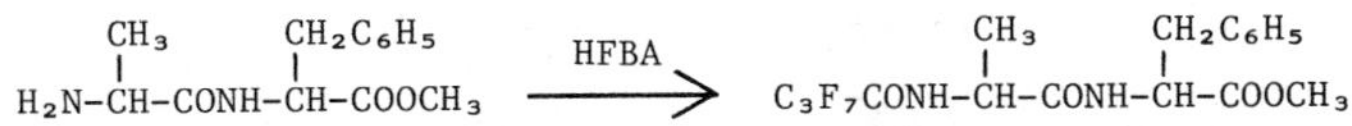

Typical Procedure. Treat peptide methyl ester with HFBA. Analyze by GC or direct inlet MS. (Further details not reported).

Comments. The HFB derivs were compared with TFA derivs (from TFAA). The HFB derivs had retention times *ca* 50% of those of the TFA derivs on Carbowax 20M and vaporized at considerably lower temp in the direct inlet. The MS fragmentation of the derivs was very similar with only the 100 amu shift for acyl containing ions of the HFB derivs.

References

[1]B. A. Andersson, *Acta Chem. Scand.* **21**, 2906 (1967).

Derivative Type. Methyl ester, *N*-heptadecanoyl, and octadecanoyl.

Reagent. Methanol (MeOH)/HCl, heptadecanoic and octadecanoic acids/ethyl chlorocarbonate/triethylamine.

Compound Type. Peptides.

Reaction

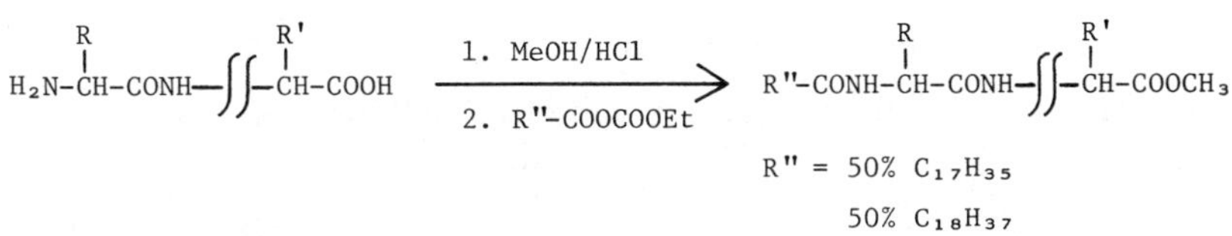

Typical Procedure. Esterify the peptide in MeOH/HCl at 25° for a few hr then evap to dryness. Prepare the mixed anhydride acylating agent by mixing 0.5 mmol of heptadecanoic acid and 0.5 mmol octadecanoic acid with 1 mmol Et_3N in 10 ml THF. Cool to −5° and add 1 mmol ethyl chlorocarbonate. After 20 min add 1 mmol of the peptide methyl ester and 0.14 ml Et_3N in 20 ml THF and stir vigorously for 2 hr while allowing to warm to room temp. Evap under vac and diss residue in $CHCl_3$. Wash with 1 *N* HCl and H_2O then analyze the deriv by MS.

Comments. The deriv was used for MS sequence determination. The use of the mixed C_{17}–C_{18} acyl group facilitated identification of acyl containing ions in the MS by giving pairs of peaks separated by 14 amu.

References

[1]E. Bricas *et al., Biochemistry* **4**, 2254 (1965).

Derivative Type. N-(3-Hydroxydecanoyl).

Reagent. *N*-(3-Hydroxydecanoyloxy)succinimide.

Compound Type. Peptides (as methyl esters).

Reaction

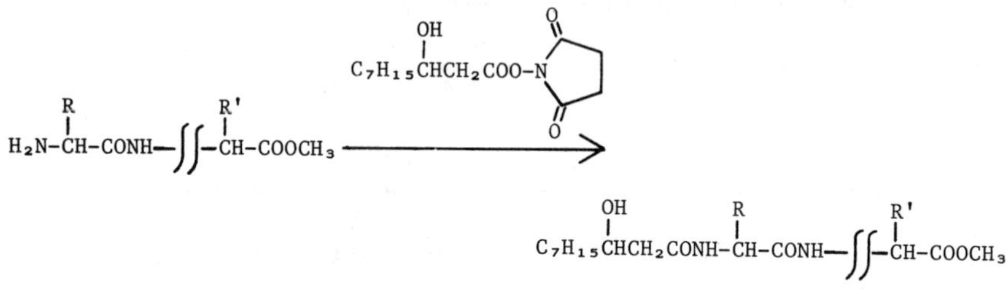

Typical Procedure. Diss 0.3 mmol of the peptide methyl ester and 0.3 mmol of the reagent in 1 ml EtOAc containing 0.5 ml DMF and 0.04 ml Et_3N. Stir 16 hr at room temp then dil with 5 ml EtOAc. Wash with 1 *N* HCl, H_2O, 1 *N* $NaHCO_3$, and H_2O. Dry over $MgSO_4$ and evap under vac. Recryst from EtOAc-1t pet ether and analyze by MS.

Comments. A series of 3-hydroxyalkanoyl and Δ^2-acyl derivs of model peptides ranging from tri- to dodecapeptides were examined for use in sequence analysis. The 3-hydroxydecanoyl and the Δ^2-decanoyl derivs were found to be superior to the earlier used acetyl and decanoyl derivs. The 3-hydroxydecanoyl was particularly useful in that it gave two series of peaks 18 amu apart due to partial dehydration. *N*-Crotonyl and *N*-cinnamoyl derivs were also prepared but their utility was highly dependent upon the nature of the peptide. Permethylation generally increased the volatility of the derivs augmenting their utility for sequence analysis.

References

[1]L. Okada *et al.*, *Tetrahedron* **30**, 1175 (1974).

Derivative Type. Methyl ester, *N*-caproyl.

Reagent. Methanol (MeOH)/HCl, caproic ethylcarbonic anhydride.

Compound Type. Peptides.

Reaction

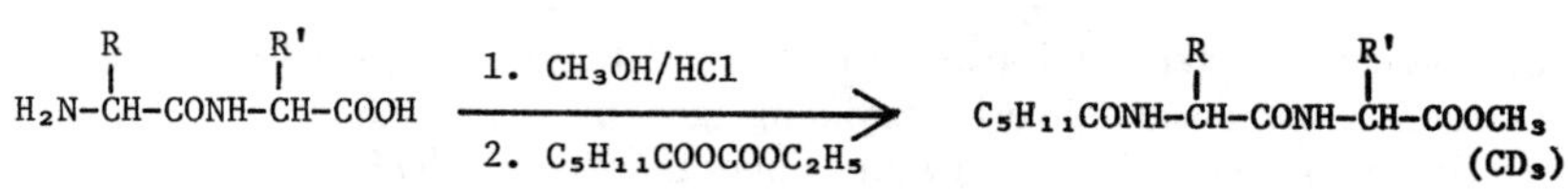

Typical Procedure (Details Not Given). Esterify the peptide with MeOH satd with HCl. Acylate by the mixed anhydride method.[2] Analyze derivs by MS.

Comments. The derivs were found to be useful for MS sequence determination. Trideuteromethyl esters (prepared with CD_3OD) were found to be useful for identification of C-terminal fragments.

References

[1]J. P. Flikweert *et al.*, *Rec. Trav. Chim.* **86**, 293 (1967).

[2]T. Wieland and H. Bernhard, *Ann. Chem.* **572**, 190 (1951).

Derivative Type. *N*-Decanoyl, methyl ester.

Reagent. *N*-Decanoyloxysuccinimide, diazomethane.

Compound Type. Peptides.

Reaction

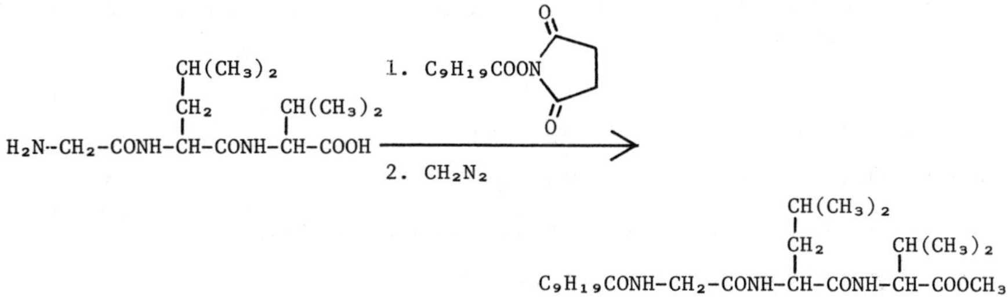

Typical Procedure (Experimental Details Not Reported). Acylate the peptide in aq soln with *N*-decanoyloxysuccinimide.[5] Esterify the acylated peptide with diazomethane. Analyze by MS.

Comments. The method was used for sequence analysis by MS.[1-4] The *N*-ethoxycarbonyl peptides were also prepared using *N*-ethoxycarbonyloxysuccinimide and *t*-butyl esters using isobutene. The method had the advantage of permitting acylation in aq soln. The derivs were subsequently used with CI-MS which had the advantage of giving sequence ions from both sides of the amide group cleavage.[6]

References

[1] A. A. Kiryushkin *et al.*, *Tetrahedron Lett* 33 (1966).
[2] A. A. Kiryushkin *et al.*, *Biochem. Biophys. Res. Commun.* **24**, 943 (1966).
[3] M. M. Shemyakin *et al.*, *Nature* **211**, 361 (1966).
[4] N. S. Wulfson *et al.*, *Tetrahedron Lett.* 39 (1966).
[5] G. W. Anderson *et al.*, *J. Am. Chem. Soc.* **86**, 1839 (1964).
[6] A. A. Kiryushkin *et al.*, *Org. Mass Spectrom.* **5**, 19 (1971).

Derivative Type. *N*-Decanoyl, methyl ester.

Reagent. *N*-Decanoyloxysuccinimide, diazomethane.

Compound Type. Peptides.

Reaction

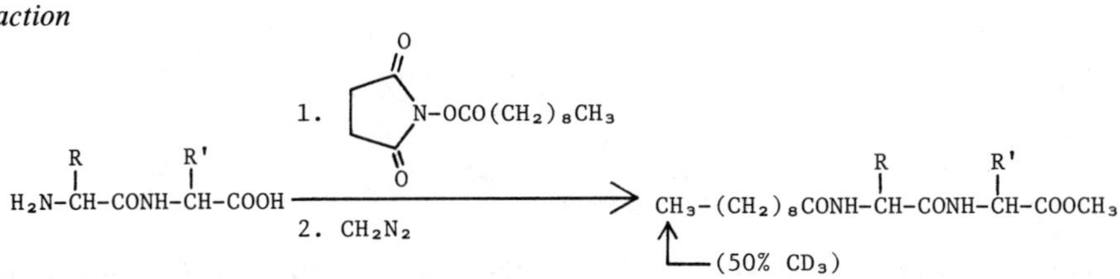

Typical Procedure. Diss 0.1 mmol peptide (e.g., 27.6 mg α-glutamyl-glutamic acid) in 1 ml warm H_2O and add 0.05 ml (0.3 mmol) Et_3N in 2 ml dimethoxyethane (DME). Add 26.9 mg (0.1 mmol) *N*-decanoyloxysuccinimide in 2 ml DME and allow to stand 16 hrs. Acidify the soln, dil with H_2O, and ext the resulting emulsion with EtOAc. Evap ext to dryness and diss residue in 1 ml anhyd MeOH. Add excess CH_2N_2 in Et_2O then evap to dryness to obtain deriv. Analyze by MS.

Comments. Peptides were derivatized with 50% d_3-methyl reagent to yield 50% d_3-derivs which facilitated MS interpretation. The preparation of the labeled reagent by deuterium reduction of sebacic hemiester and acylation of *N*-hydroxysuccinimide was given. 50%-d_3-Acetyl derivs were also prepared for the same purpose using acetic anhydride/pyridine. MS data for the derivs were given.

References

[1]J. van Heijenoort *et al*., *Tetrahedron* **23**, 3403 (1967).

Derivative Type. *N*-Succinyl, *N,O*-permethyl.

Reagent. Succinic anhydride, methylsulfinyl carbanion/methyl iodide.

Compound Type. Amino acids and peptides.

Reaction

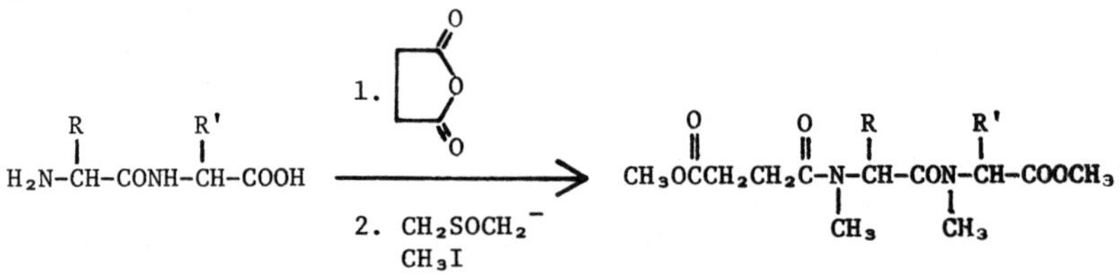

Typical Procedure. To 1 μmol of peptide add a soln of 1.1 μmol (110 μg) of succinic anhydride in 50 μl DMSO. Heat at 80° for 30 min then cool to room temp. Add approx fivefold excess of methylsulfinyl carbanion[2] (1 *M*). After 1 min add eightfold excess of CH_3I, then 1 min later quench with H_2O. Ext with $CHCl_3$, dry ext over Na_2SO_4, and evap to dryness.

Comments. The deriv gave ions characteristic of amino acid sequence. The deriv retained the advantages of the permethyl *N*-acetyl deriv but gave more abundant high mass ions.

References

[1]D. C. DeJongh *et al.*, *Biomed. Mass Spectrom.* **3**, 191 (1976).

[2]D. W. Thomas, *Biochem. Biophys. Res. Commun.* **33**, 483 (1968).

Derivative Type. Methyl ester, N-benzoyl.

Reagent. Methanol (MeOH)/HCl, benzoic ethylcarbonic anhydride.

Compound Type. Peptides.

Reaction

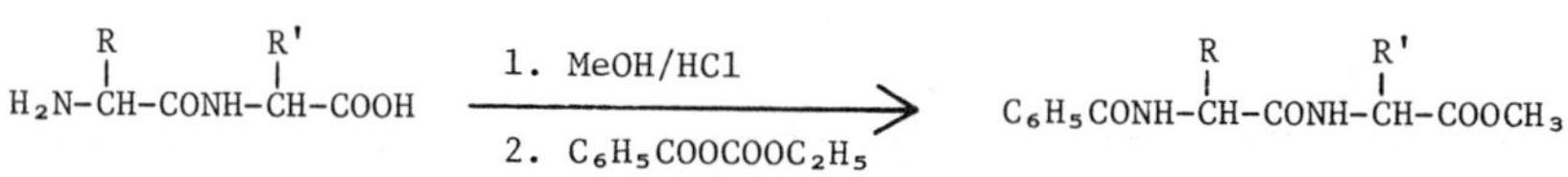

Typical Procedure (Details Not Given). Esterify the peptide with MeOH/HCl.[2] Acylate with the mixed anhydride from benzoic (or pentadeuterobenzoic) acid and ethylchlorocarbonate.[3] Analyze the deriv by MS.

Comments. The derivs were examined for use in MS sequencing of peptides but the method was troubled by partial formation of ethoxycarbonyl derivs from the mixed anhydride. This observation led to the intentional preparation of ethoxycarbonyl derivs for use in sequence analysis (q.v.).[4]

References

[1] J. P. Kamerling *et al.*, *Org. Mass Spectrom.* **1**, 345 (1968).
[2] F. E. King *et al.*, *J. Chem. Soc.* 1039 (1954).
[3] A. R. Batterby and J. C. Robinson, *J. Chem. Soc.* 259 (1955).
[4] J. P. Kamerling *et al.*, *Org. Mass Spectrom.* **1**, 351 (1968).

Derivative Type. Ethyl ester, dimethylaminobenzoyl.

Reagent. Ethanol (EtOH)/thionyl chloride, succinimido dimethylaminobenzoate.

Compound Type. Peptides.

Reaction

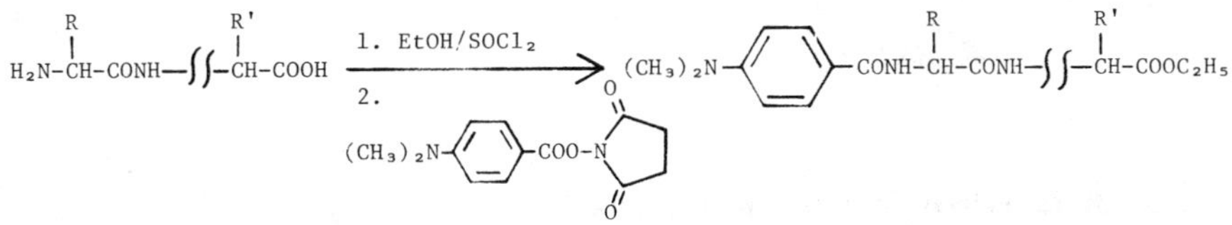

Typical Procedure. Reflux for 10 min 1–2 mg peptide in 1–2 ml EtOH containing a few drops of $SOCl_2$. Evap to dryness under vac. Add 10% Et_3N in THF and a mol equiv of succinimido dimethylaminobenzoate. Allow to react overnight at room temp. Purify the derivs by TLC on silica gel using $CHCl_3$/*sec*-BuOH/HOAc (70:30:3, v/v/v) and locate the spots by fluorescence. Elute with EtOH and analyze by direct probe MS.

Comments. Derivs were prepared from Val-Ile-Ala and Leu-Try-Ile using *o*-, *m*-, and *p*-dimethylaminobenzoates. The *m*-deriv was judged most useful in terms of MS sequence peaks. The prep of the reagents from N-hydroxysuccinimide[2] was given.

References

[1] E. Haralambidou and R. A. Day, *Org. Mass Spectrom.* **10**, 683 (1975).
[2] G. W. Anderson *et al.*, *J. Am. Chem. Soc.* **86**, 1839 (1964).

Derivative Type. β-Naphthoyl, methyl ester.

Reagent. *N*-Hydroxysuccinimido β-naphthoate (HSN), methanol (MeOH)/thionyl chloride.

Compound Type. Amino acids and peptides.

Reaction

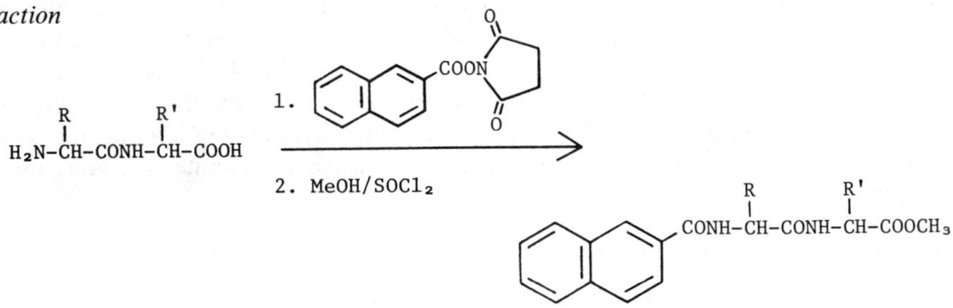

Typical Procedure. Separate amino acids or peptides by paper chromatography. Diss 60 mg HSN in 100 ml $CHCl_3$ and add 1 ml Et_3N immediately prior to use of the soln for dipping or spraying the chromatogram. Rinse chromatogram with benzene (to improve contrast) by chromatographing in this solvent overnight. Dry in an oven at 110° for a few min then locate spots with long wavelength UV. Cut out the spots and elute each with 2–3 ml of refluxing MeOH by placing the paper in a distilling column attached to a 5- or 10 ml round bottom flask. After 1 hr reflux, add 2 drops $SOCl_2$ and continue refluxing 30 min. Conc soln under vac and transfer a portion to a direct inlet probe for MS analysis.

Comments. Dimethylaminonaphthalenesulfonyl (DANS) chloride was used in a similar manner as a 100 mg/100 ml acetone soln to which 3 ml of Et_3N were added before use. The benzene rinse was not required for DANS derivs. Ethyl esters were formed in a similar manner using EtOH in place of MeOH. The derivs served both to locate the chromatogram spots and to provide sequence information in the MS. Dimethylaminobenzaldehyde and -cinnamaldehyde derivs were used for the same purpose (q.v.).

References

[1]H. Falter *et al.*, *Anal. Biochem.* **67**, 359 (1975).

5.4.4. Schiff Base Derivatives of Peptide Esters

Derivative Type. Ethyl ester, acetylacetonyl.

Reagent. Ethanol (EtOH)/HCl, acetylacetone.

Compound Type. Peptides.

Reaction

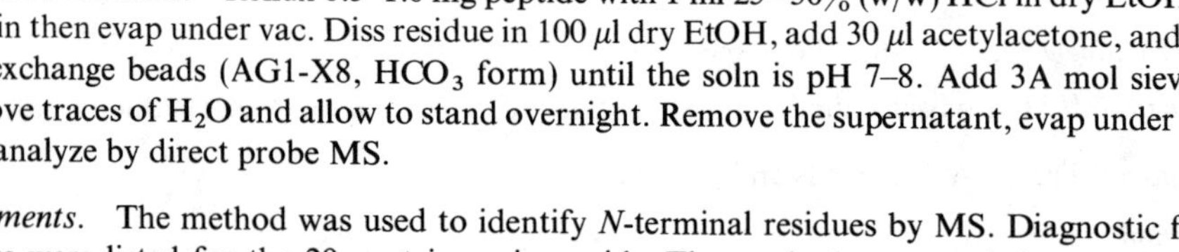

Typical Procedure. Reflux 0.5–1.0 mg peptide with 1 ml 25–30% (w/w) HCl in dry EtOH for 10 min then evap under vac. Diss residue in 100 μl dry EtOH, add 30 μl acetylacetone, and dry ion exchange beads (AG1-X8, HCO_3 form) until the soln is pH 7–8. Add 3A mol sieve to remove traces of H_2O and allow to stand overnight. Remove the supernatant, evap under vac, and analyze by direct probe MS.

Comments. The method was used to identify *N*-terminal residues by MS. Diagnostic fragments were listed for the 20 protein amino acids. The method converted Arg peptides to δ-*N*-(2-pyrimidyl)-Orn peptides and also deriv the ε amino of Lys. With Asn residues it was necessary to use CH_2N_2 for esterfication since EtOH/HCl hydrolyzed the Asn to Asp. The derivs gave readily interpretable MS for up to octapeptides containing only neutral amino acids, but the lower volatility of those containing basic amino acids limited the procedure to pentapeptides or smaller.

References

[1]V. Bacon *et al.*, *Biochem. Biophys. Res. Commun.* **37**, 878 (1969).

Derivative Type. *p*-Dimethylaminobenzylidene, methyl ester.

Reagent. *p*-Dimethylaminobenzaldehyde (DMB), methanol(MeOH)/thionyl chloride.

Compound Type. Amino acids and peptides.

Reaction

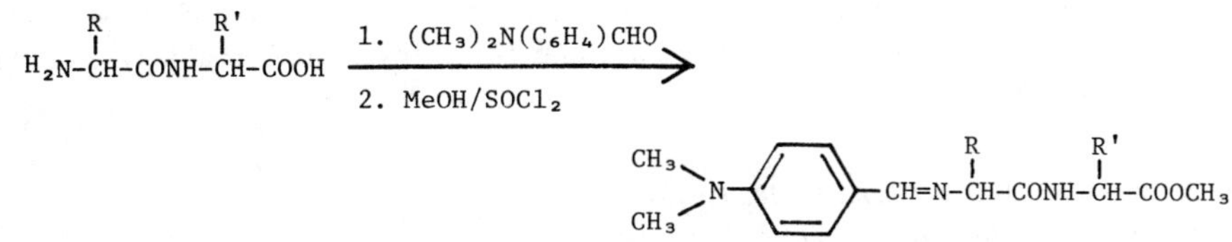

Typical Procedure. Separate amino acids or peptides by paper chromatography. Diss 100 mg DMB in 100 ml anhyd Et_2O. Add 1 ml HOAc immediately before using the soln for dipping or spraying the chromatogram. Dry the chromatogram in an oven at 105–110° for 10 min. Locate the spots with short wavelength UV and cut them out. Elute with 2–3 ml of refluxing MeOH by placing the paper in a distilling column attached to a 5- or 10-ml round bottom flask. After 1 hr reflux, add 2 drops $SOCl_2$ to the MeOH and continue refluxing 30 min. Conc soln under vac and transfer a portion to a direct inlet probe for analysis by MS.

Comments. *p*-Dimethylaminocinnamaldehyde was also used to form Schiff bases by means of 5 mg/100 ml dioxane soln to which was added 1 ml Et_3N immediately prior to use. Chromatograms were dried in an oven after dipping yielding deep red to reddish brown spots. Ethyl esters were also used and were formed in an identical procedure using EtOH in place of MeOH. The derivs not only served to locate the chromatogram spots but also provided sequence information from MS analysis. Derivs formed with dimethylaminonaphthalenesulfonyl (DANS) chloride and *N*-hydroxysuccinimido β-naphthoate were also used for the same purpose (q.v.).

References

[1]H. Falter *et al.*, *Anal. Biochem.* **67**, 359 (1975).

Derivative Type. Ethyl ester, *p*-dimethylaminobenzylidene.

Reagent. Ethanol (EtOH)/thionyl chloride, *p*-dimethylaminobenzaldehyde (DMB).

Compound Type. Peptides.

Reaction

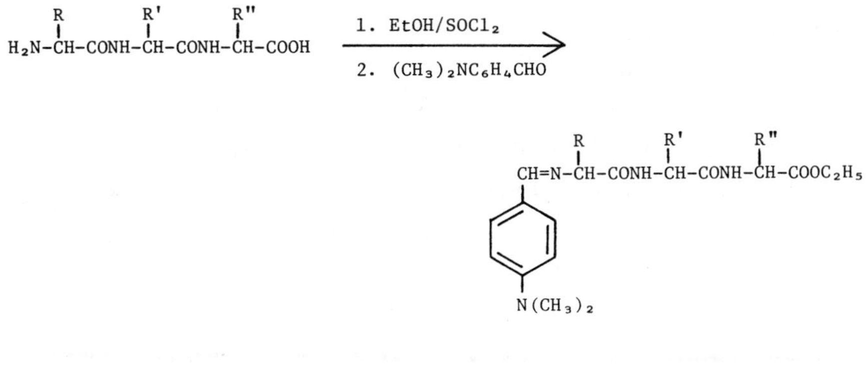

Typical Procedure. Reflux for 10 min 1–2 mg peptide in 1–2 ml EtOH containing a few drops of $SOCl_2$. Evap to dryness under vac then add 1–2 mg DMB in *ca* 1 ml HOAc. Heat 10 min at 118° then evap under vac. Diss deriv in EtOH for transfer to the probe for MS analysis.

Comments. The deriv was used for sequencing peptides by MS. The 4-dimethylamino-naphthylidene deriv prep in the same manner from 4-dimethylamino-1-naphthaldehyde gave similar results. These, along with the DANS deriv were the three most useful of 37 *N*-terminal derivs examined.[1,2] In a comparison of *o*-, *m*-, and *p*-dimethylaminobenzaldehyde derivs, the *p*- and *m*-derivs gave the most useful MS sequence peaks.[3]

References

[1]R. A. Day *et al.*, *J. Org. Chem.* **38**, 782 (1973).

[2]G. C. Patil *et al.*, *Org. Mass Spectrom.* **7**, 817 (1973).

[3]E. Haralambidou and R. A. Day, *Org. Mass Spectrom.* **10**, 683 (1975).

5.4.5. Other *N*-Derivatives of Peptide Esters

Derivative Type. Ethoxycarbonyl, methyl ester.

Reagent. Diethylpyrocarbonate, diazomethane.

Compound Type. Peptides.

Reaction

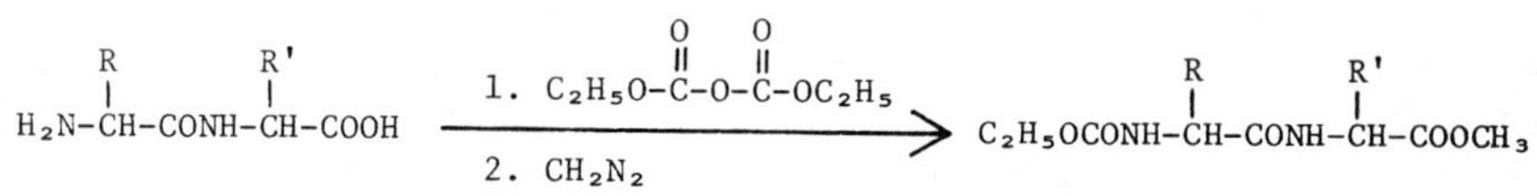

Typical Procedure. Diss 10 μmol peptide in 2 ml H_2O and add 15 μmol $NaHCO_3$ per COOH group. Add 10 μmol diethylpyrocarbonate per acylable group and shake 30–60 min at room temp. Acidify with 1 *N* HCl and ext with 2 ml EtOAc. Wash ext with 2 × 1 ml H_2O. Add CH_2N_2 in Et_2O until the yellow color persists. After 15 min, evap the soln. Purify the deriv by TLC, elute with EtOAc, and analyze by MS.

Comments. Ethoxycarbonyl derivs which were more H_2O soluble (e.g., the deriv of Gly-Ser) were isolated by lyophilization and diss in MeOH prior to esterfication. In addition to the amino group, the imidazole ring of His, the SH of Cys, and the phenolic OH of Tyr were also ethoxycarbonylated. The derivs were volatile and gave useful molecular and sequence peaks in the MS.

References

[1]J. P. Kamerling *et al.*, *Org. Mass Spectrom.* **1**, 351 (1968).

Derivative Type. 5-Dimethylamino-1-napthalenesulfonyl (DANS).

Reagent. 5-Dimethylamino-1-naphthalenesulfonyl chloride (DANS-Cl).

Compound Type. Peptides (as methyl esters).

Reaction

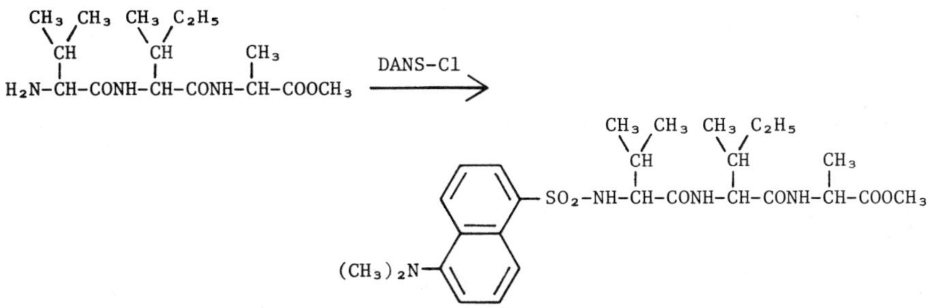

Typical Procedure. Esterfy peptide by refluxing in MeOH for 10 min after addition of $SOCl_2$. Prepare DANS derivs[2] by diss 0.5–5 nmol of peptide in 10–15 μl 0.2 *M* $NaHCO_3$ and adding an equal vol of DANS-Cl in acetone (2.5 mg/ml). Allow to react 1 hr at 37° or 2–3 hr at room temp. Analyze product by direct probe MS.

Comments. The deriv was used for sequencing of peptides by MS. Of 37 *N*-terminal derivs examined, this was one of three deemed useful for MS sequencing (the other two were the *p*-dimethylaminobenzylidine and 4-dimethylaminonaphthylidene q.v.).

References

[1] R. A. Day *et al.*, *J. Org. Chem.* **38**, 782 (1973).

[2] W. R. Gray, *Meth. Enzymol.* **11**, 142 (1967).

5.4.6. Dipeptide Derivatives

5.4.6.1 N-Acyl Dipeptide Esters

Derivative Type. Methyl ester, trifluoroacetyl.

Reagent. Methanol (MeOH)/HCl, methyl trifluoroacetate (MeOTFA).

Compound Type. Dipeptides.

Reaction

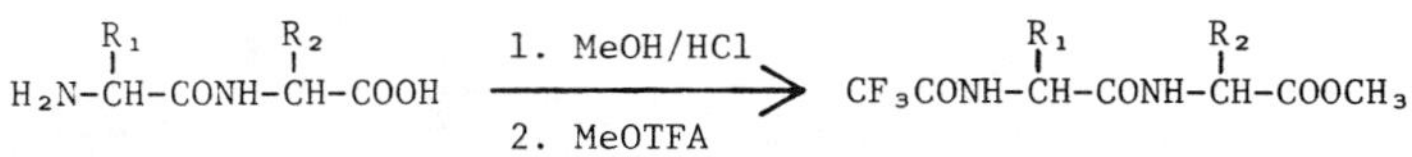

Typical Procedure. Adjust pH of aq peptide soln to 3 and dry under vac. Diss residue in 0.1 *N* HCl in MeOH and allow to stand at 20° for 24 hr. Evap soln and diss residue in dioxane/DMF (1 : 1, v/v). Add MeOTFA and Et_3N (further details not given). Analyze derivs by GC-MS.

Comments. The method was used to analyze dipeptides from dipeptidylaminopeptidase digestion mixtures. Other workers, however, prepared the derivs from a large number of dipeptides and, since not all gave volatile products, they concluded that the deriv could not be regarded as generally useful.[2]

References

[1]Y. A. Ovchinnikov and A. A. Kiryushkin, *FEBS Lett.* **21**, 300 (1972).

[2]H. Lindley and P. C. Davis, *J. Chromatogr.* **100**, 117 (1974).

Derivative Type. Methyl ester, pentafluoropropionyl.

Reagent. Methanol/HCl, pentafluoropropionic anhydride (PFPA).

Compound Type. Dipeptides.

Reaction

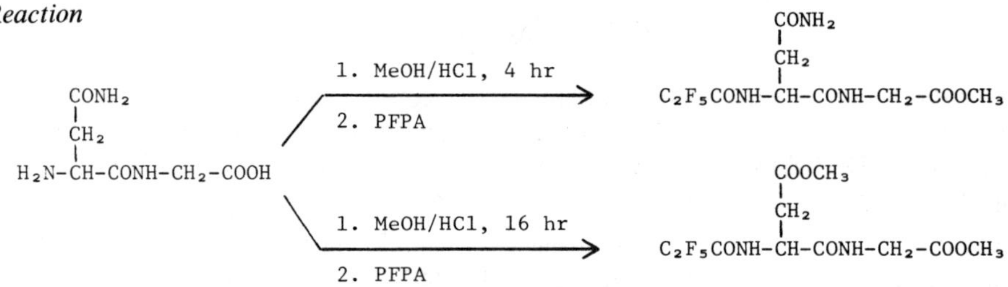

Typical Procedure. Treat peptide mixt from dipeptidyl aminopeptidase I digestion with 0.1 *N* HCl in MeOH for 4 hr at room temp or for 16 hr at 45°. Evap to dryness under vac and treat with 0.5 ml PFPA. Evap to dryness under N_2 then apply vac for 15 min. Diss residue in dry benzene or CH_2Cl_2 prior to GC or GC-MS analysis.

Comments. The method was used to differentiate Asn and Gln residues from Asp and Glu residues in peptide sequencing. The 4 hr esterfication at room temp did not hydrolyze the amide function yielding a deriv which would not chromatograph. Presence of Asn and Gln was indicated by appearance of new peaks after the 16 hr/45° esterfication.

References

[1]M. A. Young and D. M. Desiderio, *Anal. Biochem.* **70**, 110 (1976).

Derivative Type. Methyl ester, pentafluoropropionyl (PFP).

Reagent. Methanol (MeOH)/thionyl chloride, pentafluoropropionic anhydride (PFPA).

Compound Type. Dipeptides.

Reaction

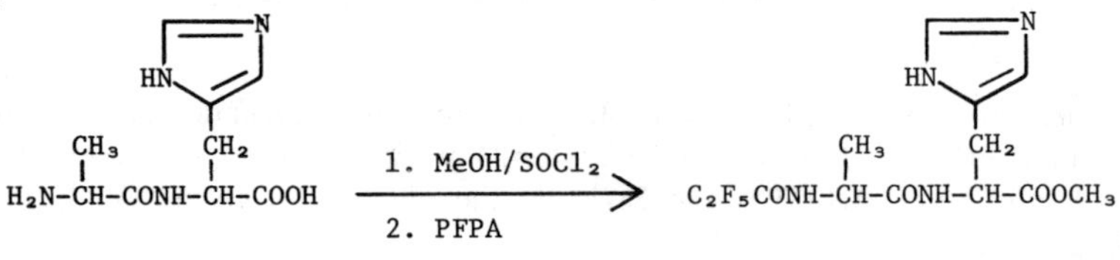

Typical Procedure. Diss lyophilized dipeptide sample in 1 ml MeOH. Cool in dry ice-acetone bath and add 0.4 ml $SOCl_2$. After several min heat at 45° for 30 min. Evap to dryness under vac. Diss residue in 200 μl PFPA and allow to react at room temp for 15 min. Evap to dryness under vac and diss residue in 200 μl dry dioxane. Analyze by GC-MS.

Comments. The method was used for peptide sequencing after hydrolysis of the peptide with dipeptidylaminopeptidase I (before and after one step of Edman degradation). Arg containing peptides were first deriv with 2,4-pentanedione (q.v.).[2] Lindley and Davis[3] prepared the methyl ester, PFP derivs (as well as TFA and acetyl derivs) of a large number of dipeptides and concluded that none of the three types of derivs could be regarded as generally useful for dipeptide analysis since none gave volatile derivs of all the dipeptides examined.

References

[1]R. M. Caprioli *et al.*, *Biochem. Biophys. Res. Commun.* **55**, 67 (1973).
[2]H. R. Morris *et al.*, *Biochem. Biophys. Res. Commun.* **51**, 247 (1973).
[3]H. Lindley and P. C. Davis, *J. Chromatogr.* **100**, 117 (1974).

5.4.6.2. Schiff Base Derivatives of Dipeptide Esters

Derivative Type. *N*-Acetylacetonyl, methyl ester.

Reagent. Trimethylphenylammonium hydroxide (TMPAH), acetylacetone.

Compound Type. Dipeptides.

Reaction

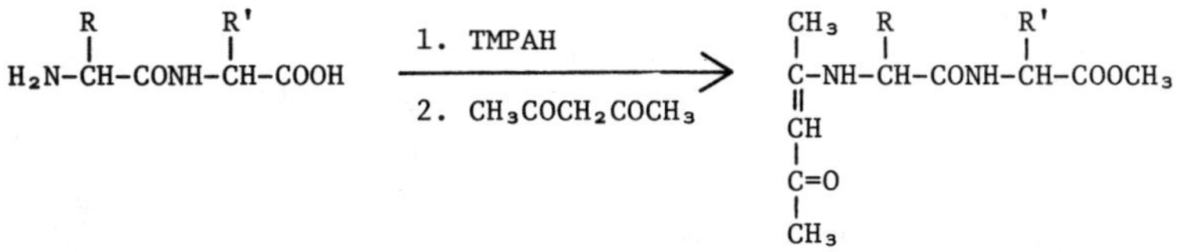

Typical Procedure. Diss *ca* 1 μmol dipeptide in 20 μl of a 0.1 *N* TMPAH soln in MeOH. Add 1 μl acetylacetone and allow mixt to stand over 3A mol sieve for at least 3 hr at room temp. Evap under N_2 and analyze residue by direct probe MS.

Comments. The method was used to analyze dipeptide mixtures without prior separation. The methyl esters were formed by pyrolysis of the ammonium salts on the direct probe. MS data (both EI and CI) were reported for 16 dipeptides.

References

[1]G. M. Schier *et al.*, *Biomed. Mass Spectrom.* **1**, 212 (1974).

Derivative Type. Acetylacetonyl, methyl ester, trimethylsilyl ether.

Reagent. Acetylacetone, diazomethane, hexamethyldisilazane (HMDS)/pyridine.

Compound Type. Dipeptides.

Reaction

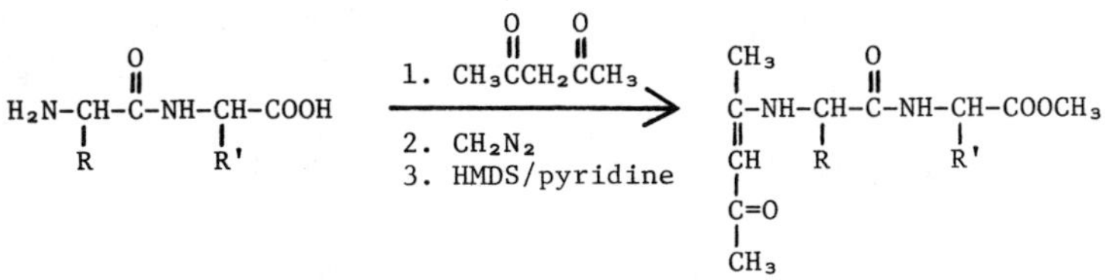

Typical Procedure. Diss mixt of dipeptides in 0.1 *M* *N*-methylmorpholine (base) in DMF/MeOH (1 : 1, v/v). Add sufficient base to yield pH approx 9. Add mol sieve and 5 μl acetylacetone per 100 μl soln (equiv to 1–2 μmol dipeptide) and heat at 100° for 2 hr. Cool soln to room temp, neutralize base with 1 *M* HCOOH in CH_2Cl_2, and blow to dryness under N_2. Diss residue in 200 μl MeOH and add excess CH_2N_2 in Et_2O. Let stand 30 min at room temp, evap solvent, diss residue in 30 μl pyridine and 15 μl HMDS, and heat to 60° for 30 min. Analyze aliquots directly by GC-MS.

Comments. The deriv was used for sequencing of peptides using dipeptide mixtures. HMDS gave TMS ethers of any hydroxyl groups present. Et_3N, TMPAH, and pyridine were also investigated as bases but gave lower yields. Trifluoroacetylacetone and hexafluoroacetylacetone were investigated but gave poor results.

References

[1]H. Frank *et al.*, *Anal. Chem.* **49**, 287 (1977).

Derivative Type. *N*-Ethoxycarbonylpropenyl, methyl ester.

Reagent. Ethyl acetoacetate, dimethyltrideuteromethylphenylammonium hydroxide (d_3-TMPAH).

Compound Type. Dipeptides.

Reaction

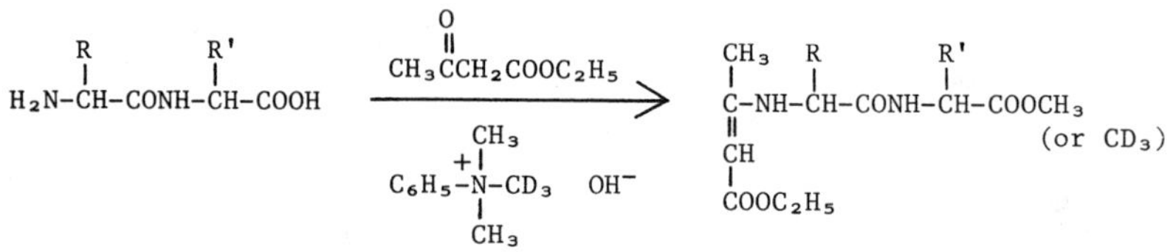

Typical Procedure. Diss the dry dipeptide mixt (0.1 μmol) in 5–10 μl of 0.1 *N* ethanolic d_3-TMPAH, 20 μl dry EtOH, and 5 μl ethyl acetoacetate in a screw cap reaction vial. Add a stick of 3A mol sieve, seal, and heat at 70° for 30 min. Remove sieve and evap solvent under vac in a capillary tube. Analyze by direct probe MS.

Comments. The method was used to identify dipeptides from dipeptidylaminopeptidase I digestion of polypeptides using CI-MS. The methylation reaction occurred on the heated probe. Lys and Arg dipeptides reacted with a second mol of acetoacetate (reaction below). Use of the partially deuterated reagent allowed determination of the number of methyl groups introduced into the molecule by virtue of the isotopic multiplet formed. The sidechain functional groups of Tyr, Asp, Glu, carboxymethyl-CySH, and the Arg deriv below were methylated in the second step. Trp and His were partially ring methylated.

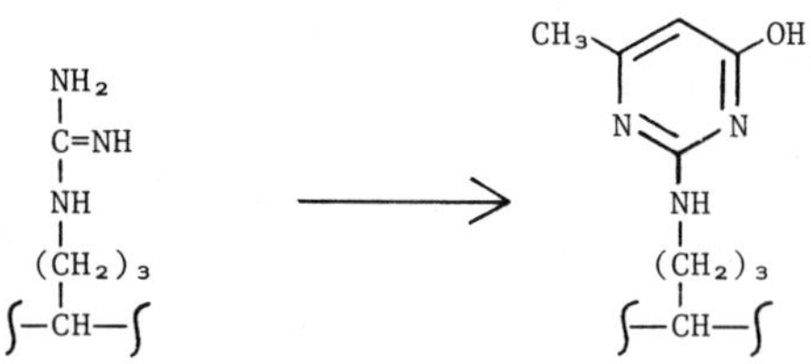

References

[1]G. M. Schier *et al.*, *Biomed. Mass Spectrom.* **3**, 32 (1976).

5.4.6.3. Other Dipeptide Derivatives

Derivative Type. Trimethylsilyl.

Reagent. Bis(trimethylsilyl)trifluoroacetamide (BSTFA).

Compound Type. Dipeptides.

Reaction

Typical Procedure. Deriv dipeptide mixt with BSTFA in CH_3CN (details not given) and analyze by GC-MS.

Comments. Sharp reproducible GC peaks were obtained for all dipeptides examined (*ca* 100 of a possible 400) except some containing His, Asn, or Gln and all containing Arg. The method was used for sequencing polypeptides using diaminopeptidase with as little as 10 nmol of sample.

References

[1]C. Krutzsch and J. J. Pisano, *Fed. Proc.* **34**, 630 (1975).

Derivative Type. Fluorescamine derivative.

Reagent. 4-Phenylspiro[furan-2(3H),1′-phthalan]-3,3′-dione (fluorescamine).

Compound Type. Amino acids and peptides.

Reaction

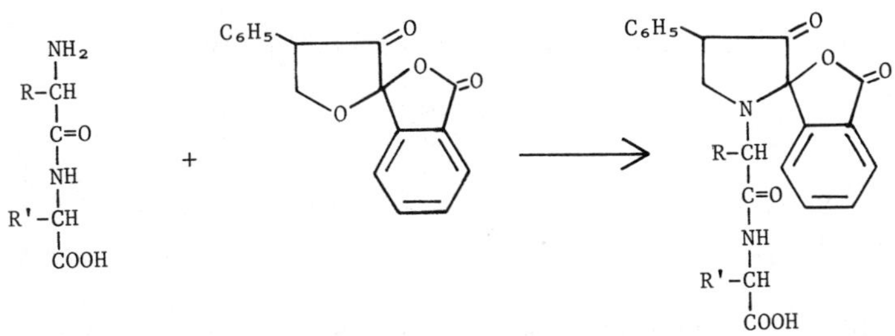

Typical Procedure. To 100 μl of a 7.2 *mM* soln of dipeptide in 0.2 *M* pyridine add an equal vol of an acetone soln of fluorescamine (0.2%). Evap in vac dessicator and analyze by direct probe MS.

Comments. The derivs were examined for MS identification of amino acids and dipeptides for use in peptide sequencing. The derivs gave identifiable spectra with their fluorescence being useful for isolation. The lactone form of the deriv which was reported likely resulted from dehydration on the direct probe.

References

[1]D. G. Pritchard *et al.*, *Biochem. Biophys. Res. Commun.* **65**, 312 (1975).

Derivative type. MDPF derivative.

Reagent. 2-Methoxy-2,4-diphenyl-3(2H)-furanone (MDPF).

Compound Type. Amino acids and peptides.

Reaction

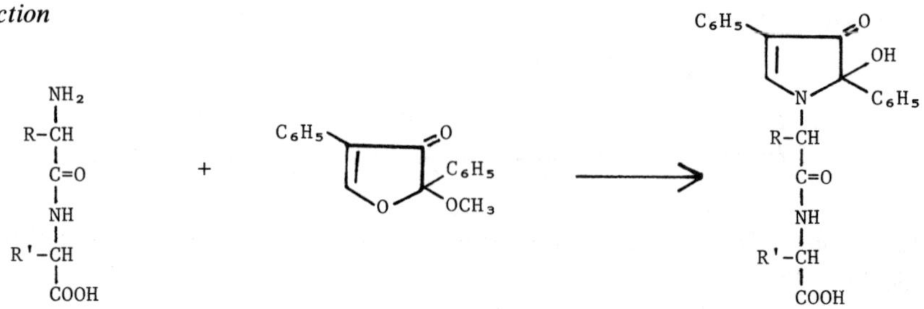

Typical Procedure. To 100 μl of a 7.2 *mM* soln of dipeptide in 0.2 *M* dimethylbenzylamine-acetate buffer add 100 μl of a 0.2% soln of MDPF in acetone. Evap to dryness in a vac dessicator and analyze by MS.

Comments. The derivs were examined for MS identification of amino acids and dipeptides for use in peptide sequencing. The derivs gave identifiable spectra with their fluorescence being useful for isolation.

References

[1]D. G. Pritchard *et al., Biochem. Biophys. Res. Commun.* **65**, 312 (1975).

Derivative Type. Oxazoline-5-one.

Reagent. Dicyclohexylcarbodiimide (DCC).

Compound Type. Dipeptides [as *N-t*-butoxycarbonyl(BOC) derivs].

Reaction

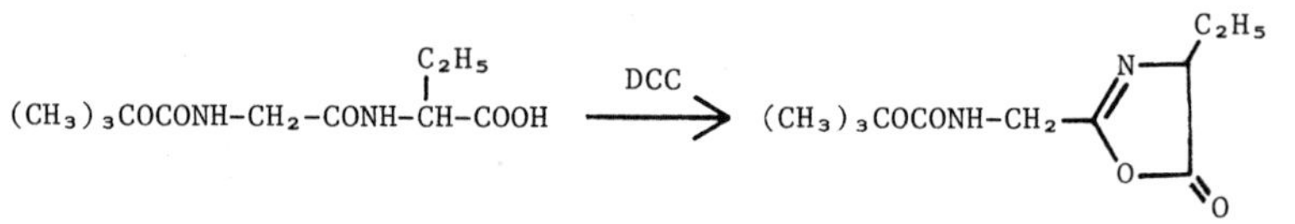

Typical Procedure. Diss 40 μmol of BOC-dipeptide in EtOAc and add 1.2 ml of a 0.1 *M* soln of DCC in EtOAc. Dil soln to 10 ml with EtOAc and analyze 1 μl by FID-GC.

Comments. The derivs gave single symmetrical GC peaks. Since the derivs could be analyzed within 1–2 min, the method competed favorably in terms of time and convenience with diazomethane esterfication.

References

[1]O. Grahl-Nielsen, *J. Chromatogr.* **93**, 229 (1974).

Derivative Type. Imidazolin-5(4)-one.

Reagent. Trifluoroacetic anhydride (TFAA).

Compound Type. Peptide amides.

Reaction

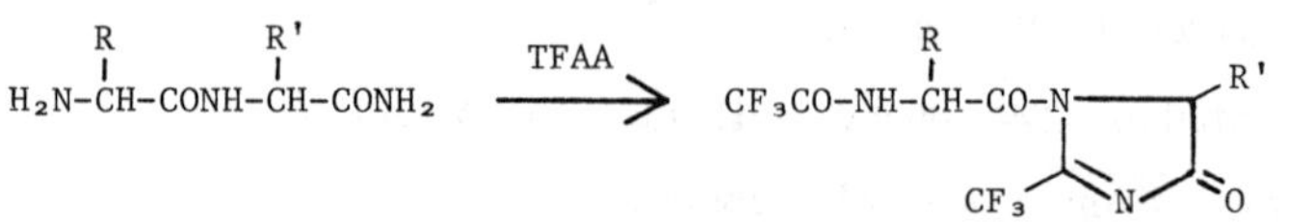

Typical Procedure. Diss 100 μg of peptide amide in 200 μl of CH_2Cl_2 and 50 μl of TFAA in a screw cap vial. After 1 hr at room temp remove excess reagent under stream of N_2. Diss residue in 50 μl of $CHCl_2$ and analyze by GC-MS or direct probe MS.

Comments. Analogous derivs were also prepared with PFPA. The MS of the derivs allowed assignment of amino acid sequence. C Terminal Pro did not allow the cyclization to occur. The method was proposed as potentially useful for sequencing polypeptides after conversion to small peptide amides by treatment of the polypeptide methyl ester with methanolic ammonia.

References

[1]C. Engelfried *et al.*, *Biomed. Mass Spectrom.* **3**, 241 (1976).

5.4.7. Derivatives for Modification of Specific Amino Acid Residues in Peptides

Derivative Type. *N*-Decanoyl, pyrimidyl ornithine peptide methyl ester.

Reagent. *N*-Decanoyloxysuccinimide, tetraethoxypropane/methanol (MeOH)/HCl.

Compound Type. Arginine containing peptides.

Reaction

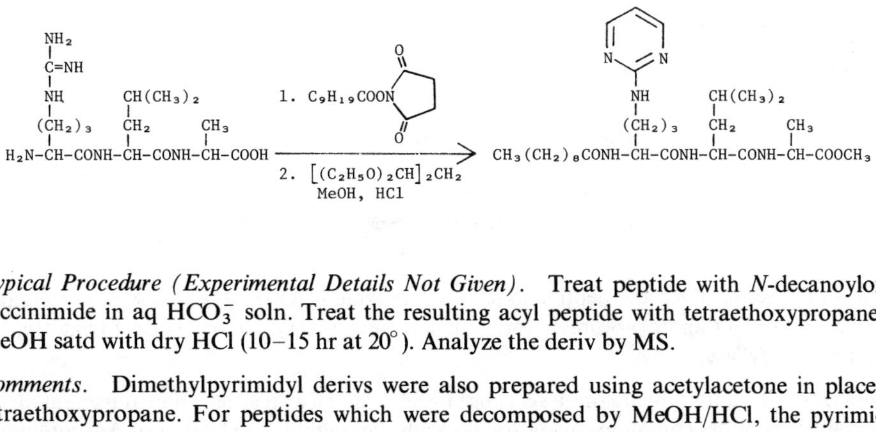

Typical Procedure (Experimental Details Not Given). Treat peptide with *N*-decanoyloxysuccinimide in aq HCO_3^- soln. Treat the resulting acyl peptide with tetraethoxypropane in MeOH satd with dry HCl (10–15 hr at 20°). Analyze the deriv by MS.

Comments. Dimethylpyrimidyl derivs were also prepared using acetylacetone in place of tetraethoxypropane. For peptides which were decomposed by MeOH/HCl, the pyrimidyl deriv was prepared with tetraethoxypropane and triethylamine followed by esterfication with diazomethane.

References

[1]M. M. Shemyakin *et al.*, *Experientia* **23**, 428 (1967).

[2]A. A. Kiryushkin *et al.*, *Tetrahedron Lett.* 33 (1966).

Derivative Type. Ornithine peptide, *N*-decanoyl, methyl ester.

Reagent. Hydrazine, *N*-decanoyloxysuccinimide, diazomethane.

Compound Type. Arginine containing peptides.

Reaction

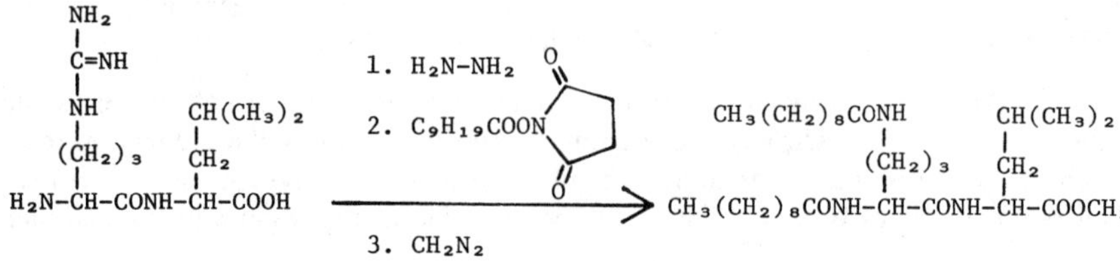

Typical Procedure (Experimental Details Not Reported). Reflux peptide with 20% aq hydrazine for 30–60 min. Treat resulting ornithine peptide with N-decanoyloxysuccinimide in aq HCO_3^- soln.[2] Esterfy the acylated ornithine peptide with CH_2N_2 in Et_2O. Analyze by MS.

Comments. The method was used to convert Arg peptides to the corresponding Orn peptide derivs for MS sequence analysis. Although the hydrazine treatment caused some peptide cleavage, it was not sufficient to interfere with the analysis.

References

[1]M. M. Shemyakin *et al.*, *Experientia* **23**, 428 (1967).
[2]A. A. Kiryushkin *et al.*, *Tetrahedron Lett.* 33 (1966).

Derivative Type. Ornithine peptides.

Reagent. Hydrazine.

Compound Type. Arginine peptides.

Reaction

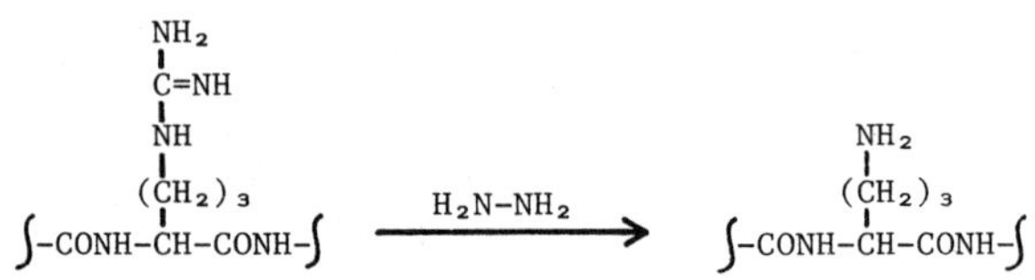

Typical Procedure. Diss the peptide in 0.5 ml hydrazine/H_2O (1 : 1, v/v) and heat 15 min at 75°. Dil with H_2O and evap under vac with warming. Acetylate and permethylate (*q.v.*)[2], then analyze by MS.

Comments. This mild hydrazinolysis caused less chain cleavage than previous methods,[3,4] especially in peptides containing acidic amino acids. Permethyl acetyl derivs of Arg containing peptides did not give molecular ions or sequence ions for the Arg residue. Conversion of Arg residues to Orn gave derivs which exhibited prominent ions for cleavage of each peptide bond. Hydrazinolysis of C terminal Arg esters yielded δ-lactams, but C terminal Arg residues with a free carboxyl gave the usual conversion to Orn.

References

[1]H. R. Morris *et al.*, *Biochem. Biophys. Res. Commun.* **51**, 247 (1973).
[2]H. R. Morris *et al.*, *Biochem. J.* **125**, 189 (1971).
[3]M. M. Shemyakin *et al.*, *Experientia* **23**, 428 (1967).
[4]D. W. Thomas *et al.*, *Biochem. Biophys. Res. Commun.* **32**, 519 (1968).

Derivative Type. Pyrimidyl ornithine peptides.

Reagent. Acetylacetone/triethylamine (Et_3N).

Compound Type. Arginine peptides.

Reaction

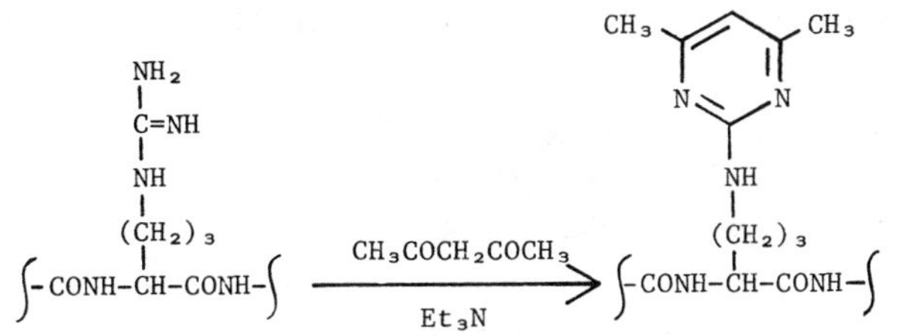

Typical Procedure. Diss the peptide in 0.1 ml H_2O and 0.2 ml EtOH containing 0.1 ml Et_3N and 0.2 ml acetylacetone. Reflux 4 hr; then evap under vac. Diss residue in 0.5 ml H_2O and acidify to pH 4 with HOAc. Heat at 100° for 10 min, cool, and ext with Et_2O to remove excess acetylacetone and its condensation dimer (2-acetyl-3,5-dimethylphenol). Acetylate[2] and permethylate (60 sec)[1] prior to MS analysis.

Comments. The volatile Et_3N buffer was used in place of the previously used $NaHCO_3$ to avoid the adverse effect of $NaHCO_3$ on subsequent permethylation.

References

[1]H. R. Morris *et al.*, *Biochem. Biophys. Res. Commun.* **51**, 247 (1973).

[2]H. R. Morris *et al.*, *Biochem. J.* **125**, 189 (1971).

Derivative Type. 4,6-Dimethylpyrimidylornithine peptide, *N*-trifluoroacetyl, permethyl.

Reagent. Acetylacetone, *S*-ethyl trifluorothioacetate, methylsulfinylcarbanion/methyl iodide.

Compound Type. Arginine containing peptides.

Reaction

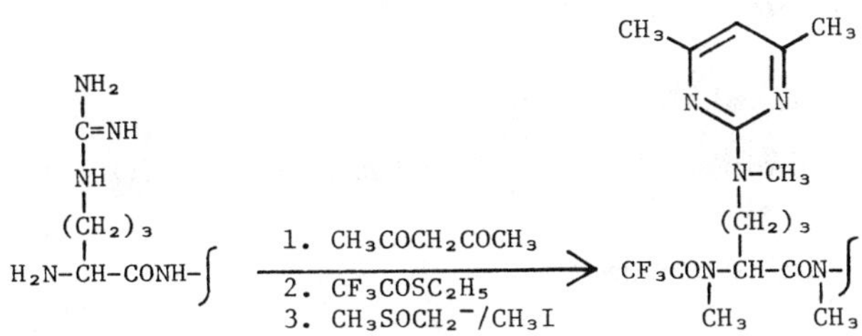

Typical Procedure.[1] To 100 nmol peptide add 10 μl 10% $NaHCO_3$, 20 μl 95% EtOH, and 20 μl acetylacetone, and heat at 100° for 4 hr. Add 1.0 ml 0.5 *M* HOAc and heat 10 min at 100° to hydrolyze Schiff bases. Ext with Et_2O to remove 3,5-dimethyl-2-acetylphenol (acetylacetone self-condensation product) and evap to dryness under vac over P_2O_5. Diss residue in 50 μl 2% Na_2CO_3 and allow to react with 5 μl $CF_3COSC_2H_5$ for 2 hr at 45°. Evap C_2H_5SH under N_2 (in a hood) and dry as above. Permethylate with methylsulfinylcarbanion.[2] Analyze by MS.

Comments. The method of Vetter-Diechtl et al.[3] was adapted for small-scale reaction of the Arg residue with acetylacetone. The Orn δ-N was difficult to methylate and some incompletely methylated material was present. The electron withdrawing TFA group may have hindered methylation of the α-N also. In applying the method to bradykinin, it was noted that the TFA deriv was more volatile than the acetyl deriv. In an attempt to further increase volatility the Arg residue was deriv with $CF_3COCH_2COCF_3$, but the MS of these derivs were inferior.

References

[1]P. A. Leclerq *et al.*, *Biochem. Biophys. Res. Commun.* **45**, 937 (1971).

[2]P. A. Leclerq and D. M. Desiderio, *Anal. Lett.* **4**, 305 (1971).

[3]H. Vetter-Diechtl *et al.*, *Experientia* **24**, 340 (1968).

Derivative Type. δ-*N*-(2-Pyrimidinyl)ornithine ethyl ester.

Reagent. 1,1,3,3-Tetraethoxypropane/ethanol (EtOH)/HCl.

Compound Type. Arginine and *N*-acyl arginine compds.

Reaction

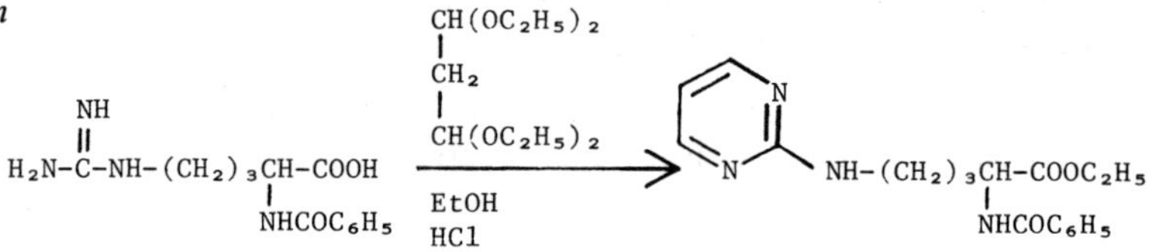

Typical Procedure. Mix 1.1 g (4 mmol) benzoylarginine with 1 ml (4.2 mmol) tetraethoxypropane in 20 ml 9.6 *N* HCl in EtOH. Stir 16 hr at 25°. Conc to a syrup at 40° then remove excess HCl by storage in a dessicator over NaOH. Add 15 ml conc NH_4OH and collect the ppt. Recryst from aq EtOH. Analyze by GC.

Comments. The method was reported as a means of selective chemical modification of Arg residues. The reagent, the ethyl acetal of malonaldehyde, reacted with the guanidine group to form the aminopyrimidine.

References

[1]T. P. King, *Biochemistry* **5**, 3454 (1966).

Derivative Type. Acetyl, CHD, methyl.

Reagent. Acetic anhydride (Ac_2O), 1,2-cyclohexanedione (CHD), methyl iodide/silver oxide.

Compound Type. Arginyl peptides

Reaction

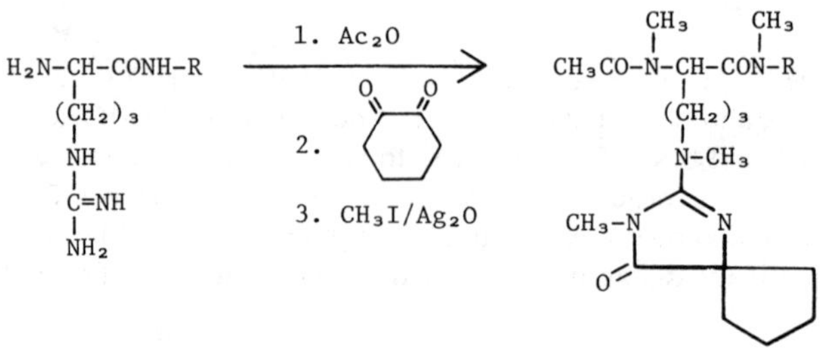

Typical Procedure. Diss peptide in 1 ml pyridine at 0° and add 0.5 ml Ac_2O. If the peptide does not diss in 1 hr add sufficient cold H_2O to diss and maintain at 0° for 1 hr more. Lyophilize and diss residue in 1 ml 0.2 N NaOH containing 10 mg CHD.[2] Allow to stand 3 hr at room temp then ext 3 × with EtOAc to remove excess CHD. Acidify and lyophilize the aq phase. Methylate[3] by diss residue in 1 ml DMF and adding 0.3 g Ag_2O and 0.5 ml CH_3I. Stir 1–2 days at room temp; then dil with 1–2 ml DMF. Centrifuge to remove white ppt then add DMF soln to 20 ml $CHCl_3$. Ext with 5% KCN, 5% $NaHCO_3$, and H_2O. Dry over Na_2SO_4 and evap to dryness under vac. Wash residue 2 × with pet ether. Diss deriv in CH_2Cl_2 for transfer to sample probe and analyze by direct inlet MS.

Comments. The acetylation could alternatively be done with MeOH/Ac_2O (4 : 1) at room temp for several hours.[4] The method was used for MS sequence analysis of Arg containing peptides. Shemyakin et al.[5] prepared CHD derivs from a large number of Arg peptides and presented a detailed discussion of their MS behavior.

References

[1]J. Lenard and P. M. Gallop, *Anal. Biochem.* **29**, 203 (1969).
[2]K. Toi *et al.*, *J. Biol. Chem.* **242**, 1036 (1967).
[3]D. C. Das *et al.*, *Biochem. Biophys. Res. Commun.* **29**, 211 (1967).
[4]D. W. Thomas *et al.*, *Biochem. Biophys. Res. Commun.* **32**, (1968).
[5]M. M. Shemyakin *et al.*, *Tetrahedron* **25**, 5785 (1969).

Derivative Type. Permethyl.

Reagent. Hydrogen peroxide, sodium hydride/dimethylformamide (DMF)/methyl iodide, potassium iodide/acetyl chloride(AcCl).

Compound Type. Methionine containing peptides (acetylated).

Reaction

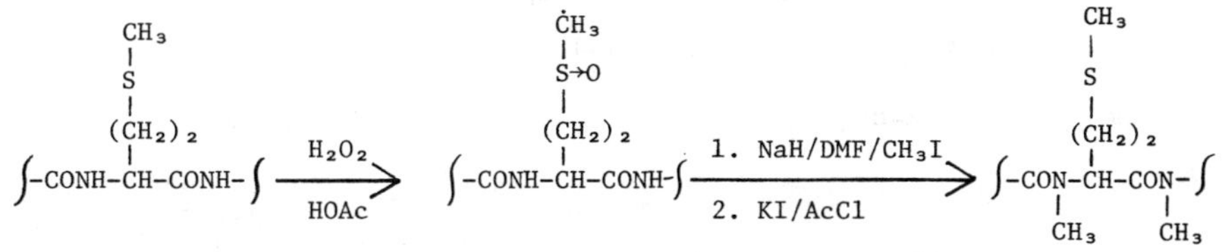

Typical Procedure. Diss 1.2 μmol acetylated peptide in 0.03 ml 0.1 M H_2O_2 in HOAc and allow to stand at room temp 1 hr. Evap under vac then permethylate with NaH and CH_3I in DMF (0.1 ml).[2] Add 0.1 ml HOAc to eliminate excess NaH then reduce the sulfoxide with 5 mg KI and 1 drop AcCl for 5 min. Add 0.1 ml aq HCl and reduce the liberated I_2 with a few drops of 0.1 N $Na_2S_2O_3$. Ext with $CHCl_3$ and wash ext 4 × with H_2O. Dry over Na_2SO_4 and evap to yield the deriv. Analyze by MS.

Comments. Met residues were protected as the sulfoxides during permethylation to prevent formation of sulfonium salts.

References

[1]P. Roepstorff *et al.*, *FEBS Lett.* **9**, 235 (1970).
[2]D. W. Thomas, *FEBS Lett.* **5**, 53 (1969).

Derivative Type. Desulfurized peptides.

Reagent. Raney nickel (RaNi).

Compound Type. Sulfur containing peptides.

Reaction

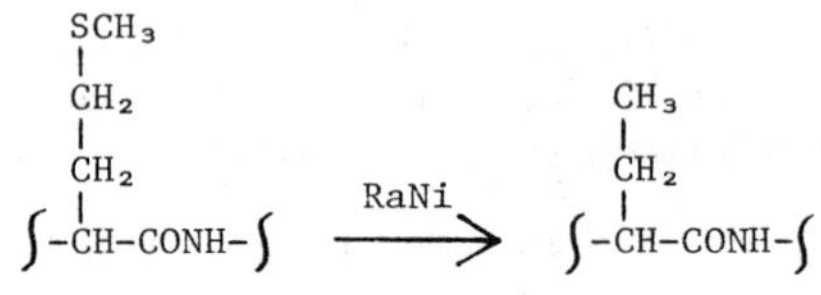

Typical Procedure. Diss 5–10 mg of peptide in H_2O(pH 7–7.5) and add RaNi (10 × the wt of peptide). Stir 3 hr at room temp then 1 hr at 40°. Filter and evap to dryness. Acetylate with Ac_2O/MeOH and esterfy with CH_2N_2.[2] Analyze by MS.

Comments. Acetyl permethyl derivs of Met containing peptides gave MS peaks delineating the sequence from the *N*-terminus up to but neither including nor beyond the Met residue. This method was used to desulfurize peptides prior to permethylation analysis to eliminate formation of sulfonium salts. Met residues were converted to α-aminobutyric acid (Abu) and CySH or Cys to Ala. The Abu residues in the reduced derivs permitted identification of Met residues since Abu differed in mass from the common aliphatic amino acids. The same technique could be applied to CySH or Cys residues, but unambiguous interpretation would be more difficult since reduction yielded Ala residues. For peptides not soluble in H_2O, an EtOH soln of the *N*-acetyl peptide was used.

References

[1] R. Toubiana *et al.*, *FEBS Lett.* **8**, 207 (1970).

[2] D. W. Thomas *et al.*, *Biochem. Biophys. Res. Commun.* **32**, 519 (1968).

Derivative Type. *N*-Ethoxycarbonyl (His ring opened), *N,O*-permethyl.

Reagent. Diethylpyrocarbonate, methylsulfinyl carbanion/methyl iodide.

Compound Type. Histidine peptides.

Reaction

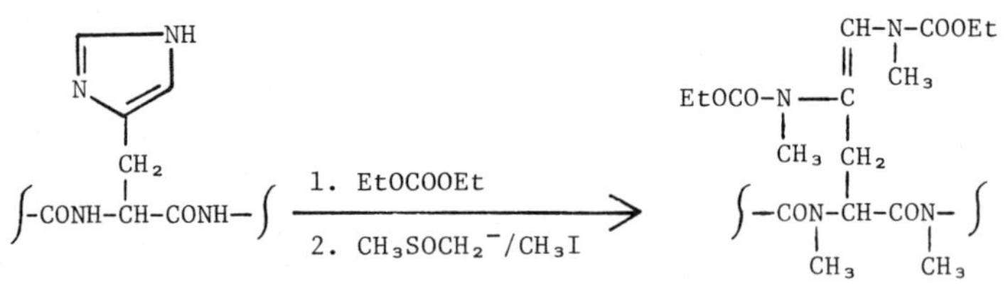

Typical Procedure. Diss 0.01 mmol peptide in 2 ml H_2O and add 0.01 mmol diethylpyrocarbonate/acylable group. Maintain pH 8.0 with a pH-stat for 30 min. Add another 4 × 1 mol equiv reagent/His residue in 1 mol portions at 30 min intervals at constant pH 8.0. Acidify to pH 1.0 with HCl and ext with EtOAc. Permethylate with methylsulfinyl carbanion/methyl iodide[2] and analyze by MS.

Comments. Without the ring opening with excess diethylpyrocarbonate, His residues formed H_2O soluble quaternary derivs in the methylation step.

References

[1]J. F. G. Vliegenthart and L. Dorland, *Biochem. J.* **117**, 31P (1970).

[2]E. Vilkas and E. Lederer, *Tetrahedron Lett.* 3089 (1968).

Derivative Type. *N*-Acetyl peptide amides, pyruvyl peptide methyl esters.

Reagent. Toluenesulfonyl chloride (TsCl), diethylamine (Et_2NH), acetic acid (AcOH)/HCl.

Compound Type. Serine containing peptides (as *N*-acetyl methyl esters).

Reaction

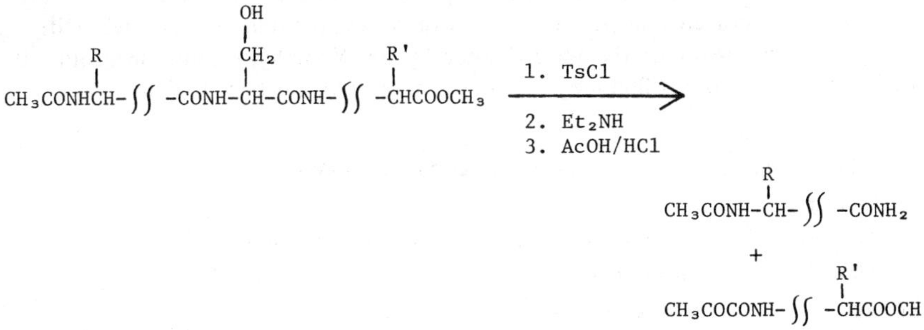

Typical Procedure. Acetylate *ca* 0.5 mg of peptide with Ac_2O and esterfy with CH_2N_2. Tosylate with excess TsCl in pyridine.[2] Treat with Et_2NH in EtOAc to eliminate the tosyl group.[2] Cleave the resulting dehydroalanine peptide by heating with HOAc/HCl. Analyze the resulting peptide mixt by direct probe MS.

Comments. The method was developed as an alternative to enzymic cleavage of peptides into smaller fragments for MS sequence analysis. The peptides were cleaved at Ser residues to yield peptide amide and pyruvyl peptide fragments. Thr underwent analogous tosylation and elimination reactions but did not undergo cleavage.

References

[1]E. Nebelin and K. Brunfeldt, in *Peptides 1976*, *Proc. 14th European Peptide Symposium*, North-Holland, Amsterdam (1977), p. 197.

[2]I. Photaki, *J. Am. Chem. Soc.* **85**, 1123 (1963).

5.4.8. Permethylated Peptide Derivatives

Derivative Type. *N*-Acetyl, permethyl.

Reagent. Acetic anhydride (Ac_2O)/methanol(MeOH), methyl iodide/silver oxide.

Compound Type. Peptides.

Reaction

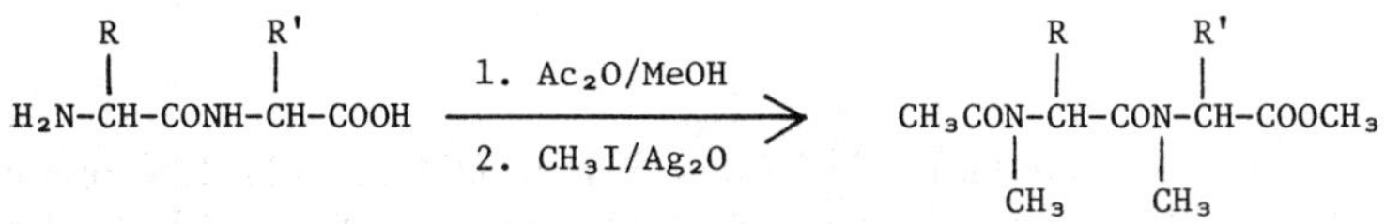

Typical Procedure. Diss peptide in 4:1 (v/v) MeOH Ac_2O and allow to stand at room temp a few hr. Evap to dryness to yield acetylated peptide. Mix 0.3–20 mg acetylated peptide with

50 mg Ag_2O, 0.3 ml DMF, and 0.2 ml CH_3I, and heat in a sealed tube at 50° for 4 hr. Allow to cool, dil with DMF, and filter. Add $CHCl_3$ to the filtrate then wash with 5% aq KCN to remove complex formed. Wash $CHCl_3$ phase repeatedly with H_2O, dry over Na_2SO_4, then evap under vac. Analyze product by direct probe MS.

Comments. The derivs were used for sequencing of peptides by MS.[1-5] For peptides already containing *N*-methyl groups, use of CD_3I permitted location of the original *N*-methyls Commercial Ag_2O was found to be satisfactory for use provided it was adequately washed (H_2O, MeOH, then Et_2O) and dried (under vac and stored over P_2O_5). Agarwal et al.[6] reported that the methylation reaction was not straightforward for peptides containing Met, Asp, or Glu in that it could give intramolecular cyclization which could lead to chain cleavage. Kiryushkin *et al.*[7] reported that permethylation of *N*-acyl peptide methyl esters with Ag_2O in DMF gave replacement of the methyl ester by an *N*-methylformamide group. It was suggested that this type of artifact may be more general than has been realized.

References

[1] B. C. Das *et al.*, *Biochem. Biophys. Res. Commun.* **29**, 211 (1967).

[2] B. C. Das *et al.*, *Nature* **217**, 547 (1968).

[3] D. W. Thomas *et al.*, *Biochem. Biophys. Res. Commun.* **32**, 199 (1968).

[4] D. W. Thomas *et al.*, *Biochem. Biophys. Res. Commun.* **32**, 519 (1968).

[5] E. Lederer, *Pure Appl. Chem.* **17**, 489 (1968).

[6] K. L. Agarwal *et al.*, *Nature* **219**, 498 (1968).

[7] A. A. Kiryushkin *et al.*, *Org. Mass Spectrom.* **5**, 19 (1971).

Derivative Type. Permethyl.

Reagent. Sodium hydride/dimethylacetamide(DMA)/methyl iodide.

Compound Type. Peptides.

Reaction

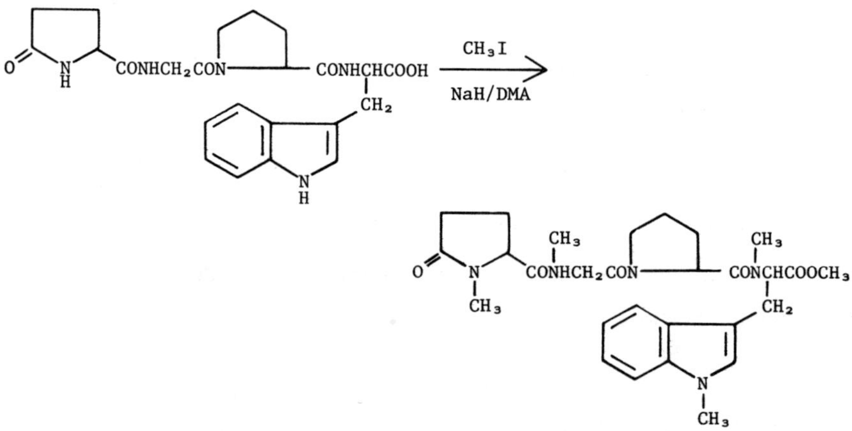

Typical Procedure. Prepare base soln by diss NaH in DMA at 120–125° (gas evolution). Cool to 20° and add to a susp of the peptide in DMA. Immediately add CH_3I. Allow to react 1 hr at 20°. Isolate deriv (details not given, presumably by addn of H_2O and ext) and analyze by MS.

Comments. The method was used to sequence one peptide segment of feline gastrin. It was reported to be cleaner than the method using NaH/DMSO/CH_3I, but some C-methylation of the Gly residue was observed.

References

[1]K. L. Agarwal *et al., J. Am. Chem. Soc.* **91**, 3096 (1969).

Derivative Type. *N*-Methyl.

Reagent. Sodium hydride/dimethylformamide (DMF), methyl iodide.

Compound Type. Peptides (as *N*-acetyl methyl esters).

Reaction

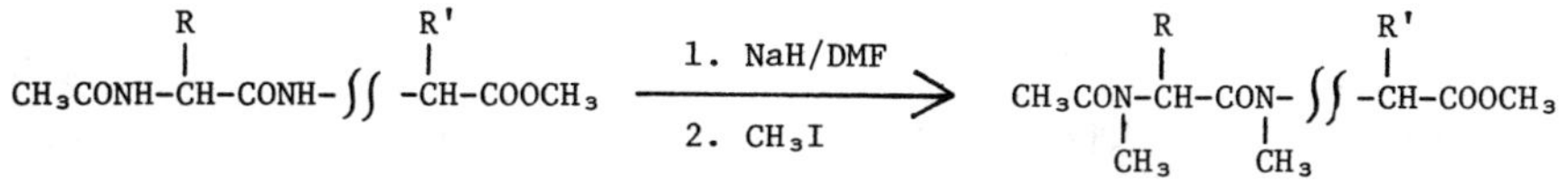

Typical Procedure. Stir 0.06 mmol Ac-Trp-Met-Asp-Phe-OMe with 3.6 mmol NaH (washed with *n*-pentane) in 2 ml DMF for 15 min at 20°. Add 3.6 mmol CH_3I and continue stirring for 20 min at 20°. Pour into 5 ml satd aq NaCl and ext with 5 ml $CHCl_3$. Wash $CHCl_3$ with 4 × 4 ml satd NaCl soln. Analyze the deriv by MS.

Comments. The method utilized milder reaction conditions than earlier methods using NaH/DMF/CH_3I, which gave sulfonium salt formation and C methylation of Gly, Asp, and Glu.[2] The method was reported to be suitable for all residues except Arg and gave no evidence of either incomplete *N*-methylation or overmethylation (no His containing examples were cited, however). Similar results were obtained with dimethyl sulfate (free of any traces of acid) and NaH in CH_3CN.

References

[1]G. Marino *et al., J. Chem. Soc. Chem. Commun.* 357 (1972).

[2]D. W. Thomas, *FEBS Lett.* **5**, 53 (1969).

Derivative Type. Acetyl, permethyl.

Reagent. Acetic anhydride (Ac_2O)/methanol(MeOH), methylsulfinylcarbanion/methyl iodide.

Compound Type. Peptides.

Reaction

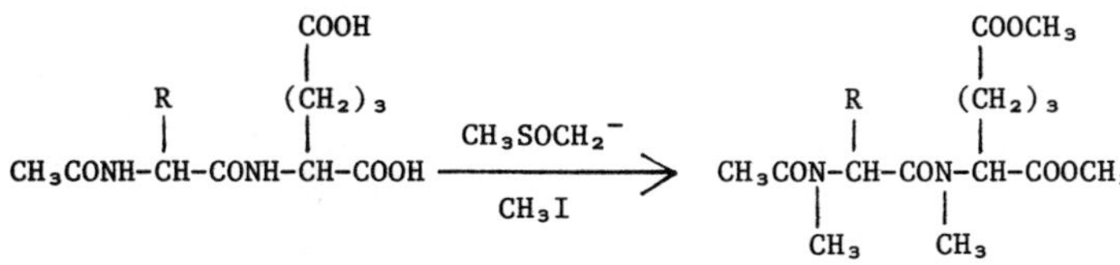

Typical Procedure. Rinse 20 mg of NaH oil dispersion 3 × with Et_2O and diss in 0.2 ml DMSO at 100° (until H_2 evolution ceases, *ca* 5 min). Add 3 mg of acetylated peptide (acetylation procedure[2]) to this soln at room temp followed by 0.3 ml CH_3I. After 1 hr, dil with H_2O and ext with $CHCl_3$. Analyze extracted product by direct probe MS.

Comments. Peptides containing Asp, Glu, or Trp residues sometimes gave by-products upon methylation with Ag_2O/CH_3I. No such products were formed by this procedure. Vilkas *et al.*[3] used a similar method for analysis of a peptidoglycolipid. They allowed the sample to react with the carbanion for 1 hr before adding the CH_3I, and then allowed the mixt to react 12 hr at room temp before quenching with H_2O. Deuteroacetyl (from d_6-Ac_2O) and deuteromethyl (from CD_3I) derivs have been used to resolve ambiguities in sequencing with acetyl, permethyl derivs.[4]

References

[1]D. W. Thomas, *Biochem. Biophys. Res. Commun.* **33**, 483 (1968).

[2]D. W. Thomas *et al.*, *Biochem. Biophys. Res. Commun.* **32**, 519 (1968).

[3]E. Vilkas *et al.*, *Tetrahedron Lett.* 3089 (1968).

[4]P. Roepstorff and K. Brunfeldt, *FEBS Lett.* **21**, 320 (1972).

Derivative Type. Acetyl, permethyl.

Reagent. Acetic anhydride (Ac_2O), methylsulfinylcarbanion/methyl iodide.

Compound Type. Peptides.

Reaction

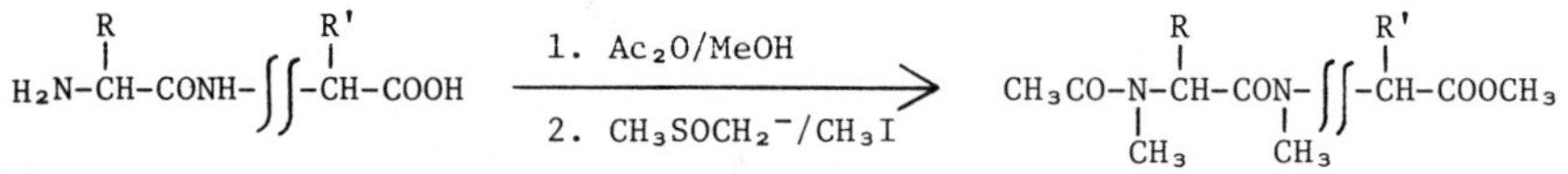

Typical Procedure. Acetylate 10–20 nmol of peptide with Ac_2O/MeOH (4 : 1, v/v) at room temp for several hr.[2] Evap to dryness under N_2. To the residue add 30 μl of methylsulfinyl carbanion soln (prepared by washing 100 mg NaH in mineral oil with Et_2O 3 × and diss in 1 ml DMSO by heating at 100° until H_2 evolution ceased) and 50 μl CH_3I.[3] After 1 hr at room temp add 2 ml H_2O and ext with 3 × 1 ml $CHCl_3$. Wash $CHCl_3$ ext with 2 × 2 ml H_2O and evap to dryness. Rinse residue 2 × with pet ether and diss in CH_2Cl_2 to transfer to sample probe. Analyze by direct inlet MS.

Comments. This procedure[1] was a modification of earlier methods[2,3] for use on smaller samples. There was evidence of considerable production of low mol wt impurities in the deriv reaction. Use of ^{3}H-CH_3I showed about twice the expected incorporation of radioactivity. The impurities however were more volatile than the peptide derivs and did not interfere with the analysis.

References

[1]J. Lenard and P. M. Gallop, *Anal. Biochem.* **34**, 286 (1970).

[2]D. W. Thomas *et al.*, *Biochem. Biophys. Res. Commun.* **32**, 199 (1968).

[3]D. W. Thomas, *Biochem. Biophys. Res. Commun.* **33**, 519 (1968).

Derivative Type. Acetyl, permethyl.

Reagent. Acetic anhydride (Ac_2O), methylsulfinylcarbanion/methyl iodide.

Compound Type. Peptides.

Reaction

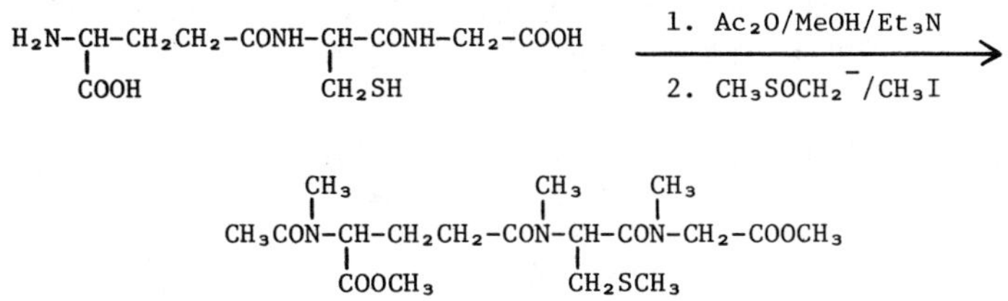

Typical Procedure. Acetylate peptide for 18 hr in MeOH (adjusted to pH 8 with Et_3N) Ac_2O (4 : 1, v/v). Reduce disulfides by treatment with 40-fold excess of mercaptoethanol in MeOH or $MeOH/H_2O$ made basic with Et_3N. Prepare 0.5 *M* methylsulfinyl carbanion by washing 200 mg NaH oil suspension 3 × with dry Et_2O and diss in 8 ml DMSO by heating at 65° under Ar until H_2 ceases. To 3–5 mg of peptide add a fivefold excess of carbanion (relative to number of peptide reactive equiv, i.e., methylation sites). Add the same number of equiv of CH_3I and allow to react 1 hr. Add 2 ml H_2O and ext with 5 ml $CHCl_3$. Wash $CHCl_3$ with 1 ml H_2O and evap to dryness. Analyze by direct probe MS.

Comments. Use of equimolar amts of CH_3I and carbanion prevented formation of cysteinyl sulfonium iodide and subsequent decomposition permitting application of the method to cysteine containing peptides. The "equimolar method" was also applied to a histidine containing peptide.[2]

References

[1]M. L. Polan *et al.*, *Biochem. Biophys. Res. Commun.* **38**, 1127 (1970).

[2]P. A. White and D. M. Desiderio, *Anal. Lett.* **4**, 141 (1971).

Derivative Type. *N*-Acetyl, *N,O,S*-permethyl.

Reagent. Acetic anhydride (Ac_2O), methylsulfinylcarbanion/methyl iodide.

Compound Type. Peptides.

Reaction

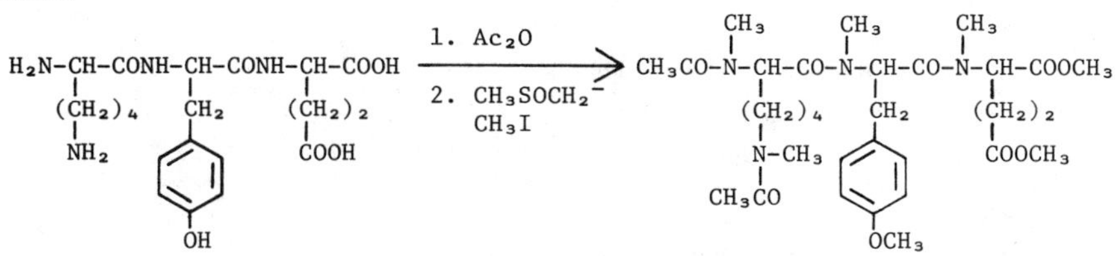

Typical Procedure. Diss 50 μg peptide in 40 μl MeOH using ultrasonic agitation. Add 10 μl Ac_2O and allow to stand at room temp for 3 hr. Remove reagents under vac. Prepare methylsulfinylcarbanion by washing 120 mg (5 mmol) of NaH in oil suspension three times with anhyd Et_2O. Add 5 ml dry DMSO and heat under N_2 until H_2 evolution ceases. Diss the acetylated peptide in 50 μl DMSO and add 8 μl of carbanion soln. After 15 min add 0.5 μl CH_3I under N_2. After 1 hr at room temp add 1 ml H_2O. Ext with 1 ml $CHCl_3$. Wash $CHCl_3$ with 3 × 1 ml H_2O. Evap under N_2 and diss residue in 10 μl $CHCl_3$. Analyze by direct probe MS.

Comments. The 3 hr time was found to be optimal for acetylation and the 1 hr time optimal for methylation. The carbanion and methyl iodide were used in 10-fold excess with respect to the number of methylation sites. These amts were approximated by use of 8 μl carbanion soln and 0.5 μl CH_3I/50 μg peptide for di- through decapeptides. Deuterated acetyl and methyl derivs were used to confirm fragmentation pathways.[2]

References

[1]P. A. Leclercq and D. M. Desiderio, Jr., *Anal. Lett.* **4**, 305 (1971).

[2]K. D. Haegele *et al.*, *Biomed. Mass Spectrom.* **1**, 175 (1974).

Derivative Type. Acetyl, permethyl.

Reagent. Acetic anhydride (Ac_2O), methylsulfinylcarbanion/methyl iodide.

Compound Type. Peptides.

Reaction

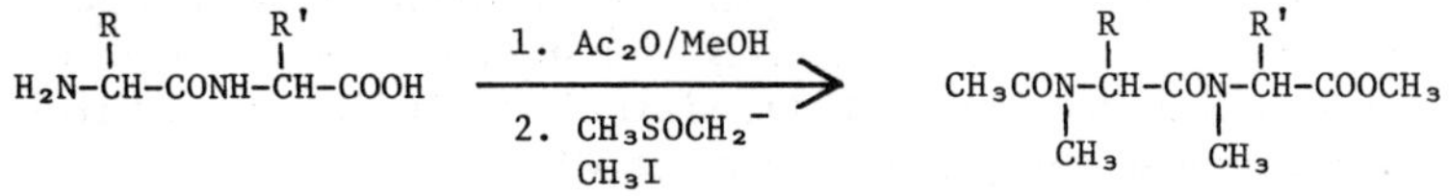

Typical Procedure. Diss the peptide in Ac_2O/MeOH (1 : 4, v/v) and allow to react at room temp for 3 hr. Evap under vac. Diss the acetylated peptide (0.1–1.0 μmol) in 100 μl dry DMSO. Add a 10-fold excess of methylsulfinyl carbanion (*ca* 1 *M*) in DMSO followed by excess CH_3I. Allow to react 1 min then dil with H_2O and ext with 1 ml $CHCl_3$. Wash the $CHCl_3$ 2 × with H_2O and evap. Analyze by MS.

Comments. The acetylation procedure acetylated only the free primary amino groups (and no hydroxyls). The basic methylation procedure[1] was modified using a 60-sec reaction time to permit analysis of His[2] and Met[3] containing peptides without formation of quaternary ammonium or sulfonium derivs. The carbanion soln was prepared by heating NaH (50 mg/ml) in DMSO at 90° for 15 min[1] with no precautions to exclude air.[2] CySH residues (analyzed as carboxymethyl CySH) were converted to dehydroalanine.[3] Arg residues required conversion to Orn or pyrimidyl Orn derivs. MS data for the derivs of 17 amino acid residues were presented in tabular form.[1]

References

[1]H. R. Morris *et al.*, *Biochem. J.* **125**, 189 (1971).

[2]H. R. Morris, *FEBS Lett.* **22**, 257 (1972).

[3]H. R. Morris *et al.*, *Biochem. Biophys. Res. Commun.* **51**, 247 (1973).

Derivative Type. Acetyl, permethyl.

Reagent. Acetic anhydride (Ac_2O), methylsulfinylcarbanion/methyl iodide.

Compound Type. Peptides.

Reaction

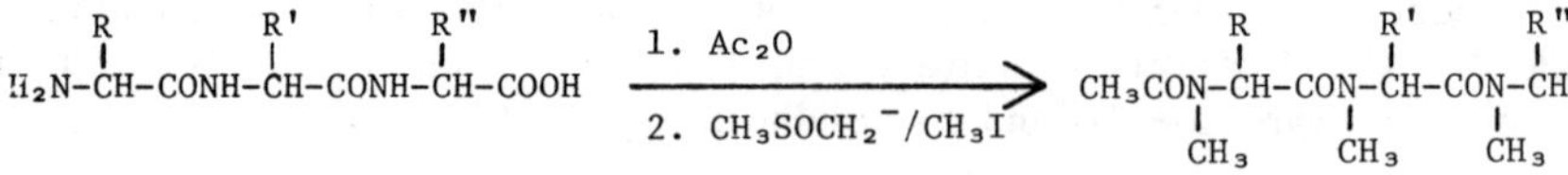

Typical Procedure. Shake 1 mg peptide with 1 ml Ac_2O and 1 ml H_2O for 2 hr at room temp. Evap to dryness under vac, add 1 ml H_2O and reevap under vac (0.1 mm) at room temp. Flush container with N_2 and seal. Wash 20 mg NaH oil dispersion (50% NaH) 3 × with hexane, dry under vac, and seal under N_2. Add 0.2 ml dry DMSO and stir 1 hr at 65°. Diss acetylated peptide in 0.4 ml dry DMSO and add to methylsulfinylcarbanion soln (cooled to room temp). Add 0.3 ml cold (0°) CH_3I immediately to the stirred soln. Stir 10 min at room temp then add 1 ml H_2O and ext with $CHCl_3$. Wash $CHCl_3$ with H_2O, evap, and analyze residue by MS.

Comments. The derivs were used for peptide sequencing by MS. The conditions described were developed to minimize C methylation.

References

[1] R. Hodges *et al.*, *Biochim. Biophys. Acta* **257**, 54 (1972).

Derivative Type. Acetyl, permethyl.

Reagent. Acetic anhydride (Ac_2O)/methanol (MeOH)/triethylamine (Et_3N), methylsulfinylcarbanion/methyl iodide.

Compound Type. Peptides.

Reaction

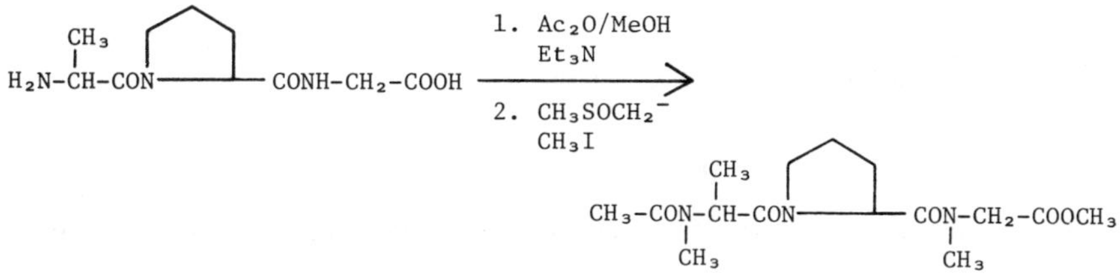

Typical Procedure. Diss 10–100 μmol peptide in 1–5 ml Ac_2O MeOH (1 : 3, v/v) to which a few drops of Et_3N have been added. Sonicate to assist dissolution; then allow to stand 3 hr at room temp. Evap to dryness under vac. Diss 1–10 μl of acetylated peptide in 0.5 ml dry DMSO. Flush container with N_2; then add 0.1 ml of *ca* 2 *M* $Na^+CH_3SOCH_2^-$ in DMSO. Mix well and allow to stand 30 min. Add 20 μl (320 μmol) CH_3I and mix. After 15 min add 1 ml ice cold H_2O and place soln in an ice bath. Quickly add another 1 ml H_2O and ext with 3 × 1 ml $CHCl_3$. Conc the combined $CHCl_3$ under N_2 to *ca* 1 ml and wash with 3 × 1 ml H_2O. Evap $CHCl_3$ to dryness under vac and analyze residue by MS.

Comments. The method was utilized in studies of the extent of C-methylation occurring in acetylation-permethylation deriv of peptides for MS analysis. C-Methylation occurred primarily on Gly residues, but other residues (Glu, Gln, Met, Tyr) were C-methylated to a lesser degree apparently dependent upon residue position and the nature of neighboring residues.

References

[1] M. Mudgett *et al.*, *Biomed. Mass Spectrom.* **2**, 254 (1975).

Derivative Type. *N*-Acetyl, d_3-permethyl.

Reagent. Acetic acid (AcOH)/acetic anhydride (Ac_2O), methylsulfinylcarbanion/d_3-methyl iodide.

Compound Type. Peptides.

Reaction

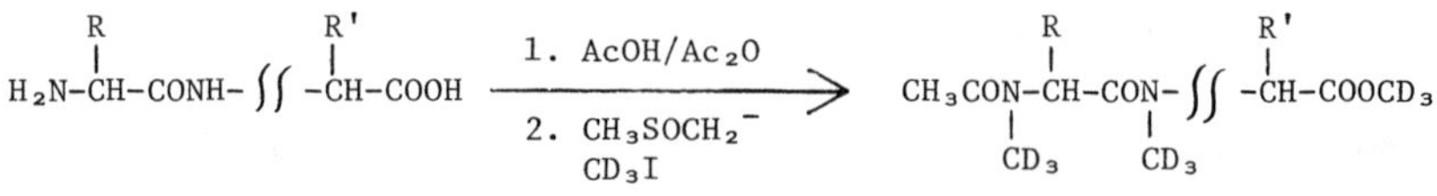

Typical Procedure. Diss 0.2 1 μmol peptide sample in 1 ml $AcOH/Ac_2O$ (1:1, v/v) and allow to stand 4 hr at room temp. Evap reagents under vac and diss residue in 0.5 ml anhyd DMSO. Add 0.1 ml of methylsulfinylcarbanion soln (prepared from 0.2 g NaH and 10 ml DMSO) followed by 0.05 ml CD_3I. Allow to react 1 hr at room temp; then add 2 ml H_2O and ext with 4×1 ml $CHCl_3$. Wash with 3×1 ml H_2O and analyze by MS.

Comments. The method was used for sequence analysis of peptides in mixts. Deuteroacetyl derivs were also prepared. The acetylation procedure acetylated all *N*-terminal and ω-amino groups. The methylation step replaced all hydrogens on heteroatoms (including S). His residues were converted to quaternary salts but these lost CD_3I thermally in the MS to yield the CD_3-His deriv. Esters were transesterfied to the CD_3 ester. Some C-methylation occurred (e.g., with Gly and Met) but the use of CD_3I permitted distinguishing the peaks so formed from those of higher homologs.

References

[1]H.-K. Wipf *et al.*, *J. Am. Chem. Soc.* **95**, 3369 (1973).

Derivative Type. Acetyl, methyl.

Reagent. Acetic anhydride (Ac_2O)/d_6-acetic anhydride, methylsulfinylcarbanion/methyl iodide.

Compound Type. Peptides.

Reaction

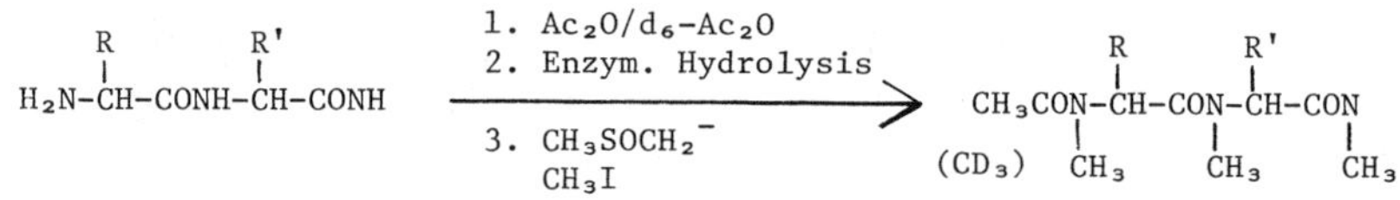

Typical Procedure. Diss 4 mg protein in 250 μl of 10% (v/v) aq pyridine. Add 200 μl pyridine and 60 μl of an equimolar mixt of Ac_2O and d_6-Ac_2O with rapid stirring. After 5 min add another 60 μl of the mixt then allow to stand 15 min at room temp. Add 2.5 ml acetone to ppt protein, allow to settle 30 min, then centrifuge. Suspend the pellet in 200 μl 0.1 *M* NH_4HCO_3 buffer (pH 8.5) and add 0.3 mg subtilisin. Incubate at 37° for 5 hr. Lyophilize in a dessicator over P_2O_5 and NaOH. Add 0.2 ml dry DMSO, flush tube with N_2, and add 0.2 ml methylsulfinyl carbanion soln (see below). After 30 min at room temp add 3×100 μl CH_3I (the first addition is exothermic). Allow to react 1 hr at room temp. Add 0.5 ml $CHCl_3$ and ext with 6×2 ml H_2O. Evap $CHCl_3$ to yield the deriv as an oil. Diss in 50 μl $CHCl_3$ for transfer to the direct probe, and analyze by MS.

Comments. The carbanion soln was prepared by washing 100 mg NaH oil dispersion (57% NaH) with 3×1 ml dry Et_2O, drying NaH under N_2, and heating under N_2 with 1 ml DMSO at 85° for 10 min. The CH_3/CD_3 groups gave distinct M, M+3 doublets in the MS which were used to determine the *N*-terminal sequence of amino acids.

References

[1]W. R. Gray and V. E. del Valle, *Biochemistry* **9**, 2134 (1970).

Derivative Type. Trifluoroacetyl, permethyl.

Reagent. Trifluoroacetic anhydride (TFAA)/trifluoroacetic acid (TFA), methylsulfinylcarbanion/methyl iodide.

Compound Type. Peptides.

Reaction

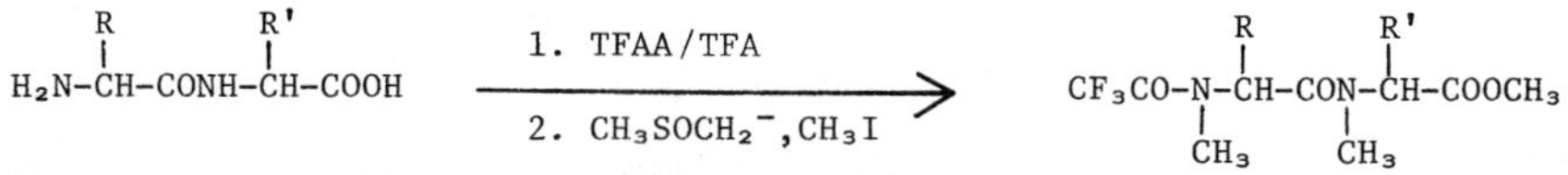

Typical Procedure. Diss peptide in TFAA/TFA (1:4, v/v). Allow to stand 1 hr at room temp; then evap under vac. Diss residue in DMSO and add slight excess of a soln of methylsulfinylcarbanion in DMSO.[2] Add excess CH_3I and allow to react at room temp 1 hr. Dil with H_2O and ext with $CHCl_3$. Wash ext with H_2O and analyze by GC-MS.

Comments. Permethyl, acetyl derivs were also examined. The derivs were used for GC-MS (both EI and CI) separation and sequencing of peptides.

References

[1]J. D. Priddle *et al., Biochem. J.* **157**, 777 (1976).
[2]H. R. Morris *et al., Biochem. J.* **125**, 189 (1971).

5.4.9. Polyamino Alcohol Derivatives

Derivative Type. Polyamine

Reagent. Lithium aluminum deuteride, thionyl chloride, lithium aluminum deuteride.

Compound Type. Peptides (acetylated ethyl esters).

Reaction

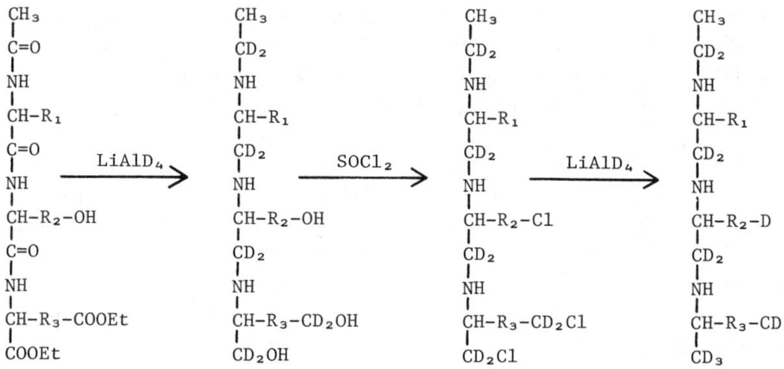

Typical Procedure. Diss 1–1.5 mg of peptide deriv in 1.5 ml THF and allow to react overnight with excess $LiAlD_4$. Decompose with a small amt of H_2O and ext 3× with Et_2O/MeOH-

(10:1, v/v). Evap to dryness and diss residue in 0.5 ml $CHCl_3$. Add 0.5 ml $SOCl_2$ and allow to stand 1.5 hr. Evap to dryness and repeat $LiAlD_4$ reduction. Analyze derivs by GC and MS.

Comments. Peptides not containing hydroxy- or dicarboxylic amino acids could be analyzed after the first reduction as the polyaminoalcohol. Polyaminoalcohols with more than one hydroxyl however did not GC well and required further reduction. The derivs permitted deduction of the sequence of small peptides from the MS.

References

[1]K. Biemann *et al.*, *J. Am. Chem. Soc.* **81**, 2274 (1959).

[2]K. Biemann and W. Vetter, *Biochem. Biophys. Res. Commun.* **3**, 578 (1960).

Derivative Type. *O*-Trimethylsilyl polyamino alcohol.

Reagent. Methanol/HCl, acetic anhydride(Ac_2O)/pyridine, Lithium aluminum deuteride, trimethylsilyldiethylamine (TMSDEA).

Compound Type. Peptides.

Reaction

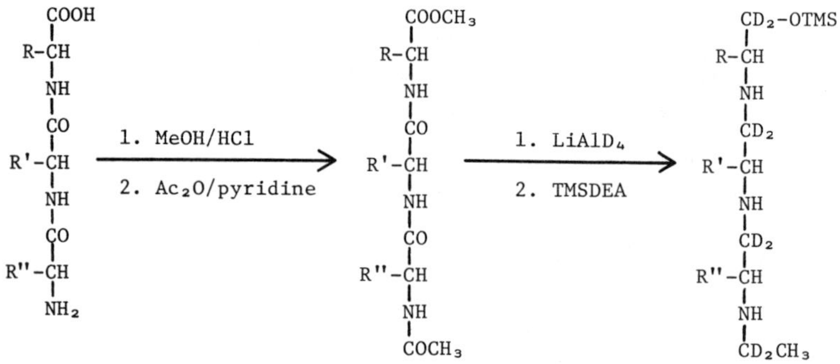

Typical Procedure. Esterfy oligopeptide mixt (from hydrolysis of *ca* 1 μmol peptide) with 2 ml MeOH satd with HCl at room temp for 1 hr. Evap to dryness under vac. Acetylate with 1 ml Ac_2O pyridine (1:1, v/v) for 45 min at room temp. Evap to dryness under vac. Transfer the residue with pyridine to a 4 ml bulb blown from 30 cm of 9 mm od pyrex tubing. Evap under vac and add 10 μl MeOH followed by 2.0 ml of 0.25 *M* $LiAlD_4$ in DME. Add a glass covered stir bar, seal the bulb, and heat with stirring for 48 hr at 95°. Open tube, pour contents into an ice cooled flash, and decompose $LiAlD_4$ with dropwise addn of 1 *N* HCl until a clear soln results.

Add 1–2 mg of 2-(2-aminoethylamino)ethanol as carrier and evap to dryness. Diss residue in 0.5 *M* oxalic acid (0.5 ml excess over calcd amt) to insure complexing of all Al and adjust pH to 6.2–7.4 with 2 *M* NH_3. Add soln to a 17 × 0.7 cm column of Bio-Rad AG1-X8 anion exchange resin (HCO_3 form) and elute with 50 ml H_2O. Evap eluate to dryness under vac and transfer to a 10 ml flask with pyridine. Evap to dryness and diss residue in 400 μl pyridine. Add 200 μl TMSDEA and heat at 55° for 50 min. Evap to dryness under vac and diss residue in 50–200 μl benzene. Analyze by GC-MS.

Comments. The derivs were used for peptide sequencing by MS. The acetylated peptide methyl esters were also reduced with diborane in THF but this method gave inferior results for polyfunctional amino acids (e.g., Lys, Thr, Asp, Met.). Arg, Gln, and Asn containing peptides were obtained in low yield. This problem was circumvented by hydrazinolysis of Arg peptides and by use of partial methanolysis during the deriv of Gln and Asn peptides.

References

[1]J. A. Kelley *et al., Biomed. Mass Spectrom.* **2**, 313 (1975).

Derivative Type. Dideuteroheptafluorobutyl amino alcohol, trimethylsilyl ether.

Reagent. Heptafluorobutyric anhydride (HFBA), lithium aluminum deuteride, trimethylsilyldiethylamine (TMSDEA).

Compound Type. Peptides (as methyl esters).

Reaction

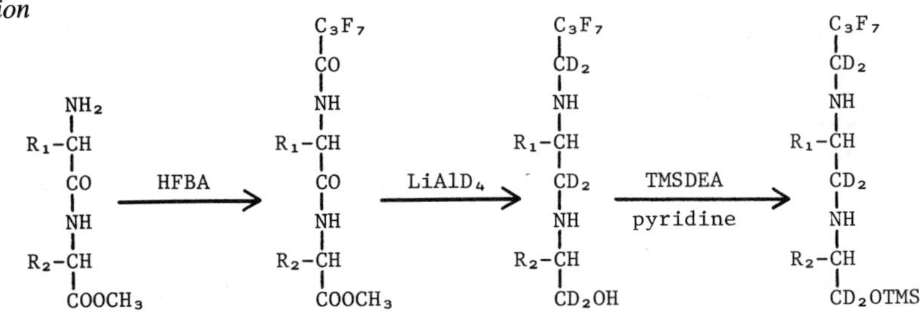

Typical Procedure.[1] Esterfy peptide mixt with 1 ml MeOH/HCl for 45 min or (if cleavage of labile peptide bonds must be avoided) with CH_2N_2 in Et_2O. Evap to dryness and add HFBA. Allow to stand at room temp 30 min (see Ref. 2 for further discussion of acylation conditions). Transfer to a 3 ml bulb (blown from 30 cm of 9 mm o.d. tubing) and evap to dryness. Add 2 ml of 1 *N* $LiAlD_4$ in DME (with ice cooling). Seal and sonicate several min (with cooling). Slowly heat to 90° with stirring and hold at 90° for 24 hr. Transfer contents to a 50 ml flask. Quench excess $LiAlD_4$ by slow addn of MeOH (ice cooling) then dil to 15 ml with MeOH. Add 10 drops H_2O (vigorous stirring) to ppt Al salts. Filter and ext residue 2× with hot MeOH. Evap combined MeOH to dryness and treat the residue with 400 μl pyridine and 200 μl TMSDEA for 1 hr at 55°. Evap to dryness and diss residue in 100 μl benzene. Analyze by GC-MS.

Comments. The derivs were found to be more volatile than previously known derivs including the corresponding products formed from the TFA (from methyl trifluoroacetate) or PFP (from PFPA) peptides. All naturally occurring amino acid residues including Arg, His, Trp, Gln, and Asn could be deriv by this method. The MS of the derivs exhibited abundant sequence ions of relatively balanced intensity. As little as 1 nmol of peptide was detectable.

References

[1]H. Nau, *Biochem. Biophys. Res. Commun.* **59**, 1088 (1974).

[2]F. Weygand and R. Geiger, *Chem. Ber.* **89**, 647 (1956).

[3]H. Nau, *Angew. Chem. Int. Ed. Eng.* **15**, 75 (1976).

[4]H. Nau and K. Biemann, *Anal. Biochem.* **73**, 139, 154, 175 (1976).

6. Aldehydes and Ketones

Many of the derivatives that have been used for GC, GC-MS, MS, and HPLC analysis of aldehydes and ketones stem from the classical carbonyl derivatives such as hydrazones and oximes. Whereas classically, these derivatives had to be isolated, purified, and analyzed, modern chromatographic and spectrometric techniques can be applied to the derivatives without prior isolation. Phenylhydrazones and dinitrophenyl hydrazones have been analyzed by GC (p. 339). The dinitrophenylhydrazones can be used with ECD to enhance detection sensitivity by 2–3 orders of magnitude over that obtained with FID. The dinitrophenylhydrazones and the 5-dimethylamino-1-naphthalenesulfonyl hydrazones have been found useful for chromogenic derivatization for HPLC analysis (p. 341). Other substituted hydrazones are also used as analytical derivatives, and the basic theme still offers potential for development of new reagents.

The oximes are also classical derivatives of carbonyl compounds. Oximes formed with hydroxylamine have been used for GC analysis of carbonyl compounds. These derivatives can exist as *syn* and *anti* isomers due to restricted rotation about the carbon nitrogen double bond. In many cases these isomers are GC separable leading to double GC peaks for each compound. The presence of multiple derivative peaks can be a significant problem for many compounds, however; the presence of an adjacent substituent (for example, a C-18 methyl in 17-ketosteroids) can sterically prevent the formation of one isomer. The weak acidity of the oxime formed with hydroxylamine has been exploited to separate the derivatives by extraction from other neutral compounds in a sample (p. 342).

Several *O*-substituted oximes have been employed analytically. For example, the *p*-nitrobenzyloxime has been used as a chromogenic derivative for HPLC analysis (p. 343). Other substituted oximes have been used to affect GC retention times for group separations (e.g., benzyloximes, p. 485) or to confer ECD sensitivity (e.g., pentafluorobenzyloximes, p. 485). These substituted oximes, applied largely to carbonyl containing compounds other than the simple aldehydes and ketones, appear in other sections.

In addition to the classical derivatives, the carbonyl compounds have been analyzed as trimethylsilyl enol ether derivatives (p. 343) and as acetals and ketals (p. 344). These derivatives, especially the TMS enol ethers, can reduce the polarity of the carbonyl group enhancing volatility for GC analysis. Since derivatization is not essential for GC analysis of most simple carbonyl compounds, volatility enhancement for these compounds is not a major goal. The diisopropyl ketal derivatives permitted GC separation of otherwise inseparable compounds (p. 344).

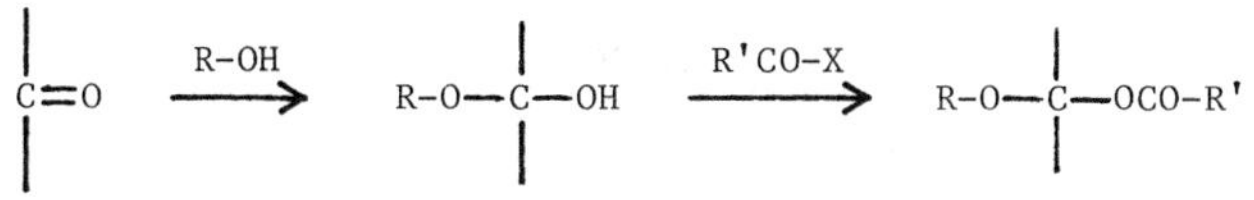

hemiacetal acyl acetal

Acyl acetal derivatives were employed in the analysis of pyridoxal (p. 344). The use of a variety of alcohol (R) and acyl (R′) groups in the preparation of these carbonyl derivatives considerably enhances the potential versatility of the acyl acetals. For example, use of the heptafluorobutyryl group permitted ECD-GC analysis of pyridoxal. It should be noted, however, that this method introduces a new asymmetric center and, in the analysis of already asymmetric compounds, could result in the formation of GC separable diastereomeric derivatives.

Other aldehydes have been identified by GC as their carboxylic acid derivatives formed by ozonolysis of their dinitrophenylhydrazones (p. 345). The dialdehyde, malonaldehyde, was converted to the hydroxypyrimidine with urea and subsequently silylated for trace analysis in tissues (p. 345). This strategy should be applicable to other 1,3-dicarbonyl compounds. For HPLC analysis of aldehydes, the heterocyclic chromophore shown on page 346 may be useful.

In summary, a large number of methods have been employed for analytical derivatization of carbonyl compounds. Not all appear in this section, since many of the methods have been applied only to compounds other than simple aldehydes and ketones. Other sections containing carbonyl derivatives include keto acids (Section 3.5, p. 196), derivatives of optical isomers (Section 9.5, p. 435), ketosteroids (Section 11.3, p. 479), prostaglandins (Section 12, p. 515), and carbohydrates (Section 13, p. 539). The reader is therefore advised to consult these sections and to utilize the indexes for additional derivatives of the carbonyl function.

6.1. HYDRAZONE DERIVATIVES

Derivative Type. Phenylhydrazone

Reagent. Phenylhydrazine

Compound Type. Carbonyl compds

Reaction

$$\text{R-CHO} \xrightarrow{C_5H_5NHNH_2} \text{R-CH=N-NH-}C_6H_5$$

Typical Procedure. Take 10 ml of 0.5 *M* phenylhydrazine HCl in 0.25 *M* aq Na_3PO_4 and ext with Et_2O. Add 10 ml MeOH to the aq soln and dil to 100 ml with H_2O. Add 5–50 μl of the carbonyl compd, shake, and allow to stand overnight at room temp. Alternatively heat 30 min at 50°. Ext with 10, 5, and 5 ml of hexane, Et_2O, or EtOAc. Evap ext to dryness under N_2 (filter N_2 stream through $NaHSO_3$ and silica gel). Diss derivs in 1, 5, or 10 ml of hexane, Et_2O, or EtOAc. Add 50–500 mg $MgSO_4$ and store at 0–5°. Analyze by GC.

Comments. Derivs were examined of 27 C_1–C_{11} carbonyl compds. The use of the derivs was reported to be easier than the use of 2,4-dinitrophenylhydrazones in that the phenylhydrazones were easier to prepare and could be analyzed on metal columns.

References

[1]J. Korolczuk *et al.*, *J. Chromatogr.* **88**, 177 (1974).

Derivative Type. 2,4-Dinitrophenylhydrazone.

Reagent. 2,4-Dinitrophenylhydrazine (DNPH).

Compound Type. Aldehydes and ketones.

Reaction

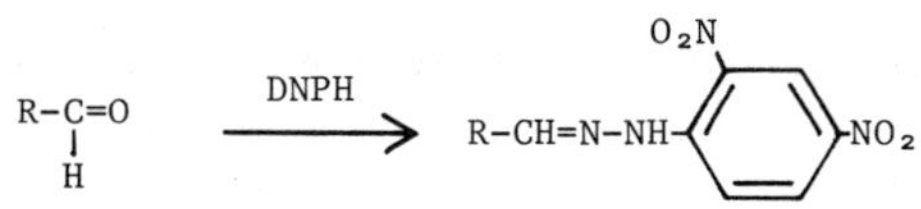

Typical Procedure. Condense the sample in a cold trap with liquid O_2 and diss in 5 ml EtOH. Add soln to a 0.1% soln of DNPH in 2 *N* HCl and allow to stand overnight at room temp. Ext with CCl_4 and evap ext under vacuum. Diss residue in 0.5 ml acetone for GC analysis.

Comments. Separation of 22 carbonyl compounds by GC was investigated. The method was applied to the analysis of aliphatic carbonyl compds in car exhaust fumes and cigarette smoke. Anthracene was used as int stand.

References

[1]Y. Hoshika and Y. Takata, *J. Chromatogr.* **120**, 379 (1976).

Derivative Type. 2,4-Dinitrophenylhydrazone.

Reagent. 2,4-Dinitrophenylhydrazine (DNPH).

Compound Type. Carbonyl compds.

Reaction

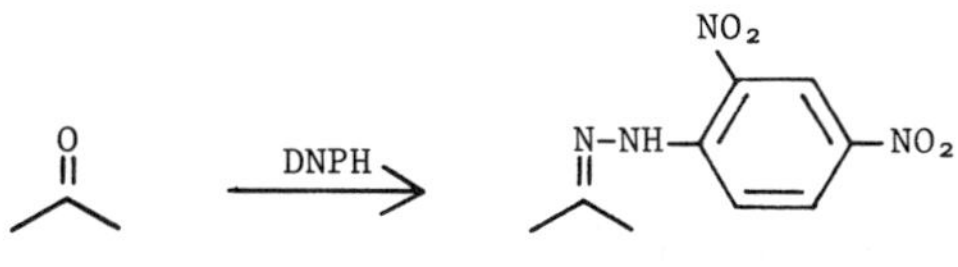

Typical Procedure. Mix 100 μl of carbonyl compd with 100 ml satd DNPH in 2 *N* aq HCl. Shake well and allow to stand at room temp overnight. Filter ppt, wash with 2 *N* HCl and H_2O, then dry in a vac dessicator over silica gel. Diss in EtOAc and analyze by GC.

Comments. Derivs of 15 flavor component carbonyl compds were examined.[1] The mixts were resolved on SE-30 or F-60, although derivs with equal carbon numbers showed some overlap. The derivs were detectable to nanogram levels by FID and showed 500 × greater sensitivity with ECD. Some derivs gave double peaks as a function of deriv solvent used; such peaks were minimized by using EtOAc. Earlier publications[2,3] reported GC of dinitrophenylhydrazones. The DNPH derivs have also been used for analysis of carbonyl compds in auto exhaust by GC[4] and HPLC[5] and for formaldehyde in cigarette smoke by HPLC.[6] Selim[7] deriv μg amts of carbonyl compds in a two phase system extracting the deriv into isooctane as soon as it was formed.

References

[1]H. Kallio *et al.*, *J. Chromatogr.* **65**, 355 (1972).

[2]R. J. Soukup *et al.*, *Anal. Chem.* **36**, 2255 (1964).

[3]E. Fedeli and M. Cirimele, *J. Chromatogr.* **15**, 435 (1964).
[4]L. J. Papa and L. P. Turner, *J. Chromatogr. Sci.* **10**, 744 (1972).
[5]L. J. Papa and L. P. Turner, *J. Chromatogr. Sci.* **10**, 747 (1972).
[6]C. T. Mansfield *et al.*, *J. Chromatogr. Sci.* **15**, 301 (1977).
[7]S. Selim, *J. Chromatogr.* **136**, 271 (1977).

Derivative Type. 2,4-Dinitrophenylhydrazone.

Reagent. 2,4-Dinitrophenylhydrazine (DNPH).

Compound Type. Aliphatic carbonyl compounds.

Reaction

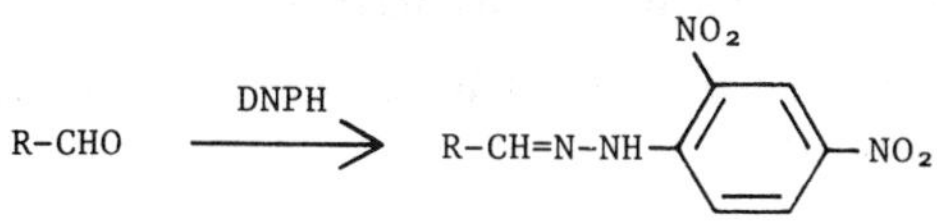

Typical Procedure.[2] Treat aq soln of carbonyl compds with a soln of 24 mg/ml DNPH in 30% aq $HClO_4$. Ext the deriv ppt with an org solvent and analyze by HPLC.

Comments. The method was used for chromogenic deriv for analysis of mixt of carbonyl compds by HPLC using 254 nm UV detection.

References

[1]M. A. Carey and A. F. Persinger, *J. Chromatogr. Sci.* **10**, 537 (1972).
[2]C. Newberg *et al.*, *Anal. Chim. Acta* **7**, 238 (1952).

Derivative Type. 5-Dimethylaminonaphthalene-1-sulfonyl (DANS) hydrazone.

Reagent. 5-Dimethylaminonaphthalene-1-sulfonyl (DANS) hydrazine.

Compound Type. Aldehydes and ketones.

Reaction

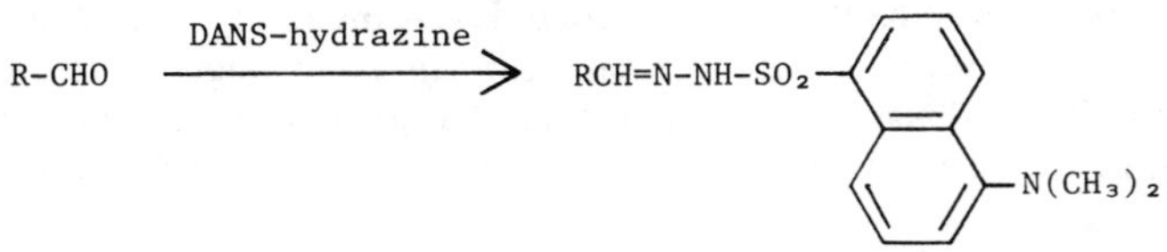

Typical Procedure. Heat the aldehyde or ketone for 15 min at 70° with 1 ml MeOH containing a twofold molar excess of DANS-hydrazine and 2 drops HOAc. Evap to dryness under vac and diss residue in 0.5 ml benzene. Analyze by HPLC.

Comments. The derivs were used for fluorescence detection HPLC (excitation max 365 nm; emission max. 530 nm). Detection limits were 1–10 ng per injection. The synthesis of the reagent from DANS-Cl and hydrazine was reported. NBD-hydrazine was also prepared from 4-chloro-7-nitrobenzo-2,1,3-oxadiazole (NBD-Cl). NBD-hydrazones were formed from aldehydes and ketones but only the aldehyde derivs were fluorescent.

References

[1]R. W. Frei and J. F. Lawrence, *J. Chromatogr.* **83**, 321 (1973).

Derivative Type. Substituted hydrazones.

Reagent. N-Aminopiperidine, N-aminohomopiperidine, pentafluorophenylhydrazine, phenylhydrazine.

Compound Type. Ketones.

Reaction

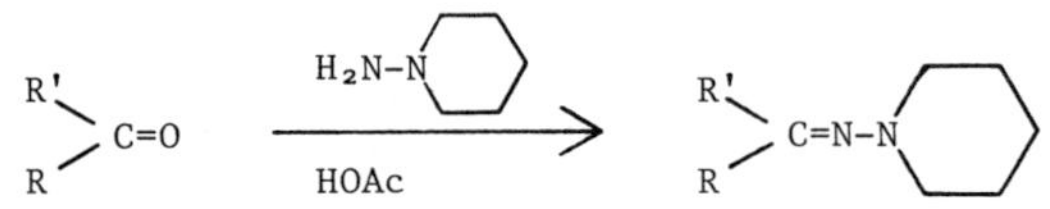

Typical Procedure. Diss compd and reagent in EtOAc containing a catalytic amt of HOAc. Let stand for 1 hr at room temp and analyze directy by GC,

Comments. The method was demonstrated for one steroidal and two aliphatic ketones. The derivs had good GC properties.

References

[1]W. J. A. VandenHeuval *et al., J. Chromatogr.* **18**, 391 (1965).

6.2. OXIME DERIVATIVES

Derivative Type. Oxime.

Reagent. Hydroxylamine.

Compound Type. Carbonyl compounds.

Reaction

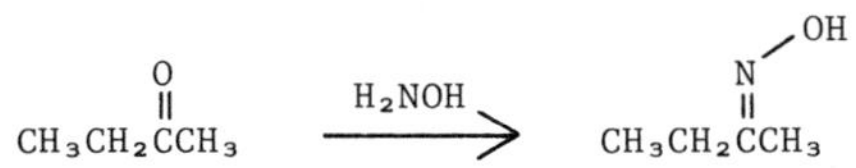

Typical Procedure. Prepare hydroxylamine soln by allowing a satd soln of NH_2OH-HCl to react with an equiv amt of KOH in a small vol of H_2O containing a half equiv amt of AcOH. Filter soln at 0° to remove KCl. Pass exhaust gas sample through the hydroxylamine soln. Alkalinize soln with solid NaOH until 1–2 *N*. Wash basic soln with pentane. Neutralize NaOH to pH 8–9 with CO_2 or AcOH. Sat soln with NaCl and ext with Et_2O. Conc ext and analyze by GC.

Comments. The weak acidity of the oximes (pK_a 10–12) and their stability in base was exploited to effect a separation of aldehydes and ketones from hydrocarbons and other neutral compounds extracted from exhaust gas. Most of the aliphatic aldoximes examined showed two GC peaks due to *syn* and *anti* isomers. The derivs decomposed on metal surfaces and were therefore analyzed on glass columns. GC data were reported for derivs of 68 carbonyl compds.

References

[1]J. W. Vogh, *Anal. Chem.* **43**, 1618 (1971).

Derivative Type. p-Nitrobenzyloxime.

Reagent. p-Nitrobenzyloxyamine hydrochloride.

Compound Type. Aldehydes and ketones.

Reaction

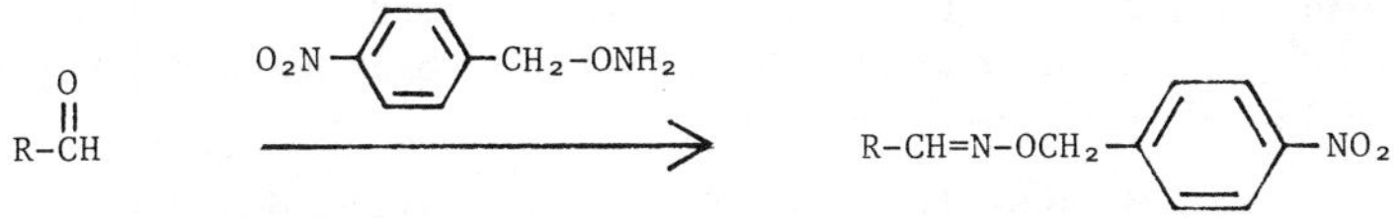

Typical Procedure. To 1–5 mg of sample in a 5 ml reaction vial add 4 ml MeOH and 40 mg *p*-nitrobenzyloxyamine·HCl. Heat at 65° for 1 hr. Cool and analyze the soln by HPLC.

Comments. The deriv was used to enhance UV adsorption for HPLC analysis.

References

[1]Regis Lab Notes, No. 17, Nov. 1974.

6.3. TRIMETHYLSILYL ENOL ETHER DERIVATIVES

Derivative Type. Trimethylsilyl.

Reagent. Ethyl trimethylsilylacetate (ETSA)/tetrabutylammonium fluoride (TBAF).

Compound Type. Ketones and alcohols.

Reaction

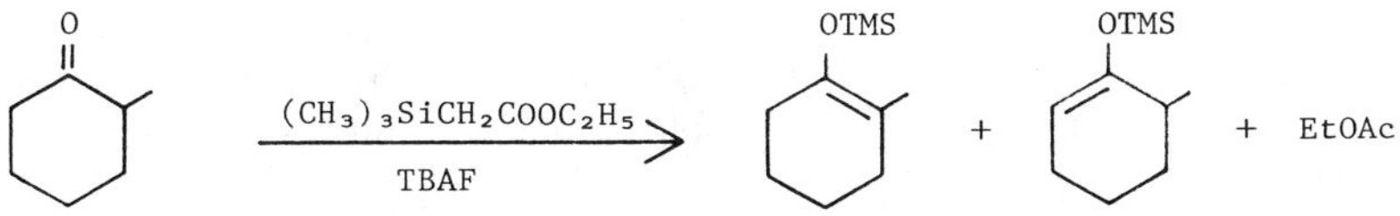

Typical Procedure. Add mixt of sample and 1.0–1.1 equiv ETSA to 0.03–0.003 equiv TBAF. The reaction takes place immediately.

Comments. The reagent was prepared by reaction of TMCS with ethyl bromoacetate in the presence of Zn dust.[2] The reagents did not significantly react with epoxides, esters or nitriles but added to aldehydes to form β-hydroxyesters. Alcohols required a larger amount of catalyst relative to ketones. The reported preparative method may be adaptable to analytical use.

References

[1]E. Nakamura *et al.*, *J. Am. Chem. Soc.* **98**, 2346 (1976).

[2]R. J. Fessenden and J. S. Fessenden, *J. Org. Chem.* **32**, 3535 (1967).

Derivative Type. Trimethylsilyl.

Reagent. Trifluoromethanesulfonic acid trimethylsilyl ester.

Compound Type. Carbonyl compounds.

Reaction

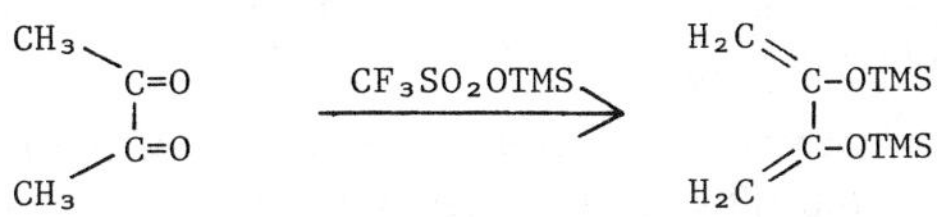

Typical Procedure (Preparative). Diss 4.3 g (0.05 mol) diacetyl and 12 g (0.12 mol) triethylamine in 100 ml abs benzene in a dry three neck flask with reflux condenser and $CaCl_2$ drying tube. With strong magnetic stirring, N_2 atmosphere, and cooling to 0–5°, add dropwise 22.2 g (0.1 mol) CF_3SO_2OTMS then heat for 1 hr. Separate the benzene phase and distill to yield 7.9 g (69%) of the TMS enol ether product.

Comments. The reagent was used to prepare TMS enol ethers on a preparative scale but could be adaptable for analytical use.

References

[1]G. Simchen and W. Kober, *Synthesis* 259 (1976).

6.4. ACETAL AND KETAL DERIVATIVES

Derivative Type. Acetals and ketals.

Reagent. Methanol (MeOH) or isopropanol (iPrOH)/toluenesulfonic acid (TsOH).

Compound Type. Aldehydes and ketones.

Reaction

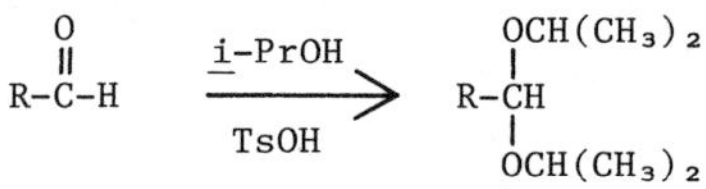

Typical Procedure. Diss 100 mg sample of aldehydes/ketones in 5 ml of the alcohol containing 10 mg TsOH. Maintain at −10° for 1 hr then add 5–10 mg of $NaOCH_3$ to stop the reaction. Analyze 10 μl aliquots by GC.

Comments. The deriv permitted GC separation of otherwise unseparated carbonyl compds. The −10° reaction temp was used to favor ketal formation.

References

[1]P. E. Manni and T. D. Streif, *J. Chromatogr. Sci.* **10**, 178 (1972).

Derivative Type. Acetal, trifluoroacetyl.

Reagent. Ethanol, *N*-methyl(bis)trifluoroacetamide (MBTFA).

Compound Type. Pyridoxal (and other B_6 vitamers).

Reaction

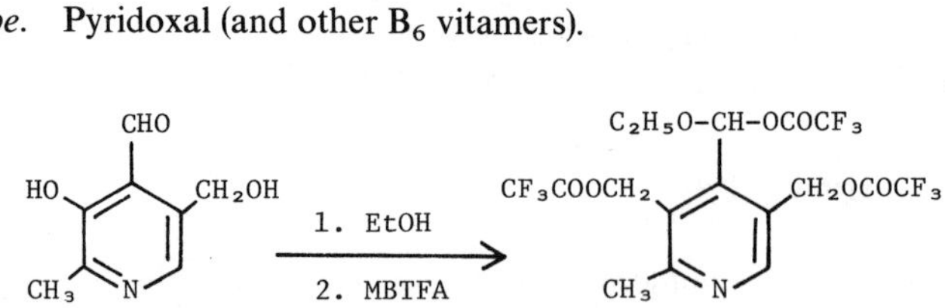

Typical Procedure. Aliquot 10–200 μl each of aq solns of pyridoxal HCl, pyridoxine HCl, pyridoxamine HCl, and desoxypyridoxine HCl into a tube and evap under N_2 at 70°. Add 40 μl abs EtOH, close tube, and reflux 15 min at 125°. Evap to dryness at 70° under N_2. Add 30 μl MBTFA, close tube, and reflux 20 min at 125°. Analyze by GC.

Comments. The method permitted simultaneous analysis of the four vitamers. The ethyl acetal, acetate deriv of pyridoxal has been prepared by refluxing with EtOH followed by treatment with Ac_2O/pyridine.[2] The methyl acetal, heptafluorobutyrate was prepared with MeOH and HFBI and analyzed by ECD-GC.[3] Treatment of pyridoxal directly with Ac_2O/pyridine yielded a diacetyl deriv which has been used for simultaneous analysis of the vitamers.[4,5]

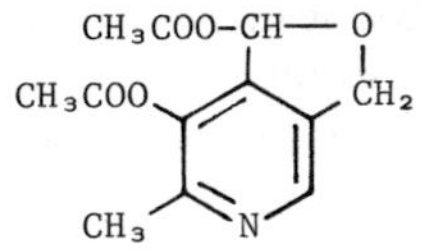

References

[1]E. M. Patzer and D. M. Kilker, *J. Chromatogr.* **135**, 489 (1977).
[2]W. Korytnyk *et al., Anal. Biochem.* **17**, 66 (1966).
[3]A. K. Williams, *J. Agr. Food Chem.* **22**, 107 (1974).
[4]A. J. Sheppard and A. R. Prosser, *Meth. Enzymol.* **18A**, 494 (1970).
[5]W. Korytnyk, *Meth. Enzymol.* **18A**, 500 (1970).

6.5. OTHER ALDEHYDE DERIVATIVES

Derivative Type. 2-Hydroxypyrimidine, trimethylsilyl.

Reagent. Urea, hexamethyldisilazane (HMDS)/trimethylchlorosilane (TMCS)/pyridine.

Compound Type. Malonaldehyde.

Reaction

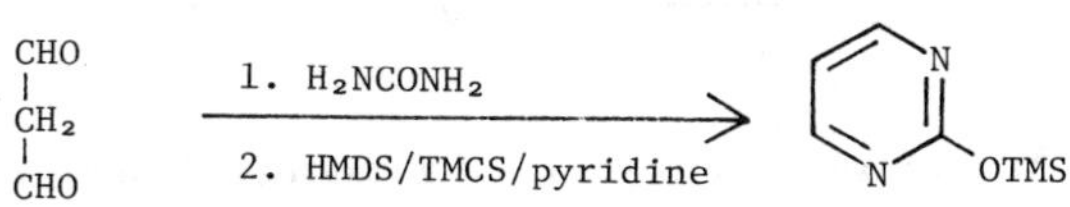

Typical Procedure. To a 0.5 m*M* aq soln of malonaldehyde add urea to a conc of 0.2 *M*. Add an equal vol of 12 *N* HCl and allow to stand at room temp 1 hr. Evap to dryness under vac at *ca* 60°. Diss residue in aq NH_4OH and evap to form the free base. Apply to a 5 ml column of Dowex 50 (H form) and elute with 20 ml 0.1 *M* HOAc and 20 ml 0.3 *M* HOAc/0.1 *M* pyridine. Evap the HOAc/pyridine eluate to dryness, treat residue with HMDS/TMCS/pyridine (2:1:7, v/v/v),[2] and reflux 1 hr. Analyze by GC-MS.

Comments. The method was used to identify microgram amts of malonaldehyde in tissue homogenates.

References

[1]M. Hamberg *et al., Anal. Biochem.* **22**, 145 (1968).
[2]Y. Sasaki and T. Hashizume, *Anal. Biochem.* **16**, 1 (1966).

Derivative Type. Carboxylic acid.

Reagent. Ozone.

Compound Type. Aldehyde 2,4-dinitrophenylhydrazones.

Reaction

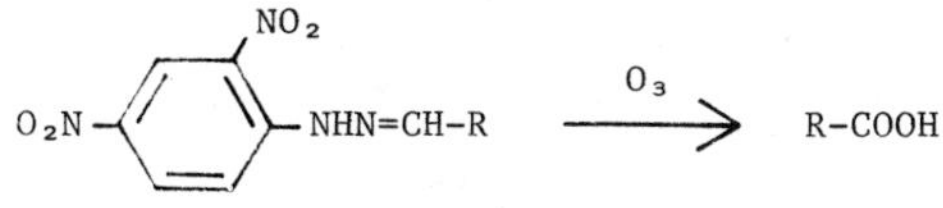

Typical Procedure. Diss sample in MeOH and cool in a dry ice/EtOH bath. Pass 2% (v/v) O_3 in O_2 through the soln at a rate of 10–20 ml/min for 30 min then pass pure O_2 for 30 min to remove residual O_3. Allow to warm to room temp then analyze by GC.

Comments. The dinitrophenylhydrazone derivs of aliphatic aldehydes were converted to the corresponding carboxylic acids to permit GC identification.[1] The method was used for analysis of aldehydes in crude spirits.[2] Ozonolysis has also been used to regenerate α-ketoacid methyl esters from the dinitrophenylhydrazones.[3]

References

[1]P. Ronkainen and S. Brummer, *J. Chromatogr.* **28**, 253 (1967).

[2]P. Ronkainen *et al.*, *J. Chromatogr.* **28**, 270 (1967).

[3]P. Ronkainen and S. Brummer, *J. Chromatogr.* **28**, 259 (1967).

Derivative Type. 6-Mercapto-3-substituted-*s*-triazolo[4,3-b]-*s*-tetrazines.

Reagent. 4-Amino-3-hydrazino-5-mercapto-1,2,4-triazole.

Compound Type. Aldehydes.

Reaction

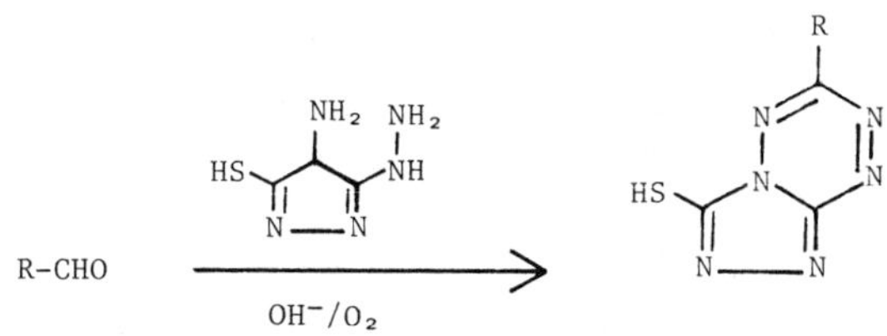

Typical Procedure. Add soln of the aldehyde to 0.1 g reagent in 10 ml 1 *N* NaOH. Aerate mixt for 30 min, dil to standard volume, and read absorbance.

Comments. This chromophore addition reaction could possibly be adaptable to HPLC (λ_{max}~ 540 nm).

References

[1]N. W. Jacobsen and R. G. Dickinson, *Anal. Chem.* **46**, 298 (1974).

7. Nitrogen Functional Groups Other than the Amino Group

This section presents the derivatives of nitrogen containing functional groups in compounds other than amines (covered in Section 2, p. 65), amino acids and peptides (Section 5, p. 242), purine and pyrimidine bases (Section 14, p. 601), barbiturates and hydantoin drugs (Section 15.3, p. 638), and benzodiazepines (Section 15.4, p. 655). Included in this section are derivatives of the *N*-hydroxylated compounds, which are of considerable analytical interest because of their role in foreign compound metabolism and in carcinogenesis, and derivatives of *N*-nitroso compounds, which are of interest largely due to their carcinogenic potential. The carbamates and ureas, also covered in this section, include many important pesticides that are of interest particularly in regard to residue analysis. Several important drugs are included among the guanidines, and other drugs are among the group of miscellaneous nitrogen compounds. This section contains a diverse array of compounds, but it includes some analytically very important compound classes.

N-*Hydroxyl Compounds.* Hydroxylamines have decreased basicity (pK_a 5–6) compared to the nonhydrolyated alkyl amines (pK_a 10–11) but are still basic enough to form salts with strong acids.[1] In addition, the hydroxyl group of the hydroxylamines is weakly acidic (the

pK_a of hydroxylamine is estimated to be 12–13).[1] Hydroxylamines have been converted to *N,O*-diacyl derivatives by treatment with trifluoroacetic anhydride (p. 350). Treatment with BSTFA gave *O*-TMS derivatives (p. 350).

hydroxylamine   hydroxamic acid   *N*-hydroxycarbamate

The hydroxyl of the hydroxamic acids is slightly more acidic (pK_a 8–9; comparable to that of phenols) and the nitrogen is still weakly basic.[2] The hydroxamic acids are sufficiently acidic to undergo methylation with diazomethane (p. 352). *O*-Propyl derivatives have been prepared with trimethylammonium hydroxide and propyl iodide (p. 350). Hydroxamic acids have also been converted to *O*-TMS derivatives (p. 351); the lack of reported structural data, however, precludes discounting the possibility of *N,O*-di-TMS derivatives in compounds with the N—H group (p. 351). *O*-Acetyl derivatives have been prepared and analyzed as isocyanates following pyrolysis in the GC (p. 351). The *N,O*-di-trifluoroacetyl derivatives, however, could be analyzed as the intact derivatives by GC-MS (p. 353). Treatment of an *N*-aryl-*N*-hydroxy-acetamide with hydrochloric acid yielded an *N*-chloro aryl amine, which was analyzed as the trifluoroacetyl derivative (p. 353). The chemically similar *N*-hydroxycarbamates have been analyzed both as *N,O*-di-TFA and *N,O*-di-TMS derivatives (p. 354).

Nitrosamines. Underivatized nitrosamines can be analyzed by GC but derivatization has been used to improve both GC behavior and detection sensitivity. After denitrosation, the resulting amines can be acylated as with other amines (p. 354), and nitrosamines can be oxidized with peracid to yield nitramine derivatives having increased ECD sensitivity (p. 355). In most cases, however, the derivatizations applied to nitrosamines are dictated by other functional groups in the molecules and not the *N*-nitroso function. Hydroxy substituted nitrosamines have been methylated (p. 355), acylated (p. 356), and silylated (p. 357). *N*-Nitrosoamino acids have been converted to methyl (p. 357) and trimethylsilyl (p. 357) esters. *N*-nitroso compounds have also been denitrosated and treated with NBD-Cl to form fluorogenic derivatives for HPLC analysis (p. 358). Denitrosated nitrosamines should also be amenable to other derivatization methods applied to amines (Section 2), but it must be noted that any nonnitrosated amine present in the sample would require prior removal or duplicate determinations (with and without the denitrosation step).

Carbamates and Ureas. Carbamates have been analyzed both as derivatives of the intact molecule and as derivatives of the products resulting from hydrolysis. The analysis of carbamate insecticides and their metabolites has been reviewed.[3] The intact carbamates are susceptible to several acylating reagents that yield *N*-acyl derivatives (p. 358) having good GC properties.

The carbamates have been converted to *N*-methyl derivatives with methyl iodide/sodium hydride (p. 363) and with trimethylphenylammonium hydroxide (p. 364). The latter reagent, however, gave *N*-methyl derivatives only with *N*-aryl carbamates; *N*-methyl carbamates gave the methyl aryl ethers. *N*-Trimethylsilyl carbamates have also been prepared (p. 364). Thiolcarbamates have been converted to the *O*-methyl carbamates by methanolysis to confirm their identification by GC (p. 365).

Hydrolysis of a carbamate yields both an amine component and an alcohol or phenol component. Acylated and alkylated derivatives of both components have been prepared for GC analysis (p. 365). Anilines derived from *N*-aryl carbamate hydrolysis have been brominated to enhance their ECD sensitivity (p. 370). Chromogenic derivatives of both components resulting from hydrolysis have likewise been prepared for HPLC analysis (p. 371).

$$R{-}O{-}\overset{\overset{\displaystyle O}{\|}}{C}{-}NH{-}R' \longrightarrow R{-}OH \;+\; H_2N{-}R'$$

The ureas have been analyzed both as derivatives of the intact urea and as derivatives of the urea hydrolysis products. The N—H groups of intact ureas can be alkylated with methyl iodide and strong base (p. 372) or with trimethylphenylammonium hydroxide (p. 373) to yield derivatives for GC analysis. Ureas can also be *N*-acylated with trifluoroacetic anhydride for GC analysis (p. 374). Thioureas give *S*-alkylation with alkyl halides yielding derivatives that can either be analyzed directly by GC or *N*-acylated (p. 375). Urea hydrolysis yields a pair of amines.

$$R{-}NH{-}\overset{\overset{\displaystyle O}{\|}}{C}{-}NH{-}R' \longrightarrow R{-}NH_2 \;+\; H_2N{-}R'$$

Derivatives of these amines have been prepared both for GC (p. 376) and HPLC (p. 377) analysis. Anilines resulting from hydrolysis of ureas have been brominated (p. 370) and converted to iodobenzenes via diazotization (p. 377) for ECD-GC analysis.

Guanidines and Isocyanates. Guanidines can be hydrolyzed by aqueous base to yield amines that have been acylated for GC analysis of several guanidine drugs (p. 377).

$$R{-}NH{-}\overset{\overset{\displaystyle NH}{\|}}{C}{-}NH_2 \longrightarrow R{-}NH_2$$

GC analysis of intact guanidines requires derivatization of the very polar guanidine group. Hexafluoroacetylacetone has been successfully employed to prepare bis(trifluoromethyl)-pyrimidine derivatives (p. 378). Biguanides have been converted to triazine derivatives by acylating agents and used for GC-MS (p. 379) and HPLC (p. 380) analysis.

Urea derivatives formed by the reaction of isocyanates with amines, have been used for GC (p. 380) and HPLC (p. 381) analyses of isocyanates.

Derivatives of miscellaneous nitrogen containing functional groups are presented in Section 7.7 (p. 381). In general, the nitrogens in these compounds are either alkylated or acylated to yield volatile derivatives with improved behavior for GC analysis. Two examples of the use of oxidation to yield derivatives for GC analysis are cited (p. 385).

References

1. P. A. S. Smith, Open-Chain Nitrogen Compounds, Vol. II, Benjamin, New York, 1966, p. 66.
2. Reference 1, p. 74.
3. H. W. Donough, *J. Chromatogr. Sci.* **13**, 212 (1975).

7.1. DERIVATIVES OF *N*-HYDROXYL COMPOUNDS

7.1.1. Derivatives of *N*-Hydroxylamines

Derivative Type. Trifluoroacetyl (TFA).

Reagent. Trifluoroacetic anhydride (TFAA).

Compound Type. *N*-Hydroxylamines.

Reaction

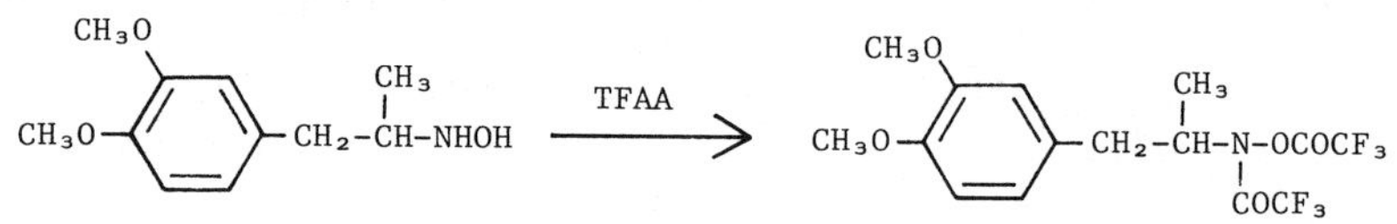

Typical Procedure. Diss 2.5 mg sample in 0.5 ml TFAA. Allow to stand 15 min then analyze by GC.

Comments. *O*-TMS derivs were also prepared using 1:1 (v/v) CH_3CN/BSTFA at room temp.[1] MS data were given for the derivs. The di-TFA deriv has been used for GC-MS-SIM quantitation in metabolism studies of 1-(2,5-dimethoxy-4-methylphenyl)-2-aminopropane.[2]

References

[1]A. H. Beckett and R. Achari, *J. Pharm. Pharmacol.* **29**, 212 (1977).
[2]J. Gal *et al.*, *J. Med. Chem.* **18**, 683 (1975).

Derivative Type. Trimethylsilyl.

Reagent. Bis(trimethylsilyl)trifluoroacetamide (BSTFA).

Compound Type. *N*-Hydroxylamines.

Reaction

Typical Procedure. Evap 1 ml of a 1 m*M* soln of the hydroxylamine free base in CH_2Cl_2 to dryness under N_2. Diss residue in 0.2 ml CH_2Cl_2 and add 20 μl BSTFA. Heat 60 min at 40–50°. Analyze by GC-MS.

Comments. Derivs were prepared from the *N*-hydroxy phenethylamine, amphetamine, phentermine, and *p*-chlorophentermine. MS data were given for the derivs.

References

[1]B. Lindeke *et al.*, *Acta Pharm. Suecica* **10**, 493 (1973).

7.1.2. Derivatives of Hydroxamic Acids

Derivative Type. *O*-Propyl.

Reagent. Tetramethylammonium hydroxide (TMAH)/propyl iodide.

Compound Type. *N*-Hydroxy-*p*-methoxyacetanilide.

Reaction

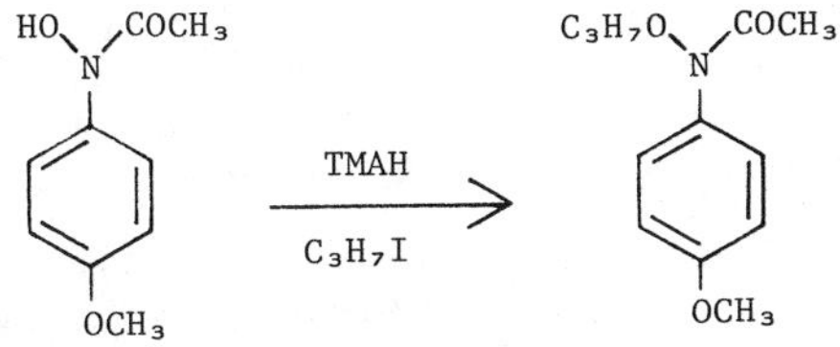

Typical Procedure. Add 4.5 ml microsomal suspension mixt to 5 ml ice cold EtOAc containing int stand and a saturating amt of NaCl. Shake, centrifuge, and sep org phase. Repeat ext with 5 ml EtOAc. Evap combined ext under vac and diss residue in 0.64 ml acetone and 0.64 ml 2% TMAH in MeOH. Mix and then add 0.40 ml *n*-propyl iodide. Allow to stand at room temp 30 min. Add 5 ml Et_2O and centrifuge. Remove supernatant and evap to dryness under vac. Diss residue in 100 μl THF and analyze by GC-MS-SIM.

Comments. *N*-Hydroxyacetanilide was used as int stand. *p*-Hydroxyacetanilide was determined simultaneously as the *O*-propyl deriv using *p*-hydroxybutyranilide as int stand. The deriv reaction worked equally well in DMSO or DMA. Other alkyl iodides could be used. The method should be generally useful for hydroxamic acids. The TMS derivs were found to be unstable.

References

[1]I. M. Kapetanovic *et al.*, *Anal. Chem.* **49**, 1843 (1977).

Derivative Type. Trimethylsilyl.

Reagent. Bis(trimethylsilyl)acetamide (BSA).

Compound Type. Hydroxamic acids.

Reaction

Typical Procedure. Mix 4 ml plasma with 4 ml pH 8 phosphate buffer and ext with 2 × 16 ml methylisobutylketone. Dry ext over Na_2SO_4 and evap under vac. Add 10 μg int stand in 20 μl dioxane. Add 40 μl BSA and allow to stand overnight at room temp. Analyze by GC.

Comments. 3-Chloro-4-propyloxyphenylacethydroxamic acid was used as int stand for quantitation of *p*-butoxyphenylacethydroxamic acid (bufexamac, shown above). No structural data were reported for the deriv, thus it may have been an *N*,*O*-di-TMS deriv.

References

[1]R. Roncucci *et al.*, *J. Chromatogr.* **57**, 410 (1971).

Derivative Type. Isocyanate.

Reagent. Acetic anhydride (Ac_2O)/pyridine.

Compound Type. Hydroxamic acids.

Reaction

$$R{-}CONH{-}OH \xrightarrow{Ac_2O} R{-}CONH{-}OCOCH_3 \xrightarrow[\Delta]{} R{-}NCO$$

Typical Procedure. Suspend up to 10 mg of the hydroxamic acid in 1 ml THF. Add 50 μl pyridine and 50 μl Ac_2O. Allow to stand 10 min at room temp then analyze by GC.

Comments. The acetylated hydroxamic acids were converted to isocyanates in the hot injector. The method was used for analysis of acyl coenzyme A compds by converting them to the hydroxamic acids and forming the isocyanate derivs.

References

[1]P. R. Vagelos *et al., Anal. Biochem.* **2**, 50 (1961).

Derivative Type. Methyl.

Reagent. Diazomethane.

Compound Type. N-Hydroxy-2-fluorenylacetamide.

Reaction

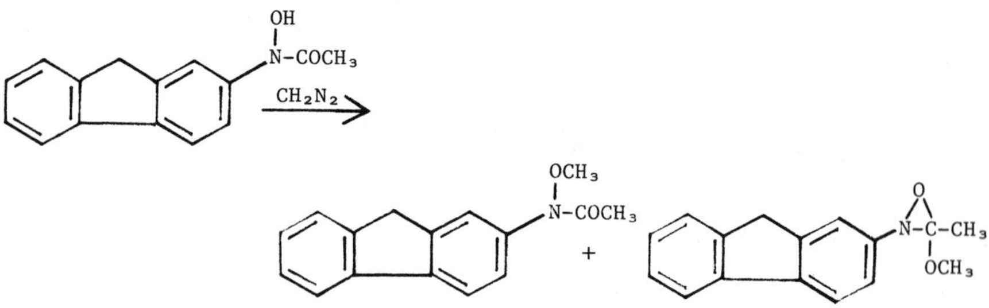

Typical Procedure. Diss a 50 μg sample in a soln of CH_2N_2 in Et_2O. After 30 min at room temp, evap to dryness and diss residue in 50 μl of $CHCl_3$. Analyze by GC-MS.

Comments. MS data for the derivs were reported.

References

[1]G. Lhoest *et al., Biomed. Mass Spectrom.* **3**, 21 (1976).

Derivative Type. Trimethylsilyl.

Reagent. Hexamethyldisilazane (HMDS)/trimethylchlorosilane (TMCS)/pyridine.

Compound Type. N-Hydroxy-2-fluorenylacetamide.

Reaction

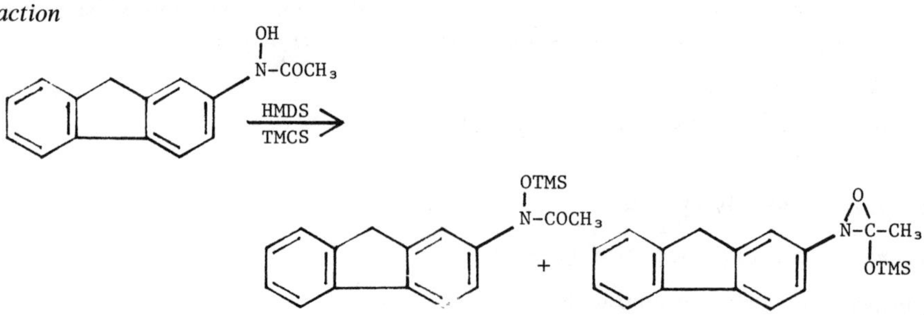

Typical Procedure. Diss 50 μg sample in a mixture of HMDS, TMCS, and pyridine (2:1:4, v/v/v). After 30 min at 40°, evap to dryness and diss residue in 50 μl CH_3CN. Analyze soln by GC-MS.

Comments. Mass spectra of the derivs were reported.[1] BSA has also been used to prepare TMS derivs of *N*-hydroxy-2-fluorenylacetamide,[2] *N*-hydroxy-2-acetamido-4-biphenyl,[3] and *N*-hydroxy-2-acetamidonaphthalene.[3]

References

[1]G. Lhoeest *et al., Biomed. Mass Spectrom.* **3**, 21 (1976).
[2]P. D. Lotlikar *et al., Biochem. J.* **136**, 1137 (1973).
[3]J. L. Radomski *et al., J. Natl. Cancer Inst.* **50**, 989 (1973).

Derivative Type. Trifluoroacetyl (TFA).

Reagent. Trifluoroacetic anhydride (TFAA).

Compound Type. *N*-Hydroxy-2-fluorenylacetamide.

Reaction

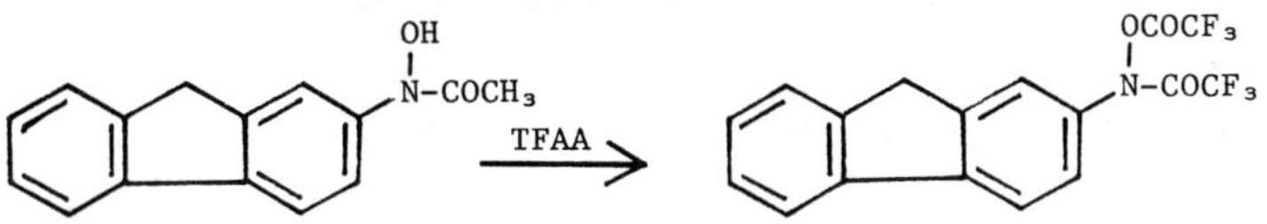

Typical Procedure. Diss 50 μg sample in 30 μl TFAA. After 30 min at 50°, evap to dryness and diss residue in 50 μl $CHCl_3$. Analyze by GC-MS.

Comments. A transacylation also replaced the acetyl group to give a di-TFA deriv. The MS fragmentation of the deriv was reported.

References

[1]G. Lhoest *et al., Biomed. Mass Spectrom.* **3**, 21 (1976).

Derivative Type. *N*-Chloro, trifluoroacetyl.

Reagent. Hydrochloric acid, trifluoroacetic anhydride (TFAA).

Compound Type. *N*-Hydroxy-2-fluorenylacetamide.

Reaction

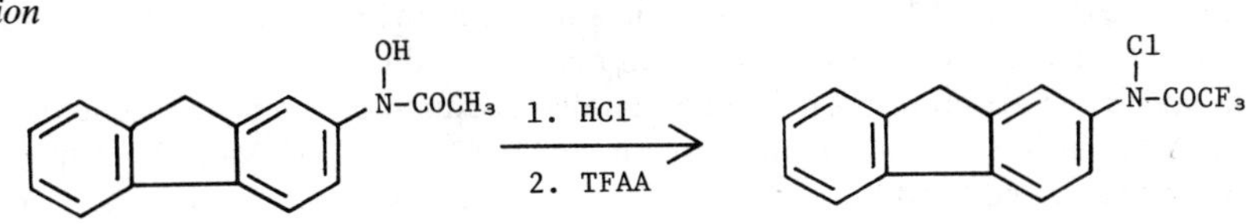

Typical Procedure. Diss 100 μg sample in 6 *N* HCl and maintain at 60° for 30 min. Basify to pH 8 with NH_4OH and ext with 2 × 1 ml $CHCl_3$. Evap ext to dryness and treat residue with 60 μl TFAA for 30 min at 50°. Evap mixt to dryness and diss residue in 100 μl $CHCl_3$. Analyze by GC-MS.

Comments. The MS fragmentation of the deriv was reported.

References

[1]G. Lhoest *et al., Biomed. Mass Spectrom.* **3**, 21 (1976).

7.1.3. Derivatives of Hydroxy Carbamates

Derivative Type. *O*-Acetyl.

Reagent. Acetic anhydride(Ac_2O)/triethylamine (Et_3N).

Compound Type. *N*-Hydroxycarbamates.

Reaction

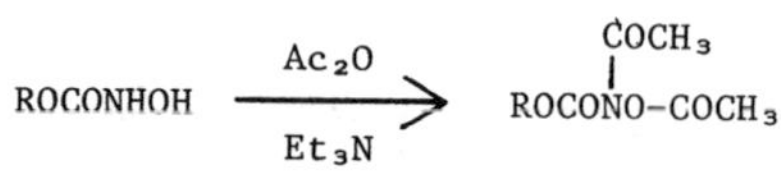

Typical Procedure. Diss sample in 0.5 ml pyridine and add 0.2 ml Ac_2O and 0.3 ml Et_3N. allow to stand 4 hr at 23° then analyze an aliquot by GC.

Comments. Derivs were prepared from a series of *N*-hydroxycarbamates where $R = C_1–C_5$. TMS derivs were also prepared using HMDS/TMCS/pyridine. The TMS derivs of the parent carbamates (i.e., nonhydroxylated) were also prepared. Although the *O*-monoacetyl derivs were initially formed, after standing 4 hr there was quant formation of the diacetyl derivs. The TMS derivs were likewise reported to be *N,O*-di-TMS.

References

[1] R. Nery, *Analyst* **94**, 130 (1969).

7.2. DERIVATIVES OF *N*-NITROSO COMPOUNDS

7.2.1. Derivatives of *N*-Nitrosamines

Derivative Type. Heptafluorobutyramide.

Reagent. Hydrogen bromide, heptafluorobutyryl chloride (HFB-Cl).

Compound Type. Nitrosamines.

Reaction

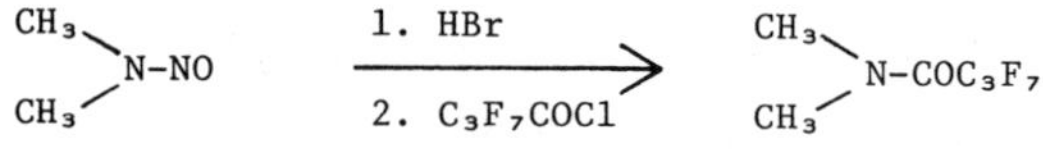

Typical Procedure. Mix 0.4 ml vol of sample and 5% HBr in CH_2Cl_2 and allow to stand 15 min. Evap to dryness under vac at 30° then remove residual HBr under N_2 stream. Diss residue in 0.5 ml MeOH and transfer to a 5 mm diam column of 200 mg SE-cellulose. Wash column with 1.5 ml MeOH; then elute with 2 ml 4 *N* HCl in 50% aq MeOH. Evap eluate to dryness under vac at 60°. Add 0.5 ml CH_2Cl_2 and 1 ml *n*-pentane. Shake, add 1 ml H_2O, shake, and remove org layer. Repeat ext of aq layer with 1 ml *n*-pentane and discard org phase. Repeat if necessary until pentane is colorless. Add to remaining aq phase 0.5 ml *n*-pentane, 1 ml 0.5% HFB-Cl in *n*-pentane and 0.5 ml 1 *M* $NaHCO_3$. Stir 15 min; then add 0.5 ml 5 *N* NaOH. Stir 5 min and discard aq layer. Add 100 mg anhyd Na_2CO_3 and stir. Transfer org phase to 2 ml vol flask and use pentane washings of Na_2CO_3 to make up to vol. Analyze by GC-MS-SIM.

Comments. The method was used to determine nitrosodimethylamine, -diethylamine, -di-*n*-propylamine, di-*n*-butylamine, -piperidine, and -pyrrolidine in food samples and infected urine using GC-MS-SIM. It was necessary to analyze nondenitrosated samples to correct for the presence of nonnitroso secondary amines in the samples.

References
[1]G. Eisenbrand *et al.*, *IARC Sci. Publ.* **14**, 65 (1976).

Derivative Type. Nitramine.

Reagent. Peroxytrifluoroacetic acid.

Compound Type. *N*-Nitrosamines.

Reaction

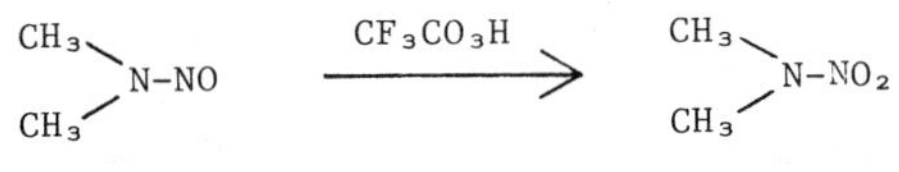

Typical Procedure.[2] To 1 ml of a 50 μg/ml soln of nitrosamine in CH_2Cl_2 add 0.1 ml peroxytrifluoroacetic acid in CH_2Cl_2 (see below). Mix then allow to stand at room temp 3.5 hr. Add 2 drops H_2O, shake gently, then allow to stand 1 min. Add 0.25 g $CaCO_3$ and, after bubbling ceases, 0.25 g anhyd Na_2SO_4. Shake gently then allow to settle. After appropriate dilution, analyze by ECD-GC.

Comments. The reagent was prepared by carefully adding 0.4 ml 85–90% (w/w) H_2O_2 to 4–5 ml CH_2Cl_2 in a 10 ml vol flask, slowly adding 2.5 ml TFAA followed by 5 min cooling in an ice bath, and dil to vol with CH_2Cl_2. The deriv was 100–1100 times more sensitive to ECD than the parent compd.

References
[1]N. P. Sen, *J. Chromatogr.* **51**, 301 (1970).
[2]J. Althorpe *et al.*, *J. Chromatogr.* **53**, 371 (1970).
[3]G. M. Telling, *J. Chromatogr.* **73**, 79 (1972).

Derivative Type. Methyl.

Reagent. Methyl iodide/sodium hydride.

Compound Type. 3-Hydroxy-1-nitrosopyrrolidine.

Reaction

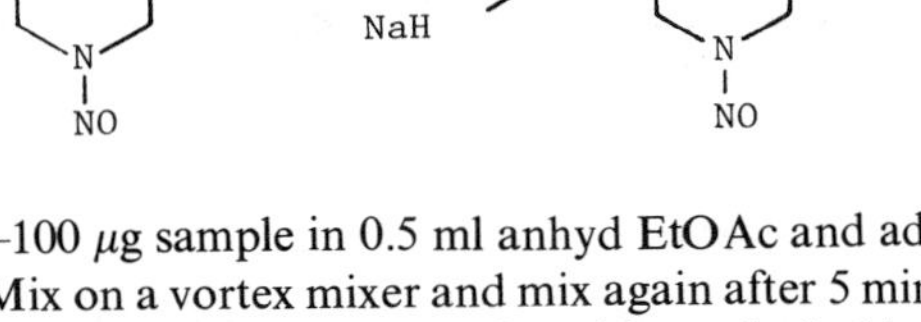

Typical Procedure. Diss 1–100 μg sample in 0.5 ml anhyd EtOAc and add 0.5 ml CH_3I and 15–20 mg powdered NaH. Mix on a vortex mixer and mix again after 5 min. Allow to stand at room temp 3 hr (or overnight). Add 1 ml H_2O dropwise with gentle shaking. Ext with 3 × 2 ml Et_2O. Dry ext by passing through Na_2SO_4 in a Pasteur pipette. Conc to 0.5–1 ml under N_2 at 30°. Analyze an aliquot by GC-MS-SIM.

Comments. The method permitted GC of the otherwise nonvolatile *N*-nitrosamine. Methylation with CH_2N_2/BF_3 or with $(CH_3O)_2SO_2$ gave lower yields of the deriv.

References

[1]N. P. Sen *et al.*, *J. Chromatogr.* **128**, 169 (1976).

Derivative Type. Trifluoroacetyl.

Reagent. *N*-Methyl-bis(trifluoroacetamide) (MBTFA).

Compound Type. 3-Hydroxy-1-nitrosopyrrolidine.

Reaction

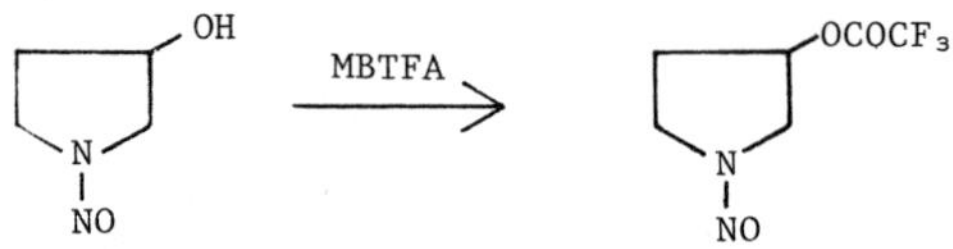

Typical Procedure. Homogenize 50 g food sample in MeOH/H_2O (9 :1, v/v) and filter through Celite. Conc 500 ml filtrate to 50 ml and ext with 50 ml *n*-pentane and 2 × 50 ml *n*-pentane-CH_2Cl_2 (1 :1, v/v). Sat aq phase with NaCl, adjust to pH 10, ext with 2 × 100 ml $HCOOC_2H_5$. Dry $HCOOC_2H_5$ ext over Na_2SO_4 and evap to 1 ml. Chromatograph on a 6 mm × 5 cm column of basic alumina covered with 2 cm of strongly acidic SE-cellulose. Elute with acetone CH_2Cl_2 (1 :1, v/v) collecting the first 20 ml. Conc the eluate to 0.2 ml. Add 0.3 ml CH_3CN and 0.1 ml MBTFA and heat in a sealed vial at 60° for 2 hr. Analyze by GC.

Comments. The method was used for detection of the nitrosamine in cured meat products to μg/kg levels.

References

[1]G. Eisenbrand *et al.*, *Proc. 2nd Int. Symp. on Nitrite in Meat Prod.*, 155 (1976).

Derivative Type. Heptafluorobutyryl (see structures).

Reagent. Heptafluorobutyric anhydride (HFBA)/pyridine.

Compound Type. *N*-Nitrosamines.

Reaction

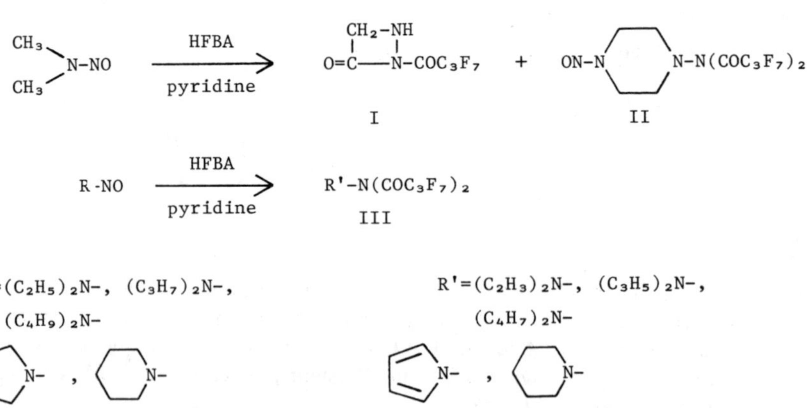

Typical Procedure. To a sample containing 1–10 μmol of each nitrosamine in 60 μl $CHCl_3$ add 1 drop (*ca* 14 μl) pyridine and 2 drops (*ca* 7 μl) HFBA. Allow to stand at room temp 1 hr.

Add 6 drops $CHCl_3$ and shake with 2 drops (*ca* 70 μl) 4% HCl. Allow to stand 3 min then discard aq phase. Add 2 drops (*ca* 70 μl) 0.1 *N* NaOH and allow to stand 30 min. Remove org phase and conc to *ca* 10 μl under an air stream. Diss residue in 100 μl Et_2O and analyze by GC.

Comments. Gough et al.[2,3] determined the structures of the derivs. In this work they used hexane in place of Et_2O in the final soln to minimize evap losses. Brooks et al.[4] used the earlier eluting dimethylnitrosamine deriv I and the derivs III to identify nitrosamines produced by urinary infection bacteria.

References

[1] J. B. Brooks *et al., Anal. Chem.* **44**, 1881 (1972).

[2] T. A. Gough *et al., Anal. Chem.* **47**, 509 (1975).

[3] T. A. Gough *et al., Anal. Chem.* **48**, 583 (1976).

[4] J. B. Brooks *et al., J. Infect. Dis.* **126**, 143 (1972).

7.2.2. Derivatives of *N*-Nitrosoamino Acids

Derivative Type. Trimethylsilyl.

Reagent. *N*-Methyl-*N*-trimethylsilyltrifluoroacetamide (MSTFA).

Compound Type. *N*-Nitrosoamino acids.

Reaction

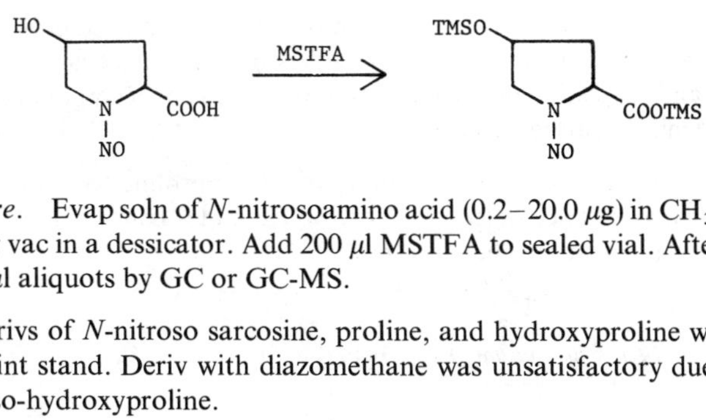

Typical Procedure. Evap soln of *N*-nitrosoamino acid (0.2–20.0 μg) in CH_3CN to dryness in 0.3 ml vial under vac in a dessicator. Add 200 μl MSTFA to sealed vial. After 30 min at room temp analyze 1 μl aliquots by GC or GC-MS.

Comments. Derivs of *N*-nitroso sarcosine, proline, and hydroxyproline were studied using myristic acid as int stand. Deriv with diazomethane was unsatisfactory due to tailing of the deriv of *N*-nitroso-hydroxyproline.

References

[1] G. Eisenbrand *et al., J. Chromatogr.* **115**, 602 (1975).

[2] G. Eisenbrand *et al., IARC Sci. Publ.* **14**, 21 (1976).

Derivative Type. Methyl ester.

Reagent. Diazomethane.

Compound Type. *N*-Nitrosoproline.

Reaction

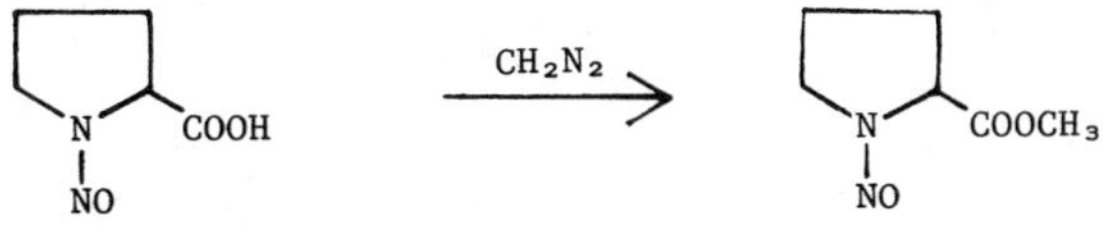

Typical Procedure. Evap aq soln of nitrosoproline to dryness under vac and diss residue in 0.5 ml MeOH. Add 2 ml of *ca* 1% CH_2N_2 in Et_2O and heat at 33° for 10 min. Repeat to insure complete deriv. Analyze by GC.

Comments. The deriv was detectable to 0.05 nmol (7 ng of the compd) using an alkali flame ionization detector.

References

[1]J. H. Wolfram *et al., J. Chromatogr.* **132**, 37 (1977).

Derivative Type. 4-Nitrobenzo-2-oxa-1,3-diazolyl (NBD).

Reagent. 7-Chloro-4-nitrobenzo-2-oxa-1,3-diazole (NBD-Cl).

Compound Type. *N*-Nitrosoproline.

Reaction

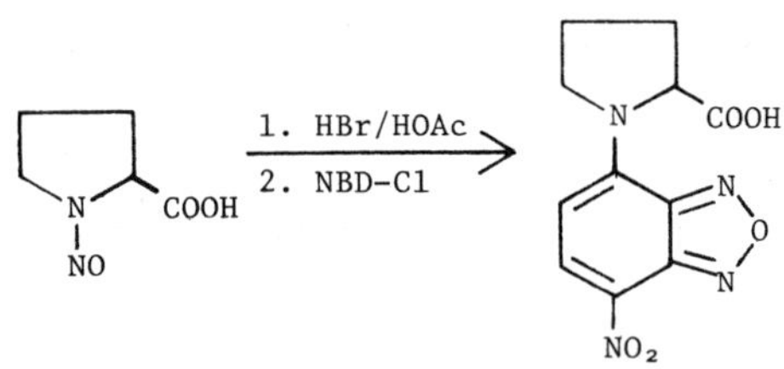

Typical Procedure. Evap under vac aq soln containing 0.1 ng nitrosoproline and diss residue in 1 ml 15% HBr in HOAc. Heat 30 min at 50°. Add 1 ml H_2O then evap to dryness under vac. Add 0.1 ml H_2O and sat soln with Na_2CO_3. Add 0.9 ml CH_3CN containing 3.0 mg NBD-Cl. heat at 50° for 30 min. Analyze by HPLC.

Comments. The deriv was used to quantitate *N*-nitrosoproline to 0.5 ng (detection limit) using HPLC with fluorescence detection. For food samples containing Pro, the Pro would have to be removed prior to the denitrosation.

References

[1]J. H. Wolfram *et al., J. Chromatogr.* **132**, 37 (1977).

7.3. DERIVATIVES OF CARBAMATES

7.3.1. Acylated Derivatives of Intact Carbamates

Derivative Type. *N*-Acetyl.

Reagent. Acetyl chloride (AcCl)/glass beads.

Compound Type. 4-Benzothienyl *N*-methylcarbamate.

Reaction

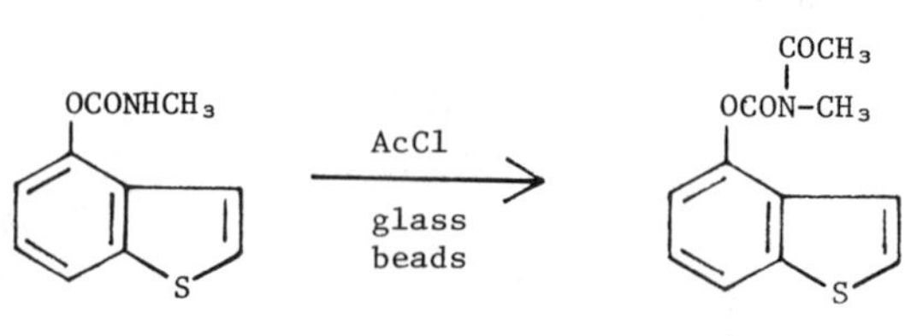

Typical Procedure. To *ca* 100 mg sample add 100 mg methyl stearate int stand, 2 ml chlorobenzene, 5 ml AcCl, and *ca* 200 mg micro glass beads (230 × 320 mesh). Reflux 4 hr, cool, and analyze a 5 μl aliquot by GC.

Comments. The method was used to quantitate the carbamate (Mobam) and its principal impurity (4-hydroxybenzothiophene).

References

[1]A. J. Epstein *et al.*, *Anal. Chem.* **39**, 721 (1967).

Derivative Type. *N*-Acetyl.

Reagent. Acetic anhydride (Ac_2O)/methanesulfonic acid (MsOH).

Compound Type. Carbamates.

Reaction

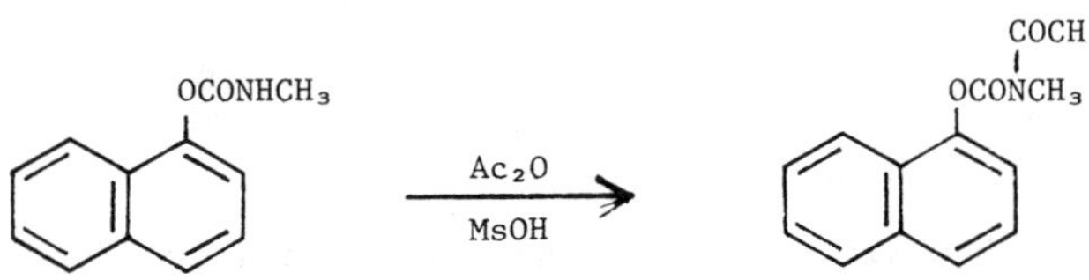

Typical Procedure. Diss 1 mg carbamate in 0.2 ml Ac_2O at 4° containing 5 mg MsOH. Allow to stand at 4° for 24 hr or raise to 97° for 30 min. Pour soln into ice H_2O and ext with $CHCl_3$. Dry over Na_2SO_4 and analyze by GC.

Comments. Use of NaOAc or pyridine as acetylation catalyst gave 30–35% 1-naphthyl acetate in addition to the desired *N*-acetyl deriv. The structure of the deriv was verified by IR and NMR.

References

[1]L. J. Sullivan *et al.*, *J. Agr. Food Chem.* **15**, 927 (1967).

Derivative Type. Trifluoroacetyl (TFA).

Reagent. Trifluoroacetic anhydride (TFAA).

Compound Type. Carbaryl and hydroxylated metabolites.

Reaction

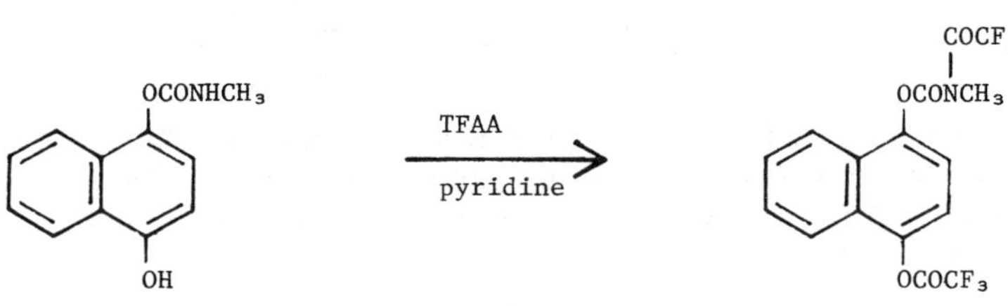

Typical Procedure. Diss 1 mg sample in 6 ml benzene. Add 0.2–0.4 ml anhyd pyridine and 0.3 ml TFAAbenzene (1 : 10, v/v). Allow to stand 15 min at 25°. Wash with 3 × 5 ml H_2O and analyze by GC.

Comments. TFA, HFB, and chloroacetyl derivs were prepared using either the anhydrides or the acyl imidazoles. Chloroacetic anhydride did not acylate the amide N. The TFA derivs were most desirable because of ease of purification, short retention time, and ECD sensitivity.

References

S. Khalifa and R. O. Mumma, *J. Agr. Food Chem.* **20**, 632 (1972).

Derivative Type. Trifluoroacetyl.

Reagent. Trifluoroacetic anhydride (TFAA)/trimethylamine (TMA).

Compound Type. 2-Methoxyethylcarbamic acid propyl ester.

Reaction

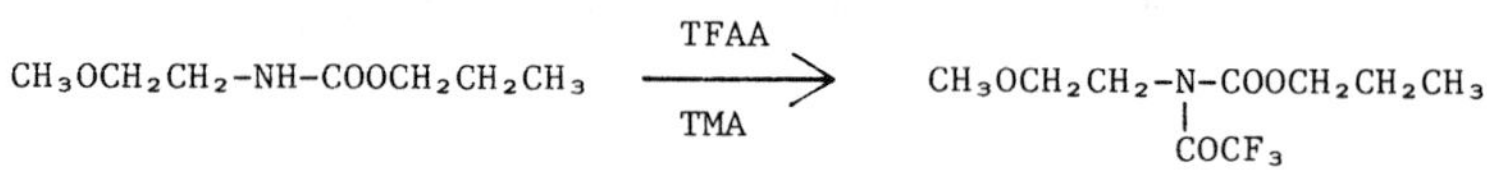

Typical Procedure. To a soln of 0.5 mg of the carbamate in 0.5 ml benzene add 20 μl TFAA and 50 μl 1.3 *M* TMA in benzene. Allow to stand at room temp 30 min then wash with 2 ml pH 7 phosphate buffer. Analyze by GC.

Comments. The method was used to quantitate the carbamate to the ng/ml range using ECD-GC. The deriv reaction was studied in benzene, hexane, and EtOAc using the TMA catalyst. The reaction was quantitative in benzene in 5 min and in hexane in 30 min but in EtOAc was incomplete even after 1.5 hr.

References

[1]B. Mellstrom and H. Ehrsson, *Acta Pharm. Suecica* **11**, 91 (1974).

Derivative Type. Trifluoroacetyl.

Reagent. Trifluoroacetic anhydride (TFAA).

Compound Type. Carbamate pesticides.

Reaction

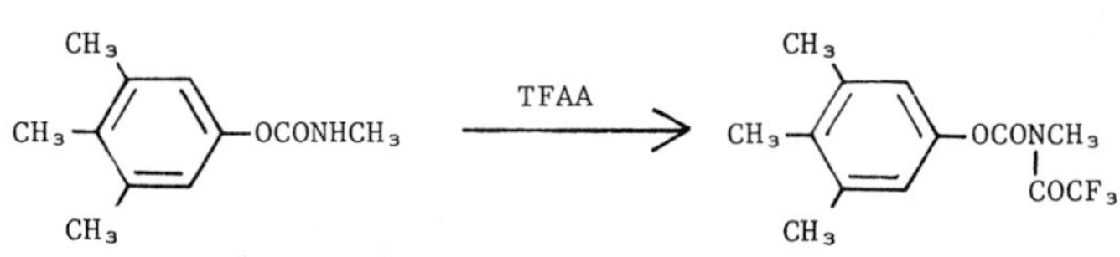

Typical Procedure. To a soln of 0.5–2.0 μg of the carbamate in 0.2 ml EtOAc add 0.2 ml TFAA and allow to react overnight at room temp (protected from light). Dil with 0.3 ml Et_2O and 4.5 ml hexane. Wash with 3×5 ml H_2O to remove excess TFAA, dry over Na_2SO_4, and analyze by ECD-GC.

Comments. The deriv was characterized by IR and elemental analysis. Ext and cleanup procedures were given for analysis of the major carbamate (shown above) of Landrin insecticide in corn samples. The method was suitable for conc down to 0.02 ppm.

References

[1]S. C. Lau and R. L. Marxmiller, *J. Agr. Food Chem.* **18**, 413 (1970).

Derivative Type. Pentafluoropropionyl (PFP).

Reagent. Pentafluoropropionic anhydride (PFPA).

Compound Type. Carbamate pesticides.

Reaction

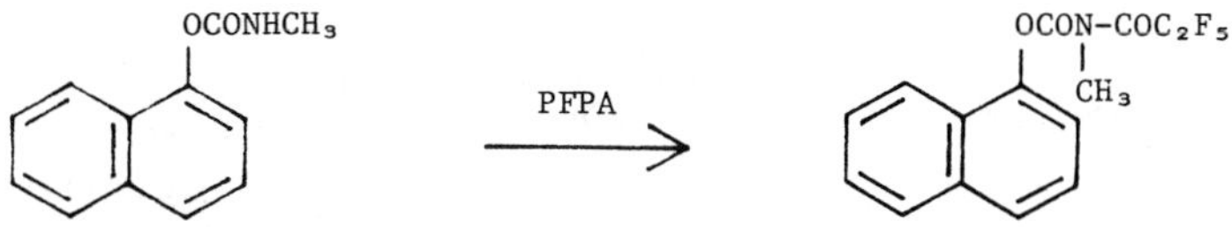

Typical Procedure. Evap sample (ext from an ethylene glycol collection from 20 m^3 air) to dryness under N_2. Add 2 ml isooctane, 1 drop redist pyridine, and 25 μl PFPA. Mix and allow to react at room temp 1 hr. Add 3 ml pH 7 phosphate buffer and mix. Add 3 ml isooctane and 0.05 ml CH_3CN. Mix then remove the bottom layer. Add 2 ml H_2O, mix, and again remove the bottom layer. Repeat H_2O wash twice. Dry soln over 0.5 g Na_2SO_4 and analyze by ECD-GC.

Comments. The method was used in a multiclass method for analysis of pesticides in air.

References

[1]J. Sherma and T. M. Shafik, *Arch. Environ. Contam. Toxicol.* **3**, 55 (1975).

Derivative Type. Heptafluorobutyryl (HFB).

Reagent. Heptafluorobutyric anhydride (HFBA)/trimethylamine (TMA).

Compound Type. Carbamate insecticides.

Reaction

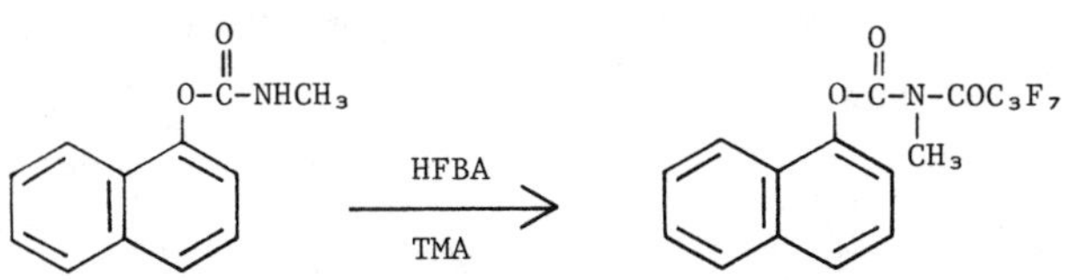

Typical Procedure. Homogenize 50 g food sample with 100 ml EtOH, filter, and transfer to a 500 ml sep funnel containing 300 ml H_2O. Ext with 3 × 50 ml $CHCl_3$. Dry ext over Na_2SO_4 and evap to dryness under vac at 30°. Transfer residue to top of a 15 × 2 cm 2% deactivated Florisil column with hexane and elute column with 250 ml 20% EtOAc in hexane. Evap eluate to dryness under vac at 30°.

To the residue add 15 μl HFBA and 0.4 ml 0.1 *M* TMA in benzene. Shake and allow to stand at room temp for 30 min. Add 4.5 ml hexane, shake, and add 10 ml H_2O. Shake, allow to separate, and analyze aliquot of hexane by GC.

Comments. Derivs of 12 carbamate insecticides were examined. TFA derivs were also studied but gave lower responses with the electrolytic conductivity detector. The method was useable to the 1 ppm level. No structural information on the derivs was reported.

References

[1]J. F. Lawrence, *J. Chromatogr.* **123**, 287 (1976).

Derivative Type. Heptafluorobutyryl.

Reagent. Heptafluorobutyric anhydride (HFBA)/pyridine.

Compound Type. Carbofuran and metabolites.

Reaction

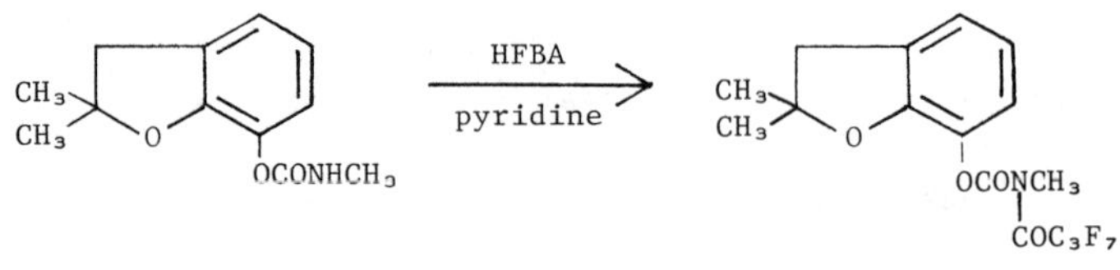

Typical Procedure. Diss the sample (up to 100 μg) in 5–10 ml benzene containing 4 drops pyridine and 0.1 ml HFBA. Allow to react overnight at room temp. Wash with 3×5 ml H_2O then dry over Na_2SO_4. Analyze by GC or GC-MS.

Comments. The method was used for analysis of carbofuran and its metabolites in crop samples. The derivs were stable in the reaction mixt for 2–3 weeks, but showed some loss in 2–3 days after the H_2O wash. Seiber[2] reported the preparation of TFA, PFP, and HFB derivs of methylcarbamate insecticides using TFAA, PFPA, and HFBA but this method, without the pyridine catalyst, failed to quant deriv the carbofuran metabs. Wong and Fisher[3] used the TFA deriv for analysis of carbofuran and its metabolites.

References

[1]R. A. Chapman and J. R. Robinson, *J. Chromatogr.* **140**, 209 (1977).

[2]J. N. Seiber, *J. Agr. Food Chem.* **20**, 445 (1972).

[3]L. Wong and F. M. Fisher, *J. Agr. Food Chem.* **23**, 317 (1975).

Derivative Type. Pentadecafluorooctanoyl (PFO).

Reagent. Pentadecafluorooctanoic anhydride (PFOA)/trimethylamine (TMA).

Compound Type. Carbofuran and linuron.

Reaction

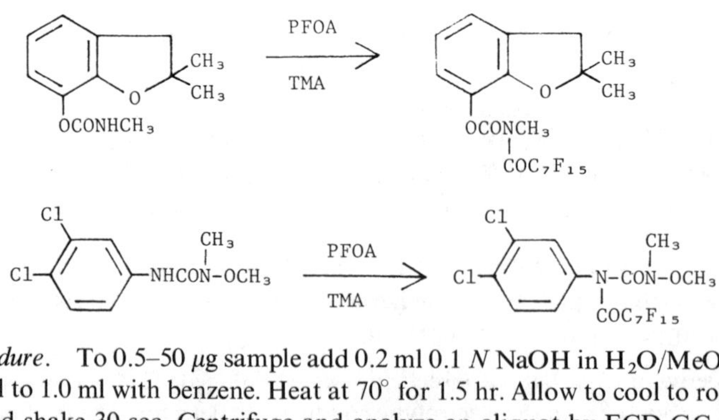

Typical Procedure. To 0.5–50 μg sample add 0.2 ml 0.1 *N* NaOH in H_2O/MeOH (3:17, v/v). PFOA and dil to 1.0 ml with benzene. Heat at 70° for 1.5 hr. Allow to cool to room temp, add 5 ml H_2O, and shake 30 sec. Centrifuge and analyze an aliquot by ECD-GC. Alternatively wash with an additional 2×5 ml H_2O and analyze by electrolytic conductivity detection GC.

Comments. Carbofuran (a carbamate) and linuron (a urea) were more difficult to deriv than the phenolic compds examined in the same study. The PFO derivs were compared with the

TFA, PFP, and HFB derivs (prepared by the same method from TFAA, PFPA, and HFBA respectively) in terms of detection sensitivity by the two detection methods. For ECD the HFB or PFO derivs were preferable. For electrolytic conductivity detection, the PFO derivs were superior. ECD sensitivities were *ca* 10–100 times greater than those by electrolytic conductivity detection.

References

[1]J. J. Ryan and J. F. Lawrence, *J. Chromatogr.* **135**, 117 (1977).

Derivative Type. Chloroacetyl.

Reagent. Chloroacetic anhydride.

Compound Type. Carbaryl.

Reaction

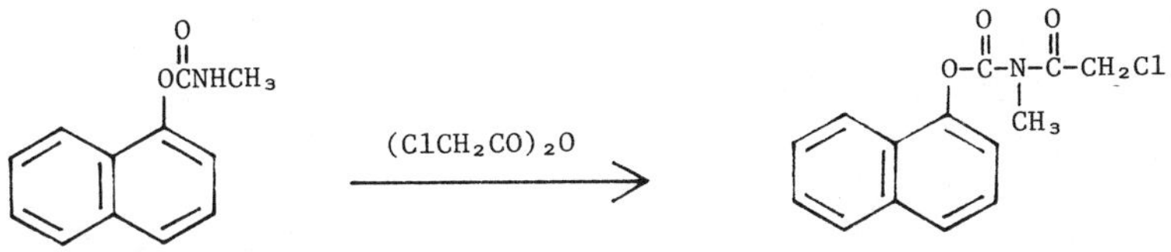

Typical Procedure (Preparative[1]). Diss 15 mmol carbaryl in 40 ml CH_2Cl_2. Add 75 mmol chloroacetic anhydride and 15 mmol pyridine and allow to stand at room temp 12 hr. Purify by recryst and chromatography on silica gel. Analyze by GC or GC-MS.

Comments. The trichloroacetyl deriv was prepared in the same manner and both derivs were also prepared from the corresponding acid chlorides. The acetyl deriv was prepared with Ac_2O/pyridine. The trichloroacetyl deriv decomposed in the GC system. The chloroacetyl deriv also underwent some thermal decomposition but was sufficiently stable for residue analysis. The acetyl deriv was thermally stable and gave adequate response (including with ECD) for residue analysis.

References

[1]E. D. Magallona *et al.*, *Arch. Environ. Contamin. Toxicol.* **5**, 177 (1977).

[2]E. D. Magallona and F. A. Gunther, *Arch. Environ. Contamin. Toxicol.* **5**, 185 (1977).

7.3.2. Other Derivatives of Intact Carbamates

Derivative Type. *N*-Methyl.

Reagent. Methyl iodide/sodium hydride.

Compound Type. Carbamate and urea herbicides.

Reaction

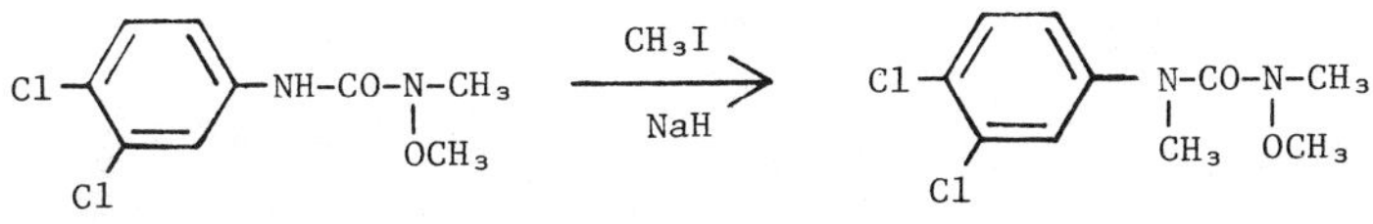

Typical Procedure. To the residue ext from a 100 g food sample add 0.5 ml DMSO, 0.5 ml benzene, 0.5 ml CH_3I, and 20–50 mg of NaH. Shake 15 min at room temp then add 3 ml hexane and shake 30 sec. Add 10 ml H_2O carefully dropwise using due care to release the H_2 formed from the NaH. After the H_2 evolution ceases, shake vigorously 30 sec then analyze an aliquot of the hexane soln by GC.

Comments. An ext method for fruit and vegetable samples was given. An electrolytic conductivity detector was used in the N mode. GC data were given for the derivs of 10 herbicides.

References

[1]J. F. Lawrence and G. W. Laver, *J. Agric. Food Chem.* **23**, 1106 (1975).

Derivative Type. *N*-Methyl.

Reagent. Trimethylphenylammonium hydroxide (TMPAH).

Compound Type. *N*-Methyl and *N*-aryl carbamates.

Reaction

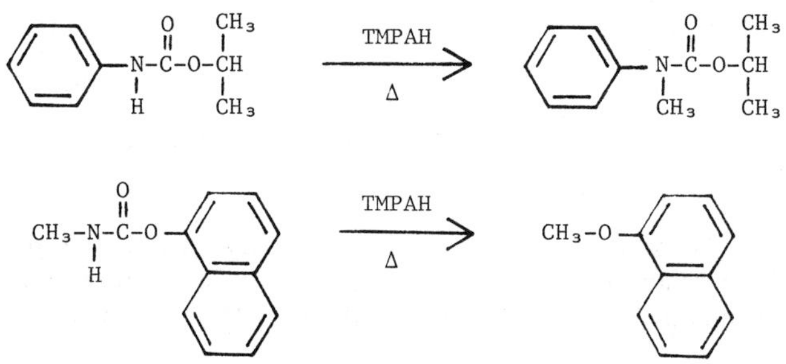

Typical Procedure. Mix soln of carbamate (0.1 mmol/10 ml MeOH) with TMPAH soln (0.2 *M* in MeOH) in TMPAH to carbamate mol ratio of 15 : 1. Inject 5 μl (0.05 μmol) of soln into GC with injector at 220°.

Comments. *N*-Aryl carbamates (e.g., propham shown above) gave the *N*-methyl deriv. *O*-aryl carbamates (e.g., carbaryl shown above) yielded the methyl aryl ether. Propoxur, carbofuran, dichlormate, chlorpropham, CEPC, and swep were also studied. Phenmedipham and monomethylmonuron gave multiple peaks.

References

[1]R. G. Wien and F. S. Tanaka, *J. Chromatogr.* **130**, 55 (1977).

Derivative Type. Trimethylsilyl.

Reagent. Hexamethyldisilazane (HMDS)/trimethylchlorosilane (TMCS).

Compound Type. Pesticidal carbamates and ureas.

Reaction

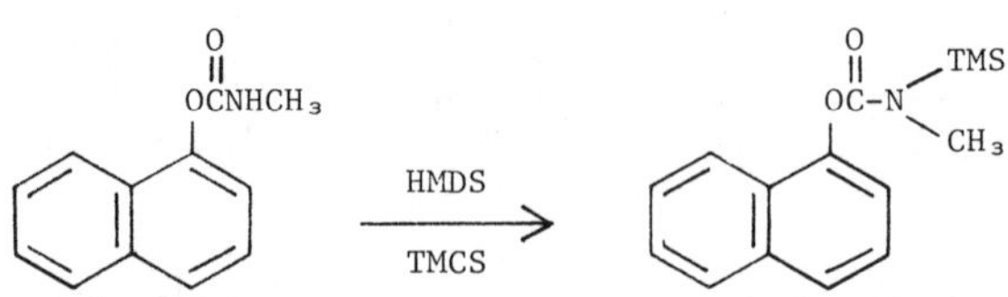

Typical Procedure. Treat sample in pyridine with HMDS and TMCS. Analyze by GC.

Comments. Experimental details for the procedure were not reported. Silylation retarded the thermal breakdown of Sevin (shown above) and yielded better GC peak symmetry. The carbamates IPC, CIPC, and Zectran and the ureas monuron, diuron, and Hercules 7531 were also examined.

References

[1]L. Fishbein and W. L. Zielinski, *J. Chromatogr.* **20**, 9 (1965).

Derivative Type. Methoxy.

Reagent. Sodium methoxide.

Compound Type. Thiolcarbamate pesticides.

Reaction

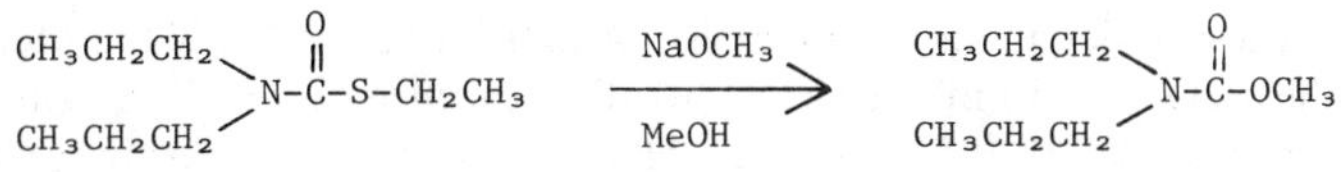

Typical Procedure. Evap sample ext to dryness under N_2 and diss residue in 0.5 ml 25% $NaOCH_3$ in MeOH. Heat at 100° for 1 hr. Cool and add 2 ml EtOAc and 5 ml H_2O. Shake 1 min and remove EtOAc. Ext aq phase with 2 × 1 ml EtOAc. Combine the three exts and conc to 0.5 ml under N_2. Analyze by GC.

Comments. The deriv was used to confirm the presence of Diallate and EPTC in corn.

References

[1]J. F. Lawrence, *J. Agric. Food Chem.* **24**, 1236 (1976).

7.3.3. Volatile Derivatives of Carbamate Hydrolysis Products

Derivative Type. Trichloroacetyl.

Reagent. Sodium hydroxide, trichloroacetyl chloride/pyridine.

Compound Type. Carbofuran and metabolites.

Reaction

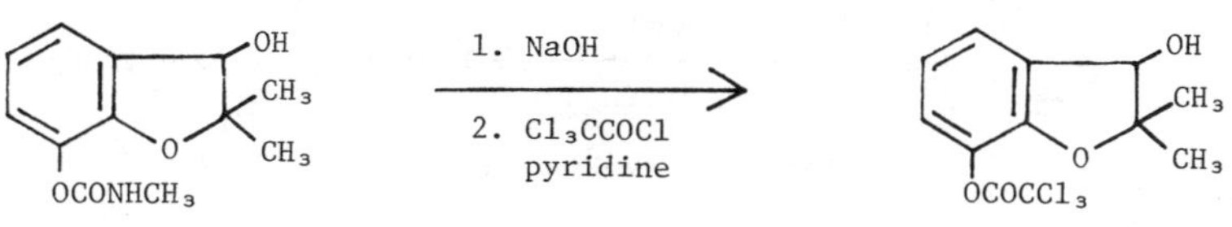

Typical Procedure. To 0.5–50 μg sample add 0.2 ml 0.1 *N* NaOH in H_2O/MeOH (3:17, v/v). Evap MeOH under air stream at 40°. Add 1 ml 0.1% (v/v) pyridine in CH_2Cl_2 and heat at 95–100° *ca* 1 min to evap CH_2Cl_2. Cool in ice and add 1 ml 1% (v/v) trichloroacetyl chloride in CH_2Cl_2. Heat again at 95–100° until CH_2Cl_2 is evapd then 1 min longer. Partition between 15 ml hexane and 5 ml H_2O. Wash hexane layer with aq $NaHCO_3$ and H_2O then dry over Na_2SO_4. Analyze by GC.

Comments. In the 3-hydroxymetabolite (shown) it appeared that only the phenolic group was derivatized since no deriv was obtained if the hydrolysis was omitted.

References

[1]L. I. Butler and L. M. McDonough, *J. Assoc. Off. Anal. Chem.* **54**, 1357 (1971).

Derivative Type. Trichloroacetyl ester.

Reagent. Sodium hydroxide, trichloroacetyl chloride.

Compound Type. Carbamate insecticides.

Reaction

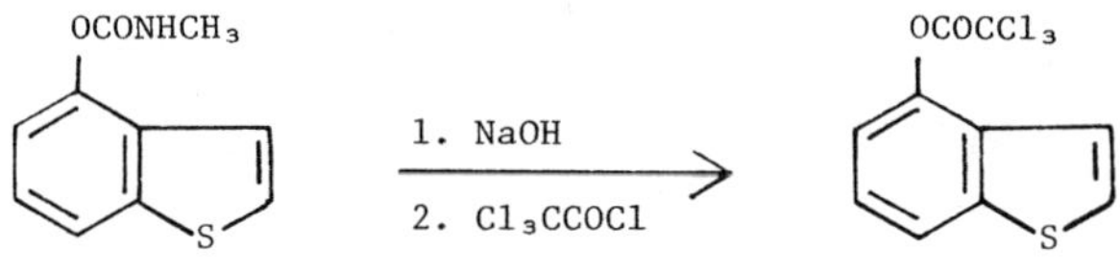

Typical Procedure. Mix 0.1–50 μg of insecticide in $CHCl_3$ with 0.1 ml mineral oil (keeper) and evap to dryness under an air stream at 40°. Add 0.2 ml 0.1 *N* NaOH in MeOH and shake 5 min. Evap the MeOH, add 1 ml 0.1% (v/v) pyridine (purified over a Na_2SO_4-Florisil column) in CH_2Cl_2, and heat at 100–105° until the CH_2Cl_2 is evapd. Cool tube in ice water and add 1 ml 1% (v/v) trichloroacetyl chloride in CH_2Cl_2. Shake briefly and heat again to remove the CH_2Cl_2. Partition the residue between 15 ml hexane and 5 ml H_2O. Wash the hexane soln with 5 ml satd $NaHCO_3$ and 5 ml H_2O. Dry over Na_2SO_4 and analyze by ECD-GC.

Comments. The preparation of the reagent from trichloroacetic acid and $SOCl_2$ in DMF was described. The method was sensitive to 0.01–0.10 ppm in crop samples.

References

[1]L. I. Butler and L. M. McDonough, *J.Agric. Food Chem.* **16**, 403 (1968).

Derivative Type. Chloroacetyl ester.

Reagent. Sodium hydroxide, chloroacetic anhydride.

Compound Type. Carbamate pesticides.

Reaction

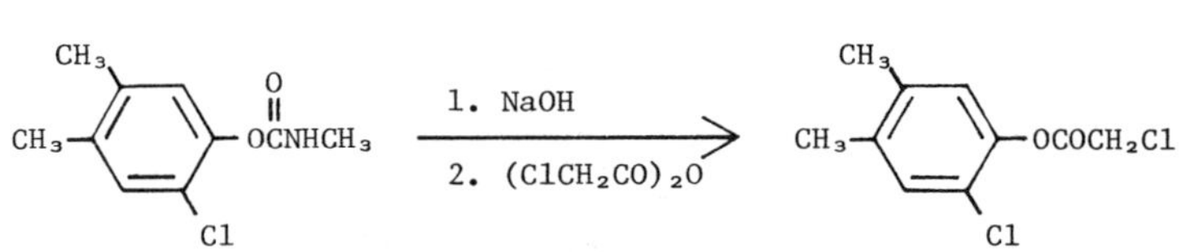

Typical Procedure. To the residue ext from 50 g vegetable matter or milk in 125 ml CH_2Cl_2 add 25 ml 0.25 *N* NaOH. Shake 3 min, then separate NaOH layer for analysis of free phenols. Evap the CH_2Cl_2 to dryness, diss residue in 50 ml 30–60° pet ether, and add 25 ml 0.25*N* NaOH. Allow hydrolysis to proceed for 18 hr. Separate the aq phase and wash with 25 ml pet ether. To the NaOH soln add 10 ml of a 5 mg/ml soln of chloroacetic anhydride in benzene and shake 3 min. Dry the benzene soln over Na_2SO_4 and analyze by ECD-GC.

Comments. The method was used for analysis of Banol (shown above), carbaryl, and 8 other methylcarbamates.

References

[1]R. L. Argauer, *J. Agr. Food Chem.* **17**, 888 (1969).

Derivative Type. *p*-Bromobenzamides.

Reagent. Sulfuric acid, *p*-bromobenzoyl chloride.

Compound Type. Carbamate pesticides.

Reaction

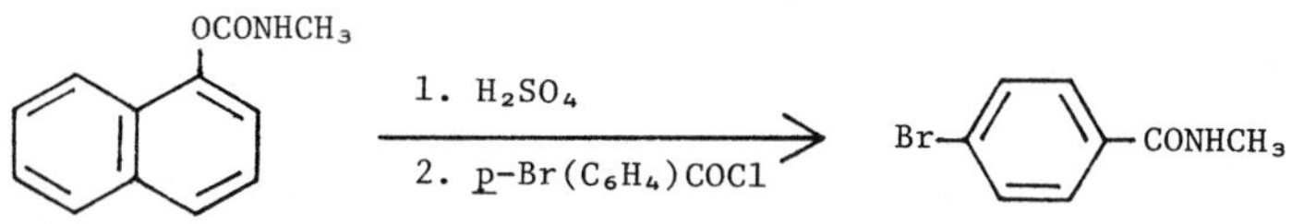

Typical Procedure. Blend a 5 g crop sample with 150 ml CH_2Cl_2, filter, and repeat ext with 100 ml CH_2Cl_2. Dry the combined ext over Na_2SO_4 and evap to dryness under an air stream. Mix residue with 0.5 ml conc H_2SO_4 and heat 2 hr at 125°. Cool to room temp and add 25 ml of a soln of *p*-bromobenzoyl chloride (1 mg/ml) in benzene followed by 10 ml aq NaOH (150 mg/ml). Shake 10 min, allow to separate, and analyze an aliquot of the org phase by ECD-GC.

Comments. The *p*-bromobenzamide derivs from *N*-methyl and *N,N*-dimethyl carbamate pesticides were compared with the 2,4-dichloro, 3,4-dichloro, and *p*-nitrobenzamides. The bromo derivs appeared to be most promising being detectable (as pure standards) to the 20 pg level. The method has been extended to *N*-aryl carbamates[2] but required use of *p*-bromobenzoyl bromide because the acid chloride would not deriv the anilines.

References

[1] R. L. Tilden and C. H. Van Middelen, *J. Agr. Food Chem.* **18**, 154 (1970).

[2] D. E. Bradway and T. Shafik, *J. Chromatogr. Sci.* **15**, 322 (1977).

Derivative Type. 2,5-Dichlorobenzenesulfonate.

Reagent. 2,5-Dichlorobenzenesulfonyl chloride.

Compound Type. Carbamate pesticides.

Reaction

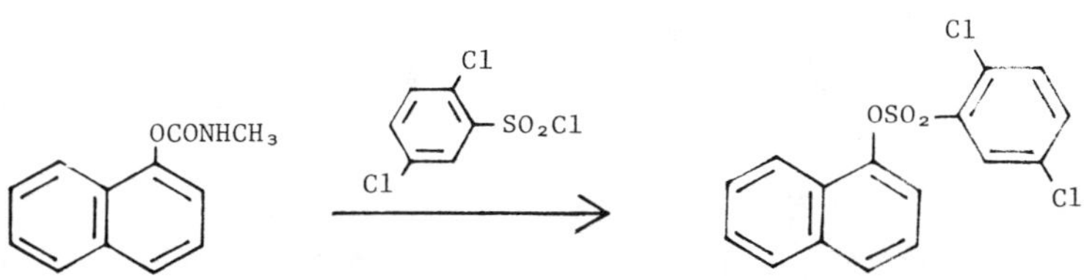

Typical Procedure. Reflux 6 g vegetable sample 1 hr with 100 ml 0.25 *N* HCl. Cool, filter, and ext with 3 × 150 ml CH_2Cl_2. Dry over Na_2SO_4 and evap to dryness under N_2. Diss residue in 10 ml benzene and wash with 3 × 3 ml 0.25 *N* NaOH at 5°. Dry over Na_2SO_4 and evap to dryness under N_2. Add 3 ml pH 8 phosphate buffer, 20 mg 2,5-dichlorobenzenesulfonyl chloride, and 3 ml acetone. Heat at 80° for 15 min. Cool, add 6 ml benzene, and remove lower aq phase. Wash benzene soln with an equal vol of H_2O, dry over Na_2SO_4, and analyze by ECD-GC.

Comments. The method was sensitive to 0.05 ppm. The 3,4-dichlo-, *p*-bromo-, and pentafluorobenzene sulfonates were also examined. All the derivs gave about the same ECD response but the 2,5-dichloro- had the longest retention time.

References

[1]H. A. Moye, *J. Agr. Food Chem.* **23**, 415 (1975).

Derivative Type. Thiophosphoryl.

Reagent. Dimethyl chlorothiophosphate.

Compound Type. Carbamate insecticides.

Reaction

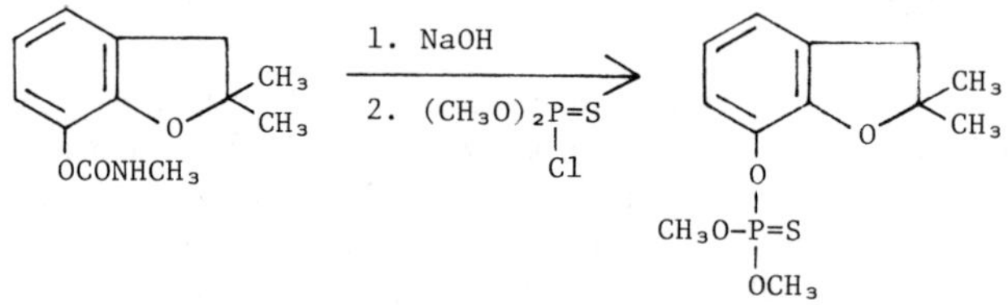

Typical Procedure. To a 250 ml flask add 4 drops diethylene glycol and 0.5 ml of a soln containing 500 μg carbamate and evap to dryness. Add 25 ml 10% NaOH and heat at 40–50° for 30 min. Add 100 ml H_2O and 25 ml 6 *N* HCl and ext with 2 × 25 ml benzene. Dry the ext over Na_2SO_4 and add 0.5 ml pyridine and 0.25 ml dimethyl chlorothiophosphate. Reflux 30 min, cool, and ext with 50 ml ice cold 5% Na_2CO_3. Dry over Na_2SO_4 and analyze by GC using a flame photometric detector.

Comments. The derivs were reported to be useful in either the phosphorus or sulfur detection mode. Methods were given for analysis of the carbamate in corn silage and milk.

References

[1]M. C. Bowman and M. Berosa, *J. Assoc. Off. Anal. Chem.* **50**, 926 (1967).

Derivative Type. 2,4-Dinitrophenyl (DNP).

Reagent. 2,4-Dinitrofluorobenzene (DNFB).

Compound Type. Carbamate insecticides.

Reaction

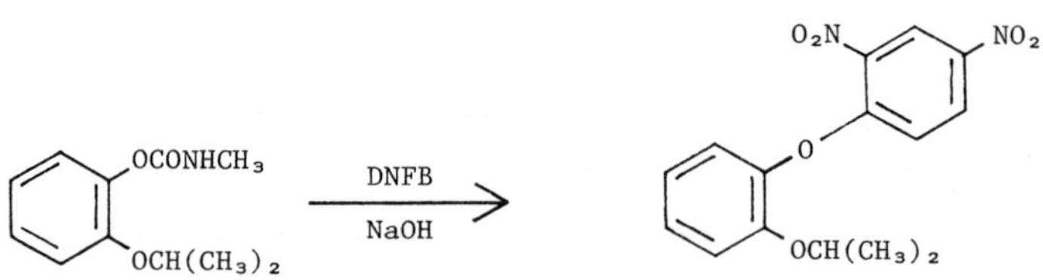

Typical Procedure. To the residue ext from 1000 ml H_2O or 50 g vegetable tissue add 0.5 ml 1% (w/v) DNFB in acetone and 10 ml buffer (see below). Heat at 50° for 30 min. Ext with 10 ml hexane and dry the ext by passing it through *ca* 5 g Na_2SO_4. Analyze 5 μl of the hexane soln by ECD-GC.

Comments. The buffer (pH 11.0) was prepared by mixing 8.2 ml 0.1 *N* NaOH and 100 ml 0.05 *M* Na_2HPO_4 and dil to 200 ml. Ext procedures were given for H_2O and vegetable tissue samples.

References

[1]I. C. Cohen *et al.*, *J. Chromatogr.* **49**, 215 (1970).

Derivative Type. 2,4-Dinitrophenyl (DNP).

Reagent. 2,4-Dinitrofluorobenzene (DNFB).

Compound Type. Carbamate insecticides.

Reaction

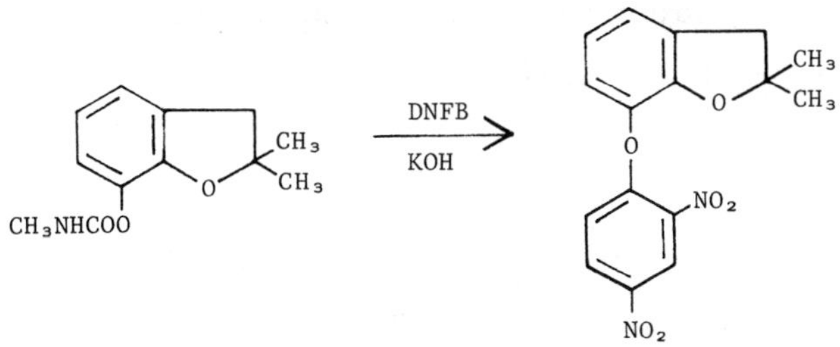

Typical Procedure. To the carbamate residue ext from 100 g crop material add 100 ml H_2O, 2 ml 0.5 *N* KOH, and 1 ml DNFB in acetone (1.5 ml/25 ml acetone). Mix 20 min then add 10 ml 5% aq $Na_2B_4O_7$ and heat on a steam bath for 20 min. Cool and ext with 5 ml isooctane. Analyze by ECD-GC.

Comments. The method was generally applicable to aromatic methyl carbamates not containing an amino substituent. The method was simpler and more specific than the earlier 2,4-dinitroaniline method.[2]

References

[1]E. R. Holden, *J. Assoc. Off. Anal. Chem.* **56**, 713 (1973).
[2]E. R. Holden, *J. Agr. Food Chem.* **17**, 56 (1969).

Derivative Type. 2,4-Dinitrophenyl methylamine.

Reagent. Sodium hydroxide, 2,4-dinitrofluorobenzene (DNFB).

Compound Type. *N*-Methylcarbamate pesticides.

Reaction

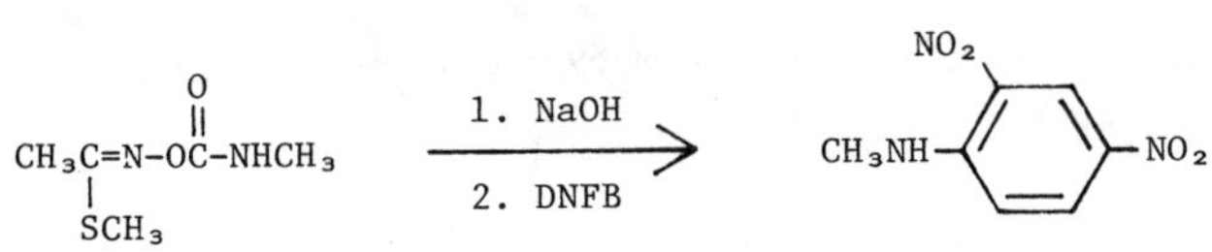

Typical Procedure. To a reference sample or residue ext add 5 ml 5% aq $Na_2B_4O_7$ and 0.5 ml 1 *N* NaOH. Heat at 82° for 10 min with shaking then add 10 μl DNFB in 1 ml dioxane and heat 10 min longer. Add 1 ml satd aq glycine soln, heat 10 min, then cool in an ice bath. Ext with 4 ml and 2 × 2 ml benzene. Wash combined benzene ext with 0.1 *N* Na_2CO_3. Wash Na_2CO_3 soln with 4 ml fresh benzene and combine benzene phases. Conc to *ca* 0.5 ml under N_2. Purify deriv on a microcolumn of silica gel[2] using benzene as eluant. Analyze by ECD-GC.

Comments. The method was applied to both the *syn* and *anti* isomers of methomyl (shown above), to carbaryl, and to Zectran (4-dimethylamino-*m*-tolyl methylcarbamate). Glycine was used to remove excess reagent.

References

[1]C. E. Mendoza and J. B. Shields, *J. Agric. Food Chem.* **22**, 255 (1974).
[2]C. E. Mendoza and J. B. Shields, *J. Agric. Food Chem.* **22**, 528 (1974).

Derivative Type. Pentafluorobenzyl (PFB).

Reagent. Potassium hydroxide, pentafluorobenzyl bromide (PFB-Br).

Compound Type. *N*-Methylcarbamate pesticides.

Reaction

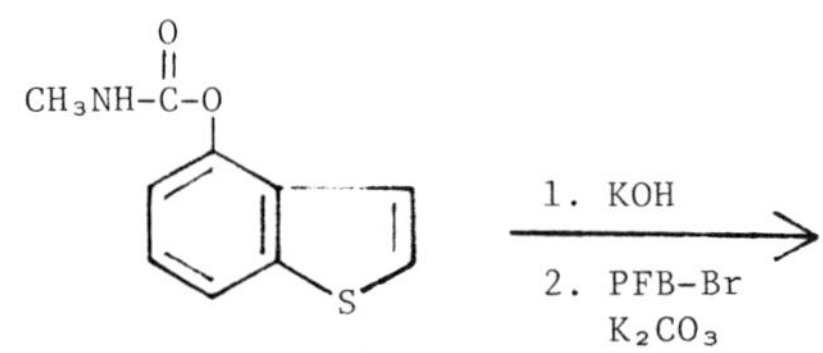

Typical Procedure. Adjust 1000 ml H_2O sample to pH 3–4 with 50% H_2SO_4 and ext with 150 then 100 ml CH_2Cl_2. Wash combined exts with 75–100 ml 0.1 *M* K_2CO_3 keeping contact time less than 5 min. Dry CH_2Cl_2 soln over Na_2SO_4 and evap to *ca* 5 ml under vac.

Add 2 ml 10% KOH in MeOH and allow to hydrolyze overnight at room temp. Transfer with 2 × 50 ml H_2O to a sep funnel and ext with 50 ml benzene. Discard benzene and adjust aq phase to pH 2 or lower with 50% H_2SO_4. Ext with 2 × 70 ml CH_2Cl_2 and wash combined exts with 2 × 70 ml H_2O. Dry CH_2Cl_2 over Na_2SO_4 and evap to 1–2 ml on rotary evaporator. Transfer with 5–6 ml acetone to a 15 ml tube and add 25 μl 5% K_2CO_3 and 100 μl 1% PFB-Br in acetone. Stopper, shake, and allow to react at room temp for 3 hr or at 60° for 30 min. Add 2 ml isoctane and evap to 1 ml at 35–40° under N_2. Repeat isooctane addition and evap to 1 ml. Transfer to a 1 g column of 3.5% H_2O-silica gel with hexane and elute with 5 ml 5% benzene in hexane. Elute with 6 ml 25% benzene in hexane (fraction 1), 8 ml 75% benzene in hexane (fraction 2) and 10 ml benzene (fraction 3). Analyze each fraction by ECD-GC.

Comments. The procedure above is modified[2] from original publication. The method was used to quantitate 0.1 μg of carbofuran and 3-ketocarbofuran, 0.5 μg carbaryl, metmercapturon, and mobam, and 1 μg propoxur per liter of H_2O.

References

[1]J. A. Coburn *et al.*, *J. Assoc. Off. Anal. Chem.* **59**, 188 (1976).
[2]J. A. Coburn, personal communication.

Derivative Type. Bromoaniline.

Reagent. Bromine/acetic acid (HOAc)/iodine/sulfuric acid.

Compound Type. *N*-Aryl carbamate and urea pesticides.

Reaction

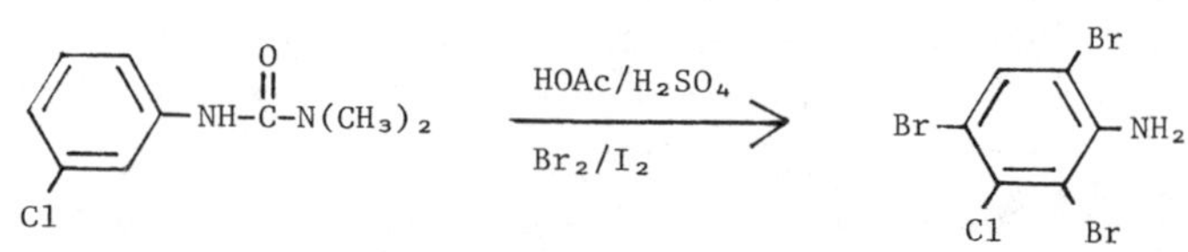

Typical Procedure. To 5 ml of a vegetable residue ext in acetone add 10 ml hexane (Skellysolve B) and dil to 100 ml with 2% aq Na_2SO_4. Shake vigorously, allow to separate, then transfer 1 ml hexane soln and evap to dryness. To the residue add 1 ml HOAc, 5 drops 9 *M* H_2SO_4, and 0.2 ml of a satd soln of I_2 in HOAc containing 5% (v/v) Br_2. Reflux in a 130° bath for 1 hr. Cool, rinse down with 4 ml H_2O, add 3 ml 50% NaOH, 1 ml hexane, and dil to 10 ml with H_2O. Analyze an aliquot of the hexane by ECD-GC.

Comments. The method was used for analysis of various carbamate and urea pesticides which yielded anilines upon hydrolysis.[1,2] A similar approach was used to analyze Sevin (*N*-methyl-1-naphthyl carbamate) as the brominated 1-naphthyl acetate.[3,4]

References

[1]W. H. Gutenmann and D. J. Lisk, *J. Agr. Food Chem.* **12**, 46 (1964).
[2]W. H. Gutenmann and D. J. Lisk, *J. Gas Chromatogr.* **4**, 424 (1977).
[3]W. H. Gutenmann and D. J. Lisk, *J. Agr. Food Chem.* **13**, 48 (1965).
[4]L. I. Butler and L. M. McDonough, *J. Assoc. Off. Anal. Chem.* **53**, 495 (1970).

7.3.4. Chromogenic Derivatives of Carbamate Hydrolysis Products

Derivative Type. 5-Dimethylaminonaphthalene-1-sulfonyl (DANS).

Reagent. 5-Dimethylaminonaphthalene-1-sulfonyl chloride (DANS-Cl).

Compound Type. Carbamate pesticides.

Reaction

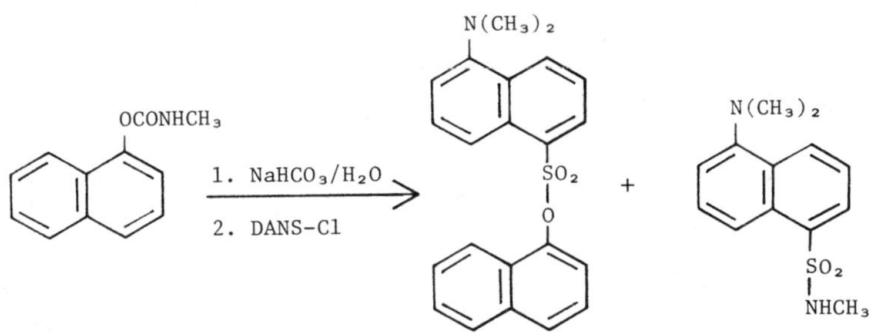

Typical Procedure. Evap 10 μl of a 0.01 μg/μl soln of the carbamate in CH_2Cl_2 to dryness by warming. Add 5 μl 0.1 *M* $NaHCO_3$ and heat at 45° for 30 min. Add 3 μl of 0.2% DANS-Cl in acetone and heat 20 min at 45°. Ext deriv into benzene and analyze by HPLC.

Comments. The fluorigenic deriv method was initially used for TLC analysis[1-6] but later adapted for HPLC.[7]

References

[1]R. W. Frei and J. F. Lawrence, *J. Chromatogr.* **61**, 174 (1971).
[2]J. F. Lawrence and R. W. Frei, *J. Chromatogr.* **66**, 93 (1972).
[3]J. F. Lawrence *et al.*, *J. Chromatogr.* **66**, 295 (1972).
[4]R. W. Frei and J. F. Lawrence, *J. Chromatogr.* **67**, 87 (1972).
[5]R. W. Frei and J. F. Lawrence, *J. Assoc. Off. Anal. Chem.* **55**, 1259 (1972).
[6]R. W. Frei and J. F. Lawrence, *J. Chromatogr.* **83**, 321 (1973).
[7]R. W. Frei *et al.*, *J. Chromatogr. Sci.* **12**, 40 (1974).

Derivative Type. Methylamine NBD derivatives.

Reagent. 4-Chloro-7-nitrobenzo-2,1,3-oxadiazole (NBD-Cl).

Compound Type. Carbamate insecticides.

Reaction

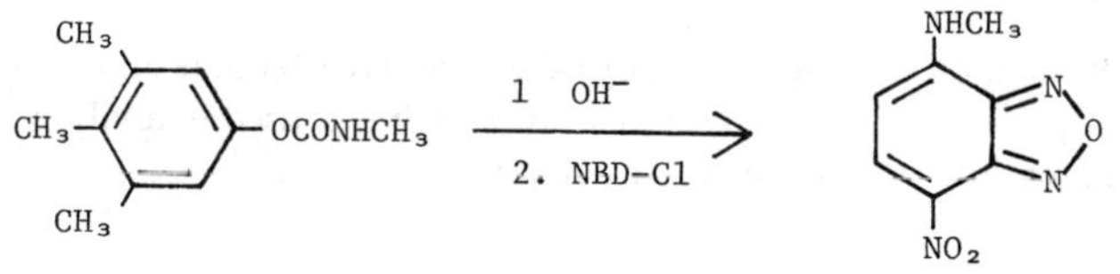

Typical Procedure. To a 10 μg sample of carbamate add 0.5 ml 1 *M* $NaHCO_3$ and 0.5 ml 1% NBD-Cl in methyl isobutyl ketone. Heat at 80° for 30 min then analyze an aliquot of the org phase by TLC.

Comments. The derivs were analyzed by TLC, but the reaction could be used for HPLC. The reagent, while not fluorescent itself, formed fluorescent derivs with alkyl amines. The reagent did not react with phenols, thiols, alcohols, or anilines. Both *N*-methyl and *N,N*-dimethyl carbamates were analyzed.

References

[1]J. F. Lawrence and R. W. Frei, *Anal. Chem.* **44**, 2046 (1972).

7.4. DERIVATIVES OF UREAS

7.4.1. *N*-Methyl Urea Derivatives

Derivative Type. Methyl.

Reagent. Sodium hydride/dimethylsulfoxide (DMSO)/methyl iodide.

Compound Type. Pesticides with reactive NH groups.

Reaction

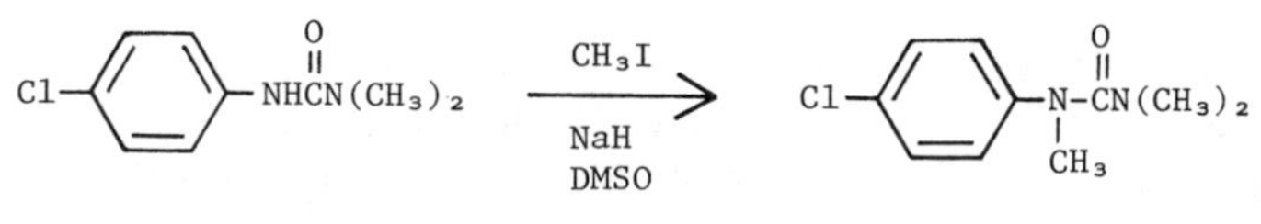

Typical Procedure. Place 10 mg NaH in a tube and wash 2 × with hexane. Add 1 ml of a 2 μg/ml soln of the pesticide followed by 0.1 ml DMSO and 0.1 ml CH_3I. Seal tube and heat at 50° for 10 min. Cool and add 1 ml H_2O. Ext 3 × with Et_2O and evap combined ext nearly to dryness under N_2. Diss in acetone and analyze by GC.

Comments. The method was used as a confirmatory test for organophosphorus and carbamate insecticides, and triazene and urea herbicides. The carbamates required milder reaction conditions (room temp, 5 min).

References

[1]R. Greenhalgh and J. Kovacicova, *J. Agr. Food Chem.* **23**, 325 (1975).

Derivative Type. Methyl.

Reagent. Methyl iodide/potassium *t*-butoxide (*t*-BuOK).

Compound Type. Trisubstituted ureas.

Reaction

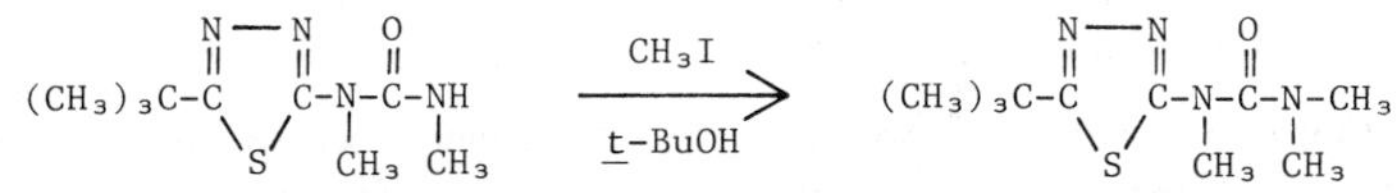

Typical Procedure. Diss 20 mg of the urea in 20 ml THF. Add 0.5–2 ml CH_3I and 200 mg *t*-BuOK. Heat at 52° for 2–16 hr. Filter and transfer THF soln to a sep funnel. Add 20 ml $CHCl_3$ and wash with 5% aq NaCl. Dry $CHCl_3$-THF phase over Na_2SO_4 and evap under vac. Diss residue in 20 ml $CHCl_3$ prior to GC analysis.

Comments. The herbicides diuron, fenuron, and monuron were also deriv by this method. The structures of the derivs were confirmed by MS.

References

[1]D. G. Saunders and L. E. Vanatta, *Anal. Chem.* **46**, 1319 (1974).

Derivative Type. *N*-Methyl.

Reagent. Dimethyl sulfoxide (DMSO)/sodium hydride/iodomethane.

Compound Type. Phenylurea herbicides.

Reaction

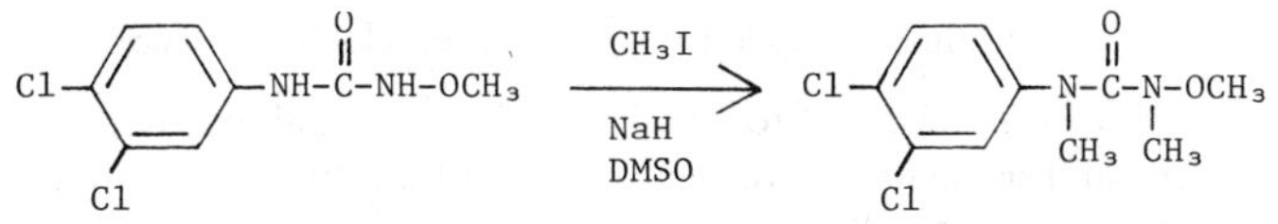

Typical Procedure. Wash 30 mg NaH oil dispersion three times with anhyd Et_2O. Dry under N_2 and suspend in 1 ml freshly dist DMSO. To 1 μmol of phenylurea in 100 μl DMSO add 200 μl of NaH suspension. Immediately add 50 μl CH_3I and stir 5–10 min. Add carefully 1 ml H_2O and ext with 3×1 ml Et_2O or lt pet ether. Dry ext over Na_2SO_4 and evap to dryness under N_2. Diss in pet ether and analyze by ECD-GC.

Comments. The method was applied to eight different phenyl ureas. Some of the compds yielded the same deriv. Such compds could be distinguished by use of trideuteromethyl derivs prepared using trideuteroiodomethane and analysis by GC-MS.

References

[1]A. Buchert and H. Lokke, *J. Chromatogr.* **115**, 682 (1975).

Derivative Type. Methyl.

Reagent. Trimethylphenylammonium hydroxide (TMPAH).

Compound Type. Substituted phenylureas.

Reaction

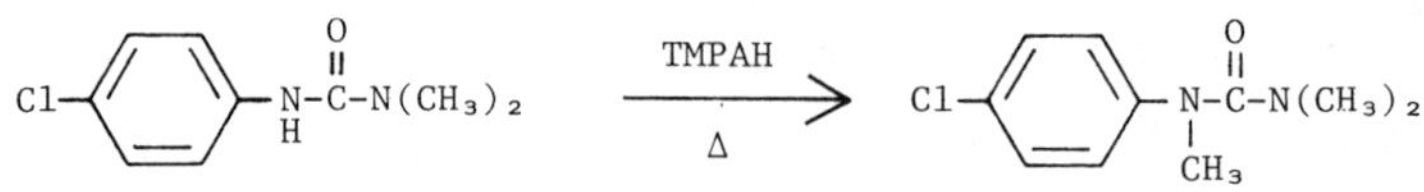

Typical Procedure. Diss phenylurea sample in 2.5 mol equiv of 0.2 *M* TMPAH in MeOH. Inject sample into GC with inlet temp 220°.

Comments. Optimum results were obtained when the reagent was used in 2.5 : 1 mol ratio and mixed with the sample prior to drawing into the injection syringe. Seven phenylurea herbicides were examined. The derivs were only measurable to *ca* 1 μg by FID-GC. Attempts to use ECD-GC were unsuccessful due to broad peaks and excessive tailing.

References

[1]F. S. Tanaka and R. G. Wien, *J. Chromatogr.* **87**, 85 (1973).

7.4.2. Other Derivatives of Intact Ureas

Derivative Type. Trifluoroacetyl.

Reagent. Trifluoroacetic anhydride.

Compound Type. Urea.

Reaction

Typical Procedure. Evap 5 μl plasma to dryness under N_2 at 70°. Add 10 μl 0.5 m*M* nitrobenzene (int stand) and 100 μl TFAA/CH_2Cl_2 (13, v/v). Mix on vortex, sonicate 1 min then allow to stand at 25° for 30 min. Analyze by GC using an electrolytic conductivity detector.

Comments. A method was reported for analysis of urea in animal tissues using the TFA deriv with ECD-GC and using extensive cleanup procedures and large (10 g) samples.[2] Use of the nitrogen specific detector allowed analysis of μg samples with no sample cleanup.[1]

References

[1]J. M. Mee, *J. Chromatogr.* **94**. 302 (1974).

[2]P. W. Miller, *J. Agric. Food Chem.* **19**, 941 (1971).

Derivative Type. Trifluoroacetyl.

Reagent. Trifluoroacetic anhydride (TFAA).

Compound Type. Trisubstituted ureas.

Reaction

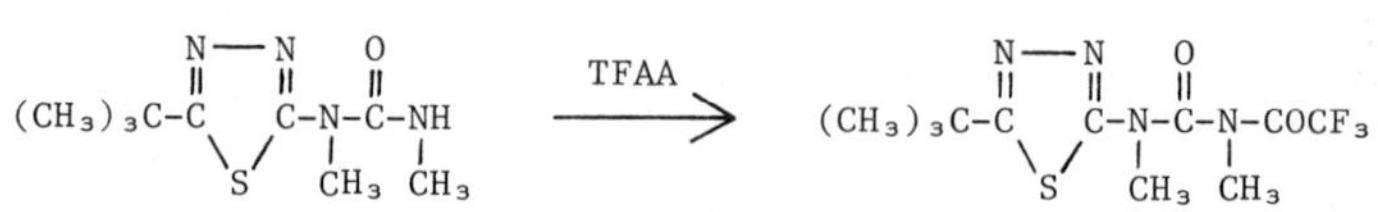

Typical Procedure. Diss 20 mg of the urea in 20 ml $CHCl_3$ and add 0.5–1 ml of TFAA. Heat at 52° for 2 hr. Remove excess reagent by partial evap under vac. Make up to 20 ml with $CHCl_3$ and analyze aliquots by GC.

Comments. Submicrogram quantities of the above herbicide were also deriv with TFAA in dry benzene. The method was also examined using Ac_2O, AcCl, HFBA, and PFPA but the TFA deriv gave the best results. TFA derivs of fenuron (3-phenyl-1,1-dimethylurea) and monuron (3-*p*-chlorophenyl-1,1-dimethylurea) were also prepared.

References

[1]D. G. Saunders and L. E. Vanatta, *Anal. Chem.* **46**, 1319 (1974).

Derivative Type. (2-(Benzylthio)-1-pentafluorobenzoyl)-2-imidazoline.

Reagent. Benzy*Reagent.* Benzyl chloride, Pentafluorobenzoyl chloride.

Compound Type. Ethylenethiourea.

Reaction

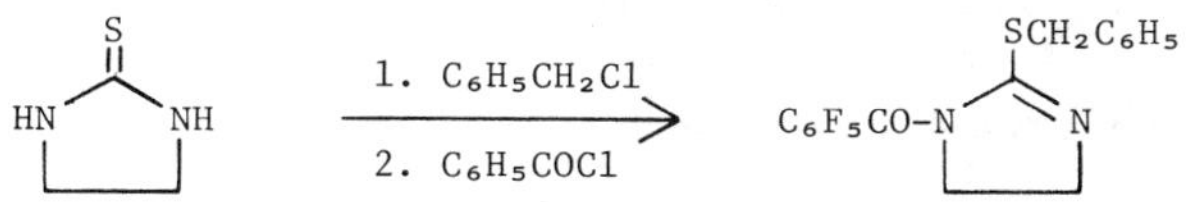

Typical Procedure. Homogenize plant sample with 10 ml MeOH, filter, and wash filter with 2 × 10 ml MeOH. Dil combined filtrate and washings to 40 ml with MeOH and reflux a 10 ml aliquot with 4 drops benzyl chloride for 30 min. Mix with 5 ml 0.2 *N* HCl and wash with 10 ml and 5 ml $CHCl_3$. Alkalinize with *ca* 10 ml 1.5 *N* KOH and ext with 10 ml $CHCl_3$. Evap $CHCl_3$ to dryness, and diss residue in 0.5 ml 1% pentafluorobenzoyl chloride in benzene and 1 drop 0.1% pyridine in benzene. Heat 15 min at 65°. Purify deriv on a column of Florisil eluting with 25 ml Et_2O/pet ether (15:85, v/v). Conc eluate to 10 ml under N_2 and analyze by ECD-GC.

Comments. The deriv gave an improved ECD response over the similar TFA deriv introduced by Newsome.[2] Nash also reported another deriv using *o*-chlorobenzyl chloride in place of benzyl chloride.[3]

References

[1]R. G. Nash, *J. Assoc. Off. Anal. Chem.* **57**, 1015 (1974).

[2]W. H. Newsome, *J. Agric Food Chem.* **20**, 967 (1972).

[3]R. G. Nash, *J. Assoc. Off. Anal. Chem.* **58**, 566 (1975).

Derivative Type. *m*-Trifluoromethylbenzyl.

Reagent. *m*-Trifluoromethylbenzyl chloride.

Compound Type. Ethylenethiourea.

Reaction

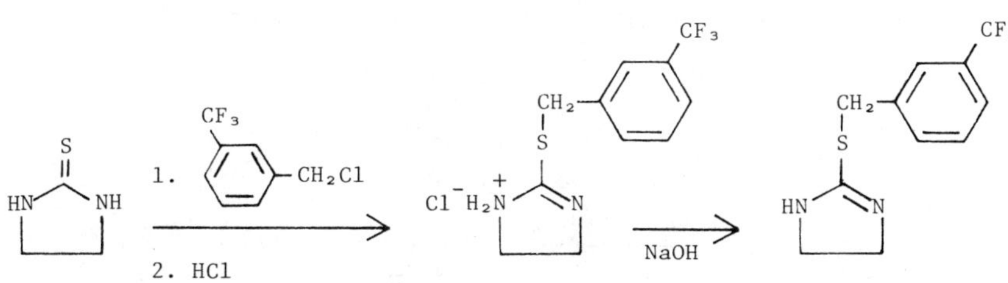

Typical Procedure. Homogenize 10 g sample of plant material with 50 ml 95% EtOH and filter. Add 4 drops of the reagent to the filtrate and reflux 30 min. Cool, add 1 drop 6 *N* HCl, and evap under vac to an aq residue. Dil with 0.5 ml H_2O and wash with 2×1 ml Et_2O. Heat aq phase at *ca* 50° to remove traces of Et_2O; then add 0.5 ml benzene and 3 drops 10% NaOH. Shake and allow layers to separate. Analyze an aliquot of the benzene layer by ECD-GC.

Comments. The *N*-TFA deriv could be formed for confirmatory purposes by addn of a few grains anhyd Na_2SO_4 and 1 drop of 20% TFAA in benzene to the final benzene ext. Similar deriv reactions have been performed on ethylenethiourea using *n*-butyl bromide[2] and benzyl bromide.[3]

References

[1] R. R. King, *J. Agric. Food Chem.* **25**, 73 (1977).
[2] J. H. Onley and G. P. Yip, *J. Assoc. Off. Anal. Chem.* **54**, 165 (1971).
[3] W. H. Newsome, *J. Agric. Food Chem.* **20**, 967 (1972).

7.4.3. Derivatives of Urea Hydrolysis Products

Derivative Type. 2,4-Dinitrophenyl.

Reagent. HCl/2,4-dinitrofluorobenzene (DNFB).

Compound Type. Urea and carbamate herbicides.

Reaction

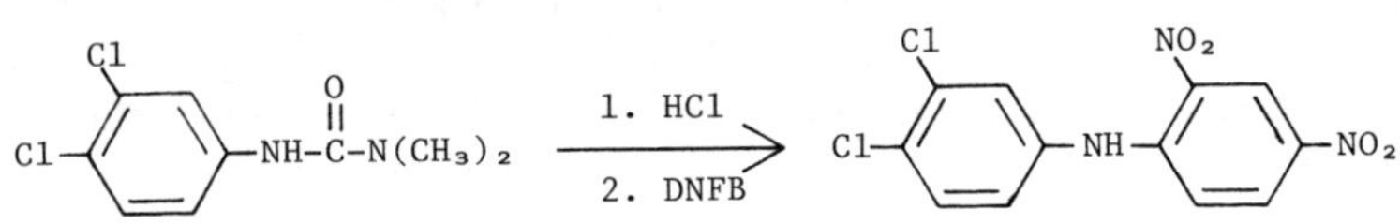

Typical Procedure. Ext 1000 ml H_2O with 3×50 ml $CHCl_3$ and dry ext by passing it through a column of Na_2SO_4. Evap to small vol and spot conc along with standards onto a 20×20 cm silica TLC plate. Develop plate with $CHCl_3$ or hexane acetone (5:1, v/v). Spray plate with 7 ml 5% v/v HCl and then with 7 ml of a 4% (w/v) soln of DNFB in acetone. Cover with a second glass plate and heat at 190° for 40 min. Cool plate and scrape the areas of interest. Elute the derivs using a short glass column and 2×5 ml acetone. Analyze by ECD-GC.

Comments. Hydrolysis and deriv were performed on the TLC plate. Hydrolysis with HI resulted in decomposition. The method was useable for levels to 0.001 ppm in water and to 0.02–0.05 ppm in soil and plant material. Data were reported for 10 herbicides.

References

[1] I. C. Cohen and B. B. Wheals, *J. Chromatogr.* **43**, 233 (1969).

Derivative Type. 5-Dimethylaminonaphthalene-1-sulfonyl (DANS).

Reagent. 5-Dimethylaminonaphthalene-1-sulfonyl chloride (DANS-Cl).

Compound Type. *N*-Phenyl urea and carbamate pesticides.

Reaction

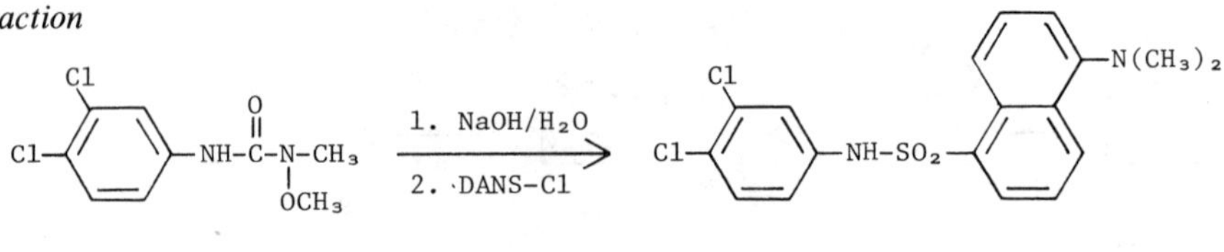

Typical Procedure. Hydrolyze 0.1 μg of the pesticide in 1 ml 2 *M* NaOH at 75° for 45 min. Add 2 drops methyl isobutyl ketone and 2 drops 0.1% DANS-Cl in acetone. Heat at 65° for 30 min. Cool, acidify with 10% HCl, and ext with 0.3 ml benzene. Analyze the ext by HPLC.

Comments. The method was used for fluorigenic deriv of pesticides for HPLC analysis. It had earlier been used in TLC analysis.[2]

References

[1] R. W. Frei and J. F. Lawrence, *J. Chromatogr.* **61**, 174 (1971).

[2] R. W. Frei *et al.*, *Analyst* **98**, 9 (1973).

Derivative Type. Iodinated benzene.

Reagent. Sodium hydroxide, nitrous acid, sodium iodide/iodine.

Compound Type. Urea pesticides.

Reaction

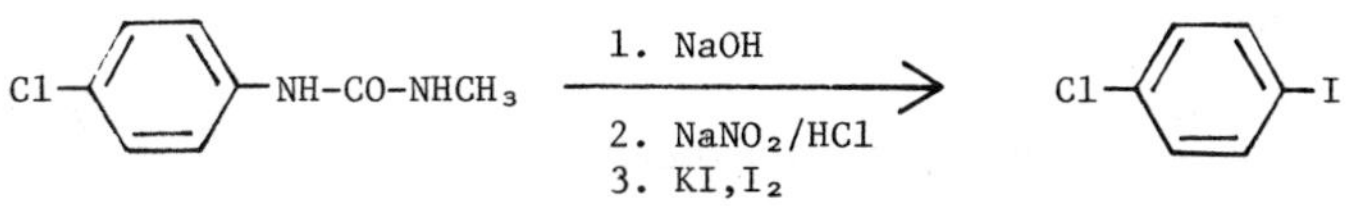

Typical Procedure. Hydrolyze soil or vegetable sample with aq NaOH and ext with hexane. Aliquot 20 ml of the hexane ext (representing 10 g of soil residues) and ext with 2 × 5 ml 1 *N* HCl. Cool the ext to 0°, add 2 ml 1% aq $NaNO_2$, and allow to stand at 0° for 25 min. Destroy excess nitrite by adding 2 ml 10% aq sulfamic acid. To the cold soln add 0.5 ml KI/I_2 soln (2.5 g I_2 in 50 ml 10% aq KI), and allow to stand at room temp for 25 min. Warm to 100° over 15 min and maintain at 100° for 5 min. Cool and reduce excess I_2 with 200 mg Na_2SO_3. Add 1.1 ml 10 *N* NaOH and ext with 20 ml hexane. Wash hexane ext with 1% $KMnO_4$ in 1 *N* NaOH and analyze by ECD-GC.

Comments. The method was developed as an alternative to the bromination procedure[2] which gave mixt of derivs.

References

[1] I. Baunok and H. Geissbuchler, *Bull. Environ. Contam. Toxicol.* **3**, 7 (1968).

[2] W. H. Gutenmann and D. J. Lisk, *J. Gas Chromatogr.* **4**, 424 (1966).

7.5. DERIVATIVES OF GUANIDINES

Derivative Type. Trifluoroacetamide.

Reagent. Potassium hydroxide, trifluoroacetic anhydride (TFAA).

Compound Type. Guanidine drugs (e.g., guanethidine).

Reaction

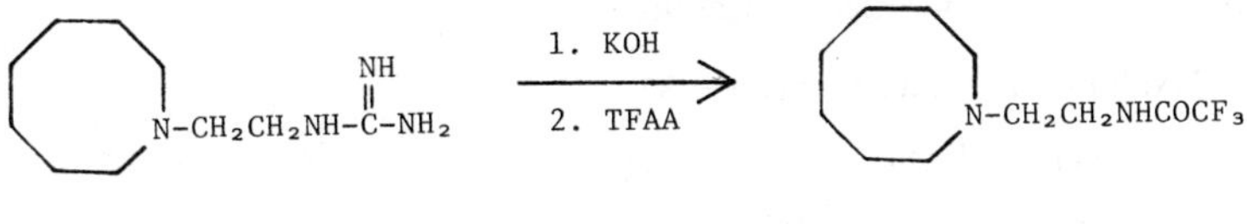

Typical Procedure. Adjust urine sample to pH 10 with 5 *N* NaOH and pipette 5 ml into 50 ml tube. Add 20 μl of int stand soln (debrisoquin, 1 mg/ml) and 20 ml of toluene and shake 10 min. Centrifuge and discard toluene. Add 0.5 ml 50% NaOH and ext 20 min with 30 ml $CHCl_3$. Centrifuge and discard aq phase. Transfer org phase to another tube containing 2 ml 0.1 *N* HCl. Shake 10 min, centrifuge, and transfer aq phase to a silanized 3.5 ml vial. Add 0.5 ml 40% KOH and 1 ml cyclohexane. Seal vial and heat 4 hr at 110° in an oil bath. Cool, mix on vortex, and centrifuge. Transfer org layer to another vial, add 20 μl HCOOH, and evap solvent under N_2 at room temp. Add 0.2 ml EtOAc and 0.2 ml TFAA. After 20 min at room temp evap solvent and excess reagent under N_2. Diss residue in 50–100 μl of EtOAc and analyze 1–5 μl by GC.

Comments. The method was used for detection of guanethidine to about 100 ng/ml by FID-GC and from 1–100 ng/ml by GC-MS. Plasma samples were first ext with 2 vol of Et_2O at pH 7.4. The method was also applicable to guanoxin, debrisoquin, and bethanidine. MS data were given for the deriv.

References

[1]J. H. Hengstmann *et al.*, *Anal. Chem.* **46**, 34 (1974).

Derivative Type. Bis(trifluoromethyl)pyrimidine.

Reagent. Hexafluoroacetylacetone.

Compound Type. Debrisoquin.

Reaction

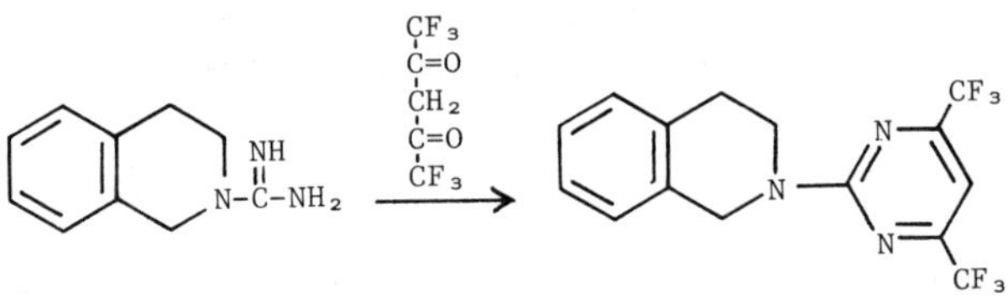

Typical Procedure. Add 10 ng d_{10}-debrisoquin int stand in 10 μl H_2O and 50 μl satd $NaHCO_3$ to 100 μl plasma. Add 50 μl hexafluoroacetylacetone in 500 μl toluene and heat at 100° for 2 hr. Remove org layer and evap to dryness under Ar stream. Diss residue in 7 μl toluene and analyze by GC-MS-SIM.

Comments. The method was used to quantitate debrisoquin and its 4-hydroxy metabolite in plasma to 1 ng/ml and 5 ng/ml, respectively.[1] The two phase system gave fewer interfering peaks than a single phase system. Acetylacetone, even in a two phase system, gave interfering peaks. The deriv had been reported earlier for GC-ECD analysis of monosubstituted guanidines.

References

[1]S. L. Malcolm and T. R. Marten, *Anal. Chem.* **48**, 807 (1976).
[2]P. Erdmansky and T. J. Goehl, *Anal. Chem.* **47**, 750 (1975).

Derivative Type. Hexafluoroacetylacetonate.

Reagent. Hexafluoroacetylacetone (HFAA).

Compound Type. Guanidines.

Reaction

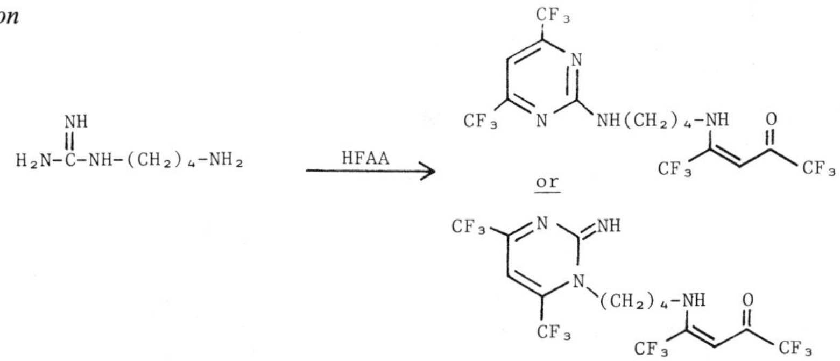

Typical Procedure. Evap in an ampoule 100 μl of a 1 mg/ml soln of the guanidine in EtOH H_2O (1 : 1, v/v) using a stream of N_2. Add 50 μl HFAA and 50 μl pyridine and seal the ampoule. Heat at 120° for 1 hr. Cool, add 1 ml Et_2O and 3 ml 3 *N* HCl. Shake, centrifuge, and analyze an aliquot of the Et_2O by ECD or alkali FID GC.

Comments. Derivs were prepared from guanidine, methylguanidine, and agmatine (shown). Full mass spectra were given but the exact structures of the derivs were not determined. Other derivs prepared included the acetyl derivs using Ac_2O, the TFA derivs using TFAA, acetylacetone derivs, trifluoroacetylacetone derivs, TMS derivs from BSTFA, and isothiocyanates using CS_2. The derivs from HFAA were found to be the most specific for the guanidines and were highly sensitive to ECD and alkali FID GC. The derivs from HFAA were also fairly resistant to hydrolysis.

References

[1]T. Kawabata *et al.*, *J. Chromatogr.* **140**, 47 (1977).

Derivative Type. 4-Trifluoromethyl-2,6-diamino-1,3,5-triazine.

Reagent. Trifluoroacetic anhydride (TFAA).

Compound Type. Phenformin.

Reaction

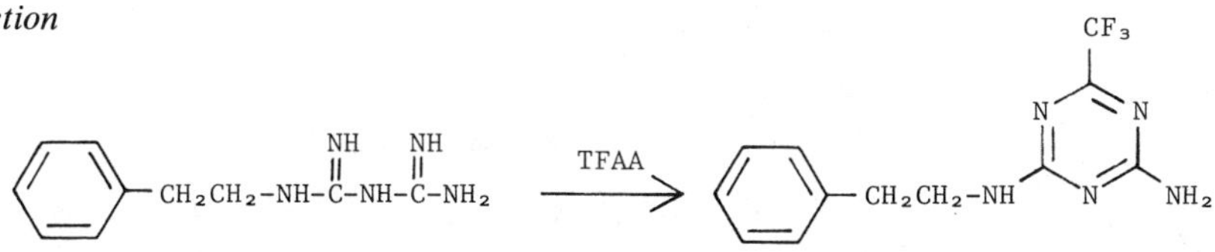

Typical Procedure. To 0.2–0.5 ml plasma add 10 ng d_2-phenformin int stand. Ppt proteins by adding 1 ml 10% trichloroacetic acid in 1 *N* HCl. Separate supernatant and basify with 10 ml 10 *N* NaOH. Ext with CH_2Cl_2 and add 50 μl TFAA to the CH_2Cl_2 ext. Heat at 50° until CH_2Cl_2 evaporates. Add 1.0 ml 1 *N* NaOH and ext mixt with 25 μl pentyl acetate. Analyze 1–2 μl aliquots by GC-MS.

Comments. The method was used to quantitate phenformin in plasma using GC-CI-MS in SIM mode with a deuterated int stand. The method should be generally applicable to biguanides.

References

[1]S. B. Matin *et al.*, *Biomed. Mass Spectrom.* **1**, 320 (1974).

Derivative Type. *s*-Triazine.

Reagent. *p*-Nitrobenzoyl chloride.

Compound Type. Metformin (1,1-dimethylbiguanide).

Reaction

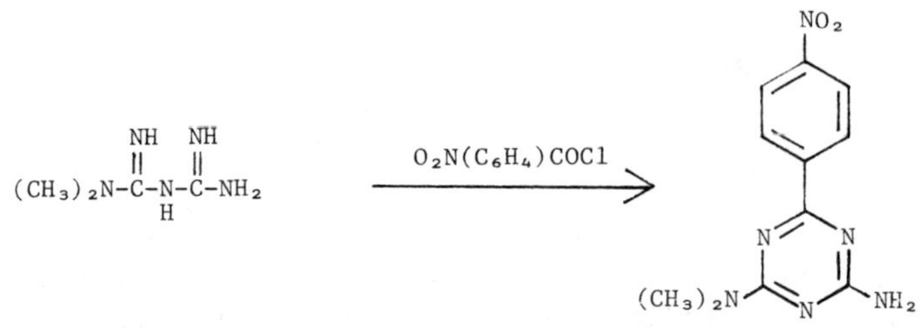

Typical Procedure. To 2 ml urine add 1 ml 20% (w/v) aq NaOH and sat with solid NaCl. Add 1 ml CH_3CN and *ca* 10 mg *p*-nitrobenzoyl chloride and allow to stand at room temp 1 hr. Add another 10 mg of reagent and, after 1 hr, analyze an aliquot of the CH_3CN soln by HPLC.

Comments. The deriv was used to impart UV absorption (monitored at 280 nm) for HPLC detection. The method above was suitable for conc to 10 μg/ml but by increasing the urine vol and decreasing the CH_3CN vol the method could be extended to conc as low as 0.2 μg/ml. MS data were given for the deriv.

References

[1]M. S. F. Ross, *J. Chromatogr.* **133**, 408 (1977).

7.6. DERIVATIVES OF ISOCYANATES

Derivative Type. Ureas.

Reagent. Benzylamine and other alkyl amines.

Compound Type. Alkyl isocyanates.

Reaction

$$\mathrm{R{-}NCO} \longrightarrow \mathrm{R{-}NH{-}CO{-}NH{-}R'}$$

Typical Procedure. To 1 ml of 0.5–10 m*M* isocyanate in $CHCl_3$ add excess benzylamine in acetone. After 15 min at room temp analyze by GC.

Comments. The derivs were detectable to 10–20 mg by FID-GC. The method was applied to dil solns of short chain alkyl (C_1–C_8) isocyanates using various alkyl amines.

References

[1]R. J. Knights *et al.*, *Anal. Biochem.* **77**, 176 (1977).

Derivative Type. *N*-*p*-Nitrobenzyl-*N*-*n*-propyl urea.

Reagent. *N*-*p*-Nitrobenzyl-*N*-*n*-propylamine.

Compound Type. Isocyanates.

Reaction

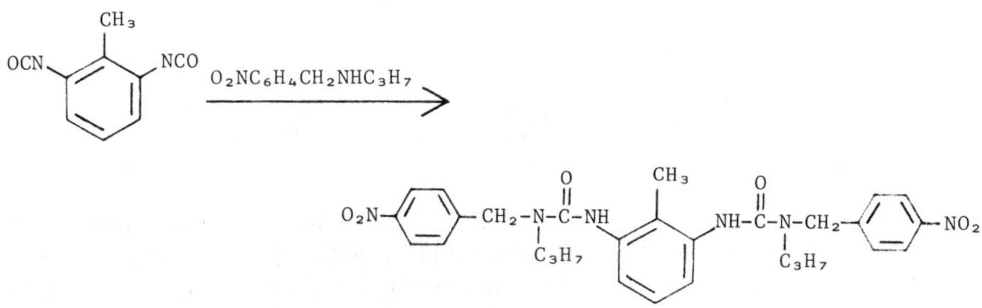

Typical Procedure. Draw air sample through a 10 ml vol of 2×10^{-3} *M* reagent soln in toluene. Evap to dryness in a vac oven at 35°. Diss residue in 1 ml CH_2Cl_2. Analyze by HPLC.

Comments. The method was used for analysis of isocyanates in working atmospheres. Toluenediisocyanate and 4,4′-diphenylmethane diisocyanate could be detected in air to 5 μg/m³.

References

[1]K. L. Dunlap *et al.*, *Anal. Chem.* **48**, 497 (1976).

7.7. DERIVATIVES OF MISCELLANEOUS NITROGEN COMPOUNDS

Derivative Type. Trifluoroacetyl (2-hydroxy-2-methyl)ethyl.

Reagent. 1,2-Epoxypropane, trifluoroacetic anhydride (TFAA).

Compound Type. Creatinine.

Reaction

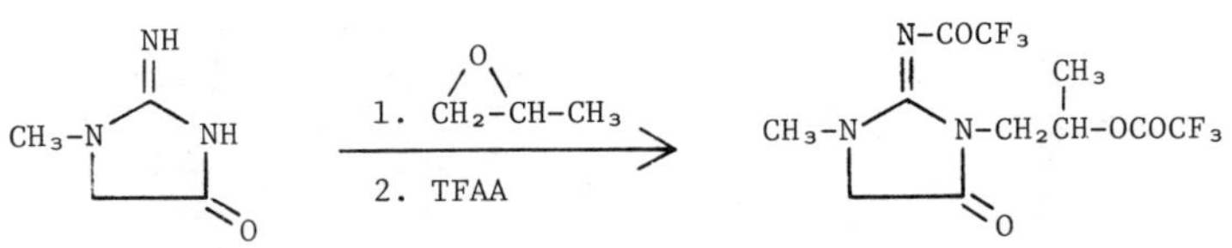

Typical Procedure. To a mixt of creatinine and $^{15}N_2$-creatinine int stand extd from serum or urine add 0.5 ml MeOH and 0.5 ml 1,2-epoxypropane. Heat at 70° for 30 min in a sealed tube. Evap to dryness and add 100 μl TFAA and 200 μl EtOAc. Heat at 37° for 30 min then evap under N_2. Diss residue in 50 μl hexane and analyze by GC-MS-SIM.

Comments. The deriv procedure was used as part of a proposed reference method for quant of creatinine in serum and urine. The creatinine was isolated from the sample by lyophilization and HPLC. The exact structure of the seriv was not ascertained; it was suggested that the *N*-TFA group may have been at the carbonyl position, although underiv creatinine did not react with TFAA.

References

[1]I. Björkhem *et al.*, *Clin. Chem.* **23**, 2114 (1977).

Derivative Type. Oxazolidinedione, *N*-methyl.

Reagent. Hydrochloric acid, diazomethane.

Compound Type. Pemoline.

Reaction

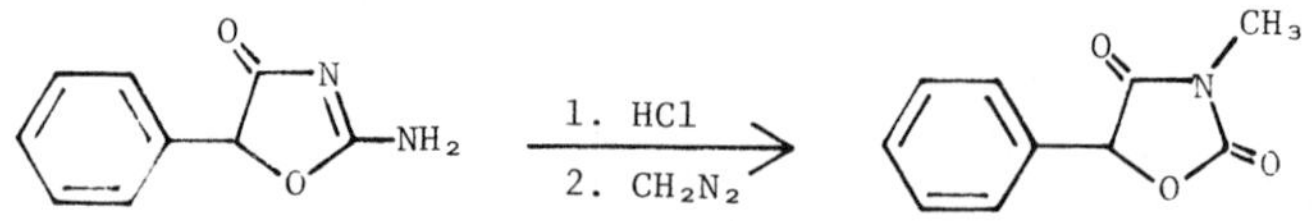

Typical Procedure.[1] Mix 5.0 ml urine and 1 ml 6 *N* HCl and allow to stand 15 min. Add 0.10 ml H_2O containing 50 μg fensuximide int stand and ext with *n*-pentane/CH_2Cl_2 (1:1, v/v). Evap the ext to dryness under N_2 at 35° then diss residue in 0.5 ml CH_2N_2 in Et_2O. After 15 min at room temp, analyze 5 μl aliquots by GC using a nitrogen selective detector.

Comments. Van Boven and Daenens[2] used essentially the same method with amobarbital as int stand. For blood samples they purified the ext on a Florisil column. Chu and Sennelo[3] utilized a similar hydrolysis for analysis of pemoline in plasma but omitted the methylation step. The plasma sample was deproteinized with sulfosalicylic acid then heated to 80° to effect conversion to the oxazolidinedione which was analyzed by ECD-GC.

References

[1]N. P. E. Vermeulen *et al.*, *J. Chromatogr.* **137**, 333 (1977).

[2]M. Van Boven and P. Daenens, *J. Chromatogr.* **134**, 415 (1977).

[3]S. Y. Chu and L. T. Sennelo, *J. Chromatogr.* **137**, 343 (1977).

Derivative Type. *N*-Methyl.

Reagent. Trimethylphenylammonium hydroxide (TMPAH).

Compound Type. Clonidine.

Reaction

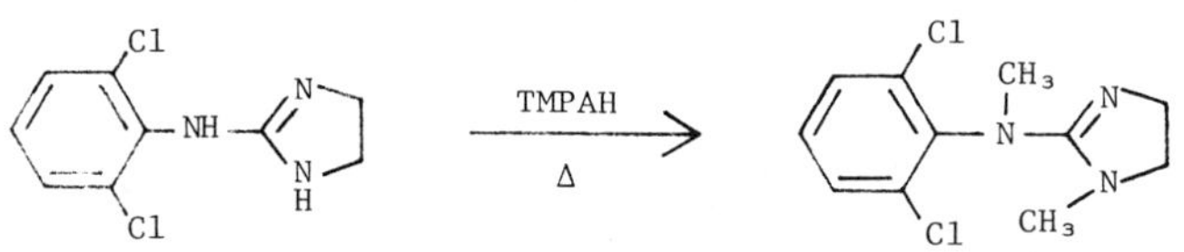

Typical Procedure. Add 400 ng d_4-clonidine int stand to 4 ml plasma, adjust to pH 10–11 with 1 ml 10% aq Na_2CO_3 and ext with 2 × 12 ml EtOAc. Combine EtOAc ext and back ext into 1 ml 0.1 *M* HCl. Adjust aq phase to pH 10 and ext with 10 ml Et_2O. Conc the ext under N_2 and diss in 0.2 *M* TMPAH in MeOH. Analyze by GC-MS-SIM.

Comments. The method was used to quantitate clonidine in plasma to 0.25 ng/ml (rel S.D. 11%). The *p*-hydroxy metabolite could be chromatographed as the trimethyl deriv formed by the same method but required CI-MS for low level (5 ng/ml) quantitation.

References

[1]G. H. Draffin *et al.*, *Adv. Mass Spectrom. Biochem. Med.* (Spectrum, New York) **2**, 389 (1977).

Derivative Type. Pentafluorobenzyl.

Reagent. Pentafluorobenzyl bromide (PFB-Br)/potassium carbonate.

Compound Type. Clonidine [2-(2,6-dichlorophenylamino)-2-imidazoline].

Reaction

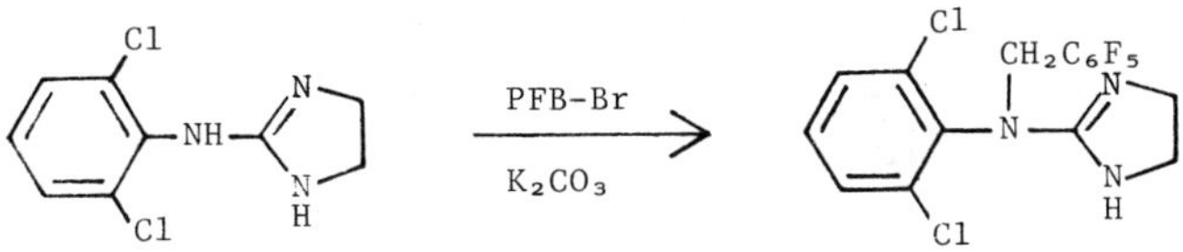

Typical Procedure.[1] To 50–500 μl plasma add 100 μl int stand soln (0.2 μg/ml in 0.1 *M* NaOH) and 0.1 *M* NaOH to make 1 ml. Ext with 2 × 3 ml cyclohexane/BuOH (9:1, v/v). Back ext the combined org phases with 0.5 ml 0.1 *N* H_2SO_4. Alkalinize the aq phase with 100 μl 10 *M* NaOH and repeat ext as above. Evap the ext under N_2 at 60°. Add 1 ml 1.0% (v/v) PFB-Br in acetone and 5–25 mg K_2CO_3 and reflux 40 min. Evap under N_2, add 1 ml 2 *N* H_2SO_4, and wash with 2 × 1 ml heptane. Alkalinize the aq phase with 250 μl 10 *M* NaOH and ext with 3 ml cyclohexane-BuOH. Evap the ext under N_2 and diss residue in 100 μl EtOAc. Analyze by ECD-GC.

Comments. The deriv was detectable to 6 pg. The 2,4-dichloro isomer was used as int stand. The structure of the deriv was confirmed by MS and the position of the PFB group on the bridge N assigned by analogy to reported alkylation studies.[2]

References

[1]P.-O. Edlund and L. K. Paalzow, *Acta Pharmacol. Toxicol.* **40**, 145 (1977).

[2]H. Stähle and K. H. Pook, *Ann. Chem.* **751**, 159 (1971).

Derivative Type. Trimethylsilyl.

Reagent. Bis(trimethylsilyl)acetamide (BSA)/trimethylchlorosilane (TMCS).

Compound Type. Cycloserine.

Reaction

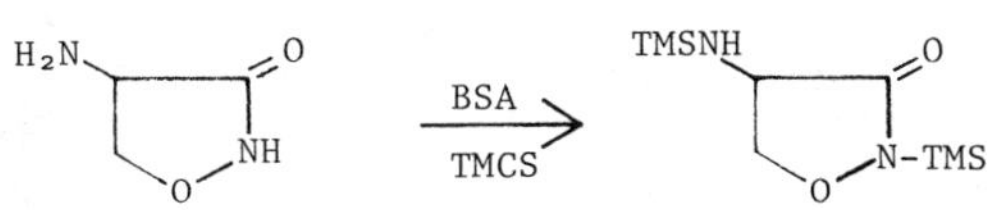

Typical Procedure. Diss sample in pyridine H_2O (2:1, v/v) to give *ca* 1 mg/ml. Allow excipients to settle and aliquot 1 ml of supernatant. Evap to dryness under an air stream. Add 100 μl BSA and 10 μl TMCS. Heat 10 min at 80°. Analyze by GC.

Comments. The structure of the deriv was confirmed by MS.

References

[1]D. L. Sondack *et al.*, *J. Pharm. Sci.* **62**, 1344 (1973).

Derivative Type. Pentafluorobenzoyl.

Reagent. Pentafluorobenzoyl chloride.

Compound Type. Thiabendazole.

Reaction

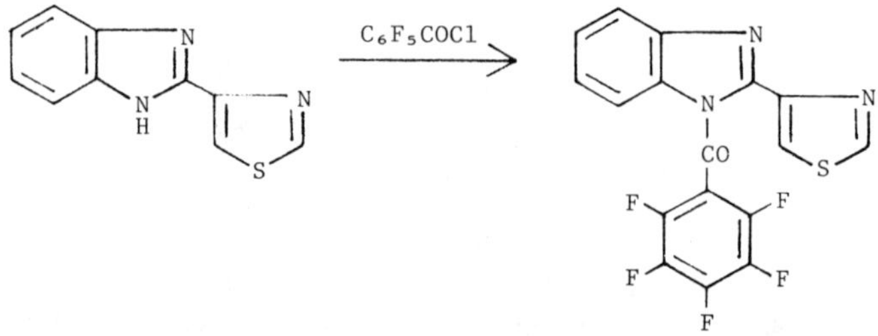

Typical Procedure. Homogenize 20 g sample with 80 ml EtOAc and 50 ml aq NaOAc (33 g/liter)-NaCl (200 g/liter) soln. Centrifuge and separate org layer. Repat ext and wash combined exts with 50 ml H_2O. Ext org phase with 2 × 50 ml 0.1 *N* HCl and adjust exts to pH 9–10 with NaOH. Ext with 2 × 40 ml EtOAc. Dry ext over Na_2SO_4 and evap to dryness. Add 0.5 ml benzene, 10 μl pentafluorobenzoyl chloride, and *ca* 2 mg anhyd Na_2CO_3. Seal tube and heat at 120° for 30 min. Evap to dryness under vac, add 0.4 ppm decachlorobiphenyl (int stand) in 1 ml acetone and analyze by GC.

Comments. The method was used for analysis of thiabendazole in fruits using ECD. The minimum detectable amt of thiabendazole was 0.01 ppm. MS data were given for the deriv.

References

[1]N. Nose *et al.*, *J. Chromatogr.* **130**, 410 (1977).

Derivative Type. Nitrile.

Reagent. Trifluoroacetic anhydride (TFAA).

Compound Type. Phenylethylmalonamide.

Reaction

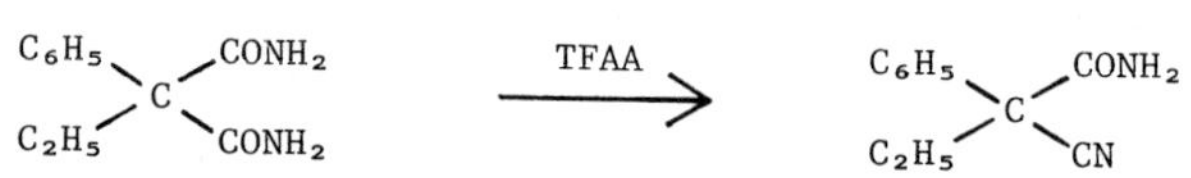

Typical Procedure. Mix 0.5 ml plasma with 0.5 ml satd $(NH_4)_2SO_4$ and ext with 10 ml 20% (v/v) EtOAc in benzene containing *p*-methylphenylethylmalonamide int stand. Evap ext under N_2 at 55°. Add 100 μl TFAA and heat 30 min at 120°. Allow to cool to room temp and evap under N_2. Diss residue in 1 ml EtOAc and analyze by ECD-GC.

Comments. MS data were given for the deriv. Flash methylation gave multiple GC peaks and trimethylsilylation gave an undesirable moisture labile deriv.

References

[1]J. E. Wallace *et al.*, *Anal. Chem.* **49**, 1969 (1977).

Derivative Type. 2,2,3,3,3-Pentafluoro-1-propyl ester, *N*-pentafluoropropionyl.

Reagent. 2,2,3,3,3-Pentafluoro-1-propanol/pentafluoropropionic anhydride (PFPA).

Compound Type. 2-Pyrrolidone-2-carboxylic acid.

Reaction

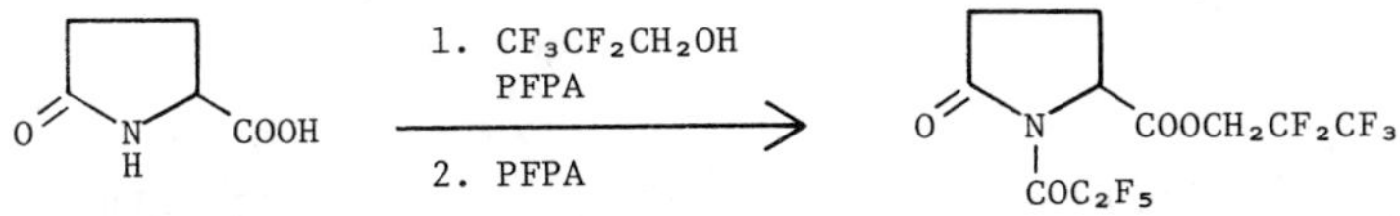

Typical Procedure.[1] To a 100 ng sample add 40 μl PFPA and 10 μl pentafluoropropanol. Heat 10 min at 75° then evap under N_2. Add 25 μl PFPA and heat at 75° for 5 min. Evap under N_2 and diss residue in 1 ml EtOAc. Analyze by ECD-GC.

Comments. The deriv was used to quantitate pyrrolidone carboxylate in urine, CSF, and tissue.[2] The compd had previously been analyzed as the methyl ester prepared with MeOH/HCl[3] or CH_2N_2[4] and measured by FID-GC, and as the *N,O*-di-TMS deriv prepared with HMDS/TMCS/pyridine.[5,6]

References

[1] E. Watson *et al.*, *Anal. Biochem.* **59**, 441 (1974).

[2] S. Wilk and M. Orlowski, *Anal. Biochem.* **69**, 100 (1975).

[3] P. Polgar and A. Meister, *Anal. Biochem.* **12**, 338 (1965).

[4] E. Jellum *et al.*, *Scand. J. Clin. Lab Invest.* **26**, 327 (1970).

[5] C. Y. Lee, *J. Assoc. Off. Anal. Chem.* **53**, 716 (1970).

[6] C. Y. Lee and T. E. Acree, *J. Agr. Food Chem.* **19**, 184 (1971).

Derivative Type. Tetrazolo[1,5-*a*]phthalazine.

Reagent. Nitrous acid.

Compound Type. Hydralazine.

Reaction

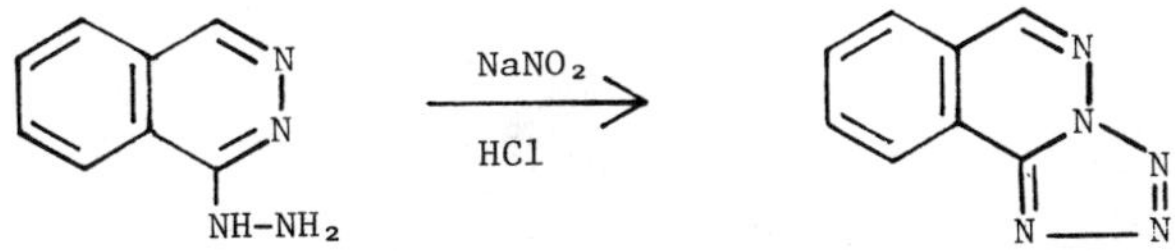

Typical Procedure. To 1 ml plasma add 50 ng 4-methylhydralazine int stand (in 0.1 ml 0.1 *N* HCl), 1 ml 2*N* HCl and 0.1 ml 50% aq $NaNO_2$. Mix and allow to react at room temp for 15 min. Adjust to pH 10 (±1) by adding 2 ml 1 *N* NaOH and 4 ml pH 10 borax buffer (0.3 *M* $Na_2B_4O_7$, 0.04 *M* NaOH). Ext with 3 ml benzene and evap ext under N_2 at 45°. Diss residue in 300–900 μl toluene and analyze by ECD-GC.

Comments. The method permitted plasma level determination to 10 ng/ml.

References

[1] D. B. Jack *et al.*, *J. Chromatogr.* **115**, 87 (1975).

Derivative Type. Pyridine analog.

Reagent. Nitrous acid.

Compound Type. 2,6-Dimethyl-4-(3-nitrophenyl)-1,4-dihydropyridine-3,5-dicarboxylic acid 3-[2-(*N*-benzyl-*N*-methylamino)]ethyl ester 5-methyl ester (YC-93).

Reaction

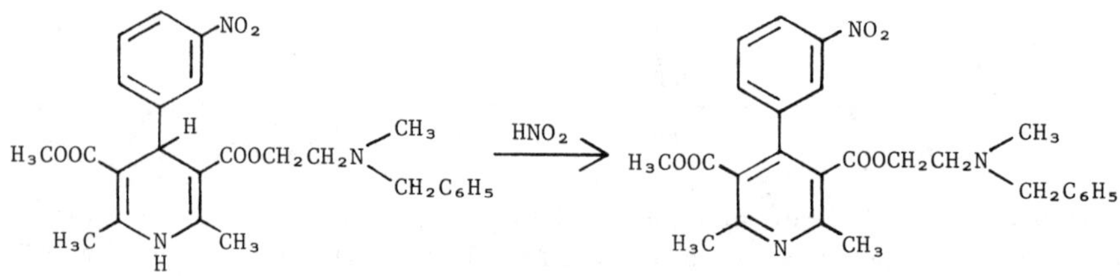

Typical Procedure. Mix 2 ml plasma, 0.5 ml int stand soln, and 0.5 ml 2 *N* NaOH. Ext with 2 × 2 ml Et_2O. Ext combined org phases with 4 ml 0.05 *N* HCl. Add 0.3 ml 1% $NaNO_2$ soln to the aq phase and heat at 45° for 1 hr. Cool and alkalinize with 0.5 ml 2 *N* NaOH. Ext with 2 × 2 ml Et_2O. Evap ext to dryness at 45° and diss residue in 50 μl benzene. Analyze 3 μl aliquots by ECD-GC.

Comments. The analogous 3-propyl (instead of ethyl) and 5-isopropyl (instead of methyl)-diester was used as int stand as a 200 ng/ml aq soln. The method was sensitive to 2–3 ng/ml. Structures of the derivs were confirmed by MS. GC of the dihydro pyridine compd (YC-93) itself gave 2 peaks due to partial conversion to the pyridine during analysis.

References

[1]S. Higuchi *et al.*, *J. Chromatogr.* **110**, 301 (1975).

8. Phospho and Sulfo Compounds

The derivatives of the phospho and sulfo compounds are used primarily to reduce the polarity of the acidic OH groups and other polar groups to impart sufficient volatility to allow GC and MS analysis. The phospho compounds consist of phosphate esters with remaining free OH groups and alkylphosphonic acids with free OH groups. The derivatization of these groups involves esterfication reactions. Both alkyl esters (formed from

alkyl phosphate   alkyl phosphonic acid

diazoalkanes, p. 388) and trimethylsilyl esters (p. 389) have been employed. A similar problem is encountered with the sulfonic acids and both alkyl (p. 394) and trimethylsilyl (p. 394) esters have been used analytically for these compounds. Sulfonic acids have also been converted to sulfonyl chlorides as volatile derivatives for GC analysis.

Organophosphate insecticides were analyzed by HPLC after hydrolysis and derivatization of the liberated phenols (p. 393).

Sulfonamides and Sulfonylureas. With the sulfonamides and sulfonylureas the objective of derivatization is to mask the polar NH groups to increase the volatility and stability of the compounds. In the sulfonamides the NH groups are sufficiently acidic to undergo alkylation by phase transfer methods using tetraalkylammonium salts. Such *N*-alkylation has been performed with methyl iodide (p. 395) and, to enhance ECD sensitivity, with pentafluorobenzyl bromide (p. 397). The conditions of the phase transfer alkylation, however, cause alteration of other sensitive structures in the molecule. For example, the succinimide portion of neosulfalepine underwent ring opening in the derivatization procedure (p. 397). Sulfonamides have also been alkylated by flash methylation using trimethylbenzylammonium hydroxide (p. 398) and trimethylphenylammonium hydroxide (p. 399), and presumably would be amenable to alkylation with other alkyl ammonium reagents.

Sulfonamides have also been *N*-acylated; both TFA and HFB derivatives were reported to have good ECD sensitivity (p. 398). Primary sulfonamides form dimethylaminomethylene derivatives with DMF dialkyl acetals and TMS derivatives with BSTFA (p. 398).

In the sulfonylureas, the sulfonamido NH is more acidic and can be alkylated with either diazomethane or dimethyl sulfate. The *N*-methyl derivatives can be analyzed directly by GC as the intact derivatives or as the sulfonamide by pyrolysis GC (p. 400). There is question regarding reported analyses of *N*-methyl sulfonamides as to whether the derivatives themselves or the pyrolysis products were actually being analyzed (p. 399). The *N*-methyl derivatives have been acylated on the other nitrogens with TFAA, PFPA, or HFBA to yield more stable derivatives (p. 401). The sulfonyl ureas have also been analyzed as dinitroaniline derivatives prepared by basic hydrolysis and treatment of the resulting alkylamines with dinitrofluorobenzene (p. 401).

Sulfate Esters. Sulfate monoesters may be methylated with diazomethane to yield volatile derivatives (p. 402). Phenolic sulfates undergo simultaneous cleavage and derivatization of the resulting phenol when treated with acylating agents (p. 402) or trimethylsilylating agents. These methods are particularly useful for analysis of sulfate esters in which the parent phenol is unstable to the normal sulfate ester hydrolysis conditions.

In addition to the methods discussed in this section, derivatization of phosphate groups are included among the carbohydrates (p. 576), in the section on nucleotides (p. 615), and in the section on indole amines (p. 110).

8.1. DERIVATIVES OF PHOSPHO COMPOUNDS

8.1.1. Alkyl Ester Derivatives

Derivative Type. Trifluoroacetyl, methyl ester.

Reagent. Trifluoroacetic anhydride (TFAA), diazomethane.

Compound Type. 2-Aminoethylphosphonic acid and 1-hydroxy-2-aminoethylphosphonic acid.

Reaction

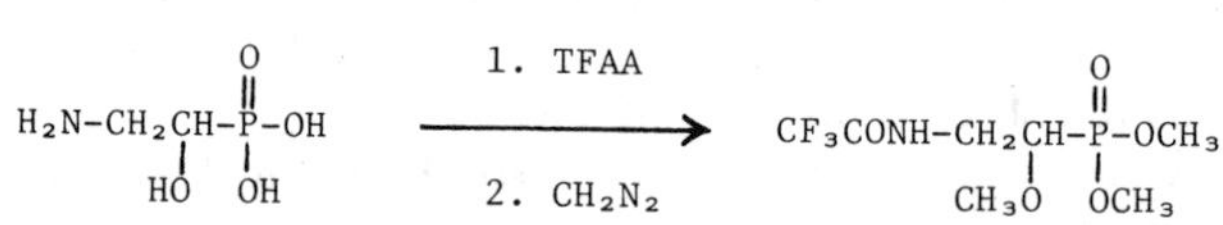

Typical Procedure. Diss sample in 20% TFAA in CH_2Cl_2. Heat at 100° for 1 hr then evap to dryness under N_2. Add CH_2N_2 in Et_2O. Evap excess CH_2N_2 and analyze by GC-MS.

Components. The TMS derivs (*N*- and *O*-TMS) were also prepared by heating 1 hr at 100° with BSTFA-1% TMCS. The compounds were identified by MS as major constituents of amoeba plasma membrane polysaccharide.

References

[1]E. D. Korn *et al.*, *J. Biol. Chem.* **248**, 2257 (1973).

Derivative Type. Trifluoroacetyl, butyl ester.

Reagents. Trifluoroacetic anhydride (TFAA)/trifluoroacetic acid (TFA), diazobutane.

Compound Type. Aminoalkylphosphonic acids.

Reaction

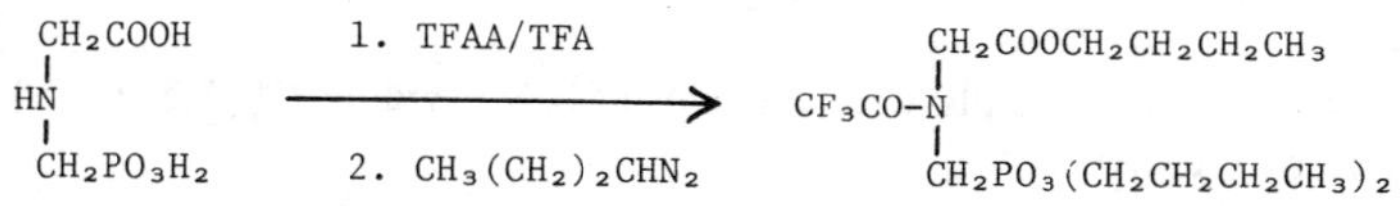

Typical Procedure. Freeze 250–500 μl of soln (containing 25–50 μg) of amino alkylphosphonic acid in 0.1 *N* NH_4HCO_3 in a vial and lyophilize. Add 2–3 mg of NH_4HCO_3 along with 1.0 ml TFAA/TFA (1:1, v/v). After 1 hr at room temp remove excess reagent under a stream of N_2. Add 0.2 ml *n*-BuOH and an excess of ethereal diazobutane.[2] After 15 min conc the sample under N_2 and analyze by GC-MS.

Comments. The generally poorly soluble (in org solvents) compds were diss by TFA. *N*-Phosphonomethyl Gly required 40° warming for trifluoroacetylation.

References

[1]M. L. Rueppel *et al.*, *Biomed. Mass Spectrom.* **3**, 28 (1976).

[2]C. W. Stanley, *J. Agric. Food Chem.* **14**, 321 (1966).

8.1.2. Trimethylsilyl Ester Derivatives

Derivative Type. Trimethylsilyl.

Reagent. Hexamethyldisilazane (HMDS)/trimethylchlorosilane (TMCS)/pyridine.

Compound Type. Pyridoxine-5-phosphate.

Reaction

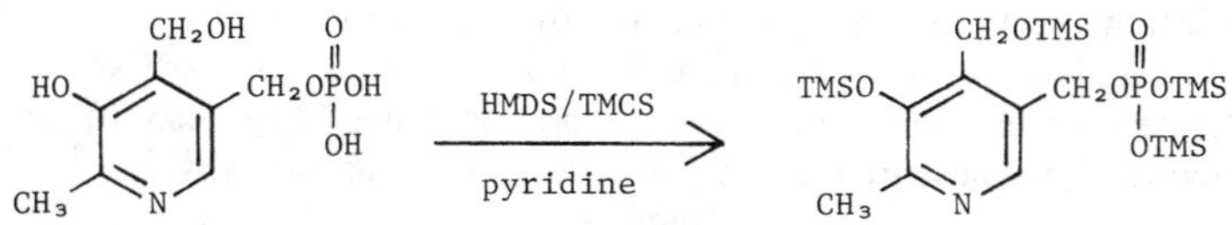

Typical Procedure. Treat sample with HMDS/TMCS/pyridine (2:1:10, v/v/v) at room temp.[2] Analyze by GC.

Comments. The derivs of pyridoxine-5-phosphate, pyridoxal-5-phosphate, and 4-deoxypyridoxine-5-phosphate each gave single GC peaks. TMS derivs of nonphosphorylated pyridoxine and its vitamers have also been used analytically.[1,3] Isopropylidene TMS derivs were also reported.[1]

References

[1]W. Korytnyk *et al.*, *Anal. Biochem.* **17**, 66 (1966).

[2]C. C. Sweeley *et al.*, *J. Am. Chem. Soc.* **85**, 2497 (1963).

[3]L. T. Sennello and C. J. Argoudelis, *Anal. Chem.* **41**, 171 (1969).

Derivative Type. Trimethylsilyl.

Reagent. Bis(trimethylsilyl)trifluoroacetamide (BSTFA)/trimethylchlorosilane (TMCS).

Compound Type. Phosphoryl and phosphono ethanolamine.

Reaction

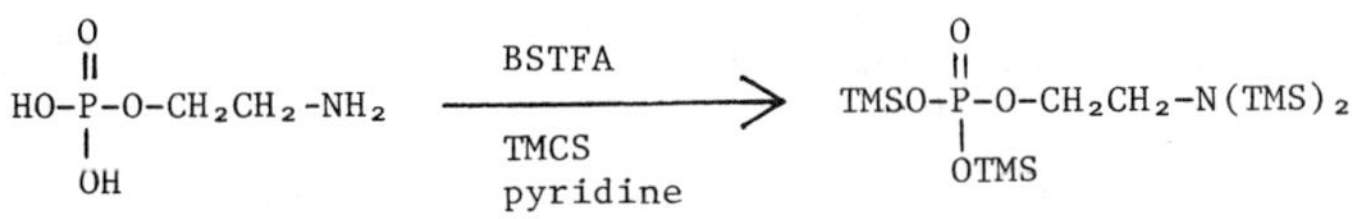

Typical Procedure. Treat sample with BSTFA/TMCS/pyridine (10:2:5, v/v/v) for 30 min at 60°. Analyze by GC-MS.

Comments. Complete MS for the derivs were given.

References

[1]K. A. Karlsson, *Biochem. Biophys. Res. Commun.* **39**, 847 (1970).

Derivative Type. Trimethylsilyl, isothiocyanate (amino compds).

Reagent. Bis(trimethylsilyl)trifluoroacetamide (BSTFA)/trimethylchlorosilane (TMCS), carbon disulfide.

Compound Type. Alkylphosphonic acids and aminoalkylphosphonic acids.

Reaction

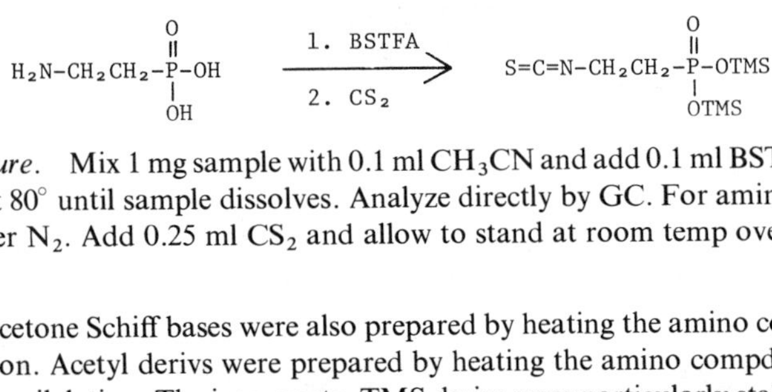

Typical Procedure. Mix 1 mg sample with 0.1 ml CH_3CN and add 0.1 ml BSTFA and 0.05 ml TMCS. Heat at 80° until sample dissolves. Analyze directly by GC. For amino compds, evap to dryness under N_2. Add 0.25 ml CS_2 and allow to stand at room temp overnight. Analyze soln by GC.

Comments. Acetone Schiff bases were also prepared by heating the amino compd in acetone prior to silylation. Acetyl derivs were prepared by heating the amino compd with Ac_2O and pyridine prior to silylation. The isocyanate, TMS derivs were particularly stable. GC and MS data for the derivs were reported.[1] Roboz *et al.*[2] used the TMS deriv (from BSTFA, 1% TMCS, in pyridine) to measure phosphonoacetic acid by GC-MS-SIM.

References

[1]D. J. Harvey and M. G. Horning, *J. Chromatogr.* **79**, 65 (1973).

[2]J. Roboz *et al., Biomed. Mass Spectrom.* **4**, 291 (1977).

Derivative Type. Trimethylsilyl.

Reagent. Bis(trimethylsilyl)trifluoroacetamide (BSTFA)/bis(trimethylsilyl)acetamide (BSA).

Compound Type. Phosphoserine and phosphothreonine.

Reaction

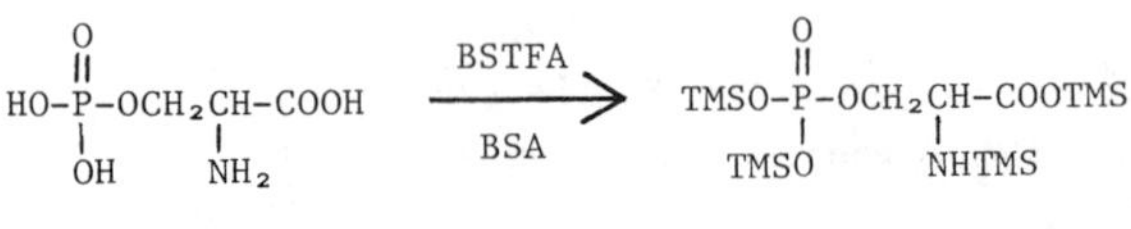

Typical Procedure. Mix a soln of 200 μg of the compd in 100 μl H_2O with 100 μl 0.1 *N* HCl and evap to dryness under N_2. Diss the residue in 80 μl CH_3CN and add 40 μl BSTFA and 40 μl BSA. Mix by sonication then allow to stand at room temp several hr. Analyze by GC-MS.

Comments. Full MS were given for the derivs.

References

[1]J.-J. Shieh and D. Desiderio, *Anal. Lett.* **10**, 831 (1977).

Derivative Type. Oxime, trimethylsilyl.

Reagent. Methoxylamine·HCl, bis(trimethylsilyl)trifluoroacetamide (BSTFA)/trimethylsilylimidazole (TMSI)/trimethylchlorosilane (TMCS).

Compound Type. Steroid phosphates.

Reaction

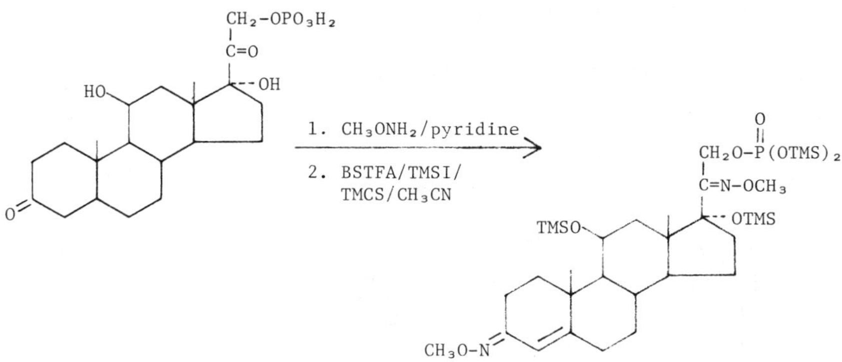

Typical Procedure. Diss 1 mg steroid phosphate in 0.2 ml pyridine and allow to react with excess CH_3ONH_2·HCl overnight at 60°. Evap under N_2 and diss residue in 0.1 ml CH_3CN, 0.1 ml BSTFA, 0.1 ml TMSI, and 0.05 ml TMCS. Heat 5 hr in a sealed tube. Analyze by GC-MS.

Comments. Steroid phosphates without keto groups were converted to TMS derivs with BSTFA/TMCS/CH_3CN (1 :5 :1, v/v/v) at 80° for 30 min. Specifically labeled d_9-TMS derivs could be prepared by on-column exchange of the more labile phosphate TMS groups. MS data were given for the derivs.

References

D. J. Harvey *et al.*, *Tetrahedron* **27**, 4231 (1971).

Derivative Type. Trimethylsilyl glycerophosphate esters.

Reagent. Bis(trimethylsilyl)trifluoroacetamide(BSTFA)/trimethylchlorosilane (TMCS).

Compound Type. Glycerophospholipids.

Reaction

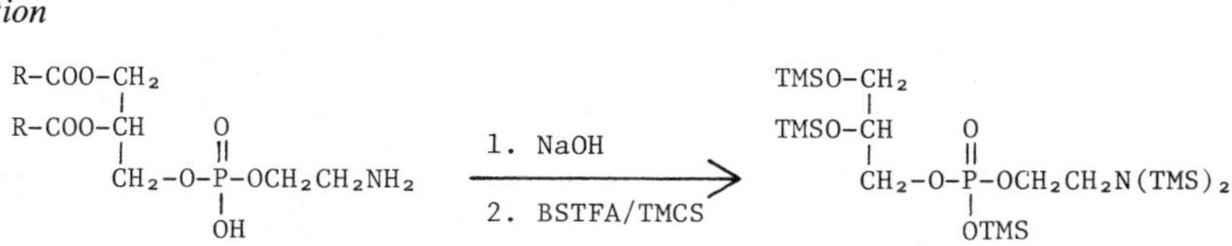

Typical Procedure. Diss 5–15 μmol of the glycerophospholipid in 1 ml $CHCl_3$/MeOH (1:4, v/v) and add 0.1 ml 1 *N* aq NaOH. Heat at 37° for 15 min; then neutralize with 0.1 ml 1 *N* HOAc. Add 2 ml $CHCl_3$/MeOH (9:1, v/v), 1 ml *i*-BuOH, and 2 ml H_2O. Mix well and centrifuge. Remove the aq phase and wash it 2× with Et_2O. Conc under vac at or below 35° then apply conc to a 1–2 g column of Dowex 50W-X4 (H form). Elute with H_2O (for the choline compd use 0.01 *M* HCl). Dry an aliquot of the eluate (corresponding to 1 mg of the glycerophosphate ester) under vac. Add 0.1 ml pyridine and 0.1 ml BSTFA containing 1% TMCS. Stir at room temp 1 hr then analyze by GC-MS or direct probe MS.

Comments. Derivs were prepared from six compds where the phosphate substituent was glycerol, ethanolamine (shown above), *N,N*-dimethylethanolamine, serine, inositol, and with no phosphate substituent. BSA gave a better yield of the penta-TMS deriv from the serine compd. Full MS for the derivs were given.

References

[1]J. H. Duncan *et al.*, *Biochemistry* **10**, 927 (1971).

Derivative Type. Trimethylsilyl diglyceride.

Reagent. Bis(trimethylsilyl)trifluoroacetamide (BSTFA).

Compound Type. Phospholipids.

Reaction

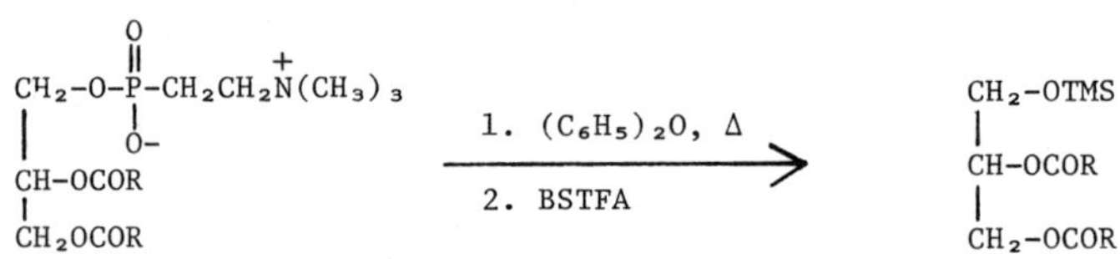

Typical Procedure. Diss 1–3 mg sample in 0.1 ml diphenyl ether and heat 5 min at 250°. Cool to room temp and add 0.1 ml BSTFA (or BSA) and allow to stand at room temp 2 hr (or heat at 60° for 30–60 min). Analyze by GC.

Comments. The same deriv was obtained by treating the sample directly with 100 μl BSA and 50 μl TMCS (room temp overnight or 60° for several hr) but higher yields were obtained with the diphenyl ether procedure. The method yielded a mixt of 1,2- and 1,3-diglyceride TMS ethers presumably due to acyl migration during the dephosphorylation step.

References

[1]M. G. Horning *et al.*, *J. Chromatogr. Sci.* **7**, 267 (1969).

Derivative Type. Trimethylsilyl deacylated polyphosphoinositides.

Reagent. Methanol (MeOH)/sodium hydroxide, bis(trimethylsilyl)trifluoroacetamide (BSTFA)/trimethylchlorosilane (TMCS).

Compound Type. Polyphosphoinositides.

Reaction

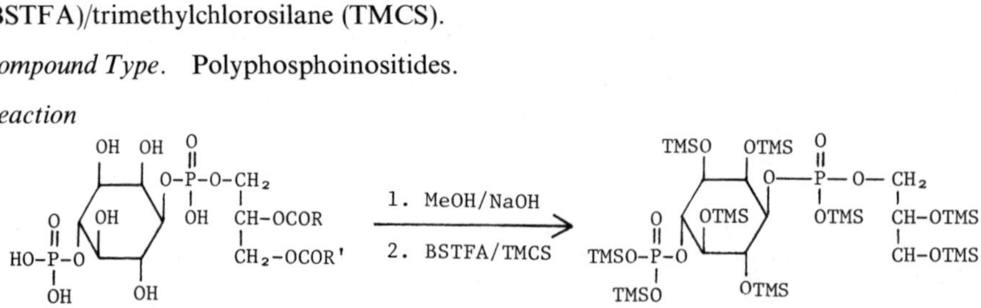

Typical Procedure. Ext lipids from brain tissue (see ref for details) and diss in $CHCl_3$ MeOH(1 :4, v/v). Add enough 1.2 *N* NaOH in $MeOH/H_2O$ (1 :1, v/v) to give 0.2 *N* in NaOH. Warm at 37° for 10 min, cool in ice, and neutralize with 1 *N* HOAc. Add 2 ml $CHCl_3$/MeOH (9 :1, v/v), 2 ml H_2O, and 1 ml *i*-BuOH then centrifuge. Convert deacylated lipids in the aq phase to the acid form on Dowex 50W then lyophilize. Diss residue in BSTFA (with 1% TMCS/pyridine (2:1, v/v) and allow to stand overnight at room temp. Analyze by GC-MS.

Comments. The method was used to identify and measure the mono-, di- (shown above), and triphosphoinositides. MS data were given for the derivs.

References

[1]T. J. Cicero and W. R. Sherman, *Biochem. Biophys. Res. Commun.* **42**, 428 (1971).

Derivative Type. Trimethylsilyl diglyceride.

Reagent. Bis(trimethylsilyl)acetamide (BSA)/trimethylchlorosilane (TMCS).

Compound Type. Phosphatidylserines.

Reaction

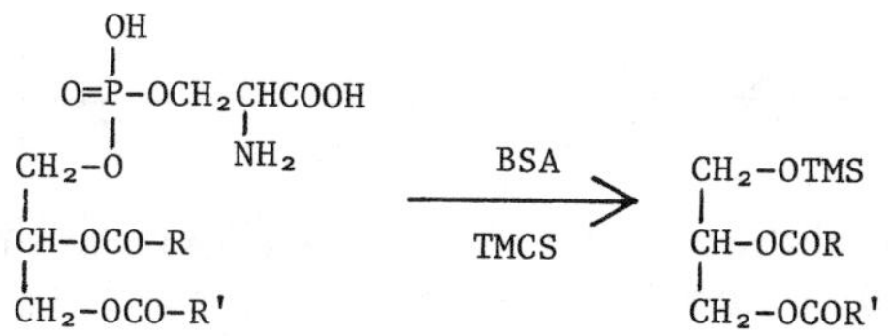

Typical Procedure. Mix 1.5 mg sample with 0.1 ml BSA and 0.05 ml TMCS. Heat at 60° for 1 hr then analyze by GC-MS.

Comments. The mechanism of the reaction was not known, but the method offered a means of studying the composition of phosphatidyl serine fractions without prior chemical or enzymic hydrolysis.

References

[1]G. Casparrini *et al.*, *Anal. Lett.* **1**, 481 (1968).

8.1.3. Derivatives of Phosphate Ester Hydrolysis Products

Derivative Type. 5-Dimethylamino-1-naphthalenesulfonyl (DANS).

Reagent. Sodium hydroxide, 5-dimethylamino-1-naphthalenesulfonyl chloride (DANS-Cl).

Compound Type. Organophosphorus insecticides.

Reaction

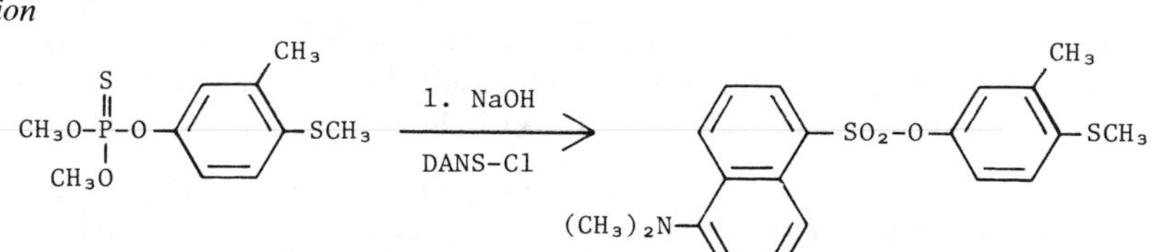

Typical Procedure. Hydrolyze sample for 45 min at 75° in 1 ml 0.5 *M* NaOH. Add 2 drops methyl isobutyl ketone and 2 drops 0.1% DANS-Cl in acetone. Shake well and heat at 65° for 30 min. Cool, acidify with 10% HCl, and ext with 0.3 ml benzene. Analyze by HPLC.

Comments. The deriv was used for fluorescence detection HPLC. The deriv reaction above is shown for *O,O*-dimethyl-*O*-[4-(methylthio)-*m*-tolyl] phosphorothiolate (Fenthion). The method was applied to five-organophosphorus insecticides with detectability to the low ng level.

References

[1] R. W. Frei and J. F. Lawrence, *J. Chromatogr.* **83**, 321 (1973).

[2] J. F. Lawrence *et al.*, *J. Chromatogr.* **121**, 343 (1976).

8.2. DERIVATIVES OF SULFONIC ACIDS

Derivative Type. Methyl ester.

Reagent. Tetramethylammonium hydroxide (TMAH).

Compound Type. Sulfonic acids.

Reaction

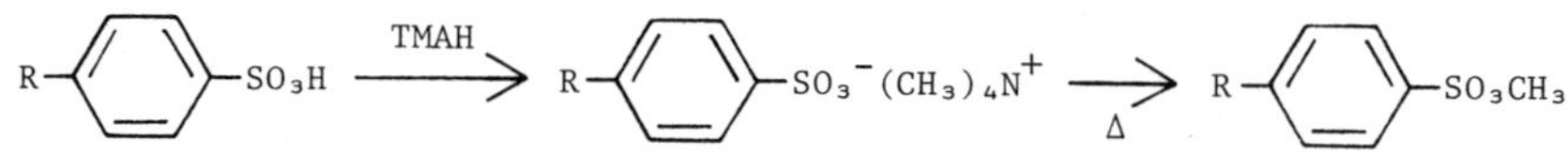

Typical Procedure. Neutralize an aq soln of the sulfonic acid with 10% aq TMAH. Evap to dryness and analyze by direct probe MS.

Comments. The methyl esters were also prepared using CH_2N_2 and gave the same MS as those from TMAH.

References

[1] A. Heywood *et al.*, *Anal. Chem.* **42**, 1272 (1970).

Derivative Type. Trimethylsilyl.

Reagent. Trimethylsilylimidazole (TMSI).

Compound Type. Sulfonic acids.

Reaction

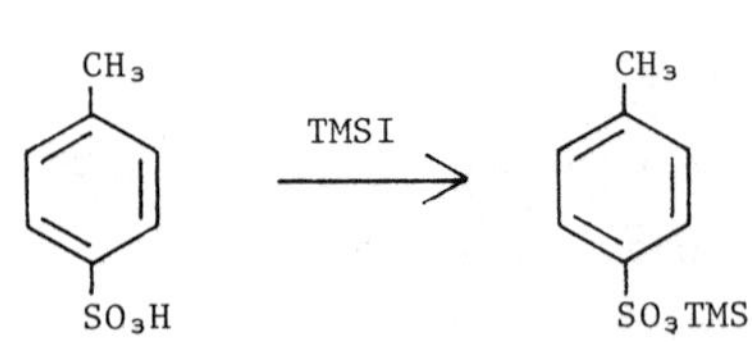

Typical Procedure. Mix 0.2–0.5 mg sulfonic acid (or the pyridinium salt) with 200–300 μl TMSI. Shake at room temp until homogeneous soln results (usually 10–15 min). Analyze 1 μl by direct probe MS allowing the excess reagent to evap in the vac lock.

Comments. Amino groups could be converted to the TFA derivs by adding to the sample in TMSI a 500 μl portion of 10% TFAI in THF. The method was also applied to a sugar sulfonic acid, its 2′,4-dinitrophenylosazone, and to cysteic acid. MS data were reported for the derivs.

References

[1] J. Eagles and M. E. Knowles, *Anal. Chem.* **43**, 1697 (1971).

Derivative Type. Sulfonyl chloride phosphoramidyl chloride.

Reagent. Phosphorus pentachloride/phosphorus oxychloride.

Compound Type. Aminobenzenesulfonic acids.

Reaction

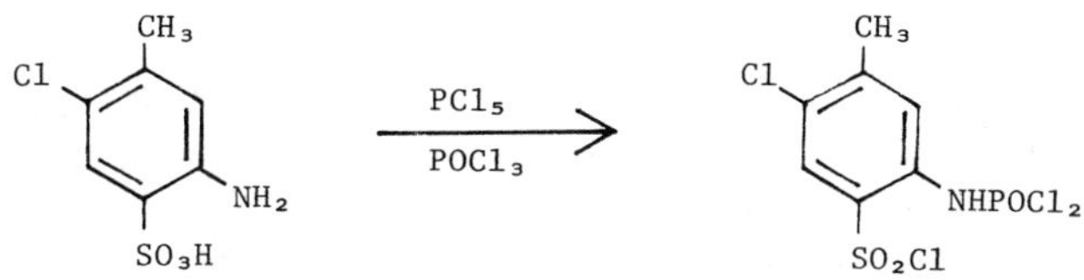

Typical Procedure. To a 0.5 g sample add carefully 3 g PCl_5 and 1 ml $POCl_3$. Attach a condenser to the sample tube and heat with stirring at 95° for 45 min. Add 10 ml toluene to diss the deriv. Analyze an aliquot of the toluene soln by GC.

Comments. GC separation of 6 aminobenzenesulfonic acids was demonstrated.

References

[1]J. S. Parsons, *J. Chromatogr. Sci.* **11**, 659 (1973).

Derivative Type. Sulfonyl chloride.

Reagent. Thionyl chloride/dimethylformamide (DMF).

Compound Type. Sulfonic acids and salts.

Reaction

$$R{-}SO_2OH \xrightarrow[\text{DMF}]{SOCl_2} R{-}SO_2Cl$$

Typical Procedure. To a 0.5 g sample of sulfonic acid or salt add 0.5 ml DMF and 20 ml $SOCl_2$. Reflux until gas is no longer evolved (several min–2 hr depending upon sample). If the sample was a salt, dil with an equal vol of CH_2Cl_2 and filter (fine sintered glass). Evap under vac and diss residue in CCl_4. Analyze by GC.

Comments. The deriv reaction was also performed using phosgene bubbled into a soln of the sample and DMF in CCl_4. Methyl ester derivs were also prepared using CH_2N_2; a large excess of reagent was required since the sulfonic acids catalyzed polymerization of CH_2N_2. It was necessary to operate the GC at subatmospheric pressure to avoid decomposition of the methyl esters.

References

[1]J. J. Kirkland, *Anal. Chem.* **32**, 1388 (1960).

8.3. DERIVATIVES OF SULFONAMIDES

8.3.1. Alkyl Derivatives of Sulfonamides

Derivative Type. Methyl.

Reagent. Methyl iodide, tetrahexylammonium sodium sulfate (THASS).

Compound Type. Chlorthalidone.

Reaction

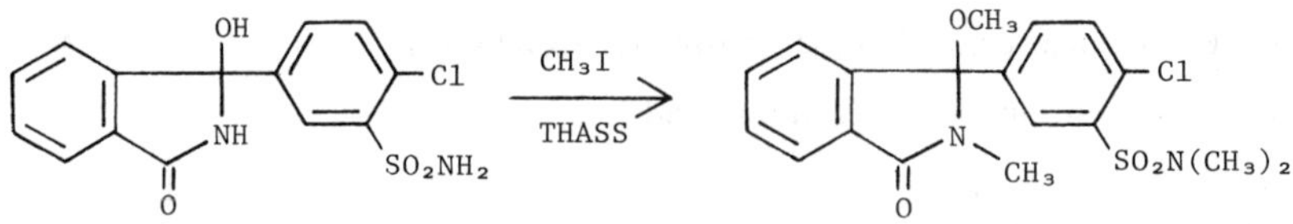

Typical Procedure. Mix 2 ml of plasma, serum, or urine with 0.2–0.3 g $NaHCO_3$ and 5 ml methyl isobutyl ketone in a 15 ml centrifuge tube. Shake for 10 min and centrifuge. Transfer 4 ml of org phase to another 15 ml tube, add 2.5 ml 0.1 *M* NaOH, and shake 10 min. Centrifuge and transfer 2 ml of the aq layer to a 15 ml screw cap tube. Add 50 μl of 0.1 *M* aq tetrahexylammonium soln and 5 ml of 0.5 *M* CH_3I in CH_2Cl_2. Shake at 50° for 20 min, centrifuge 5 min at 2500 rpm, and transfer 4 ml of the org layer to a 15 ml conical tube. Evap to dryness at room temp under N_2. Add to the residue a soln of the standard (tetraethylchlorthalidone) in hexane and place in ultrasonic bath for 5 min. Centrifuge 5 min at 2500 rpm and analyze 1–2 μl by GC.

Comments. The method was used for ECD detection down to 2 ng/ml. The THASS soln was prepared by neutralizing THA-HSO_4 with an equiv amt of NaOH.

References

[1]M. Ehrvik and K. Gustavii, *Anal. Chem.* **46**, 39 (1974).

Derivative Type. Methyl.

Reagent. Methyl iodide/tetrahexylammonium hydrogen sulfate (THA-HSO_4).

Compound Type. Furosemide.

Reaction

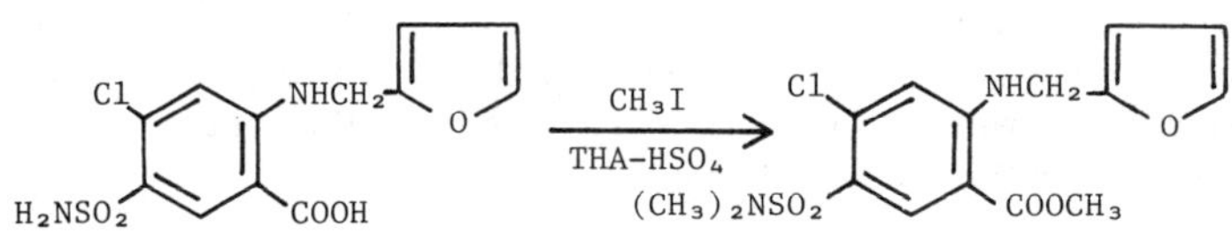

Typical Procedure. Dil 1 ml plasma with 2 ml H_2O and acidify with 0.15 ml 4 *M* HCl. Ext with 2 × 5 ml Et_2O. Evap ext to dryness under N_2. Add 2 ml 0.2 *M* NaOH, 50 μl 0.1 *M* THA-HSO_4, and 5.0 ml 0.5 *M* CH_3I in CH_2Cl_2. Shake 20 min at 50°, transfer 4.0 ml of the org phase and evap to dryness under N_2 at room temp. Add 0.2 ml hexane containing the int stand, mix, and analyze by GC.

Comments. The triethyl deriv, prepared in the same manner from ethyl iodide, was used as int stand.

References

[1]B. Lindtröm and M. Molander, *J. Chromatogr.* **101**, 219 (1974).

Derivative Type. Methyl.

Reagent. Methyl iodide/tetrahexylammonium hydrogen sulfate (THA-HSO_4).

Compound Type. Neosulfalepsine.

Reaction

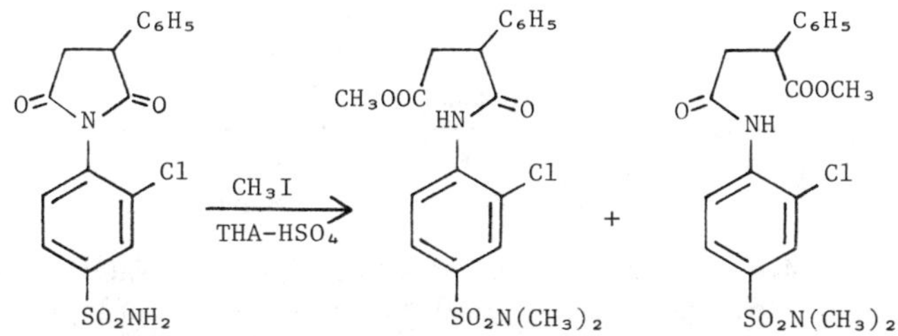

Typical Procedure. Mix 1 ml urine with 0.1 ml int stand soln, 2.0 ml 1 *N* HCl, and 2 ml pH 7 buffer (0.041 *M* Na_2HPO_3, 0.028 *M* KH_2PO_4). Ext with 2 × 3.5 ml toluene and back ext the combined toluene into 3 ml 0.1 *N* NaOH. Discard the toluene and mix the aq phase with 0.1 ml 0.05 *M* THA-HSO_4 and 2.0 ml 1 *M* CH_3I in CH_2Cl_2. Shake 40 min, separate the org phase, and evap under N_2 at 40°. Diss residue in 2 ml *n*-heptane and analyze by GC.

Comments. The deriv procedure gave ring opening of the succinimide to yield two GC separable derivs. Similar procedures were given for chlorthalidone and chlorguinaldol.

References

[1]P. H. Degen and A. Schweizer, *J. Chromatogr.* **142**, 549 (1977).

Derivative Type. Pentafluorobenzyl (PFB).

Reagent. Pentafluorobenzyl bromide (PFB-Br)/tetrabutylammonium hydrogen sulfate (TBA-HSO_4).

Compound Type. Sulfonamides.

Reaction

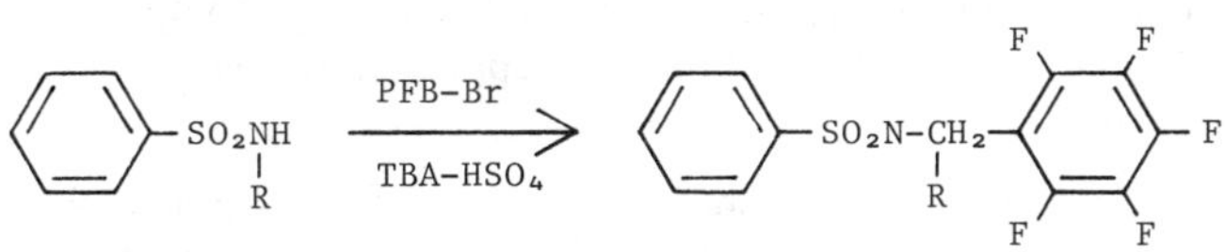

R=C_2H_5, C_6H_5

Typical Procedure. Mix 1 ml of aq 0.1 *M* TBA-HSO_4/0.2 *M* NaOH with 1 ml CH_2Cl_2 containing 2 mg of the sulfonamide and int stand. Add 20 μl PFB-Br and shake at 25° (see below for time). Separate the org layer, dil with benzene, and analyze by ECD-GC.

Comments. Benzenesulfonamide formed a di-PFB deriv completely in 20 min. *N*-Ethylbenzenesulfonamide was completely deriv in 15 min and the *N*-phenyl compd in 2 min. The shorter reaction time for the *N*-phenyl compd was presumed to be due to greater lipid character.

References

[1]O. Gyllenhaal and H. Ehrsson, *J. Chromatogr.* **107**, 327 (1975).

Derivative Type. Methyl.

Reagent. Trimethylbenzylammonium hydroxide (TMBAH).

Compound Type. Sulpyrid.

Reaction

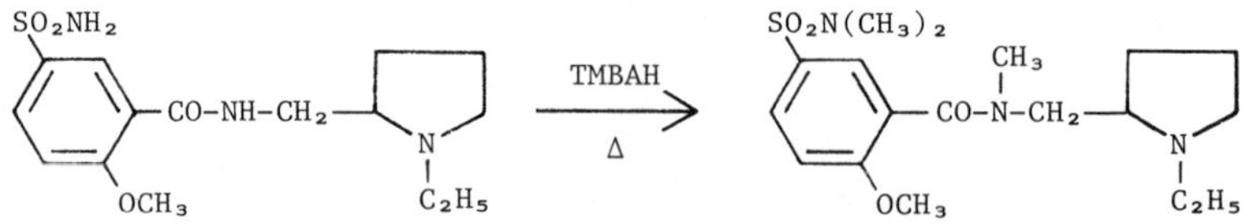

Typical Procedure. Mix 1 ml of urine or plasma with 1 ml of 0.1 *N* NaOH and ext with 2 × 5 ml EtOAc. Evap the ext to dryness under N_2 at 60° and diss the residue in 25 μl of 1 *M* TMBAH. Add 25 μl of chlorthalidone int stand (100 ng or 5 μg depending upon level being measured) in MeOH. Analyze 1–2 μl by GC-FID or GC-CI-MS-SIM.

Comments. The deriv was detectable to 200 pg by GC-CI-MS. MS data for the derivs of both compds were reported (chlorthalidone formed a tetramethyl deriv).

References

[1] A. Frigerio and C. Pantarotto, *J. Chromatogr.* **130**, 361 (1977).

Derivative Type. Trifluoroacetyl (TFA).

Reagent. Trifluoroacetic anhydride (TFAA)/trimethylamine (TMA).

Compound Type. Sulfonamides.

Reaction

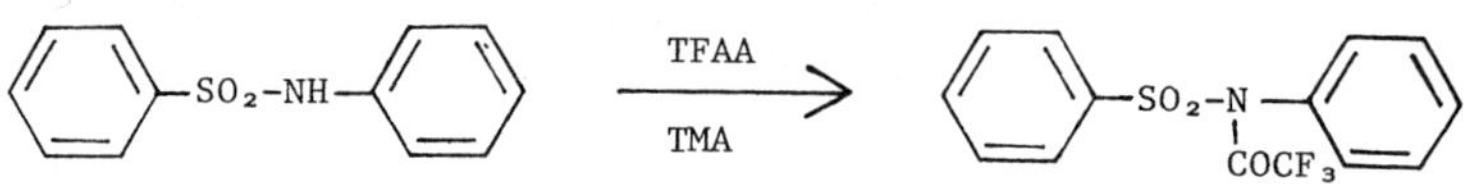

Typical Procedure. To 0.1 ml benzene containing 1.8–90 ng/ml *N*-phenylbenzenesulfonamide and 63 ng/ml 9-bromophenanthrene (int stand) add 5 μl TFAA and 10 μl 1 *M* TMA in benzene. After 20 min at 25° add 0.4 ml pH 7 phosphate buffer and mix. Analyze 2 μl of the org phase by ECD-GC.

Comments. The TFA and HFB derivs were prepared from benzenesulfonamide and *N*-ethyl- and *N*-phenylbenzenesulfonamide. Both derivs were more volatile than the parent compd and both had good ECD sensitivity.

References

[1] O. Gyllenhaal and H. Ehrsson, *J. Chromatogr.* **107**, 327 (1975).

8.3.2. Other Derivatives of Sulfonamides

Derivative Type. *N*-Dimethylaminomethylene.

Reagent. Dimethylformamide dimethylacetal.

Compound Type. Primary sulfonamides.

Reaction

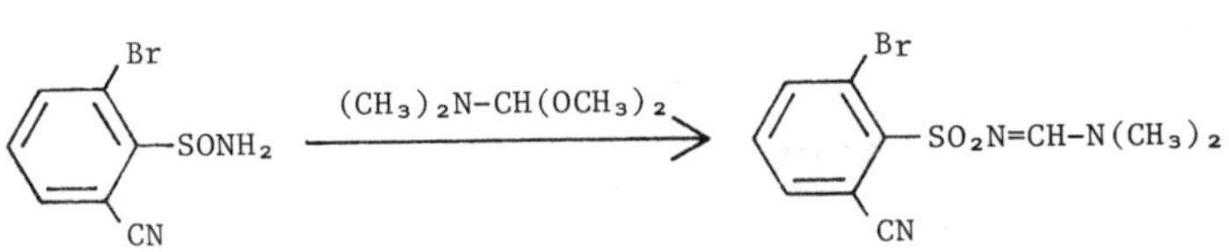

Typical Procedure. Mix 0.1–200 μg sulfonamide with 50–500 μl reagent and heat 10 min at 75–80°. Analyze by GC.

Comments. The derivs were compared with the TMS derivs (from BSTFA in pyridine) and the *N,N*-dimethyl derivs (from TMPAH in MeOH). The dimethylaminomethylene derivs had longer retention times, but their ease of preparation and lack of adsorptive properties made then attractive for GC use. The method was applied to analysis of 3-bromo-5-cyanobenzene-sulfonamide in blood using ECD-GC.

References

[1]W. J. A. VandenHeuval and V. F. Gruber, *J. Chromatogr.* **112**, 513 (1975).

8.4. DERIVATIVES OF SULFONYLUREAS

Derivative Type. Methyl

Reagent. Dimethylsulfate.

Compound Type. Tolbutamide and chlorpropamide.

Reaction

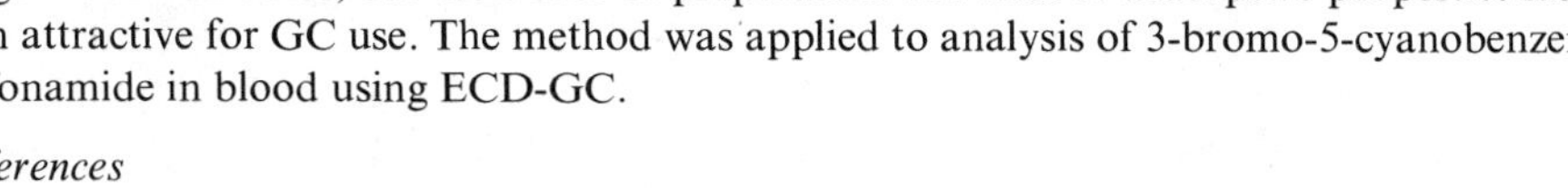

Typical Procedure. Diss the residue from evap of $CHCl_3$ ext in 1 ml K_2CO_3 in MeOH (made by mixing 1 ml of 5–10% aq K_2CO_3 in 9 ml MeOH) and add 0.1 ml dimethyl sulfate. Heat at 70° for 5 min. Evap MeOH and add 1 ml 0.2 *M* pH 5.6 acetate buffer and 5 ml heptane. Mix on vortex, centrifuge, and transfer 4 ml of heptane to another tube. Evap heptane and diss residue in 100 μl $CHCl_3$. Analyze 1–5 μl aliquots by GC.

Comments. The method was suitable for analysis of blood and other biological fluids. Tolbutamide was used as int stand for chlorpropamide and *vice versa*. Methylation was shown to occur on N by high resolution MS analysis.[2] The MS behavior of *N*-methyltolbutamide and *N*-methylacetohexamide have been thoroughly investigated, and it was noted that the MS of this class of compds required caution in interpretation due to a propensity of the compds to thermally decompose to the *N*-methylsulfonamides.[3]

References

[1]K. Sabih, *J. Pharm. Sci.* **59**, 782 (1970).

[2]K. Sabih, *Biomed. Mass Spectrom.* **1**, 163 (1974).

[3]A. D. Kossoy and J. L. Occolowitz, *Biomed. Mass Spectrom.* **5**, 123 (1978).

Derivative Type. Methyl.

Reagent. Diazomethane.

Compound Type. Sulfonylurea drugs (tolbutamide, chlorpropamide).

Reaction

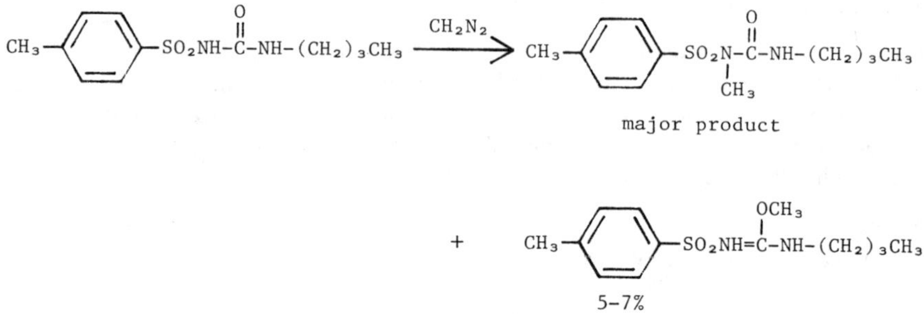

Typical Procedure. Prepare methyl deriv by treating sample of drug with excess ethereal CH_2N_2 for 30 min at room temp. Evap excess CH_2N_2 under N_2 and analyze Et_2O soln by GC.

Comments. The *N*-methyl deriv was shown to decompose in the GC injector to a temp dependent extent yielding *N*-methyl toluenesulfonamide. This decomposition was negligible when the injector was kept below 170°.[1] Matin and Knight analyzed the derivs by direct probe MS to avoid the problem of partial decomposition in the GC.[2] GC data have been reported for the methyl derivs of 15 sulfonamides prepared with CH_2N_2/Et_2O in dioxane.[3]

References

[1]K. K. Midha *et al.*, *Biomed. Mass Spectrom.* **3**, 100 (1976).

[2]S. B. Matin and J. B. Knight, *Biomed. Mass Spectrom.* **1**, 323 (1974).

[3]E. Röder and W. Stuthe, *Z. Anal. Chem.* **266**, 358 (1973).

Derivative Type. *N*-Methyl toluenesulfonamide.

Reagent. Dimethyl sulfate.

Compound Type. Tolbutamide.

Reaction

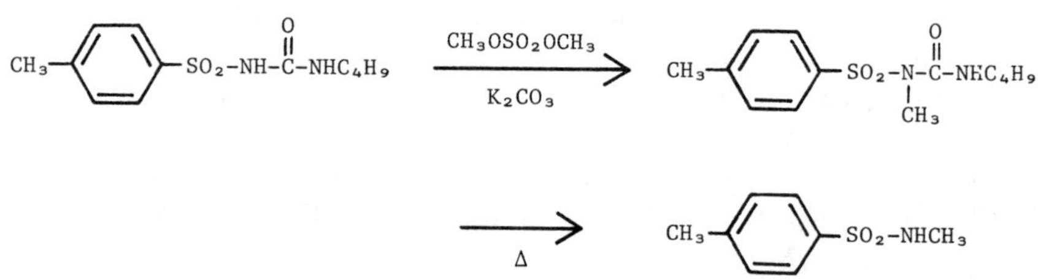

Typical Procedure. Acidify 1 ml serum with 2 ml 0.033 *M* H_3PO_4 and ext with 3 × 10 ml $CHCl_3$. Dry ext over Na_2SO_4 and partition residue between 2 ml MeOH-H_2O (9:1) and 3 ml pet ether. Discard org layer and repeat wash with 2 × 3 ml pet ether. Evap MeOH-H_2O phase to dryness under N_2 then under vac. Diss residue in 10 ml dry acetone, add 50 mg activated K_2CO_3 and 0.1 ml freshly dist dimethyl sulfate, and reflux 15 min. Conc to *ca* 1 ml and add 5 ml satd aq NaCl. Ext with 3 × 4.5 ml hexane, evap exts to dryness under N_2, and dry residue under high vac. Diss residue in 100 μl of int stand soln (0.6 mg/ml) and analyze by GC.

Comments. Pyrolysis of the deriv to *N*-methyl-*p*-toluenesulfonamide occurred in the injector portion of the column which was packed with glass wool and heated to 285°. *N*-Propyl-4-chlorobenzenesulfonamide was used as int stand. Activated K_2CO_3 was prepared by fine powdering with a mortar and pestle and heating to 125° for 3 hr followed by storage in a vac dessicator.

References

[1]D. L. Simmons *et al.*, *J. Chromatogr.* **71**, 421 (1972).

Derivative Type. *N*-Methyl, *N'*-trifluoroacetyl.

Reagent. Diazomethane, trifluoroacetic anhydride (TFAA).

Compound Type. Sulfonylureas.

Reaction

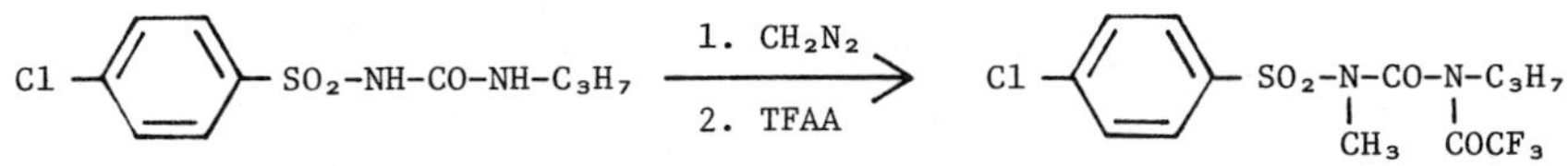

Typical Procedure. To 100 nmol of the sulfonylurea in 100 μl MeOH add 0.5 ml of an Et_2O soln of CH_2N_2. Allow to stand 5 min at room temp then evap under N_2. Diss residue in 50 μl EtOAc/pyridine (9 : 1, v/v) and add 50 μl TFAA. Heat at 65° for 30 min then evap under N_2. Diss residue in 100 μl cyclohexane and analyze by GC-MS or ECD-GC.

Comments. The PFP and HFB derivs were prepared by the same method using PFPA and HFBA. The methylation was also performed (including preparation of CD_3 derivs) with dimethyl sulfate.[2] GC and MS data were given for the derivs of five sulfonylurea drugs.

References

[1]W. E. Brazelton *et al.*, *Anal. Chem.* **48**, 1386 (1976).
[2]L. F. Prescott and D. R. Redman, *J. Pharm. Pharmacol.* **24**, 713 (1972).

Derivative Type. 2,4-Dinitroanilines.

Reagent. Aqueous sodium hydroxide, dinitrofluorobenzene (DNFB).

Compound Type. Tolbutamide and chlorpropamide.

Reaction

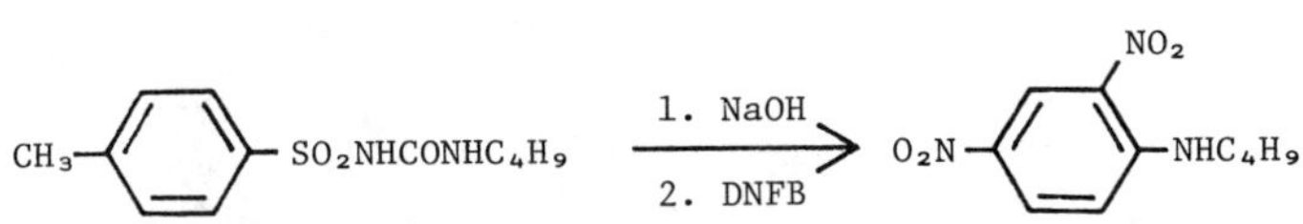

Typical Procedure. Mix 0.5 ml plasma or blood with 1 ml 2.55 *M* Na_2HPO_4 and 0.1 ml aq int stand soln, and ext with 10 ml 0.5% *i*-AmOH in hexane. Back ext 7 ml of the org phase with 1 ml 1 *N* NaOH. Acidify the aq phase with 1 ml 2 *N* HCl and ext with 2 ml pentyl acetate. Mix 1 ml of the ext with 0.1 ml 0.5% DNFB in pentyl acetate and heat 1 hr at 125°. Analyze 1–5 μl of the cooled soln by ECD-GC.

Comments. Chlorpropamide was used as int stand for tolbutamide and vice versa. For conc greater than 10 μg/ml, the deriv could be analyzed colorimetrically. The method was extended to the analysis of tolbutamide metabolites isolated by selective solvent extraction.

References

[1]S. B. Matin and M. Rowland, *J. Pharm. Pharmacol.* **25**, 186 (1973).
[2]S. B. Matin and M. Rowland, *Anal. Lett.* **6**, 865 (1973).

8.5. DERIVATIVES OF SULFATE ESTERS

Derivative Type. Methyl ester.

Reagent. Diazomethane.

Compound Type. Sugar sulfates (peracetylated).

Reaction

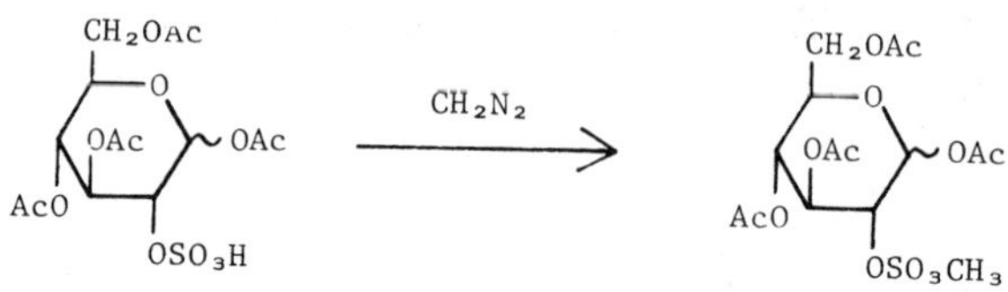

Typical Procedure (Preparative). Diss 0.1 g of the Ba salt of the acetylated sugar sulfate in 3 ml MeOH and apply to a column of Dowex 50W resin (H form). Elute with MeOH and treat eluate with CH_2N_2 in Et_2O. Allow to stand overnight in the refrigerator then purify on a column of silica gel using $CHCl_3$/acetone (97 :3, v/v). Analyze by MS.

Comments. Methylation of the monosaccharide sulfates imparted sufficient volatility for MS analysis. Acetylated sugar chlorosulfates were also prepared for MS analysis using sulfuryl chloride and pyridine.[2] MS data were reported for the derivs of eight acetylated sugar sulfates.[1]

References

[1]A. I. Kalinovski *et al.*, *Carbohydr. Res.* **58**, 473 (1977).

[2]H. J. Jennings and J. K. N. Jones, *Can. J. Chem.* **43**, 2372 (1965).

Derivative Type. Aryl acetates.

Reagent. Acetic anhydride (Ac_2O)/methanesulfonic acid.

Compound Type. Aryl sulfate esters.

Reaction

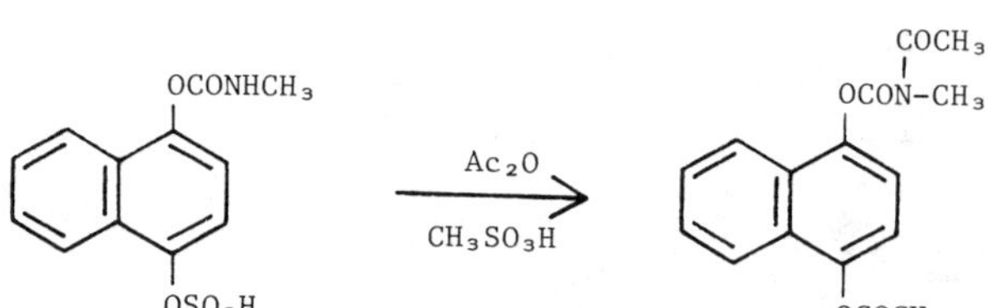

Typical Procedure. Mix 10 mg of the sulfate ester with 2 ml of 0° reagent consisting of 40 : 1 (v/v) Ac_2O/CH_3SO_3H. Heat to 100° for 30 min then cool to 0°. Add *ca* 2 g ice and ext 3 × with an equal vol of benzene. Conc the benzene ext to a small vol and analyze by GC.

Comments. The method was useful for characterization of metabolites that were unstable to the conditions normally used for hydrolysis of sulfate esters.

References

[1]G. D. Paulson and C. E. Portnoy, *J. Agr. Food Chem.* **18**, 180 (1970).

Derivative Type. Trifluoroacetyl.

Reagent. Trifluoroacetic anhydride (TFAA).

Compound Type. (3-Methoxy-4-sulfoxyphenyl)ethylene glycol (MOPEG sulfate).

Reaction

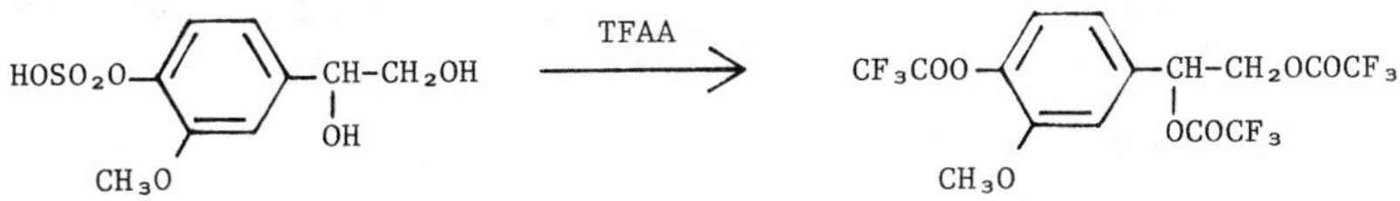

Typical Procedure. To 5 ml urine add *ca* 25 μg d_3-MOPEG sulfate in 100 μl H_2O and pass soln through a 10 × 2 cm column of Amberlite XAD-2. Wash column with 10 mol H_2O and 10 ml hexane then elute with 30 ml MeOH. Evap the MeOH under vac and diss residue in 2 ml $CHCl_3$/MeOH (1:1, v/v) containing 0.01 *M* NaCl. Apply to a 28 × 1 cm column of Sephadex LH-20 (5 g) prepared in and eluted with the same solvent. Evap the 70–100 ml effluent fraction and dry under vac over P_2O_5. Add 250 μl EtOAc and 350 μl TFAA and shake 2 hr at room temp. Centrifuge and remove the supernatant from the NaCl ppt. Evap under N_2 and diss residue in 20 μl EtO Ac/TFAA (95:5, v/v). Analyze by GC-MS-SIM.

Comments. The direct trifluoroacetylation of the sulfate conjugate eliminated the need for enzymic hydrolysis. The glucuronide conjugate was not cleaved by TFAA. A similar reaction has been observed with steroid sulfates which yielded the HFB deriv of the parent steroid upon treatment with HFBA in benzene (70°, 30 min).[2]

References

[1]S. Murray *et al.*, *J. Chromatogr.* **143**, 541 (1977).

[2]J. C. Touchstone and M. F. Dobbins, *J. Steroid Biochem.* **6**, 1389 (1975).

Derivative Type. Trimethylsilyl.

Reagent. Bis(trimethylsilyl)trifluoroacetamide (BSTFA).

Compound Type. Dopamine *O*-sulfate (3- and 4-isomers).

Reaction

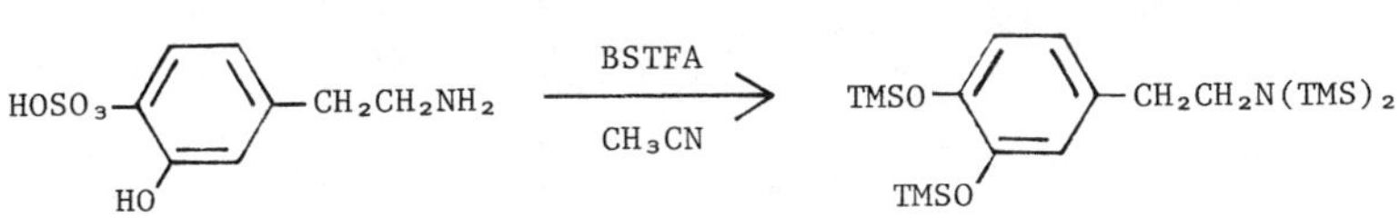

Typical Procedure. Conc 50 ml urine to 10 ml under vac and apply to a column of Dowex-1. Elute with NH_4OAc buffer and isolate eluted fractions by lyophilization. Dry the residues by addn and evap of CH_2Cl_2 at 60° under N_2. Add 50 μl BSTFA and 50 μl CH_3CN and heat 1 hr at 60°. Analyze by GC-MS.

Comments. Direct reaction of the sulfate esters with BSTFA gave hydrolysis and concomitant silylation of both phenolic groups. The tetra-TMS deriv was the major product but some tri-TMS (NHTMS) was observed. Sulfate ester cleavage and concomitant silylation has also been reported for isoetharine sulfate (BSA/pyridine, 60°, 1 hr).[2]

References

[1]R. L. Bronaugh *et al., J. Pharmacol. Exp. Ther.* **195**, 441 (1975).

[2]F. M. Williams *et al., Xenobiotica* **4**, 345 (1974).

Derivative Type. Acetyl, methyl, trimethylsilyl.

Reagent. Acetic anhydride, sodium hydroxide, dimethyl sulfate, HCl, bis(trimethylsilyl)-trifluoroacetamide (BSTFA).

Compound Type. Dopamine *O*-sulfate (3- and 4-isomers).

Reaction

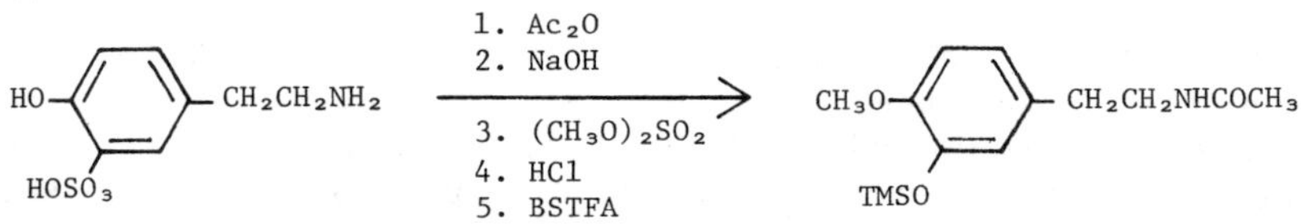

Typical Procedure. To 5–10 mg of the dopamine *O*-sulfate add 100 μl pyridine and 100 μl Ac_2O. Heat 5 min at 60° then evap under vac. Add 1 ml 0.1 *N* NaOH and heat 30 min at 60° (to deacetylate phenolic groups). Add 100 μl dimethyl sulfate and shake until soln complete. Adjust to pH 1 with 12 *N* HCl and reflux 1 hr at 95°. Ext with 3×5 ml EtOAc and dry ext over Na_2SO_4. Evap to dryness under N_2 then add CH_2Cl_2 and evap again under N_2 at 60° to remove traces of H_2O. Add 50 μl BSTFA and 50 μl CH_3CN. Heat 30 min at 60° then analyze by GC and GC-MS.

Comments. The method was used to distinguish the 3- and 4-*O*-sulfate isomers by comparison with the deriv of authentic 3-methoxytyramine.

References

[1]R. L. Bronaugh *et al., J. Pharmacol. Exp. Ther.* **195**, 441 (1975).

9. Derivatives for Chromatographic Separation of Optical Isomers

Derivatives that have been used for chromatographic separation of enantiomers are presented in this section. The majority of these derivatives have been for GC separation, but a few have been used with HPLC. Two general approaches are possible for chromatographic resolution of enantiomers: (a) the separation of the enantiomers by chromatography on an optically active stationary phase; and (b) the preparation of diastereomeric derivatives that can be separated on a nonchiral stationary phase. Only those derivatives used in the latter approach will be discussed here. The chromatographic separation of diastereomeric derivatives has been reviewed by Gil-Av and Nurok[1] and by Halpern[2]; the reader is referred in particular to the review by Gil-Av and Nurok[1] for information on mechanistic aspects of such separations. For a consideration of enantiomer resolution on optically active stationary phases the reader is referred to a review by Lochmuller and Souter.[3] Krull recently reviewed the liquid chromatographic resolution of enantiomers including the use of diastereomeric derivatives.[4]

The formation and separation of derivatives from an asymmetric sample and an asymmetric reagent are summarized in Figure 14. As can be seen from Figure 14, four possible optical isomers can result from the reaction of an asymmetric sample with an asymmetric reagent. The four derivative isomers consist of two diasteromeric pairs of chromatographically inseparable enantiomers. If an optically pure reagent is used (for example, (+)-Reagent) only two of the derivatives are formed (numbers 1 and 4) and the chromatographic separation effects a separa-

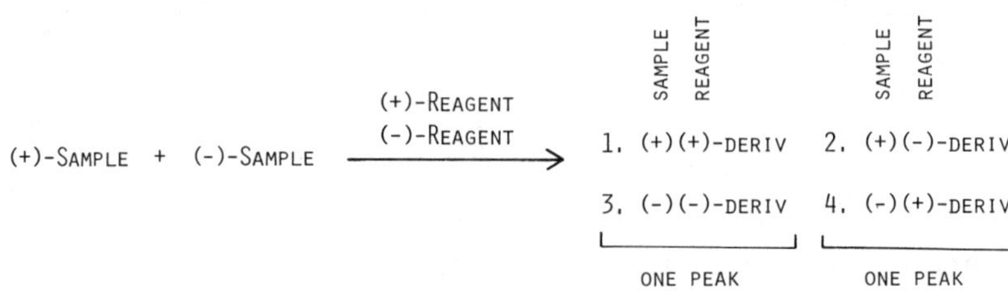

Figure 14. Products resulting from reaction of an asymmetric sample with an asymmetric reagent.

tion of the sample enantiomers. This approach derives from Pasteur's original method of enantiomer separation by fractional crystallization of diastereomeric derivatives, but utilizes the more efficient chromatographic separation process. Chromatographic separation of enantiomers via diastereomeric derivatives is used principally in determination of the enantiomeric composition of a sample compound. If an optically pure reagent is used, the enantiomeric ratio is reflected directly in the relative peak sizes. In some cases, where sufficient background information on the compound type is known, the absolute configuration of a sample compound may be determined. The technique can also be used to determine which compounds in a complex chromatogram are asymmetric. An example of this application has been reported in GC analysis of *N*-trifluoroacetyl amino acid butyl esters.[5] The use of racemic 2-butanol instead of the usual *n*-butanol yields derivatives with similar retention times, but each peak representing an asymmetric amino acid becomes a doublet due to formation of diastereomeric derivatives. With a racemic reagent, the two peaks of the doublet are of equal size regardless of the sample enantiomeric composition. As can be seen from Figure 14, if either the sample or the reagent is racemic, the two chromatographic peaks will be of equal size.

Ideally, an optimum derivative for chromatographic separation of enantiomers should be easy to prepare, and it must be stable (especially with respect to racemization) under the analysis conditions. The derivatization reaction must be essentially quantitative with either complete conservation or complete inversion of configuration of the asymmetric center and must not result in a significant kinetic resolution during the reaction. For determination of optical purity, the reagent must be optically pure, or at least of known composition. Finally, the derivative must give adequate chromatographic separation of the sample enantiomers. For maximum chromatographic separation, one wishes to maximize the "dissimilarity" between the two derivatives. Halpern[2] has summarized the factors involved in the effectiveness of a derivative for enantiomer separation. One factor is the conformational rigidity of groups attached to the asymmetric center; reagents with the asymmetric center as part of a ring system are particularly effective. Maximal differences in the size of groups attached to the asymmetric center improve separation. Minimal distance (ideally less than 3 atoms) between the asymmetric centers of the sample and reagent have the same effect. In addition, the nature of the functional group separating the asymmetric centers can influence separation. For example, it has been shown that an amide group separating the centers gives better separation than that obtained with an ester group.[6] Additional polar or polarizable groups in close proximity to the asymmetric centers and the intervening group yield improved separation presumably by enhancing interactions between these areas and the stationary phase. Such groups could also effect conformational rigidity by intramolecular effects.

Alcohols. Alcohols have been converted to diastereomeric derivatives with chiral acids and acyl reagents. The methods have included both direct esterification of chiral carboxylic acids (p. 409) and the use of acid chlorides (p. 410). Carbonate derivatives, prepared with menthylchloroformate, have been used for analysis of alcohols and hydroxylated fatty acid methyl esters (p. 412). Carbamate derivatives are formed by the reaction of alcohols with isocyanates. The isocyanate reagent prepared from optically active l-phenylethylamine has been used for analysis of enantiomeric alcohols (p. 413). By comparison to results obtained by application of these carbamate derivatives in a homologous series, the absolute configuration of the enantiomeric alcohols could be assigned. Carbamate derivatives have also been prepared by successive treatment of alcohols with phosgene and a chiral amine (p. 414). These derivatives were used for preparative HPLC but could also be used analytically.

Horeau's method for determination of absolute configuration has been adapted for microscale use (p. 414). This method utilizes the kinetic resolution of a racemic anhydride reagent in the reaction with an optically active alcohol. Determination of the enantiomeric composition of the remaining reagent via a diastereomeric amide allows assignment of the absolute configuration of the alcohol.

Amines. Amines have been converted to diastereomeric amides with several chiral acid chloride reagents for GC analysis (p. 415). Amide derivatives have also been prepared with an acylimidazole reagent (p. 417) and with a chiral acid using *N,N'*-carbonyldiimidazole as a coupling reagent (p. 417). In the latter case, the use of a pentafluorophenyl substituted acid permitted use of ECD-GC. *N*-Trifluoroacetyl-L-prolyl amides, prepared from the acid chloride reagents, were applied to the analysis of amines (p. 418) and to hydroxylamines after reduction with lithium aluminum hydride (p. 418). The *N*-pentafluorobenzoyl-L-prolyl derivatives allowed use of ECD-GC for analysis of small concentrations of enantiomeric amines (p. 419). Diastereomeric sulfonamide derivatives were obtained using a camphor sulfonyl chloride reagent (p. 419). Menthyl carbamates, prepared from menthyl chloroformate, have been used for amines and amino acid methyl esters (p. 412).

Carboxylic Acids. Enantiomeric carboxylic acids have been analyzed as diastereomeric ester and amide derivatives. Ester derivatives were prepared from (−)-menthol by acid catalyzed esterfication (p. 420). Menthyl esters were also prepared by converting the sample acid to the acid chloride with thionyl chloride (p. 420) or oxalyl chloride (p. 421) and treating the acid chloride with (−)-menthol. Diastereomeric amide derivatives have been prepared from the acid chlorides of the sample acids by reaction with chiral amines. Amines containing nitrobenzyl (p. 421) and naphthyl (p. 422) groups permitted HPLC analysis of the derivatives. Amides have also been prepared by coupling with a chiral amine using *N,N'*-dicyclohexylcarbodiimide (p. 422).

Amino Acids. Amino acids present two potential sites for reaction with a chiral reagent, and efforts have been directed toward both carboxyl derivatives and amino group derivatives. In either case, for successful GC analysis it is also necessary to derivatize the other function as well as any additional polar groups in the molecule. 2-Butyl esters were among the early derivatives examined and have been used in several studies with the amino groups derivatized as the trifluoroacetyl (p. 423) or pentafluoropropionyl (p. 425) amide. Some hydroxyl and sulfhydryl amino acids were not well resolved as the ditrifluoroacetyl derivatives but gave improved resolution after conversion to *O*- and *S*-acetyl derivatives (p. 424). Other 2-alkyl esters have also been examined (p. 425 and 426). It was observed that movement of the ester attachment toward the center of the longer chain alcohols led to decreased separations (p. 427). Menthyl esters were also among the earlier esters used in the analysis of amino acid enantiomers and gave the best separations as the *N*-trifluoroacetyl derivatives (p. 427). In a study of a series of esters, it was found that the 3,3-dimethyl-2-butyl esters, analyzed as *N*-trifluoroacetyl derivatives, gave the best separations (p. 427). In recent studies, the 3-methyl-2-

butyl esters were used as the *N*-trifluoroacetyl (p. 427) and *N*-pentafluoropropionyl (p. 428) derivatives.

Weygand *et al.*[7] reported the first GC separations of dipeptide diastereomers. The largest separation was obtained with *N*-trifluoroacetylprolylvaline methyl ester (*N*-TFA-Pro-Val-OCH_3). Other workers subsequently employed *N*-trifluoroacetyl-L-prolyl derivatives for GC separation of amino acid enantiomers as their methyl esters (p. 429). For hydroxylated amino acids, trimethylsilylation was also employed (p. 430). Sulfhydryl amino acids were converted to thiazolidines with formaldehyde prior to formation of the *N*-TFA-L-Pro methyl esters (p. 431). Thiazolidine carboxylic acid itself, used as an optically active derivatization reagent in the form of the *N*-trifluoroacetyl acid chloride, has been employed with both methyl and butyl esters of amino acids (p. 431). The converse approach, with the amino acid to be resolved forming the N-terminus of the dipeptide derivative, was used to resolve isovaline enantiomers (p. 432). The sample amino acid was converted to the *N*-trifluoroacetyl acyl chloride and used to acylate L-leucine isopropyl ester.

Nonpeptide diastereomeric amide derivatives have been utilized for chromatographic resolution. The α-methoxy-α-trifluoromethylphenylacetyl derivative of dihydroxyphenylalanine and its α-methyl analog permitted GC separation of amino acid enantiomers (p. 432). Methylation of the carboxyl and phenol groups was accomplished in one step with diazomethane. In one of the few HPLC methods reported, the optically active α-methoxy-α-methyl-1-naphthaleneacetic acid was coupled to amino acid methyl esters with dicyclohexylcarbodiimide and the derivatives used to separate the enantiomers of 17 amino acids (p. 433). A second HPLC method employed *p*-nitrobenzyl esters for detectability and *N*-camphorsulfonyl derivatives to form diastereomers (p. 433). The final subgroups of amino acid derivatives, the diketopiperazines, have been separated both as trimethylsilyl (p. 434) and trifluoroacetyl (p. 435) derivatives.

Carbonyl Derivatives. The diastereomeric ketal derivatives formed with the enantiomers of camphor and optically active 2,3-butanediol were the first nonpeptide diastereomers whose separation by GC was demonstrated (p. 435). Efforts to extend this derivative to other ketones were unsuccessful.[8] A similar ortho ester derivative prepared with the same diol has been applied to enantiomeric δ-lactones (p. 435). The ortho esters were formed using 2,3-butanediol and trimethylorthoformate with acid catalysis. A chiral hydrazone reagent, 2,2,2-trifluoromethyl-1-phenylethylhydrazine was used for resolution of ketones with an asymmetric α-carbon (p. 436). The method was not ideal, however, since ketones containing α-hydrogen could undergo racemization. A "general" technique for carbonyl compounds has yet to be devised.

References

1. E. Gil-Av and D. Nurok, *Adv. Chromatogr.* **10**, 99 (1974).
2. B. Halpern, in *Handbook of Derivatives for Chromatography*, ed. by K. Blau and G. King, Heyden, London, 1977, p. 457.
3. C. H. Lochmuller and R. W. Souter, *J. Chromatogr.* **113**, 283 (1975).
4. I. S. Krull, *Adv. Chromatogr.* **16**, 175 (1978).
5. J. K. Whelan, *J. Chromatogr.* **111**, 337 (1975).
6. B. L. Karger *et al.*, *Chem. Commun.* 625 (1969).
7. F. Weygand *et al.*, *Angew. Chem. Int. Ed.* **2**, 183 (1963).
8. W. E. Pereira *et al.*, *Austr. J. Chem.* **24**, 1103 (1971).

9.1. DERIVATIVES OF ALCOHOLS

9.1.1. Ester Derivatives

Derivative Type. L-(+)-Mandelic ester.

Reagent. L-(+)-Mandelic acid.

Compound Type. 2-Alkanols.

Reaction

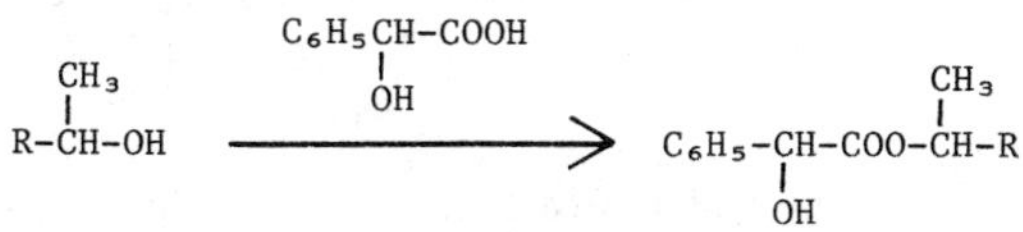

Typical Procedure.[2] Diss the alcohol sample and 1.1 mol equiv of the mandelic acid in benzene and reflux until the theoretical amt of H_2O is collected in a Dean-Stark trap. Wash soln with satd aq $NaHCO_3$ and H_2O, dry over Na_2SO_4, and analyze by GC.

Comments. GC separation of the diastereomeric esters was demonstrated for C_4-C_3 2-alkanols.[1]

References

[1]J. M. Cross *et al.*, *J. Chromatogr. Sci.* **8**, 679 (1970).

[2]H. C. Rose *et al.*, *Anal. Chem.* **38**, 469 (1966).

Derivative Type. α-Acetoxy-L-propionates.

Reagent. Lactic acid/toluenesulfonic acid (TsOH), acetyl chloride (AcCl).

Compound Type. 2-*n*-Alkanols.

Reaction

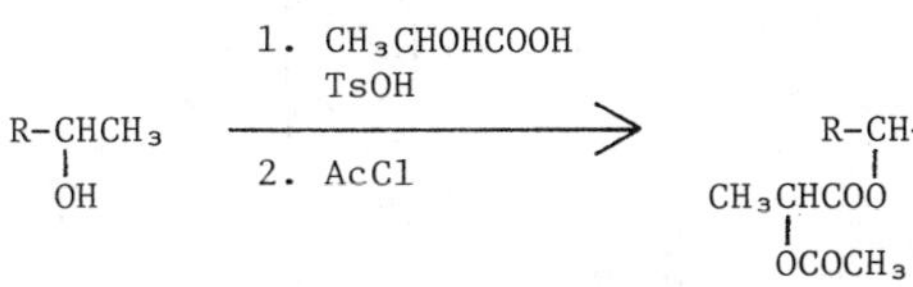

Typical Procedure. Heat the alcohol with lactic acid in benzene using TsOH as a catalyst and distil until no more H_2O is formed. Add AcCl and reflux 4 hr. Wash with 30% Na_2CO_3 and H_2O dry over Na_2SO_4, and distil under vac. Analyze the derivs by GC.

Comments. The derivs were used for GC separation of the alcohol enantiomers. α-Propionoxy and α-butyroxy derivs were also examined but did not show any significant differences in enantiomer resolution from the acetoxy derivs. In the series of C_4–C_{11} 2-*n*-alkanols it was found that the derivs of the (+)-alkanols consistently had longer retention times than those of the (−)-alkanols.[3]

References

[1]E. Gil-Av *et al.*, *J. Gas Chromatogr.* **4**, 51 (1966).

[2]E. Gil-Av and D. Nurok, *Proc. Chem. Soc.*, 146 (1962).

[3]R. Charles-Sigler *et al.*, *Israel J. Chem.* **1**, 234 (1963).

Derivative Type. 3β-Acetoxy-Δ^5-etienate.

Reagent. 3β-Acetoxy-Δ^5-etienic acid chloride/pyridine.

Compound Type. Enantiomeric alcohols.

Reaction

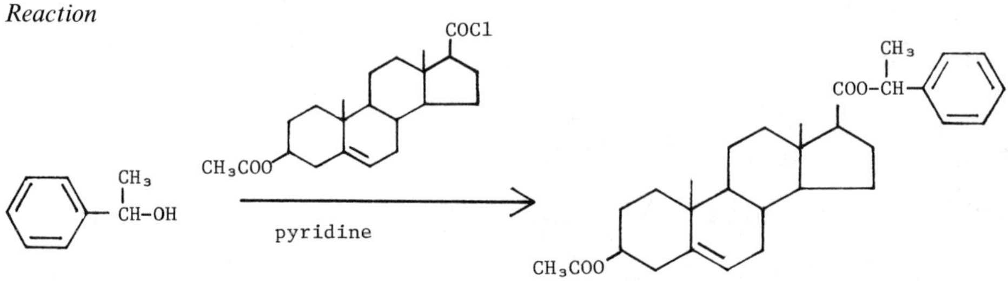

Typical Procedure. Diss 10 μmol of the carbinol, 50 μmol of 3β-acetoxy-Δ^5-etienic acid chloride, and 10 μl pyridine in 1 ml benzene and reflux 1 hr. Dil with 2 ml benzene and analyze by GC.

Comments. The method was tested on six enantiomeric alcohols and gave separations comparable to those obtained with the menthyl carbonate derivs.

References

[1]M. W. Anders and M. J. Cooper, *Anal. Chem.* **43**, 1093 (1971).

Derivative Type. Drimanoyl and chrysanthemoyl esters and amides.

Reagent. Drimanoyl chloride, chrysanthemoyl chloride.

Compound Type. Alcohols and amines.

Reaction

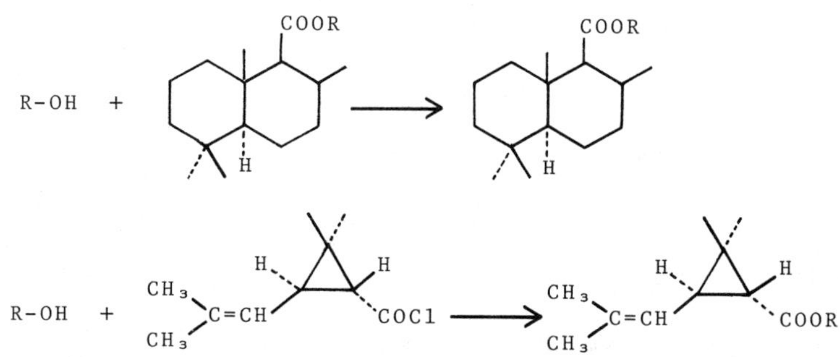

Typical Procedure. Diss 1 mg of alcohol or amine in 20 μl dry toluene and add 10 μl of a soln of the acyl chloride (3 molar proportions) in toluene. Heat at 60° (drimanoyl) or 40° (chrysanthemoyl) for 1–2 hr. Analyze directly by GC.

Comments. The drimanoyl (top) and chrysanthemoyl (bottom) derivs were used for GC resolution of enantiomeric alcohols and amines. The derivs were chosen because of substantial conformational rigidity and relatively small molecular weight increment. GC and MS data were reported for a variety of amines and alcohols.

The acid chlorides were prepared by treating 2 mg of the acid in 100 μl dry toluene with 200 μl redistilled thionyl chloride. After heating at 60° for 1 hr the excess was evapd under N_2 and the product used without further purification.

References

[1]C. J. W. Brooks *et al.*, *Anal. Chem.* **45**, 896 (1973).

Derivative Type. D-2-Phenylpropionate.

Reagent. D-2-Phenylpropionyl chloride.

Compound Type. Hydroxyacid methyl esters and alkanols.

Reaction

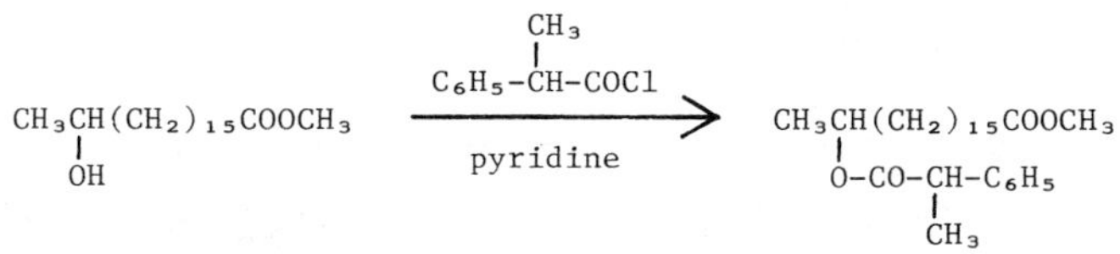

Typical Procedure. To 100 μg of hydroxy compd add 80 μl of 0.5 μmol/ml soln of D-2-phenylpropionyl chloride in benzene and 20 μl dry pyridine. Allow to stand 2 hr at room temp in a dessicator; then evap to dryness and diss residue in 100 μl $CHCl_3$. TLC on silica gel with $CHCl_3$ and elute deriv (located with UV light after spraying with 2′,7′-dichlorofluorescein) with EtOAc. Analyze by GC or GC-MS.

Comments. The deriv was used to resolve stereoisomers of 3-, ω4-, ω3, and ω2-hydroxy fatty acid methyl esters and several positional isomers of aliphatic alcohols. The *N*-(1-phenylethyl)-carbamate derivs[2] were found to be not suitable for steric analysis of 2- and 3-hydroxy acids. The synthesis of the reagent was reported.[1]

References

[1]S. Hammerström and M. Hamberg, *Anal. Biochem.* **52**, 169 (1973).

[2]M. Hamberg, *Chem. Phys.Lipids* **6**, 152 (1971).

Derivative Type. (+)-α-Methoxy-α-trifluoromethylphenylacetate.

Reagent. (+)-α-Methoxy-α-trifluoromethylphenylacetyl chloride (MTP-AcCl).

Compound Type. Enantiomeric alcohols.

Reaction

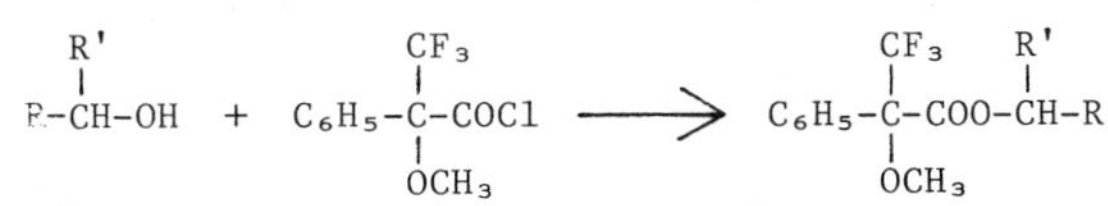

Typical Procedure.[1] Prepare the reagent by treatment of α-methoxy-α-trifluoromethyl-phenylacetic acid with $SOCl_2$ under reflux for 50 hr. Distil (54–56°, 1 mm) the MTP-AcCl and add an equimolar amt to the alcohol (0.15 mmol) in 5 drops CCl_4 and 5 drops pyridine. Allow to stand 12 hr then add 1 ml H_2O and transfer to a sep funnel with 20 ml Et_2O. Wash Et_2O soln with dil HCl, satd aq Na_2CO_3, and H_2O. Dry over $MgSO_4$ and evap to yield the deriv.

Comments. The deriv was developed for analysis of optical purity of secondary alcohols by NMR.[1,2] The deriv was also used for preparative HPLC separation of the enantiomers of an intermediate in abscisic acid synthesis[2] and thus should also be useful for analytical HPLC.

References

[1]J. A. Dale *et al.*, *J. Org. Chem.* **34**, 2543 (1969).

[2]J. A. Dale and H. S. Mosher, *J. Am. Chem. Soc.* **90**, 3732 (1968).

[3]M. Koreeda *et al.*, *J. Am. Chem. Soc.* **95**, 239 (1973).

Derivative Type. (−)-Menthoxyacetate.

Reagent. (−)-Menthoxyacetyl chloride/pyridine.

Compound Type. Benzo[a]pyrene dihydrodiol.

Reaction

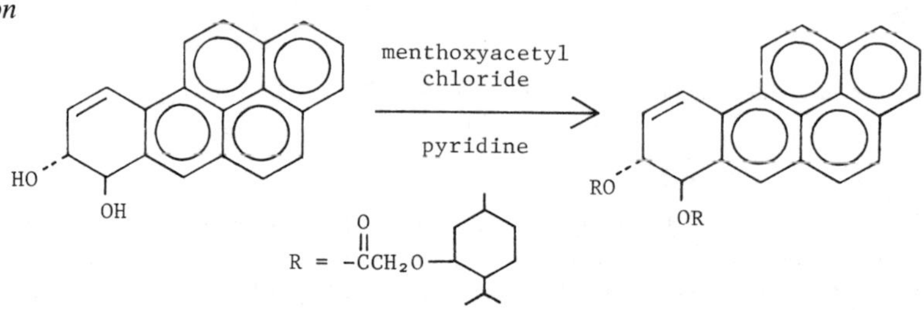

Typical Procedure.[2] Diss dihydrodiol in pyridine and add with cooling 2 mol equiv (−)-menthoxyacetyl chloride. Allow to react overnight; then pour into dil HCl. Collect the ppt, wash with H_2O, dry, and cryst from lt pet ether. Analyze by HPLC.

Comments. The deriv was used to separate the enantiomers of the dihydrodiol by HPLC. The dihydrodiols could be regenerated from the resolved derivs by treatment with NH_3 in MeOH at 0°.

References

[1]R. G. Harvey and H. Cho, *Anal. Biochem.* **80**, 540 (1977).

[2]J. Booth *et al., J. Chem. Soc.* 2808 (1950).

9.1.2. Carbonate Derivatives

Derivative Type. (−)-Menthoxycarbonyl.

Reagent. (−)-Menthylchloroformate.

Compound Type. Hydroxy fatty acid methyl esters.

Reaction

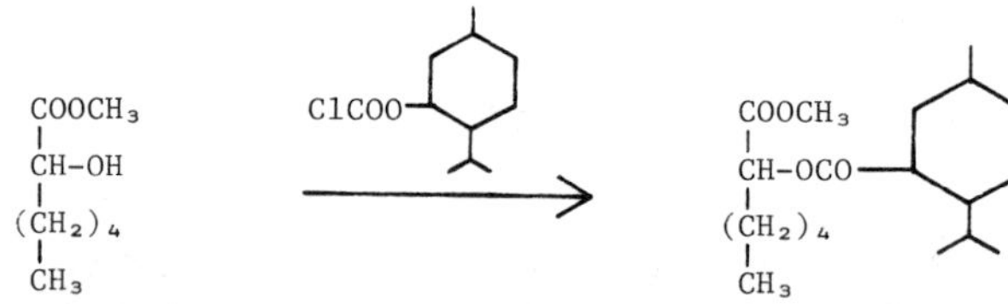

Typical Procedure.[1] To *ca* 10 μmol methyl ester in 40 μl benzene, add 12 μl pyridine and 60 μl of a 1 *M* soln of (−)-menthylchloroformate in toluene. Allow to stand 18 hr at room temp. TLC on silica gel with dioxane/benzene (3:97, v/v) and elute spots with Et_2O. Analyze by GC.

Comments. The method was used to determine the absolute configuration of hydroxylated fatty acids. Hammerstrom applied the method to 2-hydroxy acids from brain cerebrosides.[2] GC data have been reported for the derivs of 13 hydroxy acid methyl esters.[3] In addition to hydroxy acids, the reagent has been used to prepare diastereomeric derivs of enantiomeric alcohols, amines, and amino acid methyl esters.[4]

References

[1]M. Hamberg, *Anal. Biochem.* **43**, 515 (1971).

[2]S. Hammerstrom, *FEBS Lett.* **5**, 192 (1969).

[3]R. G. Arnett and P. K. Stumpf, *Anal. Biochem.* **47**, 638 (1972).
[4]J. W. Westley and B. Halpern, *J. Org. Chem.* **33**, 3978 (1968).

9.1.3. Carbamate Derivatives

Derivative Type. 1-Phenylethyl carbamate.

Reagent. R-(+)-1-Phenylethylisocyanate.

Compound Type. Enantiomeric alcohols.

Reaction

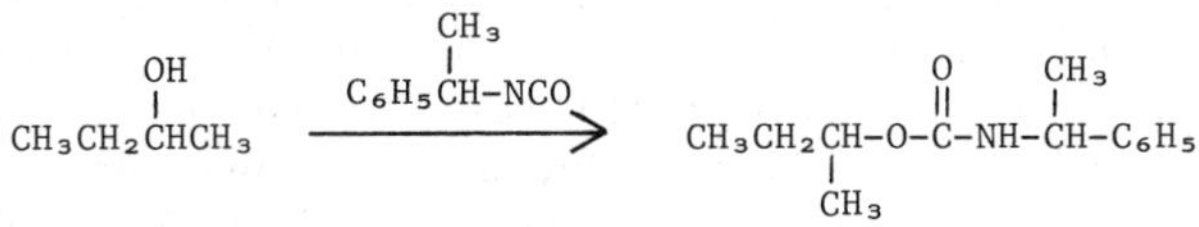

Typical Procedure. Mix 1–1.5 mg of the alcohol with 1.5 mg of the reagent and heat in a sealed tube at 110° for 7 hr. Add 0.5 ml MeOH to react with the excess reagent and analyze an aliquot of the soln by GC.

Comments. The preparation of the reagent from R-(+)-1-phenylethylamine was described. GC data for the derivs of 14 alcohols were reported. The relative retention times of the diastereomeric derivs of a homologous series of secondary alcohols could be used to assign the absolute configurations. The derivs have been used for TLC separations[2] and should also be useful for HPLC.

References

[1]W. Pereira *et al.*, *Anal. Lett.* **3**, 23 (1970).
[2]M. Freytag and K. H. Ney, *J. Chromatogr.* **41**, 473 (1969).

Derivative Type. *N*-(1-Phenylethyl)carbamate.

Reagent. R-1-Phenylethyl isocyanate.

Compound Type. ω2-Hydroxyacids (as methyl esters) and 2-alkanols.

Reaction

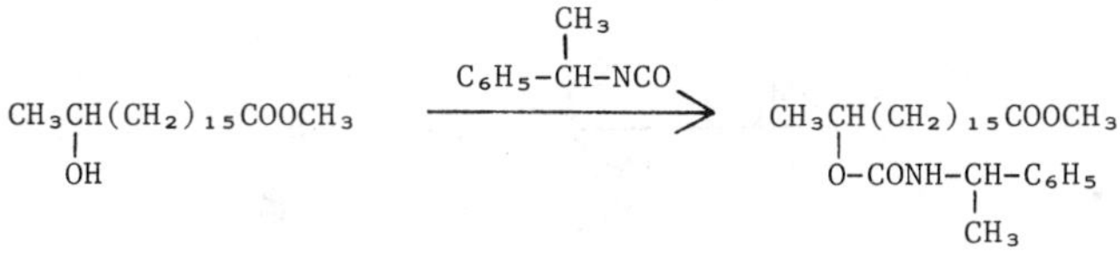

Typical Procedure. Add 0.1 ml of a 0.3 mmol/ml soln of R-1-phenylethyl isocyanate in toluene to 3–7 μmol of sample and heat at 120° for 3 hr under Ar. Evap to dryness under vac and diss residue in EtOAc. Analyze by GC.

Comments. The method was used to resolve stereoisomers of the hydroxy compds by GC resolution of the diastereomeric derivs. The preparation of the reagent was reported. The menthyl formate esters of the compds prepared from (−)-menthyl chloroformate could not be separated by GC.

References

[1]M. Hamberg, *Chem. Phys. Lipids* **6**, 152 (1971).

Derivative Type. *N*-2,2,2-Trifluoro-1-phenylethyl carbamate.

Reagent. Phosgene/2,2,2-trifluoro-1-phenylethylamine.

Compound Type. Alcohols.

Reaction

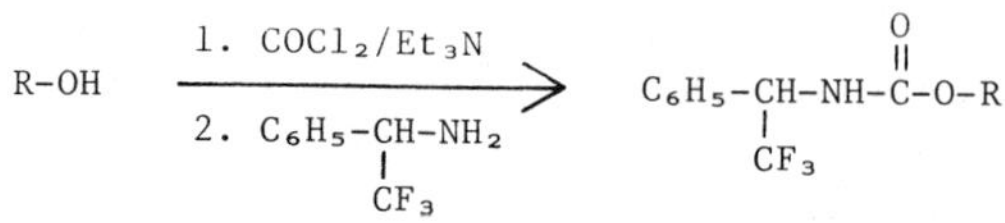

Typical Procedure.[2] To a soln of 1.47 g (15 mmol) phosgene in 15 ml dry toluene cooled to −5 add dropwise a soln of the racemic alcohol (*ca* 7.5 mmol) and Et_3N (7.5 mmol) in 25 ml dry toluene. Stir 30 min at 0° then filter under N_2 to remove Et_3N-HCl. Conc under vac at 40–50° and diss residue in 25 ml CH_2Cl_2. Add to this soln under N_2 with stirring a soln of 7.5 mmol Et_3N and 7.5 mmol 2,2,2-trifluoro-1-phenylethylamine in 25 ml CH_2Cl_2. Stir overnight at room temp then wash with 2 × 50 ml 3 *N* HCl. Dry over $MgSO_4$ and conc under vac at 30°. Separate by HPLC.

Comments. The method was used for preparative separation of enantiomers by HPLC but could be used for analytical purposes. The 2,2,2-trifluoro-1-phenylethylamine derivs showed greater chromatographic separability than the 1-phenylethylamine and 1-(1-naphthyl)-ethylamine derivs.[1]

References

[1] W. H. Pirkle and J. R. Hauske, *J. Org. Chem.* **42**, 2436 (1977).

[2] W. H. Pirkle and J. R. Hauske, *J. Org. Chem.* **42**, 1839 (1977).

9.1.4. Determination of Absolute Configuration

Derivative Type. α-Phenylbutyric ester/amide.

Reagent. (±)-α-Phenylbutyric anhydride/(+)-R-α-phenylethylamine.

Compound Type. Secondary alcohols.

Reaction

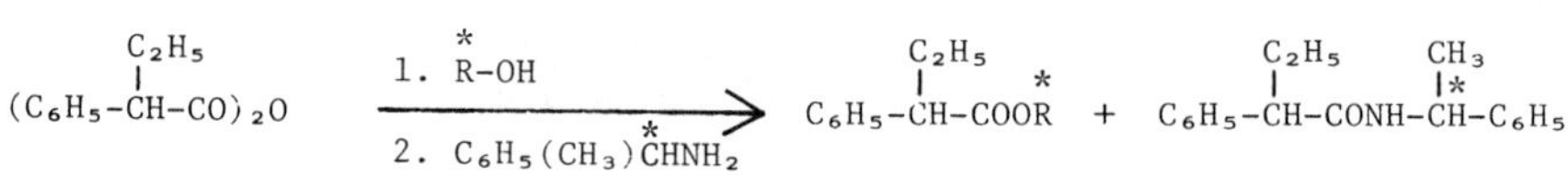

Typical Procedure. Diss 10 μmol of the alcohol in 7 μl pyridine and add 20 μmol (±)-α-phenylbutyric anhydride. Seal in vial and maintain at 40° for 1.5 hr. Carry out a parallel reaction with cyclohexanol. Add 6 μl (+)-R-α-phenylethylamine and mix thoroughly. After 15 min dil with 400 μl EtOAc and analyze by GC at 215° on a 3 m 1% OV-17 column. Determine the relative proportions of the amides of (−)-R- and (+)-S-α-phenylbutyric acid by peak areas (retention indices 2390 and 2420). Determine the percentage area of the peak from the (−)-R-acid and subtract the corresponding value from the cyclohexanol reaction. Use this value to determine the absolute configuration of the starting alcohol.

Comments. The method was a microscale adaptation of Horeau's rule[2,3] for determination of absolute configuration. The rule states that alcohols of type (I) preferentially react with

(R)-α-phenylbutyryl groups causing the remaining excess of anhydride to be deficient in this form. The acid chloride was found to be less satisfactory than the anhydride.

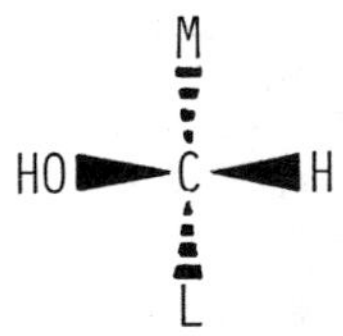

M = GROUP OF MEDIUM SIZE

L = GROUP OF LARGE SIZE

(I)

References

[1]C. J. W. Brooks and J. D. Gilbert, *Chem. Commun.* 194 (1973).

[2]A. Horeau, *Tetrahedron Lett.* 506 (1961).

[3]A. Horeau and H. B. Kagan, *Tetrahedron* **20**, 2431 (1964).

9.2. DERIVATIVES OF AMINES AND HYDROXYLAMINES

9.2.1. Amide Derivatives

Derivative Type. *O*-Methylmandelamides.

Reagent. *O*-Methylmandelyl chloride.

Compound Type. Enantiomeric amines.

Reaction

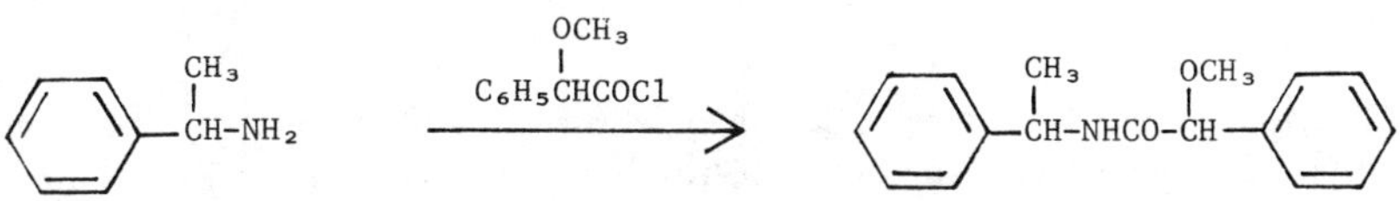

Typical Procedure. Mix 50 μl α-phenylethylamine with a 20% excess of the reagent in dioxane containing 70 μl Et_3N. Allow to react at room temp then diss the reaction mixt in Et_2O/benzene and wash with 1 *N* acid and 1 *N* base. Evap the solvent and diss in isooctane/EtOAc (4 : 1, v/v) at a conc of 10 mg/ml. Analyze by HPLC.

Comments. The reagent was prepared by refluxing (+)-(S)-*O*-methylmandelic acid with $SOCl_2$. The method was used for enantiomeric analysis of amines by HPLC. Amide derivs were also used for analysis of optically active carboxylic acids.[2,3] Feigl and Mosher[4] used the *O*-methylmandeloyl ester for GC separation of the enantiomers of 1,1,1-trifluoro-2-propanol.

References

[1]G. Helmchen and W. Strubert, *Chromatographia* **7**, 713 (1974).

[2]G. Helmchen *et al.*, *Tetrahedron Lett.* 3873 (1972).

[3]G. Helmchen *et al.*, *Tetrahedron Lett.* 1417 (1977).

[4]D. M. Feigl and H. S. Mosher, *Chem. Commun.* 615 (1965).

Derivative Type. α-Phenylbutyramide.

Reagent. α-Phenylbutyryl chloride or anhydride.

Compound Type. Enantiomeric amines and amino acid methyl esters.

Reaction

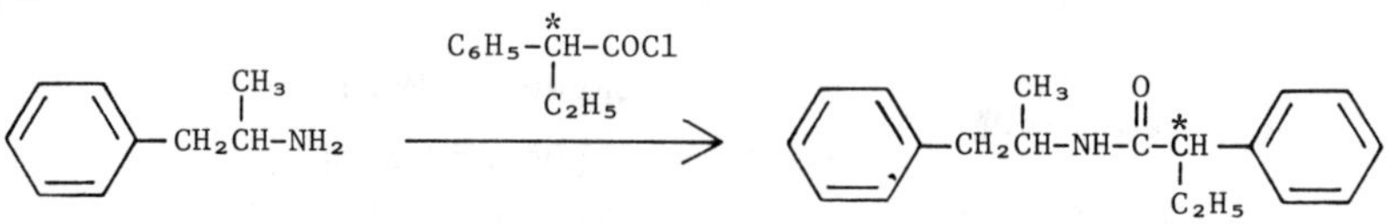

Typical Procedure. Treat 10 μmol of amine or amino acid methyl ester with an equimolar amt of the acid chloride or anhydride in 50 μl of toluene at room temp for 10 min. Dil with 1 ml EtOAc and analyze by GC or GC-MS.

Comments. The derivs were used to resolve enantiomeric amines and amino acids as diastereomeric derivs. α-Phenylpropionamide and α-chlorophenylacetamide derivs were also examined. GC and MS data were reported for six amines and five amino acids.

References

[1]M. T. Gilbert *et al., Biomed. Mass Spectrom.* **1**, 274 (1974).

Derivative Type. α-Methoxy-α-(trifluoromethyl)phenylacetyl.

Reagent. (−)-α-Methoxy-α-(trifluoromethyl)phenylacetyl chloride.

Compound Type. Enantiomeric amines.

Reaction

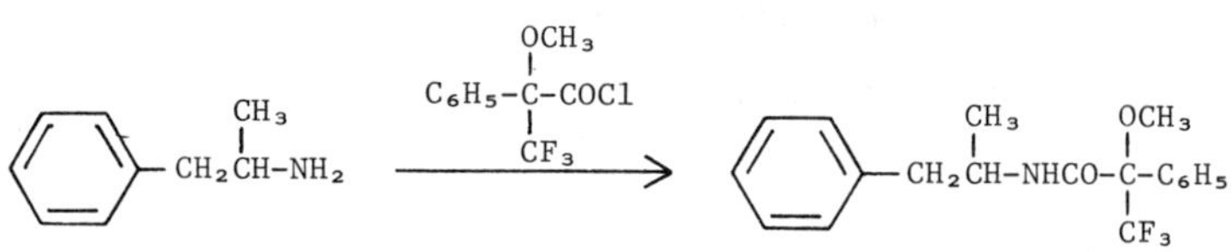

Typical Procedure. Prepare the reagent soln by refluxing for 50 hr a mixt of 1.0 g α-methoxy-α-(trifluoromethyl)phenylacetic acid in 3 ml freshly dist $SOCl_2$. Evap under vac and diss in 10 ml 1,2-dichloroethane. Mix 0.1–0.2 mg of the amine or amine salt with 50 μl 1,2-dichloroethane, 50 μl of reagent soln, and 50 μl pyridine. Heat at 70° for 30 min then cool in an ice bath. Add 1 ml 1 *N* HCl and shake. Separate the org layer, add 100 μl 1,2-dichloroethane, and wash with 0.5 ml 15% aq Na_2CO_3. Analyze by GC.

Comments. The derivs of nine amphetamine-like amines were examined; in seven cases GC conditions were found to separate the enantiomers.[1] The method was applied to an amine excreted in urine.[1] The deriv has also been used for analysis of the products of an asymmetric amine synthesis.[2] MS data for the deriv of amphetamine and several related amines have been reported.[3]

References

[1]J. Gal, *J. Pharm. Sci.* **66**, 169 (1977).

[2]D. E. Nichols *et al., J. Med. Chem.* **16**, 480 (1973).

[3]J. Gal, *Biomed. Mass Spectrom.* **5**, 32 (1978).

Derivative Type. *N*-(R)-α-Phenylbutyramide, *O*-trimethylsilyl.

Reagent. (R)-α-Phenylbutyrylimidazole, bis(trimethylsilyl)acetamide (BSA).

Compound Type. Ephedrines.

Reaction

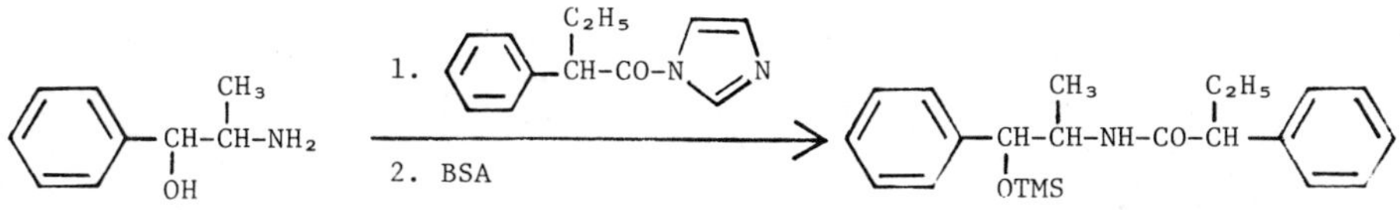

Typical Procedure. Diss 10 μmol (R)-α-phenylbutyric acid in 50 μl dry $CHCl_3$ and add 10 μmol *N,N'*-carbonyldiimidazole. After 10–15 min at room temp add 10 μmol of the amine. Allow to stand 2.5 hr at room temp then shake with 50 μl H_2O for 1 min. Centrifuge, remove $CHCl_3$ layer, and dry it over Na_2SO_4. Evap to dryness under N_2 and diss residue in BSA. Heat at 60° for 15 min then analyze by GC.

Comments. The derivs permitted GC separation of the enantiomers of ephedrine, ψ-ephedrine, and nor-ψ-ephedrine but not norephedrine on OV-1 and OV-17 columns. EI and CI MS data were reported for the derivs.

References

[1]M. T. Gilbert and C. J. W. Brooks, *Biomed. Mass Spectrom.* **4**, 226 (1977).

Derivative Type. α-Methyl-α-methoxypentafluorophenylacetamide.

Reagent. α-Methyl-α-methoxypentafluorophenylacetic acid/*N,N'*-carbonyldiimidazole (CDI).

Compound Type. β-Arylethylamines.

Reaction

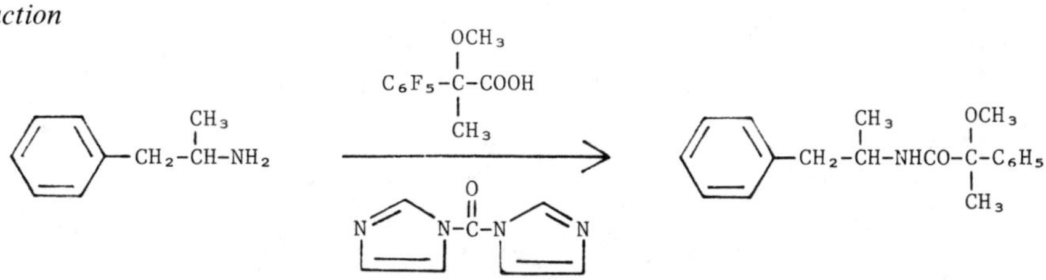

Typical Procedure. Diss 140 mg (0.52 mmol) α-methyl-α-methoxypentafluorophenylacetic acid in 5 ml anhyd THF and add 100 mg (0.62 mmol) CDI in 5 ml anhyd THF. Stir at room temp 19 hr; then add 0.52 mmol of the amine in 1 ml anhyd THF. Stir 19 hr at room temp evap under vac. Diss residue in Et_2O; wash with aq HCl, aq $NaHCO_3$, and H_2O; dry over $MgSO_4$, and evap to give a white solid deriv. Analyze by GC and NMR.

Comments. The reagent was designed for both NMR and GC analysis of enantiomeric composition of amines. The pentafluorophenyl group was chosen for ECD sensitivity and volatility. Lacking an α-hydrogen, the reagent was resistant to racemization.

References

[1]L. R. Pohl and W. I. Trager, *J. Med. Chem.* **16**, 475 (1973).

9.2.2. N-Acyl-L-prolyl Derivatives

Derivative Type. *N*-Trifluoroacetyl-L-prolyl.

Reagent. *N*-Trifluoroacetyl-L-prolyl chloride.

Compound Type. 1-Phenylisopropylamines.

Reaction

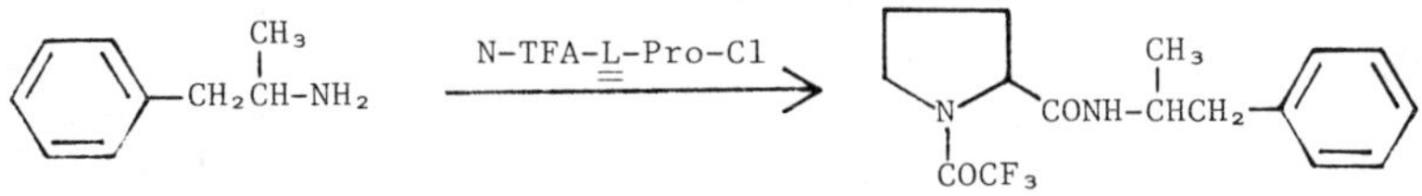

Typical Procedure. To 50 μl of a soln of the amine in $CHCl_3$ add 10 μl of a 0.1 *M* soln of *N*-TFA-L-Pro-Cl in $CHCl_3$. Allow to stand 10–15 min then analyze 1–2 μl by GC.

Comments. The method was used to separate the enantiomers of 17 ring- and *N*-alkylated amphetamines,[1] and ephedrine and pseudoephedrine and their nor- and *N*-ethyl analogs.[2] Resolution was good for all but pseudoephedrine. Gunne[3] used the deriv to separate amphetamine enantiomers in urine. The deriv has also been used for other enantiomeric amines.[4,5]

References

[1] A. H. Beckett and B. Testa, *J. Chromatogr.* **69**, 285 (1972).
[2] A. H. Beckett and B. Testa, *J. Pharm. Pharmacol.* **25**, 382 (1973).
[3] L. M. Gunne, *Biochem. Pharmacol.* **16**, 863 (1967).
[4] A. Murano, *Agr. Biol. Chem.* **37**, 981 (1973).
[5] J. C. Dabrowiak and D. W. Cooke, *Anal. Chem.* **43**, 1093 (1971).

Derivative Type. *N*-Trifluoroacetyl-L-prolyl.

Reagent. Lithium aluminum hydride, *N*-trifluoroacetyl-L-prolyl chloride.

Compound Type. Hydroxylamines.

Reaction

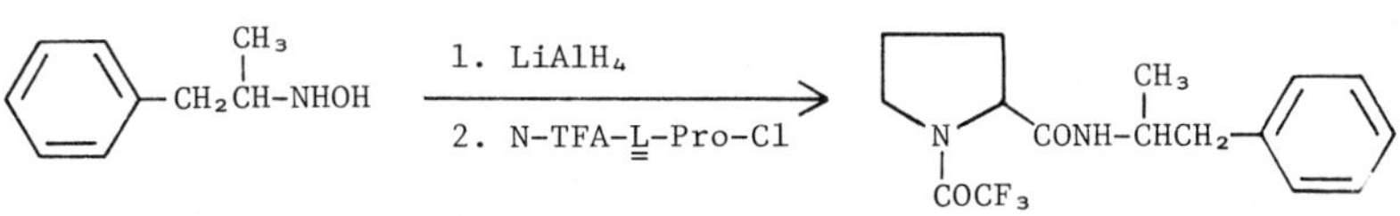

Typical Procedure. Diss 10 mg of the hydroxylamine oxalate in satd aq $NaHCO_3$ and ext with Et_2O. Dry ext over $MgSO_4$ and conc under vac. Diss residue in 5 ml Na dried Et_2O and add 10 mg $LiAlH_4$. Allow to stand 2 hr then add 2 ml H_2O. Separate the Et_2O and dry over $MgSO_4$. Evap at 40° and add 100 μl 0.1 *M* *N*-TFA-L-Pro-Cl in CH_3Cl. Analyze by GC.

Comments. The method was used to determine the optical purity of synthesized enantiomers of primary and secondary hydroxylamines. The *N*-TFA-L-Pro-Cl reagent reacted with primary hydroxylamines to give a mixt of the hydroxylamine deriv and the deriv of the precursor amine. Secondary hydroxylamines were unreactive toward the reagent. The secondary hydroxylamines (as the oxalate salts, 10 mg) were suspended in 5 ml H_2O and reduced by adding 1 ml 12.5% titanous chloride and allowing to stand 18 hr in the dark before the Et_2O extraction.

References

[1] A. H. Beckett *et al.*, *Tetrahedron* **31**, 1531 (1975).

Derivative Type. *N*-Pentafluorobenzoyl-*S*-prolyl.

Reagent. *N*-Pentafluorobenzoyl-*S*-(−)-prolyl-1-imidazole.

Compound Type. Amines.

Reaction

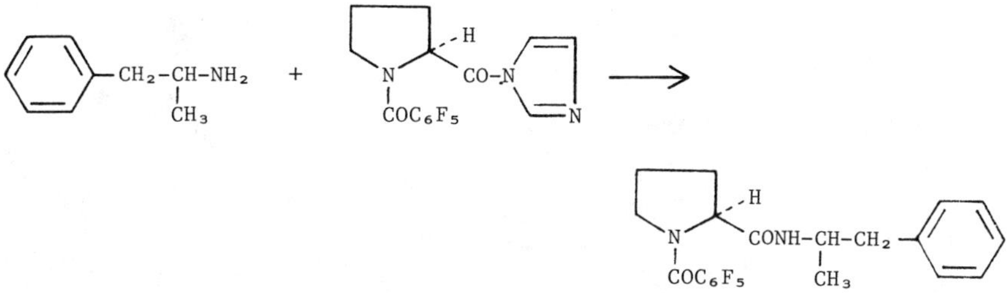

Typical Procedure. Treat 0.01–0.02 mmol amine in 1 ml benzene with an equimolar amount of the reagent in 1 ml benzene. After 30 min at room temp shake with 1 ml H_2O for 30 min. Centrifuge and analyze 1–2 μl of org phase by ECD-GC.

Comments. The deriv was developed for quantitative enantiomeric analysis of amines by ECD-GC. As little as 1 ng of the diastereomeric derivs was detectable. Of seven amines investigated, the retention times of the derivs of the levorotatory compounds were consistently shorter than those of the corresponding dextrorotatory compounds. The synthesis of the reagent was reported.[1] The method was also used with GC-CI-MS to measure the enantiomers of amphetamine in plasma and saliva.[2]

References

[1]S. B. Matin *et al.*, *J. Pharm. Sci.* **62**, 821 (1973).

[1]S. B. Matin *et al.*, *Biomed. Mass Spectrom.* **4**, 118 (1977).

9.2.3. Sulfonamide Derivatives

Derivative Type. (+)-Camphor-10-sulfonamide.

Reagent. (+)-Camphor-10-sulfonyl chloride.

Compound Type. Enantiomeric amines.

Reaction

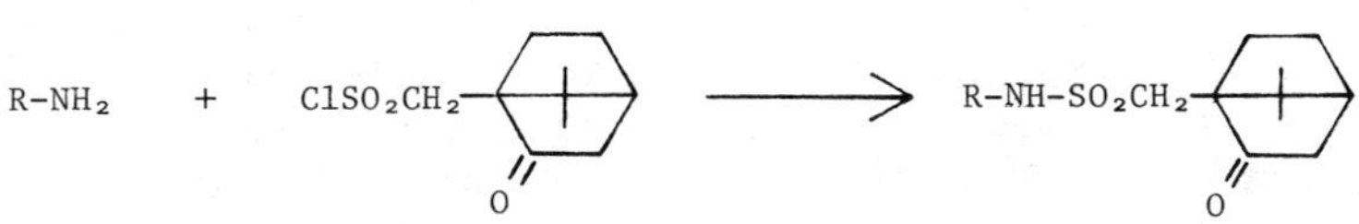

Typical Procedure. Mix the amine with an equiv amt of the sulfonyl chloride in pyridine and heat at 60° until the reaction is complete (as indicated by TLC analysis of the rxn mixt). Dil with aq HCl and ext with $CHCl_3$. Analyze by GC.

Comments. The method was demonstrated for 11 arylalkylamines with GC and NMR data reported for the derivs.

References

[1]G.-A. Hoyer *et al.*, *Tetrahedron Lett.*, 985 (1972).

9.3. DERIVATIVES OF CARBOXYLIC ACIDS

9.3.1. Ester Derivatives

Derivative Type. *O*-Acetyl, menthyl ester.

Reagent. (−)-Menthol/HCl, acetic anhydride (Ac_2O)/pyridine.

Compound Type. Lactic and glyceric acids.

Reaction

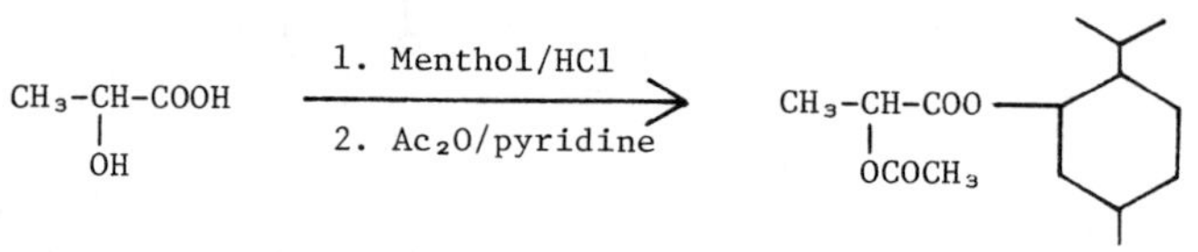

Typical Procedure. Mix 1–5 mg of dry acid with 300 mg (−)-menthol and bubble dry HCl through the soln for 2 hr at 110°. Evap under N_2 at 110° to almost dryness. Add 1 ml Ac_2O pyridine (1 :1, v/v) and heat at 100° for 30 min. Evap in the presence of abs EtOH and diss residue in $CHCl_3$. Analyze by GC.

Comments. The method was used to determine the absolute configuration of lactic and glyceric acids in urine. MS data were reported for the derivs.

References

[1] J. P. Kamerling *et al.*, *J. Chromatogr.* **143**, 117 (1977).

Derivative Type. (−)-Menthyl ester.

Reagent. Thionyl chloride, (−)-menthol.

Compound Type. Acyclic isoprenoid acids.

Reaction

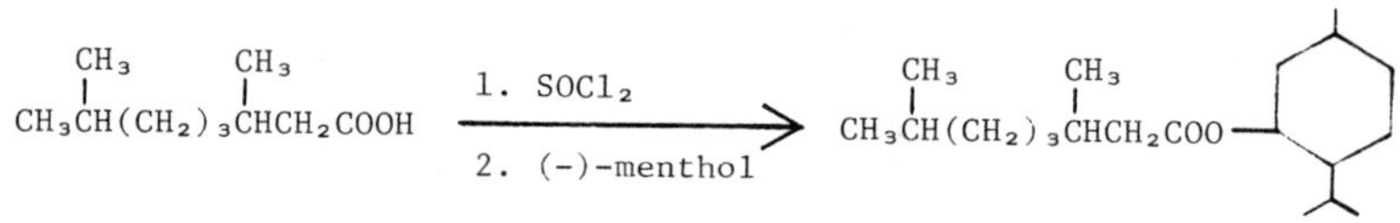

Typical Procedure. Dry sample (*ca* 5 mg) under vac over P_2O_5; then reflux 30 min with 2 ml $SOCl_2$. Evap under N_2, add toluene, and repeat evap. Add 5 mg (−)-menthol and 3 ml dry toluene containing a few drops of dry pyridine. Reflux 3 hr, allow to stand overnight, then evap to dryness. Analyze deriv by GC or GC-MS.

Comments. The derivs of a series of acyclic isoprenoid acids were examined in order to determine the stereochemistry of such acids in unknown mixt.[1] The derivs had been used to separate diastereomers of phytanic[2] and pristanic[3] acids by GC.

References.

[1] R. G. Ackman *et al.*, *J. Chromatogr. Sci.* **10**, 392 (1972).

[2] R. G. Ackman *et al.*, *J. Chromatogr.* **44**, 256 (1969).

[3] R. G. Ackman *et al.*, *Biochem. Biophys. Acta* **176**, 673 (1969).

Derivative Type. (−)-Menthyl ester.

Reagent. Oxalyl chloride, (−)-menthol.

Compound Type. Asymmetric carboxylic acids.

Reaction

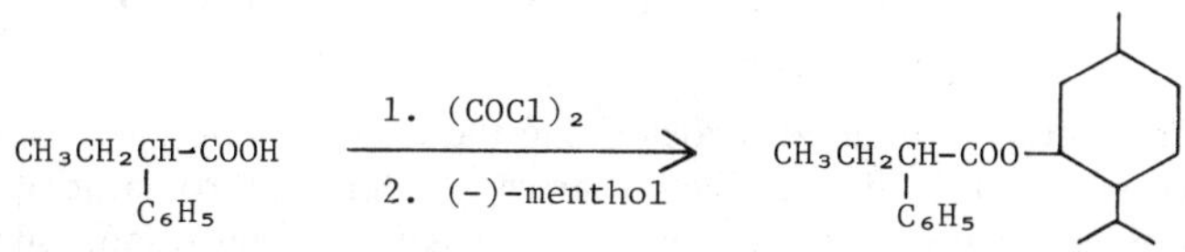

Typical Procedure. Diss 8 g α-phenylbutyric acid in 50 ml anhyd benzene and reflux with 10 g oxalyl chloride. Diss 1.8 g of the resulting acid chloride in 10 ml benzene and reflux 18 hr with excess (−)-menthol. Wash with H_2O and evap the solvent and excess menthol. Analyze by GC.

Comments. The method was used to separate enantiomers as diastereomeric derivs by GC.

References

[1]J. P. Guette and A. Horeau, *Tetrahedron Lett.* 3049 (1965).

9.3.2. Amide Derivatives

Derivative Type. *N*-α-Phenylethyl amide.

Reagent. Thionyl chloride, R-(+)-α-phenylethylamine.

Compound Type. Ibuprofen.

Reaction

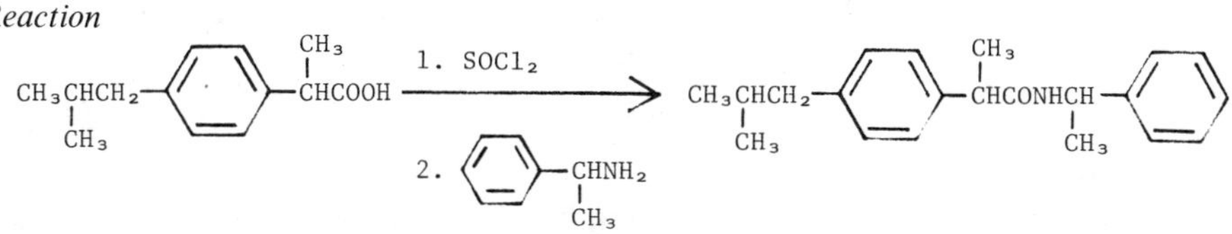

Typical Procedure. Diss the sample in toluene and treat with redistilled $SOCl_2$. Evap excess $SOCl_2$ and add a soln of chiral amine in toluene. Let stand 10 min then analyze by GC.

Comments. The deriv method was used to show that the excreted unmetabolized drug was enriched in the (+)-enantiomer after administration of racemic drug.

References

[1]C. J. W. Brooks and M. T. Gilbert, *J. Chromatogr.* **99**, 541 (1974).

Derivative Type. α-Methyl-*p*-nitrobenzylamide.

Reagent. Oxalyl chloride, (+)-R-α-methyl-*p*-nitrobenzylamine.

Compound Type. Enantiomeric acids.

Reaction

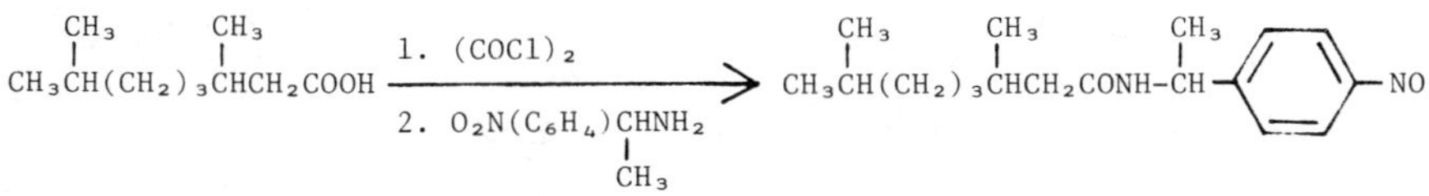

Typical Procedure.[1] Diss 0.5 mmol of the acid in 5 ml benzene and reflux with 1.5 mmol oxalyl chloride for 30 min. Evap solvent and excess reagent under vac (20 mm) at 45°. Diss crude acid chloride in 5 ml Et_2O and chill in an ice bath. Add portionwise 2.0 mmol (+)-R-α-methyl-*p*-nitrobenzylamine[3] in 3 ml Et_2O. Stir 1 hr at 0–5°; then dil with 100 ml Et_2O. Wash with 1 *N* HCl, satd $NaHCO_3$, and H_2O; then dry over $MgSO_4$. Evap Et_2O and analyze product by HPLC.

Comments. The method was used to determine the enantiomeric composition of methyl branched fatty acids, and to study the effects of structure upon the chromatographic resolution. In the adsorption chromatography system used, it was found that resolution was enhanced when the adsorbing functional group was flanked as near as possible on each side by a chiral center.

References

[1] D. Valentine *et al.*, *J. Org. Chem.* **41**, 62 (1976).
[2] C. G. Scott *et al.*, *J. Chromatogr.* **125**, 157 (1976).
[3] A. C. Cope *et al.*, *J. Am. Chem. Soc.* **92**, 1243 (1970).

Derivative Type. α-Naphthylethylamide.

Reagent. Phosphorus trichloride, α-naphthylethylamine.

Compound Type. Carboxylic acids.

Reaction

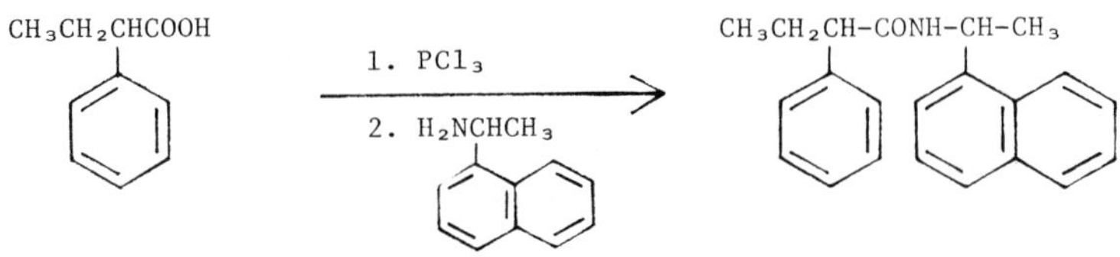

Typical Procedure. Treat the carboxylic acid with PCl_3 in benzene/pyridine to form the acid chloride then treat with excess (−)-α-naphthylethylamine (details not given). Analyze by HPLC.

Comments. The derivs were used to separate the enantiomers of several carboxylic acids containing aromatic rings including two ferrocene carboxylic acids.

References

[1] R. Eberhardt *et al.*, *Tetrahedron Lett.* 4365 (1974).

Derivative Type. *N*-methyl-*N*-(1-phenyl-2-propyl)amide.

Reagent. *S*-(+)-2-Methylamino-1-phenylpropane/dicyclohexylcarbodiimide (DCC).

Compound Type. α-Alkylphenylacetic acids.

Reaction

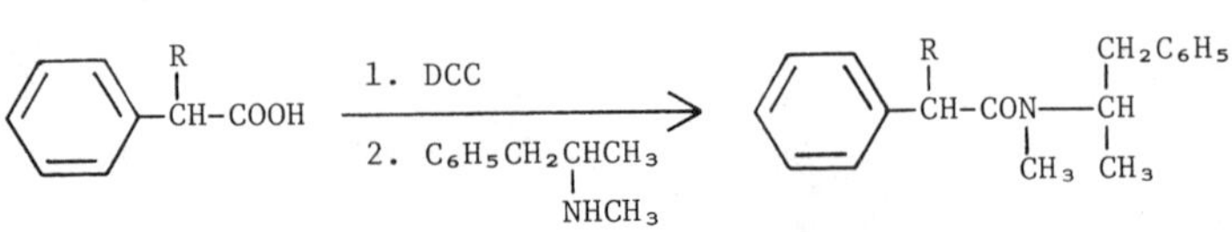

Typical Procedure. Diss 1.5 mmol α-alkylphenylacetic acid in THF and add 1 mmol DCC. Cool and add 1 mmol *S*-(+)-2-methylamino-1-phenylpropane. Stir 24 hr then dil with EtOAc and filter the dicyclohexylurea ppt. Analyze by GC.

Comments. The deriv was used for determination of the absolute configuration of α-alkylphenylacetic acids. In the series $R=C_1$ to C_4, the (+)-acid-*S*-(+)-amides consistently had shorter retention times (on 1% HI-EFF-8AP). Since (+)-hydratropic acid ($R=CH_3$) was known to have the *S*-configuration, the absolute configuration of the other (+)-acids was concluded to be *S*.

References

[1]B. Halpern and J. W. Westley, *Chem. Commun.* 237 (1967).

9.4. DERIVATIVES OF AMINO ACIDS

9.4.1. Ester Derivatives from Chiral Alcohols

Derivative Type. 2-Butyl ester, *N*-trifluoroacetyl.

Reagent. 2-Butanol (2-BuOH)/HCl, methyl trifluoroacetate (MeOTFA).

Compound Type. Amino acids.

Reaction

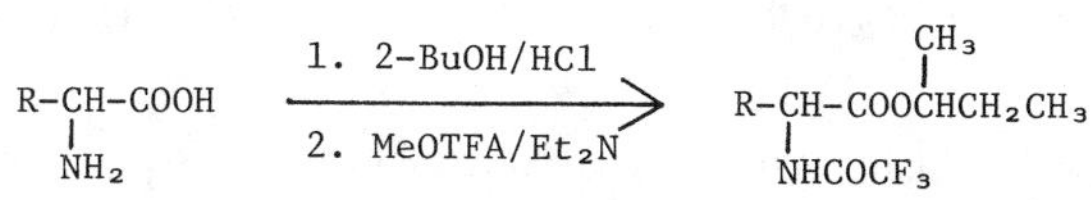

Typical Procedure. To 10 mmol amino acid add 2-BuOH (15–20 ml/g amino acid). Cool in an ice bath and bubble in dry HCl for 30–60 min. Remove from ice bath and reflux 2 hr. Evap to dryness under vac then add CH_2Cl_2 and evap to remove residual H_2O. Diss in 15 ml 2-BuOH and add 12.3 mmol Et_3N (yields ppt of Et_3N-HCl). Add 5 ml MeOTFA and allow to react at least 2 hr at room temp. Filter and evap to an oil at 60–70°. Diss in EtOAc and wash 2× with H_2O. Evap to dryness under vac then add CH_2Cl_2 and evap to remove H_2O. Analyze derivs by GC.

Comments. The derivs were used to separate the enantiomers of 10 racemic amino acids by GC.

References

[1]G. E. Pollack *et al.*, *J. Gas Chromat.* **3**, 174 (1965).

Derivative Type. 2-Butyl ester, *N*-trifluoroacetyl.

Reagent. 2-Butanol(2-BuOH)/HCl, trifluoroacetic anhydride (TFAA).

Compound Type. Amino acids.

Reaction

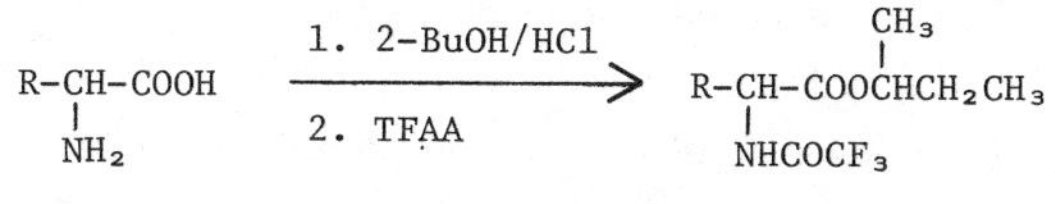

Typical Procedure. Diss up to 50 mg amino acid in 30 ml 8 *M* HCl in 2-BuOH. Reflux 1 hr protected from moisture then evap under vac at 60°. Add and evap 2 × 25 ml CH_2Cl_2 to remove H_2O. Add 5 ml anhyd CH_2Cl_2 and cool to −10° to 0°. Add dropwise with stirring 0.5 ml TFAA containing a trace of trifluoroacetic acid while maintaining the cooling (the acid and cooling were intended to prevent racemization). Stir 2 hr while allowing to warm to room temp. Evap under vac and diss residue in CH_2Cl_2 or $CHCl_3$ prior to GC analysis.

Comments. The method[1] was an extension of earlier work,[2] which utilized methyl trifluoroacetate. MeOTFA failed to *O*-acylate hydroxyamino acids. The method was applied to 21 racemic amino acids including both acidic and basic types. The method has more recently been examined in a study of analysis of Leu enantiomers.[3] In GC separation of the derivs of racemic amino acids and racemic 2-BuOH, it was reported that the earlier eluting peak consisted of LL and DD and the later peak LD and DL.[4]

References

[1]G. E. Pollack and V. I. Oyama, *J. Gas Chromatogr.* **4**, 126 (1966).

[2]G. E. Pollack *et al.*, *J. Gas Chromatogr.* **3**, 174 (1965).

[3]W. A. Bonner *et al.*, *Anal. Chem.* **46**, 2104 (1974).

[4]F. Raulin and B. N. Khare, *J. Chromatogr.* **75**, 13 (1973).

Derivative Type. 2-Butyl ester, *N*-trifluoroacetyl, *O,S*-acetyl.

Reagent. 2-Butanol(2-BuOH)/HCl, trifluoroacetic anhydride (TFAA), acetyl chloride (AcCl).

Compound Type. Hydroxyl and sulfhydryl amino acids.

Reaction

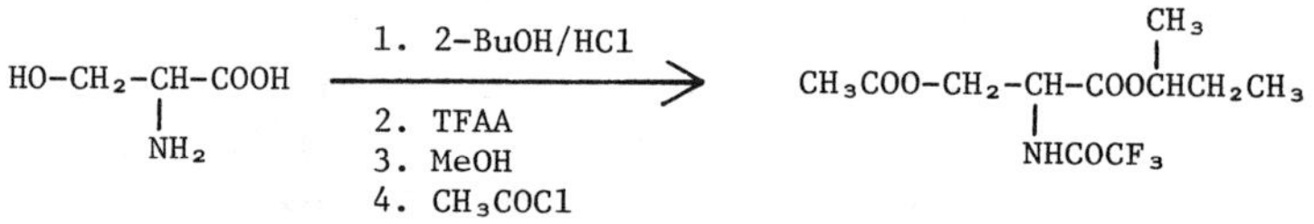

Typical Procedure. Diss 5 mg sample in 3 ml 3–4 *N* HCl in 2-BuOH. Heat at 100° for 3 hr. Evap under vac and add 1.0 ml $CHCl_2$ and add 0.2 ml TFAA. Heat at 100° for 5 min. Evap under vac and add 2 ml anhyd MeOH. Allow to stand 20 hr at room temp. Evap to dryness, add a small dry magnetic stirring bar and 1 ml anhyd CH_2Cl_2. Cool to 0–5° and add 0.05 ml anhyd pyridine with stirring. Add 0.05 ml freshly dist AcCl over 1–2 min. Allow to stand at room temp 30 min. Evap under vac and diss deriv in 1 ml anhyd alcohol free Et_2O. Filter and wash pyridine HCl ppt with Et_2O. Evap combined Et_2O to dryness and diss deriv in EtOH. Analyze by GC.

Comments. The derivs were used to separate the enantiomers of Ser, Tyr, CySH, and HO-Pro, which were not well resolved as the di-TFA esters. The resolution of Trp as the di-TFA 2-butyl ester and Asp as the *N*-TFA 2,2-dimethyl-3-butyl ester and *N*-pentafluorobenzoyl 2-butyl ester were also reported.

References

[1]G. E. Pollack and A. H. Kawauchi, *Anal. Chem.* **40**, 1356 (1968).

Derivative Type. 2-Butyl ester, *N*-pentafluoropropionyl.

Reagent. Methanol(MeOH)/HCl, 2-butanol (2-BuOH)/acetyl chloride (AcCl), pentafluoropropionic anhydride(PFPA).

Compound Type. Amino acids.

Reaction

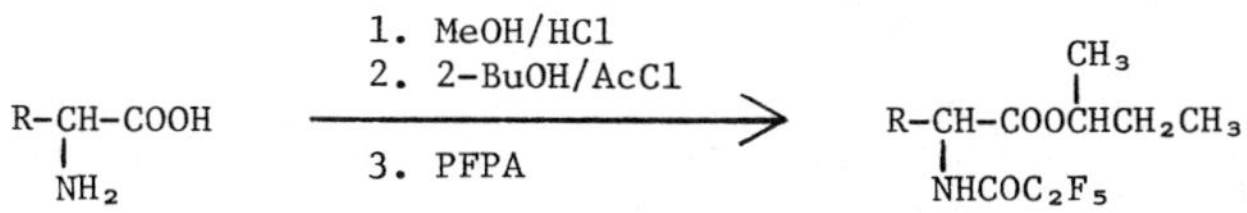

Typical Procedure. Diss 100 μg amino acid in 0.5 ml 1.25 *N* HCl in MeOH and heat 30 min at 100°. Evap the MeOH under N_2 and add 80 μl D- or L-2-BuOH and 20 μl AcCl. Heat 2 hr at 100°. Evap under N_2 and add 200 μl CH_2Cl_2 and 50 μl PFPA. Allow to stand 1 hr at room temp. Evap under N_2 and diss residue in 100 μl EtOAc. Analyze by GC.

Comments. The derivs were used to separate amino acid enantiomers. Esters were also prepared using optically active 3-methyl-2-butanol and 2-octanol. The use of optically active stationary phase permitted separation of all four optical isomers permitting the use of impure (optically) alcohol reagents.

References

[1]W. A. Koenig, *Chromatographia* **9**, 72 (1976).

Derivative Type. 2-Pentyl ester, *N*-pentafluoropropionyl (PFP).

Reagent. 2-Pentyl alcohol/HCl, pentafluoropropionic anhydride (PFPA).

Compound Type. Amino acids.

Reaction

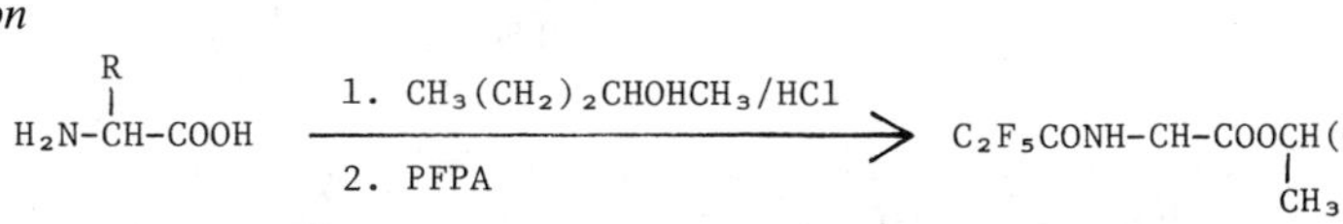

Typical Procedure. Evap 50 μl of a soln of amino acids (10 μmol each in 3 *N* HCl) to dryness under N_2 at 100°. Add 3 ml of 3–4 *N* HCl in anhyd 2-pentyl alcohol and heat at 100° for 3 hr. Evap to dryness under N_2 at 100°. Add 1.8 ml CH_2Cl_2 and 0.2 ml PFPA. Heat at 100° for 15 min then evap under vac (with ice cooling to prevent loss of volatile derivs). Diss residue in 2 ml CH_2Cl_2 and analyze 2 μl aliquots by GC.

Comments. The derivs were used for GC separation of enantiomers of amino acids. Inadequate resolution of isovaline, β-amino-*n*-butyric acid, and *N*-methylalanine and no resolution of pipecolic and β-aminoisobutyric acids were obtained using *N*-TFA R(+)-2-butyl esters. *N*-TFA, PFP, HFB, and CDF (chlorodifluoroacetyl) derivs of the 2-butyl, 2-pentyl, 2-hexyl, 2-heptyl, and 2-octyl esters were examined. The 2-pentyl and 2-hexyl PFP derivs were the most useful. The 2-heptyl and 2-octyl PFP derivs gave slightly better resolution on the average but had inconveniently long retention times.

References

[1]G. E. Pollock, *Anal. Chem.* **44**, 2368 (1972); **46**, 614 (1974).

Derivative Type. 2-Octyl ester, *N*-trifluoroacetyl.

Reagent. 2-Octanol/HCl, trifluoroacetic anhydride (TFAA).

Compound Type. Amino acids.

Reaction

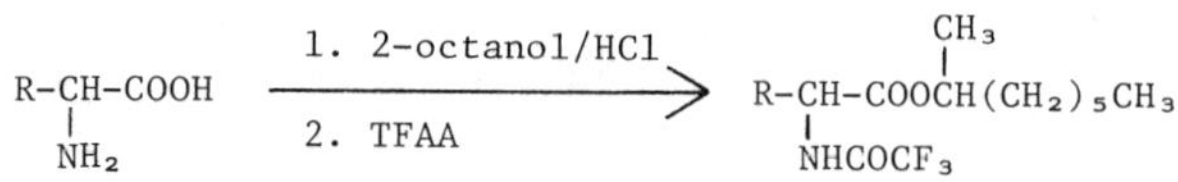

Typical Procedure. Esterfy the amino acid by diss in 2-octanol containing HCl. Evap to dryness and add CH_2Cl_2. Cool to $-20°$ and add excess TFAA with stirring. Allow to warm to room temp and allow to stand 1 hr. Analyze by GC.

Comments. The diastereomeric derivs formed were separable by GC. 2-Butanol was also used for esterfication and found to be useful for amino acids like Glu, which yield derivs with longer retention times. Derivs were prep from Ala, Val, Leu, Pro, Phe, and Glu. GC data were reported.

References

[1]E. Gil-Av *et al.*, *J. Chromatogr.* **17**, 408 (1965).

Derivative Type. Menthyl ester, *N*-trifluoroacetyl.

Reagent. (−)-Menthol/HCl, trifluoroacetic anhydride (TFAA).

Compound Type. Amino acids.

Reaction

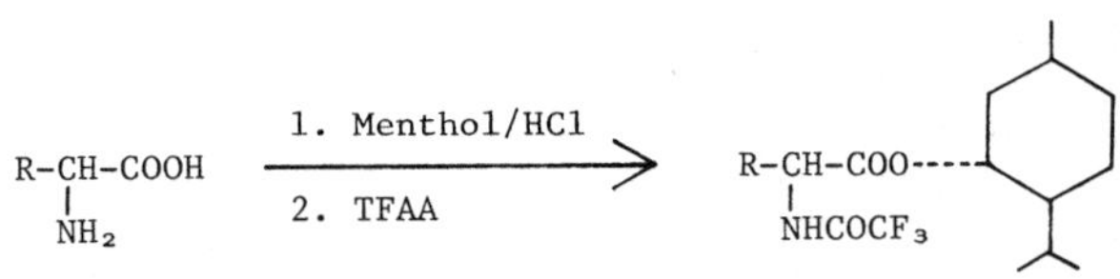

Typical Procedure. Mix 1 mg amino acid with 250 mg menthol and heat at 105° for 1–3 hr while bubbling in HCl. Remove the HCl and menthol under N_2 and add 0.5 ml TFAA. Allow to react 5 min at room temp with mixing then evap under N_2. Diss residue in 0.5 ml dry EtOAc and analyze by GC.

Comments. Acidic and basic amino acids were first converted to methyl esters with MeOH/HCl and then transesterfied with menthol as above. Hydroxy amino acids were silylated with HMDS/TMCS/pyridine after the esterfication and acylation. The method was used to resolve 19 α-amino acids and one β-amino acid by GC. Neutral and basic amino acids were well resolved on packed columns; acidic amino acids required capillary columns for complete resolution. Vitt *et al.*[2] had reported the use of the deriv for separation of the enantiomers of Val, Ala, and Leu.

References

[1]M. Hasegawa and I. Matsubara, *Anal. Biochem.* **63**, 308 (1975).
[2]S. V. Vitt *et al.*, *Tetrahedron Lett.* 2575 (1965).

Derivative Type. Menthyl ester, *N*-trifluoroacetyl.

Reagent. (−)-Menthol/*p*-toluenesulfonic acid (TsOH), trifluoroacetic anhydride (TFAA).

Compound Type. Amino acids.

Reaction

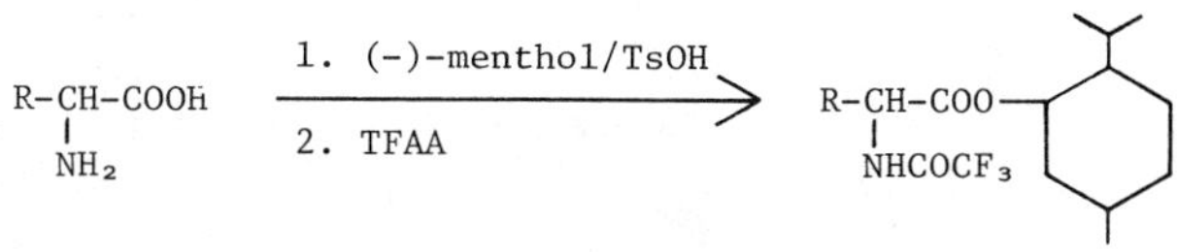

Typical Procedure. Prepare menthyl ester by refluxing amino acid with equal wts of *l*-menthol and TsOH in benzene/toluene (7:3, v/v) using azeotropic removal of H_2O with a Dean-Stark trap.[2] Acylate with 20 μl TFAA and 60 μl CH_2Cl_2 per 100 μg amino acid.[3] Heat 5 min at 150° then analyze by GC.

Comments. The derivs were used for separation of amino acid enantiomers. (−)-Borneol esters were also useful but gave smaller separation factors; (−)-isopulegol esters gave multiple GC peaks were deemed not useful. Various perfluoroacyl groups were examined but the *N*-TFA derivs showed the greatest separation factors. The *N*-TFA (−)-menthyl esters have been shown to have MS behavior suitable for GC-MS-SIM determinations.[4]

References

[1]H. Iwase, *Chem. Pharm. Bull.* **23**, 1604 (1975).

[2]K. Harada and T. Hayakawa, *Bull. Chem. Soc. Jap.* **37**, 191 (1964).

[3]D. Roach and C. W. Gehrke, *J. Chromatogr.* **44**, 269 (1969).

[4]H. Iwase and A. Murai, *Anal. Biochem.* **78**, 340 (1977).

Derivative Type. 3,3-Dimethyl-2-butyl, *N*-trifluoroacetyl.

Reagent. (+)-3,3-Dimethyl-2-butanol/HCl, trifluoroacetic anhydride (TFAA).

Compound Type. Amino acids.

Reaction

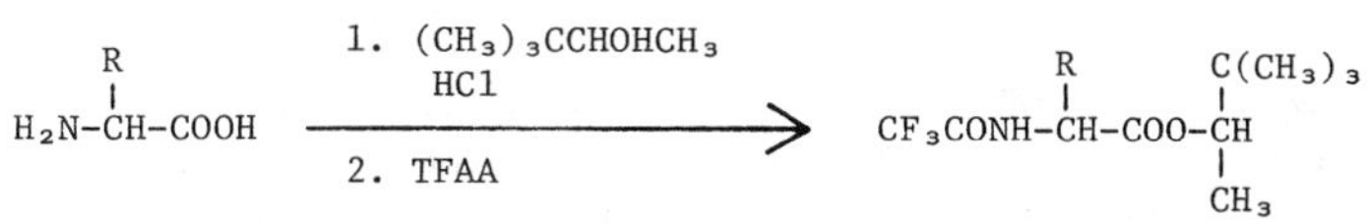

Typical Procedure. Heat 1–3 mg of the amino acid with 1–1.5 ml of the alcohol containing 1.2–1.5 m*M* HCl at 100° for 1 hr. Evap and diss the residue in 1–1.5 ml CH_2Cl_2/TFAA (4:1, v/v). Allow to stand at room temp then analyze by (packed column) GC.

Comments. 3,3-Dimethyl-2-butyl esters gave the best diastereomer resolution of several esters examined. Of 15 amino acids examined all but Asp (70%) and Pro (82%) gave resolutions of better than 93%. Arg, His, and Cys derivs were not examined. No resolution was achieved with 2-methyl-1-butyl esters. In the series 2-butyl through 2-decyl, the resolution increased with increasing alcohol size; in the larger alcohols, movement of the ester attachment to a more median position decreased resolution.

References

[1]G. S. Ayers, *J. Chromatogr.* **63**, 259 (1971).

Derivative Type. *N*-Trifluoroacetyl aspartic acid 3-methyl-2-butyl ester.

Reagent. Ozone, (+)-3-Methyl-2-butanol/HCl, trifluoroacetic anhydride.

Compound Type. Histidine enantiomers.

Reaction

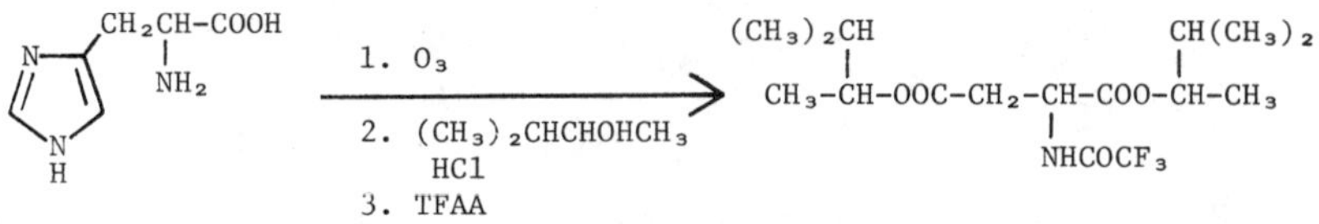

Typical Procedure. Diss 1 mg of peptide hydrolyzate in 0.5 ml 1.5 *N* HCl in MeOH. Cool to −78° and pass a 500-fold excess of O_3 into the soln over a 30 min period. Evap under N_2 and diss the residue in 100 μl 7 *N* HCl in (+)-3-methyl-2-butanol. Heat 30 min at 95° then evap under vac. Diss the residue in 200 μl CH_2Cl_2 and add 50 μl TFAA. After 30 min at room temp, evap under N_2. Diss the residue in 50 μl CH_2Cl_2 and analyze by GC.

Comments. The method was used to measure His racemization in peptide hydrolyzates. The GC separation was achieved on glass capillaries using a chiral stationary phase.

References

[1]W. Rahn *et al.*, *Z. Physiol. Chem.* **357**, 1223 (1976).

Derivative Type. (+)-3-Methyl-2-butyl ester, pentafluoropropionyl.

Reagent. (+)-3-Methyl-2-butanol/HCl, pentafluoropropionic anhydride (PFPA).

Compound Type. Amino acids.

Reaction

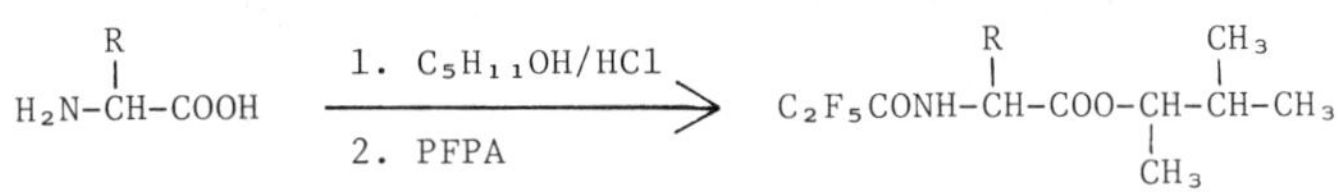

Typical Procedure. Heat 100 μg of the amino acid sample in 150 μl of 7 *N* HCl in (+)-3-methyl-2-butanol (99% optical purity) for 90 min at 100°. For basic amino acids, esterfy first with 1.25 *N* HCl in MeOH for 60 min at 100° to improve solubility. Evap under vac and diss residue in 200 μl CH_2Cl_2 and 50 μl PFPA. Allow to react 30 min at room temp then evap to dryness. Diss residue in 100 μl EtOAc and analyze by glass capillary GC.

Comments. The derivs were used for separation of enantiomers of amino acids by GC separation of the diastereomeric derivs. In order to obtain fully acylated derivs of His, Arg, and Trp it was necessary to inject the sample along with PFPA into the hot (250°) injection port.

References

[1]W. A. König *et al.*, *J. Chromatogr.* **133**, 141 (1977).

9.4.2. Amide Derivatives with Chiral Acyl Groups

Derivative Type. Methyl ester, trifluoroacetyl-L-prolyl.

Reagent. Methanol/thionyl chloride, *N*-trifluoroacetyl-L-prolyl chloride.

Compound Type. Amino acids.

Reaction

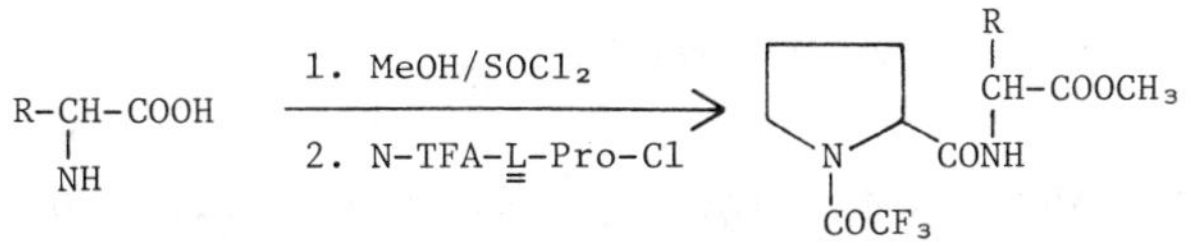

Typical Procedure. Esterfy the amino acid sample with MeOH/$SOCl_2^2$; then evap to dryness. Add excess *N*-trifluoroacetyl-L-prolyl chloride[3] in inert solvent. Cool the suspension and neutralize with Et_3N. Wash with H_2O, dry over Na_2SO_4, and analyze by GC.

Comments. The derivs were used to separate D and L enantiomers of amino acids. GC data were reported for derivs of D- and L- Ala, Val, Leu, Pro, Met, and Phe. Pro was chosen as the basis of the deriv reagent because of its resistance to racemization. Gordis[4] used the TFA-L-prolyl deriv to separate enantiomers of amphetamine.

References

[1]B. Halpern and J. W. Westley, *Biochem. Biophys. Res. Commun* **19**, 361 (1965).

[2]M. Brenner and W. Huber, *Helv. Chim. Acta* **36**, 1109 (1953).

[3]F. Weygand *et al.*, *Chem. Ber.* **90**, 1896 (1957).

[4]E. Gordis, *Biochem. Pharmacol.* **15**, 2124 (1966).

Derivative Type. Methyl ester, trifluoroacetyl-*S*-(−)-prolyl.

Reagent. Methanol (MeOH)/thionyl chloride, *N*-trifluoroacetyl-*S*-(−)-prolyl chloride.

Compound Type. Amino acids.

Reaction

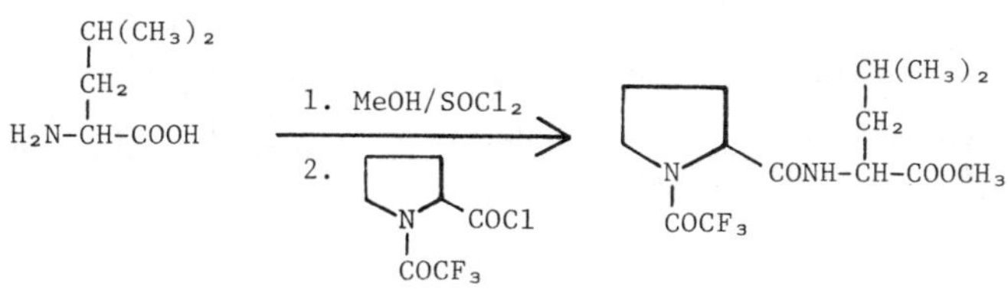

Typical Procedure. Chill 3 ml MeOH in a dry ice alcohol bath and add dropwise with stirring 1 ml $SOCl_2$. Add the soln to 1 mmol of amino acid (Leu) and heat under reflux for 30 min. Evap under vac, add 5 ml MeOH, and evap under vac; then add 5 ml benzene and evap under vac. Diss residue in 11 ml of 0.1 *M* *N*-trifluoroacetyl-*S*-(−)-prolyl chloride in CH_2Cl_2. Chill under a $CaCl_2$ drying tube in a dry ice alcohol bath then add 0.305 ml (2.2 mmol) of Et_3N in 10 ml CH_2Cl_2 dropwise with stirring over a 10 min period. Allow to warm slowly by standing in bath overnight (with $CaCl_2$ tube). Stir vigorously with 20 ml H_2O for 5 min. Sep CH_2Cl_2 layer and ext H_2O with 5 ml CH_2Cl_2. Wash combined CH_2Cl_2 with 20 ml 15% citric acid followed by 20 ml 15% $NaHCO_3$. Ext aq wash with 5 ml CH_2Cl_2 each time and add to combined CH_2Cl_2. Dry CH_2Cl_2 over Na_2SO_4 and evap under vac. Diss a portion of the product in MeOH prior to GC analysis.

Comments. The deriv was used for enantiomeric analysis by GC. The authors reported significant racemization in the use of previous methods for prep of the derivs. Some of this was found to be from racemization in the prep of the reagent. A low temp prep of the reagent with minimal racemization was reported. Racemization by Et_3N in the deriv procedure was minimized by slow addition of a measured amt of the base at low temp.

References

[1]W. A. Bonner, *J. Chromatogr. Sci.* **10**, 159 (1972).

Derivative Type. Methyl ester, *O*-trimethylsilyl, *N*-trifluoroacetyl-L-prolyl.

Reagent. Methanol (MeOH)/thionyl chloride, hexamethyldisilazane (HMDS), *N*-trifluoroacetyl-L-prolyl chloride.

Compound Type. Polyfunctional amino acids.

Reaction

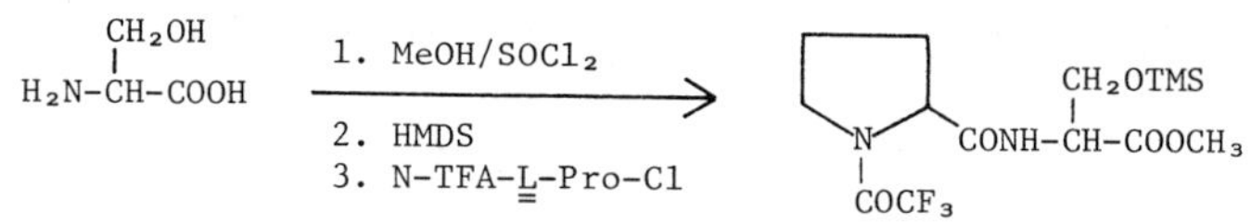

Typical Procedure. Esterfy 0.1 mmol amino acid with $MeOH/SOCl_2$.[2] Evap under vac and add 0.5 ml HMDS. Reflux until a clear soln results. Evap under vac and add 0.1 mmol *N*-TFA-L-Pro-Cl in 1 ml CH_2Cl_2. Allow to stand 1 hr then analyze by GC.

Comments. The method extended the use of *N*-TFA-L-Pro, methyl ester derivs for enantiomeric analysis to the hydroxylated amino acids. GC data for the derivs of Asp and Glu were given.

References

[1]B. Halpern and J. W. Westley, *Tetrahedron Lett.* 2283 (1966).

[2]M. Brenner and W. Huber, *Helv. Chim. Acta* **36**, 1109 (1953).

Derivative Type. Methyl ester, trimethylsilyl, *N*-trifluoroacetyl prolyl.

Reagent. Methanol(MeOH)/thionyl chloride, bis(trimethylsilyl)trifluoroacetamide (BSTFA), *N*-trifluoroacetyl prolyl chloride.

Compound Type. Threonine and allothreonine.

Reaction

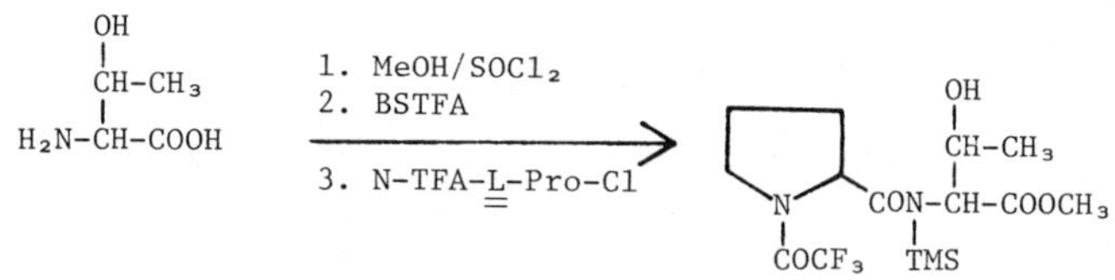

Typical Procedure. Mix 30–50 mg of amino acid sample with 6 ml $SOCl_2/MeOH$ (1:9, v/v) and reflux 4.5 hr. Cool and transfer a vol corresponding to *ca* 10 mg amino acids to another tube and evap under N_2. Add 1 ml MeOH and 3–5 drops Et_3N and evap to dryness under N_2. Diss residue in 0.5 ml CH_3CN and add 0.5 ml BSTFA. Heat on steam bath for 2 min then cool under cold H_2O stream. Add 2 ml of *N*-trifluoroacetyl R- or *S*-prolyl chloride in $CHCl_3$ (0.17 mmol/ml) and 10 drops Et_3N. Analyze 2–5 μl by GC.

Comments. The method was used to determine the relative amts of R- and *S*-threonine and R- and *S*-allothreonine in mixtures. The preparation of 99% optically pure *N*-trifluoroacetyl R- and *S*-prolyl chloride was reported. It was reported that MS data suggested the location of the TMS group on the amide-*N* rather than on the hydroxyl as would be expected. The deriv might have been a cyclic ortho amide.

References

[1]J. C. Dabrowiak and D. W. Cooke, *Anal. Chem.* **43**, 791 (1971).

Derivative Type. Thiazolidine-4-carboxylic acid, methyl ester, *N*-trifluoroacetyl-L-prolyl.

Reagent. Formaldehyde, methanol (MeOH)/thionyl chloride, *N*-trifluoroacetyl-L-prolyl chloride.

Compound Type. β-Mercapto-α-amino acids.

Reaction

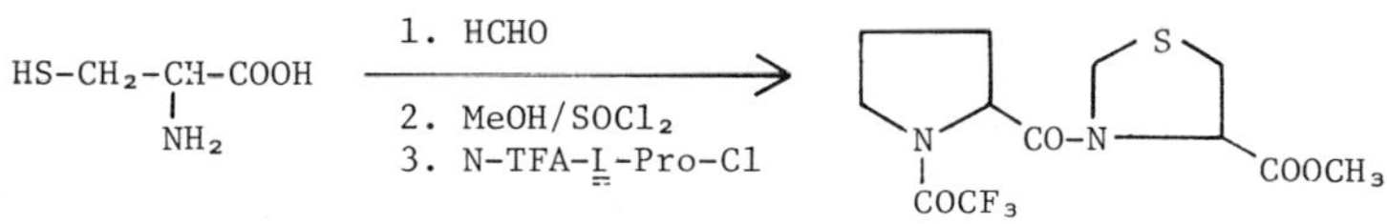

Typical Procedure. Diss 0.1 mmol sample in excess aq formaldehyde and isolate thiazolidine-4-carboxylic acid by freeze drying.[2] Esterfy with MeOH/$SOCl_2$.[3] Evap under vac and add 0.1 mmol *N*-TFA-L-Pro-Cl in 1 ml CH_2Cl_2. Allow to stand 1 hr then analyze by GC.

Comments. The method was used to separate the enantiomers of CySH and penicillamine by GC.

References

[1]B. Halpern and J. W. Westley, *Tetrahedron Lett.* 2283 (1966).
[2]S. Ratner and H. T. Clarke, *J. Am. Chem. Soc.* **59**, 200 (1957).
[3]M. Brenner and W. Huber, *Helv. Chim. Acta* **36**, 1109 (1953).

Derivative Type. Methyl ester, trifluoroacetylthiazolidine-4-carboxamide.

Reagent. Methanol/thionyl chloride, trifluoroacetylthiazolidine-4-carboxylic acid chloride.

Compound Type. Amino acids.

Reaction

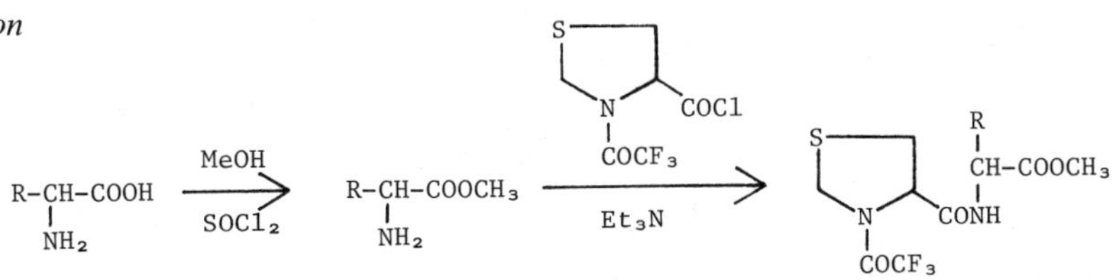

Typical Procedure. Esterfy the amino acid sample with MeOH/$SOCl_2^2$; then evap to dryness. Add an excess resolving reagent consisting of unlabeled D- and 2,2-d_2-L-trifluoroacetylthiazolidine-4-carboxylic acid chloride in inert solvent. Neutralize the suspension with Et_3N. Wash with H_2O. Analyze by GC and monitor the isotope ratios in the GC effluent by MS of collected fractions.

Comments. In the published work the fractions were collected and analyzed by MS.[1] Use of GC-MS-SIM as now available would be simpler. GC separation of diastereomers and MS detection of isotope ratios in the fractions were utilized to scan amino acid samples for optical activity. The thiazolidine reagent was chosen because of availability of both optical isomers and the ease of deuterium incorporation. Iwase[3] used the same reagent with amino acid butyl esters for separation of amino acid enantiomers.

References

[1]B. Halpern and J. W. Westley, *Biochem. Biophys. Res. Commun.* **20**, 710 (1965).
[2]M. Brenner and W. Huber, *Helv. Chim. Acta* **36**, 1109 (1953).
[3]H. Iwase, *Chem. Pharm. Bull.* **22**, 1663 (1974).

Derivative Type. *N*-Trifluoroacetyl, L-leucine isopropyl ester dipeptide.

Reagent. Trifluoroacetic anhydride (TFAA), thionyl chloride, L-leucine isopropyl ester.

Compound Type. Isovaline.

Reaction

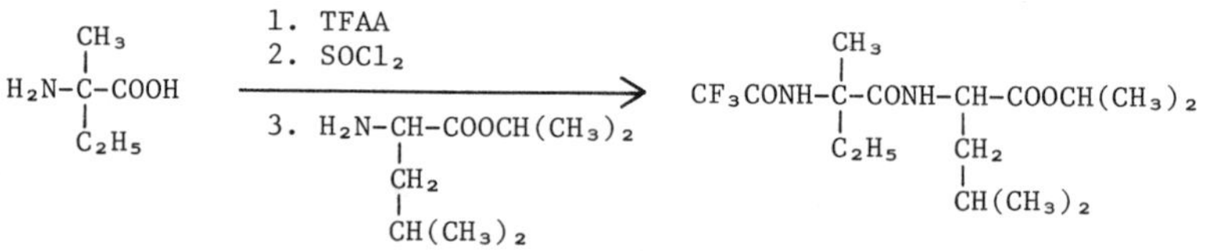

Typical Procedure. Treat suspension of DL-isovaline in CH_2Cl_2 with TFAA. Allow to stand at room temp 90 min then evap to dryness under vac. Add $SOCl_2$ and allow to stand at room temp 15 min. Evap under vac and diss residue in CH_2Cl_2. Add CH_2Cl_2 soln of acid chloride to L-leucine isopropyl ester HCl (prepared by refluxing for 3 hr a soln of L-leucine in i-PrOH satd with HCl). Chill and add dropwise with stirring 2.2 equiv of Et_3N in CH_2Cl_2. Allow to warm gradually by letting stand overnight in the dry ice alcohol bath. Add H_2O and stir vigorously. Ext with CH_2Cl_2 and wash ext with 15% citric acid soln then 15% $NaHCO_3$. Dry over Na_2SO_4 and evap under vac. Diss residue in MeOH and analyze by GC.

Comments. The deriv was used to resolve the enantiomers of isovaline by GC. The Et_3N was added slowly to the chilled soln to minimize racemization.[2]

References

[1]J. J. Flores *et al., J. Chromatogr.* **132**, 152 (1977).
[2]W. A. Bonner, *J. Chromatogr. Sci.* **10**, 159 (1972).

Derivative Type. *O*-Methyl (ether, ester), *N*-(α-methoxy-α-trifluoromethylphenylacetyl).

Reagent. Diazomethane, (*S*)-α-Methoxy-α-trifluoromethylphenylacetyl chloride.

Compound Type. Dihydroxyphenylalanine (DOPA) and α-methyldopa.

Reaction

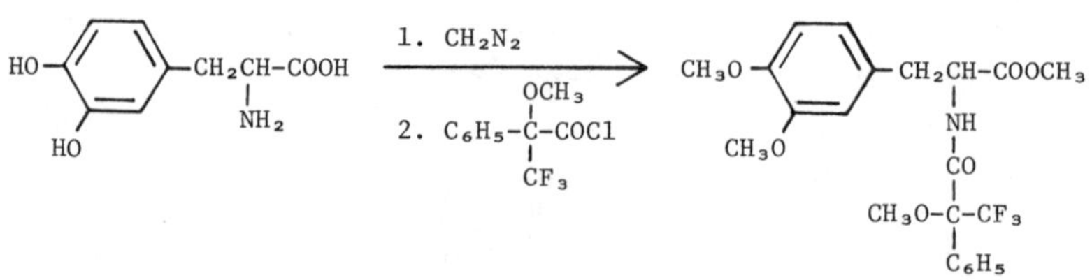

Typical Procedure. Suspend a 10 mg sample of the amino acid in 5 ml MeOH and add 5 ml of CH_2N_2 (*ca* 10-fold excess) in Et_2O. Allow to stand at room temp in the dark for 48 hr. Evap solvent and diss residue in 1 ml 1,2-dichloroethane. Acylate 0.1 ml of the soln with 0.1 ml of the acyl chloride in 1,2-dichloroethane (see below) and 0.1 ml pyridine at 70° for 30 min. Add 2 ml 1 *N* HCl, shake, and separate org layer. Wash with 0.5 ml 15% aq Na_2CO_3 then analyze by GC.

Comments. The reagent soln was prepared by refluxing (50 hr) a soln of 1 g α-methoxy-α-trifluoromethylphenylacetic acid in 3 ml freshly dist $SOCl_3$, evap under vac, and diss residue in 10 ml 1,2-dichloroethane.[2] The method was used to resolve the enantiomers by GC. *N*-Methyl and *N*,*N*-dimethyl derivs were also produced in the CH_2N_2 step, but these were not acylated by the conditions used. MS data were given for the derivs.

References

[1]J. Gal and M. M. Ames, *Anal. Biochem.* **83**, 266 (1977).

[2]J. Gal, *J. Pharm. Sci.* **66**, 169 (1977).

Derivative Type. *N*-[(−)-α-Methoxy-α-methyl-l-naphthaleneacetyl].

Reagent. (−)-α-Methoxy-α-methyl-1-naphthaleneacetic acid/*N*,*N*′-dicyclohexylcarbodiimide(DCC).

Compound Type. Amino acids (as methyl esters).

Reaction

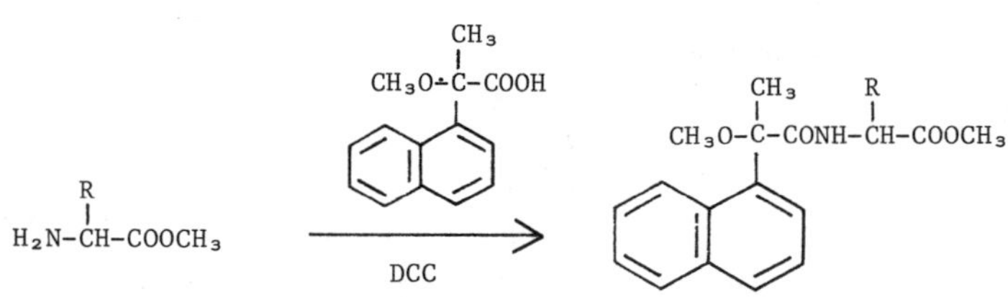

Typical Procedure. Condense the reagent with the amino acid methyl ester in pyridine in the presence of DCC (details not given). Ext the deriv with EtOAc and wash ext with H_2O. Analyze by HPLC.

Comments. HPLC resolution of the diastereomeric derivs of 17 amino acids was demonstrated. The preparation and resolution of the reagent was given.

References

[1]J. Goto *et al.*, *Chem. Pharm. Bull.* **25**, 847 (1977).

Derivative Type. *N*-(+)-10-Camphorsulfonyl, *p*-nitrobenzyl ester.

Reagent. (+)-10-Camphorsulfonyl chloride (R-SO_2Cl)/sodium hydroxide, *p*-nitrobenzyl bromide/trimethylamine (TMA).

Compound Type. Amino acids.

Reaction

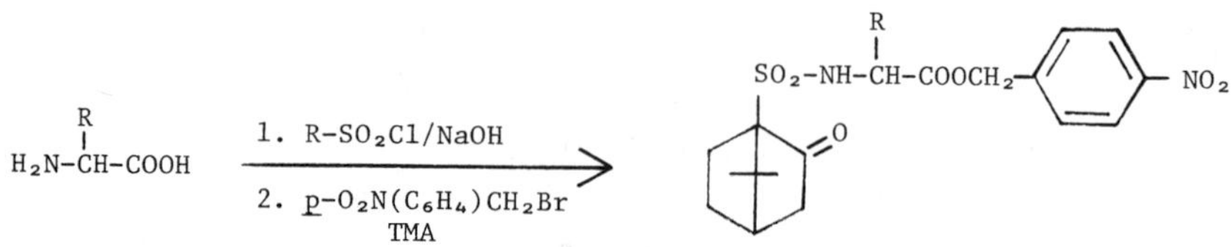

Typical Procedure. Diss 1 mmol of amino acid in a mixt of 20 ml 1 *N* NaOH and 10 ml Et_2O and, with vigorous stirring at 0°, add dropwise 30 ml of anhyd Et_2O containing 2 mmol (+)-10-camphorsulfonyl chloride. Stir 3 hr at room temp; then remove the Et_2O layer. Wash the aq phase 2 × with Et_2O, acidify with HCl, then ext with Et_2O. Dry the ext over Na_2SO_4 and evap to dryness. Diss the residue in 10 ml DMF and add 1 drop TMA and 1.1 mmol *p*-nitrobenzyl bromide. Heat 2 hr at 55° then dil with 40 ml $CHCl_3$. Wash with H_2O, dry over Na_2SO_4, and evap to dryness to yield the deriv. Diss in $CHCl_3$ for HPLC analysis.

Comments. The derivs permitted HPLC separation of the enantiomers on a silica column. The *N*-substituent added a second asymmetric center while the *p*-nitrobenzyl group imparted UV absorption. Of the amino acids examined, the L-amino acid derivs gave consistently shorter retention times than the D-isomer derivs.

References

[1]H. Furukawa *et al.*, *Chem. Pharm. Bull.* **23**, 1625 (1975).

[2]H. Furukawa *et al.*, *J. Chromatogr.* **136**, 428 (1977).

9.4.3. Diketopiperazines

Derivative Type. Trimethylsilyl.

Reagent. Bis(trimethylsilyl)acetamide.

Compound Type. Diketopiperazines.

Reaction

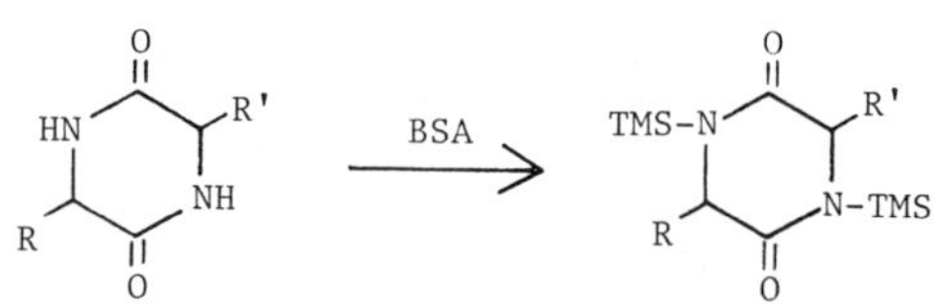

Typical Procedure. Diss 1–5 mg sample in 0.2–1.0 ml of 25% BSA in DMF. Heat 10 min at 80° then analyze by GC.

Comments. GC data were reported for the derivs of 16 diketopiperazines including some *N*-methylated compds. GC separation of diastereomers was possible.

References

A. B. Mauger, *J. Chromatogr.* **37**, 315 (1968).

Derivative Type. Trifluoroacetyl.

Reagent. Trifluoroacetic anhydride (TFAA).

Compound Type. Diketopiperazines.

Reaction

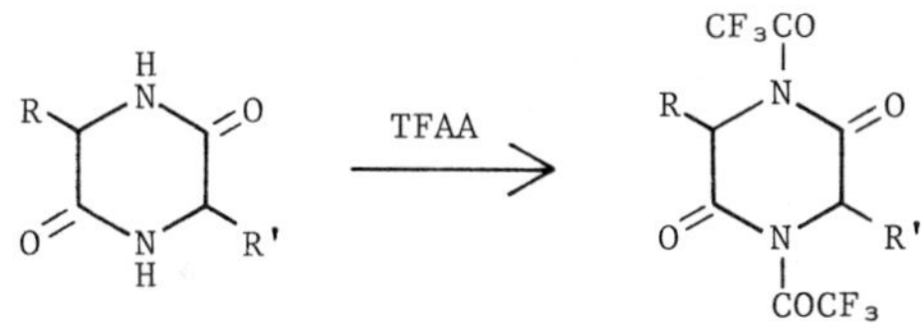

Typical Procedure. To 1–2 mg of diketopiperazine add 0.2 ml TFAA. Heat in sealed vial at 90–100° for 5–30 min depending upon solubility. Analyze by GC.

Comments. The method was used to separate diastereoisomeric diketopiperazines. Racemization was observed for the L-serine-D-phenylalanine compd but could be avoided by on-column injection.

References

[1]G. P. Slater, *J. Chromatogr.* **64**, 166 (1972).

9.5. DERIVATIVES OF CARBONYL COMPOUNDS

Derivative Type. Ketal.

Reagent. D-(−)-2,3-Butanediol/*p*-toluenesulfonic acid (TsOH).

Compound Type. Camphor.

Reaction

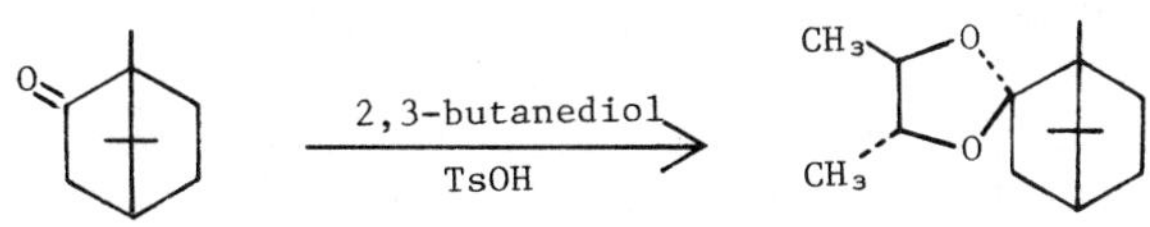

Typical Procedure. Diss 0.1 mol camphor and 0.1 mol D-(−)-2,3-butanediol in 80 ml anhyd benzene containing 75 mg TsOH. Reflux 14 hr using an H_2O separator. Wash with H_2O, dry, and dist to give the crude product. Analyze by GC.

Comments. The method was used to demonstrate GC resolution of enantiomers via diastereomeric derivs.

References

[1]J. Casanova and E. J. Corey, *Chem. Ind.*, 1664 (1961).

Derivative Type. Ortho esters.

Reagent. (−)-2R,3R-2,3-Butanediol/trimethyl orthoformate/sulfuric acid.

Compound Type. Enantiomeric δ-lactones.

Reaction

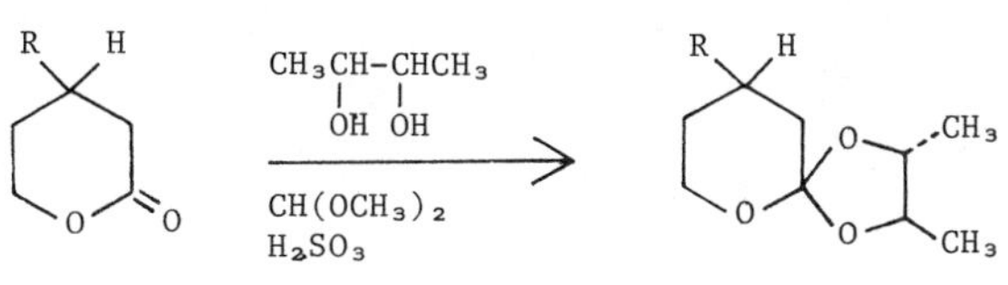

Typical Procedure. Diss 2 mmol of the lactone in 5 ml THF (dried with $LiAlH_4$) and add under N_2 270 mg (3 mmol)2,3-butanediol, 318 mg (3 mmol) trimethyl orthoformate, and 3 drops conc H_2SO_4. Stir at room temp 24 hr; then add 0.5 ml Et_3N and pour into aq $NaHCO_3$ soln. Ext 3× with benzene. Wash ext with aq $NaHCO_3$ and aq NaCl. Dry over $MgSO_4$ or Na_2SO_4 and evap to an oil. Analyze by GC.

Comments. The derivs could also be prepared by refluxing in benzene with TsOH catalyst and collecting H_2O in a Dean-Stark trap. Extension of the method to γ-lactones was not successful.

References

[1]G. Saucy *et al.*, *J. Org. Chem.* **42**, 3206 (1977).

Derivative Type. 2,2,2-Trifluoro-1-phenylethylhydrazone.

Reagent. (+)-2,2,2-Trifluoro-1-phenylethylhydrazine.

Compound Type. Asymmetric ketones.

Reaction

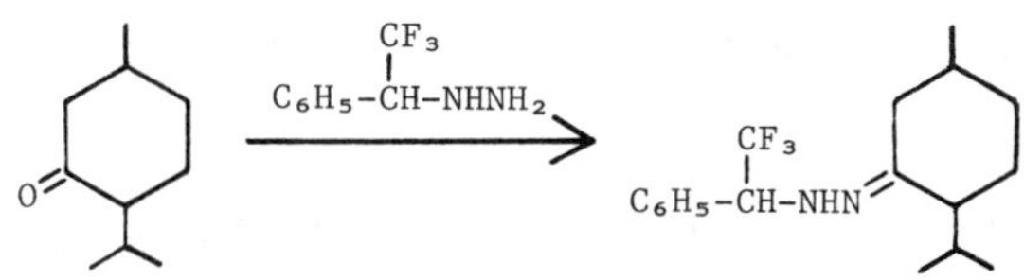

Typical Procedure. Diss the ketone and 1.2 mol equiv of the reagent in 0.5 ml of AcOH/AcONa/EtOH (1:2:100) and allow to stand at room temp 2 hr (unhindered ketones) or reflux 2 hr (hindered ketones). Analyze an aliquot of the soln by GC.

Comments. Some racemization occurred with unhindered ketones containing α-hydrogen. Sterically pure hydrazones were obtained from optically pure ketones which were hindered or which contained an asymmetric tertiary α-carbon. Diastereomeric hydrazones from β-alkyl ketones which were examined were not GC separable. The preparation and resolution of the reagent were reported.

References

[1]W. E. Periera *et al.*, *Austr. J. Chem.* **24**, 1103 (1971).

10. Fatty Lipids

The derivatives of fatty acids, either free or liberated from triglycerides were covered in Section 3. This section presents derivatives of other classes of lipids containing either fatty acid groups or other fatty alkyl chains. Derivatives of intact monoglycerides and glycerol monoethers are discussed first. These monoglyceride and monoether compounds contain two free hydroxyls that require derivatization for optimal GC performance. The hydroxyls can be converted to methyl ethers with diazomethane and boron trifluoride (p. 438) or can be trimethylsilylated (p. 439). The presence of adjacent hydroxyls make acetonide (p. 439) and boronate (p. 440) formation applicable to this group. Periodate cleavage (p. 440) and alkene formation (p. 440) have also been used for analysis of the monosubstituted glycerols.

Sphingolipids. The sphingolipids are an important class of lipids that are particularly abundant in nervous tissue. These compounds consist of a long chain base (see below), which may contain a *trans* double bond (*sphingosine*) or a saturated chain (*sphinganine*). *N*-Acylation of the base yields a *ceramide*. The acyl group may be a normal chain fatty acid or a 2-hydroxy fatty acid. Formation of a glycoside at the terminal hydroxyl yields a *cerebroside*. Alternatively, the terminal hydroxyl may be esterfied with an alcohol phosphate group such as choline phosphate in *sphingomyelin*.

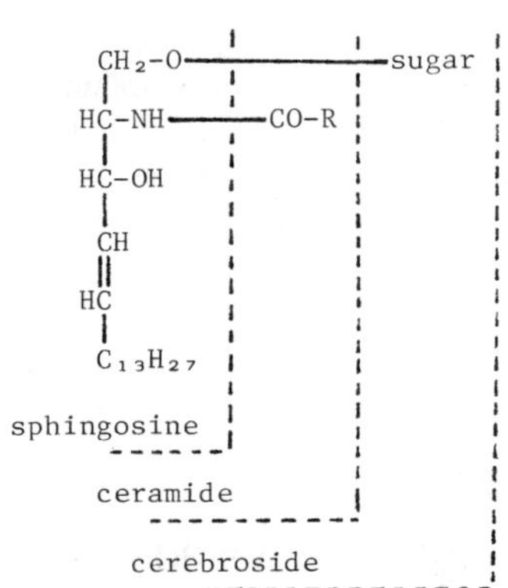

Derivatization of the sphingosine bases has employed combinations of derivatives discussed previously for hydroxyl and amino groups (p. 441). The 1,3-dihydroxy function permits formation of cyclic boronates which can undergo subsequent derivatization of the amino group (p. 441). The *N*-acylated ceramides can also form cyclic boronates (p. 444); but most studies of these compounds have employed trimethylsilyl derivatives (p. 443). In the cerebrosides the problem of sugar derivatization is present. Derivatives of these compounds have been obtained by trimethylsilylation (p. 444), acetylation (p. 446), and permethylation (p. 447) of the sugar moiety. A reduced amino "alcohol" derivative of cerebrosides, somewhat similar to the amino alcohol peptide derivatives (p. 335) has been employed for MS analysis (p. 447). Acylation of the amide nitrogen with a *p*-nitrobenzoyl group yields a chromogenic derivative suitable for HPLC analysis (p. 447). Viswanathan[1] has reviewed the derivatization and GC-MS analysis of these polar lipids.

References

[1]C. V. Viswanathan, *J. Chromatogr.* **98**, 105 (1974).

10.1. DERIVATIVES OF MONOGLYCERIDES AND GLYCEROL ETHERS

Derivative Type. Dimethoxy.

Reagent. Diazomethane/boron trifluoride.

Compound Type. Glyceryl ethers.

Reaction

```
CH2-O-(CH2)17CH3                CH2-O-(CH2)17CH3
|                   CH2N2       |
CH-OH             -------->     CH-OCH3
|                    BF3        |
CH2-OH                          CH2-OCH3
```

Typical Procedure. Hydrolyze oil sample by refluxing 2 hr with 1 *N* KOH/MeOH. Concentrate, dil with H_2O, and ext with Et_2O. Isolate glycerol ethers by chromatography on alumina. Diss a 100–150 mg sample in 1 ml of 1% $BF_3 \cdot Me_2O$ in Et_2O. Add ice cold CH_2N_2 in Et_2O dropwise until the yellow color persists. After 30 min at room temp evap excess CH_2N_2 and filter. To remove fatty acid esters, conc to approximately 1 ml and diss in 6 ml 1 *N* aq KOH. After 20 hr at room temp ext 4 × with Et_2O. Dry ext over $MgSO_4$, evap almost to dryness, and rechromatograph on alumina. Analyze the isolated dimethoxy derivs by GC-MS.

Comments. The best yield of dimethoxy ethers was obtained when there was simultaneous formation of polymethylene indicated by formation of large Et_2O insoluble aggregates. Positions of double bonds in alkyl chains were determined by oxidation with chromic acid and methylation of the resulting acids with acidic methanol.

References

[1]B. Hallgren and S. Larsson, *J. Lipid Res.* **3**, 31 (1962).

Derivative Type. Trimethylsilyl ether.

Reagent. Hexamethyldisilazane (HMDS)/trimethylchlorosilane (TMCS)/pyridine.

Compound Type. Monoacyl and monoalkyl glycerols.

Reaction

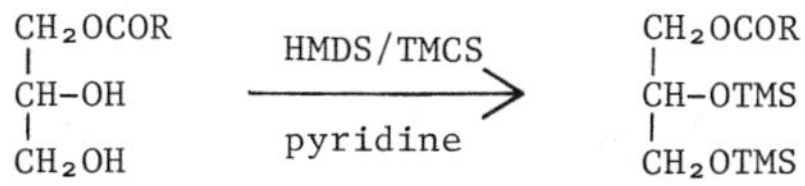

Typical Procedure. Elute TLC bands with Et_2O MeOH (80:20, v/v). Add pet ether and wash several times with H_2O. Dry over Na_2SO_4 and evap to dryness. Treat with HMDS/TMCS/pyridine (5:2:12, v/v/v) by first adding the pyridine and HMDS to the residue in a N_2 filled vial. Chill the mixt in ice, add the TMCS, and blow away the fumes with N_2. Allow to stand 60–90 min at room temp then evap to dryness under N_2. Diss the residue in pet ether and analyze by GC or GC-MS.

Comments. The deriv method did not cause isomerization. The derivs of the 1- and 2-isomers were separable by GC.[1] MS data were reported for the derivs.[2] The derivs of 1-alkyl-2-acyl-[3] and 1,2-dialkylglycerols[4] have been prepared with the same reagents (60° for 30 min) and their MS characteristics reported.

References

[1]J. J. Myher and A. Kuksis, *Lipids* **9**, 382 (1974).

[2]J. J. Myher *et al.*, *J. Lipid Res.* **15**, 586 (1974).

[3]K. Satouchi and K. Saito, *Biomed. Mass Spectrom.* **3**, 122 (1976).

[4]K. Satouchi *et al.*, *Biomed. Mass Spectrom.* **5**, 87 (1978).

Derivative Type. Acetonide (2,3-0-isopropylidene).

Reagent. Acetone/perchloric acid.

Compound Type. Glyceryl ethers.

Reaction

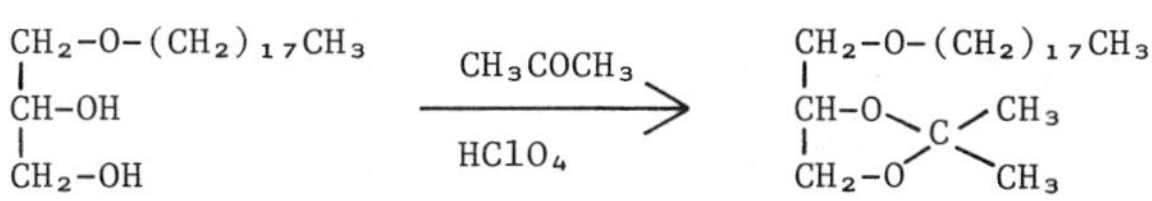

Typical Procedure. Suspend *ca* 400 mg of glycerol ether in 10 ml acetone and add 0.05 ml 12 *N* $HClO_3$. Stir 45 min (clear soln results in *ca* 10 min); then add H_2O until turbid. Ext with 4 vol Et_2O. Wash Et_2O with H_2O to remove traces of acid; then evap under vac. Diss residue in hexane and purify by column chromatography or TLC on silica gel. Analyze by GC.

Comments. The method was used in studies of the glycerol ether phospholipids of bovine erythrocytes. The acetonide derivs were separated by degree of unsaturation via chromatography of their acetoxymercurimethoxy derivs.

References

[1]D. J. Hanahan *et al.*, *Biochemistry* **2**, 630 (1963).

Derivative Type. Methaneboronate.

Reagent. Methaneboronic acid.

Compound Type. Monoglycerides and monoglycerol ethers.

Reaction

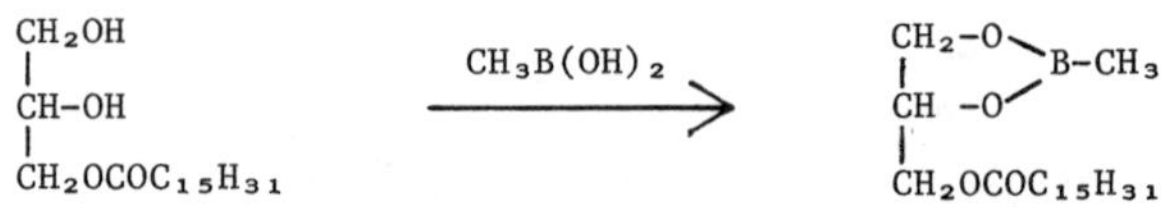

Typical Procedure. Treat sample with 1.1 mol equiv of methaneboronic acid in EtOAc. Warm if necessary to effect soln. Analyze by GC-MS.

Comments. Derivs were prepared from both 1,2- and 1,3-diols. EI and CI-MS data were reported for the derivs.

References

[1]S. J. Gaskell *et al.*, *Anal. Lett.* **9**, 325 (1976).

Derivative Type. Alkoxyacetaldehydes.

Reagent. Periodic acid, sodium arsenite/sulfuric acid.

Compound Type. Alkyl α-glycerol ethers.

Reaction

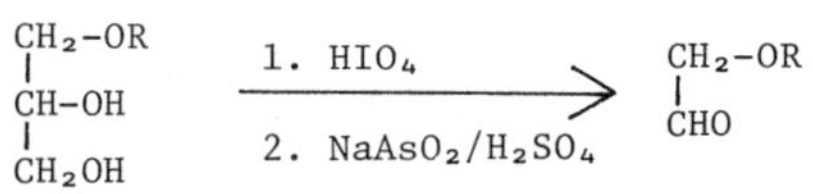

Typical Procedure. To a sample containing 0.10–0.06 μmol α-glycerol ether add 1.0 ml 5 mg/ml HIO_4 in 95% EtOH. Mix and allow to react 15 min. Add 2.0 ml H_2O, 0.7 ml 1.2 *N* $NaAsO_2$, and 0.5 ml 5 *N* H_2SO_4. Mix and allow to stand until the brown color disappears. Ext with 2 × 5 ml hexane and wash ext with 2 × 1 ml H_2O. Purify the alkoxyacetaldehydes by TLC on silica gel (hexane/Et_2O/HOAc, 80:20:1.5, v/v/v), elute with $CHCl_3$, and analyze by GC.

Comments. Batyl alcohol was used as a quantitative standard by running known amts as separate samples.

References

[1]R. A. Gelman and J. R. Gilbertson, *Anal. Biochem.* **31**, 463 (1969).

Derivative Type. Allyl ester.

Reagent. Methanesulfonyl (mesyl) chloride/pyridine, sodium iodide.

Compound Type. Monoglycerides.

Reaction

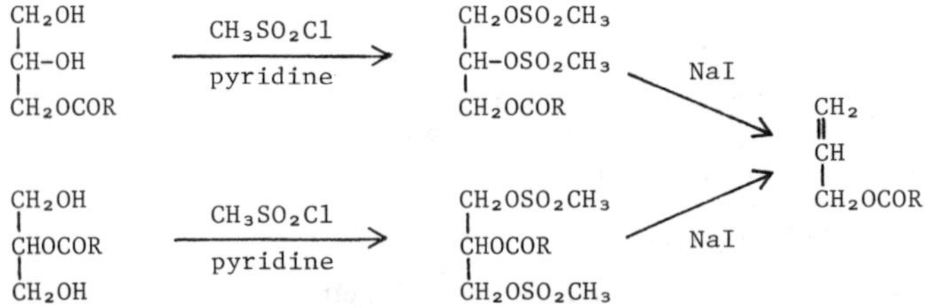

Typical Procedure. Diss 100 mg monoglyceride mixt in 1 ml anhyd alcohol free $CHCl_3$. Add 0.1 ml anhyd pyridine and cool to 0°. Add 0.1 ml mesyl chloride and allow to stand at 0° overnight. Dil with 100 ml ice cold H_2O and ext with 3 × 50 ml $CHCl_3$. Dry ext over Na_2SO_4 and evap under vac. Dry over P_2O_5; then diss in 1 ml anhyd acetone. Add 1 ml 25% NaI in acetone and heat in a sealed tube at 100° for 2 hr. Cool and wash into a sep funnel with H_2O and Et_2O. Remove free I_2 with 10% aq $Na_2S_2O_3$ and ext with 3 × 50 ml Et_2O. Dry ext over Na_2SO_4 and evap under vac. Diss residue in hexane and analyze by GC.
Comments. The procedure above gave analysis of total monoglycerides. β-Monoglycerides were determined by first oxidizing the α-monoglycerides with periodate,[2] and then analyzing as above.

References

[1]A. G. McInnes, *J. Am. Oil Chem. Soc.* **37**, 7 (1960).
[2]M. Kruty, *et al.*, *J. Am. Oil Chem. Soc.* **31**, 466 (1954).

10.2. DERIVATIVES OF SPHINGOSINE BASES

Derivative Type. Methyl boronate, *N*-acetyl or acetone Schiff base or *N,N*-dimethylaminomethylene.

Reagent. Methaneboronic acid, acetic anhydride (Ac_2O) or acetone or dimethylformamide dimethylacetal.

Compound Type. Sphingosine long chain bases.

Reaction

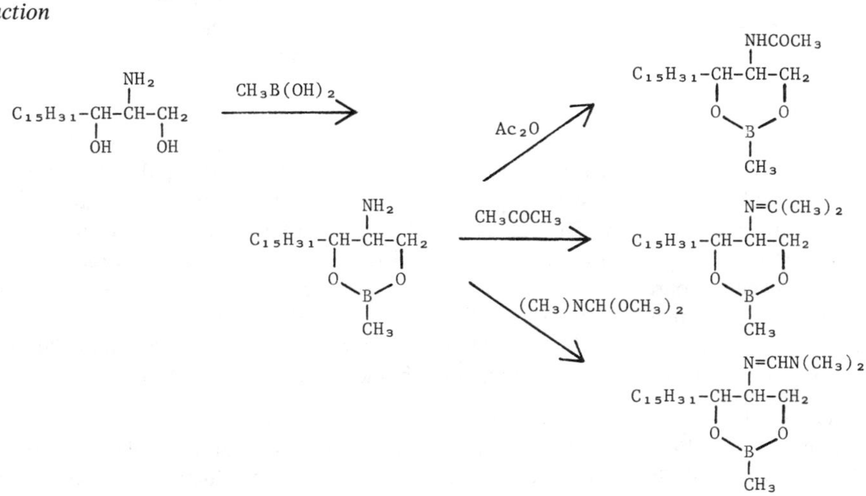

Typical Procedure. Diss sphingosine in pyridine and add an equimol amt of methaneboronic acid. Let stand 10 min at room temp then evap to dryness under N_2. Analyze by GC or deriv further. (a) *N*-Acetylation—Diss 100 μg of boronate in 100 μl of pyridine/Ac_2O(20:1). Allow to stand overnight at room temp. Evap under N_2 and diss residue in EtOAc for GC analysis. (b) Acetone Schiff base—Diss boronate in acetone and heat at 60° for 10 min. Analyze by GC. (c) *N,N*-Dimethylaminomethylene deriv—Diss 100 μg of boronate in 20 μl of a soln of 2 mmol/ml DMF dimethylacetal in pyridine. Let stand several min at room temp then evap under N_2 and diss residue in EtOAc for GC analysis.

Comments. *n*-Butyl and phenyl boronates were also prepared. Selective *N*-acetylation was also performed prior to boronate formation by diss 5 mg of sphingosine in 500 μl MeOH and 150 μl Ac_2O and allowing to stand overnight at room temp. 4-D-Hydroxysphinganine formed a bis(boronate). MS data (both EI and CI) were reported for the deriv.

References

[1]S. J. Gaskell and C. J. W. Brooks, *J. Chromatogr.* **122**, 415 (1976).

[2]S. J. Gaskell and C. J. W. Brooks, *Org. Mass Spectrom.* **12**, 651 (1977).

Derivative Type. Dihydro, trimethylsilyl.

Reagent. Hydrogen/platinum oxide, hexamethyldisilazane (HMDS)/trimethylchlorosilane (TMCS)/pyridine.

Compound Type. Sphingosines.

Reaction

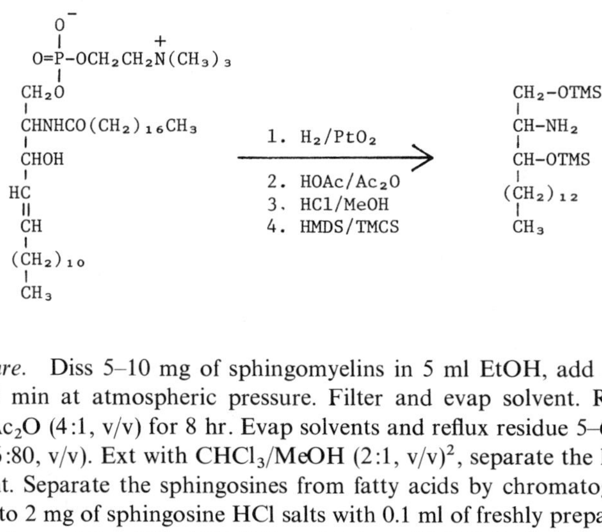

Typical Procedure. Diss 5–10 mg of sphingomyelins in 5 ml EtOH, add 5 mg PtO_2, and hydrogenate 60 min at atmospheric pressure. Filter and evap solvent. Reflux residue in 0.5 ml $HOAc/Ac_2O$ (4:1, v/v) for 8 hr. Evap solvents and reflux residue 5–6 hr in 2 ml conc HCl/MeOH (16:80, v/v). Ext with $CHCl_3$/MeOH (2:1, v/v)[2], separate the lower phase, and evap the solvent. Separate the sphingosines from fatty acids by chromatography on silicic acid.[2] Treat up to 2 mg of sphingosine HCl salts with 0.1 ml of freshly prepared reagent consisting of 1.0 ml dry pyridine, 0.2 ml HMDS, and 2 drops TMCS. Reflux 15 min at 120°. Evap under N_2 and diss residue in 5% pyridine in heptane. Analyze by GC-MS.

Comments. The sphingomyelins were reduced prior to hydrolysis to prevent formation of by-products from allylic substitution during hydrolysis. Gaver and Sweeley[3] described a methanolysis procedure which reduced the formation of by-products; they then silylated with HMDS/TMCS/ pyridine (2:1:10). Carter and Gaver[4] subsequently reported HMDS/TMCS/pyridine (13:8:10) as an improved reagent for sphingosine bases.

References

[1]K.-A. Karlsson, *Acta Chem. Scand.* **19**, 2425 (1965).

[2]K.-A. Karlsson, *Acta Chem. Scand.* **18**, 2395 (1964).

[3]R. C. Gaver and C. C. Sweeley, *J. Am. Oil Chem. Soc.* **42**, 294 (1965).

[4]H. E. Carter and R. C. Gaver, *J. Lipid Res.* **8**, 391 (1967).

10.3. DERIVATIVES OF CERAMIDES

Derivative Type. Trimethylsilyl.

Reagent. Hexamethyldisilazane (HMDS)/trimethylchlorosilane (TMCS)/pyridine.

Compound Type. Ceramides.

Reaction

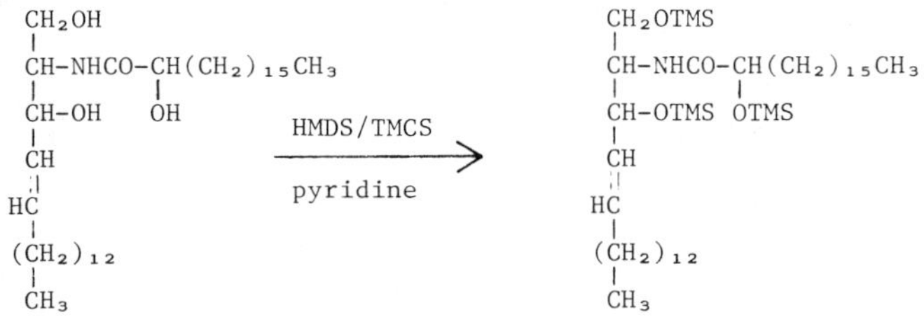

Typical Procedure. Diss a 100 μg sample in 100 μl dry pyridine, 20 μl HMDS, and 10 μl TMCS. Allow to stand 30 min at room temp. Evap to dryness and diss residue in 100 μl CS_2. Analyze by GC-MS.

Comments. Two series of ceramides (containing the bases sphingosine and sphinganine) with C_{16}–C_{25} fatty acids were examined.[1] GC separation of the nine members of each series was accomplished on an OV-1 column. 2′-Hydroxyacid ceramides were separated from nonhydroxyacid ceramides by TLC.[2] MS data were reported for a series of synthetic ceramides prepared from sphinganine and sphingosine and fatty acids ranging from C_{16}–C_{26}.[3-6] The method was applied to analysis of ceramides from human plasma sphingomyelins.[4] MS data were reported for the derivs.[1-6]

References

[1]B. Samuelsson and K. Samuelsson, *Biochem. Biophys. Acta* **164**, 421 (1968).

[2]S. Hammerstrom, *Eur. J. Biochem.* **15**, 581 (1970).

[3]B. Samuelsson and K. Samuelsson, *J. Lipid Res.* **10**, 41 (1969).

[4]B. Samuelsson and K. Samuelsson, *J. Lipid Res.* **10**, 47 (1969).

[5]S. Hammerstrom, *et al.*, *J. Lipid Res.* **11**, 150 (1970).

[6]S. Hammerstrom, *J. Lipid Res.* **11**, 175 (1970).

Derivative Type. Trimethylsilyl.

Reagent. Trimethylsilylimidazole (TMSI)/bis(trimethylsilyl)acetamide (BSA)/trimethylchlorosilane (TMCS).

Compound Type. Ceramides.

Reaction

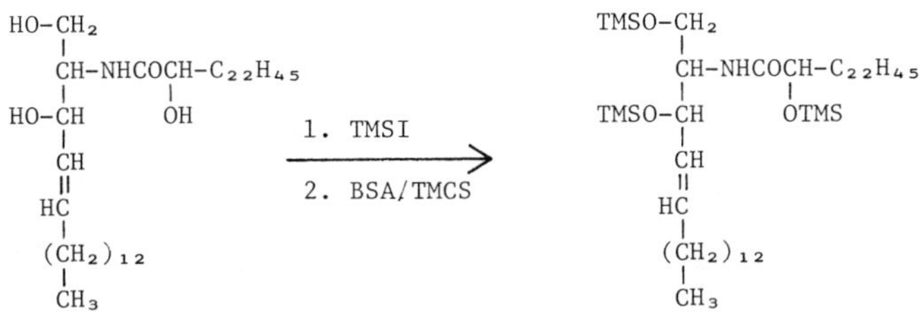

Typical Procedure. Diss 1 mg ceramide in 0.1 ml TMSI and heat 3 hr at 60°. Add 0.1 ml BSA and 0.05 ml TMCS and heat 1 hr at 60°. Analyze by GC-MS.

Comments. Derivs were prepared from ceramides with hydroxylated and nonhydroxylated acyl groups. The acyl components were further analyzed as TMS and HFB methyl esters. After transesterfication with 6% (v/v) H_2SO_4 in MeOH (15 hr at 60°), the resulting methyl esters were extd with hexane, the ext washed with 5% $NaHCO_3$, dried over $MgSO_4$, and evapd under N_2 at room temp. The resulting methyl esters were analyzed directly or converted to TMS (BSA or BSTFA, 2 hr, 60°) or HFB (HFBI, 2 hr, 60°) derivs.

References

[1]G. Casparrini *et al.*, *Chem. Phys. Lipids* **3**, 1 (1969).

Derivative Type. Methaneboronate.

Reagent. Methaneboronic acid.

Compound Type. Ceramides.

Reaction

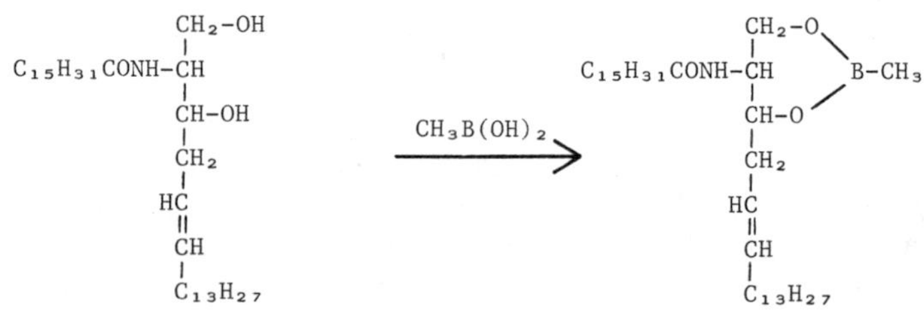

Typical Procedure. Diss sample in pyridine and add 1.1 mol equiv of methaneboronic acid. Allow to stand 10 min at room temp then analyze by GC or GC-MS.
Comments. For ceramides with α-hydroxyacyl groups, 2.2 mol equiv of the boronic acid were used forming bis-boronate derivs. GC and MS data for the derivs were reported.

References

[1]S. J. Gaskell *et al.*, *J. Chromatogr.* **126**, 591 (1976).

10.4. DERIVATIVES OF CEREBROSIDES

Derivative Type. Trimethylsilyl.

Reagent. Hexamethyldisilazane (HMDS)/trimethylchlorosilane (TMCS)/pyridine.

Compound Type. Cerebrosides.

Reaction

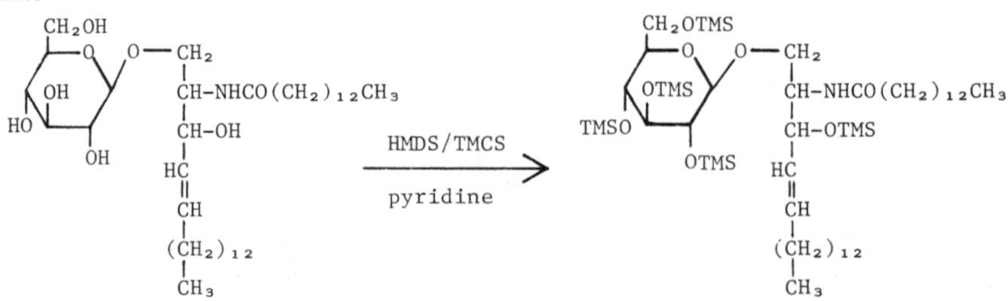

Typical Procedure. Diss sample in HMDS/TMCS/pyridine (2:1:10, v/v/v) and allow to stand at room temp at least 1 hr. Analyze by direct probe MS.

Comments. MS data were reported for the derivs of cerebrosides having a D-galactopyranose in β-glycosidic linkage to ceramides containing di- and trihydroxy long chain bases and normal or hydroxy fatty acids.[1] Other workers have also reported data on TMS cerebrosides from HMDS/TMCS/pyridine,[2] including CI-MS data.[3]

References

[1]K.-A. Karlsson *et al.*, *Chem. Phys. Lipids* **9**, 230 (1972).

[2]K. Samuelsson and B. Samuelsson, *Biochem. Biophys. Res. Commun.* **37**, 15 (1969).

[3]M. Oshima *et al.*, *Chem. Phys. Lipids* **19**, 289 (1977).

Derivative Type. Trimethylsilyl.

Reagent. Bis(trimethylsilyl)trifluoroacetamide (BSTFA)/pyridine.

Compound Type. Glycosphingolipids.

Reaction

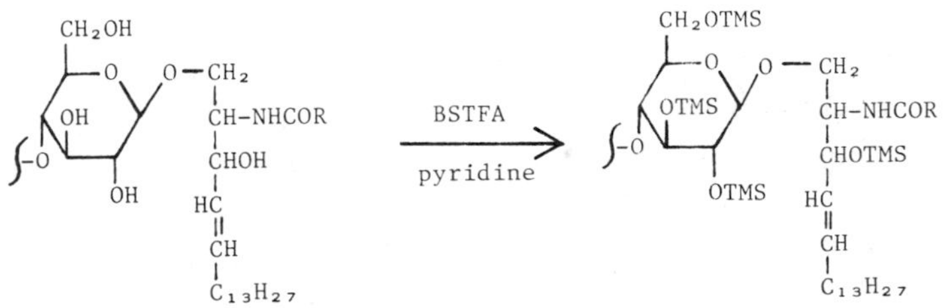

Typical Procedure. To 20–200 μg of purified glycosphingolipid add 100 μl BSTFA and 50 μl pyridine. Heat at 60° for 30 min. Evap an aliquot corresponding to 10–20 μg of sample in a probe tube and analyze by MS.

Comments. The MS data of the fully silylated intact glycosphingolipids yielded information regarding position of substitution of monosaccharide units, the nature of the sphingolipid bases, and the relative amts of constituent fatty acids.

References

[1]C. C. Sweeley and G. Dawson, *Biochem. Biophys. Res. Commun.* **37**, 6 (1969).

[2]G. Dawson and C. C. Sweeley, *J. Lipid Res.* **12**, 56 (1971).

Derivative Type. Methyl ester, trimethylsilyl.

Reagent. Methanol (MeOH)/HCl, bis(trimethylsilyl)trifluoroacetamide (BSTFA)/trimethylchlorosilane (TMCS).

Compound Type. 2-Hydroxy fatty acids in glycosphingolipids.

Reaction

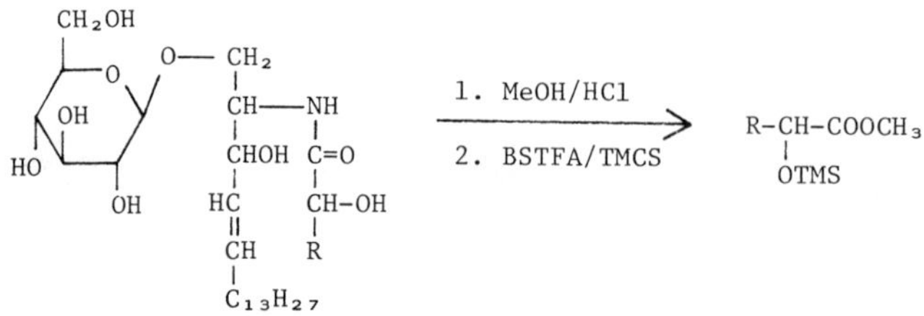

Typical Procedure. Diss 300 μg of glycosphingolipid in 1 ml 1 *N* HCl in MeOH and heat at 85° for 12 hr. Cool and ext with 3 × 2 ml hexane. Evap ext under N_2 and treat residue with 100 μl of BSTFA containing 1% TMCS. Let stand 30 min at 25°, and analyze 1–3 μl aliquots by GC-MS.

Comments. Derivs were used to identify 2-hydroxy fatty acids by use of mass chromatograms of ions characteristic of the derivs (m/e 73 for TMS derivs in general and M-59 ions for the individual compds).

References

[1]R. A. Laine *et al., Biomed. Mass Spectrom.* **1**, 10 (1974).

Derivative Type. Acetyl.

Reagent. Acetic anhydride (Ac_2O)/pyridine.

Compound Type. Cerebrosides.

Reaction

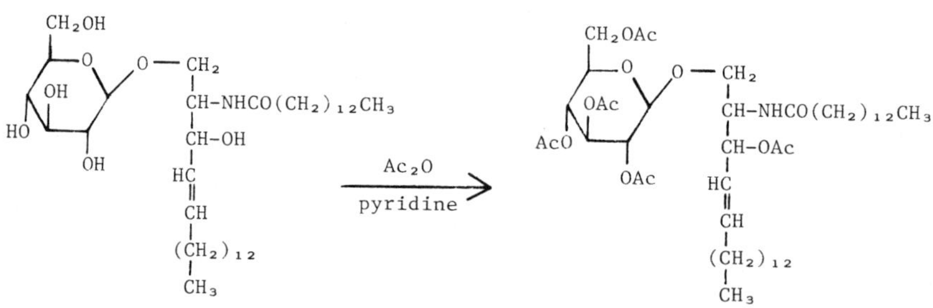

Typical Procedure. Diss sample in 1:1 (v/v) Ac_2O/pyridine and allow to stand at room temp overnight. Dil with H_2O and ext with Et_2O. Evap ext and analyze residue by direct probe MS.

Comments. *N,N*-Diacetyl derivs of glycosides without *N*-acyl groups were obtained by acetylation at 60° for 24 hr. The resulting mixt of *N*-acetyl and *N,N*-diacetyl derivs was sepd on a silica gel column. MS data were reported for seven cerebrosides.

References

[1]B. A. Andersson *et al., Chem. Phys. Lipids* **9**, 89 (1972).

Derivative Type. Permethyl, amine.

Reagent. Dimethylsulfinyl carbanion/methyl iodide, lithium aluminum hydride.

Compound Type. Glycosphingolipids.

Reaction

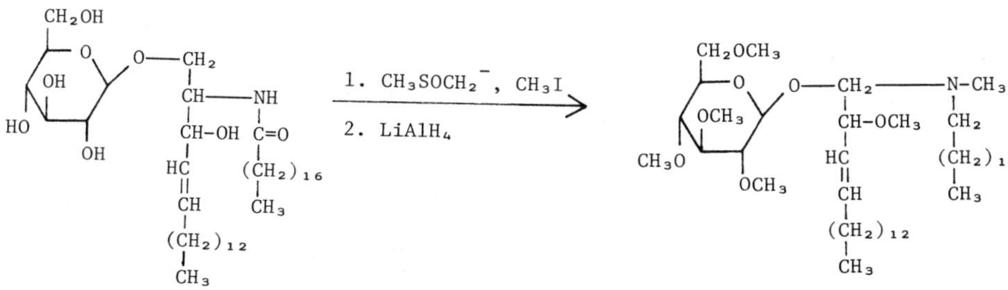

Typical Procedure. Methylate sample by the Hakomori procedure.[2] Reduce the permethyl deriv by treatment with $LiAlH_4$ in Et_2O. Purify deriv by chromatography on a silica column (detailed procedure was not given). Analyze by direct probe MS.

Comments. The reduced and nonreduced permethyl derivs were examined. Molecular weight ions (M-1) were not observed for permethyl derivs with more than four sugar units but were observed for the reduced derivs with up to six sugars.

References

[1]K.-A. Karlssen *et al.*, *Biomed Mass Spectrom.* **1**, 49 (1974).

[2]S. Hakomori, *J. Biochem.* **55**, 205 (1964).

Derivative Type. *O*-Acetyl, *N*-*p*-nitrobenzoyl.

Reagent. Acetic anhydride (Ac_2O)/pyridine, *p*-nitrobenzoyl chloride/pyridine.

Compound Type. Glycolipids.

Reaction

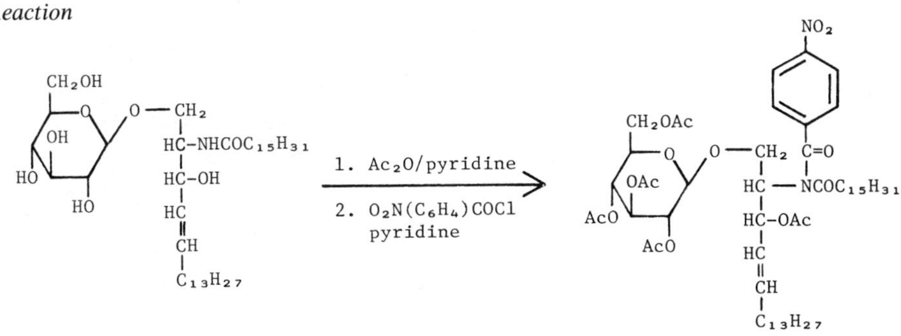

Typical Procedure. Diss 2–10 mg glycolipid in 0.5 ml Ac_2O pyridine (1:1, v/v) and heat at 80° for 2 hr. Dry in a dessicator over P_2O_5 then add 50–120 mg *p*-nitrobenzoyl chloride in 0.3 ml pyridine. Heat at 60° for 4 hr. Cool in an ice bath, add 2 ml 3% aq $NaHCO_3$, and ext with 2 ml $CHCl_3$. Wash ext with 2 × 2 ml 3% aq $NaHCO_3$ and 2 × 2 ml H_2O. Evap to dryness and diss residue in 50 μl $CHCl_3$/MeOH (1:1, v/v). Analyze by HPLC.

Comments. The deriv method was applied to glucosyl ceramide, lactosylceramide, trihexosyl ceramide, globoside, and hematoside (*N*-acetyl type) and permitted separations on the basis of fatty acid content. The deriv was an improvement over the earlier reported[2] benzoyl deriv since it had a tenfold higher UV absorption at 254 nm.

References

[1] A. Suzuki, *et al.*, *J. Biochem.* **82**, 1185 (1977).
[2] A. Suzuki *et al.*, *J. Biochem.* **80**, 1181 (1976).

11. Steroids

Derivatization of the steroids, as with many other biologically important compounds, is complicated by their multifunctional nature. The four principal polar groups encountered in steroids are alcoholic and phenolic hydroxyls, carbonyls, and carboxyls. Although some underivatized steroids can be analyzed by GC, in most cases derivatization is employed for both GC and MS analysis. For HPLC analysis, the polar groups provide "handles" for chromogenic derivatization. For details of chromatographic and mass spectrometric analysis, as well as further discussion of the derivatives employed, the reader is also referred to the recent book edited by Heftman.[1]

Silyl Derivatives. By far the most commonly employed derivatives in the steroid field are the silyl derivatives. Trimethylsilyl derivatives can be prepared from all four types of polar groups encountered in steroids, including carbonyls under forcing conditions. Most often, however, silylation is employed only for hydroxyls (both alcoholic and phenolic) with other derivatives being used for carbonyls and carboxyls. The differences in steric accessability of hydroxyls at different steroid positions, and the range of trimethylsilyl reagents available, can be exploited for selective derivatization in steroid analysis. Such selective derivatization can be advantageous in steroid structure determination, but can also be a complicating factor. For example, determination of the number of hydroxyls in an unknown, based upon the number of TMS groups added, may lead to an erroneous conclusion if the compound contains a hindered hydroxyl and insufficiently rigorous conditions have been used. Conversely, excessively rigorous conditions may produce enol ether derivatives leading likewise to an erroneous conclusion. Extended reaction times can also introduce trimethylsilyloxy groups adjacent to carbonyls[2] or lead to dehydrogenation (p. 492) in suitably predisposed structures. Thus, although trimethylsilylation is extremely useful in steroid analysis, the analyst must be aware of these potential complications. The silylation of different types of steroid hydroxyls (and carbonyls) using different trimethylsilyl reagent is summarized in Table 5.[3]

The differences in accessability between alcoholic hydroxyls (p. 469) and the differences in reactivity between alcoholic and phenolic steroids (p. 476) have been exploited to prepare mixed trimethylsilyl and other derivatives. The alcoholic and phenolic trimethylsilyl derivatives are also significantly different in stability, and the lability of the phenolic derivatives limits the use of postderivatization cleanup procedures. This problem has been circumvented in one example by first treating the phenolic group with an epoxide to form an hydroxyether. Subsequent silyation yielded a derivative with only alkyl silyl ethers (p. 478). Deuterium labeled trimethylsilyl derivatives, both d_9-TMS (p. 459) and d_3-TMS (p. 476), have been useful in MS analysis. Differences in steric accessability have permitted formation of derivatives with mixed deuterated and unlabeled trimethylsilyl groups (p. 459).

In addition to trimethylsilyl derivatives, other alkyl substituted silyl groups have also been applied to steroids. These include dimethylsilyl (p. 461), ethyl- and propyldimethylsilyl (p. 461), allyldimethylsilyl (p. 462), *t*-butyldimethylsilyl (p. 464), and cyclotetramethylene-*t*-butylsilyl (p. 465). These derivatives have been employed to improve GC separations and to enhance MS behavior. In particular, the *t*-butyl substituted silyl derivatives tend to show facile loss of the *t*-butyl group in the MS to yield intense high mass ions. The halogenomethyldimethylsilyl derivatives have been used for both GC-MS and GC-ECD analysis. In the MS, the resulting halogen isotopic doublets are useful for specific identification. The halogens also impart electron affinity with the iodomethyldimethylsilyl derivatives showing the highest ECD sensitivity. Another halogenated silyl derivative, the pentafluorophenyl dimethylsilyl, or flophemesyl, showed prominent molecular ions in the MS and also exhibited good ECD sensitivity (p. 468).

Acyl Derivatives. Steroids have also been analyzed as acyl derivatives (pp. 453 and 473). In most cases the acyl derivatives utilize halogenated acyl groups for ECD-GC analysis. A comparison of ECD sensitivity for a group of halogenated acyl derivatives is given on p. 456. These derivatives have been prepared using either acid chloride or anhydride reagents. The difference in reactivity of phenolic and alcoholic hydroxyls permits formation of mixed acylated derivatives using an acid chloride followed by an anhydride. Enol acylate derivatives can be prepared from conjugated keto groups, but the use of such derivatives is complicated by the possibility of isomeric structures (p. 493). Chromogenic acylated derivatives have also been used for HPLC analysis. Examples of such derivatives include benzoates (p. 458) and *p*-nitrobenzoates (pp. 458 and 513).

Other Hydroxysteroid Derivatives. The phenolic hydroxyl groups in steroids are sufficiently acidic to undergo methylation with dimethyl sulfate (p. 472). Such methylation of phenolic hydroxyls has been utilized extensively in the GC and GC-MS analysis of estrogens. Methylation of estrogenic steroids has also been accomplished with tetramethylammonium hydroxide

Table 5. Groups Silylated by the Different Reagents after 24 Hours at Room Temperature

Silylating agent	Group: 3α-OH; 3β-OH	HO-Ph	11β-OH	16α-OH	CH_3–C(17)–CH_2–, ····OH	CH_3–C(CH_2OH), ····OH	C(OH), ····H	C(OH), ····CH_3	20α-OH; 20β-OH	21-OH	CH_2OH–C=O, ····H	CH_3–C=O, ····OH	CH_2OH–C=O, ····OH	3-Oxo	17-Oxo	20-Oxo
HMDS	±	+	−	−	−	−	−	−	−	+	−	−	−	−	−	−
TMCS	±	+	−	−	−	−	−	−	+	±	−	−	−	−	−	−
HMDS + TMCS	+	+	−	+	−	−	+	−	+	+	−	−	−	−	−	−
BSA	+	+	−	+	−	−	+	−	+	+	−	−	−	−	−	−
BSA + TMCS	+	+	+	+	−	−	+	−	+	+	−	−	−	−	−	−
TMSDEA	+	+	−	+	−	−	+	−	+	+	−	−	−	−	−	−
TMSDEA + TMCS	+	+	+	+	−	−	+	−	+	+	−	−	−	−	−	−
BSTFA	+	+	−	+	−	−	+	−	+	+	−	−	−	−	−	−
BSTFA + TMCS	+	+	+	+	−	−	+	−	+	+	−	−	−	−	−	−
MSTFA	+	+	−	+	−	−	+	±	+	+	±	−	±	−	±	±
MSTFA + TMCS	+	+	+	+	−	−	+	−	+	+	−	−	−	−	−	−
MSTFA + KOAc	+	+	−	+	−	−	+	±	+	+	+	+	+	±	+	+
TMSI	+	+	−	+	+	+	+	−	+	+	−	−	−	−	−	−
TMSI + MSTFA	+	+	−	+	+	+	+	−	+	+	−	−	−	−	−	−

Positive reactions are indicated by +, semi-quantitative reactions by ± and negative reactions by −. Reproduced with permission from Reference 3.

(p. 473). With this method both phenolic and alcoholic hydroxyls were methylated. Application of "Hakomori" methylation to steroids, using methylsulfinyl carbanion and methyl iodide, also gave methylation of alcoholic hydroxyls (p. 470). This technique is complicated, however, by the potential for *C*-methylation adjacent to carbonyl groups (p. 470). Other derivatizations of hydroxysteroids include enzymatic dehydrogenation to ketosteroids (pp. 471 and 488) and formation of dimethylphosphinic esters for alkali flame detection (p. 479).

Oxime Derivatives. Most steroid analysis applications, especially in biological work, involve steroids that contain carbonyl and hydroxyl groups. In some cases, analysis is possible after derivatization of only the hydroxyls, but more often it is also necessary to mask the carbonyls. The most commonly used carbonyl derivatives are the oximes, which are formed with hydroxylamine or alkyloxyamines. The oxime derivatives give improved GC behavior, and different oxime substituents can be selected to alter GC separations or impart ECD sensitivity. The use of oxime derivatives may introduce new complications, however. Restricted rotation about the carbon-nitrogen double bond in oximes can lead to GC-separable *syn* and *anti* isomers yielding multiple derivative peaks. Such isomerism is not always a problem since, in many cases, steric hindrance prevents formation of one isomer leading to a single derivative peak. Another potential complication in the use of oximes is the possibility of Beckmann fission in the GC.[4] The analyst should be aware of this possible reaction (shown below) when employing oxime derivatives in GC analysis.

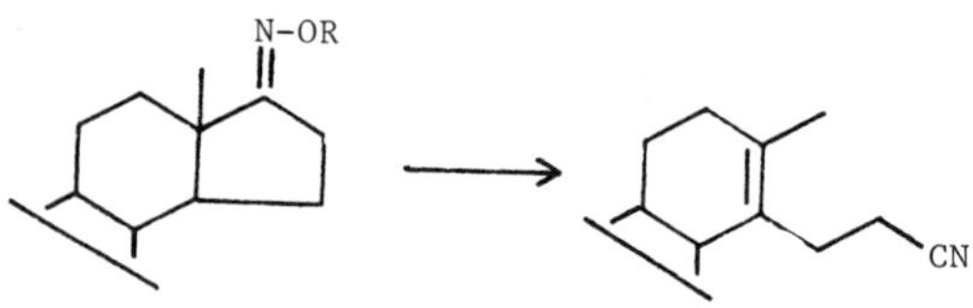

Reaction of ketosteroids with hydroxylamine and subsequent trimethylsilylation yields *O*-trimethylsilyloximes (p. 479). Use of alkoxylamines gives the corresponding O-alkyl oximes. Methoximes (p. 481) have been most commonly employed, but higher alkyl substituents (for example butyl and pentyl, p. 484, and benzyl, p. 485) permit GC separation of ketosteroids from hydroxysteroids. *O*-Pentafluorobenzyl hydroxylamine gives oxime derivatives with good ECD sensitivity (p. 485).

Hydrazone Derivatives. Dimethylhydrazone derivatives (p. 486) were among the earliest ketosteroid derivatives examined for GC, but the use of these derivatives never developed. Pentafluorophenylhydrazones have been used for ECD-GC analysis of ketosteroids (p. 487), and the chromogenic 2,4-dinitrophenylhydrazone derivatives of ketosteroids are amenable to HPLC analysis (p. 489).

Cyclic Derivatives. Cyclic ketal derivatives, also among the earliest GC derivatives examined for analysis of steroids, have been formed both from ketosteroids and from hydroxysteroids with keto reagents. Ethylene thioketals were prepared from ketosteroids with ethanedithiol (p. 494). Acetonide derivatives have been prepared from steroids having adjacent hydroxyls using acetone with acid catalysis (p. 494). 17,21-Dihydroxy-20-ketosteroids form bis(acetal) derivatives with formaldehyde (p. 502). The latter derivative is one of several which have been examined for the stabilization of the 20,21-sidechain in corticosteroids. A series of corticosteroid derivatives, both cyclic and noncyclic combinations of previously discussed derivatives, were compared for GC and MS analysis.[5] Other cyclic derivatives of corticosteroids include boronates (p. 501) and siliconides (p. 500).

Aldosterone forms an intramolecular acetal derivative under acidic conditions (p. 505). Aldosterone and other 18-oxygenated steroids form other internally cyclized derivatives due to suitably positioned functions in the molecules. Examples include lactones (p. 507) and acetal-boronates (p. 508).

Derivatives of Steroid Acids. The remaining group of the four principal polar groups found in steroids is the carboxyl. This function is encountered principally in the analysis of bile acids and is most often masked as the methyl ester formed either with diazomethane (p. 508) or methanolic hydrogen chloride (p. 510). Hexafluoroisopropyl esters of bile acids have also been used (p. 511). These bile acid derivatives employed acylation or silylation of the hydroxyl functions present.

Digitalis Steroids. The digitalis steroids, in addition to containing hydroxyls, also contain an α,β-unsaturated-λ-lactone function. This lactone can be converted to a trimethylsilyl enol ether under suitable silylation conditions (p. 512). Another method applied to this structure is reduction of the lactone to a diol and subsequent trifluoroacetylation for ECD-GC analysis (p. 512). The digitalis steroids have also been analyzed both by GC and HPLC as acylated derivatives without alteration of the lactone ring (p. 513).

References

[1]E. Heftmann (ed.), Modern Methods of Steroid Analysis, Academic Press, New York, 1973.

[2]E. M. Chambaz *et al.*, *Anal. Lett.* **1**, 749 (1968).

[3]H. Gleispach, *J. Chromatogr.* **91**, 407 (1974).

[4]J. P. Thenot and E. C. Horning, *Anal. Lett.* **4**, 683 (1971).

[5]T. A. Baille *et al.*, *Anal. Chem.* **44**, 30 (1972).

11.1. DERIVATIVES OF HYDROXYSTEROIDS AND STEROLS

11.1.1. Acyl Derivatives

Derivative Type. Trifluoroacetyl.

Reagent. Trifluoroacetic anhydride (TFAA).

Compound Type. Hydroxysteroids.

Reaction

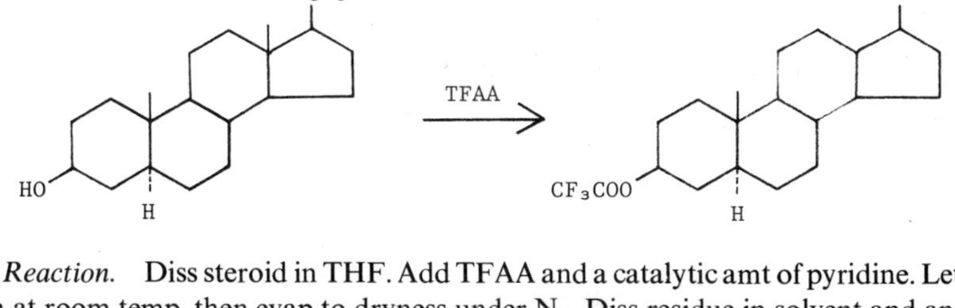

Typical Reaction. Diss steroid in THF. Add TFAA and a catalytic amt of pyridine. Let stand a few min at room temp, then evap to dryness under N_2. Diss residue in solvent and analyze by GC. (Further details of procedure were not reported.)

Comments. The deriv was found to yield a dramatic decrease in retention times relative to the free hydroxy steroid or the methyl, formyl, acetyl, or ethyl carbonate derivs. Keto groups were unaffected by the procedure.

References

[1]W. J. A. Vandenheuval, *et al., Biochim. Biophys. Acta* **48**, 596 (1961).

Derivative Type. Heptafluorobutyryl (HFB).

Reagent. Heptafluorobutyric anhydride (HFBA).

Compound Type. Testosterone.

Reaction

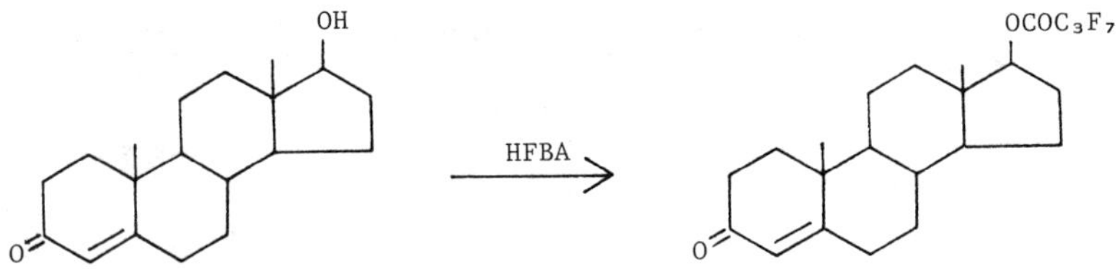

Typical Procedure. To 10 ml female plasma or 5 ml male plasma add 1,2-^{3}H-testosterone and allow to equilibrate 1 hr. Add 0.6 ml 10% NaOH and ext with 4 × 20 ml Et_2O. Wash combined ext with 10 ml 8% $NaHCO_3$ and 3 × 5 ml H_2O. Evap to dryness and transfer residue to a silica gel TLC plate with MeOH/$CHCl_3$ (1:1, v/v). Chromatograph with benzene/EtOAc (6:4, v/v), locate testosterone by radioactivity, and elute 3 × with a mixt of benzene (2 ml) and H_2O (0.2 ml). Dry eluates in a vac dessicator overnight.

Add 500 μl benzene, 5 μl pyridine, and 10 μl HFBA and heat at 70° for 1 hr. Evap to dryness under N_2 and diss residue in 3 ml benzene. Wash with 1 ml 6 *N* HCl and 3 × 1 ml H_2O. Dry over Na_2SO_4 and evap to dryness. TLC deriv on silica gel with benzene/EtOAc (9:1, v/v), locate radioactivity, and elute with 2 ml benzene and 5 drops H_2O. Evap eluate to dryness, diss residue in 0.4 ml benzene, and count 40 μl for recovery determination. Evap to dryness, diss residue in 30–40 μl benzene, and analyze by ECD-GC.

Comments. The method was used to quantitate plasma testosterone to 50 pg/ml. The conc of pyridine in the deriv reaction was critical for obtaining a high yield of mono-HFB deriv.

References

[1]I. R. Sarda et al., *Steroids* **12**, 607 (1968).

Derivative Type. 17-Heptafluorobutyryl (HFB).

Reagent. Heptafluorobutyric anhydride (HFBA).

Compound Type. Steroids.

Reaction

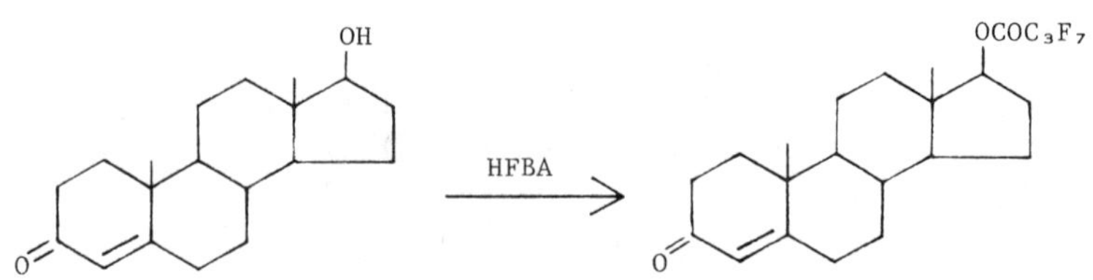

Typical Procedure (for Testosterone).[1] Dil plasma sample (0.3–1 ml male, 3–5 ml female) to 10 ml with H_2O, add 0.5 ml 25% NaOH and 1,2-^{3}H-testosterone, and ext with 3 × 15 ml Et_2O. Dry ext over Na_2SO_4, evap, and isolate testosterone by paper chromatography. Elute from the paper with MeOH, evap to dryness, and esterfy with 50 μl HFBA in 0.5 ml benzene (room temp, 1 hr). Evap under N_2 and purify deriv by TLC on silica gel (benzene/EtOAc, 9:1, v/v). Scrape testosterone deriv area, mix with 1 drop H_2O, and elute with benzene. Evap benzene and diss deriv in 30–50 μl heptane containing 0.5 ng 20-HFB 20β-hydroxyprogesterone int stand per μl. Analyze by ECD-GC.

Comments. The testosterone deriv was measurable to 20 pg. The 3-enol-17-di-HFB deriv gave a higher ECD response[2] but was too unstable to purify by TLC. The HFB deriv was also used for analysis of dihydrotestosterone and for 4-androsten-3,17-dione (after reduction to testosterone with $NaBH_4$). The deriv has also been prepared with HFBA/pyridine/benzene (10:5:500, 70°, 1 hr) for quantitation of plasma testosterone.[3] It was reported that the conc of pyridine was critical for obtaining a high yield of the mono-HFB deriv.[3] GC data were reported for HFB derivs of 16 steroids.[1]

References

[1]A. Vermeulen *et al.*, *Clin. Chim. Acta* **34**, 223 (1971).

[2]K. Nakagawa *et al.*, *Steroids* **7**, 331 (1966).

[3]I. R. Sarda *et al.*, *Steroids* **12**, 607 (1968).

Derivative Type. Pentadecafluorooctanoate.

Reagent. Pentadecafluorooctanoyl chloride/pyridine.

Compound Type. Testosterone.

Reaction

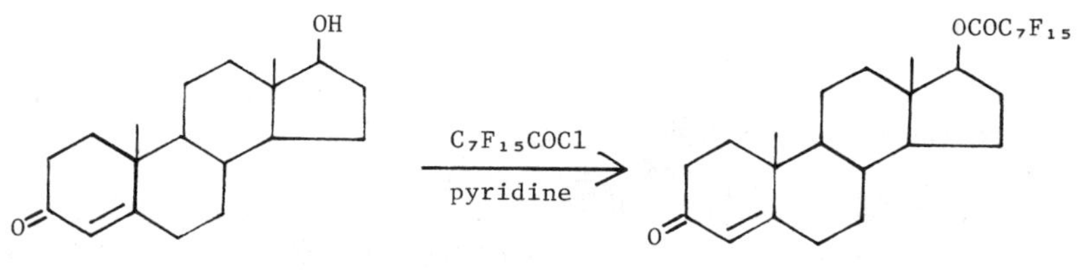

Typical Procedure (Preparative). Diss 100 mg testosterone in 100 ml benzene and add 1 ml of the acid chloride and 1 ml pyridine. Heat 30 min at 60°. Wash with 5 ml 6 *N* HCl and 50 ml 2% $NaHCO_3$. Evap the solvent and recryst the deriv from MeOH-H_2O. Analyze by FID and ECD-GC.

Comments. The deriv was compared with several other haloacyl derivs with relative responses as listed below. The diesters (3-enol-17-diacyloxy) were formed as by-products and could not be prepared in good yield. Some relative response data was also listed for derivs of cholesterol. (Table on following page).

References

[1]K. Nakagawa, *et al.*, *Steroids* **7**, 329 (1966).

	Relative Molar Response			
	Testosterone		Cholesterol	
Deriv	FID	ECD	FID	ECD
di-HFB	1.11	2.30	—	—
Perfluorooctanoyl	1.08	2.10	0.90	1.85
HFB	1.00	1.00	1.00	1.00
PFB	1.04	0.28	—	—
TFA	1.00	0.01	1.13	0.09
Chlorodifluoroacetyl	1.02	1.41	0.84	2.08
Chloroacetyl	0.99	0.20	—	—

Derivative Type. 9-H-Hexadecafluorononanoate (HFN).

Reagent. 9-H-Hexadecafluorononanoyl chloride.

Compound Type. Testosterone and other hydroxysteroids.

Reaction

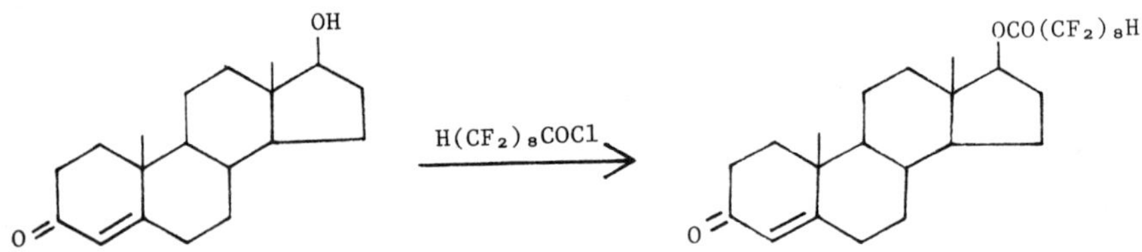

Typical Procedure. Dil 1–4 ml male plasma to 10 ml with H_2O (or use 5–10 ml female plasma). Add 0.4 ng ^{3}H-testosterone as recovery stand and 0.5 ml 1 *N* NaOH. Ext with 2 × 3 vol Et_2O. Wash ext with 0.1 vol H_2O then allow to evap overnight.

Add 0.1 ml of a 5% soln of the acid chloride in benzene and heat at 56° for 30 min. Evap to dryness then partition residue between 5 ml hexane and 1 ml 70% MeOH. Evap hexane layer and TLC residue on silica gel using benzene/EtOAc (95:5, v/v). Elute the deriv with benzene and analyze by ECD-GC.

Comments. The 11-H-eicosafluoroundecanoate (EFU) derivs were also prepared using the corresponding acid chloride. The HFN and EFU derivs of testosterone had, respectively, 2.30 and 2.55 times the ECD response of equimolar amts of the HFB deriv.

References

[1]M. A. Kirschner and J. Taylor, *Anal. Biochem.* **30**, 346 (1969).

Derivative Type. Chloroacetate.

Reagent. Chloroacetic anhydride/pyridine.

Compound Type. Testosterone.

Reaction

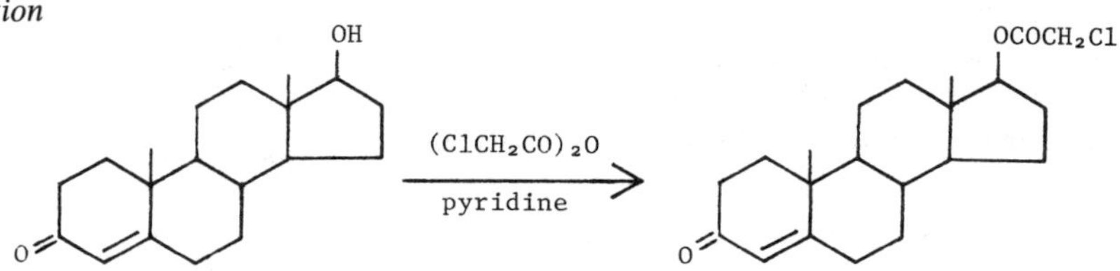

Typical Procedure. Add 0.25 ml 20% NaOH to 10 ml plasma and ext with 3 × 20 ml Et_2O. Wash exts with 2 × 5 ml H_2O and conc under N_2 at 40°. TLC on silica using cyclohexane EtOAc (1:1, v/v). Scrape testosterone area, add 3 drops H_2O to silica and elute with 3 × 1 ml MeOH.

Evap exts to dryness and add 0.5 ml of chloroacetic anhydride soln (100 mg/10 ml THF) and 0.1 ml pyridine. Allow to stand overnight in dark dessicator. Add 1 ml H_2O and ext with 3 × 1 ml EtOAc. Wash exts with 1 ml 6 *N* HCl and with 2 × 1 ml H_2O. Evap to dryness and TLC residue on silica with benzene/EtOAc (4:1, v/v). Scrape deriv area, add 4 drops H_2O, and elute with 3 × 1 ml benzene. Dry exts and diss residue in 1 ml MeOH containing 0.4 μg/ml cholesterol chloroacetate int stand. Dry, diss residue in 15 μl toluene, and analyze by GC.

Comments. GC data were given for testosterone and 7 other steroids. The method gave detection of 1 ng using ECD-GC.

References

[1]A. C. Brownie *et al., J. Clin. Endocrinol.* **24**, 1091 (1964).

Derivative Type. Chloroacetate.

Reagent. Chloroacetic anhydride.

Compound Type. Sterols.

Reaction

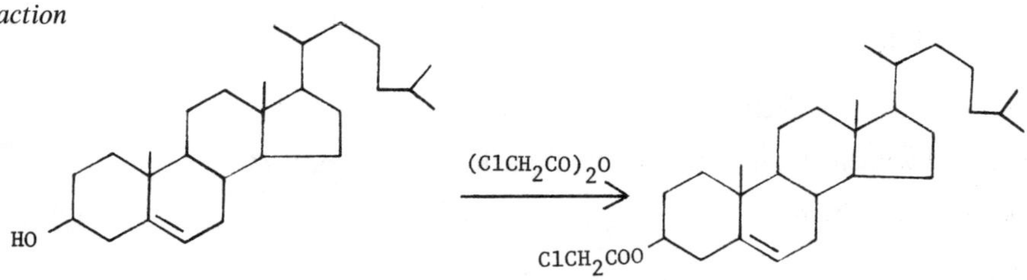

Typical Procedure (Preparative). Diss 0.25 mmol cholesterol in 5 ml THF and 0.2 ml pyridine. Add 25-fold excess of chloroacetic anhydride. Allow to stand 18 hr then dil with 5 ml H_2O. Ext with 3 × 5 ml Et_2O. Wash with 5 ml 6 *N* HCl and 2 × 5 ml H_2O. Dry over Na_2SO_4. Evap to yield product. Analyze by ECD-GC.

Comments. The ECD response of the monochloroacetate was found to be the greatest of five derivs tested. The order of response was monochloroacetate > dichloroacetate > bromoacetate > trichloroacetate > trifluoroacetate. Nakagawa *et al.*[2] prepared the chlorodifluoroacetyl (2.08), pentadecafluorooctanoyl (1.85), heptafluorobutyryl (1.00), and trifluoroacetyl (0.09) derivs of cholesterol and observed the relative molar responses to ECD indicated by the numbers in parentheses. It was reported that the heptafluorobutyryl deriv of cholesterol decomposed upon GC to form cholesta-3,5-diene, with the low ECD response being attributed to this decomposition.[3]

References

[1]R. Landowne and S. Lipsky, *Anal. Chem.* **35**, 532 (1963).
[2]K. Nakagawa *et al.*, *Steroids* **7**, 329 (1966).
[3]C. F. Poole and E. D. Morgan, *J. Chromatogr.* **90**, 380 (1974).

Derivative Type. Benzoate.

Reagent. Benzoyl chloride/pyridine.

Compound Type. Hydroxy steroids.

Reaction

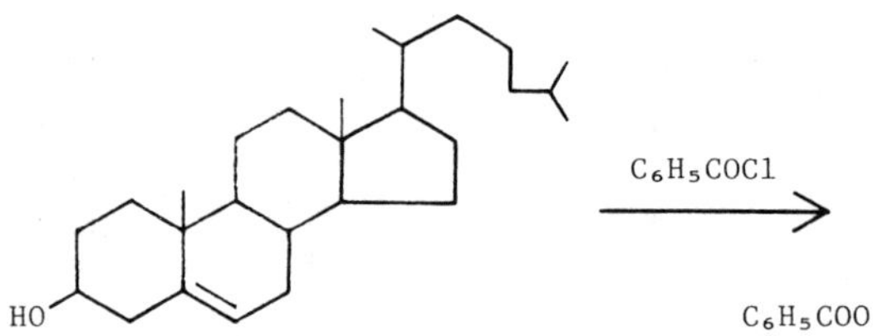

Typical Procedure. Diss steroid sample in pyridine and add 3 × mol excess of benzoyl chloride in pyridine. Shake 5 min and heat at 80° for 15 min. Partition between 0.1 *N* HCl and Et_2O. Wash Et_2O 2 × with 0.1 *N* HCl (more dil wash is required for Δ^5 steroids to prevent formation of cyclo-steroid artifacts). Analyze by HPLC.

Comments. The deriv was detectable to 1 ng (λ_{max}=230 nm). *p*-Nitrobenzoates were also examined (λ_{max}=254 nm).

References

[1]F. A. Fitzpatrick and S. Siggia, *Anal. Chem.* **45**, 2310 (1973).

11.1.2. Silyl Derivatives

Derivative Type. Trimethylsilyl.

Reagent. Hexamethyldisilazane (HMDS)/trimethylchlorosilane (TMCS).

Compound Type. Sterols.

Reaction

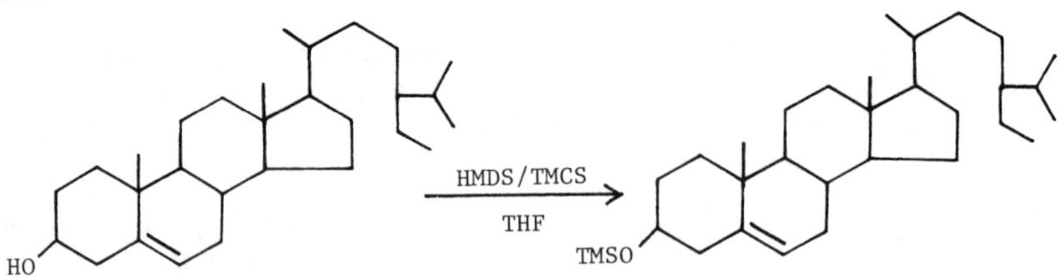

Typical Procedure.[1] Diss 25–30 mg of sterols in 1 ml anhyd THF and add 0.5 ml HMDS and 10 μl TMCS. Heat 40 min at 55°. Cool to room temp and analyze by GC.

Comments. The deriv method was a simplification of the method of Luukainen *et al.*[2] The method was used for quantitation of campesterol, β-sitosterol, and stigmasterol. TFA derivs were also prepared using the method of VandenHeuval *et al.*[3] Brooks *et al.*[4] prepared the TMS derivs of 28 sterols using HMDS-TMCS in pyridine and reported MS data for the derivs. The HMDS-TMCS reagent has also been used for analysis of 17-ketosteroids[5] and estrogens[6] in urine.

References

[1]A. Rozanski, *Anal. Chem.* **38**, 36 (1966).
[2]T. Luukainen *et al., Biochem. Biophys. Acta* **52**, 599 (1961).
[3]W. J. A. VandenHeuval *et al., Biochem. Biophys. Acta* **48**, 596 (1961).
[4]C. J. W. Brooks *et al., Lipids* **3**, 391 (1968).
[5]W. J. A. VandenHeuval *et al., Anal. Biochem.* **4**, 191 (1962).
[6]T. Luukainen *et al., Biochem. Biophys. Acta* **62**, 153 (1962).

Derivative Type. Trimethylsilyl and d_9-trimethylsilyl.

Reagent. Bis(trimethylsilyl)trifluoroacetamide (BSTFA), d_9-Trimethylsilylimidazole (d_9-TMSI).

Compound Type. Hydroxysteroids.

Reaction

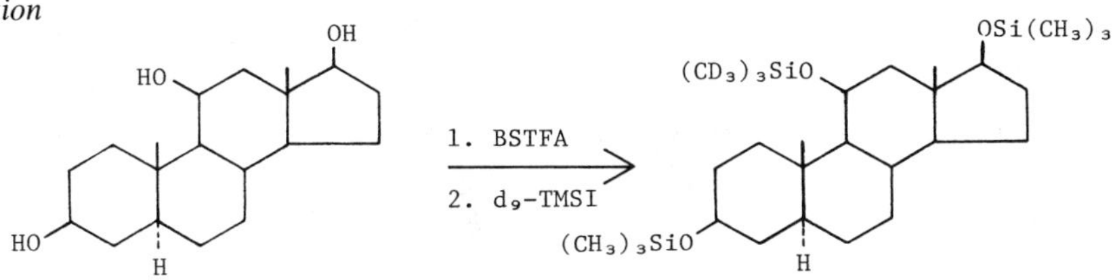

Typical Procedure. Diss 0.5 mg hydroxysteroid in 50 μl anhyd CH_3CN and add 15 μl BSTFA. Heat at 60° for 1 hr then let stand overnight at room temp. Evap to dryness under N_2 and diss residue in 50 μl cyclohexane. Add 15 μl d_9-TMSI and heat at 60°. Sample at 30 min intervals and examine by GC-MS.

Comments. The method was used to prepare derivs in which the perdeuterotrimethylsilyl group occupied the hindered position with a specificity of 90% or more. Nondeuterated TMS was introduced for 3β- and nonhindered 17β- hydroxyls and deuterated TMS for 11β- and hindered 17α- and 17β- hydroxyls. Cyclohexane was found to be the best solvent for the second step; CH_3CN and pyridine led to exchange of the previously introduced TMS groups. Ketosteroids were first converted to methoximes. It was found necessary to remove all salts from the methoximes by ext with EtOAc. Extensive information was reported on the MS characteristics of the derivs. The method was also used to prepare specifically labeled 17α, 20-di-TMSO derivs.[2]

References

[1]P. Vouros and D. J. Harvey, *Anal. Chem.* **45**, 7 (1973).
[2]P. Vouros, *J. Org. Chem.* **38**, 3555 (1973).

Derivative Type. Trimethylsilyl.

Reagent. Trimethylsilylimidazole (TMSI).

Compound Type. Ecdysones.

Reaction

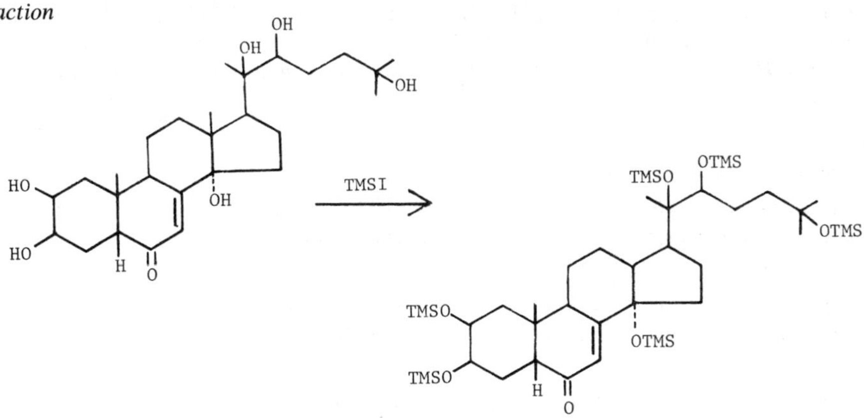

Typical Procedure. Heat sample with TMSI at 140° for *ca* 20 hr. Analyze by GC-MS.

Comments. Heating at 140° for 20 hr gave essentially quant formation of the hexa-TMS deriv of 20-hydroxyecdysone (shown). Heating 4 hr at 100° gave quant formation of the penta-TMS (2,3,20,21,25) deriv. One hr at 100° (conditions reported by other workers[2,3] to give complete silylation) yielded a mixt of the tetra-TMS (2,3,21,25) and penta-TMS derivs. The ease of silylation of the hydroxyls was in the order 2,3,22,25 > 20 ≫ 14. Addition of TMCS yielded some enol ether TMS deriv in compds lacking the 14α-hydroxyl. The derivs were unexpectedly sensitive to ECD.[4,5] Other workers have prepared TMS derivs of ecdysones with BSA[6,7] and TMSI.[8] Full MS data for the derivs were reported.[1,2]

References

[1]E. D. Morgan and C. F. Poole, *J. Chromatogr.* **116**, 333 (1976).

[2]N. Ikekawa *et al.*, *J. Chromatogr. Sci.* **10**, 233 (1972).

[3]H. Miyazaki *et al.*, *Anal. Chem.* **45**, 1164 (1973).

[4]D. W. Borst and J. D. O'Conner, *Steroids* **24**, 637 (1974).

[5]C. F. Poole *et al.*, *J. Chromatogr.* **104**, 172 (1975).

[6]M. N. Galbraith *et al.*, *Austr. J. Chem.* **22**, 1059 (1969).

[7]M. J. Thompson *et al.*, *Steroids* **16**, 85 (1970).

[8]D. S. King *et al.*, *Proc. Natl. Acad. Sci. (USA)* **71**, 793 (1974).

Derivative Type. Trimethylsilyl ethers.

Reagent. Trimethylsilylimidazole (TMSI)/bis(trimethylsilyl)acetamide (BSA)/trimethylchlorosilane (TMCS).

Compound Type. Sterols.

Reaction

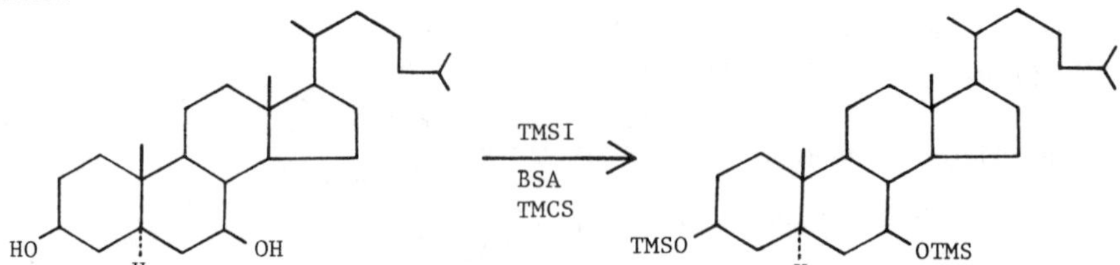

Typical Procedure. Diss sample in a mixt of TMSI, BSA, and TMCS (3:3:2, v/v/v) and heat at 60° for 60 hr in a sealed vial. Evap partially under N_2 stream, add H_2O, and ext with EtOAc. Dry ext over Na_2SO_4 and conc under vac prior to GC-MS analysis.

Comments. The method yielded complete derivatization including hindered hydroxyl groups. GC and MS data were reported for 29 sterols.

References

[1]C. J. W. Brooks *et al.*, *Biochim. Biophys. Acta* **296**, 431 (1973 .

Derivative Type. Dimethylsilyl (DMS).

Reagent. Tetramethyldisilazane (TMDS)/dimethylmonochlorosilane (DMMCS).

Compound Type. Steroids.

Reaction

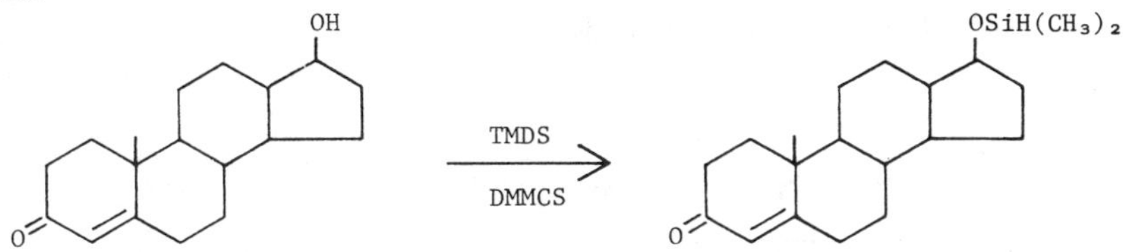

Typical Procedure. Diss 0.5–2.0 mg of steroid in 0.05 ml TMDS and 0.05 ml DMMCS. Allow to stand 3 hr at room temp. Analyze by GC.

Comments. The DMS derivs of the epimers of cholesterol, testosterone, and 4-pregnen-20β-ol-3-one were compared with the TMS and CMDMS derivs on polar and nonpolar columns. The CMDMS derivs had significantly longer retention times on both polar and nonpolar columns.

References

[1]W. J. A. VandenHeuval, *J. Chromatogr.* **27**, 85 (1967).

Derivative Type. Dimethylethylsilyl (DMES) or dimethyl-*n*-propylsilyl (DMPS).

Reagent. Dimethylethylsilyl imidazole (DMESI) or Dimethyl-*n*-propylsilylimidazole (DMPSI).

Compound Type. Hydroxysteroids.

Reaction

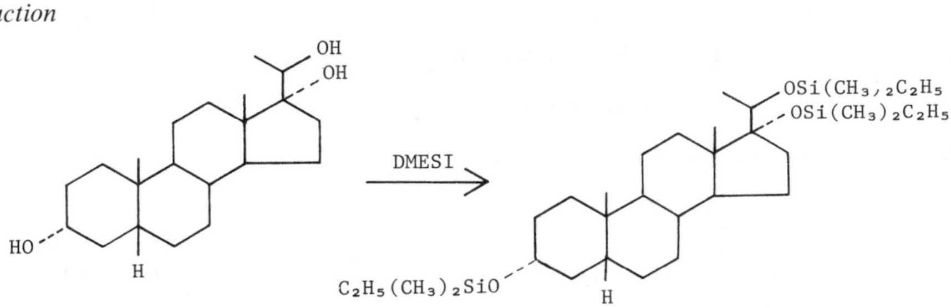

Typical Procedure. Diss 0.1–0.2 mg steroid in 100 μl DMESI (or DMPSI) and allow to stand 30 min at room temp. For steroids with the highly hindered 17α-hydroxyl group, heat at 100° for 10 hr or at 150° for 3 hr. Analyze by GC or GC-MS.

Comments. The derivs were examined as a method for determining the number of hydroxyl groups by GC methylene unit differences. DMESI and DMPSI showed comparable reactivity to TMSI. Hydroxyls (axial or equatorial) at positions 3, 7, 11, 12, 16, 17, 20, and 21 could be silylated by any of the three reagents to give TMS, DMES, or DMPS derivs. The ethyl and propyl groups were lost preferentially to methyl in the MS. GC and MS data for the 3 derivs of 44 steroids were reported.[1]

References

[1]H. Miyazaki *et al., Biomed. Mass Spectrom.* **4**, 23 (1977).

[2]H. Miyazaki *et al., Chem. Pharm. Bull.* **23**, 3033 (1975).

Derivative Type. Allyldimethylsilyl (ADMS).

Reagent. Allyldimethylchlorosilane/diethylamine (Et_2NH)/pyridine.

Compound Type. Steroids and cannabinoids.

Reaction

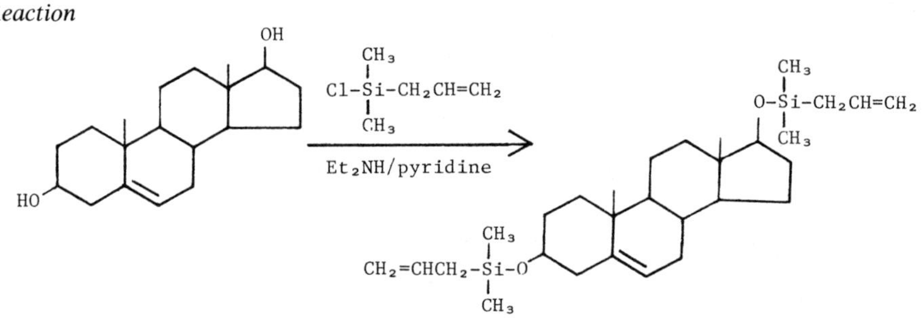

Typical Procedure. Diss 0.1 mg sample in 100 μl of allyldimethylchlorosilane/Et_2NH/pyridine (2:1:4, v/v/v). Allow to stand 5 min at room temp then analyze by GC.

Comments. Derivs were prepared from 13 mono- and dihydroxy-steroids and 11 cannabinoids. The ADMS derivs were compared with the TBDMS and the TMS derivs. The ADMS derivs contributed a smaller mass increment (98 amu) per deriv group than either the TBDMS (114 amu) or the CMDMS (106 amu) groups and exhibited shorter GC retention times than the other unsymmetrically substituted silyl derivs. The ADMS derivs, being less bulky, were also more easily prepared than the TBDMS derivs. The MS of the ADMS derivs had abundant M^+ and $[M\text{-allyl}]^+$ ions.

References

[1]D. J. Harvey, *Biomed. Mass Spectrom.* **4**, 265 (1977).

Derivative Type. Allyldimethylsilyl (ADMS).

Reagent. Allyldimethylchlorosilane/imidazole.

Compound Type. Hydroxysteroids.

Reaction

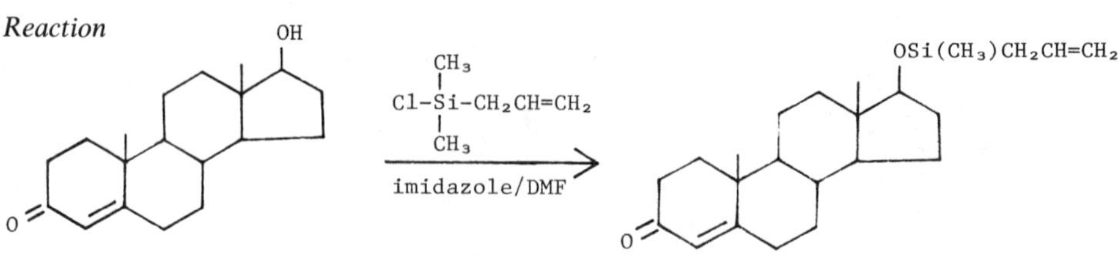

Typical Procedure. Mix steroid (<0.5 mg) with 100 μl 1 *M* allyldimethylchlorosilane in DMF and 100 μl 2 *M* imidazole in DMF. Allow to stand 1 hr at room temp. Isolate deriv by filtering through Sephadex LH-20.[2] Analyze by GC-MS.

Comments. The derivs gave intense *M*-allyl peaks in the MS. The deriv reagent was examined as a less hindered, and hence more reactive, alternative to the TBDMS reagent. The more hindered 11-hydroxy and *t*-17-hydroxy groups required longer reaction times or higher temperature (75°). At the higher temperature, enol ether formation became competitive and could not be forced to completion.

References

[1]G. Phillipou, *J. Chromatogr.* **129**, 384 (1976).

[2]R. W. Kelly and P. L. Taylor, *Anal. Chem.* **48**, 465 (1976).

Derivative Type. Allyldimethylsilyl (ADMS).

Reagent. Allyldimethylsilyl imidazole (ADMSI).

Compound Type. Steroids.

Reaction

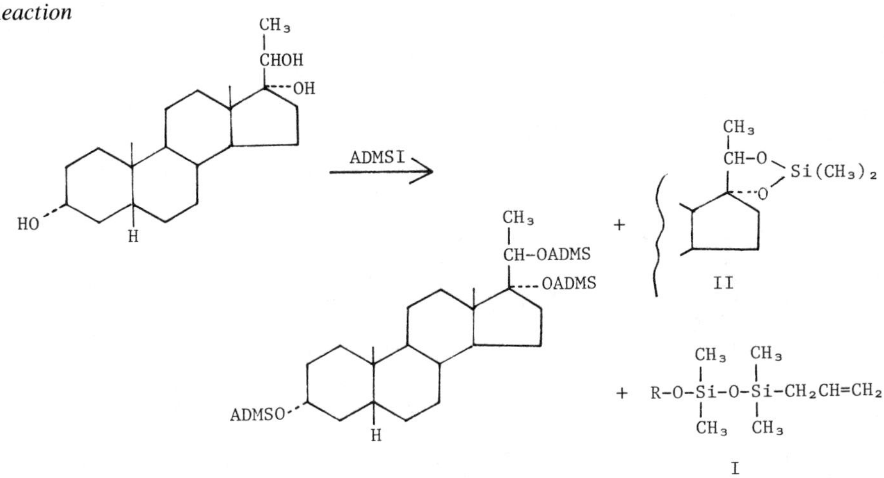

Typical Procedure.[1] Treat <1 mg steroid with 100 μl ADMSI until the reaction is complete by GC. Isolate derivs by filtration through Lipidex 5000.[2] Analyze by GC-MS.

Comments. Ketosteroids were first converted to methoximes.[3] The *in situ* generation of the ADMS reagent[4] was incompatible with the formation of methoximes leading to the preparation of the ADMSI reagent (synthesis reported[1]). The ADMS derivs were deemed unsatisfactory for steroid analysis due to competing formation of allylsiloxane derivs (I) and siliconides (II). Unhindered hydroxyls, however, were rapidly converted to ADMS derivs with minimal siloxane formation,[4] thus the derivs could be useful in some cases.

References

[1]I. A. Blair and G. Phillipou, *J. Chromatogr. Sci.* **15**, 478 (1977).

[2]J. Sjovall, *J. Steroid Biochem.* **6**, 227 (1975).

[3]J. P. Thenot and E. C. Horning, *Anal. Lett.* **5**, 21 (1972).

[4]G. Phillipou, *J. Chromatogr.* **129**, 384 (1976).

Derivative Type. *t*-Butyldimethylsilyl (TBDMS).

Reagent. *t*-Butyldimethylsilyl chloride (TBDMS-Cl)/imidazole.

Compound Type. Steroids.

Reaction

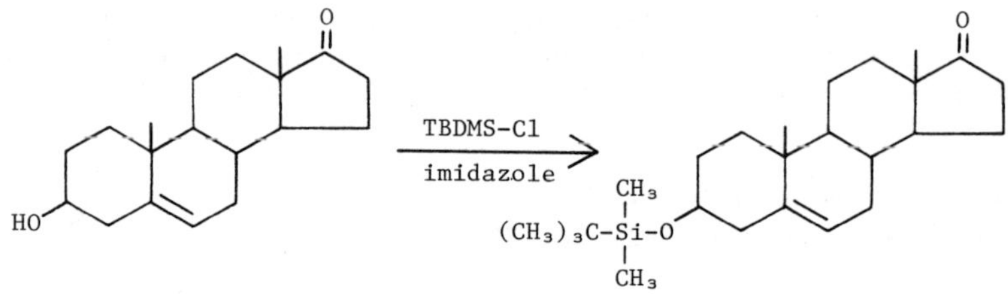

Typical Procedure. Treat steroid sample (<1 mg) with 100 μl of a soln of 2 mmol imidazole and 1 mmol TBDMS-Cl in 10 ml DMF. Heat 2 hr at 60°, cool, add H_2O, and ext deriv into lt pet ether or Et_2O. Analyze by GC or GC-MS.

Comments. A series of steroids were examined, and it was found that secondary 3α, 3β, 17α, 17β, or 20α and primary 21-hydroxy substituents were quantitatively converted to the TBDMS derivs.[1] The 20β-hydroxy group reacted slowly and 11 or tertiary-17 hydroxyls failed to react. The MS of the derivs were characterized by predominant loss of the *t*-butyl radical to give intense M-57 ions. The TBDMS deriv of testosterone has been employed to enhance the sensitivity of GC-MS-SIM measurements over that obtainable using TMS derivs.[2]

References

[1]G. Phillipou *et al.*, *Steroids* **26**, 516 (1975).

[2]D. S. Millington, *J. Steroid Biochem.* **6**, 239 (1975).

Derivative Type. *t*-Butyldimethylsilyl (TBDMS).

Reagent. *t*-Butyldimethylchlorosilane (TBDMS-Cl)/pyridine.

Compound Type. Steroids.

Reaction

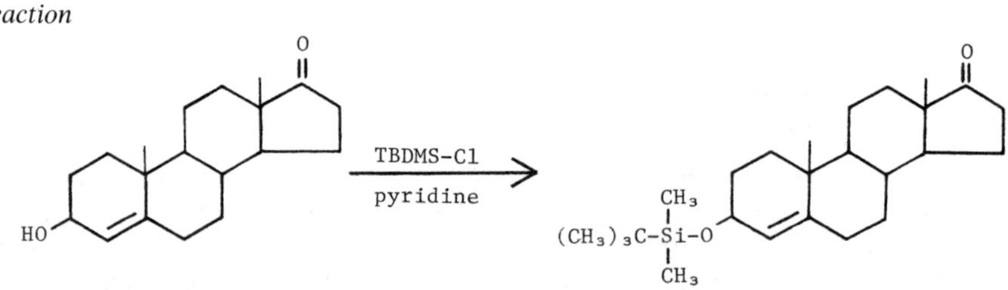

Typical Procedure. Treat steroid sample with *ca* 10 mol equiv of 1 *M* TBDMS-Cl in pyridine at room temp for 3–15 hr. Evap to dryness and diss residue in $CHCl_3$/heptane/MeOH (10:10:1). Remove excess reagent by elution through a 5 cm column of Sephadex LH-20 swollen in the eluting solvent. Analyze by GC-MS.

Comments. Conversion of nonhindered hydroxyl groups occurred quantitatively in 3–15 hr without the use of imidazole. MS data were given for the derivs of 10 steroids.

References

[1]S. J. Gaskell and C. J. W. Brooks, *Biochem. Soc. Trans.* **4**, 111 (1976).

Derivative Type. Cyclotetramethylene-*t*-butylsilyl (TMTBS).

Reagent. Cyclotetramethylene-*t*-butylchlorosilane (TMTBS-Cl)/imidazole/pyridine.

Compound Type. Steroids.

Reaction

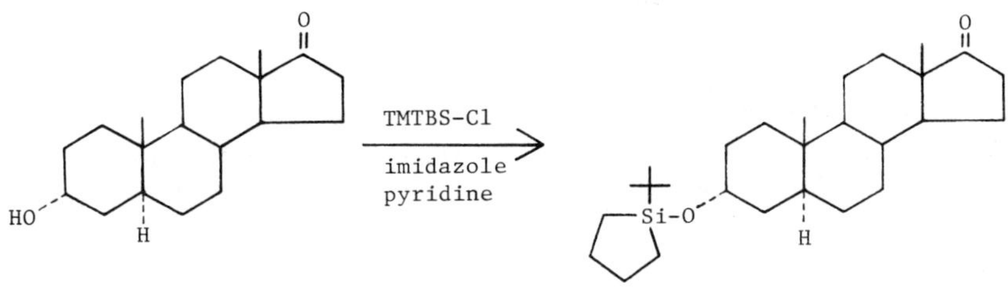

Typical Procedure. Diss sample in a soln of 1 *M* TMTBS-Cl and 1 *M* imidazole in pyridine and allow to react 1 hr (or overnight to insure complete reaction) at room temp. Analyze by GC or GC-MS.

Comments. The TMTBS and other sterically crowded trialkylsilyl derivs, cyclotetramethyleneisopropylsilyl (TMIPS) and TBDMS, prepared in a similar manner, were compared in terms of GC and MS properties with the corresponding TMS derivs. The deriv method was also carried out with DMF in place of pyridine but some of the derivs, being less soluble in DMF, pptd from the deriv mixt. The method gave rapid silylation of all primary and sterically accessable secondary hydroxyls. These included 3α, 3β, 4α, 6α, 17(secondary, α and β), and 20α hydroxyls. Most reactions were quantitative within 1 hr. Some enol ether formation was observed for 3-keto groups under forcing conditions. The chlorosilane in pyridine without imidazole was less reactive requiring overnight reaction for TBDMS and TMIPS and 80° for TMTBS, but gave no enol ether formation.

Attempts to form TMS derivs of remaining hindered hydroxyls gave silyl group interchange, especially when catalyzed by residual imidazole·HCl. Clean formation of mixed derivs required isolation of the initial deriv. This was facilitated by the high stability of the derivs (the TMIPS was somewhat more labile than the TMTBS or TMDMS and showed some hydrolysis on TLC). It was suggested that, since the sterically crowded trialkylsilyl derivs were stable to acylation conditions, acetylation or trifluoroacetylation could be the solution to complete deriv of sterically hindered hydroxyls.

The derivs showed enhanced GC separation of epimers compared to TMS derivs. The MS of the derivs were dominated by loss of the *t*-butyl (TMTBS, TBDMS) or isopropyl (TMIPS) groups to yield intense high mass ions. The stability of the derivs, the enhanced GC separations, and the favorable MS fragmentation make the derivs valuable for both qualitative and quantitative analysis.

References

[1]M. A. Quilliam and J. B. Westmore, *Anal. Chem.* **50**, 59 (1978).

[2]M. A. Quilliam and J. B. Westmore, *Steroids* **29**, 579 (1977).

[3]M. A. Quilliam and J. B. Westmore, *Steroids* **29**, 613 (1977).

Derivative Type. Chloromethyldimethylsilyl (CMDMS).

Reagent. 1,3-Bis(chloromethyl)tetramethyldisilazane/chloromethyldimethylchlorosilane.

Compound Type. Estrenols.

Reaction

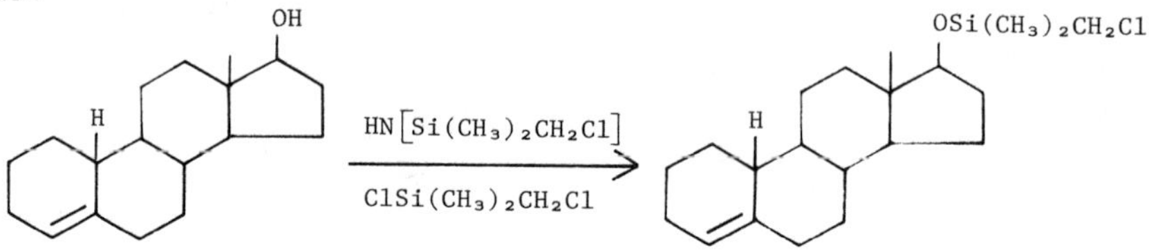

Typical Procedure. Mix 100–300 μg of steroid with 20 μl of the silazane and 5 μl of the chlorosilane. Allow to stand at room temp overnight or heat at 40° for 2 hr. Evap to dryness under N_2 and mix EtOAc with the residue. Centrifuge and analyze an aliquot of the supernatant by GC-MS-SIM.

Comments. The chlorine isotopic doublet was useful as an aid to specific detection of the deriv. The derivs were detectable down to 15 pg (CMDMS-estrenol) if ions which had no background interference were monitored.[1] The CMDMS derivs had longer retention times than the corresponding TMS derivs permitting effective GC separation of mono-, di-, and tri-TMS derivs.[2]

References

[1]C. J. W. Brooks and B. S. Middleditch, *Clin. Chim. Acta* **34**, 145 (1971).

[2]C. J. W. Brooks and B. S. Middleditch, *Anal. Lett.* **5**, 611 (1972).

Derivative Type. Chloromethyldimethylsilyl (CMDMS) ether.

Reagent. Chloromethyldimethylchlorosilane/diethylamine (Et_2NH).

Compound Type. Hydroxylated steroids.

Reaction

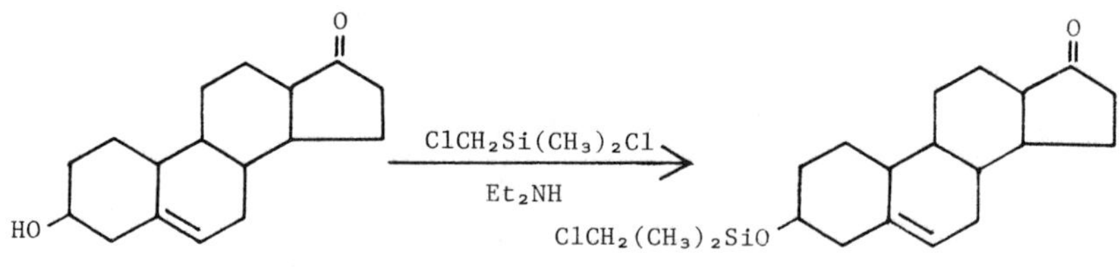

Typical Procedure.[1] Combine a soln of 0.5 ml Et_2NH in 5 ml hexane and 1 ml of chloromethyldimethylchlorosilane in 5 ml hexane in a stoppered centrifuge tube, mix, and centrifuge. Add 0.5 ml of the supernatant to the steroid sample in 0.1 ml THF. Allow to stand 1 hr at room temp. Remove volatiles in a stream of N_2, diss residue in hexane, and filter with washing through a porosity 4 filter tube. Evap soln to dryness under N_2 and redissolve in 0.2 ml hexane for GC analysis.

Comments. Plasma samples (4 ml) were dil to 10 ml, 0.2 ml 2 *N* NaOH was added, and the soln extd with 3 × 15 ml Et_2O. The ext was washed with 5 ml 2% HOAc and 5 ml H_2O, evapd to dryness under N_2, and treated as above. Bromomethyldimethylsilyl (BMDMS) ethers were also prepared from bromomethyldimethylchlorosilane. The BMDMS deriv was used to quantitate plasma testosterone by GC-MS-SIM.[2] The halomethyldimethylsilyl derivs were originally prepared for use with ECD-GC.[3,4]

References

[1]J. R. Chapman and E. Bailey, *Anal. Chem.* **45**, 1636 (1973).

[2]J. R. Chapman and E. Bailey, *J. Chromatogr.* **89**, 215 (1974).

[3]B. S. Thomas *et al.*, *Chem. Commun.* 408 (1966).

[4]C. Eaborn *et al.*, *Chem. Ind. (Lond.)* 827 (1967).

Derivative Type. Iodomethyldimethylsilyl (IDMS).

Reagent. Iodomethyldimethylchlorosilane/diethylamine (Et_2NH).

Compound Type. Steroids.

Reaction

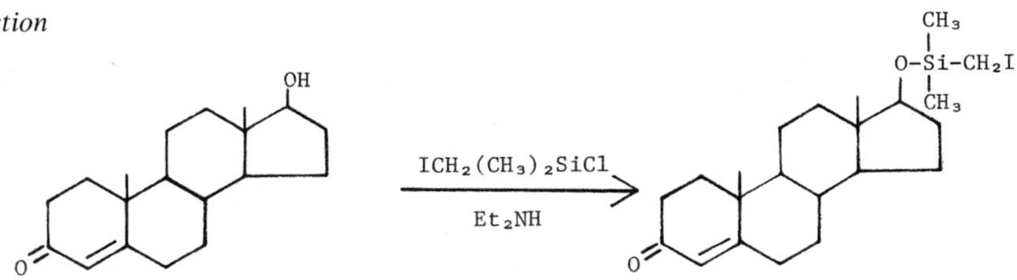

Typical Procedure. Diss the steroid in Et_2O-hexane and add a twofold excess of a mixt of equimolar amts of iodomethyldimethylchlorosilane and Et_2NH in hexane. Reflux 10 min then filter. Chromatograph the filtrate on neutral alumina with lt pet ether/acetone (95:5, v/v) and analyze the eluate by GC.

Comments. Chloromethyldimethylsilyl (CDMS) and bromomethyldimethylsilyl (BDMS) derivs were also prepared using the corresponding halogenomethyldimethylchlorosilane. The IDMS derivs were also prepared by treating the CDMS or BDMS derivs with NaI in acetone. The relative responses of the IDMS, BDMS, and CDMS derivs to ECD were on the order of 50:18:1. Retention times compared to the TMS derivs were increased 2–3× for CDMS, 5× for BMDS, and 7× for IDMS per silyl group.

References

[1]C. Eaborn *et al.*, *J. Chem. Soc.* (*C*) 2502 (1969).

Derivative Type. Iodomethyldimethylsilyl (IDMS).

Reagent. Bromomethyldimethylchlorosilane, sodium iodide.

Compound Type. Testosterone.

Reaction

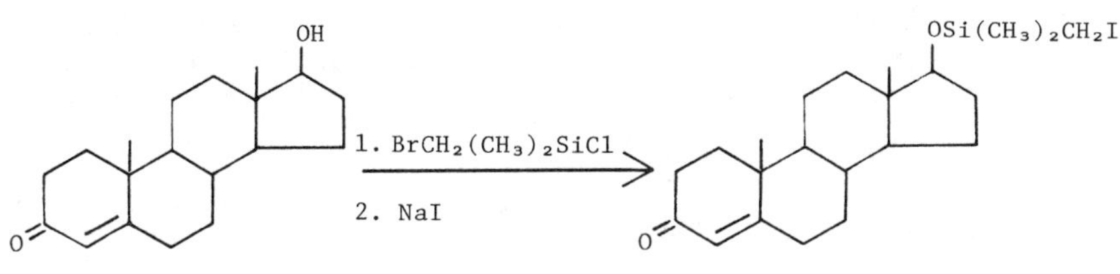

Typical Procedure. Mix 2 ml hexane, 0.2 ml bromomethyldimethylchlorosilane, and 0.1 ml Et_2NH on a Vortex mixer. Centrifuge and add 3–4 drops of the supernatant to the dried steroid ext. Allow to stand 45 min at room temp; then evap to dryness with N_2 at 37°. Add 3–4 drops of a satd soln of NaI in acetone and incubate at 37° for 30 min. Add 1 ml hexane to ppt excess NaI, centrifuge, and remove supernatant. Wash tube and ppt with another 0.5 ml hexane and combine hexane phases. Evap to dryness under N_2 and diss residue in 0.2 ml hexane. Analyze by ECD-GC.

Comments. The method was used to quantitate testosterone to less than 1 ng per 8 ml plasma. A modification of the method was used for plasma dehydroepiandrosterone.[2] The IDMS deriv was more stable than the TMS deriv.

References

[1]B. S. Thomas, *J. Chromatogr.* **56**, 37 (1971).

[2]E. Symes and B. S. Thomas, *J. Chromatogr.* **116**, 163 (1976).

Derivative Type. Pentafluorophenyldimethylsilyl (flophemesyl).

Reagent. Flophemesylamine (or flophemesyl chloride)/pyridine.

Compound Type. Steroids.

Reaction

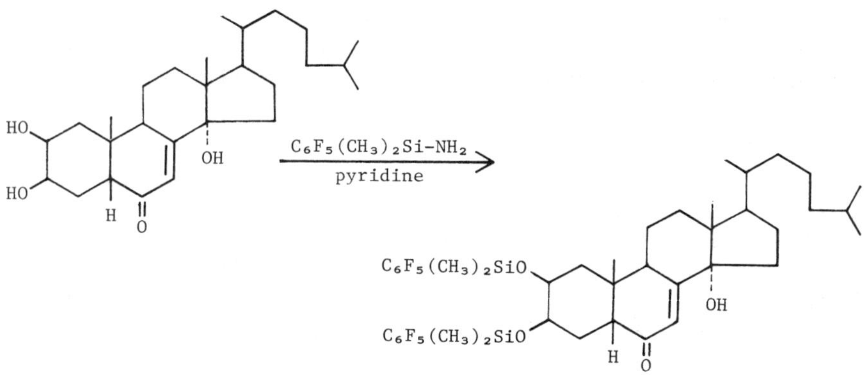

Typical Procedure. Treat sample with 1:1 (v/v) mixt of flophemesylamine and pyridine for 15 min at room temp. Analyze by GC or GC-MS.

Comments. Flophemesyl chloride in pyridine gave similar results. The deriv method formed silyl ethers of unhindered secondary hydroxyls but did not form enol ethers of enolizable ketones, e.g., steroid 3-, 6-, 11-, and 17-keto groups and 4-en-3-one and 7-en-6-one groups were unaffected even after 12 hr at 80°. Hindered hydroxyls such as in 17α-methylandrost-4-en-17β-ol-3-one, 11β-hydroxyandrost-4-ene-3,17-dione, 17α-hydroxypregn-4-ene-3,2-dione, and the 14-hydroxy in 2β,3β,14α-trihydroxy-5β-cholest-7-en-6-one (shown above) were unaffected. The flophemesylamine was most potent reagent of five evaluated; the order of reactivity in pyridine was the amine > the chloride > the diethylamine > the disilazane ≫ the imidazole.

References

[1]E. D. Morgan and C. F. Poole, *J. Chromatogr.* **104**, 351 (1975).

Derivative Type. Pentafluorophenyldimethylsilyl (flophemesyl).

Reagent. Flophemesyldiethylamine/flophemesyl chloride.

Compound Type. Steroids.

Reaction

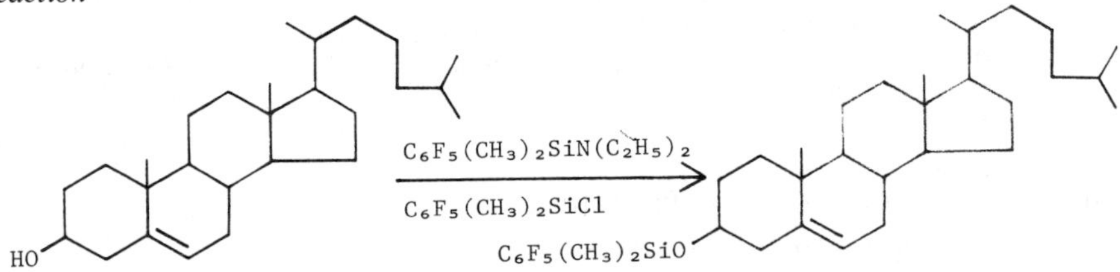

Typical Procedure.[1] Treat sample with a 10:1 (v/v) mixt of flophemesyldiethylamine flophemesyl chloride in pyridine. Let stand 15 min at room temp, then analyze by GC or GC-MS.

Comments. The use of the flophemesyl disilazane in place of the diethylamine required more vigorous conditions; cholesterol was completely derivatized in 3 hr at 65°. The reagent did not form enol ether derivs; 5α-cholestan-3-one was unaffected even after 6 hr at 80°. The reagent silylated the 17β-hydroxyl of 17α-methylandrost-4-en-17β-ol-3-one but did not affect the hindered 11β-hydroxyl of 11β-hydroxyandrost-4-ene-3,17-dione. The order of reactivity of the chloride catalyzed reagents was the diethylamine ≫ the amine > the disilazane. The deriv was compared with the chloromethyldimethylsilyl ether and found to be nearly 20 times more sensitive to ECD and more stable.[2] The corresponding trifluoropropyl- and heptafluoropentyldimethylsilyl derivs were also prepared but had less ECD sensitivity.[2] The derivs showed prominent molecular ions in the MS.[3]

References

[1] E. D. Morgan and C. F. Poole, *J. Chromatogr.* **104**, 351 (1975).

[2] E. D. Morgan and C. F. Poole, *J. Chromatogr.* **89**, 225 (1974).

[3] C. F. Poole and E. D. Morgan, *Org. Mass Spectrom.* **10**, 537 (1975).

11.1.3. Mixed Acyl and Silyl Derivatives

Derivative Type. Trimethylsilyl(TMS), heptafluorobutyryl(HFB).

Reagent. Trimethylsilylimidazole (TMSI), heptafluorobutyrylimidazole (HFBI)/heptafluorobutyric anhydride (HFBA).

Compound Type. Ecdysones.

Reaction

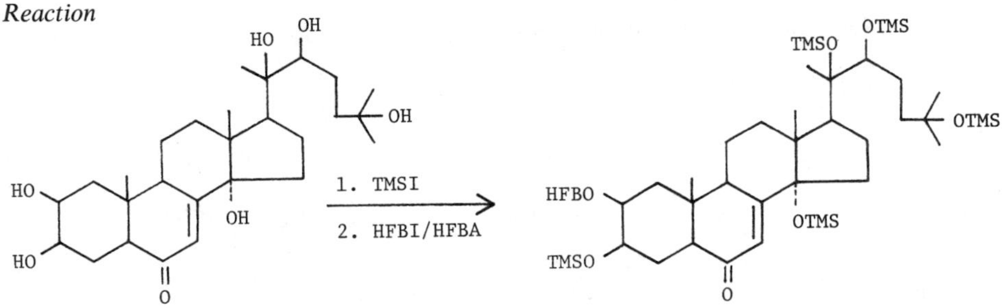

Typical Procedure. Diss 0.5 mg steroid in 20 μl TMSI and heat at 100° for 1 hr. Add 20μl HFBI and 2μl HFBA and heat at 50° for 2 hr. Add 0.2 ml benzene and wash with ice cold 0.2 ml 5% aq $NaHCO_3$ followed by 0.2 ml H_2O. Dry over Na_2SO_4 and evap to dryness. Diss residue in benzene and analyze by GC-MS or ECD-GC.

Comments. Equatorial *O*-TMS groups appeared to be more prone to exchange to *O*-HFB than did axial groups. In inokosterone the C-26 (primary) *O*-TMS was also converted to *O*-HFB. The HFB derivs permitted quantitation to pg levels by ECD-GC. The compds were also quantitated by GC-MS-SIM.

References

[1]N. Ikekawa *et al., J. Chromatogr. Sci.* **10**, 233 (1972).

[2]H. Miyazaki *et al., Anal. Chem.* **45**, 1164 (1973).

11.1.4. Other Derivatives of Hydroxylated Steroids

Derivative Type. *O*-Methyl.

Reagent. Methyl sulfinyl carbanion/methyl iodide.

Compound Type. Sterols.

Reaction

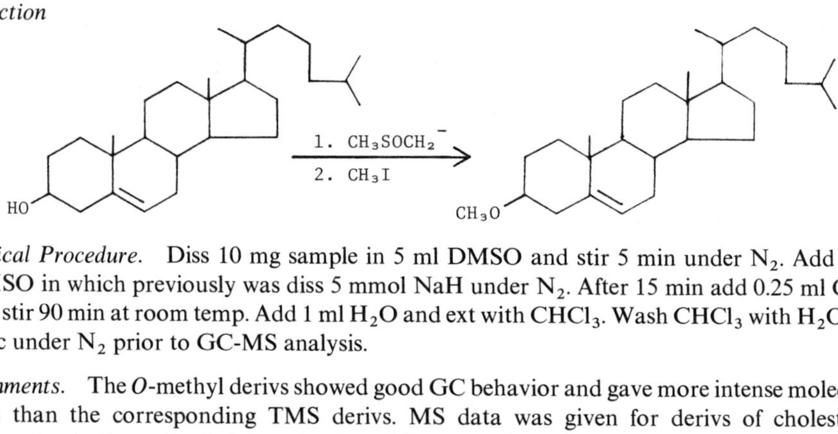

Typical Procedure. Diss 10 mg sample in 5 ml DMSO and stir 5 min under N_2. Add 1 ml DMSO in which previously was diss 5 mmol NaH under N_2. After 15 min add 0.25 ml CH_3I and stir 90 min at room temp. Add 1 ml H_2O and ext with $CHCl_3$. Wash $CHCl_3$ with H_2O and conc under N_2 prior to GC-MS analysis.

Comments. The *O*-methyl derivs showed good GC behavior and gave more intense molecular ions than the corresponding TMS derivs. MS data was given for derivs of cholesterol, campesterol, stigmasterol, desmosterol, and β-sitosterol.

References

[1]V. Mosini *et al., Adv. Mass Spectrom. Biochem. Med.*, **II**, 581 (1976).

Derivative Type. 16,16′-*O*-Permethyl.

Reagent. Methylsulfinyl carbanion/methyl iodide.

Compound Type. 17-Ketosteroids.

Reaction

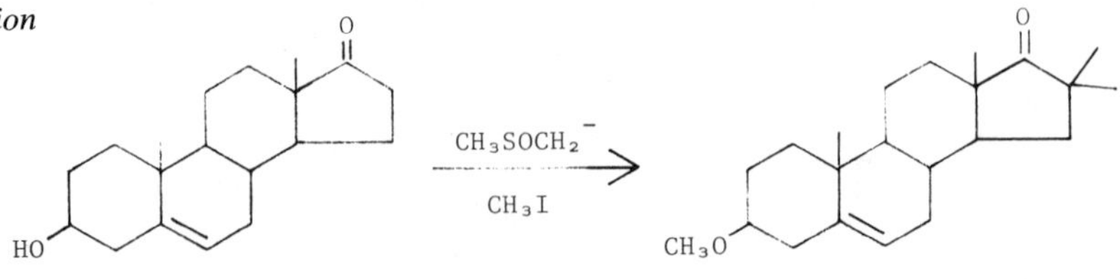

Typical Procedure. Diss sample in DMSO and add at least 5 mol equiv of methylsulfinyl carbanion soln in DMSO (prep from 120 mg, 5 mmol, NaH in 5 ml DMSO[2]) under N_2. After 15 min add the same mol equiv of CH_3I under N_2. After 1 hr add 1 ml H_2O and ext with 1 ml $CHCl_3$. Wash $CHCl_3$ with 3×1 ml H_2O. Evap to dryness under N_2 and diss in $CHCl_3$ prior to GC-MS analysis.

Comments. The permethyl derivs of a number of 3-hydroxy- and 3,11-dihydroxy-17-keto-steroids were examined and found to be useful for GC-MS analysis. 11-Ketosteroids gave mixtures of products, apparently due to greater difficulty in methylation at C-12 than at C-16.

References

[1]P. A. Leclercq, *Biomed. Mass Spectrom.* **1**, 109 (1974).
[2]P. A. Leclercq and D. M. Desiderio, *Anal. Lett.* **4**, 305 (1971).

Derivative Type. Ketosteroids.

Reagent. Cholesterol oxidase.

Compound Type. Hydroxysteroids (Δ^5-3β-ols and 5α-3β-ols).

Reaction

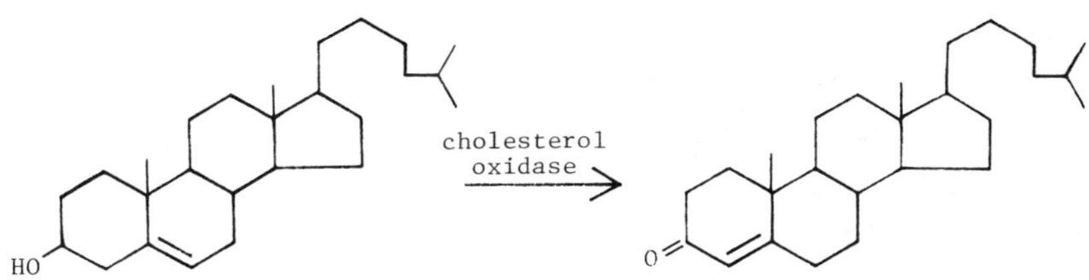

Typical Procedure. Add 0.13 μmol steroid in 250 μl isopropanol to 3 ml pH 7.0 (0.05 *M*) phosphate buffer followed by 10 μl cholesterol oxidase soln (2 I.U./ml). Incubate at 37° for 1.5 hr. Ext with 2 ml Et_2O and evap ext to dryness. Diss residue in 50 μl EtOAc for GC or GC-MS analysis.

Comments. The method was used to convert 3β-ols to 3-ketosteroids to facilitate separation and analysis of complex sterol mixtures.

References

[1]A. G. Smith and C. J. W. Brooks, *J. Chromatogr.* **101**, 373 (1974).

11.2. DERIVATIVES OF PHENOLIC STEROIDS

11.2.1. Methyl Derivatives

Derivative Type. Methyl ether.

Reagent. Dimethyl sulfate.

Compound Type. Phenolic steroids.

Reaction

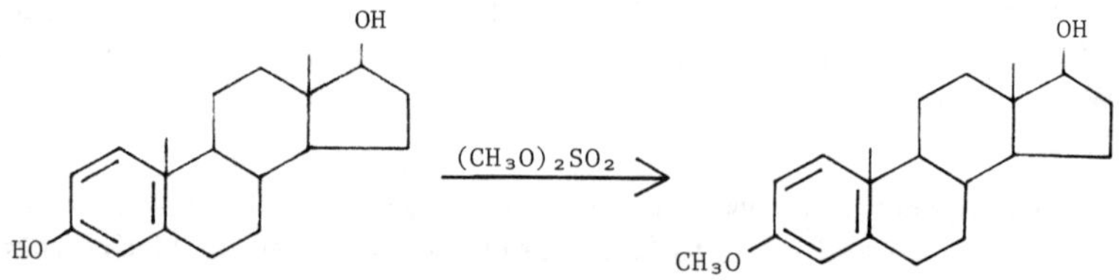

Typical Procedure. Mix estrogen sample (in dil NaOH) from 200 ml urine with 0.9 g H_3BO_3 and 4 ml 20% NaOH. Place in 37° bath and add 1 ml dimethyl sulfate. Hold at 37° for 10–30 min after soln complete then add another 1 ml dimethyl sulfate and 2 ml 20% NaOH. Hold at 37° for 20–30 min after soln complete. Ext methylated steroids with benzene.

Comments. The method was used for prep of methyl ethers for colorimetric determination but has been applied for chromatographic derivatization. More strongly alkaline conditions led to rapid hydrolysis of the dimethyl sulfate relative to the methylation rate. An ext method was reported for separating estradiol and estrone from estriol in urine.

References

[1] J. B. Brown, *Biochem. J.* **60**, 185 (1955).

Derivative Type. Methyl ether, acetate or trimethylsilyl ether.

Reagent. Dimethyl sulfate, acetic anhydride/pyridine or hexamethyldisilazane (HMDS)/ trimethylchlorosilane (TMCS).

Compound Type. Estrogens.

Reaction

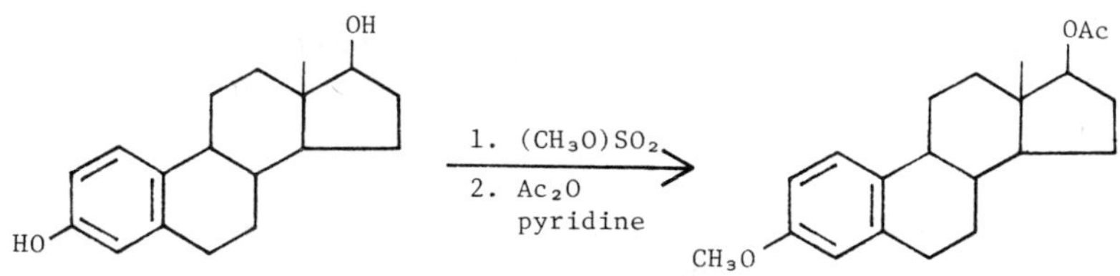

Typical Procedure.[1] Prepare an ext of urine by a modification of the Brown[2] procedure using hydrolysis, Et_2O extraction, and alkali partition. Diss ext residue in 25 ml benzene and add 25 ml 40–60° pet ether. Ext estrone, estradiol, and estriol with 2 × 25 ml 1.6% NaOH. Methylate[2] with dimethyl sulfate, ext methyl ethers with benzene, and evap ext to dryness under vac.

A. Acetylation—Treat residue with 50 μl pyridine and 130 μl Ac_2O for 1 hr at 60°. Warm with 1 ml EtOH then evap under N_2. Diss in 10–100 μl EtOH for GC injection.

B. Trimethylsilylation—Treat the residue with 250 μl of a mixt of 50 μl TMCS in 1 ml HMDS at 60° for 1.5 hr. Evap excess reagents under N_2 at 60°. Triturate residue with 5 ml hexane, centrifuge, and evap supernatant to dryness under N_2. Diss in 10–100 μl THF just prior to GC analysis.

Comments. The formation of the methyl ether served as an additional purification step and permitted increased sensitivity over a single deriv (TMS) procedure, which was also investigated.

References

[1]R. I. Cox and A. R. Bedford, *Steroids* **3**, 663 (1964).
[2]J. B. Brown, *Biochem. J.* **60**, 185 (1955).

Derivative Type. Trimethyl ether.

Reagent. Tetramethylammonium hydroxide (TMAH).

Compound Type. Estriol.

Reaction

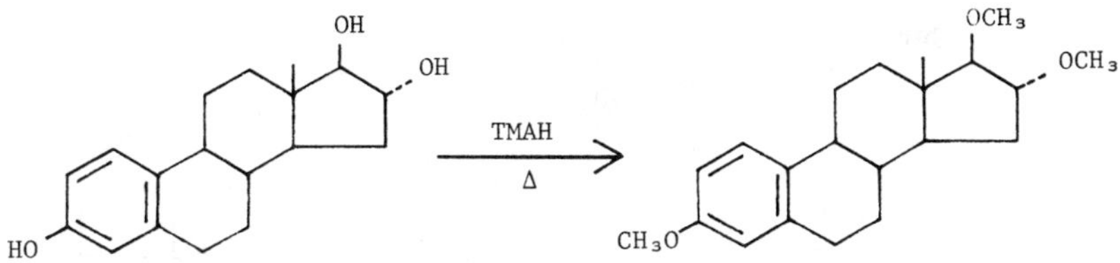

Typical Procedure. Diss the steroid in excess TMAH in MeOH and inject into the GC. Methylation occurs in the hot injector.

Comments. In preparative work (product collected by vac sublimation) it was found that use of 1 equiv of TMAH or the more weakly alkaline tetramethylammonium acetate gave only methylation of the phenolic hydroxyl while excess TMAH gave the trimethyl deriv. The use of such less basic ammonium salts might be analytically useful for specific alkylation of phenolic groups in the presence of alcohol hydroxyls. TMAH (2 *M* in MeOH) has been used to prepare a deriv for GC analysis of urinary estriol, but no data were reported on which deriv was obtained.[2]

References

[1]J. MacGee and K. G. Allen, *Steroids* **16**, 79 (1970).
[2]S. N. Millner and C. A. Taber, *Clin. Chim. Acta* **71**, 67 (1976).

11.2.2. Acyl Derivatives

Derivative Type. Acetate.

Reagent. Acetic anhydride (Ac_2O)/pyridine.

Compound Type. Estrogens.

Reaction

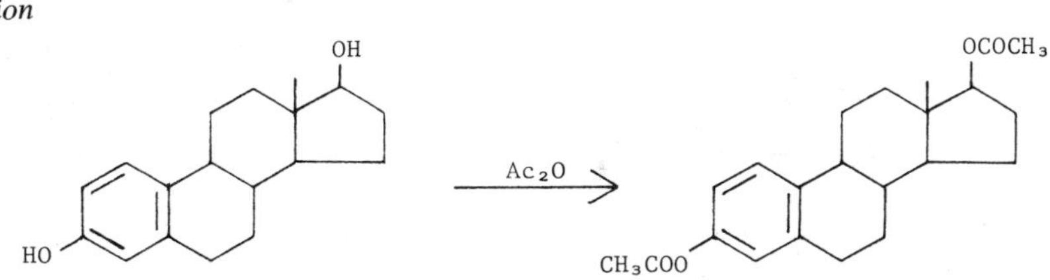

Typical Procedure. Diss sample in Ac_2O/pyridine (5:1, v/v) and heat at 68° for 1 hr. Add 5 ml H_2O with stirring then ext with 10 ml and 2×5 ml 1t pet ether. Wash combined exts with 5 ml 8% $NaHCO_3$ then with H_2O until neutral. Evap to dryness and diss residue in a small amt of pet ether for GC analysis.

Comments. The method was used for analysis of urinary estrogens using FID-GC.

References

H. Wotiz and S. Chattoraj, *Anal. Chem.* **36**, 1466 (1964).

Derivative Type. Chloroacetate.

Reagent. Chloroacetic anhydride.

Compound Type. Steroids.

Reaction

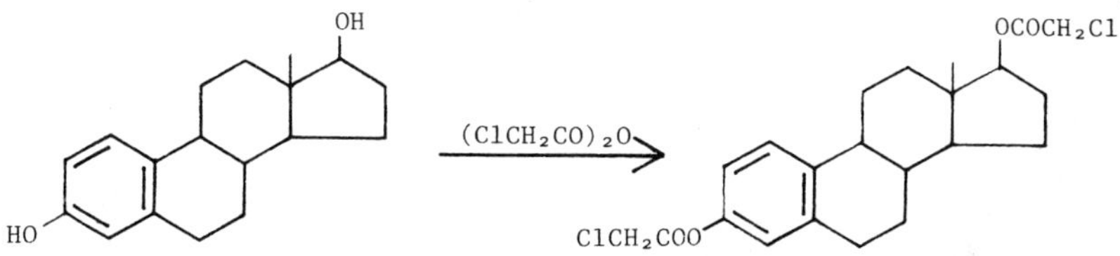

Typical Procedure. Diss 100–200 mg of steroid in 5 ml THF. Add 1 g chloroacetic anhydride and 0.2 ml pyridine. Let stand overnight in dark. Add 5 ml H_2O and ext 3× with 5 ml Et_2O. Wash ext with 5 ml 6 *N* HCl and 2× with 5 ml H_2O. Filter ext through anhyd $NaSO_4$ and evap to dryness.

Comments. The procedure was used to prepare pure derivs which were recryst from acetone. The derivs were detectable to the ng level by ECD-GC. The chloroacetate of 20β-hydroxypregn-4-en-3-one (formed by enzymic reduction) was prepared using chloroacetic anhydride in THF and pyridine (16 hr, room temp) in a method for the analysis of blood progesterone using ECD-GC.[2]

References

[1]H. J. van der Molen *et al.*, *Steroids* **6**, 195 (1965).

[2]H. J. van der Molen and D. Groen, *J. Clin. Endocrinol. Metab.* **25**, 1625 (1965).

Derivative Type. Heptafluorobutyrate.

Reagent. Heptafluorobutyric anhydride (HFBA).

Compound Type. Steroids.

Reaction

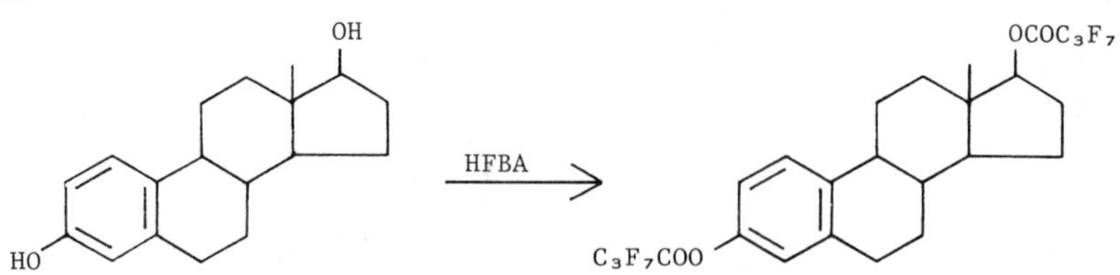

Typical Procedure. Apply stand compd or ext to stainless steel gauze with absolute EtOH and air dry. Transfer gauzes to vials (12 × 35 mm) and transfer vials to a ground glass stoppered weighing bottle containing 0.5 ml HFBA in two vials (12 × 35 mm) and P_2O_5 in a vial

(17 × 60 mm, *ca* $\frac{1}{4}$ full). Place stoppered bottle in 45° oven for 2 hr or a 35° oven for 18 hr. Elute derivs from gauzes by washing with 4 × 0.3 ml *n*-hexane. Evap eluates to dryness under N_2 and rediss in 10–50 μl prior to GC analysis.

Comments. Substitution of pyridine as deriv catalyst produced contaminants. Quantitation of estrogen derivs to the 200 pg level was reported.

References

[1]F. L. Rigby *et al.*, *Steroids* **16**, 703 (1970).

Derivative Type. Heptafluorobutyrate (HFB).

Reagent. Heptafluorobutyric anhydride (HFBA).

Compound Type. Steroids.

Reaction

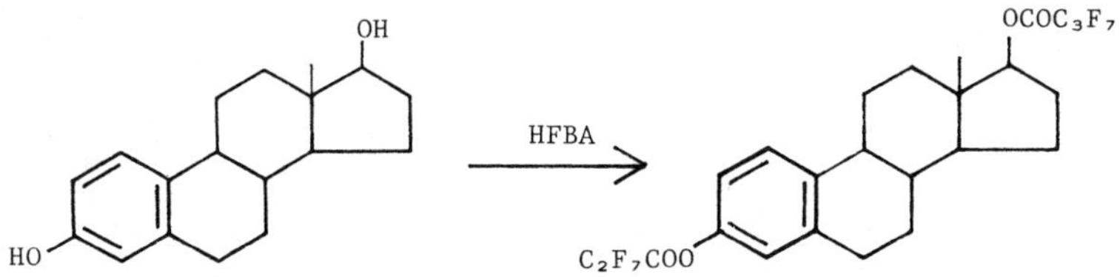

Typical Procedure. Diss compd in THF or benzene and add a large excess of HFBA. Heat at 60° for 30 min then evap under N_2. Diss residue in acetone and analyze by GC.

Comments. The derivs were detectable down to 100 pg by ECD-GC.[1] The derivs of testosterone and androsterone were listed[1] as mono-HFB derivs but no structure proof was given; similar conditions[2] gave enol-HFB derivs. The HFB deriv has been used for ECD-GC analysis of plasma estrone,[3] estradiol,[3,4] and estriol.[3]

References

[1]S. J. Clark and H. H. Wotiz, *Steroids* **2**, 535 (1963).

[2]D. Exley and J. Chamberlain, *Steroids* **10**, 509 (1967).

[3]H. H. Wotiz *et al.*, *Steroids* **10**, 127 (1967).

[4]A. K. Munson *et al.*, *Biochem. Med.* **3**, 187 (1969).

Derivative Type. 3-Pentafluorobenzoate, 17-heptafluorobutyrate.

Reagent. Pentafluorobenzoyl chloride/triethylamine (Et_3N), heptafluorobutyric anhydride (HFBA).

Compound Type. Estradiol.

Reaction

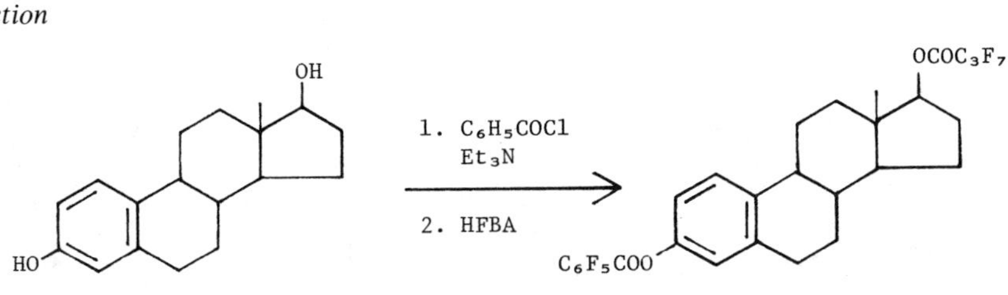

Typical Procedure. To 10^{-10}–10^{-6} g steroid add 50 μl 0.1% pentafluorobenzoyl chloride in benzene and 50 μl 2% Et_3N in benzene. Allow to react overnight in the dark; then evap under N_2. Add 20 μl 10% HFBA in hexane and heat at 60° for 30 min. Evap to dryness under N_2 and TLC deriv on silica gel with benzene/acetone (97:3, v/v) and cyclohexane/acetone (96:4, v/v). Elute deriv with EtOAc and analyze by GC.

Comments. The procedure gave specific pentafluorobenzoylation of phenolic hydroxyls allowing subsequent formation of HFB derivs of aliphatic hydroxyls.

References

[1]A. Zmigrod *et al.*, *Steroids* **15**, 635 (1970).

11.2.3. Silyl Derivatives

Derivative Type. Trimethylsilyl (TMS)/d_3-TMS.

Reagent. Trimethylsilylimidazole (TMSI)/dimethyltrideuteromethylimidazole (d_3-TMSI).

Compound Type. Hydroxysteroids.

Reaction

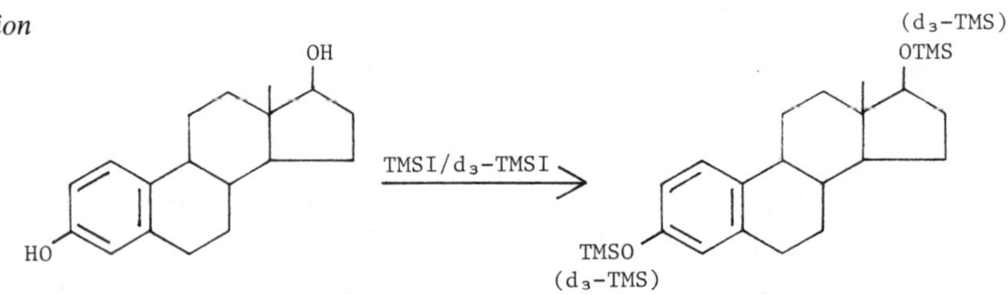

Typical Procedure. Diss 0.1–0.2 mg steroid in 100 μl of an equimolar mixt of TMSI and d_3-TMSI. Allow to stand 30 min at room temp then analyze by GC-MS or GC-MS-SIM.

Comments. The relative intensities of the M, M+3, M+6, etc. peaks in the ion clusters formed were indicative of the number of hydroxyl groups in the steroid. The reagent ratios could be adjusted for the number of hydroxyl groups to give equal intensity ion doublets (separated by 3 amu) for GC-MS-SIM detection. The synthesis of d_3-TMSI was reported. d_3-TMSI was shown to be preferable to the more readily available d_9-TMSI because of ion cluster overlap (due to M-CH_3 or M-CD_3 ions) with the latter reagent.

References

[1]H. Miyazaki *et al.*, *J. Chromatogr.* **135**, 109 (1977).

Derivative Type. Trimethylsilyl (TMS), dimethylethylsilyl (DMES).

Reagent. Trimethylsilylimidazole (TMSI), dimethylethylsilylimidazole (DMESI).

Compound Type. Hydroxysteroids.

Reaction

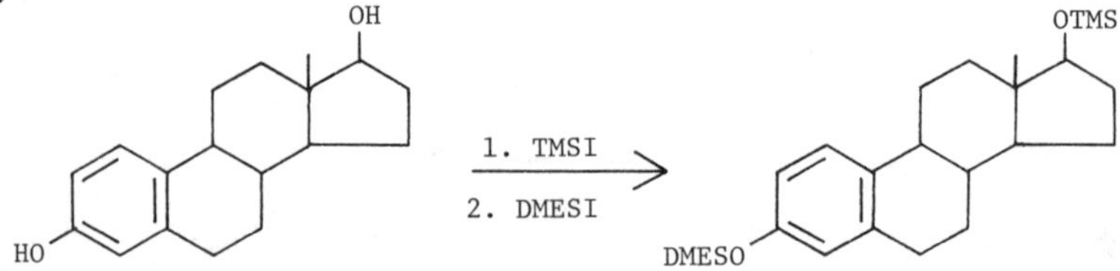

Typical Procedure. Diss 100 μg of steroid in 20 μl of TMSI (or BSTFA). Allow to stand 1 hr at room temp then evap to dryness under N_2. Diss residue in dry pyridine and inject an aliquot into the GC or GC-MS using the "sandwich" technique with DMESI containing 5% imidazole.

Comments. Phenolic TMS groups were exchanged for DMES groups but alcoholic hydroxyl TMS derivs were unchanged. The added imidazole served as a catalyst for the exchange. Diss the TMS deriv in DMESI led to some exchange of alcoholic TMS groups. The dimethylpropylsilyl (DMPS) derivs were also prepared using dimethylpropylsilylimidazole.

References

[1]H. Miyazaki *et al.*, *J. Chromatogr.* **133**, 311 (1977).

Derivative Type. Dimethylsilyl (DMS).

Reagent. Bis(dimethylsilyl)acetamide.

Compound Type. Steroids.

Reaction

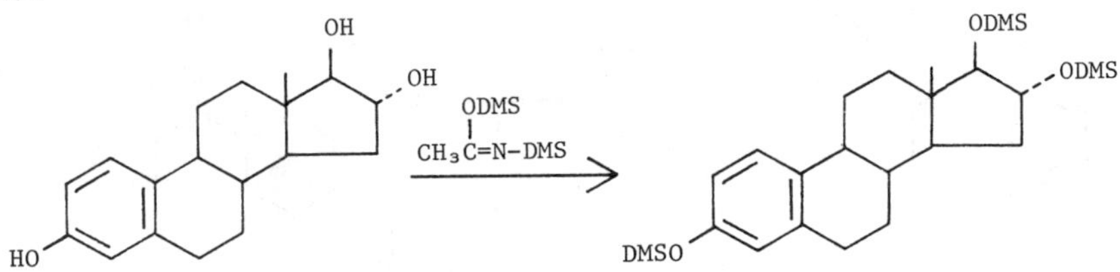

Typical Procedure. Diss 2 mg steroid in 1 ml EtOAc. Add 20 μl bis(dimethylsilyl)acetamide. Heat at 50° for 1 hr. Analyze by GC-MS.

Comments. The MS behavior of DMS derivs generally paralleled that of the TMS derivs, but the DMS derivs gave generally more intense molecular ions.

References

[1]D. H. Hunneman, in Mass Spectrometry in Biochemistry and Medicine, Raven Press, New York, NY, 1974, p. 131.

Derivative Type. *t*-Butyldimethylsilyl (TBDMS).

Reagent. *t*-Butyldimethylchlorosilane (TBDMS-Cl)/imidazole.

Compound Type. Steroids.

Reaction

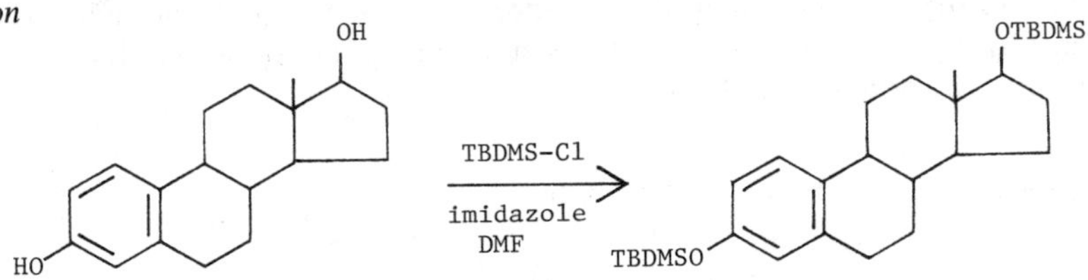

Typical Procedure (Preparative). Diss 0.1 mmol steroid in 2 ml DMF and add 1–3 mmol TBDMS-Cl and 3–5 mmol imidazole. Allow to stand at room temp 2 hr. Dil with Et_2O, wash with H_2O, and dry over Na_2SO_4. Evap to yield the deriv.

Comments. The derivs of 12 steroids were prepared for study of their hydrolysis behavior. It was found that positional isomers and some axial and equatorial isomers were sufficiently different in ease of hydrolysis that selective hydrolyses could be achieved. Such selective hydrolysis followed by a different derivatization could be exploited analytically.

References

[1]H. Hosada *et al., Chem. Pharm. Bull.* **23**, 2118 (1975).

11.2.4. Other Derivatives of Phenolic Steroids

Derivative Type. 3-*O*-[2-(Iodomethylsiloxy)propyl], 17-*O*-iodomethylsilyl.

Reagent. 1,2-Epoxypropane/sodium ethoxide (NaOEt), iodomethyldimethylchlorosilane/diethylamine.

Compound Type. Estrone and 17*β*-estradiol.

Reaction

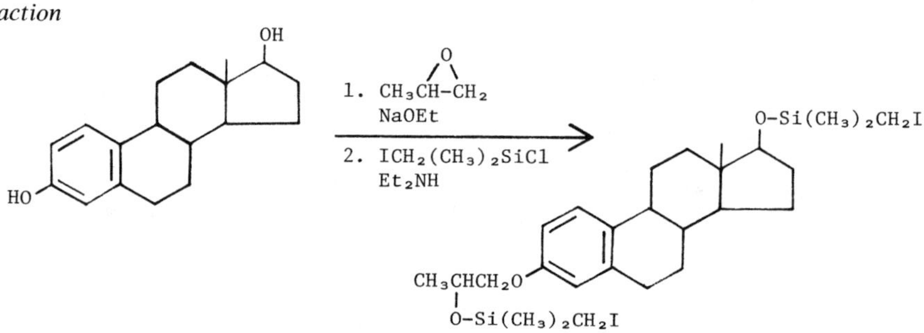

Typical Procedure. Evap an ext (containing ng quantities of estrogens) to dryness under N_2 and diss residue in 0.3 ml 0.0013 *N* NaOEt and 0.1 ml 1,2-epoxypropane. Heat 15 hr at 70°. Acidify with HOAc and ext with 2 × 5 ml $CHCl_3$. Wash ext with H_2O, dry over Na_2SO_4, and evap under N_2. Diss residue in a minimal amt of $CHCl_3$ and TLC on silica gel with benzene/MeOH (10:1, v/v). Elute the deriv with EtOAc and evap eluate to dryness under N_2. Treat the residue with iodomethyldimethylsilyl reagent (see below) at 80° for 30 min. Evap to dryness under N_2, diss residue in hexane, and purify deriv by TLC on silica gel using hexane/benzene (7:3, v/v). Analyze the purified derivs by ECD-GC.

Comments. The silylation reagent was prepared by mixing 0.1 ml of iodomethyldimethylchlorosilane in 0.5 ml dry hexane with 0.05 ml diethylamine in 0.5 ml dry hexane and removing the ppt by centrifugation. Direct silylation (without protection of the phenol as the hydroxyether) also yielded an ECD-sensitive deriv, but it was too labile to purify by TLC. A series of derivs of the estrogens were examined in an effort to maximize ECD sensitivity. The iodomethyldimethylsilyl derivs were found to be most sensitive of the group, which included HFB derivs as well as other F, Cl, and I derivs.

References

[1]D. Exley and A. Dutton, *Steroids* **14**, 575 (1969).

Derivative Type. Dimethylphosphinic ester.

Reagent. Dimethylaminodimethylphosphine/oxygen.

Compound Type. Hydroxysteroids.

Reaction

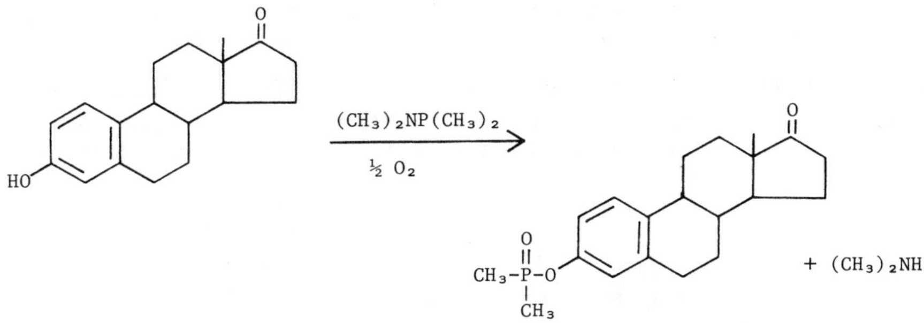

Typical Procedure.[2] Diss steroid in CH_3CN and add dimethylaminodimethylphosphine. Heat at 50–70° for 0.5–2 hr in the presence of O_2 but with careful exclusion of H_2O. Evap to dryness under vac and purify deriv by TLC on silica using acetone/$CHCl_3$ (1 : 1, v/v). Elute the deriv with dioxane and evap eluate to dryness under N_2. Diss residue in DME and analyze by GC.

Comments. The derivs could be quantitated by GC using alkali flame detection to *ca* 10 pg.[2] MS data for the derivs were reported.[4]

References

[1]W. Vogt *et al., Z. Anal. Chem.* **28**, 267 (1973).

[2]W. Vogt *et al., J. Chromatogr. Sci.* **12**, 658 (1974).

[3]W. Vogt *et al., Z. Klin. Chem.* **12**, 45 (1974).

[4]K. Jacob *et al., Biomed. Mass Spectrom.* **3**, 64 (1976).

[5]K. Jacob *et al., Tetrahedron Lett.* 1927 (1975).

11.3. DERIVATIVES OF KETOSTEROIDS

11.3.1. Oxime Derivatives

Derivative Type. Trimethylsilyloxime, trimethylsilyl.

Reagent. Hydroxylamine HCl, hexamethyldisilazane (HMDS)/trimethylchlorosilane (TMCS)/pyridine.

Compound Type. Ketosteroids.

Reaction

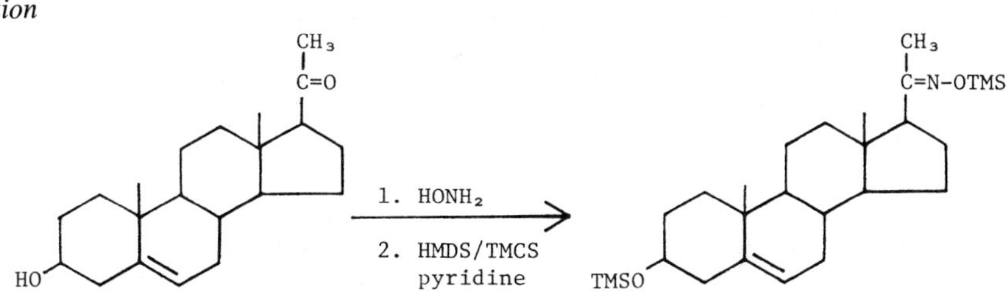

Typical Procedure. Diss 1 mg steroid and 1 mg hydroxylamine HCl in 0.1 ml pyridine and maintain at 20° overnight. Dil with 0.2 ml H_2O and ext with 1 ml EtOAc. Wash ext 2 × with

H_2O and dry over 5 A mol sieve. Analyze by GC or further deriv by evap EtOAc and adding 0.1 ml pyridine, 0.1 ml HMDS, and a trace of TMCS. Allow to stand 10 min at room temp and evap under N_2. Diss in 1 ml EtOAc and analyze 1 μl aliquots by GC.

Comments. Methoximes and ethoximes were also prepared. The derivs were shown to be useful for MS characterization of 20-oxosteroids. No indication of *syn*- and *anti*-isomers was seen. Nambara *et al.*[2] had used trimethylsilyloximes to separate isomeric 16- and 17-oxosteroids.

References

[1]C. J. W. Brooks and D. J. Harvey, *Steroids* **15**, 283 (1970).

[2]T. Nambara *et al., J. Chromatogr.* **34**, 526 (1968).

Derivative Type. Trimethylsilyloxime, trimethylsilyl.

Reagent. Hydroxylamine HCl, bis(trimethylsilyl)acetamide (BSA).

Compound Type. Testosterone.

Reaction

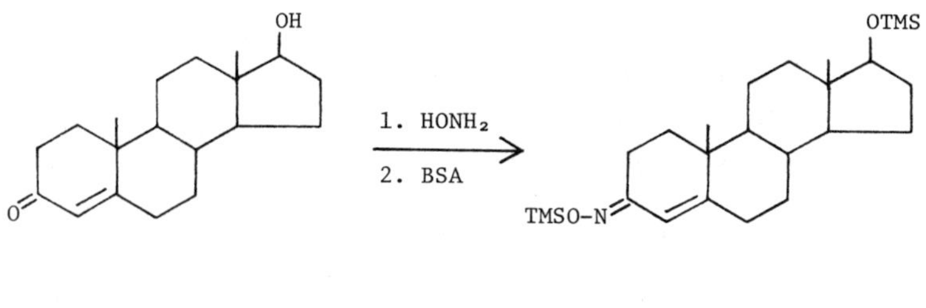

Typical Procedure. Homogenize prostatic tissue with acetone and isolate testosterone by TLC. Elute TLC area with acetone/MeOH (1:1, v/v) and evap to dryness under N_2. Diss residue in 0.5 ml EtOH for storage at $-20°$ prior to analysis.

Evap to dryness under N_2 and dessicate under vac over P_2O_5 for 1 hr. Add 30 μl of a soln of $HONH_2$-HCl (20 mg/ml) in pyridine and allow to stand overnight at 20°. Evap to dryness under N_2 and add 15 μl BSA. Allow to stand 12 hr at 20° then analyze 1.5 μl aliquots by GC-MS.

Comments. The method was used to quantitate testosterone in tissue to subnanogram levels by GC- high resolution MS-SIM. Neither TMS nor methoxime TMS derivs were satisfactory for subnanogram analysis, the latter due to formation of separable *syn-anti* isomers.

References

[1]D. S. Millington *et al., Biomed. Mass Spectrom.* **2**, 219 (1975).

Derivative Type. Methoxime.

Reagent. Methoxylamine.

Compound Type. Ketosteroids.

Reaction

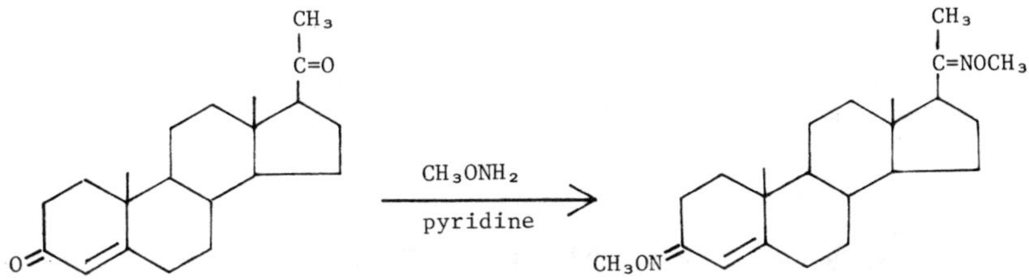

Typical Procedure. Diss equal amts of the keto steroid and methoxylamine HCl in pyridine (40 mg/ml). Allow to stand at room temp and analyze 1–2 μl aliquots by GC.

Comments. The formation of the derivs from 25 steroids having a keto group at positions 3, 11, 16, 17, or 20 were studied.[1] Only the derivs of the 3-ketosteroids studied gave two isomers (presumably *syn*- and *anti*- isomers). The 11-keto group was found to form a deriv in the absence of a C-18 methyl group. The 20-keto group was most reactive ($T_{1/2}<10$ min) with the 17-keto least reactive ($T_{1/2}>1$ hr) and the 3-keto of intermediate reactivity ($T_{1/2}=15$–20 min). Fales and Luukkainen[2] reported GC data for the derivs of 52 steroids and menadione.

References

[1]F. Dray and I. Weliky, *Anal. Biochem.* **34**, 387 (1970).

[2]H. M. Fales and T. Luukkainen, *Anal. Chem.* **37**, 955 (1965).

Derivative Type. Methoxime, trimethylsilyl ether.

Reagent. Methoxylamine, bis(trimethylsilyl)acetamide (BSA).

Compound Type. Steroids.

Reaction

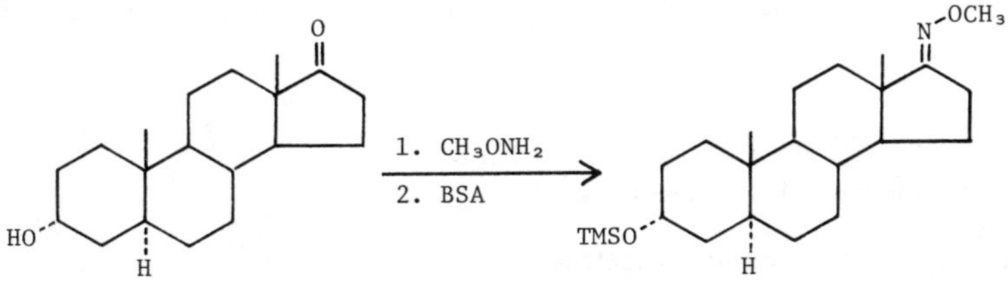

Typical Procedure

A. Stable ketones readily soluble in organic solvents—Diss 2 mg steroid in 0.5 ml pyridine. Add 8 mg methoxylamine HCl and heat 3 hr at 60°. Dil with 2 ml H_2O and ext with 3 × 5 ml benzene. Wash exts with 1 *N* HCl, H_2O, 1% $NaHCO_3$, and finally H_2O. Dry over anhyd $MgSO_4$ and evap to dryness under N_2. Proceed to C.

B. Polar steroids, e.g., corticoids—Diss 2 mg steroid in 0.5 ml pyridine. Add 8 mg methoxylamine HCl and allow to stand overnight at room temp. Dil pyridine soln with 2 ml satd NaCl and ext with 3 × 5 ml EtOAc. Wash exts with NaCl satd 1 *N* HCl, satd NaCl, NaCl satd 5% $NaHCO_3$, and finally satd NaCl. Dry over anhyd $MgSO_4$ and evap to dryness under N_2.

C. Silylation—Diss residue in 0.2 ml BSA per mg steroid and allow to stand at room temp 3–5 hr. Analyze an aliquot directly by GC or GC-MS.

Comments. Data was reported on the GC behavior of *syn*- and *anti*-isomers of the methoxime derivs of 24 steroids. Not all gave separable isomers. The 11-keto groups did not form methoximes.

References

[1]M. G. Horning *et al., Anal. Biochem.* **22**, 284 (1968).

Derivative Type. Methoxime, trimethylsilyl,

Reagent. Methoxylamine, trimethylsilylimidazole (TMSI).

Compound Type. Steroids.

Reaction

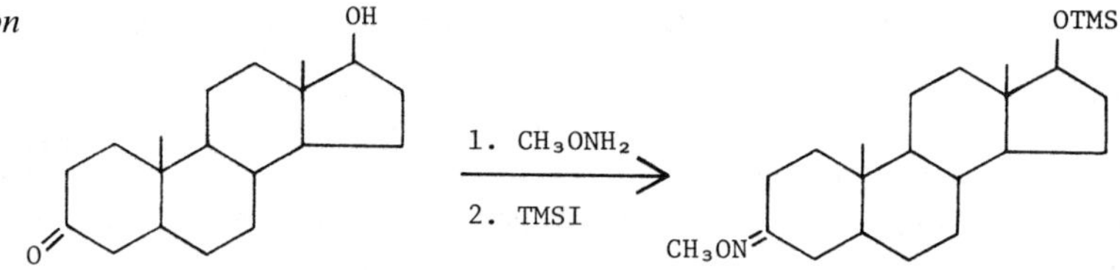

Typical Procedure. Ext 10 ml plasma dil with 10 ml saline by passage through a 4 g Amberlite XAD-2 column. Elute column with MeOAc/MeOH (2:1, v/v) and MeOH. Evap eluate and diss residue in MeOH/H_2O/$CHCl_3$ (9:1:2, v/v/v) prior to passage through a 1.5 g DEAP-LH-20 ion exchange column. Chromatograph the eluate on a 0.5 g (dry wt) column of Lipidex-5000 using the same solvent. Collect 3.5 ml of effluent and evap to dryness under N_2.

To the residue add 50 μl pyridine and 5 mg methoxylamine-HCl. Heat at 60° for 15 min; then evap pyridine. Add 50 μl TMSI. Heat at 100° for 2 hr; then add 0.5 ml hexane/pyridine/HMDS (88:1:1, v/v/v). Filter through a 0.25 g (dry wt) column of Lipidex prepared in the same solvent. Collect 2 ml of effluent, evap to dryness, and redissolve in hexane for GC-MS analysis.

Comments. The method used the deriv procedure[2] of Thenot and Horning in combination with column ext and purification procedures for analysis of neutral unconjugated steroids in plasma.

References

[1]M. Axelson and J. Sjöval, *J. Steroid Biochem.* **5**, 733 (1974).

[2]J.-P. Thenot and E. C. Horning, *Anal. Lett.* **5**, 21 (1972).

Derivative Type. Methoxime, trimethylsilyl.

Reagent. Methoxylamine, bis(trimethylsilyl)acetamide (BSA).

Compound Type. Ecdysones.

Reaction

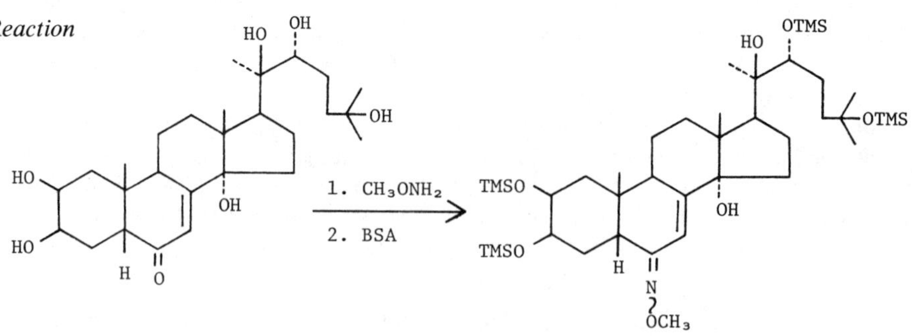

Typical Procedure. Diss 2 mg sample in 0.5 ml pyridine and add 8 mg methoxylamine-HCl. Allow to stand 100 hr at room temp. Separate the two isomeric methoximes by TLC on silica gel ($CHCl_3$/95% EtOH, 8:2, v/v). Treat each isomer with 0.5 ml BSA at room temp for 70 hr. Analyze by GC or MS.

Comments. GC and MS data were given for the derivs of ecdysterone (crustecdysone, shown above) and ecdysone.

References

[1]E. D. Morgan and A. P. Woodbridge, *Chem. Commun* 475 (1971).

Derivative Type. Methoxime, pentafluorophenyldimethylsilyl (flophemesyl).

Reagent. Methoxylamine, flophemesyldiethylamine/flophemesyl chloride.

Compound Type. Steroids.

Reaction

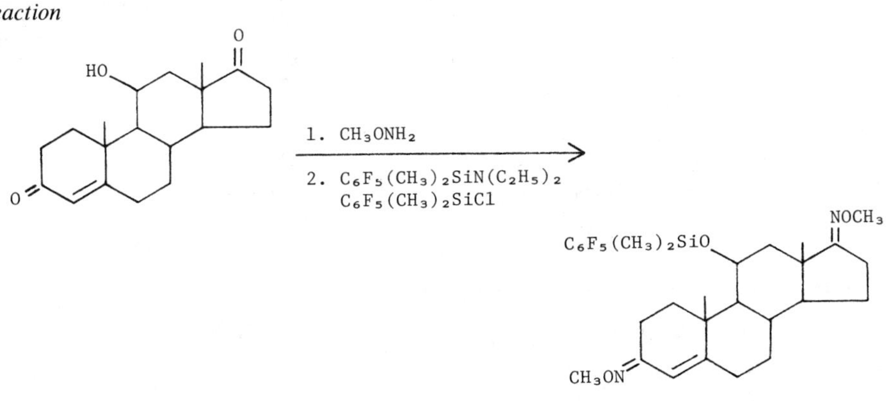

Typical Procedure. Diss steroid in anhyd pyridine and add excess methoxylamine HCl. Allow to stand at room temp overnight or heat at 65° for 3 hr. Evap to dryness under N_2. Partition residue between EtOAc and 10% aq NaCl containing 5% HCl. Wash EtOAc phase with 10% aq NaCl containing 5% $NaHCO_3$, dry over mol sieve, and evap to dryness under N_2. Diss residue in pyridine and add 1:1 (v/v) flophemesyldiethylamine/flophemesyl chloride. Heat at 60° for 3 hr. Analyze by GC or GC-MS.

Comments. Ketone groups were extensively enolized by the silylating mixt unless first protected as methoximes. The 17α-hydroxyl in 17α-hydroxypregn-4-ene-3,20,dimethoxime was silylated in 6 hr at 85° or 1 hr at 150° but 20- or 21-hydroxyls (presumably by forming the bulky silyl ether first) hindered silylation of 17-hydroxyls. Hindrance of reaction of an adjacent hydroxyl was also observed in 3β,5α,6β-trihydroxycholestane. Pyridine was a better solvent than CH_3CN, DMSO, THF, or EtOAc. Decomposition was observed in DMF.

References

[1]E. D. Morgan and C. F. Poole, *J. Chromatogr* **104**, 351 (1975).

Derivative Type. *O-sec*-Butyloxime, trimethylsilyl.

Reagent. *O-sec*-butylhydroxylamine hydrochloride, bis(trimethylsilyl)trifluoroacetamide (BSTFA).

Compound Type. Ketosteroids.

Reaction

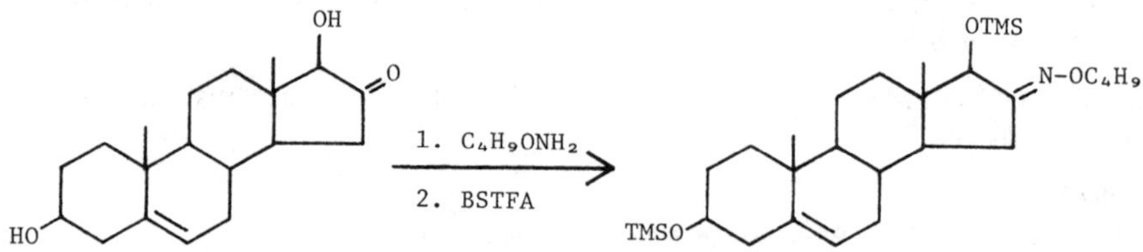

Typical Procedure. Diss 20 mg steroid in 0.5 ml pyridine containing 10 mg *O*-butylhydroxylamine HCl and heat in a conical vial at 70° overnight. Remove excess pyridine under dry N_2 and diss residue in 0.5 ml BSTFA. Heat 3 hr at 70° and inject directly into the GC.

Comments. *O*-Isopentyloximes were prepared by the same method with *O*-isopentylhydroxylamine. The *O*-butyl and *O*-pentyl oximes increased the retention times of ketosteroids and allowed their separation from hydroxysteroids. Formation of *syn*- and *anti*-isomers was observed in some cases. The MS fragmentation of the isopentyloximes was studied using d_1 and d_2-isopentylhydroxylamine.[2]

References

[1]T. A. Baille *et al., Anal. Lett.* **5**, 351 (1972).

[2]T. A. Baille *et al., Mass Spectrometry in Biochemistry and Medicine*, Raven Press, New York, NY, 1974, p. 235.

Derivative Type. *O*-Alkyl oximes.

Reagent. *O*-Alkylhydroxylamine.

Compound Type. Ketosteroids.

Reaction

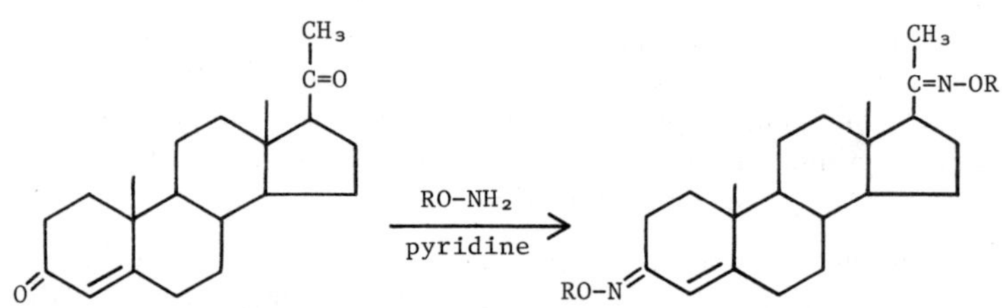

Typical Procedure. To 1 mg of steroid in 0.5 ml pyridine add 10 mg of the *O*-alkylhydroxylamine HCl. Heat at 70° for 4 hr or allow to stand at room temp overnight. Evap to dryness and diss residue in 5 ml EtOAc. Add 5 ml HCl/10% NaCl. Ext with 2 × 5 ml EtOAc and wash combined exts with 3 × 5 ml 5% $NaHCO_3$/10% HCl. Dry over $MgSO_4$ and evap under vac below 40°. Analyze the deriv by GC.

Comments. Derivs were prepared from *O*-*sec*-butyl-, *O*-isobutyl-, *O*-*n*-pentyl-, and *O*-isopentylhydroxylamines. Steroid 4-ene-3-ones and 16-ones (C-17 unsubstituted) formed separable *syn*- and *anti*-isomers. The derivs were stable and had good GC properties with retention times intermediate between methoximes and benzyloximes. The derivs permitted separation of keto from hydroxysteroids while allowing elution of diketo derivs at reasonable times and temps.

References

[1]N. Sakauchi and E. C. Horning, *Anal. Lett.* **4**, 41 (1971).

Derivative Type. *O*-Benzyloxime, trimethylsilyl ether.

Reagent. *O*-Benzylhydroxylamine, trimethylsilylimidazole (TMSI).

Compound Type. Steroids.

Reaction

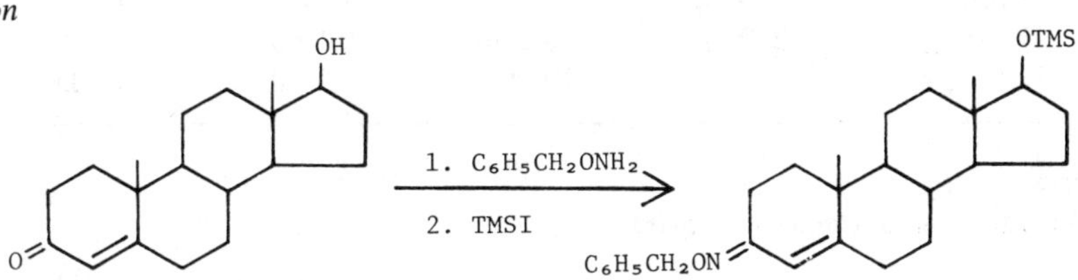

Typical Procedure. Diss 1 mg steroid in 0.5 ml of a soln of *O*-benzylhydroxylamine HCl (20–40 mg/ml) in pyridine. Heat at 60–70° for 15–20 hr then evap to dryness under N_2. (a) Trimethylsilylation of all hydroxyl groups—Add 0.5 ml TMSI or TMSI/BSA (1:1, v/v) and heat at 150° for 3 hr. (b) Trimethylsilylation of unhindered and moderately hindered hydroxyl groups—Add 0.5 ml BSA or BSTFA and heat at 100° for 1 hr. Analyze directly by GC or GC-MS.

Comments. The benzyloxime derivs eluted much later than methoxime derivs and permitted distinguishing ketosteroids from steroids containing only hydroxyl groups (or unreactive 11-keto groups). Steroid 4-ene-3-one derivs were formed as separable *syn*- and *anti*- isomers. Derivs of 16-, 17-, and 20-ones gave single GC peaks; 11-keto groups did not form derivs. GC and MS data were reported for the derivs of 16 newborn human urinary steroids.[2]

References

[1]P. G. Devaux *et al.*, *Anal. Lett.* **4**, 151 (1971).

[2]P. G. Devaux *et al.*, *Anal. Biochem.* **41**, 70 (1971).

Derivative Type. *O*-Pentafluorobenzyloxime.

Reagent. *O*-Pentafluorobenzyl hydroxylamine (PFBHA).

Compound Type. Ketosteroids.

Reaction

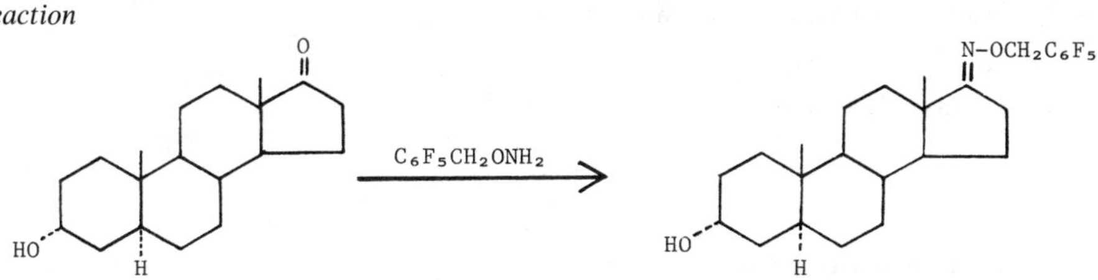

Typical Procedure. To 0.1–5 ng of steroid add 5 μg of PFBHA-HCl in 10 μl pyridine. Heat at 65° for 30 min and remove pyridine under N_2 at 40–45°. Diss residue in 25 or 100 μl of cyclohexane and wash with equal vol of 10% $HClO_4$. Analyze aliquots of cyclohexane layer by ECD-GC.

Comments. The method was used for 3, 17, and 20-monoketo and 3,17 and 3,20-diketo steroids. *Syn-* and *anti*-isomers were separated on SE-30 but gave a single peak on OV-17 or Silar 5-CP. The derivs were detectable to low pg levels using ECD-GC.

For larger quantities of sample, the following amounts of reagents were used:

Amount	PFBHA-HCl	Pyridine	Cyclohexane	Detection
1 mg	10 mg	0.2 ml	1 ml	FID
1–10 μg	250 μg	0.2 ml	0.1 ml	FID
12.5–50 ng	25 μg	10 μl	0.5 ml	ECD

References

[1]K. T. Koshy *et al.*, *J. Chromatogr. Sci.* **13**, 97 (1975).

Derivative Type. *O*-Pentafluorobenzyloxime, trimethylsilyl.

Reagent. *O*-Pentafluorobenzyl hydroxylamine(PFBHA), hexamethyldisilazane(HMDS)/trimethylchlorosilane (TMCS)/pyridine.

Compound Type. Ketosteroids.

Reaction

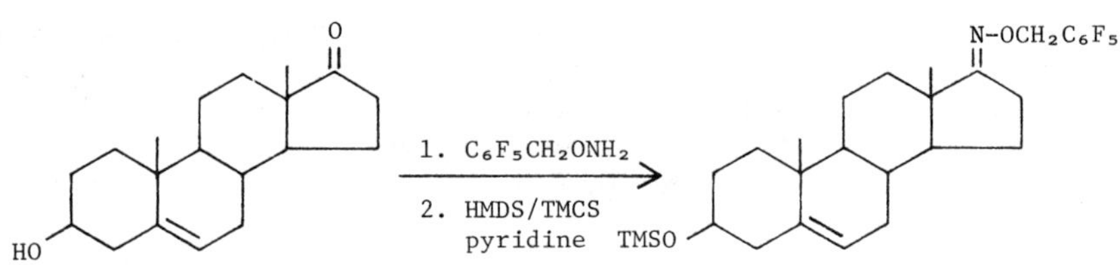

Typical Procedure. To *ca* 1 μg of 17-ketosteroid in 2 drops of pyridine add 0.2 mg PFBHA-HCl and heat at 60° for 1 hr. Dil with hexane, and wash with 1 ml H_2O, 1 ml 0.1 *N* HCl, 1 ml 0.1 *N* NaOH, and 1 ml H_2O. Evap to dryness and add 0.1 ml HMDS and 0.1 ml TMCS in 5 drops of pyridine. Allow to stand 5 min at room temp then evap to dryness. Diss residue in 1 ml hexane and analyze 2 μl aliquots by ECD-GC.

Comments. The method was used for quantitation of dehydroepiandrosterone in plasma. The synthesis of the oxime reagent was reported.

References

[1]T. Nambara *et al.*, *J. Chromatogr.* **114**, 81 (1975).

11.3.2. Hydrazone Derivatives

Derivative Type. *N,N*-Dimethylhydrazone.

Reagent. *N,N*-Dimethylhydrazine/acetic acid (HOAc).

Compound Type. Ketosteroids.

Reaction

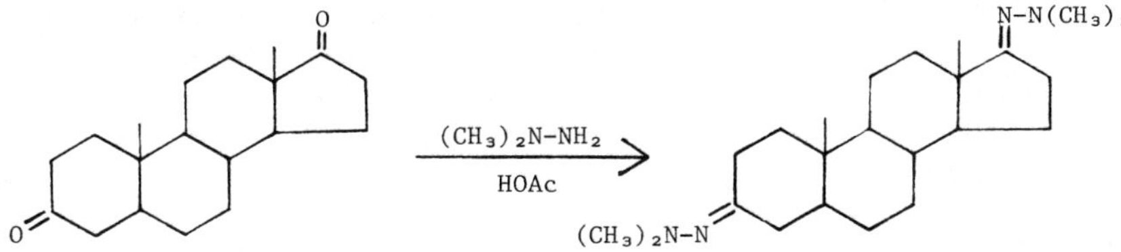

Typical Procedure. Diss 0.1–1.0 mg steroid in 0.1–0.2 ml dimethylhydrazine and add 0.05 ml HOAc. Allow to stand 2 hr at room temp then evap under N_2. Diss residue in THF and analyze by GC.

Comments. Unconjugated 3-keto groups gave quantitative derivs after 1–2 hr at room temp without the HOAc catalyst. Conjugated 3-keto groups and 16-, 17-, or 20-ones required HOAc. The 11-keto group failed to react even after 12 hr.

References

[1]W. J. A. VandenHeuval and E. C. Horning, *Biochim. Biophys. Acta* **74**, 560 (1963).

[2]E. C. Horning *et al., Rec. Progr. Hormone Res.* **19**, 57 (1963).

Derivative Type. Methyl ether, pentafluorophenylhydrazone.

Reagent. Dimethyl sulfate, pentafluorophenylhydrazine.

Compound Type. Estrone.

Reaction

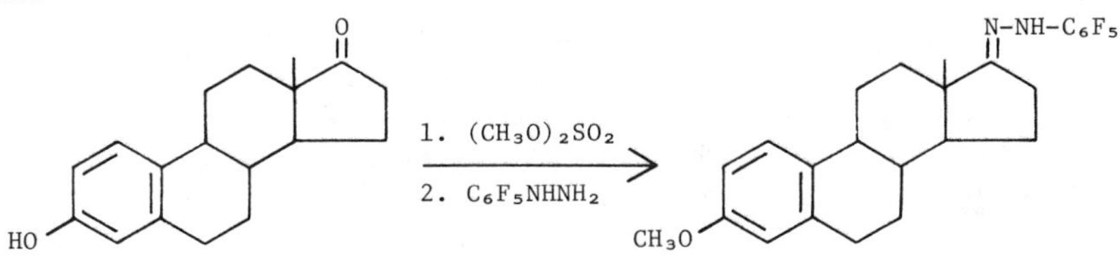

Typical Procedure. Ext 10 ml plasma with 3×20 ml Et_2O. Wash exts with 0.1 vol 8% $(NH_4)_2CO_3$ and 0.1 vol H_2O. Evap to dryness under N_2 at 40°. Diss residue in 5 ml toluene and ext with 3×2 ml 1 *N* NaOH.

To combined NaOH exts add 1.5 ml 1 *N* NaOH and 5 ml 4.5% (w/v) boric acid. Warm at 40° for 5 min then add 0.5 ml dimethyl sulfate. Shake and hold at 40° overnight. Add 2.5 ml 20% (w/v) NaOH and ext with 12 ml *n*-hexane. Wash ext with 2 ml H_2O until neutral and evap to dryness under N_2. TLC on silica gel with benzene/MeOH (99:1, v/v). Elute deriv with 6 ml EtOAc. Evap to dryness and diss residue in 0.2 ml MeOH. Add 0.2 ml of soln of 10 mg pentafluorophenylhydrazine in 9 ml MeOH and 1 ml HOAc. Allow to stand overnight at room temp. Add 1 ml H_2O and ext with 3×2 ml hexane. Wash combined hexane with 2 ml 1 *N* H_2SO_4 and then with 2 ml H_2O repeatedly to neutrality. Evap to dryness and TLC residue on silica gel with benzene. Scrape and elute deriv with 6 ml EtOAc/benzene (5:95, v/v). Wash with 0.5 ml H_2O, centrifuge, and evap org layer to dryness. Diss residue in small vol of toluene and analyze by ECD-GC.

Comments. The method was used for measurement of plasma estrone to 0.1 ng/ml. The pentafluorophenylhydrazone derivs of other steroids were also examined and found to be stable and give good GC behavior. Chloro and bromo substituted phenylhydrazines were also examined and found to be either unreactive or give unstable derivs. By measuring estradiol as estrone with this deriv, Mead *et al.*[2] determined estrone and estradiol in the same sample. The pentafluorophenylhydrazone deriv has also been used for testosterone.[3]

References

[1]J. Attal *et al., Anal. Biochem.* **20**, 394 (1967).

[2]R. A. Mead *et al., J. Chromatogr. Sci.* **7**, 554 (1969).

[3]J. Attal, *J. Endocrinol.* **85**, 280 (1969).

Derivative Type. Estrone 3-methyl ether 17-pentafluorophenylhydrazone.

Reagent. Hydroxysteroid dehydrogenase, dimethyl sulfate, pentafluorophenylhydrazine.

Compound Type. Estradiol.

Reaction

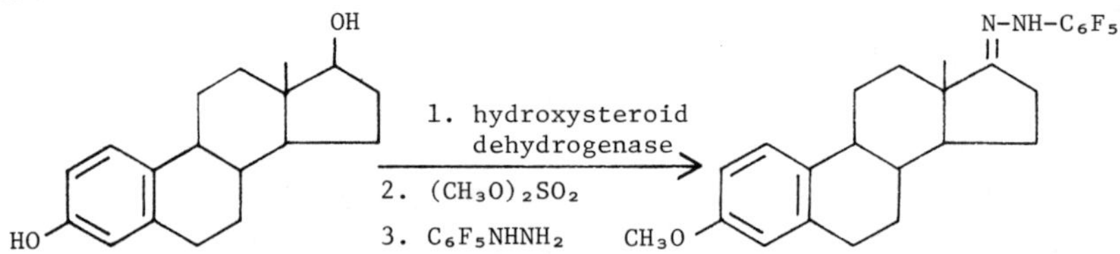

Typical Procedure. To 10 ml plasma add H^3-estradiol and ext with 6 × 15 ml Et_2O. Evap ext to dryness at 40° under N_2 and diss residue in 5 ml toluene. Ext toluene soln with 2 × 2 ml 1 *N* NaOH. Neutralize the combined NaOH ext with 8*N* HCl. Ext this aq soln with 4 × 20 ml Et_2O. Wash each Et_2O ext with 2 ml H_2O and evap the Et_2O to dryness under N_2. Purify the estradiol by TLC on silica gel with $CHCl_3$/EtOAc (5 : 1, v/v) and elute with 6 ml MeOH. Evap the eluate to dryness and incubate with hydroxysteroid dehydrogenase and cofactor at 37° for 2 hr. Ext with 3 × 5 ml Et_2O and wash ext with H_2O until neutral. Evap to dryness under N_2. Prepare the methyl ether pentafluorophenylhydrazone as for estrone[2] (q.v.) and analyze by ECD-GC.

Comments. The method was used to quantitate estradiol in plasma to as little as 50 ng/ liter.

References

[1]J. Attal and K. B. Eik-Nes, *Anal. Biochem.* **26**, 398 (1968).

[2]J. Attal *et al., Anal. Biochem.* **20**, 394 (1967).

Derivative Type. *p*-Nitrophenylhydrazone.

Reagent. *p*-Nitrophenylhydrazine.

Compound Type. Δ^4-3-Ketosteroids.

Reaction

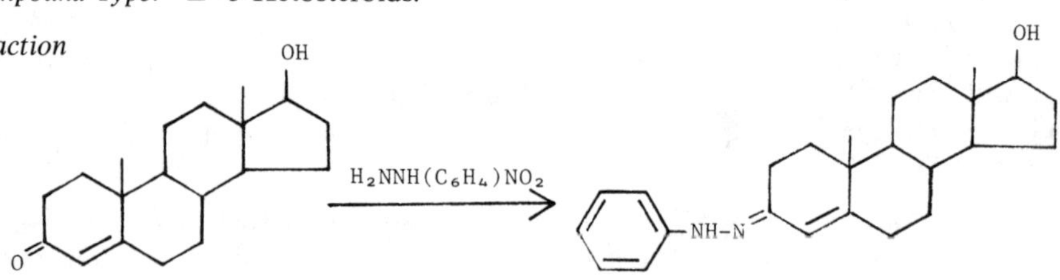

Typical Procedure. Evap steroid sample in test tube under reduced pressure with gentle warming. Diss residue in 0.5 ml MeOH. Add 0.5 ml fresh reagent (20 mg *p*-nitrophenylhydrazine diss in MeOH plus 2 ml conc HCl and MeOH to make 50 ml). Mix and allow reaction to proceed 60 min at 50°. Cool and add 0.5 ml 1 *N* NaOH and 5.0 ml DMF. Transfer to cuvette and read absorbance at 540 nm.

Comments. The method was used for colorimetric determination but could be adapted for chromogenic derivatization for HPLC.

References

[1]T. Nishina *et al.*, *Steroids* **4**, 255 (1964).

Derivative Type. 2,4-Dinitrophenylhydrazone.

Reagent. 2,4-Dinitrophenylhydrazine (DNPH).

Compound Type. Steroids.

Reaction

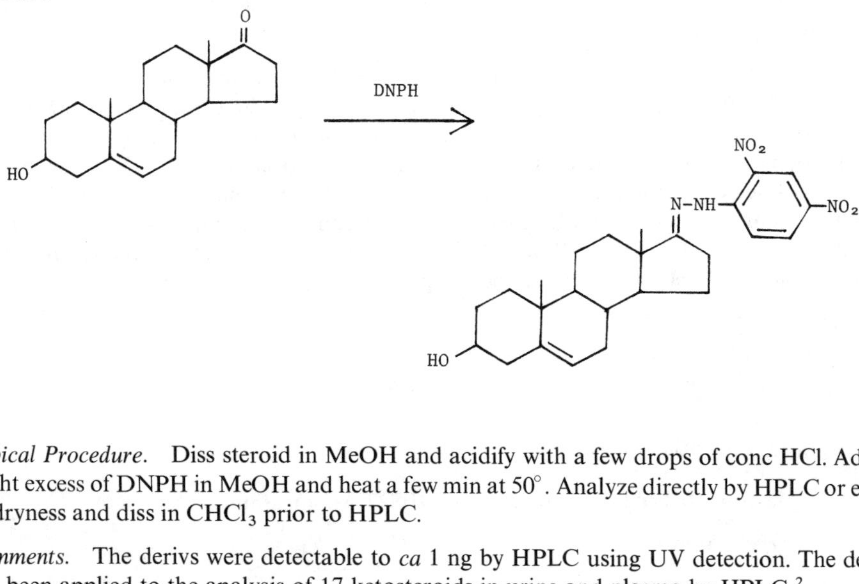

Typical Procedure. Diss steroid in MeOH and acidify with a few drops of conc HCl. Add a slight excess of DNPH in MeOH and heat a few min at 50°. Analyze directly by HPLC or evap to dryness and diss in $CHCl_3$ prior to HPLC.

Comments. The derivs were detectable to *ca* 1 ng by HPLC using UV detection. The deriv has been applied to the analysis of 17-ketosteroids in urine and plasma by HPLC.[2]

References

[1]R. A. Henry *et al.*, *J. Chromatogr. Sci.* **9**, 513 (1971).

[2]F. A. Fitzpatrick *et al.*, *Anal. Chem.* **44**, 2211 (1972).

11.3.3. Silyl Enol Ether Derivatives

Derivative Type. Trimethylsilyl.

Reagent. Hexamethyldisilazane (HMDS)/trimethylbromosilane (TMBS)/pyridine.

Compound Type. Steroids.

Reaction

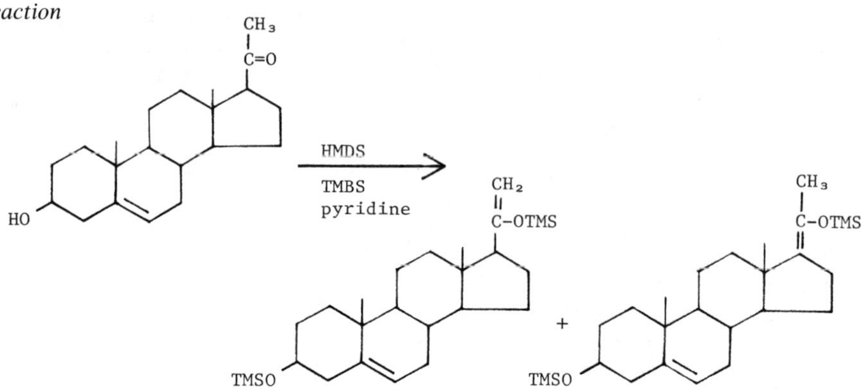

Typical Procedure. Mix 2.5 ml HMDS, 2.0 ml pyridine, and 0.5 ml TMBS. Allow ppt to settle for at least 1 hr; then add 250 μl supernatant to 50 μg steroid. Allow to react at room temp (see below for times) and then add 2 ml dry heptane. Mix, centrifuge, and remove supernatant. Evap to dryness under N_2 and diss residue in 250 μl heptane. Analyze by GC.

Comments. TMBS was found to be a more potent silylation catalyst than TMCS. Steroids with 3-, 11-, 17-, 20-, and 21-hydroxyl formed TMS derivs in 24 hr at room temp. Significant enol-OTMS deriv was formed from 17-oxo (dehydroepiandrosterone) and 20-oxo (pregnenolone) groups after 5 min and within 30 min for an 11-oxo compd (3α,20β-dihydroxy-5β-pregnan-11-one). Complete enol deriv was formed from pregnenolone in 3 hr. Incomplete enol deriv was obtained for 11-oxo and 20-oxo-17α-hydroxy compds even after 24 hr. Gaskell *et al.*[2] used BSA in place of the HMDS (BSA/TMBS/pyridine, 3:1:4, v/v/v) for preparation of enol ether derivs of Δ^4-3-ketosteroids and reported production of fewer side products.

References

[1] L. Aringer *et al.*, *Steroids* **17**, 377 (1971).

[2] S. J. Gaskell *et al.*, *Biomed. Mass Spectrom.* **2**, 148 (1975).

Derivative Type. Trimethylsilyl ethers.

Reagent. Bis(trimethylsilyl)trifluoroacetamide (BSTFA)/potassium acetate (KOAc).

Compound Type. Steroids.

Reaction

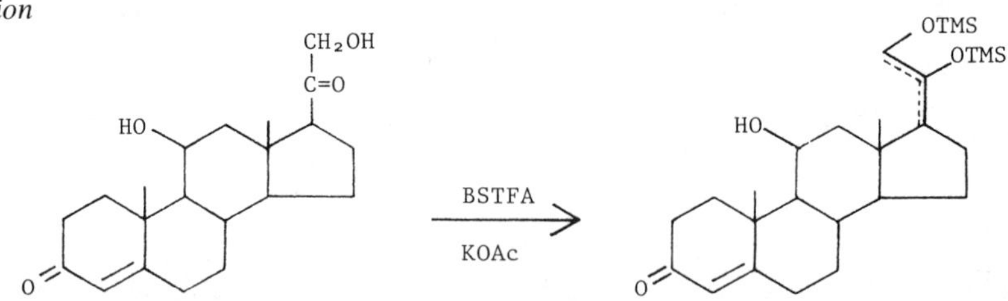

Typical Procedure. Diss 30 μg of steroid in 100 μl of BSTFA (if necessary, add CS_2 for solubility and evap under N_2 after addition of BSTFA). Add 0.5–1 mg KOAc and heat at 60° for 40 min. Evap under N_2 and diss residue in HMDS. Store in dessicator until analyzed by GC-MS.

Comments. Derivs of 28 steroids were examined. Hydroxyl groups in 3α, 3β, or 20α positions reacted completely. 17α-tertiary-hydroxyls and 11β-hydroxyls were unreactive. Partial enolization was observed for 17-keto groups, but 3-keto groups were unreactive. 20-keto groups were unreactive alone but in the presence of 21, or 17,21-hydroxyls gave the enol ether deriv. 17,20,21-trihydroxylated steroids gave mixtures of 20,21 and 17,21-di-TMS derivs.

References

[1]S. Zaraga *et al.*, *J. Steroid Biochem.* **4**, 417 (1973).

Derivative Type. Trimethylsilyl ether, trimethylsilyl enol ether.

Reagent. Trimethylsilylimidazole (TMSI)/potassium acetate.

Compound Type. Norethindrone.

Reaction

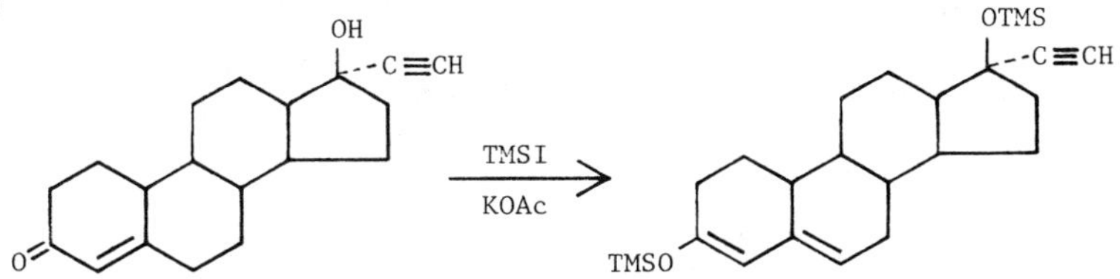

Typical Procedure. To 1 ml plasma add 40 μl norgestrel int stand soln (2 μg/ml in CH_3CN), 4 ml H_2O, 2 ml benzene, 0.05 ml EtOH, and 2.5 g $(NH_4)_2CO_3$. Mix, centrifuge, and remove benzene. Repeat with 3 × 2 ml benzene. Wash combined exts with benzene and evap under N_2. Diss residue in 0.8 ml MeOH and 0.2 ml 10% HOAc. Wash with 1 ml hexane; then evap MeOH phase to dryness under N_2.

Diss the residue in 50 μl CH_3CN, add 10 μl TMSI, and heat at 85° for 2 hr. Add 3–4 crystals of KOAc and heat for 45 min. Evap to dryness under N_2 then place tube under vac for 20 min. Diss residue in 50 μl BSTFA and analyze by GC-MS-SIM.

Comments. The 3,5(6)-diene structure shown for the deriv was one of three possible structures, the others being the 2,4-diene and the 3,5(10)-diene. Only one GC peak was observed however. The deriv was detectable to 50 pg. The HFB deriv was also examined but it gave multiple GC peaks.

References

[1]W. G. Stillwell *et al.*, *Steroids Lipids Res.* **5**, 79 (1974).

Derivative Type. Trimethylsilyl.

Reagent. Trimethylsilylimidazole (TMSI).

Compound Type. Norethynodrel.

Reaction

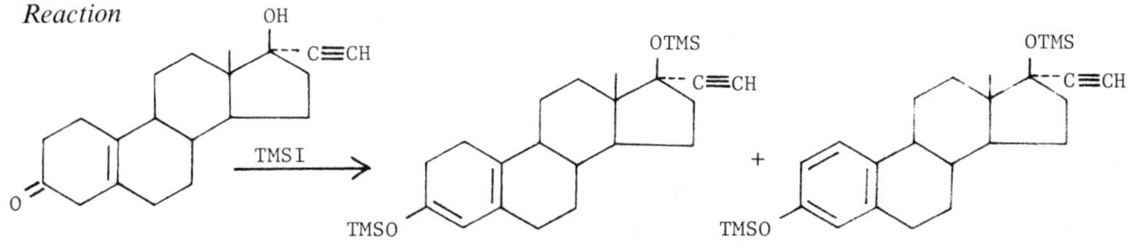

Typical Procedure. To 100 μg of the steroid add 50 μl TMSI and heat 4 hr at 100°. Analyze by GC or GC-MS.

Comments. Shorter reaction times or lower temp yielded a mixt of I and II, but heating at 100° for 4 hr or more gave only the aromatic deriv II. BSA or BSTFA at room temp gave no silylation of the compds. Norethindrone did not give an aromatic deriv after 6 hr at 100° with TMSI. The mechanism of the aromatization reaction was not determined but apparently was due to dehydrogenation rather than oxidation since the same results were obtained under N_2.

References

[1]R. M. Thompson and E. C. Horning, *Steroids Lipids Res.* **4**, 135 (1973).

11.3.4. Enol Acylate Derivatives

Derivative Type. 3-Enol heptafluorobutyrate (HFB).

Reagent. Heptafluorobutyric anhydride (HFBA).

Compound Type. 3-Oxo-Δ^4-steroids.

Reaction

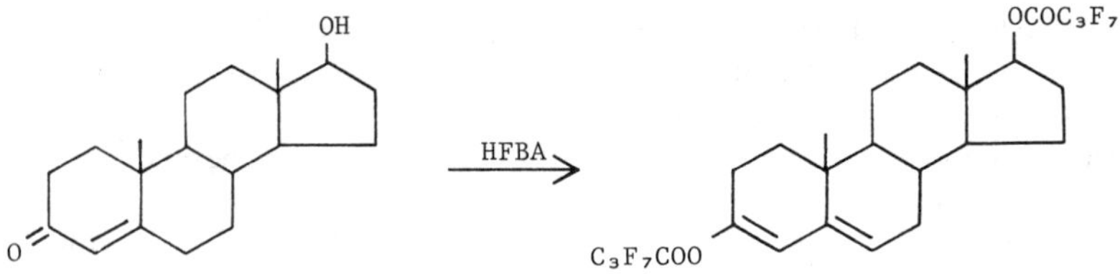

Typical Procedure. Evap steroid soln to dryness under N_2 at 53° in a small glass stoppered tube. Add 5 μl HFBA and 5 μl benzene. Heat at 70° for 30 min then place in a dessicator at room temp for 15 min. Evap to dryness under N_2 and diss residue in hexane. Analyze by ECD-GC or GC-MS.

Comments. Optimum HFBA/benzene ratios were determined for formation of mono-, di-, and triheptafluorobutyrate derivs of a series of steroids.[1] Formation of 3-enol-HFB derivs with HFBA in benzene had been reported earlier.[2,3] Unconjugated keto groups (e.g., 17-oxo) did not form enol derivs. It was demonstrated that the derivs could be purified on a Sephadex LH-20 column without decomposition.[1]

References

[1]J. R. G. Challis and R. B. Heap, *J. Chromatogr.* **50**, 228 (1970).

[2]D. Exley and J. Chamberlain, *Steroids* **10**, 509 (1967).

[3]J. Chamberlain, *J. Chromatogr.* **28**, 404 (1967).

Derivative Type. Heptafluorobutyrate.

Reagent. Heptafluorobutyric anhydride (HFBA)/pyridine.

Compound Type. Testosterone.

Reaction

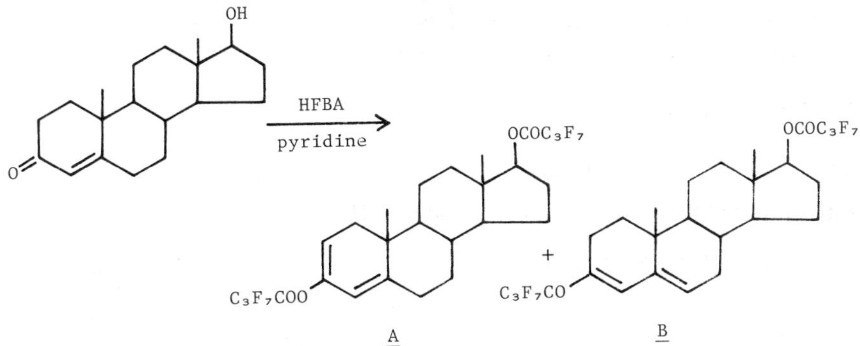

Typical Procedure. To 100 μl of a benzene soln of testosterone (5 μg/μl) and *n*-octacosane int stand (5 μg/μl) add 20 μl pyridine (or collidine) and 10 μl HFBA. Heat at 60° for 2 hr. Analyze by GC.

Comments. Deriv using pyridine or collidine gave *A* as the major product. In benzene or in acetone, without the base, isomer *B* was the major product. The reaction with pyridine was reported to be suitable for quantitative work with the deriv detectable in subnanogram amts by ECD-GC. Other workers have used the di-HFB deriv prepared in benzene without base for quantitation of plasma testosterone.[2,3] The problem of isomeric enol ether formation is a general one for 4-ene-3-one steroids.

References

[1]P. G. Devaux and E. C. Horning, *Anal. Lett.* **2**, 637 (1969).

[2]W. P. Collins *et al.*, *J. Chromatogr.* **37**, 33 (1968).

[3]D. Exley, *Biochem. J.* **107**, 285 (1968).

Derivative Type. 17β-Pentafluorobenzoate, 3-heptafluorobutyrate (HFB).

Reagent. Pentafluorobenzoic anhydride, heptafluorobutyric anhydride (HFBA).

Compound Type. Testosterone.

Reaction

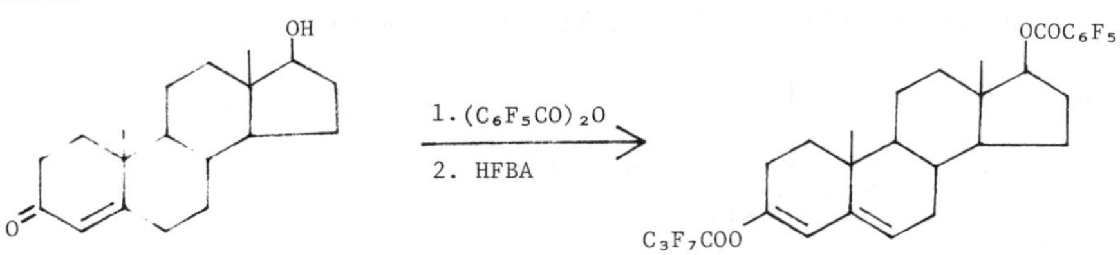

Typical Procedure. Diss 100 mg steroid in 2 ml anhyd benzene, 0.5 ml THF, and 0.05 ml pyridine. Add pentafluorobenzoic anhydride and heat overnight at 60°. Evap to dryness under N_2 and add 2 ml acetone and 0.5 ml HFBA. Allow to stand 1 hr at 25° then evap under N_2 at 70°. Diss in benzene and analyze by GC.

Comments. The deriv was shown to have higher ECD sensitivity than any previously reported deriv. The 17β-pentadecafluorooctanoate and 17β-chlorodifluoroacetate 3-HFB derivs showed nearly as high sensitivity.

References

[1]L. Dehennin *et al.*, *J. Chromatogr. Sci.* **10**, 224 (1972).
[2]L. Dehennin and R. Scholler, *Steroids* **13**, 739 (1969).

11.3.5. Thioketal Derivatives

Derivative Type. Ethylene thioketals.

Reagent. Ethanedithiol/*p*-toluenesulfonic acid (TsOH).

Compound Type. Ketosteroids.

Reaction

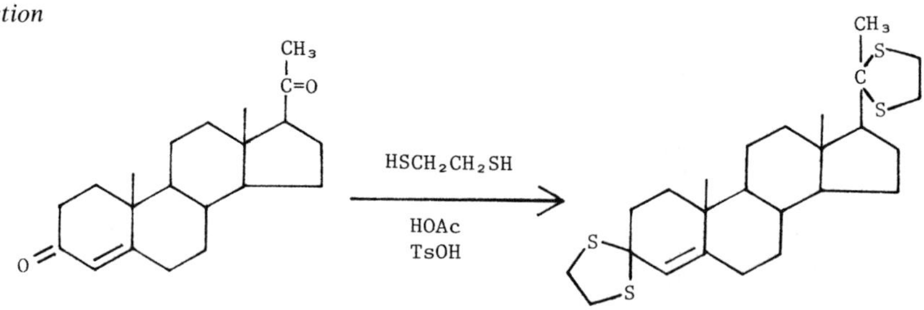

Typical Procedure. Diss 0.25–10 μg steroid in 0.1 ml reagent containing 7% (v/v) ethanedithiol and 1.7% (w/v) TsOH in HOAc. Allow to react overnight at room temp. Add 2 ml hexane and wash with 3 × 0.5 ml 0.2 *N* NaOH and 2 × with H_2O. Evap org phase under N_2 and diss residue in CS_2. Analyze by GC.

Comments. With hydroxyketosteroids, partial acetylation occurred in the ketalization procedure. These compds were therefore first acetylated in Ac_2O/pyridine (1 : 1, v/v) at room temp overnight prior to ketal formation. In both the acetylation and ketalization, the reactions could be accelerated by heating at 60°.

References

[1]A. Zmigrod and H. R. Lindner, *Steroids* **8**, 119 (1966).

11.4. DERIVATIVES OF 17,20,21-TRIHYDROXYSTEROIDS

Derivative Type. Acetonide.

Reagent. Acetone/*p*-toluenesulfonic acid (TsOH)/calcium chloride.

Compound Type. 17,20,21-Trihydroxysteroids.

Reaction

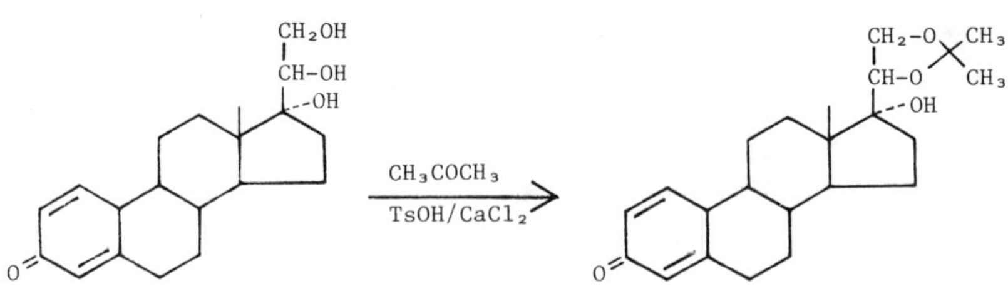

Typical Procedure. Prepare anhyd steroid by evap to dryness of an acetone-benzene soln. Diss residue in 2 ml dry acetone containing 2 mg TsOH. Add 50 mg anhyd $CaCl_2$ and shake for 4 hr. Remove $CaCl_2$ by filtration and add 2 drops of pyridine. Evap to dryness and diss residue in 20 ml benzene. Wash with 2 ml 1 *N* NaOH and 2 ml H_2O. Back ext combined washings with benzene. Dry combined benzene phases with Na_2SO_4 and evap to dryness. Redissolve residue in 1 ml benzene for GC analysis.

Comments. The deriv was used for quantitation of the following unconjugated corticosteroids in urine: 20α- and 20β-dihydrocortisol, 20-dihydrocortisone, 20α and 20β-dihydroprednisolone, and 20-dihydroprednisone.[1] The separation of epimeric cortols and cortolone in hydrolyzed urine was also demonstrated.[1] Acetonides of 17,20-diols have been prepared using acetone/$HClO_4$.[2]

References

[1]E. Bailey, *Steroids* **10**, 527 (1967).

[2]M. L. Lewbart and J. J. Schneider, *J. Org. Chem.* **34**, 3505, 3513 (1969).

Derivative Type. Trimethylsilyl ethers.

Reagents. Trimethylsilylimidazole (TMSI).

Compound Type. Hindered hydroxysteroids.

Reaction

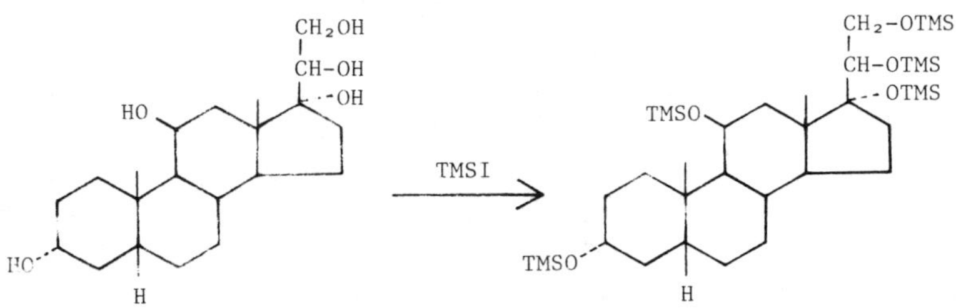

Typical Procedure. Diss 0.20 mg of steroid in 20 μl pyridine and add 200 μl TMSI. Heat at 150° for 4 hr. Analyze by GC.

Comments. Silylation rates for the hindered 17α-hydroxy group were studied using cortol as a model compd. At temp of 100° or more TMSI alone was as effective as 1:1 TMSI/BSA. Silylation with BSA TMSI TMCS (3:3:2) at 100° was more rapid than that without TMCS but at 200° the rates were nearly identical. Ketosteroids were converted to methoximes prior to silylation. GC data were reported for 20 compds.

References

[1]N. Sakauchi and E. C. Horning, *Anal. Lett.* **4**, 41 (1971).

Derivative Type. Trimethylsilyl ether.

Reagent. Trimethylsilylimidazole (TMSI)/bis(trimethylsilyl)acetamide (BSA)/trimethylchlorosilane (TMCS).

Compound Type. Steroids.

Reaction

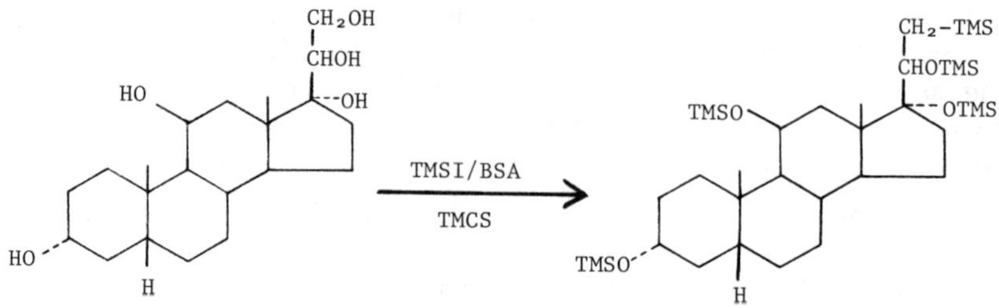

Typical Procedure. Treat steroid sample with TMSI/BSA/TMCS (3:3:2, v/v/v) at 60° for 60 hr. Analyze by GC-MS.

Comments. BSA alone would not react with 11β- or *tert*-17α-hydroxyl groups. BSA/TMCS deriv 11β-hydroxyl groups but reacted with 17α-hydroxyl groups only when C-20 was a carbonyl or methoxime. In the presence of C-20 hydroxyl, the *tert*-17α-hydroxyl did not react, presumably due to steric hindrance by the C-20 TMS ether. TMSI/BSA/TMCS at 60° yielded fully silylated derivs of steroids of the cortol, cortolone, and pregnanetriol type.[1] The conditions were also found to convert 3-, 17-, and 20- (but not 11-) keto groups to TMS enol ethers.[2]

References

[1]E. M. Chambaz and E. C. Horning, *Anal. Lett.* **1**, 201 (1967).

[2]E. M. Chambaz and E. C. Horning, *Anal. Biochem.* **30**, 7 (1969).

11.5. DERIVATIVES OF 20,21-KETOL STEROIDS

Derivative Type. 17β-Carbomethoxy.

Reagent. Periodic acid, diazomethane.

Compound Type. 20,21-Ketol steroids.

Reaction

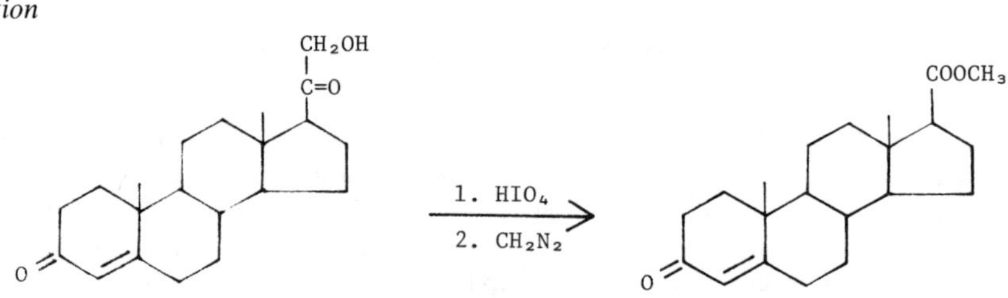

Typical Procedure. Diss steroid sample in aq EtOH or dioxane and allow to react with an equimolar quantity of HIO_4 at room temp for 15 hr. Evap under vac to remove most of the org solvent; then ext with Et_2O. Ext Et_2O with aq $NaHCO_3$. Acidify the aq soln with HCl and ext with $CHCl_3$. Evap ext and treat residue with CH_2N_2. Analyze by GC.

Comments. Aldosterone was converted to a γ-lactone which was not extd by $NaHCO_3$.

References

[1]I. Merits, *J. Lipid Res.* **3**, 126 (1962).

Derivative Type. Acetate.

Reagent. Acetic anhydride (Ac_2O)/*p*-toluenesulfonic acid (TsOH).

Compound Type. Hydroxy steroids.

Reaction

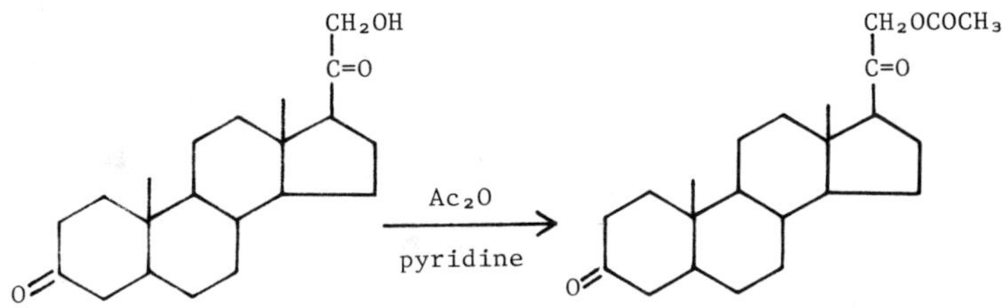

Typical Procedure. Diss 0.1–1 mg steroid in 10–50 μl of Ac_2O containing 10 mg/ml TsOH. Allow soln to evap to dryness in a vac dessicator at room temp. Ext residue with $CHCl_3$ or CH_2Cl_2 and analyze by GC.

Comments. The deriv was found to stabilize 17- and 21-hydroxy compds but the 21-acetoxy-4-ene-3,20-dione and 17,21-diacetoxy-20-one compounds suffered decomposition and losses on the GC column. 17α,21-Dihydroxy-20-ones were first acetylated at C_{21} with Ac_2O/pyridine.

References

[1]C. J. W. Brooks, *Anal. Chem.* **37**, 636 (1965).

11.6. DERIVATIVES OF 17,21-DIHYDROXY-20-KETOSTEROIDS

Derivative Type. *O*-Methyloxime, trimethylsilyl.

Reagent. Methoxyamine HCl, hexamethyldisilazane (HMDS).

Compound Type. Corticosteroids.

Reaction

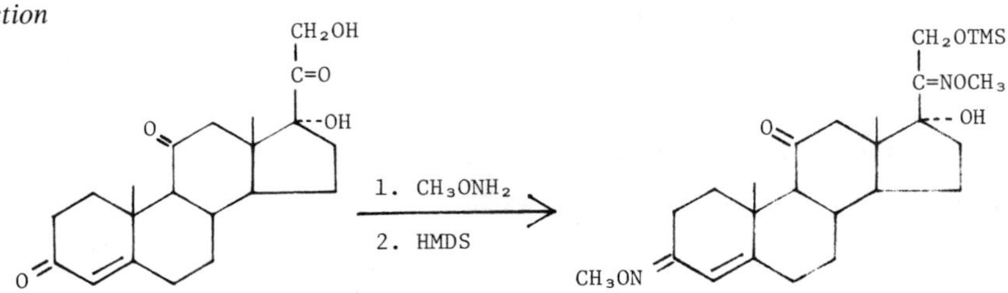

Typical Procedure. Diss sample in pyridine and add excess methoxyamine HCl. Allow to stand overnight at room temp; then add HMDS to the mixt. Allow to stand overnight again. Centrifuge to remove ppt; then analyze by GC or GC-MS.

Comments. The method did not derivatize 11β-hydroxyl, 17α-hydroxyl, or 11-keto groups. The method was reported as a means of separating C_{19} and C_{21} human urinary steroids in a single GC procedure. Hara *et al.*[2] reported preparation of an 11β-hydroxyl TMS deriv with the method. They reported that in tetrahydrocorticosteroid methoxime derivs, the 17α-hydroxyl was not converted to the TMS deriv and gave pyrolytic cleavage to the 17-ketosteroid in the GC.

References

[1]W. L. Gardiner and E. C. Horning, *Biochem. Biophys. Acta* **115**, 524 (1966).
[2]S. Hara *et al.*, *Chem. Pharm. Bull.* **14**, 1311 (1966).

Derivative Type. Methoxime, trimethylsilyl.

Reagent. Methoxylamine, bis(trimethylsilyl)acetamide (BSA)/trimethylchlorosilane (TMCS).

Compound Type. Steroids.

Reaction

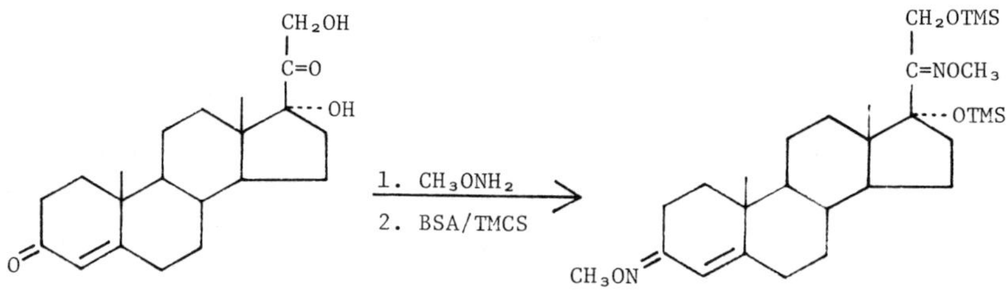

Typical Procedure. Diss 1 mg steroid in 0.1–0.2 ml pyridine and add an excess of $CH_3ONH_2 \cdot$ HCl. Heat 30 min at 65° ; then let stand at room temp 1 hr. Add 20 μl BSA and 5 μl TMCS and heat at 65° for 30 min. Allow to stand at room temp overnight. Analyze by GC.

Comments. Use of HMDS in place of BSA/TMCS gave the TMS deriv of only the 21-hydroxyl and not the 17α-hydroxyl. Engel *et al.*[2] reported a similar method using 0.4 ml satd $CH_3ONH_2 \cdot$ HCl in pyridine (75°, 30 min) followed by treatment with 0.4 ml BSA and 0.2 ml TMCS (75°, 3 hr). Methoximes were formed from 3-, 17-, and 20-keto groups (not from 11-keto) and TMS derivs from 3α, 3β, 6β, 11β, 17α, 17β, 20α, 20β, and 21-hydroxyls.[2] The derivs were purified on a column of Sephadex LH-20 with EtOAc.

References

[1]W. J. A. VandenHeuval *et al.*, *Biochim. Biophys. Acta* **144**, 691 (1967).
[2]L. L. Engel *et al.*, *Steroids* **16**, 377 (1970).

Derivative Type. Methoxime, trimethylsilyl ether.

Reagent. Methoxylamine HCl, trimethylsilyl imidazole (TMSI).

Compound Type. Corticosteroids and metabolites.

Reaction

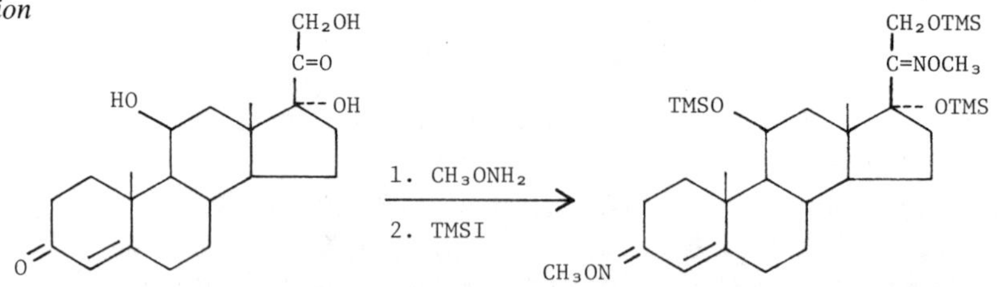

Typical Procedure.[1] To 0.1 mg steroid add 50 μl of a soln of methoxylamine HCl in pyridine (100 mg/ml). Heat at 60° for 15 min. Evap pyridine under N_2 and add 50 μl TMSI. Heat at 100° for 2 hr and analyze directly by GC or GC-MS.

Comments. Complete methoxime formation was achieved at 60° in 15 min for 3-,17-, 20-ones. It was possible to form the 11-methoxime of β-cortolone by heating at 100°. One hr at 100° was sufficient to silylate the corticosteroid 17α-hydroxyl, but α-cortol required 2 hr for complete silylation. Very little 11-enol ether was formed by this treatment.[2] The method was developed for analysis of urinary steroids including α-cortol, β-cortolone, tetrahydrocortisol, tetrahydrocortisone, and cortisone. It was found that a short section of 10% SE-30 column helped prevent decomposition of the methoxime derivs on the column.

References

[1]J.-P. Thenot and E. C. Horning, *Anal. Lett.* **5**, 21 (1972).

[2]J.-P. Thenot and E. C. Horning, *Anal. Lett.* **5**, 801 (1972).

Derivative Type. Trimethylsilyl ether, enol-trimethylsilyl ether.

Reagent. Bis(trimethylsilyl)trifluoroacetamide (BSTFA)/potassium acetate.

Compound Type. Corticosteroids.

Reaction

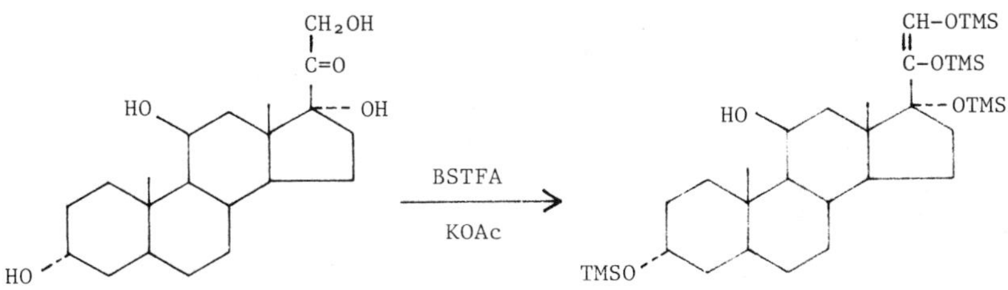

Typical Procedure. Add 5–40 mg of KOAc in MeOH soln to small tube and evap to dryness. Add 50 μg of steroid in soln and evap to dryness under N_2. Add 200 μl BSTFA, seal tube, and allow to stand at room temp overnight. Analyze aliquots directly by GC or GC-MS or evap to dryness and diss residue in hexane for analysis.

Comments. The procedure gave a thermally stable TMS deriv of the dihydroxyacetone side chain, whereas uncatalyzed trimethylsilylation did not. BSA or TMSI could be used in place of BSTFA or piperidine in place of KOAc. Hindered hydroxyl groups (e.g., 11β, 17α in cortol, and 11β in cortisol) were not trimethylsilylated under base catalysis. MS data for derivs were given.[1]

References

[1]E. M. Chambaz *et al., Anal. Chem.* **45**, 1090 (1973).

[2]E. M. Chambaz *et al., J. Steroid Biochem.* **3**, 741 (1972).

[3]E. M. Chambaz and C. Madani, *Excerpta Med. Int. Congr. Ser.* **210**, 97 (1970).

Derivative Type. Siliconide.

Reagent. Dimethyldiacetoxysilane.

Compound Type. Corticosteroids.

Reaction

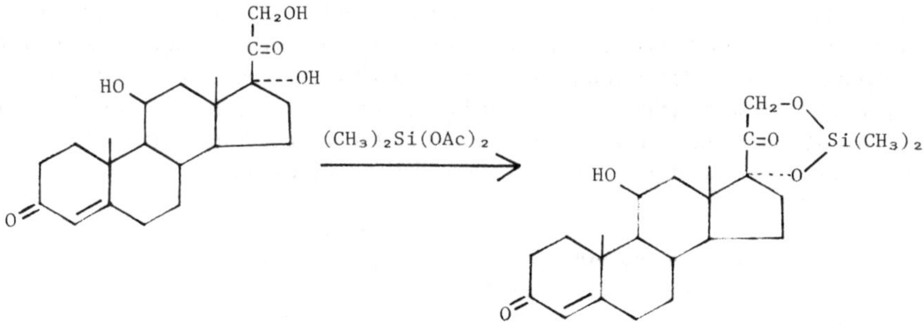

Typical Procedure. Evap steroid sample at the bottom of a 2×75 mm glass tube. Add 20 μl of a soln of 2% dimethyldiacetoxysilane and 2% Et_3N in hexane and seal the tube in a flame. Heat at 40° for 2 hr. Open tube and analyze mixt directly by GC.

Comments. The reagent was prepared by treating dimethyldichlorosilane with 2 mol of Ac_2O under reflux and distilling. The reagent also formed dimethylacetoxylsilyl derivs with unhindered hydroxyl groups. Hindered hydroxyls (such as the 11β shown above) did not react presumably due to the bulkiness of the reagent.

References

[1]R. W. Kelly, *J. Chromatogr.* **43**, 229 (1969).

Derivative Type. Siliconide.

Reagent. Dimethyldichlorosilane/pyridine.

Compound Type. Cortisone and other dihydroxy steroids.

Reaction

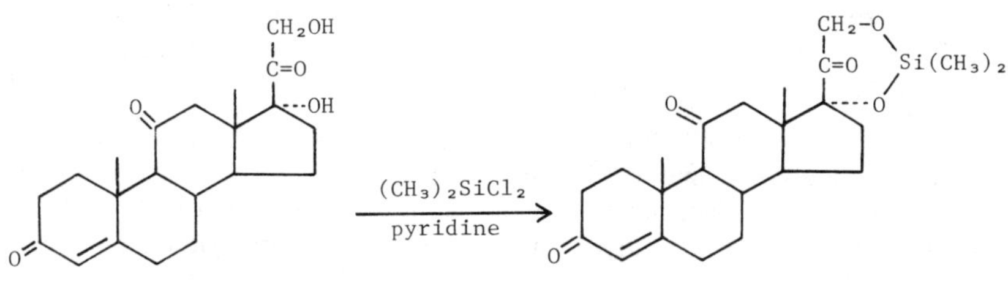

Typical Procedure. Diss 100 μg of steroid in 10 μl pyridine, 10 μl dimethyldichlorosilane, and 20 μl benzene. Seal tube and heat at 40° for 2 hr. Analyze directly by GC.

Comments. The method could not be applied to compds with other hydroxyl groups since thermally unstable derivs were produced. A siliconide deriv of a 16α,17α-diol (3β-acetoxy-16α,17α-dihydroxypregn-5-ene-20-one) prepared in a similar manner has also been reported along with MS data.[2]

References

[1]R. W. Kelly, *Steroids* **13**, 507 (1969).
[2]R. W. Kelly, *Tetrahedron Lett.* 967 (1969).

Derivative Type. Siliconide, dimethylsilyl (DMS).

Reagent. Dimethylmethoxychlorosilane/tetramethyldisilazane.

Compound Type. Corticosteroids.

Reaction.

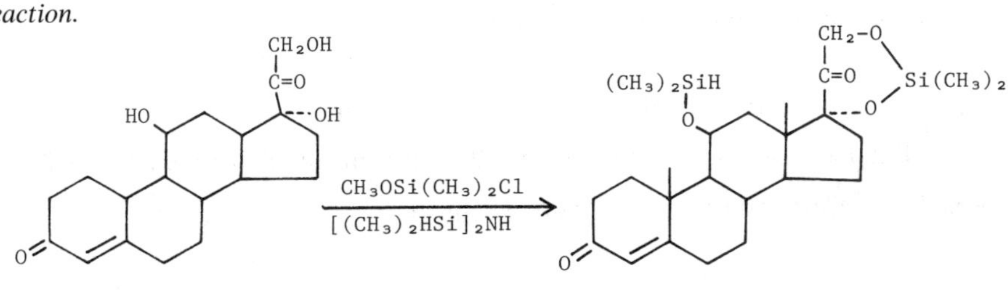

Typical Procedure. Place steroid (10 μg) soln in a 2×75 mm glass tube and evap to dryness under vac. Add 10 μl of tetramethyldisilazane and 10 μl of a 20% soln of dimethylmethoxychlorosilane in benzene. Seal tube in a flame and heat at 40° for 12 hr. Open tube and analyze mixt directly by GC. Alternatively evap to dryness, diss residue in CS_2, and analyze by GC.

Comments. Betamethasone formed a similar deriv. Compds with unhindered hydroxyl groups were not amenable to the method since unhindered hydroxyls could react with the difunctional silane to form multiple products.

References

[1] R. W. Kelly, *Steroids* **13**, 507 (1969).

Derivative Type. Alkyl boronate.

Reagent. Alkyl(methyl, *n*-butyl, *t*-butyl, cyclohexyl, phenyl)boronic acid.

Compound Type. Corticosteroids.

Reaction

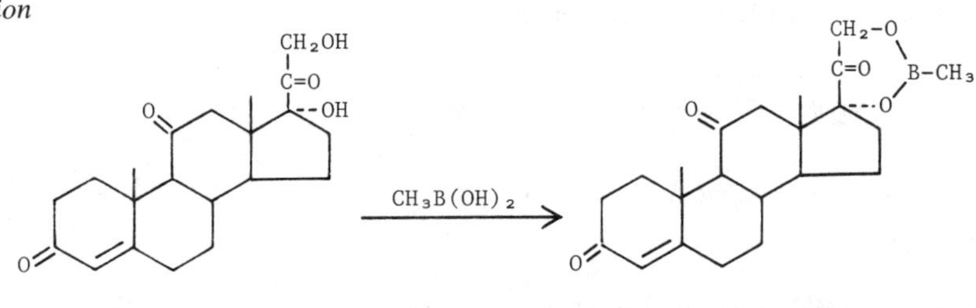

Typical Procedure. Diss 10 μmol steroid and 10 μmol of the boronic acid in 1 ml EtOAc and allow to stand at room temp 5 min. Analyze by GC or GC-MS, or evap to dryness and further deriv remaining hydroxyl or carbonyl groups.

Comments. Remaining hydroxyl groups were converted to TMS ethers by treatment with 0.1 ml pyridine, 0.1 ml HMDS, and a trace of TMCS for 5 min at room temp. The soln was evapd to dryness and the residue ext with 1 ml cyclohexane.

Remaining hydroxyl groups were acetylated by diss in 0.5 ml dry pyridine and 0.1 ml Ac_2O and allowing to stand overnight at room temp. The soln was evapd to dryness and the residue diss in 1 ml EtOAc.

17,21-Dihydroxy-20-keto steroid and 20,21-ketol boronates could not be further derivatized without destruction of the boronate unless, in the case of the dihydroxyacetone, the keto group was first protected as an oxime deriv. GC and MS data were reported for 75 boronate derivs.[1]

References

[1]C. J. W. Brooks and D. J. Harvey, *J. Chromatogr.* **54**, 193 (1971).
[2]C. J. W. Brooks and D. J. Harvey, *Biochem. J.* **114**, 15P (1969).

Derivative Type. *n*-Butylboronate, methoxime, trimethylsilyl.

Reagent. *n*-Butylboronic acid, methoxylamine HCl, bis(trimethylsilyl)acetamide (BSA).

Compound Type. Corticosteroids (tetrahydro S shown).

Reaction

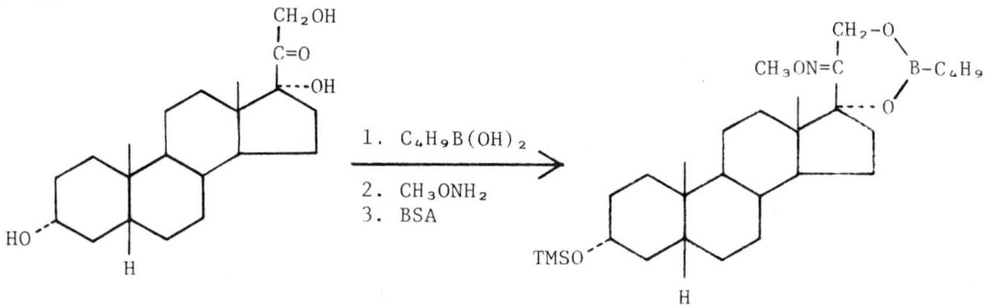

Typical Procedure. Diss 3.5 mg sample and 1.0 mg *n*-butylboronic acid in 0.5 ml pyridine. After 15 min, evap to dryness under N_2. Diss residue in 1 ml toluene and evap to dryness twice. Diss in 0.5 ml dry pyridine and add 5 mg methoxylamine HCl. After 1 hr evap to dryness under N_2 and ext residue with 4×0.5 ml CCl_4. Evap to dryness and diss residue in 0.1 ml pyridine and 0.05 ml BSA. Analyze by GC-MS.

Comments. The sample was also examined by GC-MS after each deriv step and the boronate and boronate methoxime derivs found to give satisfactory GC behavior. The derivs of the 20α and 20β epimers of 17,20,21-triols were separable by GC.

References

[1]G. M. Anthony *et al., J. Chromatogr. Sci.* **7**, 623 (1969).

Derivative Type. Bis(methylenedioxy).

Reagent. Formaldehyde.

Compound Type. 17-Hydroxycorticosteroids.

Reaction

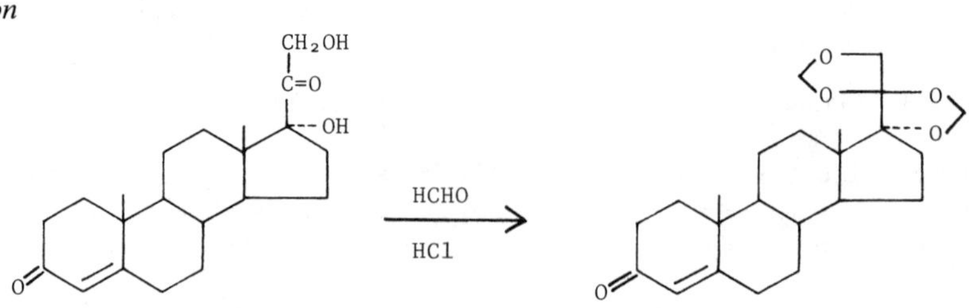

Typical Procedure. Suspend 10 mg cortisone in 5 ml $CHCl_3$ and 1.3 ml 12 *N* HCl. Add 1.3 ml

37% formaldehyde and stir at 5° for 48 hr. Wash with 1 *N* NaOH, dry over Na_2SO_4, and analyze by GC.

Comments. The method conferred thermostability to the 20,21-sidechain.

References

[1]M. A. Kirchner and H. M. Fales, *Anal. Chem.* **34**, 1548 (1962).

Derivative Type. 17-Oxosteroid.

Reagent. Sodium bismuthate.

Compound Type. Corticosteroids.

Reaction

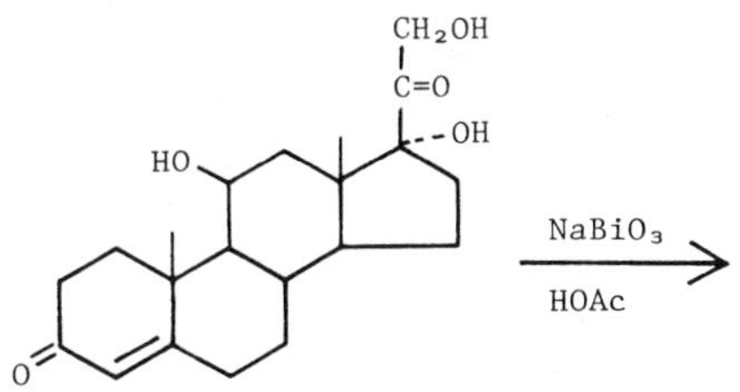

Typical Procedure. Shake steroid ext with 1 ml 15% HOAc and 25 mg $NaBiO_3$ for 1 hr in the dark. Add 2 ml satd $NaHCO_3$ and ext twice with ethylene dichloride. Wash ext with 1 *N* NaOH and H_2O then dry over Na_2SO_4. Dry ext and diss residue in MeOH. Transfer MeOH soln to siliconized glass wool and dry in stream of N_2. Analyze by GC using a solid injector.

Comments. The method was used for measurement of cortisol, cortisone, prednisolone, and prednisone in urine by GC.

References

[1]E. Bailey, *J. Endocrinol.* **28**, 131 (1964).

Derivative Type. Ketosteroids.

Reagent. Potassium borohydride, sodium metaperiodate, *t*-butyl chromate.

Compound Type. Steroids.

Reaction

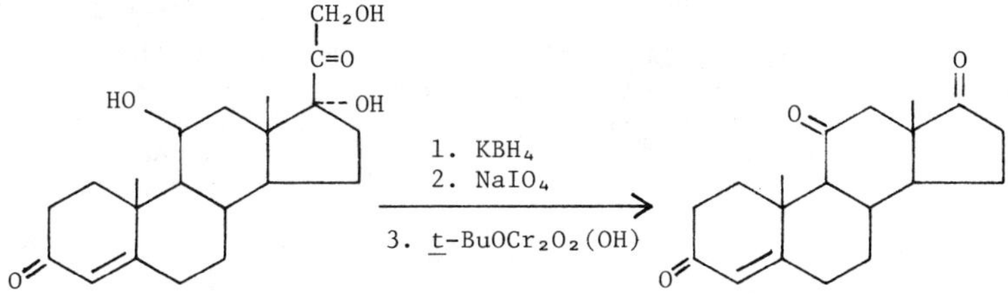

Typical Procedure. Reduce 5 ml urine with 50 mg KBH_4 overnight at room temp. Add 0.1 ml HOAc and a few min later 2 ml fresh 10% (w/v) aq $NaIO_4$. Adjust to pH 6.5–7.0 with 0.2–0.4 ml 1 *N* NaOH and shake 2 hr at room temp. Adjust to pH 9.0 with 5 *N* NaOH (0.2 ml) and hold at 37° for 15 min. Ext with 2 × 10 ml dichloroethane. Wash ext with 2 ml H_2O, 2 ml 5% (w/v)

$Na_2S_2O_4$ in 2.5 *N* NaOH, and 2 ml H_2O. Evap to dryness and separate residue into ketonic and nonketonic fractions with Girard T reagent. Oxidize each fraction separately by diss in 0.1 ml pyridine and add 0.1 ml of the chromate reagent.[3] Allow to stand overnight at room temp then add 0.2 ml fresh satd citric acid in EtOH. Shake 15 min, add 1 ml 1 *N* H_2SO_4 and ext with 3.0 ml dichloroethane. Wash with 1 ml H_2O, 1 ml 10% (w/v) $NaHSO_3$, 1 ml 1 *N* NaOH and 1 ml H_2O. Evap to dryness and analyze by GC.

Comments. The method was used to analyze the complex mixt of urinary steroids by conversion to a less complex and more readily resolvable mixt. The method involved reduction of the steroids followed by cleavage of glycols with periodate (or sodium bismuthate). The resulting products were then sepd into ketonic and nonketonic fractions which were each subsequently oxidized with *t*-butyl chromate.

References

[1]E. Menini and J. K. Norymberski, *Biochem. J.* **95**, 1 (1965).
[2]E. Menini and J. K. Norymberski, *Biochem. J.* **83**, 31P (1962).
[3]E. Menini and J. K. Norymberski, *Biochem. J.* **84**, 195 (1962).

11.7. DERIVATIVES OF 18-OXYGENATED STEROIDS

Derivative Type. Acetyl.

Reagent. Acetic anhydride (Ac_2O)/pyridine.

Compound Type. Aldosterone.

Reaction

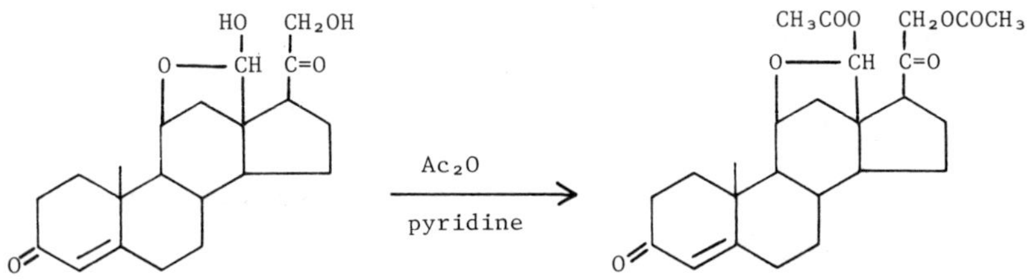

Typical Procedure. Diss 1–5 mg sample in 1 ml pyridine and 1 ml Ac_2O. Heat at 60° for 1 hr or allow to stand at room temp overnight. Analyze by GC.

Comments. Cortisone was analyzed concurrently as the 21-monoacetate.

References

[1]H. H. Wotiz *et al.*, *Biochem. Biophys. Acta* **53**, 449 (1961).

Derivative Type. Acetal, heptafluorobutyrate.

Reagent. Hydrochloric acid, heptafluorobutyric anhydride (HFBA).

Compound Type. Aldosterone.

Reaction

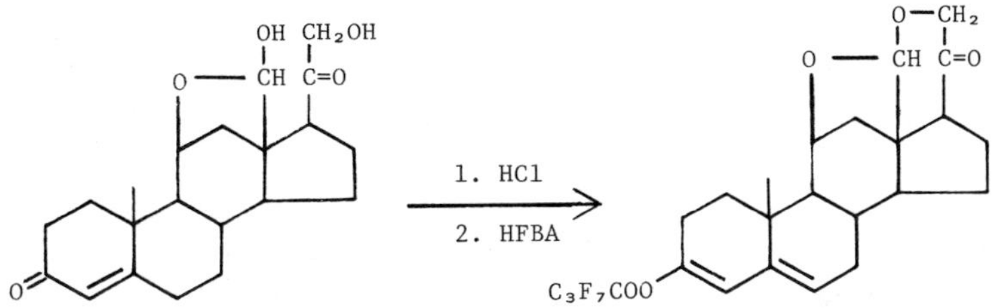

Typical Procedure. To 1 ml of urine add 0.1 μCi, 1,2-^{3}H-aldosterone (55 Ci/mmol) and 1 ml 7 *M* HCl. Heat at 60° for 3 hr. Ext with 5 ml benzene and purify by TLC on silica gel using EtOAc/cyclohexane (5:3, v/v). Treat the purified acetal with 15 μl HFBA and 15 μl dry acetone. Analyze by GC-MS.

Comments. The method was used to quantitate aldosterone in rat urine by GC-MS using $^3H_2$-aldosterone as int stand.

References

[1]L. Siekmann *et al.*, *Adv. Mass Spect. Biochem. Med.* **2**, 419 (1976).

Derivative Type. Methoxime, trimethylsilyl.

Reagent. Methoxylamine/HCl/pyridine, bis(trimethylsilyl)acetamide (BSA)/trimethylchlorosilane (TMCS).

Compound Type. 18-Hydroxycorticosteroids.

Reaction

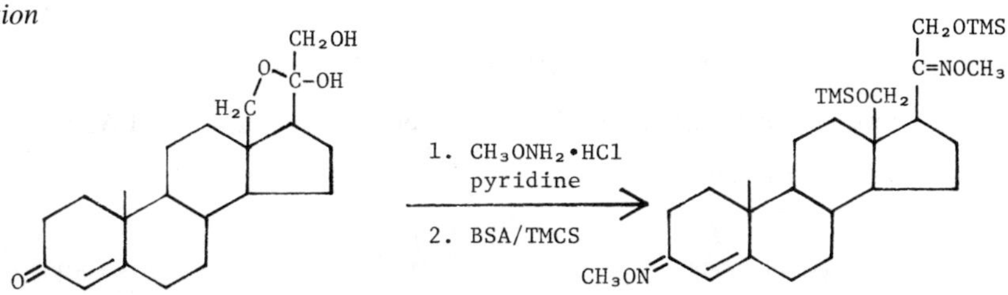

Typical Procedure. Diss 0.1 mg sample in 0.1 ml of a soln of methoxylamine·HCl (16 mg/ml) in pyridine. Heat 3 hr at 65° then evap the pyridine. Silylate with BSA/TMCS (4:1, v/v) for 16 hr at 65°. Analyze by GC.

Comments. The conditions shifted the equilibrium between the 20-ketal and 20-oxo form to give only the 20-oxo deriv. Oxime formation at room temp and subsequent silylation gave a mixt of 20-methoxime-18-OTMS and 20-TMS-ketal. The method was used for quantitation of 18-hydroxycorticosterone, 18-hydroxy-11-deoxycorticosterone, and corticosterone in rat adrenals.

References

[1]M. Prost and B. F. Maume, *J. Steroid Biochem.* **5**, 133 (1974).

Derivative Type. Methoxime, trimethylsilyl.

Reagent. Methoxylamine HCl/pyridine, trimethylsilylimidazole (TMSI).

Compound Type. 3α,18,21-Trihydroxy-5β-pregnane-11,20-dione.

Reaction

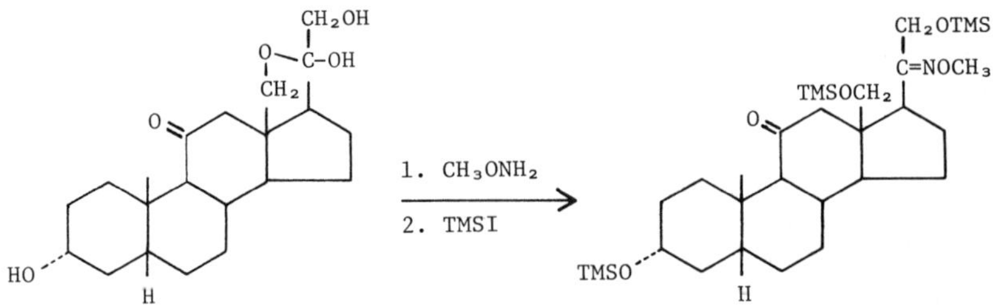

Typical Procedure. To a sample containing *ca* 10 μg of the steroid add 0.2 ml of 2% CH_3ONH_2-HCl in pyridine. Allow to stand 3 days at room temp. Add 100 μl TMSI and heat at 100° for 2 hr. Remove excess reagent by chromatography on Lipidex 5000.[2,3] Analyze deriv by GC-MS-SIM.

Comments. The 3 day reaction time for the first step was necessary for complete 20-methoxime formation since the free compd exists in the hemiketal form. Shorter times led to formation of the 3,20,21-tri-OTMS deriv. Complete 20-methoxime formation was also achieved by heating at 60° for 3 hr but with *ca* 10–15% loss of the steroid. The method was used to quantitate the compd in urine by GC-MS-SIM using 3β,17,21-trihydroxy-5α-pregnane-11,20,dione as int stand.

References

[1]C. H. L. Shackleton and J. W. Honour, *J. Steroid Biochem.* **8**, 199 (1977).

[2]Packard-Becker, Delft, Netherlands.

[3]M. Axelson and J. Sjövall, *J. Steroid Biochem.* **5**, 733 (1974).

Derivative Type. Methoxime, heptafluorobutyryl.

Reagent. Methoxyamine HCl/pyridine, heptafluorobutyric anhydride (HFBA).

Compound Type. Aldosterone.

Reaction

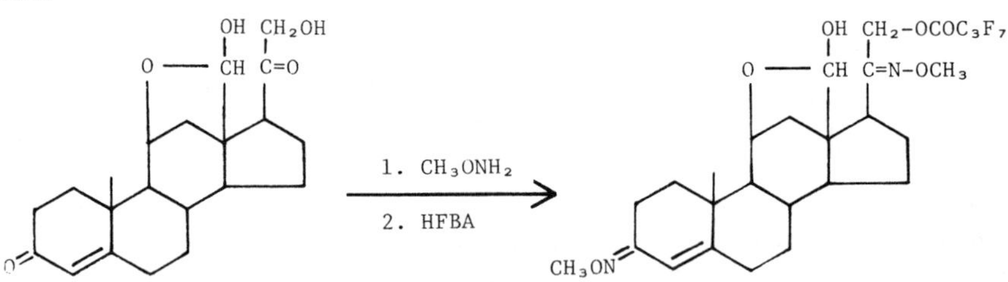

Typical Procedure. Diss 0.5–1.0 mg sample in 0.5 ml of a soln of methoxyamine HCl in pyridine (16 mg/ml) and allow to stand overnight at room temp. Dil with 2 ml satd aq NaCl and ext with 3 × 5 ml EtOAc. Wash EtOAc ext with NaCl satd 1 *N* HCl, satd NaCl, NaCl satd 5% $NaHCO_3$, and finally with satd NaCl. Dry EtOAc over $MgSO_4$ and evap to dryness. Diss residue in 0.1 ml benzene/pyridine (2:1, v/v) and add 0.2 ml HFBA. Allow to react overnight at room temp. Evap to dryness and analyze product by GC-MS or GC-ECD.

Comments. The heptafluorobutyryl deriv was also prepared by a vapor phase method by applying a 1 μg sample to a small roll of Pt gauze using CH_2Cl_2 and suspending the gauze for about 1.5 hr in a closed vessel at room temp containing HFBA in benzene (1:4, v/v).

The C-18 hemiacetal formed neither an oxime nor a heptafluorobutyrate presumably due to steric hindrance. The four isomeric derivs were eluted as a single peak on a nonselective column.

References

[1]E. C. Horning and B. F. Maume, *J. Chromatogr. Sci.* **7**, 411 (1969).

Derivative Type. γ-Lactone, chlorodifluoroacetyl.

Reagent. Periodic acid, chlorodifluoroacetic anhydride.

Compound Type. Tetrahydroaldosterone.

Reaction

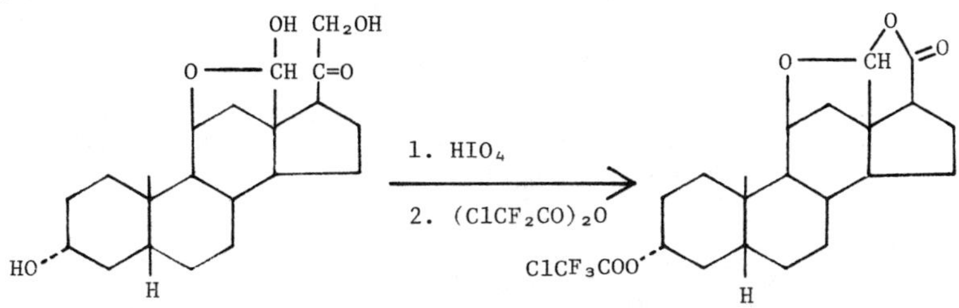

Typical Procedure. Diss a submicrogram amt of tetrahydroaldosterone in 0.1 ml of HIO_4 soln (10 mg/ml in dioxane/H_2O, 1:1, v/v) and allow to react 3 hr at room temp. Add 1 ml H_2O and ext with 10 ml $CHCl_3$. Wash ext with aq $NaHCO_3$ to remove acid then evap. Purify the γ-lactone by TLC on silica gel (cyclohexane/EtOAc, 1:1, v/v) and elute with MeOH. Evap the eluate and treat the residue with 80 μl THF and 30 μl chlorodifluoroacetic anhydride (45 min at 60°). Evap under vac and analyze the deriv by ECD-GC.

Comments. The deriv was used for quantitation of tetrahydroaldosterone in urine[1] and blood.[2]

References

[1]L. Nicholis *et al.*, *J. Clin. Endocrinol. Metab.* **28**, 547 (1968).

[2]L. F. Fabre *et al.*, *J. Chromatogr. Sci.* **7**, 632 (1969).

Derivative Type. Lactone, heptafluorobutyryl.

Reagent. Periodic acid, heptafluorobutyric anhydride (HFBA).

Compound Type. 18-Hydroxydesoxycorticosterone (18-HO-DOC) and aldosterone.

Reaction

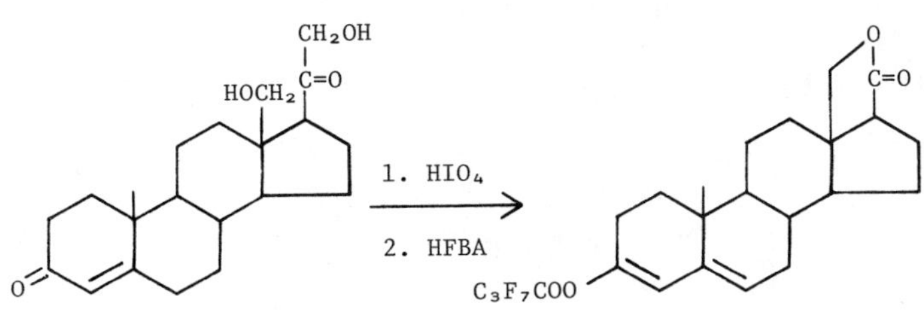

Typical Procedure. Isolate aldosterone and 18-HO-DOC fraction from 5 ml plasma by solvent extraction and paper chromatography. Evap eluate to dryness and diss in 0.1 ml EtOH. Add 0.5 ml 0.1 *M* HIO_4 containing 1% pyridine. Allow to stand in the dark at room temp for 1.5 hr. Dil with H_2O and ext with CH_2Cl_2. Wash ext with 1 ml H_2O and evap to dryness. Purify the lactone deriv by paper chromatography and elute. Diss eluted lactone in 30 μl benzene and add 30 μl HFBA. Heat at 60° for 30 min. Evap to dryness and add int stand. Analyze by ECD-GC.

Comments. The method was used in quantitation of aldosterone and 18-HO-DOC (shown) in plasma by ECD-GC. The deriv of aldosterone was used as int stand for 18-HO-DOC and *vice versa*.

References

[1]P. A. Mason and R. Fraser, *J. Endocrinol.* **64**, 277 (1975).

Derivative Type. Methaneboronate.

Reagent. Methaneboronic acid.

Compound Type. 18-Hydroxydesoxycorticosterone.

Reaction

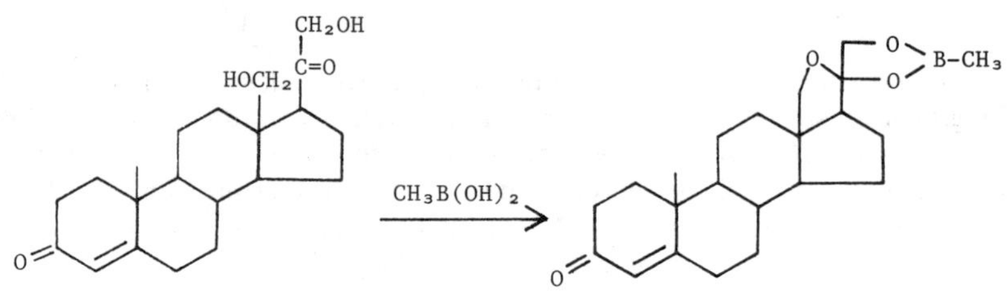

Typical Procedure.[2] Treat steroid sample with 1.1 mol equiv of methaneboronic acid in EtOAc. Warm if necessary to effect soln. Analyze by GC-MS.

Comments. The single deriv shown was obtained with methaneboronic acid as compared to mild trimethylsilylation which yielded a mixt of 3 derivs.[1]

References

[1]S. J. Gaskell *et al.*, *Anal. Lett.* **9**, 325 (1976).

[2]C. J. W. Brooks and D. J. Harvey, *J. Chromatogr.* **54**, 193 (1971).

11.8. DERIVATIVES OF BILE ACIDS

Derivative Type. Methyl ester, trimethylsilyl ether.

Reagent. Diazomethane, hexamethyldisilazane (HMDS)/trimethylchlorosilane (TMCS).

Compound Type. Bile acids.

Reaction

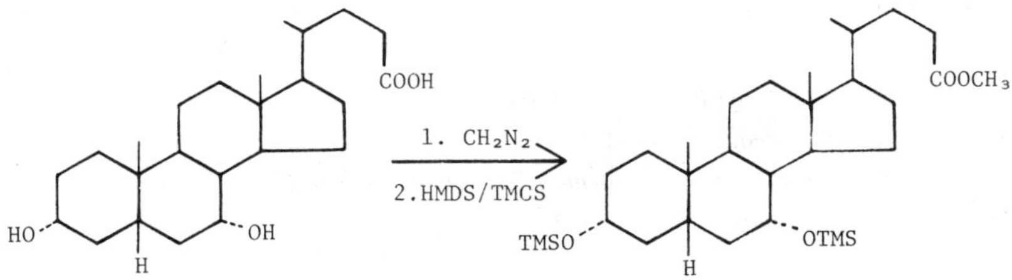

Typical Procedure. Diss sample in Et_2O/MeOH (10:1) and bubble in CH_2N_2 until a yellow color persists. Evap to dryness under N_2 and diss residue in anhyd pyridine (*ca* 1 ml/10 mg). Add 0.1 ml HMDS and 0.03 ml TMCS. Let stand at room temp at least 10 min and analyze by GC.

Comments. Use of pyridine as solvent permitted a 10 min silylation time compared with 2 hr in THF. The method was used to quantitate bile acids in rat feces.[1] Grundy *et al.*[2] used MeOH/HCl in place of CH_2N_2 in a similar analysis of fecal bile acids.

References

[1]M. Makita and W. W. Wells, *Anal. Biochem.* **5**, 523 (1963).

[2]S. M. Grundy *et al.*, *J. Lipid Res.* **6**, 397 (1965).

Derivative Type. Methyl ester, trifluoroacetate.

Reagent. Diazomethane, trifluoroacetic anhydride (TFAA).

Compound Type. Bile acids.

Reaction

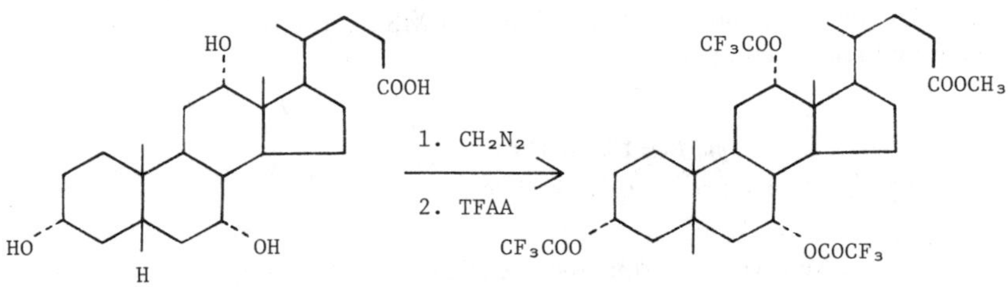

Typical Procedure. Isolate bile acids from 5–10 ml serum by ion exchange, basic hydrolysis, and Et_2O ext. Evap Et_2O ext to dryness and diss residue in 10 ml Et_2O and 1 ml MeOH. Add CH_2N_2 until yellow color persists; then allow to react 15 min. Evap to dryness and diss residue in 10 ml anhyd benzene. Apply to an alumina column. Wash column with benzene then elute with MeOH/acetone (1:9, v/v). Evap eluate to dryness under N_2. Add 0.2 ml TFAA and warm at 35° for 15 min. Evap to dryness under N_2 and diss residue in 100 μl CH_3CN. Analyze 5 μl by GC.

Comments. 3α-Trimethylsilyl methyl esters were also prepared by diss the methyl esters in 0.2 ml anhyd acetone to which was added 0.1 ml HMDS. After 3 hr at 50° (optimal cond for formation of 3α-, 3β-, 7β-, and 12β-OTMS with minimal reaction of 7α- and 12α-OH) the soln was evapd to dryness under N_2, the residue diss in CH_3CN, and the soln analyzed by GC. The methyl ester trifluoroacetates were found to be preferable for quantitation. Other workers had earlier reported the use of methyl ester trifluoroacetates of bile acids.[2,3] Kanno *et al.*[4] used the derivs for ECD-GC.

References

[1]D. H. Sandberg *et al., J. Lipid Res.* **6**, 182 (1965).
[2]J. Sjövall, *Acta Chem. Scand.* **16**, 1761 (1962).
[3]A. Kuksis and B. A. Gordon, *Can J. Biochem. Physiol.* **41**, 1355 (1963).
[4]T. Kanno *et al., J. Chromatogr. Sci.* **9**, 53 (1971).

Derivative Type. Methyl ester, acetyl.

Reagent. Diazomethane, acetic anhydride (Ac_2O)/perchloric acid.

Compound Type. Bile acids.

Reaction

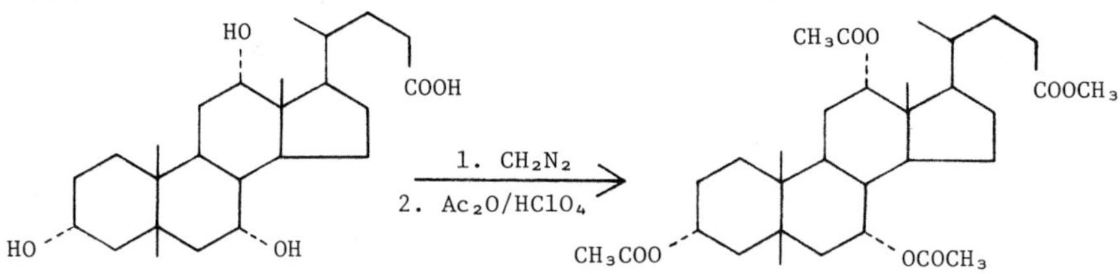

Typical Procedure. Prepare methyl esters by treatment of the bile acids with excess CH_2N_2 in Et_2O. Evap to dryness then acetylate at 0° for 15 min with a soln of 5 ml Ac_2O, 7 ml AcOH, and 1 drop $HClO_4$.[2] Add H_2O and ext with Et_2O. Evap to dryness under vac using addn and evap of added toluene to remove traces of H_2O. Diss residual derivs in acetone and analyze by GC-MS.

Comments. GC retention data and both EI and CI MS data were reported for the derivs of 18 bile acids and related compds.[1]

References

[1]P. A. Szczepanik *et al., J. Lipid Res.* **17**, 314 (1976).
[2]R. E. Roovers *et al., Clin. Chim. Acta* **19**, 449 (1968).

Derivative Type. Methyl ester, trimethylsilyl ether.

Reagent. Methanol (MeOH)/HCl, trimethylsilyl imidazole (TMSI).

Compound Type. Bile acids and cholesterol.

Reaction

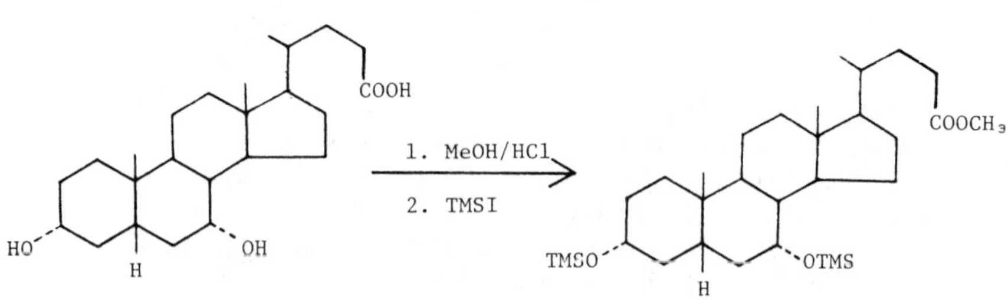

Typical Procedure. Treat the residue from the ext procedure with 5 ml of HCl satd MeOH for 3 hr at room temp. Evap to dryness and add 100 μl TMSI. After 30 min at room temp, analyze by GC-MS.

Comments. Esterfication with diazomethane gave multiple products. The persilyl derivs could not be easily separated by GC. An extraction and cleanup procedure was given for separation of the bile acids and cholesterol from a 0.3 ml bile sample into five groups: (a) free bile acids and cholesterol, (b) glycine conjugate, (c) sulfate glycine conjugate, (d) sulfate taurine conjugate and taurine conjugate, (e) taurine conjugate. Cholic, deoxycholic, and chenodeoxycholic acids and cholesterol were quantitated in each group using GC-MS and deuterium labeled standards.[1] The derivs have also been prepared using an equimolar mixt of MeOH and d_3-MeOH; the equal intensity ion cluster formed in the MS facilitated distinguishing bile acids from other endogenous substances in the GC-MS-SIM analysis of bile acids in rat serum.[2]

References

[1]H. Miyasaki *et al.*, *J. Chromatogr.* **99**, 553 (1974).

[2]M. Inoue *et al.*, *Chem. Pharm. Bull.* **22**, 1949 (1974).

Derivative Type. Hexafluoroisopropyl ester, trifluoroacetyl.

Reagent. Hexafluoroisopropanol, trifluoroacetic anhydride (TFAA).

Compound Type. Bile acids.

Reaction

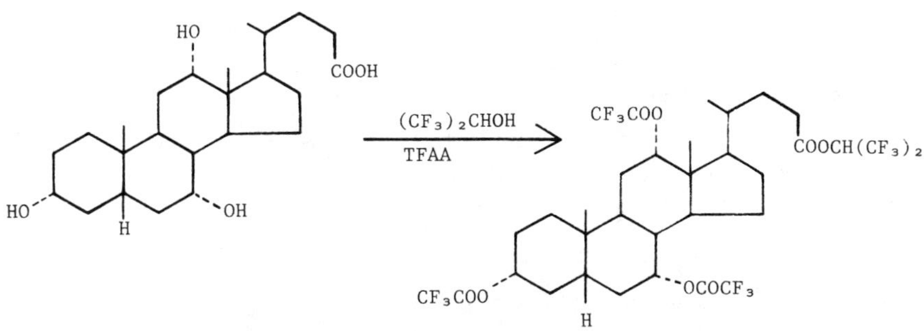

Typical Procedure. Diss bile acid sample (mg or less) in 100 μl hexafluoroisopropanol and 200 μl TFAA. Warm at 37° for 30 min. Evap under vac at room temp and diss residue in 100 μl CH_3CN. Analyze 5 μl aliquot by GC-MS.

Comments. GC and MS data were reported for five bile acids. The method was used to quantitate bile acids in human bile.

References

[1]K. Imai *et al.*, *J. Chromatogr.* **120**, 181 (1976).

11.9. DERIVATIVES OF DIGITALIS STEROIDS

Derivative Type. Trimethylsilyl.

Reagent. Trimethylsilylimidazole (TMSI)/bis(trimethylsilyl)acetamide (BSA)/trimethylchlorosilane (TMCS).

Compound Type. Cardiac aglycones.

Reaction

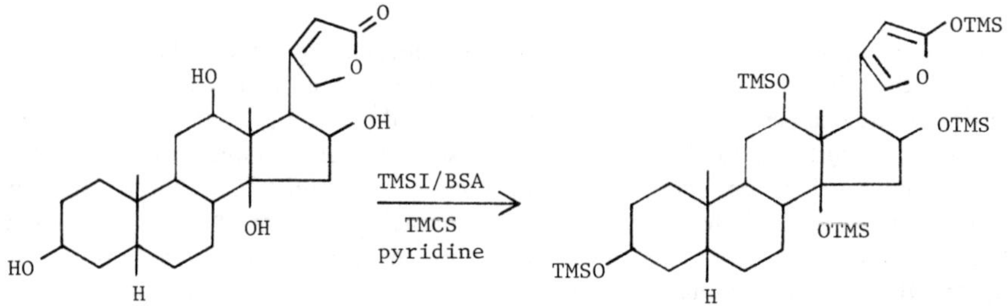

Typical Procedure. Diss 0.5 mg sample in 0.5 ml TMSI/BSA/TMCS pyridine (3:3:2:2, v/v/v/v). Heat 3 days at 60°. Analyze by GC or GC-MS.

Comments. The derivs of digoxigenin, digitoxigenin, gitoxigenin (shown above), and dihydrodigoxigenin were examined. The use of HMDS/TMCS/pyridine (4:2:4, v/v/v, at room temp) gave TMS ethers of the relatively unhindered hydroxyls (3β, 12β, 16β). Use of BSA/TMCS/pyridine (5:1:4, v/v/v, at room temp) gave also the TMS enol ether of the lactone. Use of TMSI/BSA/TMCS pyridine at 60° gave deriv of all hydroxyls including the tert 14β-hydroxyl and the enol of the lactone. MS data were reported for the derivs. GC of TMS derivs (from HMDS/TMCS in THF) of digoxigenin and digitoxigenin had been reported earlier.[2] MS studies of TMS derivs of digitalis aglycones and monoglycosides have been reported.[3]

References

[1]B. Maume *et al., Anal. Lett.* **1**, 401 (1968).

[2]R. W. Jelliffe and D. H. Blankenhorn, *J. Chromatogr.* **12**, 268 (1963).

[3]F. C. Falkner *et al., Org. Mass Spectrom.* **7**, 141 (1973).

Derivative Type. Trifluoroacetate.

Reagent. Lithium aluminum hydride, trifluoroacetic anhydride (TFAA).

Compound Type. Digitoxin and digitoxigenin.

Reaction

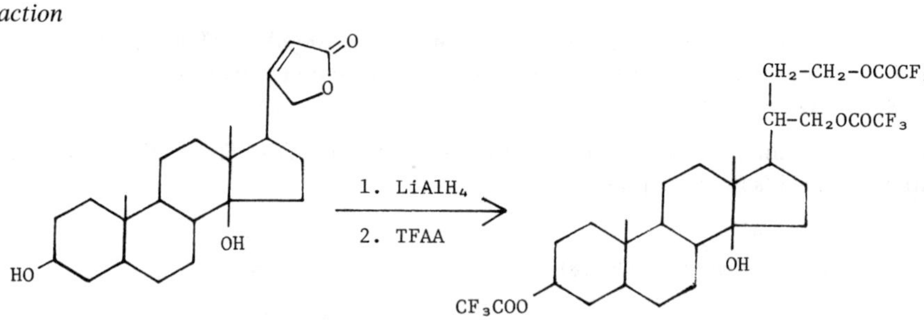

Typical Procedure (Experimental Details Not Reported). Reduce compds with $LiAlH_4$ (the paper says lithium hydride which is presumed to be a printing error). Hydrolyze with 1 *N* HCl and trifluoroacetylate products with TFAA. Analyze by ECD-GC.

Comments. The method was developed in an attempt to reduce peak broadening presumed to be due to the polar lactone ring. The ECD response for the deriv was linear from 0.5–20 ng.

References

[1]R. W. Jellife, *Circulation* **28**, 743 (1963).

Derivative Type. Heptafluorobutyryl.

Reagent. Heptafluorobutyric anhydride(HFBA).

Compound Type. Digoxin, digitoxin, and metabolites.

Reaction

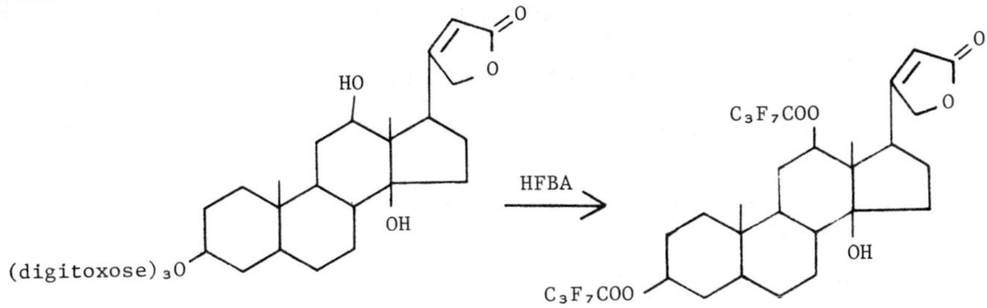

Typical Procedure. To 1 μg glycoside add 10 μl HFBA and 100 μl benzene. Heat at 90° for 20 min. Evap under N_2 and diss residue in 1 ml benzene. Analyze by GC.

Comments. The method was used for quantitation of plasma digoxin using CH_2Cl_2 ext followed by florisil column chromatography and TLC. ECD-GC permitted determination of 5–100 ng/10 ml plasma.[1]

The method was extended to analysis of submicrogram amts of digoxin, digitoxin, and metabolites in urine, feces, and tissue homogenates.[2] The compds were isolated by CH_2Cl_2 ext followed by paper chromatography and TLC. Preseparation of the compds was necessary since the deriv degraded the glycosides to the aglycone deriv.

References

[1]E. Watson and S. M. Kalman, *J. Chromatogr.* **56**, 209 (1971).

[2]E. Watson *et al.*, *J. Chromatogr.* **69**, 157 (1972).

Derivative Type. *p*-Nitrobenzoate.

Reagent. *p*-Nitrobenzoyl chloride.

Compound Type. Digitalis glycosides and aglycones.

Reaction

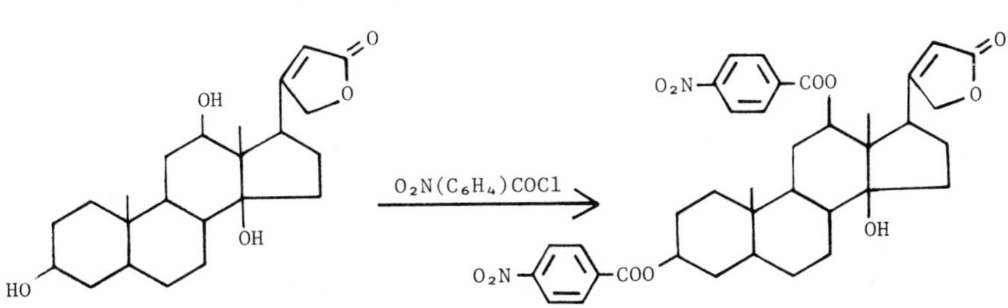

Typical Procedure

A. Aglycones—To 50 μl of a glycoside soln in pyridine (containing not more than 0.5 mg glycoside) add 150 μl of fresh reagent soln in pyridine (100 mg/ml). Shake and allow to react at room temp 10 min. Hydrolyze by adding 1 drop H_2O. After 1 min add 1 ml 5% aq $NaHCO_3$ and 2 ml $CHCl_3$. Shake and centrifuge. Remove aq phase and ext $CHCl_3$ with another 2 ml 5% aq $NaHCO_3$. Discard aq phase and wash $CHCl_3$ with 2 × 3 ml 1 *N* HCl. Analyze by HPLC.

B. Glycosides—The instability of the glycosides requires removal of pyridine prior to hydrolysis of excess reagent. After the reaction period imbed the tubes in a beaker of sand at 50°. Place the beaker in a dessicator under aspirator vac to evap pyridine (10 min). Flush tube with air or N_2 and add 2 ml 5% Na_2CO_3 containing 5 mg 4-dimethylaminopyridine. Shake or sonicate 5 min. Ext with 2 ml $CHCl_3$. Wash ext with 2 ml 5% $NaHCO_3$ and 2 × 3 ml 0.05 *N* HCl containing 5% NaCl. Analyze $CHCl_3$ soln by HPLC.

Comments. NMR of the deriv of digoxigenin (shown above) showed the compd to be the di-acyl deriv (at C-3 and C-12). The glycosides were derivatized on all free hydroxyls in the sugar moiety. The λ_{max} of the derivs was 260 nm permitting use of a 254 nm UV detector.

References

[1]F. Nachtmann *et al., Anal. Chem.* **48**, 1576 (1976).

[2]F. Nachtman *et al., J. Chromatogr.* **122**, 293 (1976).

12. Prostaglandins and Related Compounds

The study of prostaglandins and related compounds, dating from the first structure determinations[1] to the explosive growth of recent years, has relied heavily upon MS, especially GC-MS, in conjunction with analytical derivatization. The primary prostaglandins are 20 carbon unsaturated acids containing a cyclopentane ring. The prostaglandin subgroups, PGA, PGB, etc, denote the ring substitution patterns, while the subscript numbers, as in PGE_1, PGE_2, etc specify the degree of unsaturation. Subscript α or β refer to the configuration of the hydroxyl at C-9. The examples given below briefly summarize the nomenclature system:

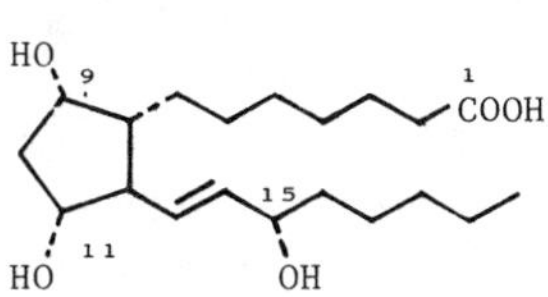

$PGF_{1\alpha}$

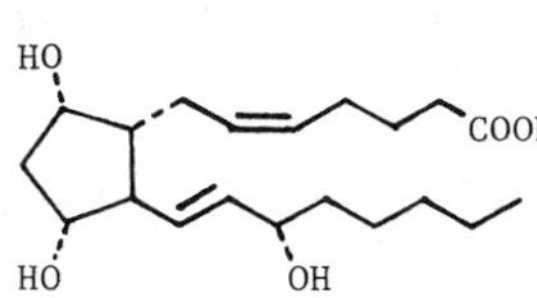

$PGF_{2\alpha}$

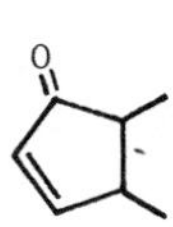

PGA

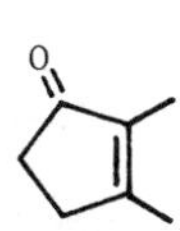

PGB

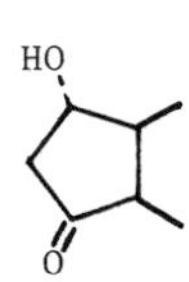

PGD

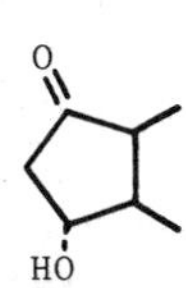

PGE

Additional C_{20} unsaturated carboxylic acid structures of the prostaglandin family include the cyclic endoperoxide PGH_2 (p. 522), PGG_2 (15-hydroperoxy-PGH_2), PGI_2 (prostacyclin, shown below), and the thromboxanes (p. 522). The multiple functionality and the lability of the prostaglandin compounds makes derivatization obligatory for GC and most MS analysis. The MS and the GC-MS of these compounds has been reviewed by Crain *et al.*,[2] Frölich,[3] and Sweetman *et al.*[4] This section includes the derivatives that have been reported for GC, MS, and LC analysis of prostaglandins and related compounds.

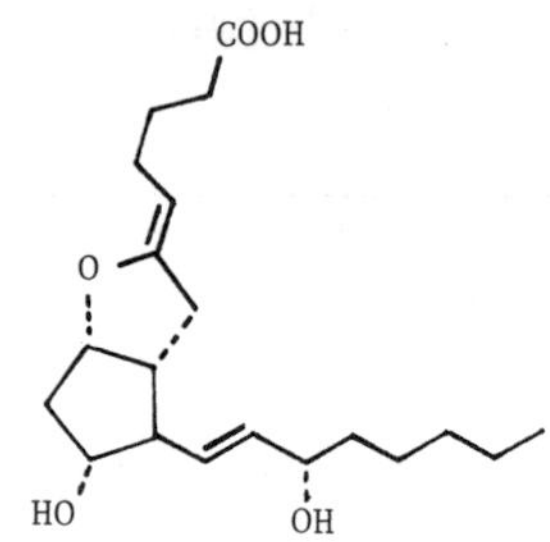

PGI_2

Methyl Ester Derivatives. The original structure determinations of PGE and PGF employed methyl ester, methyl ether derivatives formed with diazomethane under boron trifluoride catalysis (p. 517). These derivatives were quite stable, but the harsh derivatization conditions, which caused some decomposition, were soon abandoned in favor of more gentle methods. The use of diazomethane has remained the principal method of esterfication of the prostaglandins in spite of the increasing awareness of the possibility of side products (p. 524). The hydroxyl groups of the prostaglandin methyl esters have been masked by acyl or silyl groups, with acetylation (pp. 518 and 526) and trimethylsilylation (pp. 520 and 527) being the methods most widely applied. $PGF1_{\alpha}$ and $PGF_{2\alpha}$ have been analyzed by GC-MS as trifluoroacetyl methyl esters (p. 518) and by ECD-GC as heptafluorobutyryl methyl esters (p. 519).

Prostaglandin methyl esters have been trimethylsilylated with several reagents (p. 520). The methyl ester trimethylsilyl derivatives give good analytical results for the PGF compounds, but those prostaglandins containing carbonyl groups generally give better results after derivatization of the carbonyls. Trimethylsilylation methods have been reported for PGE that yield simultaneous conversion to PGB (p. 520) or that yield the PGE enol ether derivatives (p. 521). PGE has also been converted to PGB by base treatment prior to esterfication and trimethylsilylation (p. 523). Thromboxane B_2 (p. 522) and the labile endoperoxide PGH_2 (p. 522) have been analyzed by GC-MS as the methyl ester, trimethylsilyl ether derivatives.

Other silyl groups have been used with the methyl ester derivatives. *t*-Butyldimethylsilyl derivatives are particularly useful for GC-MS analysis (p. 524). The application of *t*-butyldimethylsilylation in the analysis of a PGF metabolite also employed a novel (for prostaglandins) esterfication method (p. 525). Bromomethyldimethylsilyl derivatives were examined for ECD-GC analysis (p. 526).

The best analytical results with carbonyl containing prostaglandins have generally been with methods including carbonyl derivatization. Methoximes, the most commonly employed carbonyl derivatives, have been used both with acetylation (p. 526) and trimethylsilylation (p. 527). The methoximes have the disadvantage of forming GC separable *syn-* and *anti-*isomers, but this problem has not significantly hindered their use. Benzyloxime derivatives have been used in GC-MS analysis (p. 529) and pentafluorobenzyloximes for ECD-GC analysis (p. 530).

Prostaglandin $F_{2\alpha}$ (p. 531) and thromboxane B_2 (p. 532), both of which have dihydroxy functions suitably positioned for formation of cyclic boronates, have been analyzed as methyl ester, boronate derivatives after trimethylsilylation of the C-15 hydroxyls.

Trimethylsilyl Ester Derivatives. Trimethylsilyl esters are hydrolytically less stable than the methyl esters and have been less widely employed in prostaglandin analysis. Trimethylsilylation does, however, offer the advantage of a single step derivatization of the PGF compounds (p. 532). The use of a BSTFA/piperidine mixture gave a single step derivatization of PGEs forming an enol ether derivative from the carbonyl group (p. 533). The same method applied to PGA gave a piperidine addition product (p. 533).

Trimethylsilyl ester ether derivatives have also been applied to carbonyl containing prostaglandins after oxime formations (p. 534). Methoxime and ethoxime trimethylsilyl derivatives were reported to be more stable than the trimethylsilyl derivatives alone. As with other oximes, this method formed GC separable *syn-* and *anti*-isomers.

Other Volatile Derivatives. Pentafluorobenzyl ester formation, followed by trimethylsilylation, yielded a derivative of $PGF_{2\alpha}$ suitable for ECD-GC analysis (p. 535). Pentafluorobenzyl ester pentafluorobenzyl oximes have been used for ECD-GC analysis of carbonyl containing prostaglandins and were found to give about twice the sensitivity of the methyl ester pentafluorobenzyl oximes (p. 530).

Exhaustive reduction of prostaglandins leaves the hydrocarbon skeleton either with or without the carboxyl carbon (p. 535). This derivatization has been used to screen biological samples for the presence of prostaglandin compounds.

Chromogenic Derivatives. A few examples of HPLC analysis of prostaglandins have been reported, but none of the methods yet rival GC-MS-SIM or ECD-GC in sensitivity. Chromogenic derivatives employed in these studies include *p*-nitrobenzyl esters (p. 530), *p*-nitrophenacyl esters (p. 536), *p*-bromophenacyl esters (p. 537), and *p*-nitrobenzyloximes (p. 537).

References

[1]S. Bergström *et al.*, *Acta. Chem. Scand.* **16**, 501 (1962).

[2]P. F. Crain *et al.*, *Meth. Enzymol.* **35**, 359 (1975).

[3]J. C. Frölich, in *The Prostaglandins*, Vol. 3, ed. by P. Ramwell, Plenum, New York, 1977, p. 1.

[4]B. Sweetman *et al.*, *Biomed. Mass Spectrom.* **6**, in press.

12.1. METHYL ESTER DERIVATIVES

12.1.1. Methyl Ester Methyl Ether Derivatives

Derivative Type. Methyl ester, methyl ether.

Reagent. Diazomethane/boron trifluoride methyl etherate.

Compound Type. Prostaglandins E and F.

Reaction

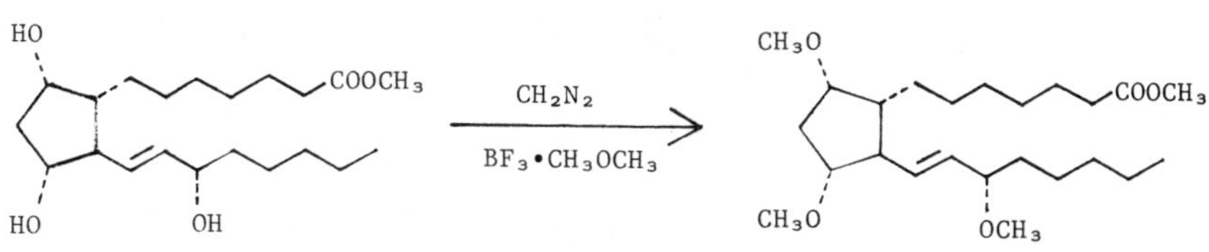

Typical Procedure.[1] Diss 5–10 mg of PGE or PGF methyl ester in 10 ml dry Et_2O and cool to $-40°$. Add 0.05 ml $BF_3{\cdot}CH_3OCH_3$ followed immediately by addition of CH_2N_2 in Et_2O

(dried over KOH and cooled to $-40°$) until the yellow color persists. Allow to react 15 min at $-40°$ and then 90 min at room temp. Filter to remove polymethylenes, wash with aq $NaHCO_3$ and H_2O, dry over Na_2SO_4, and evap under vac. Purify by preparative GC and analyze by MS.

Comments. The methyl ether derivs were used in early work on prostaglandins,[1-3] but, since the harsh conditions cause some decomposition, the method was abandoned in favor of milder deriv methods.

References

[1]S. Bergstrom *et al.*, *Acta Chem. Scand.* **16**, 969 (1962).

[2]G. Eglinton *et al.*, *Nature* **200**, 960 (1963).

[3]R. Ryhage and B. Samuelsson, *Biochem. Biophys. Res. Commun.* **19**, 279 (1965).

12.1.2. Methyl Ester, Acyl Derivatives

Derivative Type. Methyl ester, acetate.

Reagent. Diazomethane, acetic anhydride (Ac_2O)/pyridine.

Compound Type. Prostaglandins A, B, and E.

Reaction

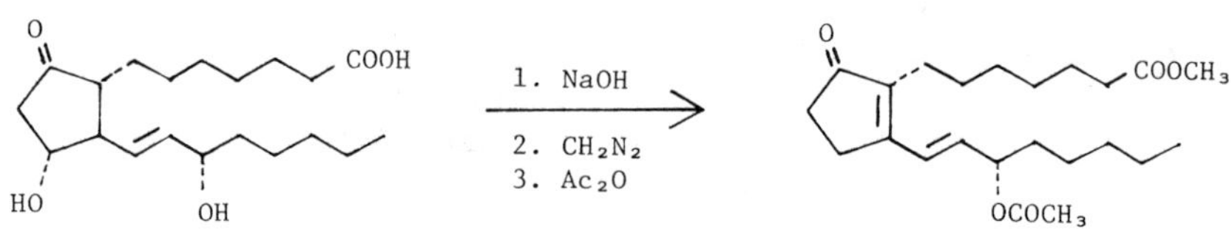

Typical Procedure. Evap PGE_1 fraction to dryness and diss in 0.5 ml EtOH. Add 0.5 ml 1 *N* aq NaOH and warm at 40° for 30 min. Cool to room temp, dil with 2.5 ml H_2O, and acidify with HCl. Ext with Et_2O, wash Et_2O with H_2O, and evap under N_2. Esterfy with CH_2N_2 in Et_2O. Evap to dryness under N_2. Treat overnight at room temp with 100 μl Ac_2O/pyridine (3:5, v/v). Add 1 ml H_2O and ext with 3×0.5 ml Et_2O. Wash Et_2O ext with H_2O and dry over Na_2SO_4. Evap to dryness under N_2 and redissolve in CH_2Cl_2. Analyze by GC-MS.

Comments. For PGA and PGB fractions, the procedure was begun at the esterfication step.

References

[1]P. W. Albro and L. Fishbein, *J. Chromatogr.* **44**, 443 (1969).

Derivative Type. Methyl ester, trifluoroacetate.

Reagent. Diazomethane, trifluoroacetic anhydride (TFAA).

Compound Type. Prostaglandins $F_{1\alpha}$ and $F_{2\alpha}$.

Reaction

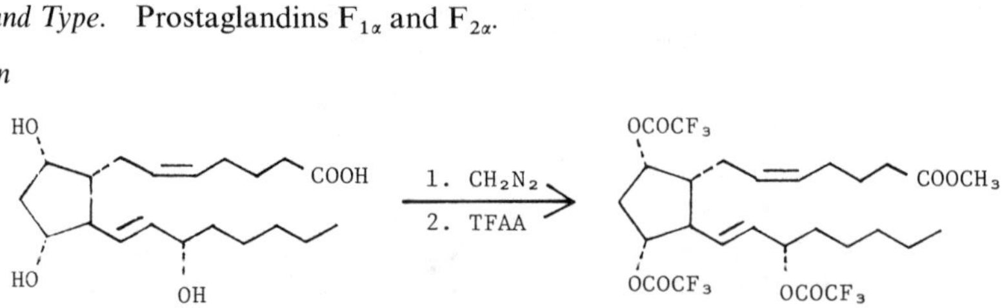

Typical Procedure. Methylate sample at room temp for 20 min using freshly prepared CH_2N_2 in Et_2O/MeOH (9:1). Add 200 μl TFAA and allow to stand for 2 hr. Remove excess reagent in a vac dessicator. Diss residue in 25 μl hexane and analyze 10 μl samples by GC-MS.

Comments. The method was used to separate, identify, and quantify $PGF_{1\alpha}$ and $PGF_{2\alpha}$ down to 10 ng.

References

[1]C. J. Thompson *et al.*, *Life Sci.* **9**, 983 (1970).

Derivative Type. Methyl ester, heptafluorobutyrate.

Reagent. Diazomethane, heptafluorobutyric anhydride (HFBA).

Compound Type. Prostaglandins $F_{1\alpha}$ and $F_{2\alpha}$.

Reaction

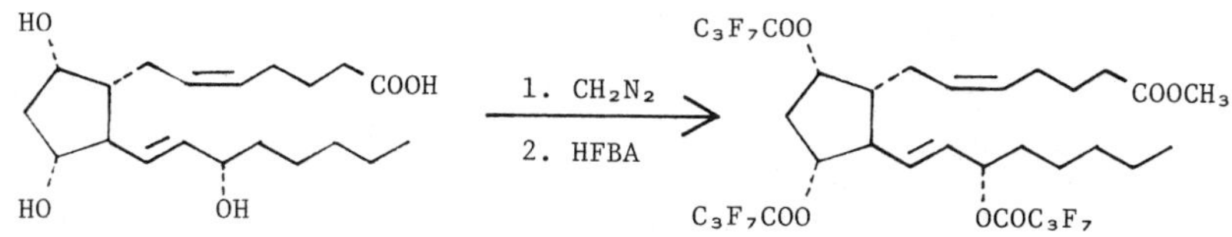

Typical Procedure.[1] Diss 0.01–100 μg of PG in 10% MeOH in Et_2O. Pass CH_2N_2 through the soln[2] for 2 min then evap to dryness. Add 10 μl HFBA. After 5–15 min evap to dryness under N_2 at 40°. Diss residue in hexane and analyze by ECD-GC.

Comments. The deriv was reported to yield multiple GC peaks presumably due to decomposition on the column.[1] Preparation of the deriv on a milligram scale and subsequent analysis by GC-MS gave only a small (<5%) second peak[3] suggesting that multiple peaks in the earlier work[1] may have been due to impurities or to deteriorated columns. The deriv was detectable down to 1 ng by GC-MS-SIM.[3] In later work using a microscale CH_2N_2 generator the deriv was reported to be detectable to less than 1 ng by ECD-GC.[4]

References

[1]M. S. Levitt *et al.*, *Prostaglandins* **1**, 121 (1972).

[2]H. Schlenk and J. L. Gellerman, *Anal. Chem.* **32**, 1412 (1960).

[3]B. S. Middleditch and D. M. Desiderio, *Prostaglandins* **2**, 195 (1972).

[4]M. J. Levitt, *Anal. Chem.* **45**, 618 (1973).

Derivative Type. Methyl ester, heptafluorobutyryl.

Reagent. Diazomethane, heptafluorobutyryl imidazole (HFBI).

Compound Type. Prostaglandins $F_{1\alpha}$ and $F_{2\alpha}$.

Reaction

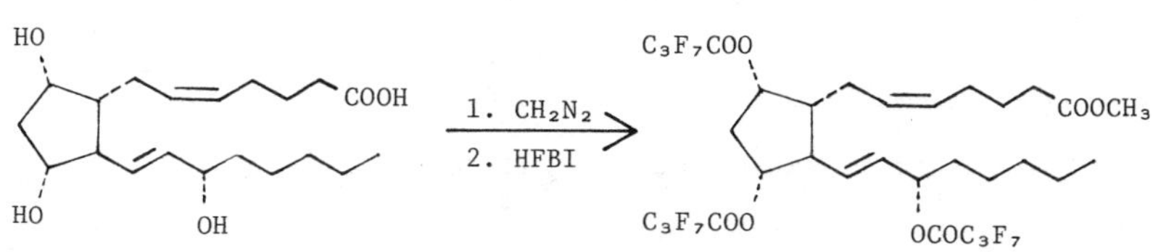

Typical Procedure. Homogenize sample with 10 vols of EtOH/Et_2O (3:1), allow to stand 30 min, filter, and evap filtrate to dryness under vac at 40°. Add 5 ml CCl_4 and 10 ml H_2O to residue, shake, and then centrifuge. Separate upper aq layer and adjust to pH 3 with 1 *N* HCl. Ext with 10 ml EtOAc, centrifuge, and separate EtOAc layer. Add 200 μl 1 *N* NH_3 soln and 2 ml MeOH and evap to dryness under vac at 40°.

Add CH_2N_2 in Et_2O/MeOH (9:1), allow to stand 10 min, then evap to dryness under N_2. Add 50 μl EtOAc and 50 μl HFBI. After 10 min evap to dryness under N_2. Diss residue in EtOAc and analyze by ECD-GC.

Comments. The method was used for analysis of PGF in seminal plasma using $PGF_{2\beta}$ as int stand. The methyl ester, trimethylsilyl derivs were also prepared using BSTFA and analyzed by FID-GC. MS data was given for the derivs.

References

[1]M. Sugiura and K. Hirano, *J. Chromatogr.* **90**, 169 (1974).

12.1.3. Methyl Ester Trimethylsilyl Derivatives

Derivative Type. Methyl ester, trimethylsilyl.

Reagent. Diazomethane, bis(trimethylsilyl)acetamide (BSA).

Compound Type. Prostaglandins A, B, E, and F.

Reaction

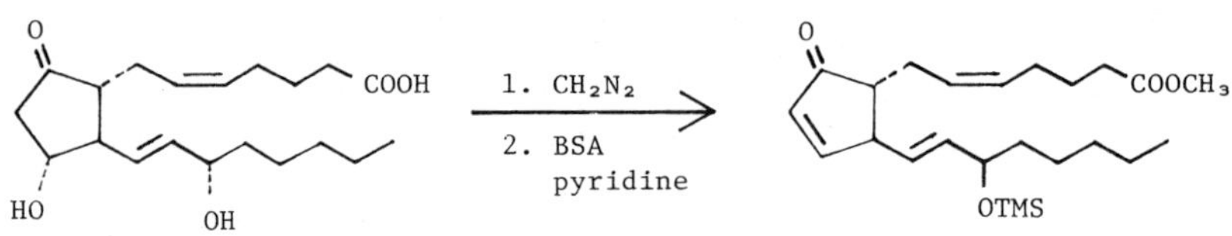

Typical Procedure. Mix ext with 20 μl MeOH and 80 μl Et_2O containing approx a 10-fold excess of CH_2N_2 and allow to react 10 min. Evap under N_2 stream. Treat with *ca* a 100-fold excess of BSA in pyridine (or MSTFA, or BSTFA) for 3–4 hr at 25°. Remove reagent under N_2 stream and redissolve in hexane. Analyze by GC-MS on 1% Dexsil 300 at 220–250°.

Comments. PGE was converted to the PGA deriv by BSA/pyridine.[1] PGA and PGB each formed methyl ester, mono-TMS derivs[1] and PGF the methyl ester, tri-TMS deriv.[2] BSTFA has been used in place of BSA for PGF.[3] An extensive study of derivs of PGA and PGB has been reported.[2]

References

[1]B. J. Sweetman *et al.*, *Prostaglandins* **3**, 75 (1973).

[2]B. J. Sweetman *et al.*, *Prostaglandins* **3**, 385 (1973).

[3]C. J. Thompson *et al.*, *Life Sci.* **9**, 983 (1970).

[4]J. T. Watson and B. J. Sweetman, *Org. Mass Spectrom.* **9**, 39 (1974).

Derivative Type. Methyl ester, trimethylsilyl.

Reagent. Diazomethane, trimethylsilylimidazole (TMSI)/piperidine.

Compound Type. Prostaglandins E_1, E_2, $F_{1\alpha}$, $F_{2\alpha}$.

Reaction

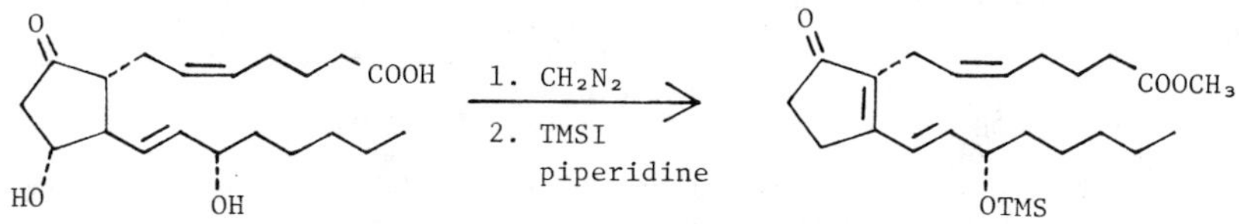

Typical Procedure. Diss sample in 0.3 ml of MeOH and treat with freshly prepared CH_2N_2 in Et_2O. Evap to dryness under N_2 and treat with 1:1 TMSI/piperidine (20 μl/10 μg). For 1–500 ng diss residue in 50 μl benzene and add 3 μl of TMSI/piperidine soln. Analyze by GC-MS.

Comments. Either deuterated PG or arachidic acid was used as int stand. The method was demonstrated down to the picomole level. PGE was converted cleanly to PGB methyl ester, mono-TMS. PGF formed the tri-TMS, methyl ester.

References

S. Nicosia and G. Galli, *Anal. Biochem.* **61**, 192 (1974).

Derivative Type. Methyl ester, trimethylsilyl.

Reagent. Diazomethane, bis(trimethylsilyl)acetamide (BSA)/trimethylchlorosilane (TMCS)/ trimethylsilylimidazole (TMSI).

Compound Type. Prostaglandins E_2 and $F_{2\alpha}$.

Reaction

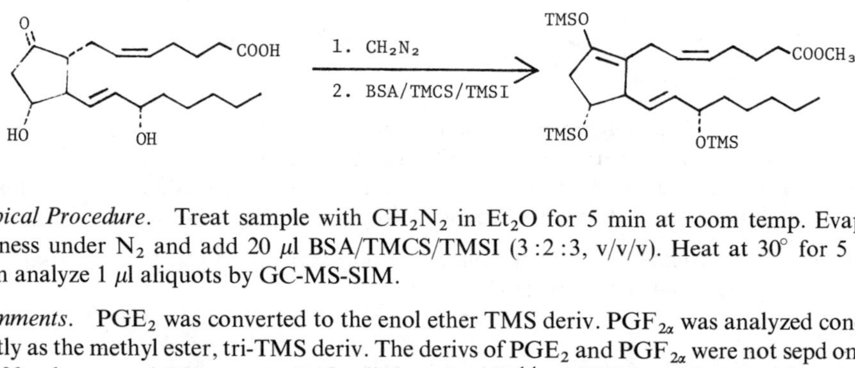

Typical Procedure. Treat sample with CH_2N_2 in Et_2O for 5 min at room temp. Evap to dryness under N_2 and add 20 μl BSA/TMCS/TMSI (3:2:3, v/v/v). Heat at 30° for 5 min then analyze 1 μl aliquots by GC-MS-SIM.

Comments. PGE_2 was converted to the enol ether TMS deriv. $PGF_{2\alpha}$ was analyzed concurrently as the methyl ester, tri-TMS deriv. The derivs of PGE_2 and $PGF_{2\alpha}$ were not sepd on the SE-30 column used. Measurement of as little as 5×10^{-14} g of $PGF_{2\alpha}$ with a precision (SE) of $\pm 10\%$ was reported. The detection limit for PGE_2 was 5×10^{-13} g with SE 15%.

References

[1]B. W. Wilson *et al.*, *Adv. Mass Spectrom. Biochem. Med.* **2**, 487 (1976).

[2]B. W. Wilson *et al.*, *Agents and Actions* **6**, 176 (1976).

Derivative Type. Methyl ester, trimethylsilyl ether.

Reagent. Diazomethane, hexamethyldisilazane (HMDS)/trimethylchlorosilane (TMCS).

Compound Type. Unsaturated hydroxy acids.

Reaction

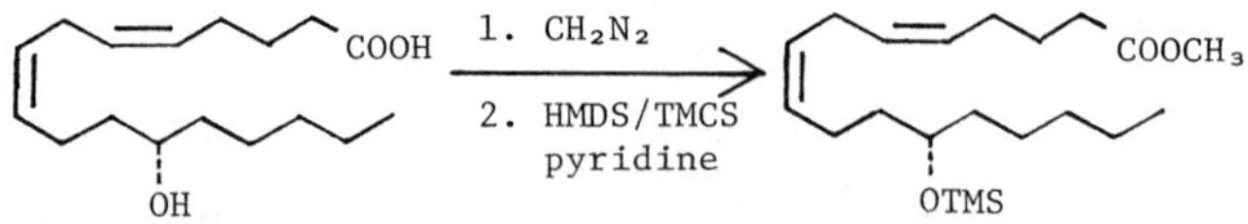

Typical Procedure. To 1–2 ml of platelet suspension add 4.48 nmol deuterated int stand in 10 ml 90% EtOH. Dil with H_2O, acidify to pH 3 and ext 2× with Et_2O. Evap ext and esterfy with CH_2N_2. TLC on silica with org layer from EtOAc/2,2,4-trimethylpentane/H_2O (75 : 75 : 100, v/v/v). Elute sample areas with Et_2O. Evap eluates to dryness and treat with HMDS/TMCS/pyridine (actual TMS reagents were not stated; HMDS/TMCS was inferred from earlier publications). Analyze by GC-MS-SIM.

Comments. The method was used for analysis of transformation products of arachidonic acid in platelets by GC-MS with deuterated int stand.

References

[1]M. Hamberg *et al., Proc. Natl. Acad. Sci. USA* **71**, 3824 (1974).

Derivative Type. Methyl ester, trimethylsilyl.

Reagent. Diazomethane, hexamethyldisilazane (HMDS)/trimethylchlorosilane (TMCS)/ pyridine.

Compound Type. Thromboxane B_2.

Reaction

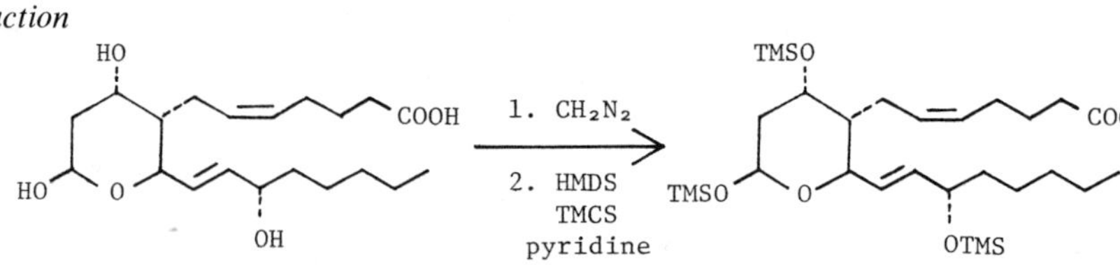

Typical Procedure. Add deuterated int stand to sample, acidify to pH 3, and ext 2× with Et_2O. Evap ext and esterfy with CH_2N_2. Purify ester by TLC on silica with org phase from EtOAc/2,2,4-trimethylpentane/H_2O (75:75:100, v/v/v) and elute with EtOAc. Evap eluate to dryness and treat with HMDS/TMCS/pyridine. Analyze by GC-MS-SIM.

Comments. The method was used to quantitate thromboxane B_2 in platelet suspensions by GC-MS-SIM using a deuterated int stand.

References

[1]M. Hamberg *et al., Proc. Natl. Acad. Sci. USA* **71**, 3824 (1974).

Derivative Type. Trimethylsilyl.

Reagent. Bis(trimethylsilyl)trifluoroacetamide (BSTFA).

Compound Type. Prostaglandin H_2 methyl ester.

Reaction

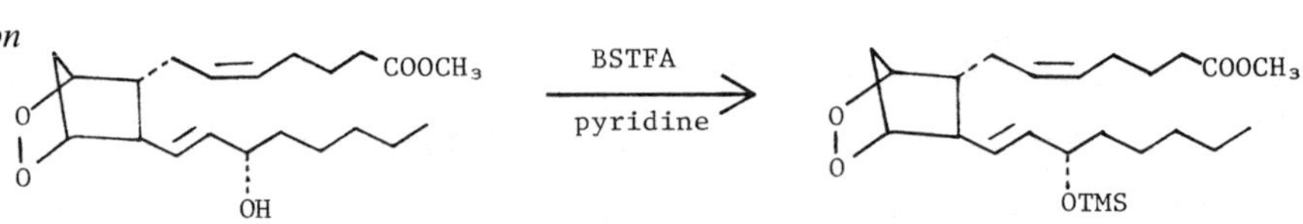

Typical Procedure. Prepare deriv by treatment of PGH_2 methyl ester with BSTFA in pyridine (details not reported) and analyze by GC-MS.

Comments. The method was used to compare chemically synthesized PGH_2 with biosynthetic material. PGH_2 had previously been analyzed by MS (direct probe) as the methyl ester.[2]

References

[1]R. A. Johnson *et al.*, *J. Am. Chem. Soc.* **99**, 7738 (1977).
[2]R. R. Gorman *et al.*, *Prostaglandins* **13**, 1043 (1977).

Derivative Type. Prostaglandin B, methyl ester, trimethylsilyl.

Reagent. Potassium hydroxide, diazomethane, bis(trimethylsilyl)acetamide (BSA).

Compound Type. Prostaglandin E.

Reaction

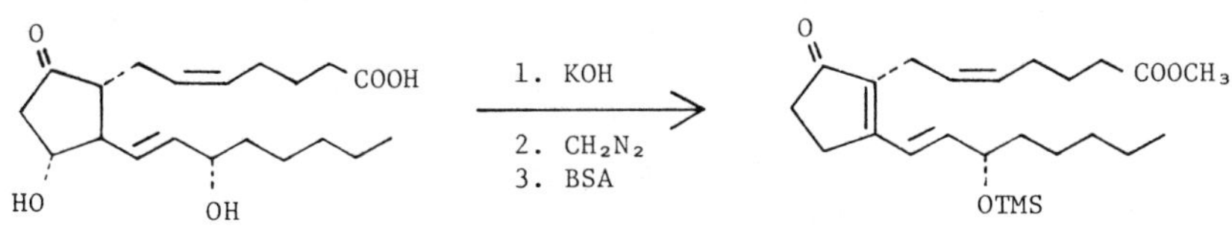

Typical Procedure.[1] Add 0.1 ml 0.5 *N* KOH in MeOH to the dry PGE sample and allow to react 5 min at 20°. Add 0.05 ml 2 *M* citric acid and 0.3 ml H_2O. Ext twice with an equal vol Et_2O. Evap the ext under N_2 and diss residue in 0.1 ml Et_2O/MeOH (9:1, v/v). Add excess CH_2N_2 in Et_2O and allow to react 2 min at 20°. Evap under N_2 and add 10 μl BSA. After 15 min at 20° evap under N_2. Diss residue in heptane and analyze by ECD-GC.

Comments. The method was used to determine nanogram amts of PGE_1 and PGE_2 in tissues and biological fluids. ω-Homo-PGE_1 and ω-nor-PGE_2 were used as int stand. Korteweg *et al.*[5] reported detection to the low picogram level using ECD-GC with glass capillary columns.

References

[1]G. H. Jouvenaz *et al.*, *Biochim. Biophys. Acta* **202**, 231 (1970).
[2]G. H. Jouvenaz *et al.*, *Prostaglandins* **3**, 175 (1973).
[3]E. Anggard and C. E. Jonsson, *Acta Physiol. Scand.* **81**, 440 (1971).
[4]K. Green and B. Samuelsson, *Eur. J. Biochem.* **22**, 391 (1971).
[5]M. Korteweg *et al.*, *Prostaglandins* **13**, 1221 (1977).

Derivative Type. Pyrazoline adduct, methyl ester, trimethylsilyl.

Reagent. Diazomethane, bis(trimethylsilyl)acetamide (BSA).

Compound Type. 15-Oxo-prostaglandin $F_{2\alpha}$.

Reaction

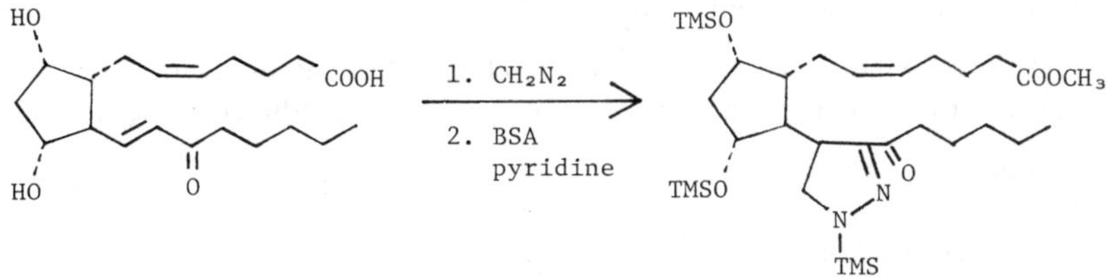

Typical Procedure. Diss a 20 μg sample in Et_2O containing excess CH_2N_2. Allow to react for several min then evap under N_2. Add 20 μl BSA and 20 μl pyridine and heat at 60° for 10 min. Evap 2 μl of the soln in a probe tube and analyze by MS.

Comments. It was shown that pyrazoline adduct formation was a serious side reaction in the esterfication of 15-oxo-$PGF_{2\alpha}$. The pyrazoline adduct did not give a GC peak.

References

[1]A. F. Conkerill *et al.*, *Biomed. Mass Spectrom.* **4**, 187 (1977).

12.1.4. Methyl Ester, Other Silyl Derivatives

Derivative Type. *t*-Butyldimethylsilyl ether.

Reagent. *t*-Butyldimethylsilyl chloride (TBDMS-Cl).

Compound Type. Prostaglandins (as methyl esters) and steroids.

Reaction

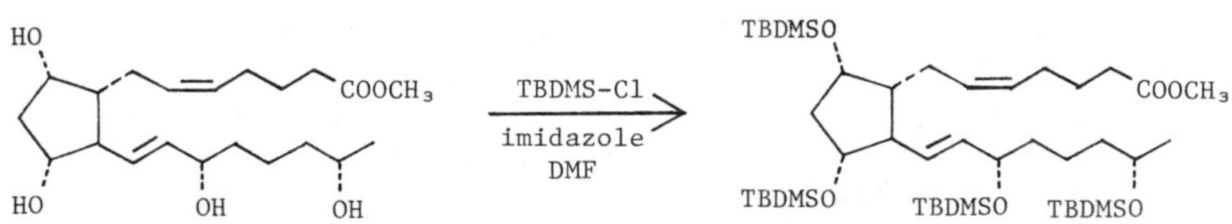

Typical Procedure. Evap sample in the bottom of a 2 × 60 mm glass tube. Add 50 μl of 2 *M* imidazole in dry DMF and 50 μl of TBDMS-Cl (2 *M* in DMF). Seal tube and heat at 100° for 1 hr. Place contents of tube on top of 0.5 × 30 cm column of Sephadex LH-20 swollen in heptane/EtOAc (3:1). Elute column with 4 ml of same solvent and evap eluate to dryness to yield deriv. Analyze by GC-MS.

Comments. The method gave high yields of derivs of submicrogram quantities of prostaglandins and steroids. The method was also utilized with mixed derivs including cyclic boronates and oximes.

References

[1]R. W. Kelley and P. L. Taylor, *Anal. Chem.* **48**, 465 (1976).

Derivative Type. Methyl ester, *t*-butyldimethylsilyl (TBDMS).

Reagent. Diazomethane, *t*-butyldimethylchlorosilane (TBDMS-Cl)/imidazole/dimethylformamide (DMF).

Compound Type. Thromboxane B_2.

Reaction

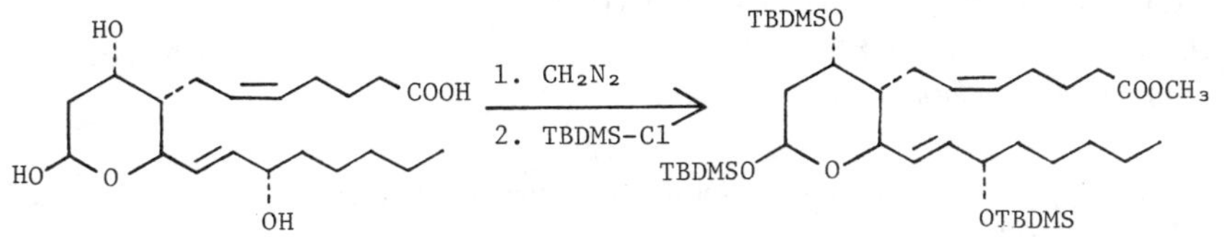

Typical Procedure. Prepare the methyl ester by treatment with CH_2N_2 in Et_2O and evap under N_2. To 50 μg of the ester add 25 μl TBDMS-Cl/imidazole/DMF (1:1:6, w/w/w) and heat overnight at 85° under N_2. Evap the DMF under vac and purify the deriv on a 5 × 0.5 cm column of Sephadex LH-20 eluting with 2 ml $CHCl_3$/hexane/MeOH (10:10:1, v/v/v). Evap eluate and diss deriv in 25 μl EtOAc prior to GC or GC-MS-SIM analysis.

Comments. Dil of the TBDMS reagent with pyridine (dist from KOH) gave only partial deriv (C-9 hydroxyl underiv). Although TMS, methane (and butane) boronate, methyl ester derivs were successfully prepared, attempts to form the corresponding TBDMS deriv were unsuccessful. MS data were given for the deriv.

References

[1] A. G. Smith *et al.*, *J. Chromatogr.* **142**, 553 (1977).

Derivative Type. Methyl ester, *t*-butyldimethylsilyl (TBDMS) ether.

Reagent. Tetramethylammonium hydroxide (TMAH)/methyl iodide, *t*-butyldimethylsilyl chloride (TBDMS-Cl)/imidazole.

Compound Type. 5α,7α-Dihydroxy-11-oxotetranorprostane-1,16-dioic acid (PGF metabolite).

Reaction

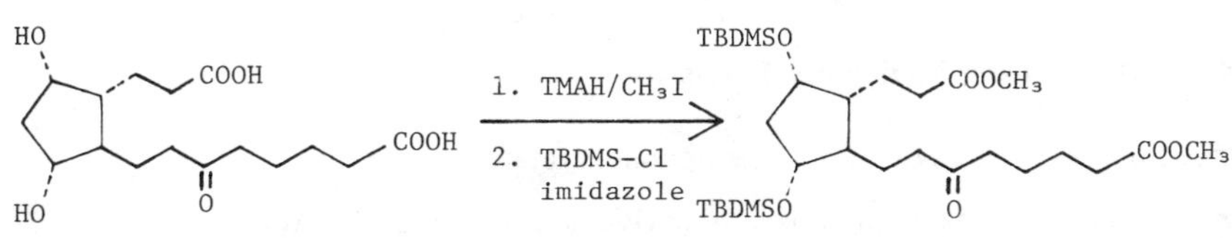

Typical Procedure.[1] To 10 ml of urine add 200 ng of the 4β,6α,6β-$^2H_3$-5β-^3H PGF metabolite int stand as the dimethyl ester. Add 2 ml 5 *M* NaOH and allow to stand overnight. Acidify to pH 3 and ext with 2 g column of Amberlite XAD-2. Elute column with MeOH and esterfy eluted material with 2% TMAH in MeOH and CH_3I.[2] Treat the resulting mixt of dimethyl ester and methyl ester, δ-lactone with 0.1 *M* pH 10 $Na_2B_4O_7$ buffer overnight to hydrolyze the C-1 methyl ester. Wash soln with CH_2Cl_2, acidify aq soln to pH 2, and allow to stand 1 hr (converts mono ester to δ-lactone). Adjust to pH 8 and ext with CH_2Cl_2. Evap ext and esterfy as above.[2] Treat diester with TBDMS-Cl/imidazole/DMF. Purify deriv by TLC (assumed to be on silica) using EtOAc/heptane (2:3, v/v). Analyze by GC-MS-SIM.

Comments. The methylation, selective C-1 hydrolysis, and extraction sequence afforded considerable purification in lieu of a chromatographic step. The method was used to quantitate the metabolite to 1 ng/ml.

References

[1]A. R. Brash *et al.*, *Biochem. Soc. Trans.* **4**, 706 (1976).
[2]H. Greeley, *J. Chromatogr.* **88**, 229 (1974).
[3]A. R. Brash *et al.*, *Biochem. Med.* **16**, 77 (1976).

Derivative Type. Methyl ester, bromomethyldimethylsilyl ether.

Reagent. Diazomethane, bromomethyldimethylsilyldiethylamine (BMDMS-DEA).

Compound Type. Prostaglandins F.

Reaction

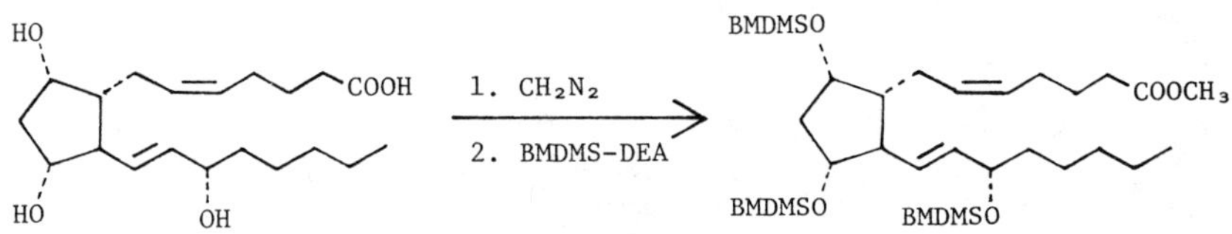

Typical Procedure. Diss sample in 0.1 ml Et_2O/MeOH (9:1, v/v) and add excess CH_2N_2 in Et_2O. Allow to react 2 min at 20°; then evap under N_2. Add 0.1 ml silylating reagent (see below) and allow to stand 1 hr at room temp. Evap under N_2 and diss residue in 0.5 ml heptane. Centrifuge and decant heptane soln from the ppt of $Et_2NH{\cdot}HCl$. Conc the heptane soln and analyze by ECD-GC.

Comments. The reagent was prepared by diss 0.1 ml dry Et_2NH in 1 ml dry Et_2O and adding it slowly to a soln of 0.2 ml bromomethyldimethylsilyl chloride in 1 ml dry Et_2O. The resulting mixt was centrifuged and the supernatant used as the reagent. The method was used to determine nanogram amts of $PGF_{1\alpha}$ and $PGF_{2\alpha}$ in tissues and amniotic fluid.

References

[1]G. H. Jouvenaz *et al.*, *Biochim. Biophys. Acta* **202**, 231 (1970).
[2]G. H. Jouvenaz *et al.*, *Prostaglandins* **3**, 175 (1973).

12.1.5. Methyl Ester Methoxime Derivatives

Derivative Type. Methyl ester, methoxime, acetate.

Reagent. Diazomethane, methoxyamine HCl, acetic anhydride (Ac_2O).

Compound Type. Prostaglandins.

Reaction

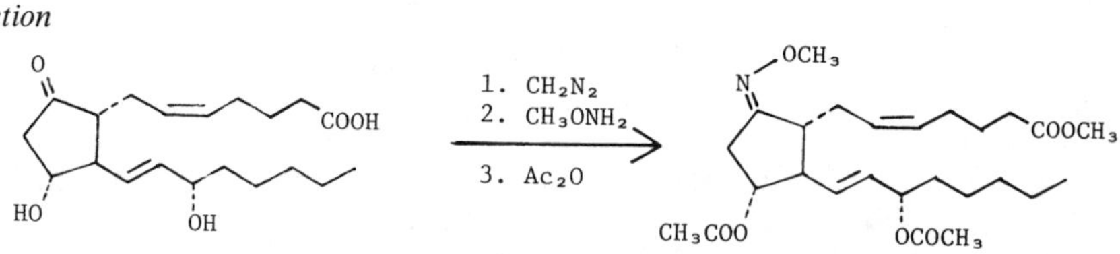

Typical Procedure. Diss 100–200 μg of PG in 0.3 ml MeOH and treat with CH_2N_2 in Et_2O. Evap to dryness and diss residue in 0.2 ml pyridine containing 5 mg CH_3ONH_2-HCl. Allow to stand overnight in a stoppered tube in a dessicator. Evap to dryness and ext residue with 3 × 0.5 ml Et_2O. Evap Et_2O ext to dryness and diss residue in 0.3 ml pyridine and 0.3 ml Ac_2O.

Allow to stand overnight at room temp. Add ice, acidify, and ext 3 × with Et_2O. Wash Et_2O soln with sm vol of 0.2 *N* HCl, 10% aq $NaHCO_3$, then H_2O until neutral. Evap to dryness and diss residue in CS_2. Analyze by GC-MS.

Comments. Deuterated derivs were prepared with CD_3ONH_2, d_6-Ac_2O, and by treatment of the methyl esters with 0.5 *N* $NaOCD_3$ in CD_3OH. Methoxime TMS derivs were also prepared. The methyl ester, methoxime, acetate derivs were used to quantitate PGE_2 and $PGF_{2\alpha}$ (omitting the methoxime step) by GC-MS-SIM using deuterated int stand.[2,3]

References

[1]K. Green, *Chem. Phys. Lipids* **3**, 254 (1969).

[2]U. Axen *et al.*, *Biochem. Biophys. Res. Commun.* **45**, 519 (1971).

[3]K. Green *et al.*, *Anal. Biochem.* **54**, 434 (1973).

Derivative Type. Methoxime, methyl ester, trimethylsilyl ether.

Reagent. Methoxylamine HCl, diazomethane, bis(trimethylsilyl)acetamide (BSA).

Compound Type. Prostaglandins and metabolites.

Reaction

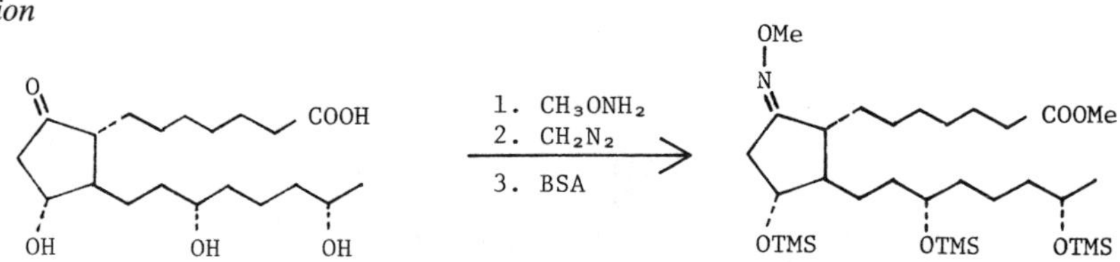

Typical Procedure. Mix semen sample with 4 vols 1.5 *M* pyridine acetate buffer (pH 5) containing 5 mg/ml methoxylamine HCl and place in ultrasonic bath for 40 min. Ext with Et_2O/EtOAc (4:1). Evap ext and esterfy with CH_2N_2. Treat with BSA then analyze an aliquot by GC-MS.

Comments. Oximation of the unextracted semen prevented degradation of β-ketol system to PGA or PGB. TBDMS ethers allowed separation of 19-hydroxy PGE_1 and PGE_2.

References

[1]P. L. Taylor and R. W. Kelly, *Nature* **250**, 665 (1974).

Derivative Type. Methyl ester, methoxime, trimethylsilyl ether.

Reagent. Diazomethane, methoxyamine HCl, bis(trimethylsilyl)acetamide (BSA).

Compound Type. Prostaglandin E metabolite (7α-hydroxy-5,11-diketotetranorprostane-1,16-dioic acid).

Reaction

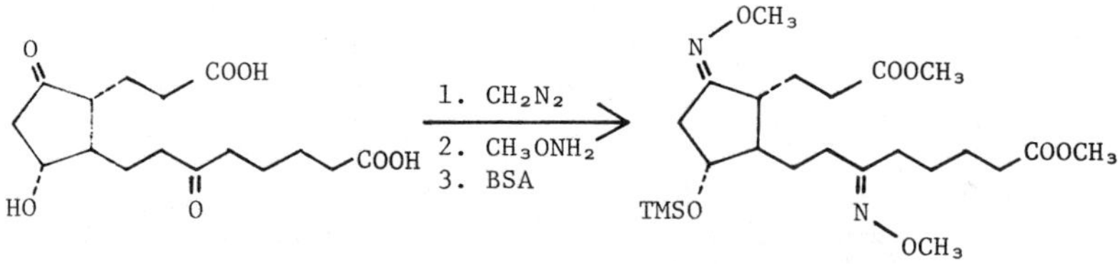

Typical Procedure. Adjust 10 ml urine to pH 4.0 and pass through a 6×1.8 cm column of XAD-2 resin. Wash column with 300 ml H_2O and elute with 90 ml EtOH. Evap EtOH and diss residue in 0.75 ml MeOH. Treat with excess CH_2N_2 in Et_2O for 10 to 15 min. Evap to dryness and diss residue in 1.5 ml pyridine. Add 12.5 mg methoxyamine HCl and let stand for 18 hr at room temp. Evap pyridine under reduced pressure and transfer with 3×1 ml MeOH and 3×1 ml H_2O into a separatory funnel. Ext $3\times$ with Et_2O. Add int stand to Et_2O extracts (1.0 μg, 45,000 dpm). Purify by reverse phase partition chromatography on Hyflo Super-Cel.[2] Purify material from radioactive peak by TLC.[2] Elute plate area containing major isomer of methoxime with Et_2O. Evap eluate to dryness and treat with 5 μl pyridine and 5 μl BSA. Leave standing overnight at room temp, and analyze by GC-MS-SIM.

Comments. The method was used to quantitate down to a few ng/ml using bis-d_3-methoxime-2-^{3}H-dimethyl ester as int stand.[1] Hamberg prepared the methoxime before the esterfication in applying a similar procedure to analysis of the PGE metabolite[2] and the analogous PGF metabolite.[3]

References

[1]H. W. Seyberth *et al.*, *Prostaglandins* **11**, 381 (1976).
[2]M. Hamberg, *Biochem. Biophys. Res. Commun.* **49**, 720 (1972).
[3]M. Hamberg, *Anal. Biochem.* **55**, 368 (1973).

Derivative Type. Methyl ester, methoxime, trimethylsilyl.

Reagent. Diazomethane, methoxylamine HCl, bis(trimethylsilyl)trifluoroacetamide (BSTFA).

Compound Type. Thromboxane B_2.

Reaction

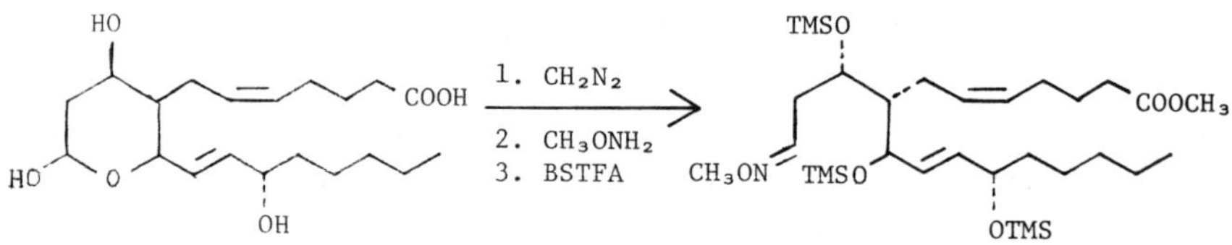

Typical Procedure.[1] Prepare the methyl ester by treatment with CH_2N_2 in Et_2O and evap under N_2. Diss 50 μg of the ester in 50 μl pyridine containing 5 mol equiv of methoxylamine · HCl and heat 3 hr at 85°. Add 10 μl BSTFA and heat 3 hr at 85°. Evap and diss the deriv in 25 μl EtOAc prior to GC or GC-MS-SIM analysis.

Comments. The *n*-propyloxime was also prepared in a similar manner. The methoxime deriv gave 2 GC peaks, a major (later eluting) and minor, presumably due to *syn*- and *anti*-isomers. Use of a large excess of the reagent and long reaction times gave only the later eluting isomer. The *n*-propyloxime deriv gave a single broad peak. MS data were given. The methoxime, methyl ester, TMS deriv was used in the original structure determination[2] and in other later studies of thromboxane B_2.[3-5]

References

[1]A. G. Smith *et al.*, *J. Chromatogr.* **142**, 533 (1977).
[2]M. Hamberg and B. Samuelsson, *Proc. Natl. Acad. Sci. USA* **71**, 3400 (1974).

[3]W. Dawson *et al.*, *Nature* **262**, 699 (1976).
[4]P. Falardeau *et al.*, *Biochim. Biophys. Acta* **441**, 193 (1976).
[5]W.-C. Chang *et al.*, *Prostaglandins* **13**, 17 (1977).

Derivative Type. Methyl ester, methoxime, trimethylsilyl.

Reagent. Diazomethane, methoxylamine HCl, TMS reagent not given.

Compound Type. 6-Keto-prostaglandin $F_{1\alpha}$ (6K-$PGF_{1\alpha}$).

Reaction

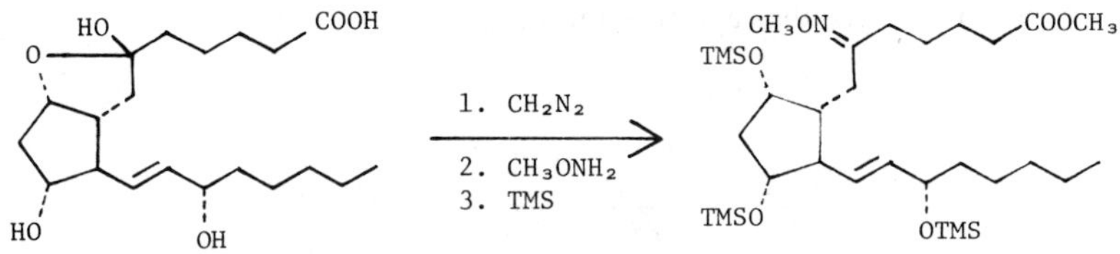

Typical Procedure. Homogenize tissue sample in 20 vol pH 7.4 (0.05 *M*) KH_2PO_4-NaOH buffer and ext with 5 vol EtOH. Add 1 μg d_5-6K-$PGF_{1\alpha}$ int stand and methylate entire ext with CH_2N_2 in Et_2O. Purify methyl ester by TLC on silica gel using $CHCl_3$/MeOH/HOAc/H_2O (90 :9 :1 :0.65, v/v/v/v). Elute the ester (R_f=0.45) with MeOH and convert to the methoxime with methoxylamine HCl in pyridine. Convert to the TMS ether and analyze by GC-MS-SIM.

Comments. The method was used to quantitate 6K-$PGF_{1\alpha}$ in rat stomach fundus.[1] The methyl ester, *n*-butylboronate, trimethylsilyl deriv was also prepared.[2] MS data for the derivs were reported.[1–4]

References

[1]C. R. Pace-Asciak, *Anal. Biochem.* **81**, 251 (1977).
[2]C. R. Pace-Asciak, *J. Am. Chem. Soc.* **98**, 2348 (1976).
[3]A. F. Cockerill *et al.*, *Biomed. Mass Spectrom.* **4**, 358 (1977).
[4]C. R. Pace-Asciak, *Biochim. Biophys. Acta* **487**, 495 (1977).

12.1.6. Methyl Ester Benzyloxime Derivatives

Derivative Type. Benzyloxime, trimethylsilyl ether, methyl ester.

Reagent. Benzylhydroxylamine-HCl, TMS reagent not given, diazomethane.

Compound Type. Prostaglandin E_2.

Reaction

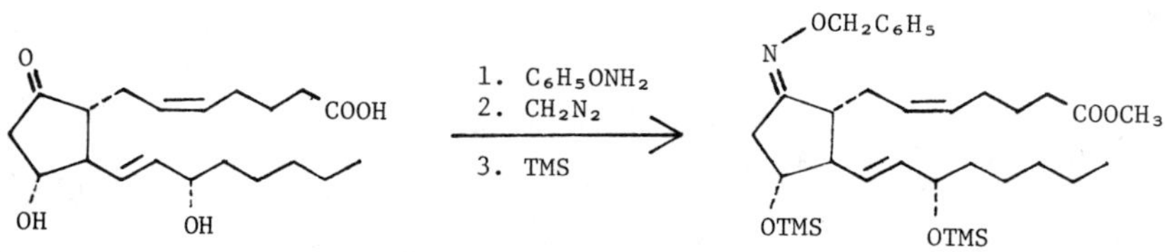

Typical Procedure. Treat sample with benzylhydroxylamine,[2] treat with CH_2N_2, then trimethylsilylate. Analyze by GC-MS-SIM.

Comments. The method was used to quantitate PGE_2 using d_4-PGE_2 as int stand.

References

[1]L. Baczynskyj *et al.*, *Anal. Chem.* **45**, 479 (1973).

[2]P. G. Devaux *et al.*, *Anal. Biochem.* **41**, 70 (1971).

Derivative Type. Methyl ester, pentafluorobenzyloxime, trimethylsilyl ether.

Reagent. Diazomethane, pentafluorobenzylhydroxylamine (PFHBA), bis(trimethylsilyl)-acetamide (BSA).

Compound Type. Prostaglandins and thromboxanes.

Reaction

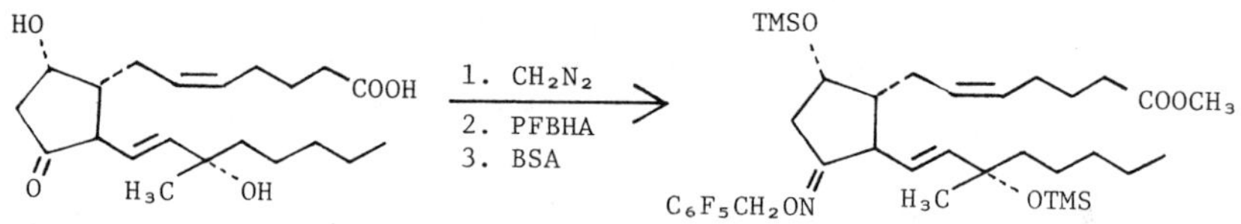

Typical Procedure. Prepare methyl ester by treating with excess CH_2N_2 in Et_2O for 5 min. Evap and diss residue (10 μg or more) in 0.2 ml anhyd pyridine containing a 10-fold mol excess of PFBHA HCl. Heat at 40° for 2 hr or allow to stand at room temp overnight. Evap under N_2 and mix residue with 0.5 ml 0.9% NaCl soln. Ext with 2 × 2 ml hexane. Dry over Na_2SO_4 and evap to dryness. Treat residue with 100 μl BSA for 1 hr at 40°. Dil with hexane and analyze by ECD-GC.

Comments. The deriv was used to quantitate compds to picogram levels by ECD-GC. Pentafluorobenzyl ester,[2] pentafluorobenzyloximes were also examined and found to give about twice the sensitivity of the methyl ester derivs when retention times were adjusted to be the same. Thromboxane B_2 formed the pentafluorobenzyl ester poorly.

References

[1]F. A. Fitzpatrick *et al.*, *Anal. Chem.* **49**, 1032 (1977).

[2]A. J. Wickrema Sinha *et al.*, *J. Pharm. Sci.* **62**, 1428 (1973).

Derivative Type. Methyl ester, pentafluorobenzyl oxime, trimethylsilyl ether.

Reagent. Diazomethane, pentafluorobenzylhydroxylamine (PFBHA), bis(trimethylsilyl)-acetamide (BSA).

Compound Type. Thromboxane B_2, prostaglandins A, B, D, and E.

Reaction

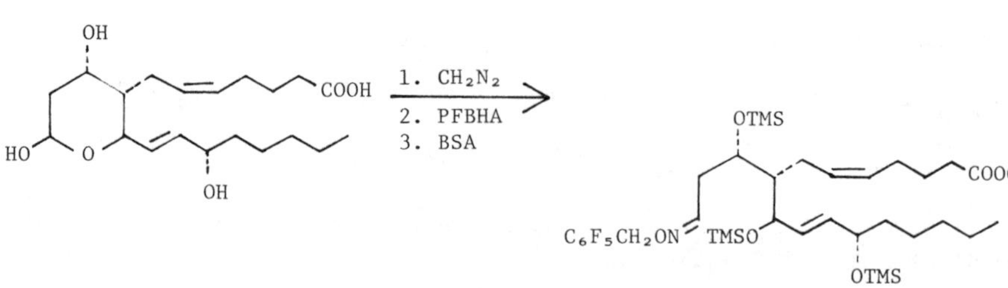

Typical Procedure. Quench platelet aggregation sample (1 ml) with 1 ml acetone, centrifuge, and transfer supernatant to another tube containing 0.1 ml 2 *M* citric acid. Ext with 3 × 5 ml Et_2O. Evap ext to dryness under N_2.

Treat residue with excess CH_2N_2 in Et_2O for 5 min and evap to dryness under N_2. Diss residue in 1 ml pyridine containing 1.0 mg PFBHA. Heat at 40° for 2 hr to allow to stand overnight at room temp. Evap to dryness under N_2, mix residue with 0.5 ml 0.9% NaCl, and ext with 2 × 2 ml hexane. Dry over Na_2SO_4, conc, and transfer to 0.5 ml glass stoppered tube. Evap to dryness and treat residue with 0.1 ml BSA and 0.01 ml DMF for 1 hr at 40°. Analyze 1 μl aliquots by ECD-GC.

Comments. The method was used to quantitate PGA_2, PGE_2, and PGD_2 (unseparated), PGB_2, and TxB_2 (2 peaks, *syn*- and *anti*-isomers). Recovery of TxB_2 was shown to be 90% but the rel SD for submicrogram quantities was ±11 to ±42%. MS data were given for the TxB_2 deriv.

References

[1]F. A. Fitzpatrick *et al.*, *Prostaglandins* **13**, 201 (1977).

12.1.7. Methyl Ester Boronate Derivatives

Derivative Type. Methyl ester, butyl boronate (9,11), trimethylsilyl ether (15).

Reagent. Diazomethane, butaneboronic acid, bis(trimethylsilyl)trifluoroacetamide (BSTFA).

Compound Type. Prostaglandin $F_{2\alpha}$.

Reaction

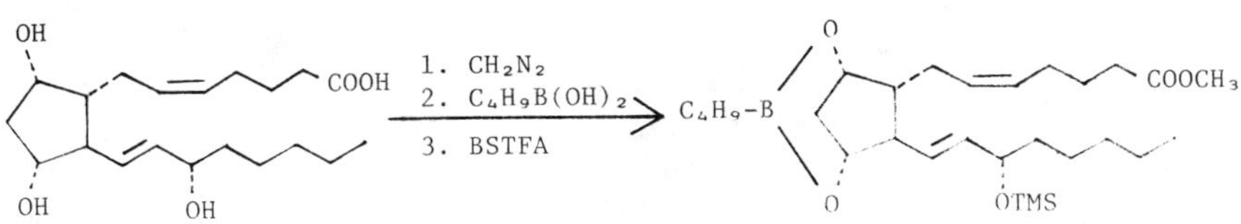

Typical Procedure. Add 20 μl of 3.8 ng/μl d_4-$PGF_{2\alpha}$ to 5 ml plasma. Acidify to pH 3.0 and ext with 20 ml Et_2O/EtOAc (4:1). Wash Et_2O layer to neutrality and evap under vac at 45° Treat *ca* 1/3 of the extract with 25 μl CH_2N_2 soln (1% in Et_2O/MeOH, 2:1) for 10 min. Evap to dryness and add 30 μl of butaneboronic acid (5 mg/ml in acetone benzene, 2:1). Heat at 60° for 20 min, then evap in vac oven. Add 25 μl of BSTFA (20% in hexane). After 2 hr, evap under vac. Add 7 μl 5% BSTFA soln and seal tube. Analyze the total tube contents by GC-MS (1% Dexsil 300 at 240°).

Comments. The method was used to quantitate $PGF_{2\alpha}$ using a deuterated int stand.[1,2] The method was also used to identify 19-hydroxy PGFs.[3] CI-MS data have been reported for PGF butyl boronate derivs.[4]

References

[1]R. W. Kelley, *Anal. Chem.* **45**, 2079 (1973).

[2]C. Pace-Asciak and L. S. Wolf, *J. Chromatogr.* **56**, 129 (1971).

[3]P. L. Taylor and R. W. Kelley, *FEBS Lett.* **57**, 22 (1975).

[4]A. G. Smith and C. J. W. Brooks, *Biomed. Mass Spectrom.* **4**, 258 (1977).

Derivative Type. Methyl ester, methaneboronate, trimethylsilyl.

Reagent. Diazomethane, methaneboronic acid/pyridine, bis(trimethylsilyl)trifluoroacetamide (BSTFA).

Compound Type. Thromboxane B_2.

Reaction

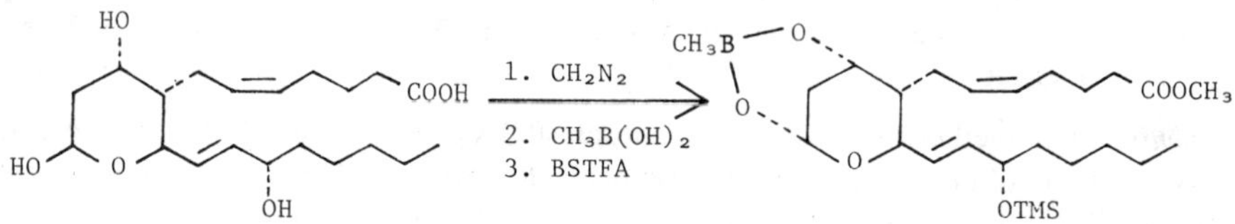

Typical Procedure. Prepare the methyl ester by treatment with CH_2N_2 in Et_2O and evap under N_2. Diss 50 μg of the ester in 25 μl pyridine containing 5 mol equiv of methaneboronic acid and heat 30 min at 85°. Add 10 μl BSTFA and heat 30 min at 85°. Evap and diss deriv in 25 μl EtOAc prior to GC or GC-MS-SIM analysis.

Comments. The deriv gave a poor FID response (less than 25% of the tri-TMS deriv) and a badly tailing solvent front. The *n*-butaneboronate gave a poorer response, and no peaks at all were observed with cyclohexane- and benzeneboronic acids. The methaneboronate, methyl ester, di-TMS deriv was also prepared from the borohydride reduction ($NaBH_4$ in 2-PrOH/MeOH, 1 :1, v/v) product of thromboxane B_2. MS data were given for the derivs.

References

[1]A. G. Smith *et al.*, *J. Chromatogr.* **142**, 533 (1977).

12.2. TRIMETHYLSILYL ESTER DERIVATIVES

12.2.1. Trimethylsilyl Ester Ether Derivatives

Derivative Type. Trimethylsilyl.

Reagent. Bis(trimethylsilyl)trifluoroacetamide (BSTFA)/trimethylchlorosilane (TMCS).

Compound Type. Prostaglandins F_1, F_2.

Reaction

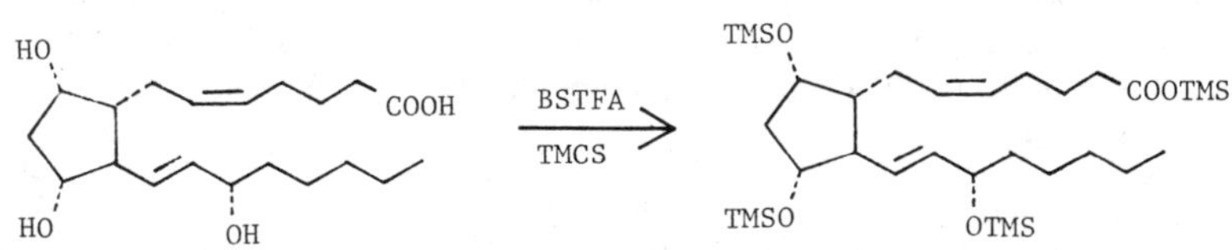

Typical Procedure. Allow mixt of 0.1 ml BSTFA, 0.05 ml TMCS, and 0.5 mg PG to stand at room temp for 2 hr. Analyze 0.5–3 μl by GC.

Comments. MS data for the derivs and deuterated analogs have been reported.[2]

References

[1]F. Vane and M. G. Horning, *Anal. Lett.* **2**, 357 (1969).
[2]B. S. Middleditch and D. M. Desiderio, *Anal. Biochem.* **55**, 509 (1973).

Derivative Type. Trimethylsilyl.

Reagent. Bis(trimethylsilyl)trifluoroacetamide (BSTFA)/piperidine.

Compound Type. Prostaglandins E_1 and E_2.

Reaction

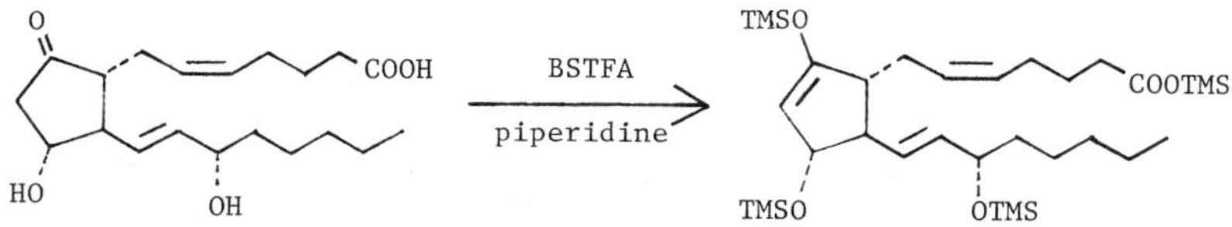

Typical Procedure. Diss 1–10 μg of PGE in 40 μl of BSTFA/piperidine (1 : 1, v/v) and heat at 60° for 1 hr. Analyze by GC-MS.

Comments. MS data were reported to support the deriv structure shown.[1] Application of the method to a mixt of PGA, PGB, PGE, and PGF permitted GC separation of the four members of the 1-series on OV-17 and OV-225 and of the four members of the 2-series on OV-225. The derivs of PGE_2 and PGF_2 were not resolved on OV-17.[2]

References

[1] J. Rosello *et al.*, *Biomed. Mass Spectrom.* **4**, 237 (1977).
[2] J. Rosello *et al.*, *J. Chromatogr.* **122**, 469 (1976).
[3] J. Rosello *et al.*, *J. Chromatogr.* **130**, 65 (1977).

Derivative Type. Trimethylsilyl, 11-piperidyl.

Reagent. Bis(trimethylsilyl)trifluoroacetamide (BSTFA)/piperidine.

Compound Type. Prostaglandins A_1 and A_2.

Reaction

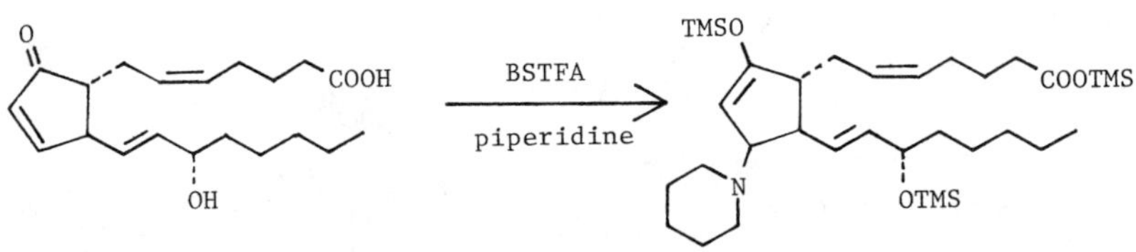

Typical Procedure. Diss 1–10 μg of PGA in 40 μl of BSTFA/piperidine (1 :1, v/v) and heat at 60° for 1 hr. Analyze by GC-MS.

Comments. MS data were reported[1] to support the 11-piperidyl structure shown rather than the earlier postulated[3] 9-piperidyl structure. No piperidine addition was seen with PGB. In the absence of piperidine, PGA formed a di-TMS deriv like that from PGB.

References

[1] J. Rosello *et al.*, *Biomed. Mass Spectrom.* **4**, 237 (1977).
[2] J. Rosello *et al.*, *J. Chromatogr.* **122**, 469 (1976).
[3] J. Rosello *et al.*, *J. Chromatogr.* **130**, 65 (1977).

12.2.2. Methoxime Trimethylsilyl Ester Ether Derivatives

Derivative Type. Methoxime, trimethylsilyl ether, trimethylsilyl ester.

Reagent. Methoxylamine HCl, bis(trimethylsilyl)trifluoroacetamide (BSTFA).

Compound Type. Prostaglandins A, B, and E.

Reaction

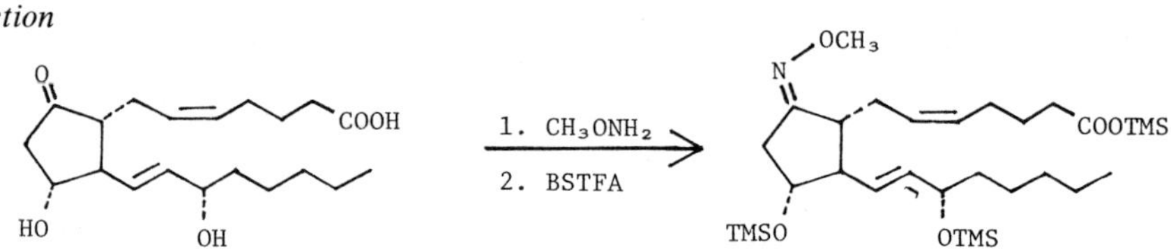

Typical Procedure. Diss 0.5 mg prostaglandin in 0.5 ml pyridine and add 5 mg methoxylamine HCl. Heat at 60° for 2–3 hr; then remove pyridine under N_2. Add 0.1 ml BSTFA. After 1–2 hr at room temp analyze 0.5–3.0 μl aliquots by GC-MS.

Comments. Prostaglandins A_1, E_1, and E_2 gave double peaks presumably due to *syn*- and *anti*-isomers. PGB_1 and 8-*iso*-PGE_1 each gave single peaks. MS data for the derivs was given. The d_3-methoxime, prepared from d_3-methoxylamine,[2] was used as an int stand to quant PGE_1 by GC-MS-SIM.[3]

References

[1]F. Vane and M. G. Horning, *Anal. Lett.* **2**, 357 (1969).
[2]M. Hamberg and B. Samuelsson, *J. Am. Chem. Soc.* **91**, 2177 (1969).
[3]B. Samuelsson *et al.*, *Anal. Biochem.* **38**, 301 (1970).

Derivative Type. Methoxime (MO), trimethylsilyl.

Reagent. Methoxylamine HCl, bis(trimethylsilyl)trifluoroacetamide (BSTFA).

Compound Type. Prostaglandins.

Reaction

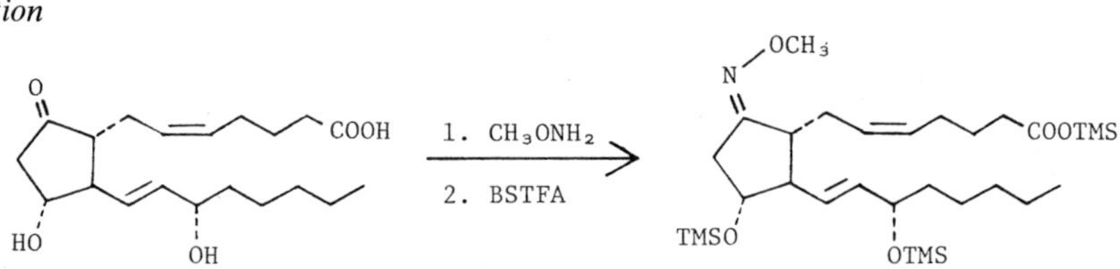

Typical Procedure. Diss 100 μg of prostaglandin in 250 μl pyridine and add 1 mg CH_3ONH_2-HCl. Allow to react at 50° overnight then evap to dryness under N_2. Add 100 μl BSTFA and heat at 50° for 3 hr. Analyze by GC-MS.

Comments. The MO-TMS derivs of PG's A, B, D, and E of the 1- and 2- series were prepared and compared with TMS and ethoxime (EO)-TMS derivs.[1–3] TMS derivs of $PGF_{1\alpha}$, $PGF_{2\alpha}$, and $PGF_{2\beta}$ were also prepared omitting the MO step. The MO-TMS and EO-TMS derivs of the PGEs and PGDs were more stable than the TMS derivs. MS data for the d_9-TMS derivs and the d_9-TMS ether, d_0-TMS ester (prepared by on-column TMS ester interchange with TMSI were reported.[2] A thorough discussion of the MS behavior of the derivs was subsequently reported.[4]

References,

[1]B. S. Middleditch and D. M. Desiderio, *Prostaglandins* **2**, 115 (1972).

[2]B. S. Middleditch and D. M. Desiderio, *Prostaglandins* **4**, 31 (1973).

[3]B. S. Middleditch and D. M. Desiderio, *Prostaglandins* **4**, 459 (1973).

[4]B. S. Middleditch and D. M. Desiderio, *J. Org. Chem.* **38**, 2204 (1973).

12.3. OTHER VOLATILE PROSTAGLANDIN DERIVATIVES

Derivative Type. Pentafluorobenzyl ester, trimethylsilyl ether.

Reagent. Pentafluorobenzyl bromide (PFB-Br), bis(trimethylsilyl)acetamide (BSA).

Compound Type. Prostaglandin $F_{2\alpha}$.

Reaction

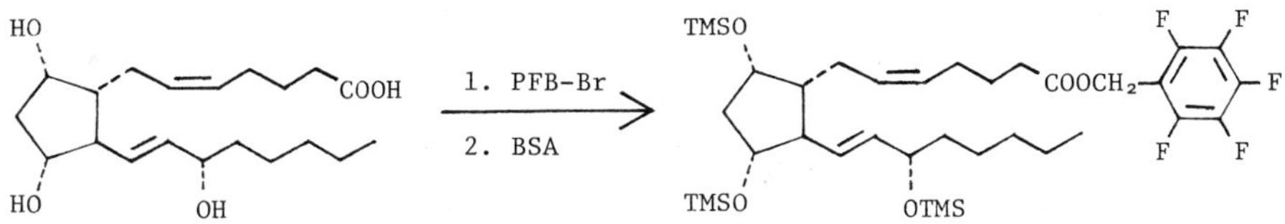

Typical Procedure.[1] Diss a submicrogram sample of $PGF_{2\alpha}$ in 10 μl of a soln of PFB-Br (5.8 mg, 0.022 mmol) in CH_3CN and 10 μl of a soln of diisopropylethylamine (0.886 mg, 0.007 mmol) in CH_3CN. Heat at 40° for 5 min and evap to dryness under N_2. Repeat the above procedure on the residue and again evap to dryness under N_2. Diss the residue in 10 μl BSA and, after 5 min at room temp, analyze by ECD-GC.

Comments. The deriv was detectable down to 12.5 pg.

References

[1]J. A. F. Wickramasinghe *et al.*, *J. Pharm. Sci.* **62**, 1428 (1973).

[2]J. A. F. Wickramasinghe and R. S. Shaw, *Biochem. J.* **141**, 179 (1974).

Derivative Type. Carbon skeleton.

Reagent. Hydrogen/palladium on calcium carbonate.

Compound Type. Prostanoids.

Reaction

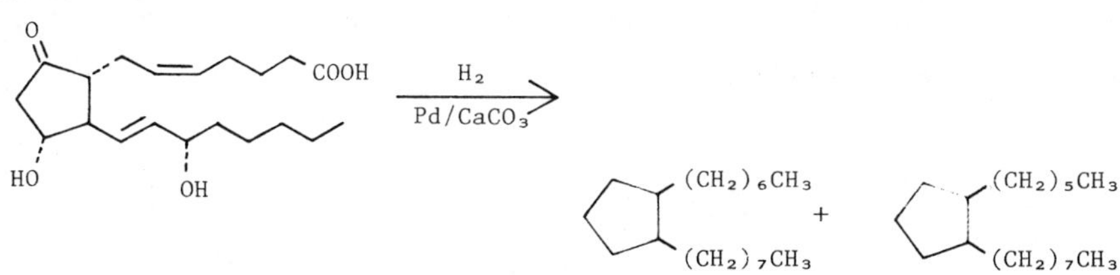

Typical Procedure. Mix sample (50–100 mg for unprocessed biological sample, 20–50 mg for extracts, 1–20 mg for pure compds) with 1 g 5% $Pd/CaCO_3$. Grind solid samples with catalyst in mortar; mix organic ext with catalyst and dry under vac; mix aq sample with catalyst and freeze dry. Place sample mixt in oven tube connected to solid CO_2 cooled condensing tube. Purge tube with inert gas then heat sample mixt in a stream of H_2 (10 ml/min) from 25° to 330° at 10°/min then hold at 330° for 1 hr. Cool and rinse product from condenser tube with 2 ml $CHCl_2$. Filter through cotton to remove H_2O and stir with 2 ml 95–98% H_2SO_4 for 18 hr. Separate org layer and analyze by GC-MS.

Comments. The method was applied to screening biological samples for prostanoids. Authentic PGE_2 gave predominantly hydrocarbons resulting from initial decarboxylation with small amounts of the parent hydrocarbons (*cis*- and *trans*-isomers).

References

[1]I. T. Harrison *et al.*, *Chem. Ind.* 219 (1976).

12.4. CHROMOGENIC PROSTAGLANDIN DERIVATIVES

Derivative Type. *p*-Nitrobenzyl ester.

Reagent. *p*-Nitrobenzyl bromide/diisopropylethylamine.

Compound Type. Prostaglandins A_2, E_2, $F_{2\alpha}$.

Reaction

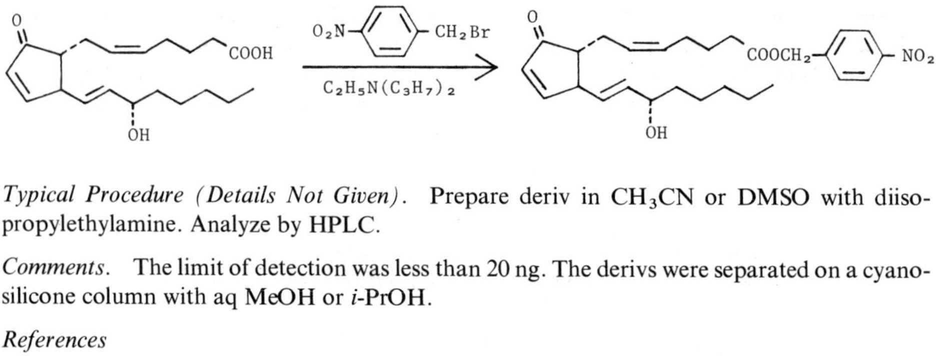

Typical Procedure (Details Not Given). Prepare deriv in CH_3CN or DMSO with diisopropylethylamine. Analyze by HPLC.

Comments. The limit of detection was less than 20 ng. The derivs were separated on a cyanosilicone column with aq MeOH or *i*-PrOH.

References

[1]W. Morozowich, *A. Ph. A. Acad. Pharm. Sci. Abstr.*, p. 69, April 1972.

Derivative Type. *p*-Nitrophenacyl esters.

Reagent. *p*-Nitrophenacyl bromide/diisopropylethylamine.

Compound Type. Prostaglandins.

Reaction

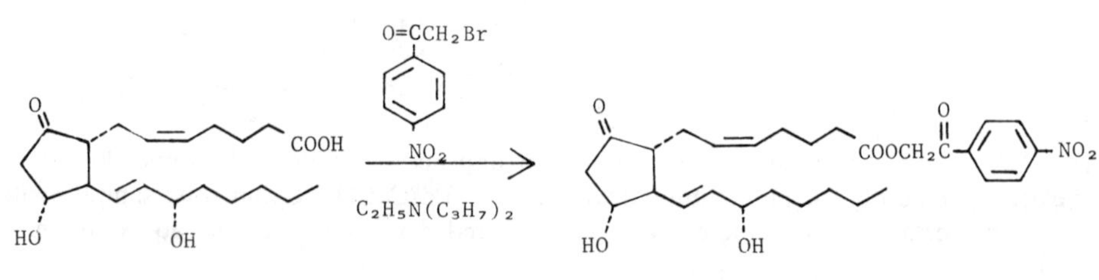

Typical Procedure. Treat PGE_2 samples (0.4–1.13 mg) in 1 ml acetonitrile with 9.60 mg of *p*-nitrophenacyl bromide and 2.9 μl of diisopropylethylamine. After 15 min at room temp remove aliquots and inject into HPLC.

Comments. The method was used for chromogenic deriv of 22 prostaglandins and analogs for HPLC analysis. The derivs were detectable by UV to about 1 ng. Merritt and Bronson[2,3] have used the derivs for separations of prostaglandins on silver loaded HPLC columns.

References

[1]W. Morozowich and S. L. Douglas, *Prostaglandins* **10**, 19 (1975).
[2]M. V. Merritt and G. E. Bronson, *Anal. Chem.* **48**, 1851 (1976).
[3]M. V. Merritt and G. E. Bronson, *Anal. Biochem.* **80**, 392 (1977).

Derivative Type. *p*-Bromophenacyl ester.

Reagent. *p*-Bromophenacyl bromide/diisopropylethylamine.

Compound Type. Prostaglandins (PGD_2, PGE_2, $PGF_{2\alpha}$).

Reaction

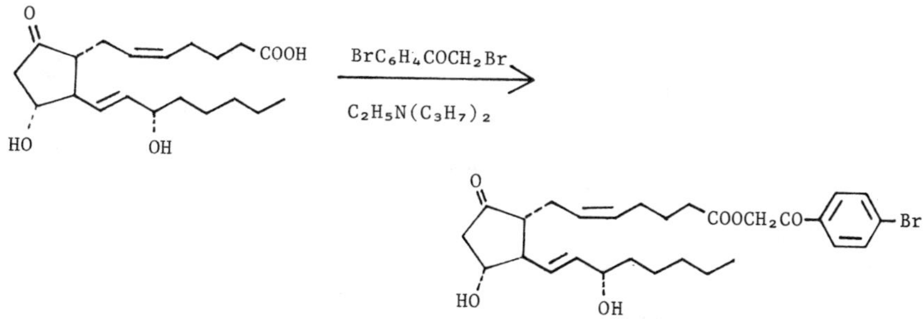

Typical Procedure.[1] Diss 500 μg or more of PG in 1.0 ml anhyd CH_3CN containing a threefold mol excess of *p*-bromophenacyl bromide. Add 2 μl diisopropylethylamine and allow to stand 1 hr at 25°. Analyze by HPLC.

Comments. For quantities of 1–100 μg the vol of CH_3CN was reduced to 0.1–0.2 ml and the amine to 0.5 μl. The threefold mol excess of deriv reagent was maintained. The method was used to quantitate the PGs to less than 3 μg using an alkylsilane reverse phase column and 254 nm detection. Attempts to separate the three PGs as *p*-nitrophenacyl esters were unsuccessful. The method was used to quantitate PGs in incubation mixtures.[2] A similar method had been reported earlier.[3]

References

[1]F. A. Fitzpatrick, *Anal. Chem.* **48**, 499 (1976).
[2]F. A. Fitzpatrick, *J. Pharm. Sci.* **65**, 1609 (1976).
[3]W. Morozowich and S. Douglas, *Prostaglandins* **10**, 19 (1975).

Derivative Type. Methyl ester, *p*-nitrobenzyloxime.

Reagent. Diazomethane, *p*-nitrobenzylhydroxylamine HCl (PNBA).

Compound Type. Prostaglandins and thromboxanes.

Reaction

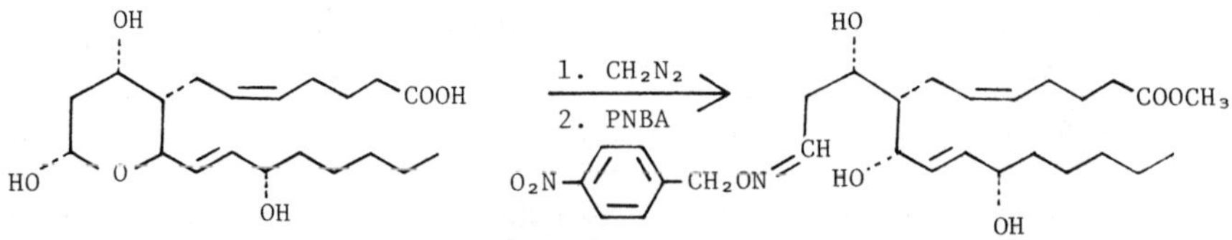

Typical Procedure. Prepare methyl ester by treating with excess CH_2N_2 in Et_2O for 5 min. To 10 μg of ester add 0.2 ml anhyd pyridine containing a 10-fold mol excess of *p*-nitrobenzylhydroxylamine HCl. Heat at 40° for 2 hr or allow to stand overnight at room temp. Evap under N_2 and diss deriv in MeOH for HPLC analysis.

Comments. The deriv was used for HPLC analysis of prostaglandins and thromboxanes. Detection limits ranged from 25 to 225 ng for the 11 compds examined. Pentafluorobenzyl esters and oximes were also examined but showed lower HPLC detection sensitivities.

References

[1]F. A. Fitzpatrick *et al.*, *Anal. Chem.* **49**, 1032 (1977).

13. Carbohydrates

The first application of GC to carbohydrates was reported in 1958 and described efforts to separate the fully methylated methyl glycosides of some monosaccharides.[1] Since that beginning, carbohydrates have been subjected to GC analysis using a variety of derivatives. The turning point toward widespread application of GC to carbohydrates was the publication in 1963 by Sweeley, Bentley, Makita, and Wells of the use of trimethylsilyl derivatives for GC analysis of carbohydrates including mono- to tetrasaccharides.[2] This paper has become a "citation classic" having been cited 1258 times in the period through 1975.[3] In spite of the numerous other derivatives that have been applied to carbohydrates the trimethylsilyl derivatives remain the most widely utilized. The analysis of carbohydrates by GC, and the necessary derivatives for such analysis, have been the subject of several reviews.[4–10]

The initially used permethyl derivatives of monosaccharide methyl glucosides had relatively long retention times. In an effort to find more volatile derivatives, the monosaccharides were reduced with sodium borohydride and acetylated to give alditol acetate derivatives[11] (p. 545). The alditol acetates had good GC properties and, with subsequent improvements in GC columns, allowed separation of relatively complex mixtures. The GC separation of various methylated alditols as the acetate derivatives[12] facilitated the use of "methylation analysis" of polysaccharides (p. 586). This technique, involving methylation of the free hydroxyl groups in the polysaccharide followed by hydrolysis and analysis of the resulting partially methylated monosaccharides, allows study of the sugar composition and positions of glycosidic linkages in polysaccharides.[13]

Although the alditol acetates had favorable GC properties, the trimethylsilyl derivatives of carbohydrates were more volatile and became the derivative of choice for most carbohydrate analysis (p. 550). Substitution of the trifluoroacetyl group for the acetyl in carbohydrate acetate derivatives yields a derivative more volatile than either the acetyl or trimethylsilyl derivative. The trifluoroacetates have experienced relatively little use compared to the other carbohydrate derivatives, but application of more recently developed reagents (e.g., MBTFA,[14] p. 550) could lead to wider use. The volatility of the TMS and TFA derivatives have permitted GC of oligosaccharides (p. 581) as well as the monomers. Oligosaccarides as large as an octaose have been separated as the TMS derivatives.[15] Dimethylsilyl (DMS) derivatives (p. 555) are more volatile than the corresponding TMS derivatives[16] but few applications of DMS derivs have been reported. The DMS derivs were used in the analysis of cyclodextrins containing up to eight anhydroglucose units.[17]

Despite their widespread use, these carbohydrate derivatives have limitations: a propensity to yield multiple derivatives or, when a single derivative is formed, the failure to yield a unique derivative. Sugars with anomeric centers can yield mixtures of alpha- and beta-anomeric derivatives. For example, glucose can yield two TMS derivatives.

The possibility of both pyranose and furanose forms of sugars increases the number of potential derivatives. For example, galactose forms four different TMS derivatives due to pyranose, furanose, and anomeric forms.[18]

Since the sugar acids can exist in either open chain form or as 1,4- or 1,5-lactones, the combinations of lactones and anomers yield even more possibilities for multiple derivatives with the sugar acids.

One solution to the problem of multiple derivatives has been to reduce sugar carbonyl

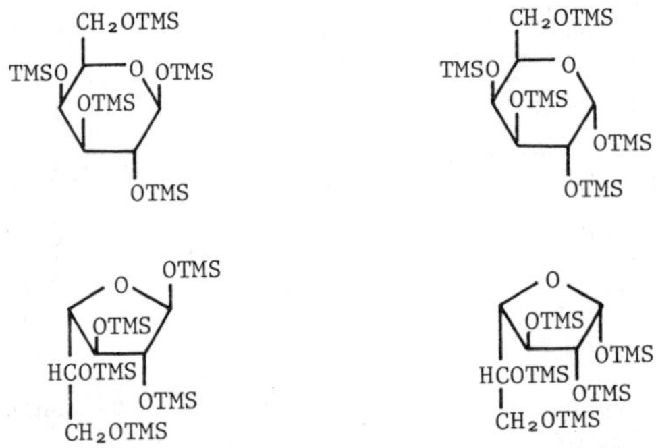

groups to hydroxyls as, for example, in the formation of alditol acetate derivatives of aldoses. While this approach does indeed eliminate the possibility of anomeric derivatives, the reduction creates new problems. Since the 1-position is no longer unique in the alditols, different aldoses can yield the same alditol upon reduction. For example, both D-arabinose and D-lyxose yield the same alditol D-arabitol. The same holds for the hexoses although in this case two D-aldoses can yield enantiomeric alditols. Thus D-glucose and D-gulose yield, respectively, D- and L-glucitol. The derivatives of these enantiomeric alditols are not GC separable (on the normal nonasymmetric phases); hence, for GC analysis, the two aldoses yield the same derivative. The problem is further complicated when the reduction approach is applied to the ketoses.

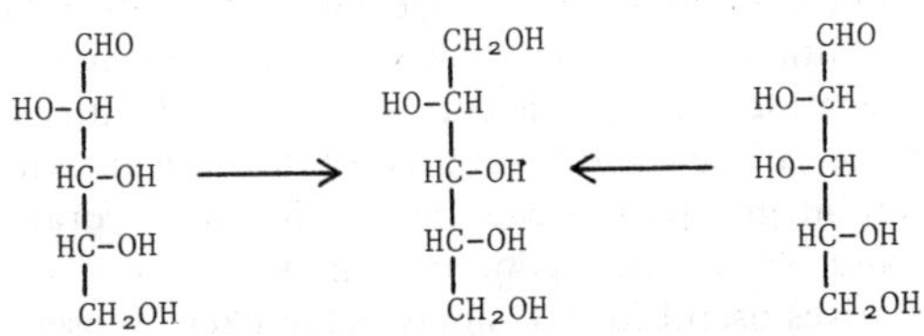

D-arabinose   D-arabitol   D-lyxose

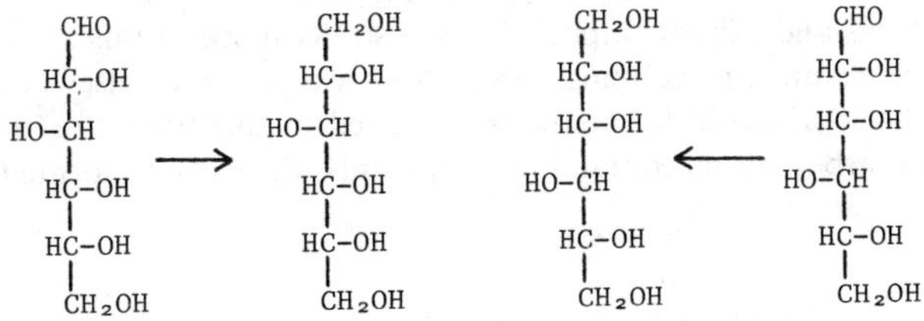

D-glucose   D-glucitol   L-glucitol   D-gulose

Reduction of an aldose yields a primary alcohol, but reduction of a ketose yields a secondary alcohol that, in the sugars, introduces another asymmetric center. Chemical reduction of a ketose therefore yields a mixture of two alditols. Thus there is not only the problem of two ketoses (e.g., fructose and sorbose) yielding the same alditol (glucitol), but each ketose yields a mixture of two alcohols.

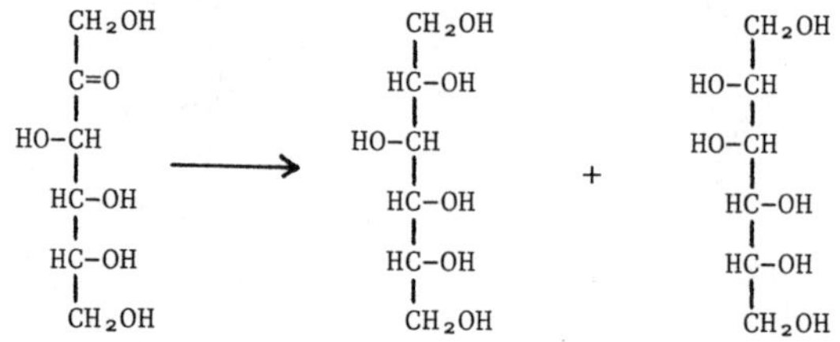

D-fructose D-glucitol D-mannitol

In summary, the reduction approach to eliminating anomeric derivatives solves one problem but creates new ones.

Other approaches that have been taken to eliminate the anomeric center include the formation of oxime derivatives of the carbonyl. This approach is not entirely satisfactory because the oximes can exist as GC separable *syn-* and *anti-*isomers. Perhaps the closest approach to a solution has been the formation of aldononitrile acetates (Wohl derivatives) from aldoses (p. 558). These derivatives preserve the unique 1-position with the formation of a nonisomeric functional group (nitrile). Thus each aldose gives a single unique derivative. Unfortunately, since the Wohl derivatives do not give volatile products with ketoses, these derivatives do not completely solve the problem of obtaining a single unique derivative for each of the monosaccharides. The ultimate solution may lie, not in new derivatization methods, but in high resolution (capillary) GC to separate the multiple derivatives formed with existing methods.

The low volatility of the carbohydrates that necessitates derivatization for GC analysis also makes derivatization an integral part of MS analysis of carbohydrates. The MS analysis of carbohydrates[10,19,20] and carbohydrate containing natural products[21] has been reviewed. In general the same derivatives useful for GC analysis are likewise useful for MS and GC-MS analysis. HPLC analysis of carbohydrates has largely utilized postcolumn derivatization,[22] although *p*-nitrobenzoates (p. 559) have been formed prior to HPLC. Derivatization in HPLC analysis is relatively new, and it is likely that other existing chromogenic and fluorogenic derivatives are applicable to carbohydrates.

Derivatives that have been applied to analysis of glycoside conjugates (p. 590), particularly glucuronides of drugs and other compounds are also discussed in this section. Some glycosides, however (for example glycolipids, see p. 444), are presented elsewhere. The reader is reminded that other derivatization methods discussed in this book are also applicable to carbohydrates and carbohydrate containing compounds; these can be located via the indexes.

References

[1] A. G. McInnes *et al.*, *J. Chromatogr.* **1**, 556 (1958).

[2] C. C. Sweeley *et al.*, *J. Am. Chem. Soc.* **85**, 2497 (1963).

[3] C. C. Sweeley, *Curr. Cont.* **1977** (43), 9 (1977).

[4] J. M. Berry, *Adv. Chromatogr.* **2**, 271 (1966).

[5] C. C. Sweeley *et al.*, *Meth. Enzymol* **8**, 95 (1966).

[6] J. H. Soneker, in *Biomedical Applications of Gas Chromatography*, Vol. 2, ed. by H. A. Szymanski, Plenum, New York, NY, 1968, p. 87.

[7] J. R. Clamp *et al.*, *Meth. Biochem. Anal.* **19**, 229 (1971).

[8] K. M. Brobst, *Meth. Carbohydr. Chem.* **6**, 3 (1972).

[9] R. A. Laine *et al.*, *Meth. Enzymol.* **28B**, 159 (1972).

[10]G. G. S. Dutton, *Adv. Carbohydr. Chem. Biochem.* **28**, 11 (1973); **30**, 9 (1974).
[11]C. T. Bishop and F. P. Cooper, *Can. J. Chem.* **38**, 388 (1960).
[12]H. Björndal *et al.*, *Acta Chem. Scand.* **21**, 1801 (1967).
[13]H. Björndal *et al.*, *Angew Chem. Int. Ed.* **9**, 610 (1970).
[14]J. E. Sullivan and L. R. Schewe, *J. Chromatogr. Sci.* **15**, 196 (1977).
[15]J. B. Beadle, *J. Agr. Food Chem.* **17**, 904 (1969).
[16]W. R. Supina *et al.*, *J. Am. Oil Chem. Soc.* **44**, 74 (1967).
[17]J. B. Beadle, *J. Chromatogr.* **42**, 201 (1969).
[18]T. E. Acree *et al.*, *Carbohydr. Res.* **6**, 498 (1968).
[19]N. K. Kochetkov and O. S. Chizhov, *Adv. Carbohydr. Chem.* **21**, 39 (1966).
[20]R. Radford and D. C. DeJongh, in *Biomedical Applications of Mass Spectrometry*, ed. by G. R. Waller, Wiley, New York, NY, 1972, p. 313.
[21]S. Hanessian, *Meth. Biochem. Anal.* **19**, 105 (1971).
[22]J. F. Lawrence and R. W. Frei, *Chemical Derivatization in Liquid Chromatography*, Elsevier, New York, NY, 1976, p. 131.

13.1. DERIVATIVES OF ALDOSES

13.1.1. Methyl Derivatives

Derivative Type. Permethyl.

Reagent. Methyl iodide/silver oxide.

Compound Type. Mono-, di-, and trisaccharides.

Reaction

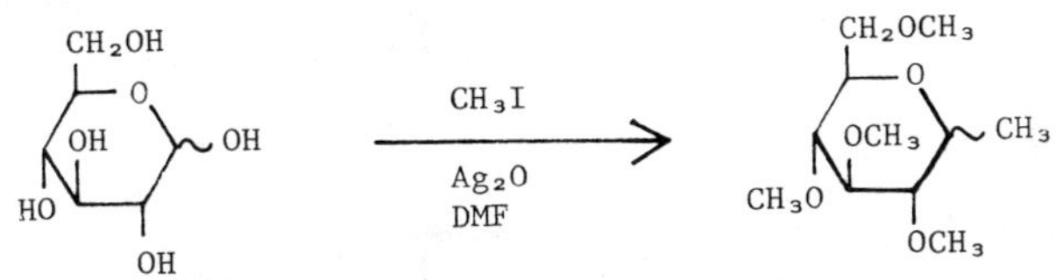

Typical Procedure. Diss the sugar in 10–25 times its wt of DMF with protection from moisture. Heat if necessary to diss. Add 3 equiv CH_3I per OH group and 2 equiv Ag_2O per OH at room temp. Stir at room temp 16 hr then filter and wash salts with DMF. Evap to a small vol under vac at 100°. Add $CHCl_3$ and centrifuge to remove ppt. Wash $CHCl_3$ several times with H_2O. Analyze by GC.

Comments. The deriv method was based upon the method of Kuhn *et al.*,[3] which had also been used in GC analysis by others.[4,5] The method gave anomeric methyl glycosides that were generally GC separable. The MS behavior of the derivs, studied with the aid of deuteromethyl derivs, has been described.[6]

References

[1]M. Gee and H. G. Walker, *Anal. Chem.* **34**, 650 (1962).
[2]H. G. Walker *et al.*, *J. Org. Chem.* **27**, 2100 (1962).
[3]R. Kuhn *et al.*, *Angew. Chem.* **67**, 32 (1955).
[4]A. G. McInnes *et al.*, *J. Chromatogr.* **1**, 556 (1958).
[5]H. W. Kircher *et al.*, *Anal. Chem.* **32**, 1103 (1960).
[6]N. K. Kochetkov and O. S. Chizhov, *Tetrahedron* **21**, 2029 (1965).

Derivative Type. Permethyl.

Reagent. Methylsulfinylcarbanion/methyl iodide.

Compound Type. Carbohydrates.

Reaction

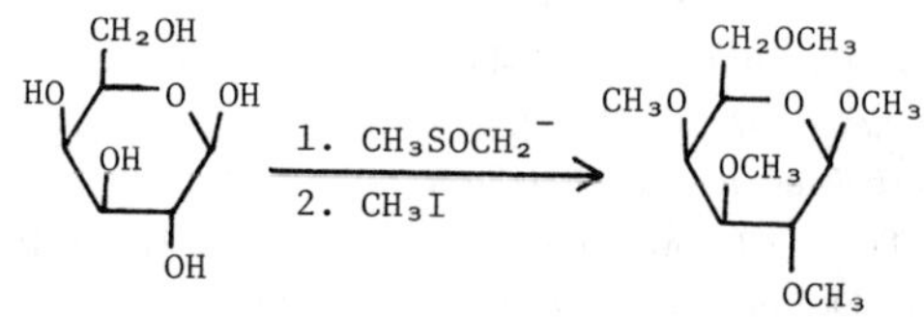

Typical Procedure. Diss carbohydrate sample in sm amt DMSO and add an amt of methylsulfinylcarbanion equivalent to the hydroxyl content of the sample. (The methylsulfinylcarbanion is prepared by diss a weighed amt of NaH in DMSO.[2]) Stir under N_2 for 10 min then add excess CH_3I. Stir under N_2 for 20 min, dil with H_2O, and ext with $CHCl_3$. Wash ext with H_2O and evap to dryness. Diss residue in Et_2O-pet ether and wash with H_2O to remove remaining DMSO. Evap to yield cryst deriv.

Comments. Reference 1 was the original report of Hakomori permethylation of carbohydrates using methylsulfinyl carbanion. The method has been applied to monosaccharides, oligosaccharides, and polysaccharides. MS data for the permethyl derivs of 25 oligosaccharides have been reported.[3]

References

[1]S. Hakomori, *J. Biochem.* **55**, 205 (1964).

[2]E. J. Corey and M. Chaykovsky, *J. Am. Chem. Soc.* **84**, 886 (1962).

[3]J. Moor and E. S. Waight, *Biomed. Mass Spectrom.* **2**, 36 (1975).

Derivative Type. O-Methyl.

Reagent. Potassium/liquid ammonia, methyl iodide.

Compound Type. Carbohydrates.

Reaction

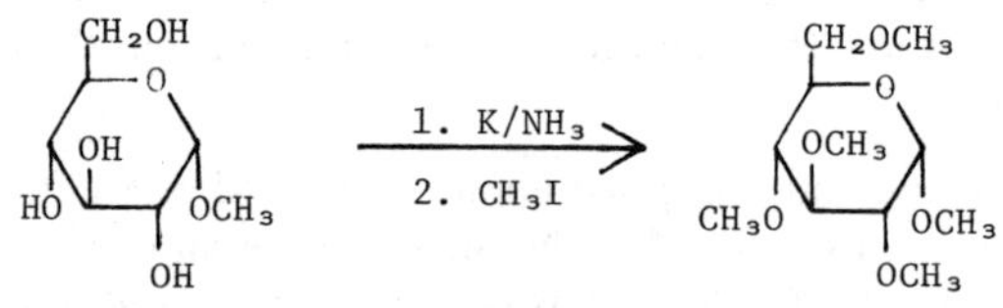

Typical Procedure. Diss carbohydrate in liquid NH_3 and add to a soln of K in liquid ammonia. Add a 5% excess of CH_3I and allow to stand until the NH_3 evaporates. Ext the residue with $CHCl_3$ and evap the ext to yield the permethyl deriv.

Comments. Sugars containing free (or potential) aldehyde or ketone groups were converted to amines by K/NH_3. The alkylations could also be carried out with KNH_2 in NH_3. The formation of KNH_2 in the K/NH_3 soln was catalyzed by addition of Fe_2O_3 (a rusty nail). This method of alkylation has apparently not been examined for analytical use but might be useful.

References

[1]I. E. Muskat, *J. Am. Chem. Soc.* **56**, 693 (1934).

[2]I. E. Muskat, *J. Am. Chem. Soc.* **56**, 2449 (1934).

13.1.2. Acetyl Derivatives

Derivative Type. Acetate.

Reagent. Acetic anhydride (Ac_2O)/pyridine.

Compound Type. Partially methylated pentoses and hexoses.

Reaction

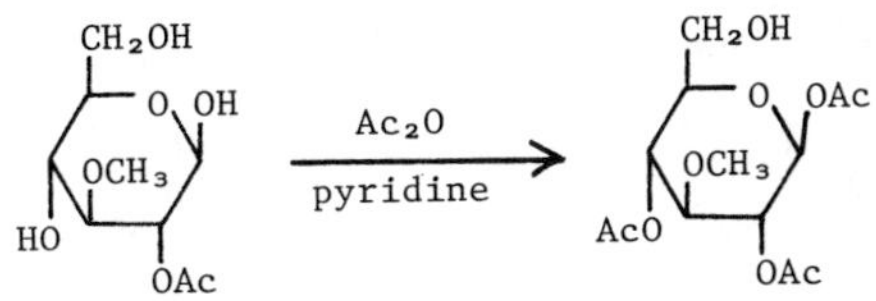

Typical Procedure. Mix 1 mg sugar with 30 μl pyridine and 15 μl Ac_2O and seal in an ampoule. Heat on a steam bath 1 hr. Purify the deriv by preparative GC. Analyze by MS.

Comments. Derivs of various mono-, di-, tri-, and tetra-*O*-methyl derivs of xylose, glucose, and mannose were studied.[1] Nonmethylated sugars have also been studied as cyclic[2] and acyclic[3] peracetate derivs. GC data were reported for several sugar peracetates.[4]

References

[1]D. C. DeJongh and K. Biemann, *J. Am. Chem. Soc.* **85**, 2289 (1963).

[2]K. Biemann *et al.*, *J. Am. Chem. Soc.* **85**, 1763 (1963).

[3]D. C. DeJongh, *J. Org. Chem.* **30**, 453 (1965).

[4]W. J. A. VandenHeuval and E. C. Horning, *Biochem. Biophys. Res. Commun.* **4**, 399 (1961).

Derivative Type. Alditol acetate.

Reagent. Sodium borohydride, acetic anhydride (Ac_2O).

Compound Type. Sugars.

Reaction

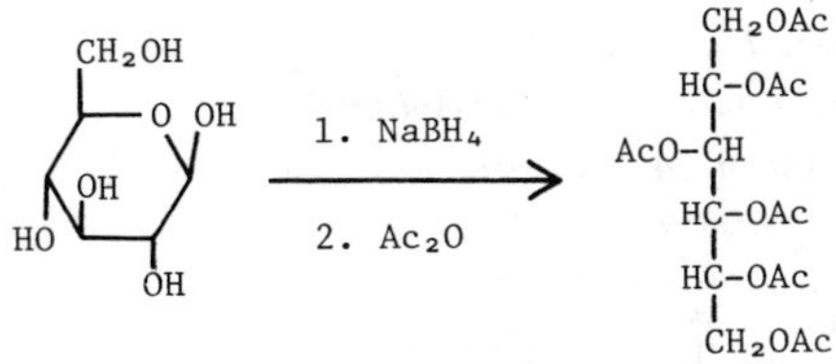

Typical Procedure. Mix 5 ml of sugar sample soln (0.01 *M*) with 5 ml of 0.01 *M* glucose (stand), and 40 mg $NaBH_4$ and allow to stand 1 hr. Add a few drops of HOAc then evap to a syrup. Add 3×3 ml 1% HCl in MeOH and evap each under vac (20 Torr, 40°) to remove the borate as methyl borate. Diss the residue in 5 ml Ac_2O to which 0.12 ml conc H_2SO_4 had been added. Heat at 80° for 1 hr. Pour the cooled soln into a previously frozen mixt of 5 ml pyridine and 10 ml H_2O. After 10 min ext with $CHCl_3$. Wash ext with H_2O, 1 *N* H_2SO_4, 5% $NaHCO_3$, and H_2O. Dry over Na_2SO_4 and evap to a syrup. Diss in dry $CHCl_3$ and analyze by GC.

Comments. The derivs of the hexoses, the pentoses, five deoxy sugars and erythrose were separated as their alditol acetates.[1] GC data were reported for the derivs of 10 alditols.[2] TMS alditol derivs were also examined but adequate GC separation of the derivs was not achieved.[2]

References

[1]D. H. Shaw and G. W. Moss, *J. Chromatogr.* **41**, 350 (1967).

[2]J. S. Sawardeker *et al.*, *Anal. Chem.* **37**, 1602 (1965).

Derivative Type. Alditol acetate.

Reagent. Sodium borohydride, acetic anhydride (Ac_2O).

Compound Type. Neutral sugars.

Reaction

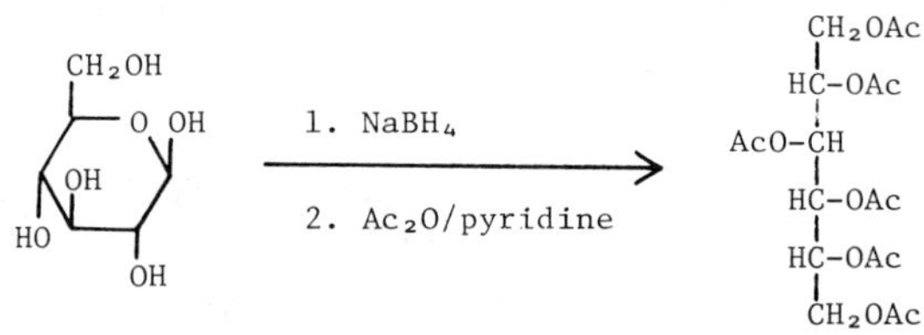

Typical Procedure.[4] Diss sugar sample in 0.5 ml H_2O and add $NaBH_4$ (5 mg/mg sugar) in 0.1 ml H_2O dropwise at 0°. Store overnight at 4° then destroy excess $NaBH_4$ by addn of HOAc. Pass soln through a column of Dowex 50 (H form) to remove Na. Collect effluent and washings and evap to dryness under vac. Add 3 ml anhyd MeOH, heat at 50° for 30 sec, then evap to dryness at room temp (to remove borate as methyl borate). Repeat twice. Add 300 μl pyridine and 200 μl Ac_2O and reflux (100° bath) 20 min. Evap under vac over NaOH and H_2SO_4. Diss residue in pyridine and analyze by GC.

Comments. The deriv has been used for analysis of sugars in many sample types, e.g., glycoproteins,[4,7] soil hydrolysates,[5] and wood pulp.[6] Sorbitol has been analyzed as the hexaacetate.[3]

References

[1]S. W. Gunner *et al.*, *Chem. Ind* 255 (1961).

[2]S. W. Gunner *et al.*, *Can. J. Chem.* **39**, 1892 (1961).

[3]J. A. Hause *et al.*, *Anal. Chem.* **34**, 1567 (1962).

[4]J. H. Kim *et al.*, *Anal. Biochem.* **20**, 258 (1967).

[5]J. M. Oades, *J. Chromatogr.* **28**, 246 (1967).

[6]E. P. Crowell and B. B. Burnett, *Anal. Chem.* **39**, 121 (1967).

[7]W. F. Lehnhardt and R. J. Winzler, *J. Chromatogr.* **34**, 471 (1968).

Derivative Type. Benzylidene, *O*-acetyl.

Reagent. Benzaldehyde/zinc chloride, acetic anhydride (Ac_2O)/pyridine.

Compound Type. Hexopyranosides.

Reaction

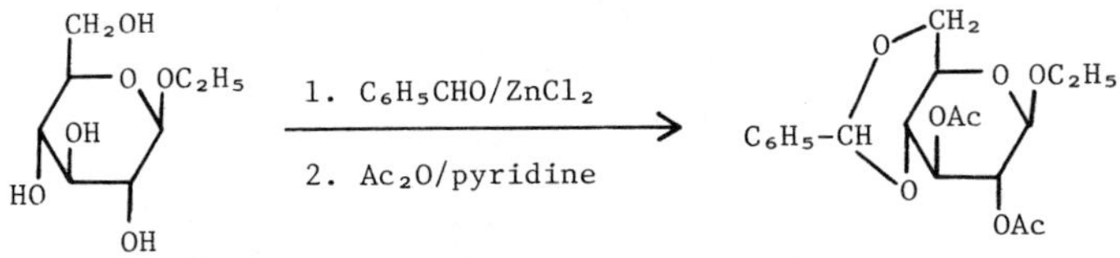

Typical Procedure. Prepare the benzylidene derivs by stirring a 4.8 g sample in benzaldehyde (30 ml) with zinc chloride (1 g) for 3 days at room temp. Add 300 ml $CHCl_3$ and wash with 2×50 ml ice H_2O and 50 ml aq $NaHCO_3$. Evap and treat the residue with Ac_2O/pyridine (1 :1, v/v) at room temp overnight. Add to an equal vol of ice H_2O and an equal vol of dil H_2SO_4. Ext with $CHCl_3$ and wash ext with aq $NaHCO_3$. Dry over Na_2SO_4 and evap. Analyze by MS.

Comments. Acetylation was also performed with *N*-acetyl imidazole, the preparation of which was given. Extensive MS data were reported for a large number of the derivs.

References

[1]C. Bosso *et al., Org. Mass Spectrom.* **12**, 493 (1977).

Derivative Type. Butaneboronate, acetate.

Reagent. Butaneboronic acid, acetic anhydride (Ac_2O).

Compound Type. Carbohydrates.

Reaction

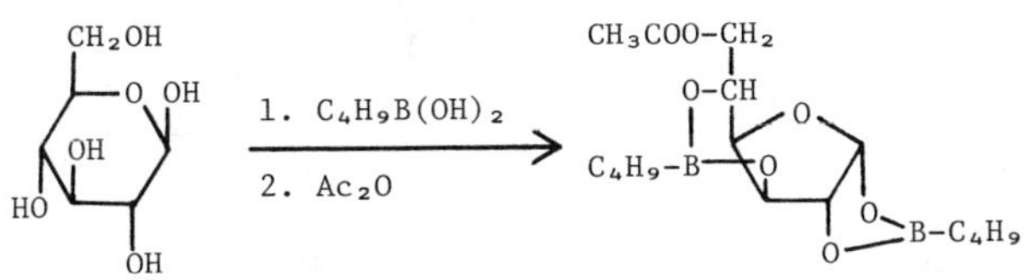

Typical Procedure. Suspend 2 mg (0.01 mmol) of hexose in 1 ml dry pyridine and add 10 mg (0.1 mmol) butaneboronic acid. Shake 2 hr then add 50 μl (0.5 mmol) Ac_2O. After 1 hr, analyze by GC-MS.

Comments. The derivs gave intense M-C_4H_9 ions in which the charge was stabilized by interaction of the acetate groups with the boron atom. The corresponding methane-boronate derivs were also examined. The above "equilibrium boroacetylation" procedure gave single or multiple products depending upon the structure of the hexose. Preferential 6-acetylation was obtained by treating the hexose (5 mg) with a soln (1 ml) of 50% AcOH at 100° for 12 hr. The soln of 6-acetyl deriv was evapd to dryness and the residue treated with the boronic acid in pyridine.

References

[1]J. Wiecko and W. R. Sherman, *J. Am. Chem. Soc.* **98**, 7631 (1976).

13.1.3. Trifluoroacetyl Derivatives

Derivative Type. Trifluoroacetyl.

Reagent. Trifluoroacetic anhydride (TFAA)/sodium trifluoroacetate.

Compound Type. Carbohydrates.

Reaction

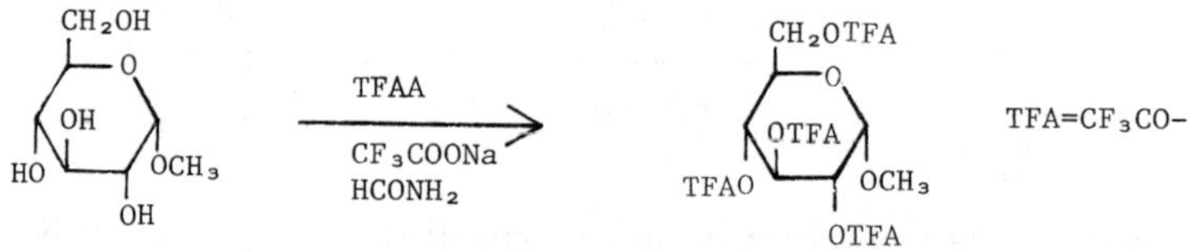

Typical Procedure. Diss 1 mg of the sugar and 0.3 mg sodium trifluoroacetate in 50 μl formamide with heating. Cool to room temp and add 0.2 ml TFAA with cooling in ice. Allow to stand 10 min at room temp then analyze by GC.

Comments. The method was applied to a variety of carbohydrates including mono- and disaccharides, glycosides, amino sugars, and cyclitols.

References

[1]T. Ueno *et al*., *Agr. Biol. Chem*. **31**, 1189 (1967).

Derivative Type. Trifluoroacetyl (TFA).

Reagent. Trifluoroacetic anhydride (TFAA)/sodium trifluoroacetate.

Compound Type. Carbohydrates.

Reaction

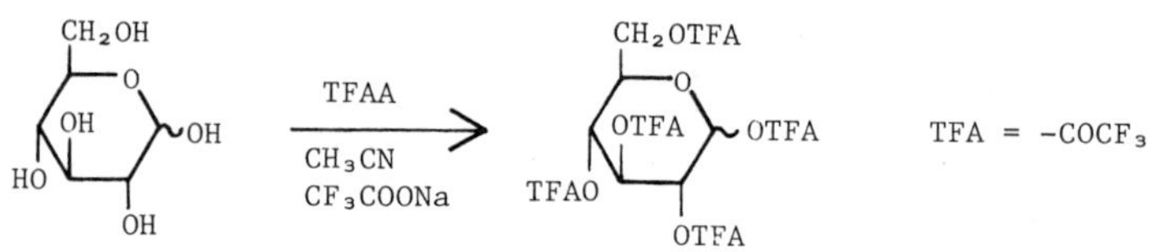

Typical Procedure. Diss 4–5 mg of the sugar in 0.2 ml CH_3CN/TFAA (1:1, v/v) containing 10% (w/v) sodium trifluoroacetate and heat at 50–60°. Analyze by GC.

Comments. The method was applicable to mono-, di-, and trisaccharides including amino-sugars and aminocyclitols.

References

[1]M. Vilkas *et al*., *Tetrahedron Lett*. **14**, 1441 (1966).

Derivative Type. Alditol trifluoroacetate.

Reagent. Sodium borohydride, trifluoroacetic anhydride.

Compound Type. Aldoses.

Reaction

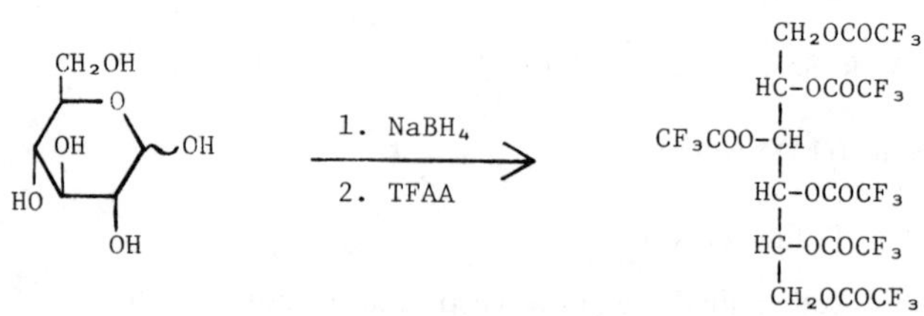

Typical Procedure.[1] To 0.5 ml of an aq soln of 10–500 μg aldoses add 0.5 ml 1% aq $NaBH_4$. Allow to stand 30 min at room temp then destroy excess borohydride by adding 0.5 ml Amberlite CG-120 (H form). Remove the resin by filtration then evap to dryness. Add 1 ml MeOH and evap to remove borate as methyl borate. Repeat 2–3 times then dry under vac. Add 0.1 ml EtOAc and 0.1 ml TFAA and allow to stand at room temp 30 min. Analyze 1–2 μl by GC.

Comments. The TFA derivs of 11 alditols that constitute all of the theoretically possible tetritols, pentitols, and hexitols were separated by GC.[2] The derivs were stable in the presence of excess TFAA but were quite moisture labile when isolated.[1] The TFA derivs have been used for analysis of sugar alcohols in urine by GC-MS.[3]

References

[1]T. Imanari *et al.*, *Chem. Pharm. Bull.* **17**, 1967 (1969).
[2]M. Matsui *et al.*, *Chem. Pharm. Bull.* **16**, 1383 (1968).
[3]H. Haga *et al.*, *Chem. Pharm. Bull.* **20**, 1805 (1972).

Derivative Type. Trifluoroacetyl (TFA).

Reagent. Trifluoroacetic anhydride (TFAA).

Compound Type. Monosaccharides.

Reaction

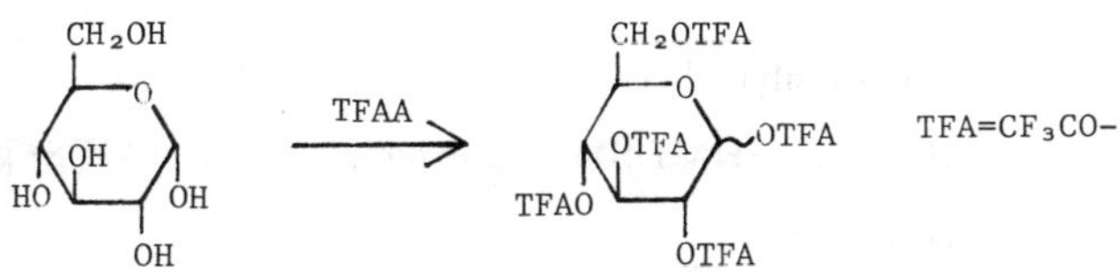

Typical Procedure.[1] To a sample containing μg amts of each monosaccharide add 100 μl TFAA and 100 μl CH_2Cl_2, seal vial under N_2, and heat at 130° for 2 hr. Evap to 10 μl under N_2 using heat from an IR lamp. Add 100 μl dry *n*-hexane containing 1% TFAA and again evap to 10 μl. Add another 100 μl hexane-TFAA and analyze by ECD-GC.

Comments. The method was applicable to aldoses, ketoses, deoxy sugars and amino sugars. GC separation of all the anomeric derivs of nine monosaccharides was achieved on a glass capillary column.[1] The method was used for analysis of sugars in seawater.[1] HFB and PFP derivs were also prepared and gave higher ECD responses, but losses of the sugar derivs were encountered in the evap of excess reagents.[1] Zanetta *et al.*[2] used the same method with heating at 150° for 5 min. König *et al.*[3] prepared the derivs with $CHCl_2$/TFAA/pyridine (13:6:1, v/v/v) by heating 3 hr at 20° and reported on their MS fragmentation.

References

[1]G. Eklund *et al.*, *J. Chromatogr.* **142**, 575 (1977).
[2]J. P. Zanetta *et al.*, *J. Chromatogr.* **69**, 291 (1972).
[3]W. A. König *et al.*, *Chem. Ber.* **106**, 1905 (1973).

Derivative Type. Trifluoroacetyl.

Reagent. *N*-Methylbis(trifluoroacetamide)(MBTFA).

Compound Type. Carbohydrates (mono-, di-, tri- and tetrasaccharides).

Reaction

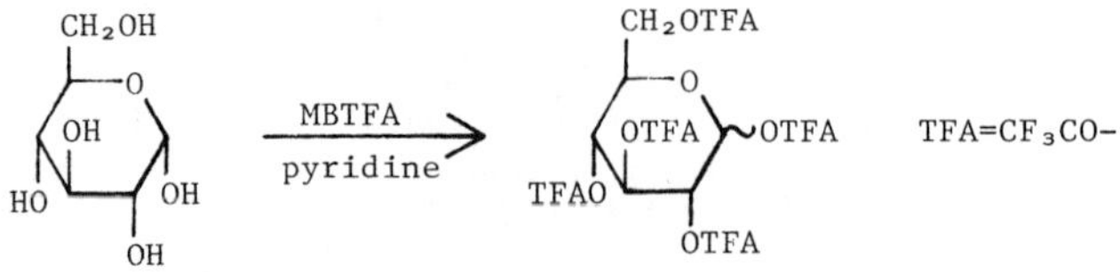

Typical Procedure. To 5–10 mg sample add 0.5 ml MBTFA and 0.5 ml pyridine. Cap vial and heat at 65° for 1 hr with occasional shaking. Analyze 1 μl aliquot by GC.

Comments. Attempts to deriv sugars with TFAA and TFAI yielded multiple and irreproducible peaks. Single peaks were obtained with MBTFA. The derivs were sufficiently volatile (more volatile than TMS) to permit separation of mono- through tetrasaccharides in a single (15 min) GC run. The structures of the derivs were not established but the peaks were identical to those of the less volatile members of the pairs of major derivs obtained with TFAA or TFAI.

References

[1]J. E. Sullivan and L. R. Schewe, *J. Chromatogr. Sci.* **15**, 196 (1977).

13.1.4. Trimethylsilyl Derivatives

Derivative Type. Trimethylsilyl ether.

Reagent. Hexamethyldisilazane (HMDS)/trimethylchlorosilane (TMCS)/pyridine.

Compound Type. Carbohydrates.

Reaction

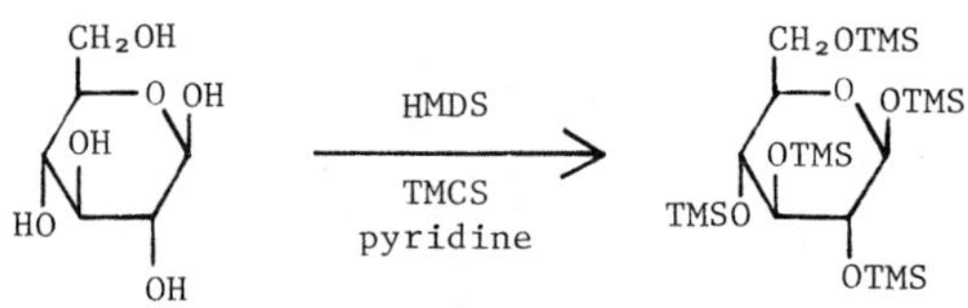

Typical Procedure. Treat 10 mg of carbohydrate with 1 ml anhyd pyridine (stored over KOH), 0.2 ml HMDS, and 0.1 ml TMCS. Shake 30 sec then allow to stand at room temp 5 min or more before GC analysis.

Comments. The fine ppt (NH_4Cl) which formed was not removed and did not interfere with the GC analysis. Nearly 100 carbohydrates and related compds were examined ranging from C_2 (glycoaldehyde) to C_{24} (stachyose).[2] The MS fragmentations of the derivs have been studied using deuterated sugars.[3] Partridge and Weiss reported a method for ext of the derivs to remove the badly tailing pyridine peak.[4]

References

[1]R. Bentley *et al.*, *Biochem. Biophys. Res. Commun.* **11**, 14 (1963).
[2]C. C. Sweeley *et al.*, *J. Am. Chem. Soc.* **85**, 2497 (1963).
[3]D. C. DeJongh *et al.*, *J. Am. Chem. Soc.* **91**, 1728 (1969).
[4]R. D. Partridge and A. H. Weiss, *J. Chromatogr. Sci.* **8**, 553 (1970).

Derivative Type. Trimethylsilyl.

Reagent. Hexamethyldisilazane (HMDS)/trimethylchlorosilane (TMCS)/dimethylsulfoxide (DMSO).

Compound Type. Monosaccharides.

Reaction

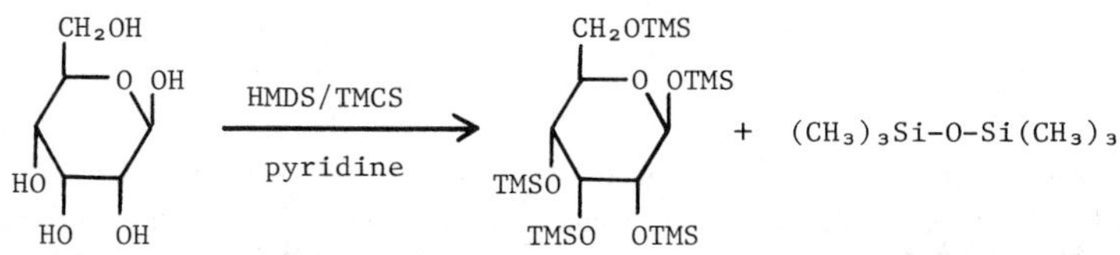

Typical Procedure. Evap aq soln of monosaccharide (0.2–6.0 mg) under vac at 40°. Diss residue in 1 ml DMSO and add 0.3 ml HMDS and 0.2 ml TMCS. Mix and allow to stand at room temp at least 10 min. Centrifuge to separate phases and analyze an aliquot of the upper phase by GC.

Comments. The monosaccharide derivs partitioned preferentially into the upper phase of the hexamethyldisiloxane formed. Use of DMF in place of DMSO gave similar results, but 16–20 hr were required for the upper phase to form. The hexamethyldisiloxane phase could be washed with H_2O to remove residual solvent without extensive loss of the deriv.

References

[1]W. C. Ellis, *J. Chromatogr.* **41**, 325 (1969).

Derivative Type. Trimethylsilyl.

Reagent. Trimethylsilylimidazole (TMSI).

Compound Type. Carbohydrates.

Reaction

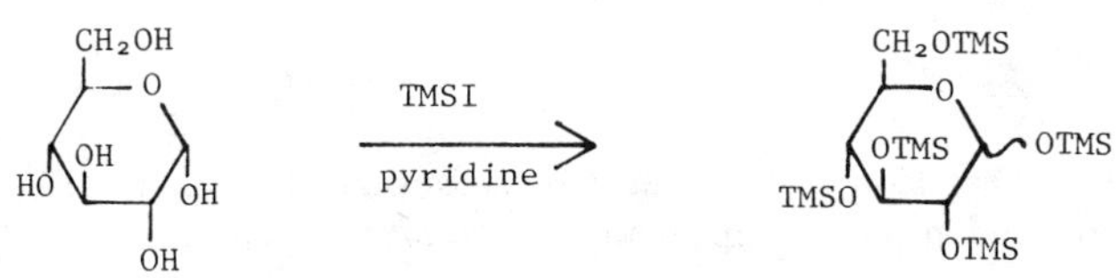

Typical Procedure. Diss 5 mg sample in 100 μl 1.5 *M* TMSI in pyridine. After 1 hr analyze by GC.

Comments. The method was applied to identification of urinary carbohydrates separated by anion exchange chromatography. Sodium borate and boric acid in the buffer used were removed before deriv by ion exchange (for Na) and addition and evap of MeOH (to remove borate as the methyl ester). GC data were given for the derivs of 37 C_3–C_{18} carbohydrates.

References

[1]W. C. Butts and R. L. Jolley, *Clin. Chem.* **16**, 722 (1970).

Derivative Type. Trimethylsilyl.

Reagent. Hexamethyldisilazane (HMDS)/trimethylchlorosilane (TMCS)/pyridine.

Compound Type. Carbohydrates and polyols.

Reaction

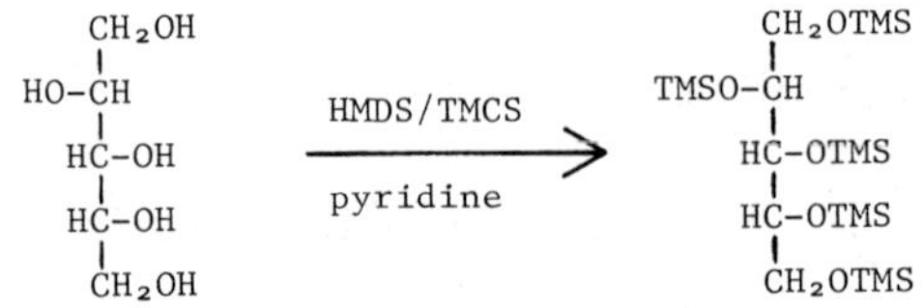

Typical Procedure. Add a 0.1 ml aq sample of carbohydrates or polyols to 2.0 ml HMDS/TMCS/pyridine (2:1:4, v/v/v) and allow to react 3 hr at 35–40°. Add 1 ml $CHCl_3$, cool in ice, and add 3 ml 3 *N* HCl. Shake then remove aq layer. Wash $CHCl_3$ with 2×0.1 ml ice H_2O. Analyze an aliquot of the $CHCl_3$ soln by GC.

Comments. The method permitted direct analysis of aq solns obviating the need for freeze drying. Sufficient TMS reagents were used to react with all of the H_2O in the sample.

References

A. H. Weiss and H. Tambawala, *J. Chromatogr. Sci.* **10**, 120 (1972).

Derivative Type. Trimethylsilyl.

Reagent. Trimethylsilylimidazole (TMSI).

Compound Type. Carbohydrates (mono-, di-, and trisaccharides).

Reaction

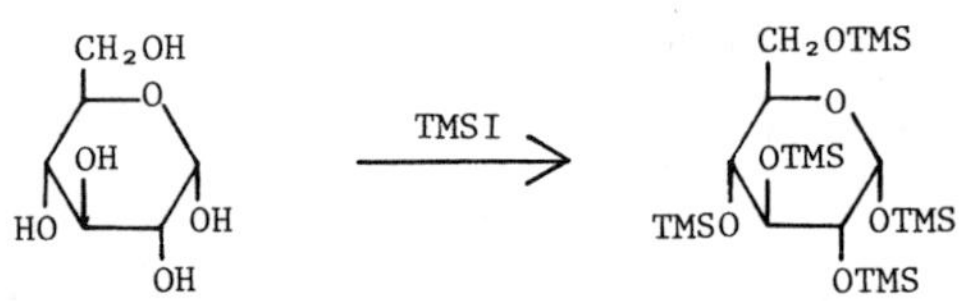

Typical Procedure. Diss 15 mg of the sugar or aq sugar soln in 1 ml 21% (w/v) TMSI in pyridine. Allow to stand at room temp 15 min then analyze by GC.

Comments. The method was developed for deriv of wet carbohydrate samples. A 10% aq soln of the trisaccharide raffinose was completely silylated in 15 min at room temp. The reagent also caused minimal anomerization.

References

[1]G. D. Brittain *et al.*, *Recent Advances in Gas Chromatography*, ed. by I. I. Dromsky and J. A. Perry, Dekker, New York, NY, 1971, p. 223.

Derivative Type. Oxime, trimethylsilyl.

Reagent. Hydroxylamine·HCl, trimethylsilylimidazole (TMSI).

Compound Type. Sugars.

Reaction

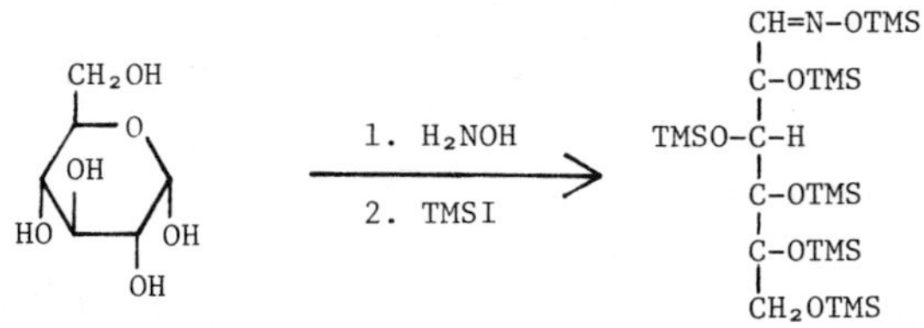

Typical Procedure. To 10–15 mg of sugar sample add 1.0 ml of "STOX," seal, and heat for 30 min at 70–75°. Cool and add 1.0 ml TMSI, shake for 30 sec and let stand for 30 min. Analyze by GC.

Comments. "STOX" oxime int stand/reagent contains 25 mg/ml hydroxylamine·HCl and 6 mg/ml phenyl β-D-glucopyranoside (int stand) in pyridine. Alternatively, 1.0 ml HMDS and 0.1 ml TFA could be used for trimethylsilylation.[2] Anderle *et al.*[3] used HMDS/TMCS for prep of the TMS derivs of triose and tetrose oximes. The derivs yielded two GC peaks due to *syn* and *anti* isomers.

References

[1] *Pierce Previews*, Sept. 1974, Pierce Chemical Co., Rockford, Ill.

[2] K. M. Brobst and C. E. Lott, *Cereal Chem.* **43**, 35 (1966).

[3] D. Anderle *et al.*, *Anal. Chem*, **49**, 137 (1977).

Derivative Type. Methoxime, trimethylsilyl.

Reagent. Methoxylamine HCl, bis(trimethylsilyl)trifluoroacetamide.

Compound Type. Carbohydrates.

Reaction

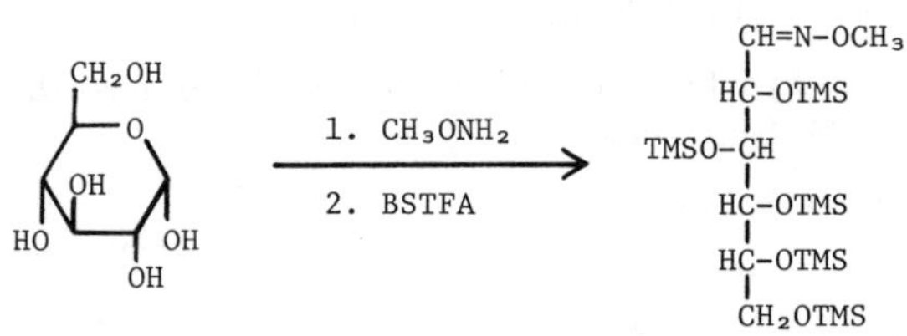

Typical Procedure. Diss 1 mg methoxylamine HCl and 1 mg hexose in 50 μl pyridine and allow to react 2 hr at 80°. Add 50 μl BSTFA and allow to react 15 min at 80°. Analyze by GC-MS.

Comments. MS data were reported for derivs of glucose and deuterated analogs.[1] The methoxime group, by stabilizing ions containing that group, yielded mass spectral peaks permitting analysis of substitution on each carbon. MS data were subsequently reported for the derivs of C_3–C_7 aldoses, partially methylated aldoses, deoxyaldoses, and ketoses.[2]

References

[1] R. A. Laine and C. C. Sweeley, *Anal. Biochem.* **43**, 533 (1971).

[2] R. A. Laine and C. C. Sweeley, *Carbohydr. Res.* **27**, 199 (1973).

Derivative Type. Methaneboronate, trimethylsilyl ether.

Reagent. Methaneboronic acid, bis(trimethylsilyl)trifluoroacetemide (BSTFA)/trimethylchlorosilane (TMCS).

Compound Type. Carbohydrates.

Reaction

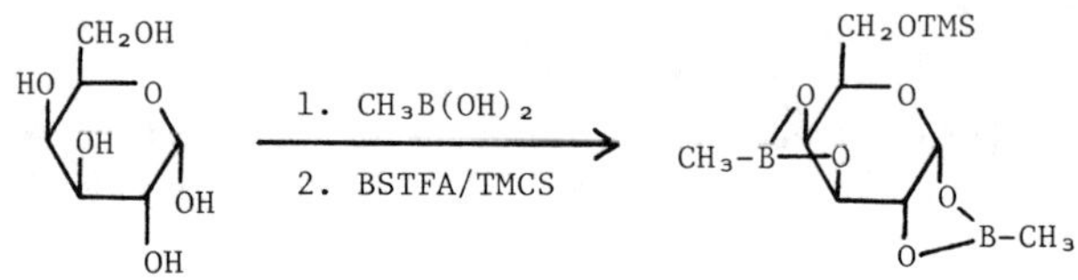

Typical Procedure. Evap a soln of *ca* 200 μg carbohydrate to dryness under N_2. Remove traces of H_2O azeotropically with CH_2Cl_2. Add *ca* 20 mol equiv of methaneboronic acid in pyridine and heat at 110° for 30 min. Add 50 μl BSTFA/TMCS (1:1, v/v) and heat at 110° for 10 min. Analyze by GC or GC-MS.

Comments. MS data were reported for methane-, benzene-, and butaneboronate TMS derivs of a series of isomeric pentoses, hexoses, 6-deoxyhexoses and 2-acetamido-2-deoxyhexoses.[1] The derivs were also prepared using butaneboronic acid in pyridine at 100° followed by HMDS/TMCS[2,3] or BSA.[2] Boronate derivs of carbohydrates have been examined by GC using alkali flame ionization detection.[4]

References

[1]V. N. Reinhold *et al.*, *Carbohyd. Res.* **37**, 203 (1974).

[2]P. J. Wood *et al.*, *Carbohyd. Res.* **42**, 1 (1975).

[3]P. J. Wood and I. R. Siddiqui, *Carbohyd. Res.* **19**, 283 (1971).

[4]R. Greenhalgh and P. J. Wood, *J. Chromatogr.* **82**, 410 (1973).

Derivative Type. *O*-Trimethylsilyl-aldono-1,4-lactones.

Reagent. Hypoiodite (or hypobromite), hydrochloric acid, hexamethyldisilazane (HMDS)/trimethylchlorosilane (TMCS)/pyridine.

Compound Type. Neutral glycoses.

Reaction

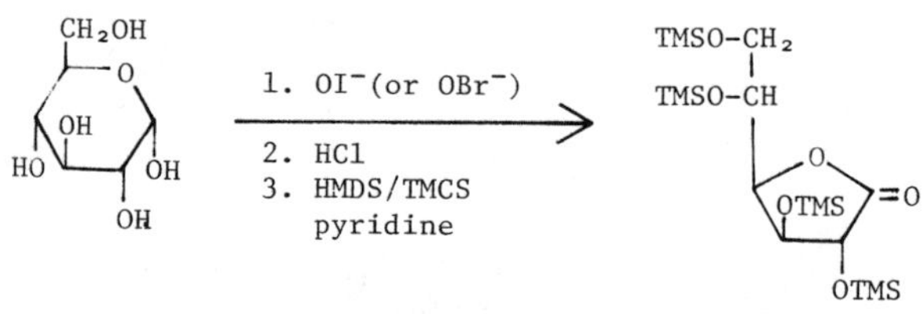

Typical Procedure. Diss 10 mg aldoses in 1 ml H_2O and add 4 ml 0.1 *N* iodine soln followed by 6 ml 0.1 *N* NaOH (or add 0.2 ml Br_2 and 100 mg $BaCO_3$). Allow to react 24 hrs at room temp then pass through a 6 ml column of Rexyn 101 (H form) ion exchange resin. Ext combined eluate and column washings (H_2O) with 6 × 10 ml $CHCl_3$ to remove I_2. (For samples oxidized with OBr^-, filter through sintered glass; then pass through column.) Evap aq soln to dryness under vac, add 0.15 ml 2 *N* HCl, and again evap under vac. To the residue add 0.2 ml HMDS,

0.2 ml TMCS, and 1.5 ml pyridine. Shake 20 min at room temp; then evap under vac below 35°. Diss residue in $CHCl_3$ (EtOH free) and analyze by GC.

Comments. The method was used for analysis of aldopentoses, aldohexoses, and aldoheptoses. The method was developed as an alternative to use of aldose *O*-acetyl, *O*-methyl, or *O*-TMS methyl glycoside derivs (which yield anomeric mixtures) or the reduced glycitol derivs (where two glycoses can give the same deriv).

References

[1]I. M. Morrison and M. B. Perry, *Can. J. Biochem.* **44**, 1115 (1966).

13.1.5. Dimethylsilyl Derivatives

Derivative Type. Dimethylsilyl (DMS).

Reagent. Tetramethyldisilazane (TMDS)/dimethylmonochlorosilane (DMMCS)/pyridine.

Compound Type. Aldoses and alditols.

Reaction

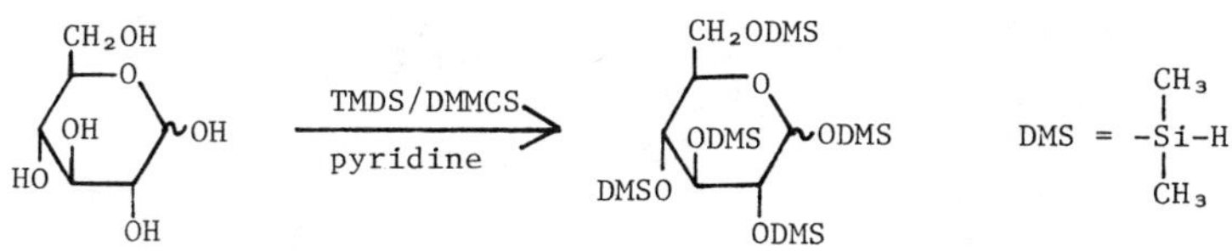

Typical Procedure. Diss *ca* 50 mg of the sample compd in 1 ml TMDS/DMMCS/pyridine (3 :1 :9, v/v/v) and allow to react 10–30 min. Analyze by GC.

Comments. The DMS derivs gave shorter retention times than the corresponding TMS derivs and, in some cases, allowed separations not possible with TMS derivs (the converse was also true).[1] The DMS derivs have been used for analysis of cyclodextrins containing 6, 7, and 8 anhydroglucose units.[2] The fully silylated TMS derivs could not be prepared presumably due to steric hindrance.

References

[1]W. R. Supina *et al., J. Am. Oil Chem. Soc.* **44**, 74 (1967).

[2]J. B. Beadle, *J. Chromatogr.* **42**, 201 (1969).

13.1.6. Acetal and Ketal Derivatives

Derivative Type. *O*-Isopropylidine.

Reagent. Acetone/cupric sulfate.

Compound Type. Pentoses and hexoses.

Reaction

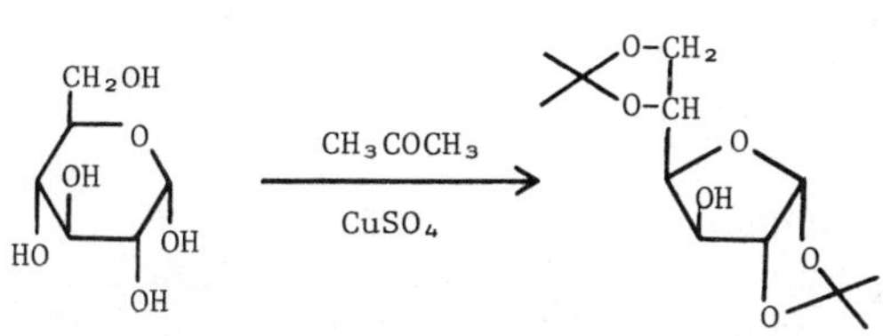

Typical Procedure. Mix 10 mg sugar, 50 mg cupric sulfate, and 0.8 ml acetone and seal in an ampoule. Heat on a steam bath for 18 hr. Remove the acetone layer and make it slightly basic with K_2CO_3. Evap to dryness and purify the residual deriv by preparative GC. Analyze by MS.

Comments. The MS of *O*-isopropylidine derivs of glucose, galactose, mannose, fructose, sorbose, arabinose, fucose, xylose, ribose, and lyxose were studied. Deuterated derivs were also prepared with d_6-acetone.

References

[1]D. C. DeJongh and K. Biemann, *J. Am. Chem. Soc.* **86**, 67 (1964).

[2]A. Buchs *et al.*, *Helv. Chim. Acta* **57**, 1333 (1974).

Derivative Type. Isopropylidene.

Reagent. Acetone/sulfuric acid.

Compound Type. Aldoses.

Reaction

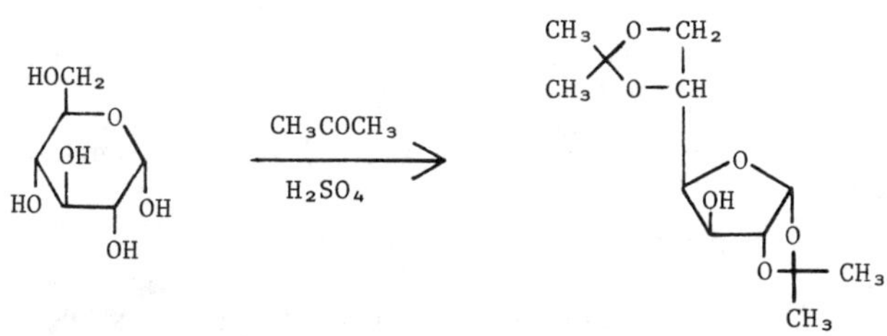

Typical Procedure. To 1–2 mg of sugars add 1.5 ml 1% (v/v) H_2SO_4 in acetone and shake 2 hr. Neutralize with solid $NaHCO_3$ and analyze by GC.

Comments. The common aldoses, with the exception of ribose, gave predominantly single derivs. GC data were reported for the derivs of eight aldoses. The method was used for analysis of monosaccharide composition of soil.

References

[1]S. Morgenlie, *Carbohydr. Res.* **41**, 285 (1975).

Derivative Type. Isopropylidene.

Reagent. 2,2-Dimethoxypropane/*p*-toluenesulfonic acid (TsOH).

Compound Type. D-Glucose.

Reaction

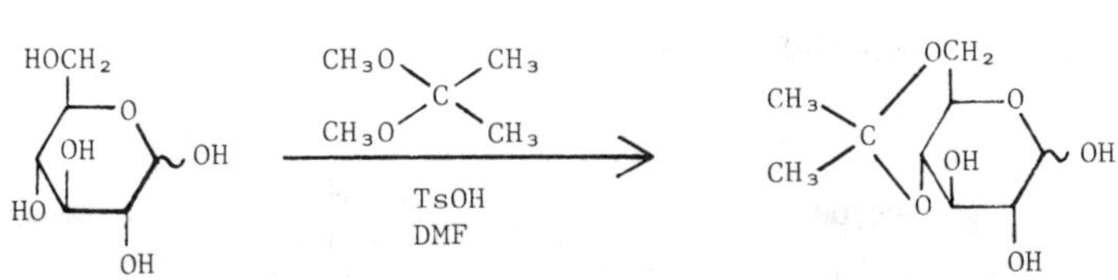

Typical Procedure (Preparative). Diss 4.0 g (22.2 mmol) D-glucose in 50 ml DMF. Add 85 mg TsOH and 8 ml (64.6 mmol) 2,2-dimethoxypropane. Stir 2 hr at room temp then add Amberlite IRA-410 (OH form) ion exchange resin to remove the TsOH. Filter and wash resin with MeOH. Evap under vac to a syrup and chromatograph on silica with $CHCl_3$ then $CHCl_3$/MeOH (100 : 1, 50 : 1, 30 : 1) with the major product (80%) shown above eluting at 30 : 1.

Comments. This preparative reaction could be adaptable for analytical use. Acetonation at 95° yielded multiple products.

References

[1]M. Kiso and A. Hasegawa, *Carbohydr. Res.* **52**, 87 (1976).

Derivative Type. Acetal, methyl ether.

Reagent. Methyl vinyl ether/toluenesulfonic acid(TsOH), methyl iodide/silver oxide.

Compound Type. Sugars.

Reaction

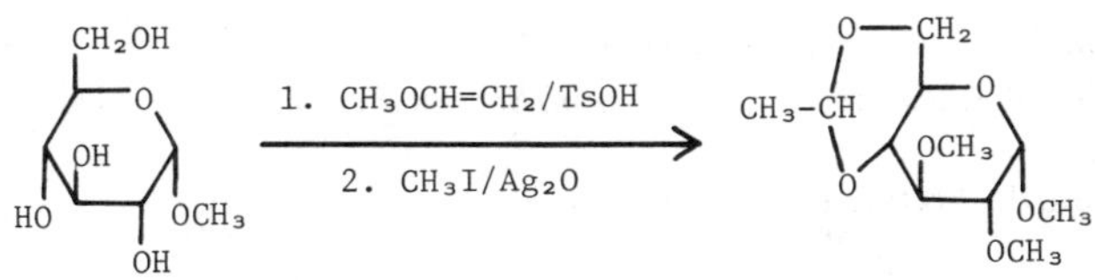

Typical Procedure.[1]Mix 15 g methyl α-D-glucopyranoside with 30 ml DMF over 2 g Drierite. Add 1 mol equiv methyl vinyl ether and 80 mg TsOH and stir at room temp several hr. Filter, add 100 mg $NaHCO_3$ and 2 ml H_2O, and conc to a syrup. Shake with Et_2O and filter. Conc filtrate to a syrup. Diss syrup in 100 ml CH_3I and add 25 g Ag_2O. Shake overnight at room temp. Repeat methylation until IR shows negligible HO peak.

Comments. The acetal formation was also performed with chloroethyl, ethyl, cyclohexyl, and isopropyl vinyl ethers.[1] The reaction with methyl vinyl ether has been used in analysis of acetylated sugar residues with free C-4 and C-6 hydroxyls.[2] The method is potentially useful for analytical deriv of other 1,3-dihydroxy compds.

References

[1]M. L. Wolfram *et al.*, *J. Org. Chem.* **33**, 1067 (1968).

[2]H. Bjorndal *et al.*, *Angew. Chem. Int. Ed.* **9**, 610 (1970).

Derivative Type. Diethyl dithioacetal (or ketal), peracetate.

Reagent. Ethanethiol/HCl, acetic anhydride (Ac_2O)/pyridine.

Compound Type. Monosaccharides.

Reaction

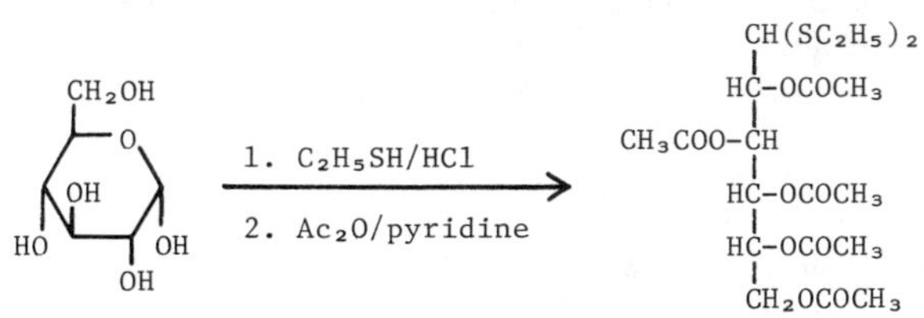

Typical Procedure. Prepare the diethyl dithioacetal (ketal)[2] by diss 1 g aldose (ketose) in 1 ml conc HCl and cooling to 0° in a tightly sealed glass stoppered flask. Add 1 ml ethanethiol and shake 10 min. Allow to warm to room temp yielding a clear soln out of which the product begins to crystallize. After 5 min, cool again to 0°, open flask, and add 2 ml H_2O. Cool to 0° and collect the crystalline products. Wash with cold H_2O and recryst from H_2O. Acetylate with Ac_2O and pyridine at room temp. Analyze by MS.

Comments. The derivs of hexoses, pentoses, and deoxyhexoses were examined.[1] From the MS it was possible to distinguish aldoses from ketoses and to determine deoxy positions. Cyclic ethylene dithioacetal peracetates were prepared in a similar manner using 1,2-ethanedithiol.[3] The MS of nonacetylated dithioacetals of sugars[4] and deoxysugars[5] have also been studied.

References

[1]D. C. DeJongh and K. Biemann, *J. Am. Chem. Soc.* **86**, 3149 (1964).

[2]H. Zinner, *Chem. Ber.* **84**, 780 (1951).

[3]D. C. DeJongh, *J. Am. Chem. Soc.* **86**, 4027 (1964).

[4]D. C. DeJongh, *J. Org. Chem.* **30**, 1563 (1965).

[5]D. C. DeJongh and S. Hanessian, *J. Am. Chem. Soc.* **88**, 3114 (1966).

13.1.7. Aldononitrile Acetate Derivatives

Derivative Type. Aldononitrile acetate.

Reagent. Hydroxylamine·HCl/sodium acetate (NaOAc), acetic anhydride (Ac_2O).

Compound Type. Aldoses.

Reaction

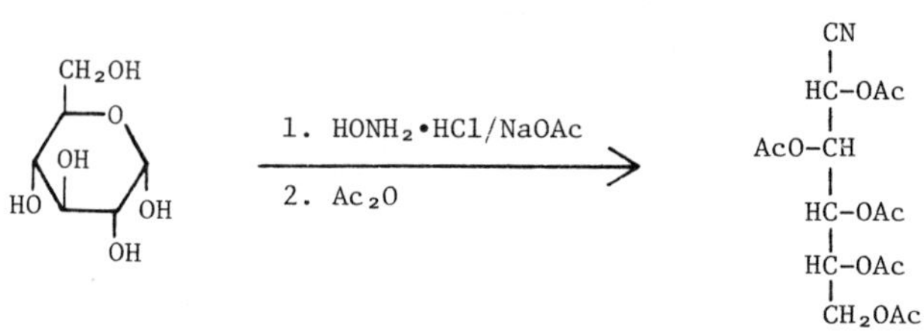

Typical Procedure. Diss 1 mg aldose, 1 mg $H_2NOH \cdot HCl$, and 2.5 mg dry NaOAc in 200 μl dry MeOH and heat at 60° for 1 hr. Evap under N_2 and add 200 μl Ac_2O and heat at 120° for 1 hr. Analyze by GC.

Comments. The aldononitrile acetate derivs (Wohl derivs) gave a single deriv for each aldose (in contrast, e.g., to TMS derivs that give anomeric pairs) yet retained the terminal dissymetry (in contrast, e.g., to reduced derivs for which two aldoses yield the same deriv). Sodium acetate was found to give a faster reaction than pyridine. The method has been used for simultaneous analysis of polyols and aldoses in urine[2] and in lens tissue.[3]

References

[1]J. Szafranek *et al., Anal. Lett.* **6**, 479 (1973).

[2]C. D. Pfaffenberger *et al., Anal. Biochem.* **63**, 501 (1975).

[3]C. D. Pfaffenberger *et al., J. Chromatogr.* **126**, 535 (1976).

Derivative Type. Aldononitrile acetate.

Reagent. Hydroxylamine·HCl/pyridine, acetic anhydride (Ac_2O).

Compound Type. Aldoses and partially methylated aldoses.

Reaction

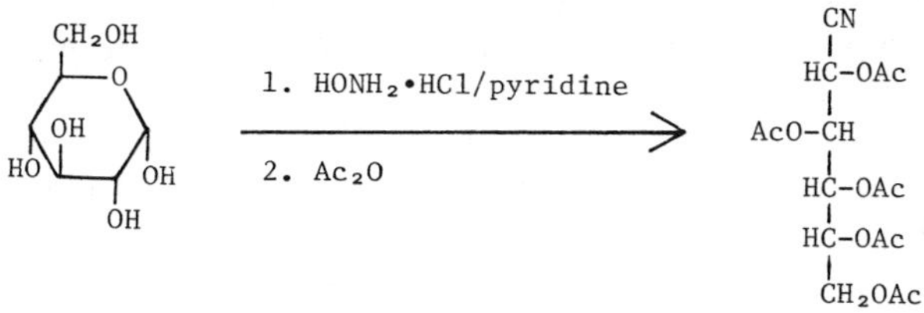

Typical Procedure. Diss a mixt of 1 mg each of 12 monosaccharides in 0.6 ml pyridine and add 12 mg dry hydroxylamine·HCl. Heat in a sealed ampoule for 30 min at 90°. Add 1.8 ml Ac_2O and reseal the ampoule. Heat another 30 min, then evap under vac at 40°. Diss residue in 0.1 ml dry $CHCl_3$ and analyze by GC.

Comments. In contrast to the TMS aldononitriles, which were not very stable,[2] the aldononitrile acetates were quite stable and could be stored several months at room temp in a vac dessicator.[1] GC data have been reported for the derivs of eight partially methylated xyloses.[3] The method was also applied to pentoses, hexoses, 6-deoxyhexoses, and their methyl ethers with MS data reported for the derivs.[4] This report[4] also stated that reaction times could be reduced to 15 min. Seymour *et al.* prepared the derivs of all the possible methyl ethers of mannose[5] and used the data in analysis of mannans.[6]

References

[1]R. Varma *et al.*, *J. Chromatogr.* **77**, 222 (1973).
[2]R. Varma and D. French, *Carbohydr. Res.* **25**, 71 (1972).
[3]D. G. Lance and J. K. N. Jones, *Can. J. Chem.* **45**, 1995 (1967).
[4]B. A. Dmitriev *et al.*, *Carbohyd. Res.* **19**, 432 (1971).
[5]F. R. Seymour *et al.*, *Carbohyd. Res.* **44**, 181 (1975).
[6]F. R. Seymour *et al.*, *Carbohyd. Res.* **48**, 225 (1976).

13.1.8. Chromogenic Derivatives

Derivative Type. *p*-Nitrobenzoate.

Reagent. *p*-Nitrobenzoyl chloride.

Compound Type. Carbohydrates.

Reaction

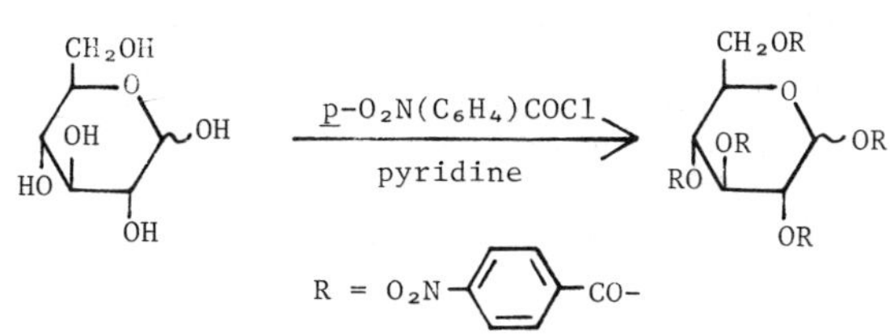

Typical Procedure. Diss up to 0.5 mg carbohydrate in 50 μl pyridine and add 150 μl of *p*-nitrobenzoyl chloride (100 mg/ml) in pyridine. Allow to react 10 min at room temp then evap the pyridine in a vac dessicator at 50° (use a beaker of sand at 50°). Add 2 ml 5% Na_2CO_3 containing 5 mg 4-dimethylaminopyridine (as catalyst) and mix 5 min. Ext with 2 ml $CHCl_3$. Wash ext with 2 ml 5% $NaHCO_3$ and 2×3 ml 0.05 *N* HCl containing 5% NaCl. Analyze by HPLC.

Comments. The method was applicable to mono-, di-, and trisaccharides except fructose for which all steps had to be carried out at room temp (20–22°). The methyl group of methyl glycosides was stable to the procedure. Anomerizations in pyridine for slowly mutarotating sugars could be minimized by diss the sample in pyridine immediately before the deriv formation. Derivs of 12 sugars and 11 digitalis glycosides were examined and HPLC separation of 7 digitalis glycoside derivs was reported.[2]

References

[1]F. Nachtmann and K. W. Budna, *J. Chromatogr.* **136**, 279 (1977).
[2]F. Nachtmann, *Z. Anal. Chem.* **282**, 209 (1976).

13.2. DERIVATIVES OF SUGAR ALCOHOLS

13.2.1. Acetyl Derivatives

Derivative Type. Acetyl.

Reagent. Acetic anhydride (Ac_2O)/Dowex 50X4-200(H form).

Compound Type. Inositols (as TMS derivs).

Reaction

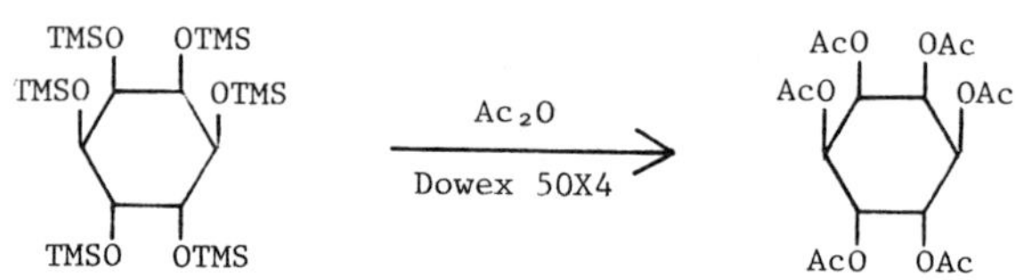

Typical Procedure. Diss 100 μg TMS-inositols (prepared with HMDS/TMS/pyridine) in 40 μl Ac_2O and add a few beads of Dowex 50X4-200 (H form). Heat 2 hr at 70° then analyze by GC-MS.

Comments. The hexa-*O*-TMS and hexa-*O*-acetyl derivs of the various inositol diastereomers were examined by MS. The TMS derivs gave more variation in the MS for the different diastereomers than did the acetyl derivs. The direct conversion of TMS derivs to acetyl derivs could be generally useful for analysis of small amts of natural products allowing the obtaining of MS data on two types of derivs without any required purification.

References

[1]W. R. Sherman *et al.*, *Org. Mass Spectrom.* **3**, 829 (1970).

Derivative Type. Acetyl.

Reagent. Acetic anhydride (Ac_2O)/sulfuric acid.

Compound Type. Glycitols.

Reaction

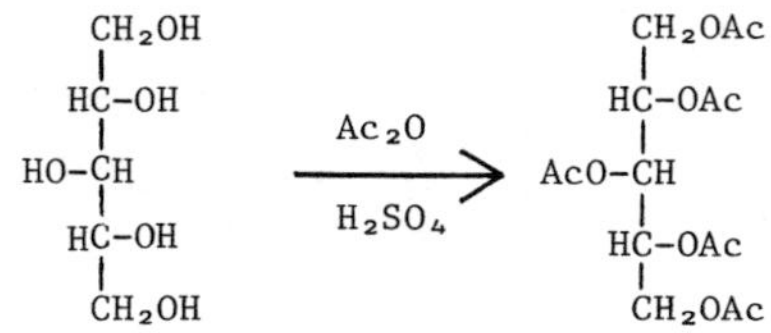

Typical Procedure. Diss *ca* 30 mg heptitol in 4 ml Ac_2O containing 0.1 ml H_2SO_4 and heat 4 hr at 50°. Pour into 25 ml ice and H_2O containing 3 ml pyridine. After 30 min, ext with 5 ml $CHCl_3$. Wash ext with H_2O, 2 *N* H_2SO_4, H_2O, dil $NaHCO_3$, and H_2O. Dry over Na_2SO_4 and analyze by GC.

Comments. The method was used for analysis of glycitols produced by a yeast.

References

[1]H. Onishi and M. B. Perry, *Can. J. Microbiol.* **11**, 929 (1965).

13.2.2. Trifluoroacetyl Derivatives

Derivative Type. Trifluoroacetyl

Reagent. Trifluoroacetic anhydride (TFAA)/sodium trifluoroacetate.

Compound Type. Alditols.

Reaction

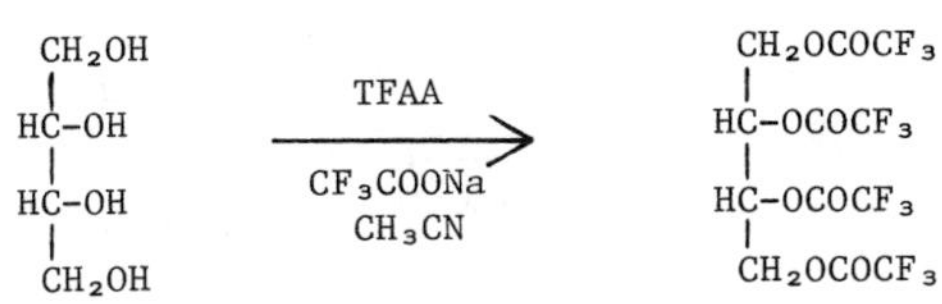

Typical Procedure. Diss 1–3 mg of the alditol in 0.1 ml 10% sodium trifluoroacetate in CH_3CN and add 0.1 ml TFAA. Heat 10–15 min in a sealed tube at 60° then evap to dryness. Analyze by MS.

Comments. The deriv method was based upon that of Vilkas *et al.*[2] MS data were reported for the derivs of a series of alditols and it was found that the derivs were useful for determining the location of a deoxy function.

References

[1]O. S. Chizhov *et al., Org. Mass Spectrom.* **2**, 947 (1969).
[2]M. Vilkas et al., *Tetrahedron Lett.* 1441 (1966).

13.2.3. Trimethylsilyl Derivatives

Derivative Type. Trimethylsilyl.

Reagent. Hexamethyldisilazane (HMDS)/trimethylchlorosilane (TMCS)/pyridine.

Compound Type. Inositols.

Reaction

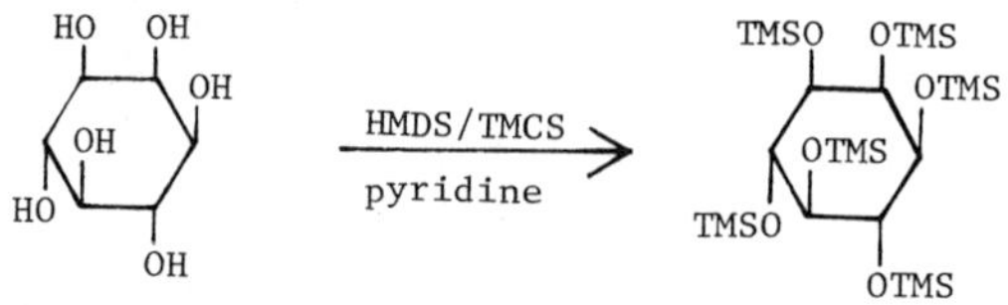

Typical Procedure. Diss sample in HMDS/TMCS/pyridine (2:1:17, v/v/v) and allow to react 48 hr at room temp. Analyze by GC-MS.

Comments. The method was used for analysis of inositols in tissues.[1] Other workers have also reported GC separations of inositols as TMS derivs from HMDS/TMCS/pyridine.[2,3]

References

[1]W. R. Sherman *et al., Biochim. Biophys. Acta* **158**, 197 (1968).

[2]R. N. Roberts *et al., Anal. Biochem.* **10**, 282 (1965).

[3]Y. C. Lee and C. E. Ballou, *J. Chromatogr.* **18**, 147 (1965).

13.2.4. Boronate Derivatives

Derivative Type. Butaneboronate.

Reagent. Butaneboronic acid.

Compound Type. Carbohydrates.

Reaction

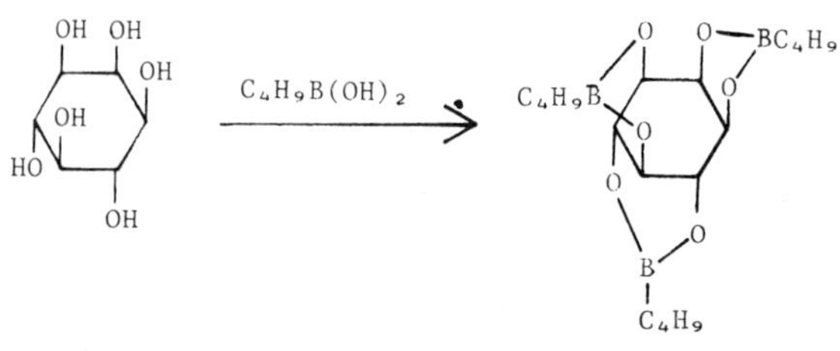

Typical Procedure. Diss 1 mg sugar in 1 ml pyridine containing 5 mg butaneboronic acid. Heat if necessary to diss the sugar. Inject an aliquot of the soln into the GC.

Comments. The derivs of the aldohexoses showed overlapping peaks but the cyclitols and alditols gave single sharp peaks. Compared to acetylation, both the deriv time and the analysis time were much shorter. MS data were reported showing the deriv of *myo*-inositol (shown above) to be the tris-butaneboronate. Some commercial butaneboronic acid was found to be contaminated with isobutaneboronic acid which gave spurious GC peaks.

References

[1]F. Eisenberg, *Carbohydr. Res.* **19**, 135 (1971).

[2]F. Eisenberg, *Meth. Enzymology* **28B**, 168 (1972).

Derivative Type. Butaneboronate.

Reagent. *n*-Butaneboronic acid.

Compound Type. Sorbitol and mannitol.

Reaction

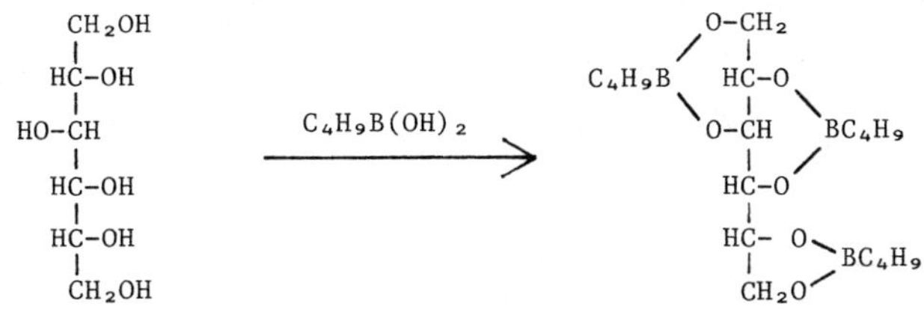

Typical Procedure. Dry a sample containing *ca* 1 mg of sorbitol or mannitol in a vac dessicator or vac oven. Add 1 ml of a 10 mg/ml soln of *n*-butaneboronic acid in pyridine and containing the int stand (methyl nonadecanoate[1] or *m*-diphenylbenzene[2]). Analyze an aliquot by GC.

Comments. The derivs of sorbitol, mannitol, and galactitol were separated on a Silar-5CP column.[1] The methods were developed for analysis of sorbitol and mannitol in pharmaceutical preparations. The exact structures of the tris-butaneboronate derivs were not reported. McKinley and Weigel[3] prepared a series of phenylboronate derivs of diols and polyols and reported MS evidence indicative of formation of six-membered rings in the deriv.

References

[1]M. P. Rabinowitz *et al.*, *J. Pharm. Sci.* **63**, 1601 (1974).

[2]D. L. Sondack, *J. Pharm. Sci.* **64**, 128 (1975).

[3]I. R. McKinley and H. Weigel, *Chem. Commun.* 1051 (1972).

13.3. DERIVATIVES OF AMINO SUGARS

13.3.1. Acetyl Trimethylsilyl Derivatives

Derivative Type. *N*-Acetyl, *O*-trimethylsilyl.

Reagent. Acetic anhydride (Ac_2O), hexamethyldisilazane (HMDS)/trimethylchlorosilane (TMCS).

Compound Type. Hexosamines.

Reaction

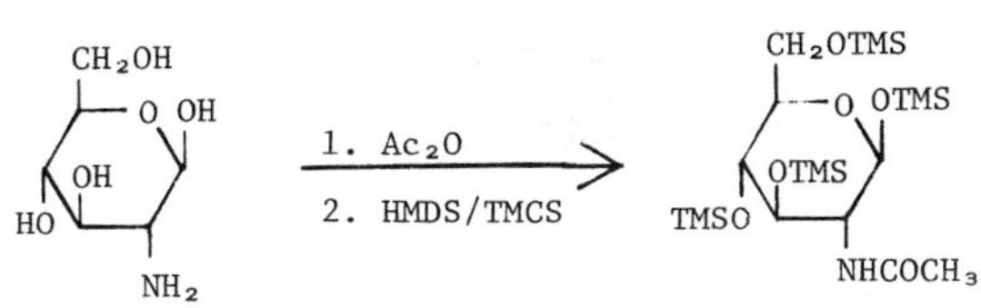

Typical Procedure.[1] Diss 2–10 mg of hexosamine-HCl and 2–5 mg of 2-amino-2-deoxy-D-glucitol-HCl (int stand) in 1 ml H_2O and 0.1 ml MeOH. Add 0.7 g dry powdered Dowex-1 (CO_3 form) ion exchange resin and 0.1 ml Ac_2O. Shake at room temp 20 min then filter through glass wool. Evap to dryness under vac at 50°. Add 1 ml pyridine, 0.1 ml HMDS, and 0.1 ml TMCS. Shake 15 min at room temp, centrifuge, and analyze an aliquot of the supernatant by GC.

Comments. Attempts to prepare a nonacetylated GC deriv using HMDS/TMCS were unsuccessful,[1] but Stimson[3] prepared TMS derivs of hexosamines with BSTFA. Clamp *et al.*[4] used the *N*-acetyl, *O*-TMS derivs for analysis of monosaccharide residues in glycoproteins after methanolysis. MS studies of *O*-TMS derivs of *N*-acetyl amino sugars and disaccharides have been reported.[5]

References

[1]M. B. Perry, *Can. J. Biochem.* **42**, 451 (1964).

[2]J. M. Richey *et al.*, *Anal. Biochem.* **9**, 272 (1964).

[3]W. H. Stimson, *FEBS Lett.* **13**, 17 (1971).

[4]J. R. Clamp *et al.*, *Biochem. Biophys. Acta* **148**, 342 (1967).

[5]P. L. Coduti and C. A. Bush, *Anal. Biochem.* **78**, 21 (1977).

Derivative Type. *N*-Acetyl, *O*-trimethylsilyl.

Reagent. Acetic anhydride (Ac_2O)/hexamethyldisilazane (HMDS)/trimethylchlorosilane (TMCS).

Compound Type. Hexosamines.

Reaction

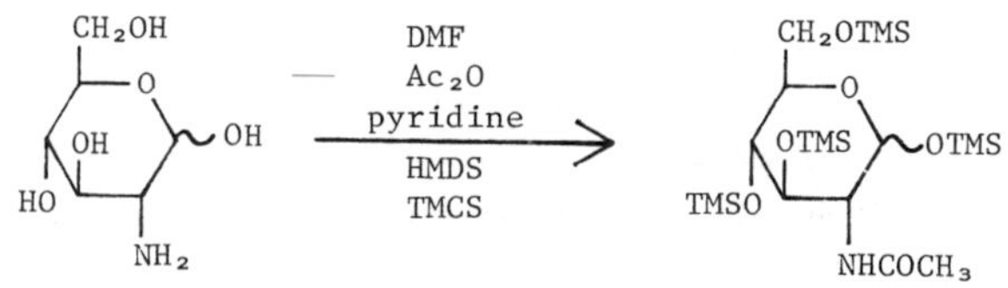

Typical Procedure. Diss sample in 50 μl DMF and add 40 μl pyridine, 10 μl Ac_2O, 50 μl HMDS, and 30 μl TMCS. Shake well then allow to stand 30 min at room temp. Analyze by GC.

Comments. The method was used to analyze reduced chito-oligosaccharides after HCl hydrolysis.

References

[1]S. Hara and Y. Matsushima, *J. Biochem.* **71**, 907 (1972).

Derivative Type. *N*-Acetyl, *O*-trimethylsilyl.

Reagent. Acetic anhydride (Ac_2O)/pyridine, bis(trimethylsilyl)trifluoroacetamide (BSTFA).

Compound Type. Amino sugar methyl glycosides.

Reaction

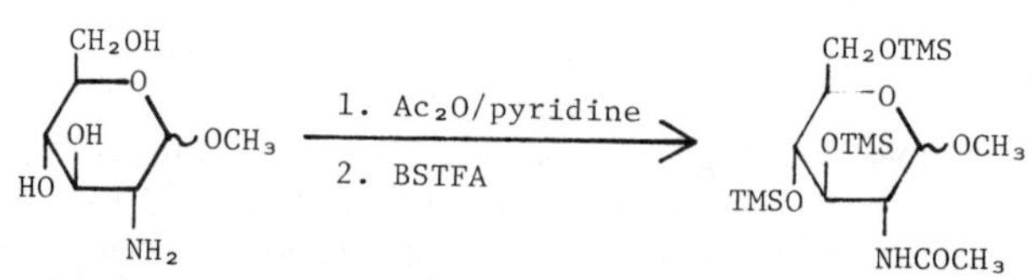

Typical Procedure. Diss 0.5 ml of 1.5 *N* HCl in MeOH methanolysate of glycopeptide in 0.15 ml redist anhyd pyridine. Add 0.10 ml AcCl and mix immediately. After 30 min at room temp dry the sample under vac over KOH then over P_2O_5 for 4 hr. Diss the dry residue in 50 μl anhyd pyridine and add 50 μl BSTFA. After 30 min at room temp, analyze by GC.

Comments. Pyridine was used to both neutralize the acidic methanolysate and to catalyze the acetylation. The method was reported to be superior to the use of Ag_2CO_3 which gave decomposition problems with some samples.

References

[1]J. R. Etchison and J. J. Holland, *Anal. Biochem.* **66**, 87 (1975).

13.3.2. Trifluoroacetyl Derivatives

Derivative Type. Trifluoroacetyl.

Reagent. Trifluoroacetic anhydride.

Compound Type. Hexosamines.

Reaction

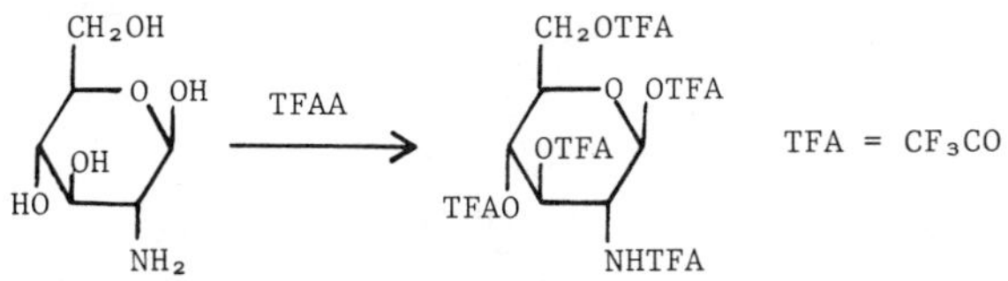

Typical Procedure. To a 50 nmol sample of hexosamine add 5–10 μl of 0.5 m*M* nitrobenzene (int stand) in toluene. Add 100 μl of $TFAA/CH_2Cl_2$ (1:3, v/v), mix well, then heat at 100° for 15 min. Analyze by GC.

Comments. Derivs were prepared from glucosamine and galactosamine. Anomeric derivs of the latter compd were GC separable. An electrolytic conductivity detector was used.

References

[1]J. M. L. Mee, *J. Chromatogr.* **94**, 298 (1974).

Derivative Type. Trifluoroacetyl hexosaminols.

Reagent. Sodium borohydride, trifluoroacetic anhydride.

Compound Type. Hexosamines.

Reaction

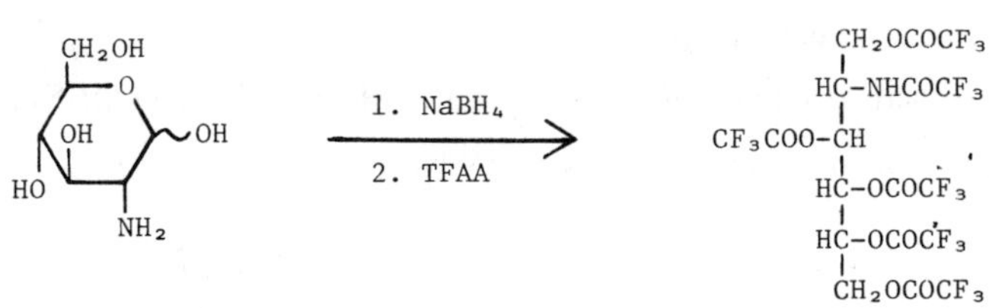

Typical Procedure. To 0.5 ml of an aq soln of hexosamines (100–500 μg) add 0.5 ml 1% aq $NaBH_4$. Allow to stand 30 min at room temp then add 2 drops 0.5 *N* HCl in MeOH. Evap to dryness under vac, add 2 ml MeOH, and evap again. Add 1 ml MeOH and remove NaCl ppt. Evap and diss residue in 0.5 ml H_2O. Pass through a DEAE-Sephadex (borate form) column and wash with 10 ml H_2O. Elute amino alcohols with 2 ml 0.5 *N* HCl in MeOH and evap eluate to dryness. Remove boric acid as above by adding 2 ml MeOH and evap. Diss residue in

0.2 ml THF and add 0.1 ml TFAA at 0°. Allow to stand 10 min at room temp then analyze by GC.

Comments. GC data were given for the TFA derivs of glucosaminol, galactosaminol, and mannosaminol.

References

[1]Z. Tamura *et al.*, *Chem. Pharm. Bull.* **16**, 1864 (1968).

13.3.3. Trimethylsilyl Derivatives

Derivative Type. Trimethylsilyl.

Reagent. Hexamethyldisilazane (HMDS)/trimethylchlorosilane (TMCS)/pyridine/bis(trimethylsilyl)acetamide (BSA).

Compound Type. 2-Amino-2-deoxyhexoses.

Reaction

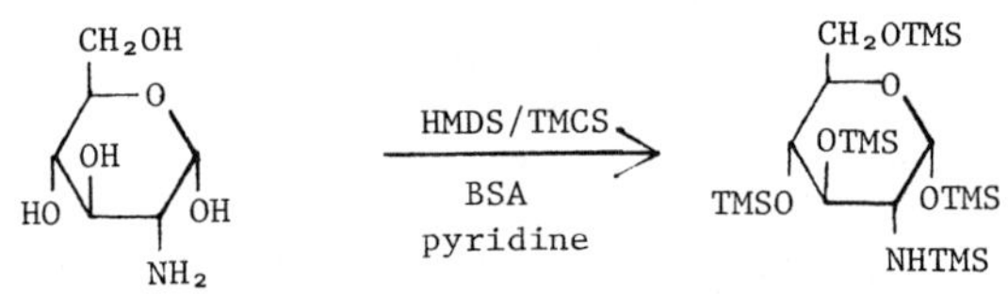

Typical Procedure.[1] Treat aminosugar with HMDS/TMCS/pyridine/BSA (1:0.5:10:1) for 30 min at room temp. Analyze by GC-MS.

Comments. BSA and pyridine alone gave multiple peaks even after heating several hr at 70° HMDS/TMCS/pyridine alone gave only *O*-TMS derivs. Ext of the deriv into hexane led to hydrolysis of the labile *N*-TMS. Full MS were given for the derivs of 12 amino hexoses.[1] Other workers have used HMDS/DMF (1:5, v/v; 100°, 2 min) for the preparation of TMS derivs of hexosamines.[3]

References

[1]J. Kärkkäinen and R. Vihko, *Carbohydr. Res.* **10**, 113 (1969).

[2]J. Kärkkäinen *et al.*, *J. Chromatogr.* **20**, 457 (1965).

[3]B. Radhakrishnamurphy *et al.*, *Anal. Biochem.* **17**, 545 (1966).

Derivative Type. Trimethylsilyl.

Reagent. Bis(trimethylsilyl)acetamide(BSA)/trimethylchlorosilane (TMCS)/pyridine.

Compound Type. Hexosamines in glycosaminoglycans.

Reaction

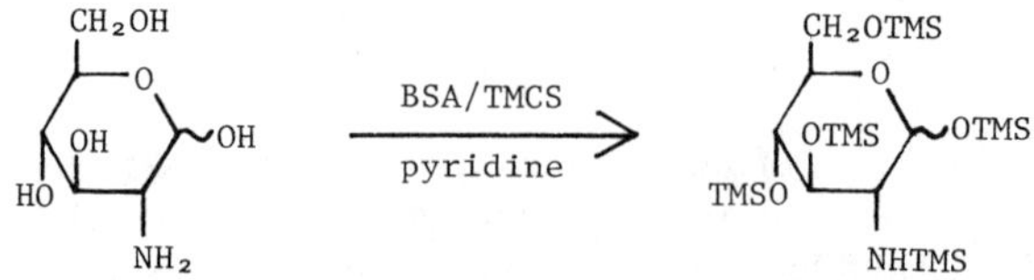

Typical Procedure. Hydrolyze a 1 mg sample of glycosaminoglycan in 0.3 ml 1 *M* HCl for 3 hr at 105° or 0.3 ml 6 *M* HCl for 16 hr at 105°. Lyophilize and add 0.2 ml BSA/TMCS/pyridine (100 :1 :400, v/v/v). Analyze by GC.

Comments. Several conditions were examined for silylation of the hexosamines as summarized below:

Method	Reagents	Conditions	Number of TMS Groups
Ia	HMDS/TMCS/pyridine (2:1:10)	r.t., 20 min	4
Ia′	Ia fresh reagents	r.t., 20 min	4, 5
Ib	Ia + 1 vol BSA	r.t., 20 min	5
IIa	BSA/pyridine (1:4)	anal. immediately	5
IIb	BSA/TMCS/pyridine (100:1:400)	anal. immediately	5
III	HMDS/DMF (1:5)	105°, 3 min	4
IV	TMSI/pyridine (1:4)	60°, 20 min	4
Va	BSA/TMCS/CH_3CN (100:1:400)	60°, 10 min	6
Vb	Same as Va	60°, 5 hr	6
VI	BSTFA/TMCS/pyridine (100:1:400)	60°, 3 hr	4, 5

Two anomeric forms were observable for each degree of silylation. It was possible to convert the samples to essentially single anomers by allowing the more stable anomer to crystallize from aq soln before deriv. Methods II-b (to give mono-*N*-TMS) and V-a (to give di-*N*-TMS) were deemed most useful analytically from the standpoint of separability of derivs.

References

[1] R. E. Hurst, *Carbohydr. Res.* **30**, 143 (1973).

Derivative Type. Trimethylsilyl (alditol or substituted tetrahydrofuran).

Reagent. Nitrous acid, sodium borohydride, hexamethyldisilazane (HMDS)/trimethylchlorosilane (TMCS)/pyridine.

Compound Type. Amino sugars.

Reaction

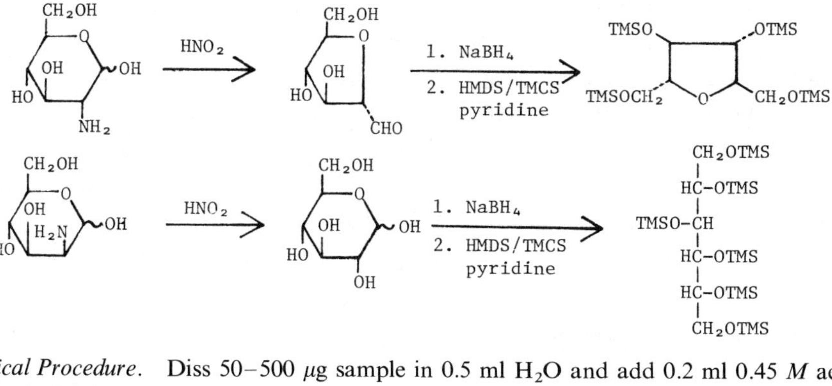

Typical Procedure. Diss 50–500 μg sample in 0.5 ml H_2O and add 0.2 ml 0.45 *M* acetate buffer (pH 3.6) and 0.2 ml 0.4% $NaNO_2$ soln. Allow to stand 5 hr at room temp. Adjust to pH 8–10 with 2% NaOH, cool in ice, and add 0.2 ml 1.3% $NaBH_4$. Allow to stand 2 hr at room temp. Add Dowex 50 X8 (H form, 50–100 mesh) to adjust to pH 2–3. Filter resin and evap soln under vac. Add MeOH and evap (4×) to remove borate. Diss residue in 50 μl pyridine, 50 μl HMDS, and 30 μl TMCS. Mix 1 min then analyze by GC.

Comments. The method was used for analysis of 2-amino-2-deoxy hexoses. The configuration of the 2-amino group determined whether the nitrous deamination yielded an intramolecular reaction product or the hydrolysis product. GC data were given for derivs of glucosamine, galactosamine, mannosamine, muramic acid, and 3-*O*-methylglucosamine.

References

[1]S. Hase and Y. Matsushima, *J. Biochem.* **66**, 57 (1969).

13.3.4. Other Derivatives of Amino Sugars

Derivative Type. Aminoglycitol acetate.

Reagent. Sodium borohydride, acetic anhydride (Ac_2O).

Compound Type. 2-Amino-2-deoxyhexoses.

Reaction

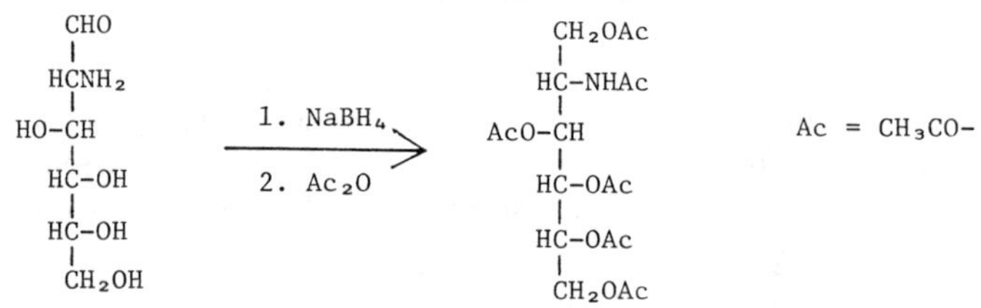

Typical Procedure. Diss 0.5–5 mg 2-acetamido-2-deoxyhexose or 2-amino-2-deoxyhexose in 0.5 ml H_2O and reduce with 10 mg $NaBH_4$. After 2 hr at room temp destroy the excess $NaBH_4$ with HOAc. Evap to dryness under vac below 50°. Add and evap 4×3 ml MeOH containing 1 drop HOAc. Add 1.0 ml Ac_2O and heat at 135° for 2 hr. Analyze 1 μl aliquots by GC.

Comments. GC data were reported for the derivs of eight 2-amino-2-deoxyhexoses.[1] Each deriv gave a single GC peak. The deriv has been used for simultaneous analysis of neutral and amino sugars.[2] Jones *et al.*[3] reported GC separations of peracetylated glycosamines.

References

[1]M. B. Perry and A. C. Webb, *Can. J. Biochem.* **46**, 1163 (1968).
[2]L. J. Griggs *et al.*, *Anal. Biochem.* **43**, 369 (1971).
[3]H. G. Jones *et al.*, *Can. J. Chem.* **40**, 1559 (1962).

Derivative Type. *N*-Carboethoxy, *O*-trimethylsilyl.

Reagent. Ethyl chloroformate/sodium bicarbonate, hexamethyldisilazane (HMDS)/trimethylchlorosilane (TMCS)/pyridine.

Compound Type. Amino sugars.

Reaction

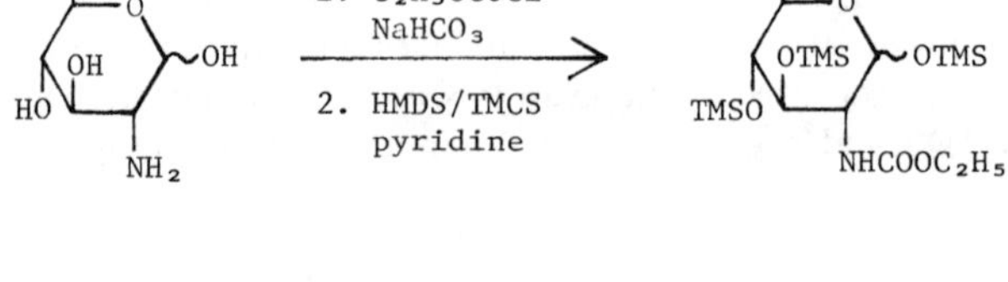

Typical Procedure. To 5 ml of a 1 m*M* soln of amino sugar add 0.3 ml satd aq $NaHCO_3$ and 25 μl ethylchloroformate. Allow to stand 1 hr then deionize by passing over a 2 cm column of "De-acidite" G (7–9% crosslinking, 100–200 mesh) over which was laid 2 cm of Dowex 50W (X12, 200–400 mesh, H form). Discard the first 2.5 ml and collect the following 2 ml of eluate. Evap the eluate to dryness in a vac dessicator over NaOH. Add 0.4 ml HMDS/TMCS/ pyridine (3 : 2 : 10, v/v/v) and allow to stand at room temp 30 min. Analyze an aliquot by GC.

Comments. The method was used for simultaneous analysis of neutral sugars and amino sugars in mucopolysaccharide hydrolyzates.

References

[1]M. D. G. Oates and J. Schrager, *J. Chromatogr.* **28**, 232 (1967).

Derivative Type. Dithioacetal.

Reagent. Ethane thiol/HCl.

Compound Type. Acetamido deoxy sugars.

Reaction

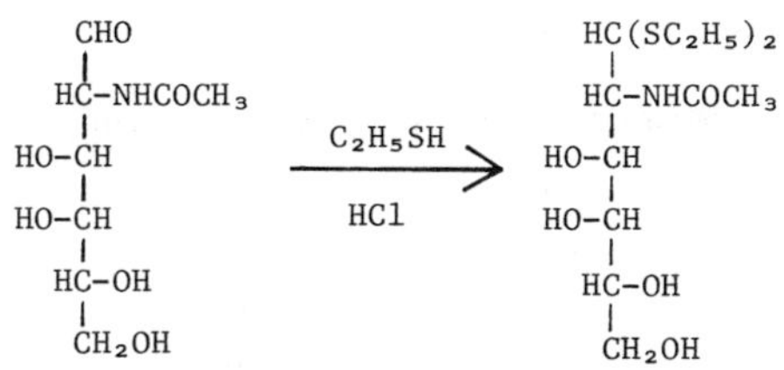

Typical Procedure.[3] Mix 10 g 2-acetamido-2-deoxy galactose with 40 ml 35% HCl and 40 ml ethane thiol. Stir 24 hr at 0° then neutralize with excess basic lead carbonate. Add 500 ml H_2O and collect ppt. Wash ppt with H_2O and recryst from EtOH-$CHCl_3$-Et_2O. Analyze by MS.

Comments. MS of the derivs of a variety of deoxy acetamido hexoses and pentoses were studied.

References

[1]D. C. DeJongh and S. Hanessian, *J. Am. Chem. Soc.* **87**, 1408 (1965).

[2]D. C. DeJongh and S. Hanessian, *J. Am. Chem. Soc.* **87**, 3744 (1965).

[3]M. L. Wolfram and Z. Yosizawa, *J. Am. Chem. Soc.* **81**, 3474 (1959).

13.4. DERIVATIVES OF SUGAR ACIDS

13.4.1. Trimethylsilyl Derivatives

Derivative Type. Trimethylsilyl.

Reagent. Hexamethyldisilazane (HMDS)/trimethylchlorosilane (TMCS).

Compound Type. Hexuronic acids.

Reaction

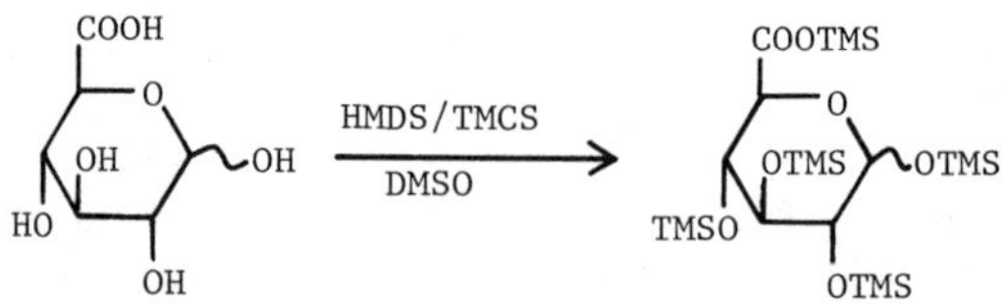

Typical Procedure. Mix 100 μl of HMDS and 25 μl TMCS and add 250 μl of a 1 mg/ml soln of hexuronic acid in DMSO. Analyze an aliquot of the upper layer by GC.

Comments. The uronic acids were more soluble in DMSO than in the more commonly used pyridine. Silylation occured immediately with the deriv partitioning predominantly into the upper (HMDS) phase. Methoxime TMS derivs were also prepared by treatment of the uronic acid (1 mg) with 1 ml 1% methoxylamine HCl in pyridine followed by addition of 200 μl HMDS and 100 μl TMCS and heating at 37° for 1 hr.

References

[1]J. F. Kennedy *et al.*, *Carbohydr. Res.* **49**, 243 (1976).

Derivative Type. Trimethylsilyl.

Reagent. Bis(trimethylsilyl)trifluoroacetamide (BSTFA)/trimethylchlorosilane (TMCS)/pyridine.

Compound Type. Sugar acids.

Reaction

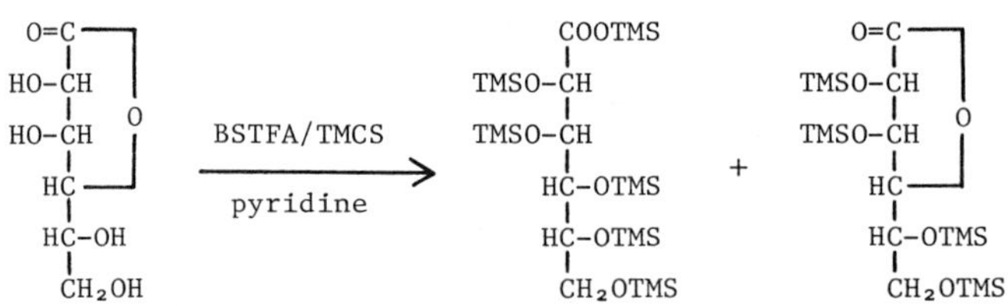

Typical Procedure. Diss 1 mg of the sugar acid (or lactone) in 100 μl dry pyridine and add 85 μl BSTFA and 15 μl TMCS. Heat at 60° for 1 hr then analyze by GC.

Comments. The method was applied to mixts of aldonic, deoxyaldonic, hexuronic, and hexaric acids. The compds yielded mixts of TMS lactones and linear TMS-ether TMS-esters. The derivs were separated on a 60 m SE-30 glass capillary column. Kennedy *et al.* prepared TMS derivs of hexuronic acids with HMDS/TMCS[2] and studied their structures by NMR and MS.[3]

References

[1]J. Szafranek *et al.*, *J. Chromatogr.* **88**, 149 (1974).
[2]J. F. Kennedy *et al.*, *Carbohydr. Res.* **49**, 243 (1976).
[3]J. F. Kennedy *et al.*, *Carbohydr. Res.* **57**, 205 (1977).

Derivative Type. Trimethylsilyl.

Reagent. Bis(trimethylsilyl)trifluoroacetamide (BSTFA)/trimethylchlorosilane (TMCS).

Compound Type. Heptono-1,4-lactone.

Reaction

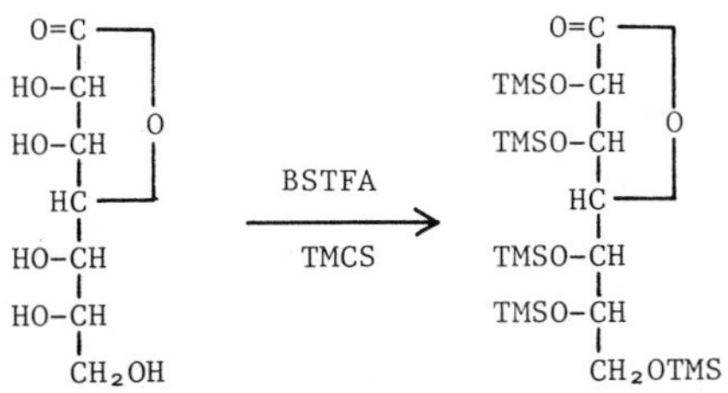

Typical Procedure. Diss 10 mg sample in 0.2 ml CH_3CN. Add 0.2 ml BSTFA containing 1% TMCS. Heat at 80° for 2 hr. Analyze by GC.

Comments. The derivs were prepared for use in analysis of aldoheptoses in bacterial cell walls by hydrolysis and mild oxidation to yield heptono-1,4-lactones.[2]

References

[1]M. B. Perry *et al.*, *J. Chromatogr.* **44**, 614 (1969).

[2]G. A. Adams *et al.*, *Can. J. Microbiol.* **13**, 1605 (1967).

Derivative Type. 2,3,5,6-Tetra-*O*-trimethylsilyl-aldono-1,4-lactone.

Reagent. Sodium borohydride, hexamethyldisilazane (HMDS)/trimethylchlorosilane (TMCS)/pyridine.

Compound Type. Hexuronic acids.

Reaction

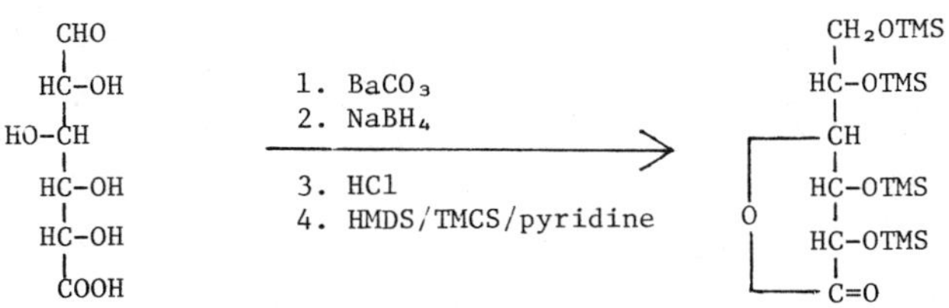

Typical Procedure. Diss mixt of hexuronic acids (1–10 mg) in 2 ml H_2O and convert to the salts by stirring with 150 mg $BaCO_3$ for 10 min at 65°. Filter and reduce with 50 mg $NaBH_4$ for 1 hr at room temp. Add 2 g Rexyn 101 (H form) ion exchange resin to destroy excess $NaBH_4$ then pass through a *ca* 6 ml column of the same resin. Evap eluate and column washings to dryness under vac and remove boric acid by addition and evap of 5 × 20 ml MeOH. Diss in 0.5 ml conc HCl and evap to dryness under vac. Add HMDS/TMCS/pyridine (1:1:10, v/v/v) and shake 15 min at room temp. Evap to dryness under vac. Diss residue in $CHCl_3$, centrifuge, and analyze by GC.

Comments. The method was used for analysis of hexuronic acids in polysaccharides by acid hydrolysis of the polymers in the presence of Pt and H_2, which converted the liberated hexuronic acids to their stable aldonic acid derivs. The aldonic acids were then analyzed as above.

References

[1]M. B. Perry and R. K. Hulyalkar, *Can. J. Biochem.* **43**, 573 (1965).

Derivative Type. Trimethylsilyl.

Reagent. *N*-Trimethylsilylacetamide.

Compound Type. Ascorbic acid.

Reaction

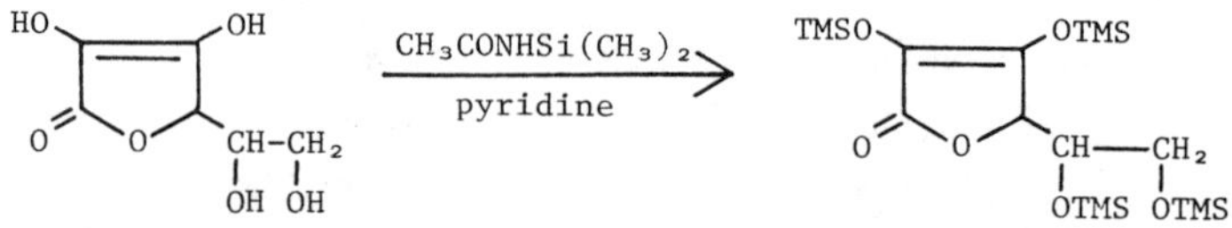

Typical Procedure. Diss a 50 mg sample of ascorbic acid in 10 ml pyridine and add 50 mg octadecane int stand. Add 1.5 g *N*-trimethylsilylacetamide, flush bottle with N_2 or Ar, and seal. Allow to stand at least 4 hr at room temp. Analyze by GC.

Comments. MS of the deriv showed it to contain four TMS groups. HMDS formed a mixt of the tetra-TMS and a lesser subst deriv. HMDS/TMCS has, however, been used in a method for analysis of ascorbic acid in tissues.[2]

References

[1]M. Vecchi and K. Kaiser, *J. Chromatogr.* **26**, 22 (1967).

[2]J. H. Allison and M. A. Stewart, *Anal. Biochem.* **43**, 401 (1971).

Derivative Type. Oxime, trimethylsilyl.

Reagent. Sodium hydroxide, hydroxylamine, bis(trimethylsilyl)trifluoroacetamide (BSTFA)/ trimethylchlorosilane (TMCS).

Compound Type. Sugars and sugar acids.

Reaction

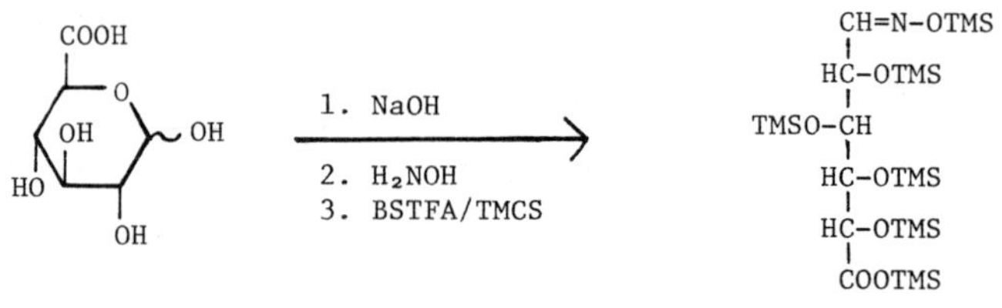

Typical Procedure. Diss 0.1–10 mg sample in 2–5 ml H_2O and adjust pH to 8.5 with 0.05 *N* NaOH. Stir 4 hr at room temp maintaining at pH 8.5 with an automatic titrator. Adjust to pH 7 with 0.01 *N* HCl and evap under vac at 35°. Dry by adding and evap 2×2 ml CH_2Cl_2 or by drying over P_2O_5 in a vac dessicator. Diss residue in 1–2 ml dry pyridine and add *ca* an equal amt of $H_2NOH \cdot HCl$. Sonicate at 30° for 2 hr. Calculate the required amt of TMS reagents from the wt of dry sample (*a* mg), $H_2NOH \cdot HCl$ (*b* mg), and residual H_2O (*c* mg). Add BSTFA ($10a + 10b + 25c + 100$ μl) followed by half this volume of TMCS. Agitate 2 hr then evap under vac at 30°. Diss residue in 0.5–1 ml anhyd Et_2O containing 0.5% BSTFA. Analyze by GC or GC-MS.

Comments. The oximes formed *syn*- and *anti*-isomers, but it was possible to choose stationary phases on which they were not separated. GC retention data were given for the derivs of an extensive list of aldonic and deoxyaldonic acids, aldoses, ketoses, and uronic acids.

References

[1]G. Petersson, *Carbohydr. Res.* **33**, 47 (1974).

13.4.2. Other Derivatives of Sugar Acids

Derivative Type. Aldonic acid, butaneboronate.

Reagent. Butane boronic acid.

Compound Type. Glucuronic and iduronic acids.

Reaction

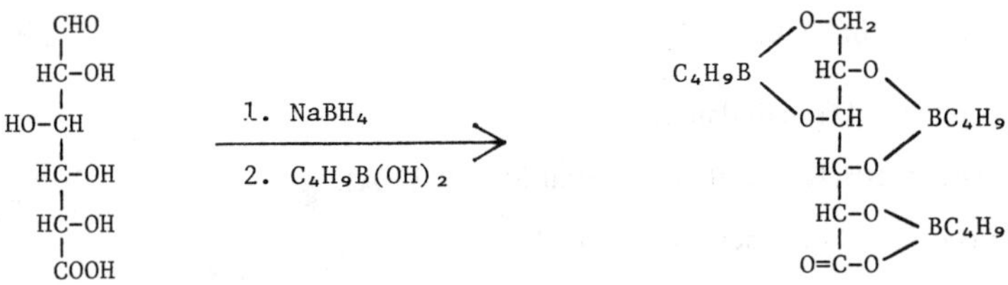

Typical Procedure. Adjust a 1 ml sample to neutrality with a slight excess of solid $Ba(CO_3)_2$ and warm at 60° for 15 min. Add 10 μg mannitol int stand then chill the sample in ice. Add *ca* 5 mg solid $NaBH_4$ and allow to react 1 hr in the cold. Add a few beads of sulfonic acid resin (until bubbling stops) to destroy excess $NaBH_4$. Filter through a small column of the same resin then evap the eluate to dryness at 30–35°. Add MeOH and evap twice to remove boric acid. Diss residue in 0.5 ml conc HCl and evap to dryness at 20–25° (converts mixt of aldonic acids and lactones to 1,4-aldonolactones[2]). Dry under high vac at 100° for 5 min. Add 1 ml 0.5% butaneboronic acid in pyridine and heat at 100° for 10 min. Evap to *ca* 50 μl under N_2 at 60° and analyze by GC.

Comments. For samples with large solid residues, the pyridine was removed completely and the deriv extd with toluene. The exact structures of the derivs were not given but it was reported that MS data indicated tris-boronate derivs.

References

[1]F. Eisenberg, *Anal. Biochem.* **60**, 181 (1974).

[2]M. B. Perry and R. K. Hulyalkar, *Can. J. Biochem.* **43**, 573 (1965).

Derivative Type. Acetyl.

Reagent. Acetic anhydride (Ac_2O)/pyridine.

Compound Type. Methyl (methyl O-methyl-α-D-glucopyranosid)uronates.

Reaction

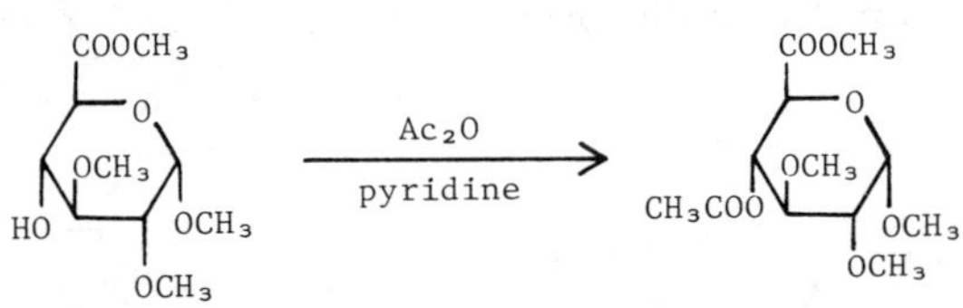

Typical Procedure. Diss sample in excess Ac_2O in pyridine and allow to stand at room temp overnight. Evap under vac and diss residue in $CHCl_3$. Analyze by GC.

Comments. Eight compds of varying degree and position of methylation were examined underiv and as acetyl, TFA, and TMS derivs. Only the acetyl derivs gave adequate GC separation of all the compds.[1] The MS behavior of the acetyl derivs has been studied.[2]

References

[1]D. Anderle and P. Kovac, *J. Chromatogr.* **91**, 463 (1974).
[2]V. Kovacik *et al.*, *Carbohydr. Res.* **54**, 23 (1977).

13.4.3. Derivatives of Sialic Acids

Derivative Type. Trimethylsilyl.

Reagent. Bis(trimethylsilyl)trifluoroacetamide (BSTFA).

Compound Type. *N*-Acetylneuraminic acid.

Reaction

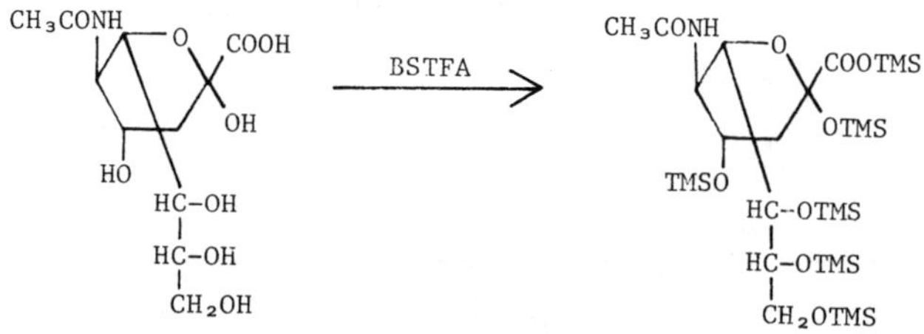

Typical Procedure. Hydrolyze 10 mg protein in 2 ml 0.05 *N* HCl for 1.5 hr at 80°. Lyophilize and ext residue with 2 ml warm (60°) MeOH containing *trans*-stilbene int stand. Evap under vac and add 1 ml CH_3CN and 0.5 ml BSTFA. Heat at 125° for 2 hr. Analyze by GC.

Comments. The method was successfully applied to the determination of N-acetylneuraminic acid in casein (*ca* 2%).

References

[1]D. A. Craven and C. W. Gehrke, *J. Chromatogr.* **37**, 414 (1968).

Derivative Type. Methyl ester, trimethylsilyl.

Reagent. Diazomethane, hexamethyldisilazane (HMDS)/trimethylchlorosilane (TMCS)/pyridine.

Compound Type. *N*-Acetylneuraminic acid and derivs.

Reaction

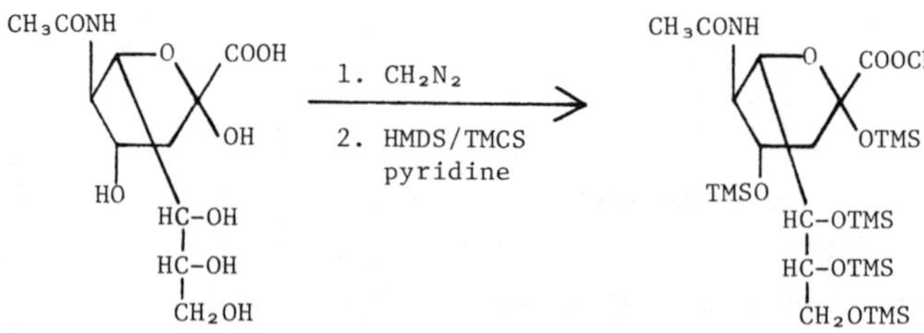

Typical Procedure. Diss 50–100 μg sample in 0.5 ml MeOH and add CH_2N_2 until a faint yellow color is obtained. Evap immediately and diss residue in 1 ml pyridine. Add 0.2 ml HMDS and 0.1 ml TMCS. Allow to stand 2 hr at room temp then add 2 ml $CHCl_3$ and 2 ml H_2O. Separate $CHCl_3$ and dry over Na_2SO_4. Analyze by MS.

Comments. The method was applied to *N*-acetylneuraminic acid,[1] glycoside derivs,[1] and *O*-acetylated derivs.[2] Deuterated derivs were used to clarify MS fragmentation patterns.

References

[1] J. P. Kamerling and F. G. Vliegenthart, *Carbohyd. Res.* **33**, 297 (1974).
[2] J. P. Kamerling *et al.*, *Carbohyd. Res.* **41**, 7 (1975).

Derivative Type. Methyl ester, methyl glycoside, peracetyl.

Reagent. Methanol(MeOH)/H^+, acetic anhydride (Ac_2O)/pyridine.

Compound Type. Sialic acids.

Reaction

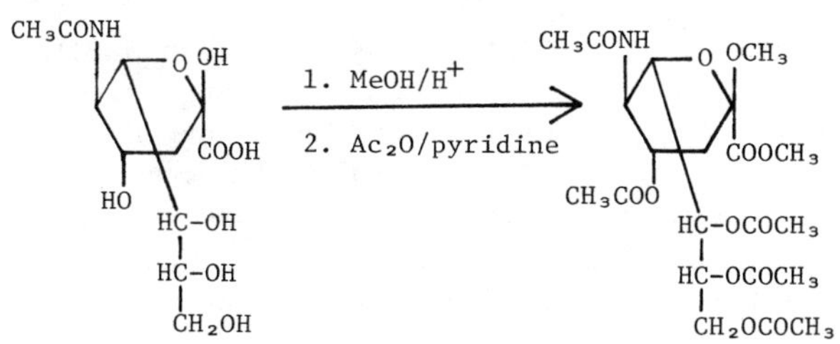

Typical Procedure. Prepare the methyl glycoside, methyl ester by boiling the sialic acid in MeOH in the presence of either Dowex-1 (H form) or 0.1 *M* HCl. Isolate by column chromatography on Dowex-50 (AcO form). Acetylate by heating with Ac_2O/pyridine (1:1, v/v) and analyze by GC or MS.

Comments. Deuterated methyl derivs were prepared using 0.1 *M* HCl in CD_3OD. The derivs were used in a study of the MS properties of sialic acids and in the identification of a new sialic acid, 8-*O*-methyl-*N*-acetylneuraminic acid. The peracetyl derivs were chosen as being more stable and more convenient than the TMS derivs.

References

[1] N. K. Kochetkov *et al.*, *Carbohydr. Res.* **27**, 5 (1973).

Derivative Type. Permethyl, *O*-acetyl.

Reagent. Methyl sulfate/sodium hydroxide, methyl iodide/sodium hydroxide, methyl iodide/methylsulfinyl carbanion, methanol/HCl, acetic anhydride (Ac_2O).

Compound Type. Sialic acid containing polysaccharides.

Reaction

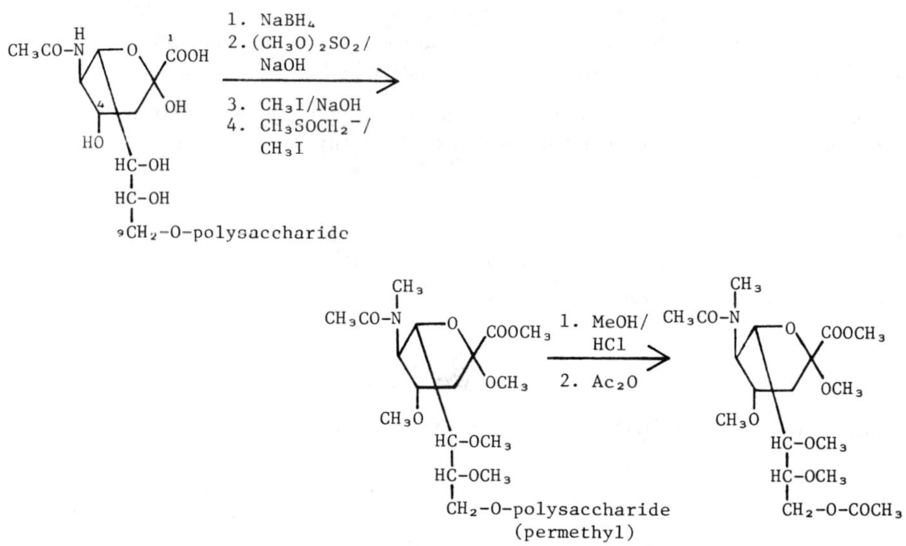

Typical Procedure. Reduce 100 mg of the polysaccharide with $NaBH_4$ to prevent alkaline degradation during methylation. Methylate the reduced sample with $(CH_3O)_2SO_2/NaOH$ (Haworth methylation).[2] Diss the partially methylated sample in THF/H_2O (4:1, v/v) and repeat methylation, then methylate with $CH_3I/THF/NaOH$.[3] Finally methylate with methylsulfinylcarbanion/CH_3I in DMSO.[4] Treat methylated sample with 0.1 *M* HCl in MeOH at 100° for 3 hr (some samples require 12 hr). Re-*N*-acetylate the methanolysis product with Ac_2O. *O*-Acetylate with Ac_2O/pyridine. Alternatively, trimethylsilylate with TMSI. Analyze by GC-MS.

Comments. The method was studied as a means of determining the position of linkages to sialic acid residues in biopolymers. The multistep methylation was presumably employed because studies of model compds showed extensive degradation upon direct treatment with the Hakomori method. MS data were reported for reference compds which would result from C-4, C-8, and C-9 linked sialic acid residues

References

[1] A. K. Bhattacharjee and H. J. Jennings, *Carbohyd. Res.* **51**, 253 (1976).

[2] E. L. Hirst and E. Percival, *Meth. Carbohyd. Chem.* **2**, 146 (1963).

[3] E. L. Falconer and G. A. Adams, *Can. J. Chem.* **34**, 338 (1956).

[4] S. Hakomori, *J. Biochem. (Tokyo)* **55**, 205 (1964).

13.5. DERIVATIVES OF SUGAR PHOSPHATES

13.5.1. Trimethylsilyl Derivatives

Derivative Type. Trimethylsilyl.

Reagent. Trimethylsilylimidazole (TMSI).

Compound Type. Sugar phosphates and related compounds.

Reaction

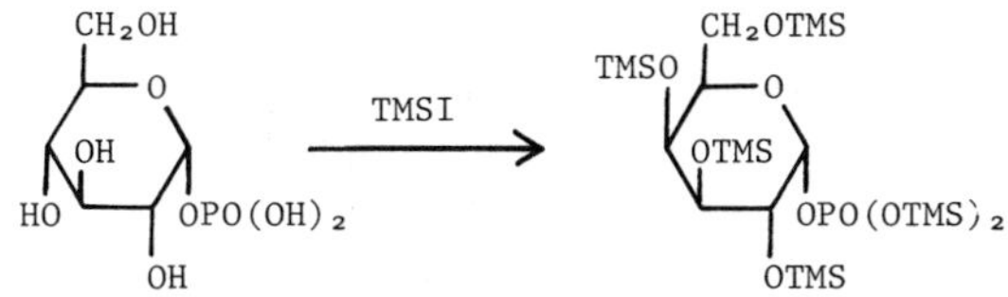

Typical Procedure. Diss 1 mg of the sugar phosphate or its cyclohexylammonium salt in 0.1 ml TMSI and 0.1 ml CH_3CN. Allow to stand at room temp for 5 min then analyze 1 μl by GC or GC-MS.

Comments. All phosphates except pentose- and hexose-1-phosphates and 1-phosphoaldur-onic acids could also be deriv with 0.1 ml BSTFA, 0.05 ml TMCS, and 0.1 ml CH_3CN at 80° for 10 min.

Derivs were prep of aldose-1-, -4-, -5-, and -6-phosphates, ketose diphosphates, phospho-enolpyruvate, and various polyol phosphates. GC data were reported for derivs of 43 sugar and polyol phosphates. The structures of the derivs were confirmed by MS. The derivs of aldose diphosphates were too unstable to GC. TMS-oxime derivs were prep in an effort to reduce the number of GC peaks due to α- and β-furanose and -pyranose structures and to increase stability but it was found that the GC separation of the derivs of isomeric hexoses was poor.

References

[1]D. J. Harvey and M. G. Horning, *J. Chromatogr.* **76**, 51 (1973).

Derivative Type. Trimethylsilyl.

Reagent. Bis(trimethylsilyl)acetamide (BSA)/trimethylchlorosilane (TMCS)/pyridine.

Compound Type. Sugar phosphates.

Reaction

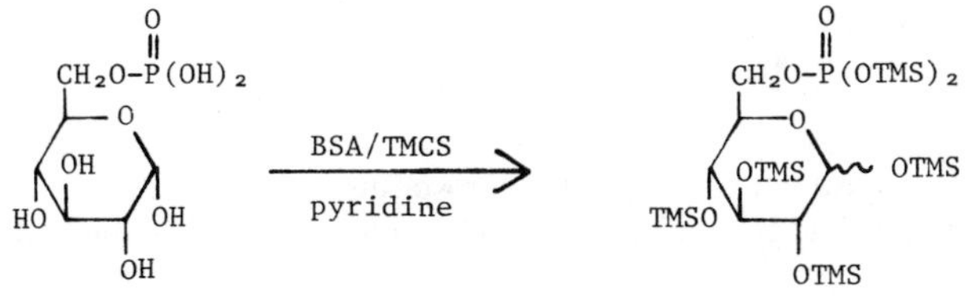

Typical Procedure. Mix 2 μmol of the sugar phosphate (as the Na or cyclohexylammonium salt) with 50 μl BSA, 20 μl TMCS, and 30 μl pyridine (dried over 5A sieve). Allow to stand 1 hr at room temp then analyze by GC-MS.

Comments. The MS behavior of five sugar phosphate TMS derivs was studied. Deuterated derivs were prepared with d_{18}-BSA. The method was also used in the analysis of *myo*-inositol-1-phosphate.[2]

References

[1]M. Zinbo and W. R. Sherman, *J. Am. Chem. Soc.* **92**, 2105 (1970).
[2]W. R. Sherman *et al., J. Biol. Chem.* **244**, 5703 (1969).

Derivative Type. Trimethylsilyl.

Reagent. Bis(trimethylsilyl)acetamide (BSA).

Compound Type. Sugar phosphates and other urinary acids.

Reaction

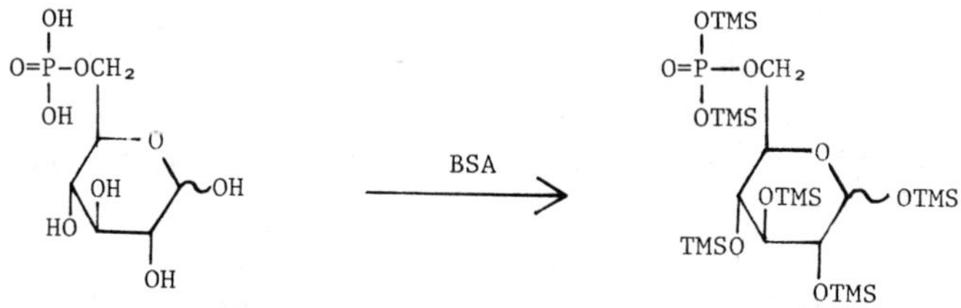

Typical Procedure. Diss 1–2 mg sample in 0.2 ml BSA (if necessary add 0.2 ml pyridine for solubility). Allow to stand several hr then analyze by GC.

Comments. The derivs of a series of acids found in urine were examined. These included aliphatic and aromatic carboxylic acids, glucuronides, and sugar phosphates. The use of BSA was found to be less destructive than HMDS/TMCS. The acids were extracted from urine by solvent ext or with DEAE-Sephadex.

References

[1]M. G. Horning *et al.*, *J. Gas Chromatogr.* **5**, 297 (1967).

Derivative Type. Trimethylsilyl.

Reagent. Bis(trimethylsilyl)trifluoroacetamide (BSTFA)/trimethylchlorosilane (TMCS).

Compound Type. Glycerophosphoric acids.

Reaction

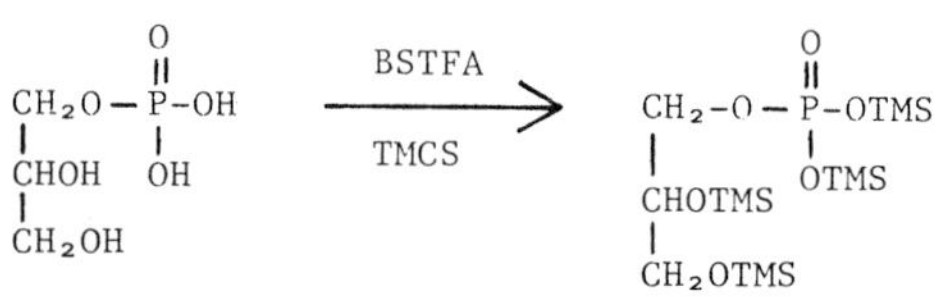

Typical Procedure. Heat 1 mg sample (as the sodium or cyclohexylammonium salt) with 100 μl BSTFA and 50 μl TMCS in 100 μl CH_3CN at 80° for 30 min in a sealed tube. Analyze by GC-MS.

Comments. The d_9-TMS analogs (all TMS groups labeled) were obtained by heating the sample with d_{18}-BSA and d_9-TMCS. The derivs containing d_9-TMS on the phosphate only were obtained by first saturating the GC column with 20 : 1 mixt of d_{18}-BSA and d_9-TMCS. The unlabeled TMS deriv (1 μl) was then injected followed in 30 sec by 10 μl of the d_{18}-BSA/d_9-TMCS mixt. The derivs containing d_9-TMS on the nonphosphate hydroxyls only were obtained by injecting the perdeutero deriv onto a column satd with BSA-TMCS and then injecting 10 μl of the BSA-TMCS mixt. An extensive discussion of the MS behavior of the derivs of seven glycerophosphates was given.

References

[1]D. J. Harvey *et al.*, *J. Chem. Soc. Perkin.* **I**, 1074 (1972).

Derivative Type. Trimethylsilyl.

Reagent. Bis(trimethylsilyl)trifluoroacetamide (BSTFA)/trimethylchlorosilane (TMCS).

Compound Type. Glycerophosphates.

Reaction

Typical Procedure. Diss 1 mg sample in 0.1 ml CH_3CN. Add 0.1 ml BSTFA and 0.05 ml TMCS. Analyze by GC or GC-MS.

Comments. Derivs were prepared from the two glycerol monophosphates, the two glyceric acid monophosphates, and glycerol-1,3-diphosphate, and each deriv gave a single GC peak. MS data were reported for the derivs.

References

[1]D. J. Harvey *et al.*, *Anal. Lett.* **3**, 489 (1970).

13.5.2. Methyl Ester Derivatives

Derivative Type. Methyl ester, trimethylsilyl.

Reagent. Diazomethane, hexamethyldisilazane (HMDS)/trimethylchlorosilane (TMCS)/ pyridine.

Compound Type. Sugar phosphates.

Reaction

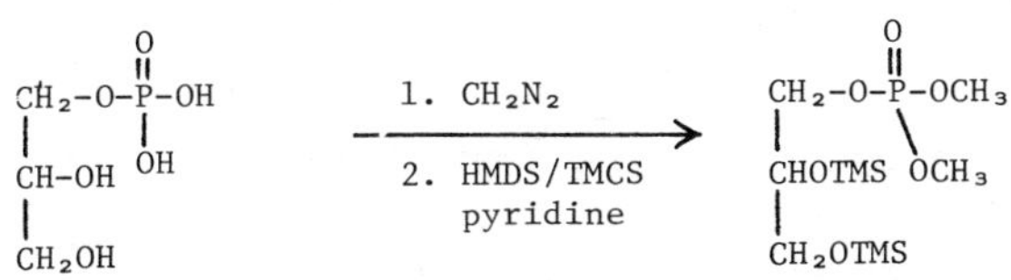

Typical Procedure. Diss a 5 mg sample in 1–2 ml MeOH and dil with 1–2 ml Et_2O. Pass CH_2N_2 into the soln until the yellow color persists then evap under vac. Diss residue in 1 ml pyridine and add 0.2 ml HMDS and 0.1 ml TMCS. Allow to stand at room temp then analyze by GC.

Comments. The phosphate residue was methylated because attempts to form phosphate TMS derivs with the silylation reagent were unsuccessful. GC data were given for the derivs of both glycerol monophosphates and eight hexose phosphates.

References

[1]W. W. Wells *et al.*, *Biochim. Biophys. Acta* **82**, 408 (1964).

Derivative Type. Methyl ester, *O-t*-Butyldimethylsilyl (TBDMS).

Reagent. Diazomethane, *t*-butyldimethylchlorosilane (TBDMS-Cl)/imidazole.

Compound Type. α-Phosphoroglycerol.

Reaction

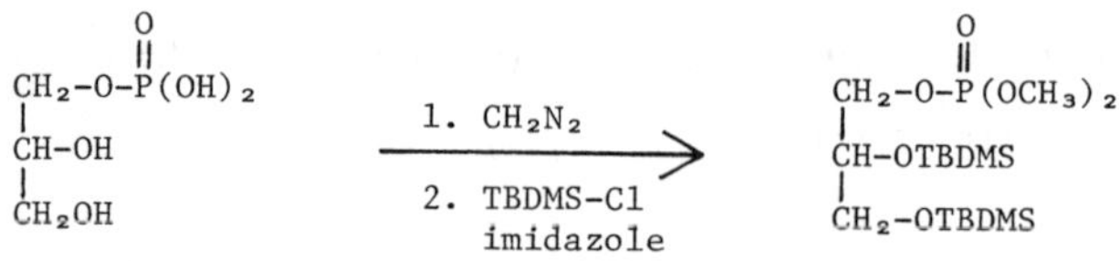

Typical Procedure. Diss sample in 1 ml MeOH and add CH_2N_2 in Et_2O dropwise until the yellow color persists. Evap to dryness under vac and add 20 μl 1 *M* TBDMS-Cl and 2.5 *M* imidazole in DMF. Allow to stand 15 min at room temp then analyze 1 μl aliquots by GC-MS-SIM.

Comments. The method was used to measure ^{18}O abundance in α-phosphoroglycerol[1] and in xylopyranosides.[2] Kraska *et al.* reported MS data for a series of carbohydrate TBDMS derivs.[3]

References

[1]E. J. Heron and R. M. Caprioli, *Biochemistry* **13**, 4371 (1974).

[2]W. D. Hitz *et al.*, *Carbohydr. Res.* **46**, 195 (1976).

[3]B. Kraska *et al.*, *Carbohydr. Res.* **36**, 398 (1974).

Derivative Type. Methyl, methaneboronate.

Reagent. Diazomethane, methaneboronic acid.

Compound Type. Sugar phosphates.

Reaction

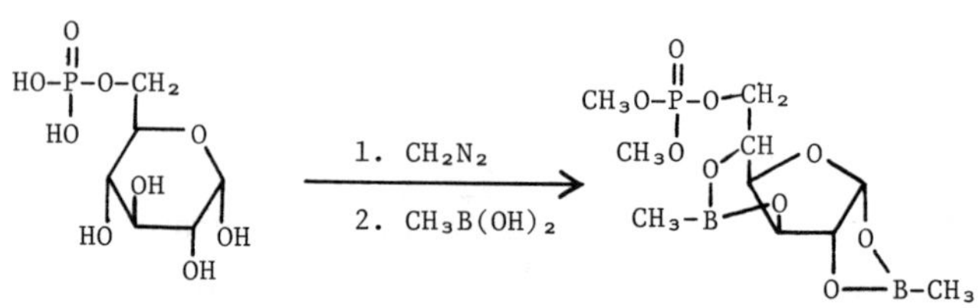

Typical Procedure. Diss 5 mg hexose phosphate disodium salt in 1 ml H_2O and pass through a small column of Dowex 50W (H form). Lyophilize the eluate and diss the residue in a few ml MeOH. Add CH_2N_2 in Et_2O until the yellow color persists. Evap to dryness and diss residue in 1 ml dry pyridine. Add 10 mg methaneboronic acid. After a few min, analyze by GC-MS.

Comments. Butaneboronate derivs were prepared in the same manner. Derivs of glucose, galactose, mannose, fructose, and methyl gluconate phosphates and deuterated analogs were studied.

References

[1]J. Wiecko and W. R. Sherman, *Org. Mass Spectrom.* **10**, 1007 (1975).

13.6. DERIVATIVES OF INTACT OLIGOSACCHARIDES

13.6.1. Trimethylsilyl Derivatives

Derivative Type. Trimethylsilyl.

Reagent. Hexamethyldisilazane (HMDS)/trimethylchlorosilane (TMCS)/pyridine.

Compound Type. Oligosaccharides.

Reaction

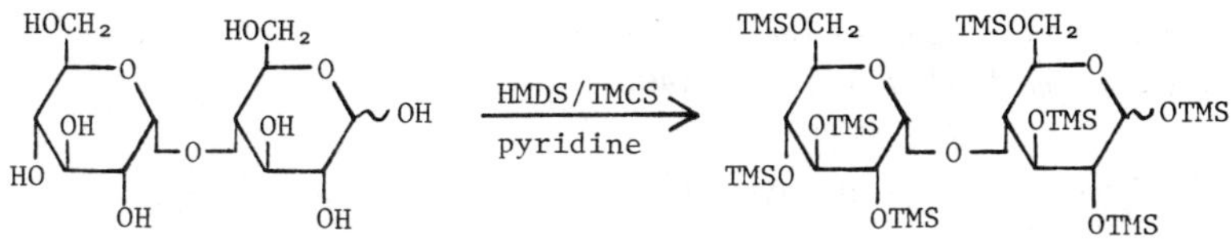

Typical Procedure. Diss 0.5 mg oligosaccharide in 1 ml pyridine and add 0.2 ml HMDS and 0.1 ml TMCS. Allow to react 2 hr at room temp (disaccharides) or 3 hr at 70° (higher oligosaccharides). Add 2 ml hexane and 2 ml H_2O, mix well, and discard H_2O layer. Wash hexane with 1 ml H_2O, dry over Na_2SO_4, and evap under vac. Diss residue in 1 ml hexane and analyze by MS.

Comments. When the oligosaccharide did not diss easily in pyridine, a small drop of H_2O was added before the pyridine. The HMDS addition was then increased to 0.6 ml and the TMCS to 0.3 ml. The MS behavior of 18 TMS disaccharides containing only aldohexoses and connected via all possible linkages, 3 aldopentose containing disaccharides, and four TMS trisaccharides composed of aldohexoses were studied.[1] The method was also applied to disaccharides containing 2-acetamido-2-deoxyglycoses[2] and fructose.[3] Bhatti *et al.*[4] used the reagents to prepare derivs of an extensive series of mono- through pentasaccharides.

References

[1] J. P. Kamerling *et al.*, *Tetrahedron* **27**, 4275 (1971).
[2] J. P. Kamerling *et al.*, *Tetrahedron* **27**, 4749 (1971).
[3] J. P. Kamerling *et al.*, *Tetrahedron Lett.* 2367 (1971).
[4] T. Bhatti *et al.*, *Biochem. Biophys. Acta* **222**, 339 (1970).

Derivative Type. Trimethylsilyl.

Reagent. Hexamethyldisilazane (HMDS)/trifluoroacetic acid (TFA)/pyridine.

Compound Type. Sugars.

Reaction

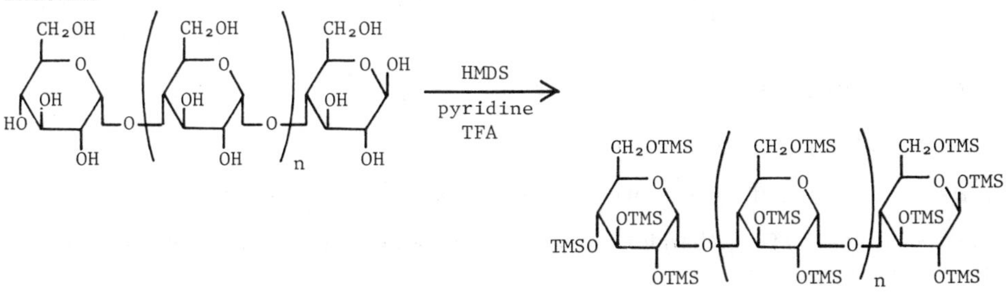

Typical Procedure. Diss 60–70 mg of 80% solids syrup in 1.0 ml pyridine. Add 0.9 ml HMDS and 0.1 ml TFA. Shake vigorously 30 sec then let stand for 15 min with occasional shaking. analyze a 5 μl aliquot by GC.

Comments. The method was developed to avoid the necessity for dry samples and was applicable for use in the presence of up to 40 mg H_2O. The ammonium trifluoroacetate by product was soluble in pyridine giving a clear soln. In the original work the deriv was used to analyze mono- through tetrasaccharides in corn syrup.[1] Beadle[2] used the method to analyze maltooligosaccharides up to the octaose from starch hydrolyzates.

References

[1]K. M. Brobst and C. E. Lott, *Cereal Chem.* **43**, 35 (1966).

[2]J. B. Beadle, *J. Agr. Food Chem.* **17**, 904 (1969).

Derivative Type. Oxime, trimethylsilyl.

Reagent. Hydroxylamine·HCl/pyridine, hexamethyldisilazane (HMDS)/trifluoroacetic acid (TFA)/pyridine.

Compound Type. Disaccharides.

Reaction

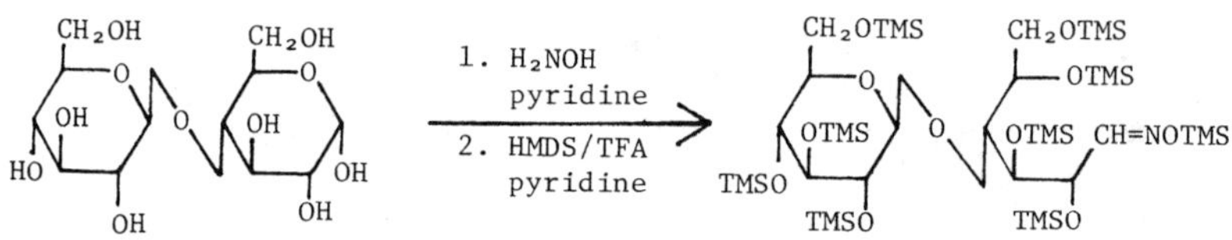

Typical Procedure.[2] Diss *ca* 5 mg sugar in 1 ml $H_2NOH \cdot HCl$ soln (25 mg/ml) in pyridine. Heat 30 min at 70–80° then evap to dryness at 45°. Remove traces of H_2O by adding 1 drop benzene and evap to dryness. Diss residue in 0.25 ml anhyd pyridine and add 225 μl HMDS and 25 μl TFA.[3] Allow to react 30 min at room temp then analyze by GC.

Comments. The derivs, which were quite unstable, generally gave multiple GC peaks due to isomeric forms. The derivs of lactose, maltose, gentiobiose, turanose, and lactulose, however, each gave a single main GC peak. GC data were reported for the derivs of 10 disaccharides.

References

[1]T. Toba and S. Adachi, *J. Chromatogr.* **135**, 411 (1977).

[2]B. S. Mason and H. T. Slover, *J. Agr. Food Chem.* **19**, 551 (1971).

[3]K. M. Brobst and C. E. Lott, *Cereal Chem.* **43**, 35 (1966).

13.6.2. Permethyl Alditols

Derivative Type. Permethyl alditol.

Reagent. Sodium borohydride, methylsulfinylcarbanion/methyl iodide.

Compound Type. Trisaccharide, α-D-mannopyranoside-(1→3)-β-D-mannopyranoside-(1→4)-2-acetamido-2-deoxy-D-glucose.

Reaction

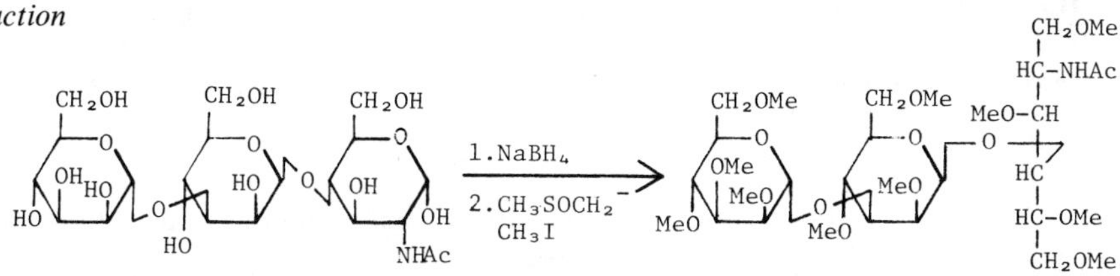

Typical Procedure. Diss the sample and 200 μg of isomaltotetraose (int stand) in 2 ml H_2O and reduce with 10 mg $NaBH_4$ for 2 hr. Adjust to pH 3 by adding Dowex 50 (H form) then filter and evap to dryness. Remove boric acid by adding and evap 3 × 2 ml MeOH. Diss residue in 2 ml DMSO and add 2 ml 2 *M* methylsulfinylcarbanion in DMSO. Sonicate 30 min, freeze, and add 1 ml CH_3I. Sonicate thawed mixt 30 min, pour into 10 ml H_2O and ext with 3 × 5 ml $CHCl_3$. Wash ext with 3 × 5 ml H_2O and evap to dryness. Diss residue in 20 μl acetone and analyze 1 μl aliquots by GC-MS-SIM.

Comments. The method was used to quantitate the trisaccharide in urine of humans with mannosidosis. The sample was isolated from the urine on a Sephadex G-25 column.

References

[1] A. Lundblad *et al.*, *Biomed. Mass Spectrom.* **2**, 285 (1975).

Derivative Type. Permethyl alditol.

Reagent. Sodium borodeuteride, methylsulfinylcarbanion/methyl iodide.

Compound Type. Disaccharides and trisaccharides.

Reaction

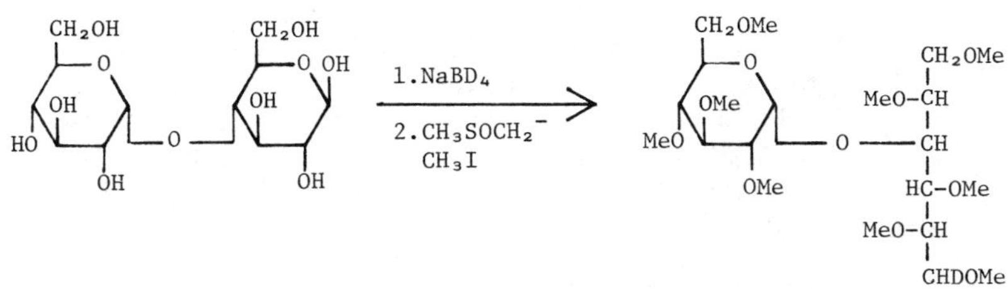

Typical Procedure. Reduce the disaccharide sample with sodium borodeuteride.[2] Diss 0.2 mg of the resulting alditol in 0.7 ml dry DMSO. Add under N_2 0.7 ml 2 *M* methylsulfinylcarbanion in DMSO. Sonicate 30 min then add 0.7 ml CH_3I with ice cooling. Sonicate 20 min. Add 5 ml $CHCl_3$. Remove $CHCl_3$ phase, wash 2 × with H_2O, dry over Na_2SO_4, and evap under N_2. Diss in acetone and analyze by GC or GC-MS.

Comments. Deuterium reduction permitted differentiation of (1→3)- and (1→4)- linked alditols. Disaccharides containing *N*-acetyl amino sugars yielded *N*-methyl-*N*-acetyl derivs. Monosaccharide components were identified by methanolysis of the disaccharide derivs followed by GC analysis. The method was also used for trisaccharides.[3]

References

[1] J. Kärkkäinen, *Carbohydr. Re.* **14**, 27 (1970).

[2] J. Kärkkäinen, *Carbohydr. Res.* **11**, 247 (1969).

[3] J. Kärkkäinen, *Carbohydr. Res.* **17**, 1 (1971).

13.7. DERIVATIVES OF POLYSACCHARIDE CONSTITUENTS

13.7.1. Alditol Dervatives

Derivative Type. Alditol acetate.

Reagent. Propionic anhydride/pyridine, diazomethane, lithium borohydride, sulfuric acid, sodium borohydride, acetic anhydride (Ac_2O)/pyridine.

Compound Type. 4-*O*-Methyl-D-glucuronic and other uronic acids (in polysaccharides).

Reaction

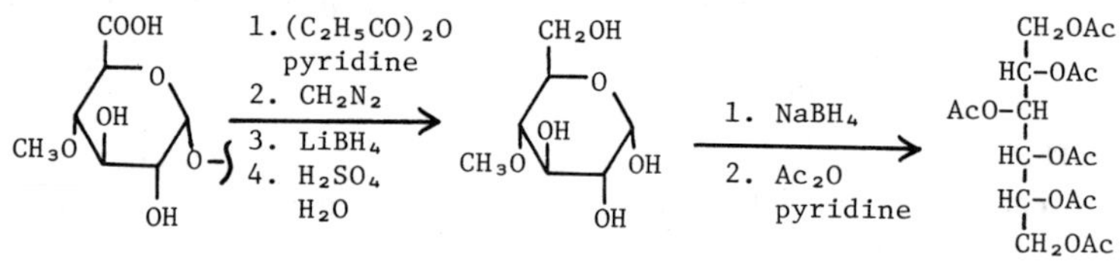

Typical Procedure. Diss 5 g powdered gum sample in 60 ml DMF, 45 ml pyridine, and 30 ml propionic anhydride and allow to react 2 days. Pour soln into 1000 ml 2% HCl, filter ppt, wash with cold H_2O, and dry under vac. Retreat 6 g of the product with 80 ml pyridine and 12 ml propionic anhydride for 3 days to yield 6.8 g of propionated acid. Diss 500 mg in 15 ml THF, and add to a soln of CH_2N_2 in Et_2O (50 ml) at $-73°$. Stir at $-73°$ for 1 hr then pour into 50 ml lt pet ether (30–60°). Diss 450 mg of the ppt in 50 ml THF, bring to reflux, and add dropwise 1 g $LiBH_4$ in 25 ml THF over 90 min. Reflux 18 hr, cool in ice, and add 25 ml H_2O over 2 hr. Dialyze 24 hr then recover polysaccharide by freeze drying.

Hydrolyze 100 mg in 10 ml 1 *N* H_2SO_4 (6 hr, 100°), neutralize with $BaCO_3$, and deionize with Amberlite IR-120 (H form) resin. Reduce 75 mg of the monosaccharide mixt with 100 mg $NaBH_4$ in 50 ml H_2O (12 hr). Pass through Amberlite IR-120 (H^+ form), distill with MeOH, and acetylate with Ac_2O/pyridine (10 ml, 1:1, v/v) for 1 hr at 100°. Evap to dryness, diss in $CHCl_3$, and analyze by GC.

Comments. The method permitted simultaneous estimation of 4-*O*-methyl-D-glucuronic, glucuronic, and galacturonic acids in polysaccharides. The method gave better recovery of uronic acid than an earlier method[2] employing esterfication with propylene oxide and $NaBH_4$ reduction. The MS of 4-*O*-methyl-D-glucitol pentaacetate was given in the latter paper.[2]

References

[1]G. G. S. Dutton and S. Kabir, *Anal. Lett.* **4**, 95 (1971).

[2]E. Sjöström *et al.*, *Acta Chem. Scand.* **23**, 3610 (1969).

Derivative Type. Alditol trifluoroacetate.

Reagent. Sodium borohydride, trifluoroacetic anhydride (TFAA).

Compound Type. Oligosaccharides.

Reaction

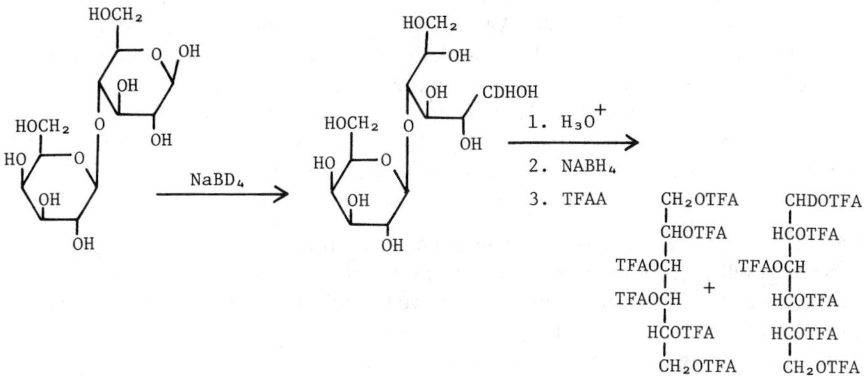

Typical Procedure. Treat 1 μg–1 mg of oligosaccharide with $NaBD_4$ in D_2O (sample conc *ca* 0.5%) in a sealed tube for 2 hr. Destroy excess $NaBD_4$ by adding Dowex 1 × 8 (50–100 mesh). Remove ion exchanger by filtration and wash with water. Combine filtrate and washings and evap to dryness. Hydrolyze residue with 1 *N* H_2SO_4 at 100°. Remove acid by addition of $BaCO_3$ and filtration and centrifugation of $BaSO_4$. Reduce hydrolyzate with $NaBH_4$ as above. After destruction of excess $NaBH_4$ with ion exchanger, evap filtrate to dryness and treat with MeOH to remove H_3BO_3. Diss residue in H_2O and transfer aliquot to reaction vial. Dry and add 0.1 ml TFAA and 0.1 ml EtOAc. After 15 min at 80°, cool and analyze by GC-MS.

Comments. Alditol trifluoroacetates from mannose, glucose, and galactose residues in oligosaccharides were determined by GC-MS. The deuterium reduction allowed identification of the reducing end group simultaneously.

References

[1] A. Kamei *et al.*, *Chem. Pharm. Bull.* **24**, 1108 (1976).

Derivative Type. Methylated alditol acetate.

Reagent. Methyl vinyl ether/toluene sulfonic acid (TsOH), sodium methoxide (NaOMe), methyl iodide/silver oxide, sulfuric acid, sodium borohydride, acetic anhydride.

Compound Type. Acetylated polysaccharides.

Reaction

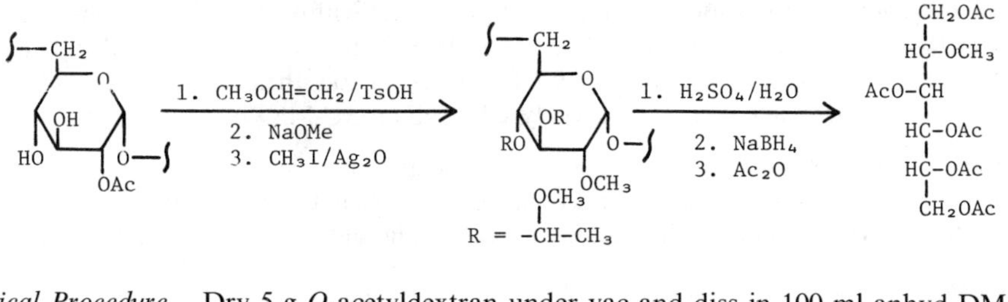

Typical Procedure. Dry 5 g *O*-acetyldextran under vac and diss in 100 ml anhyd DMSO. Add 185 mg TsOH and maintain at 15°. With stirring add 15 g methyl vinyl ether (condensed at −10°) portionwise. Stir 3.5 hr at 15° then place on a 25 × 7 cm column of Sephadex LH-20 and elute with anhyd acetone. Evap the product eluate and diss 5 g of the amorphous residue in 100 ml 0.2% NaOMe in MeOH. Reflux 4 hr then place on an 80 × 4 cm column of Sephadex LH-20 and elute with MeOH. Conc the product eluate to a red syrup (3.3 g) and diss in 50 ml DMF. Add 5 ml CH_3I and 5 g Ag_2O. Allow to react 20 hr. Filter and wash salts with 25 ml DMF. Pool filtrate and washings and repeat methylation twice. Carefully add 200 ml benzene to the resulting filtrate, filter ppt, and evap benzene from the filtrate. Place the remaining DMF soln on a 80 × 4 cm Sephadex LH-20 column and elute with MeOH. Evap product eluate to a lt yellow syrup. Hydrolyze in 50 ml 2 *N* H_2SO_4 for 5.5 hr at 100°. Deionize soln with IR-4B resin (free base form) and conc to a syrup. Reduce with $NaBH_4$ and acetylate with Ac_2O.[2,3] Analyze the resulting partially methylated alditol acetates by GC.

Comments. The method was used to determine the positions of acetylation in polysaccharides. The originally acetylated positions were converted to the *O*-methyl in the final derivs. This method was a shorter and simpler alternative to the earlier similar method in which the free hydroxyls were first treated with phenyl isocyanate to form phenyl carbamoyl derivs.[4,5] In addition to forming *O*-(1-methoxyethyl) derivs, methyl vinyl ether also forms cyclic ethylidene acetals from suitably positioned 1,3-dihydroxy functions (see p. 557).[6]

References

[1]A. N. DeBelder and B. Norrman, *Carbohyd. Res.* **8**, 1 (1968).
[2]H. Bjorndal *et al.*, *Acta Chem. Scand.* **21**, 1801 (1967).
[3]A. C. Richardson, *Carbohydr. Res.* **4**, 415 (1967).
[4]H. O. Bouveng *et al.*, *Acta Chem. Scand.* **14**, 742 (1960).
[5]H. O. Bouveng, *Acta Chem. Scand.* **15**, 96 (1961).
[6]M. L. Wolfram *et al.*, *J. Org. Chem.* **33**, 1067 (1968).

Derivative Type. Methylated alditol acetate.

Reagent. Methylsulfinylcarbanion/methyl iodide, formic acid/water, sodium borohydride, acetic anhydride.

Compound Type. Polysaccharides.

Reaction

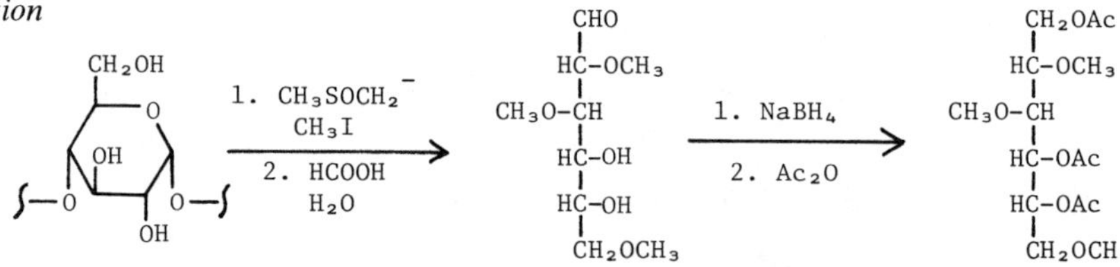

Typical Procedure. Methylate polysaccharide in DMSO with methylsulfinylcarbanion and CH_3I. After several hr pour mixt into H_2O and isolate methylated polysaccharide by dialysis or solvent ext. Hydrolyze by heating with 90% HCOOH at 100° then evap to dryness. Reduce with $NaBH_4$ and acetylate with Ac_2O. Analyze by GC or GC-MS.

Comments. The method was used to determine the sugar composition of polysaccharides and the positions of glycosidic linkages (the acetylated positions in the final deriv). Ref 1 discusses methylation analysis of polysaccharides in general and lists alternative methylation methods.

References

[1]H. Björndal *et al.*, *Angew. Chem. Int. Ed.* **9**, 610 (1970).
[2]H. Björndal *et al.*, *Acta Chem. Scand.* **21**, 1801 (1967).
[3]H. Björndal *et al.*, *Carbohydr. Res.* **5**, 433 (1967).

Derivative Type. *O*-Acetyl-penta-*O*-methylhexitol.

Reagent. Sodium borohydride, methylsulfinylcarbanion/methyl iodide, formic acid/sulfuric acid, acetic anhydride/pyridine.

Compound Type. Reducing-end sugar residues of polysaccharides.

Reaction

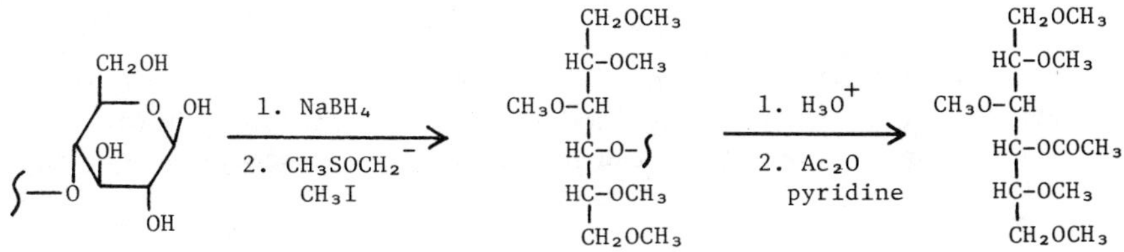

Typical Procedure. Diss 20 mg polysaccharide in 2 ml H_2O containing 40 mg $NaBH_4$ and allow to react at 35° for 16 hr. Destroy excess $NaBH_4$ with 6 *N* HOAc and ppt reduced polysaccharide by adding 8 ml EtOH. Desalt by repeated dispersions in 80% EtOH followed by centrifugations then lyophilize. Diss 10 mg of reduced polysaccharide in 1 ml DMSO and add 1 ml 2 *M* methylsulfinylcarbanion in DMSO. Stir under N_2 at room temp 4 hr then add 0.5 ml CH_3I dropwise at 18–20° with ice cooling. Dil with H_2O and dialyze against H_2O for 2 days. Ext with $CHCl_3$ and evap ext.

Heat residue with 90% HCOOH at 100° for 2 hr then with 1 *N* H_2SO_4 under vac at 100° for 6 hr. Add 0.1 ml int stand soln containing 20 μg 2,3,4,6-tetra-O-methyl-D-glucitol. Neutralize with $Ba(OH)_2$ and centrifuge. Remove Ba on a 0.2 ml column of Dowex 50 × 8 (H form). Remove reducing sugars on a 0.2–1 ml column of Dowex 1 × 8 (OH form).[2,3] Evap the aq soln to a small vol and again treat with Dowex 50 × 8 as above. Evap effluent to dryness and remove traces of boric acid with repeated evap of added MeOH. Acetylate with 0.1 ml Ac_2O and 0.1 ml pyridine at 100° for 2 hr. Add 0.2 ml H_2O and evap under vac. Analyze by GC-MS.

Comments. The method was used to identify reducing-end sugar residues in polysaccharides. *O*-Ethyl derivs were prepared by the same method using ethyl iodide. Sodium borodeuteride reduction was used to distinguish 3- and 4-mono-*O*-acetyl-penta-*O*-alkylhexitols which have the same substitution pattern. GC and MS data were reported for the various derivs of D-glucitol and D-mannitol.

References

[1]H. Yamaguchi and K. Okamoto, *J. Biochem. (Tokyo)* **82**, 511 (1977).

[2]H. Yamaguchi *et al.*, *J. Biochem. (Tokyo)* **79**, 299 (1976).

[3]H. Yamaguchi and K. Makino, *J. Biochem. (Tokyo)* **81**, 563 (1977).

Derivative Type. Ethylated alditol acetate.

Reagent. Methylsulfinylcarbanion/ethyl iodide, trifluoroacetic acid/water, sodium borodeuteride, acetic anhydride.

Compound Type. Polysaccharides.

Reaction

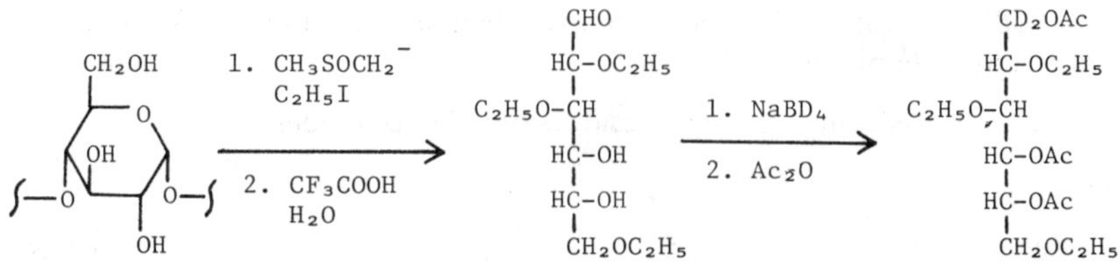

Typical Procedure. Diss 1–6 mg of carbohydrate (dried over P_2O_5 for at least 10 hr at 60° under vac) in 1.0 ml dry DMSO under N_2. Add 100 μl of 2.0 *M* methylsulfinylcarbanion soln.[2] Stir 15 min for soluble carbohydrates and several hr if the carbohydrate is only partially soluble. Add 22 μl C_2H_5I and stir 2 hr. Add a second 100 μl of 2.0 *M* carbanion soln followed by 22 μl of C_2H_5I as above and stir 2 hr. Sep ethylated polysaccharide from DMSO and other reagents by either dialysis or chromatography on Sephadex LH-20. Hydrolyze by heating with 1 ml 2.0 *M* CF_3COOH for 2 hr at 121° in a sealed tube. Evap to dryness under an air stream at 50° taking care to remove samples from heat as soon as they are dry to prevent decomposition. Reduce the freed partially ethylated aldoses to the alditols with 10 mg $NaBD_4$ in 1 ml 1 *M* NH_3 at room temp for 2 hr. Decompose excess $NaBD_4$ by dropwise addn of HOAc until H_2 evolution ceases. Evap to dryness under an air stream at 50°. Remove borate by adding and evap 10% HOAc in MeOH (2 ×) and MeOH (8 ×). Acetylate by heating in a sealed tube with Ac_2O at 121° for 3 hr. Analyze by GC or GC-MS.

Comments. The derivs were explored as an alternative to methylation analysis to enable GC resolution of some sugars which are not separated well by GC of the methyl derivs. Sodium borodeuteride was used in the aldose to alditol reductions to facilitate MS interpretation. Borate (from hydrolysis of $NaBD_4$) was removed because it has been shown to interfere with acetylation.[4] MS data has been tabulated for identification of the derivs.[3]

References

[1]D. P. Sweet *et al.*, *Carbohyd. Res.* **40**, 199, 217 (1975).

[2]P. A. Sandford and H. E. Conrad, *Biochemistry* **5**, 1508 (1971).

[3]D. P. Sweet *et al.*, *Biomed. Mass Spectrom.* **1**, 263 (1974).

[4]P. Albersheim *et al.*, *Carbohyd. Res.* **5**, 340 (1967).

13.7.2. Other Derivatives of Polysaccharide Constituents

Derivative Type. Methyl glycoside, *N*-acetyl, trimethylsilyl ether.

Reagent. Methanol/HCl, acetic anhydride (Ac_2O)/pyridine, trimethylsilylimidazole (TMSI)/pyridine.

Compound Type. Polysaccharide constituents.

Reaction

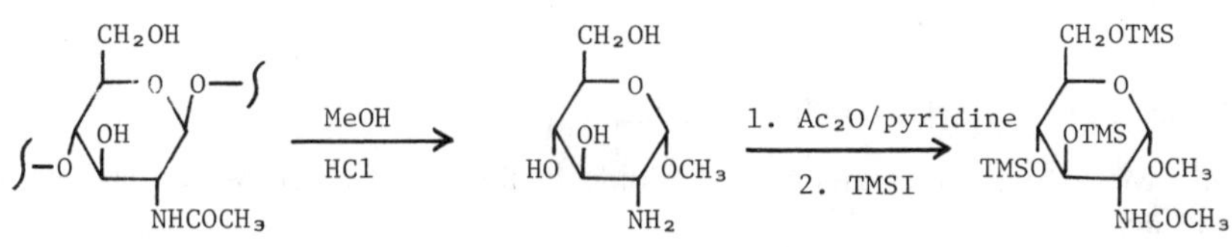

Typical Procedure. Place sample in 1 ml glass ampule and dry over P_2O_5 in a vac dessicator.

Add 0.5 ml 1.5 *N* HCl in anhyd MeOH, seal under N_2, and heat at 80° for 24 hr. Cool ampule, open, and add 0.15 ml pyridine and 0.1 ml Ac_2O. Seal and allow to stand at room temp for 30 min. Evap to dryness in vac dessicator over KOH. Add 50 μl of int stand soln and evap to dryness over P_2O_5. Add 100 μl of 1.5 *M* TMSI in pyridine. Analyze by GC.

Comments. The method was simpler and involved fewer transfers than similar previous methods. The int stand soln contained 2 m*M* arabitol, mannitol, and perseitol. The methanolysis gave multiple but characteristic peaks for each sugar. Partial O-acetylation occurred during the *N*-acetylation step.

References

[1]D. G. Pritchard and C. W. Todd, *J. Chromatogr.* **133**, 133 (1977).

Derivative Type. *N*-Acetyl, trimethylsilyl.

Reagent. Hydrochloric acid, acetic anhydride (Ac_2O)/acetone/Dowex 1, hexamethyldisilazane (HMDS)/trimethylchlorosilane (TMCS)/pyridine.

Compound Type. Amino sugars (in polysaccharides).

Reaction

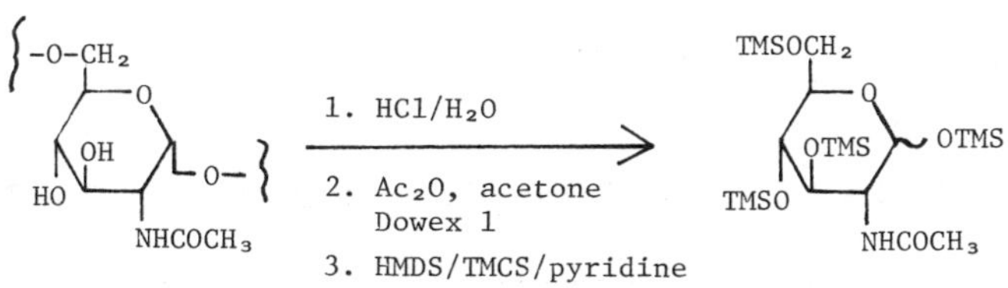

Typical Procedure. Mix polysaccharide sample with 1.25 ml 2 *M* HCl and heat 3 hr at 100°. Cool to room temp, add 3.75 ml H_2O and transfer 1–1.5 ml at a time to a 1 × 4 cm column of Dowex 1 (CO_3 form). Rinse hydrolysis tube and pipette with 3 × 2 ml H_2O and add to the column. Discard the first 2.5 ml of eluate and continue elution with H_2O collecting the next 12 ml.

Mix 5 ml of the eluate with 5 ml 0.6% (v/v) Ac_2O in acetone and transfer this mixt to a 1 × 12 cm column of the same resin [prewashed with 1 : 1 (v/v) aq acetone]. Elute with H_2O discarding the first 9 ml and collecting the next 17 ml. Add mannitol int stand and evap eluate to dryness under vac over P_2O_5. Treat the residue (1–2 mg) with 0.5 ml HMDS/TMCS/pyridine (2:1:10, v/v/v) at room temp overnight. Analyze by GC.

Comments. The novel acetylation procedure required the presence of acetone; the reaction would not go to completion in pure aq soln. MeOH was less effective than acetone in promoting acetylation.

References

[1]G. A. Levvy *et al., Biochim. Biophys. Acta* **222**, 333 (1970).

Derivative Type. Diethyl dithioacetal, trimethylsilyl.

Reagent. Ethanethiol (EtSH)/trifluoroacetic acid (TFA), hexamethyldisilazane (HMDS)/trimethylchlorosilane (TMCS).

Compound Type. Aldehydes (periodate oxidation products of pyranose sugars).

Reaction

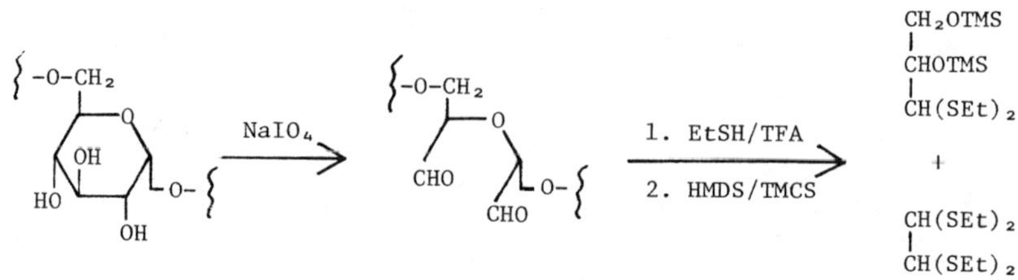

Typical Procedure. Diss 1–100 μmol aldehyde sample in 20 μl EtSH/TFA (10:1, v/v) and allow to react 30 min at 25°. Add 50 μl of a soln of D-xylitol int stand in pyridine followed by 100 μl HMDS and 50 μl TMCS. Heat 30 min at 50°, centrifuge, then analyze an aliquot of the supernatant by GC.

Comments. The method was developed for analysis of the component aldehydes formed on oxidation of carbohydrates with periodate. The EtSH/TFA both catalyzed release of the component aldehydes and formed the dithioacetal derivs. Unlike oxime and hydrazone type derivs, the dithioacetals had no steric isomers simplifying the GC analysis.

References

[1]S. Honda *et al.*, *Anal. Chem.* **50**, 55 (1978).

13.8. DERIVATIVES OF GLUCURONIDES AND OTHER GLYCOSIDES

13.8.1. Permethyl Derivatives of Glucuronides

Derivative Type. Permethyl.

Reagent. Methylsulfinylcarbanion/methyl iodide.

Compound Type. Steroid glucuronides.

Reaction

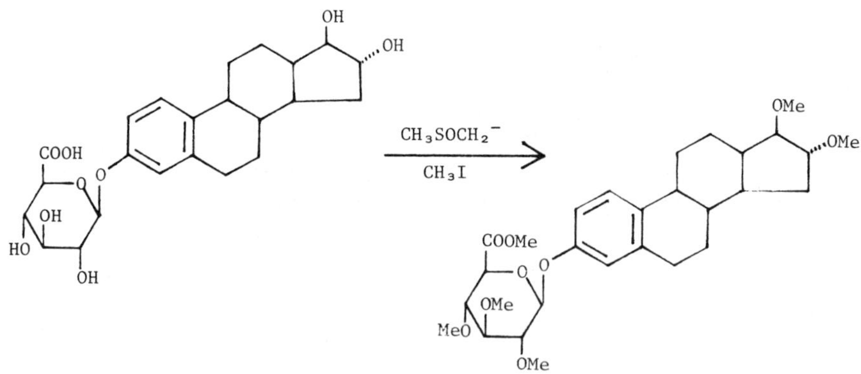

Typical Procedure. Diss 100 μg steroid glucuronide in 100 μl dry DMSO and add 10 μl or more 1 *M* methylsulfinylcarbanion in DMSO.[3] Allow to stand 15 min then add excess CH_3I. Allow to stand 1 hr then add 1 ml H_2O. Ext with 1 ml CH_2Cl_2 and wash ext with 3 × 1 ml H_2O. Evap to dryness under N_2 and diss residue in 100 μl $CHCl_3$. Analyze by GC-MS.

Comments. Permethyl derivs of the glucuronides were obtained with a tenfold or greater excess of carbanion whereas the aglycones required a thirtyfold or greater excess.[2] Steroids containing an α,β-unsatd ketone function underwent additional methylation.[1]

References

[1]R. M. Thompson and D. M. Desiderio, *Biochem. Biophys. Res. Commun.* **48**, 1303 (1972).

[2]R. M. Thompson, *J. Steroid Biochem.* **7**, 845 (1976).

[3]P. A. Leclerq and D. M. Desiderio, *Anal. Lett.* **4**, 305 (1971).

Derivative Type. Permethyl.

Reagent. Methylsulfinylcarbanion/methyl iodide.

Compound Type. Drug glucuronides.

Reaction

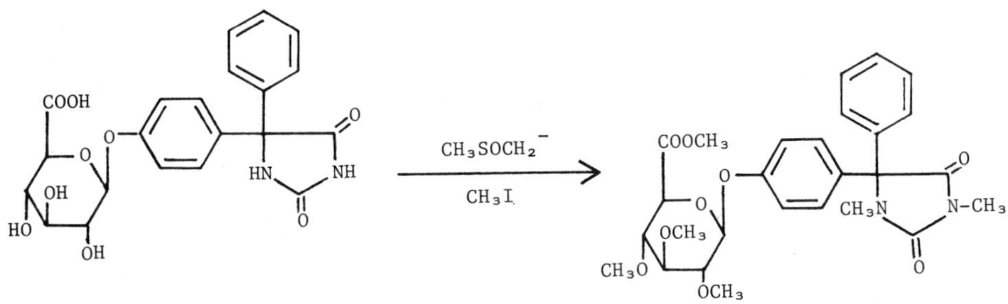

Typical Procedure. Evap a 20 μl aliquot of bile to dryness under N_2. Diss residue in 100 μl of DMSO (distilled over CaH_2). Flush vial with N_2 and add 25 μl 1 *M* methylsulfinylcarbanion in DMSO.[3] After 15 min add 10 μl of CH_3I or CD_3I. After 1 hr add 1 ml H_2O. Ext product into 1 ml $CHCl_3$ and wash with 2×1 ml H_2O. Evap $CHCl_3$ to dryness and redissolve in 10 μl $CHCl_3$ for GC-MS analysis.

Comments. Glucuronides were identified in bile by presence of characteristic ions at m/e 101 and 201.[1,2] MS data were given including data for pertrideuteromethyl derivs.[1,2] The method has been used in studies of the metabolism of diphenylhydantoin,[4,5] phenyramidol,[6] methocarbamol,[7] carbamazepine,[8] and menadione.[9] Unconjugated barbiturates have been shown to undergo partial ring opening to give multiple derivs by this permethylation method.[10,11] The deriv method has been extended to the analysis of drug metabolites, sugars, and acids in urine.[12]

References

[1]R. M. Thompson *et al.*, *Drug Metab. Disp.* **1**, 489 (1973).

[2]R. M. Thompson and N. Gerber, *J. Chromatogr.* **124**, 321 (1976).

[3]P. A. Leclerq and D. M. Desiderio, *Anal. Lett.* **4**, 305 (1971).

[4]N. Gerber *et al.*, *Res. Commun. Chem. Pathol. Pharmacol.* **6**, 499 (1973).

[5]R. M. Thompson *et al.*, *Drug Metab. Disp.* **4**, 349 (1976).

[6]N. Gerber *et al.*, *Drug Metab. Disp.* **2**, 140 (1974).

[7]R. M. Thompson *et al.*, *Xenobiotica* **5**, 145 (1975).

[8]J. E. Bauer *et al.*, *Experientia* **32**, 1032 (1976).

[9]R. M. Thompson *et al.*, *Res. Commun. Chem. Pathol. Pharmacol.* **4**, 543 (1972).

[10]R. M. Thompson and D. M. Desiderio, *Org. Mass Spectrom.* **7**, 989 (1973).
[11]R. M. Thompson, *J. Pharm. Sci.* **65**, 288 (1976).
[12]R. M. Thompson, *Res. Commun. Chem. Pathol. Pharmacol.* **16**, 145 (1977).

Derivative Type. Permethyl.

Reagent. Methylsulfinylcarbanion/methyl iodide.

Compound Type. Diphenylhydantoin *N*-3 glucuronide.

Reaction

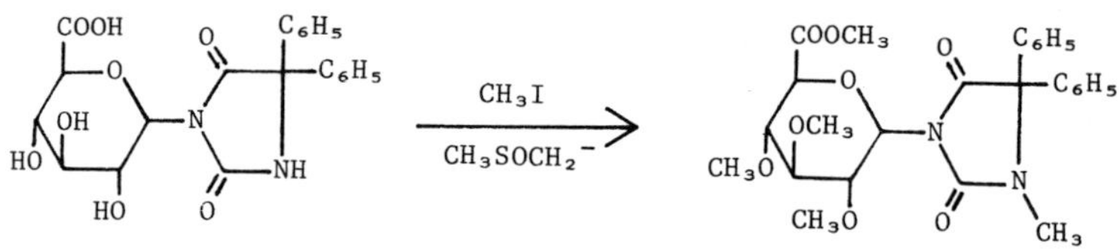

Typical Procedure. Evap 50 μl bile to dryness under N_2 and diss residue in 100 μl dry DMSO. Add 50 μl of 1 *M* methylsulfinylcarbanion in DMSO under N_2 and mix on Vortex 1–2 min. Add 10 μl CH_3I and mix 1 min. Quench with 1.5 ml H_2O and ext with 0.25 ml $CHCl_3$. Wash ext with 2 × 1.5 ml H_2O. Dil $CHCl_3$ soln to 50 μl and analyze 3 μl aliquots by GC-MS.

Comments. Various CD_3 derivs were also prepared and the MS data used to support the proposed *N*-glucuronide structure. This example illustrates the applicability of permethylation to *N*-glucuronides.

References

[1]R. G. Smith *et al., Biomed. Mass Spectrom.* **4**, 275 (1977).

13.8.2. Trimethylsilyl Derivatives of Glucuronides

Derivative Type. Trimethylsilyl.

Reagent. Bis(trimethylsilyl)trifluoroacetamide (BSTFA)/trimethylchlorosilane (TMCS).

Compound Type. Glucuronide conjugates.

Reaction

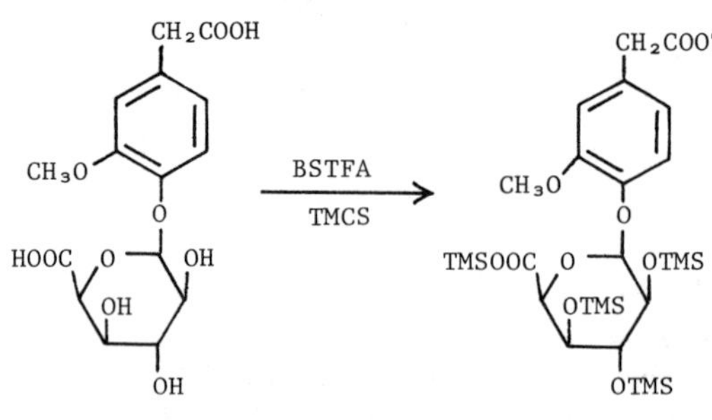

Typical Procedure.[2,3] Diss 1–20 μg sample in 50–100 μl pyridine. Add 50–100 μl BSTFA containing 1% TMCS. Heat at 50° overnight. Analyze by GC-MS.

Comments. MS data were reported for a series of ether and ester glucuronides including morphine glucuronide.[1] The ether glucuronides gave intact TMS derivs. The ester glucuronides apparently underwent cleavage in the deriv reaction go give the TMS aglycone and TMS-glucopyrano-(6→1)-lactone.

References

[1]J. E. Mrochek and W. T. Rainey, *Anal. Biochem.* **57**, 173 (1974).
[2]J. E. Mrochek *et al.*, *Clin. Chem.* **17**, 72 (1971).
[3]W. C. Butts, *Anal. Biochem.* **46**, 187 (1972).

Derivative Type. Trimethylsilyl.

Reagent. Bis(trimethylsilyl)acetamide (BSA)/bis(trimethylsilyl)trifluoroacetamide (BSTFA).

Compound Type. Glucuronides of phenylethanolamine drugs.

Reaction

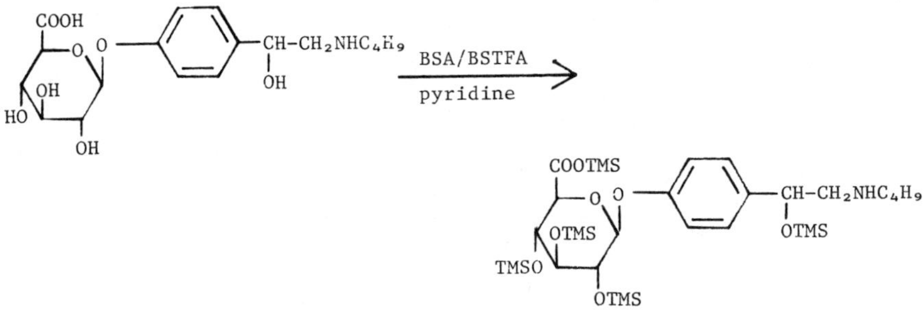

Typical Procedure. Diss 1 mg of the glucuronide in 40 μl pyridine and 40 μl BSA/BSTFA (1 :1, v/v). Warm slightly then allow to stand at room temp 5 hr. Analyze by GC-MS.

Comments. MS data were reported for a series of glucuronide derivs.

References

[1]K. H. Pook *et al.*, *J. Pharm. Sci.* **65**, 1513 (1976).

Derivative Type. Trimethylsilyl.

Reagent. *N*-Methyl-*N*-trimethylsilyl trifluoroacetamide (MSTFA)/trimethylchlorosilane (TMCS).

Compound Type. Steroid glucuronides.

Reaction

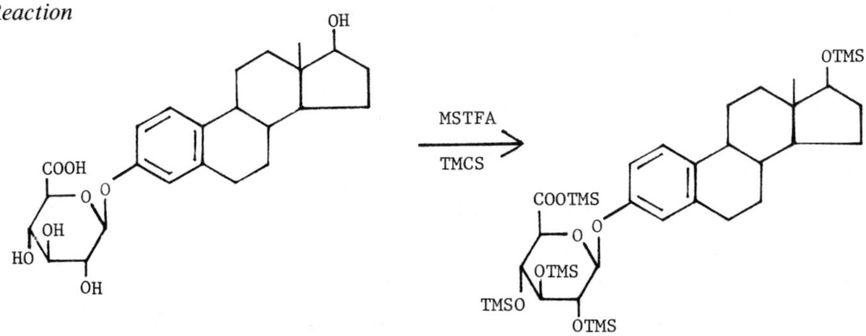

Typical Procedure. Diss steroid containing sample (100–200 μg) in 95 μl MSTFA and 5 μl TMCS. Heat 2 hr at 80°. Evap and analyze deriv by direct probe MS.

Comments. The derivs were also prepared using *N*-methyl-*N*-trimethylsilyl heptafluorobutyramide in place of MSTFA. Full MS were given for the derivs of 13 steroid glucuronides. The MS of the derivs permitted distinguishing "aliphatic" (alcohol) from "aromatic" (phenol) glucuronides.

References

[1]B. Spiegelhalder *et al.*, *J. Steroid Biochem.* **7**, 749 (1976).

Derivative Type. Methyl ester, trimethylsilyl ether.

Reagent. Diazomethane, hexamethyldisilazane (HMDS)/trimethylchlorosilane (TMCS).

Compound Type. Glucuronide conjugates.

Reaction

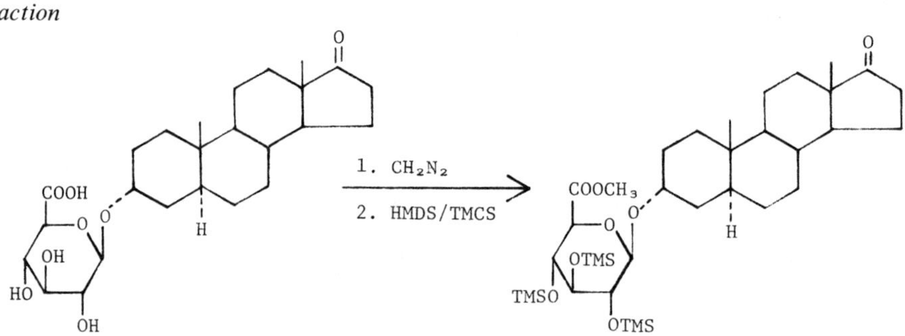

Typical Procedure. Ext 20 ml urine with 2 × 20 ml $CHCl_3$ and discard $CHCl_3$. Add 20 ml satd NaCl and 20 ml aq 3 *M* pyridinium sulfate, and ext with 2 × 60 ml $CHCl_3$. Evap ext and diss residue in minimum amt of $Et_2O/CHCl_3$ (1:1). Add ethereal diazomethane. Evap soln to dryness and TLC residue on silica gel G using sequentially benzene/acetone (8:2) and benzene acetone (1:1). After second development elute glucuronides ($R_f \approx 0.3$) with $CHCl_3$/MeOH (1:1) and MeOH/pyridine (1:1). Evap combined eluates and treat with 0.15 ml pyridine, 1.10 ml HMDS, and 0.05 ml TMCS. Allow to stand overnight and analyze soln directly by GC-MS.

Comments. The method was used to identify glucuronide conjugates of pregnanediol, androsterone, and etiocholanolone in human pregnancy urine.[1] BSA has been used in place of HMDS/TMCS for steroid glucuronides[2,3] and phenolic glucuronides.[3] GC separation of the derivs of the glucuronides of androstane-17-one-3α-ol and -3β-ol has been demonstrated.[4] MS data were reported for the derivs.

References

[1]P. I. Jaakomaki, *Biochim. Biophys. Acta* **137**, 216 (1967).

[2]E. C. Horning *et al.*, *J. Gas Chromatogr.* **5**, 283 (1967).

[3]S. Billets *et al.*, *J. Med. Chem.* **16**, 30 (1973).

[4]W. J. A. VandenHeuval, *J. Chromatogr.* **28**, 406 (1967).

Derivative Type. *n*-Propyl ester, trimethylsilyl ether.

Reagent. *n*-Propanol (*n*-PrOH)/HCl, trimethylsilylimidazole (TMSI).

Compound Type. Estrogen glucuronides.

Reaction

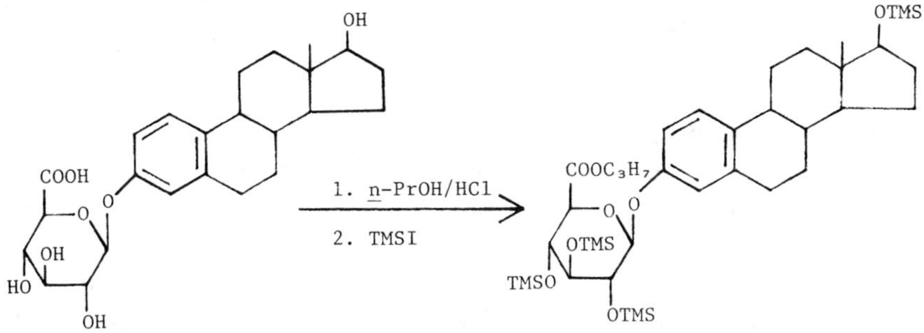

Typical Procedure. Convert 300 μg of the steroid glucuronide sodium salt to the free acid by passage through an ion exchange column (Amberlite GC-120). Evap the eluate to dryness under reduced pressure and diss the residue in HCl-*n*-PrOH (5% w/v) soln. After 30 min at room temp evap to dryness and treat the residue with TMSI for 30 min at room temp. Analyze by GC-MS.

Comments. The deriv was used for mass spectral characterization of 9 estrogen glucuronides. d_9-Trimethylsilyl derivs were also prepared using d_9-TMSI.

References

[1]H. Miyasaki *et al*., *Biomed. Mass Spectrom*. **3**, 55 (1976).

13.8.3. Other Derivatives of Glucuronides

Derivative Type. Methyl ester, *O*-acetyl.

Reagent. Diazomethane, acetic anhydride (Ac_2O)/pyridine.

Compound Type. Glucuronides.

Reaction

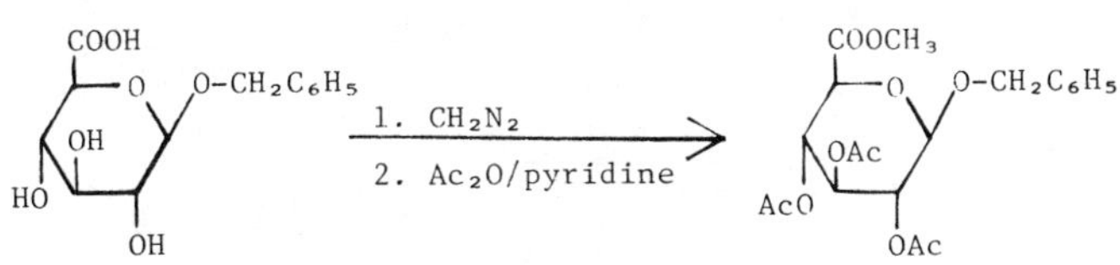

Typical Procedure.[1] Diss 0.5–1 mg of glucuronide in 1 ml MeOH and add CH_2N_2 in Et_2O until the yellow color persists. Evap to dryness and acetylate residue with Ac_2O pyridine (1 : 1, v/v). Evap and diss residue in acetone prior to GC analysis.

Comments. Trimethylsilyl derivs (prepared with HMDS/TMCS/pyridine) and permethyl derivs (from NaH/CH_3I in DMF or DMSO) were also examined for various types of glucuronides including ether-, ester-, *N*-, *S*-glucuronides. None of the derivs however were deemed suitable by the authors[1] for quantitation of glucuronides in urine.

References

[1]T. Imanari and Z. Tamura, *Chem. Pharm. Bull*. **15**, 1677 (1967).
[2]Z. Tamura and T. Imanari, *Chem. Pharm. Bull*. **12**, 1386 (1964).

Derivative Type. Acetyl, methyl ester.

Reagent. Acetic anhydride(Ac_2O)/methanesulfonic acid (MsOH), diazomethane.

Compound Type. Glucuronides.

Reaction

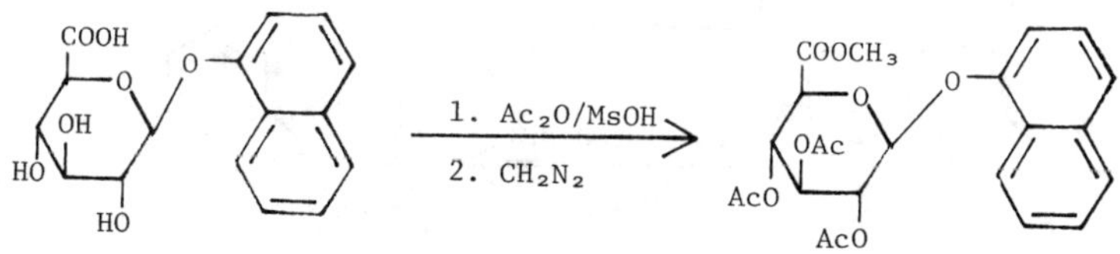

Typical Procedure. Diss glucuronide sample in 1 ml cold Ac_2O containing 50 μl MsOH. Allow to stand at room temp 24 hr then pour into ice H_2O and ext with $CHCl_3$. Dry ext over Na_2SO_4 and evap under vac. Methylate with excess CH_2N_2 in Et_2O then evap under vac. Diss residue in $CHCl_3$ and analyze by GC.

Comments. TMS methyl esters (from CH_2N_2 and HMDS/TMCS) and TMS ether, TMS esters (from HMDS/TMCS) were also examined but the acetyl methyl ester derivs were deemed superior for quant deriv of glucuronides in crude mixt.[1] MS data for the derivs have been reported.[2]

References

[1]J. B. Knaak *et al.*, *J. Agr. Food Chem.* **15**, 605 (1967).

[2]G. D. Paulson *et al.*, *Anal. Chem.* **45**, 21 (1973).

Derivative Type. Methyl ester, trifluoroacetate (TFA).

Reagent. Diazomethane, trifluoroacetic anhydride (TFAA).

Compound Type. Glucuronides.

Reaction

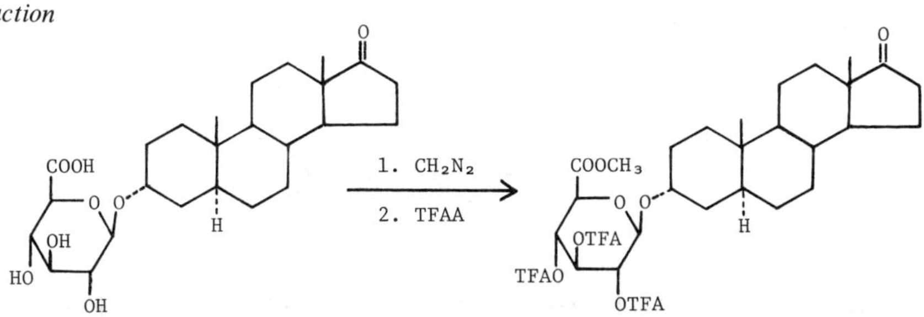

Typical Procedure. To 100 μl of 1 mg/ml soln of glucuronide in MeOH, add soln of CH_2N_2 in Et_2O until color persists. After 15 min at room temp evap to dryness under N_2 at 50°. Add 200 μl EtOAc and 25 μl TFAA. Mix on Vortex mixer for 15 min at room temp and inject soln into GC.

Comments. The method was also demonstrated for 1-naphthol glucuronide. Removal of excess TFAA by evap or ext resulted in rapid decomposition of the deriv.

References

[1]H. Ehrsson *et al.*, *J. Chromatogr.* **101**, 206 (1974).

13.8.4. Derivatives of Other Glycosides

Derivative Type. Trifluoroacetyl (TFA).

Reagent. Trifluoroacetylimidazole (TFAI).

Compound Type. Sterol glycosides.

Reaction

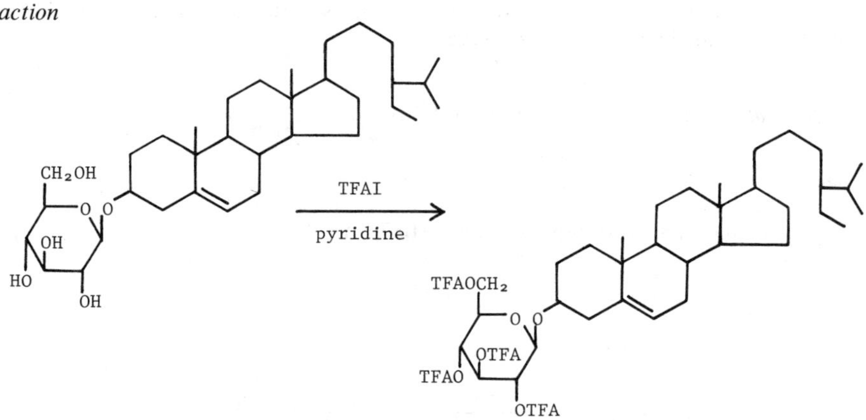

Typical Procedure. Prepare TFA derivs by treating sterol glycoside with TFAI in pyridine soln at room temp. Analyze soln directly by GC-MS.

Comments. HFB derivs were also prepared using HFBI and TMS derivs using BSA. The TFA derivs were judged most useful for GC-MS analysis.[1] The TMS derivs have also been prepared using BSTFA/CH_3CN (1:1, v/v) at 110° for 30 min.[2]

References

[1]B. A. Knights, *Anal. Lett.* **6**, 495 (1973).

[2]R. A. Laine and A. D. Elbein, *Biochemistry* **10**, 2547 (1971).

13.9. DERIVATIVES OF MISCELLANEOUS CARBOHYDRATE COMPOUNDS

Derivative Type. Permethyl.

Reagent. Potassium *t*-butoxide (*t*-BuOK)/methyl iodide.

Compound Type. OH and NH compounds.

Reaction

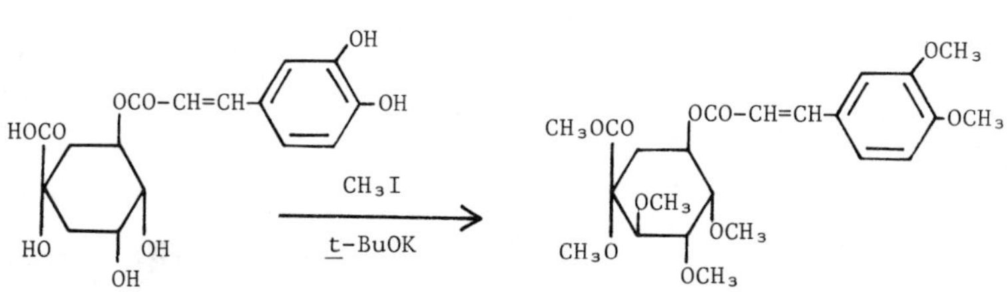

Typical Procedure. Diss up to 5 mg sample in 3 drops of dry DMSO. Add 1 ml 5.6% (w/v) soln of *t*-BuOK in DMSO (soln prep by heating to 70°). Add 0.1 ml CH_3I and allow to stand at room temp 1 hr. Add 1 ml 1 *M* AcOH and ext with 10 ml $CHCl_3$. Wash ext with 4 × 10 ml H_2O and evap to dryness under vac. Analyze by direct probe MS.

Comments. The method was examined as a simpler alternative to the use of NaH for permethylation. In the series of compds examined there were no differences in results from those obtained using NaH in DMSO. Some acetylated oxygens were rearranged to methyl ethers and some N and C methylations were observed.

References

[1] J. Eagles *et al.*, *Biomed. Mass Spectrom.* **1**, 43 (1974).

Derivative Type. Trimethylsilyl.

Reagent. Bis(trimethylsilyl)trifluoroacetamide (BSTFA).

Compound Type. Glycals.

Reaction

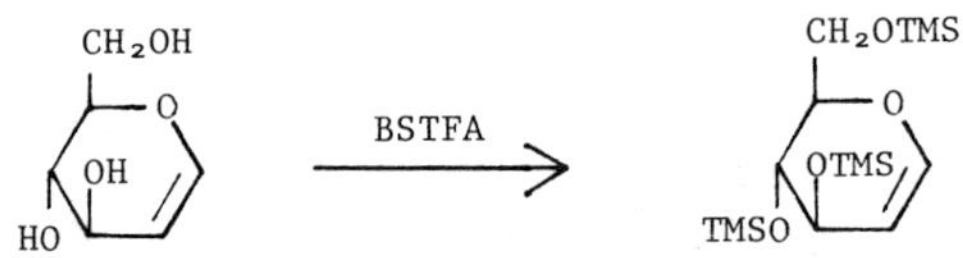

Typical Procedure. Diss the glycal in BSTFA and allow to react at room temp. Analyze by GC.

Comments. Derivs were prepared from D-glucal and D-xylal. Acetyl and specifically deuterated acetyl derivs were also prepared. The MS fragmentation of the derivs was discussed extensively.

References

[1] J. Szafranek *et al.*, *Carbohydr. Res.* **56**, 11 (1977).

14. Nucleotides, Their Constituents, and Related Compounds

The nucleic acids and their constituents constitute an important area of biochemical analysis. The compounds have been analyzed using various types of chromatography, but their highly polar nature necessitates derivatization for GC analysis. In the first MS studies of nucleosides and their free bases, the compounds were analyzed without prior derivatization.[1] Subsequent developments in methodology for the formation of more volatile derivatives have facilitated MS analysis of this group of compounds and have even permitted MS analysis of dinucleotides. In contrast to the need for derivatization of the compounds for GC and MS analysis, there has been little activity in the area of chromogenic derivatization, since the heterocyclic bases already possess UV absorption adequate for detection in HPLC analysis. This section presents derivatives which have been applied to nucleosides, the purine and pyrimidine bases, and structurally related compounds.

Purine and Pyrimidine Bases. The first derivatives for GC analysis of the nucleic acid bases were the *N*-methyl derivatives reported by MacGee in 1964.[2] This method, utilizing tetramethylammonium flash methylation, was applied to the determination of base ratios in nucleic acids (p. 601). A similar method with trimethylphenylammonium hydroxide has been used for plasma level analysis of 5-fluorouracil (p. 601). Phase transfer alkylation has been applied to 6-mercaptopurine for GC-MS analysis (p. 602).

The alkylated derivatives have been largely supplanted by trimethylsilyl derivatives for analysis of purine and pyrimidine bases. Hashizume and Sasaki reported analysis of TMS derivatives prepared using HMDS/TMCS/pyridine (p. 608). Gehrke and co-workers reported an improved method for preparing TMS derivatives of the purine and pyrimidine bases using

BSA in acetonitrile (p. 602). The synthesis of the silylating reagent BSTFA[3] in 1968 lead to even more improved results for nucleic base analysis (p. 603). BSTFA appears still to be the reagent of choice, although it has been reported that the trimethylsilyl amine reagents TMSDMA and TMSDEA give a more rapid reaction than either BSA or BSTFA; TMSI was ineffective for silylation of the bases (p. 604). The antimetabolite base 5-fluorouracil has also been analyzed as the TMS derivative prepared with BSTFA (p. 604).

Nucleosides. Miles and Fales, in the first report of GC analysis of nucleosides (1962),[4] reported use of acetyl, methyl, and isopropylidene derivatives to achieve needed volatility. Other investigators have subsequently studied permethyl derivatives of nucleosides prepared with methyl iodide/silver oxide (p. 605), methyl iodide/methylsulfinyl carbanion (p. 606), and trimethylphenylammonium hydroxide (p. 618). Most of the work with nucleosides, however, has involved silyl derivatives. GC analysis of TMS derivatives of nucleosides was first reported by Hancock and Coleman in 1965 using HMDS/TMCS/pyridine to prepare the derivatives (p. 608). Hashizume and Sasaki also employed HMDS/TMCS/pyridine in the preparation of TMS derivatives of nucleoside (p. 608). Subsequent studies have shown BSTFA to yield better results and most of the more recent work has utilized this reagent (p. 609). Problems with multiple derivative formation from those compounds containing amino groups (e.g., cytidine, adenosine, guanosine) have complicated work with the TMS derivatives. Cytidine in particular has been troublesome but improved results have been obtained with methoxime TMS derivatives (p. 612) or with trifluoroacetyl TMS derivatives (p. 611).

The "sterically crowded" trialkylsilyl derivatives have been shown to be useful both for GC and MS analysis of nucleosides (p. 614). These derivatives are more stable with respect to hydrolysis than the TMS derivatives and have favorable MS behavior. Using a two step procedure, it has been possible to prepare mixed derivatives with the *t*-butyldimethylsilyl group specifically on the 5′-hydroxyl and TMS, trifluoroacetyl, or acetyl at other positions (p. 614). Trifluoroacetyl derivatives (p. 615), phenylboronate TMS derivatives (p. 619), and isopropylidene TMS derivatives (p. 613) have also been used for nucleoside analysis. 7-Methylpurine nucleosides undergo oxidation at C-8 during trimethylsilylation giving an 8-oxo derivative specific for this type of nucleoside (p. 612).

Hattox and McCloskey[5] studied the relationship between structure and GC behavior of 32 nucleoside trimethylsilyl derivatives and used methylene unit values to express correlations of structure with retention times on OV-17 and SE-30. Dolhum and Wiebers[6] and Hignite[7] have summarized the MS behavior of nucleoside and nucleotide derivatives.

Nucleotides. Nucleotides have been analyzed primarily as the TMS derivatives (p. 615), but recently permethyl derivatives have been prepared using trimethylphenylammonium hydroxide (p. 618). Trimethylsilylation has been extended to dinucleotides enabling the use of MS for nucleic acid sequencing (p. 619). The determination of the base sequence within the dinucleotide is facilitated by formation of an asymmetric derivative using phenylboronic acid followed by TMS formation (p. 619). The boronate then identifies the 5′-nucleoside in the dinucleotide derivative. The use of periodate cleavage between the 2′- and 3′-positions has been examined for use as a method of end group analysis of nucleic acids by MS (p. 618).

References

1. K. Biemann and J. A. McCloskey, *J. Am. Chem. Soc.* **84**, 2005 (1962).
2. J. MacGee, *Fed. Proc.* **23**, 2575 (1964).
3. D. L. Stalling *et al.*, *Biochem. Biophys. Res. Commun.* **31**, 616 (1968).
4. H. T. Miles and H. M. Fales, *Anal. Chem.* **34**, 860 (1962).
5. S. E. Hattox and J. A. McCloskey, *Anal. Chem.* **46**, 1378 (1974).

6. J. J. Dolhun and J. L. Wiebers, *Org. Mass Spectrom.* **3**, 669 (1970).
7. C. Hignite, in *Biochemical Applications of Mass Spectrometry*, ed. by G. R. Waller, Wiley Interscience, New York, NY, 1972, p. 429.

14.1. DERIVATIVES OF PURINE AND PYRIMIDINE BASES

14.1.1. Methylated Derivatives

Derivative Type. Methyl.

Reagent. Tetramethylammonium hydroxide (TMAH).

Compound Type. Purine and pyrimidine bases.

Reaction

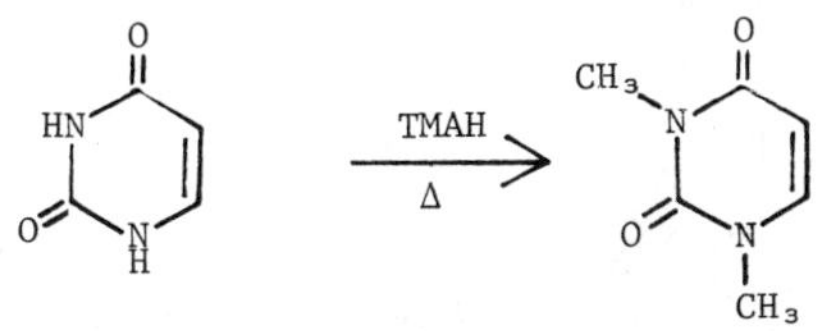

Typical Procedure. Diss the mixt of bases isolated from hydrolysis of 0.5 mg nucleic acid in 100 μl 1 *M* TMAH in EtOH. Analyze 5 μl of the soln by GC with the injector port at 360°.

Comments. The method was used to determine base ratios in nucleic acids. Uracil and thymine were shown to yield 1,3-dimethyl derivs. Xanthine, theobromine, and theophylline all yielded caffeine. The derivs of adenine, guanine, cytosine, and other purines and pyrimidines were not structurally characterized but all gave characteristic derivs suitable for analysis. The procedure was not applicable to the minor methylated bases. The ethyl and propyl derivs of adenine have been prepared from the corresponding tetraalkylammonium hydroxides,[2] thus higher homologs of the reagent could be useful for analysis of the minor methylated bases.

References

[1]J. MacGee, *Anal. Biochem.* **14**, 305 (1966).
[2]T. C. Meyers and L. Zeleznick, *J. Org. Chem.* **28**, 2087 (1963).

Derivative Type. Methyl.

Reagent. Trimethylphenylammonium hydroxide (TMPAH).

Compound Type. 5-Fluorouracil.

Reaction

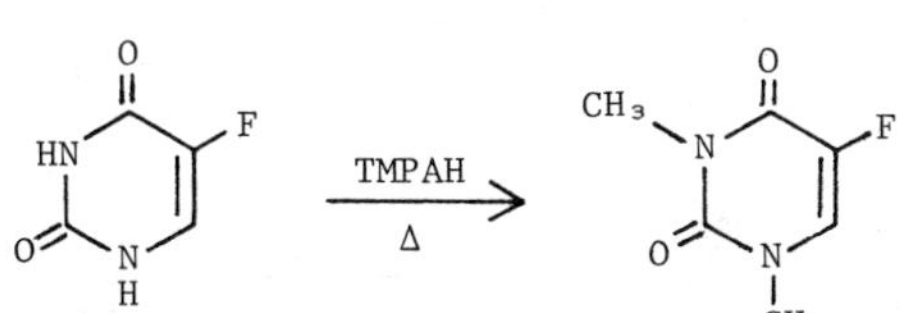

Typical Procedure. Dil 1 ml plasma to 4 ml with pH 6.8 phosphate buffer (to give final buffer conc of 0.25 *M*). Heat at 100° for 5 min and centrifuge. Wash supernatant with an equal vol of Et_2O then ext 2 ml of washed supernatant with 30 ml of 16% *n*-PrOH in Et_2O. Remove

25 ml of the ext and add to it 1.5 ml H_2O (to yield 88° *n*-PrOH-H_2O azeotrope). Evap at 70° under N_2. Diss residue in 2.5 ml MeOH and transfer 2.0 ml to a tube containing 200 ng thymine int stand. Evap to dryness under N_2 and add 30 μl 0.2 *M* TMPAH in MeOH. Analyze 0.5 μl by GC.

Comments. MS data were given for the derivs. The method permitted quantitation to 10 ng/ml in plasma.

References

[1]B. L. Hillcoat *et al.*, *Brit. J. Clin. Pharm.* **3**, 135 (1976).

Derivative Type. Methyl.

Reagent. Methyl iodide/tetrahexylammonium hydroxide (THAH).

Compound Type. 6-Mercaptopurine.

Reaction

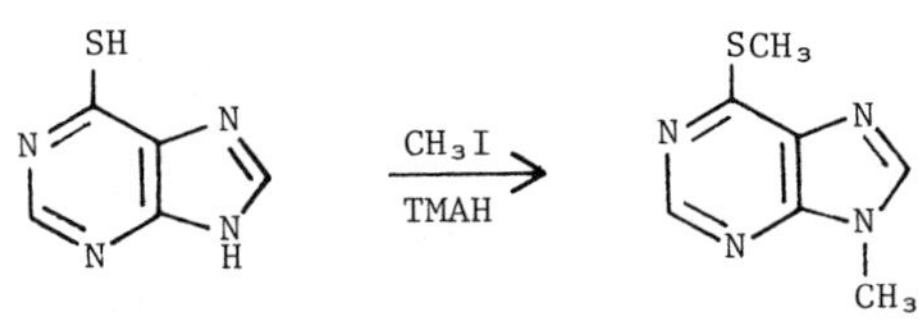

Typical Procedure. Dil 1.0 ml plasma with 3.0 ml H_2O and acidify with 1 ml 0.67 *N* H_2SO_4. Add 1 ml 10% (w/v) Na_2WO_4 soln. Centrifuge to remove proteins. Add 1 ml 3 *N* NaOH to the supernatant and wash with 4 ml Et_2O.

Treat aq phase with 100 μl 0.1 *M* THAH in MeOH and shake 5 min with 5 ml 0.5 *M* CH_3I in CH_2Cl_2. Centrifuge and separate org phase. Add 633 ng of 6-methyl-N^9-perdeuterio-methyl-6-mercaptopurine int stand. Dry over Na_2SO_4 and evap to dryness. Diss residue in 50 μl toluene and analyze by GC-MS.

Comments. The method was used to quantitate 6-mercaptopurine in plasma by GC-MS-SIM. The limit of detection was 20 ng/ml.

References

[1]J. M. Rosenfeld *et al.*, *Anal. Chem.* **49**, 725 (1977).

14.1.2. Trimethylsilyl Derivatives

Derivative Type. Trimethylsilyl.

Reagent. Bis(trimethylsilyl)acetamide (BSA).

Compound Type. Purine and pyrimidine bases.

Reaction

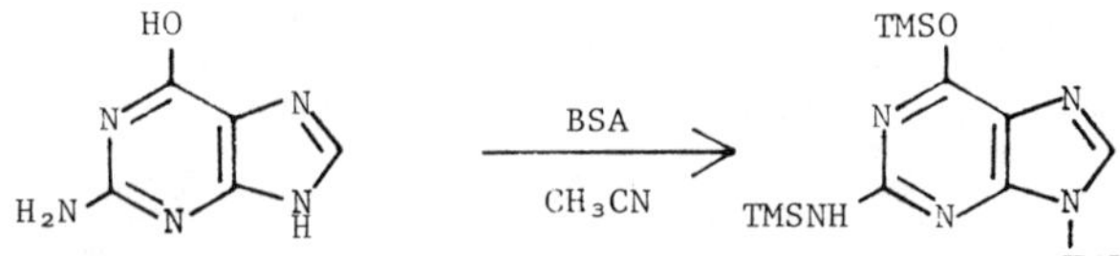

Typical Procedure. Diss the base sample in 100 mol equiv BSA/CH_3CN (1:3, v/v) and heat in a sealed vial at 150° for 45 min. Cool to room temp and analyze by GC.

Comments. The 3 : 1 ratio of CH_3CN : BSA was necessary for complete miscibility. Cytosine and 5-methyl cytosine each gave two GC peaks. More extreme conditions did not force the reaction to a single product. The method has been used for base analysis of μg quant of RNA after enzymic hydrolysis.[3]

References

[1]C. W. Gehrke *et al.*, *Biochem. Biophys. Res. Commun.* **28**, 869 (1967).
[2]C. W. Gehrke and C. D. Ruyle, *J. Chromatogr.* **38**, 473 (1968).
[3]M. Jacobson *et al.*, *Anal. Biochem.* **25**, 363 (1968).

Derivative Type. Trimethylsilyl.

Reagent. Bis(trimethylsilyl)trifluoroacetamide (BSTFA).

Compound Type. Purine and pyrimidine bases.

Reaction

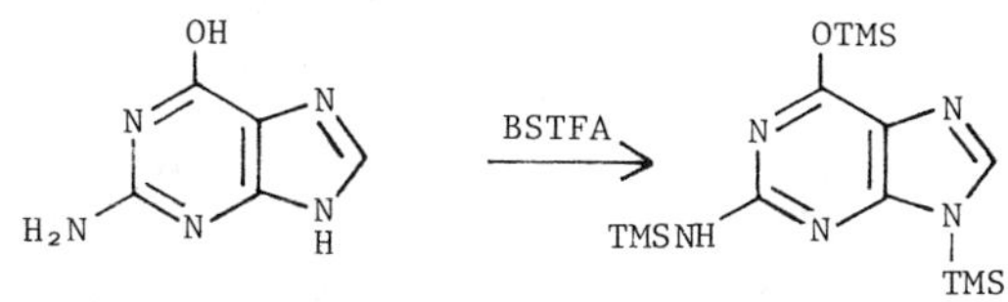

Typical Procedure. Diss 500 ng of base in 50 μl BSTFA, 25 μl CH_3CN, and 25 μl dichloroethane and heat in a closed tube at 150° for 30 min. Analyze by GC.

Comments. The derivs were prep in CH_3CN alone for larger samples (10 μg) but the mixed solvent was necessary to prevent smaller amts of derivs being obscured in the solvent peak. BSA and BSA/TMCS were also used to prep derivs of purine and pyrimidine bases[2] but BSTFA was found to be best for low level analysis. The method has been used for base analysis in urine[3] and nucleic acid hydrolyzates.[4,5] Miller *et al.*[6] have published a systematic study of the TMS derivs of purine and pyrimidine bases and reported GC retention and FID response data.

References

[1]C. W. Gehrke and D. B. Lakings, *J. Chromatogr.* **61**, 45 (1971).
[2]C. W. Gehrke and C. D. Ruyle, *J. Chromatogr.* **38**, 473 (1968).
[3]D. B. Lakings *et al.*, *J. Chromatogr.* **116**, 69 (1976).
[4]D. B. Lakings and C. W. Gehrke, *J. Chromatogr.* **62**, 347 (1971).
[5]D. B. Lakings and C. W. Gehrke, *Clin. Chem.* **18**, 810 (1972).
[6]V. Miller *et al.*, *J. Chromatogr.* **119**, 355 (1976).

Derivative Type. Trimethylsilyl.

Reagent. Bis(trimethylsilyl)trifluoroacetamide (BSTFA)/trimethylchlorosilane (TMCS).

Compound Type. Pyrimidine and purine bases.

Reaction

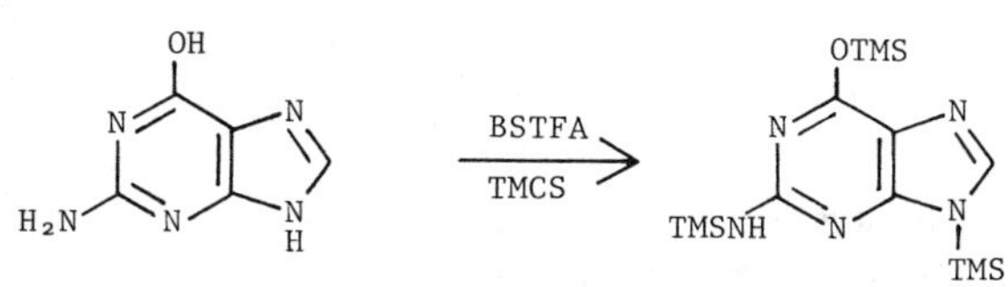

Typical Procedure. Diss 0.5–1.5 mg of base in BSTFA to a conc of *ca* 1 μg/μl. Add 1% TMCS and heat at 100° for 1–2 hr. Analyze by GC-MS.

Comments. The same procedure was used with BSA in place of BSTFA. A third procedure utilized BSA/CH_3CN (1:3, v/v) containing 1% TMCS for 1 hr at room temp. The latter procedure was also used for deuterated derivs using d_{18}-BSA and d_9-TMCS. MS data were reported for the derivs of 20 bases.

References

[1]E. White *et al.*, *J. Org. Chem.* **37**, 430 (1972).

Derivative Type. Trimethylsilyl.

Reagent. Trimethylsilyldiethylamine (TMSDEA)/trimethylchlorosilane (TMCS).

Compound Type. Guanine and hypoxanthine.

Reaction

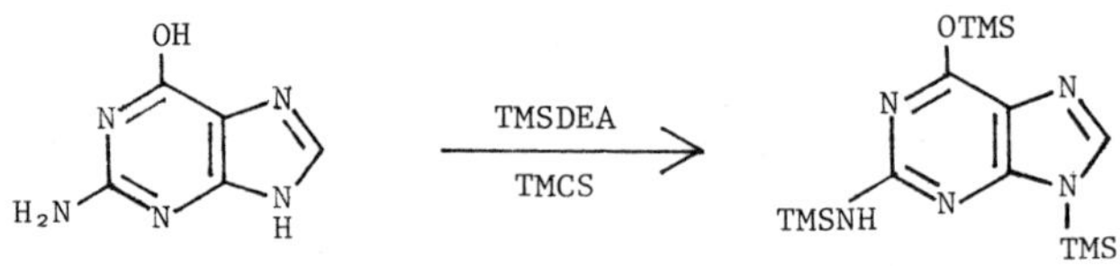

Typical Procedure. Diss 5 mg of each base in 1 ml pyridine and 1 ml TMSDEA containing a few drops of TMCS. Allow to stand 1 hr at room temp. Analyze by GC.

Comments. Deriv of guanine and hypoxanthine was examined using TMSDEA, TMSDMA, BSA, BSTFA, MSA, and TMSI. The TMS amines reacted more rapidly than the TMS amides (except for BSA which reacted rapidly, 30 min, with guanine). TMSI was ineffective.

References

[1]H. Iwase *et al.*, *J. Chromatogr.* **106**, 213 (1975).

Derivative Type. Trimethylsilyl.

Reagent. Bis(trimethylsilyl)trifluoroacetamide (BSTFA)/trimethylchlorosilane (TMCS).

Compound Type. 5-Fluorouracil.

Reaction

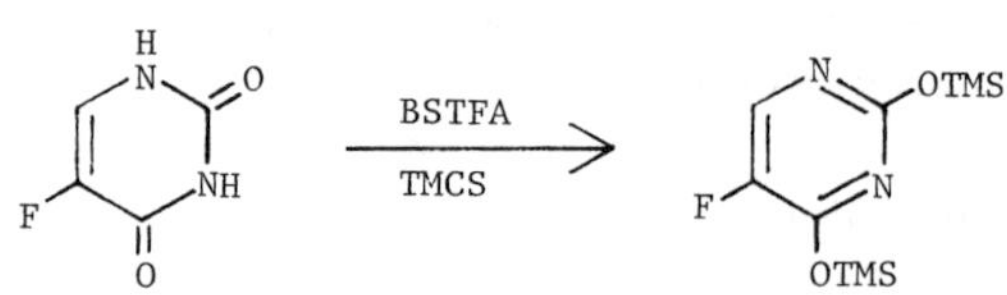

Typical Procedure. Adjust plasma or urine to pH 6 by addition of a known vol of pH 4.7, 0.5 *M* acetate buffer. Mix 1.0 ml of this sample with 0.4 g anhyd Na_2SO_4 and ext with 15 ml 16% *n*-PrOH in Et_2O. To an aliquot of the ext add 25 μl of a 16 μg/ml soln of anthracene (int stand) in pyridine and evap under N_2. Add 50 μl pyridine and 50 μl BSTFA containing 1% TMCS. Allow to react 5 min at room temp then analyze by GC.

Comments. The method was sensitive to 0.2 μg/ml in plasma. Use of ECD in place of FID gave only a twofold increase in sensitivity. Windheuser *et al.*[2] isolated 5-fluorouracil from plasma by dialysis and formed the TMS deriv with BSA. Finn and Sadee[3] prepared the deriv with BSTFA (without TMCS) in pyridine for GC-MS-SIM analysis. The actual structure of the deriv (*O*-TMS or *N*-TMS) was not established.

References

[1]J. L. Cohen and P. B. Brennan, *J. Pharm. Sci.* **62**, 572 (1973).

[2]J. J. Windheuser *et al., J. Pharm. Sci.* **61**, 301 (1972).

[3]C. Finn and W. Sadee, *Cancer Chemother. Rep.* **59**, 279 (1975).

14.2. DERIVATIVES OF NUCLEOSIDES AND RELATED COMPOUNDS

14.2.1. Methylated Derivatives

Derivative Type. Methyl.

Reagent. Methyl iodide/silver oxide.

Compound Type. Nucleosides.

Reaction

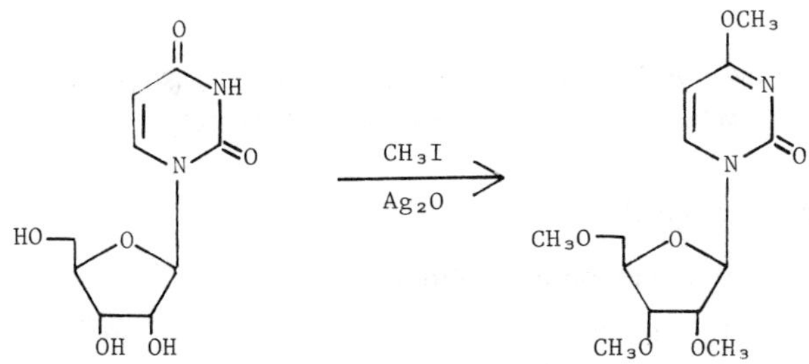

Typical Procedure.[2] Suspend a 2.8 g sample in 15 ml dry MeOH and add 2.5 g Ag_2O and 2.5 ml CH_3I. Shake vigorously at 25° for 12 hr. Add another 2.5 g Ag_2O and 2.5 ml CH_3I, shake another 12 hr, then repeat addition and 12 hr shaking. Filter and ext the filtered salts with 60 ml hot MeOH. Combine the filtrate and ext and evap under 35° to yield the methylated deriv as a syrup. Analyze by MS.

Comments. Derivs were prepared from adenosine, uridine, cytidine, and guanosine, and MS data were reported. The methylation of adenosine did not go to completion and gave a mixt of 2 derivs having 1 and 2 methyl groups on the base. Attempts to methylate 5′-mononucleotides by the procedure were unsatisfactory due to production of deriv mixts which were difficult to purify.

References

[1]J. J. Dolhun, and J. L. Wiebers, *Org. Mass Spectrom.* **3**, 669 (1970).

[2]A. S. Anderson *et al., J. Chem. Soc.* 369 (1952).

Derivative Type. Permethyl.

Reagent. Methylsulfinylcarbanion/methyl iodide.

Compound Type. Nucleosides.

Reaction

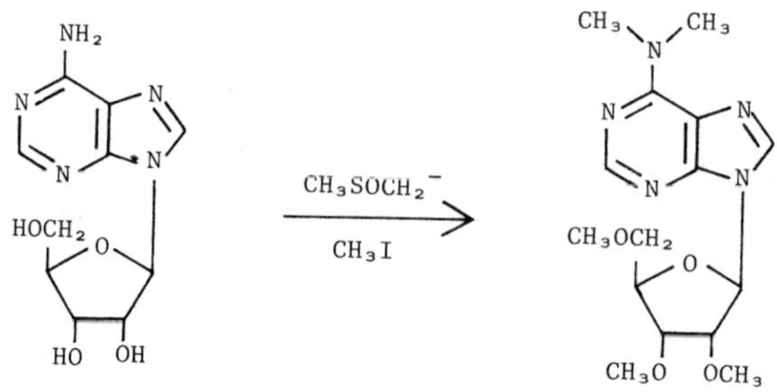

Typical Procedure. Diss 2–100 μg of nucleoside in 80 μl DMSO. Add tenfold excess (in terms of sample "reactive equivalents") of carbanion soln (prepared according to Ref. 2). After 30 min add an equimolar amt of CH_3I and allow to stand 90 min. Add 1 ml H_2O and ext with 1 ml $CHCl_3$. Wash $CHCl_3$ with 3 × 1 ml H_2O and evap to dryness under vac. Diss residue in $CHCl_3$ and analyze by GC-MS.

Comments. The derivs were used for MS analysis of nucleosides. CD_3I and other alkyl iodides were also used.

References

[1]D. L. von Minden and J. A. McCloskey, *J. Am. Chem. Soc.* **95**, 7480 (1973).

[2]M. L. Polan *et al.*, *Biochem. Biophys. Res. Commun.* **38**, 1127 (1970).

Derivative Type. Permethyl.

Reagent. Methylsulfinylcarbanion/methyl iodide.

Compound Type. Nucleosides.

Reaction

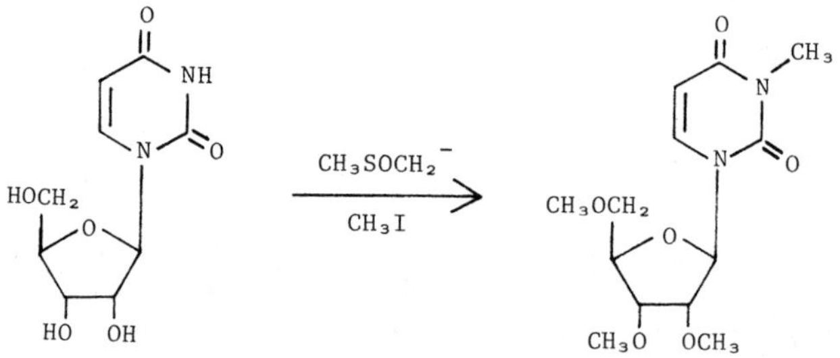

Typical Procedure. Diss 50 μg sample in 100 μl DMSO and add 50 μl of dil methylsulfinylcarbanion soln (see below). After 1 min add 100 μl CH_3I, mix, and allow to stand 1 hr. Add 5 ml H_2O and 25 μg of *N,O*-permethyl-2′-deoxyuridine int stand in DMSO. Ext with 5 ml $CHCl_3$. Wash ext with 3 × 5 ml H_2O, dry over Na_2SO_4, and evap to dryness under N_2 at 60°. Diss residue in 100 μl $CHCl_3$ and analyze 1 μl aliquots by GC.

Comments. The methylsulfinylcarbanion reagent was prepared by heating 6 g *t*-BuOK in 100 ml dry DMSO at 70° for 2 hr. The soln was filtered and stored at −18° in 10 ml aliquots as stock soln. The stock soln was dil 4 × with DMSO for use. The dil soln was tested prior to use by mixing 50 μl with 100 μl of a soln of 1 mg triphenylmethane in DMSO. A red color was indicative of good reagent.

GC behavior was studied for the derivs of 10 pyrimidine and 3 purine nucleosides on 13 stationary phases. The number of methyl groups introduced was determined by MS. UV data supported the location of the methyl group at N-3 rather than O^4 in the pyrimidine derivs.

References

[1] A. P. DeLeenheer and C. F. Gelijkens, *Anal. Chem.* **48**, 2203 (1976).

Derivative Type. Methyl.

Reagent. Methylsulfinylcarbanion/methyl iodide.

Compound Type. Dihydrouracil nucleosides.

Reaction

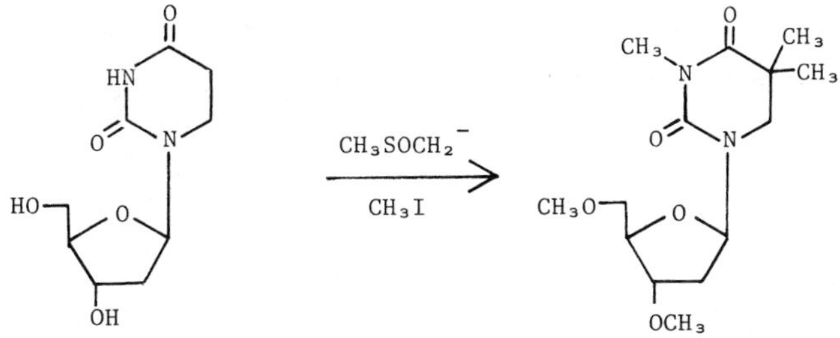

Typical Procedure. Diss 20–50 μg of the compd in 80 μl DMSO. Add a tenfold excess (on a "reactive equiv" basis) of methylsulfinylcarbanion soln (prepared from NaH and DMSO[2]). After 30 min add an equimolar amt of CH_3I and allow to stand 90 min. Add 1 ml H_2O and ext with 1 ml $CHCl_3$. Wash ext with 3×1 ml H_2O and analyze by GC-MS.

Comments. d_3-Methyl derivs were formed using CD_3I. The formation of 3,5,5-trimethyl derivs of the base was unique to the 5,6-dihydropyrimidine system and provided a sensitive MS test for this 5,6-dihydro function. The structures of the derivs were confirmed by CI and EI-MS and NMR.

References

[1] R. P. Panzica *et al.*, *Biochim. Biophys. Acta* **331**, 147 (1973).

[2] M. L. Polan *et al.*, *Biochem. Biophys. Res. Commun.* **38**, 1127 (1970).

14.2.2. Trimethylsilyl Derivatives

Derivative Type. Trimethylsilyl.

Reagent. Hexamethyldisilazane (HMDS)/trimethylchlorosilane (TMCS)/pyridine.

Compound Type. Nucleosides.

Reaction

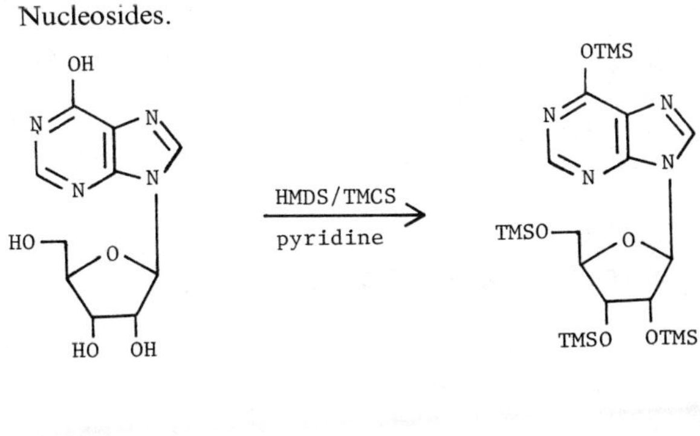

Typical Procedure. Diss 10–20 mg nucleoside in 1 ml pyridine and add 0.2 ml HMDS and 0.2 ml TMCS. Allow to stand at room temp 5 min then analyze by GC.

Comments. GC data were reported for the derivs of 12 nucleosides.[1] Two GC peaks were obtained for adenosine, deoxyadenosine, cytidine, and deoxyguanosine presumably due to derivs with and without *N*-TMS groups.[1] Dimethylsilyl, triethylsilyl, tri-*n*-butylsilyl, and triphenylsilyl derivs of adenosine were prepared using the corresponding chlorosilanes.[2]

References

[1]R. L. Hancock and D. L. Coleman, *Anal. Biochem.* **10**, 365 (1965).

[2]R. L. Hancock, *J. Gas Chromatogr.* **6**, 431 (1968).

Derivative Type. Trimethylsilyl.

Reagent. Hexamethyldisilazane (HMDS)/trimethylchlorosilane (TMCS)/pyridine.

Compound Type. Nucleosides and bases.

Reaction

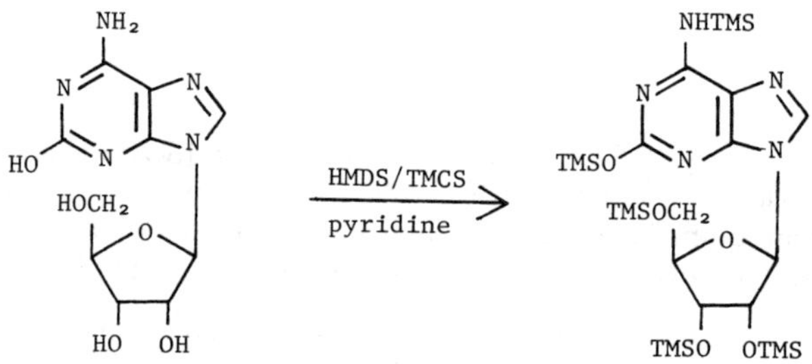

Typical Procedure. Diss 10–20 mg of nucleoside or base and 15 mg of phenanthrene int stand in 0.7 ml pyridine, 0.2 ml HMDS, and 0.1 ml TMCS. Reflux 1 hr under anhyd conditions. Cool to room temp and analyze by GC.

Comments. The TMS groups on the base moiety were more prone to hydrolysis than those on the sugar group. Efforts to isolate the fully trimethylsilylated derivs were unsuccessful due to hydrolysis of the TMS groups on the bases by atmospheric moisture. The isolable partially silylated nucleosides (2′,3′,5′-tri-TMS) and deoxynucleosides (3′,5′-di-TMS) were characterized by IR and NMR. GC data were reported for 18 nucleosides and bases.[1] The deriv has been used for analysis of the base and phosphorus composition of nucleic acids after chemical hydrolysis.[2]

References

[1]Y. Sasaki and T. Hashizume, *Anal. Biochem.* **16**, 1 (1966).

[2]T. Hashizume and T. Sasaki, *Anal. Biochem.* **24**, 232 (1968).

Derivative Type. Trimethylsilyl.

Reagent. Bis(trimethylsilyl)trifluoroacetamide (BSTFA).

Compound Type. Nucleosides.

Reaction

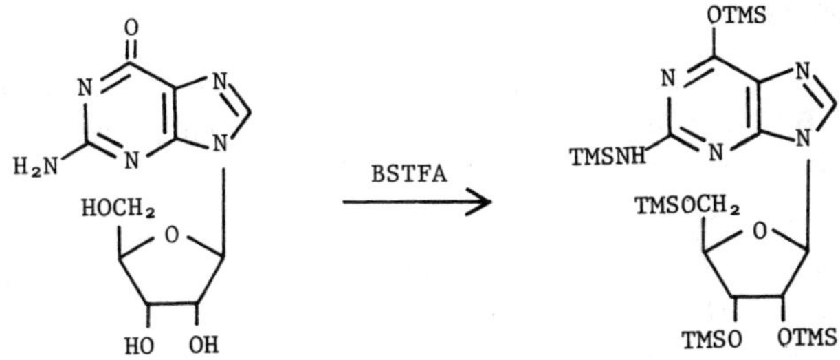

Typical Procedure. Diss nucleoside standards (1 μmol each) in 0.5 ml CH_3CN int stand containing 250 μg pyrene and add 225 μmol (0.24 ml) BSTFA per μmol of nucleoside. Dil to 2.5 ml with CH_3CN and maintain at 150° for 15 min. Analyze 5 μl aliquots by GC.

Comments. BSTFA was compared with BSA and TMSI and found to give superior yields of the derivs. Acetonitrile was the best solvent of several investigated. Solubility was found to be a primary factor in incomplete deriv. The presence of H_2O significantly affected the extent of deriv and the stability of the products. Addition of 1% TMCS had no significant effect on the deriv with BSTFA. In earlier work[2] it was observed that 2′-deoxyribonucleosides yielded peaks for the respective bases with mol excesses of BSTFA greater than 500. No single deriv was obtained with 2′-deoxycytidine. The minimum detectable amts for the derivs were 5–10 ng using FID. The effects on derivatization of different substances (salts, reaction by-products) present in nucleic acid hydrolysis mixtures were studied.[3] It was found that these substances do interfere with deriv formation and should be removed prior to derivatization. Hancock[4] had previously reported the preparation of derivs of 16 nucleosides with BSTFA in pyridine (room temp), but some gave multiple peaks, presumably due to partial formation of *N*-TMS derivs.

References

[1]C. W. Gehrke and A. B. Patel, *J. Chromatogr.* **130**, 103 (1977).

[2]C. W. Gehrke and A. B. Patel, *J. Chromatogr.* **123**, 335 (1976).

[3]A. B. Patel and C. W. Gehrke, *J. Chromatogr.* **130**, 115 (1977).

[4]R. L. Hancock, *J. Chromatogr. Sci.* **7**, 366 (1969).

Derivative Type. Trimethylsilyl.

Reagent. Bis(trimethylsilyl)trifluoroacetamide (BSTFA)/trimethylchlorosilane (TMCS).

Compound Type. Cyclonucleosides.

Reaction

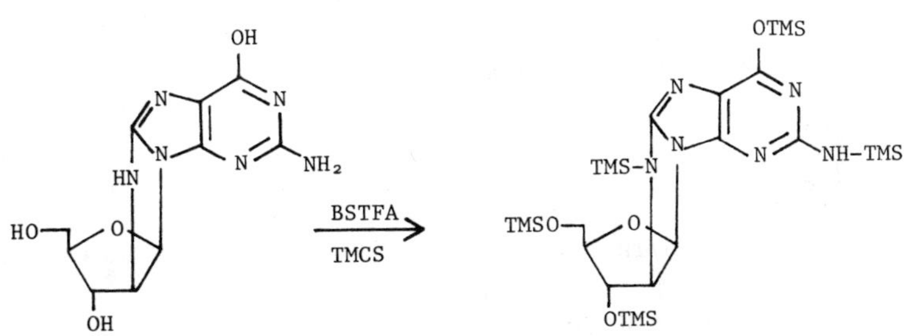

Typical Procedure. Mix 0.5 mg sample with 100 μl BSTFA and 15 μl TMCS. Allow to stand overnight at room temp. Analyze by MS.

Comments. MS data were reported for a series or purine O, S, and NH linked anhydro nucleosides.[1-3] BSA/TMCS/pyridine (30:1:30, v/v/v; room temp 10–30 min, then 100 5–10 min) has also been used for cyclonucieosides.[4]

References

[1]D. C. K. Lin *et al.*, *J. Org. Chem.* **38**, 1118 (1973).

[2]J. B. Westmore *et al.*, *Org. Mass Spectrom.* **6**, 1243 (1972).

[3]J. B. Westmore *et al.*, *Org. Mass Spectrom.* **7**, 317 (1973).

[4]S. Tsuboyama and J. A. McCloskey, *J. Org. Chem.* **37**, 166 (1972).

Derivative Type. Trimethylsilyl.

Reagent. Bis(trimethylsilyl)trifluoroacetamide (BSTFA).

Compound Type. Methylated nucleosides.

Reaction

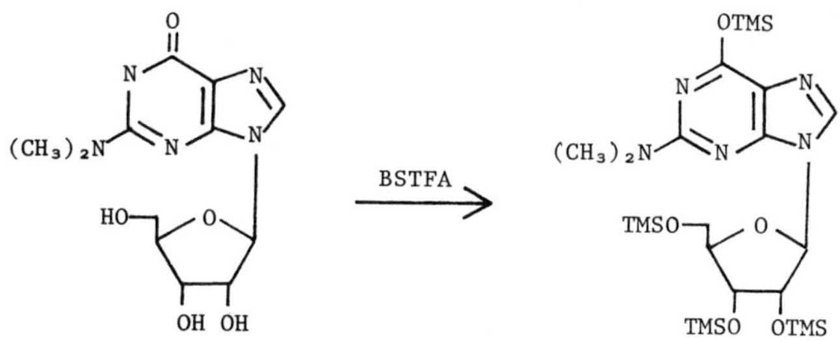

Typical Procedure. Isolate nucleosides from 1.0 ml urine by adsorption on charcoal and elute with pyridine-EtOH (see ref for detailed procedure). Evap eluate to dryness at 75° under N_2 and add *n*-tetratriacontane and *n*-octacosane int stands. Add 0.5 ml BSTFA/CH_3CN (1:1, v/v) and heat at 150° for 15 min. Analyze by GC.

Comments. The method was used for analysis of N^2,N^2-dimethylguanosine and 1-methylinosine in urine. Pseudouridine was isolated by a similar procedure and deriv with BSTFA/CH_2Cl_2/CH_3CN (2:1:1, v/v/v) at 150° for 15 min.

References

[1]S. Y. Chang *et al.*, *J. Lab. Clin. Med.* **83**, 816 (1974).

Derivative Type. Trimethylsilyl.

Reagent. Bis(trimethylsilyl)trifluoroacetamide (BSTFA)/trimethylchlorosilane (TMCS).

Compound Type. Cytokinin glycosides.

Reaction

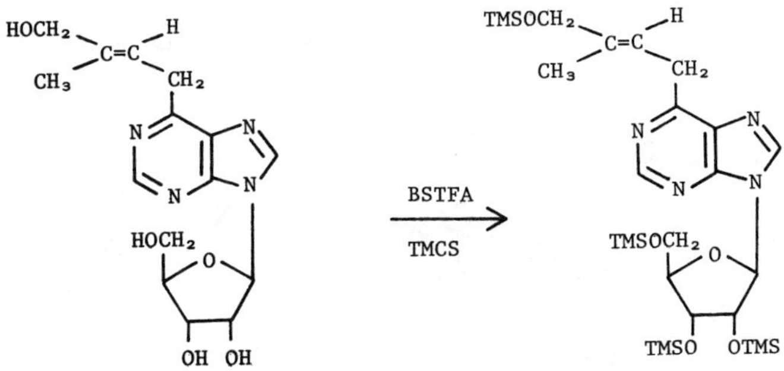

Typical Procedure. Diss 10–100 μg of sample in 10–20 μl pyridine and add 100 μl of BSTFA containing 1% TMCS. Warm at 60° for 30–60 min then analyze by GC-MS.

Comments. The method was used in MS studies of glycosides of the cytokinins zeatin and 6-benzylaminopurine. Permethyl derivs were also studied but the TMS derivs showed more significant isomer differences and appeared less prone to multiple product formation.

References

[1]J. K. MacLeod *et al.*, *J. Org. Chem.* **41**, 3959 (1976).

14.2.3. Mixed Trimethylsilyl Derivatives

Derivative Type. *O*-Trimethylsilyl, N^4-trifluoroacetyl.

Reagent. Bis(trimethylsilyl)trifluoroacetamide (BSTFA)/trimethylchlorosilane (TMCS), trifluoroacetylimidazole (TFAI).

Compound Type. Cytidine.

Reaction

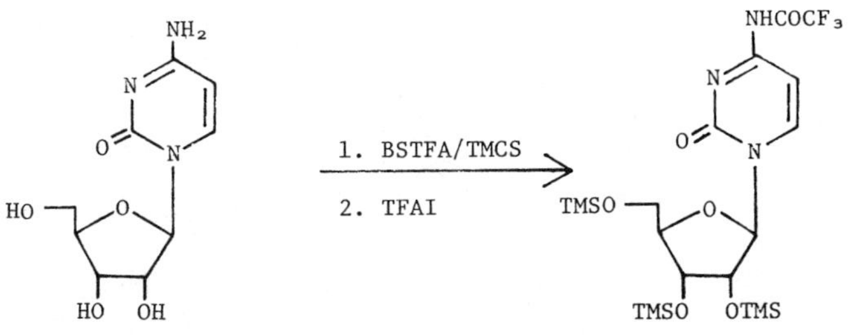

Typical Procedure. To a 50 μg sample add 50 μl BSTFA containing 1% TMCS and 5 μl dry pyridine. Seal tube and heat at 100° for 1 hr. Cool and add 0.5 μl TFAI. Mix and allow to stand at room temp 5–10 min. Analyze by GC-MS.

Comments. The deriv showed improved GC behavior over the TMS deriv which eluted later and with a very broad peak shape. The unsatisfactory GC behavior of cytidine-$(TMS)_4$ was presumed to be due to charge separated resonance forms involving O^2. The electron withdrawing effect of the TFA group opposed the formation of these charge separated forms.

References

[1]U. I. Krahmer *et al.*, *Anal. Biochem.* **82**, 217 (1977).

Derivative Type. Methoxime, trimethylsilyl.

Reagent. Methoxylamine HCl, bis(trimethylsilyl)trifluoroacetamide (BSTFA)/trimethylchlorosilane (TMCS).

Compound Type. Cytidine and deoxycytidine.

Reaction

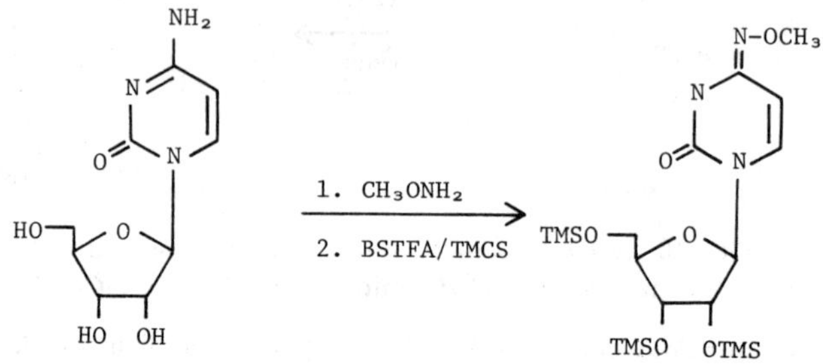

Typical Procedure. To 1–2 mg sample add 1 ml pyridine containing 15–20 mg methoxylamine HCl. Heat at 75° for 3 hr then evap to dryness under N_2. Add 0.2 ml pyridine and 0.3 ml BSTFA containing 1% TMCS. Heat at 75° for 3 hr for ribonucleosides and overnight for deoxyribonucleosides. Analyze by GC.

Comments. The methoxime, TMS derivs showed better GC properties than the TMS derivs. Other nucleosides were not affected by the methoxime treatment except that deoxyadenosine and deoxyguanosine were quantitatively hydrolyzed to adenine and guanine.

References

[1]W. C. Butts, *J. Chromatogr. Sci.* **8**, 474 (1970).

Derivative Type. Trimethylsilyl, 8-oxo.

Reagent. Bis(trimethylsilyl)trifluoroacetamide(BSTFA)/trimethylchlorosilane (TMCS).

Compound Type. 7-Methylpurine nucleosides.

Reaction

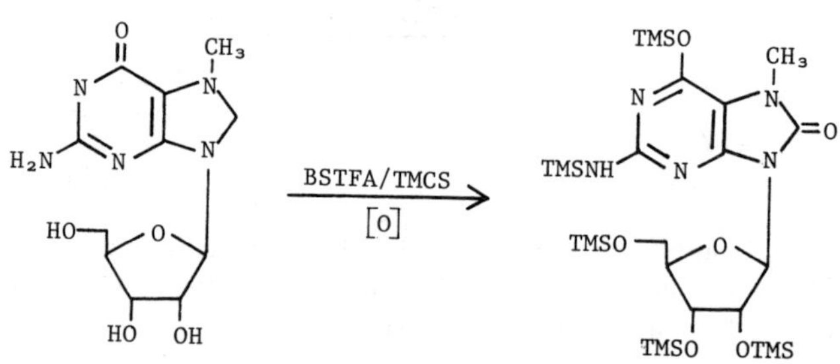

Typical Procedure. Diss 0.2 μmol of the nucleoside in 80 μl BSTFA and 1 μl TMCS, heat 2 hr at 100°, then heat 30 min at 150°. Analyze by GC-MS.

Comments. Preparation of the deriv in an ^{18}O enriched atmosphere showed the source of O in the oxo group to be diss O_2. Various silylation reagents were examined with the BSTFA/TMCS (80 : 1) giving the highest yield (>25%) of the 8-oxo deriv. The oxygen incorporation did not occur with other nucleosides or with the base 7-methylguanine.

References

[1]D. L. von Minden *et al.*, *Anal. Biochem.* **50**, 110 (1972).

Derivative Type. Isopropylidene, trimethylsilyl.

Reagent. Acetone/cupric sulfate, trimethylsilylimidazole (TMSI).

Compound Type. Nucleosides.

Reaction

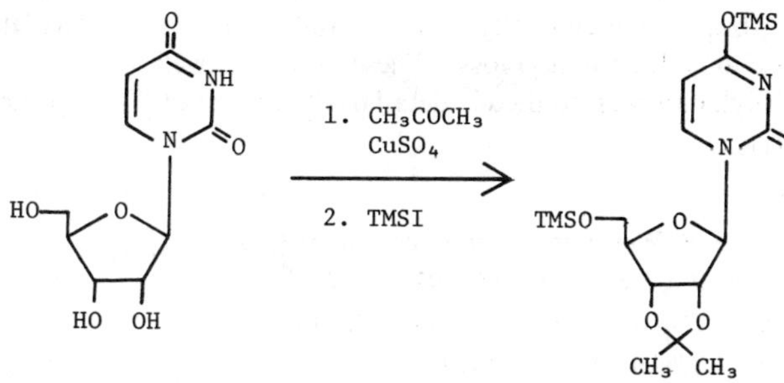

Typical Procedure. Prepare isopropylidene deriv by heating nucleoside sample with acetone in the presence of $CuSO_4$.[2] Separate the acetone layer and evap. For 2 mg isopropylidene nucleoside, add 0.2 ml CH_3CN and 0.4 ml TMSI. Heat at 60° for 3 hr. Analyze by MS.

Comments. Acetate and trifluoroacetate derivs of isopropylidene nucleosides were also prepared using acetyl- and trifluoroacetylimidazole respectively in pyridine (80°, 3 hr).

References

[1]J. J. Dolhun and J. L. Wiebers, *Org. Mass Spectrom.* **3**, 669 (1970).

[2]J. A. McCloskey and M. J. McClelland, *J. Am. Chem. Soc.* **87**, 5090 (1965).

14.2.4. Other Silyl Derivatives

Derivative Type. *t*-Butyldimethylsilyl (TBDMS).

Reagent. *t*-Butyldimethylchlorosilane (TBDMS-Cl)/imidazole.

Compound Type. Nucleosides.

Reaction

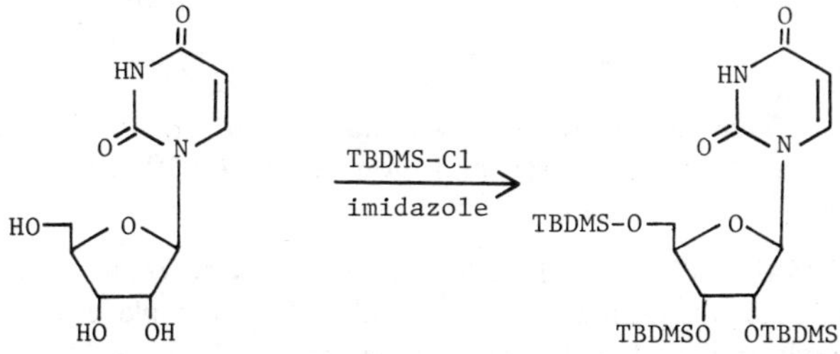

Typical Procedure. To 1–20 μmol of nucleoside add 10–100 μl of a soln of 1 *M* TBDMS-Cl and 1 *M* imidazole in DMF or pyridine. Heat at 80° for 1 hr then analyze by GC or GC-MS.

Comments. Heating at 80° for 1 hr was required to effect quant silylation of uridine (shown above). Reaction at room temp yielded predominantly a mixt of 2′,5′- and 3′,5′-bis(*O*-TBDMS) derivs. With adjacent hydroxyls, deriv of one hydroxyl was relatively rapid but yielded steric hindrance of the other. TBDMS-imidazole (TBDMSI) was prepared[2] and found to be a much less reactive silyl donor, but was specific for hydroxyl groups. Addition of the chlorosilane to a soln of TBDMSI in pyridine greatly increased the reactivity. The reagent solns containing the chlorosilane gave some *N*-silylation of amino groups. Adenosine and deoxyadenosine showed some silylation of the base amino groups, but could not be quant silylated even with heating. The MS of the derivs were characterized by intense high mass ions due to loss of the *t*-butyl group. Other sterically crowded trialkylsilyl derivs (TMTBS, TMIPS) likewise showed intense ions due to alkyl loss.[1] These derivs, as well as the triisopropylsilyl and methyldiisopropylsilyl, had been studied in detail by Ogilvie, *et al.*[4-7] as protecting groups for nucleoside synthesis.

References

[1]M. A. Quilliam and J. B. Westmore, *Anal. Chem.* **50**, 59 (1978).
[2]M. A. Quilliam *et al.*, *J. Chromatogr.* **105**, 297 (1975).
[3]M. A. Quilliam *et al.*, *Biomed. Mass Spectrom.* **1**, 78 (1974).
[4]K. K. Ogilvie, *Can. J. Chem.* **51**, 3799 (1973).
[5]K. K. Ogilvie and D. J. Iwacha, *Tetrahedron Lett.* 317 (1973).
[6]K. K. Ogilvie *et al.*, *Tetrahedron Lett.* 2861 (1974).
[7]K. K. Ogilvie *et al.*, *Tetrahedron Lett.* 2865 (1974).

Derivative Type. 5′-*t*-Butyldimethylsilyl (TBDMS), 3′-trimethylsilyl (TMS).

Reagent. *t*-Butyldimethylsilylimidazole (TBDMSI), trimethylsilylimidazole (TMSI).

Compound Type. Deoxynucleosides.

Reaction

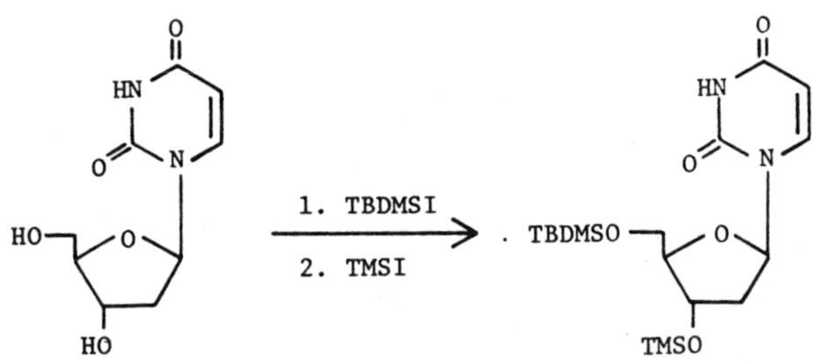

Typical Procedure. Treat deoxynucleoside with an excess of TBDMSI (1 *M* soln in pyridine) at room temp for 24 hr. Add excess TMSI and allow to react for 10 min at room temp. Analyze by GC or GC-MS.

Comments. TBDMSI (and also bis-TBDMS-acetamide) were found to be nearly specific for the 5′-hydroxyl of deoxynucleosides. BSTFA could be used in place of TMSI, but was not specific for hydroxyls and reacted also with amines and enols. TFA derivs could also be prepared in the second step by use of TFAI or TFAA. With TFAA the excess reagent had to be removed before analysis. Second step acetylation could be achieved with Ac_2O (excess reagent likewise had to be removed before analysis) but acetylimidazole reacted too slowly to be useful.

References

[1]M. A. Quilliam *et al.*, *J. Chromatogr.* **105**, 297 (1975).

[2]K. K. Ogilvie *et al.*, *Tetrahedron Lett.* 2865 (1974).

14.2.5. Other Derivatives

Derivative Type. Trifluoroacetyl.

Reagent. Trifluoroacetic anhydride (TFAA).

Compound Type. Nucleosides.

Reaction

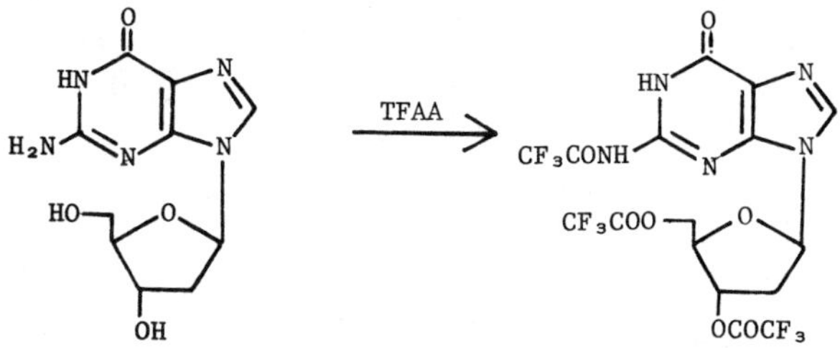

Typical Procedure. Mix 50–100 μg of nucleosides with 100 μl CH_2Cl_2 and 7 μl TFAA and heat in a sealed vial at 100° for 30 min. Analyze by GC-MS or direct probe MS.

Comments. Guanine nucleosides showed better deriv formation after heating 5 min at 150°. This procedure was not used generally however because some cleavage of the glycosidic bond occurred. The procedure (100°, 30 min) acylated all ribose and deoxyribose hydroxyls and all amino groups with active hydrogens. Enolizable carbonyls in general did not react. MS data were given for the derivs of 22 nucleosides.[1] The MS of 23 adenosine analogs (differing in the sugar moiety) were also examined.[2] The orientation of hydroxyl groups could not be determined by systematic analysis of the spectra but the data did give indication of the sugar ring size.

References

[1]W. A. Koenig *et al.*, *Biochemistry* **10**, 3968 (1971).

[2]W. A. König *et al.*, *Chem. Ber.* **105**, 262 (1972).

14.3. DERIVATIVES OF MONONUCLEOTIDES AND DINUCLEOTIDES

14.3.1. Trimethylsilyl Derivatives of Mononucleotides

Derivative Type. Trimethylsilyl.

Reagent. Hexamethyldisilazane (HMDS)/trimethylchlorosilane (TMCS).

Compound Type. Nucleotides.

Reaction

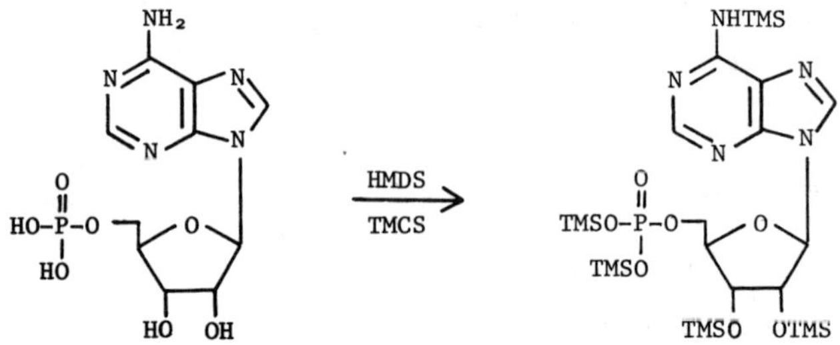

Typical Procedure. Suspend 10 mg of the nucleotide or Na salt in 0.1 ml anhyd pyridine and add 100 μl HMDS and 50 μl TMCS. Reflux 1 hr under anhyd conditions then analyze by GC.

Comments. Derivs were prepared from eight nucleotides including three 3′(2′) nucleotides. All gave single GC peaks except the deriv of 3′(2′)-UMP which gave two peaks. No GC peaks were obtained for TMS cytidylate. Hancock[2] used HMDS/TMCS/pyridine to prepare adenosine phosphate derivs.

References

[1]T. Hashizume and Y. Sasaki, *Anal. Biochem.* **15**, 199 (1966).

[2]R. L. Hancock, *J. Gas Chromatogr.* **4**, 363 (1966).

Derivative Type. Trimethylsilyl.

Reagent. Bis(trimethylsilyl)trifluoroacetamide (BSTFA)/trimethylchlorosilane (TMCS).

Compound Type. Nucleotides.

Reaction

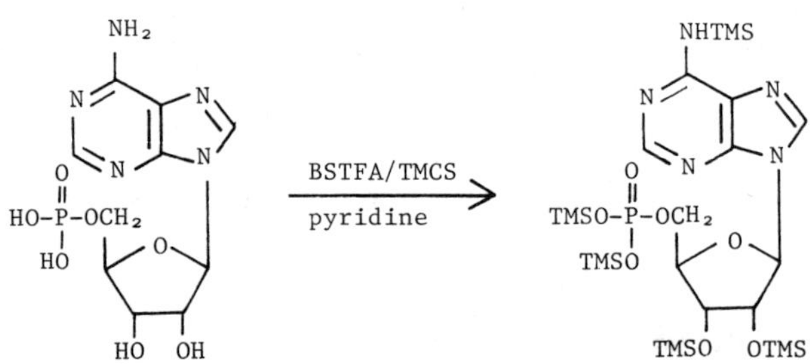

Typical Procedure. Diss 30 μg of the nucleotide or its salt in 30 μl pyridine, 30 μl BSTFA, and 1 μl TMCS. Seal and allow to stand at room temp 30 min or heat to 150°. Analyze by GC-MS.

Comments. This procedure was most generally satisfactory for all nucleotides except those of cytidine and cyclic AMP.

References

[1]A. M. Lawson *et al., J. Am. Chem. Soc.* **93**, 1014 (1971).

Derivative Type. Trimethylsilyl.

Reagent. Bis(trimethylsilyl)trifluoroacetamide (BSTFA)/trimethylchlorosilane (TMCS).

Compound Type. Adenosine-3′,5′-cyclic monophosphate (cyclic AMP).

Reaction

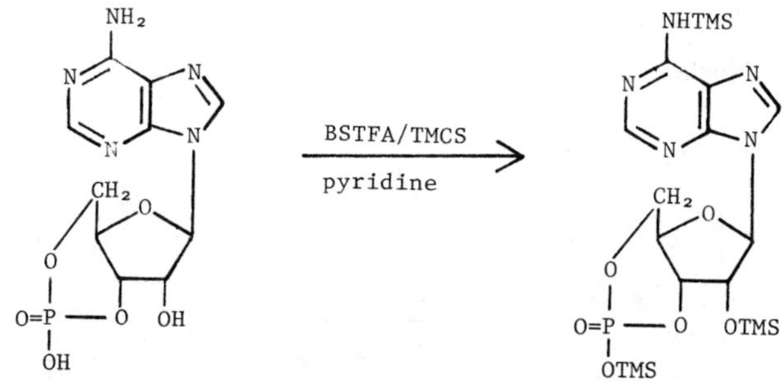

Typical Procedure. Diss 50 μg of the acid or its disodium or diammonium salt in pyridine/BSTFA/TMCS (9:90:1, v/v/v) in a sealed vial. Heat at 100° for 3 hr or until no $(TMS)_2$ deriv remains. Analyze by GC-MS.

Comments. This procedure was better for cyclic AMP than the pyridine/BSTFA/TMCS (30:30:1, v/v/v) which was generally used for other nucleotides.

References

[1] A. M. Lawson *et al., J. Am. Chem. Soc.* **93**, 1014 (1971).

Derivative Type. Trimethylsilyl.

Reagent. Bis(trimethylsilyl)acetamide (BSA)/trimethylchlorosilane (TMCS).

Compound Type. Cytidine monophosphate (CMP).

Reaction

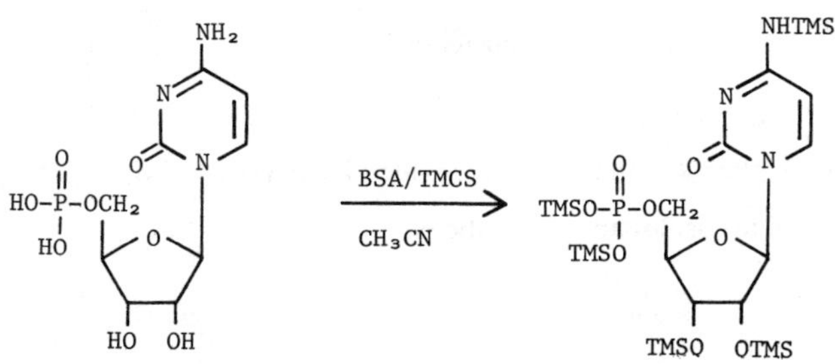

Typical Procedure. Heat 200 μg of the acid or its salt with 35 μl BSA and 5 μl TMCS in 16 μl CH_3CN at 70° for 60–90 min. Analyze by GC-MS.

Comments. This method was the optimum conditions for deriv of CMP or 2′-deoxy-CMP. The yield from this deriv method was only 20–30% of that obtained with the other major nucleotides. An alternate method[2] using BSA/TMCS without CH_3CN gave the $(TMS)_4$ deriv which could not be introduced via the GC but was amenable to direct probe MS.

References

[1] A. M. Lawson *et al., J. Am. Chem. Soc.* **93**, 1014 (1971).

[2] J. A. McCloskey *et al., J. Am. Chem. Soc.* **90**, 4182 (1968).

Derivative Type. Trimethylsilyl ring opened derivatives.

Reagent. Sodium periodate, Potassium borohydride, bis(trimethylsilyl)trifluoroacetamide (BSTFA)/trimethylchlorosilane (TMCS).

Compound Type. Nucleotides.

Reaction

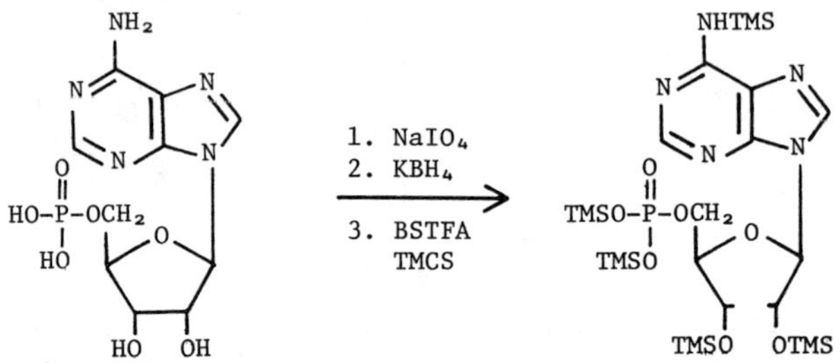

Typical Procedure. Mis 25 μg of nucleotide with 200 μl of 0.01 *M* NaOAc and 10 μl of 0.1 *M* $NaIO_4$ and allow to stand 30 min in the dark. To the resulting 2′,3′-dialdehyde soln add 10 μl of 0.15 *M* ethylene glycol. After 20 min add 10 μl 0.2 *M* KBH_4. Allow to stand 30 min then evap under vac and dry over P_2O_5. Add 30 μl pyridine, 30 μl BSTFA, and 1 μl TMCS. After 30 min at room temp, analyze by GC-MS.

Comments. The oxidative ring cleavage-tritium labeling has been used to identify terminal nucleotides in nucleic acids.[2] This deriv was therefore examined as a potential method for end group analysis by MS.

References

[1] A. M. Lawson *et al.*, *J. Am. Chem. Soc.* **93**, 1014 (1971).

[2] U. L. RajBhandary, *J. Biol. Chem.* **243**, 556 (1968).

14.3.2. Permethyl Derivatives of Mononucleotides

Derivative Type. Permethyl.

Reagent. Trimethylphenylammonium hydroxide (TMPAH).

Compound Type. Nucleosides and nucleotides.

Reaction

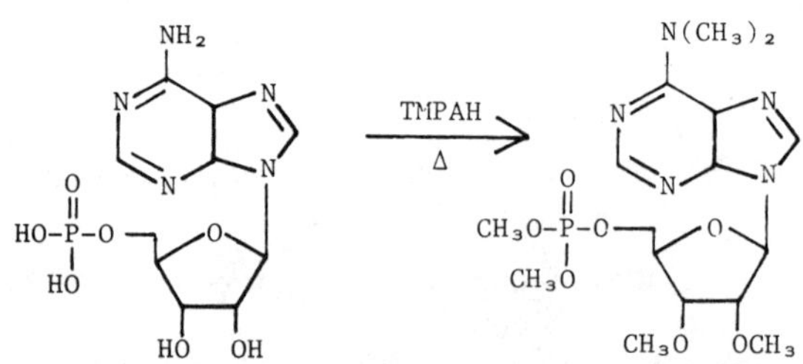

Typical Procedure.[1] Diss the sample in an excess of 1% TMPAH in MeOH. Analyze an aliquot of the soln by GC-MS with the injector port at 300°.

Comments. Single deriv peaks were obtained for the six nucleosides (adenosine, guanosine, uridine, cytidine, 6-mercaptopurine riboside, and 6-methylmercaptopurine riboside) and one nucleotide (AMP) studied.[1] Pettit *et al.*[2] applied the method to 5′-UMP, 5′-TMP, and 2′-,3′-, and 5′-AMP and reported MS data for the derivs.

References

[1]K. M. Baker, *Adv. Mass Spectrom. Biochem. Med.* **1**, 231 (1976).
[2]G. R. Pettit *et al.*, *Biomed. Mass Spectrom.* **5**, 153 (1978).

14.3.3. Derivatives of Dinucleotides

Derivative Type. Trimethylsilyl.

Reagent. Bis(trimethylsilyl)trifluoroacetamide (BSTFA)/trimethylchlorosilane (TMCS).

Compound Type. Dinucleotides.

Reaction

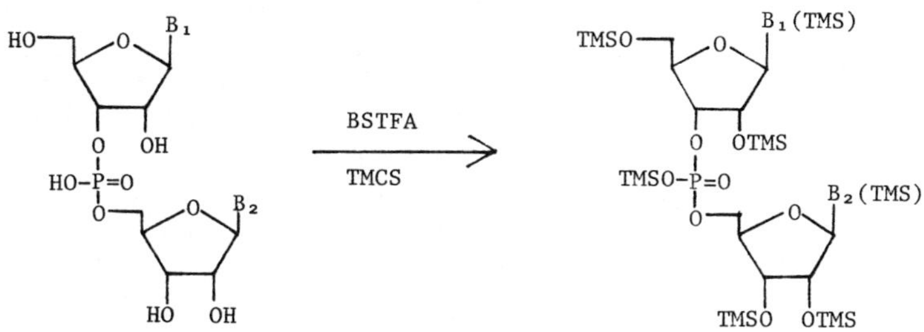

Typical Procedure. Diss 500 μg dinucleotide in 50 μl pyridine and add 100 μl BSTFA and 20 μl TMCS. Allow to stand 3 hr at room temp or heat 30 min at 60°. Evap 10–20 μl of the soln on a direct probe capillary (under vac) and analyze by MS.

Comments. MS data were given for the derivs of 16 dinucleotides.

References

[1]D. F. Hunt *et al.*, *Biochem. Biophys. Res. Commun.* **33**, 378 (1968).

Derivative Type. Phenylboronate, trimethylsilyl.

Reagent. Benzeneboronic acid, Trimethylsilylimidazole (TMSI).

Compound Type. Nucleosides and dinucleoside phosphates.

Reaction

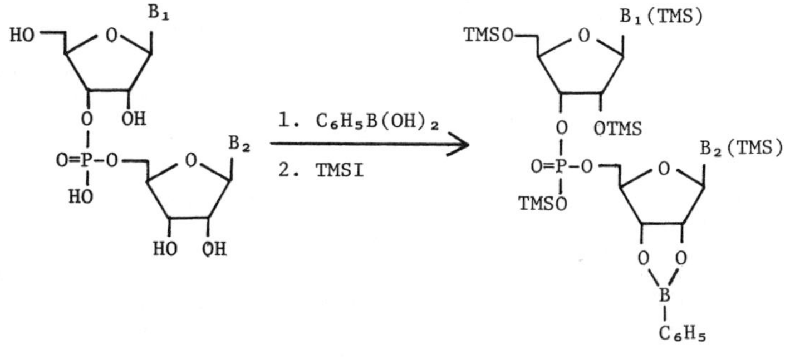

Typical Procedure. Mix 2 mg nucleoside or dinucleoside phosphate with an equimolar amt of benzene-boronic acid in dry pyridine (DMF for guanosine) and heat 2 hr at 120° in a sealed tube. Evap to dryness and wash residue several times with Et_2O. Recryst from acetone. Treat with TMSI and analyze by direct probe MS.

Comments. Derivs were prepared from uridine, adenosine, cytidine, guanosine, and six dinucleoside phosphates. The unsymmetrical dinucleoside deriv permitted determination of the order of the bases. The phenylboronate preparation was adapted to mg scale from a published procedure.[2] The MS behavior of the derivs, as well as methyl, acetyl, TMS, TFA, and isopropylidene derivs has been described.[3]

References

[1] J. J. Dolhun and J. L. Wiebers, *J. Am. Chem. Soc.* **91**, 7755 (1969).

[2] A. M. Yurkevich *et al.*, *Tetrahedron* **25**, 477 (1969).

[3] J. J. Dolhun and J. L. Wiebers, *Org. Mass Spectrom.* **3**, 669 (1970).

15. Drugs

Drug derivatives have already been discussed in several of the preceding sections. This section includes derivatives of several classes of drugs which do not conveniently fall under earlier classification. For the most part, the derivatives in these examples are simply drug applications of previously discussed derivatization methods. Drugs present a wide array of functional groups to be derivatized and often include multiple types of functional groups within a given compound. Examples of derivatization of drugs, therefore, offer valuable information on the scope of a derivatization method and the compatability of the method with other structures present. Most of the derivatives listed in this section have been used for gas chromatography. The derivatization of drugs for gas chromatographic analysis has been reviewed by Ahuja.[1]

Alkaloids. The importance of morphine and related compounds in medicine has stimulated much interest in the analysis of these compounds. Morphine itself contains both phenolic and alcoholic hydroxyls and a tertiary amino group. Acylation (p. 623), alkylation (p. 626), and silylation (p. 627) methods for this class of compounds have been reported for GC analysis. These hydroxyl group derivatives give satisfactory GC performance without any derivatization of the amino group. For very low level analysis of pentazocine, the adverse effects of the amino group were minimized by coinjection of another amino compound to compete for the column adsorption sites (p. 626). The difference in reactivity of the phenol and alcohol groups allows preparation of mixed alkyl and acyl derivatives. This approach was utilized to yield a derivative that avoided interfering impurities in the analysis (p. 626). Chromogenic derivatization of the phenolic group of morphine type compounds has been reported for HPLC analysis (p. 628).

Cocaine is a tertiary amine containing two ester groups. The compound can be analyzed by GC underivatized, but enhanced sensitivity has been obtained by reduction to a dihydroxy derivative and formation of the electron capturing diheptafluorobutyrate (p. 628). The cocaine hydrolysis products, ecgonine and benzoylecgonine, contain free carboxyl groups and require esterfication for GC analysis. Benzoylecgonine was analyzed concurrently with cocaine as the propyl ester derivative (p. 629). Halogenated esters of both benzoylecgonine (p. 629) and ecgonine (p. 630) have been used for ECD-GC analysis.

The catechol alkaloids, apomorphine and tetrahydroberberine, were analyzed by GC as heptafluorobutyryl (p. 630) and trimethylsilyl (p. 631) derivatives.

Antibiotics. The antibiotics present a diverse array of functional groups that must be derivatized for successful GC or MS analysis. Most applications have employed trimethylsilyl derivatives. A series of methods for preparation of these derivatives are presented beginning on p. 632. Some of the antibiotics contain so many functional groups that trimethylsilylation yields a derivative with an inconveniently high molecular weight for mass spectral analysis. Permethylation has been employed to yield derivatives with a smaller mass increment (p. 636). The acetyl derivative was used for analysis of clindamycin (p. 637) and several penicillins could be analyzed by GC simply as methyl esters (p. 637). The zwitterionic cephaloridine was cleaved by base treatment and quantitated by analysis of liberated pyridine (p. 638).

Barbiturates, Hydantoins, and Related Compounds. Underivatized barbiturates and hydantoins can, in many cases, be analyzed by GC, but in most cases the compounds are derivatized to improve separation and peak shape..Derivatization of this group of compounds principally involves masking the polar amide and imide N-H groups. The acidic hydrogens may be replaced by various substituents with alkylation being most commonly used. Methylation, the most widely studied alkylation, has been accomplished with dimethyl sulfate (p. 638), methyl iodide and base (p. 639), diazomethane (p. 641), and quaternary ammonium reagents (p. 642). Each type of reagent has disadvantages. For example, diazomethane can yield isomer mixtures (p. 642) and the quaternary ammonium reagents can yield ring opening (p. 644). A "buffering" method has been reported to minimize the latter problem (p. 644). Higher alkyl derivatives have been examined to improve GC separation and to improve derivative stability (p. 646). Pentafluorobenzyl derivatives permit application of ECD-GC for increased sensitivity (p. 650).

A few applications of acyl (p. 652) and trimethylsilyl (p. 653) derivatives have been reported for barbiturates and related compounds. The alkylating agent dimethylformamide dimethyl acetal, instead of forming *N*-methyl derivatives, gave what appeared to be dimethyl ketals with barbiturates and some related compounds (p. 654). Barbiturates have been analyzed by HPLC as the fluorogenic DANS derivatives (p. 654).

Benzodiazepines. The benzodiazepines also contain acidic amide nitrogens that require replacement for optimum GC behavior. Phase transfer methylation has been applied to several members of this compound class (p. 655). An alternative approach to benzodiazepine analysis employs hydrolysis to substituted benzophenone derivatives (p. 656). These electron capturing derivatives have been applied to the analysis of several benzodiazepines.

Cannabinoids. Analysis of cannabinoids by GC involves derivatizations primarily of phenolic groups with some compounds also containing alcohol hydroxyl or carboxyl groups. Several investigators have examined silylated derivatives for both GC and GC-MS analysis (p. 657). An uncommon trimethylsilylacetyl derivative was demonstrated on tetrahydrocannabinol and shown to be stable in water-methanol solution (p. 658). Other derivatives used for GC analysis of cannabinoids include methyl (p. 660) and trifluoroacetyl (p. 660). Diethyl phosphate derivatives were used with flame photometric detection (p. 661). Cannabinolic acids formed cyclic boronates suitable for GC or GC-MS analysis (p. 661).

Diphenylmethanes. Electron capturing benzophenone derivatives were discussed in conjunction with benzodiazepine analysis. Benzophenone derivatives can also be formed from diphenylmethane drugs by oxidation. Early studies employed chromic acid oxidation (p. 662), but subsequent reports indicated improved results with barium peroxide (p. 663). A related compound containing a phenyl pyridyl methane group was oxidized to a "benzophenone" type derivative with potassium permanganate (p. 664).

References

1. S. Ahuja, *J. Pharm. Sci.* **65**, 163 (1976).

15.1. ALKALOIDS

15.1.1. Morphine and Related Compounds

Derivative Type. Acetyl.

Reagent. Acetic anhydride (Ac_2O).

Compound Type. Morphine.

Reaction

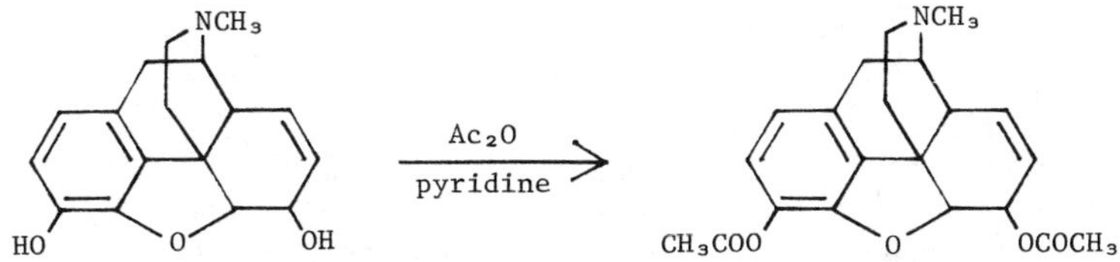

Typical Procedure. Treat ext (see below) with 0.75 ml Ac_2O and 0.15 ml pyridine for 1 hr at 60°. Add MeOH (to remove excess Ac_2O as MeOAc) and evap under vac at 60°. Diss deriv in 0.5 ml *n*-BuOAc containing 100 μg/ml estradiol-17β diacetate int stand. Analyze by GC.

Comments. Ext procedures were given for 10 ml samples of bile, urine, blood, and liver tissue.

References

[1] J. E. Wallace *et al.*, *Clin. Chim. Acta* **36**, 85 (1972).

Derivative Type. Trifluoroacetyl.

Reagent. Trifluoroacetic anhydride (TFAA).

Compound Type. Morphine.

Reaction

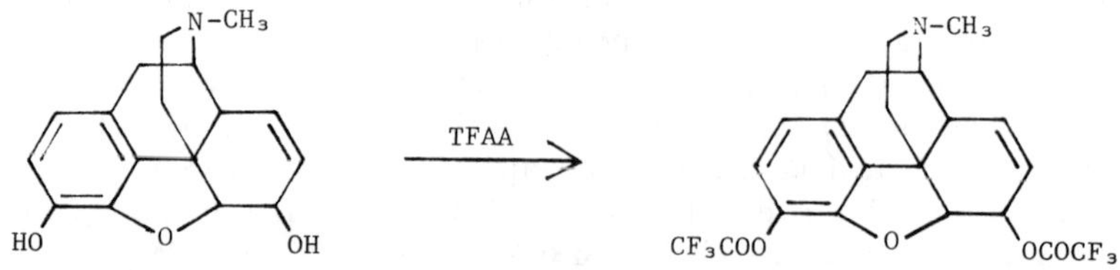

Typical Procedure. Mix 1–2 ml serum with 2 ml pH 8.9 borate buffer and ext with 25 ml 10% *i*-BuOH in $CHCl_3$. Wash ext with 2×5 ml pH 10.1 phosphate buffer. Add 5 ml 0.5 *N* HCl and shake 3 min. Remove 4.5 ml of the HCl soln and adjust to pH 8.7±0.2 with solid Na_2CO_3. Ext with 5 ml 10% *i*-PrOH in EtOAc. Transfer 4.5 ml of the ext to another tube containing 0.1 ml 2 μg/ml soln of nalorphine int stand in EtOAc. Evap to dryness under a stream of filtered air at 50°. Add 0.1 ml EtOAc and 0.1 ml TFAA. Heat 20 min at 50° then evap to dryness as before. Diss residue in 0.1 ml EtOAc and analyze by ECD-GC.

Comments. The deriv was detectable to less than 0.1 ng injected. The method permitted serum quantitation down to 25 ng/ml.

References

[1]J. E. Wallace *et al., Anal. Chem.* **46**, 2107 (1974).

Derivative Type. Pentafluoropropionyl.

Reagent. Pentafluoropropionic anhydride (PFPA).

Compound Type. Morphine.

Reaction

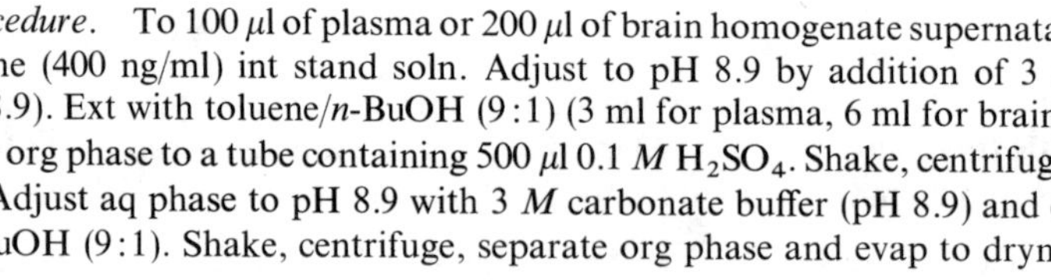

Typical Procedure. To 100 μl of plasma or 200 μl of brain homogenate supernatant add 100 μl of nalorphine (400 ng/ml) int stand soln. Adjust to pH 8.9 by addition of 3 *M* carbonate buffer (pH 8.9). Ext with toluene/*n*-BuOH (9:1) (3 ml for plasma, 6 ml for brain). Centrifuge and transfer org phase to a tube containing 500 μl 0.1 *M* H_2SO_4. Shake, centrifuge and discard org phase. Adjust aq phase to pH 8.9 with 3 *M* carbonate buffer (pH 8.9) and ext with 3 ml toluene/*n*-BuOH (9:1). Shake, centrifuge, separate org phase and evap to dryness under N_2 at 70°.

Diss residue in 50 μl dry EtOAc and add 100 μl PFPA. Seal and heat at 60° for 30 min. Remove excess reagent under N_2 and diss residue in 50 μl dry EtOAc. Analyze 1–2 μl by GC.

Comments. The method was used to quantitate morphine to 500 pg/ml in plasma and to 100 pg/30 mg in brain tissue. The addition of 10% *n*-BuOH to the ext solvent reduced adsorption losses.

References

[1]B. Dahlström and L. Paalzow, *J. Pharm. Pharmacol.* **27**, 172 (1975).

Derivative Type. Heptafluorobutyryl (HFB).

Reagent. Heptafluorobutyric anhydride (HFBA).

Compound Type. Codeine, morphine, norcodeine, and normorphine.

Reaction

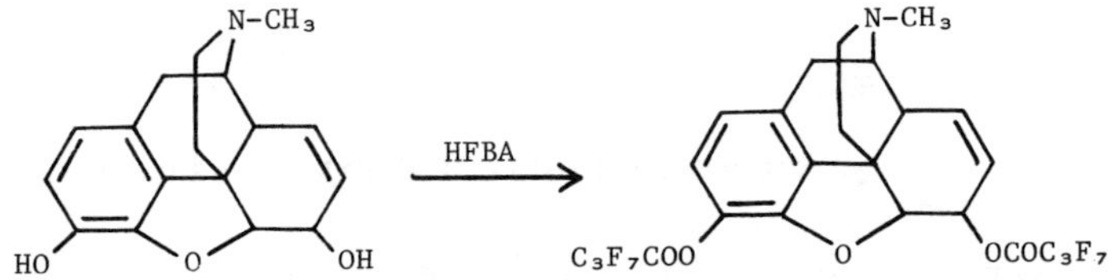

Typical Procedure. Add deuterated int stand to 0.5 ml plasma and ppt proteins by addition of 6 vol of acetone. Evap the supernatant and hydrolyze the residue in 1 ml 6 *N* HCl (30 min, 100°). Add 1 ml $Na_2B_4O_7$-HCl buffer (0.04 *M*, pH 8.85) and adjust to pH 8.85 with 8 *N* NaOH. Ext 20 min with 4 ml *i*-AmOH/$CHCl_3$ (1:3, v/v) then evap ext to dryness under N_2 at 80–90°. To the residue add 100 μl of 40% HFBA in Et_2O. Allow to react 30 min at room temp, evap under N_2, and diss residue in 100 μl $CHCl_3$. Analyze by GC-MS-SIM.

Comments. The procedure was also used for analysis of urine. TFA derivs were prepared by the same method using 40% TFAA in CH_2Cl_2. The TFA deriv proved preferable for GC-MS analysis due to interferences with the HFB deriv.

References

[1]W. O. R. Ebbighausen *et al., Adv. Biochem. Psychopharmacol.* **7**, 135 (1973).
[2]W. O. R. Ebbighausen *et al., J. Pharm. Sci.* **62**, 146 (1973).

Derivative Type. Heptafluorobutyryl (HFB).

Reagent. Heptafluorobutyric anhydride (HFBA).

Compound Type. Naloxone and naltrexone.

Reaction

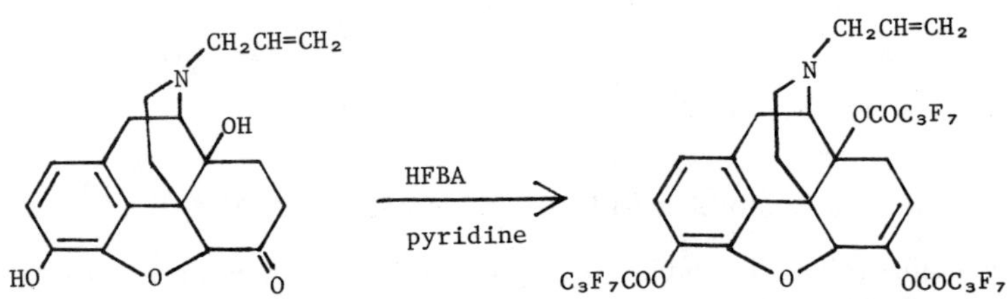

Typical Procedure. To submicrogram samples of naloxone or naltrexone in 15 × 125 mm glass culture tubes add 25 μl of HFBA and 50 μl of 1% pyridine in benzene. Seal the tubes and heat in a 70° oil bath (50 mm immersion to permit refluxing in the upper part of the tube) for 2 hr. Cool tubes in an ice bath. Immediately before analysis wash with 5 ml satd aq $Na_2B_4O_7$ and analyze 1–3 μl of the org phase by ECD-GC.

Comments. The PFP and TFA derivs were prepared using PFPA and TFAA in the same procedure. The HFB derivs were more stable than the PFP that were more stable than the TFA. Each compd was used as the int stand for the other. Use of larger conc of pyridine led to ECD interferences. Omission of pyridine gave di-HFB (underiv keto group). Structures of the derivs were confirmed by MS and IR. Derivs were also prepared from cyclazocine, pentazocine, morphine, and oxymorphone and their GC behavior examined to insure that these compds would not interfere in the analysis.

References

[1]R. A. Sams and L. Malspeis, *J. Chromatogr.* **125**, 409 (1976).

Derivative Type. Pentafluorobenzyl.

Reagent. Pentafluorobenzyl bromide (PFB-Br)/tetrabutylammonium hydrogen sulfate (TBA-HSO_4).

Compound Type. Pentazocine.

Reaction

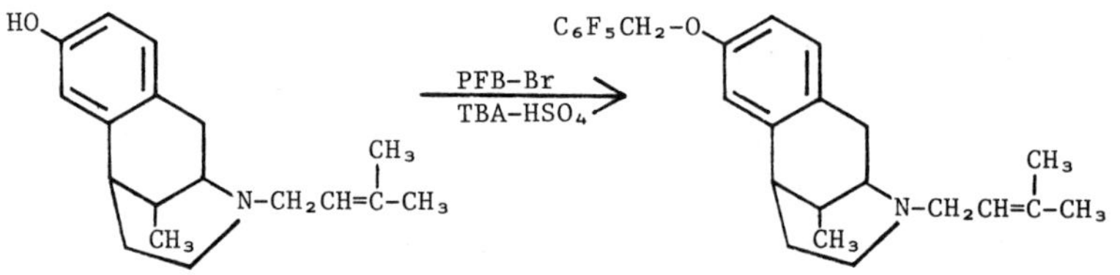

Typical Procedure. Mix 0.5 ml plasma with 0.1 ml pH 10.5 carbonate buffer. Ext with 3.0 ml benzene. Transfer 2 ml of benzene to another tube and ext with 0.5 ml 0.1 *M* H_2SO_4 containing ketobemidone int stand (10 ng/ml).

Mix aq phase with 0.1 ml 12 *M* NaOH, 0.1 ml 0.5 *M* TBA-HSO_4, and 0.2 ml CH_2Cl_2 soln of PFB-Br (0.4%, v/v). Shake 10 min and discard aq phase. Mix CH_2Cl_2 phase with 0.2 ml *n*-heptane and 0.5 ml 0.5 *M* H_2SO_4. Centrifuge and remove org phase by evap under N_2. Wash aq phase with 4×0.2 ml *n*-heptane. Add 0.1 ml 6 *M* NaOH and ext with 0.1 ml *n*-heptane. Analyze 1–3 μl of heptane by ECD-GC.

Comments. The method was used to quantitate pentazocine in plasma to 1 ng/ml by ECD-GC Samples were "bracketed" in the syringe with a soln of desipramine base to reduce adsorption due to the tertiary amine group.

References

H. Brotell *et al.*, *J. Chromatogr.* **78**, 293 (1973).

Derivative Type. Pentafluorobenzyl(PFB), trifluoroacetyl (TFA).

Reagent. Pentafluorobenzyl bromide (PFB-Br)/tetrabutylammonium hydroxide (TBAH), trifluoroacetic anhydride (TFAA).

Compound Type. Morphine and related compds.

Reaction

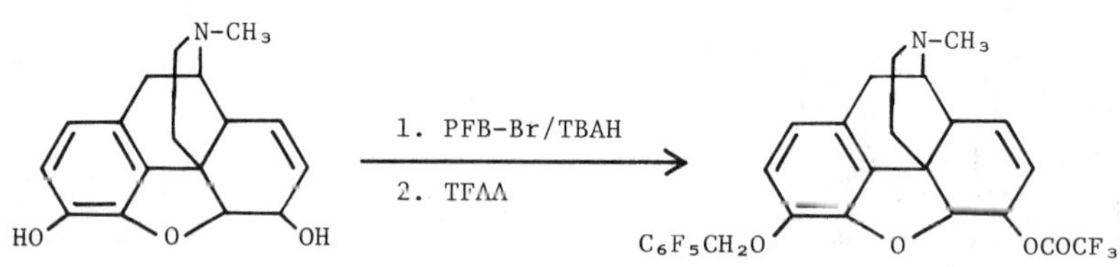

Typical Procedure. To 1 ml plasma add 600 ng d_3-morphine int stand, 50 μl 4 *M* NaOH, 250 μl TBAH, 1 ml EtOAc, and 20 μl PFB-Br. Shake 30 min at 22°. Remove the EtOAc and mix with 1 ml 0.05 *M* H_2SO_4. Remove aq phase and adjust to pH 14 with 4 *M* NaOH. Ext with 1 ml EtOAc, wash ext with H_2O, and evap to dryness at 75° under N_2. Add 50 μl benzene/MeOH (1 :4, v/v) and evap to remove traces of H_2O. Add 50 μl benzene and 25 μl TFAA and heat 15 min at 75°. Analyze by GC-MS-SIM.

Comments. PFB TFA derivs were prepared from morphine and nalorphine; PFB derivs were prepared from pentazocine, levorphanol, and levallorphan. Use of the PFB,TFA deriv avoided some of the long retention time impurities seen in using the di-TFA deriv.

References

[1]W. J. Cole *et al.*, *J. Chromatogr.* **136**, 409 (1977).

Derivative Type. Trimethylsilyl.

Reagent. Bis(trimethylsilyl)trifluoroacetamide (BSTFA)/trimethylchlorosilane (TMCS).

Compound Type. Morphine.

Reaction

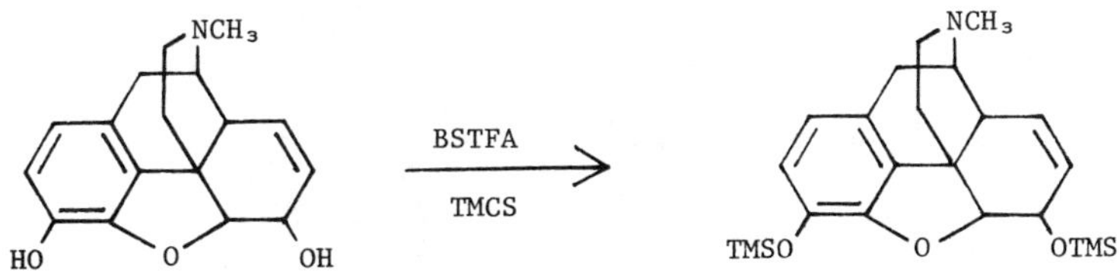

Typical Procedure.[1] Mix 0.1–1.0 ml plasma or CSF with 250 mg $NaHCO_3$ and adjust vol to 2.0 ml with H_2O. Ext with 5.5 ml EtOAc containing 10% (v/v) *i*-PrOH. Remove 5.0 ml of the org phase and back ext into 1.2 ml 1 *N* HCl. Evap 1.0 ml of the aq phase to dryness under vac at 35°. Add 50 μl of 10 μg/ml tetraphenylethylene (int stand) in MeOH and evap. Add 25 μl BSTFA containing 1% TMCS. After 20–30 min at room temp, analyze by GC.

Comments. The method was sensitive down to *ca* 25 ng/sample. Sine *et al.*[2] also used BSTFA/TMCS for determination of morphine and other alkaloids in urine. The TMS deriv has also been prepared using BSA[3,4] and HMDS/TMCS.[5,6,7]

References

[1]G. R. Wilkinson and E. L. Way, *Biochem. Pharmacol.* **18**, 1435 (1969).

[2]H. E. Sine *et al.*, *Chin. Chem.* **19**, 340 (1973).

[3]N. Ikekawa *et al.*, *Anal. Biochem.* **28**, 156 (1969).

[4]F. Fish and W. D. C. Wilson, *J. Chromatogr.* **40**, 164 (1969).

[5]E. Brochmann-Hanssen and A. B. Svendsen, *J. Pharm. Sci.* **51**, 1095 (1962).

[6]E. Brochmann-Hanssen and A. B. Svendsen, *J. Pharm. Sci.* **52**, 1134 (1963).

[7]G. E. Martin and J. S. Swinehart, *Anal. Chem.* **38**, 1789 (1966).

Derivative Type. 5-Dimethylamino-1-naphthalene sulfonyl (DANS).

Reagent. 5-Dimethylamino-1-naphthalene sulfonyl chloride (DANS-Cl).

Compound Type. Alkaloids.

Reaction

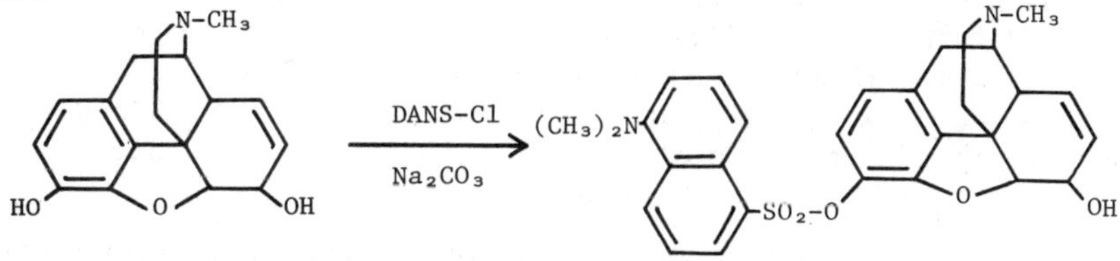

Typical Procedure. To 10 μl of aq alkaloid soln (containing not more than 30 nmol reactive groups) add 50 μl 0.1% DANS-Cl in acetone and 50 μl 0.1 *M* Na_2CO_3. Heat 20 min at 45° in the dark then cool to room temp and ext with 0.5 ml benzene. Analyze an aliquot of the benzene ext by HPLC.

Comments. The method was developed for analysis of cephaeline, codeine, emetine, ephedrine, morphine, narcotine, and other alkaloids in complex pharmaceutical dosage forms (codeine and narcotine did not form DANS derivs). The method was applied directly to syrups and aq slurries of capsule contents, but employed a TLC cleanup before HPLC. The structures of the derivs were confirmed by NMR. UV absorption and fluorescence data were reported for the derivs.

References

[1]R. W. Frei *et al.*, *J. Chromatogr.* **116**, 365 (1976).

15.1.2. Cocaine and Related Compounds

Derivative Type. Heptafluorobutyryl (HFB).

Reagent. Lithium aluminum hydride, heptafluorobutyric anhydride (HFBA).

Compound Type. Cocaine.

Reaction

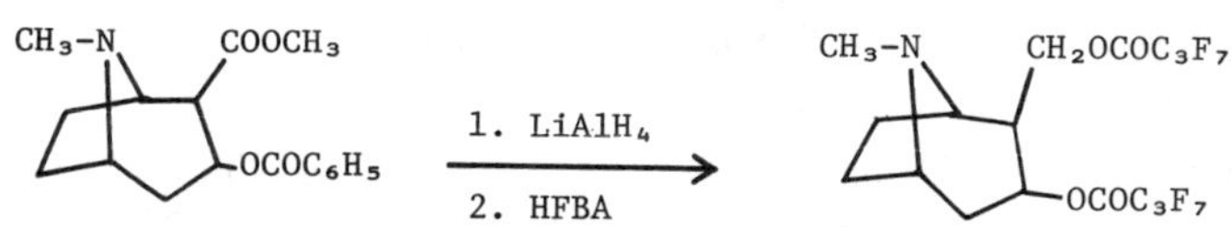

Typical Procedure. Mix 5 ml urine with 1 ml satd aq $Na_2B_4O_7$ and ext with 2 ml cyclohexane. Separate the cyclohexane phase and add to it 50 μl satd $LiAlH_4$ in Et_2O. Allow to react 3 min; then add 50 μl H_2O and shake. Add 50 μl HFBA and allow to react 3–5 min at room temp. Wash with 6 ml satd $Na_2B_4O_7$ and analyze an aliquot of the org phase by ECD-GC.

Comments. The PFP deriv was prepared by the same method using PFPA. GC data were given for the two derivs, as well as the HFB and PFP derivs of amphetamine and methamphetamine.

References

[1]J. W. Blake *et al.*, *Anal. Chem.* **46**, 288 (1974).

Derivative Type. Propyl ester.

Reagent. 1-Iodopropane/tetramethylammonium hydroxide (TMAH)/trimethylphenylammonium hydroxide (TMPAH).

Compound Type. Cocaine and benzoylecgonine.

Reaction

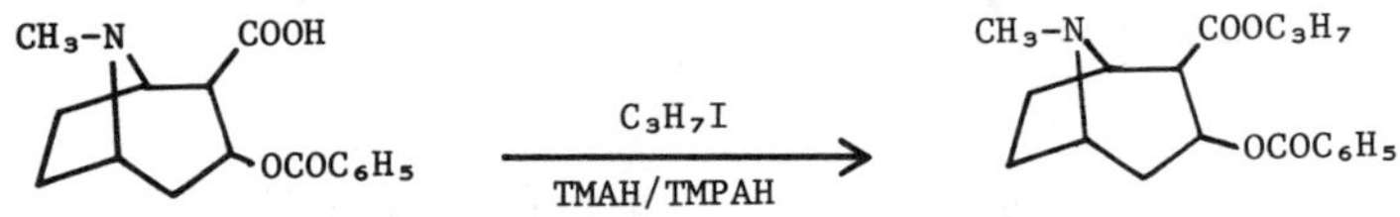

Typical Procedure. Mix 5 ml urine with 0.2 ml pH 9.6 Na_2CO_3-$NaHCO_3$ buffer (0.75 *M*) and ext with 25 ml 25% v/v EtOH in $CHCl_3$ containing 25 μg int stand. Evap ext to dryness under N_2 at 60°. Add 0.5 ml of the base (see below). Mix and after 1 min add 10 μl C_3H_7I. Heat at 50° for 15 min. Add 5 ml 0.2 *N* H_2SO_4 and wash with 2 × 5 ml *n*-BuOAc. Add 2 ml of the carbonate buffer and ext with 2 ml $CHCl_3$. Evap ext under N_2 at 60° and diss residue in 20 μl $CHCl_3$. Analyze by GC.

Comments. The base was composed of 1 part 25% TMAH in MeOH and 20 parts 0.1 *M* TMPAH in MeOH in 200 parts DMF (v/v). The mixed base was used because TMPAH alone required a long (>1 hr) reaction time and TMAH alone caused hydrolysis of the sample compd and int stand. *n*-Pentylbenzoylecgonine was used as the int stand. The method was used for simultaneous determination of cocaine and benzoylecgonine in urine.

References

[1]D. L. von Minden and N. A. D'Amato, *Anal. Chem.* **49**, 1974 (1977).

Derivative Type. Pentafluorobenzyl ester.

Reagent. Pentafluorobenzyl bromide (PFB-Br)/pyridine.

Compound Type. Benzoylecgonine.

Reaction

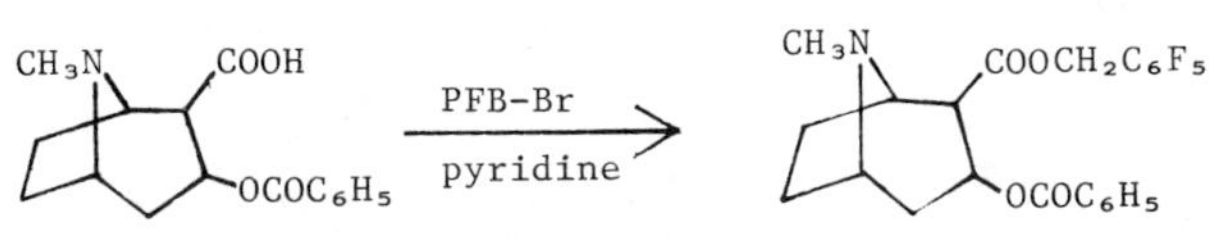

Typical Procedure. To 0.5 ml plasma add 0.1 ml 50 ng/ml soln of chlorproethazine HCl (int stand) in MeOH and basify with 0.5 ml pH 9.5 Na_2CO_3-$NaHCO_3$ buffer. Ext with 10 ml 20% (v/v) EtOH in $CHCl_3$. Evap ext to dryness and diss residue in 0.2 ml PFB-Br (5 μl/10 ml) in CH_2Cl_2 and 5 μl 1% (v/v) pyridine in benzene. Heat 1 hr at 80°; then evap solvents and excess reagents. Partition residue between 0.2 ml 0.15 *M* NH_4OH and 2 ml benzene. Transfer benzene to another tube containing 0.5 ml 0.03 *N* H_2SO_4. Mix well then discard the benzene layer. Add 5 mg $NaHCO_3$ and ext with 2 ml benzene. Evap the benzene ext to dryness and diss residue in 20 μl MeOH. Analyze by ECD-GC.

Comments. The method permitted analysis of benzoylecgonine in plasma to 5 ng/ml. The method was unusual in that it used a post derivatization back extraction cleanup step. Benzoyl ecgonine has also been analyzed as the TMS deriv from BSTFA.[2]

References

[1]M. J. Kogan *et al.*, *Anal. Chem.* **49**, 1965 (1977).

[2]A. P. Graffeo *et al.*, *J. Chromatogr.* **126**, 717 (1976).

Derivative Type. 2-Chloroethyl ester.

Reagent. 2-Chloroethanol/boron trichloride.

Compound Type. Ecgonine.

Reaction

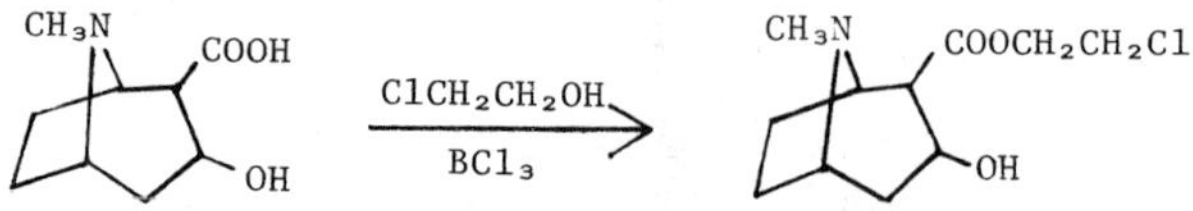

Typical Procedure. Add 10 ml 0.1 g/ml BCl_3 in 2-chloroethanol to 1 ml urine. Heat at 100° for 1 hr. Cool and add 10 ml ice cold H_2O. Ext with several 40 ml portions of Et_2O and discard ext. Make the aq soln basic to litmus with NH_4OH and ext with 4 × 30 ml Et_2O. Evap exts under an air stream and diss residue in Et_2O containing aldrin (40 ng/ml) int stand. Analyze by ECD-GC.

Comments. The method permitted identification of as little as 10 μg in 1 ml urine. It was suggested that further deriv of the hydroxyl to a PFP or HFB deriv would give a deriv much more sensitive to ECD.

References.

[1]J. M. Moore, *Clin. Chem.* **21**, 1538 (1975).

15.1.3. Other Alkaloids

Derivative Type. Heptafluorobutyryl.

Reagent. Heptafluorobutyric anhydride (HFBA).

Compound Type. Apomorphine.

Reaction

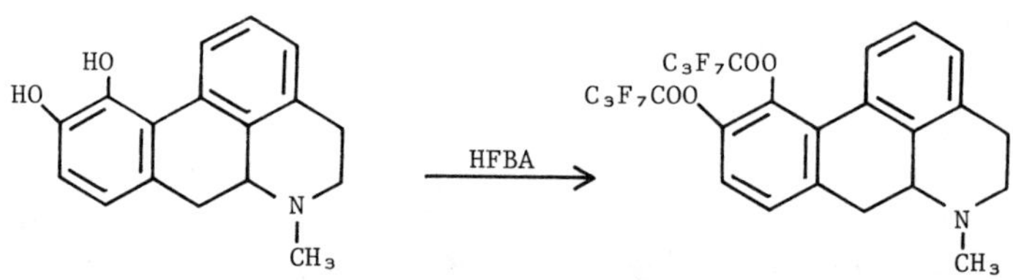

Typical Procedure. To 2 ml plasma add 0.2 ml of 0.5 *M* pH 7.0 phosphate-citrate buffer and ext with 4 ml and 2 ml of 4:1 (v/v) benzene/EtOAc. Evap ext to dryness and diss residue in 0.5 ml benzene. Add 10 μl 5% Et_3N in benzene and 40 μl HFBA and heat at 60° for 40 min. Wash with 10 ml satd aq $Na_2B_4O_7$ (pH 9.0). Analyze an aliquot of the benzene phase by ECD-GC.

Comments. The method was sensitive to 5 ng/ml in plasma. The use of EtOAc for ext or pyridine as catalyst gave higher blanks. MS data for the deriv were reported and were claimed to be consistent with the structure shown.

References

[1]J. R. Miller *et al*., *Res. Commun. Chem. Pathol. Pharmacol.* **15**, 447 (1976).

Derivative Type. Trimethylsilyl.

Reagent. Bis(trimethylsilyl)acetamide (BSA).

Compound Type. Apomorphine.

Reaction

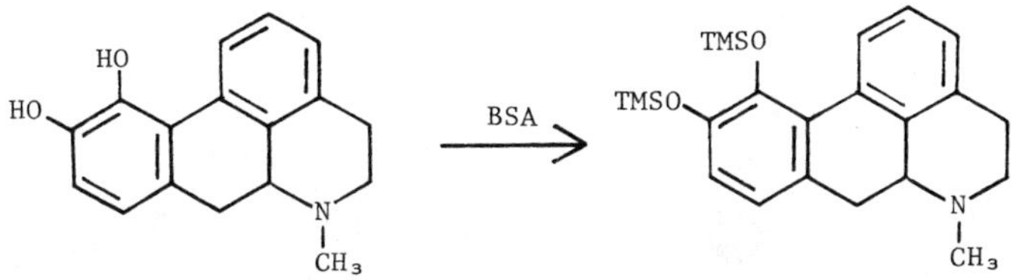

Typical Procedure. Adjust a 10 ml urine sample to pH 7.00 ± 0.05 with 10 *N* NaOH or 5% HCl and ext with 4×5 ml EtOAc. Combine ext, evap to dryness, and treat with three successive 0.3 ml portions of BSA. Adjust volume to 1 ml with BSA and gently shake 20 min. Analyze 2–5 μl by GC.

Comments. Freeze dried feces samples were analyzed by the same method after powdered samples were hydrolyzed for 30 min at 90–100° in 5 ml 4.8 *N* HCl.

References

[1]R. V. Smith and A. W. Stocklinski, *Anal. Chem.* **47**, 1321 (1975).

Derivative Type. Trimethylsilyl ether.

Reagent. Hexamethyldisilazane (HMDS)/trimethylchlorosilane (TMCS).

Compound Type. Tetrahydroberberine alkaloids.

Reaction

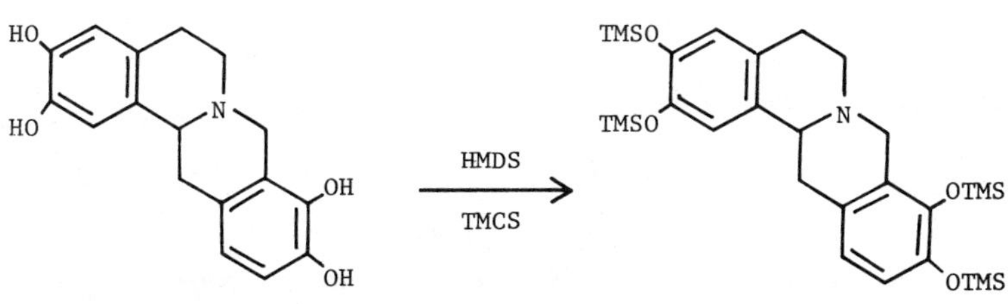

Typical Procedure. Treat urine (70 ml) with glucuronidase and sulfatase at pH 5 then buffer with 5 g NaH_2PO_4 and adjust to pH 7.0 with NaOH. Ext with 4 vol toluene/*i*-AmOH (3:2, v/v). Ext org phase with one tenth vol. 0.1 *N* HCl.

Evap 1 ml aliquot of HCl to dryness and treat residue with 100 μl of HMDS/TMCS (9:1, v/v) for 8 hr at 65°. Analyze by GC-MS.

Comments. The compds were also analyzed in tissue homogenates.

References

[1]J. L. Cashaw *et al.*, *J. Chromatogr.* **99**, 567 (1974).

15.2. ANTIBIOTICS

15.2.1. Trimethylsilyl Derivatives

Derivative Type. Trimethylsilyl.

Reagent. Hexamethyldisilazane (HMDS).

Compound Type. Penicillins.

Reaction

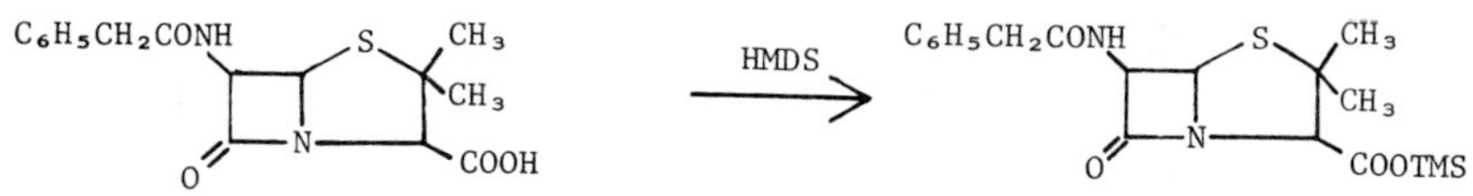

Typical Procedure. Mix 2 ml of 20 mg/ml aq soln of the penicillin with 2 ml pH 2.2 $(NH_4)_2SO_4$ buffer and ext with 8 ml $CHCl_3$. Mix 2 ml of the ext with 2 ml of HMDS/pyridine (1:1, v/v) containing 0.375 mg/ml 5-α-cholestane (or 5-α-cholestan-3-one) int stand. Allow to stand at room temp 10 min then analyze by GC.

Comments. Derivs were prepared from seven penicillins and their structures confirmed by MS. Deriv of penicillins G and V, phenethicillin, and methicillin was complete in 10 min; oxacillin, cloxacillin, and dicloxacillin required up to 60 min.

References

[1]C. Hishta *et al., Anal. Chem.* **43**, 1530 (1971).

Derivative Type. Trimethylsilyl.

Reagent. Hexamethyldisilazane (HMDS)/trimethylchlorosilane (TMCS)/trimethylsilylimidazole (TMSI)/bis(trimethylsilyl)acetamide (BSA).

Compound Type. Ampicillin.

Reaction

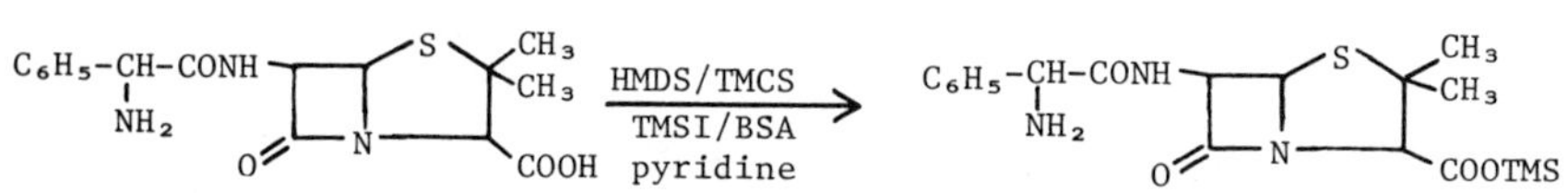

Typical Procedure. Prepare sample by diss capsule contents in MeOH and evap an aliquot to dryness or by evap an aliquot of an aq prep to dryness. Diss residue in 0.3 ml pyridine, 0.3 ml HMDS, 0.3 ml TMCS, 0.1 ml TMSI, and 0.05 ml BSA. Allow to stand 2 hr at room temp then analyze by GC.

Comments. The method gave a single mono-TMS product from either ampicillin or its sodium salt. Use of BSA in pyridine or CH_3CN gave multiple peaks due to formation of *N*-TMS derivs. HMDS/TMSI/pyridine (5:0.5:3) gave a single mono-TMS deriv with ampicillin but not with the salt. The method was used to quantitate ampicillin in pharmaceutical preparations using α-cholestane as int stand. MS data for the derivs were reported.

References

[1]H.-L. Wu *et al., J. Chromatogr.* **137**, 127 (1977).

Derivative Type. Trimethylsilyl.

Reagent. Hexamethyldisilazane (HMDS)/trimethylchlorosilane (TMCS)/pyridine.

Compound Type. Lincomycin.

Reaction

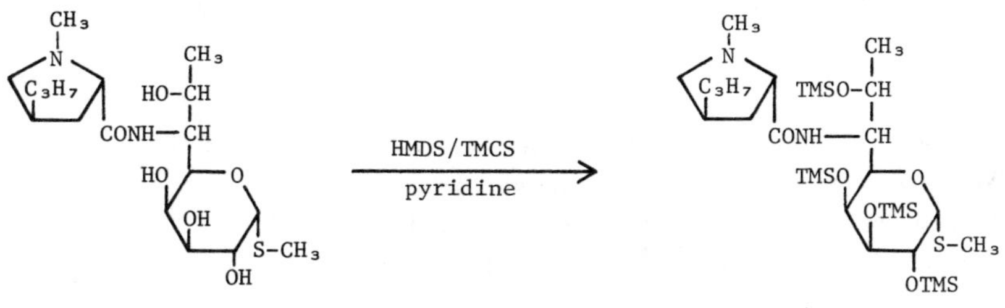

Typical Procedure. Diss 10 mg sample in 1 ml pyridine and add 200 μl HMDS and 100 μl TMCS. Allow to react 30 min at room temp then analyze by GC.

Comments. The procedure was suitable for analysis of bulk drug, pharmaceutical preparations, and fermentation samples. GC separation of lincomycin analogs and drug and excipients in a pharmaceutical preparation was demonstrated.

References

[1]R. L. Houtman *et al.*, *J. Pharm. Sci.* **57**, 693 (1968).

Derivative Type. Trimethylsilyl.

Reagent. Trimethylsilyldiethylamine (TMSDEA).

Compound Type. Neomycins B and C.

Reaction

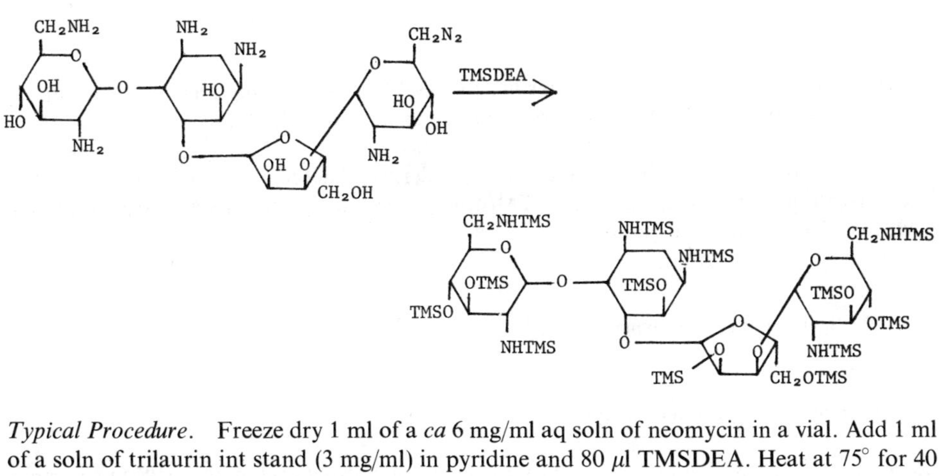

Typical Procedure. Freeze dry 1 ml of a *ca* 6 mg/ml aq soln of neomycin in a vial. Add 1 ml of a soln of trilaurin int stand (3 mg/ml) in pyridine and 80 μl TMSDEA. Heat at 75° for 40 min. Analyze by GC or GC-MS.

Comments. The derivs of the two neomycins could be separated and quantitated on an OV-1 column. The derivs were very moisture labile. Although a mol ion was not observed, MS data for the derivs suggested that all N and O sites were converted to the TMS deriv as shown. MS data for similarly prepared derivs of neamine, neobiosamine B, neosamine B, and deoxystreptamine (one and two ring fragments of neomycin) likewise showed silylation of each O and N carrying replaceable H. The use of TRI-SIL Z[3] in place of pyridine in the int stand soln was reported to give more stable derivs. The method was subsequently adapted for the analysis of kanamycin and paromycin.[4]

References

[1]K. Tsuji and J. H. Robertson, *Anal. Chem.* **41**, 1332 (1969).

[2]B. Van Giessen and K. Tsuji, *J. Pharm. Sci.* **60**, 1069 (1971).

[3]TMSI (1.5 meq/ml) in pyridine, Pierce Chemical Co., Rockford, Ill.

[4]K. Tsuji and J. H. Robertson, *Anal. Chem.* **42**, 1661 (1970).

Derivative Type. Trimethylsilyl.

Reagent. Hexamethyldisilazane (HMDS).

Compound Type. Spectinomycin.

Reaction

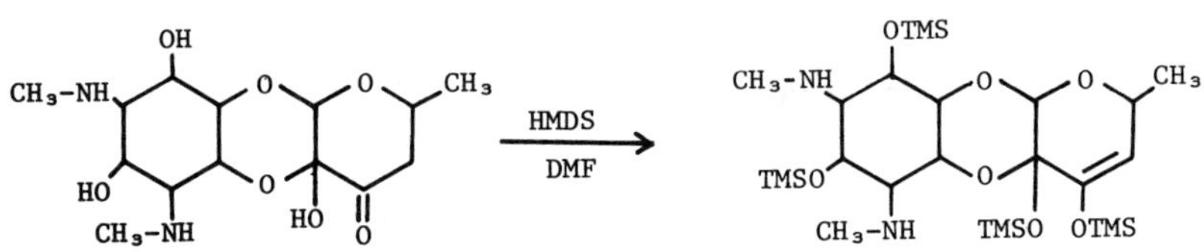

Typical Procedure. Diss 2–5 mg sample in 1 ml DMF containing 2 mg/ml triphenylantimony int stand. Add 0.1 ml HMDS. Allow to stand 1 hr then analyze by GC.

Comments. MS data for the deriv and for the derivs of actinamine, dihydro-, and tetrahydrospectinomycin suggested four *O*-TMS groups in the spectinomycin deriv. BSA and TMSI yielded multiple GC peaks.

References

[1]L. W. Brown and P. B. Bowman, *J. Chromatogr. Sci.* **12**, 373 (1974).

Derivative Type. Trimethylsilyl.

Reagent. Trimethylsilylimidazole (TMSI)/bis(trimethylsilyl)acetamide (BSA)/trimethylchlorosilane (TMCS).

Compound Type. Erythromycins.

Reaction

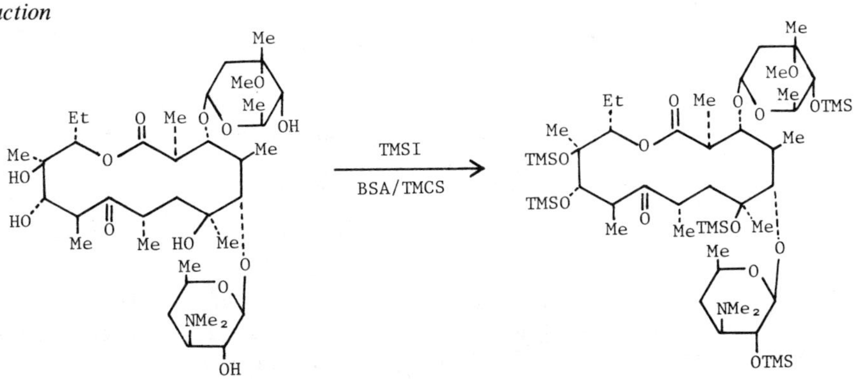

Typical Procedure. Add 5 ml TMCS to 5 ml BSA followed by 2 ml TMSI (a ppt will form if the reagents are not mixed in this order). Add 10 ml of this mixt to 10 ml pyridine containing 4.4 mg/ml 1,3-dimyristin int stand. To 10 mg of erythromycin powder add 1 ml of the reagent-int stand soln. Heat at 75° for 24 hr. Analyze by GC.

Comments. MS analysis of the derivs indicated that all active hydrogens of the hydroxy groups were silylated.

References

[1]K. Tsuji and J. H. Robertson, *Anal. Chem.* **43**, 818 (1971).

[2]J. H. Robertson and K. Tsuji, *J. Pharm. Sci.* **61**, 1633 (1972).

Derivative Type. Trimethylsilyl.

Reagent. Bis(trimethylsilyl)acetamide (BSA).

Compound Type. Chloramphenicol.

Reaction

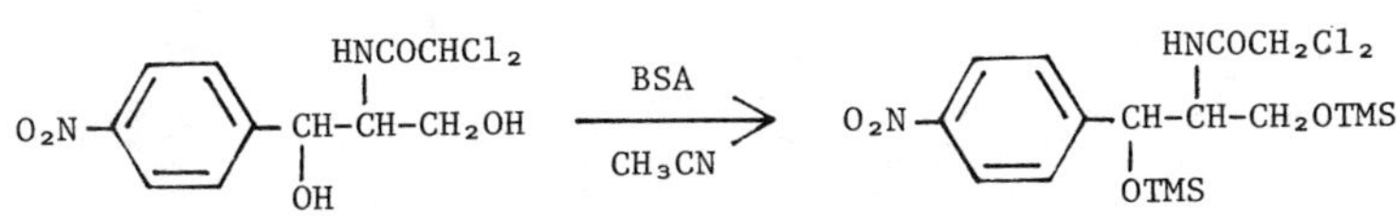

Typical Procedure. To a sample containing *ca* 10 mg chloramphenicol add 0.5 ml of 6% (v/v) BSA in CH_3CN containing 20 mg/ml *m*-phenylene dibenzoate int stand. Mix on a vortex mixer until soln is complete then analyze 1 μl by GC.

Comments. Presence of hydroxylic solvents (e.g., MeOH) gave inversion of configuration to yield a second GC peak due to a diastereomeric deriv. The deriv structure was confirmed by UV, IR, NMR, and MS data.[2]

References

[1]M. Margosis, *J. Chromatogr.* **47**, 341 (1970).

[2]M. Margosis, *J. Pharm. Sci.* **59**, 501 (1970).

Derivative Type. Trimethylsilyl.

Reagent. Bis(trimethylsilyl)acetamide (BSA)/trimethylchlorosilane (TMCS)/pyridine.

Compound Type. Tetracyclines.

Reaction

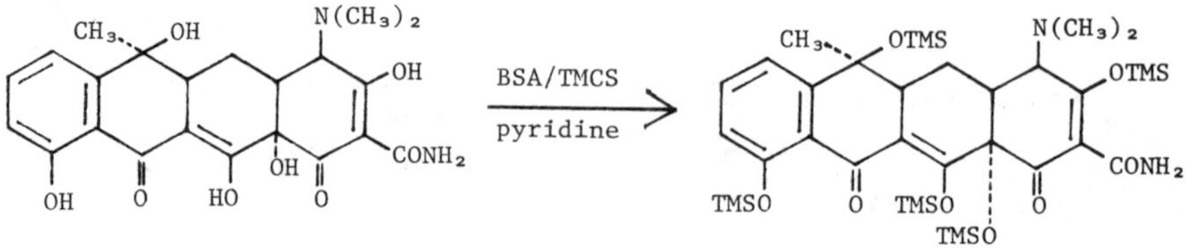

Typical Procedure. Prepare fresh reagent by mixing 5 ml BSA and 5 ml TMCS with 10 ml pyridine. Add 1.5 μl/ml trioctanoin int stand. To 10 mg tetracycline HCl powder add 1 ml of the reagent-int stand soln. Allow to react 24 hr at room temp. Analyze by GC.

Comments. MS data were given for the deriv. Attempted use of TMSDEA gave a mixt of TMS_5 and TMS_6 derivs.

References

[1]K. Tsuji and J. H. Robertson, *Anal. Chem.* **45**, 2137 (1973).

15.2.2. Other Derivatives of Antibiotics

Derivative Type. Permethyl.

Reagent. Methylsulfinyl carbanion/methyl iodide.

Compound Type. Polyene macrolides.

Reaction

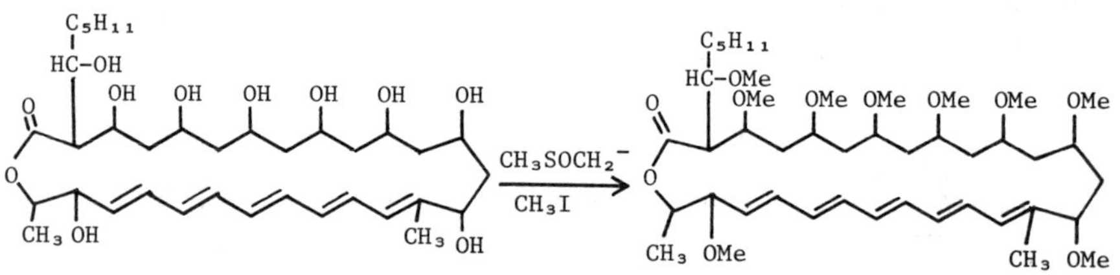

Typical Procedure. Treat the sample with a fivefold mol excess (relative to the number of deriv sites) of methylsulfinyl carbanion soln.[2] Add the same number of equiv of CH_3I and allow to react 1 hr at room temp. Add H_2O and ext with $CHCl_3$. Wash ext with H_2O and evap to dryness. Analyze residue by MS.

Comments. The method was applied to the nine isomers of the Filipin complex. MS of the TMS derivs of the Filipin complex have been reported[3] but the mass increment of each TMS group leads to an inconveniently high mol wt deriv. This application was a demonstration of the advantage of permethylation in terms of mass increment for deriv of polyfunctional compds.

References

[1]P. A. White and D. M. Desiderio, *Anal. Letters* **3**, 499 (1970).

[2]M. L. Polan *et al.*, *Biochem. Biophys. Res. Commun.* **38**, 1127 (1970).

[3]B. T. Golding *et al.*, *Tetrahedron Lett.* **37**, 2615 (1964).

Derivative Type. Acetyl.

Reagent. Acetic anhydride (Ac_2O)/pyridine/4-(dimethylamino)pyridine (DMAP).

Compound Type. Clindamycin palmitate.

Reaction

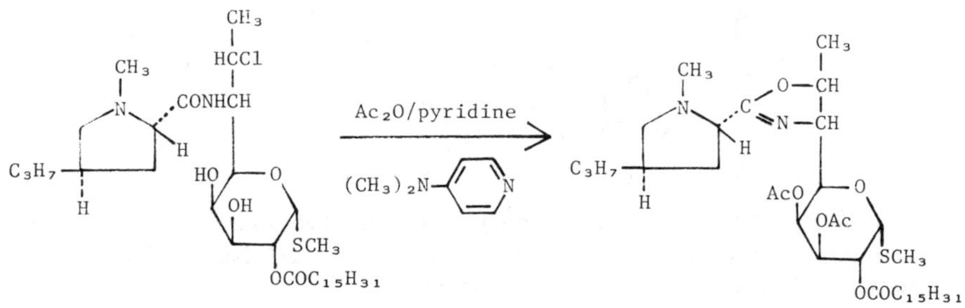

Typical Procedure. Diss *ca* 75 mg of clindamycin palmitate HCl in 5 ml H_2O, add 1 ml 30% aq Na_2CO_3, and ext with 5 ml $CHCl_3$ containing 5 mg/ml trilaurin int stand. To 1 ml of the ext add 1 ml of a 15 mg/ml soln of DMAP in pyridine and 1 ml Ac_2O. Allow to react 30 min at room temp then analyze by GC.

Comments. Use of 4-(dimethylamino)pyridine as catalyst gave quantitative acetylation in 30 min at room temp. Without the catalyst, acetylation was not complete even after 2 hr at 95°. Brown showed that the acetyl deriv lost HCl on the GC column.[2] TFA derivs have also been prepared.[2]

References

[1]E. L. Rowe and S. M. Machkovech, *J. Pharm. Sci.* **66**, 273 (1977).

[2]L. W. Brown, *J. Pharm. Sci.* **63**, 1597 (1974).

Derivative Type. Methyl ester.

Reagent. Diazomethane.

Compound Type. Penicillins and penicillanic acids.

Reaction

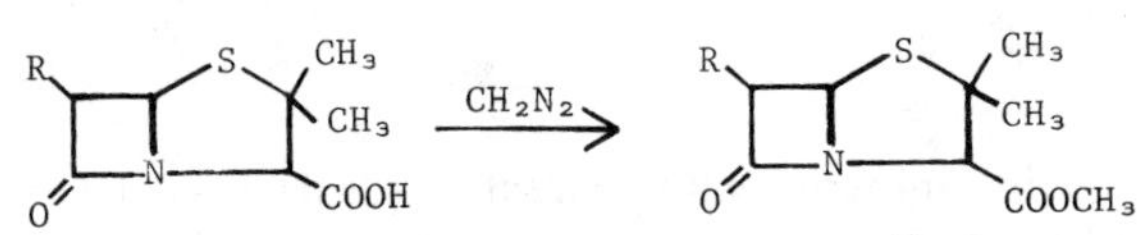

Typical Procedure. Diss sample in a soln of CH_2N_2 in Et_2O. Evap excess CH_2N_2 and analyze by GC.

Comments. GC retention data were given for derivs of four penicillanic acids and four penicillins. The side chain D- and L-isomers of α-phenoxyethyl- and 3,4-dichloro-α-methylbenzyl-penicillin were GC separable.

References

[1]E. Evrard *et al.*, *Nature* **201**, 1124 (1964).

Derivative Type. Pyridine.

Reagent. Sodium hydroxide.

Compound Type. Cephaloridine.

Reaction

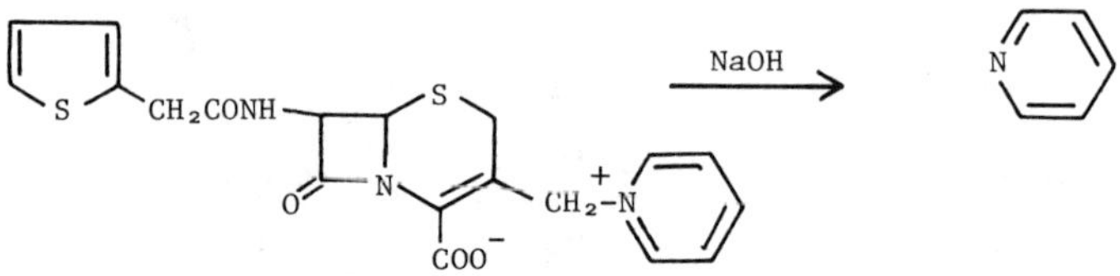

Typical Procedure. Mix 0.5–5 ml (depending upon suspected conc) serum with 1.0 ml of 35 μg/ml aq soln α-picoline int stand, 0.5 ml CH_2Cl_2, and 3 ml 5 *N* NaOH. Shake 5 min and discard aq phase. Analyze 10 μl of the CH_2Cl_2 by GC.

Comments. Although not strictly a deriv reaction, this method was included because of the paucity of GC methods available for the β-lactam antibiotics. The method was useable for therapeutic serum conc of 5–80 μg/ml.

References

[1]P. W. Mullen *et al.*, *Res. Commun. Chem. Pathol. Pharmacol.* **7**, 85 (1974).

15.3. BARBITURATES, HYDANTOINS, AND RELATED COMPOUNDS

15.3.1. Methyl Derivatives

Derivative Type. Methyl.

Reagent. Dimethyl sulfate.

Compound Type. Barbiturates.

Reaction

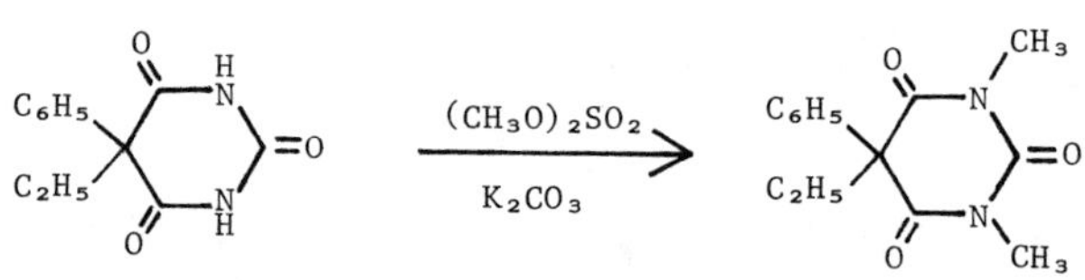

Typical Procedure. Mix 2 ml serum with 2 ml satd aq NaH_2PO_4 and ext with 10 ml Et_2O. Transfer 7 ml of the Et_2O to another tube and evap under an air stream. Diss the residue in 2 ml of 10% (v/v) H_2O in MeOH satd with K_2CO_3. Add 0.1 ml dimethyl sulfate and heat 4 min at 135°. Evap MeOH at 135° under an air stream. Add 1 ml 1 *M* pH 6 acetate buffer and ext with 1.5 ml heptane. Evap the heptane ext to dryness and diss residue in 100 μl acetone containing 50 ng methyl palmitate int stand. Analyze by GC.

Comments. GC data were reported for the derivs of amobarbital, secobarbital, and phenobarbital.

References

[1]H. F. Martin and J. L. Driscoll, *Anal. Chem.* **38**, 345 (1966).

Derivative Type. Methyl.

Reagent. Dimethyl sulfate/potassium carbonate.

Compound Type. Barbiturates.

Reaction

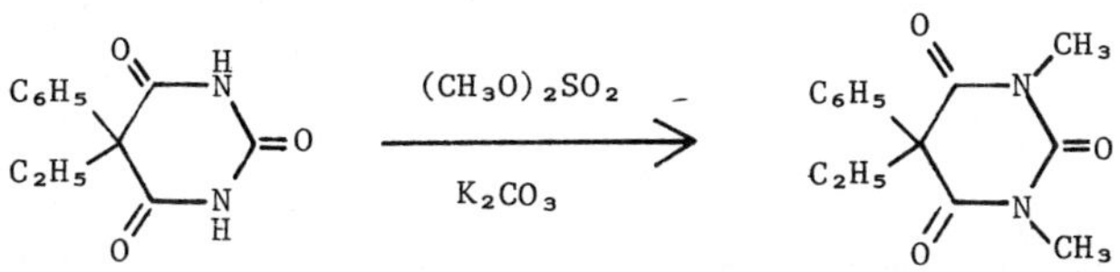

Typical Procedure. To 1 ml serum, tissue homogenate, or aq soln add 1 ml satd K_2CO_3. Carefully add 0.2 ml dimethyl sulfate and heat 4–5 min on a steam bath. Cool in an ice bath and ext with 1 ml *n*-heptane containing methyl palmitate int stand on a vortex mixer. Analyze 1 μl of the heptane phase by GC.

Comments. The rapid procedure permitted barbiturate analysis in 15 min. Minimum detectable amts were 10–50 ng.

References

[1]J. T. Stewart *et al., Anal. Lett.* **2**, 449 (1969).

Derivative Type. Methyl.

Reagent. Methyl iodide/potassium carbonate.

Compound Type. Barbiturates.

Reaction

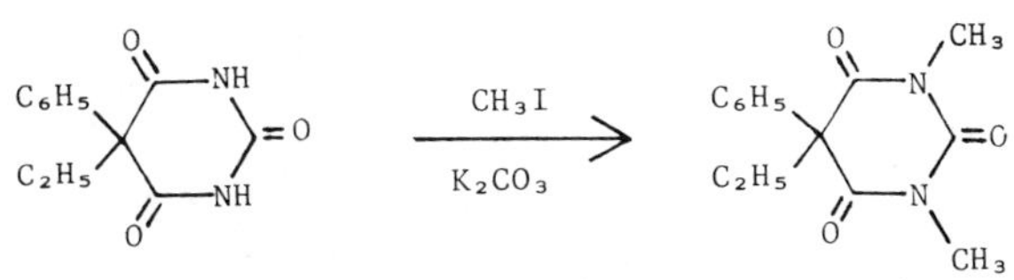

Typical Procedure. Mix 100–200 μl of barbiturate soln in acetone with 3–1000 mol equiv of CH_3I. Add 3–5 mg K_2CO_3 and reflux 30 min. Analyze soln directly by GC.

Comments. The method was reported to be useful for quantitative work in the microgram and nanogram range. The derivs of 13 barbiturates were prepared. The method was also used with chloromethyl methyl ether, dimethyl sulfate, diethyl sulfate, allyl iodide, and benzyl iodide.[2] The methoxy derivs of the barbiturates showed enhanced separation properties compared to the methyl derivs.

References

[1]W. Dünges and E. Bergheim-Irps, *Anal. Lett.* **6**, 185 (1973).

[2]W. Dünges, *Chromatographia* **6**, 196 (1973).

Derivative Type. Dimethyl.

Reagent. Methyl iodide/sodium hydroxide.

Compound Type. Diphenylhydantoin (DPH).

Reaction

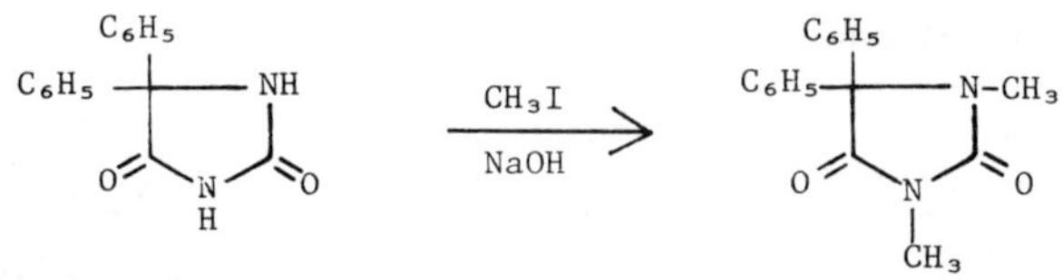

Typical Procedure. To a sample of up to 3 μg add 100 μl pH 13 (0.47 *M*) buffer (see below) and 600 μl of a 4% (v/v) soln of CH_3I in acetone. Shake 1 hr at room temp. Remove supernatant org layer and evap to dryness under N_2 at room temp. Diss residue in 200 μl $CHCl_3$ and analyze by GC.

Comments. The buffer was prepared by diss 37.28 g KCl and 18.84 g NaOH in 1000 ml H_2O. The method was used for analysis of DPH in 100 μl plasma samples using 5-(*p*-methylphenyl)-5-phenylhydantoin (MPPH) as int stand.

References

[1]J. Gordos *et al.*, *J. Chromatogr.* **143**, 171 (1977).

Derivative Type. Methyl.

Reagent. Tetrahexylammonium hydrogen sulfate (THA-HSO_4)/methyl iodide.

Compound Type. Barbiturates.

Reaction

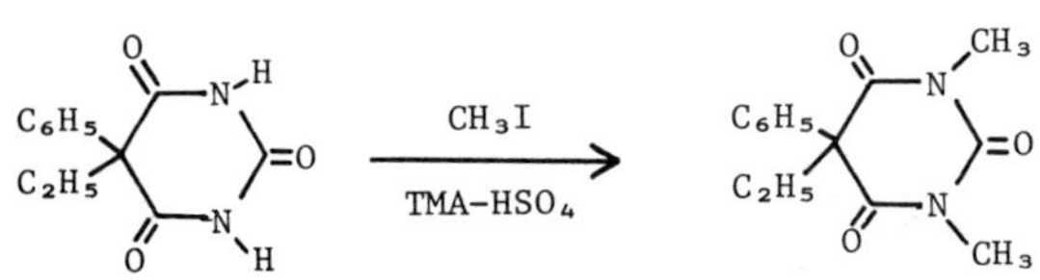

Typical Procedure. Mix 2.5 μg of barbiturate with 0.5 ml 0.01 *M* THA-HSO_4 in pH 10 phosphate buffer. Add 0.5 ml CS_2 and 10 μl CH_3I. Shake 5 min at room temp and centrifuge. Inject 3–4 μl into GC.

Comments. The tetrabutyl and tetrapentyl ammonium salts were also examined but the tetrahexyl gave the most rapid deriv formation. CH_2Cl_2 gave a badly tailing solvent front compared to CS_2.

References

[1]H. Ehrsson, *Anal. Chem.* **46**, 922 (1974).

Derivative Type. Methoxy.

Reagent. Diazomethane.

Compound Type. Diphenylhydantoin.

Reaction

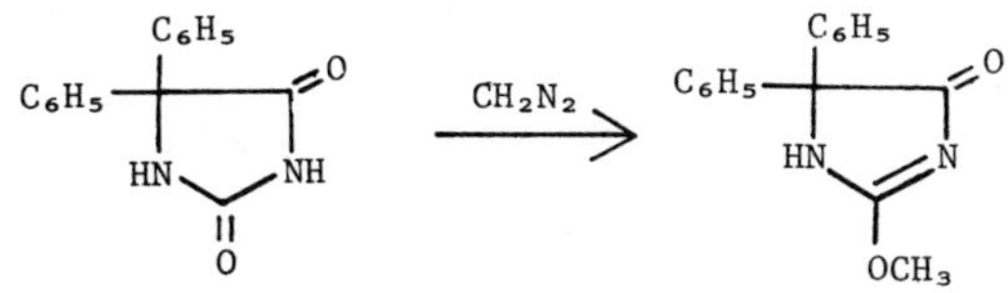

Typical Procedure. Acidify 0.5–1.0 ml serum with 0.5 ml 0.5 *N* HCl and ext with 10 ml $CHCl_3$. Evap 5 ml of the $CHCl_3$ ext to dryness under N_2. Add CH_2N_2 in Et_2O and allow to stand until N_2 evolution stops. Evap under N_2 and diss residue in 100 μl CS_2. Analyze by GC.

Comments. The $CHCl_3$ was freshly dist to eliminate formation of interfering peaks. Androsterone acetate was used as int stand. The structure of the deriv was based upon IR evidence.

References

[1]D. H. Sandberg *et al., Anal. Chem.* **40**, 736 (1968).

[2]J. J. Von G. Grimmer and H. Schäfer, *Arzn. Forschung* **19**, 1287 (1969).

Derivative Type. *N,N′*-Dimethyl.

Reagent. Diazomethane.

Compound Type. Phenobarbital.

Reaction

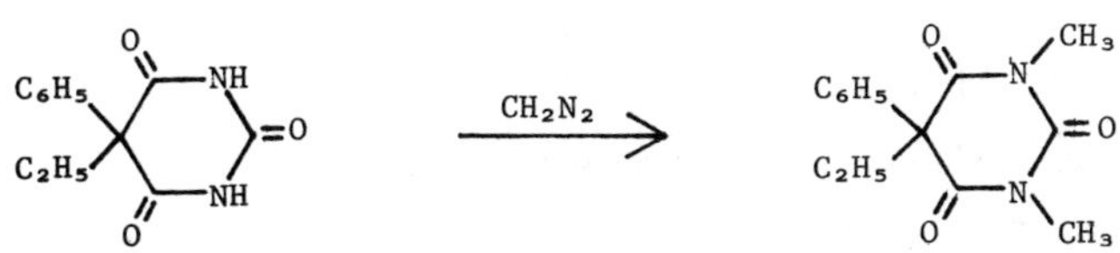

Typical Procedure. Saturate sample (100–200 μl plasma, 5–7 ml urine, 0.5–1.0 ml milk) with powdered $(NH_4)_2CO_3$ and ext with EtOAc. Evap to dryness under N_2 and diss residue in 0.5 ml MeOH. Add excess CH_2N_2 in Et_2O. Allow to stand 15 min at room temp and evap to dryness under N_2. Analyze residue by GC-MS-SIM.

Comments. 2,4,5-^{13}C-Phenobarbital was used as int stand. Deriv with diazoethane was also demonstrated and allowed distinguishing the derivs of phenobarbital and mephobarbital. Diphenylhydantoin was converted to the mono-*N*-methyl deriv. Cook *et al.*[2] reported the preparation of derivs of 11 barbiturates; thiopental gave multiple GC peaks. Gilbert *et al.*[3] reported MS data on barbiturate derivs prepared with diazomethane.

References

[1]M. G. Horning *et al., J. Chromatogr. Sci.* **12**, 630 (1974).

[2]J. G. H. Cook *et al., J. Chromatogr.* **6**, 182 (1961).

[3]J. N. T. Gilbert *et al., J. Pharm. Pharmacol.* **22**, 897 (1970).

Derivative Type. Dimethyl.

Reagent. Diazomethane.

Compound Type. Barbiturates.

Reaction

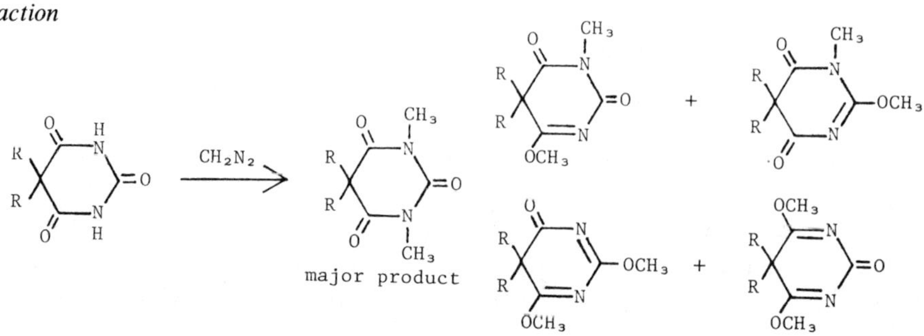

Typical Procedure. Diss 1 mg barbiturate in 1 ml MeOH and add excess CH_2N_2 in Et_2O. Allow to react 15 min then evap to dryness under N_2. Diss residue in EtOAc and analyze by GC-MS.

Comments. Diethyl derivs were also prepared in EtOH soln from diazoethane introduced by the method of Schlenk and Gellerman.[2] The side products in CH_2N_2 deriv of barbiturates which account for 10–15% were shown to be isomeric dimethyl derivs resulting from tautomerization. These side products were formed in larger amts from diazoethane.

References

[1]D. J. Harvey *et al.*, *Biomed. Mass Spectrom.* **1**, 340 (1974).

[2]H. Schlenk and J. L. Gellerman, *Anal. Chem.* **32**, 1412 (1960).

Derivative Type. Methyl.

Reagent. Tetramethylammonium hydroxide (TMAH).

Compound Type. Barbiturates.

Reaction

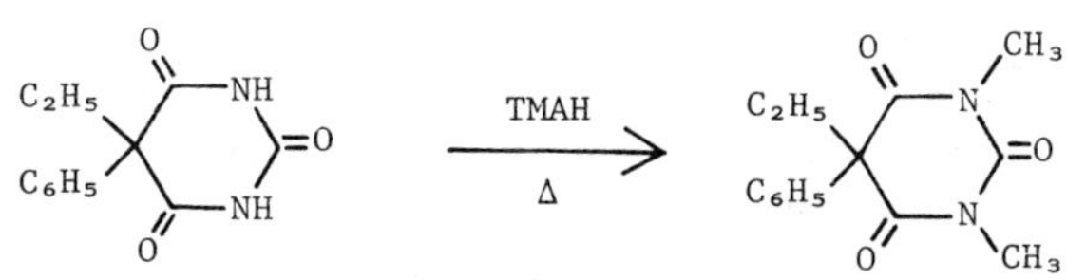

Typical Procedure. Add 0.1 *M* TMAH in MeOH to barbituric acid (1 μl/μg). Inject soln into GC with injector at *ca* 240°.

Comments. Eighteen barbiturates were examined with a maximum of 13 being resolvable in a single GC run.[1] It was suggested that the unsatd barbiturates might be resolvable from the satd analogs by prior treatment with Br_2 in CCl_4 to give the dibromo deriv.[1] The method has been used for barbiturate analysis in 100 μl plasma samples using GC-MS.[2]

References

[1]G. W. Stevenson, *Anal. Chem.* **38**, 1948 (1966).

[2]G. H. Draffan *et al.*, *J. Chromatogr.* **75**, 45 (1973).

Derivative Type. Methyl.

Reagent. Tetramethylammonium hydroxide (TMAH).

Compound Type. Anticonvulsant drugs.

Reaction

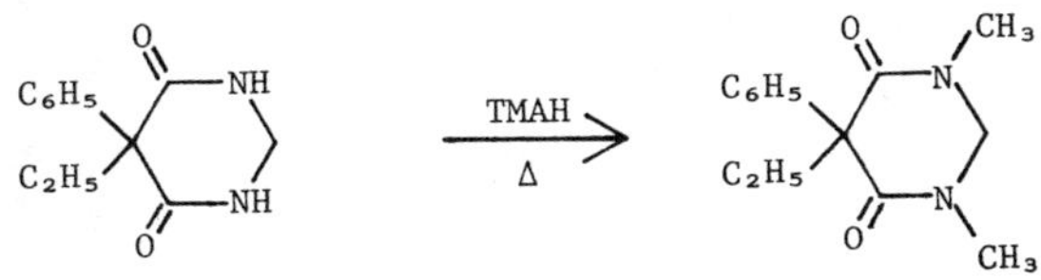

Typical Procedure. To 1 ml serum add 0.5 ml 0.3 *M* phosphate buffer and 5 ml int stand soln. Mix well and separate the toluene phase. Add 25 μl 2 *M* TMAH in MeOH. Mix well then centrifuge. Inject 1 μl of the lower TMAH phase into the GC.

Comments. Two int stand solns were used. Soln A contained 30 μg 5-ethyl-5-tolylbarbituric acid and 15 μg 5-*p*-methylphenyl-5-phenylhydantoin/5 ml toluene. Soln B was soln A plus 50 μg α-dimethyl-β-methylsuccinimide. Soln B was substituted for soln A when ethosuximide was to be measured. The method was used to quantitate ethosuximide, metharbital, methsuximide, phensuximide, mephytoin, mephobarbital, phenobarbital, primidone (shown above) and diphenylhydantoin. A micro adaptation using 200 μl serum samples was also reported. The use of TMAH for analysis of plasma diphenylhydantoin had been reported earlier.[2]

References

[1]E. B. Solow *et al.*, *J. Chromatogr. Sci.* **12**, 256 (1974).
[2]J. MacGee, *Anal. Chem.* **42**, 421 (1970).

Derivative Type. Methyl.

Reagent. Trimethylphenylammonium hydroxide (TMPAH).

Compound Type. Barbiturates, phenolic alkaloids, and dimethylxanthines.

Reaction

Typical Procedure. Diss the sample compd(s) in 0.1 *M* TMPAH in MeOH (*ca* twofold excess). Inject 1 μl of soln (containing *ca* 0.2–1 μg of each sample component) into the GC with the injector at 250°.

Comments. The method permitted GC separation of 12 barbiturates. The dimethylxanthines theophylline and theobromine were both converted to caffeine by the deriv method. The phenolic alkaloids morphine, codamine, laudanine, and reticuline gave complete methylation of phenolic hydroxyl groups without reaction of secondary alcohol groups. An injector temp of 275° was used for the xanthines and the phenolic alkaloids. Tetramethylammonium hydroxide was also examined but gave incomplete methylation. TMPAH has also been used in the analysis of diphenylhydantoin[2,3] and its *p*-hydroxylated metabolite,[2] primidone,[3] and phenobarbital.[3] Baty and Robinson[4] used TMPAH in their analysis of diphenylhydantoin by GC-MS-SIM. Triethylphenylammonium hydroxide has been used in the same deriv reaction to distinguish methylated drugs from their demethylated metabolites.[5]

References

[1]E. Brochmann-Hansen and T. O. Oke, *J. Pharm. Sci.* **58**, 370 (1969).
[2]R. H. Hammer *et al.*, *J. Pharm. Sci.* **60**, 327 (1971).
[3]H. J. Kupferberg, *Clin. Chim. Acta* **29**, 283 (1970).
[4]J. D. Baty and P. R. Robinson, *Biomed. Mass Spectrom.* **4**, 36 (1977).
[5]P. Friel and A. S. Troupin, *Clin. Chem.* **21**, 751 (1975).

Derivative Type. Methyl.

Reagent. Trimethylphenylammonium hydroxide (TMPAH).

Compound Type. Barbiturates.

Reaction

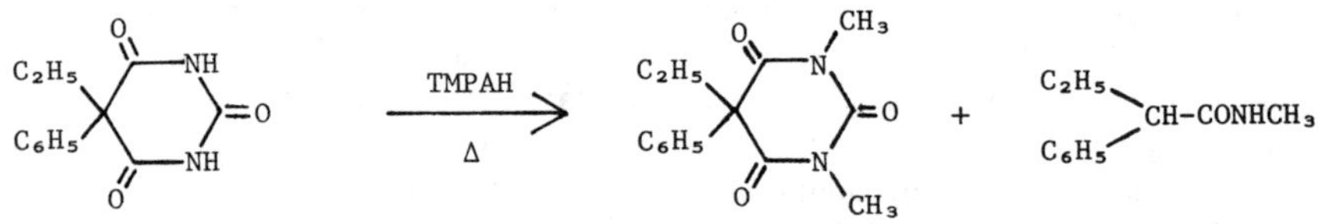

Typical Procedure. Mix 1 ml blood, serum, or plasma with 1 ml acetate buffer (see below) and ext with 5 ml Et_2O/toluene (1:1, v/v). To 4.5 ml of the ext add 0.2 ml 25% TMPAH in MeOH and mix well. Analyze 1 μl of the lower TMPAH phase by GC.

Comments. The buffer was prepared by diss 12 g NaOAc in 60 ml 1 *N* HOAc and dil to 1000 ml with H_2O. The method was used for analysis of 19 barbiturates, glutethimide, and diphenylhydantoin. GC and MS data for the 21 derivs were reported.[1] An early eluting side product formed from phenobarbital was identified as *N*-methyl-α-phenylbutyramide.[2] Use of a high TMPAH conc (1.8 *M*) was reported to give a reproducible amt of the side product yielding a better quantitation method.[2]

References

[1]R. F. Skinner *et al.*, *Anal. Chem.* **45**, 574 (1973).
[2]R. Osiewicz *et al.*, *J. Chromatogr.* **88**, 157 (1974).

Derivative Type. Methyl.

Reagent. Trimethylphenylammonium hydroxide (TMPAH).

Compound Type. Barbiturates and hydantoins.

Reaction

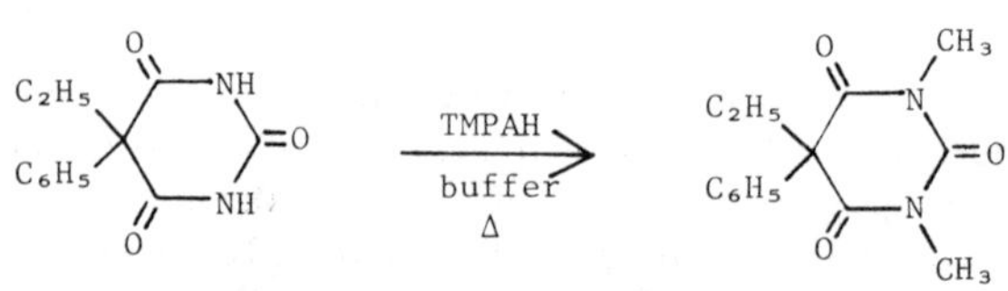

Typical Procedure. Mix 1 ml plasma with 0.5 ml 1 *M* H_3PO_4 and ext with 5 ml toluene. Sep the ext and mix with 100 μl of 0.2 *M* TMPAH and 20 μl H_2O. Centrifuge and remove the lower aq phase. Mix with 20 μl of "acid buffer" (see below) and inject 1.5 μl (slowly, 6 sec) into the GC.

Comments. The buffer was prepared by diss 16.0 g citric acid and 4.0 ml 85% H_3PO_4 in 100 ml H_2O and adding 7.0 g H_3BO_3 and 340 ml 2 *N* NaOH. This soln was dil to 500 ml and 400 ml of it mixed with 2.5 g NaF and adj to pH 9.0 with conc HCl. The resulting pH 9.0 soln was dil to 500 ml. The "acid buffer" was prep by adjusting the pH of 20 ml of this soln to 4.0 with 6 *N* HCl immediately before use.

The method prevented the formation of *N*-methyl-α-phenylbutyramide by hydrolytic cleavage of phenobarbital. The same result was achieved by adding 20 μl DMSO and 15 μl CH_3I to 50 μl of the TMPAH soln before injection.

References

[1]W. J. Serfontein and L. S. DeVilliers, *J. Chromatogr.* **130**, 342 (1977).

Derivative Type. d_3-Methyl.

Reagent. d_9-Trimethylphenylammonium hydroxide (d_9-TMPAH).

Compound Type. Phenobarbital and mephobarbital.

Reaction

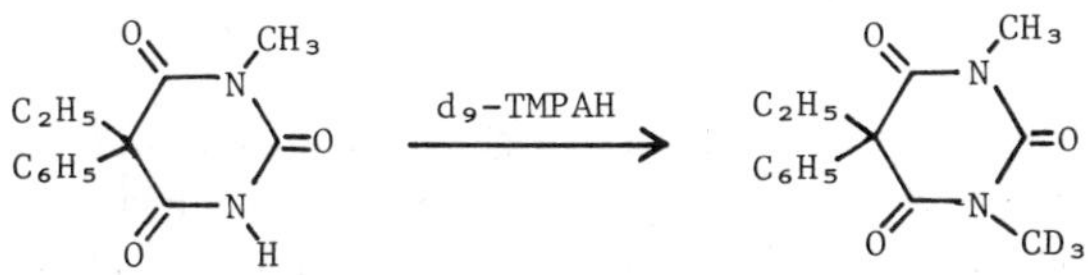

Typical Procedure. Ext 2.5 ml serum with 12.5 ml CH_2Cl_2. Adjust pH of serum to 9.0 with solid $NaHCO_3$ and repeat extraction. Dry combined ext over Na_2SO_4 and conc on a rotary evaporator and finally under N_2 to 50 μl. Mix 1 μl with 4 μl of d_9-TMAH soln (*ca* 0.5 *M* in EtOH) and inject into the GC-MS.

Comments. Use of the deuterated reagent allowed differentiation between phenobarbital and mephobarbital present in the same sample since both gave the same deriv with unlabeled reagent. The synthesis of d_9-TMPAH from aniline and d_3-methyl iodide was given.

References

[1]L. K. Wong *et al.*, *J. Chromatogr.* **116**, 321 (1976).

Derivative Type. Methyl.

Reagent. Lecithin.

Compound Type. Barbiturates and hydantoins.

Reaction

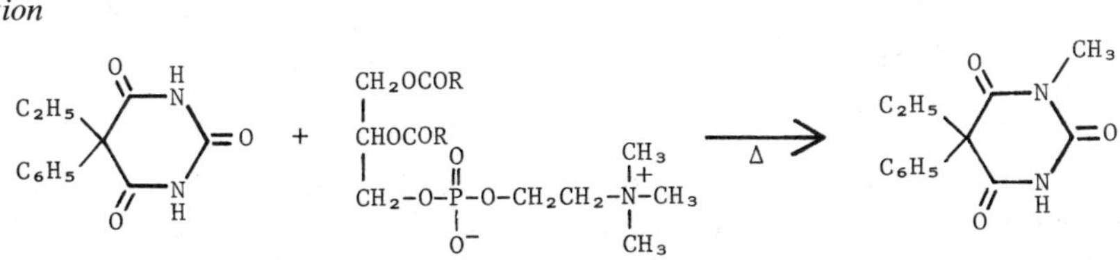

Typical Procedure. Diss 2 mg phenobarbital in 0.5 ml CH_2Cl_2 and mix 1 μl with 1 μl of lecithin soln (25 mg/ml benzene). Take up soln in 10 μl syringe and inject into GC-MS.

Comments. The ability of lecithin to methylate phenobarbital by the above procedure was demonstrated to substantiate the theory that lecithin in serum is responsible for methylated artifacts seen in GC analysis of serum extracts.

References

[1]L. K. Wong *et al.*, *J. Chromatogr.* **116**, 321 (1976).

15.3.2. Other Alkyl Derivatives

Derivative Type. Ethyl.

Reagent. Tetraethylammonium hydroxide (TEAH).

Compound Type. Barbiturates, glutethimide.

Reaction

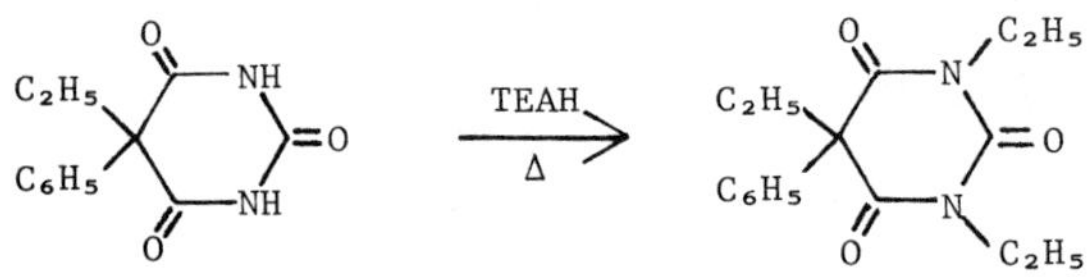

Typical Procedure. To 1 ml whole blood, plasma, or serum, add int stand, 0.5 ml 1 *M* H_3PO_4, and 5 ml toluene. Shake, centrifuge and remove toluene layer. Add 10 μl of 40% aq TEAH to the toluene. Shake and centrifuge. Inject 1 μl of TEAH phase slowly (10 sec) into GC.

Comments. The alkylation reaction occurred in the GC injector. Slow injection gave better formation. Kananen *et al.*[2] used a modification of the method with tetramethylammonium hydroxide for barbiturate analysis.

References

[1]J. MacGee, *Clin. Chem.* **17**, 587 (1971).

[2]G. Kananen et al., *J. Chromatogr. Sci.* **10**, 283 (1972).

Derivative Type. Propyl.

Reagent. Tetrapropylammonium hydroxide (TPrAH).

Compound Type. Theophylline.

Reaction

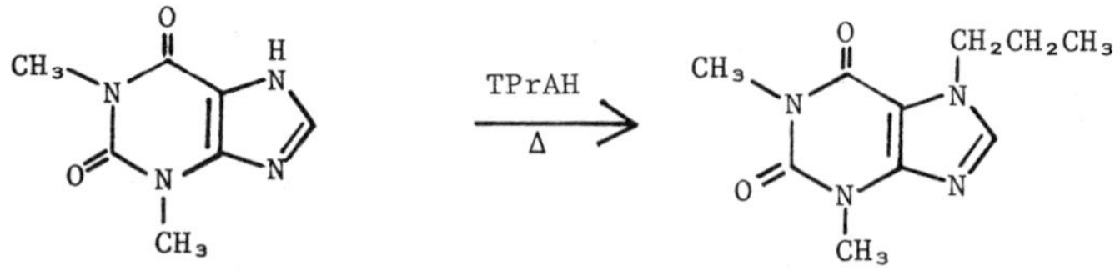

Typical Procedure. Ext 1 ml plasma or saliva with 10 ml $Et_2O/CH_2Cl_2/i\text{-}PrOH$ (6:4:1, v/v/v). Centrifuge and cool in dry ice/acetone bath. Decant upper org layer into another tube containing 2 ml 1 *N* NaOH. Mix well, centrifuge, cool, and discard upper org layer. Acidify aq layer with 0.4 ml 8 *M* H_3PO_4, mix with 1 ml satd aq Na_2SO_4, and ext with 2×10 ml of the org solvent mixt. Add int stand and evap to dryness under N_2 at 40°. Diss residue in 25 μl TPrAH. Analyze an aliquot by GC with injector port at 265°.

Comments. Thiobarbital or fluoranthene were used as int stand. The thiobarbital was unstable to base and could not be stored over 30 min. GC data were given for derivs of 3-methylxanthine, 1,3-dimethyluric acid, and 1-methyluric acid. The theophylline deriv gave a GC peak distinct from other xanthines, theophylline metabolites, or barbiturates.

References

[1]V. P. Shah and S. Riegelman, *J. Pharm. Sci.* **63**, 1283 (1974).

Derivative Type. *N*-Butyl.

Reagent. Iodobutane/tetramethylammonium hydroxide (TMAH).

Compound Type. Ethosuximide.

Reaction

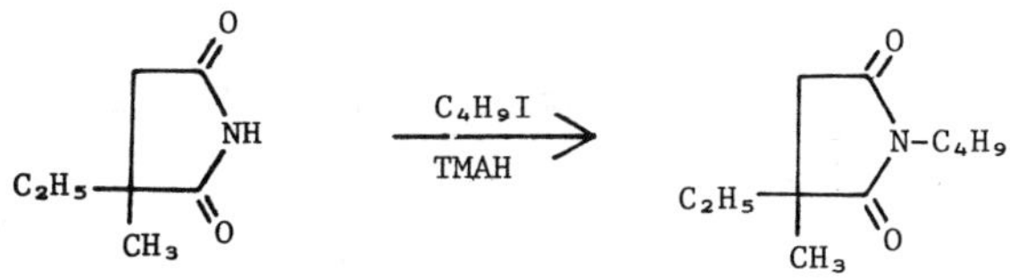

Typical Procedure. Mix 0.5 ml plasma with 1.0 ml 1 *M* pH 6.8 phosphate buffer and ext with 7.0 ml toluene containing 5 μg/ml α,α-dimethyl-β-methylsuccinimide int stand. Sep the toluene phase and ext with 0.5 ml 0.5 *M* NH_4OH. Acidify the aq phase with 0.1 ml 5.7 *M* H_3PO_4 and ext with 5.0 ml CH_2Cl_2. Dry ext over 0.5 g Na_2SO_4 and evap under N_2. Add 40 μl DMA, 10 μl TMAH (2.4 g/100 ml MeOH), and 10 μl iodobutane. Mix and let stand 10 min. Centrifuge and analyze an aliquot of the supernatant by GC.

Comments. The method yielded a stable deriv with better GC properties than that obtained by methylation, on-column alkylation with tetrabutylammonium hydroxide, or silylation.

References

[1]C. J. Least *et al.*, *Clin. Chim. Acta* **60**, 285 (1975).

Derivative Type. Butyl.

Reagent. 1-Iodobutane/tetramethylammonium hydroxide (TMAH).

Compound Type. Theophylline.

Reaction

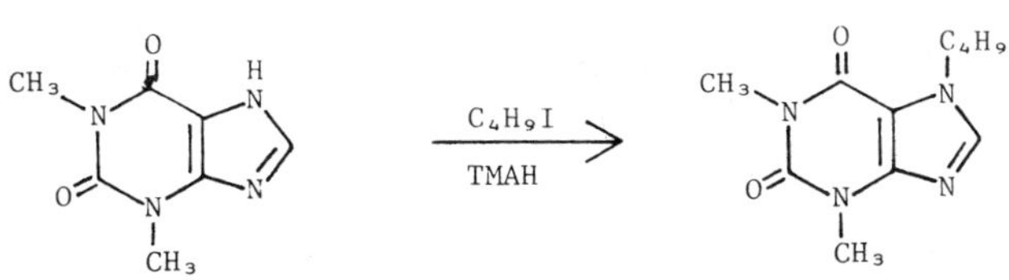

Typical Procedure. To 2 ml of EDTA treated blood add 1 ml aq soln of 3-isobutyl-1-methylxanthine (int stand) (20 mg/liter) and 10 drops HOAc. Ext with 10 ml $CHCl_3$ and evap ext to dryness under N_2 at 60°. Diss residue in 40 μl DMA and add 5 μl TMAH soln (240 mg/ml in MeOH). Mix and add 10 μl C_4H_9I. Mix 30 sec and centrifuge to remove ppt. Analyze 1 μl by GC-MS.

Comments. Samples of 0.2 ml were used by decreasing the HOAc and $CHCl_3$ volumes 10-fold and halving the deriv reagents. The method was an adaptation of Greeley's method for barbiturates.[2] Johnson *et al.*[3] used the method for analysis of saliva and blood. Dechtiaruk *et al.*[4] reported an improved procedure where they used 1-iodopentane and 3-isobutyl-1-methylxanthine as int stand to simultaneously determine theophylline, theobromine, and phenobarbital.

References

[1]M. Sheehan *et al.*, *Clin. Chem.* **23**, 64 (1977).

[2]R. H. Greeley, *Clin. Chem.* **20**, 192 (1974).

[3]G. F. Johnson *et al.*, *Clin. Chem.* **21**, 144 (1975).

[4]W. Dechtiaruk *et al.*, *Clin. Chem.* **21**, 1038 (1975).

Derivative Type. Butyl.

Reagent. 1-Iodobutane/tetramethylammonium hydroxide (TMAH).

Compound Type. Barbiturates.

Reaction

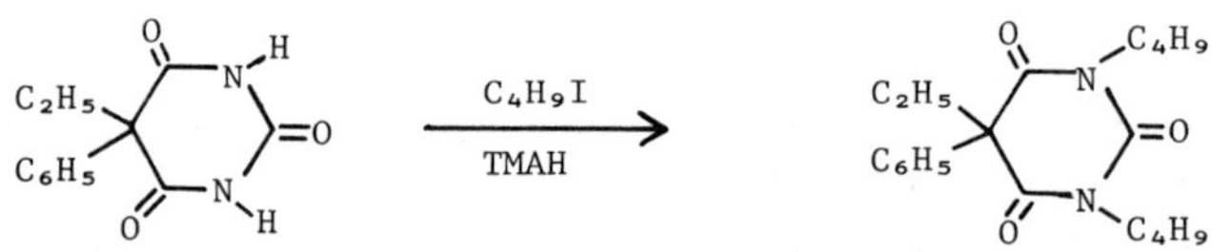

Typical Procedure. To 5.0 mg phenobarbital add 50 μl of a 24% (v/v) soln of TMAH in MeOH, 0.95 ml MeOH, and 4.0 ml *N,N*-dimethylacetamide. Mix and add 100 μl of 1-iodobutane. Mix thoroughly (tetramethylammonium iodide precipitates) for 3 min then analyze by GC.

Comments. Methyl, ethyl, 1-butyl, and 2-butyl derivs were examined as well as the use of TMPAH as the base. The 2-butyl derivs gave multiple products. The 1-butyl derivs were found to give the best separation of the 16 barbiturates studied.

References

[1]R. H. Greeley, *Clin. Chem.* **20**, 192 (1974).

Derivative Type. *n*-Butyl.

Reagent. Tetrabutylammonium hydroxide (TBAH).

Compound Type. Xanthines and barbiturates.

Reaction

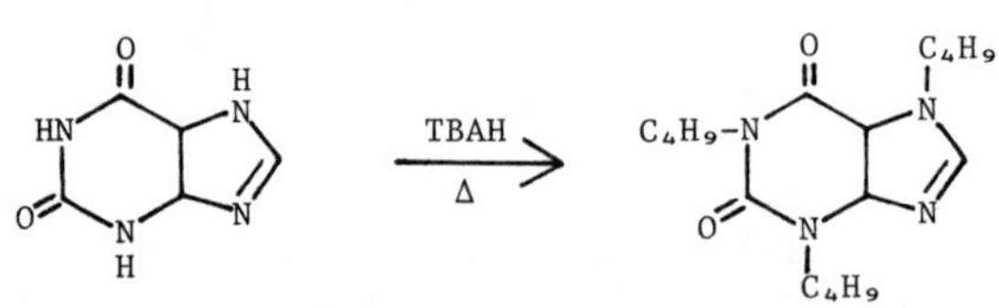

Typical Procedure. To a 1 ml sample soln in MeOH (containing not more than 1.5 μequiv N—H) add 7 μl 25% TBAH in MeOH. Inject 1 μl into the GC with the injector at least 270°.

Comments. Structural isomers, e.g., theophylline and theobromine, which gave identical derivs upon methylation could be resolved by GC as the butyl derivs.[1] Hooper *et al.*[2] used the same reagent for simultaneous analysis of methobarbital and phenobarbital.

References

[1] M. Kowblansky *et al.*, *J. Chromatogr.* **76**, 467 (1973).
[2] W. D. Hooper *et al.*, *J. Chromatogr.* **110**, 206 (1975).

Derivative Type. *n*-Butyl.

Reagent. Dimethylformamide (DMF) di-*n*-butyl acetal.

Compound Type. Theophylline and probenecid.

Reaction

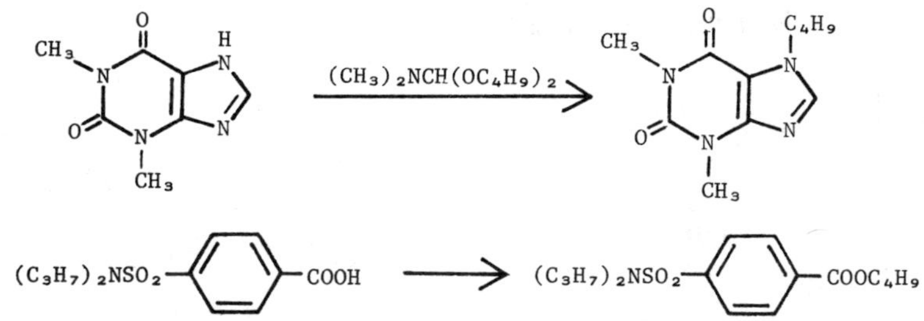

Typical Procedure. Dil 0.2–1.0 ml plasma with pH 4.0 phosphate buffer to 2.0 ml and mix with 2.0 g cellulose. Transfer to the top of a chromatographic column packed with a mixt of 4.5 g cellulose and 3.0 ml pH 7.0 phosphate buffer. Elute with $CHCl_3$ discarding the first 15 ml and collecting the following 30 ml. Evap the eluate to dryness and diss residue in 60 μl of int stand soln. Add 40 μl DMF di-*n*-butyl acetal and allow to stand at room temp 5 min. Analyze 1 μl by GC.

Comments. The method was used to quantitate theophylline and probenecid in plasma and rat brain tissue using probenecid propyl ester as int stand. The int stand soln was prepared by heating 10 mg probenecid in 5 ml HCl satd PrOH at 60° for 60 min, evap to dryness, and diss the residue in 50 ml acetone. Methyl, ethyl, propyl, and *t*-butyl derivs were also examined but the *n*-butyl deriv gave the best separation of theophylline, theobromine, and caffeine.

References

[1] A. Arbin and P. Edlund, *Acta Pharm. Suecica* **11**, 249 (1974).

Derivative Type. Hexyl.

Reagent. Tetrahexylammonium hydroxide (THAH).

Compound Type. Diphenylhydantoin (DPH) and phenobarbital (PB).

Reaction

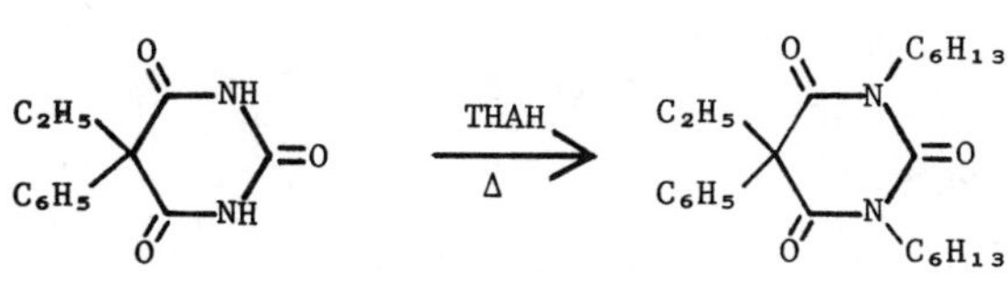

Typical Procedure. To 1 ml serum add 1 ml 1 *N* HCl and 7.0 ml ETBA (4.5 μg/ml)-MPPH (1.5 μg/ml) int stand soln in $CHCl_3$. Shake, centrifuge, and separate $CHCl_3$ phase. Add 5 ml 0.45 *N* NaOH to the $CHCl_3$ and shake for 1 min. Discard $CHCl_3$ phase and add 1 ml 6 *N* HCl to the aq phase. Shake, add 10 ml $CHCl_3$, shake, centrifuge, and discard aq phase. Transfer 7 ml of the $CHCl_3$ to another tube and evap to dryness under N_2 at 35–40°. Diss residue in 25 μl 0.2 *M* THAH in MeOH and inject 1–2 μl into GC.

Comments. The method was used for simultaneous determination of diphenylhydantoin and phenobarbital by isothermal GC. 5-(*p*-Methylphenyl)-5-phenylhydantoin (MPPH) and 5-ethyl-5-*p*-tolylbarbituric acid (ETBA) were used as internal standards. The lower limit of detection was about 1 μg/ml. Hexyl and heptyl derivs were prepared from 12 barbiturates using the tetraalkylammonium hydroxides.[2]

References

[1]T. J. Giovanniello and J. Pecci, *Clin. Chim. Acta* **67**, 7 (1976).

[2]T. J. Giovanniello and J. Pecci, *Clin. Chem.* **23**, 2154 (1977).

Derivative Type. Benzyl.

Reagent. Benzyl iodide/potassium carbonate.

Compound Type. Barbiturates.

Reaction

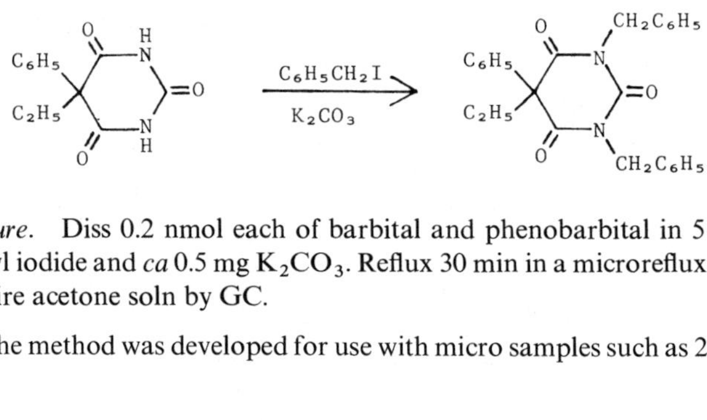

Typical Procedure. Diss 0.2 nmol each of barbital and phenobarbital in 5 μl acetone. Add 100 nmol benzyl iodide and *ca* 0.5 mg K_2CO_3. Reflux 30 min in a microrefluxer apparatus and analyze the entire acetone soln by GC.

Comments. The method was developed for use with micro samples such as 20 μl "fingerstick" blood samples.

References

[1]W. Dünges, *Anal. Chem.* **45**, 963 (1973).

Derivative Type. Pentafluorobenzyl.

Reagent. Pentafluorobenzyl bromide (PFB-Br)/tetrapentylammonium hydroxide (TPAH).

Compound Type. Theophylline.

Reaction

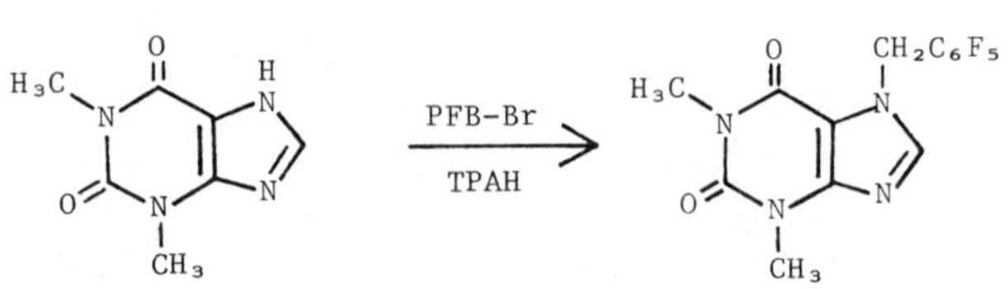

Typical Procedure. Mix 100 μl plasma (containing 50–1000 ng theophylline) with 100 μl int stand soln (2 μg/ml theobromine in pH 7.0 phosphate buffer), 300 μl pH 7.0 phosphate buffer, and 0.5 g cellulose. Pack mixt into a chromatographic column and elute with 10 ml CH_2Cl_2. Shake eluate with 500 μl 2 *M* NaOH for 10 min and centrifuge. Discard org phase and wash aq phase with 2 × 4 ml benzene.

Mix aq phase with 500 μl 0.14 *M* TPAH, 5.0 ml CH_2Cl_2, and 10 μl PFB-Br. Shake 10 min and centrifuge. Sep CH_2Cl_2 phase and evap under N_2. Add 500 μl heptane and 1.5 ml 4 *M* HCl. Shake 10 min, centrifuge, and discard org layer. Add 0.5 ml 15 *M* NaOH and ext with 2 × 3.0 ml cyclohexane. Evap ext under N_2 to *ca* 3 ml. Analyze 2 μl by ECD-GC.

Comments. This extractive alkylation method permitted quantitation to about 5 ng. The method was also applied to rat brain tissue. Use of tetrabutylammonium gave incomplete deriv of the theobromine int stand.

References

[1]A. Arbin and P. Edlund, *Acta Pharm. Suecica* **12**, 119 (1975).

Derivative Type. Pentafluorobenzyl.

Reagent. Pentafluorobenzyl bromide (PFB-Br)/tetrabutylammonium hydrogen sulfate (TBA-HSO_4).

Compound Type. Phenobarbital.

Reaction

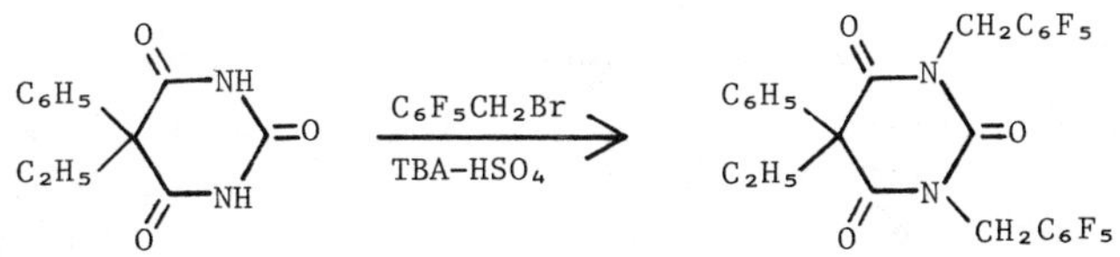

Typical Procedure. Mix 100 μl saliva with 10 μl of 0.1 *M* Na_3PO_4. Add 0.5 ml of 0.1 *M* TBA-HSO_4 in pH 9 phosphate buffer and 500 μl of 0.1 *M* PFB-Br in CH_2Cl_2 (containing 150 ng/ml heptabarbital int stand). Shake 20 min then ext with 3.0 ml *n*-heptane. Analyze 1 μl of the heptane phase by ECD-GC.

Comments. Use of a precolumn venting system permitted direct GC analysis of the reaction soln. The method permitted detection of 60 ng phenobarbital in 100 μl saliva.

References

[1]O. Gyllenhaal *et al., J. Chromatogr.* **122**, 471 (1976).

Derivative Type. Pentafluorobenzyl.

Reagent. Pentafluorobenzyl bromide (PFB-Br)/triethylamine (Et_3N).

Compound Type. Barbituric acids, diphenylhydantoin.

Reaction

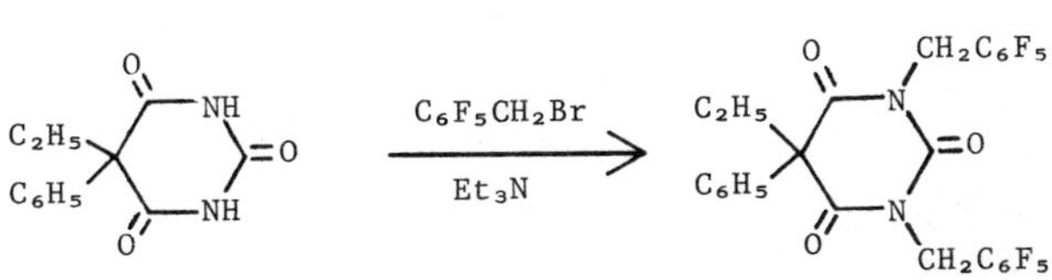

Typical Procedure. Diss 5 μg sample in 100 μl MeOH containing 5 μg 9-bromophenanthrene int stand. Add 50 μmol Et_3N and 1 μl (5 μmol) PFB-Br. Heat at 50° for 60 min. Analyze by FID-GC. (Analysis of reaction mixt by ECD-GC would require a cleanup procedure).

Comments. Derivs (as pure compds) were detectable to the pg level by pulsed ECD-GC. Use of Na_2CO_3 as catalyst caused decomposition of barbituric acids and diphenylhydantoin.

References

[1]T. Walle, *J. Chromatogr.* **114**, 345 (1975).

15.3.3. Acyl, Silyl, and Other Derivatives

Derivative Type. Trifluoroacetyl.

Reagent. Trifluoroacetic anhydride (TFAA).

Compound Type. Amides.

Reaction

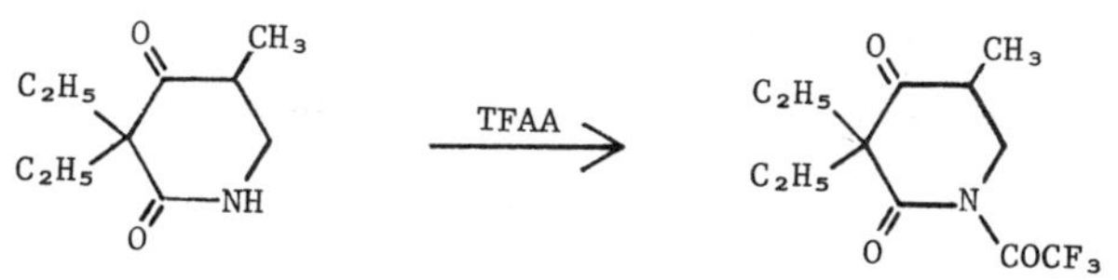

Typical Procedure. To 1 mg of the amide in 0.5 ml hexane (or benzene, EtOAc, or CS_2) add 25 μl TFAA and allow to react 15 min at room temp. Analyze an aliquot by GC.

Comments. The reaction was quantitative in less than 15 min in the four solvents examined. The amide derivs were susceptible to hydrolysis by traces of H_2O. Derivs were prepared from methylprylon (shown above), piperidione, phenacetin, acetaminophen, and *N-n*-decylacetamide. Pentafluorobenzoyl derivs were also prepared using 200 μl of 1.4 *M* TMA in hexane and 25 μl pentafluorobenzoyl chloride per 0.2 mg amide in 0.5 ml hexane (2 hr at room temp). The reaction was slower than with TFAA, and without the TMA the reaction was incomplete. The pentafluorobenzoyl derivs were considerably more stable than the TFA derivs. The pentafluorobenzoyl derivs gave good ECD response but the TFA derivs were too labile to survive the dilution necessary for ECD examination.

References

[1]H. Ehrsson and B. Mellström, *Acta Pharm. Suecica* **9**, 107 (1972).

Derivative Type. Pentafluorobenzoyl.

Reagent. Pentafluorobenzoyl chloride.

Compound Type. Theophylline.

Reaction

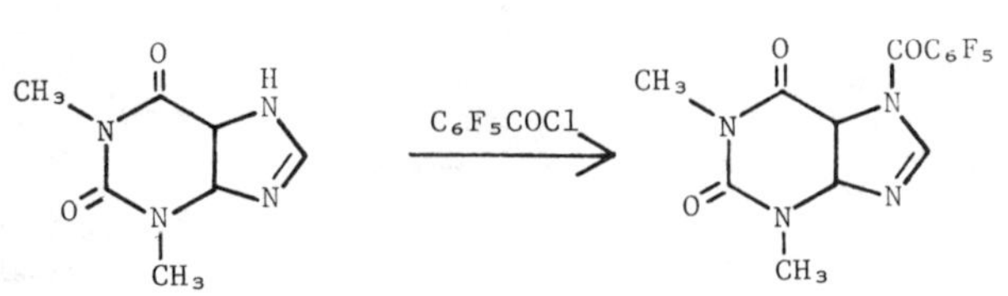

Typical Procedure. To 1 ml plasma add 100 μl of int stand soln (12.5 μg/100 μl EtOH) and *ca* 0.80 g $(NH_4)_2SO_4$ (alternatively 2 ml of satd aq soln). Vortex for 15 sec and add 10 ml of toluene/EtOAc (8:2, v/v). Shake 15 min, centrifuge and sep org phase. Evap to dryness under N_2 at 90°. Cool to room temp and add 25 μl C_6F_5COCl. Seal and heat at 90° for 1 hr. Evap to dryness under N_2 and diss residue in 5 ml EtOAc. Analyze 2–5 μl by ECD-GC.

Comments. The method was used to quantitate theophylline in plasma to 10 ng/ml using 3-isobutyl-1-methylxanthine as int stand.[1] Essentially the same method was used to quantitate primidone in plasma using *p*-methyl primidone as int stand.[2]

References

[1]H. A. Schwertner *et al.*, *Anal. Chem.* **48**, 1875 (1976).
[2]J. E. Wallace *et al.*, *Anal. Chem.* **49**, 903 (1977).

Derivative Type. Trimethylsilyl.

Reagent. Bis(trimethylsilyl)acetamide (BSA).

Compound Type. Diphenylhydantoin and its *p*-hydroxy metabolite.

Reaction

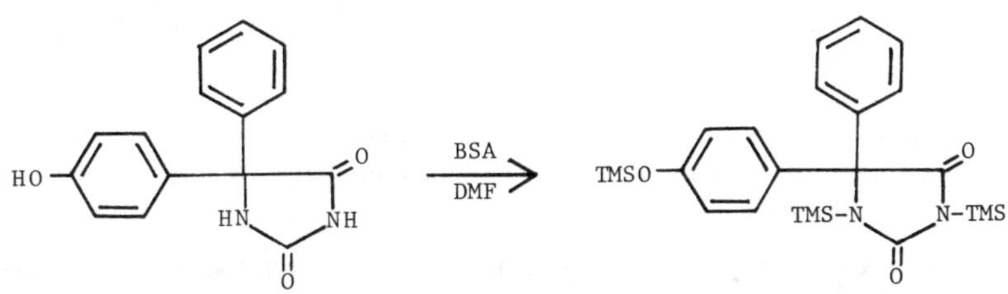

Typical Procedure. Ext 1 ml of glucuronidase treated plasma with 10 ml CH_2Cl_2 and dry ext over Na_2SO_4. Evap 80% of ext to dryness under N_2 and add 50 μl BSA. Heat for 30 min at 80° and inject 1–2 μl into the GC-MS.

Comments. The method was used for quantitating the hydroxylated metabolite in plasma by GC-MS-SIM using a deuterated int stand. BSA in DMF (5:1, v/v; room temp, 10 min) had previously been used for analysis of the drug and metabolite by GC.[2]

References

[1]J. D. Baty and P. R. Robinson, *Biomed. Mass Spectrom.* **4**, 36 (1977).
[2]T. Chang and A. J. Glazko, *J. Lab. Clin. Med.* **75**, 145 (1970).

Derivative Type. *N*-Trimethylsilyl.

Reagent. Bis(trimethylsilyl)acetamide (BSA)/trimethylchlorosilane (TMCS).

Compound Type. Ethosuximide.

Reaction

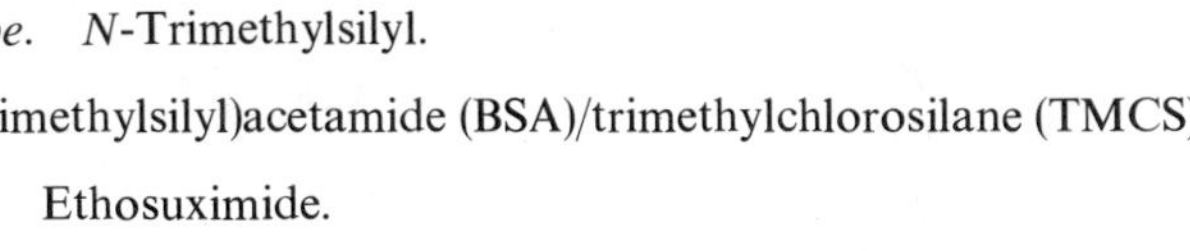

Typical Procedure. Saturate sample (100–200 μl plasma, 5–7 ml urine, 0.5–1.0 ml milk) with $(NH_4)_2CO_3$ and ext with EtOAc. Add 100 μl piperidine to the ext and evap to dryness under N_2. Diss residue in 10 μl pyridine and add 10 μl BSA and 5 μl TMCS. Heat at 60° for 1 hr then analyze by GC-MS-SIM.

Comments. The piperidine salt was formed to prevent loss of the volatile drug during evap of the solvent.

References

[1]M. G. Horning *et al.*, *J. Chromatogr. Sci.* **12**, 630 (1974).

Derivative Type. Dimethyl ketal.

Reagent. Dimethylformamide (DMF) dimethyl acetal.

Compound Type. Barbiturates and related compds.

Reaction

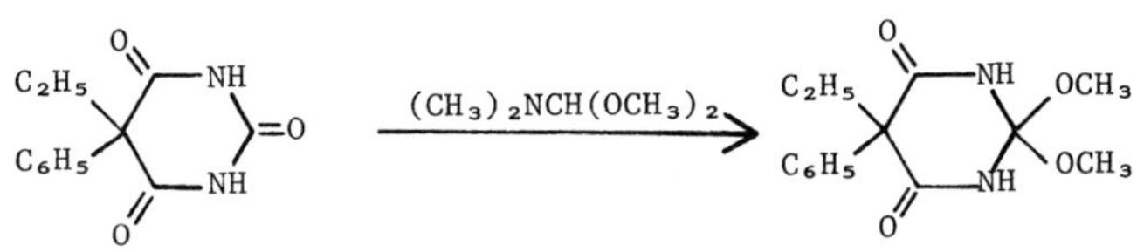

Typical Procedure. Diss 5–40 mg of sample in 1 ml reagent and warm on a steam bath for 10 min. Add int stand soln (40 mg/ml dibutyl phthalate in $CHCl_3$) and dil to 5 ml with DMF. Analyze 3 μl aliquots by GC.

Comments. NMR and MS data were given to support the ketal structure assigned to the deriv. For glutethimide, *n*-eicosane int stand was added to the reagent/sample soln and it was analyzed without dilution. The glutethimide deriv was subject to solvent induced reversibility, and methyprylon formed the deriv only with difficulty.

References

[1]V. S. Venturella *et al.*, *J. Pharm. Sci.* **62**, 662 (1973).

Derivative Type. 5-Dimethylaminonaphthalene-l-sulfonyl (DANS).

Reagent. 5-Dimethylaminonaphthalene-l-sulfonyl chloride (DANS-Cl).

Compound Type. Barbiturates.

Reaction

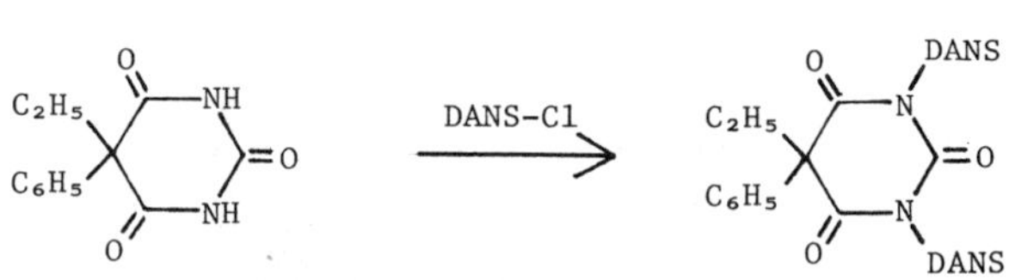

Typical Procedure. Ext a 20 μl blood sample by stirring with 3 × 20 μl acetone/Et_2O (1:1, v/v). Add 5 μl EtOAc to org phase and dry over 4 A mol sieve. Evap Et_2O and EtOAc stepwise under controlled conditions[2] to a vol of 5 μl. Add 15 μg DANS-Cl and 2 mg H_2CO_3 and reflux 2 hr. Dil mixt to 100 μl and inject 5 μl aliquots into the HPLC.

Comments. The fluorescent deriv was used for HPLC analysis (excitation 360 nm; emission measured at 250 nm). The detection limit was 0.3–0.5 ng/ml eluate.

References

[1]W. Dünges, *J. Chromatogr. Sci*, **12**, 655 (1974).

[2]K. Beyerman *et al.*, *Z. Anal. Chem.* **251**, 289 (1970).

15.4. BENZODIAZEPINES

15.4.1. Methyl Derivatives

Derivative Type. *N*-Methyl.

Reagent. Methyl iodide/tetrabutylammonium hydrogen sulfate (TBA-HSO_4).

Compound Type. Benzodiazepines.

Reaction

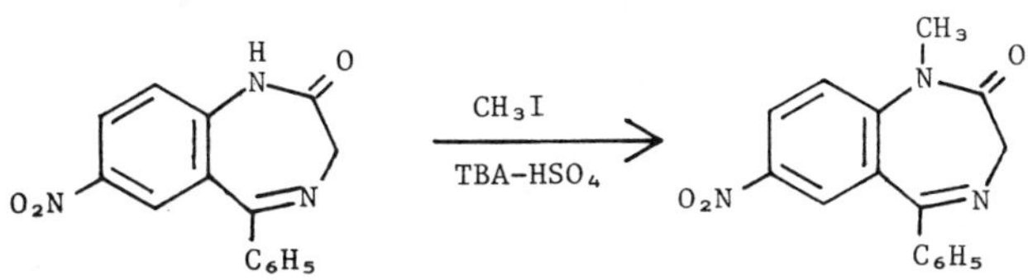

Typical Procedure. Mix 1 ml whole blood with 2 ml pH 9.0 borate buffer and heat *ca* 12 sec on a boiling H_2O bath. Mix on Vortex mixer 5 sec then repeat heating (6 sec) and mixing (2–3 sec) twice. Cool, add 6 ml benzene/CH_2Cl_2 (9:1, v/v), and shake 5 min. Centrifuge, remove 5.5 ml supernatant, and add to it 50 μl CH_3I and 0.50 ml fresh 0.025 *M* TBA-HSO_4 in 0.1 *N* NaOH. Shake 10 min at room temp, centrifuge, and discard aq phase. Wash with 2 ml 0.1 *N* HCl followed by 2 ml H_2O. Evap to dryness under N_2 then under vac. Diss residue in 100 μl benzene/acetone/MeOH (80:15:5, v/v/v) and analyze by ECD-GC.

Comments. The method was applied to nitrazepam (shown above), clonazepam, and (with slight modification) demoxepam. Analytical procedures were reported for 20 benzodiazepine compounds.[1] A similar method had been reported earlier for nitrazepam and employed griseofulvin as int stand.[2] Use of the tetrapropylammonium salt gave a slower reaction.[2]

References

[1]J. A. F. de Silva *et al.*, *Anal. Chem.* **48**, 10 (1976).

[2]H. Ehrsson and A. Tilly, *Anal. Lett.* **6**, 197 (1973).

Derivative Type. N_1,O_3-Dimethyl.

Reagent. Methyl iodide/tetrahexylammonium hydrogen sulfate (THA-HSO_4).

Compound Type. Oxazepam.

Reagent

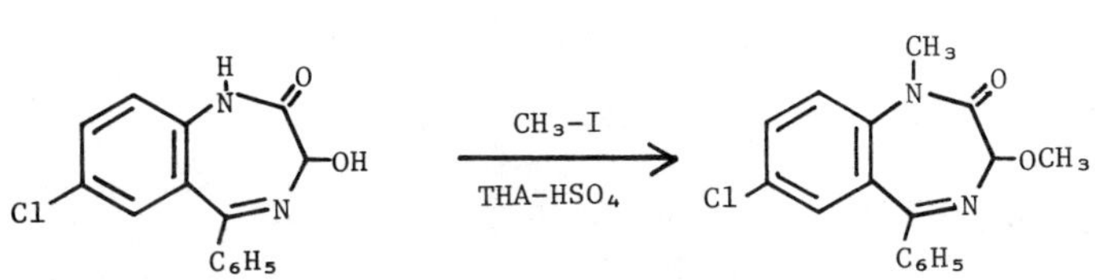

Typical Procedure. Dil a serum sample ($\leqslant$ 2 ml and containing 5–50 ng oxazepam) with 4 ml pH 7.4 phosphate buffer. Add 50 μl int stand soln (500 ng/ml lorazepam in CH_2Cl_2) and ext with 8 ml CH_2Cl_2. Centrifuge and transfer org layer to another tube by filtering it through silanized glass wool in a Pasteur pipette. Evap to dryness and add 1.5 ml 4 *M* CH_3I in CH_2Cl_2 and 0.5 ml 0.01 *M* aq THA soln (pH 13). Shake 5 min then transfer org phase to another tube via a glass wool filter as before. Evap to dryness and add 1 ml hot satd aq Ag_2SO_4 soln. Mix 10 sec on Vortex mixer, add 0.1 ml toluene, and mix again 10 sec. Analyze 1–2 μl of the org phase by ECD-GC.

Comments. The diethyl deriv was prepared by the same method using C_2H_5I and a reaction time of 15 min. The THA soln was prep by diss 4.51 g THA-HSO_4 in 1000 ml 0.1 *M* NaOH. The deriv could be converted to an isomeric form (structure not given) having a shorter GC retention time.

References

[1]J. Vessman *et al.*, *Anal. Chem.* **49**, 1545 (1977).

15.4.2. Hydrolysis Product Derivatives

Derivative Type. Benzophenones.

Reagent. Sulfuric acid/hydrochloric acid.

Compound Type. 7-Nitro-1,4-benzodiazepin-2-ones (Clonazepam and flunitrazepam).

Reaction

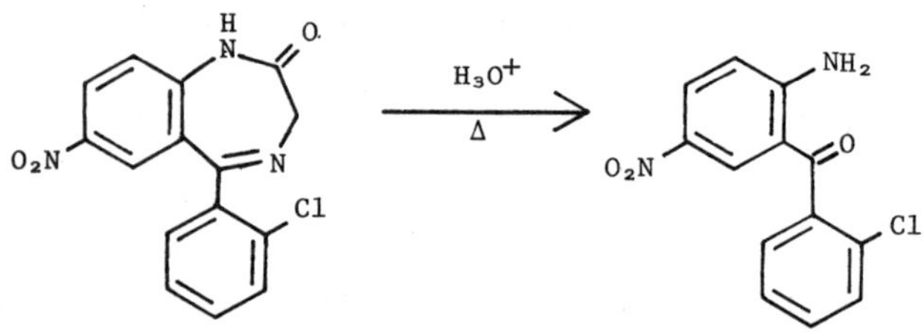

Typical Procedure (Clonazepam). Mix 2 ml blood or 5 ml urine with 5 ml pH 9.0 borate buffer and ext with 10 ml Et_2O. Centrifuge, remove the ext, and repeat ext with 10 ml Et_2O. Back ext the combined Et_2O into 5.0 ml 4 *N* HCl/4 *N* H_2SO_4 (95:5, v/v). Discard the Et_2O and wash the acid phase with 2×15 ml Et_2O. Heat the acid phase at 100° for 1 hr. Cool in ice, add 1 drop bromthymol blue indicator, and carefully add 3.5 ml 6.0 *N* NaOH dropwise until the blue end point is reached. Ext with 2×10 ml Et_2O. Evap ext to dryness under vac and diss residue in 100 μl 20% acetone in *n*-hexane. Analyze by ECD-GC.

Comments. Hydrolysis in HCl alone (especially "old" HCl with increased O_2 content) gave a chlorinated side product. The neutralization had to be done carefully with cooling to prevent cyclization to the 9-acridanone. Similar methods were applied to diazepam,[2] oxazepam,[3] and nitrazepam.[4]

References

[1]J. A. F. de Silva *et al.*, *J. Pharm. Sci.* **63**, 520 (1974).

[2]J. A. F. de Silva *et al.*, *Anal. Chem.* **36**, 2099 (1964).

[3]J. Vessman *et al.*, *Acta Pharm. Suecica* **9**, 447 (1972).

[4]G. P. Beharrell *et al.*, *J. Chromatogr.* **70**, 45 (1972).

Derivative Type. Trifluoroacetyl aminobenzophenone.

Reagent. Hydrochloric acid, trifluoroacetic anhydride (TFAA)/pyridine.

Compound Type. Pivoxazepam.

Reaction

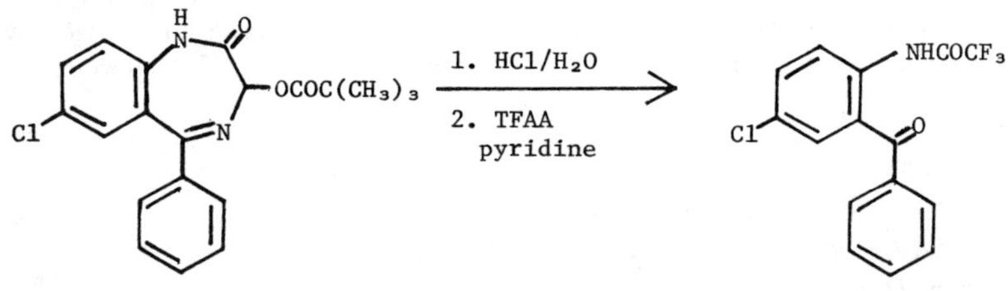

Typical Procedure. Ext 3 ml plasma with 9 ml benzene containing 2′-chloropivoxazepam int stand. Wash 4 ml of the benzene with 3 × 5 ml 1 *N* HCl then evap 3.5 ml of the washed benzene ext to dryness. Diss residue in 40 μl abs EtOH, add 5 ml 6 *N* HCl, and heat at 100° for 1 hr. Neutralize with NaOH, adjust to pH 7.4 with 1 ml phosphate buffer, and ext with 2 × 10 ml Et_2O. Evap ext to dryness and add 10 μl TFAA and 25 μl 0.05% pyridine in benzene. Heat 10 min at 60° then evap to dryness. Diss residue in 100 μl hexane/acetone (1:1, v/v) and analyze 2 μl aliquots by ECD-GC or GC-MS-SIM.

Comments. The method permitted detection to 5–10 ng/ml by ECD-GC and 0.5–0.8 ng/ml by GC-MS-SIM.

References

[1]M. C. Sanchez *et al.*, *J. Chromatogr.* **126**, 601 (1976).

15.5. CANNABINOIDS

15.5.1. Silylated Derivatives

Derivative Type. Trialkylsilyl(methyl, ethyl, *n*-propyl, *n*-butyl, *n*-hexyl).

Reagent. Trialkylchlorosilane/diethylamine (DEA)/pyridine.

Compound Type. Cannabinoids.

Reaction

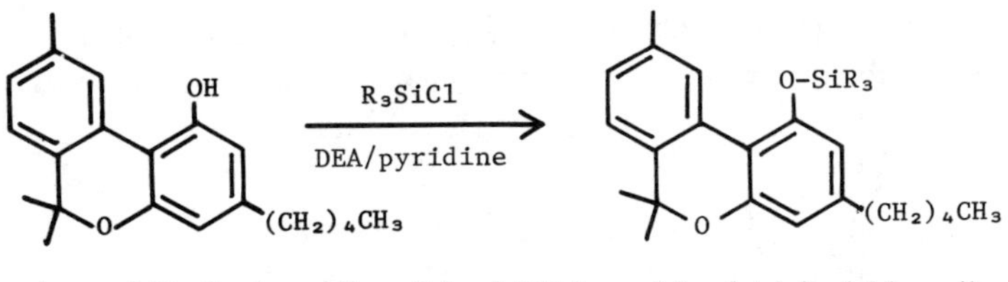

Typical Procedure. Mix 2 ml pyridine, 0.5 ml DEA, and 1 ml trialkylchlorosilane. Cool in ice water then centrifuge. Treat the cannabinoid sample with 100 μl of the supernatant for 30 min at room temp. Analyze by GC-MS.

Prepare TMS derivs by treating sample with BSTFA (40 μl), TMCS (20 μl), and CH_3CN (40 μl) for 30 min at room temp. Analyze by GC-MS.

Comments. The derivs were used to achieve separation of monohydroxy- from dihydroxy-cannabinoids; the TMS derivs were poorly resolved but increasing the alkyl length to *n*-butyl gave complete fractionation of the two groups. GC data were given for the 5 derivs of 11 cannabinoids.[1] The homologous derivs were also used in the analysis of the diphenylpropanoids of nutmeg[2] and the butyl homologs of cannabinoids.[3] Data were reported for the derivs of a series of aliphatic mono- and di-alcohols.[1] The MS behavior of the derivs of some cannabinoids, steroids, and alcohols has been reported.[4] The spectra are characterized by abundant *M*-alkyl ions but have generally fewer diagnostic peaks than the TMS derivs.

References

[1]D. J. Harvey and W. D. M. Paton, *J. Chromatogr.* **109**, 73 (1975).

[2]D. J. Harvey, *J. Chromatogr.* **110**, 91 (1975).

[3]D. J. Harvey, *J. Pharm. Pharmacol.* **28**, 280 (1976).

[4]D. J. Harvey, *Org. Mass Spectrom.* **12**, 473 (1977).

Derivative Type. *O*-Ethyl, *O*-trimethylsilyl.

Reagent. Tetrahexylammonium hydroxide(THAH)/ethyl iodide, Bis(trimethylsilyl)trifluoroacetamide (BSTFA)/trimethylchlorosilane (TMCS).

Compound Type. 11-Hydroxy-Δ^9-tetrahydrocannabinol (11-OH-Δ^9-THC).

Reaction

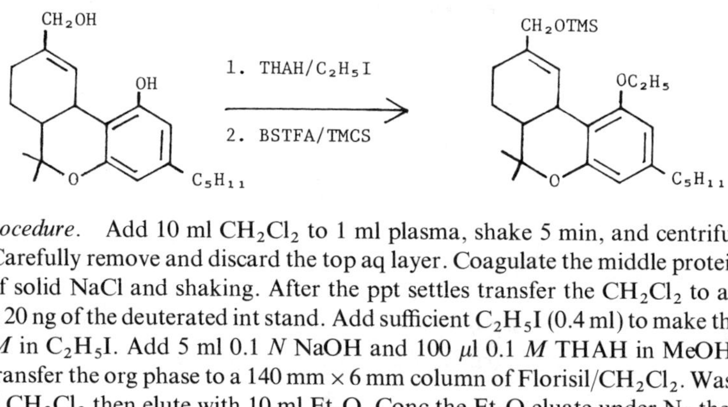

Typical Procedure. Add 10 ml CH_2Cl_2 to 1 ml plasma, shake 5 min, and centrifuge (yields 3 layers). Carefully remove and discard the top aq layer. Coagulate the middle protein layer by addition of solid NaCl and shaking. After the ppt settles transfer the CH_2Cl_2 to a new tube containing 20 ng of the deuterated int stand. Add sufficient C_2H_5I (0.4 ml) to make the CH_2Cl_2 plase 0.5 *M* in C_2H_5I. Add 5 ml 0.1 *N* NaOH and 100 μl 0.1 *M* THAH in MeOH. Shake 5 min then transfer the org phase to a 140 mm × 6 mm column of Florisil/CH_2Cl_2. Wash column with 10 ml CH_2Cl_2 then elute with 10 ml Et_2O. Conc the Et_2O eluate under N_2 then transfer to 5 ml vial. Evap to dryness under N_2. Diss residue in 20 μl BSTFA/TMCS (9:1, v/v). After 5 min analyze 1–3 μl aliquots by GC-MS-SIM.

Comments. 1-OC_2D_5-11-OH-Δ^9-THC was used as int stand. The ethyl deriv was used because of the possibility of endogenous *O*-methyl compound. The compd was detectable to 3 ng/ml by the method.

References

[1]J. M. Rosenfeld and V. Y. Taguchi, *Anal. Chem.* **48**, 726 (1976).

Derivative Type. Trimethylsilyl acetate.

Reagent. Trimethylsilyl ketene (TMS-ketene).

Compound Type. Amines, alcohols, phenols (e.g., tetrahydrocannabinol).

Reaction

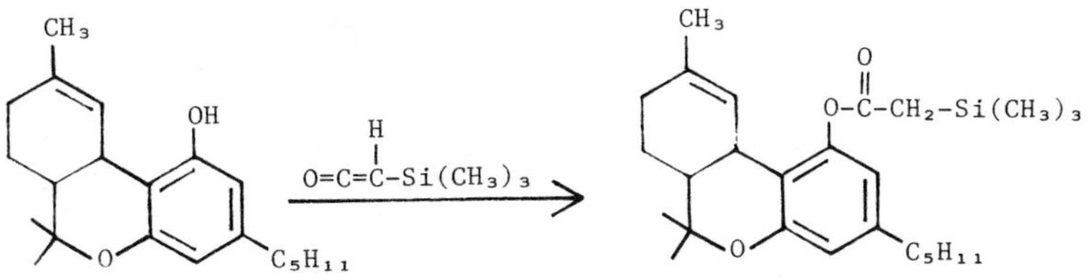

Typical Procedure.[1] Add excess TMS-ketene (10 mmol) in 100 μl CCl_4 to 10 μmol cannabinol and allow soln to stand in a sealed tube for 24 hr at 25°. Evap solvent in a stream of N_2 to give 98% yield of deriv. Analyze by GC, GC-MS, or HPLC.

Comments. The reagent was prepared by the following sequence[2]:

$$HC{\equiv}C{-}OEt \xrightarrow[2.\ TMS-Cl]{1.\ MeLi} TMS{-}C{\equiv}C{-}OEt \xrightarrow[120^\circ]{} TMS{-}CH{=}C{=}O$$

TMS-ketene was reported to acylate hindered alcohols and amines.[2] The acylation could be catalyzed by $BF_3 \cdot Et_2O$.[2] The derivs showed no evidence of hydrolysis in the MeOH-H_2O HPLC solvent.[1]

References

[1]E. E. Knaus *et al.*, *J. Chromatogr. Sci.* **14**, 525 (1976).
[2]R. A. Ruden, *J. Org. Chem.* **39**, 3607 (1974).

Derivative Type. *t*-Butyldimethylsilyl (TBDMS) ether.

Reagent. *t*-Butyldimethylsilyl chloride (TBDMS-Cl)/imidazole.

Compound Type. Cannabinoids.

Reaction

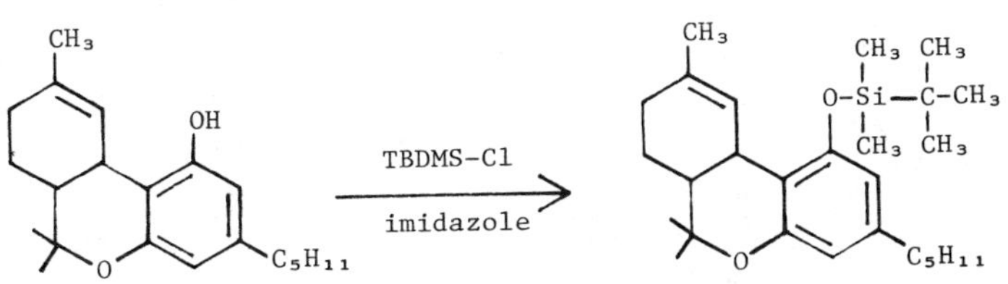

Typical Procedure. Mix 1 mmol of tetrahydrocannabinol, 4 mmol imidazole, and 2 mmol TBDMS-Cl in 0.5 ml DMF and warm on steam bath 1 hr. Add 2 ml 10% aq NaOH and ext with 3 × 2 ml pet ether. Evap solvent to give quant yield of TBDMS-O-THC. Analyze by GC or GC-MS.

Comments. The deriv was stable to moisture, base, and even steam. When the cannabinoid contained two reactive functional groups, the reagent quantities were increased by 50%. Cannabinolic acids gave TBDMS esters. MS data were given for the derivs.

References

[1]E. E. Knaus *et al.*, *J. Chromatogr. Sci.* **14**, 525 (1976).

15.5.2. Other Derivatives of Cannabinoids

Derivative Type. *O*-Methyl ether (Δ^9-THC-OCH_3).

Reagent. Trimethylphenylammonium hydroxide (TMPAH).

Compound Type. Δ^9-Tetrahydrocannabinol (Δ^9-THC).

Reaction

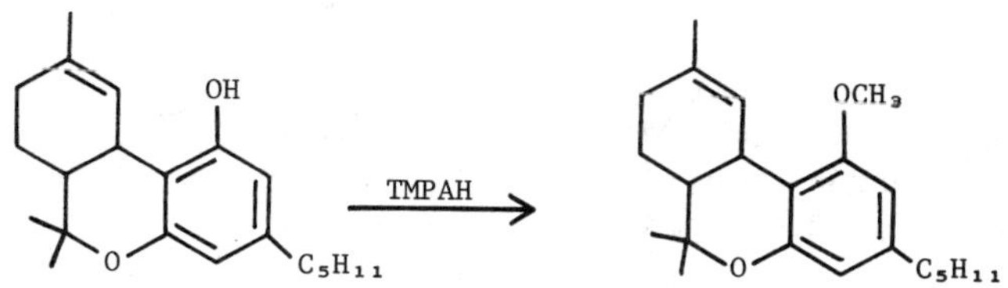

Typical Procedure. Ext 2 ml plasma with 4 × 2 ml hexane containing 5% (v/v) *i*-AmOH. Ext combined hexane with 4 × 1 ml modified Claisens alkali (see below). Acidify the combined alkali ext with 0.2 ml conc HCl and diss the KCl ppt by adding 1 ml H_2O. Ext with hexane and evap hexane ext to dryness under N_2. Add int stand (1.15 μg) soln and evap. Diss the residues in 50 μl 0.2 *M* TMPAH in MeOH. Analyze 1 μl aliquots by GC-MS-SIM.

Comments. The modified Claisens alkali was prepared by diss 17.5 g KOH (1/2 normal amt) in 25 ml H_2O, cooling, and adding 100 ml MeOH.[2] Δ^9-THC-OCD_3 was used as int stand. The synthesis of the standard was reported. Purification of Δ^9-THC was effected by the selective extractability of lipid soluble phenols from hexane by Claisens alkali. The method was used for quantitation in the range of 1–200 ng/ml plasma.

References

[1]J. J. Rosenfeld *et al.*, *Anal. Chem.* **46**, 2232 (1974).

[2]L. F. Fieser and M. Fieser, *Reagents for Organic Synthesis*, Wiley, New York, NY, 1967, p. 153.

Derivative Type. Trifluoroacetyl (TFA).

Reagent. Trifluoroacetic anhydride (TFAA)/pyridine.

Compound Type. Δ^1-Tetrahydrocannabinol (Δ^1-THC).

Reaction

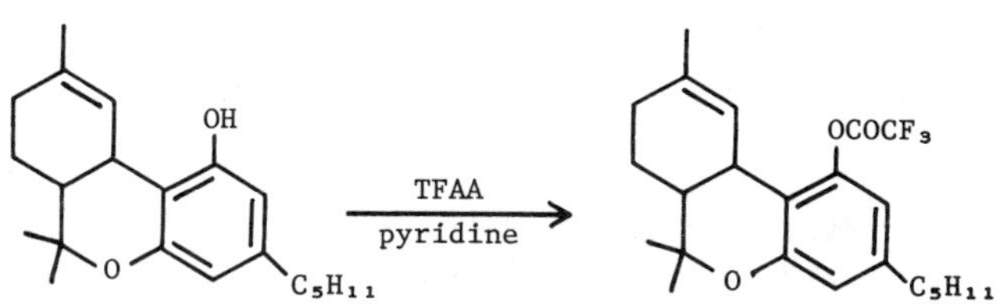

Typical Procedure. Combine TFAA, pyridine, and Δ^1-THC in the ratio 15 : 3 : 1 (w/w/w) in isooctane soln with the Δ^1-THC conc 0.2–0.7 mg/ml. Mix then add BSTFA in an amt 1.5 times the wt of pyridine (to remove the ppt of pyridinium trifluoroacetate). Analyze by GC.

Comments. The TMS ether deriv of Δ^1-THC was also prepared using BSTFA/pyridine/trifluoroacetic acid/Δ^1-THC (120:60:2:1, w/w/w/w in isooctane. Δ^1-THC 0.2–0.7 mg/ml). The trifluoroacetic acid was added to promote silylation of the sterically hindered phenol.[2]

It was reported that neither the TFA nor the TMS derivs gave any advantage over Δ^1-THC itself for GC quantitation down to 50 ng injections.

References

[1]N. E. Hoffman and R. Y. Yang, *Anal. Lett.* **5**, 7 (1972).
[2]N. E. Hoffman and K. A. Petranetz, *Anal. Lett.* **5**, 589 (1972).

Derivative Type. Diethyl phosphate.

Reagent. Sodium diethylphosphorochloridate.

Compound Type. Cannabinols.

Reaction

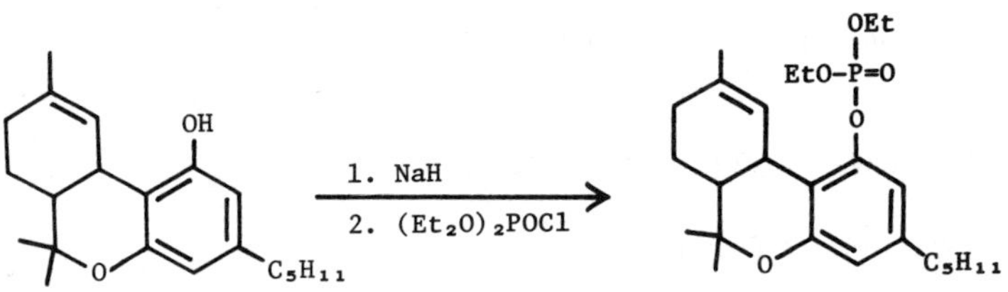

Typical Procedure. Ext 10 ml blood or plasma with 40 ml lt pet ether by shaking for 7 min. Add 10 ml 1 *N* HCl and continue shaking 7 min. Sep org phase and evap to *ca* 5 ml under N_2 warming. Transfer to 10 ml sep funnel and wash with 3×5 ml 1 *N* NaOH, 2×5 ml 1 *N* HCl, and once with H_2O. Dry over anhyd Na_2SO_4 and evap under N_2. Diss residue in 0.3 ml dry benzene/*t*-BuOH (9:1, v/v) and add to *ca* 300 mg NaH pellets in tube arranged to deliver directly into another tube containing 2 drops of redistilled diethyl phosphorochloridate. After 2 min drop soln onto reagent and maintain the mixt at 50° for 30 min. Evap benzene under stream of N_2 at <45° and add 1 ml 5% Na_2CO_3 at 45°. Shake and allow to stand in 45° bath 1 min. Add 1 ml hexane, shake, then remove hexane phase. Evap to 50 μl and analyze 5–10 μl aliquots by GC.

Comments. The deriv was used for GC analysis with flame photometric detector. The practical assay limit for Δ^9-THC was <1 ng/ml if >10 ml plasma was used.

References

[1]N. K. McCallum, *J. Chromatogr. Sci.* **11**, 509 (1973).

Derivative Type. Alkylboronate.

Reagent. Methyl- or *n*-butylboronic acid.

Compound Type. Cannabinolic acids.

Reaction

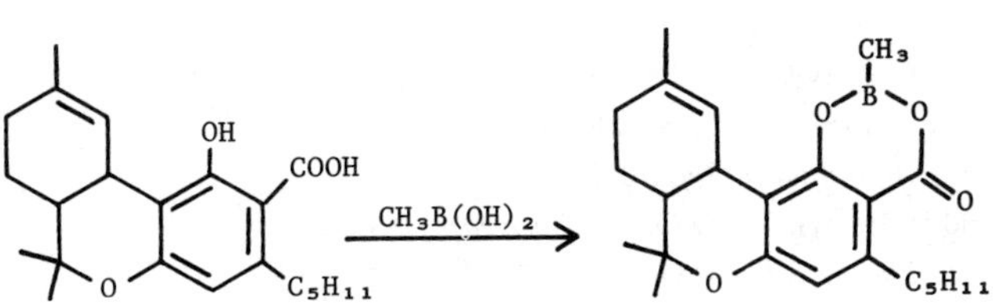

Typical Procedure. Ext 50 mg sample of cannabis with EtOAc by crushing in a mortar then allowing to stand 1 hr at room temp. Filter then evap ext to dryness under N_2. Diss resin in 0.1 ml pyridine and 0.1 ml benzene. Add 1 mg of alkylboronic acid and allow to stand at room temp 30 min. Evap to dryness under N_2 and diss residue in 0.1 ml pyridine and 0.1 ml benzene. In the case of the volatile methylboronic acid, add another 0.5 mg of the acid. After another 30 min at room temp, evap to dryness under N_2. Diss residue in 0.1 ml CH_3CN and add 0.1 ml BSTFA and 0.05 ml TMCS to convert unreacted hydroxyls and carboxyls to TMS derivs. After 30 min at room temp, analyze 1 μl aliquots by GC-MS.

Comments. GC and MS data were reported for the alkylboronate derivs of five cannabinol acids. Data were given for TMS derivs of these acids and other cannabinoids. Phenylboronate derivs for two of the acids were also examined.

References

[1]D. J. Harvey, *Biomed. Mass Spectrom.* **4**, 88 (1977).

15.6. DIPHENYLMETHANE TYPE DRUGS

Derivative Type. Benzophenone.

Reagent. Chromic acid.

Compound Type. Diphenylmethane compounds.

Reaction

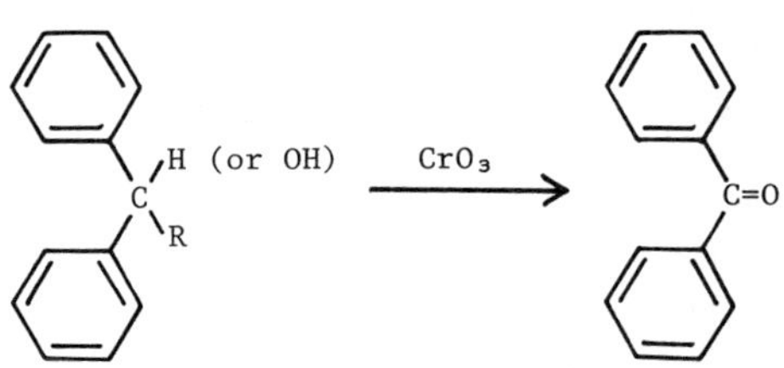

Typical Procedure. Mix 6.5 μg of compd in 1.0 ml HOAc with 9.0 ml oxidation reagent (334 mg CrO_3 in 100 ml of 6% (w/v) H_2SO_4 in HOAc) in a glass stoppered tube. Shake gently for 1 hr. Transfer 1.0 ml to another tube containing 5.0 ml 5 *M* NaOH and 4 ml H_2O. Add 1.0 ml int stand (*p*-methylbenzophenone) soln (160 ng/ml) in heptane. Shake, centrifuge, and remove heptane layer. Dry over Na_2SO_4 and analyze by GC.

Comments. The method yielded an ECD sensitive deriv. The reactions of 48 diphenylmethane compds were studied. The method had a potential problem of selectivity for drug analysis since some metabolites could give the same deriv.

References

[1]J. Vessman *et al.*, *Acta Pharm. Suecica* **7**, 373 (1970).

[2]J. Vessman and P. Hartvig, *Acta Pharm. Suecica* **8**, 235 (1971).

[3]J. Vessman, *Acta Pharm. Suecica* **8**, 251 (1971).

[4]P. Hartvig *et al.*, *Acta Pharm. Suecica* **9**, 269 (1972).

[5]J. Vessman and P. Hartvig, *Acta Pharm. Suecica* **9**, 463 (1972).

[6]P. Hartvig, *Acta Pharm. Suecica* **11**, 109 (1974).

Derivative Type. Benzophenone.

Reagent. Barium peroxide.

Compound Type. Emepronium bromide.

Reaction

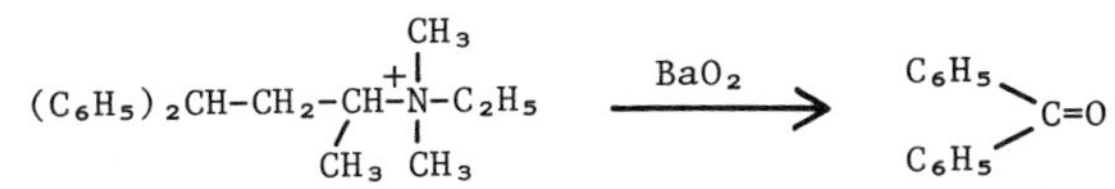

Typical Procedure. To a serum sample containing 0.2–4 ng of emepronium in a centrifuge tube add 1 ml of a 9 ng/ml int stand soln, 1 ml of 0.5 *M* $NaClO_4$, H_2O to 5 ml and 1 ml CH_2Cl_2. Shake 30 min and centrifuge. Transfer CH_2Cl_2 to another tube and evap to dryness under N_2. To the residue add 0.25 ml of 3.6 *M* H_2SO_4 and *ca* 20 mg of BaO_2. Add 50 μl heptane and heat in a 114° sand bath (sand up to level of water phase). Cool, sep heptane phase, and wash with 3 drops 1 *M* NaOH. Analyze 10–20 μl of the heptane phase by GC.

Comments. This method gave increased sensitivity over the earlier method which employed chromic acid oxidation.[2] The small ext and reaction vol permitted use in the 0.2–8 ng/ml range.

References

[1]P. Hartvig *et al.*, *J. Pharm. Sci.* **65**, 1707 (1976).

[2]J. Vessman *et al.*, *Acta Pharm. Suecica* **7**, 363 (1970).

Derivative Type. Benzophenone.

Reagent. Barium peroxide/sulfuric acid.

Compound Type. Methadone.

Reaction

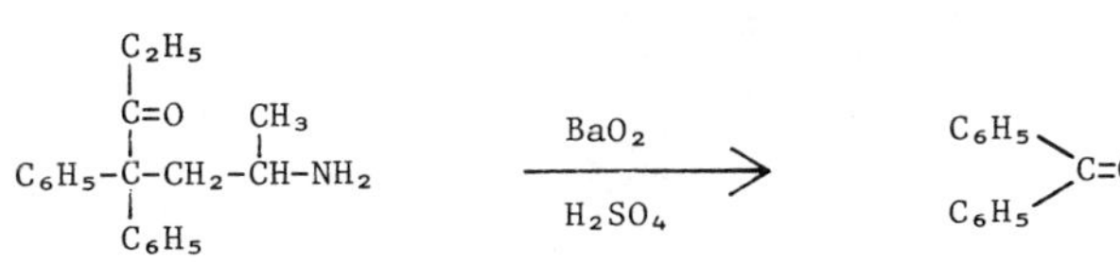

Typical Procedure. Mix an aliquot of serum corresponding to 16–120 ng of methadone (not more than 2 ml) with 3.0 ml of int stand soln, 1 ml of 1 *M* NaOH, and H_2O to 5 ml. Ext with 5 ml *n*-heptane for 30 min. Sep org phase and ext it with 2 ml 0.1 *M* H_3PO_4 for 30 min. Transfer aq phase to a tube containing 100 mg BaO_2 and 1 ml 10.8 *M* H_2SO_4. Add 0.5 ml *n*-heptane, attach an air condenser, and heat in a glycerol bath at 116° for 15 min. Sep the org phase and wash it with 2 ml 0.1 *M* NaOH. Analyze 2–5 μl of the org phase by ECD-GC.

Comments. The lower limit for determination in serum was 4 ng/ml. The int stand soln was prepared by diss 4-(4-chlorophenyl)-4-phenyl-2-dimethylaminobutane in 0.1 *M* H_3PO_4 and diluting to 49 ng/ml. The int stand was converted in the procedure to 4-chlorobenzophenone.

References

[1]P. Hartvig and B. Naslund, *J. Chromatogr.* **111**, 347 (1975).

Derivative Type. *p*-Bromophenyl-2-pyridyl ketone.

Reagent. Potassium permanganate.

Compound Type. Brompheniramine.

Reaction

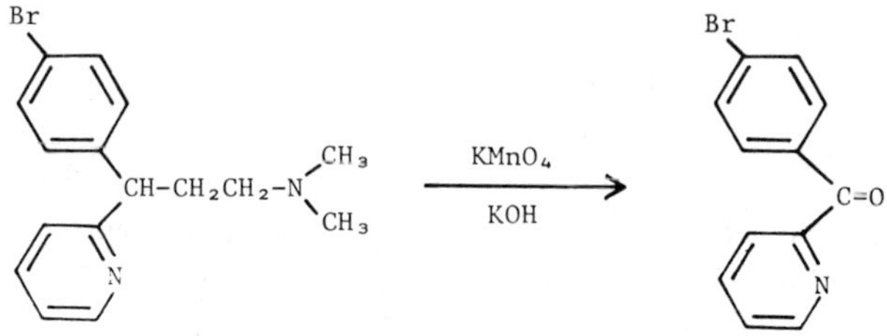

Typical Procedure. Mix 5.0 ml blood, 0.5 ml of a soln containing 100 ng of chlorphenir-amine int stand, and 0.5 ml 10 *N* KOH. Heat in boiling H_2O bath 20 min. Cool and ext with 100 ml Et_2O. Centrifuge, sep Et_2O, and mix with 1 ml 0.1 *N* H_3PO_4. Evap Et_2O to *ca* 10 ml under N_2. Mix on Vortex mixer 1 min then sep the aq phase. Ext the Et_2O with another 0.5 ml 0.1 *N* H_3PO_4 and combine the acid ext. Alkalinize with 3–4 drops 10 *N* KOH, sat with NaCl, and ext with 2 ml and 2×1 ml Et_2O. Place Et_2O ext on a 1.3 g Celite column buffered with pH 5.4 phosphate. Elute with 15 ml Et_2O (equilibrated with pH 5.4 phosphate buffer). Ext eluate with 1 ml 0.1 *N* H_3PO_4.

Heat ext under vac to expel Et_2O, cool, and add 0.25 ml 2 *N* KOH and 3 ml 1% $KMnO_4$. Heat at 100° for 15 min with an air condenser. Cool and ext soln with 0.20 ml isooctane. Analyze 10–20 μl by ECD-GC.

Comments. The method was sensitive to less than 10 ng/ml in blood. The column chromato-graphy was used to separate unchanged drug from demethylated metabolites that would yield the same seriv.

References

[1] R. B. Bruce *et al.*, *Anal. Chem.* **40**, 1246 (1968).

Part III Appendices and Indexes

Appendix A
Compositions of Brand Name Reagents

Authors occasionally report analytical work with brand name reagents without identifying the composition of the reagents. To facilitate identification of the constituents of such reagents, and to make the reader aware of available premixed reagents, this list of brand name reagents and their compositions is presented. Complete addresses of sources appear in Appendix B.

Reagent Name	Composition	Source
p-Bromophenacyl-8®	*p*-Bromophenacyl bromide (0.1 mmol/ml) and crown ether (0.005 mmol/ml) in CH_3CN	Pierce
Butyl-8® Concentrate	DMF di-*n*-butyl acetal	Pierce
n-Butyl-8®	DMF di-*n*-butyl acetal (2 meq/ml) in pyridine	Pierce
CHROMOTAG®	A series of chromogenic reagents for HPLC: DNBC; 3,5-Dinitrobenzoyl chloride PBPB; *p*-Bromophenacyl bromide PNBA; *p*-Nitrobenzyloxyamine HCl PNBDI; *O*-*p*-Nitrobenzyl-*N*,*N*′-diisopropylisourea PNBPA; *p*-Nitrobenzyl-*n*-propylamine HCl SNPA; *N*-Succinimidyl *p*-nitrophenyl acetate	Regis
Deutero-Regisil® (RD-1)	d_{18}-BSTFA	Regis
Diazald®	*N*-methyl-*N*-nitroso-*p*-toluene-sulfonamide	Aldrich
DI-SIL-PREP®	TMDS/DMMCS/pyridine (3 : 1 : 9, v/v/v)	Applied Science
DMF-SIL-PREP®	HMDS/TMCS/DMF (40 : 1 : 40, v/v/v)	Applied Science
Esterate-CM®	DMF dimethyl acetal	Supelco
Esterate-M®	DMF dimethyl acetal (2 meq/ml) in pyridine	Supelco
Eston-F_3®	2,2,2-Trifluoroethanol with 10% (w/w) BCl_3	Supelco
Ethyl-8®	DMF diethyl acetal (2 meq/ml) in pyridine	Pierce
Ethyl-8® Concentrate	DMF diethyl acetal	Pierce
FLOROX®	*O*-(Pentafluorobenzyl)hydroxylamine HCl (2.5 mg/ml) in pyridine	Pierce
Fluorōpa®	*o*-Phthalaldehyde	Durrum (sold by Pierce)

Reagent Name	Composition	Source
FLUOROTAG®	A series of fluorogenic reagents for HPLC: BMC; 4-Bromomethyl-7-methoxycoumarin NBD-chloride; 7-chloro-4-nitrobenzyl-2-oxa-1,3-diazole Dansyl chloride; 5-Dimethylaminonaphthalene-1-sulfonyl chloride Dansyl hydrazine; 5-Dimethylaminonaphthalene-1-sulfonyl hydrazine	Regis
Fluram®	Fluorescamine	Roche (sold by Pierce)
Hydrox-Sil®	HMDS/TMCS/pyridine (2 : 1 : 10, v/v/v)	Regis
Hydrox-Sil AQ®	TMSI (21% w/v; 1.5 meq/ml) in pyridine	Regis
MethElute®	Trimethylphenylammonium hydroxide (TMPAH, 0.2 *M*) in MeOH	Pierce
METH-PREP I®	0.2 *N* aq (*m*-trifluoromethylphenyl)-trimethylammonium hydroxide	Applied Science
METH-PREP II®	0.2 *N* (*m*-trifluoromethylphenyl)trimethylammonium hydroxide in MeOH	Applied Science
Methyl-8®	DMF dimethyl acetal (2 meq/ml) in pyridine	Pierce
Methyl-8® Concentrate	DMF dimethyl acetal	Pierce
MOX®	methoxyamine-HCl (2%) in pyridine	Pierce
p-Nitrobenzyl-8®	*O*-*p*-Nitrobenzyl-*N*,*N'*-diisopropyl isourea (0.1 *M*) in CH_2Cl_2	Pierce
Phenacyl-8®	Phenacyl bromide (0.1 mmol/ml) and crown ether (0.005 mmol/ml) in CH_3CN	Pierce
Power SIL-PREP®	TMSI/BSA/TMCS (3 : 3 : 2, v/v/v)	Applied Science
Propyl-8®	DMF di-*n*-propyl acetal (2 meq/ml) in pyridine	Pierce
Propyl-8® Concentrate	DMF di-*n*-propyl acetal	Pierce
Regisil®	BSTFA	Regis
Regisil® RC-2	BSTFA with 1% TMCS	Regis
Regisil® RC-3	BSTFA with 10% TMCS	Regis
SIL-PREP®	HMDS/TMCS/pyridine (3 : 1 : 9, v/v/v)	Applied Science
STOX®	methoxyamine-HCl (25 mg/ml) and phenyl-*β*-D-glucopyranoside (6 mg/ml) in pyridine	Pierce
SYLON® BFT	BSTFA/TMCS (5 : 1, v/v)	Supelco
SYLON®-BT	BSA/TMCS (5:1, v/v)	Supelco
SYLON®-BTZ	BSA/TMCS/TMSI (3 : 2 : 3, v/v/v)	Supelco
SYLON® CT	DMCS in toluene	Supelco
SYLON® HT	HMDS/TMCS (3 : 1, v/v)	Supelco
SYLON® HTP	HMDS/TMCS/pyridine (3 : 1 : 9, v/v/v)	Supelco
SYLON® TP	TMSI (1.5 mmol/ml) in pyridine	Supelco
Tri-Deuter-8®	deuterium labeled DMF dimethyl acetal for preparation of d_3-methyl esters	Pierce
TRI-SIL®	HMDS/TMCS/pyridine (2 : 1 : 10, v/v/v)	Pierce
TRI-SIL®/BSA	BSA (2.5 meq/ml) in DMF	Pierce

Reagent Name	Composition	Source
Formula D		
TRI-SIL®/BSA	BSA (2.5 meq/ml) in pyridine	Pierce
Formula P		
TRI-SIL® BT	BSA/TMCS (1 : 5, v/v)	Pierce
TRI-SIL® Concentrate	HMDS/TMCS (2 : 1, v/v)	Pierce
TRI-SIL® DMF	TRI-SIL Concentrate, DMF (equal parts, mixed as needed)	Pierce
TRI-SIL® TBT	TMSI/BSA/TMCS (3 : 3 : 2, v/v/v)	Pierce
TRI-SIL® Z	TMSI (1.5 meq/ml) in pyridine	Pierce
X-Tract Elute®	Trimethylphenylammonium hydroxide (TMPAH) (2 *M*) in MeOH	Pierce

Appendix B

Suppliers of Reagents and Apparatus for Analytical Derivatization

United States

Aldrich Chemical Co., Inc.
940 West Saint Paul Avenue
Milwaukee, Wisconsin 53233

Alltech Associates, Inc.
202 Campus Drive
Arlington Heights, Illinois 60004

Analabs, Inc.
80 Republic Drive
North Haven, Connecticut 06473

Applied Science Laboratories, Inc.
P.O. Box 440
State College, Pennsylvania 16801

Durrum Chemical Corp.
1228 Titan Way
Sunnyvale, California 94086

Organomation Associates
P.O. Box 5
Turnpike Station
Schrewsbury, Massachusetts 01545

PCR Research Chemicals, Inc.
P.O. Box 1778
Gainesville, Florida 32602

Pierce Chemical Company
P.O. Box 117
Rockford, Illinois 61105

RFR Corporation
1 Main Street
Hope, Rhode Island, 02831

Regis Chemical Company
8210 Austin Avenue
Morton Grove, Illinois 60053

Reliance Glass Works, Inc.
17 Gateway Road
Bensenville, Illinois 60106

Supelco, Inc.
Supelco Park
Bellefonte, Pennsylvania 16823

Wheaton Scientific
1000 North Tenth Street
Millville, New Jersey 08332

Argentina

Sirex
Libertad 836
Buenos Aires, Argentina

Australia

Laboratory Supply Pty., Ltd.
48-52 Sydenham Road
Marrickville, N.S.W., Australia

Quantum Labs
237 Milton Road
Milton, Quantum, 4064
Brisbane, Australia

Belgium

Etablissements Mornard
Frans Smoldersstraat 18
B-1940 Sint Stevens Woluwe
Belgium

Brazil

Acatec Comercio Rep. Ltd.
Rue Augusta, 1351, 7° Andar
01305—Sao Paulo
Brazil

Chile

Sirex
J.M. De La Barra 507 (7°)
Santiago, Chile

England

Phase Separations
Deeside Industrial Estate
Queensferry
CLWYD
Great Britain

Pierce and Warriner (U.K.) Ltd.
Chester, Cheshire CH1 4EF
England

France

Interchim
Depot de la Loue
B.P. 15
03103 Montlucon, France

Lab Eurobio
20, Bld, Saint-Germain
Paris, V France

Touzart & Matignon
8, Rue Eugene Henaff
94400 Vitry-Sur-Seine
France

Touzart & Matignon
3, Rue Amyot
75005 Paris, France

German Federal Republic

Aug, Hedinger KG
7 Stuttgart-Wangen
Heiligenwiesen 26
German Federal Republic

Macherey-Nagel & Co.
D-516 Duren
Werkstrabe 6-8
German Federal Republic

Serva International
Chemie und Handels-GmbH & Co.
Karl-Benz-Strasse 7, Postfach 10 52 60.
D-6900 Heidelberg 1
German Federal Republic

Holland

Pierce Eurochemic B.V.
Box 1151
Rotterdam, Holland

Iran

Tektrom Co. Ltd.
2nd Alley—Pakestan Ave, Abbasabad Ave.
P.O. Box 14/1975
Tehran, Iran

Italy

B.D.H. Italia S.p.A.
Via Ernesto Breda, 142
20126 Milano, Italy

Japan

Chemco Scientific Co., Ltd.
Kyo-Building, 5th Floor, No. 165
2-Chome, Kyomachi-Bari
Nishi-Ku, Osaka 550, Japan

Kokusan Chemical Works, Ltd.
3-9 Honcho-Nihonbashi
Chuo-ku
Tokyo, Japan

Netherlands

Brunschwig Chemie B.V.
P.O. Box 70213, Butaanweg 8
Amsterdam-1003, Netherlands

New Zealand

Lab Supply Pierce, N.Z. Ltd.
Box 64-049, Birkenhead South
Auckland 10, New Zealand

Smith-Biolab, Ltd.
39 Woodwide Ave, Northcote
Auckland, New Zealand

Poland

Polskie Odczynniki Chemiczne
ul. Sowinskiego 11
44-1015 Gliwice, Poland

South Africa

Coulter Electronics SA
84 Halfway House
1685 TVL
Republic South Africa

Spain

Xpectrix
Carretera Sardanyola, 73-75
San Cugat del valles
Barcelona, Spain

Sweden

Labkemi AB
Frofastegatan 111
S-421 31 Vastra Frolunda
Sweden

Switzerland

Chemie Brunschwig AG
Belchenstrasse 12
4054 Basel
Switzerland

Supelco, S.A.
3 Rte. De Celigny
1299 Crans, Switzerland

Technosa SA
Chemin Chamblandes 47
CH-1009 Pully-Lausanne, Switzerland

Derivative and Reagent Index

This index lists the derivative types and the reagents used in the methods sections of Part II (references to derivatives discussed in the introductory sections are found in the general subject index). Those derivative types having numerous entries are subdivided according to type of compound derivatized. Under each derivative type are listed the various reagents used to prepare that derivative and page references. Page numbers in bold print indicate the experimental procedure is given. Page numbers in regular print indicate the derivative is mentioned in the comments. The format is as follows:

DERIVATIVE TYPE
Type of compound
Reagent, 00, 00, 00
Reagent, 00, 00, 00

In those instances of numerous entries for a derivative type all using the same reagent, the reagent is listed first followed by the subdivisions according to compound type:

DERIVATIVE TYPE
Reagent
Type of compound, 00, 00, 00
Type of compound, 00, 00, 00

The abbreviation "N.S." indicates that the reagent is not stated in the text.

Author Index

The author index includes the names of all authors whose work is cited including those cited as "et al." Numbers in parentheses indicate the reference number on the page where necessary to avoid ambiquity in "et al." citations.

General Subject Index